The Mathematics of Frobenius in Context

Sources and Studies in the History of Mathematics and Physical Sciences

For further volumes:
http://www.springer.com/series/4142

Thomas Hawkins

The Mathematics of Frobenius in Context

A Journey Through 18th to 20th Century Mathematics

Springer

Thomas Hawkins
Department of Mathematics & Statistics
Boston University
Boston, MA, USA

ISBN 978-1-4614-6332-0 ISBN 978-1-4614-6333-7 (eBook)
DOI 10.1007/978-1-4614-6333-7
Springer New York Heidelberg Dordrecht London

Library of Congress Control Number: 2013941038

Printed on acid-free paper

Springer is part of Springer Science+Business Media (www.springer.com)

Preface

This book grew out of my research on the history of mathematics over the past 40 years. Time and again, the path of my investigation led me to the consideration of work by Frobenius that played an important role in the historical development I was attempting to understand. I finally decided it would be appropriate to bring these research experiences together into a book on the mathematics of Frobenius, especially since little has been written about him despite the fact that he made many important contributions to present-day mathematics—as suggested by the many theorems and concepts that bear his name.

Initially, the focus of the book was dictated by my earlier research experiences and interests involving Frobenius. These had all involved his work on the theory and application of linear algebra, including the application involved in his creation of the theory of group characters and representations; and so the initial working title was "Frobenius and the History of Linear Algebra." As the reader will see, much of Frobenius' work did indeed involve linear algebra somewhere along the way; but I began to realize that to focus exclusively on this aspect of his work would present a distorted picture of his mathematical activities and their significance, as well as of the sources of his inspiration and the reasons so much of his work has become a part of basic mathematics. His creation of representation theory may have been his most important achievement, but he also did work of lasting significance in many other areas of mathematics. Frobenius was an algebraist at heart, but he looked for interesting problems of an essentially algebraic or formal nature in a broad spectrum of areas of nineteenth- and early twentieth-century mathematics. To do him and his work justice, the scope of the book had to be broadened into a more well-rounded intellectual biography; and that is what I have attempted. Whence the first part of the title of the book: "The Mathematics of Frobenius."

The second part of the title also requires clarification. I have attempted to present the mathematics of Frobenius "in context" in two senses. The first of these involves providing the reader with the historical background necessary to understand why Frobenius undertook to solve a particular problem and to appreciate, by seeing what had been done before, the magnitude of his achievement, as well as what he owed to his predecessors. In addition to the backgrounds peculiar to various particular problems, it was also necessary to say something about Frobenius' educational background, namely his training in the Berlin school of mathematics presided over by Weierstrass (Frobenius' dissertation advisor and principal supporter), Kronecker, and Kummer. Of particular importance is the work done by Weierstrass and Kronecker on the theory of equivalence (respectively, congruence) of families of bilinear (respectively quadratic) forms. As we shall see, from their work Frobenius learned both theorems and concomitant disciplinary ideals that together motivated and informed much of his early work. In addition, from Kummer's groundbreaking work on ideal complex numbers, and Kronecker's interest in extending it, as well as from Dedekind's quite different extension by means of his theory of algebraic numbers and ideals, Frobenius acquired the background that motivated much of his work on the theory of numbers and abstract group theory. Thus considerable attention is given to these arithmetic developments.

I have also attempted to present Frobenius' mathematics "in context" in the sense that I have sought to trace the various ways in which his work was subsequently applied, developed, and ultimately incorporated into present-day mathematics. By presenting the mathematics of Frobenius in context in both these senses, my hope is that the reader will come away not only with an enriched appreciation of Frobenius' work but also with a glimpse of the broad swath of diverse and important strands of eighteenth- to twentieth-century mathematics that results from the contextual approach and that ranges from the work of Lagrange and Laplace on terrestrial and celestial mechanics in the last decades of the eighteenth century, which involved them with the theory of systems of linear differential equations, to the theory of complex abelian varieties in the mid-twentieth century. This is the "Journey through Eighteenth- to Twentieth-Century Mathematics" of the subtitle.

The book has been divided into three parts. Part I is devoted to an overview of Frobenius' entire mathematical career and thus serves as an introduction to the main body of the book. Here, within the context of Frobenius' educational and professional career, his contributions to mathematics and the attendant backgrounds are briefly sketched and their subsequent impact on the development of mathematics indicated. It is my hope that the reader will come away from Part I with a broad sense of Frobenius' many contributions to mathematics, of the institutional and personal connections that affected his work, of the broad scope and progression of his mathematical interests, and of the ways in which his work has been incorporated into present-day mathematics. Of course, in order to gain more than just a vague sense, in order to fully appreciate what Frobenius accomplished, how it grew out of or was motivated by earlier work, and how it has affected present-day mathematics, a reading of the chapters in Parts II and III is necessary. The two chapters that form Part II deal with the development of linear algebra up to and including the work of Weierstrass and Kronecker and are essential background for all that is to follow. The chapters of Part III deal in depth with Frobenius' major works, a subset of the works discussed in Part I. These chapters range over many areas of mathematics and can be read independently of one another, with little loss of continuity thanks to the overview provided by Part I. Thus, for example, a reader particularly interested in Frobenius' arithmetic work could turn next to Chapters 8 and 9, where this work is treated. Readers wishing to know more about his work on group characters and representations could start with Chapter 12. I have provided a detailed table of contents to guide readers to those parts of Frobenius' work of special interest to them.

In addition to a detailed table of contents, I have provided an extensive index that will enable readers to look for a specific topic that may not be included in the table of contents. The index can also be used to find the meaning of unfamiliar terms, such as "Dedekind characters," the "containment theorem of Frobenius," or "winter semester" in German universities. If several pages are given for an entry, the page number containing the explanation of the term is given in boldface. The index is also helpful for tracking down various recurring themes of the book, such as "generic reasoning," "disciplinary ideals of Berlin school" (also found under Kronecker, who articulated them), and "multiple discoveries" involving Frobenius. By the latter term I mean instances in which Frobenius and one or more mathematicians independently made essentially the same discovery or developed similar ideas. As the index shows, Frobenius was involved in many instances of multiple discovery. The entry for Frobenius is particularly extensive and should prove useful in locating various sorts of information about him, such as a listing of all the evaluations Weierstrass made of him and his work, as well as all evaluations Frobenius made of other mathematicians or mathematical theories. In addition, there is a listing of all mathematicians influenced by Frobenius and a listing of all mathematicians who influenced him in the broad sense that includes mathematicians who provided him with useful ideas and results, as well as mathematicians whose work, due to deficiencies, motivated him to develop a theory that removed them.

My interest in Frobenius began circa 1970 with an attempt to reconstruct the origins of his remarkable theory of group characters [266]. I knew that he had been in correspondence with Dedekind, who had introduced him to group determinants. Important excerpts of Dedekind's side of the correspondence had been published by E. Noether in Dedekind's collected works [119, pp. 414–442], but Frobenius' side of the correspondence seemed lost to posterity until the year after my paper [266] appeared, when Clark Kimberling announced his fortuitous discovery of the Dedekind–Frobenius correspondence [339, §8], which runs to over 300 pages. At my request he kindly provided me with a copy of the correspondence, which showed that my reconstruction of how Frobenius had created his theory of group characters needed to be significantly modified. The result was my paper [268], which quoted extensively (in translation) from the correspondence. Much of that material is incorporated into this book. In addition, the correspondence during 1882 has proved enlightening in discussing Frobenius' work on density theorems, which was done in 1880 but not published until 1896. By the time I investigated Frobenius' work on density theorems, two unpublished transcriptions of the correspondence had been made, the first by the late Walter Kaufmann-Bühler, and the second, building upon the first, by Ralf Haubrich. They kindly sent me copies of drafts of their transcriptions, which greatly facilitated a careful reading of the entire correspondence. The Dedekind–Frobenius correspondence was initially housed at the Clifford Memorial Library of the University of Evansville, and I am grateful to the library for permission to use the correspondence in my publications. In 1995, the correspondence was moved to its present location in the archives of the library of the Technical University at Braunschweig, the institution where Dedekind spent almost all of his mathematical career.[1] All of the citations from Frobenius' letters are from this archival source. The citations from Dedekind's letters that are printed in his collected works are so indicated by footnotes.

Besides the individuals and institutions mentioned above, I am indebted to many others who, at one point or other during the past 40 years, assisted me with some aspect of my work on Frobenius. Although I am sure I have now forgotten some, I do wish to express my gratitude to those I have remembered: Armand Borel, Keith Conrad, Harold Edwards, Walter Feit, Jeremy Gray, Rob Gross, Walter Ledermann, Franz Lemmermeyer, Peter Neuman, Wilfried Parys, Klaus Peters, Peter Roquette, Michael Rosen, David Rowe, Yvonne Schetz, Hans Schneider, Shlomo Sternberg, and Dan Weiner. I am also grateful to the NSF Science and Technology Studies program for providing the financial support that enabled me to initiate my efforts to write a book on Frobenius.[2] My greatest debt of all is to Jean-Pierre Serre. To begin with, shortly before my interests turned toward Frobenius, he took on the burden of editing Frobenius' mathematical works for publication. Frobenius' *Mathematische Abhandlungen* appeared in 1968 [232] and has facilitated my study of his work ever since. In addition, throughout my career as historian of mathematics he has encouraged my efforts and generously given his time to critically evaluate and respond with many helpful suggestions to drafts of various parts of my work. His contributions to the writing of this book in particular have been manifold. Some of these are reflected in the index but many are not. Among the latter, I would mention that the decision to transform my book from "Frobenius and the History of Linear Algebra" into a book that attempts to deal with all of Frobenius' major mathematical contributions was sparked by his remark, on hearing of my plans to write the first sort of book, that I should really look into Frobenius' work on theta functions, since C.L. Siegel had told him that Frobenius had done important work in this area. (That Siegel was right can be seen from Chapter 11.) Initially I dismissed Serre's suggestion of a more inclusive work on the grounds of personal inadequacy, but his suggestion remained in the back of my mind and eventually led to the following book, imperfect as it may prove to be.

Finally, I wish to express my gratitude to David Kramer, whose scrupulous and informed copyediting of the book has resulted in many significant improvements.

Boston, MA, USA Thomas Hawkins

[1]Frobenius' letters to Dedekind are archived under the reference Universitätsarchiv Braunschweig G 98 : 10. G 98. Frobenius' letters are under G 98 : 10.

[2] Through grant SES-0312697.

Contents

Part I
Overview of Frobenius' Career and Mathematics

Chapter 1
A Berlin Education

Ferdinand Georg Frobenius was born in Berlin on 26 October 1849.[1] He was a descendant of a family stemming from Thüringen, a former state in central Germany and later a part of East Germany. Georg Ludwig Frobenius (1566–1645), a prominent Hamburg publisher of scientific works, including those written by himself on philology, mathematics, and astronomy, was one of his ancestors. His father, Christian Ferdinand, was a Lutheran pastor, and his mother, Christiane Elisabeth Friedrich, was the daughter of a master clothmaker.

Frobenius grew up in Berlin and attended high school there at the Joachimstalische Gymnasium, where he distinguished himself as one of its most outstanding students [22, p. 190]. He began his university studies in Göttingen, where he enrolled for the "summer semester" of 1867. In German universities the summer semester runs from about mid-April to mid-July and the winter semester from about mid-October through February. Thus the winter semester corresponds roughly to the fall semester in an American university and the summer semester to the spring semester. Frobenius took two courses in analysis at Göttingen and a course in physics given by the well-known physicist Wilhelm Weber. His primary interest was already in mathematics, and it is likely that his intention from the outset was to do what he ended up doing: enrolling for the winter semester 1867 at the University of Berlin and pursuing his study of mathematics there for six semesters through completion of his doctorate. This would have been a reasonable course of action because at the time, the University of Berlin was the leading center for mathematics in Germany and one of the major centers for mathematics in the world.

---

[1] The following biographical details about Frobenius are drawn from [4, 22, 553].

T. Hawkins, *The Mathematics of Frobenius in Context*, Sources and Studies in the History of Mathematics and Physical Sciences, DOI 10.1007/978-1-4614-6333-7_1,

1.1 Student Years: 1867–1870

In his invaluable study of mathematics in Berlin [22], K.-R. Biermann has characterized the period 1855–1892, which he calls the "Kummer–Weierstrass–Kronecker era," as a historical high point for instruction and research in mathematics at the university, what one who was there (Adolf Kneser) described as the "heroic period" [22, p. 75]. Indeed, thanks largely to the concerted efforts of the above three mathematicians, a bona fide school of mathematics emerged during these years. Frobenius became a devoted member of this school, and his choice of research problems and the manner in which he approached them was colored by his experiences at Berlin, which resonated with his own mathematical proclivities. Figure 1.1 shows a youthful Frobenius as he may have looked during his student and postdoctoral years at the University.

A closer look at the Berlin school is now in order. It will be helpful to begin by indicating the three principal categories of lecturers at a German university. The highest category was that of *ordentlicher Professor*, which I have translated as "full professor." The full professors had substantial salaries and powers within the university, e.g., to recommend new faculty appointments and to direct doctoral dissertations. Then there was the category of "extraordinary professor" (*ausserordentlicher Professor*), which I have translated as assistant professor, since the salary was modest and came with no powers within the university. The lowest category consisted of the "private lecturers" (*Privatdozenten*), which I have translated as "instructors." How someone with a doctorate became an instructor by writing a

Fig. 1.1 Frobenius as he may have looked during the early years of his career, when he absorbed and began applying the teachings of the Berlin school of mathematics. Image courtesy of ETH-Bibliothek Zürich, and located in its Image Archive

Habilitationsschrift is indicated in the next section. There were other lecturers outside these categories as well; an example is provided by Kronecker, as will be seen below.

At Berlin, the two full professors of mathematics at the university in 1867 were Eduard Kummer (1810–1893) and Karl Weierstrass (1815–1897). After impressing Jacobi and Dirichlet with work in analysis, Kummer had achieved more widespread fame for his groundbreaking theory of ideal complex integers associated to the ordinary complex integers of $\mathbb{Z}[\omega_p]$, ω_p a primitive pth root of unity for the prime number p (see Section 9.1.2). He had begun his work on this theory in 1846, but by 1867, when Frobenius arrived, his research interests had shifted to geometry (ray systems). As for Weierstrass, he had gone from the relative obscurity of a high-school mathematics teacher into the mathematical limelight by virtue of his solution, in 1854, to the Jacobi inversion problem for hyperelliptic integrals in any number of variables (see Section 10.1). Kummer's first administrative accomplishment after his appointment to a Berlin professorship in 1855 was to pull the strings necessary to get Weierstrass to Berlin the following year.

Mention should also be made of Carl Borchardt (1817–1880) [22, pp. 61–62]. Borchardt had obtained his doctorate under Jacobi's direction at the University of Königsberg (1843) and subsequently became his friend. Jacobi was in poor health at the time and Borchardt accompanied him to Italy, where (in Rome) they met Jacob Steiner (1796–1863) and P.G. Lejeune Dirichlet (1805–1859), who were then both full professors at the University of Berlin. Borchardt became an instructor at the university in 1848. In 1855, Dirichlet, about to leave for a professorship at Göttingen, persuaded Borchardt, who was independently wealthy, to take over the editorship of "Crelle's Journal"—the *Journal für die reine und angewandte Mathematik*—which had been founded in 1829 by August Crelle, who had edited it until his death in 1855. Borchardt also became a member of the Berlin Academy of Sciences in 1855, at the recommendation of Dirichlet, perhaps in part as a reward for assuming the task of editing *Crelle's Journal*, which was in effect the journal of the Berlin school of mathematics. That is, members of the academy, such as Kronecker, Kummer, and Weierstrass, tended to publish in the proceedings of the academy, but their students aspired to publication in *Crelle's Journal*. For example, from 1871 until 1893, when Frobenius became a Berlin professor and member of the academy, virtually all of his mathematical output was published in that journal.

Borchardt remained editor until his death in 1880, and during his tenure the quality of papers accepted increased. After over a decade as editor he had considered retiring, but when in 1869 Alfred Clebsch, then at the University of Göttingen, founded the rival *Mathematische Annalen*, Borchardt, sensing the competitive challenge, decided to continue as editor. Borchardt also became a close friend to Weierstrass, whom Borchardt had made a point of meeting in 1854, after Weierstrass had solved the Jacobi inversion problem for hyperelliptic integrals. Borchardt was one of the few people whom Weierstrass addressed with the familiar "*Du*" form.

By 1857 Weierstrass had solved the inversion problem for general abelian integrals, but he held back his results because he had discovered that Riemann had already published his quite different solution to the problem that same year.

Fig. 1.2 Weierstrass, through his work, teaching, and support was to exert the greatest influence on Frobenius of any mathematician. Of his many doctoral students, Frobenius was to become the most accomplished. Photo courtesy of Mathematisches Forschungsinstitut, Oberwolfach, Germany

Weierstrass then set for himself the goal of understanding Riemann's results and their relation to his own. Riemann's solution was couched in terms of what are now called Riemann surfaces and were not as rigorously founded as Weierstrass wished. His next goal was to develop his own solution in the light of his understanding of Riemann's results. This he did not do in print but, gradually, through the medium of his lectures on abelian integrals and functions at Berlin. Frobenius probably attended Weierstrass' lectures on this subject as given in the summer semester 1869.[2] Some important aspects of Weierstrass' lectures are discussed in Section 11.1 because of their relevance to Frobenius' work. Part of the foundations for the theory of elliptic and abelian functions was supplied by the general theory of functions of one or more complex variables, which Weierstrass also developed in lectures. Although Weierstrass' primary area of research was complex analysis, he also thought about and occasionally published in other areas. These included the theory of the transformation of quadratic and bilinear forms, to which he devoted two important papers (1858, 1868) that were to prove of great consequence to Frobenius (as indicated below). Weierstrass, who became Frobenius' mentor and dissertation advisor, is pictured in Fig. 1.2.

---

[2]A listing of the semesters in which Weierstrass lectured on abelian integrals and functions is given in vol. 3 of his *Mathematische Werke*, the volume containing of a version of these lectures.

The third member of the Berlin mathematical triumvirate was Leopold Kronecker (1823–1891). Although Kronecker had studied mathematics at Berlin and obtained a doctorate under the direction of Dirichlet in 1845, he did not pursue a traditional academic career thereafter. Instead he managed the considerable wealth of his family while at the same time indulging his interests in mathematics. These interests were shaped more by Kummer than by Dirichlet. Kummer had known Kronecker since his high-school days at the gymnasium in Liegnitz (now Legnica, in Poland), where he was then teaching, and the bond between them had developed into a friendship based on a mutual interest in mathematics. As Frobenius wrote many years later, "In so far as Kronecker was a student, he was Kummer's student ..." [202, p. 710]. Indeed, one of Kronecker's goals was to extend a version of Kummer's theory of ideal numbers from the context of $\mathbb{Q}(\omega_p)$ to that of far more general fields.[3]

Kronecker's interest in algebra, and in particular in algebraic equations, had been stimulated by the posthumous publication in 1846 of Évariste Galois' "Memoir on the conditions for the solvability of equations by radicals" [239]. Kronecker, however, did not find Galois' group-theoretic conditions for solvability satisfying [613, pp. 121ff.]. Galois had not described the nature of a polynomial $f(x)$ solvable by radicals; he had only characterized the associated (Galois) group G by the property (stated here in modern terms) of the existence of a chain of subgroups $G \supset G_1 \supset G_2 \supset \cdots \supset G_k = 1$ such that G_i is a normal subgroup of G_{i-1} and the factor group G_{i-1}/G_i is abelian. Kronecker wanted to characterize the equations solvable by radicals in a manner relating more directly to their coefficients and roots. Galois' results showed that it sufficed to do this for abelian equations, i.e., polynomial equations with abelian Galois group. Kronecker obtained many results of considerable interest, including what has become known as the Kronecker–Weber theorem: if $f(x) \in \mathbb{Q}[x]$ is abelian, then all its roots are rational functions of roots of unity.[4]

By the time Frobenius arrived in Berlin, Kronecker, who is pictured in Fig. 1.3, had been residing there for a dozen years and was very much engaged in the mathematical life of the university. He was a member of the Berlin Academy of Sciences, and at Kummer's suggestion, he claimed the right accorded members of presenting lectures at the university. Indeed, along with Kummer and Weierstrass, he was involved in the 2-year lecture cycle that had been devised to provide students, within that time span, with a broad and up-to-date education in what were deemed the main areas of mathematics. Thus, although Kronecker was not a full professor, and so could not officially direct doctoral theses or run the Berlin mathematics

---

[3] According to Frobenius [202, p. 712], in 1858 Kronecker sent a (now lost) manuscript containing some such extension to Dirichlet and Kummer. In 1882, on the occasion of the fiftieth anniversary of Kummer's doctorate, he presented a sketch of his ideas as they stood then [363], but they were difficult to follow. Edwards [150] has given a possible reconstruction of what Kronecker had in mind.

[4] See Frobenius' discussion of this and related results [202, pp. 712–713].

Fig. 1.3 Through his publications Kronecker was to exert a strong influence on Frobenius, who found in Kronecker's sketchy communications many problems to investigate and resolve. Photo courtesy of Mathematisches Forschungsinsitut, Oberwolfach, Germany

seminar, he played an important role in the educational program of the Berlin school. Reflecting his primary interests in algebra and the theory of numbers, Kronecker's lectures covered such areas as the theory of algebraic equations, number theory, the theory of determinants (including much that would now be classified as linear algebra), and the theory of simple and multiple integrals [22, p. 81]. It was probably through Kronecker that the young Frobenius learned about Galois' memoir [239], which he seems to have carefully studied along with the related works by Abel. Indeed, as we shall see in the following section, Frobenius' first paper beyond his dissertation involved the ideas of Abel and Galois, albeit interpreted within the context of Weierstrassian complex function theory. Kronecker was to exert a major influence on the directions taken by Frobenius' research.

Kummer's lectures covered analytic geometry, mechanics, the theory of surfaces, and number theory. His polished lectures were relatively easy to follow and never ventured beyond well-established theories, unlike both those of Kronecker, whose lectures were very difficult to follow, and those of Weierstrass, whose lectures were challenging but more accessible, the result of an ongoing effort to present the material in a rigorous and appropriate manner. In addition to abelian integrals and functions, Weierstrass' lectures were on such topics as the foundations of the theory of analytic functions, elliptic integrals and functions, the calculus of variations, and applications of elliptic functions to problems in geometry and mechanics. In his lecture cycle Weierstrass strove to present a rigorous development of mathematical analysis that took nothing for granted [22, p. 77]. Although other courses in mathematics were taught by assistant professors, instructors, and other special lecturers, Frobenius managed to take all his mathematics courses from Kummer,

Weierstrass, and Kronecker.[5] In addition to courses in mathematics, he took courses in physics and philosophy.

Kronecker's role within the Berlin school was not at all limited to teaching. The Berlin Academy of Sciences afforded him a forum for mathematical discourse with his colleagues. As Frobenius later wrote,

> Nothing gave him more pleasure than to share his views on mathematical works or problems with his colleagues. Insatiable in his craving for scientific discourse, he could hold forth on his ideas until deep into the night with those who listened with intelligence and comprehension; and he who could not be convinced could be certain that already the next morning he would find a write-up of the matter discussed [202, p. 711].

Although in later years, relations between Kronecker and Weierstrass became strained to the breaking point [22, pp. 100ff.], during the period 1868–1875, when Frobenius was in Berlin, the situation was quite different. Even after years of strained relations, Weierstrass retained vivid, fond memories of the early years when he and Kronecker freely exchanged mathematical ideas [202, p. 711]. Several examples of such mutual interaction will be given in the following chapters because they inspired work by Frobenius. By far the most important example of their mutual interaction involved the theory of the transformation of quadratic and bilinear forms, the subject of Part II.

In Part II, I set the stage for, and then present, the work of Weierstrass and Kronecker that informed the Berlin school's approach to what would now be classified as linear algebra. As we shall see, much of Frobenius' work in the early years after his departure from Berlin in 1874 involved, in one way or another, Berlin style linear algebra. In developing and applying linear algebra to diverse mathematical problems drawn from analysis and arithmetic, Frobenius made extensive use of Weierstrass' theory of elementary divisors and the disciplinary ideals implicit in it and made explicit by Kronecker. Weierstrass' theory originated in his interest in the eighteenth-century discussion by Lagrange and Laplace of mechanical problems leading to a system of linear differential equations, which in modern notation would be expressed by $B\ddot{y}+Ay=0$, where A,B are symmetric matrices, B is positive definite, and $y = \begin{pmatrix} y_1 & \cdots & y_n \end{pmatrix}^t$. Lagrange and Laplace had inherited the method of algebraic analysis that had revolutionized seventeenth-century mathematics. They perfected the method and made brilliant applications of it to terrestrial and celestial mechanics. In their hands, mathematical analysis became both elegant and general in its scope, but it retained the tendency to reason with algebraic symbols as if they had "general" values, a tendency that obscured the possibility of special relations or singularities that need to be considered in order to attain truly general results. In this way they came to believe that the above system of equations would yield stable solutions only if it were assumed, in addition, that the roots of $f(s) = \det(sB - A)$ are real and distinct. Building upon work of Cauchy, Dirichlet, and Jacobi, Weierstrass showed in 1858 that the additional

---

[5]This is clear from the vita at the end of his doctoral dissertation [171, p. 34], as is the fact that his other courses at Berlin were in physics and philosophy.

assumptions were unnecessary, that by means of rigorous, nongeneric reasoning it could be established that (in the language of matrices introduced later by Frobenius) a real nonsingular matrix P exists such that $P^tBP = I$ and $P^tAP = D$, where D is a diagonal matrix. From this result the correct form for the solutions to $B\ddot{y} + Ay = 0$ then followed, and their stability was established.

Weierstrass presented his results in a paper published by the Berlin Academy in 1858. He had couched his results in the language of the transformation of quadratic forms so as to relate them to the work of Cauchy and Jacobi, which had been motivated by the principal axes theorems of mechanics and the theory of quadric surfaces. Cauchy had been the earliest critic of the generic reasoning of the eighteenth century, whereas Jacobi continued, despite Cauchy's example, to pursue his characteristically elegant form of algebra on the generic level. Jacobi's work on the simultaneous transformation of pairs of quadratic forms into sums of squared terms raised the question as to when, in nongeneric terms, two pairs of quadratic or bilinear forms can be transformed into one another. Weierstrass discovered that the method he had employed in his 1858 paper could be generalized to apply to all nonsingular pairs of bilinear forms so as to give necessary and sufficient conditions for the transformation of one nonsingular pair into another. The result was his theory of elementary divisors, which he presented to the Berlin Academy in a paper of 1868. It will be helpful to briefly explain Weierstrass' theory at this point so as to make the later sections of Part I intelligible to readers unfamiliar with the language of elementary divisors.

Following Frobenius (Section 7.5), I will identify pairs of bilinear forms $\Phi = x^tBy$ and $\Psi = x^tAy$ with their coefficient matrices B,A. Such a pair can be simultaneously transformed into another pair $\tilde{B},\tilde{A}$ if nonsingular matrices P,Q exist such that $PAQ = \tilde{A}$ and $PBQ = \tilde{B}$. In this case, the pairs (B,A) and $(\tilde{B},\tilde{A})$ are said to be *equivalent*. A pair (B,A) is *nonsingular* if $\det B \neq 0$. Weierstrass' main theorem was that two nonsingular pairs are equivalent if and only if they have the same "elementary divisors." To understand what elementary divisors are, observe that pairs (B,A) and $(\tilde{B},\tilde{A})$ are clearly equivalent precisely when the matrix families $sB - A$ and $s\tilde{B} - \tilde{A}$, s a complex variable, are equivalent. By means of determinant-theoretic considerations, Weierstrass introduced a sequence of polynomials $E_n(s),\ldots,E_1(s)$ associated to $sB - A$, which, thanks to Frobenius (Chapter 8), can be seen to be the *invariant factors* of $sB - A$ with respect to the polynomial ring $\mathbb{C}[s]$. That is, the Smith normal form of $sB - A$ over $\mathbb{C}[s]$ is the diagonal matrix with $E_n(s),\ldots,E_1(s)$ down the diagonal. They satisfy $E_i(s) \mid E_{i+1}(s)$ and $\det(sB - A) = \prod_{i=1}^{n} E_i(s)$. The $E_i(s)$ of course factor into linear factors over $\mathbb{C}$, and so Weierstrass wrote $E_i(s) = \prod_{j=1}^{k} (s - a_j)^{m_{ij}}$, where $a_1,\ldots,a_k$ denote the distinct roots of $\varphi(s) = \det(sB - A)$ and m_{ij} is a nonnegative integer. The factors $(s - a_i)^{m_{ij}}$ with $m_{ij} > 0$ are Weierstrass' elementary divisors. They are thus the powers of the distinct prime factors of each invariant factor $E_i(s)$ in the polynomial ring $\mathbb{C}[s]$. In order to prove his main theorem, Weierstrass showed that given $sB - A$ (with $\det B \neq 0$), P,Q may be determined such that $W = P(sB - A)Q$ has a simple form from which its elementary divisors can be immediately ascertained. The matrix W is essentially the same as the familiar Jordan canonical form of $sB - A$, which was introduced independently by Camille Jordan at about the same time. The determinant of each "Jordan block" of W equals an elementary divisor $(s-a_i)^{m_{ij}}$ of $sB-A$ with m_{ij} giving the dimension of the block.

For a corollary to his theory, Weierstrass returned to the context of his paper of 1858: pairs (B,A) of symmetric matrices; but he was now able to replace his earlier hypothesis that B is definite with the weaker one that B is nonsingular. His corollary was that two such symmetric pairs, (B,A) and $(\tilde{B},\tilde{A})$, are *congruent* in the sense that a nonsingular P exists such that $P^tBP=\tilde{B}$ and $P^tAP=\tilde{A}$ if and only if $sB-A$ and $s\tilde{B}-\tilde{A}$ have the same elementary divisors. His corollary provided a rigorous counterpoint to Jacobi's generic theorem that a symmetric pair is congruent to a pair of diagonal matrices. It should also be noted (as Frobenius did) that Weierstrass' theory also provided necessary and sufficient conditions that two matrices A and $\tilde{A}$ be *similar* in the sense that $A=S^{-1}\tilde{A}S$ for some nonsingular matrix S. It is only necessary to apply Weierstrass' main theorem to pairs with $B=\tilde{B}=I$.

Kronecker was not a member of the Berlin Academy in 1858 when Weierstrass presented his paper on pairs of quadratic forms, and he was apparently unfamiliar with it when, in 1866, while working with Weierstrass' encouragement on the problem of a viable generalization to abelian functions of the notion of elliptic functions admitting a complex multiplication, he became interested in the problem of the congruent transformation of the special families $sB-B^t$, where B is any $2n\times 2n$ matrix, into a normal form. This problem had emerged from his method of attacking the complex multiplication problem and seems to have overshadowed the original problem in his mind. He dealt with it in the generic spirit of Jacobi, but when he became aware of Weierstrass' paper of 1858 and his subsequent theory of elementary divisors, he abandoned the generic approach and took upon himself the highly nontrivial task of extending Weierstrass' theory to singular pairs (A,B), i.e., the problem of determining a set of invariants for *any* matrix pair (B,A) that would provide necessary and sufficient conditions for two such pairs to be equivalent. (The invariant factors $E_i(s)$ alone are insufficient to this end.)

Kronecker worked on this problem during 1868–1874 (see Section 5.6), while Frobenius was in Berlin, and he succeeded in solving it, although he only sketched his solution for symmetric pairs (see Theorem 5.13) and then, returning to his work of 1866, he developed from scratch the necessary and sufficient conditions for $sB+B^t$ and $s\tilde{B}+\tilde{B}^t$ to be congruent: $P^t(sB-B^t)P=s\tilde{B}+\tilde{B}^t$, $\det P\neq 0$. It was in these communications to the Berlin Academy that Kronecker—embroiled in a quarrel with Camille Jordan stemming from the latter's criticism of Weierstrass' theory of elementary divisors—explicitly articulated two disciplinary ideals that were implicit in Weierstrass' and his own work on the transformation of forms. Frobenius knew of this work and these ideals, and as we shall see (especially in Chapters 6 and 7), they served to motivate and inform his choice of research problems, the solutions to which involved the creation of mathematics of lasting significance (as indicated in subsequent chapters of Part I).

Weierstrass' paper on elementary divisors appeared in the proceedings of the Berlin Academy during Frobenius' second year at the university, although it is unlikely that he was then aware of it. Weierstrass, however, had already become aware of him. One of the customs at Berlin was for one of the full professors to pose a mathematical prize problem. In 1868, during Frobenius' second semester at Berlin, it was Weierstrass who posed the problem. Seven students submitted solutions, and one of them was Frobenius. He did not win the prize, but he did receive an honorable mention and a prize of 50 thaler [22, p. 88].[6] What had impressed Weierstrass was the uncommon facility for mathematical calculations his solution displayed [22, p. 190], a talent his other teachers had also observed. Whether the 18-year-old also had a talent for independent, original mathematical thought, however, remained to be seen. It was in the Berlin Mathematics Seminar that this talent seems to have first revealed itself. The Berlin Mathematics Seminar, the first seminar in pure mathematics in a German university, had been instituted by Kummer and Weierstrass in order to help students learn to think independently about the mathematics they had been learning [22, p. 72]. The seminar was open only to a limited number of pre- and postdoctoral students who were deemed qualified to benefit from it. No doubt Frobenius' solution to the prize problem was a major factor in his acceptance into the seminar at the beginning of his second year. In fact, he participated in the seminar for four semesters [171, p. 34] and so throughout his final 2 years as a student.

It was in the seminar that Weierstrass realized that Frobenius was much more than a mindless calculator. As he explained in 1872[7]:

> At first he [Frobenius] attracted the attention of his teachers by virtue of his extraordinary facility with mathematical calculations, which enabled him, already as a second-semester student, to solve a prize question posited by the faculty. However, it soon became clear that he possessed to a high degree the mental aptitude and capability necessary for original mathematical research. As a member of the mathematics seminar he produced various works, which would have been worthy of publication and were certainly not inferior in value to many recent publications. In the seminar, when it came to scientific matters, he always proved himself to be an independent thinker, although he was otherwise unassuming and almost childlike in manner. What was dictated to him in lectures he zealously made his own, but rather than being content with that, he always used what he had learned to determine his own scientific endeavors.

Weierstrass had further opportunity to observe Frobenius' talents as the director of his doctoral dissertation.

Incidentally, Frobenius' choice of Weierstrass over Kummer as dissertation advisor is not surprising. During the 9-year period 1867–1874 when Frobenius was in Berlin, 13 doctoral degrees were awarded, and eight of them were done under Weierstrass' direction. During this period, Weierstrass' area of research was no doubt perceived as more in the mainstream. Indicative of this is the fact that Lazarus Fuchs (1833–1902), who had received his doctorate under Kummer's

---

[6]In 1875 Frobenius published a paper [176] that was an outgrowth of the work he had done on the prize problem.

[7]The occasion was a proposal to the Prussian minister of culture of a new associate professorship in mathematics, with Frobenius as the choice to fill the new position. The entire document is given by Biermann [22, pp. 189ff.]; the quotation below is from pp. 190–191.

direction in 1858, created quite a stir in Berlin in the period 1865–1868, when, after attending Weierstrass' lectures on abelian integrals and functions in 1863, he had applied Weierstrass' treatment of algebraic differential equations to an important class of linear homogeneous differential equations, thereby displaying the capacity for independent mathematical research needed to be habilitated as an instructor (*Privatdozent*) at the university. Eventually, Frobenius was drawn into the enterprise of developing Fuchs' theory, but not until after he had completed his doctoral dissertation and received his doctorate.

The subject of Frobenius' dissertation appears to have been of his own devising. Its starting point was the Cauchy integral formula

$$f(z) = \frac{1}{2\pi i}\int_{C_\rho} \frac{f(w)}{w-z}dw.$$

For f analytic in a suitable region including the circle C_ρ defined by $|w| = \rho$, it was well known that the integral formula could be used to derive the Laurent expansion of f by expanding the kernel $\frac{1}{w-z}$ in a geometric series and integrating $\frac{f(w)}{w-z}$ term by term. Frobenius' idea was to consider other expansions of this kernel, suitably chosen so that term-by-term integration of the uniformly convergent expansion would yield series expansions for $f(z)$ of the form $\sum c_n F_n(z)$ with the coefficients c_n given by integrals analogous to those that occur in the Laurent expansion. Frobenius' dissertation revealed a mathematician with a broad knowledge of complex analysis and the related theory of hypergeometric series and differential equations combined with an extraordinary ability to manipulate complicated analytical expressions so as to achieve interesting, original, and conclusive results, although these results were not really part of the mainstream mathematics of his day.[8]

Frobenius officially obtained his doctorate on 28 July 1870, and he had clearly impressed the Berlin faculty.[9] Weierstrass described his dissertation as "a first-rate work," one that was "distinguished by thoroughgoing studies" and "outstanding" in its form of presentation. It left Weierstrass convinced that its author "possesses a definite talent for independent research." Frobenius' oral doctoral examination (on 23 June 1870) was equally impressive. According to the records, Weierstrass asked the candidate questions on the theory of abelian functions and integrals and their theoretical basis in complex function theory—material that was central to Weierstrass' cycle of lectures at the university—and Frobenius showed himself to be "completely familiar with this difficult theory"— and was even able to present with detailed exactitude both complicated proofs and derivations, much to the satisfaction of the examining committee. Kummer then took over the questioning,

[8] A good description of Frobenius' impressive results can be found in Meyer Hamburger's review [259] of the German version of the Latin dissertation that Frobenius published three months later [172].

[9] See [22, p. 85], where the quotations below about Frobenius' dissertation and doctoral examination are given in the original German.

which covered questions about the application of the theory of elliptic functions in number theory and mechanics and ended with some questions about geometric problems. Here too the committee found the candidate well informed throughout. Unlike a typical mathematics doctoral candidate nowadays, Frobenius was also questioned in philosophy (by Harms) and physics (by Dove). He was judged to have a "deep and thorough" understanding of Kant's *Critique of Pure Reason* and to have satisfactorily answered Dove's questions relating to phenomena in the theory of heat. His overall performance earned him a pass summa cum laude.

A doctoral candidate had to submit three theses, one of which he would be asked to defend at his oral doctoral examination. Frobenius' three theses [171, p. 34] were (in English translation from the Latin[10]): "(1) Kant did not support his thesis concerning time and space by sufficiently weighty arguments; (2) The theory of definite integrals ought to precede the treatment of differential calculus; (3) It is better for the elements of higher analysis to be taught in the schools than the elements of the more recent synthetic geometry." Judging by the committee's remarks about Frobenius' understanding of Kant's *Critique of Pure Reason*, it would seem that the committee chose the first thesis. The other two theses are of interest because they were essentially pedagogical in nature. They indicate Frobenius' early concern for the teaching of mathematics—what should be taught and how it should be presented. As we shall see, Frobenius proved to be an excellent teacher, and his creative mathematical work was always characterized by a concern for a clear and rigorous presentation in a form he deemed the most appropriate. These pedagogical tendencies of his mathematical output were probably reinforced by his exposure to Weierstrass' lectures. I believe they were one of the reasons why Frobenius' work proved to be influential.

1.2 Postdoctoral Years: 1870–1874

After obtaining his doctorate, as was the custom, Frobenius took, and passed, the examination required to become a secondary school teacher.[11] He also received an invitation from the University of Freiburg to habilitate there so as to become an instructor. To become an instructor, a further published proof of independent original mathematical work—called a *Habilitationsschrift*—was required. Undoubtedly on the basis of a glowing report from Weierstrass, Freiburg was offering him a generous remuneration and the promise of rapid advancement; but Frobenius declined the offer due to family matters. Instead, he spent a probationary year teaching at the Joachimstalische Gymnasium in Berlin, where he had himself studied, and he proved to be an excellent high-school teacher. An experienced schoolmaster who carefully observed him during that year reported that he possessed an unmistakable

[10] I am grateful to my colleague Dan Weiner for supplying the translations.

[11] The information in this paragraph is drawn from a document published by Biermann [22, p. 191].

inborn pedagogical talent. Having thus passed his probationary year with great success, he was given a regular teaching position at another high school in Berlin (the *Sophienrealschule*).

Despite his teaching duties, Frobenius sought to pursue a career as a mathematician. As a first step, within three months of passing his doctoral examination, he submitted a German language version of his dissertation to *Crelle's Journal* [172]. That it was accepted for publication was certainly due to the influence of Weierstrass, who praised the dissertation for its "many new ideas and results" [22, p. 191]. Still, the dissertation topic was not part of mainstream mathematics; it was something of a mathematical dead end. Frobenius seemed to realize this, for he never returned to the subject. Instead, he turned to a subject that had recently become of considerable interest in Berlin due to the work of Lazarus Fuchs (1833–1902) mentioned above: the application of the new complex function theory to the study of linear differential equations.

Stimulated by the landmark papers of Gauss (1812) and Riemann [495] on the hypergeometric differential equation, Fuchs combined Weierstrassian power series techniques and Weierstrass' theory of algebraic differential equations with the monodromy method introduced by Riemann to study, in groundbreaking papers of 1865–1868, linear homogeneous differential equations.[12] Fuchs had initially published his results in 1865, in the proceedings of the *Gewerbeschule* [236] (later to become the Berlin Technical Institute), where he was teaching. Weierstrass was impressed with Fuchs' results, seeing in them proof of the latter's capability for independent, original mathematical work [22, p. 94]. As a result, in 1866, Fuchs was appointed as an instructor at the university, his 1865 paper [236], with a version published in *Crelle's Journal* in 1866 [237], serving as *Habilitationsschrift*. He remained in that position until 1868 (the year after Frobenius arrived in Berlin), when he left for a professorship at the University of Greifswald. The instructorship vacated by Fuchs was filled by L. Wilhelm Thomé (1841–1910), who had received his doctorate under Weierstrass' direction in 1865.[13] Thomé was regarded an "an indispensable replacement for Fuchs" [22, p. 95], and in 1870, he was made an assistant professor. Thomé was indeed a replacement for Fuchs, since beginning in 1872, he began publishing papers related to Fuchs' theory and its generalization. Frobenius was familiar with the work of Fuchs and Thomé, and it was to Fuchs' theory that he decided to turn for a new direction in research.

Fuchs had studied linear differential equations of the form

$$L(y) = y^{(n)} + q_1(z)y^{(n-1)} + \cdots + q_n(z)y = 0, \tag{1.1}$$

[12] In discussing Fuchs' work, as well as the related work of Frobenius, I have drawn upon Gray's more definitive account [255, Chs. II–III].

[13] Thomé should not be confused with Carl Johannes Thomae (1840–1921), who also worked in complex function theory but had received his doctorate from Göttingen in 1864 and then spent two semesters attending Weierstrass' lectures in Berlin before becoming an instructor in Göttingen in 1866. In 1874 he became a full professor at the University of Freiburg, where he spent the rest of his career.

where $y^{(k)} = d^k y/dz^k$, and the coefficient functions $q_i(z)$ are meromorphic in a simply connected region of the complex plane and have at most a finite number of poles there. Thus the total number of singular points of the coefficients is finite. As Fuchs showed, the singular points of the coefficient functions are the only possible points of singularity of the solutions. One of his main achievements was to characterize those equations (1.1) with the property that in a neighborhood of a singular point $z = a$ of the coefficients, all solutions, when multiplied by $(z-a)^\rho$, for some complex number ρ, remain bounded. These later became known as *linear differential equations of the Fuchsian class*. Fuchs was able to establish a fundamental set of solutions in a neighborhood of such a singular point.

Fuchs made an observation in his paper [237, §6] that seems to have piqued Frobenius' interest in his theory. Fuchs observed that due to earlier work of Puiseux, it followed that the class of linear homogeneous differential equations all of whose solutions are algebraic was contained within the class of differential equations he had studied. Although Fuchs did not mention it, his observation suggested the problem of characterizing those differential equations of the Fuchsian class for which all solutions are algebraic functions. By 1871, Frobenius was thinking about this problem [173, p. 65]. One reason may have been due to work of Kummer's former student H.A. Schwarz (Ph.D., 1864), who since 1869 was a professor at the Zurich Polytechnic (now the Eidgenössische Technische Hochschule Zürich). In August 1871, Schwarz announced that he had solved the problem of determining the hypergeometric differential equations for which all solutions are algebraic functions. The solution involved some beautiful mathematics, and Weierstrass and his circle in Berlin were no doubt discussing Schwarz's work.[14] The hypergeometric equations were special second-order equations of the Fuchsian class.

Schwarz's work may have encouraged Frobenius to think about the analogous but far more formidable problem for nth-order equations of Fuchsian type. I believe this general problem appealed to him because he found it analogous to the problem solved by Galois in his "Memoir on the conditions for solvability of equations by radicals" [239]: characterize those polynomial equations $f(x) = 0$ that can be solved "algebraically," i.e., by means of radicals. The problem implied by Fuchs' paper was to characterize those linear differential equations $L(y) = y^{(n)} + q_1(z)y^{(n-1)} + \cdots + q_n(z)y = 0$ that can be *integrated algebraically* (as Frobenius later put it [177]) in the sense that all solutions are algebraic functions. Such differential equations even resembled polynomials with the kth powers x^k of the unknown being replaced by a kth derivative $y^{(k)}$ of the unknown function y. And the problem of characterizing the algebraically integrable ones seems analogous to the problem solved by Galois, namely to characterize those polynomial equations that are algebraically solvable in the sense of solvable by radicals.

It was natural for Frobenius, an algebraist at heart, but a student of Weierstrass as well, to look to Galois' theory for function-theoretic analogues of what he regarded as the most important elements of Galois work—the theory of the Galois group

[14] For an account of Schwarz's work see [255, pp. 70–77].

associated to a polynomial and the concept of the irreducibility of a polynomial (with respect to a field of known coefficients) [172, p. 65]—in order to deal with the problem of characterizing algebraically integrable equations $L(y) = 0$. Frobenius and—independently and quite differently—Sophus Lie were among the earliest mathematicians to consider how to apply Galois' ideas to differential equations.[15]

The starting point of the theory of groups in Galois' work was his construction of what later became known as a *Galois resolvent* V associated to a polynomial $f(x)$ of degree n with "known" coefficients and no multiple roots. Expressed using modern terminology, the assumption is that $f(x) \in \mathbb{K}[x]$, where $\mathbb{K} \supset \mathbb{Q}$ is the field of "known quantities." Let $a_1, \ldots, a_n$ denote the roots of $f(x)$ and $\mathbb{L} = \mathbb{K}(a_1, \ldots, a_n)$ the associated splitting field. Galois began by sketching a proof that constants $c_1, \ldots, c_n$ can be chosen from $\mathbb{K}$ in various ways, including as integers, such that $V = c_1 a_1 + \cdots + c_n a_n$ takes on $n!$ distinct numerical values when the roots $a_1, \ldots, a_n$ are subjected to all $n!$ possible permutations. Using this property of V, he was able to show that every root of $f(x)$ is a rational function of V with known coefficients, i.e., as we would put it, $\mathbb{L} = \mathbb{K}(V)$. Galois used V to define a set G of permutations of the roots $a_1, \ldots, a_n$ defining what he called the "group of the equation $f(x) = 0$." This group corresponds to the Galois group in the modern sense, although Galois' definition of it was complicated.[16]

Getting back to Frobenius, he realized that Galois' construction of the Galois resolvent V could be extended to the context of the above-mentioned problem suggested by Fuchs' paper. That is, if $L(y) = 0$ is a differential equation of the Fuchsian class with singular point $z = a \in \mathbb{C}$ and if all its solutions are algebraic functions of z, then, Frobenius realized, this means that a polynomial in y,

$$f(y,z) = a_m(z)y^m + \cdots + a_1(z)y + a_0(z), \tag{1.2}$$

of degree $m \geq n$, n the order of $L(y) = 0$, exists with rational functions $a_k(z)$ as coefficients such that in accordance with Weierstrass' theory, the m roots $y_k(z)$ defined locally by power series in $z - a$ and satisfying $f(y_k(z), z) = 0$ in a neighborhood of $z = a$ are all solutions to $L(y) = 0$; furthermore, n of these

---

[15] Regarding Lie, see [276, Ch. 1]. For an accessible exposition of the modern approach to applying Galois' ideas to the theory of differential equations of the Fuchsian class, see [377].

[16] Let $g(x) \in \mathbb{K}[x]$ denote the minimal polynomial of V with $m = \deg g(x)$. Then since $\mathbb{K}(a_1, \ldots, a_n) = \mathbb{K}[V]$, each root a_i of $f(x)$ is uniquely expressible as a polynomial in V, $a_i = \phi_i(V)$, where $\phi(x) \in \mathbb{K}[x]$ has degree at most $m-1$. If $V', V'', \ldots, V^{(m-1)}$ are the other roots of the minimal polynomial $g(x)$, then G consists of m permutations $\sigma_1, \ldots, \sigma_m$ of $a_1, \ldots, a_n$ with σ_k the mapping that takes the root $a_i = \phi_i(V)$ to $\phi_i(V^{(k)})$, which is also a root $a_{i'}$ of $f(x)$, so $\sigma_k : a_i \to a_{i'}$ for $i = 1, \ldots, n$ and $k = 0, \ldots, m-1$. In the nineteenth century, permutations in the sense of mappings of a finite set of symbols were called "substitutions." In the example at hand, σ_k substituted the arrangement (or permutation) $a_{1'}, \ldots, a_{n'}$ for the original arrangement $a_1, \ldots, a_n$. Readers interested in a more detailed and historically accurate portrayal of Galois' ideas, including a detailed working out of Galois' sketchy remarks about the construction and properties of V, should consult Edwards' lucid exposition of Galois' memoir [148], which includes as appendix an annotated English translation of the memoir.

roots form a fundamental set of solutions. This means that every (local) solution to $L(y) = 0$ is a linear combination of these n roots, and so a fortiori, every solution is a linear combination of all m roots $y_k(x)$. At this point, Galois' construction of the Galois resolvent V became relevant, for Frobenius realized that it could be imitated to determine constants $c_1, \dots, c_m$ such that $v(z) = c_1 y_1(z) + \cdots + c_m y_m(z)$ transforms into $m!$ distinct functions under all $m!$ permutations of $y_1(z), \dots, y_m(z)$. From this property he was able to conclude, by analogy with Galois' argument, that every root $y_k(x)$ of $f(x,y)$ is expressible as a rational function of $v(x)$. Clearly $v(x)$ is also a solution to $L(y) = 0$, being a linear combination of solutions. Furthermore, every local solution of $L(y) = 0$ is a linear combination of the $y_k(z)$, and such a linear combination, like the $y_k(x)$ themselves, will be a rational function of v. To sum up, Frobenius had discovered, thanks to Galois' work, the following necessary condition for all solutions to the Fuchsian class equation $L(y) = 0$ to be algebraic functions of z:

Proposition 1.1. *If every solution to $L(y) = 0$ is algebraic, then it is necessary that a solution $y = v(x)$ exist such that every solution to $L(y) = 0$ is a rational function of v.*

Frobenius first stated the above proposition in a paper of 1875 [177, §1], when he had obtained conditions on $L(y)$ under which the converse of the above proposition holds (see below), but given his early familiarity with Galois' work, it seems very likely that he knew Proposition 1.1 already in 1871 but kept it to himself, since it did not amount to much without a converse.

In 1871, unable yet to find a converse to Proposition 1.1, Frobenius wrote up a paper in which he showed how to express Galois' group-theoretic notions in terms of notions from complex function theory. This was no doubt something he had done in the course of seeking to apply Galois' group-based notions and theorems to differential equations. His stated goal in writing up his results for publication was to clothe the "abstract theory" of Galois in the "comfortable geometrical robes of analysis" so that it would gain in comprehensibility and intuitiveness [173, p. 65]. I doubt that Frobenius himself had difficulty with Galois' "abstract" approach; rather he hoped to make Galois' important ideas more accessible to Weierstrass' circle of function theorists. The paper was submitted to *Crelle's Journal* in October 1871 [173], a year after the German version of his dissertation had been submitted to that journal. It was his first paper since his dissertation. He stressed that it did not contain new results but "only seeks to shed light on a known theory from a new viewpoint" [173, p. 65].

Here is a brief summary of Frobenius' viewpoint. Galois' introduction of the group associated to a polynomial equation had been complicated, and the group property, namely, that the permutations constituting the group had the property that the composition of any two is again a member of the group, was not immediately clear. Frobenius realized that the complex analytic setting made possible a more straightforward—and from the Berlin perspective, more natural—definition of the group of an equation from which the group property, as well as other properties, followed readily. Thus he considered a polynomial in two complex variables y

and z like the one in (1.2) associated to his version of the Galois resolvent: $f(y,z) = a_n(z)y^n + a_{n-1}(z)y^{n-1} + \cdots + a_0(z)$, where the coefficients $a_k(z)$ are rational functions of z. (In more familiar terms, $f \in \mathbb{K}[y]$, where $\mathbb{K} = \mathbb{C}(z)$.) To define the analogue of Galois' group associated to a polynomial, he proceeded as follows. Without loss of generality assume that the coefficients $a_k(z)$ are polynomials in z. Also assume that $f(y,z)$ has no quadratic divisors. Say that $z = z_0$ is a singular point of $f(y,z)$ if the polynomial $f(y,z_0)$ has multiple roots or if $f(y,z_0)$ has "an infinite root," by which Frobenius seems to have meant that $a_n(z_0) = 0$.[17]

The permutations that define the analogue of the Galois group of $f(y,z)$ are then defined as follows. Fix a nonsingular point $z = a$. Then by a theorem of Weierstrass, a neighborhood of a exists in which convergent power series in $z - a$ define analytic functions $y_1(z), \ldots, y_n(z)$ satisfying $f(y_i(z),z) = 0$, $i = 1, \ldots, n$, in a neighborhood of a. They are therefore (locally defined) roots of $f(y,z)$. Now consider a closed curve γ that begins and ends at $z = a$ and passes through no singular points of $f(y,z)$. Then if each of $y_1(z), \ldots, y_n(z)$ is analytically continued around γ in the sense of Weierstrass, each function $y_i(z)$ returns to $z = a$ as another power series in $z - a$ defining some root $y_{i'}(z)$ of $f(y,z)$. Furthermore, it followed from Weierstrass' theory that $y_{1'}(z), \ldots, y_{n'}(z)$ is a permutation of $y_1(z), \ldots, y_n(z)$. In other words, the mapping $S_\gamma : y_i(z) \to y_{i'}(z)$ defines a permutation of the roots, as function elements $y_i(z)$, and so as well a permutation of indices $i \to i'$. Thus as γ runs through all such curves γ, a finite set G of distinct permutations $i \to i'$ of $1, \ldots, n$ is generated by the S_γ. The permutations of $G \subset S_n$ are Frobenius' analogue of the Galois group of a polynomial equation. By contrast with Galois' original approach, it is easy to show that G is closed under composition: if S_{γ_1} and S_{γ_2} define elements of G, then so does the composite $S_{\gamma_1} \circ S_{\gamma_2}$, since it is the permutation corresponding to the closed curve $\gamma_2 + \gamma_1$ (curve γ_2 followed by γ_1). It turns out that the permutations $i \to i'$ associated with G are independent of the choice of the nonsingular point a [173, p. 71].

Galois had also introduced the now-familiar notion of adjoining a root of an auxiliary polynomial equation to the field of known constants and had considered the manner in which adjunction might reduce the group of the equation to a subgroup [239, Props. II–III]. Frobenius gave a function-theoretic version of adjunction [173, p. 71] with corresponding reduction properties [173, §§ 5–6]. He ended by saying, "it still remains to establish that this definition [of G] agrees with the one given by Galois" [173, p. 82]. Galois had asserted (albeit without explicit field-theoretic terminology) that given $f(x) \in \mathbb{K}[x]$ without multiple roots and with $\mathbb{L}$ a splitting field of $\mathbb{K}$ with respect to $f(x)$, his group of permutations of the roots $a_1, \ldots, a_m$ of $f(x)$ was characterized by the following two properties: (1) if a rational function of $a_1, \ldots, a_n$ is left invariant under all the permutations of his group, then it is a known quantity, i.e., is in $\mathbb{K}$; (2) any rational function of $a_1, \ldots, a_n$ that is a known quantity is left invariant under all the permutations of his group [239,

[17]Perhaps what he meant is illustrated by the example $f(y,z) = z^n y^n - 1 = 0$. In this case, $y^n = 1/z^n$, and so letting $z \to z_0 = 0$, $y^n \to \infty$. Of course, $f(y,0) = -1$ is a polynomial of degree zero with no roots. What Frobenius meant more generally was perhaps that at singular points $z = z_0$, $a_n(z_0) = 0$.

Prop. I]. In other words, a group of permutations of $a_1,\ldots,a_n$ is the Galois group if and only if it satisfies (1) and (2). This is the origin of the modern definition of the Galois group as Aut $(\mathbb{L},\mathbb{K})$.

With this proposition in mind—but in the special case in which $\mathbb{K}=\mathbb{C}(z)$, the field of rational functions in z—Frobenius said that the agreement of his definition with Galois' was established by his Theorem IX [173, p. 82, IX]: "Every rational function of the roots of an equation that is unchanged by the substitutions of $\ldots[G]\ldots$ can be expressed rationally by means of known quantities, and every rational function of the roots that can be expressed rationally by means of known quantities will not be changed by the substitutions of $\ldots[G]\ldots$." This theorem is expressed in language very similar to that used by Galois, but the meaning is different.[18] In his Theorem IX, "a rational function of the roots of an equation" means $R(y_1(z),\ldots,y_n(z))$, where $R\in\mathbb{C}(u_1,\ldots,u_n)$ and the $y_k(z)$ are defined in some neighborhood $N(a)$ of $z=a$. To say that $R(y_1(z),\ldots,y_n(z))$ "can be expressed rationally by means of known quantities" means that $R(y_1(z),\ldots,y_n(z))$ is equal to a rational function of z in $N(a)$. The converse part means that if $R(y_1(z),\ldots,y_n(z))$ agrees with a rational function of z in $N(a)$, then $R(y_1(z),\ldots,y_n(z)))$ is left unchanged by the permutations of G. Frobenius went no further to justify his claim that his permutation group is consequently identical to that defined by Galois, but he probably had in mind Weierstrass' monodromy theorem, combined with the identity theorem, as the basis for his claims.[19]

This first postdoctoral paper by Frobenius was more in the mathematical mainstream of the times than his dissertation. This is reflected in the fact that his "new viewpoint" was not as novel as he had thought. For the first time—but hardly the last—Frobenius was not the only mathematician to have a similar idea. Unbeknownst to him, this mode of definition had already been used to define what was called the monodromy group of $f(y,z)$ in the post-Galois French literature. Most notably, Jordan in his 1870 *Traité des substitutions* [322, pp. 277–278]—a work whose contents were still evidently unfamiliar to Frobenius in October 1871—defined the monodromy group as follows (with $F(x,k)$ playing the role of $f(y,z)$ above):

---

[18] In what follows, I focus on the group G and omit the subgroups G^* that result by adjunction of function elements, although Theorem IX applies more generally to G^*.

[19] Assume that $f(y,z)\in\mathbb{K}[y]$, $\mathbb{K}=\mathbb{C}(z)$, is irreducible and that $a_1,\ldots,a_m$ are the singular points a, meaning the points at which $f(y,a)$ has a multiple root or has degree in y less than n, i.e., $a_n(a)=0$. By the identity theorem, the singular points are finite in number, say $a_1,\ldots,a_m$. Let Γ denote a non-self-intersecting polygonal line joining $a_1,\ldots,a_m,\infty$, and set $D=\mathbb{C}\sim\Gamma$. Then D is an open, connected, and simply connected set, and by the Weierstrass monodromy theorem each locally defined root $y_j(z)$ has an extension $Y_j(z)$ to D that is single-valued and analytic. (See, e.g., [348, pp. 126–127].) The identity theorem implies that $f(z,Y_j(z))=0$ throughout D. That same theorem implies that Frobenius' Theorem IX holds with the $y_j(z)$ replaced by the $Y_j(z)$. Thus if $\mathbb{L}$ is the field of all meromorphic functions defined on D and expressible rationally in terms of $Y_1,\ldots,Y_n$, then $\mathbb{L}\supset\mathbb{K}=\mathbb{C}(z)$ and $\mathbb{L}=\mathbb{K}(Y_1,\ldots,Y_n)$ is a splitting field for $f(y,z)$ over $\mathbb{K}$. Furthermore, by Frobenius' Theorem IX (as extended to $Y_1,\ldots,Y_n$), his group G can be identified with Aut$(\mathbb{L},\mathbb{K})$. Thus G is a bona fide Galois group.

> Having given an initial value to k ... suppose it is made to vary according to some law. The roots $x_1, x_2, \ldots$ of the equation will vary continually with k, and if, at the end of the operation, k takes again its initial value k_0, the final values of $x_1, x_2, \ldots$ will satisfy the equation $F(x,k) = 0$: except for their order, they are the same as the initial values. Thus ... the result of the operation will be represented by a certain permutation[20] S of its roots.
>
> If the law of variation of k is modified in all possible manners, one will obtain diverse permutations $S, S_1, \ldots$, which evidently form a group H: if in making k vary in a certain manner the permutation S is obtained, then the permutation S_1 is obtained by making k vary in another manner; the permutation SS_1 will be obtained by submitting it successively to these two modes of variation.

The group H was called the monodromy group. Jordan assumed that $F(x,k) \in \mathbb{K}[y]$, where $\mathbb{K} = \mathbb{Q}(k)$ (rather than $\mathbb{K} = \mathbb{C}(k)$, as with Frobenius). However, the idea is a nonrigorous version of what Frobenius did using Weierstrass' theory. The one notable difference was the different choice of $\mathbb{K}$. As a result, H is a subgroup of the Galois group G for Jordan. Jordan illustrated this with the example $F(x,k) = x^2 - 2k^2$, which has roots $r_1 = \sqrt{2}k$ and $r_2 = -\sqrt{2}k$, which remain unchanged about closed curves, so that H consists solely of the identity permutation. On the other hand, since $\pm\sqrt{2} \notin \mathbb{Q}$, it follows that $G = \mathrm{Aut}(\mathbb{L}, \mathbb{K})$, $\mathbb{L} = \mathbb{K}(\sqrt{2})$, has order two. Jordan showed that in general, H is a normal subgroup of G [322, pp. 278–279], and his proof implies that when $\mathbb{K}$ is taken as $\mathbb{C}(k)$, as in Frobenius' paper, then H is the Galois group of the equation.

In defining the group of the equation $f(x,y) = 0$, Frobenius was thus simply, but unwittingly, giving a much more rigorous presentation, based upon the techniques of Weierstrassian complex function theory—including the theory of analytic continuation—of the monodromy group of the equation, which in his setting ($\mathbb{K} = \mathbb{C}(z)$) coincided with the Galois group of the equation. He did show that for this particular type of Galois group, many of its properties could be derived in ways more intelligible to function theorists than those of Galois' far more general (and "abstract") approach, where $\mathbb{K}$ could be, in effect, any known field of characteristic zero. He had also digested Cauchy's important memoir of 1844 on transitive and multiply transitive groups of permutations [77] and proved several theorems using these notions about the Galois group of the equation [173, §4].

Frobenius' paper is of historical interest because it reveals his early interest in algebra, group theory in particular, and its application within other areas of interest—in this case complex function theory. As we shall see, Frobenius was at heart an algebraist, and a hallmark of his mathematical modus operandi was to perceive problems of an essentially algebraic nature within diverse mathematical contexts. The above discussed paper was, as he himself said, a by-product of his efforts to bring Galois' seminal ideas to bear on Fuchs' theory of linear homogeneous differential equations, presumably as in Proposition 1.1 above.

---

[20] The French word is *substitution*, which was used in the nineteenth century for permutations in the modern sense of mappings, as indicated in the earlier footnote on Galois' definition of the group associated to a polynomial equation.

By the end of July 1872, 2 years after he had obtained his doctorate, Frobenius had published just two papers, both in *Crelle's Journal*: the German version of his doctoral dissertation and the above paper on Galois groups as monodromy groups. It is doubtful that the paper on Galois groups was a sufficient display of original independent work to serve as a *Habilitationsschrift*. Like his doctoral dissertation, it displayed a great mastery of the mathematical material at hand and led to interesting results. In the case of the dissertation, the results were, however, not connected with other theories in a fruitful way. In the case of the Galois group paper, the results were admittedly already known in a more general form, and the method of deriving them was not really as novel as Frobenius had imagined. Certainly the paper does not compare favorably with Fuchs' *Habilitationsschrift*—his 1866 paper on linear homogeneous differential equations.

Nonetheless, Weierstrass had great faith in Frobenius' promise as a mathematician and great respect for his capabilities as a teacher. This is documented in a proposal dated 22 June 1872 to the minister of culture, drawn up by Weierstrass and signed by many members of the Philosophical Faculty, to create a new assistant professorship in mathematics, due to the steadily increasing number of students wishing to study mathematics at the university, and to appoint Frobenius to fill it—even though he was not yet habilitated, i.e., was not yet an instructor at the university, and was only 24 years old.[21]

In the proposal, Weierstrass described Frobenius' two published papers as reflecting favorably on his "capability for independent research," and he also noted with approval that these publications were characterized by an admirable "clarity and skillfulness" of presentation. To establish Frobenius' teaching ability, Weierstrass referred to the favorable report by the experienced teacher who had observed him during his trial year at the Joachimstalische Gymnasium and had reported (according to Weierstrass) that Frobenius possessed "an unmistakable inborn didactic talent" and had a characteristic way of expressing ideas that enabled him "to stimulate an entire class and keep them in suspense." Because Weierstrass and his colleagues were recommending for a professorship "a very young man" who had never been a university instructor, Weierstrass felt it necessary to say more about their candidate. After a careful survey of all the young instructors who were in a position to accept a call to Berlin at a reasonable salary, Weierstrass explained, they found no one about whom they could say with the same degree of conviction as with Frobenius "that with the objective we have presently in mind, he is the most suitable person. In talent he towers above them all ...; and in terms of basic knowledge and solid education he takes second place to no one." Weierstrass admitted that Frobenius was not as well known by virtue of his (limited) publications as some of the others, but he discounted this fact on two grounds: it had been only 2 years since he had obtained his doctorate and, more importantly, his restraint in publishing was a point in his favor in the opinion of Weierstrass and others who viewed with regret the

---

[21]The document is transcribed in full by Biermann [22, pp. 189–192]; see also his discussion of it [22, pp. 95–96].

uncontrollable excess of publication that was taking place in mathematics. Quality, not quantity, Weierstrass implied, was what counted, and he evidently believed that, given some time, Frobenius would establish himself as a first-rate mathematician.

On 27 March 1874 Weierstrass and his colleagues were granted permission to offer a new assistant professorship to Frobenius. By that time, he was well on his way to justifying Weierstrass' optimistic view of his potential as a mathematician. Let us now consider what he had achieved since his 1872 paper on Galois groups as monodromy groups. It was in the opening lines of that paper that Frobenius had expressed the view that the two key elements of Galois' theory were the group concept and the concept of polynomial irreducibility. In the paper itself he had developed a function-theoretic analogue of the Galois group, and at the same time, as I suggested above, he had used an analogue of the Galois resolvent construction to obtain the unpublished Proposition 1.1. Six months later, he submitted a paper to *Crelle's Journal* proposing a suitable notion of irreducibility for linear differential equations, with the resulting properties developed for equations of the Fuchsian class [175], but which could also be suitably extended to more general linear homogeneous differential equations.

At that time (1873), most mathematicians working in complex function theory and its application to differential equations would probably have thought of any polynomial of degree greater than one as reducible in the sense that it can be factored into a product of linear factors with complex coefficients. Frobenius, however, had read Galois for whom the idea of reducibility was predicated on the assumption of factoring polynomials using only known coefficients. Thus in Galois' theory, and again using modern terminology, $f(x) \in \mathbb{K}[x]$ ($\mathbb{K} \subset \mathbb{C}$) is reducible if $f(x) = g(x)h(x)$, with $g, h \in \mathbb{K}[x]$ having degrees strictly between 0 and $n = \deg f$. Otherwise, it is irreducible. Galois' characterization of $f(x)$ being irreducible can be seen to be equivalent to the following: $(*)$ no polynomial $g(x) \in \mathbb{K}[x]$ exists with $\deg g < \deg f$ and $g(a) = f(a) = 0$ for some $a \in \mathbb{C}$.[22] Frobenius realized this equivalence, and, more importantly, saw that it led to a viable analogue for differential equations: an equation $L(y) = 0$ of order n and of the Fuchsian class is said to be *irreducible* if there is no such equation $M(y) = 0$ of order less than n that has a solution in common with $L(y) = 0$. Otherwise, $L(y) = 0$ is said to be *reducible* [175, pp. 107–108].

Frobenius' paper on irreducible differential equations was his first publication introducing a significant and entirely new concept, and in his paper he developed its implications with a thoroughness and clarity that reflected his considerable mathematical and pedagogical talents and that was to become a hallmark of his work. Starting with the proposition that if $L(y) = 0$ has a solution in common with

---

[22]Suppose f is irreducible over $\mathbb{K}$ in the usual sense and that $(*)$ fails to hold. Then $g \in \mathbb{K}[x]$ exists with $\deg g < \deg f$ and $g(a) = f(a) = 0$. But $(f,g)_{\mathbb{K}} = 1$, so $p, q \in \mathbb{K}[x]$ exist such that $p(x)f(x) + q(x)g(x) = 1$, and setting $x = a$ implies $0 = 1$. Conversely, if f satisfies $(*)$, it cannot be reducible, for then $f(x) = g(x)h(x)$, where g, h have degrees less than n. If $a \in \mathbb{C}$ is a root of f, then $0 = f(a) = g(a)h(a)$ implies without loss of generality that $g(a) = 0$, contrary to $(*)$.

an irreducible equation $M(y) = 0$, then all solutions to $M(y) = 0$ are solutions to $L(y) = 0$ [175, §3, III], he established a succession of propositions culminating in the fact that if $L(y) = 0$ is reducible, so that it has a solution in common with an irreducible equation $M(y) = 0$ of lesser order, then $N(y) = 0$ exists such that $L(y) = N(M(y))$ and so, with operator composition viewed as multiplication, $L(y)$ does contain $M(y)$ as a factor [175, p. 128]. In addition to these results, he determined necessary and sufficient conditions on both Gauss' and Riemann's hypergeometric differential equations for them to be irreducible [175, §5]. These results are described in some detail by Gray [255, pp. 57–59], who also describes how Frobenius used his theory of irreducibility "to clarify three aspects of the theory of linear differential equations: the behavior of solutions under analytic continuation; the nature of accidental singular points; the occurrence of solutions with essential singularities when the equation is not of the Fuchsian type" [255, p. 59]. This last area in which Frobenius utilized his theory of irreducibility was motivated by Thomé's papers (1873–1874) on the properties of equations not of Fuchsian type; Frobenius' paper was published in 1875 [178].

Incidentally, in this paper [178, p. 243] Frobenius also presented what he called a "reciprocity theorem," which, he explained, he and Thomé had discovered independently and applied in 1873 ([175, p. 133], [562, p. 277]) but which Frobenius now (in 1875) expressed in the following elegant form: If A and B are linear differential operators and A^* and B^* their respective adjoints, then the adjoint of the composite AB is B^*A^*. This result has become known as the Frobenius reciprocity theorem. Only the elegant form of statement was uniquely due to Frobenius, but the fact that his name alone is attached to the theorem is a reflection of the clarity, rigor, and elegance of his manner of mathematical expression, a quality not shared by the prolix Thomé.[23]

As will be seen in Section 16.3, many years later, Frobenius' paper on irreducible equations was to give rise to a line of development in the theory of linear differential operators that together with Frobenius' rational theory of elementary divisors (Sections 8.6 and 16.2) was to lead, through the work of Alfred Loewy and his student Wolfgang Krull in the twentieth century, to the module-theoretic approach to the theory of canonical matrix forms that one finds in van der Waerden's *Moderne Algebra* (1931) and in many present-day texts.

So far, I have considered that part of Frobenius' work relating to Fuchs' theory of differential equations that was guided by the notions he took from Galois' theory of polynomial equations. In the course of familiarizing himself with the details of Fuchs' work, however, Frobenius also discovered in the early months of 1873 that Fuchs' theory itself could be developed in a much simpler manner. To explain this, it is first necessary to say something more about Fuchs' theory. As mentioned earlier, Fuchs had been able to characterize those homogeneous linear differential equations (1.1) with the property that in a neighborhood of a singular point $x = a \in \mathbb{C}$ of the

[23]Frobenius submitted his paper [175] slightly earlier than Thomé submitted his [562]—24 April 1873 versus 7 May 1873—but I doubt this is the reason the theorem bears Frobenius' name alone.

coefficients, all solutions, when multiplied by a suitable power of $(x-a)$, remain bounded. (These are the linear differential equations of the Fuchsian class.) One of Fuchs' major achievements was to determine a fundamental set of solutions in a neighborhood of such a singular point.

When the singular point is $a=0$, these equations can be expressed locally in the form

$$L(y) = x^n y^{(n)} + x^{n-1} p_1(x) y^{(n-1)} + \cdots + p_n(x) y = 0, \tag{1.3}$$

where x is a complex variable and the functions $p_i(x)$ are analytic at $x=0$, i.e., analytic in a neighborhood of $x=0$ [238, p. 360]. Fuchs' method of determining a fundamental set of solutions to (1.3) in a neighborhood of $x=0$ was Riemann's monodromy method that Fuchs had developed in the context of the more general equations (1.1). If $y_1,\ldots,y_n$ is a fundamental set of solutions to (1.3) in a neighborhood of $a=0$, and if these solutions are analytically continued around all the singular points, the result is another fundamental set of solutions $\tilde{y}_1,\ldots,\tilde{y}_n$, related to the first by

$$\tilde{y}_i = \sum_{j=1}^{n} \alpha_{ij} y_j, \tag{1.4}$$

where the α_{ij} are constants with $\det(\alpha_{ij}) \neq 0$ [237, §3]. Central to Fuchs' analysis was what he called the "fundamental equation," namely

$$F(\omega) = \det(A - \omega I) = 0, \quad A = (\alpha_{ij}). \tag{1.5}$$

Fuchs expressed a root ω of this equation in the form $\omega = \mathrm{e}^{2\pi i\rho}$. He called ρ an *index* associated to ω. Clearly there are infinitely many indices associated to ω, but any two differ by some integer.

Fuchs realized that a linear change of fundamental set, $y^* = Ky$ in modern notation, where $y = \begin{pmatrix} y_1 & \cdots & y_n \end{pmatrix}^t$, corresponds to the initial fundamental set, $K = (k_{ij})$, and so on, changes A into $A' = K^{-1}AK$ [237, p. 133]. This was a few years before the introduction of the canonical forms of Weierstrass (Section 5.4) and Jordan (Section 5.5), and Fuchs contented himself with the observation that if $\omega_1,\ldots,\omega_m$ were the distinct roots of the fundamental equation, with ω_j having multiplicity μ_j, then K could be chosen such that $A' = K^{-1}AK = B_1 \oplus \cdots \oplus B_m$, where B_j is a $\mu_j \times \mu_j$ lower triangular matrix with ω_j down the diagonal. The associated fundamental set $y_1^*,\ldots,y_n^*$ thus included, corresponding to a root ω of multiplicity μ, μ solutions $y_1^*,\ldots,y_\mu^*$, so that $\tilde{y}_1^* = \omega y_1^*$ and, for $\alpha > 1$, $\tilde{y}_\alpha^* = \omega y_\alpha^*$ plus a linear combination of $y_1^*,\ldots,y_{\alpha-1}^*$. Consider now the differential equation (1.3). A function with the property that $\tilde{y}_1^* = \omega y_1^*$ is easily seen to be $y_1^* = x^\rho \varphi(x)$, where $\mathrm{e}^{2\pi i\rho} = \omega$ and $\varphi(x)$ is analytic at $x=0$. Likewise, a function with the property that for $\alpha > 1$, $\tilde{y}_\alpha^* = \omega y_\alpha^*$ plus a linear combination of $y_1^*,\ldots,y_{\alpha-1}^*$ is $y_\alpha^* = x^\rho [b_{\alpha,1} + b_{\alpha,2} \log x + \cdots + b_{\alpha,\alpha-1} (\log x)^{\alpha-1}]$. Fuchs showed that corresponding to each distinct root ω of the fundamental equations, solutions of the above sort actually exist, and that the totality of such solutions corresponding to the m distinct roots ω form a fundamental set of solutions to (1.3).

Fuchs also showed how to determine indices ρ from the coefficients of the differential equation (1.3). In this connection he observed that $L[x^\rho \varphi] = x^\rho \tilde{L}[\varphi]$, where

$$\tilde{L}[\varphi] = x^n \varphi^{(n)}(x) + x^{n-1}\tilde{p}_1(x)\varphi^{(n-1)}(x) + \cdots + \tilde{p}_n(x)\varphi(x), \tag{1.6}$$

and the coefficients $\tilde{p}_j(x)$, which are expressible in terms of $p_1,\ldots,p_n$, are consequently also analytic at $x = 0$. Thus $y = x^\rho \varphi$ is a solution to $L(y) = 0$ if and only if $\tilde{L}[\varphi] = 0$. Now $\tilde{L}[\varphi]$ as given by (1.6) is analytic in a neighborhood of $x = 0$, and its value at $x = 0$ is easily seen from (1.6) to be $\tilde{p}_n(0)\varphi(0)$. Thus in order that $\tilde{L}[\varphi] = 0$, it is necessary that $\tilde{p}_n(0) = 0$ (since $\varphi(0) \neq 0$). If $\tilde{p}_n(x)$ is expressed in terms of $p_1,\ldots,p_n$, the condition $\tilde{p}(0) = 0$ becomes, if we take $p_0(x) \equiv 1$,

$$I(\rho) \stackrel{\text{def}}{=} p_n(0) + \sum_{k=1}^{n} \rho(\rho-1)\cdots(\rho-k+1)p_{n-k}(0) = 0. \tag{1.7}$$

Fuchs called $I(\rho) = 0$, which is a polynomial equation in ρ of degree n, the "fundamental determining equation," but it has become known (thanks to Cayley) as the *indicial equation*. Thus if a solution to (1.3) of the above form $y = x^\rho \varphi(x)$ exists, ρ must be a root of the indicial equation.

Fuchs divided the roots of the indicial equation into groups. The groups form what would now be called equivalence classes under the relation $\rho \sim \rho'$ if $\rho - \rho'$ is an integer. In other words, $\rho \sim \rho'$ precisely when $\omega = e^{2\pi i\rho}$ and $\omega' = e^{2\pi i\rho'}$ are the same root of the fundamental equation (1.5). The roots in a given equivalence class may be ordered as $\rho_0,\ldots,\rho_\mu$ so that the real parts decrease as the index of the root increases from 0 to μ, i.e., for $\alpha < \beta$, $\rho_\alpha - \rho_\beta$ is a nonnegative integer. Fuchs began by showing that a solution of the form $y = z^{\rho_0}\varphi(x)$ exists in a neighborhood of $x = 0$, where $\varphi(x) = \sum_{\nu=0}^{\infty} g_\nu x^\nu$. He did this using a recurrence relation of the form $g_{k+1} = A_1 g_1 + \cdots + A_k g_k$ to introduce a second nth-order linear differential equation with a convergent solution of the form $y = \sum_{k=0}^{\infty} c_k z^k$ and such that $|g_k| \leq |c_k|$, thereby establishing the convergence of the series defining the solution $y_0 = x^{\rho_0}\varphi(x)$ to (1.3) [237, pp. 148–152]. To obtain the further solutions to (1.3) containing logarithms, Fuchs went through the lengthy reduction procedure introduced by Weierstrass in his theory of algebraic differential equations. It involved consideration of a succession of differential equations of increasingly lower orders to obtain the various logarithmic solutions that went along with $y_0 = x^{\rho_0}\varphi(x)$.

It was at this point in Fuchs' theory that Frobenius asked whether, since Weierstrass' method was devised to deal with all algebraic differential equations, a simpler method might not be found for the special case of the linear differential equations (1.3) treated by Fuchs [174, p. 85]. He found such a method and submitted a paper containing it to *Crelle's Journal* in April 1873 [174], a week before submitting his paper on irreducible equations [175].

Frobenius' idea was to work directly with Fuchs' equation (1.3) and, if $L(y) = 0$ denotes it, to consider what L does to a function of two complex variables x and ρ of the form

$$g(x,\rho) = x^\rho \left[\sum_{\nu=0}^{\infty} g_\nu x^\nu\right] = \sum_{\nu=0}^{\infty} g_\nu x^{\rho+\nu}, \tag{1.8}$$

where the g_ν are constants and $g_0 \neq 0$. Now, $L[g(x,\rho)] = \sum_\nu g_\nu L[x^{\rho+\nu}]$ and $L[x^\rho] = x^\rho \sum_{\nu=0}^{\infty} f(x,\rho)$, where (again with $p_0(x) \equiv 1$)

$$f(x,\rho) = p_n(x) + \sum_{k=1}^{n} \rho(\rho-1)\cdots(\rho-k+1)p_{n-k}(x). \tag{1.9}$$

It then follows readily from the form of the differential equation (1.3) that

$$L[g(x,\rho)] = \sum_{\nu=0}^{\infty} g_\nu f(x,\rho+\nu). \tag{1.10}$$

Since the functions $p_1,\ldots,p_n$ are analytic at $x = 0$, the same is true of $f(x,\rho)$ by virtue of (1.9), and so we may express it as a convergent power series

$$f(x,\rho) = \sum_{\mu=0}^{\infty} f_\mu(\rho)x^\mu; \tag{1.11}$$

and (1.9) shows that the coefficients $f_\mu(\rho)$ are polynomials in ρ of degree at most n. By substituting (1.11) in (1.10) and rearranging, we have

$$L[g(x,\rho)] = \sum_{\nu=0}^{\infty} [g_\nu f_0(\rho+\nu) + g_{\nu-1} f_1(\rho+\nu-1) + \cdots + g_0 f_\nu(\rho)].$$

As a first conclusion, Frobenius could thus observe from this expression that if $g(x,\rho)$ converges and is a solution to (1.3), then necessarily the following recurrence formulas hold:

$$\begin{aligned} 0 &= g_0 f_0(\rho), \\ 0 &= g_1 f_0(\rho+\nu) + g_0 f_1(\rho), \\ &\cdots \\ 0 &= g_\nu f_0(\rho+\nu) + g_{\nu-1} f_1(\rho+\nu-1) + \cdots + g_0 f_\nu(\rho). \end{aligned} \tag{1.12}$$

From the first of these equations, since $g_0 \neq 0$, it followed that ρ must be a root of $f_0(\rho) = 0$, and since the identity $f_0(\rho) = f(0,\rho)$ follows from (1.11), equation (1.9) shows that $f_0(\rho) = 0$ is precisely Fuchs' indicial equation (1.7). Thus for $g(x,\rho)$ to be a solution, ρ must be a root of the indicial equation.

Frobenius now introduced a new idea [174, pp. 87ff.]. Rather then think of the g_k as constants that depend on the choice of the root ρ_k, think of them as functions of the complex variable ρ. If, for the moment, we consider $g_0(\rho)$ arbitrary, then the recurrence relations (1.12) determine the functions $g_k(\rho)$ recursively in terms of $g_0(\rho)$ and the functions $f_\nu(\rho)$. Frobenius chose $g_0(\rho)$ in such a way that if $g(x,\rho)$ as defined by (1.8) is replaced $g(x,\rho)$ as defined by

$$g(x,\rho) = \sum_{\nu=0}^{\infty} g_\nu(\rho) z^{\rho+\nu}, \tag{1.13}$$

then the recurrence relations may be used to show that the series in (1.13) converges for $(x,\rho) \in D \times U$, where D is a disk $|z| < R$ and U is a bounded open subset of complex numbers ρ containing all the roots of the indicial equation. Moreover, the convergence is uniform on closed subsets of $D \times U$. From this it followed that

$$L[g(x,\rho)] = f_0(\rho) g_0(\rho) x^\rho, \tag{1.14}$$

which shows first of all that for every root ρ_k of the indicial equation, $g(x,\rho_k)$ is a solution.

Actually, the uniform convergence in (x,ρ) in (1.13) showed much more. Let $\rho_0, \rho_1, \ldots, \rho_\mu$ denote one of Fuchs' groups in his ordering such that $\alpha < \beta$ means that $\rho_\rho - g_\beta$ is a nonnegative integer. Some of these roots may be equal (as Fuchs realized). Frobenius therefore denoted the distinct roots in the group by $\rho_0, \rho_\alpha, \rho_\beta, \rho_\gamma, \ldots$. Thus $\rho_0 = \rho_1 = \cdots = \rho_{\alpha-1}$, and ρ_0 has multiplicity α as a root of the indicial equation $f_0(\rho)$. Likewise, ρ_α is a root of $f_0(\rho)$ of multiplicity $\beta - \alpha$, ρ_β a root of multiplicity $\gamma - \beta$, and so on. However, $g_0(\rho)$ had been chosen such that $f_0(\rho) g_0(\rho)$ has ρ_0 as a zero of multiplicity α, ρ_α as a zero of multiplicity β, ρ_β as a zero of multiplicity γ, and so on [174, p. 91 (8)]. From this it followed that if ρ_κ is any one of the μ roots in the group, it is a zero of $f_0(\rho) g_0(\rho) x^\rho$ of multiplicity at least $\kappa + 1$, and so $\partial^\kappa [f_0(\rho) g_0(\rho) x^\rho] / \partial \rho^\kappa = 0$. This means that if

$$g^{(\kappa)}(x,\rho) = \frac{\partial^\kappa g(x,\rho)}{\partial \rho^\kappa},$$

then $g^{(\kappa)}(x,\rho_k)$ is a solution to (1.3), since, thanks to the uniform convergence in x and ρ, differentiation with respect to x and ρ may be interchanged, and so for $\rho = \rho_\kappa$,

$$L[g^{(\kappa)}(x,\rho)] = \frac{\partial^\kappa}{\partial \rho^\kappa} \{L[g(x,\rho)]\} = \frac{\partial^\kappa \{f_0(\rho) g_0(\rho) x^\rho\}}{\partial \rho^\kappa} = 0.$$

Since the series (1.13) defining $g(x,\rho)$ may be differentiated with respect to ρ term by term, the result of doing this κ times is (by Leibniz's rule for the κ-fold derivative of a product)

$$g^{(\kappa)}(x,\rho_\kappa) = x^{\rho_\kappa} \sum_{\nu=1}^{\infty} \left[\sum_{j=0}^{\kappa} \binom{\kappa}{j} (\log x)^j g_\nu^{(\kappa-j)}(\rho_\kappa) \right] x^\nu. \tag{1.15}$$

In this manner Frobenius obtained Fuchs' solutions containing logarithms without the need to consider monodromy, the resulting fundamental equation (1.5), the related quasicanonical form $A' = K^{-1}AK = B_1 \oplus \cdots \oplus B_m$, and the succession of lower-order differential equations required by Weierstrass' method for algebraic differential equations.

In effect, Frobenius had shown that Fuchs' original approach, via monodromy considerations, could be dispensed with. Frobenius thereby infused Fuchs' theory with a simplicity and clarity that it had originally lacked. This is no more clearly seen than in Fuchs' efforts to distinguish those differential equations $L(y) = 0$ of the form (1.3) that have logarithms in their solutions from those that do not [238, pp. 373–378]. In 1826, Cauchy had already made a study of the differential equations (1.3) in the special case in which all the coefficients $p_i(x)$ are constant: $p_i(x) = p_i(0)$ [70]. Let $L_0(y) = 0$ denote the differential equation (1.3) in this special case. Since the indicial equation (1.7) of $L(y) = 0$ depends only on the values of the $p_i(x)$ at $x = 0$, $L_0(y) = 0$ has the same indicial equation. Cauchy's solution implied that there are no logarithms in the solutions to $L_0(y) = 0$ precisely when the indicial equation has no multiple roots. Fuchs' idea was to extend this result to the differential equations $L(y) = 0$ of (1.3). This required developing a succession of complicated theorems [238, I–V, pp. 371–375] leading to complicated conditions [238, I–II, pp. 376–378] under which $L(y) = 0$ had no logarithms in the solutions associated to a group of roots $\rho_1, \ldots, \rho_\mu$ if and only if this was true for $L_0(y) = 0$ [238, pp. 371–378]. The upshot of this rather tortuous line of reasoning was the theorem that the differential equation $L(y) = 0$ of (1.3) has no logarithms in its solutions if and only if the associated indicial equation has no multiple roots. In slightly more than a single page, Frobenius proved this theorem using the properties of his function $g(x,\rho)$ and the formulas (1.15) [174, §4].

As Gray has pointed out, "Frobenius was quite scrupulous in acknowledging Fuchs' work, but as his simpler methods drove out those of Fuchs, the comparison became blurred until he is sometimes remembered more for the results than the methods ..." [255, p. 57]. As in his doctoral thesis, here again we see Frobenius' facility for elegant calculations on display, albeit here applied to a theory of general interest to mathematicians. Here he showed how he could quickly master a theory and devise a clearer and simpler mathematical rendition of it. No doubt Weierstrass was duly impressed by the same "clarity and skillfulness" of presentation that he had already noted in Frobenius' first two publications (when proposing him for an assistant professorship in 1872).

By the beginning of 1875, Frobenius had finally succeeded in establishing a partial converse to Proposition 1.1—a sufficient condition for an equation of Fuchsian type $L(y) = 0$ to be algebraically integrable [176]. Recall that he knew that a necessary condition for $L(y) = 0$ to be algebraically integrable was that all solutions can be expressed as rational functions of a single solution. To explore the extent to which this rationality condition might also be sufficient, he assumed that a transcendental solution to $L(y) = 0$ exists under the rationality hypotheses. He showed that the rationality and transcendental hypotheses together imply that $L(y) = 0$ has a transcendental solution that (by considering several distinct cases) is also a solution to an equation of order at most 2. This does lead to a contradiction if it is assumed that $L(y) = 0$ is irreducible of order greater than 2, for then by definition, it cannot share a solution with an equation of lower order. In this way, Frobenius obtained the following partial converse to Proposition 1.1: *Suppose that* $L(y) = 0$ *is irreducible and of order greater than two. Then it is algebraically integrable if and only if every solution is a rational function of a single solution.*

The same year that Frobenius obtained the above general theorem, Fuchs independently published a paper determining all second-order equations of the Fuchsian class that are algebraically integrable. Fuchs' work, which was not wholly satisfactory, was then taken up and perfected by Felix Klein and Paul Gordan, so that all algebraically integrable equations of order two could be precisely tabulated.[24] This work lacked the generality of Frobenius' theorem, but it motivated Camille Jordan to develop methods for dealing with nth-order equations. As in the case $n = 2$, an algebraically integrable nth-order equation has the property that its associated group of Fuchs' monodromy transformations (1.4), which is a subgroup of $\mathrm{GL}(n, \mathbb{C})$, must be finite. The focus then shifted to group theory—the determination of all finite subgroups of $\mathrm{GL}(n, \mathbb{C})$. In 1878, Jordan completely solved this problem for $n = 2, 3$ and proved a general theorem about the structure of finite subgroups of $\mathrm{GL}(n, \mathbb{C})$; but as Gray remarked after describing Frobenius' contribution, "not much else was done with differential equations of order greater than 2 all of whose solutions are algebraic, for Jordan's work indicated how technically complicated it could become" [255, p. 96].[25]

Frobenius began his university teaching career with the 1874 summer semester. He gave lectures on determinants, integral calculus, and synthetic geometry.[26] He actually remained in the position for only one semester. He had become engaged to Auguste Lehmann, and then he was offered a full professorship at the Zurich Polytechnic Institute, which later became the Federal Technical Institute in Zurich (i.e., *Eidgenössische Technische Hochschule (ETH), Zürich*). The full professorship made marriage financially possible, and he decided to accept the position. Weierstrass was dismayed by this turn of events. No doubt frustrated that Frobenius was giving up the position that he had worked so hard to create for his promising student, a position that would have kept him at one of the most vibrant centers for mathematical research, he deemed Frobenius' fiancee "thoroughly unsuitable"

[24] See Table 3.2 in Gray's book [255, p. 87].

[25] Chapter III of Gray's book [255] is devoted to all the work done on Fuchs-type equations that can be integrated algebraically.

[26] Stenographic notes of these lectures were reproduced by the university. Copies are located in the Bibliothek Mathematik und Geschichte der Naturwissenschaften at the University of Hamburg.

for him and blamed her for persuading him to give up his position in Berlin.[27] In October 1875, Frobenius assumed his new position at the Zurich Polytechnic, and the following year, he married Auguste Lehmann. They had one child, a son Otto, shortly thereafter. There is no evidence at hand to judge Weierstrass' opinion about the suitability of Frobenius' bride, but as the following chapter shows, Weierstrass' negative opinion of Frobenius' move to Zurich, although quite understandable from his own perspective, turned out to be off the mark. Although Frobenius did indeed miss the Berlin mathematical environment, his separation from it seems to have encouraged him to turn to problems outside the purview of the Berlin mathematicians.

[27] Weierstrass' negative opinion of Frobenius' fiancée and of his decision to leave Berlin for Zurich is contained in a letter to Sonya Kovalevskaya dated 23 September 1875 [28, p. 219].

Chapter 2
Professor at the Zurich Polytechnic: 1874–1892

By the time the 26-year-old Frobenius arrived at the Zurich Polytechnic Institute, a tradition had already been established whereby a professorship there served as a springboard for promising young German mathematicians en route to a professorship back in Germany.[1] This tradition was unwittingly initiated by Richard Dedekind (1831–1916) in 1858, not long after the polytechnic was founded. As we shall see in Chapters 8 and 9 and in Chapters 12 and 13, Dedekind ranks with Frobenius' teachers, Weierstrass and Kronecker, as having had a major influence on the directions taken by his mathematics. Some more information about Dedekind is thus in order.

Dedekind had received his doctorate from the University of Göttingen in 1854. Gauss had been his thesis director, but the mathematicians who most influenced Dedekind were Dirichlet and Riemann. Dedekind was an instructor at Göttingen in 1855, when Dirichlet left Berlin for Göttingen to become Gauss' successor. Dedekind had come to Göttingen in 1850 as a 19-year-old.[2] There he met another student, Bernhard Riemann (1826–1866), with whom he became close friends. Riemann had begun his studies at Göttingen in 1846, but when he decided to concentrate on mathematics, he left Göttingen for Berlin, where he made the acquaintance of Jacobi and Dirichlet. At Göttingen, Gauss, who disliked teaching, did not teach any advanced mathematics courses. Riemann returned to Göttingen in 1849, after the intellectual environment there had improved somewhat due to the return of the physicist Wilhelm Weber, and in 1851, he submitted his now famous doctoral dissertation on complex analysis. The following year, Dedekind received his doctorate with a more modest dissertation on Eulerian integrals written under Gauss' direction.

---

[1] In discussing mathematics at the polytechnic, I have drawn upon [168].

[2] The following biographical details are drawn from the accounts by Biermann [21] and Freudenthal [169].

T. Hawkins, *The Mathematics of Frobenius in Context*, Sources and Studies in the History of Mathematics and Physical Sciences, DOI 10.1007/978-1-4614-6333-7_2,

In the summer of 1854, both Riemann and Dedekind qualified as lecturers at the university. When Dirichlet arrived in 1855, Dedekind attended his lectures on the theory of numbers, potential theory, definite integrals, and partial differential equations, as well as Riemann's lectures on abelian and elliptic functions. His education in advanced mathematics began in earnest in 1855, and Dirichlet and Riemann were his teachers. Dedekind later recalled that Dirichlet had made a "new man" of him, having expanded his scholarly and personal horizons. When the polytechnic, under the impetus of a new president, began searching for a research-oriented mathematician, Dirichlet was consulted. He provided an extensive list of names but with Riemann and Dedekind mentioned most favorably. Although Dirichlet ranked Riemann above Dedekind as a mathematician, the latter was offered the position over Riemann because he was regarded as a more effective teacher at the levels required at the polytechnic. In 1862, Dedekind left the polytechnic for a professorship at the polytechnic institute in his home town of Braunschweig. There he remained until his death, declining professorships at more prestigious German universities such as Göttingen. Frobenius met him when he returned to the Zurich Polytechnic for a visit in 1880, and in 1882, they began a correspondence over arithmetic matters (Section 9.3.3). The picture of Dedekind in Fig. 2.1 probably indicates how he looked at the time they began corresponding. The correspondence continued for many years, and (in 1896) Dedekind's letters provided Frobenius with a concept (that of a group determinant) that was unknown in the mathematical literature and that, in Frobenius' hands, led to his creation of the theory of group characters and representations, as indicated briefly in Chapter 3 and in vivid detail (thanks to the Dedekind–Frobenius correspondence) in Chapters 12 and 13.

Dirichlet had intended to publish his lectures on the theory of numbers as a book, but he died before he was able to accomplish this himself. Although Dirichlet never wrote out his lectures in any detail, Dedekind fortunately had done so, and in 1863, the first edition of Dirichlet's *Vorlesungen über Zahlentheorie* appeared under Dedekind's editorship. There is no doubt that Dedekind put a considerable amount of time and effort into this publication; for the bare outline presented by Dirichlet in the lecture room had to be augmented, and this required an intimate knowledge of Dirichlet's work. The result was a treatise that bore the marks of brilliance of both mathematicians. To the basic text itself, which was based largely on the lectures during the winter term of 1856–1857, Dedekind appended a number of supplements, some of them based on papers by Dirichlet and others presenting his own original work such as Supplement X of the second edition of 1871 [137], in which he first presented his theory of ideals.[3] Dedekind's theory, especially as expounded in his 1877 monograph [113] and in the third (1879) edition of Dirichlet's lectures, influenced Frobenius' work in several ways, as we shall see.

Dedekind's successor at the polytechnic was E.B. Christoffel (1829–1900), who was joined the following year (1863) by F. Prym, the first mathematician to fill a

[3]See the writings of Edwards [145, 146] on the development of the theory of ideals.

Fig. 2.1 Dedekind was to become, along with Weierstrass and Kronecker, a major influence on Frobenius. The photograph shows him as he may have looked in 1882, when he began corresponding with Frobenius. Photo courtesy of Mathematisches Institut, Oberwolfach, Germany

newly created second full professorship at the polytechnic. Prym was also German and a student of Riemann but had received his doctorate from Berlin in 1863 with a dissertation on Riemann-style complex analysis. The conditions supportive of mathematical research at the polytechnic improved considerably during the ensuing years. For example, Christoffel, who had been Dirichlet's student in Berlin and later (1859–1862) was an instructor there during the Kummer–Weierstrass–Kronecker era, was instrumental in instituting a mathematics seminar at the polytechnic that was closely modeled on the Berlin mathematics seminar. In 1868 and 1869, respectively, Christoffel and Prym left the polytechnic for professorships in Germany. The more prestigious position held by Christoffel was given in 1869 to H.A. Schwarz—already mentioned in the Section 1.2 for his outstanding work characterizing algebraically integrable hypergeometric equations—and Prym's position went in 1870 to Heinrich Weber (1842–1913). Weber had received his doctorate from the University of Heidelberg in 1863 after studying also at Leipzig and Königsberg. His primary interests were in number theory and algebra, including collaborative work with Dedekind, as well as work aimed at clarifying and developing ideas of both Dedekind and Kronecker—work that drew him to Frobenius' attention, as will be seen in Chapters 9 and 12–15.

Frobenius was never a colleague of Weber's because both Weber and Schwarz left the polytechnic for professorships in Germany as Frobenius was arriving. Frobenius was given the "Christoffel" professorship that Schwarz had vacated. Unlike his German predecessors at the polytechnic, Frobenius remained there for 18 years, until he was called back to Berlin. His years at the polytechnic proved to be quite productive. He found it very difficult to be separated from the Berlin mathematicians,[4] and yet that separation seems to have encouraged him to break off his work on linear differential equations and to move on to problems dictated more by his own mathematical tastes. These problems were certainly informed by his training at Berlin, as we shall see, but were generally on matters not being actively pursued there.

Weber's position seems to have been left unfilled for a while. This may have been due to several changes in the mathematics faculty. First of all, the Swiss mathematician C.F. Geiser (1843–1934), who had taught for many years at the polytechnic, was appointed to an assistant professorship in 1869 and then in 1873 to a full professorship. Geiser was Frobenius' colleague throughout the latter's tenure in Zurich, and they became friends.[5] For much of that time (1881–1887, 1891–1892) Geiser served as director of the polytechnic. His speciality was that of his great uncle, Jakob Steiner, namely synthetic algebraic geometry, and he had obtained his doctorate from the university in Bern, Switzerland, under the direction of Steiner's student Ludwig Schläfli. He may have helped turn Frobenius' interest in the direction of algebraic geometry (see below), although they never collaborated.[6]

Fortunately for Frobenius, Ludwig Stickelberger (1850–1936), whose primary interests were in algebra and number theory and whom Frobenius had known in Berlin, accepted a teaching position at the polytechnic in 1874. Stickelberger, who was Swiss, had also been Weierstrass' student and had obtained his doctorate in 1874. As a student, Stickelberger had been on Frobenius' 1870 doctoral examination committee,[7] and Frobenius was then on Stickelberger's committee in 1874. Stickelberger's dissertation, which involved an application of Weierstrass' theory of elementary divisors, contained ideas that Frobenius was to utilize in a fundamental way soon thereafter in his work on the problem of Pfaff (Section 6.2). A portrait of Stickelberger in later years is given in Fig. 2.2.

The problem of Pfaff, or at least that aspect of it that interested Frobenius, was at the interface of analysis and algebra, just as Christoffel's work (1868) on what was to become Riemannian geometry had been. Christoffel had been concerned with the transformation of quadratic differential forms, whereas the problem of Pfaff involved the transformation of linear differential forms, or 1-forms as they are now often called. In both problems these transformations are given by analytic functions of several variables, not linear transformations, so the problem considered by Christoffel and the problem of Pfaff represented analytical analogues

[4]In a letter of 24 January 1895, now back in Berlin, Frobenius wrote to Dedekind regarding the latter's reluctance to leave Braunschweig: "Wenn Sie schreiben, an einer Universität würden Sie wahrscheinlich noch nicht zurückgetreten sein, so bedauere ich sehr, dass wir die Berliner Universität, oder doch wenigstens ihre mathematische Abtheilung nicht nach Braunschweig verlegen können, wie einst mein Freund Geiser bedauerte, sie nicht nach Zürich verlegen zu können, als er sah, wie schwer mir das Scheiden wurde."

[5]See the quotation in the last footnote.

[6]Frobenius does cite one of Geiser's papers in one of his algebraic-geometric works [195, p. 382].

[7]The Berlin doctoral committees included three "adversaries" who were either students, recent PhDs, or junior faculty members.

Fig. 2.2 Ludwig Stickelberger was Frobenius' friend and collaborator at the Zurich Polytechnic until his departure in 1879 for the University of Freiburg, where he remained for the rest of his career. From 1897 on, Stickelberger had Alfred Loewy as colleague, and undoubtedly helped foster Loewy's appreciation of Frobenius' work, which he used to great effect (Sections 16.3.1–16.3.3). This photograph is located in the Image Archive, ETH-Bibliothek Zürich

of the problem of the linear transformation of quadratic or bilinear forms that had concerned Weierstrass and Kronecker. Indeed, Christoffel, who had done important work on the transformation of pairs of Hermitian symmetric forms (Section 5.2) that generalized results of Weierstrass, attacked his analytical problem in the same spirit as Weierstrass and Kronecker had attacked theirs: determine in- or covariants that characterize the equivalence classes of quadratic differential forms, and the latter problem was to be investigated first on the algebraic or, as we might now say, tangent space, level. Frobenius was familiar with Christoffel's work, and given his affinity for algebra, it encouraged him to deal with the problem of Pfaff in the same way, thereby engaging him on the research level with linear algebra. Further motivation came from Clebsch's work on the problem of Pfaff, which, despite preliminary claims to the contrary, had succeeded only on the generic level, thereby presenting Frobenius with the challenge of providing a Berlin style solution to the problem. Frobenius succeeded in doing this in a paper submitted in 1876, his first paper written in Zurich.

As we shall see in Chapter 6, Frobenius' method of solution involved him with linear algebra, which he developed elegantly, and led to his well-known integrability theorem (for systems of Pfaffians or 1-forms). Central to his method was the bilinear covariant of a 1-form that Lipschitz, who, like Christoffel, was interested in the transformation of quadratic differential forms, had mentioned in passing. As we shall see in Section 6.6, Frobenius' emphasis on the importance of the bilinear covariant provided Élie Cartan with the key idea behind his exterior calculus of differential forms, which he went on to apply to many problems besides the problem of Pfaff. Through the work of Cartan, Frobenius' paper on the problem of Pfaff had a significant impact on the development of present-day mathematics. I have devoted the final section of Chapter 6 to summarizing and characterizing the nature of Frobenius' work on the problem of Pfaff because it turns out to be paradigmatic of Frobenius' subsequent work in diverse areas of mathematics and suggests a principal reason why his work has had a considerable impact on present-day mathematics, as the reader will see in the chapters that follow Chapter 6.

Frobenius' next major paper was submitted to *Crelle's Journal* in May 1877 [181], and so 11 months after his paper on the problem of Pfaff. It was motivated by a problem posed and solved generically by Hermite and inspired by a special case of a problem solved earlier by Cayley. I call it the Cayley–Hermite problem, and Frobenius' nongeneric solution to the problem is given in Chapter 7. The problem inspired Frobenius to develop matrix algebra in order to facilitate its solution. As we shall see, the idea of matrix algebra was suggested by the problem and was introduced independently in that connection by Cayley (1858), Laguerre (1867), and Frobenius (1877). Both Cayley and Laguerre reasoned on the generic level and developed matrix algebra accordingly, thereby creating a symbolism that was not equipped to deal correctly and rigorously on a nongeneric level with linear-algebraic problems. Frobenius, by contrast, succeeded in fusing matrix algebra with the work of Weierstrass and Kronecker on the transformation of families of bilinear forms, so as to fashion a new mathematical tool of considerable power, capable of rigorously solving problems on the nongeneric level, such as the Cayley–Hermite problem (Section 7.5.4) and a related, more general problem suggested by Rosanes (Section 7.5.3).

Frobenius' paper constituted a veritable treatise on the theory and application of matrix algebra that present-day mathematicians can read with ease due to the striking modernity of his approach to mathematical reasoning. Here we find for the first time the notion of the minimal polynomial of a matrix and a proof of its many properties (Section 7.5.1), including the fact that the minimal polynomial of a matrix A divides its characteristic polynomial $p(r) = \det(rI - A)$. An immediate consequence was the so-called Cayley–Hamilton theorem that $p(A) = 0$, a theorem that, from a formal standpoint, appears to be true but could be proved by Cayley only for 2×2 and 3×3 matrices by direct computation. As an application of the minimal polynomial, Frobenius proved the theorem that still bears his name: the only real division algebras are $\mathbb{R}$, $\mathbb{C}$, and the quaternions (Section 7.5.6).

In his later work, Frobenius made use of his brand of matrix algebra whenever it was deemed critical to the solution of a problem. For example, in 1883, he used it in this way to give a definitive solution to a problem posed by Kronecker concerning abelian functions admitting a complex multiplication (see below). He also used it in 1896 to solve two matrix congruence problems, which made possible a major simplification of Weierstrass' theory of elementary divisors, as well as Kronecker's generalization of it to singular families, and also a major simplification of the latter's theory of families of the form $x^t(A+A^t)y$ (Section 16.1). These simplifications had the effect of dispensing with many determinant-based arguments in favor of far simpler matrix algebra, including the theorem that matrices with nonzero determinant always possess a square root. Also, when he was creating the theory of group characters and representations, one of his theorems on matrix algebra led him in 1899 to his theory of the primitive idempotents of a group algebra (Section 15.2). Eventually, Frobenius' many accomplishments using matrix algebra gained the audience and influence they deserved (Section 16.1.6).

After submitting his paper on the Cayley–Hermite problem in May 1877, Frobenius submitted two others based on his earlier work. These are not discussed in Part III but are worth a brief mention here. The first, submitted in June 1877 [183] applied the analytical classification theorem (Theorem 6.7) from his paper on the problem of Pfaff to the study of Pfaffian equations $\omega = a_1(x)dx_1 + \cdots + a_n(x)dx_n$ with the property that the coefficient functions $a_i(x)$ are homogeneous functions of a fixed degree g in $x_1, \ldots, x_n$, a type of equation Euler had discussed by examples in three variables. The second, submitted in February 1878 [184], applied results from his paper on the Cayley–Hermite problem to a problem suggested by Kronecker's study of families of bilinear forms with coefficient matrices of the special type $rA - A^t$. If $\varphi(r) = \det(rA - A^t) \not\equiv 0$, then if P is any nonsingular transformation such that $P^t(rA - A^t)P = rB - B^t$, it follows, taking determinants, that $p^2 = \det(rB - B^t)/\det(rA - A^t)$, where $p = \det P$. This shows that p^2 is independent of the choice of P and only the sign of the square root in $p = \sqrt{\det(rA - A^t)}/\sqrt{\det(rB - B^t)}$ can possibly depend on P. Thus the first question is, can the sign vary? It is easily seen that this question can be reduced to the same question when $A = B$, i.e., when $P^tAP = A$, so that P is what I call a Rosanes transformation in discussing Frobenius' paper on the Cayley–Hermite problem (Section 7.5.3). Using results and ideas from that paper, Frobenius quickly proved that when $\varphi(1) \neq 0$, the sign of the square root never changes [184, p. 458, III]. But what is that sign? More to the point, can it be determined without finding a P such that $P^tAP = B$? Frobenius showed that it could. Since $A - A^t$ is skew-symmetric with $\det(A - A^t) = \varphi(1) \neq 0$, he realized from his work on the problem of Pfaff that $\det(A - A^t) = \mathrm{Pf}(A - A^t)^2$, where $\mathrm{Pf}(A - A^t)$ denotes the Pfaffian of $A - A^t$, a polynomial in the coefficients a_{ij} of A. The end result was that $p = \mathrm{Pf}(B - B^t)/\mathrm{Pf}(A - A^t)$ [184, p. 458, (2)]. When $\varphi(1) = 0$ but $rA - A^t$ has no elementary divisors of the form $(r-1)^{2k+1}$, Frobenius was able to obtain, with considerable computational skill and effort, an analogous result with "skew invariants" that generalized the Pfaffians. He also obtained similar results for families $rA - B$, where A and B are symmetric, although the mathematics in this case was different in nature and more difficult.

In both these papers, the results from his earlier papers simply formed the starting point for new mathematical developments and results, but these developments, although containing thorough analyses based on new ideas and masterly calculations, were in a sense old-fashioned; they did not open up entirely new mathematical vistas for exploration and application. While he was working on these papers, however, Frobenius already had in mind a quite different and largely unexplored mathematical domain in which to develop his penchant for linear-algebraic reasoning. The domain was the theory of numbers, an area of mathematics deemed especially important at Berlin, where Kummer and Kronecker were known for their many arithmetic contributions. Frobenius' education thus included a thorough understanding of the principal achievements in this area, which he supplemented by reading the masters.

The first arithmetic line of research Frobenius pursued seems to have been inspired by his careful study of Gauss' *Disquisitiones Arithmeticae* of 1801 [244]. Gauss had pointed out that the arithmetic theory of binary quadratic forms was but a part of a vast and potentially fertile theory of forms in any number of variables, which he left for his successors to explore. In papers submitted in April 1878 and January 1879 [182, 185], Frobenius sought to make a contribution to this theory by considering bilinear forms in two sets of variables $x_1, \ldots, x_m$ and $y_1, \ldots, y_n$, viz., $F(x,y) = x^t Ay$, where A is $m \times n$ and has integers as coefficients (Chapter 8).

Frobenius focused on two problems that were motivated by Gauss' theory in the binary and ternary cases. Expressed in terms of matrix algebra rather than forms, the first problem was to determine necessary and sufficient conditions that two $m \times n$ integral matrices A and B be equivalent in the sense that integral square matrices P and Q exist with determinants ± 1—what are now called unimodular matrices—so that $A = PBQ$. This problem is of course analogous to the problem that Weierstrass had solved by means of his theory of elementary divisors. Indeed, Frobenius, realized that if $d_n, \ldots, d_1, d_0$ denotes the analogue of the sequence introduced by Weierstrass in order to define the invariant factors $E_i(s)$ (Section 1.1), so that d_i is the greatest common divisor of all $i \times i$ minors of A for $i = 0, \ldots, n-1$ and $d_0 = 1$, then (as in Weierstrass' theory) $d_{i-1} \mid d_i$ for all $i \leq r = \operatorname{rank} A$ and so $e_i = d_{i+1}/d_i$ is an integer. (For $i > r$, of course, $d_i = 0$, and Frobenius defined $e_i = 0$ for $i > r$.) It also follows, as in Weierstrass' theory, that if A and B are equivalent in the above sense, then they must have the same e_i—invariant factors as we called them in discussing Weierstrass' theory. Frobenius' first problem thus boiled down to establishing the converse. He did this by proving that for any integral A, unimodular P and Q can be determined such that $PAQ = N$, where N is the $m \times n$ "diagonal" matrix with the invariant factors of A, $e_1, \ldots, e_r, 0, \ldots, 0$, down the diagonal.

Nowadays N is usually called the Smith normal form, because unbeknownst to Frobenius, in 1861, H.J.S. Smith had proved the above-stated theorem in the special case in which A is square and has nonzero determinant.[8] Both Frobenius and Smith used their normal forms to study systems of linear equations and congruences, although Frobenius expressed his results in a more satisfying and general form (Section 8.3.1). Smith's work was motivated by the desire for results about solutions to systems of the above type, and he pursued application of his results no further. Frobenius, however, found several other important applications for his results. For example, for skew-symmetric integral matrices, he used his results to obtain a more

[8] As indicated in Section 8.5, Smith's extension of his result to rectangular matrices was different from Frobenius' above result and done only for the generic case in which A has full rank.

informative normal form (Section 8.3.2), which proved useful a few years later in his work on the general theory of theta functions (Section 11.3).

Frobenius' most important application of his solution to the first problem was his rational theory of elementary divisors (Section 8.6). For many years before 1879, he had been dissatisfied with the way in which Weierstrass' theory was established. That is, given two families $sA - B$, $sC - D$ of $n \times n$ matrices, it is possible to determine by rational operations—i.e., operations taking place within the field $\mathbb{Q}(a_{ij}, b_{ij}, c_{ij}, d_{ij})$—the invariant factors of the two families and so whether they are equivalent. Furthermore, granted that they are equivalent, it was possible to determine rationally a P and Q such that $P(sA - B)Q = sB - C$. However, Weierstrass' proof of equivalence was not rational, since it employed a canonical form that required knowing the roots of the characteristic polynomial. Once he had established his theorem on the normal form, Frobenius realized that the reasoning behind it could be repeated with the integers replaced by polynomials with coefficients from any known field, not just the field of complex numbers (so as to yield a rational justification of Weierstrass' theory), but also the algebraic number fields then being studied by Dedekind, as well as the finite fields introduced by Galois. Although it was not essential to his theory, Frobenius also introduced the now-familiar rational canonical form to show that for any set of invariant factors (over any given field), there is a family $sA - B$ with precisely those invariant factors.

Frobenius' rational theory of elementary divisors had the effect of greatly diminishing the role played by the theory of determinants in the development of elementary divisor theory. Indeed, his theory was a major factor behind the subsequent development of the theory of matrices and linear transformations over abstract fields, as will be seen in Sections 16.2–16.3. In fact, in conjunction with his rational theory of elementary divisors, Frobenius' early paper introducing the concept of irreducibility into linear differential equations (discussed above in Section 1.2) and a related paper by one of his Berlin doctoral students (Landau) played a considerable role (through the intermediary of Alfred Loewy, Wolfgang Krull's mentor) in bringing about the now-familiar module-theoretic approach to elementary divisor theory first found in van der Waerden's *Moderne Algebra* (Section 16.3).

Frobenius devoted his second paper [185] exclusively to the second problem, which was inspired by Gauss' notion of a binary quadratic form $G(x) = x^t Bx$ being "contained in" the form $F(y) = y^t Ay$ when a nonsingular integral linear transformation $y = Px$ exists such that $G(x) = F(y)$. (Hence $B = P^t AP$.) This means that every integer represented by G can be represented by F but not conversely, since P is not assumed unimodular. Thus in general, the integers representable by G are contained within the totality of those representable by F. However, if F is also contained in G, then, since all forms are assumed nonsingular, it follows that P must be unimodular and so the two forms are equivalent and represent the same integers. That was as far as Gauss went with these matters. Frobenius, however, considered what would happen if for any two bilinear forms A and B with integral $m \times n$ coefficient matrices, B were said to be contained in A if integral, possibly singular, P and Q existed such that $B = PAQ$. (In view of $B = PAQ$, Frobenius' successors spoke of B as a *multiple* of A.) This definition is much weaker than Gauss', although it still implies that all the integers representable by $G(u,v) = u^t Bv$ are contained among the integers representable by $F(x,y) = x^t Ay$. This then raises the following question: if G is contained in F and F is contained in G, are they equivalent? This is no longer the trivial question it was in Gauss' theory. It is remarkable that Frobenius was able to give an affirmative answer. He did so (with considerable effort) by solving the following analogue of the first problem: determine necessary and sufficient conditions that G be contained in F, i.e., that B be a multiple of A in the sense that $B = PAQ$ for some P and Q. The solution was given by what I have called his containment theorem (Theorem 8.16): B is a multiple of A in the above sense if and only if $\operatorname{rank} B \leq \operatorname{rank} A$ and every invariant factor of B is a multiple of the corresponding invariant factor of A. Frobenius made use of this theorem in developing his generalized theory of theta functions (Section 11.3).

Frobenius' arithmetic interests during 1878–1880 were not limited to the arithmetic theory of bilinear forms. We have already seen in Section 1.2 that early on (1872), Frobenius had a keen interest in and appreciation for Galois' theory of equations and, in particular, for Galois' notion of a group. Galois' groups were groups of permutations, but other sorts of groups were emerging from the theory of numbers. As will be seen in Section 9.1, implicit in Gauss' *Disquisitiones Arithmeticae* are, e.g., the multiplicative groups of integers modulo a prime p, $(\mathbb{Z}/p\mathbb{Z})^\times$, or relatively prime to a nonprime integer m, $(\mathbb{Z}/m\mathbb{Z})^\times$, as well as several groups of equivalence classes of binary forms under the highly nontrivial operation of Gaussian "composition" of such forms. In addition, Kummer's work on ideal numbers in cyclotomic number fields, and Dedekind's later work on ideals in algebraic number fields, led to what are now called ideal class groups. All these groups are finite and commutative, but before Dedekind (1871) they were not regarded as groups in the sense that the objects under consideration were not regarded as falling under the notion of what Galois had called a group.

However, by 1869 it was pointed out by Ernst Schering that a problem that arose in connection with all these groups—the problem (as we would now say) of expressing them as a direct product of cyclic subgroups—could be solved for all of them by the same line of reasoning, which Schering was the first to devise. Schering's work prompted Kronecker in 1870 to formulate an explicitly abstract set of rules for a finite number of unspecified "elements" that can now be seen to make the elements into a finite abelian group. For such elements Kronecker presented what amounts to an abstracted version of Schering's proof. He did not, however, expressly conceive of what he was doing as part of a more general theory that would embrace as well the groups of Galois' theory and the theory of finite permutation groups that Cauchy had developed independently of connections to Galois' theory in the 1840s. It was Dedekind, in his Supplement X to the second edition (1871) of Dirichlet's lectures on the theory of numbers, who emphasized that the above-mentioned developments in number theory were simply an aspect of a general theory of abstract finite groups.

The observations of Dedekind inspired Frobenius, encouraged in this regard by Kronecker's own support for an abstract viewpoint, to deal, in collaboration with Stickelberger, with that part of Dedekind's envisioned general theory having to do with the decomposition of a finite abelian group as a direct product of cyclic subgroups. Schering had proved, in effect, that if $\mathfrak{G}$ is any finite abelian group, then cyclic subgroups $\mathfrak{C}_{e_1}, \ldots, \mathfrak{C}_{e_N}$ can be determined with orders $e_1, \ldots, e_N$, respectively, such that $e_{i+1} \mid e_i$ and $\mathfrak{G} = \mathfrak{C}_{e_1} \times \cdots \times \mathfrak{C}_{e_N}$. This is of course the existence part of one version of the fundamental theorem of finite abelian groups. Frobenius and Stickelberger were interested in the extent to which such a direct product decomposition was unique. Motivated by analogy with Dedekind's theory of ideal factorization into powers of prime ideals, they also proved that $\mathfrak{G}$ could be factored as a direct product of cyclic subgroups of prime power orders and considered the uniqueness question for this factorization as well. Nowadays, their conclusions can be interpreted as implying uniqueness up to isomorphism for these two factorizations, but this notion was not yet commonplace, and so what Frobenius and Stickelberger did was to prove, e.g., that the e_i were group invariants. All this, and much more, was contained in a paper they submitted in July 1878 to *Crelle's Journal*. The paper runs to 56 pages in Frobenius' collected works and represents, in effect, the first monograph on the theory of finite abelian groups.

Frobenius' interest in arithmetic applications of the theory of groups was kept alive as the result of a paper Kronecker published in 1880 (Section 9.3). Kronecker was interested in criteria for the irreducibility of polynomials with integer coefficients, and in the 1880 paper he posited a theorem that enabled him to formulate the notion of the (analytical) density of a set of primes and to use it to obtain a new irreducibility criterion. In the course of trying to understand Kronecker's sketchy paper, Frobenius discovered a connection between the density of the primes p for which a given polynomial $f(x) \in \mathbb{Z}[x]$ with nonzero discriminant has a factorization of the form $f(x) \equiv \varphi_1(x) \cdots \varphi_r(x) \pmod{p}$, with the $\varphi_i(x)$ irreducible mod p and of degree f_i, and the structure of the Galois group $\mathfrak{G}$ of $f(x)$. Specifically, if $\mathfrak{G}$ is regarded as a subgroup of the symmetric group $\mathfrak{S}_n$, then the above density is equal to the fraction of $g \in \mathfrak{G}$ that belong to the conjugacy class of $\mathfrak{S}_n$ that consists of all permutations expressible as the product of r disjoint cycles of respective lengths $f_1, \ldots, f_r$.

Thanks to work of Dedekind, Frobenius was able to translate the above result into one within the context of Dedekind's theory of ideals: if $\mathbb{K}$ is an extension of $\mathbb{Q}$ of degree n, and $\mathfrak{o}_{\mathbb{K}}$ denotes the algebraic integers in $\mathbb{K}$, then the density of all primes p for which $p\mathfrak{o}_{\mathbb{K}} = \mathfrak{p}_1 \cdots \mathfrak{p}_r$, where $\mathfrak{p}_i$ is prime ideal of degree f_i, is the fraction of $g \in \mathfrak{G}$ that belong to the conjugacy class of $\mathfrak{S}_n$ of all permutations expressible as the product of r disjoint cycles of respective lengths $f_1, \ldots, f_r$, where now $\mathfrak{G}$ is the Galois group of an irreducible integral polynomial of degree n. In the process of translating his theorem into one in Dedekind's theory and investigating a related conjecture, Frobenius was led to introduce the now familiar concept of a Frobenius automorphism. The concept and properties of Frobenius automorphisms led him to attempt to adapt his proof techniques to prove a different sort of density theorem. It became known as Chebotarev's density theorem because Frobenius did not succeed in proving it—although he evidently believed it true—and had to settle for a lesser theorem, which is now customarily called the Frobenius density theorem.

Although Frobenius did all the above-described work in 1880, for a combination of reasons (discussed in Section 9.3.6), it did not appear in print until 1896. He did, however, initiate a correspondence with Dedekind in 1882 regarding the above-mentioned conjecture. (He had met Dedekind in 1880 when the latter visited the Zurich Polytechnic.) The correspondence with Dedekind, which continued sporadically for many years, turned out to be very consequential for Frobenius, because it was through the correspondence that he learned about a problem involving group determinants that Dedekind had toyed with, and which in Frobenius' hands led him to his creation of his theory of group characters and representations of finite groups, which is discussed in the following chapter. The work in 1880 also forced Frobenius to delve more deeply into the properties of noncommutative finite groups; his first-mentioned theorem about densities led him to explore various counting techniques in such groups as well as properties of a new group-theoretic notion, namely what are now called double cosets (Section 9.4).

With his work on density theorems in limbo, Frobenius turned to several other areas of mathematics to work on problems of interest to him. The majority of these areas fell within the broad framework of the theories of elliptic and abelian functions, which was the focus of Weierstrass' research interests, and so it is not entirely surprising to find his student Frobenius also displaying an interest in problems associated to these theories. Indeed, during his years in Zurich, Frobenius submitted 27 papers for publication, and 14 of them involved either elliptic functions, abelian functions, or the allied theories of theta functions. The 4-year period 1879–1883 turned out to be especially fertile; 9 of the 14 works were produced then, including all the ones discussed in detail in later chapters by virtue of their importance.

Chapter 10 is devoted to two problems solved by Frobenius that grew out of the Berlin reaction to an important 1855 paper by Hermite on the transformation of abelian functions in two complex variables. (There the reader will find an exposition of all that is needed about elliptic and abelian functions in order to understand the historical developments under discussion.) Weierstrass, who was interested in the theory of abelian functions in any number of variables, asked Kronecker to look into the possibility of extending to g variables certain fundamental parts of Hermite's theory that lay on the boundary between arithmetic and algebra. In the relatively simple case of $g = 2$ variables, Hermite had not paused to justify several of the assertions and assumptions he was making in developing his theory, and presumably Weierstrass wanted Kronecker not only to formulate Hermite's theory for any number g of variables but also to justify the assertions and assumptions underlying it. Kronecker wrote up his conclusions and gave them to Weierstrass but only made brief passing references to his results in his own publications, which prompted Heinrich Weber to attempt to flesh out Kronecker's results in a paper of 1878. It was probably Weber's 1878 paper that called Frobenius' attention to these matters, and, in particular, to the two problems referred to in Chapter 10 as Hermite's abelian matrix problem and Kronecker's complex multiplication problem.

Although Kronecker had made important contributions related to Hermite's problem, his efforts at completely solving it left much to be desired, as was also the case with Weber's attempts to clarify and develop Kronecker's ideas. Frobenius realized that the line of reasoning he had used to prove his theorem on the Smith–Frobenius normal form could be appropriately modified to provide a different approach to Hermite's problem that had the advantage of yielding a completely general and satisfactory solution. Frobenius submitted his results for publication in May 1879, and no doubt began to think about the other problem discussed by Weber, namely, Kronecker's complex multiplication problem. That problem was a byproduct of Kronecker's attempt, with Weierstrass' encouragement, to formulate a generalization to abelian functions of the notion of an elliptic function admitting complex multiplication. Kronecker used the theoretical apparatus of Hermite's theory to give such a formulation in terms of Hermite's abelian matrices (defined in Section 10.2.1), but he was unable to formulate conditions sufficient to guarantee that an abelian matrix would give rise to a complex multiplication, even though he assumed that the characteristic roots of the abelian matrix were distinct. This then became Kronecker's complex multiplication problem: to determine necessary and sufficient conditions on any abelian matrix that it determine a complex multiplication. However, Kronecker also made comments on what happens when multiple roots are present that suggested another problem: if an abelian matrix gives rise to a complex multiplication, the associated canonical period system need not be unique. Determine necessary and sufficient conditions for uniqueness.

Both of the Kronecker problems were solved generally and definitively by Frobenius by January 1883. The matrix algebra he had developed in response to the Cayley–Hermite problem played a key role in his solution. It was also here for the first time that unitary matrices were introduced and their key properties established. Although Frobenius' approach to complex multiplication, like that of Kronecker, was essentially algebraic, other mathematicians found geometric applications for his results (Section 10.7). In particular, even though Humbert (1899–1900) was to find Kronecker's notion of complex multiplication far too limiting for the purposes of algebraic geometry—the algebraic-geometric study of abelian varieties with complex multiplications—Frobenius' work, even though based on Kronecker's limited notion of complex multiplication, played a significant role in the theory of abelian varieties with complex multiplication, especially through the work of G. Scorza and S. Lefschetz.

In 1879, Stickelberger, Frobenius' friend and collaborator, left the Zurich Polytechnic, where he did not hold any sort of a professorship, for an assistant professorship at the University of Freiburg, in Germany. Stickelberger remained there for the rest of his career. Also in Freiburg, Stickelberger had as a colleague (starting in 1897) Alfred Loewy, who was a great admirer of Frobenius' rational theory of elementary divisors, an admiration that may have been encouraged by Stickelberger. As mentioned above, Loewy was instrumental in bringing about, through the impact of his work on that of his student Wolfgang Krull, the module-theoretic approach to rational elementary divisor theory, i.e., the theory of canonical matrix forms for linear transformations over any field.

Stickelberger's departure was a real loss for Frobenius, who, as noted at the beginning of this section, missed the company of Berlin mathematicians.

Fig. 2.3 Friedrich Schottky replaced Stickelberger as Frobenius' friend and colleague at the Zurich Polytechnic. This photo was taken in 1886 while they were together there and is located in the Image Archive, ETH-Bibliothek Zürich. Later, Frobenius managed to get Schottky to join him in Berlin, but at a great institutional and personal cost, as indicated in Chapter 3

Stickelberger's loss, however, was compensated by the appointment in 1882 of another Weierstrass student, Friedrich Schottky (1851–1935), as full professor at the polytechnic. (See Fig. 2.3.) He and Frobenius had overlapped at Berlin, where Schottky had impressed Weierstrass with an outstanding doctoral dissertation (1875) on the conformal mapping of multiply connected regions.[9]

Schottky, then an instructor at the University of Breslau, was recommended for the Zurich position by Weierstrass who regarded him as "the most gifted for deeper mathematical speculation" among prospective young German mathematicians, although he added a caveat: "Schottky is a peculiar person. He is somewhat of a dreamer and not very adept in practical life" [168, p. 43]. Weierstrass expanded on the basis for these remarks in a letter of 7 May 1875 to his confidante, Sonya Kovalevskaya[10]:

> On Christmas eve he was suddenly arrested and led away to the barracks to serve a 3-year term as a common soldier, for "he had forgotten" to register in time for the 1-year term as a volunteer (as every student does). Fortunately, he proved to be so useless as a soldier that he was discharged as unsuitable after 6 weeks. Thus he could return to his dissertation. He then signed up for the examination without [presenting the requisite] certificates and without knowing anything about the necessary formalities. As rector I had to cancel his name from the register because neither had he attended lectures nor were his whereabouts in Berlin known.

---

[9]According to Frobenius' recollections in 1902 [22, p. 210].

[10]The letter is transcribed in [28]; I have followed the English translation in [29, p. 80].

Such idiosyncrasies did not seem to bother Frobenius, himself somewhat unworldly, and he welcomed Schottky as a talented mathematician and came to know him as a friend. As for Schottky, the period before coming to Zurich was a fallow one for him, but the stimulation of contact with Frobenius seems to have inspired him to publish important work on abelian functions (e.g., on the Schottky problem) while in Zurich[11] and may be part of the reason Schottky once referred to Frobenius as "an irreplaceable friend" [22, p. 132].[12]

Virtually all Schottky's work was on complex analysis, including, in particular, the theories of elliptic, abelian, and theta functions, which Frobenius had also begun to explore a few years before Schottky arrived. Frobenius must have been delighted to have as a colleague a mathematician who was characterized by Weierstrass as a deep thinker and who, in addition, was a source of information and sounding board about branches of complex analysis of current interest to him as well. In particular, Schottky's 1880 book on abelian functions in three variables [519] was frequently cited by Frobenius in his own work and encouraged Frobenius' research into what he called "Jacobian functions," which are theta functions in the modern sense of that term. This important work by Frobenius, which he submitted for publication in December 1883, is the subject of Chapter 11.

In Frobenius' time, theta functions were conceived in a more limited sense that was related to their origin in the Jacobi inversion problem for abelian integrals and the resulting special abelian functions that emerged as solutions. Much of Weierstrass' lectures were devoted to the inversion problem for abelian integrals and so to these special theta functions. However, Weierstrass also realized that the infinite series defining these special theta functions had a form that could be considered independently of whether or not the data substituted into this form originated from abelian integrals. And so he devoted a small portion of his lectures to this more general class of theta functions, which, however, was still not coextensive with the modern class of theta functions. In his monograph of 1880, Schottky presented Weierstrass' theory of general theta functions more or less as presented in Weierstrass' lectures while Schottky was in Berlin but with a few differences of conception, which, I suggest, helped pique Frobenius' interest in further generalizing Weierstrass' theory. I say "helped pique" because there were other sources of inspiration as well, which served to suggest the theoretical basis for such a generalization and its theoretical goals.

In unpublished documents only hinted at in publications, Weierstrass also considered the question whether all abelian functions, not just those that arise from the inversion of abelian integrals, can be expressed rationally in terms of his general theta functions (Section 11.2). An apparent impasse to an affirmative answer was surmounted by Weierstrass' discovery in 1870 that if Ω is a $(g \times 2g)$

---

[11]See [29, p. 80], as well as Frobenius' remarks [22, p. 213].

[12]The fact that they never wrote any joint papers may be a reflection of their differing mathematical orientations: Frobenius was primarily interested in algebraic aspects, whereas Schottky's approach to function theory was Riemannian in spirit, although combined with Weierstrassian rigor [170].

period matrix for abelian functions, then necessarily there are constraints on Ω: there must exist a skew-symmetric integral matrix L such that (I) $\Omega L \Omega^t = 0$ and (II) the Hermitian matrix $i\Omega L \Omega^h$ is positive definite.[13] Weierstrass communicated his discovery to Adolf Hurwitz, who had spent several years in Berlin studying with him, because it was relevant to a problem Hurwitz was working on. In 1883, when Hurwitz published the solution to his problem, he also made the mathematical community, including Frobenius, aware for the first time of Weierstrass' theorem about conditions (I)–(II).

Hurwitz's paper and Schottky's presentation of Weierstrass' general theory of theta functions were two of the three major sources of inspiration for Frobenius' theory of Jacobian functions. The third was the realization that if Jacobian functions with period matrix Ω and secondary period matrix H, and so of type (Ω, H), exist, then the skew-symmetric matrix $K = \Omega^t \mathrm{H} - \mathrm{H}^t \Omega$ must have integer coefficients—and so fall within the purview of his recently developed arithmetic theory bilinear forms. Among the results of Frobenius' theory, in which K played the leading role, was a theorem analogous to Weierstrass': if Jacobian functions of type (Ω, H) exist, then necessarily, there exists a skew-symmetric integral matrix L for which Weierstrass' conditions (I) and (II) hold. Frobenius also generalized to Jacobian functions a theorem that Weierstrass had established for his more special theta functions: the number of linearly independent Jacobian functions of a given type (Ω, H) is $\ell = \sqrt{\det K} = \mathrm{Pf}(K)$, where $\mathrm{Pf}(K)$ is the Pfaffian of K. A third theorem of Weierstrass' that Frobenius thought he had generalized was, in generalized form, that any $g+1$ abelian functions $f_1, \ldots, f_{g+1}$ in g variables $z_1, \ldots, z_g$ with period matrix Ω satisfy a polynomial equation $P(f_1, \ldots, f_{g+1}) = 0$ for all values of $z_1, \ldots, z_g$. Frobenius seemed to think that this result followed from his theorem that any $g+2$ Jacobian functions $\varphi_1, \ldots, \varphi_{g+2}$ in g variables of the same type (Ω, H) and characteristic satisfy a homogeneous polynomial equation $\mathrm{H}(\varphi_1, \ldots, \varphi_{g+2}) = 0$ for all $z_1, \ldots, z_g$. Weierstrass' theorem in the above generalized form would indeed follow had it been known at the time that every abelian function with period matrix Ω is the quotient of Jacobian functions. For reasons suggested at the end of Section 11.3, Frobenius seems to have thought, mistakenly, that Weierstrass had proved such a result.

The theorem that every abelian function is the quotient of Jacobian functions was later proved by Paul Appell for $g = 2$ variables and then by Poincaré for any number of variables (Section 11.4.1). It was the Austrian mathematician Wilhelm Wirtinger who first realized the significance of the Appell–Poincaré theorem when combined with Frobenius' theory of Jacobian functions, which was unknown to Appell and Poincaré when they did their work: abelian functions with period matrix Ω exist if and only if Jacobian functions of some type (Ω, H) exist if and only if Ω satisfies Weierstrass' conditions (I) and (II). Wirtinger's observations

---

[13] These conditions are usually called Riemann's conditions, because unbeknownst to Weierstrass and most mathematicians, Riemann had discovered these conditions in a special case a decade earlier (see Section 11.4.1).

implied that the foundations of the theory of abelian functions could be built upon Frobenius' theory of Jacobian functions and general function-theoretic theorems as the attendant tools. But such an implication was not appealing to Wirtinger, who preferred the traditional approach, suitably refined and augmented, of the theory of abelian integrals.

It was not until the late 1920s that someone (Lefschetz) regarded the implications of Wirtinger's observations as of potential merit—if only the theorem of Poincaré could be established more simply using purely function-theoretic results, as had been the case with Appell's proof for $g = 2$. This finally occurred in the 1940s (Section 11.4.3). The first to do so was Fabio Conforto, followed by Frobenius' "mathematical grandson" Carl Ludwig Siegel.[14] Both Conforto and Siegel, drawing upon the Franco-Italian tradition, included abelian varieties within their foundational framework, although their methods were what would now be described as classical. Independently of their work, Frobenius' theory also became fundamental to the modern approach to abelian varieties thanks to a 1949 paper by André Weil, who was perhaps the first to equate "theta functions" with Frobenius' "Jacobian functions." Weil's greatest achievement in his paper was to show how Poincaré's theorem, which still required relatively long function-theoretic proofs on the part of both Conforto and Siegel, could be given a much shorter proof by applying ideas underlying recent proofs of de Rham's theorems. Whereas Poincaré had disliked Frobenius' purely algebraic theory (Section 11.4.3), Weil found Frobenius' algebraic study of theta functions appealing, and after translating Frobenius' main results into his own conceptual framework, he concluded that "The majority of known results on abelian functions and varieties (in the 'classical case' where the field of constants is the complex field) can be deduced very easily from the preceding and knowledge of the cohomology ring of the torus" [598, p. 421]. The approach envisioned by Weil has now become preponderant, and so Frobenius' algebraic theory of Jacobian functions, coupled with Poincaré's theorem (with Weil's proof), also lives on as well in the modern treatment of complex abelian varieties.

Frobenius' paper on Jacobian functions was not his first to deal with theta functions. A trio of earlier works on theta functions with integral characteristics were initiated by Weber's above-mentioned 1879 paper. (This trio is discussed in Section 12.4 for reasons indicated below.) The overall objective of Weber's paper had been to generalize the results of Hermite's important 1855 paper from $g = 2$ to any number g of variables. As we have already seen, parts of Weber's paper had inspired Frobenius' solution to Hermite's abelian matrix problem and to Kronecker's complex multiplication problem. Weber's paper also prompted Frobenius' first paper on theta functions (1880) because in another part of that paper

[14] Siegel (1896–1981) had attended Frobenius' lectures on number theory in Berlin before being drafted into the German army in 1917, and as a consequence decided to pursue a career in number theory rather than astronomy (see his personal recollections about Frobenius [232, pp. iv–vi]). Because he refused military service, he was sent to a psychiatric institute, where the father of Frobenius' former doctoral student Edmund Landau helped him to endure the ordeal. After the war, he received his doctorate under Landau's supervision at the University of Göttingen.

Weber considered the way theta functions with integral characteristics transform under the transformation defined by an abelian matrix of order n. Hermite had proved a result along these lines showing that $2g = 4$ theta functions satisfying relations due to Göpel transform into nth-degree polynomials in the transformed theta functions *with the same characteristics*. This result was not easy to generalize directly, and Weber did not attempt it. Instead, he proved the less specific result that all 2^{2g} theta functions with integral characteristics transform into polynomials of degree n in the 2^{2g} transformed theta functions. Frobenius sought to generalize Hermite's more specific result to g variables. He saw a way to do this by means of an addition theorem for theta functions that generalized a theorem due to Hermite. This required generalizing Göpel's relations to $g > 2$ variables, which in turn required Frobenius to focus on the characteristics of theta functions.

In the context of the work of Hermite and Weber, characteristics are specified by 2^{2g} matrices with integer coefficients $A = \begin{pmatrix} a_1 & \cdots & a_g \\ b_1 & \cdots & b_g \end{pmatrix}$ considered modulo 2. If we define a "multiplication" by

$$AA' = \begin{pmatrix} a_1 + a_1' & \cdots & a_g + a_g' \\ b_1 + b_1' & \cdots & b_g + b_g' \end{pmatrix} \pmod{2},$$

then the characteristics form a primary abelian group $\mathfrak{C}_g$ of order 2^{2g} and rank $2g$, as Frobenius pointed out [187, p. 14n] with a reference to his recent joint paper with Stickelberger on abelian groups [187, p. 14n]. General theorems on the structure of finite abelian groups, such as the various forms of the fundamental theorem, however, were not relevant to the problem at hand; instead, what was required were complex relations among sets of group elements (such as those indicated in Section 12.4). In addition to successfully generalizing Hermite's addition theorem and his theorem on the transformation of theta functions with characteristics satisfying Göpel's relations, Frobenius devoted the second half of his paper to applying his addition theorem to reformulate, extend, and complete work done by Hermann Stahl (1880), one of Weierstrass' students (PhD, 1882). Frobenius submitted this paper for publication in February 1880, and 2 months later he submitted another with the title "On groups of theta characteristics." Here he considered subgroups of $\mathfrak{C}_g$ classified by their rank and developed their theory so as to obtain thereby "a sharper insight into the essence of the formulas" that Stahl, Max Noether, and Prym had obtained in various publications [191, p. 130]—a characteristically Frobenian enterprise.

The final paper in the trilogy was submitted a month after the second. What is particularly significant about this paper is that whereas in the earlier ones, Frobenius had focused on the nontrivial innovations needed to generalize or rework known results, here he ventured out in an entirely new direction and introduced a homogeneous polynomial $F[x_R]$ in $h = [\mathfrak{C}_g : 1]$ variables $x_A, x_B, \ldots, x_R, \ldots$, one variable for each $R \in \mathfrak{C}_g$, that was defined by means of a determinant (Section 12.4). He focused on how this polynomial factors subject to various conditions on the variables. Sixteen years later, when he learned from Dedekind about the latter's notion of a group determinant and the problem of its factorization (Sections 12.2–12.3),

Frobenius immediately saw the analogy with his polynomial $F[x_R]$ and employed several of the techniques used on $F[x_R]$ in his solution of the group determinant problem, which involved a generalization of Dedekind's notion of a group character and ultimately led to his theory of group characters and representations.

Frobenius' trilogy on theta functions with characteristics was less about the theory of groups than it was about the numerous and highly complex relations among sets of elements in the characteristic group $\mathfrak{C}_g$ and the complicated formulas they make possible. I think Frobenius had these works especially in mind when he wrote in 1893:

> In the theory of theta functions it is easy to set up an arbitrarily large number of relations, but the difficulty begins when it comes to finding a way out of this labyrinth of formulas.
>
> Consideration of that mass of formulas seems to have a withering effect upon the mathematical imagination. Many a distinguished researcher who, through tenacious perseverance, has advanced the theory of theta functions in two, three, or four variables, has, after an outstanding demonstration of brilliant analytical talent, grown silent either for a long time or forever. I have attempted to overcome this paralysis of the mathematical creative powers by time and again seeking renewal at the fountain of youth of arithmetic [202, pp. 575–576].

In this passage, by "arithmetic" Frobenius did not mean the theory of numbers per se—his arithmetic period was over by 1882—but rather the theory of abstract finite groups, and especially nonabelian ones, that had begun to engage him in the midst of his work on density theorems.

Armed with techniques he had learned in that enterprise (Section 9.4), he began to venture out into the theory of finite groups without any application to number theory in mind. In March 1884, he submitted a little paper on Sylow's 1872 theorem that if a prime power p^ν divides the order of a permutation group, then it contains a subgroup of order p^ν. As Frobenius wrote, all previous proofs of this theorem "drag the symmetric group" into the argument, even though it is "completely foreign to the content of Sylow's theorem" [193, p. 301]. Frobenius presented a brief proof of the theorem as stated for an abstract finite group, and his proof is still one of the standard ones. This little paper was his first dealing with the theory of finite groups without any connection whatsoever to arithmetic applications. He followed this 2 years later (December 1886) with a lengthy paper on the theory of double cosets of abstract groups, in part a translation from permutation groups to abstract groups of what he had learned in working on density theorems. He also used the resultant theory to give proofs of abstract versions of all of Sylow's main theorems.

In addition to the above works on group theory, which appeared during his final years at the Zurich Polytechnic (1886–1891), Frobenius published six other papers that will be passed over here. Coming from the pen of Frobenius, they all contained new ideas and insights and clever calculations, but to my knowledge they do not warrant closer attention in the following chapters.[15]

---

[15]The papers in question are numbers 34, 37, 38, 39, 40, and 41 in Frobenius' *Abhandlungen*. My knowledge in the areas covered is limited, and I would be delighted to learn that I have underestimated the significance of some of these papers.

Chapter 3
Berlin Professor: 1892–1917

During Frobenius' initial years in Berlin (1867–1875), the mathematical leaders, Kummer, Weierstrass, and Kronecker, had worked together in personal and intellectual harmony as illustrated in Chapter 5; but during the 1880s, personal and philosophical differences between Kronecker and Weierstrass emerged. Weierstrass saw Kronecker's intuitionist views on mathematics as a threat to his own life's work in analysis, which was based on foundations rejected by Kronecker. Weierstrass' concerns were not merely a figment of his imagination. For example, in 1885, Kronecker wrote to H.A. Schwarz, declaring that his successors would finish what he had begun and "they will also recognize the incorrectness of all those conclusions with which so-called analysis currently operates."[1] Schwarz immediately shared the letter with his friend Weierstrass, and as a result, relations between Schwarz and Kronecker were completely severed. Weierstrass' distress was so great that despite increasingly bad health, he entertained a plan to move to Switzerland and devote his remaining energy exclusively to mathematical research [20, pp. 211–213]. The main reason why he did not carry out this plan was his realization that with Kronecker in power at the university—he had taken over Kummer's professorship in 1883—it would be impossible for his successor to be chosen from among those he deemed worthy, such as Schwarz, who was then a full professor in Göttingen. Fuchs had been a full professor in Berlin since 1884 but did not prove to be a staunch supporter of Weierstrass vis-à-vis Kronecker.

Weierstrass' health continued to deteriorate, but he held on to his professorship nonetheless. The situation changed unexpectedly on 29 December 1891 with the sudden death of Kronecker. Now Weierstrass was in a position to influence the choice of successors for both himself and Kronecker. The faculty committee, which consisted of the dean of the faculty (a philologist) and Fuchs, recommended Schwarz and Frobenius as the respective successors of Weierstrass and Kronecker.

---

[1]Letter of 14 November 1885 quoted in [22, p. 101] and translated by me. Much of the following information about the circumstances surrounding Frobenius' appointment are drawn from [22].

T. Hawkins, *The Mathematics of Frobenius in Context*, Sources and Studies in the History of Mathematics and Physical Sciences, DOI 10.1007/978-1-4614-6333-7_3,

Both recommendations were approved, and so in 1892, Frobenius became full professor at Berlin, with Fuchs and Schwarz as his colleagues.

The committee's report (dated 8 February 1892) appears to be in the handwriting of the dean and was probably influenced in terms of its content by Weierstrass as much or more so than by Fuchs, especially in its recommendation of Frobenius. It is worth quoting at length because of its perceptive characterization of Frobenius as a mathematician.[2]

> Professor Frobenius (b. 1849) stands above all the other mathematicians under consideration, not only by virtue of the extraordinary fruitfulness of his mathematical productivity, but above all by virtue of the universality of his mathematical talent. There is scarcely a branch of mathematics that he has not made the subject of his research. *Almost every one of his writings provides a formally polished representation of the material underlying a problem, upon which he then constructs an improved or extended solution to it.* Although in Mr. Frobenius the university would on the whole have a teacher who could hold lectures in almost any branch of mathematics by virtue of his own researches, above all the curriculum of the university would gain an important addition, which is based on Mr. Frobenius' special aptitude. In writings that are full of achievements he has by preference made the formal parts of analysis and the algebraic theory of forms the object of his studies and seems accordingly to have a special calling to represent these indispensable branches of science. In addition, Mr. Frobenius, who earlier had been active at our university as assistant professor, is an outstanding teacher.

The italics in the quotation have been added because the highlighted sentence is a particularly apt characterization of Frobenius' major mathematical publications up to that point in time. (Figure 3.1 portrays Frobenius as he may have looked at the time.)

The recommendation of Frobenius for membership in the Berlin Academy of Sciences, which more or less came with a full professorship at Berlin, is quite similar and confirms Weierstrass' role, since he was one of the signatories (along with Fuchs and the distinguished physicist and physiologist Hermann von Helmholtz). Indeed, the characteristics of Frobenius' work that are singled out for praise are characteristics that Weierstrass expressly advocated and aspired to in his own work. Written 9 months after the above-quoted document, it is worth quoting in part because it further develops the above-italicized characterization of the nature of Frobenius' work. The parts of the quotation within square brackets are additions in the hand of Fuchs, who probably composed them after discussing the matter with Weierstrass. What follows is my translation of a portion of the document.[3]

> [In all the disciplines that he treats, he turns to the formal foundations, and in the great majority of his works, starting from these foundations and in a characteristically original manner, he constructs anew an entire discipline from a unifying viewpoint, making it possible to see previously existing results of the discipline in an entirely new light, to fill previously existing gaps, and to create a formal foundation that provides an outstanding basis for subsequent investigations. Each of his larger works could justifiably be characterized as a little compendium for the discipline in question.]

---

[2]The entire document is transcribed in Biermann [22, pp. 206–209].

[3]The entire document is transcribed in [19, pp. 61–63].

Fig. 3.1 Frobenius as he may have looked circa 1896 when, at age 47 and in his fifth year as Berlin professor, he began creating his theory of group characters and representations. Photo courtesy of Institut Mittag-Leffler, Djursholm, Sweden

> All of his works are distinguished by a sure mastery of the material, and all contain either new results or known results in a new form. Mr. Frobenius is thus an outstanding stylist, who writes clearly and understandably without ever attempting to mislead by empty rhetoric.

As indicated already in Section 1.1, I believe that this characterization of much of Frobenius' mathematical work is right on the mark and reveals one of the principal reasons why it has had such widespread influence on the subsequent developments leading to present-day mathematics. This characterization of his work and the concomitant historical implications are confirmed in the following chapters that deal in greater depth with some of his writings and their role in subsequent developments. It is also from these chapters that the reader will gain an appreciation of the nature of Frobenius' remarkable and atypical genius as a mathematician.

By "the formal parts of analysis" the university report presumably referred to the algebraically oriented work on Fuchs' theory, on the problem of Pfaff, and on the theory of abelian and theta functions; and by "the algebraic theory of forms" it meant Frobenius' work on the linear algebra of bilinear forms, which he had also applied, e.g., to number theory and the theory of abelian functions. The entire quotation is a very accurate representation of Frobenius' mathematical activity up to 1892 and illustrates the impossibility of predicting the future on the basis of the past. Frobenius' predilection for the formal, algebraic aspects of all mathematics remained a constant during his years as Berlin professor, as did his penchant for writing pellucid "compendia," but the theory of finite groups, which is understandably not mentioned in the committee report, was quickly to become an increasingly major focus of his research during this period.

We saw in the previous chapter that Frobenius had begun dabbling in the theory of abstract finite groups while still in Zurich, but in Berlin, the intensity of his research in this area increased. It began with a paper of 1893 [200] related to the subject of his earlier paper of 1887 [193], namely Sylow's work of 1872. As Frobenius explained in the introduction to [200], Sylow's paper showed that simply knowing the order of a group, and hence the prime factorization of that order, one can draw far-reaching conclusions about the constitution of the group. If the order of a group $\mathfrak{G}$ is $\prod_{i=1}^{k} p_i^{e_i}$, then Sylow's theorem proves the existence of a subgroup of order $p_i^{e_i}$ for every i. Sylow had also proved that a group of order p^e is always solvable. In [200] Frobenius proved what he aptly regarded as a counterpoint to this theorem. That is, Sylow had proved that a group $\mathfrak{G}$ is solvable if its order is a product of *identical* primes, and Frobenius now proved the solvability of $\mathfrak{G}$ if its order is a product of *distinct* primes.

Not long thereafter, he began writing another paper, which appeared early in 1895 with the title "On finite groups" [204]. Here once again we see his penchant for reworking a subject systematically from a new point of view. Here the new viewpoint was supplied by his concept of a *complex*, which reflects Dedekind's influence on his work. (The content of [204] is described in Section 12.1.) He also published a third paper on finite groups in 1895 entitled "On solvable groups II" [206]. Using the results of "On finite groups" [204], he was able to generalize theorems and prove conjectures contained in his 1893 paper [200] mentioned earlier. At this time, there was a growing interest among mathematicians (e.g., Hölder, Burnside, Cole) in two problems: determining what groups are solvable and what groups are simple. Solvable groups were of interest in Galois theory, since the polynomials with solvable Galois groups were the ones solvable by radicals. Simple groups were of interest because the factor groups in a composition series of any finite group are simple, and thus the problem of classifying all simple groups was regarded as a significant step toward the classification of all finite groups.

Among the results Frobenius obtained in [206] relating to these problems I will mention two: (1) If $p < q < r$ are primes with $(p,q,r) \neq (2,3,5)$, and if the order of $\mathfrak{G}$ is of the form p^2qr^c, then $\mathfrak{G}$ is solvable [206, p. 692]; (2) (conjectured by Frobenius in [200]) among all groups whose order is a product of five (not necessarily distinct) primes, there are only three (nonabelian) groups that are simple, namely the groups of proper linear fractional transformations mod p for $p = 7, 11, 13$. These are now called the projective special linear groups $\mathrm{PSL}_2(p)$. As Frobenius and his contemporaries conceived of them, for a fixed prime p, $\mathrm{PSL}_2(p)$ consists of all linear fractional (or projective) transformations

$$w \equiv \frac{az+b}{cz+d} \pmod{p}, \quad ad - bc \equiv 1 \pmod{p}, \tag{3.1}$$

where a,b,c,d are integers. It was known that for groups whose order is a product of at most four primes, the only nonabelian simple group is $\mathrm{PSL}_2(5)$, which has order 60 and was known to be isomorphic to the alternating group $\mathfrak{A}_5$ and to Klein's icosahedral group. Thus Frobenius' result (2) was an extension of what was known at the time about nonabelian simple groups.

In January 1895, Frobenius had submitted his paper "On finite groups" [204] for publication in the proceedings of the Berlin Academy of Sciences, where he now published most of his work, and he was probably working on the group-theoretic results to appear in the two papers [205, 206] discussed above. Thus preoccupied with the theory of finite groups, he received a fateful letter from Dedekind dated 19 January 1895. Dedekind wrote to Frobenius about Kronecker. One of Frobenius' first duties as Kronecker's successor had been to write, in 1893, the memorial essay on the life and work of the late Kronecker for the Berlin Academy of Sciences [202]. Dedekind wrote to Frobenius to suggest that a letter he had received from Kronecker in 1880 was of sufficient mathematical interest to warrant publication in the proceedings of the academy. (This is the letter containing Kronecker's "Jugendtraum theorem" on abelian extensions of imaginary quadratic fields [149, p. 30].)

On 24 January 1895, Frobenius responded with a long, friendly letter. Besides expressing his agreement with Dedekind's suggestion, he touched on many matters of common interest: a quarrel Dedekind had with Hilbert, Weierstrass' failing health, the reactions of the Frobenius family to their new surroundings in Berlin, and so on. One passing remark turned out to be consequential. As mentioned above, Frobenius knew from his correspondence with Dedekind in the 1880s that the latter had used group-theoretic ideas in his work on prime factorizations of ideals in algebraic number fields. Since his forthcoming paper "On finite groups" was written from the abstract point of view of complexes—a viewpoint that Frobenius could reasonably expect Dedekind to wholeheartedly approve—it was only natural to call it to Dedekind's attention, and so he wrote, "I am curious what you will say about my work on the theory of groups that I will present next to the academy. I mean, I know you have concerned yourself with the subject but I do not know how far you have gone into it."

The correspondence with Dedekind that ensued is discussed and quoted at length in Chapters 12–13 because it documents how Dedekind provided Frobenius with a problem and a related theorem that eventually led him, starting in 1896, to his creation of the theory of group characters and representations for finite groups. The problem was the factorization of a determinant associated to a finite group $\mathfrak{H}$, namely, $\Theta = \det(x_{PQ^{-1}})$, where $\mathfrak{H} = \{E, A, B, \ldots\}$ and $x_E, x_A, x_B, \ldots$ are $h = (\mathfrak{H} : 1)$ independent complex variables. Thus Θ is a homogeneous polynomial of degree h in h variables. The related theorem was that when $\mathfrak{H}$ is abelian, the group determinant factors into linear factors: $\Theta = \prod_{\lambda=1}^{h} \left[\sum_{R\in\mathfrak{H}} \chi^{(\lambda)}(R)x_R\right]$, where the $\chi^{(\lambda)}$ denote what Dedekind called the characters of $\mathfrak{H}$. Dedekind introduced this notion as a generalization of seemingly disparate considerations that underlay arithmetic work of fundamental importance by Gauss and Dirichlet. By Dedekind's definition $\chi : \mathfrak{H} \to \mathbb{C} \sim \{0\}$ is a character on an abelian group $\mathfrak{H}$ if for all $R, S \in \mathfrak{H}$ one has $\chi(RS) = \chi(R)\chi(S)$. Today, these are called first-degree or linear characters.

Computation of Θ for several nonabelian groups showed Dedekind that the prime factorization of Θ in $\mathbb{C}[x_E, x_A, x_B, \ldots]$ involves some nonlinear prime factors, and further computations suggested to Dedekind the problem of determining an algebra, related to the group algebra of $\mathfrak{H}$, such that Θ factors into linear factors if coefficients from this algebra are allowed. He communicated this problem and the above-mentioned theorem to Frobenius and gave him permission to do with them whatever he wanted.

Frobenius was fascinated by Dedekind's group determinant problem, but he preferred posing it in a different, more straightforward manner: if $\Theta = \prod_{\lambda=1}^{l} \Phi_\lambda^{e_\lambda}$ is the prime factorization of Θ in $\mathbb{C}[x_E, x_A, x_B, \ldots]$, how are l, e_λ, and $f_\lambda = \deg \Phi_\lambda$ related to the structure of the underlying group $\mathfrak{H}$? Frobenius' letters to Dedekind amount to enlightening progress reports on his many-sided attacks on this problem. Out of these efforts came the fact that if $\Phi_\lambda = x_E^{f_\lambda} + \psi^{(\lambda)}(A) x_E^{f_\lambda - 1} x_A + \cdots$, then the functions $\psi^{(\lambda)} : \mathfrak{H} \to \mathbb{C}$, with $\psi^{(\lambda)}(E)$ set equal to f, satisfy remarkable orthogonality relations, for which Frobenius found many uses and so began to think the functions $\psi^{(\lambda)}$ especially important. Notice that if $\mathfrak{H}$ is abelian, Dedekind's above-mentioned theorem implies that $f_\lambda = 1$ for all λ and that $\psi^{(\lambda)}$ coincides with one of Dedekind's characters. Thus the functions could be regarded as generalizations of Dedekind's characters to nonabelian groups, and for this reason, as the properties of the $\psi^{(\lambda)}$ proved increasingly useful, Frobenius decided to call them characters as well. Computed examples suggested that l always equals the number k of conjugacy classes of $\mathfrak{H}$, and when Frobenius finally observed that the functions $\psi^{(\lambda)}$ remain constant on these classes, he was able to use the orthogonality relations to prove that $l = k$ (Section 13.2). The computed examples also suggested that $e_\lambda = f_\lambda$, so that $h = \deg \Theta = \sum_{\lambda=1}^{k} f_\lambda^2$. Frobenius eventually proved $e_\lambda = f_\lambda$ but it took him several months to succeed and required what he jokingly called the "principle of the horse trade" (for reasons indicated in Section 13.4).

While Frobenius was thinking about why e_λ and f_λ must be equal, a technique of variable specialization he had first introduced in his work on theta functions with characteristics (see Section 12.4) led him to a remarkable discovery: Specialize the variables $x_E, x_A, x_B, \ldots$ by setting $x_P = x_Q$ if P and Q are conjugate. Then Θ changes from a polynomial in h variables to a polynomial Θ^* in k variables, and the prime factorization $\Theta = \prod_{\lambda=1}^{k} \Phi_\lambda^{f_\lambda}$ becomes $\Theta^* = \prod_{\lambda=1}^{k} \left[\Phi_\lambda^*\right]^{f_\lambda}$, where, as he showed,

$$\Phi_\lambda^* = \left(\frac{1}{f_\lambda} \sum_{R \in \mathfrak{H}} \psi^{(\lambda)}(R^{-1}) x_R^* \right)^{f_\lambda} = \left(\frac{1}{f_\lambda} \sum_{\rho=1}^{k} h_\rho \psi_{\rho'}^{(\lambda)} x_\rho \right)^{f_\lambda}.$$

Here ρ' denotes the conjugacy class of R^{-1} for R in the ρth class, and $x_1, \ldots, x_k$ denote the distinct specialized variables with x_ρ the variable for all x_S with S conjugate to R and R in the ρth conjugacy class. Frobenius was very happy with this result because it showed that Θ^* was a product of linear factors with his generalized characters $\psi^{(\lambda)}$ providing the coefficients—a satisfying generalization of Dedekind's theorem.

At the same time, as a result of returning reluctantly to Dedekind's "group algebra" approach to the factorization of Θ, he discovered, by means of a theorem on matrix algebra he had proved in 1877, that if $A = (a_{\alpha,\beta})$ is the $k \times k$ matrix with $a_{\alpha,\beta} = \sum_{\gamma=1}^{k} \frac{h_{\alpha\beta\gamma}}{h_\alpha} x_\gamma$, where h_α denotes the number of elements in the αth conjugacy class and $h_{\alpha\beta\gamma}$ denotes the number of distinct solutions (A,B,C) to $ABC = E$ with A,B,C in the αth, βth, and γth conjugacy classes, respectively, then

$$\det A = \prod_{\lambda=1}^{k} \left(\frac{1}{f_\lambda} \sum_{\alpha=1}^{k} h_\alpha \psi_{\alpha'}^{(\lambda)} x_\alpha \right).$$

The significance of this formula was that the matrix A is defined completely in terms of constants $h_\alpha, h_{\alpha\beta\gamma}$ directly related to the structure of the group $\mathfrak{H}$, and so the characters $\psi^{(\lambda)}$ are directly related to $\mathfrak{H}$. This enabled him to write a paper "On group characters" [211] using the $\det A$ formula to define the characters $\psi^{(\lambda)}$ by means of matrix algebra and pure group theory (Section 13.3). In [211], however, the characters $\psi^{(\lambda)}$ were defined only up to a "temporarily undetermined" constant f_λ, and the connection with the prime factors Φ_λ and the problem of factoring Θ, on which he was still working, trying to show $e_\lambda = f_\lambda$, was never mentioned.

After he had proved $e_\lambda = f_\lambda$ and published his paper on the factorization of Θ in December 1896 [212], Frobenius discovered, once again prompted by some observations by Dedekind, that he could translate his theory of the factorization of the group determinant into the language of matrix representations of the group $\mathfrak{H}$, by which he meant a group homomorphism $\sigma : \mathfrak{H} \to \mathrm{GL}(n,\mathbb{C})$ for some n called the degree of σ (Section 13.5). Then if ρ denotes the left regular representation of $\mathfrak{H}$ and we set $\rho(x) = \sum_{R\in\mathfrak{H}} \rho(R)x_R$, it turns out that $\rho(x) = (x_{PQ^{-1}})$, so that $\Theta = \det\rho(x)$. Furthermore, the factorization $\Theta = \prod_{\lambda=1}^{k} \Phi_\lambda^{f_\lambda}$, $f_\lambda = \deg\Phi_\lambda$, translates into the theorem that a nonsingular matrix L exists such that (to use some modern language) $L^{-1}\rho(x)L$ is a matrix direct sum of k irreducible representations σ_λ of $\mathfrak{H}$ with each σ_λ occurring $f_\lambda = \deg\sigma_\lambda$ times in the direct sum. In Frobenius' terminology ρ is thus *equivalent* to the above direct sum of the σ_λ. Moreover, if we set $\sigma_\lambda(x) = \sum_{R\in\mathfrak{H}} x_R\sigma_\lambda(R)$, then $\det\sigma_\lambda(x) = \Phi_\lambda$. Furthermore, the generalized character $\psi^{(\lambda)}$ is simply the trace function of σ_λ: $\psi^{(\lambda)}(R) = \mathrm{tr}\,\sigma_\lambda(R)$. These results were published in 1897. Two years later, Frobenius published a sequel (discussed in Section 15.2) in which he presented a more general complete reducibility theorem than the one above, which relates only to the regular representation ρ. In the sequel he proved that any representation of $\mathfrak{H}$ is equivalent to a direct sum of the above irreducible representations σ_λ, so that in particular, the σ_λ are the only irreducible representations of $\mathfrak{H}$.

Even before he had published the above sequel, Frobenius, an expert at calculation, began thinking of devising theoretical means to facilitate the computation of character tables for groups of large order. The most powerful tool he created was his theory of induced characters and representations (1898), including the reciprocity theorem that still bears his name (Section 15.1). He used this theory to determine the character tables for the general symmetric and alternating groups in 1900–1901. He also sought to apply his theory of group characters to major problems of the day in the theory of finite groups: determine which groups are simple and which are solvable. One of his most important results was his 1901 result (Theorem 15.4), proved using his theory of induced characters, that if $\mathfrak{H}$ has the properties that make it what is nowadays called a Frobenius group, then it contains a normal subgroup (now called the Frobenius kernel of $\mathfrak{H}$) and so is not simple. A proof of Frobenius' Theorem 15.4 that does not use character theory has yet to be found, and all proofs are essentially variants of the one Frobenius gave.

Two papers from 1901–1902 by the British mathematician Alfred Young caused Frobenius to return to the study of the representations of the symmetric group $\mathfrak{S}_n$ (Section 15.2). Young's papers, although presented as contributions to the theory of invariants, were really about the structure of the group algebra of $\mathfrak{S}_n$ and, as Frobenius could see, had implications for the characters and representations of $\mathfrak{S}_n$. This was because in his second paper on matrix representations of groups (1899) Frobenius had introduced his theory of primitive characteristic units of a group, which were analogous to the irreducible characters of a group but more fundamental in that they led Frobenius to a remarkable theorem giving an explicit formula for the irreducible representation associated to a primitive characteristic unit (Theorem 15.2).[4] When he read Young's papers, Frobenius realized that Young had unwittingly focused on the primitive characteristics of $\mathfrak{S}_n$, which Young associated to the tableaux that now bear his name, although he had not managed to fully determine the numerical coefficients that occur in his formulas defining them. Frobenius showed how to do this using the characters of $\mathfrak{S}_n$ (see Theorem 15.3 and what follows). It was Frobenius' reformulation and completion of Young's results that led Hermann Weyl in 1925 to see how the "Young symmetrizers" introduced by Frobenius may also be used to determine the irreducible representations of the special linear group.

Frobenius was very fortunate that Dedekind decided to tell him about group determinants, because this notion had not been introduced in print by any mathematician, nor does it seem to have been "in the air" at the time. Otherwise, it is unlikely that Frobenius would be known today as the creator of the theory of group characters and representations. This is not to say that the theory would have remained undiscovered for a long time. On the contrary, three lines of mathematical investigation—all of which Frobenius was known to dislike—were leading to essentially the same theory as Frobenius had begun to explore: (1) the theory of noncommutative "hypercomplex number systems" (linear associative algebras over $\mathbb{C}$); (2) Lie's theory of continuous groups; and (3) Felix Klein's research program on a generalized Galois theory. These matters are discussed in Chapter 14. The first

---

[4]In the module-theoretic approach to representation theory, primitive characteristic units correspond to the primitive idempotents of the group algebra. The left ideal generated by such an idempotent induces by left multiplication the corresponding irreducible representation of the group algebra.

line was pursued in 1893 by the Estonian mathematician Theodor Molien, who was also influenced by line (2) in the form of Wilhelm Killing's groundbreaking work on Lie algebras (Section 14.2). Then, in 1897, under the influence of line (3), Molien applied his results to the group algebra of a finite group in 1897 to independently obtain some of Frobenius' basic results about group representations and characters. Molien's theory of hypercomplex number systems was also further investigated by Élie Cartan, who had begun his career by reworking and correcting Killing's work on Lie algebras. Line (2) led William Burnside to the brink of discovering the beginnings of Frobenius' theory when he learned of Frobenius' work (Section 14.3). The experience turned Burnside into the earliest exponent of Frobenius' theory as a valuable tool for the study of finite groups and led to a friendly rivalry between the two mathematicians that was characterized by many instances of independent rediscovery of each other's results about finite groups, as indicated in Section 15.4. Line (2) also led Heinrich Maschke to inadvertently rediscover Frobenius' general complete reducibility theorem in the sense that this theorem was implicit in the reasoning he used to solve a problem about finite groups of linear transformations (Section 14.4).

Although Frobenius himself did not think that the theory of hypercomplex number systems should be regarded as a basic mathematical tool, he was impressed by Molien's results and felt compelled to develop them more rigorously using the tools of the theory of determinants. He accomplished this in two papers from 1903 (Section 15.3) and then proceeded, in another paper from 1903, to apply the resultant theory to deal with results of Cartan that went beyond Molien's and involved the determinants associated to the left and right representations of a hypercomplex system, which can differ when the system is not semisimple. Cartan gave formulas relating the prime factorizations of these two determinants, which always have the same prime factors but to differing powers. His formulas involved integers c_{ij} that have since become known as Cartan invariants. Later, they played a fundamental role in Brauer's modular theory of representations, as will be seen in Section 15.6. Frobenius applied the results of his reworking of Cartan's results to obtain necessary and sufficient conditions that the two above determinants have the same prime factorization. The systems characterized by Frobenius' conditions were termed Frobenius algebras by Brauer, and that is how they are still known today.

By 1903, Frobenius' major contributions to the representation theory of groups and algebras had been made, although several noteworthy contributions still lay ahead. In 1904, he published a paper on the characters of multiply transitive permutation groups (discussed at the end of Section 15.2) containing many interesting results. Among other things, he introduced a new way to present the character table of the symmetric group $\mathfrak{S}_n$ and then applied it to determine the character tables of the two 5-fold transitive Mathieu groups $\mathfrak{M}_{12}$ and $\mathfrak{M}_{24}$, some of the earliest examples of what are now called sporadic simple groups. Along the way, he showed that $\mathfrak{M}_{24}$ contains a subgroup isomorphic to $\mathfrak{M}_{12}$, something that Mathieu had not realized. As a consequence, it followed that isomorphic copies of all the Mathieu groups are contained in $\mathfrak{M}_{24}$, a subgroup of $\mathfrak{S}_{24}$ of order 244,823,040. His last

Fig. 3.2 Frobenius as he looked in his later years. Original is located in the Porträtsammlung, Universitätsbibliothek der Humboldt-Universität zu Berlin and is used with permission

paper having anything to do with representation theory was published in 1907 and involved a generalization of a theorem in the theory of groups made possible using characters [227].

By that time, he was 58 years old and his health was deteriorating. (The photograph displayed in Fig. 3.2 was perhaps taken at about this time.) Most of his publications after 1907 involved tying up mathematical loose ends and reworking the results of others—something that Frobenius had done with brilliance since the outset of his career. Many of them had to do with the theory of numbers, a subject he taught at the university. In one work, published in 1912, however, Frobenius produced another, his final, masterpiece. It had to do with the remarkable theory of nonnegative matrices, which he had developed masterfully in response to the equally remarkable discoveries of a young mathematician named Oskar Perron. These matters are discussed in Chapter 17. The resultant theory (sometimes referred to generically as Perron–Frobenius theory), although motivated by purely mathematical interests, later provided the mathematical foundation for a broad spectrum of applications to such diverse fields as probability theory, numerical analysis, economics, dynamical programming, and demography. Section 17.3 is devoted to the first such application of the theory, which was to the probabilistic theory of Markov chains.

The above discussion of Frobenius' years as full professor at the University of Berlin has been focused on the progression of his mathematical research. I now turn to the institutional context within which he worked.[5] The year 1892, when Frobenius at age 43 returned to Berlin as full professor, marked the transition from one era of the Berlin school of mathematics to another. During 1855–1892, the school had been led by Kummer, Weierstrass, and Kronecker, and its golden years more or less began with the period 1867–1875, when Frobenius was there. When Frobenius returned to Berlin as full professor, Kummer and Kronecker were dead and Weierstrass was retired. The other full professors of mathematics were Fuchs, who was appointed in 1884, and Schwarz, who had come at the same time as Frobenius. At this time also, Kurt Hensel (1861–1941), who was Kronecker's student (Ph.D., 1884) was made an assistant professor.

During his first years as Berlin professor, the presence of Weierstrass, to whom Frobenius was devoted, must have been a comfort. As Frobenius explained to Dedekind in a letter of 24 January 1895,

> Weierstrass' condition is always the same, not better, but fortunately, also not worse. He does not leave his room, spends the entire day, and often also the entire night, sitting in his armchair, gets his morphine injections with great regularity, bears his affliction like a hero, is still always mentally lively, and is happy when his friends visit quite often. He is not in great pain, and so still knows how to chat interestingly and to tell all kinds of stories.

By September 1896, however, Weierstrass' condition had worsened. After not visiting him for 5 weeks (due to an extended summer vacation), Frobenius was greatly troubled by Weierstrass' "considerably increased frailty."[6] Regarding a visit a week later, Frobenius wrote to Dedekind (6 September 1896): "Last night I was with Weierstrass. He slept almost the entire time (after an injection), and I conversed with his sister, who counts on him for little and hence is very, very lonely." Weierstrass died a few months later, in February 1897, while Frobenius was in the midst of developing his new theory of group characters and representations.

In 1902, Fuchs died unexpectedly, just short of his 69th birthday. A faculty committee was quickly formed to come up with the customary list of three candidates for the open full professorship, ordered according to the preferences of the committee. The committee of course included the remaining two full professors of mathematics, Frobenius and Schwarz, as well as the dean (a philosopher) and four other full professors, one of whom was the theoretical physicist Max Planck. Presumably the opinions of the mathematicians were given more weight in the deliberations, and the memorandum containing the list of candidates and the rationale behind their choice and ordering was written by a mathematician, in this case Frobenius.[7] The memorandum was sent to the Prussian minister of culture, who had the power to accept or reject the recommendation of the memorandum. At this time, the minister was Friedrich Althoff (1839–1908), a powerful and intimidating

---

[5]For further details on Frobenius' role as professor in Berlin see Chapter 6 of Biermann's important study of mathematics at the University of Berlin [22], which also contains several documents written by Frobenius.

[6]Letter to Dedekind dated 4 September 1896.

[7]Biermann has included this important document in its entirety in his book [22, pp. 209–211].

official who was on good terms with Felix Klein, whose organizational talents and ideas about private financing of educational projects he appreciated. Ever since the Kummer–Kronecker–Weierstrass era, Klein had been held in low esteem by the Berliners, who watched in dismay as Klein moved to Göttingen in 1886 and began creating a school of mathematics there that by 1902 was threatening to eclipse Berlin as the foremost center for mathematics in Germany.[8]

It was not Klein's own mathematical talent that was responsible for this turn of events; rather it was his organizational ability and his keen eye for promising new mathematical talent. In 1895, he managed to get David Hilbert a full professorship at Göttingen. In 1888 Hilbert had proved his famous finite basis theorem in the theory of invariants, and Klein was quick to perceive the enormous originality and talent of the 26-year-old Hilbert. Having established his finite basis theorem, Hilbert began working in the theory of numbers, and in 1897 published his now famous "*Zahlbericht*," which contained a brilliant synthesis of the work of Kummer, Kronecker, and Dedekind that Hilbert used to introduce entirely new ideas, ideas that ultimately led to the creation of class field theory (see Section 15.6.3). Next he turned to work on the foundations of geometry, and here, too, introduced highly original ideas that culminated in his book *Grundlagen der Geometrie* (1899). In 1900, it was awarded the Steiner Prize of the Berlin Academy of Sciences. Given his achievements and relative youth, Hilbert was the obvious first choice to replace Fuchs, and this is what the committee as a whole decided, with Schottky, still a full professor at the University of Marburg, as second choice and Otto Hölder as the third choice.

Although the memorandum spoke for the entire committee, its content, in its specifics, expressed the views of the two mathematicians on the committee and especially the views of its author, Frobenius. Frobenius' statement that in terms of mathematical accomplishments, Schottky was Hilbert's equal [22, p. 210], reflects an underassessment of Hilbert's achievements, due, it would seem, to an inadequate familiarity with some of his work,[9] and a corresponding overassessment of Schottky's achievements, due no doubt to his friendship with Schottky, his colleague for 10 years in Zurich, and also to the fact that he was one of Weierstrass' top students, and so someone who would help preserve the traditions of the Weierstrass era. When Frobenius wrote that "in many respects" Schottky would

---

[8]On Klein's role in making Göttingen into the leading center for pure and applied mathematics in Germany, see the informative account by Rowe [510].

[9]Frobenius' degree of familiarity with Hilbert's massive *Zahlbericht*, published only 5 years earlier, was probably not great. He had done no research on algebraic number theory since 1880 when he worked on density theorems; and although he taught the subject at Berlin, he probably presented it more or less in accordance with Dedekind's rendition of the theory. (Cf. Frobenius' 1909 letter to H. Weber [230].) This would account for his one-sentence evaluation of the *Zahlbericht* as an "outstanding" report that had succeeded in "filling numerous gaps in earlier developments" of the theory of algebraic numbers [22, p. 210]. Evidently, Frobenius did not realize that Hilbert had done much more than fill gaps, that he had introduced many new and fertile ideas into the theory. Of course, it is much easier to realize this in hindsight, after the creation of class field theory.

better complement himself and Schwarz than would Hilbert [22, p. 210], one cannot help but wonder whether that "complementarity" had something to do with the fact that with the inclusion of Schottky, all three full professors would be from the Weierstrass school. According to the memorandum, Hilbert was chosen over Schottky for two reasons, reasons that were no doubt especially compelling for the nonmathematician members of the committee: (1) Hilbert was highly successful as a teacher, drawing many students to him and to his lectures, whereas Schottky was not; (2) Schottky had published very little during the past 10 years, whereas much was rightly to be expected from Hilbert.

Although Frobenius was supporting the choice of Hilbert with his head, he was evidently supporting the choice of Schottky with his heart. No doubt Althoff could read this between the lines. If Althoff had the interests of mathematics at Berlin uppermost in his mind, he would have pressured Hilbert to accept the call to Berlin, but he did not. The interests of his friend Klein and what he could accomplish at Göttingen, given what he had already accomplished, seem to have had a higher priority; thus Althoff put no pressure on Hilbert to accept, and when a new assistant professorship in mathematics at Göttingen was proposed by Klein and Hilbert, evidently as a reason for Hilbert to stay, Althoff astonished them by offering instead a new full professorship.[10] That professorship was offered to Hilbert's close friend Minkowski, and Hilbert chose to remain at Göttingen. With the addition of Minkowski, Göttingen, like Berlin, now had three full professorships in mathematics, but this did not put Göttingen on a par with Berlin; in reality Göttingen now surpassed Berlin as the leading German center for mathematics.

When Althoff communicated the bad news about Hilbert to Frobenius' committee, he also rejected offering the Berlin position to the other proposed candidates, Schottky (a poor teacher) or Hölder (too sickly), which was the first time the mathematicians recommended by the faculty had been flatly rejected [22, p. 131]. But that was not all. Althoff proposed his own list of candidates, clearly assembled in consultation with Klein: Friedrich Schur and Friedrich Engel, who worked on matters related to Klein's friend Lie's theory of transformation groups, Hans von Mangoldt, a Berlin Ph.D., whose principal achievement was to have given complete proofs of two theorems only partially proved by Riemann in his celebrated 1859 paper on the prime number theorem, and Carl Runge, another Berlin Ph.D. with considerable talent but who had switched from mathematics to physics circa 1887. (In 1904, Klein managed to get Runge a full professorship in applied mathematics at Göttingen.)

None of Althoff's proposed candidates were worthy of a full professorship in pure mathematics at Berlin. In fact, in terms of achievements in pure mathematics, Schottky stood far above them all. No doubt under the urging of Frobenius, who

[10] I am grateful to David Rowe for calling these facts to my attention. Many of them are contained in a draft of a letter dated 24 June 1902 that Klein sent to Althoff. The letter is located in the archives of the Niedersächsische Staats- und Universitätsbibliothek, Göttingen (Cod. Ms. F. Klein I D, 33–34).

must have been infuriated by Althoff's Klein-inspired candidate recommendations, the committee sent a second memorandum to Althoff, also written by Frobenius, with Schottky, H. Weber, and Kurt Hensel (then an assistant professor at Berlin) as first, second, and third choices.[11] Although Frobenius' arguments on behalf of Schottky were specious and his arguments against Althoff's proposed candidates lacking in tact, for whatever reason, the ministry relented and Schottky was appointed as Fuchs' replacement. What appeared as a victory for Frobenius, however, was just the opposite. At Berlin, as could have been predicted, Schottky did relatively little to enhance the prestige of mathematics there or to draw mathematics students to the university by virtue of his teaching. His appointment contributed to the widening gap between Göttingen and Berlin [22, p. 133]. Weber, who was 60, would probably have been a better temporary choice. Furthermore, Frobenius had lost all credibility with the Ministry of Culture, and, as we shall now see, this hurt his more justifiable efforts to obtain a full professorship at Berlin for his prize student, Issai Schur.

During his 25 years as Berlin professor, Frobenius had a total of ten doctoral students. Two of them are worth mentioning here.[12] His fourth student was Edmund Landau (1877–1938). Both Landau's doctoral dissertation (1899) and his subsequent work leading up to his habilitation to qualify for a Berlin instructorship (1901) were focused on the analytic number theory surrounding the Riemann hypothesis. Although Frobenius approved Landau's appointment as instructor, he felt that Landau's work was too narrowly focused and encouraged him to broaden his research interests [22, p. 128]. Landau seems to have taken Frobenius' advice, for in 1902 he published a paper further developing ideas in Frobenius' 1873 paper [175] on irreducible linear differential equations (discussed above in Section 1.2). Landau's extension of Frobenius' ideas and results played a role in the developments leading up to the work of Krull and the module-theoretic approach to elementary divisor theory (Section 16.3.1). Landau was a stimulating and popular lecturer at Berlin, and during his years there as instructor (1901–1909), he introduced new areas into the curriculum such as set theory and integral equations.[13] In 1904, Schwarz wrote a memorandum to the ministry proposing Landau for an assistant professorship [22, pp. 219–221], but the ministry did not approve it. Again in 1908 Landau was proposed (this time by Frobenius) for such a position and again the proposal was rejected [22, pp. 137–139]. The following year, however, Landau was offered, and accepted, the full professorship at Göttingen vacated by the death of Minkowski. Althoff had passed away by then, but the ministry was still in Klein's pocket,[14] and Göttingen, not Berlin, was the new center of gravity for German mathematics [22, p. 138].

[11] This document can also be found in Biermann's book [22, pp. 211–216].

[12] A third talented Frobenius student was Robert Remak, whose doctoral dissertation (1911) is discussed in Section 16.3.2 because of its influence on Krull.

[13] The following discussion of Landau is drawn from Biermann's book [22, pp. 136–139].

[14] See in this connection Rowe's quotation from a letter of 1904 to Klein from a ministry official [510, p. 197].

Fig. 3.3 Schur as he may have looked during his first Berlin period. Frobenius rightly perceived Schur's brilliance as a mathematician and envisioned him as his successor at Berlin. Photograph courtesy of Mathematisches Forschungsinstitut, Oberwolfach, Germany

The only one of Frobenius' doctoral students who elicited exalted praise from him was Issai Schur (1875–1941), his fifth doctoral student (Fig. 3.3). Starting with his doctoral thesis (1901), much of Schur's work for many years was related to Frobenius' theory of group characters and representations and involved brilliant and far-reaching extensions and reformulations of the theory, as can be seen in Section 15.5, where his work is discussed at length. From the outset, Frobenius fully appreciated Schur's extraordinary mathematical talent, which was focused on an algebraic approach to mathematical problems that was similar in spirit to his own. Before long, he envisioned Schur as his successor at Berlin,[15] but as we shall now see, Frobenius did not live to see his hopes for Schur realized.

Despite his extraordinary mathematical talents and accomplishments, Schur remained an instructor at Berlin for 10 years (1903–1913), without receiving any outside offers [22, p. 139]. Finally, in 1913, he was offered, and accepted, an assistant professorship at the University of Bonn. Given his considerable mathematical achievements by then, it is somewhat surprising and regrettable that he did not receive an offer of a full professorship.[16] Perhaps this had something to do with what Frobenius characterized as Schur's only fault: his "extreme modesty" [22, p. 225].

---

[15]This expectation is expressed in a letter of 1913; see [167, pp. 12–15].

[16]Similar sentiments were expressed by E. Steinitz in a letter to Schur dated 13 May 1913 [407, pp. lxii–lxiii].

But it may have had more to do with the ministry's antagonism toward Frobenius and its distrust of his words of praise, which in the case of Schottky had been exaggerated and misleading.

Frobenius did manage to get Schur back to Berlin to fill the assistant professorship that had opened up due to the death of Johannes Knoblauch. In a postcard dated 15 December 1915, Frobenius sent the good news to Schur[17]:

> Dear Friend,
>
> On Friday, the 10th of December, I went to the ministry and there Naumann promised to accept your conditions and to write to you the next day. Now I did not want to anticipate him, and yet wanted to be the first to congratulate you. Therefore I chose this droll and apparently purely sentimental form. From the tone of your reply I believe I may infer that you will be pleased to come here. ... Perhaps you will get in touch with Knopp and ask him to find a suitable apartment even now. I understand there are 65000 empty ones. If you (or your wife) would like to visit me, please let me know in advance. I am still not particularly well, despite the many Sugodin that I swallow. Well, enjoy your last Christmas in Bonn as much as possible in these hard times. With most cordial greetings from house to house.
>
> Your old friend
> Frobenius

As this card reminds us, not only was Frobenius in poor health by this time due to his heart condition, but Germany was undergoing the attendant "hard times" of the First World War (whence Frobenius' caustic joke about 65,000 vacant apartments).

Schur's position in Berlin was to commence in the spring of 1916. Before then, Frobenius wrote Schur on 19 February 1916 to express his sympathy at Schur's loss of his mother. "I know how much you were attached to your mother although you lived far apart. Basically it is only after this blow, which no one is spared, that one becomes a grown man who stands on his own two feet. The terrible circumstances of our time make this loss especially painful." The letter then proceeded to give further information about Frobenius' health and circumstances:

> I have been considerably better for the past eight days. Since then the albumen has vanished completely. My appetite is excellent and I sleep well. The three attacks that I suffered last year were not asthma but angina pectoris. Perhaps they were complicated by some bronchial asthma.
>
> My doctor is a rather odd gentleman, who proceeds rather slowly and tries out one thing after another. Thus 4 weeks ago he prescribed the oxygen baths twice a week, which do me a lot of good. Two weeks ago he gave me Théocine, which contains a lot of digitalis and seems to have worked well.

---

[17]This quotation and the following one are taken from the essay "The life of Issai Schur through letters and other documents" by W. Ledermann and P. Neumann [331, pp. lxiii–lxvi], which includes transcriptions of the original German as well as the quoted translations, which I have followed except for minor modifications. The published correspondence is part of a collection of 15 letters and postcards that Frobenius sent to Schur between 1901 and 1917. They were supplied by Susan Abelin, Schur's great-granddaughter. The original documents are in the possession of Hilde Abelin, Schur's daughter. I am grateful to Susan Abelin for granting me permission to obtain a copy of these letters and to Walter Ledermann for initially calling their existence to my attention and for supplying me with copies.

> I have been able to give my lectures without interruption. Of course I have to take care of myself. My colleagues have taken over the seminar from me. Even in the summer I cannot think about holding it in addition to my lectures.
>
> ...
>
> I have received a great deal of sympathy during my illness. But I am very glad to know that so old and valued a friend as you will again be nearby. For in one's old age the band of friends thins out considerably.

Although by the spring of 1916 Schur had moved to Berlin, Frobenius' health continued to decline, and he sometimes communicated with Schur on mathematical matters via postcard from his home in Charlottenburg.[18]

In March 1916, Schwarz, whose mental powers were deteriorating, resigned. His position was filled by the faculty committee's first choice, Erhardt Schmidt (1876–1959), one of Hilbert's students, who was also an analyst but of a different sort from Schwarz, having specialized in Hilbert's theory of integral equations. Despite his poor health, Frobenius composed most of the memorandum (dated 9 March 1917 [22, pp. 224–226]), which had Schur as the second choice and gave Frobenius many opportunities to promote his favorite, e.g., "the many-sided Schur is to Landau [now a Göttingen full professor] as genius is to talent" [22, p. 224]. True as these words were, they no doubt irritated the ministry. This memorandum contained Frobenius' final words to the ministry in praise of Schur, for he passed away on 3 August 1917, shortly before Schmidt took up his duties as Schwarz's replacement.

With the appointment of Schmidt, mathematics at Berlin began its transition to a new era with new faculty and a new organization that, like Göttingen under Klein and Hilbert, exhibited greater tolerance for applied mathematics. The death of Frobenius opened up a chair in mathematics. Schwarz (as emeritus professor) and Schottky put forward Schur as Frobenius' replacement, in accordance, as they stressed, with Frobenius' wishes but also "by virtue of his penetrating mind, uncommon knowledge of mathematical research in all areas, including geometric ones—and not least of all on account of his noble character" [22, p. 143]. However, both the physicist Max Planck and Schmidt were against proposing Schur as the exclusive first choice; they argued that Schur and Constantin Carathéodory (1873–1950) be proposed on an equal footing as the top-ranking choices. Carathéodory, who was Schmidt's close friend [15, p. 105], had been born in Berlin and had studied at the university for a while before going to Göttingen, where he obtained his doctorate in 1904 under the direction of Minkowski. Since 1913 he had been full professor at Göttingen, replacing Klein.

Schmidt wrote the report to the ministry on the faculty recommendations and pointed out that due to his areas of research, "Schur is suited like no other to fill the gap left by the death of Frobenius. Indeed, it was Frobenius' own parting wish that Schur be his successor" [22, p. 227]. However, since Schur had already been put forward unsuccessfully three times for a full professorship, Schmidt did not evaluate

[18] Schur's Nachlass contains four such postcards, two from the second half of 1916 and two from the first half of 1917.

Schur's mathematics but simply referred to the earlier recommendations (written by Frobenius, whose opinions the ministry distrusted) and went on to extol at great length the virtues of Carathéodory. Given Schmidt's report and the bad relations that had developed between Frobenius and the ministry, it is not entirely surprising that the latter chose Carathéodory.

Carathéodory commenced his duties at Berlin in the fall of 1918, but remained there only 15 months. His ancestry was Greek, and when the Greek government invited him to oversee the creation of a new university in Greece, he agreed out of a sense of patriotic duty. Again, a Berlin professorship became available, and one would think Schur would have obtained it, but he did not. Instead, at the end of 1919, he was given a "personal full professorship," which amounted to an unchaired full-professorial title with the compensation of an assistant professor [22, p.152, n.1]. Schmidt, who had more or less assumed the leadership of the Berlin mathematicians, next focused on setting up a full professorship in applied mathematics, probably influenced by his experiences at Göttingen, where Hilbert and Klein had advocated—and practiced—applying mathematics, especially to physics.[19] He finally accomplished this with the appointment of Richard von Mises (1883–1953) in the spring of 1920. (As we will see in Section 17.3.2, in 1931 von Mises applied Frobenius' theory of nonnegative matrices to deal with a problem in the foundations of statistical mechanics.)

It was not until Schottky retired in 1921 that Schur was finally made a chaired full professor at Berlin. At last the algebraic tradition that Frobenius had established at Berlin could be continued by his chosen successor! During his 15 years as full professor at Berlin—his tenure being cut short by the anti-Semitic policies of the Nazi regime—Schur had many doctoral students (22 compared with Frobenius' total of 10 over 25 years), students who were devoted to him and to his style of algebra, which they continued to develop.[20]

Among the several distinguished mathematicians who emerged from the Schur school of algebra at Berlin was Richard Brauer (1901–1977), whose work in particular involved important extensions of Frobenius' and Schur's work related to group characters and representations. These are discussed in Section 15.6. Here, by way of closing this overview of Frobenius' work and its influence, I will just mention the induction theorem that Brauer proved in 1946. It showed that every Frobenius character χ on a finite group $\mathfrak{G}$ is expressible in the form $\chi = \sum_{i=1}^{m} n_i \hat{\chi}_i$, where the n_i are ordinary integers and $\hat{\chi}_i$ is induced, in the sense of Frobenius' theory of induced characters, from one of Dedekind's (linear) characters χ_i on an "elementary subgroup" of $\mathfrak{G}$ (Theorem 15.13). Brauer's theorem showed that Frobenius' characters were even more intimately connected with those that had been introduced earlier by Dedekind than either of them had realized.

---

[19]See in this connection Section 3 of Chapter 9 of my book [276].

[20]This can be seen, e.g., in the recent book *Studies in Memory of Issai Schur* [331].

Part II
Berlin-Style Linear Algebra

Chapter 4
The Paradigm: Weierstrass' Memoir of 1858

In presenting an overview of Frobenius' career, I indicated that an important component in his education at Berlin involved the work of Kronecker and Weierstrass on the classification of families of quadratic and bilinear forms and the disciplinary ideals their work embodied. Both Weierstrass' theory of elementary divisors and Kronecker's generalization of it were inspired by a paper that Weierstrass published in 1858 (Section 4.6). The purpose of this chapter is to sketch the developments that motivated Weierstrass' work as well as those that provided the means for him to establish it. The former line of development arose from the mathematical analysis of a discrete system of masses oscillating near a point of stable equilibrium. The latter line of development goes back to the mechanics of a rotating rigid body and the existence of principal axes with respect to which the product moments of inertia vanish.

For an appreciation of what is to follow it will be helpful to keep in mind the following special case of a theorem from Weierstrass' 1858 paper, which I state here in modern matrix notation. *If* $\Phi(y) = y^t By$ *and* $\Psi(y) = y^t Ay$, $y = \begin{pmatrix} y_1 & \cdots & y_n \end{pmatrix}^t$, *are quadratic forms in variables* $y_1, \ldots, y_n$ *(so that A and B are symmetric) and if* $\Phi(y) > 0$ *for all* $y \neq 0$ *(so* Φ*, B are positive definite), then all the roots of the polynomial* $P(\lambda) = \det(\lambda B - A)$ *are real. Furthermore, a nonsingular linear change of variables* $y = Lz$ *can be determined such that, expressed in the variables* $z_1, \ldots, z_n$, $\Phi(z) = z_1^2 + \cdots + z_n^2$ *and* $\Psi(z) = \lambda_1 z_1^2 + \cdots + \lambda_n z_n^2$*, where* $\lambda_1, \ldots, \lambda_n$ *are the roots of* $P(\lambda)$*, each listed as many times as its multiplicity.* An easy consequence of this theorem is that if $B\ddot{y} + Ay = 0$ denotes a linear system of second-order differential equations in the unknown functions $y = y(t)$, and if A and B are symmetric and positive definite, then all solutions to this system are stable in the sense that they remain bounded as $t \to \infty$. This is because Weierstrass' theorem says that $L^t BL = I$ and $L^t AL = D$, D being the diagonal matrix with $\lambda_1, \ldots, \lambda_n$ down the diagonal; and so $y(t)$ is a solution of the given system if and only if $z(t) = L^{-1}y(t)$ is a solution to the "decoupled" system $\ddot{z} + Dz = 0$, i.e., $\ddot{z}_j + \lambda_j z_j = 0$, $j = 1, \ldots, n$. The additional hypothesis that A is also positive definite implies that the roots λ_j of $P(\lambda)$ are not only real but actually positive. Thus all solutions to the decoupled system can be

T. Hawkins, *The Mathematics of Frobenius in Context*, Sources and Studies in the History of Mathematics and Physical Sciences, DOI 10.1007/978-1-4614-6333-7_4,

expressed in the form $z_j = \delta_j \sin(\sqrt{\lambda_j} \cdot t + \varepsilon_j)$, $j = 1, \ldots, n$, for suitable choice of the constants δ_j, ε_j (so as to meet any initial conditions $z(0) = a$, $\dot{z}(0) = b$). Since all these solutions remain bounded as $t \to \infty$, the same is true of all solutions to $B\ddot{y} + Ay = 0$ because they are all of the form $y(t) = Lz(t)$, i.e., each function $y_j(t)$ is a linear combination of $z_1(t), \ldots, z_n(t)$.

4.1 Generic Reasoning

As we shall see, the challenge faced by Weierstrass in 1858 was to go beyond a mode of reasoning that was commonplace among eighteenth- and early nineteenth-century mathematicians, a mode of reasoning that I will refer to as generic reasoning. The purpose of this brief section is to explain the meaning of this term and the historical circumstances that produced it.

The geometers of the eighteenth century had inherited two magnificent achievements from the previous century. One was an entirely new way of doing mathematics, the method of symbolic analysis, which had been introduced by Viète and Descartes and further developed by Newton and Leibniz. The second great achievement was Newton's *Philosophiae naturalis principia mathematica* (1687). Like most great works, the *Principia* left many problems to clarify and resolve. Furthermore, although Newton had earlier been an enthusiastic exponent of Descartes' method of symbolic analysis, by the time he composed the *Principia*, he preferred the less controversial geometric mode of presentation by means of a type of synthetic infinitesimal geometry peculiar to him.

It was the continental mathematicians of the eighteenth century, people such as Euler, d'Alembert, Lagrange, and Laplace who perfected the analytic methods of their predecessors and demonstrated their power by systematically and successfully applying them to the problems in terrestrial and celestial mechanics suggested by Newton's *Principia*. Two works that summed up and symbolized these accomplishments were Lagrange's *Mécanique analytique* of 1788 [391] and Laplace's *Traité de mécanique céleste* [402] of 1799. These works represented unequivocal triumphs for the new method of analysis. Lagrange and Laplace were well aware of this fact, which seems to have induced them to place unlimited confidence in the method of analysis.

Thus in the preface to *Mécanique analytique*, Lagrange [391] boasted that his book contained neither geometric figures nor geometric constructions, that the methods expounded by him involved only "algebraic" operations subjected to a "systematic and uniform progression," so that mechanics becomes a new branch of analysis. And Laplace, in the course of contrasting the geometric methods of Newton with his own, wrote that although algebraic analysis necessitates consideration of "abstract combinations" far removed from the particulars of the problem at hand,

> by abandoning oneself to the operations of Analysis ... one is led, by the generality of this method and by the inestimable advantage of transforming the reasoning into mechanical procedures, to results often inaccessible to synthesis. Such is the fruitfulness of Analysis that it suffices to translate particular truths into this language in order to see emerge from their very expression a multitude of new and unexpected truths [403].

Being understandably preoccupied with the power of the method of symbolic analysis, Lagrange and Laplace tended to overlook certain concomitant difficulties. As the above quotations indicate, the great virtue of analysis was its generality and the resulting uniform, even mindlessly mechanical, nature of its procedures. The generality of the method of analysis had been viewed as its great virtue ever since its inception. Thus Viète stressed that the new method of analysis "does not employ its logic on numbers—which was the tediousness of the ancient analysts—but uses its logic through a logistic which in a new way has to do with species. This logistic is much more successful and powerful than the numerical one for comparing magnitudes with one another in equations ..." [571, pp. 321–322]. In the new analysis one reckons with species by means of "signs" (symbols) that do not represent specific magnitudes, but an entire species of magnitudes. Analysis became a method for reasoning with, manipulating, expressions involving symbols with "general" values, and a tendency developed to think almost exclusively in terms of the "general" case with little, if any, attention given to potential difficulties or inaccuracies that might be caused by assigning certain specific values to the symbols. Such reasoning with "general" expressions I shall refer to for the sake of brevity as *generic reasoning*. As we shall see, it was not only commonplace in the eighteenth century but it continued to have practitioners during the first half of the nineteenth century, despite Cauchy's warnings of its dangers.

4.2 Stability of Solutions to $B\ddot{y} + Ay = 0$

4.2.1 Lagrange

A good illustration of the pitfalls of generic reasoning is presented by the method of integrating systems of second-order linear differential equations that d'Alembert introduced in the late 1740s and 1750s and that Lagrange perfected in a memoir of 1766 [387] and then utilized in his *Mécanique analytique* (1788) to analyze small oscillations of discrete conservative mechanical systems near a stable equilibrium point [391, Pt. II, §VI]. Early in his career (1759–1762) he had considered several examples of such systems, but now he proposed to develop the general theory with the formal elegance that was characteristic of his work.

It was here that he introduced what are now called the generalized coordinates $q_1, \ldots, q_n$ of the mechanical system in order to make the mathematical formulation more elegant.[1] The equilibrium point is taken to be at $\mathbf{q} = (q_1 \cdots q_n)^t = 0$. In a neighborhood of $q = 0$, the kinetic energy (T) and potential energy (V) are then given by

$$T = \frac{1}{2}\Big[\sum_{j,k=1}^{n} b_{jk}\dot{q}_j\dot{q}_k\Big], \quad V = a_0 + \frac{1}{2}\Big[\sum_{j,k=1}^{n} a_{jk}q_jq_k\Big] + \cdots, \tag{4.1}$$

where a_0 is a constant and "$\cdots$" indicates the higher-order terms of the series expansion of $V = V(q_1,\ldots,q_n)$ about $q = 0$. Here the a_{jk} and b_{jk} are constants satisfying $a_{kj} = a_{jk}$, $b_{kj} = b_{jk}$. Actually, Lagrange only introduced these constants for $j \le k$ in his expressions for T and V [391, pp. 372–375].

By neglecting the higher-order terms of V (since the q_j are assumed very small) and by applying the equations of motion to T and the truncated V, Lagrange obtained the system of linear differential equations $B\ddot{q} + Aq = 0$. Since $T(\dot{q}_1,\ldots,\dot{q}_n) = (1/2)\dot{q}^tB\dot{q}$ is the kinetic energy $\sum mv^2$ of the system, Lagrange realized that for any real numbers $x_1,\ldots,x_n$ that are not all zero, $T(x_1,\ldots,x_n) > 0$ [391, p. 384]. In terms of the above matrix notation this means that B is positive definite. Furthermore, because $q = 0$ corresponds to a stable equilibrium, the potential function V is assumed to have a strict local minimum there, which means that the quadratic part of V, viz., $V_2 = (1/2)q^tAq$, is taken to be strictly positive for all $q \neq 0$. In other words, the hypothesis is that A is also positive definite.

Given this setup, Lagrange then proceeded to integrate the system $B\ddot{q} + Aq = 0$ by the methods he had developed in 1766.[2] Thus one begins by considering the possibility of solutions of a certain form, and in *Mécanique Analytique*, Lagrange chose the form $q = \xi(t)v$, where $\xi = C\sin(\sqrt{\lambda}\cdot t + \varepsilon)$, and the components $v_1,\ldots,v_n$ of v are arbitrary constants. Substitution of such an expression into $B\ddot{q} + Aq = 0$ gives $\ddot{\xi}Bv + \xi Av = 0$, and since $\ddot{\xi} = -\lambda\xi$, this reduces to a system of linear equations in the unknowns $v_1,\ldots,v_n$ constituting v and in λ, which corresponds in matrix notation to $(\lambda B - A)v = 0$. Since there are n equations and actually only the n unknowns $v_2/v_1,\ldots,v_n/v_1,\lambda$, the first $n-1$ may be eliminated so as to arrive at an nth-degree polynomial equation $P(\lambda) = 0$ satisfied by λ. This polynomial, which was called a resultant by Laplace, would today be denoted by $P(\lambda) = \det(\lambda B - A)$.[3] Since the reasoning is generic, the roots $\lambda_1,\ldots,\lambda_n$ of this equation are assumed to be distinct. If $v^{(j)} = \left(v_1^{(j)} \cdots v_n^{(j)}\right)^t$ is a nonzero solution to $(\lambda_j B - A)v = 0$ for $j = 1,\ldots,n$, then $q^{(j)} = \sin(\sqrt{\lambda_j}\cdot t + \varepsilon_j)v^{(j)}$ is a solution to $B\ddot{q} + Aq = 0$ and every solution is expressible as a linear combination of these [391, p. 377]:

[1] Lagrange used the notation $\xi, \psi, \varphi, \ldots$ for $q_1,\ldots,q_n$.

[2] These methods were an elegant and general version of those introduced earlier by d'Alembert [269, pp. 3ff.].

[3] The development of the theory of determinants is discussed below in Section 4.3.

$$q=\sum_{j=1}^{n} C_j \sin\left(\sqrt{\lambda_j}\cdot t+\varepsilon_j\right) v^{(j)}. \tag{4.2}$$

I will refer to this as *the generic solution to* $B\ddot{q}+Aq=0$.

In order for the generic solution to represent the behavior of a mechanical system oscillating close to a stable equilibrium, $q(t)$ must remain bounded as $t\to\infty$. In view of the form of the solution, this means that the roots λ_j must be real and positive; otherwise, $\sqrt{\lambda_j}=\alpha+i\beta$, with $\beta\neq 0$, and the expression for $\sin(\sqrt{\lambda_j}+\varepsilon_i)$ will involve exponentials $e^{\pm\beta t}$, one of which becomes infinite as $t\to\infty$.[4] There is no evidence that Lagrange suspected that the reality of the λ_j might be connected, mathematically, with the symmetry of the coefficient systems A and B. (He did, however, realize that the positive definiteness of T and V_2 implies that *if* the λ_j are assumed to be real, then they must be positive.)[5] In sum, on the basis of the form of the generic solution, Lagrange concluded that the roots λ_j must be real if the solutions are to be meaningful mechanically.

That was not all. He concluded not only that the roots λ_j must be real and positive but that they must also be distinct if the solution (4.2) is to be mechanically meaningful. He referred in this connection to the "known methods" he had first developed in the above-mentioned memoir of 1766 [387, pp. 520ff.]. There he had considered the integration of a system $\ddot{y}+My=0$, where the coefficient system $M=(m_{jk})$ is completely arbitrary, i.e., it is not assumed that $m_{kj}=m_{jk}$. This includes the systems $B\ddot{y}+Ay=0$ of *Mécanique analytique*, because they may always be rewritten in the form $\ddot{y}+My=0$ with $M=B^{-1}A$. Lagrange gave an elegant version of the generic solution to $\ddot{y}+My=0$ that replaced the arbitrary constants C_j with explicit formulas involving a given set of initial conditions $y(0)=y_0$, $\dot{y}(0)=\dot{y}_0$ [387, p. 526].

Perhaps because his generic formula of the solution to $\ddot{y}+My=0$ was being developed with an eye toward applications to specific examples of discrete mechanical systems oscillating near a stable equilibrium, e.g., a weightless string loaded with a finite number of equidistant masses and swinging from one end [387, pp. 534ff.], Lagrange considered what would happen if the positive roots λ_j were not all distinct, in which case his generic solution would become indeterminate. To deal with the simplest case, he considered what happens if $\lambda_1=\lambda_2$. It might seem that he was about to free himself from the generic mode of reasoning, but that was not so. What he did was to assume that $\sqrt{\lambda_2}=\sqrt{\lambda_1}+\omega$, where ω is a "vanishing quantity" [387, p. 528] and $\lambda_2,\dots,\lambda_n$ are all distinct. What this meant was that initially it would be assumed that $\omega\neq 0$, so that the formula for the generic solution holds. This formula was then manipulated until it was in a form that remained determinate when ω is

---

[4]Lagrange realized that $\sin\theta=(e^{i\theta}-e^{-i\theta})/(2i)$. He paid no attention to the possibility that when $\sqrt{\lambda_j}=\alpha+i\beta$, with $\beta\neq 0$, corresponding solutions could be complex-valued.

[5]He reasoned correctly that if λ_j is real, then there is a real $v^{(j)}\neq 0$ such that $Av^{(j)}=\lambda_j Bv^{(j)}$, and so $\lambda_j=[v^{(j)}]^t\lambda_j Bv^{(j)}/[v^{(j)}]^t Bv^{(j)}=[v^{(j)}]^t Av^{(j)}/[v^{(j)}]^t Bv^{(j)}=V_2(v^{(j)})/T(v^{(j)})$ is the quotient of two positive quantities and so is positive [391, p. 384].

set equal to 0. The manipulated formula with $\omega = 0$ then contains terms that involve $t\sin(\sqrt{\lambda_1} \cdot t + \varepsilon_1)$, which does not remain bounded as $t \to \infty$. And so Lagrange concluded that multiple roots are incompatible with stability. The reasoning leading to this conclusion, however, was generic. For "general values" of the coefficients m_{jk}, the manipulated formula with $\omega = 0$ does indeed involve the above term with a nonzero coefficient, i.e., a coefficient that is not identically zero as a function of the m_{jk}. No doubt Lagrange realized that for specific singular numerical values of the m_{jk} this coefficient might vanish, but thinking generically, he glossed over the possibility that for m_{jk} satisfying a relation such as $m_{kj} = m_{jk}$ but otherwise variable, it could be that the coefficient of $t\sin(\sqrt{\lambda_1} \cdot t + \varepsilon_1)$ vanishes, thereby yielding a solution compatible with stability.

Lagrange's blindness to such possibilities is illustrated by his application of his method of integration to the above-mentioned swinging string, loaded with n equally spaced weights. In this case the system of coefficients is the symmetric matrix

$$M = \frac{1}{a}\begin{pmatrix} 1 & -1 & 0 & 0 & 0 & \cdots & 0 & 0 \\ -1 & 3 & -2 & 0 & 0 & \cdots & 0 & 0 \\ 0 & -2 & 5 & -3 & 0 & \cdots & 0 & 0 \\ 0 & 0 & -3 & 7 & -4 & \cdots & 0 & 0 \\ \cdots & \cdots & \cdots & \cdots & \cdots & \cdots & \cdots & \cdots \\ 0 & 0 & 0 & 0 & 0 & \cdots & -(n-1) & 2n-1 \end{pmatrix},$$

where a denotes the distance between the weights [387, pp. 535ff.]. Lagrange expressed the resultant polynomial as $P(\lambda) = \det(\lambda I + M)$ (rather than $P(\lambda) = \det(\lambda I - M)$ as above), so that the above generic reasoning implies that the roots of this polynomial must be real, negative, and distinct. Lagrange even computed the coefficients of $P(\lambda)$ [387, p. 536], but of course it was another matter altogether to deduce directly from the equation of the polynomial, the nature of its roots. Thus he explained that

> although it would be difficult, perhaps impossible, to determine the roots of the equation $P = 0$ in general, one can nevertheless be assured by the very nature of the problem that these roots are all real, unequal, and negative; for otherwise, the values of $\ldots [y_1(t), y_2(t), y_3(t), \ldots] \ldots$ could increase to infinity, which would be absurd [387, p. 538].

This passage indicates that Lagrange never doubted that his mathematical model of motion, based on $\ddot{y} + My = 0$, faithfully represented the motion of the swinging string as $t \to \infty$. There is no evidence of an awareness of the possibility that the symmetry of the coefficients might make such systems exceptions to his tacitly generic conclusions about the incompatibility of multiple roots and stability and so afford a mathematical means to justify the reasonableness of his mathematical model.

Twenty-two years later, in *Mécanique analytique*, Lagrange attempted a more mathematical justification of the boundedness of the solutions to $B\ddot{q} + Aq = 0$. From (4.1) we have that $V = a_0 + V_2 + R$, where $V_2 = \frac{1}{2}\sum_{j,k=1}^{n} a_{jk}q_jq_k$ and R denotes the terms in the expansion of V in (4.1) that are of third and higher degrees.

The hypothesis of a stable equilibrium meant that $|q_j(t)| \ll 1$ and so the higher-order terms comprising R were very small relative to the initial terms $a_0 + V_2$. Lagrange took them to be negligible and so expressed the principle of conservation of energy ($T + V = \text{const}$) as $T + a_0 + V_2 = \text{const}$. From $T + a_0 + V_2 = \text{const}$, it then follows that for any time t, if $q(t)$ represents the state of the mechanical system at time t, then

$$T(\dot{q}(t)) + a_0 + V_2(q(t)) = T(\dot{q}(0)) + H + V_2(q(0)),$$

and so, since T and V_2 are nonnegative,

$$0 \leq V_2(q(t)) \leq T(\dot{q}(t) + V_2(q(t)) = T(\dot{q}(0)) + V_2(q(0)). \tag{4.3}$$

Thus $V_2(q(t))$ remains bounded for all t, and Lagrange concluded that the same would therefore be true for the $q_i(t)$; and so "the roots of the equation in [λ] will necessarily be all real, positive, and unequal" [391, p. 385]. The quoted conclusion shows that in (4.3), Lagrange was tacitly replacing $q(t)$, the actual state of the mechanical system at time t, by a solution to $B\ddot{q} + Aq = 0$, and so once again taking it for granted that the solutions to these equations faithfully represent the oscillatory behavior of the system, thereby essentially assuming what was to be proved.[6]

Lagrange's above "proof" did show an effort to bring the mathematical properties of A and B into the picture, namely the property of positive definiteness. However, the symmetry of A and B entered the picture only in the sense that T and V_2 are quadratic forms. It was Laplace who first discovered the relevance of symmetry conditions to stability in the context of the mechanical system comprising the solar system.

4.2.2 Laplace

While Lagrange was studying systems of linear differential equations in the 1760s, he was also working on problems in celestial mechanics. In 1764, he won the prize of the Paris Academy for a memoir on the libration of the moon, and in 1766, he again won a prize for a memoir on the satellites of Jupiter. In the same year, he left Turin to fill the position at the Berlin Academy vacated by Euler. At Berlin, Lagrange continued to work on problems in celestial mechanics, and in 1774, he sent off an important memoir [389] to the Paris Academy on the secular perturbations of the planetary orbits.

When the manuscript of Lagrange's memoir [389] arrived in Paris, it was read with great interest by Laplace, who was then a young man of 24 just

---

[6] As we shall see in Section 4.6, the boundedness of solutions to $B\ddot{q} + Aq = 0$ does follow from a theorem published by Dirichlet in 1846.

beginning to establish himself as an outstanding mathematical astronomer. Well before Lagrange's memoir appeared in print in 1778, Laplace had composed two of his own that were published in 1775 and 1776 [398, 399] and in which he called attention to Lagrange's methods and extended them to obtain analogous results for the other orbital parameters, including the following differential equations for the planetary eccentricities $e_j(t) = \sqrt{h_j(t)^2 + \ell_j(t)^2}$:

$$\dot{h}_j + \sum_{k=1}^{n} c_{jk}\ell_k = 0, \quad \dot{\ell}_j - \sum_{j=1}^{n} c_{jk}h_j = 0, \quad i = 1,\dots,n, \tag{4.4}$$

where n is the number of planets. The coefficients c_{jk} are given by formulas that depend on the masses and mean solar distances of the jth and kth planets. Thus, although Laplace had the planetary system of the sun in mind, he developed the mathematics more generally for any number of planets and with unspecified masses and mean solar distances.

In matrix notation with $C = (c_{jk})$, $h = \left(h_1 \ \cdots \ h_n\right)^t$, and $l = \left(l_1 \ \cdots \ l_n\right)^t$, the above equations are $\dot{h} = -Cl$ and $\dot{l} = Ch$. Therefore, $\ddot{h} = -C\dot{l} = -C^2h$, and likewise $\ddot{l} = -C^2l$. In other words, h and l are solutions to

$$\ddot{y} + C^2y = 0. \tag{4.5}$$

Thus h and l are solutions to the system $\ddot{y} + My = 0$, $M = C^2$, the type of system studied by Lagrange in his 1766 memoir [387], with $M = C^2$. Laplace accepted the conclusions of that work. Thus he accepted Lagrange's claim that in order for the solutions to $\ddot{y} + C^2y = 0$ to remain bounded as $t \to \infty$, it must be that the roots of $P(\lambda) = \det(\lambda I - C^2)$ are real, positive, and distinct. Laplace chose to work with the coefficient system C and so introduced the polynomial $Q(\rho) = \det(\rho I - C)$. If $\rho_1,\dots,\rho_n$ are the roots of $Q(\rho)$, then $\rho_1^2,\dots,\rho_n^2$ are the roots of $P(\lambda)$, and so Laplace concluded that $\rho_1,\dots,\rho_n$ must be real and distinct if his solutions $h(t), l(t)$ to $\dot{h} = -Cl, \dot{l} = Ch$ were to remain bounded as $t \to \infty$.

While Laplace accepted Lagrange's conclusions about the connection between the boundedness of solutions to $\dot{h} = -Cl, \dot{l} = Ch$ and the nature of the roots of $Q(\rho) = \det(\rho I - C)$, he did not accept Lagrange's assumption that the stability of a mechanical system, such as the planetary system was assumed to be, automatically implies that the solutions to an associated linear system of differential equations, derived by taking approximations, are bounded as $t \to \infty$. Suppose $h(t), l(t)$ are a solution to the differential equations (4.4), so that $\dot{h} = -Cl$ and $\dot{l} = Ch$, and set $e_j(t) = \sqrt{h_j(t)^2 + l_j(t)^2}$. Laplace had reason to believe that he could prove that for all t, one has

$$\sum_{j=1}^{n} m_j a_j^{1/2} e_j(t)^2 = \text{const}, \tag{4.6}$$

where m_j is the mass of the jth planet and a_j its mean solar distance. If (4.6) were true, then all the $e_j(t)$ would have to remain bounded as $t \to \infty$ and so this would also be true for $h(t)$ and $l(t)$. A first attempt at a proof [400, pp. 89–91] was flawed; it suffered from the same sort of deficiencies that occurred in Lagrange's proof of the analogous relation (4.3).

Some time between 1787 and 1789, Laplace discovered how to establish (4.6) rigorously. He was so pleased with his discovery that he devoted a memoir to the matter [401].[7] The key was a relation satisfied by the coefficients c_{jk} by virtue of their definition in terms of the planetary masses and mean solar distances:

$$m_j a_j^{1/2} c_{jk} = m_k a_k^{1/2} c_{kj}. \tag{4.7}$$

Evidently showing that (4.6) holds is equivalent to showing that the derivative of the left-hand side of (4.6) is identically zero. Since $e_j(t)^2 = h_j(t)^2 + l_j(t)^2$, the derivative of the left-hand side is

$$2\sum_{j=1}^{n} m_j a_j^{1/2}[h_j\dot{h}_j + l_j\dot{l}_j], \tag{4.8}$$

and since h_j, l_j are solutions to (4.4), we have

$$\dot{h}_j = \sum_{k=1}^{n} c_{jk} l_k \quad \text{and} \quad \dot{l}_j = -\sum_{k=1}^{n} c_{jk} h_k.$$

Substituting these expressions in (4.8) transforms it into

$$2\Big[\sum_{j,k=1}^{n} m_j a_j^{1/2} c_{jk} h_j l_k - \sum_{j,k=1}^{n} m_j a_j^{1/2} c_{jk} l_j h_k\Big]. \tag{4.9}$$

The symmetry relation $m_j a_j^{1/2} c_{jk} = m_k a_k^{1/2} c_{kj}$ then implies that (4.9) is identically zero, whence (4.6) follows.

"I thought," Laplace proudly added—perhaps with Lagrange in mind—"that it would be pleasing to see the same equations [i.e., (4.6)] result directly from the differential equations which determine the secular variations of the orbits" [401, p. 300]. He was fully aware of the generality of his reasoning, i.e., that it was independent of the number n of planets and the positive values assigned to their masses m_j and mean solar distances a_j. In fact, his reasoning clearly remains valid for any system of coefficients $C = (c_{jk})$ for which positive constants $d_1, \ldots, d_n$ exist such that

$$d_j c_{jk} = d_k c_{kj}, \tag{4.10}$$

[7] The memoir was included in the memoirs of the Paris Academy for the year 1787, which was not published until 1789. It is possible that Laplace submitted it after 1787.

and hence, in particular, for all symmetric C (all $d_j = 1$). Laplace reproduced his proof in even greater detail in the first volume (1799) of his *Mécanique céleste* [402, pp. 318ff.].

Expressed in matrix notation, what Laplace did was to produce a diagonal matrix D with $d_{jj} = m_j a_j > 0$ such that his symmetry relation (4.7) holds, i.e., $DC = C^t D$, which is what (4.10) asserts. This implies that DC and DC^2 are symmetric.[8] Thus the differential equation $\ddot{y} + C^2 y = 0$ is equivalent to $D\ddot{y} + DC^2 y = 0$. Here D and DC^2 are symmetric and D is positive definite.

This was the first time the symmetry of the coefficients of a system of linear differential equations was seen to be mathematically relevant to the boundedness of the solutions. Of course, Laplace thought he had also proved that his symmetry relations imply that the roots of $Q(\rho) = \det(\rho I - C)$ must be real and unequal, and this conclusion remained unquestioned for some time. For a mathematician who accepted this result, a natural question to ask would be whether the properties of the roots of $Q(\rho) = \det(\rho I - C)$ or $P(\lambda) = \det(\lambda D - DC^2)$ might be deduced directly from the associated symmetry properties, i.e., without the need to introduce the solutions to the associated system of differential equations. More generally, thinking of Lagrange's *Mécanique analytique*, could the symmetry of A and B and the positive definiteness of B be used to show directly that the roots of $P(\lambda) = \det(\lambda B - A)$ are real and distinct? As we shall now see, Charles Sturm (1803–1855) was such a mathematician.

4.2.3 Sturm

Sturm was born and raised in Geneva, but by his early twenties he had moved permanently to Paris. In 1829 he was appointed as an editor of the *Bulletin des sciences mathématiques*, an abstracting journal published by the Baron de Férussac.[9] Sturm used his newly acquired editorial privilege to present in the *Bulletin* abstracts of memoirs he was submitting to the Paris Academy of Science. Many of these memoirs were not published, but the abstracts provide us with an idea of the progression of his research.[10] In the abstract [554] of a memoir submitted to the academy on 23 May 1829, we find what is now known as Sturm's theorem, as well as allusions to some generalizations. Sturm had discovered that if $p(r)$ is an nth-degree polynomial with real coefficients, then the number of distinct real roots it has in an interval $[a,b]$ can be calculated using certain sequences of

---

[8] To see that $DC = C^t D$ implies that DC^2 is symmetric, observe that $DC = C^t D$ means $C^t = DCD^{-1}$, and so $(DC^2)^t = (C^t)^2 D = (DCD^{-1})^2 D = (DC^2 D^{-1})D = DC^2$.

[9] For details about Sturm's life and work, as well as reprints of many of his publications, see [556].

[10] A list of the memoirs Sturm presented to the academy at various dates in 1829 together with the corresponding abstract in the *Bulletin* is given by Bôcher [27, p. 2], who focuses on Sturm's work on the theory of equations.

polynomials $p_0, p_1, \ldots, p_m$, where always $p_0 = p$, the given polynomial, $\deg p_{j+1} < \deg p_j$, and $m \leq n$. These sequences are now called Sturm sequences. Sturm's theorem says that if $\sigma(c)$ denotes the number of changes of sign in the sequence $p_0(c), p_1(c), \ldots, p_m(c)$ (terms $p_j(c) = 0$ being ignored), then the number of distinct real roots of $p(r)$ in $[a,b]$ is $\sigma(a) - \sigma(b)$. In his abstract, Sturm defined what is now known as the canonical Sturm sequence, but he also indicated that other sequences could be used to the same end.

Sturm was familiar with the above-discussed work of Lagrange and Laplace. All of it, he realized, involved the integration of linear systems of the form $B\ddot{y}+Ay=0$, with $B = (b_{jk})$ and $A = (a_{jk})$ symmetric and with at least B positive definite. On 27 July 1829 he presented a memoir to the academy in which, judging by his lengthy abstract [555], he applied his theorem to give a purely algebraic proof, without resort to differential equations and their solutions, that the roots of the polynomial $p(r) = \det(rB+A)$ are all real and distinct [555, p. 315].

Sturm explained his ideas for the case of $n = 5$ variables "for greater simplicity" [555, p. 314]. This was helpful, because Sturm's algebraic tools for dealing with systems of equations such as $(rB+A)v = 0$ were the same elimination-theoretic ones used by Lagrange and Laplace. As we shall see in the next section, Cauchy had laid the foundations for the now-familiar theory of determinants in a paper published in 1815, but in 1829, this work was still not widely known. The sequence $p_0(r), \ldots, p_5(r)$ introduced by Sturm is easy to define using the theory of determinants: $p_5(r)$ is any positive constant, e.g., $p_5 = 1$, and $p_4(r)$ is the principal minor determinant of $rB+A$ obtained by deleting the last four rows and columns of $rB+A$, i.e., $p_4(r) = rb_{11}+a_{11}$. Likewise, $p_3(r)$ is the principal minor of $rB+A$ obtained by deleting its last three rows and columns, i.e.,

$$p_3(r) = \begin{vmatrix} rb_{11}+a_{11} & rb_{12}+a_{12} \\ rb_{21}+a_{21} & rb_{22}+a_{22} \end{vmatrix};$$

and so on for $p_2(r)$ and $p_1(r)$, and, of course, $p_0(r) = p(r) = \det(rB+A)$.

To define these polynomials elimination-theoretically, as Sturm did, was more complicated. I repeat Sturm's definitions to convey an idea of the great simplification afforded by Cauchy's theory of determinants. Think of $(rG+K)v = 0$ as a system of linear equations in the unknowns $v_1, \ldots, v_5$ that constitute v. Then $p_4(r)$ is the coefficient of v_1 in the first equation. Now proceed to eliminate v_1 by solving the first equation for v_1 to get $v_1 = -[\sum_{j=2}^{5}(rb_{1j}+a_{1j})v_j]/p_4(r)$. This is then substituted into the second equation, which then becomes a linear equation in the four unknowns $v_2, \ldots, v_5$. This equation is multiplied through by $p_4(r)$, so that the coefficients of $v_2, \ldots, v_5$ are now all quadratic polynomials in r. Then $p_3(r)$ is defined to be the coefficient of v_2 in this equation. Next this equation is solved for v_2 and the result substituted in the third equation, which then becomes a linear equation in the three unknowns $v_3, \ldots, v_5$. When this equation is multiplied by $p_3(r)$, the coefficients of $v_3, \ldots, v_5$ become cubic polynomials in r, and $p_2(r)$ is by definition the coefficient of v_3. And so on.

Sturm realized that since B is assumed positive definite, the leading coefficients of the polynomials $p_j(r)$ are all positive [555, p. 317]. (Expressed in the language of determinants, the leading coefficient of $p_j(r)$ is the principal minor determinant of B obtained by deleting the last j rows and columns, and these principal minors are positive if and only if B is positive definite [240, v. 1, p. 306].) This means that $\lim_{r\to+\infty} p_j(r) = +\infty$ for all $j < 5$. Thus we may choose b such that $p_j(b) > 0$ for all j, which means that $\sigma(b) = 0$. Likewise, $\lim_{r\to-\infty} p_j(r)$ will be $+\infty$ or $-\infty$, depending on whether the degree of $p_j(r)$ is even or odd, respectively. This means that $a < b$ may be chosen such that $p_5(a), p_3(a), p_1(a)$ are positive and $p_4(a), p_2(a), p_0(a)$ are all negative; and so $\sigma(a) = 5$. This means that $\sigma(a) - \sigma(b) = 5$, and Sturm invoked his theorem to conclude that $p_0(r) = \det(rB + A)$ has five real and distinct roots.

Sturm was, of course, reasoning generically. It is easy to come up with counterexamples, e.g., $B = I_5$ (the identity matrix) and A the matrix with all $a_{jk} = 1$, for which $\det(rB + A) = r^4(r+5)$, so that there are only two real roots. Sturm himself realized the generic nature of his reasoning: "I must add that some of these theorems could be subject to exception if 2 or more of the consecutive functions among L, M, N, P, Q [$= p_4, p_3, p_2, p_1, p_0$] were to have common factors,[11] something that does not occur as long as the coefficients ... [of A and B] ... are indeterminate" [555, pp. 320]. Regarding coefficients as indeterminate was the hallmark of generic reasoning.

Sturm's generic "theorem" on the roots of $\det(rB + A)$ led him to some suggestive results about quadratic functions $\Phi(v) = v^tBv$ and $\Psi(v) = v^tAv$, which are correct under Sturm's assumption that all the roots of $p(r) = \det(rB + A)$ are real and distinct [555, p. 320ff.]. Suppose that $\lambda_1, \ldots, \lambda_5$ are the five distinct real roots of $p(r) = \det(rB + A)$, so that $e_j \neq 0$ exist for which $(B + \lambda_j A)e_j = 0$. Consider Φ and Ψ as bilinear functions $\Phi(v,w) = v^tBw$ and $\Psi(v,w) = v^tAw$. Then for $j \neq k$, $\Phi(e_j, e_k) = \Psi(e_j, e_k) = 0$. Since the λ_j are assumed real and distinct, this is a correct result.[12] Sturm realized that it implies that the change of variables $v = Px$, where P is the matrix whose columns are $e_1, \ldots, e_5$, transforms the two quadratic functions $\Phi(v,v)$ and $\Psi(v,v)$ into sums of squares: $\Phi(x,x) = \sum_{j=1}^5 \Phi(e_j,e_j)x_j^2$, $\Psi(x,x) = \sum_{j=1}^5 \Psi(e_j,e_j)x_j^2$.[13]

---

[11] Here Sturm was perhaps alluding to an important defining property of what are now called generalized Sturm sequences $p_0, p_1, \ldots, p_m$ on $[a,b]$, namely that if $c \in (a,b)$ is such that $p_j(c) = 0$ for some j between 1 and $m-1$, then $p_{j-1}(c)p_{j+1}(c) < 0$. In the above-mentioned example, $p_0, \ldots, p_4 = r^4(r+5), r^3(r+4), r^2(r+3), r(r+1), r+1$, so that $c = 0$ fails to have this property with respect to any interval $[a,b]$ containing it because $p_0, \ldots, p_3$ all have r as a common factor.

[12] This can be seen as follows. We have that $Ae_j = -\lambda_j Be_j$ for all j and $A^t = A$, $B^t = B$. Thus $-\lambda_k(e_j^tBe_k) = e_j^tAe_k = (Ae_j)^te_k = -\lambda_j e_j^tBe_k$, and since $\lambda_i \neq \lambda_j$, it must be that $e^tBe_k = 0$, and so also $e_j^tAe_k = 0$.

[13] In matrix form: $P^tBP = D_B$ and $P^tAP = D_A$, where D_B, D_A are diagonal matrices with, respectively, $e_1^tBe_1, \ldots, e_5^tBe_5$ and $e_1^tAe_1, \ldots, e_5^tAe_5$ down the diagonal. This follows immediately from $e_j^tBe_k = e_j^tAe_k = 0$ for $j \neq k$.

The reasoning leading to this result is clearly the same for any number of variables. It must be kept in mind, however, that for Sturm, what had been proved was a vague generic theorem, namely that "in general," two quadratic functions, one of which is positive definite, can be simultaneously transformed into sums of square terms, i.e., as long as the coefficients a_{jk} and b_{jk} remain indeterminate. It was based on the generic theorem that all the roots of $p(r) = \det(rB + A)$ are real and distinct. As we have seen, Sturm realized that for certain specific values of the coefficients, his proof of that theorem might fail. That failure meant that even the reality of all the roots of $p(r)$ had not been proved for all choices of coefficients; and it remained unclear for exactly which coefficients it had been proved. Thus the theorem about transforming quadratic functions Φ, Ψ with Φ positive definite might fail as well. This dilemma was an intrinsic feature of generic reasoning. Weierstrass' theorem stated at the beginning of the chapter, of course, removed all the ambiguity and uncertainty: $p(r)$ always has all its roots real, although they need not be distinct. The functions Φ and Ψ can always be transformed simultaneously into sums of square terms, although Sturm's formulas for Φ, Ψ as sums of squares need not hold. Weierstrass' theorem was nongeneric.What gave him the courage to reject generic reasoning, to believe that it was possible to give nongeneric proofs? The answer, as we shall now see, is to be found in the work Cauchy.

After stating the above result, Sturm added by way of conclusion that the formulas he had derived leading up to the simultaneous transformation of Φ and Ψ into sums of squares "include as very special cases those on the transformation of coordinates and those by means of which the surfaces of second degree that have a center are related to their principal axes" [555, p. 322]. It is possible that he was made mindful of this very special application of his results by Cauchy, who had generalized to equations in n variables the mathematics behind the determination of the principal axes of a central quadric surface. According to Cauchy [72, p. 195], he had met Sturm sometime before Sturm submitted the memoir he described in his abstract [555]. They discovered that they had arrived at similar results, and presumably by agreement, they both submitted memoirs to the Paris Academy of Science on the same day. To my knowledge, neither was published by the academy, although, as we have seen, Sturm managed to publish an extract in Férussac's *Bulletin* in 1829. Cauchy, however, was able to do more. In 1829, he published a full-length account of his results—possibly identical to the memoir submitted to the academy—in his own private journal, *Exercises de Mathématiques*. Sturm's results, being buried away and only sketched in an abstracting journal, were not well known and had little impact on subsequent developments, whereas Cauchy's paper had a major impact. That was not solely because he was more famous and had published his work in detail. It was also because he introduced a powerful new algebraic tool—the theory of determinants more or less as we now know it—and because he rejected the prevalent generic mode of reasoning. These two features of Cauchy's paper are why it was to have a great influence on Weierstrass.

4.3 Cauchy's Theory of Determinants

Soon after its founding in 1794, the École Polytechnique in Paris became a breeding ground for talented engineers, scientists, and mathematicians. Among its most distinguished students was Augustin Cauchy (1789–1857), who gained admission in 1805 at age 16. Cauchy remained at the École two years, then pursued more practical engineering studies elsewhere, and worked as an engineer for several years. But by 1812, when Cauchy submitted a lengthy essay [68] on the theory of determinants to the *Journal de l'École Polytechnique*, his interests had turned toward an academic career in mathematics.

As we have seen in the previous sections, the homogeneous polynomial formed from the coefficients of a square matrix that is now called the determinant of the matrix had been considered by several mathematicians before Cauchy. Cauchy himself referred to Cramer (1750), Bézout (1764), Vandermonde (1771), and Laplace (1772), who introduced the term "resultant" for this special polynomial.[14] In what follows I will use the term "resultant" when referring to the eighteenth-century notion and attendant theory. Resultants had been introduced in connection with the consideration of systems of linear equations, the number of equations being equal to the number of unknowns. Three main properties relating to them had been discovered. Since Cramer it was known that for such a system of inhomogeneous equations, viz., $Ax = b$ with A an $n \times n$ matrix and x, b, $n \times 1$ column matrices, the (generic) solution was expressible in terms of resultants, each unknown being a ratio of such. Nowadays this result is frequently referred to as Cramer's rule; it can be expressed in present-day notation as $x_i = \det[A_i(b)]/\det A$, $i = 1, \ldots, n$, where $A_i(b)$ denotes the matrix A except that its ith column has been replaced by b. It was also known that the condition for a homogeneous system $Ax = 0$ to have a nontrivial solution was that the resultant of the coefficient system be zero. Much attention was given in the eighteenth century to the matter of calculating resultants, and in this connection, Laplace introduced what are now known as the Laplace expansions.

To gain a greater appreciation of the manner in which Cauchy transformed the theory of resultants, it is worth indicating how Laplace introduced the eponymous expansions.[15] The context was homogeneous systems of equations. As noted above, it was necessary to compute the resultant of the coefficient system and set it equal to zero in order to have the equation of condition for a nontrivial solution. For systems of equations with $n = 2, 3, 4, 5, 6$ equations, Laplace gave detailed step-by-step recipes for how to do this. He began with a system of two equations, which he wrote as

$$0 = {}^1a.\mu + {}^1b.\mu'; \quad 0 = {}^2a.\mu + {}^2b.\mu'.$$

---

[14] For further details on the early use of determinants, including exact references, see the account by Muir [449]. It contains many extensive quotations from original sources that enable the reader to appreciate the theoretical coherence and concomitant notational innovations introduced by Cauchy.

[15] The following is based on Muir's account [449, v. 1, pp. 27–32].

He then gave a recipe for computing the resultant so as to get the equation of condition $0 = +{}^1a.{}^2b - {}^1b.{}^2a$, namely that the resultant of the coefficient system must vanish. Next he considered a system of three equations with

$$\begin{matrix} {}^1a & {}^1b & {}^1c \\ {}^2a & {}^2b & {}^2c \\ {}^3a & {}^3b & {}^3c \end{matrix}$$

as its coefficient system. What then follows is a lengthy recipe for computing the resultant in a special form, so that the equation of condition becomes

$$0 = ({}^1a.{}^2b - {}^1b.{}^2a){}^3c - ({}^1a.{}^3b - {}^1b.{}^3a){}^2c + ({}^2a.{}^3b - {}^2b.{}^3a){}^1c. \tag{4.11}$$

We can now recognize this as the Laplace expansion up the third column of the above coefficient system. For Laplace, it represented the first step in an algorithm for reducing resultants of any degree to those of degree at most two. Thus for a system of four equations, the resultant is expressed as a sum of six products of second-degree resultants. And for five equations, the algorithm expresses the resultant as a sum of triple products, each product consisting of two degree-two resultants and one of degree one (i.e., a coefficient). Laplace realized that there was an analogue of (4.11) for a degree-four resultant, but this for him again played a computational role analogous to (4.11), where the new goal was to express resultants as sums of products of resultants of degree at most three. Thus, e.g., his algorithm led to an expression for a resultant of degree five as a sum of products of second- and third-degree resultants.

By contrast, for Cauchy the Laplace expansions were not so much a computational tool as the theoretical basis for enriching the theory of resultants along lines suggested by his reading of Gauss' *Disquisitiones Arithmeticae* (1801). Cauchy also introduced a superior notation for $n \times n$ coefficient systems, which he wrote in the form

$$\left\{\begin{matrix} a_{1,1}, & a_{1,2}, & \cdots & a_{1,n}, \\ a_{2,1}, & a_{2,2}, & \cdots & a_{2,n}, \\ \cdots & \cdots & \cdots & \cdots \\ a_{n,1}, & a_{n,2}, & \cdots & a_{n,n}, \end{matrix}\right.$$

and denoted for brevity by $(a_{1,n})$. His notation was thus very close to modern double-index notation, and I will use the notation $A = (a_{ij})$ in what follows.

Rather than speaking of the resultant of the system $A = (a_{ij})$, Cauchy spoke of its *determinant*. In this way he acknowledged his debt to Gauss, who, in the part of *Disquisitiones Arithmeticae* that inspired Cauchy, developed the theory of the representation of integers by means of quadratic forms in two or three variables [244, art. 153ff., art. 266ff.]. Gauss never spoke of "resultants," although they appeared in his work for the 2×2 and 3×3 coefficient systems corresponding to quadratic forms and to linear changes of variable in these forms. The negatives of the resultants of the systems associated to quadratic forms he called determinants. As we shall see below, the reason Cauchy honored Gauss with his choice of nomenclature was that Gauss was the first to recognize and utilize the now-familiar multiplicative property of determinants $\det(AB) = \det A \cdot \det B$; it was this property, and the prospect of its ramifications and generalizations, that led Cauchy to devote a lengthy essay to the properties of resultants.

Cauchy began by directly defining the determinant of an $n \times n$ coefficient system $A = (a_{ij})$. This he did in an interesting but now unfamiliar manner [68, pp. 113ff.]. Before defining a determinant, Cauchy had devoted considerable space to a preliminary exposition of permutations of $1, 2, \ldots, n$, and he realized that his definition of the determinant of $A = (a_{ij})$ implied the now-familiar formula

$$\det A = \sum_{\sigma} (\operatorname{sgn} \sigma) a_{\sigma(1)1} a_{\sigma(2)2} \cdots a_{\sigma(n)n}, \tag{4.12}$$

where $\sum_{\sigma}$ means that the permutation

$$\sigma = \begin{pmatrix} 1 & 2 & \cdots & n \\ \sigma(1) & \sigma(2) & \cdots & \sigma(n) \end{pmatrix}$$

runs through all $n!$ permutations (which he called "substitutions"). He even showed by examples how to compute $\operatorname{sgn} \sigma = \pm 1$ by factoring σ into a product of disjoint cycles (called "circular substitutions").

Using his superior notation and (4.12), Cauchy derived the Laplace cofactor expansions. In the now-familiar way, he characterized what is now called the (i, j) minor determinant of $A = (a_{ij})$ as the determinant of the $(n-1) \times (n-1)$ coefficient system obtained from A by "suppressing" the horizontal line i and the vertical line j on which a_{ij} is located [68, p. 126]. I will denote this minor by $M_{ij}(A)$. Cauchy had no notation for it because what was important to him was the coefficient (now called the (i, j) cofactor)

$$b_{ij} = (-1)^{(i+j)} M_{ij}(A), \tag{4.13}$$

which arises in the Laplace expansions.[16] Laplace's column expansion (4.11) then takes the general form (for column μ)

$$a_{1\mu} b_{1\mu} + a_{2\mu} b_{2\mu} + \cdots + a_{n\mu} b_{n\mu} = D_n,$$

where D_n is Cauchy's notation for the determinant of A. Cauchy also deduced the corresponding row expansion formula

$$a_{\mu 1} b_{\mu 1} + a_{\mu 2} b_{\mu 2} + \cdots + a_{\mu n} b_{\mu n} = D_n. \tag{4.14}$$

[16] Cauchy did not explicitly introduce the expression $(-1)^{(i+j)}$ for the sign of the minor.

To both expansion formulas he added, for reasons to be indicated below, a companion formula that followed from the well-known fact that a coefficient system with two identical rows (or columns) has zero determinant. Thus if for $\nu \neq \mu$ the μth row of A is replaced by the νth, the resulting system will have zero determinant. This yields a companion expansion, which he wrote together with (4.14), namely

$$a_{\nu 1}b_{\mu 1} + a_{\nu 2}b_{\mu 2} + \cdots + a_{\nu n}b_{\mu n} = 0 \quad (\nu \neq \mu). \tag{4.15}$$

For Cauchy, (4.14) and (4.15) were equally important, for they said something about the system $B = (b_{ij})$, which Cauchy introduced and called the *adjoint* of the system A [68, p. 126]. This term as well Cauchy took from Gauss. In present-day terms, Cauchy's adjoint B is just the the matrix of cofactors of A, which I will denote by $B = \mathrm{Cof}(A)$. Nowadays, the adjoint (or adjunct) of A is the transpose of Cauchy's adjoint. The reason for the discrepancy is indicated below.

Cauchy honored Gauss by using his terms "determinant" and "adjoint" for the following reason. Gauss had pointed out that if a binary or ternary form f is transformed into f' by $x = Py$ and if, in turn, f' is transformed into f'' by $y = Qz$, then f'' may be regarded as obtained from f by a single linear transformation $x = Rz$ i.e., $x = Py = P(Qz) = Rz$. For ternary forms, Gauss wrote down the coefficient system defining R [244, art. 159]. If the systems P and Q are written in Cauchy's double index notation, then Gauss' formula for the resultant substitution translates immediately into the following: given substitutions $P = (p_{ij})$ and $Q = (q_{ij})$, then the substitution $R = (r_{ij})$ resulting from the composition of P and Q (in that order) is given by

$$r_{ij} = \sum_{k=1}^{n} p_{ik}q_{kj} \tag{4.16}$$

with $n = 3$. Undoubtedly, Gauss realized that the relation $\det R = \det P \cdot \det Q$ was true for $n > 3$ as well, but he had no need of such a relation in *Disquisitiones Arithmeticae*. For Cauchy, the validity of $\det R = \det P \cdot \det Q$ for systems of any dimension n was what made determinants and adjoints interesting.

Instead of Gauss' definition (4.16) of the composition of two linear substitutions, however, Cauchy defined a system $R = (r_{ij})$ to be the result of the composition of P followed by Q if [68, p. 138, eqn. (30)]

$$r_{ij} = \sum_{k=1}^{n} q_{jk}p_{ik}. \tag{4.17}$$

This means that R is the matrix defined by $R = PQ^t$, and Cauchy seems to have added the transpositional twist so that the expansion formulas (4.14) and (4.15) would say that the system $(\det A)I$ is the result of the composition of A followed by its adjoint $B = \mathrm{Cof}(A)$, rather than the now-familiar $A \cdot [\mathrm{Cof}(A)]^t = (\det A)I$.

Having given his definition of the composition of systems, Cauchy proceeded to prove "a very remarkable theorem," which he stated as follows [68, p. 142]:

Theorem 4.1 (Cauchy's product theorem). *"When a system of quantities is determined ... [by (4.17)] ... from two other systems, the determinant of the resultant system is always equal to the product of the determinants of two composing systems."*

Since systems and their transposes have equal determinants, as Cauchy had observed, his product theorem is equivalent to the customary one implied by Gauss' treatment of binary and ternary forms.

The first application that Cauchy made of his product theorem was to establish a generalization of what he had seen in Gauss' *Disquisitiones* in the case $n = 3$ (and with A symmetric): the determinant of the adjoint of an $n \times n$ system A equals $(\det A)^{n-1}$ [68, p. 142], i.e., in more modern notation,

$$\det(\mathrm{Cof}(A)) = (\det A)^{n-1}. \tag{4.18}$$

His second application was to generalize another of Gauss' results about adjoints of ternary quadratic forms. Gauss had proved that if F is the adjoint of the form f with determinant $D \neq 0$, then the adjoint of F is the form Df [244, art. 267]. Cauchy perceived that the correct generalization of this result to $n \times n$ systems $A = (a_{ij})$ with $\det A \neq 0$ was the following: if B is the adjoint of A, then the adjoint of B is the system $C = (c_{ij})$, where $c_{ij} = (\det A)^{n-2} a_{ij}$ [68, p. 142], i.e., in more familiar notation,

$$\text{If } B = \mathrm{Cof}(A) \text{ then } \mathrm{Cof}(B) = (\det A)^{n-2} A. \tag{4.19}$$

The new results about adjoints given in (4.18) and (4.19) were generalizations of relations Gauss had discovered. Cauchy also ventured onto entirely new ground with his theory of derived systems [68, pp. 153–169], which had as its goal a generalization of the above multiplication theorem. A brief description of the elements of this theory is warranted, because one of its consequences was a theorem that was basic to Weierstrass' formulation of his theory of elementary divisors, the backbone of Berlin-style linear algebra.

Given an $n \times n$ system $A = (a_{ij})$, let p denote an integer satisfying $1 \leq p \leq n-1$. Let $\{i_1, \ldots, i_p\}$ denote a set consisting of p of the integers $1, \ldots, n$. If $\{j_1, \ldots, j_p\}$ is another such set, consider the $p \times p$ system obtained from A by considering the coefficients a_{ij} of A that are on rows $i_1, \ldots, i_p$ and on columns $j_1, \ldots, j_p$. Form the determinant of this system, which I will denote by $m(i_1, \ldots, i_p; j_1, \ldots, j_p)$. The $m(i_1, \ldots, i_p; j_1, \ldots, j_p)$ are what became known as degree-p minor determinants. They first became prominent in Cauchy's essay. There are $P = \binom{n}{p}$ distinct sets $\{i_1, \ldots, i_p\}$ and hence P^2 determinants $m(i_1, \ldots, i_p; j_1, \ldots, j_p)$. Cauchy formed them into a $P \times P$ system as follows. Determine an ordering of the P sets $\{i_1, \ldots, i_p\}$ such that

$$\{i_1,\ldots,i_p\} \prec \{j_1,\ldots,j_p\} \text{ if } i_1 \times \cdots \times i_p < j_1 \times \cdots \times j_p. \tag{4.20}$$

For some choices of n and p this stipulation completely determines the ordering, while for others, it does not, since distinct sets can give the same product. In all cases, however, the first set will be $\{1,\ldots,p\}$ and the last will be $\{n-p+1,\ldots,n\}$. Following Cauchy, let us use the notation $\{i_1,\ldots,i_p\} = (\mu)$ to mean that $\{i_1,\ldots,i_p\}$ is the μth set in the ordering. Define a $P \times P$ system $A^{(p)} = (a^{(p)}_{\mu\nu})$ by declaring

$$a^{(p)}_{\mu\nu} = \pm m(i_1,\ldots,i_p; j_1,\ldots,j_p), \tag{4.21}$$

where $\{i_1,\ldots,i_p\} = (\mu)$, $\{j_1,\ldots,j_p\} = (\nu)$, and the $\pm$ sign has yet to be determined.

The ordering and the choice of signs are subject to some additional compatibility conditions between the ordering and signs for $A^{(p)}$ and the "complementary" system $A^{(n-p)}$, both of which are $P \times P$ with $P = \binom{n}{p}$ [68, pp. 156–157]. For example, for $p = 1$ the sets $\{i_1,\ldots,i_p\}$ consist of a single number μ, and so (4.20) completely determines the ordering: $\{\mu\} = (\mu)$. Also $m(\mu;\nu) = a_{\mu\nu}$, so if in (4.21) we take all $\pm$ signs as $+$, then $A^{(1)} = A$. Now let $p = n-1$ and consider the complementary derived system $A^{(n-1)}$. Again the above ordering is completely determined by (4.17), since $i_1 \times \cdots \times i_{n-1} = n!/\mu$, where μ is the integer missing from $\{i_1,\ldots,i_{n-1}\}$. Thus the smaller μ is, the larger $\{i_1,\ldots,i_{n-1}\}$ is in the ordering. This means that $\{i_1,\ldots,i_{n-1}\} = (n-\mu+1)$, where μ is the missing integer. Thus $a^{(n-1)}_{n-\mu+1,n-\nu+1}$ is $\pm$ the degree-$(n-1)$ minor determinant of A that corresponds to striking out row μ and column ν. In this case, Cauchy's rule on the $\pm$ signs in (4.21) makes

$$a^{(n-1)}_{n-\mu+1,n-\nu+1} = [\mathrm{Cof}(A)]_{\mu\nu}.$$

The derived system $A^{(n-1)}$ is thus just a permuted form of the adjoint system of A.[17] Cauchy's theory of derived systems can be regarded as a vast generalization of the theory of adjoint systems.

After developing many interesting properties relating to complementary systems $A^{(p)}$ and $A^{(n-p)}$, including, for example, $(\det A)^P = \det A^{(p)} \cdot \det A^{(n-p)}$ [68, p. 161], which generalizes (4.18), Cauchy turned to $n \times n$ systems A, B, C related by his composition rule (4.17), i.e., $C = AB^t$ in modern notation. I will refer to this relation by saying that *C is the composite of A with B*. Before introducing derived systems, Cauchy had shown that when C is the composite of A and B, then the adjoint of C is the composite of the adjoint of A with the adjoint of B [68, p. 149, eqn. (44)],

[17] If σ denotes the permutation of $1,\ldots,n$ defined by $\sigma(\mu) = n-\mu+1$, then $A^{(n-1)}$ is obtained from Adj A by permuting the latter's rows according to σ and then permuting the columns of the resulting system by σ. That is, if P_σ is the corresponding $n \times n$ permutation matrix (obtained from I_n by permuting its rows according to σ), then $A^{(n-1)} = P_\sigma \,\mathrm{Adj}\, A P_\sigma$.

and he realized (in view of how adjoints and $(n-1)$st derived systems are related) that the derived system $C^{(n-1)}$ is the composite of the derived systems $A^{(n-1)}$ and $B^{(n-1)}$ [68, pp. 165–166].[18] He was able to generalize this relation to pth derived systems [68, pp. 164–165]:

Theorem 4.2 (Cauchy's second product theorem). *If C is the composite of A and B, then the $\pm$ signs in the pth derived systems may be chosen so that $C^{(p)}$ is the composite of $A^{(p)}$ and $B^{(p)}$.*

Theorem 4.2 states that

$$c_{ij}^{(p)} = \sum_{k=1}^{P} a_{ik}^{(p)} b_{kj}^{(p)}, \quad P = \binom{n}{p}, \tag{4.22}$$

and since $c_{ij}^{(p)}$, $a_{ik}^{(p)}$, and $b_{kj}^{(p)}$ are all $\pm$ a $p \times p$ minor determinant of C, A, and B, respectively, (4.22) implies the following:

Corollary 4.3. *If C is the composite of A and B then any $p \times p$ minor of C is a sum of signed products of $p \times p$ minors of A with $p \times p$ minors of B.*

It should be clear that the corollary remains true whether "C is the composite of A and B" is understood in the sense of Cauchy ($C = AB^t$) or in the sense of Gauss ($C = AB$).

Despite the clear and elegant presentation of Cauchy's essay and the wealth of new theorems established, it does not seem to have been known for many years after it appeared in 1815. The first mathematician who mastered its contents was apparently Carl Gustav Jacobi (1804–1851) [449, v. 1, p. 254]. Although Jacobi may have discovered Cauchy's essay as early as 1827 [449, v. 1, p. 178], it was, as we shall see, Cauchy's generalization of the principal axes theorem (1829) that moved Jacobi to apply and develop Cauchy's theory of determinants. In this way, determinants became a principal tool in the study of the transformation of quadratic and bilinear forms—the context of linear algebra in Berlin. Jacobi himself published an essay on determinants in 1841 [311], together with a companion essay on what are now called Jacobian determinants [312]. By the 1850s, textbooks on determinants were appearing in English [546], Italian [40], French [43], and German [12, 42]. By mid-century, Cauchy's theory as expounded and embellished by Jacobi had become a part of mainstream mathematics.

By the time it was part of the mainstream, the idiosyncrasies of Cauchy's original presentation (his definition of composition and the concomitant definition of the adjoint) were gone. Thus the composition of $n \times n$ coefficient systems $A = (a_{ij})$ and $B = (b_{ij})$ was understood as resulting in the system $C = AB$. The adjoint of A was understood as the transpose of the matrix $\mathrm{Cof}(A)$ of cofactors of A. Throughout this book I will use the notation

---

[18] That $\mathrm{Adj}\, A[\mathrm{Adj}\, B]^t = \mathrm{Adj}\, C$ implies $A^{(n-1)}[B^{(n-1)}]^t = C^{(n-1)}$ follows readily from the previously explained relation $A^{(n-1)} = P_\sigma \,\mathrm{Adj}\, A P_\sigma$, since $\sigma^{-1} = \sigma$, and so $P_\sigma^{-1} = P_\sigma$.

$$\mathrm{Adj}\, A = [\mathrm{Cof}(A)]^t \tag{4.23}$$

for the adjoint of A. Cauchy's version of the Laplace expansions then took the form that corresponds in modern matrix notation to

$$A \cdot \mathrm{Adj}\, A = (\det A)I. \tag{4.24}$$

Finally, the corollary to Cauchy's second multiplication theorem implies that *if* $C = AB$, *then every* $j \times j$ *minor of* C *is a sum of terms, each of which is* $\pm$ *a* $j \times j$ *minor of* A *multiplied by a* $j \times j$ *minor of* B. The above summary of facts from the theory of determinants is what needs to be kept in mind in this and in the following chapters.

4.4 Cauchy and the Principal Axes Theorem

Cauchy became an adjunct professor at the École Polytechnique in 1815 and a full professor the following year. His essay on determinants already displayed the great talent he had for expounding an enriched version of a mathematical subject, and this talent can be seen in the many books he published based on his lectures at the École. In particular, Cauchy's teaching duties gave him the opportunity to develop the elements of analysis in what he deemed to be a more rigorous fashion than his eighteenth-century predecessors. The several books that he published based on these lectures encouraged mathematicians such as Dirichlet and Weierstrass in their efforts to develop analysis more rigorously. Among other things, Cauchy was critical of the generic reasoning of his predecessors. Thus in the preface to his *Cours d'Analyse* of 1821 [69], he insisted that mathematical analysis must not be based on arguments "drawn from the generality of algebra," arguments that "tend to attribute an indefinite scope to algebraic formulas, whereas in reality the majority hold true only under certain conditions and for certain values of the variables involved." In what follows, we shall see how Cauchy sought to adhere to this stipulation in treating the principal axes theorem.

4.4.1 The three-dimensional principal axes theorem

Under the guidance of Monge, the "application of algebra to geometry"—analytic geometry—had become a vital component of the educational program of the École Polytechnique ([31, ch. 7], [560]). As the outline of 1802 prepared by Monge and Hachette [447] indicates, a central part of the study of analytic geometry was the classification of quadric surfaces, i.e., surfaces defined by a quadratic equation in three variables. In 1748, Euler had initiated the study of these surfaces in an appendix to the second volume of *Introductio in analysin infinitorum* [156].

The classification is based on the fact that a central quadric surface, i.e., one with an equation of the form

$$\sum_{i,j=1}^{3} a_{ij}x_ix_j = M, \tag{4.25}$$

can be expressed in the form

$$\sum_{i=1}^{3} L_iy_i^2 = M \tag{4.26}$$

with respect to a suitably chosen rectangular coordinate system. Thus it was necessary to prove that a quadratic form in three variables can be transformed into a sum of square terms by a change of rectangular coordinates.

Euler had not bothered to provide a proof in the sketch he presented, although he briefly indicated the approach [156, Art. 114] later developed into a proof by Hachette and Poisson in 1802 [447]. The idea is to make a rotational change of variables of the form given by Euler in his discussion of quadric surfaces, namely

$$\begin{aligned} x &= t\cos\xi + u\sin\xi\cos\eta - v\sin\xi\sin\eta, \\ y &= -t\sin\xi + u\cos\xi\cos\eta - v\cos\xi\sin\eta, \\ z &= u\sin\eta + +v\cos\eta, \end{aligned} \tag{4.27}$$

and to write down the equations that state that the tu and uv terms of the resulting quadratic in t, u, and v should vanish. With suitable manipulation these equations yield a cubic equation in the tangent of one of the angles in (4.27). Since a cubic equation with real coefficients must have at least one real root, it followed that the transformation (4.27) could be chosen such that the tu and uv vanish. A further orthogonal transformation can then be chosen to remove the tv term.

The Hachette–Poisson approach to the principal axes theorem became the standard one in the early nineteenth century. Nowadays, this theorem is linked to the eighteenth-century "eigenvalue problem" of determining λ and x such that $Ax = \lambda x$, where $A = (a_{ij})$ is the coefficient matrix of the quadratic form M of (4.25); but this connection is lacking in the Hachette–Poisson approach. In fact, it was only later that J. Bret, a former student at the École, observed in 1812 [38] that the cubic having the L_i of (4.26) as its zeros can be expressed in terms of the original coefficients a_{ij} of (4.25) as

$$f(x) = x^3 - (a_{11} + a_{22} + a_{33})x^2 + \cdots .$$

The cubic $f(x)$ is the characteristic polynomial associated with $Ax = \lambda x$, and Bret's observation, together with the proof of Hachette and Poisson, implies that its roots must always be real. Bret, however, did not regard $f(x)$ as a determinant or in terms of an eigenvalue problem. Although Hachette's subsequent exposition of quadric surfaces in 1813 [257, p. 1152] did make a connection between the transformation of the equation for a quadric surface and an eigenvalue problem, the connection remained incidental until the work of Cauchy. For example, it is missing in Biot's popular exposition of quadric surfaces published in 1826 [24].

Cauchy was not the first to emphasize the connection between the principal axes theorem and an eigenvalue problem. That honor belongs to Lagrange, whose treatment of the matter proved inspirational to Cauchy. Lagrange's detailed treatment of the principal axes theorem was motivated by mechanics. Rotational motion had first been successfully treated by Euler in 1765 [157, 158] on the basis of his discovery that a rigid body possesses three mutually perpendicular principal axes (Euler's term) with respect to which the product moments of inertia vanish. Their existence was established by determining the axes for which the moment of inertia takes an extreme value, the moment of inertia being expressed in terms of the angles the axes make with a fixed coordinate system. Euler's introduction of principal axes and his resolution of the problem marked an important advance in mechanics, which was another Eulerian triumph for the analytic, as opposed to the geometric, approach. On the other hand, Euler's work was not analytic in the extreme sense practiced by Lagrange, and we find Lagrange presenting a new analysis of the problem in a memoir presented in 1773 [388].

Lagrange's stated objection to Euler's solution concerned the latter's use of the principal axes as a starting point. The problem, he felt, should be considered "in itself" and resolved "directly and independently" of the properties of the principal axes, which should follow as consequences of the analysis rather than be its starting point. What Lagrange wished to do (and Euler had not done) was to reduce the mechanical problem to analysis in the sense later articulated in the preface to *Mécanique analytique* [391], when he wrote

> I have set myself the problem of reducing this science, and the art of solving the problems pertaining to it, to general formulas, the simple development of which gives all the equations necessary for the solution of each problem .... No figures will be found in this work. The methods I expound in it require neither constructions nor geometric or mechanical reasoning, but only algebraic operations subjected to a systematic and uniform progression. Those who like analysis will be pleased to see mechanics become a new branch of it and will be grateful to me for having thus extended its domain.

In the same spirit as the above quotation, Lagrange wrote that "The merit of my solution, if it has one, thus consists solely in the Analysis I employ..." [388, p. 580]. In particular, in Lagrange's approach, the role formerly played by the principal axes is taken over by a purely algebraic theorem. Stated in present-day terms, it is that a quadratic form in three variables can be transformed into a sum of square terms by an orthogonal transformation of the variables.

In discussing Lagrange's solution to the problem, I will follow the slightly different treatment in *Mécanique analytique* [391, Pt. II, §VI], for it was probably the version that Cauchy read.[19] After deducing, from general principles, differential

[19]The treatment in the second (1815) and subsequent editions is essentially the same as in the first. See the bibliography for section numbers in later editions.

equations describing the rotational motion of a rigid body, Lagrange observed that they can be integrated by making a linear change of variables

$$\begin{aligned} p &= p'x + p''y + p'''z, \\ q &= q'x + q''y + q'''z, \\ r &= r'x + r''y + r'''z, \end{aligned} \tag{4.28}$$

which transforms the quadratic function

$$T = \frac{1}{2}\left(Ap^2 + Bq^2 + Cr^2\right) - Fqr - Gpr - Hpq$$

into a sum of squares

$$T = \frac{1}{2}(\alpha x^2 + \beta y^2 + \gamma z^2) \tag{4.29}$$

in such a manner that

$$p^2 + q^2 + r^2 = x^2 + y^2 + z^2.$$

Lagrange observed that the above relationship implies that the coefficients of the linear transformation (4.28) must satisfy the further relationships

$$\begin{aligned} (p')^2 + (q')^2 + (r')^2 &= 1, \quad p'p'' + q'q'' + r'r'' = 0, \\ (p'')^2 + (q'')^2 + (r'')^2 &= 1, \quad p'p''' + q'q''' + r'r''' = 0, \\ (p''')^2 + (q''')^2 + (r''')^2 &= 1, \quad p''p''' + q''q''' + r''r''' = 0. \end{aligned} \tag{4.30}$$

Such a linear transformation we recognize now as an orthogonal transformation. Of course, the coefficients of Euler's rotational transformation (4.27) satisfy the orthogonality relations (4.30), but the form (4.28)/(4.30) is not only more formally symmetric and general but also more suggestive. That is, in Lagrange's form it is immediately clear how to extend the definition of this type of transformation from $n = 3$ to any number $n > 3$ of variables.

Lagrange showed that to prove the existence of the transformation (4.28) satisfying (4.30), one can consider the problem of solving $Ax = \lambda x$, where A is the matrix of coefficients determined by the quadratic function T. The characteristic roots, he showed, are $\lambda = \alpha, \beta, \gamma$ of (4.29); and the coefficients of the transformation (4.28) are obtained from the corresponding characteristic vectors. (Throughout this book I will use the term "characteristic vector" rather than "eigenvector.") The possibility of transforming T into (4.29) thus depended on the reality of the characteristic roots. In this case, Lagrange could not make use of the physical context, but since the characteristic equation was cubic, he was able to establish the reality algebraically. The idea of his proof [391, pp. 239–240], which was limited to the generic case, is easy to explain in present-day terms. Assume that not all the characteristic roots are real. Then two of the roots will be complex conjugates, say $\beta = p+qi$ and $\gamma = p-qi$, and this means the last two "columns" of the orthogonal transformation (4.28) (being characteristic vectors for β and γ, respectively) will also be complex conjugates. But then the orthogonality relation for these columns, namely $p''p''' + q''q''' + r''r''' = 0$, takes the form $|p''|^2 + |q''|^2 + |r''|^2 = 0$, which is absurd.

Cauchy's exposition of the theory of quadric surfaces was contained in his *Leçons sur l'application du calcul infinitésimale à la géométrie* [71, pp. 244ff.], which was published in 1826. There the influence of Lagrange's treatment of the principal axes theorem is evident, for unlike his contemporaries, Cauchy focused on the consideration of an eigenvalue problem, its associated characteristic equation, and the need to prove the reality of all its roots. Cauchy also realized that Lagrange's formulation of the principal axes theorem could be generalized to $n > 3$ variables. Thus in November 1826, he presented a paper to the Academy of Sciences "On the equation that has for its roots the principal moments of inertia of a solid body and on various equations of the same type" [73]. In order to justify, by means of an application, the consideration of a characteristic equation $f(s) = \det(A - sI)$ corresponding to an $n \times n$ symmetric coefficient system $A = (a_{ij})$ with $n > 3$, Cauchy pointed to the following problem. Let

$$\Psi = \sum_{i,j=1}^{n} a_{ij}x_i x_j \quad \text{and} \quad \Phi = x_1^2 + \cdots + x_n^2,$$

and consider the problem of determining the maximum or minimum values of the quotient Ψ/Φ. Then the determination of these values, he announced, "will depend on an equation of degree n for which all the roots are real" [73, p. 80]. As Cauchy showed in a paper of 1829, which gives the details to support the brief announcements of his 1826 paper, the nth-degree equation is the characteristic polynomial associated to the eigenvalue problem $Ae = \lambda e$ to which the above extremum problem leads, and the desired extreme values are among the associated characteristic roots [72, pp. 174–176]. He also announced the generalization of the principal axes theorem to the n-variable quadratic form Ψ. Let us now turn to the 1829 paper and Cauchy's proofs.

4.4.2 The n-dimensional principal axes theorem (1829)

Cauchy's above-mentioned paper of 1829 was published in his own mathematical journal, *Exercises de mathématiques*, and appeared before his brief announcement, which was not published until 1830. It contains the proofs of two theorems that served to generalize the principal axes theorem to n variables. The first posits the reality of the roots of the characteristic equation associated to a symmetric coefficient system $A = (a_{ij})$:

Theorem 4.4 (Cauchy's reality theorem). *Let $\Psi = \sum_{i=1}^{n} a_{ij}x_i x_j$ denote a homogeneous function of degree two and $f(s) = \det(A - sI)$ the associated nth-degree polynomial. Then all the roots of $f(s)$ are real.*

Nowadays, the proof of the reality of the roots of a real symmetric (or Hermitian symmetric) matrix follows in a line or two from the consideration of Hermitian inner products.[20] In Cauchy's time, however, inner products (real or Hermitian) were not common mathematical notions. Dazzled by the brilliance of the new theory of determinants, mathematicians overlooked simple inner product considerations and focused instead on reasoning utilizing determinants. As a result, Cauchy's determinant-based proof, although nongeneric and perfectly rigorous, is much longer than modern ones. The interested reader will find it sketched below in Section 4.4.3. Others may proceed to Section 4.5 with no loss of continuity.

The idea behind Cauchy's proof was the same as Lagrange had used in dealing with the three-variable case: show that the assumption of a complex root—and therefore of a pair of complex conjugate roots—leads, by means of the orthogonality relations for the associated conjugate characteristic vectors, to a contradiction. Lagrange's own generic proof was already considerably more complicated than the modern proof, and Cauchy had to deal with a characteristic polynomial of arbitrary degree n. Given Cauchy's approach to the proof and the tools at his disposal, his proof is remarkably succinct and elegant. Indeed, as we shall see, it made a great impression on his contemporaries, especially Jacobi, since it suggested that the elegance of Lagrangian algebra could be extended to n variables by utilizing the theory of determinants that Cauchy had developed. Furthermore, Cauchy's proof was completely rigorous, i.e., he was able to avoid the pitfalls of generic reasoning, although his proof must be read carefully to appreciate this fact. It was thus the first valid proof that the characteristic roots of a symmetric matrix must be real.

Cauchy's second theorem, his generalization of the principal axes theorem of mechanics and analytic geometry, may be stated as follows:

Theorem 4.5 (Cauchy's principal axes theorem). *Given a homogeneous function $\Psi = \sum_{i,j=1}^{n} a_{ij}x_i x_j$ of second degree, there exists a linear change of variables $x = Py$ such that $\sum_{i=1}^{n} x_i^2 = \sum_{i=1}^{n} y_i^2$ and $\Psi = \sum_{i=1}^{n} s_i y_i^2$, where $s_1,\dots,s_n$ are the roots of $f(s) = \det(A - sI) = 0$.*

[20] If A is symmetric and $Ae = \lambda e$, where $e \neq 0$, to show that λ must be real, consider the Hermitian inner product $(e, Ae) = e^t\overline{(Ae)}$. On the one hand, $(e, Ae) = \bar{\lambda}(e^t\bar{e})$. On the other hand, $(e, Ae) = (e^tA)\bar{e} = (Ae)^t\bar{e} = \lambda(e^t\bar{e})$. Since $e^t\bar{e} > 0$, it must be that $\bar{\lambda} = \lambda$. The same type of Hermitian inner product argument can be used to show that the roots of $\det(\lambda B - A)$ must all be real when A, B are symmetric and B is positive definite. Such an inner product argument was first used by Christoffel in 1864 (Section 5.2).

Cauchy began by proving the theorem assuming that the roots $s_1,\ldots,s_n$ are distinct. To this end, he used the fact, established in the proof of the above reality theorem, that if s and s' are distinct characteristic roots and if $e = \begin{pmatrix} e_1 & \cdots & e_n \end{pmatrix}^t$ and $f = \begin{pmatrix} f_1 & \cdots & f_n \end{pmatrix}^t$ are corresponding characteristic vectors, then

$$e_1 f_1 + \cdots + e_n f_n = 0.$$

It then follows that if $p_i = \begin{pmatrix} p_{1i} & \cdots & p_{ni} \end{pmatrix}^t$ is a characteristic vector for s_i chosen such that $p_{1i}^2 + \cdots + p_{ni}^2 = 1$, then the coefficient system $P = (p_{ij})$, which has the coefficients of p_i as its ith column, has the orthogonality properties summed up in $P^t P = I$, from which it then follows that the variable change $x = Py$ satisfies the conditions of Theorem 4.5 [72, pp. 192–194].

In order to extend Theorem 4.5 to the case in which the characteristic equation has multiple roots, Cauchy used another theorem he had proved in his paper [72, p. 187], which is based on the following considerations. In the $n \times n$ symmetric array A, cross out the first row and the first column so as to obtain an $(n-1) \times (n-1)$ symmetric array, and let $R = R(s)$ denote its characteristic polynomial. (Thus $R(s)$ is the $(1,1)$ minor determinant of $A - sI$.) By the reality theorem, its roots are real. Hence they may be ordered as $r_1 \leq r_2 \leq \cdots \leq r_{n-1}$. Cauchy proved that the roots of $f(s) = \det(A - sI)$, suitably ordered, satisfy

$$s_1 \leq r_1 \leq s_2 \leq r_2 \leq \cdots \leq s_{n-1} \leq r_{n-1} \leq s_n.$$

For any symmetric coefficient system A he then defined

$$K = K(A) = \prod_{i=1}^{n-1} f(r_i), \quad f(s) = \det(A - sI).$$

It follows that if $K(A) \neq 0$, then none of the roots r_i of $R(s)$ are roots of $f(s)$. This means that the roots of $f(s)$ must be distinct, since if $s_{k-1} = s_k$, then, since $s_{k-1} \leq r_k \leq s_k$, we must have $f(r_k) = f(s_k) = 0$ and so $K(A) = 0$.

The above considerations show that for the symmetric systems A satisfying $K(A) \neq 0$, Cauchy's above proof of Theorem 4.5 holds, since the characteristic roots of such A are all distinct. Now K is, as Cauchy observed, a symmetric function of the roots $r_1, r_2, \ldots, r_{n-1}$ of $R = 0$, and by the properties of such symmetric functions it follows that K is a polynomial in the coefficients a_{ij} of A. Hence "in general," $K \neq 0$. To cover the nongeneric case $K = 0$, Cauchy simply remarked that if the a_{ij} satisfy the condition $K = 0$, then

> it would suffice, in order to make it cease to hold, to attribute to one of the coefficients $[a_{ij}]$ in question an infinitely small increment ε; and since ε could be made to converge to zero without ... [Theorem 4.5] ... ceasing to hold, it is clear that it still holds at the moment when ε vanishes [72, p. 195].

Cauchy thus ended up resorting to the type of limit-infinitesimal reasoning that Lagrange had applied (with unfortunate results) when trying to see how multiple roots affect the solution of a system of linear differential equations (Section 4.2.1). However, Cauchy's argument was much clearer than Lagrange's, for it brings the coefficients a_{ij} into the picture, whereas Lagrange's formulas obscured their dependency on the a_{ij}. In fact, Cauchy's proof can be made rigorous by applying the multidimensional Bolzano–Weierstrass theorem.[21] At the time, however, such a compactness argument was not an established technique, and Weierstrass himself deemed Cauchy's proof of Theorem 4.5 to be convincing only for the case in which A has distinct characteristic roots. This was one of the motivating considerations behind his paper of 1858.

4.4.3 Cauchy's proof of his reality theorem

To prove his reality theorem (Theorem 4.4), Cauchy's idea was to show that if $S = f(s) = \det(A - sI) = 0$ has a complex root, and hence a pair of complex conjugate roots, then so does $R = f_{11}(s) = 0$, where $f_{11}(s)$ denotes the minor of $A - sI$ corresponding to the $(1,1)$ entry. (Cauchy described R as the characteristic polynomial of the $(n-1)\times(n-1)$ symmetric system obtained from A by deleting its first row and column.) If $R = 0$ has a pair of complex conjugate roots, then the same reasoning implies that if $Q = 0$ denotes the characteristic polynomial of the symmetric system obtained from A by deleting the first two rows and columns, then $Q = 0$ has a pair of complex conjugate roots. Continuing in this manner, we eventually arrive at the absurd conclusion that $\det(a_{nn} - s) = a_{nn} - s = 0$ has a pair of complex conjugate roots. The heart of Cauchy's proof thus involved showing that if $S = 0$ has a pair of complex conjugate roots, then so does $R = 0$. Let us consider how he did this.

The first step was the following now-familiar lemma:

Lemma 4.6. *If A is symmetric and $Ax = s_1x$, $Ay = s_2y$, with $s_1 \neq s_2$, then*

$$x_1y_1 + \cdots + x_ny_n = 0.$$

Nowadays, this lemma is a simple consequence of inner product considerations, but Cauchy's proof was longer and in the style of Lagrangian algebra [72, pp. 177–178].

---

[21] Since $K(A)$ is a polynomial in the $d = n(n+1)/2$ variables a_{ij}, $i \leq j$, that determine A and since $K(A)$ is not identically zero, it follows that $K(A) \neq 0$ on an open dense subset of $\mathbb{R}^d$. Thus given an A with multiple roots, there is a sequence $A_n \to A$ for which $K(A_n) \neq 0$ for all n. For each such A_n, Theorem 4.5 holds by virtue of Cauchy's proof in the case of distinct characteristic roots. Thus for each A_n, there exists an orthogonal matrix P_n such that $D_n = P_n^t A_n P_n$ is diagonal. Now, the totality of all orthogonal $n \times n$ matrices P forms a compact subset of $\mathbb{R}^\ell$, $\ell = n^2$. Thus the sequence P_n has a convergent subsequence $P_{n_j} \to P$, where P is orthogonal. Thus $P^t AP = \lim_{j\to\infty} P_{n_j}^t A_{n_j} P_{n_j} = \lim_{j\to\infty} D_{n_j}$ is a limit of diagonal matrices and hence is diagonal. (I am grateful to Richard Beals for pointing out this compactness argument to me many years ago.)

By eliminating a_{11} from the first equations of the two systems $Ax = s_1x$ and $Ay = s_2y$, and then a_{22} from the second equations of each system, and so on, Cauchy obtained n equations, which when added gave $(s_2 - s_1)(x_1y_1 + \cdots + x_ny_n) = 0$, from which the lemma follows.

For the remainder of the proof [72, pp. 178–180], assume that s_1 is a complex characteristic root of A. The goal is to show that this assumption implies that $R(s_1) = 0$, so that the reasoning indicated above applies and leads to a contradiction. Let $x = (x_1 \cdots x_n)^t \neq 0$ be such that $Ax = s_1x$ and without loss of generality that $x_1 \neq 0$.[22] If the first equation in the system $(A - s_1I)x = 0$ is ignored, what is left can be interpreted as an inhomogeneous system of $n-1$ equations of the form $(B - s_1I_{n-1})z = -x_1c$, where B is the $(n-1) \times (n-1)$ symmetric matrix obtained from A by deleting row and column one, whence $\det(B - s_1I_{n-1}) = R(s_1)$, $z = (x_2 \cdots x_n)^t$, and $c = (a_{21} \cdots a_{n1})^t$. Still under the hypothesis that s_1 is a complex root, we now show that the hypothesis $R(s_1) \neq 0$ leads to a contradiction. Assuming that hypothesis, suppose first that $c = 0$. Then since A is symmetric, its $(1,2)$ through $(1,n)$ entries are also all 0's, and so $S(s_1) = \det(A - s_1)$ factors as $(a_{11} - s_1)R(s_1)$. Since $R(s_1) \neq 0$ but $S(s_1) = 0$, we must have that $s_1 = a_{11}$ is real, contrary to the hypothesis that s_1 is complex. Thus the hypothesis $R(s_1) \neq 0$ requires that $c \neq 0$, and the solution to $(B - s_1I_{n-1})z = -x_1c$ is given by Cramer's rule. The result is

$$x_i = -x_1[f_{1i}(s_1)/R(s_1)] \quad \text{for} \quad i = 2, \ldots, n, \tag{4.31}$$

where $f_{1i}(s_1)$ denotes the $(1,i)$ minor of $A - s_1I$.

Following Cauchy, let us introduce a more symmetric notation,

$$X_1(s) = f_{11}(s) = R(s), \quad X_i(s) = -f_{1i}(s), \quad i = 2, \ldots, n.$$

Then (4.31) implies that if $R(s_1) \neq 0$, then

$$x_1^{-1}X_1(s_1)x = (X_1(s_1) \cdots X_n(s_1))^t \stackrel{\text{def}}{=} V(s_1)$$

is a characteristic vector for the characteristic root $s = s_1$ of A. Now consider the complex conjugate $s_2 = \bar{s}_1$. Clearly $R(s_2) = R(\bar{s}_1) = \overline{R(s_1)} \neq 0$, and since evidently, $X_i(\bar{s}_1) = \overline{X_i(s_1)}$ for all i, we can conclude by the above reasoning that $V(\bar{s}_1) = \overline{V(s_1)}$ is a characteristic vector for characteristic value $\bar{s}_1$. Finally, since $\bar{s}_1 \neq s_1$ by hypothesis, Lemma 4.6 applies, and since $V(\bar{s}_1) = \overline{V(s_1)}$, the orthogonality relation in the lemma for the roots s_1 and $s_2 = \bar{s}_1$ becomes

$$|X_1(s_1)|^2 + \cdots + |X_n(s_1)|^2 = 0,$$

[22] Since $x \neq 0$, $x_k \neq 0$ for some k. If $k \neq 1$, use the (k,k) minors of A and $A - s_1I$ in lieu of the $(1,1)$ minors and modify the following proof accordingly, deleting the kth row of $(A - s_1I)x = 0$, etc.

which implies in particular that $R(s_1) = X_1(s_1) = 0$ and of course $R(\bar{s}_1) = \overline{R(s_1)} = 0$ as well. Thus if $S = 0$ has s_1 as a complex root, then so does $R = 0$. In accordance with the above-described proof outline, this completes Cauchy's proof. It is a clever nontrivial extension of Lagrange's proof idea to the nongeneric case and any number of variables. Today the proof is obsolete, but historically, it reveals how Cauchy used his theory of determinants to surpass his illustrious predecessor and impress his successors.

4.5 A Very Remarkable Property

In 1834, Jacobi presented an elegant determinant-theoretic derivation of Cauchy's principal axes theorem, and in 1840, Cauchy published a method for integrating systems of linear differential equations, based on his calculus of residues, that was valid whether or not the associated characteristic polynomials possessed multiple roots. Hidden between the lines of both their works was the same property. In the case of Jacobi's work, his generic formula for the orthogonal transformation of the principal axes theorem could be seen, by a critical eye such as Weierstrass', to be generally valid, provided this property was known to hold. In the case of Cauchy's work, his formulas for the solutions to the differential equations indicated that the powers of t that always occurred in the solutions of Lagrange and Laplace when multiple roots were present would, in fact, not occur, provided the same property held. The proof of this property became the key to Weierstrass' proof of the theorem stated at the beginning of the chapter.

4.5.1 *Jacobi's generic formula*

Not long after Jacobi became aware of Cauchy's essay on determinants, Cauchy published his proof of his principal axes theorem (Theorem 4.5) in his paper of 1829. As we saw, Cauchy had introduced the notion of an orthogonal transformation of n variables, a notion that Jacobi had already studied in the case $n = 3$ in 1827. Inspired by Cauchy's paper, Jacobi turned to the theory and application of n-variable orthogonal transformations in a paper of 1834 published in *Crelle's Journal* [310] and made considerable use of Cauchy's theory of determinants. Thus, for example, he showed that if $x = Py$ is any orthogonal transformation, i.e., any transformation with coefficient matrix $P = (p_{ij})$ satisfying the equations that correspond to $P^tP = I$, then $\det P = \pm 1$ [310, p. 201]. And for the orthogonal transformation $x = Py$ that takes a quadratic form $\Psi = \sum_{i,j=1}^{n} a_{ij}x_jx_j$ into $\Psi = \sum_{k=1}^{n} s_k y_k^2$, he obtained the following elegant formula relating the coefficients p_{ij} to the characteristic polynomial $f(s) = \det(A - sI)$ [310, p. 212, eqn. (36)]:

$$p_{ik}p_{jk} = -\frac{b_{ij}(s_k)}{f'(s_k)}. \tag{4.32}$$

Here $b_{ij}(s)$ is the (i,j) coefficient of $\mathrm{Adj}\,(A-sI)$, the adjoint of $A-sI$. (See (4.23) for the definition of the adjoint of a matrix.) If we take $i=j$ in (4.32), we get $p_{ik}^2 = -b_{ii}(s_k)/f'(s_k)$, and so $p_{ik} = \pm\sqrt{-b_{ii}(s_k)/f'(s_k)}$. A correct choice of $\pm$ signs can be determined using (4.32) with $i \neq j$, and so Jacobi's formula determines the desired orthogonal transformation P.

Jacobi's reasoning was exclusively on the generic level. Suppose we consider his formula (4.32) in the spirit of Cauchy's more critical approach to algebra. If some root $s_k = a$ of the characteristic polynomial $f(s)$ is a root of multiplicity $m > 1$, then $f'(a) = 0$, and (4.32) has a zero in the denominator, thereby throwing its meaning into question when multiple roots exist. Some attention was paid to the validity of Jacobi's formula in the case of a multiple root $s = a$ by V.A. Lebesgue in 1837 [404]. The purpose of Lebesgue's paper was an expository synthesis of Cauchy's 1829 memoir on quadratic forms, the related papers by Sturm (Section 4.2.3), and Jacobi's paper of 1834. In a brief section on the nongeneric case, Lebesgue showed, using the symmetry properties of the coefficients, that if $s = a$ is a multiple root of $f(s)$, then $b_{ij}(a) = 0$ for all i and j. Thus formula (4.32) "can occur in the form $0/0$ but not in the form $m/0$" [404, p. 352]. Apparently satisfied with this observation, Lebesgue chose to pursue the matter no further.

The $0/0$ form is, of course, not in itself reassuring. However, if $s_k = a$ is a root of $f(s) = \det(A-sI)$ of multiplicity $m > 1$, then $s = a$ is a root of $f'(s)$ of multiplicity $m-1$. Thus Jacobi's formula would remain meaningful provided $(s-a)^{m-1}$ divides $b_{ij}(s)$ for all (i,j). In that case, $b_{ij}(s) = (s-a)^{m-1}b^*_{ij}(s)$ and $f'(s) = (s-a)^{m-1}g(s)$, with $g(a) \neq 0$, and so for all s with $|s-a| > 0$ sufficiently small, $b_{ij}(s)/f'(s) = b^*_{ij}(s)/g(s)$ has a removable singularity at $s = a$, and $b_{ij}(a)/f'(a) = b^*_{ij}(a)/g(a)$ is meaningful. In sum, Jacobi's formula would retain its meaning whenever the following property could be shown to hold:

Property 4.7. If $s = a$ is a root of multiplicity $m > 1$ of $f(s) = \det(A-sI)$, then $(s-a)^{m-1}$ divides every cofactor of $A-sI$.

Thus if all symmetric matrices A could be shown to satisfy this property, Jacobi's formulas would lose their generic character and would yield a proof of Cauchy's principal axes theorem that—by contrast with Cauchy's more dubious justification of the existence of P in the presence of multiple roots—would be perfectly rigorous.

4.5.2 *Cauchy's method of integration*

When Cauchy composed his memoir of 1829 on the transformation of quadratic forms, he was not particularly interested in systems of linear differential equations with constant coefficients. His paper was motivated by the principal axes theorem of mechanics and the theory of quadric surfaces. In the 1830s, however, he became increasingly interested in the problem of deriving the properties of light from a theory of the small vibrations of a solid elastic medium [605, Ch. 5]. The groundbreaking work of Fresnel in the 1820s was undoubtedly a factor behind his interest in this approach to the theory of light.

In developing the mathematics of his approach, Cauchy was naturally led to systems of linear differential equations, which in the simplest cases could be assumed to have constant coefficients [74]. At first, Cauchy limited his treatment to the case of distinct characteristic roots while briefly alluding to the "very simple" Lagrangian method of handling the case of multiple roots [74, pp. 211–212, 220]. A few months later, he revised his opinion about the simplicity of that method in a lengthy paper published in the *Comptes rendus* of the Academy of Sciences [75] and reprinted the following year (1840) in his *Exercises d'Analyse* [76]. He now portrayed Lagrange's method as one designed for distinct roots of the associated characteristic equation—a term he here introduced for the first time—that Lagrange was then "forced" to modify to cover multiple roots [76, p. 76]. Indeed, as we saw in Section 4.2.1, Lagrange had modified his elegant solution formula to cover only the case of one double root. Cauchy could afford to be critical because he had discovered a new method of expressing the solutions that was valid for multiple roots as well as for distinct roots—a method he felt would be useful "not only to geometers but to physicists as well" [76, p. 76].

Cauchy's method utilized his calculus of residues—a calculus for which he was constantly seeking new applications.[23] For the sake of simplicity, I will indicate the method as applied to the first-order system of n linear differential equations

$$\dot{y} + Ay = 0, \quad y(0) = c_0, \tag{4.33}$$

which Cauchy himself considered first and in the greatest detail and generality. Cauchy introduced the equation $f(s) = \det(sI + A) = 0$, which he again referred to as the "characteristic equation" [76, p. 80]. Let $s_1, \ldots, s_k$ denote the distinct roots of $f(s)$. Then Cauchy's solution to (4.33) [76, p. 81, eq. (11)], expressed here using matrix notation is $y(t) = R(t)c_0$, where

$$R(t) = \sum_{j=1}^{k} \mathrm{Res}_{s=s_j} \left[\mathrm{e}^{st} \frac{\mathrm{Adj}\,(sI + A)}{f(s)} \right]. \tag{4.34}$$

Since $R(0) = I_n$ [76, p. 82, eq. (14)], the initial condition is satisfied.

Cauchy did not use any determinant-theoretic language in this paper, undoubtedly because it was still too novel for a readership that he hoped would include some physicists. Indeed, both here and in his 1829 paper on the generalized principal axes

[23] The pages of Cauchy's private journal, *Exercises de mathématique* (1826–1830), are filled with such applications, including an application to the solution of the nth-order differential equation with constant coefficients [71].

theorem [72], he never expressly invoked the concepts and results of his memoir on determinants [68], even though he was clearly thinking in terms of them, as is clear, e.g., from the proof of his reality theorem (Section 4.4.3). For readers like Weierstrass who were acquainted with the Cauchy–Jacobi theory of determinants, however, the connection of the coefficients of $R(t)$ with the adjoint of the system $sI+A$ would have been clear.

Cauchy introduced his formula $y(t)=R(t)c_0$ because it covered all possible cases of multiple roots. Let us consider what the formula implies if $s_k=a$ is a root of multiplicity m of $f(s)=\det(sI+A)$, so that $f(s)=(s-a)^m f_0(s)$ and $f_0(a)\neq 0$. If $b_{ij}(s)$ denotes the (i,j) coefficient of $\mathrm{Adj}\,(sI+A)$, then $s=a$ cannot be a root of multiplicity m or greater for all (i,j). For if that were the case, then $(sI+A)\,\mathrm{Adj}\,(sI+A)=f(s)I$ could be rewritten as

$$(sI+A)\frac{\mathrm{Adj}\,(sI+A)}{(s-a)^m}=f_0(s)I, \tag{4.35}$$

and $L=\lim_{s\to a}\frac{\mathrm{Adj}\,(sI+A)}{(s-a)^m}$ would exist, so that letting $s\to a$ in (4.35) would yield $(aI+A)L=f_0(a)I$, and so, taking determinants, $f(a)\det L=[f_0(a)]^n$ would follow from Cauchy's multiplication theorem and would imply incorrectly that $f(a)\neq 0$. Thus (i,j) exists such that $b_{ij}(s)$ will have $s=a$ as a zero of multiplicity $m'<m$. This means that $\varphi(s)=b_{ij}(s)/f(s)$ will have a pole of order $p=m-m'>0$ at $s=a$, and the same will be true of $\psi(s)=e^{ts}\varphi(s)$.

Now according to Cauchy's own formula [71, p. 28], the residue at a pole $s=a$ of order p is given by

$$\mathrm{Res}_{s=a}\big[\psi(s)\big]=\lim_{s\to a}\frac{1}{(p-1)!}\frac{d^{p-1}}{ds^{p-1}}\Big[(s-a)^p\psi(s)\Big].$$

If $p\geq 2$ and if $\varphi(s)=\frac{a_p}{(s-a)^p}+\cdots+\frac{a_1}{(s-a)}+\cdots$ denotes the Laurent expansion of φ about $s=a$, then by the above formula, the residue of $\psi(s)$ equals, upon calculation,

$$\frac{1}{(p-1)!}\big[a_{-p}t^{p-1}+\cdots+(p-1)!a_{-2}t+(p-1)!a_{-1}\big]e^{at}.$$

Since $a_{-p}\neq 0$, it follows that the (i,j) coefficient of $R(t)$ will have powers of t in it. Thus even if all the characteristic roots are purely imaginary, so that $a=i\theta$ and $e^{at}=e^{i\theta t}$ remains bounded as $t\to\infty$, this will not be true of the (i,j) coefficient of $R(t)$. Thus if $b_{ij}(s)/f(s)$ has a pole of order $p=m-m'\geq 2$ at $s=a$, not all solutions $y=R(t)c_0$ will remain bounded. This is analogous to what Lagrange had discovered in studying $\ddot{y}+My=0$.

Lagrange's erroneous conclusion about the necessary occurrence of powers of t when multiple roots are present was based on generic reasoning and concomitant generic formulas. Cauchy's nongeneric formula for the solutions to $\dot{y}+Ay=0$ made it possible to see that, in fact, no powers of t occur in $R(t)$ provided that for all (i,j),

$b_{ij}(s)/f(s)$ has either no pole at a characteristic root $s = a$ or a pole of order $p = 1$. Clearly $b_{ij}(s)/f(s)$ has a pole of order $p = m - m' = 1$ at $s = a$ precisely when $b_{ij}(s)$ has $s = a$ as a zero of multiplicity $m' = m - 1$. Thus the requirement that $(s-a)^{m-1}$ divide $b_{ij}(s)$ for all (i, j) implies either no pole (because $(s-a)^m$ also divides $b_{ij}(s)$) or a pole of order 1 at $s = a$. This requirement is, of course, that of Property 4.7: *if* $f(s) = \det(sI + A)$ *has* $s = a$ *as a root of multiplicity* $m > 1$, *then* $(s-a)^{m-1}$ *divides all cofactors of* $sI + A$. Thus, thanks to Cauchy's nongeneric formulas, it follows that *the solutions,* $y = R(t)c_0$, *to* $\dot{y} + Ay = 0$ *never contain powers of* t *if and only if Property 4.7 holds.*

After describing a method for expediting calculation of the solution $y(t) = R(t)c_0$ to $\dot{y} + Ay = 0$, as well as inhomogeneous first-order systems, Cauchy showed how in general to integrate a linear system of order $\ell \geq 2$ by reducing the problem to that of solving a larger first-order system. Applied to a second-order $n \times n$ system $\ddot{y} + Ay = 0$, Cauchy's idea was to replace it by a $2n \times 2n$ first-order system by introducing auxiliary unknown functions $z_1, \ldots, z_n$ so that the second-order system is replaced by $\dot{z} + Ay = 0$, $\dot{y} - z = 0$, i.e., by $\dot{w} + \tilde{A}w = 0$ with

$$w = \begin{pmatrix} y \\ z \end{pmatrix} \quad \text{and} \quad \tilde{A} = \begin{pmatrix} 0 & -I \\ A & 0 \end{pmatrix}.$$

This leads to the characteristic polynomial $F(s) = \det(sI + \tilde{A}) = \det(s^2 I + A) = f(s^2)$,[24] and it is not difficult to see that Property 4.7 must still hold for $f(s) = \det(sI + A)$ if no powers of t occur in the solutions to $\dot{w} + \tilde{A}w = 0$ and so in the solutions to $\ddot{y} + Ay = 0$.

Weierstrass was familiar with Cauchy's method of integration, as well as with Jacobi's reworking of the proof of Cauchy's principal axes theorem.

4.6 Weierstrass' Memoir of 1858

Weierstrass' interest in the transformation of quadratic forms, which lay outside his main area of research, may have been stimulated through his friendship with Carl Borchardt (1817–1880), the editor of *Crelle's Journal* since 1856. Borchardt had written his doctoral dissertation under Jacobi's supervision and was primarily interested in the theory and application of determinants, a subject that Jacobi had popularized after acquiring an interest in it through Cauchy's 1829 memoir. Borchardt and Weierstrass first met in Braunsberg, East Prussia, in 1854, when

---

[24] It can be seen by means of elementary row operations that

$$F(s) = \det \begin{pmatrix} s^2 I + A & sI \\ 0 & I \end{pmatrix} = \det(s^2 I + A).$$

Weierstrass was still a gymnasium professor. An account of that meeting was given by Weierstrass in a letter to H.A. Schwarz shortly after Borchardt's death in 1880:

> At the time he was still a Privatdozent [in Berlin] but had already drawn to himself the attention of colleagues through two solid works. I still recall with great joy the pair of days we passed together. The pleasure they provided me can only be grasped by one who knows that from 1841 until then I had not spoken a mathematical word with anyone [140, p. 158].

One of Borchardt's "solid works" was his paper on "A new property of the equation by means of which the secular perturbations of the planets are determined" [30] in which he used determinants to give a new, direct proof of the reality of the characteristic roots of a symmetric matrix. It seems likely that through that meeting and their subsequent contact in Berlin, where Weierstrass moved the following year, Weierstrass became interested in the theory of determinants and in the work of Cauchy and Jacobi on its application to quadratic forms. (He had already studied, and been impressed by, Laplace's *Mécanique Céleste* during his years as a student in Bonn.) Weierstrass adopted the elegant methods and notation of Jacobi but infused them with his critical approach to mathematics.

In a memoir presented to the Berlin Academy in 1858 [587], Weierstrass proposed to study the problem of simultaneously transforming two quadratic forms $\Psi(x) = \sum_{i,j=1}^{n} a_{ij}x_ix_j$ and $\Phi(x) = \sum_{ij=1}^{n} b_{ij}x_ix_j$, where Φ is assumed to have the property that $\Phi(x) \neq 0$ for all $x \neq 0$. In other words, $A = (a_{ij})$ and $B = (b_{ij})$ are symmetric matrices and B is strictly definite. Cauchy's principal axes theorem (Theorem 4.5) concerned the special case in which $B = I$, i.e., $\Phi = \sum_{k=1}^{n} x_k^2$, and the conclusion was that a linear transformation $x = Py$ exists such that $\Phi = \sum_{k=1}^{n} y_k^2$ (hence P is orthogonal) and $\Psi = \sum_{i=1}^{n} s_k y_k^2$, where the s_k are the roots of $f(s) = \det(sI - A)$. Weierstrass proposed to show that in the more general situation a linear transformation $x = Py$ exists such that $\Phi = \pm\sum_{k=1}^{n} y_k^2$ (depending on whether Φ is positive or negative definite) and $\Psi = \sum_{i=1}^{n} s_k y_k^2$, where the s_k are the roots of $f(s) = \det(sB - A)$.

According to Weierstrass, the above simultaneous transformation of Φ and Ψ "is one of the most interesting and important algebraic problems arising in diverse investigations" [587, p. 233]. The proofs given by Cauchy, Jacobi, and others, he admitted, left nothing to be desired as long as the roots s_k are all unequal. But

> it does not appear that special attention has been given to peculiar circumstances that arise when the roots of the equation $f(s) = 0$ are not all different; and the difficulties which they present—of which I was made aware by a question to be discussed more fully later—were not properly cleared up. I also at first believed that this would not be possible without extensive discussions in view of the large number of different cases that can occur. I was all the more pleased to find that the solution to the problem given by the above-named mathematicians could be modified in such a way that it does not at all matter whether some of the quantities $s_1, s_2, \ldots, s_n$ are equal [587, p. 234].

The "question to be discussed more fully later" that alerted Weierstrass to the issues involved with multiple roots was the question of multiple roots and stability in mechanical problems (Section 4.2), which he turned to in the final section of his paper. Indeed, I suggest that consideration of this question in the light of a paper by Dirichlet on stable equilibria not only convinced him that Lagrange and Laplace were wrong in insisting that multiple roots were incompatible with stability but also indicated to him the possibility of proving this—as well as the above-proposed theorem on the transformation of quadratic forms—by a slight modification of the generic approach.

After discussing the claim of Lagrange and Laplace about the incompatibility of multiple roots and stability, Weierstrass wrote: "But it is unfounded. To be convinced of this it is only necessary to recall Dirichlet's proof of the fundamental theorem of this theory [of small oscillations]" [587, p. 244]. He was referring to a paper published by Dirichlet in 1846 [136], which became an appendix to the third edition (1853) of Lagrange's *Mécanique analytique*. Dirichlet was one of the earliest German mathematicians to accept Cauchy's more critical approach to analysis. He had also acquired an interest in mathematical physics through his contact with Fourier in Paris in the 1820s. Combining his interest in mathematical physics with Cauchy's critical approach to analysis, he produced several papers dealing with the mathematical foundations of physics, such as his celebrated paper of 1829 on the convergence of Fourier series.[25] In the 1846 paper, Dirichlet turned his critical eye upon a key proposition in Lagrange's *Mécanique analytique*: A state of equilibrium in a conservative mechanical system is stable if the potential function assumes a strict maximum value.[26] He pointed out that Lagrange's proof was decidedly circular and replaced it with a rigorous one that utilizes Cauchy's conception of continuity.

In the above quotation, Weierstrass referred to Dirichlet's proof, rather than to his theorem. What was it about Dirichlet's proof that proved so convincing to Weierstrass? Although Weierstrass did not explain, I would suggest it was that Dirichlet's proof [136, p. 7] established the following general theorem: *Suppose $T(y) = T(y_1,\ldots,y_n)$ and $V(y) = V(y_1,\ldots,y_n)$ are continuous functions with the following properties: (1) $V(0) = 0$ is a local maximum of V in the strict sense that $\tau > 0$ exists such that $V(y) < 0$ for $0 < |y_i| < \tau$, $i = 1,\ldots,n$; (2) $T(y) > 0$ for all $y \neq 0$ and $T(0) = 0$; (3) $y(t) = (y_1(t),\ldots,y_n(t))$ is a curve with continuous derivative $\dot{y}(t)$ with the property that $T(\dot{y}(t)) = V(y(t)) + C$ for all t and some constant C. Then for a given $\varepsilon > 0$ there is a $\delta > 0$ such that if the curve $y(t)$ is such that there is a "time" $t = t_0$ for which $|y_i(t_0)| \leq \delta$ and $|\dot{y}_i(t_0)| \leq \delta$ for all $i = 1,\ldots,n$, then $|y_i(t)| \leq \varepsilon$ for all t and all i.*

For Dirichlet, T and V were the kinetic and potential energy functions expressed in Lagrange's generalized coordinates $q_1,\ldots,q_n$ and with $T - V = C$ holding by virtue of the principle of energy conservation. His proof showed that if $q(t)$ gives the generalized coordinates at time t of a mechanical system for which $q = 0$ is an equilibrium point at which V has a maximum (taken to be 0), then the equilibrium is stable, i.e., at some time t_0, the system is sufficiently close to the equilibrium point

[25] For more details about this aspect of Dirichlet's mathematics, see [265, 271].

[26] Lagrange's potential function is the negative of Dirichlet's, so that Lagrange spoke of a minimum [392, Pt. I, §III Art. V].

and the kinetic energy is sufficiently small, then the system will remain arbitrarily close to the equilibrium point indefinitely. This was the fundamental theorem of the theory of small oscillations.

Dirichlet's proof, however, established the above-stated general result. It therefore applies to any two quadratic forms $T = y^t By$ and $V = y^t Ay$ that are positive and negative definite, respectively, and to any solution curve $y(t)$ of the system of associated differential equations, $B\ddot{y} = Ay$, since such a curve $y(t)$ is easily seen to satisfy (3) above, namely that $T(\dot{y}(t)) - V(y(t)) = C$ for all t.[27] The theorem implicit in Dirichlet's proof thus implies that for any $\varepsilon > 0$ (no matter whether large or small), if the initial conditions satisfy $|y_i(t_0)| \leq \delta$, $|\dot{y}_i(t_0)| \leq \delta$ for the δ of Dirichlet's theorem, but are otherwise arbitrary, then all solutions to $B\ddot{y} = Ay$ with such initial conditions will be bounded: $|y_i(t)| \leq \varepsilon$ for all t and all i. As Weierstrass said, this shows that Lagrange's claim that multiple roots can produce unbounded solutions is unfounded: symmetric A, B certainly exist with A, B respectively negative and positive definite such that $f(s) = \det(sB - A)$ has multiple roots, but the solutions to $B\ddot{y} = Ay$ satisfying the above initial conditions are nonetheless bounded by virtue of Dirichlet's proof. In fact, it then follows from Cauchy's method of integrating $B\ddot{y} = Ay$ (Section 4.5.2), which Weierstrass also knew [587, p. 245], that all solutions to $B\ddot{y} = Ay$ must be bounded.

Because multiple roots can exist, Weierstrass could see that Lagrange's method of integrating $B\ddot{y} = Ay$ was inadequate, since his formulas were based on the assumption of distinct roots. His familiarity with the work on the principal axes theorem by Cauchy and Jacobi enabled him to see what I pointed out in the introductory remarks to this chapter: a correct method of integrating $B\ddot{y} = Ay$ would be an easy consequence of the following generalization of the principal axes theorem: given the quadratic forms $\Phi(y) = y^t By$ and $\Psi(y) = y^t Ay$, with Φ positive definite, (1) the (not necessarily distinct) roots $s_1, \ldots, s_n$ of $f(s) = \det(sB - A)$ are real; and (2) there is a nonsingular linear variable change $y = Lz$ such that in the variables $z_1, \ldots, z_n$, $\Phi = z_1^2 + \cdots + z_n^2$ and $\Psi = s_1 z_1^2 + \cdots + s_n z_n^2$. Weierstrass' idea was to prove this sort of a theorem.

He was undoubtedly encouraged by the fact that Cauchy had given a rigorous proof of (1) in the special case $B = I$. As for (2), here Weierstrass perceived the theoretical significance of Property 4.7: If $f(s) = \det(sI - A)$ has $s = a$ as a root of multiplicity m, then $(s - a)^{m-1}$ must divide every cofactor of $sI - A$. In the special case $B = I$, if Property 4.7 could be proved true, Jacobi's generic proof of the existence of L would become rigorous (Section 4.5.1). In fact, Weierstrass would have realized that Property 4.7 does hold for $f(s) = \det(sB - A)$ whenever A and B are respectively negative and positive definite. This conclusion follows by combining the implications of Dirichlet's proof with Cauchy's method of integration, which together imply that all solutions to $B\ddot{y} = Ay$ are bounded, and

[27] To see this, consider $\varphi(t) = T(\dot{y}(t)) - V(y(t)) = \dot{y}^t B\dot{y} - y^t Ay$. Then $\varphi'(t) = \ddot{y}^t B\dot{y} + \dot{y}^t B\ddot{y} - \dot{y}^t Ay - y^t A\dot{y} = (\ddot{y}^t B - y^t A)\dot{y} + \dot{y}^t(B\ddot{y} - Ay) = (B\ddot{y} - Ay)^t\dot{y} + \dot{y}^t(B\ddot{y} - Ay) = 0$, since $B\ddot{y} = Ay$. Thus $\varphi(t) = C$, and (3) is satisfied.

so by Cauchy's formula for the solutions, no powers of t can occur in the solutions, and so (as indicated in Section 4.5.2) Property 4.7 must hold.

Weierstrass discovered that he could show that Property 4.7 must hold for any symmetric A and B when B is either positive or negative definite. The discovery of this "very remarkable condition" [587, p. 240] assumed a critically important role in his quest for a proof of the above sort of generalized principal axes theorem and later became the inspiration for his theory of elementary divisors. (In discussing the work of Jacobi and Cauchy, I referred to it as a remarkable property, and it is indexed under "Weierstrass' remarkable property" because he first noted its significance.) For future reference I will state it here as a lemma:

Lemma 4.8. *If A and B are symmetric and B is strictly definite, then if $s = a$ is a root of multiplicity m of $f(s) = \det(sB - A)$, it follows that $(s-a)^{m-1}$ divides every cofactor of $sB - A$.*

Weierstrass' proof of this lemma becomes easy to understand if we express it in terms of matrix algebra. Frobenius, who was the first to develop the symbolic algebra of matrices in a substantial and rigorous manner in 1878 (Chapter 7), realized this.[28] Indeed, he extracted from the matrix version of Weierstrass' proof, which involves Laurent expansions of matrix functions, a basic tool for developing the theory of matrices. For these reasons, I will use matrix algebra in sketching Weierstrass' proof.

Let $L(s) = sB - A$ and $M(s) = [f(s)]^{-1}\,\mathrm{Adj}\,(sB - A)$. Since A, B are both symmetric, so are $L(s)$ and $M(s)$. (From the viewpoint of Frobenius' matrix algebra, $M(s) = L(s)^{-1}$.) The entries of $M(s)$ are the rational functions $\beta_{ij}(s)/f(s)$, where $\beta_{ij}(s)$ denotes the (i,j) cofactor of $sB - A$. Each of these rational functions has a Laurent series expansion about $s = a$, viz.,

$$\beta_{ij}(s)/f(s) = G_{ij}(s-a)^{\ell_{ij}} + G'_{ij}(s-a)^{\ell_{ij}+1} + \cdots, \quad G_{ij} \neq 0,$$

where in general, the ℓ_{ij} will be negative integers. Weierstrass had given a proof that the roots of $f(s)$ are all real. Thus a is real, and so all the above Laurent coefficients are real. Let ℓ denote the minimum of the integers ℓ_{ij}. Then (to now follow Frobenius) the above Laurent expansions can be written in matrix form as

$$M(s) = (z-a)^{\ell} H + (z-a)^{\ell+1} H' + \cdots, \quad H \neq 0. \tag{4.36}$$

Here $H, H', \ldots$ are all constant real matrices, which are symmetric, since $M(s)$ is. That $H \neq 0$ follows from the definition of ℓ, which implies that there is at least one term $\beta_{ij}(s)/f(s)$ with $\ell_{ij} = \ell$ and so the (i,j) entry of H is $G_{ij} \neq 0$. (In other words, $-\ell$ is the maximal order of a pole of a coefficient of $M(s)$ at $z = a$.) Note that if $\ell \geq -1$, then $\beta_{ij}(s)/f(s)$ has either a simple pole at $s = a$ or a removable singularity. Thus if $\ell \geq -1$, $(s-a)^{m-1}$ must divide all cofactors $\beta_{ij}(s)$ of $sB - A$. In other words,

[28] See in this connection the letter Frobenius wrote to Weierstrass in November 1881 [207].

if it can be proved that $\ell \geq -1$, then Condition 4.7 holds and Weierstrass' lemma is proved.

Weierstrass' proof idea was to consider the Laurent expansions of the derivatives $[\beta_{ij}(s)/f(s)]'$, which follow from (4.36) by term-by-term differentiation with respect to s:

$$M'(s) = \ell(s-a)^{\ell-1}H + \cdots. \tag{4.37}$$

Here, and in the remainder of the proof, "$\cdots$" denotes a sum of terms involving greater powers of $(s-a)$. The (i,j) entry of the right-hand side of (4.37) gives the Laurent expansion of $[M'(s)]_{ij}$. These Laurent expansions are unique, and Weierstrass' idea was to determine these expansions in another way in the hope of gaining more information about ℓ. To this end, he used the fundamental identity $L(s)\,\mathrm{Adj}\,L(s) = f(s)I$, which implies the identity $L(s)M(s) = I$. Differentiation of this identity with respect to s gives $L'(s)M(s) + L(s)M'(s) = 0$, and since $L'(s) = (sB-A)' = B$, we get $M'(s) = -M(s)BM(s)$. Substitution of (4.36) in this equation gives (since $[M(s)]^t = M(s)$)

$$M'(s) = -[(s-a)^{\ell}H + \cdots]^t B[(s-a)^{\ell}H + \cdots] = -(s-a)^{2\ell}H^tBH + \cdots. \tag{4.38}$$

Now, the (i,j) coefficient of H^tBH is $h_i^tBh_j$, where $h_1,\ldots,h_n$ are the columns of H. Thus the diagonal entries are $h_i^tBh_i = \Phi(h_i) \neq 0$ by virtue of the definiteness of Φ and the fact that h_i is real. The Laurent expansion of a diagonal term is therefore $[M'(s)]_{ii} = -\Phi(h_i)(s-a)^{2\ell} + \cdots$. Now, (4.37) states that the Laurent expansion of $[M'(s)]_{ii}$ has the form $K(s-a)^{\ell'} + \cdots$, where $\ell' \geq \ell - 1$. By uniqueness of the expansion we know that $\ell' = 2\ell$, and so $2\ell = \ell' \geq \ell - 1$, i.e., $\ell \geq -1$, and Lemma 4.8 is proved.

Using Lemma 4.8 together with the sort of determinant-theoretic reasoning he found in the papers of Cauchy and Jacobi, Weierstrass gave a proof of the following generalization of the principal axes theorem.

Theorem 4.9. *Let $\Phi = \sum b_{ij}x_ix_j$ and $\Psi = \sum a_{ij}x_ix_j$ be quadratic forms with Φ strictly definite. Then (i) $f(s) = \det(sB-A)$ is a polynomial of degree n whose roots are all real; (ii) if $s_1,\ldots,s_n$ denote the roots, each listed as many times as its multiplicity, then there exists a nonsingular linear transformation $x = Py$ such that in the y-variables $\Phi = \varepsilon[\sum_{i=1}^n y_i^2]$ and $\Psi = \varepsilon\sum_{i=1}^n s_iy_i^2$, where ε is $+1$ or -1 according to whether Φ is positive or negative definite, respectively.*

Expressed in modern matrix notation, Weierstrass' theorem asserts the existence of a nonsingular P such that $P^tAP = \varepsilon I$ and $P^tBP = \varepsilon D$, where D is the diagonal matrix with entries $s_1,\ldots,s_n$. In the papers of Cauchy and Jacobi, $B = I$, and so in this special case, $\Phi = x^tx = y^tP^tPy = y^ty$, i.e., $P^tP = I$ and P is an orthogonal linear transformation. This shows that Cauchy's two theorems—his reality theorem (Theorem 4.4) and his principal axes theorem (Theorem 4.5)—follow as special cases of Weierstrass' theorem.

There is no need to go into the details of Weierstrass' proof, except to point out one glaring, rather ironic, weak point: his proof that the roots of $f(s) = \det(sB - A)$ are always real. First, in Section 2 of [587], under the additional hypothesis that $f(s)$ has no multiple roots, he proved that the transformation P of Theorem 4.9 exists, and he did it by means of Jacobi's "elegant ... formulas" [587, p. 237] for the coefficients of P, namely (4.32), which here take the form

$$p_{ik}p_{jk} = \beta_{ij}(s_k)/f'(s_k), \tag{4.39}$$

where $\beta_{ij}(s)$ is the (i,j) cofactor of $sA - B$ [587, p. 237]. Then in Section 4.3 he used the existence of P to show that in the case of no multiple roots, the roots must all be real. He did this by showing that if complex roots exist, then Φ cannot be definite. The third step—mentioned only in passing and a bit vaguely in the midst of step four, the proof of Lemma 4.8—was to argue that even in the presence of multiple roots, the roots must all be real. Weierstrass did this by using a limit-infinitesimal argument akin to the one Cauchy had used to extend his principal axes theorem (Theorem 4.5) from the generic case to the case of multiple roots. Thus Weierstrass argued that "if one of ... [the roots] ... be imaginary, then by means of an infinitely small variation of the coefficients $[a_{ij}]$ of Ψ one could arrange it so that the roots of the equation $f(s) = 0$ are all different but that [at least] one of them remains imaginary, which according to what was previously proved cannot occur" [587, pp. 240–241]. Given that Weierstrass had rejected Cauchy's use of the same sort of reasoning (as indicated by the quotation at the beginning of this section), it is remarkable that he himself utilized it! It was evidently the best he could do, and the reality of the roots was required (as noted) in the proof of Lemma 4.8. As we will see in the next chapter, a few years after Weierstrass published his paper, Christoffel gave a very simple, purely algebraic and more general reality proof—with symmetry replaced by the more general property of Hermitian symmetry.

The remainder of Weierstrass' paper was devoted to the application of Theorem 4.9 to the integration of $B\ddot{y} = Ay$ with A and B respectively negative and positive definite. In this connection, he went much further than what I indicated in the introduction to this chapter. He wrote down an explicit formula for the solution in terms of arbitrary initial conditions $y_i(0) = p_i$ and $\dot{y}_i(0) = q_i$ and the characteristic roots $s_k = -\rho_k^2$ of $f(s) = \det(sB - A)$, which, he noted, "agrees with the one that Cauchy derived by means of his calculus of residues" [587, p. 245]. Weierstrass' formula was comparable in elegance to the one Lagrange had given in 1766 under the assumption of distinct roots, but it was also completely general.

Weierstrass emphasized that Lagrange's "erroneous view" of the link between multiple roots and unstable solutions to systems of linear differential equations was the result of generic reasoning [587, p. 244]. That is, as he explained, when the coefficients in such a system are completely arbitrary, then when there is a multiple characteristic root "in general," polynomials in t occur as factors in the solution functions (as Cauchy's formula (4.34) shows), but it is not at all evident when these polynomials reduce to constants, i.e., it is not at all clear from the generic approach that, as Weierstrass was able to show, this reduction to constant polynomials occurs when A and B have the above symmetry and definiteness properties but in all other respects can have arbitrary coefficients.

To Weierstrass' Berlin colleagues and students, the message of his 1858 paper was clear: Generic reasoning is unacceptable due to its lack of rigor, but this does not mean that the elegance of generic reasoning and the resulting theorems must be completely abandoned due to the need for "extensive discussions in view of the large number of different cases that can occur." On the contrary, new mathematical tools (e.g., the theory of determinants and Laurent expansions) make it possible to avoid a case-by-case analysis and to achieve truly general results in a manner that retains the elegance of the generic approach. Of course, this same message can be read in Cauchy's 1829 paper on the principal axes theorem and in his method of integrating linear systems of differential equations—and indeed, Weierstrass himself read that message in those works. But in Berlin, as we shall see in the next chapter, Weierstrass' paper of 1858 became the paradigm of a rigorous approach to algebraic problems, based on disciplinary ideals that Kronecker was the first to articulate.

Chapter 5
Further Development of the Paradigm: 1858–1874

5.1 Weierstrass' Unpublished Theory

In his paper of 1858, Weierstrass had considered pairs of real quadratic forms $\Phi = x^t Bx$, $\Psi = x^t Ax$, with Φ definite, because his goal was to establish a generalization of the principal axes theorem for them that would provide the basis for a nongeneric treatment of $B\ddot{y} = Ay$. He realized, however, that it was possible to simultaneously transform more general pairs of quadratic forms Φ, Ψ into sums of square terms. Indeed, Jacobi, in his above-discussed paper of 1834, had already done so.

After presenting his elegant version of Cauchy's principal axes theorem, Jacobi proceeded to generalize it as follows [310, pp. 247ff.]. Suppose that $\Psi = \sum_{i,j} a_{ij}x_ix_j$ and $\Phi = \sum_{i,j} b_{ij}x_ix_j$ are two quadratic forms, written such that $a_{ij} = a_{ji}$ and $b_{ij} = b_{ji}$. Determine a linear transformation of the variables $x = Py$ such that

$$\Phi = g_1y_1^2 + \cdots + g_ny_n^2 \quad \text{and} \quad \Psi = h_1y_1^2 + \cdots + h_ny_n^2. \tag{5.1}$$

In matrix form, $\Phi = x^tAx$, $\Psi = x^tBx$ with A, B symmetric, and (5.1) states that $P^tAP = G$ and $P^tBP = H$, where G and H are diagonal matrices with the g_k and h_k, respectively, as diagonal entries. The problem is thus to determine a nonsingular linear transformation that simultaneously "diagonalizes" A and B in the above sense. This was the sort of problem that Sturm had considered in 1829 (Section 4.2.3), although in 1834, Jacobi was evidently unfamiliar with Sturm's work, which is not surprising given its obscure mode of publication. The problem Jacobi posed was much more general than Sturm's, because Jacobi made no assumptions about the positive (or negative) definiteness of either Ψ or Φ.

Jacobi tacitly formulated and solved the problem on the generic level, making effective use of his elegant notation. He began by assuming that the desired transformation P exists. Although Jacobi had no matrix notation at his disposal, what he did can be summarized succinctly using it. If $p_1, \ldots, p_n$ denote the columns of P, so that $P = \begin{pmatrix} p_1 & \cdots & p_n \end{pmatrix}$, then since $[P^tAP]_{jk} = p_j^tAp_k$ and $[P^tBP]_{jk} = p_j^tBp_k$, $P^tAP = G$ and $P^tBP = H$ are equivalent to

T. Hawkins, *The Mathematics of Frobenius in Context*, Sources and Studies in the History of Mathematics and Physical Sciences, DOI 10.1007/978-1-4614-6333-7_5,

$$p_j^t A p_k = \delta_{jk} g_k \quad \text{and} \quad p_j^t B p_k = \delta_{jk} h_k, \quad j,k = 1,\ldots,n. \tag{5.2}$$

These equalities imply that for all $j,k = 1,\ldots,n$,

$$p_j^t(h_k A - g_k B)p_k = h_k(p_j^t A p_k) - g_k(p_j^t B p_k) = 0.$$

This equality must hold for any fixed k and for all $j = 1,\ldots,n$, i.e., as we would say it today, the vector $(h_k A - g_k B)p_k$ must be orthogonal to the n linearly independent vectors $p_1,\ldots,p_n$. This means that for each k, $(h_k A - g_k B)p_k = 0$, as Jacobi concluded (without disclosing his reasoning). This means, of course, that $\det(h_k A - g_k B) = 0$ for all pairs (h_k, g_k). Jacobi was thus led to a generalization of the procedure of Lagrange and Laplace, albeit now articulated using Cauchy's theory of determinants: first determine the roots $(h_1, g_1),\ldots,(h_n, g_n)$ of $f(u,v) = \det(uA - vB) = 0$. These roots are determined only up to a constant factor $\lambda \neq 0$, i.e., (h_k, g_k) and $(\lambda h_k, \lambda g_k)$ are not regarded as distinct roots. Because Jacobi was reasoning generically, he tacitly assumed that $f(u,v) = 0$ had n distinct roots (h_k, g_k) in the above sense. In the same spirit he also wrote the expressions g_k/h_k and h_k/g_k [310, p. 249], which require g_k and h_k to be nonzero. If $p_k \neq 0$ is any solution such that $(h_k A - g_k B)p_k = 0$, it follows readily that the relations (5.2) hold for $j \neq k$, and so the variable change $x = Py$, with $P = \begin{pmatrix} p_1 & \cdots & p_n \end{pmatrix}$, transforms Φ and Ψ into sums of squares $\Phi = \sum_{k=1}^n \tilde{g}_k y_k^2$, $\Psi = \sum_{k=1}^n \tilde{h}_h y_k^2$, where $\tilde{g}_k = p_j^t A p_j$, $\tilde{h}_k = p_j^t B p_j$. It is easily seen that $\tilde{g}_k / g_k = \tilde{h}_k / h_k$, which means that each p_k can be suitably chosen as a multiple μ_k of the initially chosen p_k so as to get $\tilde{g}_k = g_k$ and $\tilde{h}_k = h_k$.

Jacobi's solution to the above-stated problem can thus be summarized as the following nongeneric theorem.

Theorem 5.1. *If $f(u,v) = \det(uA - vB)$ has n distinct roots (h_k, g_k), $k = 1,\ldots,n$, then for each k, $p_k \neq 0$ can be chosen such that $(h_k A - g_k B)p_k = 0$, and if $P = \begin{pmatrix} p_1 & \cdots & p_n \end{pmatrix}$, then by virtue of the variable change $x = Py$, $\Phi(y) = \sum_{k=1}^n g_k y_k^2$ and $\Psi(y) = \sum_{k=1}^n h_k y_k^2$.*

Although for the purpose of later reference I have stated Jacobi's result in a precise, nongeneric form, it should be kept in mind that Jacobi's own formulation, being in the generic style, was stated ambiguously, so that the exact extent of validity is left to the reader. Readily producible 2×2 examples show that the assumption of distinct roots is essential when the positive (or negative) definite hypothesis is not made. On the other hand, Jacobi's conclusions go through if $(g_k, 0)$ and/or $(0, h_{k^*})$ are among the (distinct) roots; and so such roots are allowed in the above statement of Jacobi's theorem.

Jacobi's generic theorem went far beyond Sturm's, because no hypothesis of definiteness was made. Indeed, it is possible for A and/or B to be singular and still $f(u,v) = \det(uA - vB)$ has distinct roots. Also without any definiteness hypothesis, the roots (g_k, h_k) of $f(u,v)$ are generally complex even if the coefficients of A and B are assumed to be real. Another characteristic of generic reasoning is that the precise range of values of the symbols being used is never mentioned. All of

Jacobi's reasoning goes through if the coefficients of A and B are also complex, and Jacobi certainly must have realized this. Thus he in effect inaugurated the study of the simultaneous transformation of quadratic forms over the field of complex numbers.

For a mathematician such as Weierstrass, who was not content with generic reasoning, Jacobi's theorem must have raised in his mind the question of what can be said when not all the roots are distinct. In his memoir of 1858, he had given a precise and satisfying answer under the hypothesis that one of Φ, Ψ is definite. But what about when Φ and Ψ are indefinite? It would have been easy for Weierstrass to give examples in which $f(u,v) = \det(uA - vB)$ has multiple roots and yet the conclusion of Jacobi's theorem still holds.[1]

The possibility of symmetric A, B such that $\det(uA - vB)$ has multiple roots but the conclusion of Jacobi's theorem does not hold must have seemed even more likely, although how does one prove the impossibility of a P for which $P^tAP = G$ and $P^tBP = H$? Weierstrass may have thought about it in the following manner. If G and H are diagonal matrices, consider $f(s) = \det(sG - H)$. It is easy to see that $sG - H$ has the property presented in Weierstrass' Lemma 4.8: if $s = a$ is a zero of multiplicity m for $f(s)$, then it must be a zero of multiplicity at least $m - 1$ for each cofactor of $sG - H$ and hence for each minor $f_{jk}(s)$. (In other words, for any j and k, the poles of $f_{jk}(s)/f(s)$ are all simple.) Then if $P^tAP = G$ and $P^tBP = H$, so $P^t(sA - B)P = sG - H$, it follows from Cauchy's first multiplication theorem that if $s = a$ is a zero of multiplicity m for $F(s) = \det(sA - B)$, it is a zero of multiplicity m for $f(s) = \det(sG - H) = (\det P)^2 F(s)$. Since $sG - H$ has the property of Lemma 4.8, it follows that $(s-a)^{m-1}$ divides each minor $f_{jk}(s)$. But then Corollary 4.3 to Cauchy's second multiplication theorem states that every minor $F_{\alpha\beta}(s)$ of $sB - A$ is a linear combination of the minors $f_{jk}(s)$; and since all $f_{jk}(s)$ are divisible by $(s-a)^{m-1}$, this must be true of $F_{\alpha\beta}(s)$. In other words, *a necessary condition that a nonsingular P exist such that $P^t(sA - B)P = sG - H$ is that $sB - A$ have the property of Weierstrass' Lemma 4.8.* Examples of $sB - A$ that do not have this property are relatively easy to manufacture.[2] Also, an obvious—but more difficult—question would be whether the property in Lemma 4.8 is also sufficient for the conclusion of Jacobi's theorem to hold.

Jacobi's penchant for elegant generalization did not stop with the diagonalization of pairs of quadratic forms. He seems to have been the first to consider the transformation of *bilinear* forms, i.e., functions of the form

---

[1]For example, let P be any 3×3 orthogonal matrix, and take $A = PD_1P^t$, $B = PD_2P^t$, where $D_1 = \text{Diag. Matrix}(1,1,-1)$ and $D_2 = \text{Diag. Matrix}(1,1,-3)$. Then A and B are symmetric and indefinite. Obviously, since $P^t = P^{-1}$, we have $P^tAP = D_1$ and $P^tBP = D_2$, even though $f(u,v) = \det(uA - vB) = (u-2v)^2(u-3v)$ has $(2,1)$ as a double root.

[2]If $B = \begin{pmatrix} 1 & 1 & 0 \\ 1 & 2 & 1 \\ 0 & 1 & 0 \end{pmatrix}$ and $A = \begin{pmatrix} 1 & 1 & 1 \\ 1 & 1 & 1 \\ 1 & 1 & 1 \end{pmatrix}$, then $f(s) = \det(sB - A) = -s^3$, so $s = 0$ is a root of multiplicity $m = 3$, but s^2 does not divide all minors of $sB - A$, e.g., $s^2 - f_{11}(s) = -s(1-s)$.

$$\Psi(x,y) = \sum_{i,j=1}^{n} a_{ij}x_i y_j, \tag{5.3}$$

where now a_{ij} and a_{ji} are generally different and presumably may take on complex values. Although Jacobi never published his results on bilinear forms—his health began to fail in 1842—they were published posthumously by Borchardt in 1857 [314].

In this work Jacobi considered the problem of determining two linear transformations, $x = Pu$ and $y = Qv$, such that the bilinear form $\Psi(x,y)$ of (5.3) becomes in the new variables

$$\Psi(u,v) = g_1 u_1 v_1 + \cdots + g_n u_n v_n. \tag{5.4}$$

In matrix form this is $P^t AQ = G$, where G is the diagonal matrix with entries $g_1,\ldots,g_n$. As Jacobi probably realized, for any $\Psi(x,y)$ with actual numerical coefficients a_{ij}, such a transformation is always possible by means of elementary row and column operations. But he was thinking of $\Psi(x,y)$ with the a_{ij} regarded as algebraic symbols; he was seeking a generic transformation. Making essential use of the theory of determinants, he derived elegant determinant formulas for $u = P^{-1}x$, $v = Q^{-1}y$, viz.,

$$u_m = \begin{vmatrix} a_{1,1} & \cdots & a_{m-1,1} & \frac{\partial\Psi}{\partial y_1} \\ a_{1,2} & \cdots & a_{m-1,2} & \frac{\partial\Psi}{\partial y_2} \\ \cdots & \cdots & \cdots & \cdots \\ a_{1,m} & \cdots & a_{m-1,m} & \frac{\partial\Psi}{\partial y_m} \end{vmatrix}, \quad v_m = \begin{vmatrix} a_{1,1} & \cdots & a_{1,m-1} & \frac{\partial\Psi}{\partial x_1} \\ a_{2,1} & \cdots & a_{2,m-1} & \frac{\partial\Psi}{\partial x_2} \\ \cdots & \cdots & \cdots & \cdots \\ a_{m,1} & \cdots & a_{m,m-1} & \frac{\partial\Psi}{\partial x_m} \end{vmatrix},$$

so that

$$\Psi = \frac{1}{p_1}u_1 v_1 + \frac{1}{p_1 p_2}u_2 v_2 + \cdots + \frac{1}{p_{n-1}p_n}u_n v_n, \tag{5.5}$$

where p_i is the principal $i \times i$ minor determinant of A obtained by deleting the last $n-i$ rows and columns [314, p. 589]. As Jacobi pointed out, u_m and v_m involve only the variables $x_m,\ldots,x_n$ and $y_m,\ldots,y_n$, respectively.[3] (Thus P^{-1}, Q^{-1}, P, Q are all upper triangular.)

Jacobi's theorem is manifestly generic. For example, since $p_1 = a_{1,1}$, if $a_{1,1} = 0$, the diagonal form for Ψ becomes meaningless. Nonetheless, Jacobi's above-described paper [314], along with his paper of 1834 [310] and his 1841 essay on determinants [311], constituted some of the principal primary sources upon which Weierstrass and Kronecker were to draw in their researches on a nongeneric theory

[3]It is easily seen that for all $i < m$, one has $\partial u_m/\partial x_i = \partial v_m/\partial y_i = 0$. For example, $\partial u_m/\partial x_i = \sum_{\sigma\in S_m} \operatorname{sgn}(\sigma) a_{1,\sigma(1)}\cdots a_{i,\sigma(i)}\cdots a_{i,\sigma(m)} = 0$, since it represents the determinant of a matrix with ith and mth rows equal.

of the transformation of families of quadratic and bilinear forms. In particular, Kronecker called the transformation of Ψ given by (5.5) the "Jacobi transformation" [358, p. 390] and stressed the important role it played in Weierstrass' work [358, p. 391] as well as his own, since Jacobi's determinant-theoretic reasoning leading to it could be made generally applicable [358, p. 395]. Thus, on the one hand, Jacobi's work by virtue of his generic theorems posed the challenge of a nongeneric theory of the transformation of families of quadratic and bilinear forms $s\Phi + \Psi$ into one another and, in particular, into especially simple types. On the other hand, his work held out the promise that the theory of determinants might offer a powerful tool to carry out such a program.

According to Frobenius [202, pp. 719–720], Weierstrass did in fact go on in 1858 to consider the more general problem of when the one-parameter family $s\Phi(x,y) + \Psi(x,y)$ of bilinear forms is "equivalent" to another such family $s\tilde{\Phi}(u,v) + \tilde{\Psi}(u,v)$ in the sense that nonsingular linear transformations $x = Hu$, $y = Kv$ exist that transform $s\Phi(x,y) + \Psi(x,y)$ into $s\tilde{\Phi}(u,v) + \tilde{\Psi}(u,v)$. Furthermore,

> Weierstrass solved this problem in the year 1858 by transforming the family of forms into an aggregate of elementary families which did not admit a further decomposition. Since the manuscript of his work was lost on a trip, however, he became so disgusted with the consideration of the subject that he first returned to it ten years later but then pursued the investigation in a completely different, entirely direct manner.

Frobenius' remarks indicate that Weierstrass had indeed taken up the challenge posed by Jacobi's generic treatment of the transformation of families of quadratic and bilinear forms, although he was perhaps not satisfied with the indirect approach he had pursued. Also in 1858 and the following years, Weierstrass' principal concern was to understand Riemann's solution, in 1857, to the Jacobi inversion problem (Chapter 10) so as to compare it with his own (unpublished) solution, an enterprise that took him a decade to complete. That concern, together with health problems, make it understandable that Weierstrass did not at once attempt to reconstruct his solution to the problem of the equivalence of families of bilinear forms. Although we do not know what prompted Weierstrass, circa 1868, to take up the problem once again, the papers of Christoffel (1864) and Kronecker (1866) discussed in the following two sections seem plausible candidates for renewing Weierstrass' interest in the problem and ultimately leading him to discover the direct approach of his theory of elementary divisors (Section 5.4).

5.2 Christoffel and Hermitian Symmetry

Weierstrass' paper of 1858 was the culmination of a line of development involving mechanics and analytic geometry that led to the consideration of symmetric coefficient systems and quadratic forms. About the time his paper appeared, developments in the theory of numbers and in mathematical optics led independently to the consideration of a different type of symmetry, now called Hermitian symmetry.

In this section, I consider these developments and their culmination in a paper of 1864 [91] by E.B. Christoffel (1829–1900), who generalized some of the results in Weierstrass' paper to the case in which Φ and Ψ are what are now called Hermitian symmetric forms. A few years later, Christoffel published another paper on the general theory of bilinear forms. For reasons indicated below, Weierstrass did not find these papers satisfying, although they certainly reflected a growing interest in the subject of quadratic and bilinear forms.

The property of Hermitian symmetry seems to have been noted first by Charles Hermite (1822–1901) in 1855 [290] in the course of the following arithmetic considerations. The introduction of complex quantities into the theory of numbers was initiated by Gauss, who showed in his 1832 work on biquadratic residues [246] that many arithmetic notions carry over to what are now called Gaussian integers, i.e., complex numbers $a+bi$, where a,b are ordinary integers. It was natural to ask whether the arithmetic theory of forms could also be extended to Gaussian integers. An answer was given by Dirichlet in 1842 [135]. He showed that many of Gauss' results can be extended to the representation of Gaussian integers by forms $f = \alpha z^2 + 2\beta zw + \gamma w^2$, where $\alpha, \beta, \gamma, z, w$ are Gaussian integers and equivalence is defined using linear substitutions with Gaussian integers as coefficients. In this manner, the idea of considering forms with complex coefficients was introduced into the theory of numbers. Hermite, however, showed that there was another, less straightforward, mode of generalization. That is, he indicated that many of the propositions in Gauss' *Disquisitiones Arithmeticae* [244] have analogues for the special type of forms

$$f = \alpha zz + \beta z\overline{w} + \overline{\beta}\overline{z}w + \gamma w\overline{w},$$

where α and γ are real and β, z, w are Gaussian integers.

Hermite explained that he was led to study such forms through his interest in the representation of an integer as a sum of four squares [289, pp. 258–259]. The fact that every integer can be expressed as the sum of four squares had already been proved by Lagrange in 1770, but in the nineteenth century, new proofs of this result were of interest, especially if they could shed light on the number of representations of an integer as a sum of four squares [126, v. 2, 168]. A Hermitian form is especially suitable for consideration of the four-square theorem, since, as Hermite stressed, it is actually a real quadratic form in the four variables x, y, u, v, where $z = x + iy$ and $w = u + iv$. Perhaps Hermite was led to introduce Hermitian forms by observing that a sum of four squares can be regarded as a sum of two Gaussian integer "squares," i.e., $x^2 + y^2 + u^2 + v^2 = z\overline{z} + w\overline{w}$. In any case, he obtained the following proof that every odd integer m is representable as a sum of four squares [289, pp. 258–259]. Consider the form

$$f = mz\overline{z} + (a+bi)z\overline{w} + (a-bi)\overline{z}w + \left[\frac{a^2+b^2}{m}\right] w\overline{w},$$

where $a^2+b^2+1\equiv 0 \pmod{m}$. (Hermite had proved that such a and b exist in connection with another proof of the four-square theorem for odd integers [286].) The form f represents m when $z=1$ and $w=0$. Furthermore, since its determinant is $+1$, it followed from the arithmetic theory as developed by Hermite that f is arithmetically equivalent to $Z\overline{Z}+W\overline{W}$. Hence m is representable by this form and therefore as a sum of four squares.

Although Hermite's discussion of Hermitian forms in [289] was limited to the case of two variables, he was familiar with Cauchy's paper on the transformation of quadratic forms [72] and in his 1855 paper expressed his interest in the n-variable theory, especially in Cauchy's theorem that the characteristic polynomial of a real symmetric A has the "remarkable property" that all its roots are real [290, p. 479]. Hermite realized that Cauchy's reality theorem could be used to conclude that the roots of $f(s)=\det(A-sI)$ remain real when A is Hermitian symmetric. Judging by his brief remarks, what he had in mind was that if $\Phi(z)=\bar{z}^t Az$ and if $z=x+iy$ and $A=S+iT$, then $\bar{A}^t=A$ means that S is symmetric and T is skew-symmetric ($T^t=-T$), and $\Phi=(\bar{x}^t-i\bar{y}^t)(S+iT)(x+iy)$ is real-valued and equals $x^t Sx-x^t Ty+y^t Tx+y^t Sy$. In other words, Φ can be regarded as a real quadratic form in $2n$ variables $x_1,\ldots,x_n,y_1,\ldots,y_n$. The coefficient matrix of Φ so conceived is (in block form) the symmetric matrix $M=\begin{pmatrix} S & -T \\ T & S \end{pmatrix}$. By Cauchy's theorem, the roots of $F(s)=\det(M-sI)$ are all real, but, and this was Hermite's point, $F(s)=f(s)^2$, where $f(s)=\det(A-sI)$, with $A=S+iT$ Hermitian. Thus all the roots of $f(s)$ are real; Hermite's proof is perfectly rigorous, since Cauchy's was.

Alfred Clebsch (1833–1872) also considered Hermitian symmetric systems in a paper of 1860 [95], and it is likely that his work impressed Christoffel more than that of Hermite, for Clebsch was investigating Cauchy's theory of light when he made the discovery. Cauchy's method of integrating the differential equation derived from his model was first to obtain particular solutions of a specific form. (See Cauchy [78].) This, in turn, required consideration of a system of equations $Cx=\lambda x$, where the coefficients of C are in general complex. (Cauchy made extensive use of complex numbers in his work on optics.) In working out the details of Cauchy's theory in a special case, Clebsch ended up with a 3×3 matrix of coefficients of the form $C=(c_{jk})$, where c_{jk} is the complex number $a_{jk}+ib_{jk}$ and $a_{kj}=a_{jk}$, $b_{kj}=-b_{jk}$. Expressed in matrix notation, $C=A+iB$, where the matrices A and B have real coefficients, A is symmetric, and B is skew-symmetric ($B^t=-B$). Hence as noted above, C is Hermitian symmetric.

Clebsch apparently did not know of Hermite's above-mentioned paper of 1855, for he never mentioned Hermite, but he also discovered that the characteristic equation of C "has remarkable properties whereby it and equation (5.4) [$Cx=\lambda x$] appear as extensions of those equations that arise in the theory of secular perturbations and in so many other investigations" [95, p. 326]. What Clebsch meant was that if $C=A+iB$ is $n\times n$ with A real and symmetric and B real and skew-symmetric, then all the characteristic roots of C are real. He gave a proof of his claim, and had it been valid, it would have represented the first elementary reality proof in the sense that—unlike the earlier proofs of Cauchy and Borchardt [30]—it did not utilize the theory of determinants but only elementary algebraic considerations of what would nowadays be interpreted in terms of inner products. But the proof contained a lacuna that I cannot see how to fill.[4]

Clebsch does not seem to have realized the lacuna, for in a paper of 1863 [98], he cited his reality theorem with the claim that he had proved it. Taking $A = 0$ in that theorem, he observed that it states that the roots of iB are real and hence that the roots of the skew-symmetric matrix $B = (1/i)(iB)$ are purely imaginary. Noting that it is easy to prove this directly, Clebsch gave the proof—and this time it contained no gaps. It seems to be the earliest example of an elementary proof involving the nature of the roots of a matrix with a symmetry property, and it probably encouraged Christoffel to seek a comparable elementary proof of the reality of the roots of $f(s) = \det(sA - B)$, where A and B are Hermitian symmetric and A is definite.[5] Clebsch's proof is thus of some historical interest.[6]

Christoffel was in a position to be familiar with both Weierstrass' 1858 paper and the investigations of Clebsch. He had been a student in the mid-1850s at the University of Berlin, where he attended Dirichlet's lectures on mathematical physics and was so impressed that he resolved to concentrate his own research in that area

---

[4]The proof started as follows. If λ is a characteristic root of $C = A + iB$, then $z = p + iq \neq 0$ exists such that $\lambda z = Cz = (A + iB)z$. From this, Clebsch concluded without any proof [95, p. 327, eqn. (14)] that

$$\lambda p = Ap - Bq \quad \text{and} \quad \lambda q = Aq + Bq. \tag{5.6}$$

These equalities follow readily by taking real and imaginary parts if λ, p, and q are assumed real, but Clebsch assumed that (5.6) held without these reality assumptions. I have been unable to see how his assumption can be justified. Using (5.6), Clebsch correctly proved by what would now be translated into inner product considerations that λ must be real [95, pp. 327–329].

[5]Although he may have been encouraged by the precedent of Clebsch's proof, Christoffel's actual proof seems to owe more to observations made by Lagrange, as indicated below.

[6]Clebsch began his proof by considering any two characteristic roots λ, λ' of B. His proof follows directly from the fact that nonzero e and e' exist for which $Be = \lambda e$ and $Be' = \lambda e'$. He observed that these equations imply that

$$\lambda \sum_j e_j e'_j = \sum_{jk} b_{jk} e'_j e_k \quad \text{and} \quad \lambda' \sum_j e_j e'_j = \sum_{jk} b_{jk} e_j e'_k, \tag{5.7}$$

where the e_j and e'_j are the respective components of e and e'. In more familiar notation, the first equation in (5.7) is $\lambda(e \cdot e') = (Be \cdot e')$, and the second is $\lambda(e \cdot e') = (e \cdot \lambda' e') = (e \cdot Be')$, where $(e \cdot f) = e^t f$ is the usual real inner product. Adding these two equations and invoking the skew-symmetry of B, he concluded that

$$(\lambda + \lambda')(e \cdot e') = 0. \tag{5.8}$$

That is, he realized that by virtue of skew-symmetry, the right-hand sides of the equations in (5.7) are negatives of one another. Indeed, in matrix notation the right-hand side of the first equation is $Be \cdot e'$, which is the same as $e \cdot B^t e' = -(e \cdot Be')$, the negative of the right hand side of the second equation. To complete the proof, Clebsch took $\lambda' = \bar{\lambda}$. Since B is real, taking conjugates in $Be = \lambda e$ yields $B\bar{e} = \bar{\lambda}\bar{e}$, and so he took $e' = \bar{e}$. In this case (5.8) becomes $(\lambda + \bar{\lambda})(e \cdot \bar{e}) = 0$, whence $\lambda + \bar{\lambda} = 0$ and λ is a pure imaginary number.

and regarded Dirichlet as his mentor.[7] His doctoral dissertation (1856) [90] was on a problem in the theory of electricity. For the following three years, family considerations brought Christoffel back to his home town, where he continued his study of mathematical physics and in particular became acquainted with Cauchy's theory of light. Christoffel returned to the University of Berlin as instructor in 1859, where Weierstrass had become assistant professor. He remained in Berlin until 1862, when, on the recommendation of Kummer and Weierstrass, he became professor at the Polytechnic in Zurich. As noted in Chapter 2, Christoffel was one of Frobenius' predecessors in Zurich, and he did much to improve conditions there for mathematical research.

In 1864, while in Zurich, Christoffel published his work related to Cauchy's theory of light in two back-to-back papers in *Crelle's Journal* [91, 92], now edited by Christoffel's former teacher Borchardt. The second paper [92] contained the mathematical physics, and the first supplied the mathematics to justify it. The second was concerned with Cauchy's mechanical model for light propagation. It led to a system of linear differential equations $\ddot{y} = Cy$. The coefficients of C were not necessarily constant, but Cauchy had limited his attention to that case, except for a few hints as to how to proceed more generally. Christoffel therefore set himself the task of dealing more fully with the case of nonconstant coefficients. In [92] he sought to justify Cauchy's brief remarks about the integration of $\ddot{y} = Cy$ when C is not constant. Using the idea that the solutions to $\ddot{y} = Cy$ should resemble averages, he argued that the solution of $\ddot{y} = Cy$ can be reduced to solving a system with constant coefficients of the form $B\ddot{y} = Ay$, where B is diagonal with positive entries and the coefficients of A are, in general, complex. Christoffel showed that the coefficients of A have the symmetry property called to attention by Hermite and Clebsch. He also realized that $\Phi(z) = \bar{z}^t Bz$, with B as above, has the property that $\Phi(z) > 0$ for any complex $z = x + iy \neq 0$, and so considered the more general case in which B is not necessarily diagonal but simply Hermitian symmetric with the "Hermitian" definiteness property that $\Phi(z) = \bar{z}^t Bz \neq 0$ for all $z \neq 0$.

The problem of integrating $B\ddot{y} = Ay$ with A, B Hermitian symmetric and B positive definite was therefore analogous to the problem studied by Weierstrass in 1858. Weierstrass had used the transformation $x = Py$ of his Theorem 4.9 to obtain his elegant formula for the solutions to $B\ddot{y} = Ay$ when A and B are real symmetric and B is positive definite. Christoffel saw that he could do the same with his system $B\ddot{y} = Ay$ by proving the following generalizations of Weierstrass' results, which he presented in the first paper [91, pp. 159–160, III–IV].

Theorem 5.2 (Christoffel). *Let $\Phi(z) = \sum b_{ij} z_i \bar{z}_j$ and $\Psi(z) = \sum a_{ij} z_i \bar{z}_j$ have the symmetry property $a_{ji} = \bar{a}_{ij}$, $b_{ji} = \bar{b}_{ij}$, and let Φ be definite. Then (1) the roots of $f(s) = \det(sB - A)$ are all real; and (2) for any $(n-1) \times (n-1)$ minor $f_{ij}(s)$ of $f(s)$, the poles of $f_{ij}(s)/f(s)$ are all simple.*

---

[7] For biographical details and documents concerning Christoffel and his work see [249, 319].

By virtue of (1) and (2), the analogue of the Weierstrass transformation $x = Py$ of Theorem 4.9 existed and was used by Christoffel in his second paper to integrate the system $B\ddot{y} = Ay$ that occurred there [92, pp. 159–160].

Christoffel's proof of (1) was apparently the first completely rigorous yet brief *and elementary* proof of this theorem—even when specialized to $A = I$ and B real and symmetric, since Cauchy's proof was not elementary. Expressed in modern notation, Christoffel's proof proceeds as follows [91, p. 131]. Since $f(0) = \det(-A)$, $s = 0$ can be a root of $f(s)$, but since it is real, it is only necessary to show that every root $s = \sigma \neq 0$ is real. Since $f(\sigma) = \det(\sigma B - A) = 0$, there is a $u \neq 0$ in $\mathbb{C}^n$ such that $\sigma Bu = Au$. Thus for every $v \in \mathbb{C}^n$, we have

$$\sigma v^t Bu = v^t Au.$$

Take $v = \bar{u}$. Then the above becomes (with $\bar{u}^t = u^h$, the Hermitian transpose of u)

$$\sigma \Phi(u) = \sigma u^h Bu = u^h Au = \Psi(u).$$

Since $\Phi(u) \neq 0$ by definiteness, we may write the above as $\sigma = \Psi(u)/\Phi(u)$. Now, as Hermite had shown, for any $u \in \mathbb{C}^n$, $\Psi(u)$ and $\Phi(u)$ are real numbers. (Christoffel proved this by showing that $\Phi(u)$ and $\Psi(u)$ are equal to their complex conjugates.) Thus σ is the quotient of two real numbers and hence is real, and so all the roots of $f(s)$ must be real. Since, in present-day terms, $u^h Au$ and $u^h Bu$ are the (Hermitian) inner products (u, Au) and (u, Bu), Christoffel's simple proof is in effect carried out by means of inner product reasoning.

Lagrange in his *Mécanique Analytique* of 1788 had already considered the relation $\sigma = \Psi(u)/\Phi(u)$ in conjunction with the question of the reality of σ (Section 4.2.1), but he had done so within the context of real, symmetric A and B; and because it was unclear whether the components of u were real, he sought other means to establish the reality of σ. Weierstrass would also have been familiar with these considerations by Lagrange, yet he too did not envision a viable reality proof based on such considerations. Thus he gave the unsatisfactory reality proof indicated in Section 4.6. Once the context is generalized to Hermitian symmetric forms, however, the question whether $u \in \mathbb{R}^n$ becomes irrelevant, since, as noted, Hermitian forms are always real-valued. It would therefore seem that Christoffel's proof, which may have been suggested by Lagrange's remarks, illustrates Hilbert's later dictum that when a problem is formulated within an appropriately general context, it can become quite easy to solve.

Weierstrass, however, was not satisfied with Christoffel's paper. His less than favorable opinion may have been colored by his personal dislike of Christoffel. The three years (1859–1862) that Weierstrass and Christoffel were together in Berlin had been preceded by a three-year period (1856–1859) during which Christoffel lived in isolation in his hometown in order to be near his ailing mother. There he assiduously studied mathematics and mathematical physics and became something of a recluse. According to his biographer Geiser [249, p. vi], who was his colleague for many years in Zurich, the period of self-study transformed Christoffel into an highly independent thinker, an intellectual trait no doubt fostered as well by his personality, for he was shy to the point of being antisocial.[8] He was also given to bouts of irascibility during which he treated even his friends with harshness and mistrust and made many outlandish derogatory remarks about mathematicians and mathematical theories. Weierstrass once referred to him as a "queer chap"[9] and likened him to a harmless version of Kronecker—the Kronecker of 1885, who was critical of Weierstrass' approach to analysis, not the Kronecker of the 1860s and 1870s, who (as we shall see) staunchly defended Weierstrass' work on quadratic and bilinear forms and sought to complement it with work of his own. As for Christoffel, his emphasis upon intellectual independence seems to have produced a somewhat negative view of the lectures of Kummer and Weierstrass, because as excellent as he admitted they might be in many respects, he felt they were not conducive to a complete study of a subject [319, p. 21n 9(§7)]. Here Christoffel, like Lie and Klein a few years later, seems to have objected to the exclusive presentation of mathematics from the Berlin point of view. In the case of Weierstrass' lectures on complex analysis, this meant exclusion of the quite different viewpoint of Riemann, whose publications Christoffel greatly admired.

That Weierstrass' dislike of Christoffel colored his appreciation of his mathematics is suggested by a postscript to his 1858 paper that he published in 1879 [591]. We saw in Section 4.6 that the weak point of the 1858 paper was Weierstrass' proof that the roots of $f(s) = \det(sA - B)$ are real when A and B are real symmetric and A is definite. He had to prove it first for the case of distinct roots and then use a hand-waving limit-infinitesimal argument to push it through in the case of multiple roots. The purpose of the 1879 postscript was to give a more satisfactory proof. Although he referred to reality proofs given by others—Cauchy, Jacobi, Borchardt, Sylvester, Kronecker—no mention whatsoever is made of Christoffel's paper of 1864 [91] with its simple, rigorous reality proof for a more general class of forms. Toward the end of his life, when editing his papers for appearance in his collected works, Weierstrass added a footnote to his 1879 paper in which he expressed his regret for omitting a reference to Christoffel's paper and suggested that he had probably done so because Christoffel's proof of part (2) of his Theorem 5.2 was "neither completely algebraic nor free of objections and so seemed unsatisfactory to me" [591, p. 140n]. Given that Weierstrass cited the likes of Jacobi and Sylvester, who reasoned generically, his explanation doesn't seem convincing, especially since Christoffel's proof of part (1) of Theorem 5.2 was what was relevant to Weierstrass' 1879 postscript. Indeed, as Weierstrass also admitted in the footnote: "I should have pointed out that his proof of the first theorem [i.e., part (1)] is not only completely correct but also leaves nothing to be desired in the way of simplicity."

---

[8]Geiser [249, p. vi] suggests that this personality trait also developed during the solitary years 1856–1859.

[9]In a letter of 24 March 1885 to Sonya Kovalevskaya Weierstrass referred to Christoffel as a "wunderlicher Kauz" [441, pp. 194–195]. The letter is also contained in [28].

It is possible that Christoffel's interest in generalizing Weierstrass' results about quadratic forms might have induced him to consider the transformation of complex bilinear forms. In fact, Christoffel treated forms with coefficients a_{ij} satisfying $\bar{a}_{ji} = a_{ij}$ (so that generally $a_{ij} \neq a_{ji}$) as bilinear forms $\varphi = \sum_{i,j=1}^{n} a_{ij}u_i v_j$, which he denoted by a capital letter Φ when u_k and v_k are taken to be complex conjugates $u_k = x_k + iy_k$, $v_k = x_k - iy_k$ [91, p. 129]. Such a possibility may have helped encourage Weierstrass to attempt to recreate his lost theory of the equivalence of pairs of complex bilinear forms before Christoffel had the opportunity to render an "unsatisfactory" treatment. As it turned out, Christoffel did go on to study the transformation of bilinear forms in a paper of 1868 [93] but in terms of the theory of invariants rather than along the lines of the problem posed, solved, and lost by Weierstrass.[10] When in 1868, Dedekind published Riemann's habilitation lecture hinting at what was to become n-dimensional Riemannian geometry, Christoffel applied himself to the theory of the analytic transformation of quadratic differential forms [94]. This work at the interface of linear algebra and geometry proved inspirational to Frobenius, whose work on the problem of Pfaff (Chapter 6) was likewise at the interface of linear algebra and analysis.

5.3 Kronecker on Complex Multiplication and Bilinear Forms

In 1866, Kronecker published a paper in the proceedings of the Berlin Academy, "On bilinear forms," [353] that was probably a major stimulus for Weierstrass to redevelop and publish (in 1868) his lost work on the transformation of bilinear forms, what became known as his theory of elementary divisors. Such mathematical give and take was characteristic of Weierstrass' relations with Kronecker during the 1860s and 1870s. Indeed, Kronecker's 1866 paper was in turn inspired by arithmetic-algebraic questions relating to the theory of abelian and theta functions proposed to him by Weierstrass in 1858 (and discussed at length in Sections 10.2 and 10.3). Here it will suffice to say that Kronecker looked into the matter, which involved certain problems implicit in Hermite's theory of the transformation of abelian functions in two variables and in its extension to any number g of variables, and sent a report to Weierstrass that indicated how to resolve the g-variable problems. (One of the problems was later resolved more definitively by Frobenius; see Section 10.4.)

[10] That is, given a bilinear form $F = x^t Ay$, a nonsingular linear transformation $x = Px'$, $y = Py'$ induces a transformation $T_P : A \to A'$ of the coefficients of F, namely what we can now write as $T_P(A) = P^t AP$. An example of a (relative) invariant is $I_{\lambda,\mu}(A) = \det(\lambda A + \mu A^t)$ for any fixed values of λ, μ, since by Cauchy's product theorem, $I_{\lambda,\mu}(A') = (\det P)^2 I_{\lambda,\mu}(A)$. As we shall see in the next section, Kronecker introduced the determinant $I_{\lambda,\mu}(A)$ in 1866, as Christoffel realized.

The involvement with Hermite's theory suggested to Kronecker a way to generalize the notion of an elliptic function admitting complex multiplication to that of abelian functions. This matter is discussed in detail in Section 10.5, since Frobenius later also gave a definitive solution to the problem posed by Kronecker's treatment of complex multiplication (Section 10.6). Here it is enough to know that Kronecker's study of complex multiplication led to the following linear algebraic problem, which for ease of comprehension I will describe using matrix symbolism, although Kronecker did not. Indeed, such notation was not in common use in 1866, as we shall see in Chapter 7. Let

$$\tilde{A} = \begin{pmatrix} \mathrm{A} & \mathrm{B} \\ \Gamma & \Delta \end{pmatrix} \tag{5.9}$$

denote a $2g \times 2g$ matrix of integers, partitioned into four $g \times g$ blocks as indicated. The matrix $\tilde{A}$ is also assumed to have the following property:

$$\tilde{A}^t J \tilde{A} = nJ, \quad \text{where} \quad n \in \mathbb{Z}^+ \quad \text{and} \quad J = \begin{pmatrix} 0 & I_g \\ -I_g & 0 \end{pmatrix}. \tag{5.10}$$

The question of the existence of a complex multiplication in Kronecker's sense centered on the following problem.

Problem 5.3. *With $\tilde{A}$ as in (5.9) and satisfying (5.10), determine when there exists a $g \times g$ complex symmetric matrix T with* $\operatorname{Im} T$ *positive definite and satisfying the relation*

$$\mathrm{B} + T\Delta - \mathrm{A}T - T\Gamma T = 0. \tag{5.11}$$

As Kronecker explained, in the course of investigating this problem he was led to the "general investigation of those transformations of bilinear forms in $2n$ variables x and y for which the coefficient system for [the transformation of] both variable systems is the same" [353, p. 146]. The general investigation of the transformation of bilinear forms seems to have fascinated Kronecker more than the above motivating problem, and so it is not surprising that he entitled his paper simply "On bilinear forms."

The manner in which Kronecker related the above problem to an investigation of the transformation of bilinear forms was quite ingenious, although too much of a digression to include here. Here it will suffice to say that if $x = \begin{pmatrix} x_1 & \cdots & x_{2g} \end{pmatrix}^t$ and $y = \begin{pmatrix} y_1 & \cdots & y_{2g} \end{pmatrix}^t$ and we consider the bilinear form

$$\Phi(x,y) = x^t \tilde{B} y, \quad \text{where} \quad \tilde{B} = \begin{pmatrix} \mathrm{B} & -\mathrm{A} \\ \Delta & -\Gamma \end{pmatrix} = -J\tilde{A}, \tag{5.12}$$

then Kronecker's solution required finding a simultaneous nonsingular linear transformation of both variable sets, viz., $x = Qz$, $y = Qw$, such that the bilinear form

$\Phi(x,y) = x^t\tilde{B}y$, when expressed in the z,w variables as $\Phi(z,w) = z^t\tilde{C}w$, so that $Q^t\tilde{B}Q = \tilde{C}$, has coefficient matrix $\tilde{C}$ of the form $\tilde{C} = \begin{pmatrix} 0 & * \\ * & * \end{pmatrix}$. As the simplest type of bilinear form with this property Kronecker singled out z^tNw, where

$$N = \begin{pmatrix} 0 & D_1 \\ D_2 & 0 \end{pmatrix}, \tag{5.13}$$

and D_1, D_2 are diagonal matrices with entries $\lambda_1,\ldots,\lambda_g$ and $\lambda_{g+1},\ldots,\lambda_{2g}$, respectively, along the diagonals. It is tacitly assumed that none of the diagonal entries are zero. Kronecker called z^tNw the normal form and proposed to consider the following more general problem.

Problem 5.4. *Given any nonsingular bilinear form $x^t\tilde{B}y$ in an even number of variables, is there a nonsingular linear transformation of the type $x = Qz$, $y = Qw$, i.e., the same linear transformation is applied to both sets of variables, such that $x^t\tilde{B}y = z^tNw$, i.e., $Q^t\tilde{B}Q = N$?*

Thus in this problem $\tilde{B}$ is no longer assumed to have the special form $\tilde{B} = -J\tilde{A}$ specified by (5.12); it can be any $2g \times 2g$ matrix with complex coefficients and nonvanishing determinant. (When $\tilde{B} = -J\tilde{A}$, $\det\tilde{B} = \det\tilde{A} \neq 0$ by virtue of (5.10).)

Kronecker chose the name "normal form" for z^tNw

> because *every* bilinear form can be transformed into it. This reduction of the bilinear form ... into the given normal form is of the greatest significance, because not only is the above question ... resolved thereby, but also the general transformation of any bilinear form into another is thereby obtained [353, p. 148].

This passage conveys Kronecker's manifest enthusiasm for the theory of the transformation of bilinear forms. His problem is reminiscent of the problem solved by Weierstrass in his paper of 1858, but I suspect that Kronecker was not aware of this paper, for, as we have seen, Weierstrass' point of view was that the goal was to obtain results that transcended the limits of the generic case of distinct characteristic roots. By contrast, in 1866, Kronecker tacitly shared Jacobi's view that the generic case was of primary interest. This is reflected in his above claim that "*every* bilinear form can be transformed into" the normal form—a generic assertion. As we shall now see, in justifying his claim, Kronecker explicitly limited his attention to the generic case.

Kronecker's starting point in dealing with his transformation problem was that if there is a Q that transforms $x^t\tilde{B}y$ into z^tNw, then it also takes the transposed form $x^t\tilde{B}^ty$ into the transposed normal form z^tN^tw. (In matrix form this is immediately clear, since $Q^t\tilde{B}Q = N$ implies by transposition that $Q^t\tilde{B}^tQ = N^t$.) Thus the transformation of a bilinear form into the normal form is equivalent to the transformation of the *family* of forms $x^t(u\tilde{B}+v\tilde{B}^t)y$ into $z^t(uN+vN^t)w$, i.e., $Q^t\tilde{B}Q = N$ is equivalent to $Q^t(u\tilde{B}+v\tilde{B}^t)Q = uN+vN^t$ for all u,v. Taking determinants in the last equation and using the product theorem gives $(\det Q)^2\det(u\tilde{B}+vB^t) = \det(uN+vN^t)$. Since $Q^t\tilde{B}Q = N$ implies $(\det Q)^2\det\tilde{B} = \det N$, Kronecker wrote the determinant relation in the form

$$\det N \cdot \det(u\tilde{B} + v\tilde{B}^t) = \det \tilde{B} \cdot \det(uN + vN^t). \tag{5.14}$$

Thus (5.14) is a necessary condition for the transformation problem to have a solution, and Kronecker realized it was also a sufficient condition when it is assumed that $f(u,v) = \det(u\tilde{B} + v\tilde{B}^t)$ factors into distinct linear factors, i.e., no linear factor $au + bv$ is a constant multiple of another. For future reference, this result will be stated as a formal theorem.

Theorem 5.5. *Let $\tilde{B}$ be $2g \times 2g$ with $\det \tilde{B} \neq 0$ and assume that $f(u,v) = \det(u\tilde{B} + v\tilde{B}^t)$ factors into distinct linear factors. Then a nonsingular linear transformation $x = Qz$, $y = Qw$ exists such that $x^t \tilde{B} y = z^t N w$ (i.e., $Q^t \tilde{B} Q = N$), where N is the normal form (5.13), if and only if (5.14) holds, i.e., if and only if $f(u,v)$ and $\det(uN + vN^t)$ have the same linear factors.*

Clearly, if (5.14) holds, then $f(u,v)$ and $\det(uN + vN^t)$ must have the same linear factors; and if they have the same linear factors, they differ by a constant factor, which can be evaluated by setting $u = 1$ and $v = 0$, yielding (5.14). As Kronecker noted [353, p. 157], the linear factors of $f(u,v)$ are unequal precisely when the roots of $\varphi(r) = \det(r\tilde{B} - \tilde{B}^t)$ are distinct.

After indicating the proof of Theorem 5.5, Kronecker applied it to Problem 5.3, using the fact that $\tilde{B}$ has the specific form $\tilde{B} = -J\tilde{A}$ given in (5.12). For future reference, his result may be summed up as follows [353, pp. 155–156].

Theorem 5.6. *Let $\tilde{A} = \begin{pmatrix} \mathrm{A} & \mathrm{B} \\ \Gamma & \Delta \end{pmatrix}$ be a $2g \times 2g$ integral matrix satisfying $\tilde{A}^t J \tilde{A} = nJ$ for some $n \in \mathbb{Z}^+$. Set $\tilde{B} = -J\tilde{A}$ and assume that $\varphi(r) = \det(r\tilde{B} - \tilde{B}^t)$ has distinct roots. Then there exists a complex symmetric $g \times g$ matrix T that satisfies (5.11), viz., $\mathrm{B} + T\Delta - \mathrm{A}T - T\Gamma T = 0$.*

What the omitted part of Kronecker's reasoning showed was that if Theorem 5.5 is applied to $\tilde{B} = -J\tilde{A}$, then if $Q = \begin{pmatrix} Q_1 & Q_2 \\ Q_3 & Q_4 \end{pmatrix}$ is the $2g \times 2g$ matrix posited by Theorem 5.5 for this choice of $\tilde{B}$, with Q partitioned into $g \times g$ matrices, then $T = Q_3 Q_1^{-1}$ is complex symmetric and satisfies (5.11).

As Kronecker realized, Theorem 5.6 did not provide a solution to Problem 5.3. That is, suppose $\tilde{A}$ is as in that problem and has the additional property that the roots of $\varphi(r) = \det(r\tilde{B} - \tilde{B}^t)$ are all distinct. Then Theorem 5.6 implies that a complex symmetric T exists such that $\mathrm{B} + T\Delta - \mathrm{A}T - T\Gamma T = 0$, but it does not show that T has the requisite property of having an imaginary part that is positive definite. Indeed, as Kronecker stressed [353, p. 157], examples of $\tilde{A}$'s can be given that satisfy the conditions of the above theorem and yet there is no associated T with positive definite imaginary part. As we shall see in Section 10.6, Frobenius completely resolved Problem 5.3 in 1882. It was one of several research projects on Frobenius' part that were motivated by the work of Kronecker. In dealing with such problems, Frobenius was able to draw on the theory of quadratic and bilinear forms created by Weierstrass and Kronecker in the period 1868–1874. This is the subject of the ensuing sections, and it would seem that Kronecker's above-discussed 1866 paper, which he tellingly chose to entitle "On bilinear forms" rather than "On the transformation of abelian and theta functions"—its ostensible subject—was a principal source of motivation for those developments.

5.4 Weierstrass' Theory of Elementary Divisors

Weierstrass was present at the 15 October 1866 session of the Berlin Academy when Kronecker presented his paper on "On bilinear forms" [353],[11] and he must have found Kronecker's enthusiasm for the theory of the transformation of families of bilinear forms contagious. Of course, Kronecker had remained on the generic level of distinct characteristic roots, and Weierstrass must have been reminded of his old nongeneric, but lost, study of such forms. At some point he must have informed Kronecker of this old work, and perhaps also referred Kronecker to his paper of 1858 [587], which, judging by the generic character of Kronecker's 1866 paper, was not familiar to him at that time.[12] Sometime between 1866 and 1868 Weierstrass and Kronecker discussed the following problem.

Problem 5.7. *Given two bilinear forms $\Gamma(x,y) = x^t Py$ and $\Delta(x,y) = x^t Qy$, consider the family $u\Gamma(x,y) + v\Delta(x,y)$, where u, v are complex parameters. If $u\tilde{\Gamma}(u,v) + v\tilde{\Delta}(u,v)$ is another such family, where $\tilde{\Gamma}(u,v) = u^t\tilde{P}v$ and $\tilde{\Delta}(u,v) = u^t\tilde{Q}v$, determine necessary and sufficient conditions for these two families to be equivalent in the sense that nonsingular variable changes $x = Hu$, $y = Kv$ exist that transform the family $u\Delta(x,y) + v\Psi(x,y)$ into the family $u\tilde{\Gamma}(u,v) + u\tilde{\Delta}(u,v)$.*

Of course, equivalence of the two families is the same as saying that the pair $\Gamma(x,y)$, $\Delta(x,y)$ can be simultaneously transformed into the pair $\tilde{\Gamma}(u,v)$, $\tilde{\Delta}(u,v)$. In terms of the coefficient matrices, equivalence means that $H^tPK = \tilde{P}$ and $H^tQK = \tilde{Q}$. This problem was a broad generalization of the problem Weierstrass had posed and resolved in his 1858 paper. In the special case in which $P = \tilde{P} = I$, equivalence means $H^tK = I$ and $H^tQK = \tilde{Q}$, so that $H^t = K^{-1}$, and equivalence means in this case that $K^{-1}QK = \tilde{Q}$, i.e., that Q and $\tilde{Q}$ are similar, a term introduced by Frobenius in 1878 (Section 7.5).

There is no record of the conversations between Weierstrass and Kronecker concerning Problem 5.7, but there is enough documentary evidence to indicate

[11] At the same session, Weierstrass summarized one of his papers. See p. 612 of *Monatsberichte der Akademie der Wiss. zu Berlin 1866* (Berlin, 1867).

[12] Although Kronecker moved to Berlin as an independent scholar in 1855, it was not until 1861 that he became a member of the Berlin Academy. Thus when Weierstrass presented his paper [587] to the academy, Kronecker was probably not in attendance.

that the two mathematicians decided to join forces in working on it. It is easily seen that nonzero P, Q exist for which $D(u,v) = \det(uP + vQ)$ is identically zero.[13] The families $uP + vQ$ with $D(u,v) \equiv 0$ were called singular families. It is not clear whether Weierstrass had considered singular families in his lost work, but now he proposed to work on Problem 5.7 for nonsingular families, no doubt attempting to rederive his old lost results. Kronecker would focus on the formidable singular case.[14] Weierstrass had a distinct advantage over Kronecker, in that he had already solved his part of Problem 5.7 once before. As a result, he was the first to present his part of the solution to the Berlin Academy in a paper "Toward the theory of quadratic and bilinear forms" [588], read at the 18 May 1868 session. (Kronecker provided an addendum communicating some thoughts germane to the singular case.)[15]

In a footnote, Weierstrass characterized his paper as "a reworking and further development" of his 1858 memoir.[16] As he had done in 1858, and as Kronecker had done in his 1866 paper, Weierstrass focused on the characteristic polynomial $D_n(u,v) = \det(uP + vQ)$ in formulating his necessary and sufficient conditions for equivalence. Since $D_n(u,v) \not\equiv 0$, it is a homogeneous polynomial in u, v of degree n, where n is the dimension of the (square) matrices P, Q. If $D_n(g,h) \neq 0$, the matrix $A = gP + hQ$ has the nozero determinant $D_n(g,h)$. Since $(g,h) \neq (0,0)$, it is always possible to determine g', h' such that $gh' - hg' = 1$. Let $B = g'P + h'Q$, and set $u = gs - g'$, $v = hs - h'$, so that u, v are parametrized by s. Then if $\Phi = x^t Ay$ and $\Psi = x^t By$, it follows that $u\Gamma + v\Delta = s\Phi - \Psi$, and we are back in the notational framework of the 1858 paper, except that now Φ and Ψ are bilinear rather than quadratic forms and the hypothesis that Φ is definite has been replaced by the hypothesis that Φ is nonsingular—$\Phi(x,y) = 0$ for all x and some y means $y = 0$.

Although Weierstrass articulated the conceptual apparatus underlying his necessary and sufficient conditions for equivalence in terms of $u\Gamma + v\Delta$, to prove that his conditions were indeed necessary and sufficient, he worked with the associated one-parameter family $s\Phi - \Psi$. For this reason and because it was in the form $s\Phi - \Psi$ that Frobenius and most other mathematicians utilized Weierstrass' results, I will expound them for $s\Phi - \Psi = x^t(sA - B)y$. This will also bring out more clearly the connection with the remarkable property of 1858 discovered by Weierstrass and

---

[13]Simple examples with $D(u,v) \equiv 0$ with $P, Q \neq 0$ are given by $P = \begin{pmatrix} a^2 & ab \\ ab & b^2 \end{pmatrix}$ and $Q = \begin{pmatrix} ac & bc \\ ad & bd \end{pmatrix}$.

[14]Weierstrass explained in a footnote added to the version of his paper that appeared in his collected works (*Werke* 2 (1902), p. 19n) that "This case [$D(u,v) \equiv 0$] was not treated by me because I knew that Mr. Kronecker would investigate it thoroughly. (See the relevant works of Mr. K in the monthly reports [*Monatsberichten*] of the academy.)" The "relevant works" of Kronecker are considered in the next section.

[15]"Hr. Kronecker knüpfte an den versehenden Vortrag folgende Bemerkungen an: ...," *Monatsberichte der Akademie der Wiss. zu Berlin 1868*, pp. 339–346. Reprinted in Kronecker's collected works with the title "Über Schaaren quadratischer Formen" [354].

[16]The footnote occurs on the first page of [588] but is omitted from the edited version that Weierstrass included in his collected works.

embodied in his Lemma 4.8: If $s = a$ is a root of multiplicity m of $f(s) = \det(sA - B)$, then $(s-a)^{m-1}$ divides every degree-$(n-1)$ minor of $sA - B$.

Since $f(s)$ has $\det A \neq 0$ as the coefficient of s^n, it is always of degree n. Let $D_n(s) = (\det A)^{-1} f(s)$, so that the coefficient of s^n in $D_n(s)$ is 1. Let

$$D_n(s) = \prod_{i=1}^{k} (s - a_i)^{m_i}$$

denote the factorization of $D_n(s)$ into distinct factors, so that $a_1, \ldots, a_k$ are the distinct roots of $D_n(s)$ and $f(s)$ and m_i is the multiplicity of a_i. Then if the above remarkable property holds, then $(s-a_i)^{m_i - 1}$ divides every degree-$(n-1)$ minor of $sA - B$. However, for reasons indicated in Section 4.5.2, it cannot be that $(s-a)^{m_i}$ divides every degree-$(n-1)$ minor. This means that if $D_{n-1}(s)$ denotes the polynomial greatest common divisor of all the degree-$(n-1)$ minors of $sA - B$, its factorization must be

$$D_{n-1}(s) = \prod_{i=1}^{k} (s - a)^{m_i - 1}.$$

Thus

$$E_n(s) = D_n(s)/D_{n-1}(s) = \prod_{i=1}^{k} (s - a_i). \tag{5.15}$$

Weierstrass' remarkable property is thus equivalent to the fact that $E_n(s)$ factors into distinct linear factors, one for each distinct root of $f(s)$.

The above characterization of the remarkable property segues into Weierstrass' notion of elementary divisors. For any nonsingular family $s\Phi - \Psi$, let $D_n(s)$ and $D_{n-1}(s)$ be defined as above, and in general, let $D_{n-i}(s)$ denote the polynomial greatest common divisor of all the degree-$(n-i)$ minors of $sA - B$. It follows by considering Laplace expansions that $D_{n-i-1}(s)$ divides $D_{n-i}(s)$ for all i, and so $E_{n-i}(s) = D_{n-i}(s)/D_{n-i-1}(s)$ is always a polynomial for all $i < n-1$. This is also true for $E_1(s) = D_1(s)/D_0(s)$ by defining $D_0(s) = 1$. I will refer to the series of polynomials $D_n, \ldots, D_0$ as the *Weierstrass series* or, more briefly, the *W-series* associated to $s\Phi - \Psi$ (or $sA - B$), since he was the first to introduce it. Nowadays, the polynomials $E_n, \ldots, E_1$ would be referred to as the invariant factors of $sA - B$.[17]

The reason Weierstrass should think to consider the W-series in connection with his equivalence problem is not hard to find. The corollary to Cauchy's second multiplication theorem (Corollary 4.3) says that if $s\Phi - \Psi$ and $s\tilde{\Phi} - \tilde{\Psi}$ are equivalent, so $H^t(sA - B)K = s\tilde{A} - \tilde{B}$ and $sA - B = (H^t)^{-1}(s\tilde{A} - \tilde{B})K^{-1}$, then

[17] Unbeknownst to Weierstrass, arithmetic analogues of the W-series and invariant factors had been introduced a few years earlier (1861) by H.J.S. Smith. His work and its relation to Frobenius' rational version of Weierstrass' theory of elementary divisors are discussed in Chapter 8.

every degree-$(n-i)$ minor of $s\tilde{A}-\tilde{B}$ is a linear combination of the corresponding minors of $sA-B$ and vice versa. Thus if $(s-a)^{\mu}$ divides all degree-$(n-i)$ minors of $sA-B$, it divides all degree-$(n-i)$ minors of $s\tilde{A}-\tilde{B}$ and vice versa. That is to say, equivalent families $s\Phi-\Psi$ and $s\tilde{\Phi}-\tilde{\Psi}$ always have the same W-series and so the same invariant factors. The identity of W-series (and so invariant factors) was thus a necessary condition for equivalence, and this suggests the question whether it is also sufficient. Weierstrass' solution to the above equivalence problem involved showing that this was indeed the case.

As we shall see, Weierstrass' sufficiency proof suggested introducing the notion of what he called the "elementary divisors" of $s\Phi-\Psi$. If $a_1,\ldots,a_k$ denote the distinct roots of $\det(sA-B)$, so that $D_n(s)=\prod_{i=1}^{k}(s-a_i)^{m_i}$, then for all $j=1,\ldots,n$, the factor $E_j=D_j/D_{j-1}$ has a corresponding factorization

$$E_j(s)=\prod_{i=1}^{k}(s-a_i)^{e_{ji}}.$$

Here, of course, some of the exponents e_{ji} may be zero. The factors $(s-a)^{e_{ji}}$ with $e_{ji}>0$ Weierstrass named the *elementary divisors* of $s\Phi-\Psi$. Since $D_n=(D_n/D_{n-1})(D_{n-1}/D_{n-2})\cdots(D_1/D_0)=E_nE_{n-1}\cdots E_1$, it follows that $D_n(s)$ is the product of its elementary divisors.

To see the connection between elementary divisors and more familiar notions, consider the 7×7 example $A=I_7$ and $B=B_3(a)\oplus B_2(a)\oplus B_2(a)$, where $B_k(a)$ denotes the $k\times k$ Jordan block matrix with a down the diagonal, and 1 along the superdiagonal. The W-series for $sI-B$ is

$$\{D_7,D_6,D_5,D_4,D_3,\ldots\}=\{(s-a)^7,(s-a)^4,(s-a)^2,1,1,\ldots\},$$

and so the corresponding series of invariant factors $E_i=D_i/D_{i-1}$ is

$$\{E_7,E_6,E_5,E_4,\ldots\}=\{(s-a)^3,(s-a)^2,(s-a)^2,1,\ldots\}.$$

The elementary divisors are therefore $(s-a)^3$, $(s-a)^2$, and $(s-a)^2$. Each elementary divisor is thus the characteristic polynomial of one of the Jordan blocks. As we shall see, this sort of a connection is at the heart of Weierstrass' solution to Problem 5.7, i.e., his proof of the following theorem.

Theorem 5.8 (Weierstrass' elementary divisor theorem). *Two nonsingular families of bilinear forms $s\Phi(x,y)-\Psi(x,y)$ and $s\tilde{\Phi}(X,Y)-\tilde{\Psi}(X,Y)$ are equivalent in the sense that nonsingular linear transformations $x=HX$, $y=KY$ exist such that $s\Phi(x,y)-\Psi(x,y)=s\Phi(X,Y)-\tilde{\Psi}(X,Y)$ if and only if $s\Phi(x,y)-\Psi(x,y)$ and $s\tilde{\Phi}(X,Y)-\tilde{\Psi}(X,Y)$ have the same W-series and hence the same elementary divisors.*

In 1858, Weierstrass had proved his Theorem 4.9 by showing that the family of quadratic forms $s\Phi(x)-\Psi(x)$ with Φ definite could be transformed into what

Kronecker might have referred to as the "normal form" $s\tilde{\Phi}(X) - \tilde{\Psi}(X)$, with $\tilde{\Phi} = \varepsilon(X_1^2 + \cdots + X_n^2)$, $\tilde{\Psi} = \varepsilon(a_1X_1^2 + \cdots + a_nX_n^2)$, and $\varepsilon = \pm 1$, depending on whether Φ is positive or negative definite. The existence of the transformation had depended on the remarkable property of Lemma 4.8, which, in 1868, could be translated into the fact that the elementary divisors of $s\Phi - \Psi$ were all linear, i.e., were $(s-a_1),\ldots,(s-a_k)$, where $a_1,\ldots,a_k$ are the distinct roots of $\det(sA-B)$. To prove Theorem 5.8, Weierstrass introduced the following canonical form associated to given elementary divisors $(s-a_i)^{f_i}$, $i=1,\ldots,r$. Since $D_n(s)$ is the product of its elementary divisors, $D_n(s) = \prod_{i=1}^r (s-a_i)^{f_i}$, and so $n = f_1 + \cdots + f_r$. The n variables constituting X and Y, respectively, are then each divided into r groups of, respectively, $f_1,\ldots,f_r$ variables. Let X_{ij}, Y_{ij}, $j=1,\ldots,f_i$, denote the variables in the ith group. Then the canonical form is

$$s\tilde{\Phi}(X,Y) - \tilde{\Psi}(X,Y) = \sum_{i=1}^{r} s\tilde{\Phi}_i(X,Y) - \tilde{\Psi}_i(X,Y), \tag{5.16}$$

where $s\tilde{\Phi}_i(X,Y) - \tilde{\Psi}_i(X,Y)$ depends only on the variables X_{ij}, Y_{ij}, $j=1,\ldots,f_i$, of the ith group. If we let $X_i = \begin{pmatrix} X_{i1} & \cdots & X_{if_i}\end{pmatrix}^t$ and $Y_i = \begin{pmatrix} Y_{i1} & \cdots & Y_{if_i}\end{pmatrix}^t$, then in matrix form, $\tilde{\Phi}_i = X_i^t J_{f_i} Y_i$ and $\tilde{\Psi}_i = X_i^t W_{f_i}(a_i) Y_i$, where

$$J_{f_i} = \begin{pmatrix} 0 & \cdots & 0 & 1 \\ 0 & \cdots & 1 & 0 \\ \vdots & \ddots & \vdots & \vdots \\ 1 & \cdots & 0 & 0 \end{pmatrix}, \quad W_{f_i}(a_i) = \begin{pmatrix} 0 & \cdots & 0 & 1 & a_i \\ 0 & \cdots & 1 & a_i & 0 \\ \vdots & \ddots & \vdots & \vdots & \\ a_i & \cdots & 0 & 0 & 0 \end{pmatrix}. \tag{5.17}$$

It is easily verified that the sole elementary divisor of $sJ_{e_i} - W_{e_i}(a_i)$ is $(s-a_i)^{f_i}$, and that since the matrix of $s\tilde{\Phi} - \tilde{\Psi}$ has the block form

$$\tilde{W}(s) = \begin{pmatrix} sJ_{f_1} - W_{f_1}(a_1) & \cdots & 0 \\ \vdots & \ddots & \vdots \\ 0 & \cdots & sJ_{f_r} - W_{f_r}(a_r) \end{pmatrix}, \tag{5.18}$$

its elementary divisors are precisely the elementary divisors $(s-a_i)^{f_i}$, $i=1,\ldots,r$.

I will refer to the canonical form $s\tilde{\Phi}(X,Y) - \tilde{\Psi}(X,Y)$ described by (5.16)–(5.18) as the *Weierstrass canonical form*. It is analogous to the more familiar Jordan canonical form and is equivalent to it.[18] As its name suggests, this form was

[18] If $B_{e_i}(a_i)$ denotes the Jordan block corresponding to $W_{e_i}(a_i)$, then $J_{e_i}P_{e_i} = I_{e_i}$ and $W_{e_i}(a_i)P_{e_i} = B_{e_i}(a_i)$, where P_{e_i} is the permutation matrix corresponding to the permutation

$$\begin{pmatrix} 1 & 2 & \cdots & e_i \\ e_i & e_i - 1 & \cdots & 1 \end{pmatrix}.$$

introduced independently in 1871 by Camille Jordan, who had introduced a mod p version in 1870 (Section 5.5). Note, however, that the Weierstrass canonical form matrix (5.18) is symmetric, whereas the corresponding Jordan canonical form is not. Weierstrass may have chosen a symmetric canonical form so as to be able to deal with families of quadratic forms, as in Corollary 5.10 below.

The difficult direction in Weierstrass' proof of Theorem 5.8 involved showing, by highly nontrivial reasoning based on considerations of determinants [including the Jacobi transformation (5.5)] and Laurent expansions, that if $s\Phi(x,y) - \Psi(x,y)$ is any nonsingular family of bilinear forms with elementary divisors $(s-a_i)^{f_i}$, $i = 1,\ldots,r$, then with X and Y as defined above, H and K may be determined such that $x = HX$, $y = KY$ transforms $s\Phi(x,y) - \Psi(x,y)$ into the above canonical form $s\tilde{\Phi}(X,Y) - \tilde{\Psi}(X,Y)$. In matrix form this is the assertion that $H^t(sA-B)K = \tilde{W}(s)$, where $\tilde{W}(s)$ is the Weierstrass canonical form matrix given in (5.18). It then follows immediately that any two nonsingular families with the same elementary divisors can be transformed into the same canonical form and so into each other.

Weierstrass deduced several corollaries from his theorem. For example, he considered the question suggested by Jacobi's paper of 1834 and its implicit Theorem 5.1: When can $\Phi(x,y)$ and $\Psi(x,y)$ be transformed, respectively, into $\Phi^* = \sum_{i=1}^n g_i X_i Y_i$ and $\Psi^* = \sum_{i=1}^n h_i X_i Y_i$? To apply Theorem 5.8 it was necessary to assume that $\Phi(x,y)$ is nonsingular, which implies that $g_i \neq 0$ for all i. Thus the above transformation is possible if and only if the transformation into $\Phi^{**} = \sum_{i=1}^n X_i Y_i$, $\Psi^{**} = \sum_{i=1}^n a_i X_i Y_i$, $a_i = (h_i/g_i)$, is possible. Since the elementary divisors of $s\Phi^{**} - \Psi^{**}$ are easily seen to be linear, the desired transformation is possible if and only if all the elementary divisors of $s\Phi - \Psi$ are linear. This was the heyday of the theory of determinants, however, and so Weierstrass sought a simple determinant characterization of this situation. He could have simply pointed out that a necessary and sufficient condition for the transformation is his remarkable property (if $s = a$ is a root of multiplicity m in $\det(s\Phi - \Psi)$, then $(s-a)^{m-1}$ divides a minor of degree $n-1$). But he preferred a more sophisticated version, which was based on the fact that if the factorization of $D_i(s)$ into powers of distinct linear factors is $D_i(s) = (s-a)^{\mu}\cdots$, then that of $D_{i-1}(s)$ is of the form $D_{i-1} = (s-a)^{\mu'}\cdots$, where $\mu' < \mu$. When all the elementary divisors are linear, however, it must be that $\mu' = \mu - 1$. This means that $D_{n-i} = (s-a)^{m-(n-i)}\cdots$ for $n-i < m$, and so one obtains the following [588, p. 37]:

Corollary 5.9. *Given bilinear forms $\Phi(x,y)$ and $\Psi(x,y)$, with $\Phi(x,y)$ nonsingular, nonsingular transformations $x = HX$, $y = KY$ can be determined such that $\Phi(x,y)$ and $\Psi(x,y)$ are transformed, respectively, into $\Phi^* = \sum_{i=1}^n g_i X_i Y_i$ and $\Psi^* = \sum_{i=1}^n h_i X_i Y_i$ if and only if the following condition holds: if $s = a$ is a root of $\det(s\Phi - \Psi)$ of multiplicity $m > 1$, then every minor of degree $n-m+1$ is divisible by $(s-a)$.*

Thus if $P = P_{e_1} \oplus \cdots \oplus P_{e_r}$, then the nonsingular transformations $X = X'$, $Y = PY'$ take Weierstrass' canonical form $s\tilde{\Phi} - \tilde{\Psi}$ into the bilinear form $(X')^t(sI_n - J)Y'$, where $J = B_{e_1}(a_1) \oplus \cdots \oplus B_{e_r}(a_r)$ is the Jordan canonical form.

By such a corollary, Weierstrass showed how a (generalized) version of a generic theorem of Jacobi's could be made nongeneric and rigorous by means of an easily stated condition.

Weierstrass also considered the special case in which Φ and Ψ are quadratic (with possibly complex coefficients) [588, §5]. He observed [588, p. 326] that in this case, his formulas for the coefficients of the transformations H and K show that $H = K$. Expressed in matrix form, this means that $H^t(sA - B)H = \tilde{W}(s)$ when $sA - B$ is symmetric. (This is why $\tilde{W}(s)$ needs to be symmetric: when $sA - B$ is symmetric, so is $H^t(sA - B)H$.) As a result he obtained the following corollary.

Corollary 5.10 (Weierstrass' corollary on quadratic forms). *I. Two quadratic forms $\Phi(x), \Psi(x)$ with Φ not singular can be simultaneously transformed into $\Phi^*(X)$ and $\Psi^*(X)$ with Φ^* nonsingular if and only if $s\Phi - \Psi$ and $s\Phi^* - \Psi^*$ have the same elementary divisors. II. Two quadratic forms $\Phi(x), \Psi(x)$ with Φ not singular can be simultaneously transformed into $\Phi^* = \sum_{i=1}^n g_i X_i^2$ and $\Psi^* = \sum_{i=1}^n h_i X_i^2$ with all $g_i \neq 0$ if and only if all the elementary divisors of $s\Phi - \Psi$ are linear.*

As Weierstrass pointed out, a sufficient condition for the simultaneous transformation posited in Part II above had been given by him in his Theorem 4.9 of 1858, namely that Φ and Ψ be real with Φ strictly definite. Part II must have appeared to him a very satisfying, nontrivial generalization of that result.[19]

5.5 The Canonical Form of Camille Jordan

In France, Camille Jordan (1838–1921) was led to what is now called the Jordan canonical form through his efforts to develop the profound but sketchy ideas on the solvability of equations that Galois had left to posterity.[20] The aspect of Galois' work that was the most fundamental and that has received the greatest attention by historians is his discovery that every polynomial equation has associated with it a group of permutations and that the solvability of the equation by radicals depends on whether the associated group possesses (in modern terminology) a composition series with factor groups that are abelian. But Galois also devoted considerable attention to the problem of using his criterion for solvability to determine when various types of equations are solvable by radicals. His approach

---

[19]The reasoning that led Weierstrass to his formulas for H and K was based on an assumption that was later seen to be far from obvious when $sA - B$ is symmetric, so that the conclusion that $H = K$ when the forms are quadratic—and so the proof of Corollary 5.10—was seen to contain a gap. On the efforts to rework Weierstrass' theory so that Corollary 5.10 followed, including the important role played by Frobenius, see Section 16.1.

[20]Nineteenth-century mathematicians developed Galois' work as it was known through his collected works as published posthumously in the *Journal de mathématiques pures et appliquées* in 1846 [239].

involved representing the permutation group of the equation, i.e., what is now called its Galois group, by what he called linear substitutions.

For example, Galois sought to determine necessary and sufficient conditions for an irreducible equation of prime degree p to be solvable. He showed that when the equation is solvable, its permutation group $\mathfrak{G}$ has the following form: Let $x_0, x_1, \ldots, x_{p-1}$ denote the p roots of the equation. Then the permutations of the x_i that constitute $\mathfrak{G}$ will be of the form $x_k \to x_{k'}$, where $k' \equiv ak + b \pmod{p}$, $a \not\equiv 0 \pmod{p}$. Hence the permutations of $\mathfrak{G}$ can be represented by linear substitutions $k' \equiv ak + b \pmod{p}$. Galois also sought to characterize those primitive equations that are solvable [239, pp. 26–27, 51–61], and in this connection he made two important discoveries (asserted without proof): (1) the degrees of such solvable equations are of the form p^n, p prime; (2) the corresponding group $\mathfrak{G}$ of permutations must have the following form analogous to that for a solvable equation of prime degree: If the p^n roots of the equation are written in the form $a_{k_1 \ldots k_n}$, then the permutations $a_{k_1 \ldots k_n} \to a_{k'_1 \ldots k'_n}$ that constitute $\mathfrak{G}$ are such that

$$k'_i \equiv m_{i1}k_1 + \cdots + m_{in}k_n + a_i \pmod{p}, \quad i = 1, 2, \ldots, n, \tag{5.19}$$

where $\det(m_{ij}) \not\equiv 0 \pmod{p}$. Galois observed that all linear substitutions of type (5.19) form a group, so that the problem was to determine all maximal solvable subgroups of this group.

Galois went no further with the general problem, but for $n = 2$, he made some additional observations of interest. He pointed out that the "circular substitutions" $k'_i \equiv k_i + a_i \pmod{p}$ form (in modern terminology) a normal subgroup. Thus he was led to consider the factor group

$$k'_i \equiv m_{i1}k_1 + m_{i2}k_2 \pmod{p}, \quad i = 1, 2. \tag{5.20}$$

Since the group of substitutions (5.20) is not primitive, it did not figure prominently in Galois' analysis. Instead, he focused on the factor group defined by

$$k' \equiv \frac{m_{11}k + m_{12}}{m_{21}k + m_{22}} \pmod{p}, \tag{5.21}$$

i.e., the projective linear group now denoted by $\mathrm{PSL}_2(p)$.

During the period 1846–1866 some attempts were made to clarify Galois' ideas and establish his assertions, but significant progress was first made by Jordan. Prior to Jordan, Galois' successors, following his lead, concentrated on projective linear substitutions (5.21). Jordan, however, preferred to make the consideration of homogeneous linear substitutions (5.20) fundamental, and in a paper of 1867 [320, pp. 132–133], he indicated their important role in the problem of determining all the irreducible equations of a given degree that are solvable by radicals. In connection with this problem he sought, in a paper of 1868 [321], to determine the solvable subgroups of the group of linear substitutions in two variables (5.20). Jordan showed that, contrary to Galois' opinion, there exist three general types of solvable subgroups. To do it he used the fact that by a linear change of variables, a linear substitution S could be put in one of a limited number of "canonical forms" depending on the nature of the roots of $\det(S - kI) \equiv 0 \pmod{p}$ [321, pp. 111–113]. His method of constructing solvable subgroups was to build them up from their composition series, and this involved determining all linear substitutions that commute with a given substitution S. To this end, he introduced the possible canonical forms for S.

In 1870, Jordan published his *Traité des substitutions* [322], which was largely a compendium and exposition of his own work during the 1860s. The study of linear homogeneous substitutions of the form (5.20) naturally occupied a prominent position in the *Traité* [322, Ch. II]. Among other things, Jordan generalized his canonical forms from 2 to n variables (or indices as he continued to term them) [322, pp. 114ff.]. That is, he showed that if S is a linear homogeneous substitution with defining equations (in anachronistic matrix notation) $x' \equiv Ax \pmod{p}$, then a suitable linear change of variables $x = Py$ can be determined such that the equations of S are $y' \equiv Jy \pmod{p}$, where J is the canonical form. This of course means that $J \equiv P^{-1}AP \pmod{p}$. The only difference between Jordan's J and the familiar Jordan canonical form is that the coefficients in the Jordan blocks corresponding to the root k are written as

$$\begin{array}{ccccc} k & 0 & 0 & 0 & \cdots \\ k & k & 0 & 0 & \cdots \\ 0 & k & k & 0 & \cdots \\ \cdots & \cdots & \cdots & \cdots & \cdots \end{array}$$

where $k \not\equiv 0 \pmod{p}$, since $\det S \not\equiv 0 \pmod{p}$. Using his canonical form, Jordan determined the linear substitutions which commute with a given substitution [322, pp. 128ff.]. He also considered the problem of determining all substitutions S, T with the same canonical form and showed that they are related by $S \equiv U^{-1}TU \pmod{p}$ for some substitution U [322, pp. 136ff.].

When Jordan wrote his *Traité*, he was unaware that Weierstrass had introduced essentially the same canonical form for ordinary complex coefficients in 1868 [588]. Jordan was himself stimulated to introduce his canonical form within the context of complex numbers by the appearance of a note in the *Comptes rendus* of 1870 by the engineer Yvon Villarceau [616]. The object of the note was to call attention to the fact that Lagrange was mistaken when he claimed that multiple roots always produce unstable solutions in the associated linear differential equations. Weierstrass, of course, had clarified this matter in his 1858 memoir [587], but Villarceau was unaware of it. His discovery of Lagrange's error was based on a detailed analysis of $d^2y/dt = Ay$ in the case of two variables.

At the conclusion of his note, Villarceau raised the question of when a system $d^nx/dt^n = Ax$ can be resolved into separate equations each of which can be directly integrated. Since any such system can be reduced to a system of the form $dy/dt = By$, Villarceau studied this case for two variables and announced that "it was easy to establish that if, when the characteristic equation has equal roots, these equations can nevertheless be integrated by means of exponential and trigonometric functions unaffected by algebraic factors containing the independent variable t, the proposed system resolves into two equations that can be integrated separately" [616, p. 766]. Villarceau had raised some interesting questions, although he evidently did not possess the algebraic tools necessary to deal with the case of n variables. Jordan responded with a note in 1871 [323] in which he pointed out that the question raised by Villarceau can be easily resolved in general "by a procedure identical to that which we have used in our Traité des substitutions to reduce a linear substitution to its canonical form" [323, p. 313].

Thus, Jordan showed that $dx/dt = Ax$ could be transformed by a linear change of variables into $dy/dt = Jy$, in which form the equations separate into subsystems

$$\frac{dy_1}{dt} = \sigma y_1, \quad \frac{dy_2}{dt} = \sigma y_2 + y_1, \quad \ldots, \quad \frac{dy_r}{dt} = \sigma y_r + y_{r-1}, \tag{5.22}$$

which can be integrated directly to yield $y_r = \psi(t)e^{\sigma t}$, $y_{r-1} = \psi'(t)e^{\sigma t}, \ldots, y_1 = \psi^{(r-1)}(t)e^{\sigma t}$, where $\psi(t)$ is any polynomial of degree $r-1$.[21] Hence Villarceau's conclusion for two variables was generally true: the solutions involve no algebraic factors $\psi(t), \psi'(t), \ldots$ if and only if the system reduces to equations $dy_i/dt = \sigma_i y_i$, $i = 1, 2, \ldots, n$. Jordan included this method of solution in the second edition (1887) of his *Cours d'analyse* [330, pp. 173–175]. As the form of the subsystem (5.22) indicates, in formulating his canonical form over the complex numbers, Jordan expressed the Jordan blocks that make up his canonical form J in the now familiar form

$$\begin{pmatrix} \sigma & 1 & 0 & \cdots & 0 \\ 0 & \sigma & 1 & \cdots & 0 \\ \vdots & \vdots & \vdots & \ddots & \vdots \\ 0 & 0 & 0 & \cdots & \sigma \end{pmatrix}.$$

5.6 Singular Families and Disciplinary Ideals: Kronecker's Memoirs

As we saw in Section 5.4, Kronecker had taken on the problem of solving the equivalence problem (Problem 5.7) in the singular case, and in an addendum [354] to Weierstrass' 1868 paper on his theory of elementary divisors, had already presented some ideas germane to its solution. Except for these preliminary results however,

[21] The same result on the integration of $A\,dx/dt = Bx$ was communicated by Weierstrass to the Berlin Academy in 1875, but apparently first published in his *Werke* [591]. No reference was made to Jordan's note [323].

his more definitive results were discovered shortly before and during a polemical exchange with Camille Jordan. As a consequence, many of Kronecker's discoveries were first presented in the January–March monthly proceedings (*Monatsberichte*) of the Berlin Academy interspersed with criticisms of Jordan's work. In what follows, I have not attempted an exhaustive description of this polemic,[22] but it cannot be entirely ignored, since it prompted Kronecker to articulate two important disciplinary ideals epitomized by Weierstrass' two papers on quadratic and bilinear forms and adhered to not only by Kronecker in his work on singular families but also by Frobenius. Frobenius, who was still in Berlin in 1874, was familiar with these papers by Kronecker, and the disciplinary ideals that he articulated in the midst of the controversy with Jordan not only informed all of Frobenius' work on linear algebra but, in particular, provided the motivation for his important early work on the problem of Pfaff (Chapter 6) and on the Cayley–Hermite problem (Chapter 7). Thus in what follows, the emphasis will be on Kronecker's theorems about singular families and the disciplinary ideals they epitomized, with the polemic with Jordan described only in so far as it provides the historical background needed for understanding Kronecker's pronouncements.

5.6.1 Singular families of quadratic forms 1868–1874

Kronecker's above-mentioned addendum [354] was divided into two parts. In the first, he focused on Weierstrass' Theorem 4.9 from his 1858 paper, i.e., on the simultaneous diagonalization of two real quadratic forms, one of which is definite. Kronecker had discovered a simple algorithm for the reduction of quadratic forms, which provided a quite different proof of Weierstrass' theorem. The second part contained Kronecker's first, tentative, steps toward the goal of generalizing the theory of Weierstrass' 1868 paper so as to cover the case of singular families. Ideally, such a generalization would provide a complete set of invariants to play the role of the elementary divisors in Weierstrass' theory so as to provide necessary and sufficient conditions for two families, singular or not, to be equivalent.

Kronecker focused his attention on families of quadratic forms $u\Phi(x)+v\Psi(x)$, $\Phi = x^tAx$, $\Psi = x^tBx$ (A,B symmetric)—the context of Weierstrass' 1858 paper [587]. He apparently did this because from Kronecker's viewpoint, the study of the transformation of quadratic forms $\Phi = x^tAx$ by linear transformations $x = HX$, which changes the coefficient matrix from A to $\tilde{A} = H^tAH$, was more general than that of the transformation of bilinear forms $\Psi = x^tBy$ under $x = HX$, $y = KY$, which changes the coefficient matrix from B to $\tilde{B} = H^tBK$ [356, p. 352]. This was because given the bilinear form $\Psi = x^tBy$, we may consider the associated quadratic form

[22] For those interested in following it, here is a list of the sources in chronological order: Jordan [324], Kronecker [356, 357], Jordan (before seeing [357]) [325], Kronecker [358, 360], Jordan [326].

in $2n$ variables $z = \left(x_1 \ldots x_n\ x_{n+1} \ldots x_{2n}\right)^t$ given by $\Phi = z^t Az$, where A is the symmetric matrix

$$A = \begin{pmatrix} 0 & B \\ B^t & 0 \end{pmatrix}.$$

If Φ is transformed by $z = PZ$, where

$$P = \begin{pmatrix} H & 0 \\ 0 & K \end{pmatrix},$$

then $z = PZ$ means that $x = HX$ and $y = KY$, and the matrix of Φ in the Z coordinates is

$$\tilde{A} = P^t AP = \begin{pmatrix} 0 & H^t BK \\ (H^t BK)^t & 0 \end{pmatrix} = \begin{pmatrix} 0 & \tilde{B} \\ \tilde{B}^t & 0 \end{pmatrix},$$

so $\tilde{B} = H^t BK$. Thus the study of the transformation of bilinear forms can be regarded as a special case of the study of the transformation of quadratic forms in an even number $2n$ of variables with special linear transformations $z = PZ$ that act separately on the first n and last n variables. With this viewpoint in mind, Kronecker focused his attention on singular families of quadratic forms.

When a family of quadratic forms is singular, so that $\det(uA + vB) \equiv 0$, the rank[23] of $uA + vB$ is less than n for all u and v, and so the rows of $uA + vB$ are linearly dependent. Kronecker observed that this means that linear dependency relations of the form

$$\sum_{i=1}^{n} \alpha_i(u,v)\mathrm{Row}_i(uA + vB) \equiv 0$$

will exist. In other words, if $uA + vB$ is regarded as a matrix with coefficients from $\mathbb{C}[u,v]$, then its rows are linearly dependent, and so a vector $v = v(u,v)$ exists for which

$$v^t (uA + vB) = 0, \quad v = \left(\alpha_1(u,v) \ \ldots \ \alpha_n(u,v)\right)^t, \tag{5.23}$$

and there will be $\mu = n - r$ linearly independent solutions, where r denotes the rank of $uA + vB$ over $\mathbb{C}[u,v]$. Kronecker observed that the coefficients $\alpha_i(u,v)$ corresponding to a particular solution v can be taken as homogeneous polynomials of the same degree.

[23]The term "rank" was introduced by Frobenius in 1877, but the notion (without a name) was in existence much earlier. See Section 6.3.

By way of a simple illustration, consider the quadratic family

$$uA+vB=\begin{pmatrix} 0 & 0 & u+v & 2u+v \\ 0 & 0 & u+v & 2u+v \\ u+v & u+v & 0 & 0 \\ 2u+v & 2u+v & 0 & 0 \end{pmatrix}, \tag{5.24}$$

which has rank two. Thus (5.23) will have $\mu = n - r = 2$ linearly independent solutions. Solving (5.23) in this case yields, by the usual elimination methods, the general solution $v = \big(a \ -a \ b \ -b(u+v)/(2u+v)\big)$, where a and b are free variables. Two homogeneous linearly independent solutions are obtained by taking, e.g., $a = 1$, $b = 0$ and $a = 0$, $b = 2u + v$, respectively, to get

$$v_1 = \big(1 \ -1 \ 0 \ 0\big)^t, \quad v_2 = \big(0 \ 0 \ 2u+v \ -(u+v)\big)^t,$$

a homogenous vector of degree $m_1 = 0$ and one of degree $m_2 = 1$.

In his 1868 paper, Kronecker focused exclusively on m_1, the minimal degree of a homogeneous vector solution to (5.23). In the above example, $m_1 = 0$, but in general, m_1 can be any nonnegative integer. For example, for the family in $n = 5$ variables $u(x_1x_2 + x_3x_4) + v(x_2x_3 + x_4x_5)$, the rank is $r = 4$, and (5.23) has $\mu = 5 - 4 = 1$ linearly independent solution, which can be expressed in homogeneous form as $v = \big(v^2 \ 0 \ -uv \ 0 \ u^2\big)^t$. The nonzero coefficients have degree 2, and so $m_1 = 2$.

For singular quadratic families $\mathcal{Q} = x^t(uA + vB)x$, Kronecker's main result in his paper of 1868 was that if the above minimal degree of homogeneity m_1 is nonzero, which means that (5.23) has no solutions $v \neq 0$ with constant components, then a variable change $x = Hz$ is possible such that in the new variables, $\mathcal{Q}$ is transformed into a reduced form [354, p. 174], but this did not bring with it the sort of generalization of Weierstrass' theorem envisioned above.

According to Kronecker, shortly after presenting his 1868 paper he saw how to combine its result with Weierstrass' theory so as to obtain an extension to singular families $\mathcal{Q}$ in n variables of rank $n - 1$ [356, pp. 354–355]. In this connection he introduced a series of invariants that he called the "series of determining classes" [356, p. 352], which is simply the Weierstrass series (or W-series), $\{D_n(u,v), \ldots, D_1(u,v)\}$, except now extended to singular families of quadratic forms $\mathcal{Q} = x^t(uA + vB)x$. Thus $D_n(u,v) = \det(uA + vB)$, which might now be zero. Likewise, $D_{n-1}(u,v)$ is the polynomial greatest common divisor of all degree-$(n-1)$ minors of $uA + vB$, and so on down to $D_1(u,v)$. For example, if $uA + vB$ is the singular family defined in (5.24), the W-series is $\{0, 0, 1, 1\}$. For the other example given above, viz., $uA + vB = u(x_1x_2 + x_3x_4) + v(x_2x_3 + x_4x_5)$, the W-series is $\{0, 1, 1, 1, 1\}$.

Here, just as in Weierstrass' theory, the $D_i(u,v)$ are invariants under variable transformations $x = HX$, and $D_{i-1}(u,v)$ is a divisor of $D_i(u,v)$. For nonsingular families, the W-series completely determines the elementary divisors, since the latter arise from the factorization over $\mathbb{C}$ of the invariant factors $E_i = D_i/D_{i-1}$, and so the W-series likewise forms a complete set of invariants. Kronecker discovered that this is still true for singular families of rank $r = n - 1$, i.e., he had discovered the following theorem [356, pp. 353–354, (A)].

Theorem 5.11. *If $\mathcal{Q}(x)$ and $\tilde{\mathcal{Q}}(X)$ are two families of quadratic forms with W-series containing at most one zero term, then a nonsingular transformation $x = HX$ exists such that $\mathcal{Q}(x) = \tilde{\mathcal{Q}}(X)$ if and only if their W-series are identical.*

For singular families of lower rank he was also able to push their reduction further but had not yet established the existence of a complete set of invariants. Kronecker communicated these results to Weierstrass, but had no intention of publishing them, since they seemed straightforward consequences of the results in his and Weierstrass' papers of 1868. Of course, they were also far removed from a definitive solution to the problem of extending Weierstrass' theory to arbitrary singular families.

For singular families of rank $r = n - \mu$ with $\mu > 1$, it is not difficult to see using examples known to Kronecker at the time that the W-series is no longer a complete set of invariants. For example, for $m \geq 1$, let $\mathcal{E}_{2m+1}$ denote Kronecker's "elementary family" in $2m+1$ variables

$$\mathcal{E}_{2m+1} = u(x_1x_2 + \cdots + x_{2m-1}x_{2m}) + v(x_2x_3 + \cdots + x_{2m}x_{2m+1}). \tag{5.25}$$

Then $\mathcal{Q}_1 = \mathcal{E}_3 \oplus \mathcal{E}_7$ and $\mathcal{Q}_2 = \mathcal{E}_5 \oplus \mathcal{E}_5$ are two families in 10 variables with the same W-series $\{0, 0, 1, 1, \ldots, 1\}$. However, for $\mathcal{Q}_1$, the minimal degree of homogeneity is $m_1 = 1$, whereas for $\mathcal{Q}_2$ we have $m_1 = 2$. This shows that $\mathcal{Q}_1$ and $\mathcal{Q}_2$ cannot be congruent, since it is easy to see that the degree m of any homogeneous vector solution to (5.23) remains invariant under a linear transformation $x = HX$ of variables.[24] Judging by his later remarks [357, p. 379] he realized this when he wrote his 1868 paper.

During the summer of 1868, Kronecker sought to extend the reductive algorithm he had used to obtain the reduction in his 1868 paper to more general families of quadratic forms, but he did not succeed [356, p. 355]. Then, toward the end of 1873, and so more than five years later, a more general line of research suggested to him how to achieve the extension of his reduction method that had eluded him in 1868. The new insight involved a reduction procedure based on a partition of variables into groups, which were to be treated collectively. He spoke to Kummer about his new discoveries and had the idea of making the effort necessary to present a lengthy, detailed essay describing his reduction procedure and the concomitant theorem to which I now turn.

The notion of the minimal degree of homogeneity m_1 of a singular quadratic family that Kronecker had introduced in 1868 can be further developed to produce

[24] If $\mathcal{Q}_1 = x^t(uA + vB)x$ and $\mathcal{Q}_2 = X^t(u\tilde{A} + v\tilde{B})X$ were equal for $x = HX$, so $H^t(uA + vB)H = uA' + vB'$, then if $v_1 = v(u, v)$ is a solution to $v^t(uA + vB) = 0$ with components homogeneous of degree m, it follows that $v_2 = H^{-1}v_1$ is also homogeneous of degree m and $v_2(u\tilde{A} + v\tilde{B}) = 0$.

a second series of invariants for such families. Suppose $\mathcal{Q} = x^t(uA+vB)x$ has rank $r = n-\mu$. Then (5.23), viz., $v^t(uA+vB) = 0$, has μ linearly independent solutions $v = v(u,v)$ over $\mathbb{C}[u,v]$. As illustrated by the example at (5.24), any solution can be multiplied by a homogeneous polynomial so as to make its nonzero components homogeneous polynomials of the same degree d. Then m_1 was defined as the minimal value of d. Let v_1 denote a homogeneous solution to $v^t(uA+vB) = 0$ of degree $d = m_1$. When $\mu > 1$, we may next consider the degrees of all homogeneous solutions v to $v^t(uA+vB) = 0$ that are not multiples of v_1 and set m_2 equal to the minimum of such degrees. Let v_2 denote a vector of degree m_2. If $\mu > 2$ we may consider the degrees of all homogeneous solutions to $v^t(uA+vB) = 0$ that are not linear combinations of v_1 and v_2 and define m_3 as the minimum of these degrees, and so on. Clearly $m_1 \leq m_2 \leq \cdots \leq m_\mu$. (Expressed in more familiar terms, the vectors $v_1,\ldots,v_\mu$ form a basis for the null space of $(uA+vB)^t$ over $\mathbb{C}[u,v]$, which by symmetry is the same as the null space of $uA+vB$.)[25] This then leads to the definition of what I will refer to as the *Kronecker series* (or *K-series*) of a singular family:

Definition 5.12. Given a singular family $\mathcal{Q} = x^t(uA+vB)x$ in n variables and of rank $r = n-\mu$, the μ integers $\{m_1,m_2,\ldots,m_\mu\}$ will be called the K-series of $\mathcal{Q}$.[26]

As with m_1, all the integers m_i in a K-series are invariants with respect to linear transformations $x = HX$ of the family, as Kronecker realized. For example, given the family in (5.24), the form of the vectors v_1 and v_2 given there shows that the K-series is $\{1,2\}$. More generally, for $m_1 \leq m_2 \leq \cdots \leq m_\mu$, the direct sum

$$\mathcal{Q} = \mathcal{E}_{2m_1+1} \oplus \cdots \oplus \mathcal{E}_{2m_\mu+1},$$

where $\mathcal{E}_{2m+1}$ denotes Kronecker's elementary family (5.25), has $\{m_1,m_2,\ldots,m_\mu\}$ as its K-series. (The W-series consists of μ 0's followed by all 1's.)

The theorem that Kronecker had discovered with the aid of his improved reduction procedure—although not proved in complete detail—was that together the W- and K-series form a complete set of invariants:

Theorem 5.13 (Kronecker). *Two families of quadratic forms are transformable into one another if and only if they have the same W- and K-series.*

For families of quadratic forms, this theorem does indeed constitute a definitive solution to the problem of extending Weierstrass' theory to the singular case. For nonsingular families, there is no K-series, and the theorem reduces to Weierstrass' Theorem 5.8 as specialized to quadratic forms in part I of Corollary 5.10 with the

---

[25]The `NullSpace` command in Mathematica provides a basis that is easily converted into Kronecker's basis $v_1,\ldots,v_\mu$.

[26]The K-series is independent of its mode of construction. For example, as defined above using vectors (which Kronecker did not employ), the numbers $m_1,\ldots,m_\mu$ are independent of the choice of vectors $v_1,v_2,\ldots$ used to define them. See [240, v. 2, p. 38].

W-series replacing the elementary divisors; and for arbitrary families, the K-series provided the additional invariants needed to cover the singular case.

5.6.2 The first disciplinary ideal

Kronecker's plan to write an extensive work on his improved method of reduction and the concomitant Theorem 5.13 never materialized, however, due to a note "On bilinear polynomials" by Camille Jordan that appeared in the 22 December 1873 *Comptes rendus* of the of the Paris Academy of Sciences [324]. In the note, Jordan reported on the results of a memoir of the same title he had submitted in August 1873 to the *Journal de mathématiques pures et appliquées* (founded by Liouville in 1836). The memoir was motivated in large part by Weierstrass' 1868 paper on his theory of elementary divisors: by the challenge of obtaining an extension of Weierstrass' results to the singular case and by the challenge of "providing a simpler reduction procedure than the synthetic method" adopted by Weierstrass.[27] Although Kronecker's 1868 addendum to Weierstrass' paper contained a reduction, it fell far short of extending Weierstrass' theory to the singular case, and during the ensuing five years, Kronecker published nothing more that would have indicated he was still working on it.

Jordan explained in his announcement [324] that his memoir dealt with three problems: Given a bilinear form—or "polynomial," as Jordan called it—$\Phi = x^t By$, (1) reduce Φ to a "simple canonical form" by two orthogonal substitutions $x = SX$, $y = TY$ (so that $\tilde{B} = S^t BT = S^{-1}BT$); (2) reduce Φ to a "simple canonical form" by applying a single nonsingular (but not necessarily orthogonal) transformation to both the x- and y-variables—say $x = HX$, $y = HY$ (so that $\tilde{B} = H^t BH$); (3) to reduce to a "canonical form" two bilinear forms Φ and Ψ by arbitrary nonsingular linear transformations H and K of the x- and y-variables, respectively (so that $\tilde{B} = H^t BK$) [324, p. 7]. Jordan went on to state that whereas the first problem was, he believed, new, the second had already been treated by Kronecker in his 1866 paper [353], and the third has already been treated by Weierstrass in his 1868 paper, although

> the solutions given by these eminent geometers from Berlin are incomplete in that they have left aside certain exceptional cases, which are, nevertheless, not without interest. Moreover, their analysis is quite difficult to follow, especially that of Mr. Weierstrass. The new methods that we propose are, by contrast, extremely simple and admit no exceptions [324, p. 7].

The exceptional case left aside by Kronecker in 1866 was of course the nongeneric case in which $\det(uA + vA^t)$ has repeated linear factors, and Weierstrass' omitted exceptional case involved the singular families that had been left for Kronecker to study.

---

[27] According to Jordan's note of 2 March 1874 [325, p. 13].

One can imagine Kronecker's reaction to Jordan's note. In the first place, Jordan's reference to the lack of generality of Kronecker's 1866 work must have stung all the more, since as we have seen, through Weierstrass' work, Kronecker had been converted to the view that the goal in the theory of forms should be general, nongeneric results. Jordan was correct to criticize Kronecker's 1866 paper, and Kronecker knew it! As for Weierstrass' 1868 paper, the paradigm for Kronecker's own subsequent efforts, Jordan's criticism that Weierstrass had not considered the exceptional case of singular families was also legitimate; and that criticism, too, must have stung Kronecker, since he had taken on the task of resolving the singular case, although—five years later—he had still not published a solution. Stung by these legitimate criticisms, Kronecker turned his attention to that part of Jordan's criticism asserting the difficulty and attendant lack of simplicity of Weierstrass' method of proof. Weierstrass' method was based on complicated transformations involving determinants that had been used earlier by Jacobi.[28] To Kronecker, who, like Weierstrass and Jacobi, was accustomed to intricate linear-algebraic reasoning with determinant-based constructs, Weierstrass' procedure seemed "completely transparent" [356, p. 370], but Jordan's style of doing linear algebra was relatively free of determinants. Most present-day mathematicians would find Weierstrass' paper as difficult to follow as Jordan did.

Equally irking and suspicious to Kronecker must have been Jordan's claim to have discovered new methods that were "by contrast extremely simple and admit no exceptions." After all, he had been stuck on a reduction procedure that seemed similar to Jordan's—in so far as it could be understood from Jordan's brief note—and Kronecker's solution to this impasse—a technique of variable grouping—was far from obvious, i.e., not straightforward [358, pp. 404–405]. Finally, Kronecker must have been struck by the fact that Jordan only claimed to have obtained what he deemed to be simple canonical forms; he did not use them to specify a complete set of invariants for the equivalence classes implicit in each of the three problems, as Weierstrass had done and as he himself had done in his recently discovered (but not yet published) Theorem 5.13.

Kronecker's response to Jordan's note was presented in the 19 January 1874 proceedings of the Berlin Academy. Jordan's full memoir in the *Journal de mathématiques pures et appliquées* had not yet appeared, and so Kronecker was not in a position to comment on the simplicity and generality of Jordan's methods. Instead, he focused on what was manifest from Jordan's note: that he had simply found what he deemed simple canonical forms without justifying their theoretical significance by using them, as Weierstrass had done, to determine a complete set of invariants defined in terms of the original (nonsingular) family and characterizing equivalence, namely his elementary divisors (or equivalently, his W-series).

Kronecker's elaboration of this point constitutes what I will call his first disciplinary ideal [356, pp. 367–368]:

---

[28] Kronecker later called these transformations "Jacobi transformations," as indicated in Section 5.1 and below in Section 5.6.4.

> In fact, the expression "canonical form" or "simple canonical form" ... has no generally accepted meaning and in and of itself represents a concept devoid of objective content. No doubt someone who is faced with the question of the simultaneous transformation of two bilinear forms may, as an initial vague goal of his efforts, have in mind finding general and simple expressions to which both forms are to be simultaneously reduced. But a "problem" in the serious and rigorous meaning which justifiably attends the word in scientific discourse certainly may not refer to such a vague endeavor. In retrospect, after such general expressions have been found, the designation of them as canonical forms may at best be motivated by their generality and simplicity. But if one does not wish to remain with the purely formal viewpoint, which frequently comes to the fore in recent algebra—certainly not to the benefit of true knowledge—then one must not neglect the justification of the posited canonical form on the basis of intrinsic grounds. In reality, the so-called canonical or normal forms are usually determined merely by the tendency of the investigation and hence are only regarded as the means, not the goal of the investigation. In particular, this is always much in evidence when algebraic work is performed in the service of another mathematical discipline, from which it obtains its starting point and goal. But, of course, algebra itself can also supply sufficient inducement for positing canonical forms; and thus, e.g., in the two works by Mr. Weierstrass and myself cited by Mr. Jordan the motives leading to the introduction of certain normal forms are clearly and distinctly emphasized. In the case of Weierstrass, the "peculiar" simultaneous transformation of two bilinear forms, which is given in formula (44) on p. 319 of the oft-mentioned work, expressly serves to establish the agreement of the elementary divisors as a sufficient condition for the transformability of two form-pairs [into one another].

Thus the proper motive for introducing a normal form is theoretical, e.g., in Weierstrass' case to establish the difficult sufficiency part of his above-mentioned theorem on the equivalence of nonsingular families of bilinear forms, thereby establishing the elementary divisors as a complete set of invariants for the equivalence classes formed by nonsingular families. These invariants formed the intrinsic grounds or basis for equivalence. The canonical form was simply a means to this end. This was a criticism of Jordan because he had not shown how his canonical forms lead to a complete set of invariants for the equivalence classes corresponding to his three problems.

Jordan responded to Kronecker's criticisms, which included more than the above-mentioned one, in a note presented at the 2 March 1874 proceedings of the Paris Academy [325]. He agreed with Kronecker that a canonical form is not justified by simplicity alone, and then, focusing on the third problem, he claimed that he could prove that the identity of the canonical forms of two families $u\Phi + v\Psi$ and $u\tilde{\Phi} + v\tilde{\Psi}$ was both necessary and sufficient for their equivalence [325, p. 14]. In his reply in the 19 March 1874 proceedings of the Berlin Academy, Kronecker suggested that Jordan had not fully understood the above-quoted passage, and so he proceeded to elaborate on the disciplinary ideal it contained. Admitting that Jordan's necessary and sufficient condition for equivalence was correct, Kronecker went on to characterize it as insubstantial[29] because in articulating necessary and sufficient conditions for equivalence,

[29] "zu dürftigen Inhalts" [358, p. 382].

> it is not a question of a practical procedure for deciding the equivalence of given systems of forms; rather it is a question of a theoretical criterion for equivalence that is as closely linked as possible to the coefficients of the given forms, i.e., a question of positing a complete system of "invariants" in the higher sense of that word [358, pp. 382–383].

In other words, the invariants constituting the W-series and the K-series associated to a given family $uA + vB$ can be calculated directly from the coefficients of A and B, and if this calculation is performed for two given families, their equivalence or inequivalence is known a priori from a comparison of these invariants, without the need to reduce each family to a canonical form. Thus, as Kronecker continued, the reduction to a canonical form is certainly a necessary first step, but it is then necessary to use that form to determine the associated invariants, thereby proceeding from a purely formal conception of a canonical form to a loftier conception that leads to a complete system of invariants.

The disciplinary ideal articulated here by Kronecker made explicit what was implicit in Weierstrass' paper on elementary divisors, as well as in Kronecker's Theorem 5.13 on the equivalence of singular families of quadratic forms, which he stated in a hastily written appendix to his note of 19 January 1874.[30] It was certainly a disciplinary ideal of Berlin-style linear algebra. Frobenius, who was still in Berlin in 1874, was familiar with the above-quoted passages, and as we shall see, they formed the motivation for Frobenius' work on the problem of Pfaff, the subject of the next chapter.

5.6.3 The second disciplinary ideal

By the time of his note of 16 March 1874, Jordan's detailed memoir had appeared in the *Journal de mathématiques pures et appliquées* [327], and Kronecker had studied it with a critical eye toward the former's claim of new methods that are "extremely simple and admit no exceptions" [324, p. 7]. He pointed out that in his reduction procedure, Jordan had utilized linear transformations given by formulas with denominators that could vanish [358, p. 402], thereby throwing into question the generality of his methods. On a more comprehensive level, Kronecker questioned whether Jordan's simple methods could suffice without the more complicated considerations behind his own method of grouping variables, and in this connection he pointed to Section 12 of Jordan's paper [327], where the lack of a method of variable grouping, Kronecker correctly claimed, vitiated the claimed generality of Jordan's method.[31]

---

[30]The appendix arrived too late to be published in the 19 January proceedings and so appeared in the proceedings of 16 February 1874 [357]. See Section V [357, pp. 378–381]. A complete proof was of course not given, and it is unclear whether Kronecker had already worked out all the details. See in this connection Section 5.6.5 below.

[31]For Kronecker's criticisms, see [358, pp. 406–408]. At first, Jordan did not correctly understand the significance of the criticism and dismissed it, but in 1881, while working on his lectures at the

Jordan's emphasis on formal simplicity and his claims of simple methods apparently reminded Kronecker of the attitude underlying the generic reasoning that had dominated linear algebra prior to Weierstrass' paper of 1858 and prompted him to articulate a second disciplinary ideal of the Berlin school:

> It is common—especially in algebraic questions—to encounter essentially new difficulties when one breaks away from those cases that are customarily designated as general. As soon as one penetrates beneath the surface of the so-called generality, which excludes every particularity, into the true generality, which comprises all singularities, the real difficulties of the investigation are usually first encountered, but at the same time, also the wealth of new viewpoints and phenomena that are contained in its depths [358, p. 405].

That is the general disciplinary ideal, and its history as an achievement of the Berlin school then follows as Kronecker continued [358, pp. 405–406]:

> This has proved to be the case in the few algebraic questions that have been completely resolved in all their details, namely in the theory of families of quadratic forms, the main features of which have been developed above. As long as no one dared to drop the assumption that the determinant contains only unequal factors, the well-known question of the simultaneous transformation of two quadratic forms ... led only to extremely inadequate results, and the true viewpoint of the investigation went entirely unrecognized. By dropping that assumption, Weierstrass' work of 1858 already led to a deeper insight, namely to a complete resolution of the case in which only simple elementary divisors occur. The general introduction of the concept of elementary divisors, however, first occurred in Weierstrass' work of 1868, and an entirely new light was shed on the theory of arbitrary families of forms, albeit with nonvanishing determinants. When I did away with this last limitation and developed the more general concept of elementary families from that of elementary divisors, the wealth of newly arising algebraic forms was infused with the greatest clarity, and by virtue of this complete treatment of the subject, a most valuable insight into the theory of higher invariants, conceived in their true generality, was attained.

This disciplinary ideal, which goes hand in hand with the first, was also accepted by Frobenius and guided and informed his research on the problems in linear algebra discussed in the following chapters, especially his work on the Hermite–Cayley problem (Chapter 7).

5.6.4 *Bilinear families* $x^t(\mathbf{uA}+\mathbf{vA^t})y$ *revisited*

As we have seen (Section 5.3), in 1866, Kronecker investigated the congruence of bilinear forms of the special type $x^t(uA+vA^t)y$, $X^t(u\tilde{A}+v\tilde{A}^t)Y$, congruence meaning that the former family is transformed into the latter by $x = HX$, $y = HY$, i.e., that $H^t(uA+vA^t)H = u\tilde{A}+v\tilde{A}^t$. Not only was his Theorem 5.5 limited to distinct linear factors, the number of variables was assumed to be even and $\det A$ assumed

Collège de France, he realized the import and validity of Kronecker's criticism. In a note in the proceedings of the Paris Academy [329], he acknowledged his mistake and graciously attributed the first completely general reduction procedure to Kronecker.

nonzero, because that was the case in the application of the theorem to the problem about the complex multiplication of abelian functions that he was considering.

In his note of 22 December 1873 [324], Jordan had pointed out that Kronecker had not provided a general solution to this congruence problem in 1866. Given Jordan's criticism and Kronecker's two disciplinary ideals, it is not surprising that in a memoir presented to the Berlin Academy for the proceedings of 23 April 1874 [359], we find him providing a solution to the congruence problem commensurate with those ideals.

Kronecker's solution was along the following lines. Consider an arbitrary $n \times n$ family of the form $uA + vA^t$. Then its associated W-series $\{D_n, \ldots, D_1\}$ and K-series $\{m_1, \ldots, m_\mu\}$ are invariant under congruence transformations. Since D_i divides D_{i+1}, the quotients $E_i = D_{i+1}/D_i$ are homogeneous polynomials in u, v, and Kronecker used their properties to define a table of nonnegative integers denoted by (J) [359, p. 471], which are invariants under conjugate transformations $x = HX$, $y = HY$ and "completely replace" the W- and K-series [359, p. 470]. Kronecker's idea was to show that (J) provides a complete set of invariants for bilinear forms Φ with respect to "conjugate transformations," i.e., when x and y are subject to the same linear transformation. To this end, he showed how, by means of conjugate transformations, to reduce Φ to a direct sum (an "aggregate" in Kronecker's terminology) of "elementary" bilinear forms of four distinct types [359, p. 463, VII]. This was accomplished using what Kronecker called "Jacobi transformations" [359, p. 426] in honor of Jacobi, who had used them in [314] to establish the generic theorem discussed in Section 5.1. After developing the properties of such transformations in the first section, he used the results to obtain, in the second section, the above-mentioned decomposition of Φ into elementary forms. In the third section he then showed that two such decompositions that were not identical had different (J) tables. This meant that two bilinear forms $\Phi(x,y) = x^t Ay$ and $\tilde{\Phi}(X,Y) = X^t \tilde{A} Y$ are conjugate if and only if they have the same table (J), and the same is then true for the families corresponding to $uA + vA^t$ and $u\tilde{A} + v\tilde{A}^t$ [359, pp. 472, 476]. Kronecker had now solved his problem from 1866 in complete generality and so in accordance with the two disciplinary ideals that had been inspired by Weierstrass' papers of 1858 and 1868.

Kronecker also emphasized another immediate implication of his theorem that related to one of Weierstrass' results. Recall from Section 5.4 that in order to specialize his theory of elementary divisors to families of quadratic forms $x^t(uA + vB)x$, $A^t = A$, $B^t = B$, Weierstrass had to show that two such families are equivalent in the sense of $H^t(uA + vB)K = u\tilde{A} + v\tilde{B}$ for nonsingular H, K if and only if they are congruent, i.e., if and only if $P^t(uA + vB)P = u\tilde{A} + v\tilde{B}$ for some nonsingular P. The nontrivial part of the theorem is of course to show that equivalence implies congruence. Kronecker observed that his theorem implied an analogous result for families of bilinear forms of the special type $uA + vA^t$. For if $uA + vA^t$ and $u\tilde{A} + v\tilde{A}^t$ are equivalent in the sense of Weierstrass, then they will have the same W- and K-series and so the same table (J). By Kronecker's above theorem, this means that $uA + vA^t$ and $u\tilde{A} + v\tilde{A}^t$ must be congruent. For future reference, Kronecker's observation will be stated as a theorem.

Theorem 5.14. *Two families of bilinear forms of the special type* $x^t(uA+vA^t)y$, $X^t(u\tilde{A}+v\tilde{A}^t)Y$ *are equivalent, i.e.,*

$$H^t(uA+vA^t)K = u\tilde{A}+v\tilde{A}^t, \quad H,K \text{ nonsingular},$$

if and only if they are congruent, i.e.,

$$P^t(uA+vA^t)P = u\tilde{A}+v\tilde{A}^t, \quad P \text{ nonsingular}.$$

Although the above theorem was an immediate consequence of Kronecker's proof that the table (J) provides a complete set of invariants for $uA+vA^t$ under congruent transformations, that proof was itself highly nontrivial, involving as it did most of the 54 pages of his paper. The analogous above-mentioned theorem of Weierstrass also involved a lengthy proof, which like Kronecker's made heavy use of determinants and which—at about this time (1874)—was discovered to contain a gap (Section 16.1.1). I mention this because Frobenius discovered that by means of his symbolic algebra of matrices (Section 7.5), a byproduct of a research program motivated by Kronecker's second disciplinary ideal, both theorems could be deduced immediately from the lemma that every nonsingular matrix has a square root (Sections 16.1.2–16.1.3), thereby bypassing the lengthy determinant-based considerations that had been used by his mentors.

5.6.5 Generalization of Weierstrass' theory

Probably due to Kronecker's preoccupation with the more challenging case of singular families of quadratic forms and his polemic with Jordan, it was not until November 1874 that he turned to the task of utilizing his reductive procedures and related results as outlined in the academy's proceedings for January–April to sketch out a bona fide generalization of Weierstrass' theory of elementary divisors, i.e., a theory that would specify a complete set of invariants and concomitant canonical form characterizing the equivalence classes of all families of bilinear forms $x(uA+vB)y$, singular as well as nonsingular, under nonsingular transformations of the form $x=HX$ and $y=KY$. The manuscript that Kronecker produced in 1874, however, was held back from publication until 1890, because, as he explained then [367, p. 140], while composing it he decided that the "analytic-algebraic" techniques he was employing were unsatisfactory. He decided to publish the manuscript in 1890 because in the meantime he had succeeded in developing an arithmetic approach to the theory, which he expected to present soon to the academy, and he wanted to provide the basis for a comparison of the two approaches.[32] By way of concluding

[32]Kronecker died in December 1891, without, apparently, ever having published his promised arithmetic theory. During December–January 1890–1891, he did publish a version of his algebraic

the discussion of Berlin linear algebra in the period 1858–1874, I will briefly describe Kronecker's generalization, since it also epitomizes the disciplinary ideals of the Berlin school that he had articulated.

The obvious candidate for a complete set of invariants are the W- and K-series that Kronecker had introduced in dealing with families of quadratic forms. The definition of the W-series $\{D_n(u,v),\ldots,D_1(u,v)\}$ remains the same as in the quadratic case. The K-series $\{m_1,\ldots,m_\mu\}$ as defined for quadratic forms (Definition 5.12) also provides a set of invariants within the context of bilinear families, but now a further set of invariants is needed. Recall that the K-series $\{m_1,\ldots,m_\mu\}$ is defined in terms of the solutions to $\mathbf{v}^t(uA+vB)=\mathbf{0}$, and reflects the linear dependency relations among the *rows* of $uA+vB$ over $\mathbb{C}[u,v]$. For families of quadratic forms, the linear dependency relations among the columns of $uA+vB$, which are given by solutions to $(uA+vB)w=0$, are exactly the same due to symmetry. For families of bilinear forms, this is not the case, and a second K-series $\{\overline{m}_1,\ldots,\overline{m}_\mu\}$ must be defined in the same manner as the first but with respect to the solutions to $(uA+vB)w=0$ [367, p. 150]. I will refer to $\{m_1,\ldots,m_\mu\}$ as the *row K-series* and to $\{\overline{m}_1,\ldots,\overline{m}_\mu\}$ as the *column K-series*. Kronecker showed that the W-series, together with the row and column K-series, do in fact constitute a complete set of invariants, i.e., that any two families of bilinear forms are equivalent in the sense of Weierstrass if and only if the have the same W-series and the same row and column K-series.[33] He had indeed completed what Weierstrass had begun with his theory of elementary divisors, thereby further vindicating the disciplinary ideals Kronecker had articulated.[34]

---

theory as it applies to quadratic forms [368, 369], again for the purpose of comparison with his forthcoming arithmetic theory.

[33] Although Kronecker's reasoning implied the above-stated result, he focused instead on a table of integer invariants denoted by $(\mathfrak{J})$ [367, p. 151] derivable from the above W- and K-series and analogous to the table (J) he had introduced in his memoir on the special type of bilinear family $uA+vA^t$ discussed in Section 5.6.4.

[34] Kronecker also generalized Weierstrass' theory in another direction. He developed the entire theory outlined above for forms involving r x-variables and s y-variables, and so the matrices involved are $r\times s$. When the matrices are not square the lengths of the row and column K-series are generally different. Kronecker's theory was elaborated by Muth in 1899 [450, pp. 93–133]. In the twentieth century, various approaches and refinements were introduced by Dickson [128], Turnbull [565], and Ledermann [405]. In his comprehensive treatise on the theory of matrices, Gantmacher devoted a chapter to Kronecker's theory [240, V. 2, Ch. XII].

Part III
The Mathematics of Frobenius

Chapter 6
The Problem of Pfaff

Having now discussed at length the nature and development of linear algebra at Berlin during Frobenius' years there, we next turn to Frobenius' first major paper in which Berlin linear algebra and its concomitant disciplinary ideals played a role.[1] This was also his first paper from Zurich that reflects a break with his work while in Berlin, where it was focused on the theory of ordinary differential equations. The new direction involved what was called the problem of Pfaff. The problem was at the interface of analysis (total differential equations) and algebra and, as perceived by Frobenius, was analogous to the problem of the transformation of quadratic and bilinear forms as treated by Weierstrass and Kronecker. As we shall see, Pfaff's problem had been around for many years, but work by Clebsch and Natani in the 1860s had revived interest in it. Frobenius' work was clearly motivated by Clebsch's treatment of the problem and the issues it suggested vis à vis the disciplinary ideals of the Berlin school. Frobenius' paper on the problem of Pfaff, which was submitted to *Crelle's Journal* in September 1876, marked him as a mathematician of far-ranging ability. His analytic classification theorem (Theorem 6.7) and integrability theorem (Theorem 6.11) have become basic results, and his overall approach by means of the bilinear covariant was to have a great influence on Élie Cartan (Section 6.6), especially as regards his exterior calculus of differential forms and its applications.

6.1 Mathematical Preliminaries and Caveats

Before entering into the diverse mathematical treatments of the problem of Pfaff, it will be helpful to make some general comments on the nature of the mathematical reasoning in this period. The theory involves functions $f = f(x_1,\ldots,x_n)$ of any

[1]This chapter draws extensively on my paper [277].

T. Hawkins, *The Mathematics of Frobenius in Context*, Sources and Studies in the History of Mathematics and Physical Sciences, DOI 10.1007/978-1-4614-6333-7_6,

number of variables whose properties are not explicitly specified. It is not even clear whether the variables are assumed to be real or are allowed to be complex, although it does seem that the latter possibility is the operative one, since occasionally telltale expressions such as $\log(-1)$ occur.[2] In general, mathematicians in this period tended to regard variables as complex rather than real [255, p. 29]. It is taken for granted that partial derivatives of these functions exist, and frequent use is made of the equality of mixed partial derivatives, e.g., $\partial^2 f/\partial x_i \partial x_j = \partial^2 f/\partial x_j \partial x_i$. Also, the inverse function theorem and the implicit function theorem are applied whenever needed. In the case of Frobenius, who had been trained in a school emphasizing mathematical rigor, he never applied these theorems without first showing that the requisite Jacobian determinant does not vanish, but even Frobenius never expressly points out that these theorems are local in nature. It seems likely to me that Frobenius regarded the functions under consideration as complex-analytic functions of complex variables $x_1, \ldots, x_n$, but continued the tradition of not being explicit about such assumptions.[3] Whether he was fully aware of the local nature of his results is far less certain, but the reader should understand them as local results valid in a neighborhood of any point satisfying the specified conditions. It was not until the twentieth century that the distinction between local and global results began to be taken seriously by mathematicians.[4]

In the previous chapters we have seen that the Berlin school stressed the importance of going beyond the generic case in dealing with algebraic matters. As applied, e.g., to a matrix $A = (a_{ij})$, this meant not thinking of the coefficients a_{ij} as symbols or variables, so that ("in general") A has full rank. As a student of that school, Frobenius was careful to base his reasoning on the rank of A, which is not presumed to be maximal. In the problem of Pfaff, however, matrices arise whose coefficients are functions of $x = (x_1, \ldots, x_n)$, so that $A = A(x)$ is likewise a function of these variables. Nonetheless, Frobenius spoke of *the* rank of $A(x)$ without any clarification, whereas (as he certainly realized) it can vary with x in the type of matrices that occur in the theory. For example, if $x = (x_1, x_2)$ and

$$A(x) = \begin{pmatrix} 0 & x_1 - 1 \\ -(x_1 - 1) & 0 \end{pmatrix},$$

then A has rank 2 for points $x = (x_1, x_2)$ with $x_1 \neq 1$ but rank 0 at points with $x_1 = 1$. By *the* rank of these matrices Frobenius meant their maximal rank, so that in the above example the rank of $A(x)$ is 2.

For every matrix $A(x)$, the points at which it has maximal rank r are the points of $\mathbb{C}^n$ that do not lie on the manifold of points x such that all degree-r minors

---

[2] For example, in a key paper by Clebsch [96, pp. 210–212] discussed in Section 6.3.

[3] This assumption is explicitly made in the 1900 treatise on Pfaff's problem by von Weber [575, Ch. 2, §1], although the local nature of the results is glossed over.

[4] How this occurred in the theory of Lie groups is described in [275]. See also "local vs. global viewpoint" in the index of [276].

of $A(x)$ vanish. Assuming, e.g., that all functions are complex-valued and analytic, which appears to be Frobenius' tacit assumption, the points of maximal rank form an open, dense subset G of $\mathbb{C}^n$. It is the points in G that Frobenius was tacitly considering. For purposes of reference I will follow Cartan's book of 1945 [66, p. 45] and refer to them as the *generic points* of $\mathbb{C}^n$.

In what follows, I will present the deliberations of the various mathematicians involved more or less as they did, and so these preliminary remarks should be kept in mind. In the case of Frobenius' main analytic theorems, namely Theorems 6.7 and 6.11, besides stating them as he did, I have given my interpretation of them as the local theorems that his reasoning implies. As proofs of local theorems, Frobenius' reasoning is rigorous in the sense that the necessary details for a proof by present-day standards can be filled in. I do not think this is a coincidence. Based on my study of a variety of Frobenius' papers on diverse subjects, I would venture to say that he was very careful to present his mathematics in a clear and precise manner. I believe that he himself could have filled in the omitted details in his proofs. However, since this was not the custom in the theory of partial differential and Pfaffian equations at the time, he omitted them, content to focus on the algebraic aspects of the problem, which were his primary interest.

6.2 The Problem of Pfaff

The problem that became known as Pfaff's problem had its origins in the theory of first-order partial differential equations, which as a general theory began with Lagrange.[5] Although other eighteenth-century mathematicians such as Euler had studied various special types of first-order partial differential equations, Lagrange was primarily responsible for initiating the general theory of such equations, which I will express with the notation

$$F(x_1,\ldots,x_m,z,p_1,\ldots,p_m)=0, \quad p_i=\frac{\partial z}{\partial x_i}, \quad i=1,\ldots,m. \tag{6.1}$$

Here z is singled out as the dependent variable, and the goal is to obtain a general (or complete) solution $z=\varphi(x_1,\ldots,x_m,C_1,\ldots,C_m)$, where the C_i are arbitrary constants. Here by "obtain a solution" I mean to show how such a solution can be obtained by means of solutions to one or more systems of ordinary differential equations. This is what it meant to integrate a partial differential equation throughout the eighteenth century and up to the time period of Frobenius. As Lagrange put it, "the art of the calculus of partial derivatives is known to consist in nothing more than reducing this calculus to that of ordinary derivatives, and a partial differential

[5] The following introductory remarks are based on the more extensive discussion in Section 2 of my paper [273].

equation is regarded as integrated when its integral depends on nothing more than that of one or more ordinary differential equations" [390, p. 625].

For linear first-order equations, Lagrange showed how to do this for any number m of variables. He was, however, less successful in dealing with nonlinear equations and was able to integrate any first-order partial differential equation (6.1) only when $m = 2$, i.e., in the case of two independent variables. The integration of nonlinear equations with $m > 2$ was first achieved in 1815 by Johann Friedrich Pfaff (1765–1825), a professor of mathematics at the University of Halle. Pfaff's bold and brilliant idea was to consider the more general problem of "integrating," in a sense to be discussed, a total differential equation

$$\omega = a_1(x)dx_1 + \cdots + a_n(x)dx_n = 0 \tag{6.2}$$

in any number of variables $x = (x_1, \ldots, x_n)$ [471]. The reason he sought to deal with the integration of total differential equations (in the sense explained below) was that he had discovered that the integration of any first-order partial differential equation in m variables can be reduced to the integration of a total differential equation in $n = 2m$ variables.[6] Thus, by solving the more general problem of integrating (6.2), he obtained as a special case what had eluded Lagrange and won thereby the praise of Gauss, who described Pfaff's result as "a beautiful extension of the integral calculus" [245, p. 1026].

At the time Pfaff wrote his memoir, there was no consensus on what it meant to integrate (6.2) even for $n = 3$. Pfaff observed [471, p. 6] that Euler had expressed the view that it makes sense to speak of the integration of $\omega = 0$ only when $M\omega$ is exact for some nonvanishing factor $M = M(x_1, x_2, x_3)$. This means that if $M\omega = d\Phi$, or equivalently that $\omega = Nd\Phi$ with $N = 1/M$, then the equation $\Phi(x_1, x_2, x_3) = C$ represents an integral of $\omega = 0$ in the sense that for all x_1, x_2, x_3 satisfying $\Phi(x_1, x_2, x_3) = C$, one has $d\Phi = 0$, and so $\omega = Nd\Phi = 0$ for these x_1, x_2, x_3. In geometric terms, the integral $\Phi(x_1, x_2, x_3) = C$ defines a surface with the property that all the vectors $dx = (dx_1, dx_2, dx_3)$ in the tangent plane to the surface at a point $P = (x_1, x_2, x_3)$ lying on it are perpendicular to the vector (a_1, a_2, a_3) evaluated at P, i.e., $a_1dx_1 + a_2dx_2 + a_3dx_3 = 0$.

Pfaff pointed out that Monge had disagreed with Euler and stated that two simultaneous equations $\Phi = C_1$, $\Psi = C_2$ could also be regarded as an integral of $\omega = 0$. That is, viewed geometrically, the simultaneous equations $\Phi = C_1$ and $\Psi = C_2$ define a curve as the intersection of two surfaces. The equation (6.2) stipulates that the tangent space at each point x of a solution curve should consist of vectors $dx = (dx_1, \ldots, dx_3)$ orthogonal to $a = (a_1, \ldots, a_3)$. If this is true of the tangent to the above curve, then from Monge's viewpoint, $\Phi = C_1, \Phi = C_2$ would constitute a solution to (6.2).

Monge's viewpoint, which Pfaff accepted, can be stated for any number of variables $x_1, \ldots, x_n$ as follows. A system of $k < n$ simultaneous equations

[6] A description of this reduction is given in [273, §2].

$$\Phi_i(x_1,\ldots,x_n) = C_i, \quad i = 1,\ldots,k, \tag{6.3}$$

is an integral of $\omega = a_1dx_1 + \cdots + a_ndx_n = 0$ if (1) the Φ_i are functionally independent in the sense that the $k \times n$ Jacobian matrix

$$\partial(\Phi_1,\ldots,\Phi_k)/\partial(x_1,\ldots,x_n)$$

has full rank k; (2) for the points satisfying the constraints imposed by (6.3), it follows that $\omega = 0$. That is, if (by the implicit function theorem) we express (6.3) in the form $x_i = \Psi_i(t_1,\ldots,t_d)$, $i = 1,\ldots n$, where $d = n - k$ and $t_1,\ldots,t_d$ denotes d of the variables $x_1,\ldots,x_n$, then setting $x_i = \Psi_i(t_1,\ldots,t_d)$ and $dx_i = \sum_{j=1}^{d}(\partial\Psi_i/\partial x_j)dt_j$ in the expression for ω makes $\omega = 0$. Viewed geometrically, the solution (6.3) represents a manifold of dimension $d = n - k$ with the property that all vectors $dx = (dx_1,\ldots,dx_n)$ in its tangent space at a point x are orthogonal to $a = (a_1(x),\ldots,a_n(x))$. Of course, as indicated in Section 6.1, all this needs to be understood locally.

In general, it is not at all clear how to determine an integral in this sense for a given equation $\omega = 0$. Pfaff's idea was that an integral of $\omega = 0$ is immediate when ω has a simple form. Consider, for example, $n = 3$ and

$$\omega = dx_1 + x_2dx_3 = 0. \tag{6.4}$$

Then it turns out that $\omega = 0$ does not have an integral in Euler's sense, i.e., $\omega \neq Nd\Phi$.[7] On the other hand, the form of ω is so simple it is clear that the simultaneous equations $x_1 = C_1$, $x_3 = C_3$ define a (straight line) curve on which $dx_1 = dx_3 = 0$, and so $\omega = 0$ there. In other words, $x_1 = C_1$, $x_3 = C_3$ represents an integral of (6.4).

Pfaff therefore posed to himself the problem of showing that a suitable variable change $x_i = f_i(y_1,\ldots,y_n)$ always exists such that when ω is expressed in the variables $y_1,\ldots,y_n$, it has a simple form that makes determining an integral of $\omega = 0$ an easy matter. Pfaff's solution to this problem was the subject of his memoir of 1815. In solving his problem, Pfaff reasoned generically in the sense explained in Section 4.1 and illustrated throughout Chapter 4. The generic theorem implicit in Pfaff's memoir may be stated as follows using the sort of index notation introduced later by Jacobi.

Theorem 6.1 (Pfaff's theorem). *There exists "in general" a change of variables $x_i = f_i(y_1,\ldots,y_n)$, $i = 1,\ldots,n$, such that*

$$\omega = a_1(x)dx_1 + \cdots + a_n(x)dx_n$$

[7]The necessary condition for $\omega = Nd\Phi$ was given in an elegant form by Jacobi and is displayed below in (6.6). The Pfaffian equation (6.4) fails to satisfy this condition.

transforms into an expression involving m differentials,

$$\omega = b_1(y)dy_1 + \cdots + b_m(y)dy_m, \tag{6.5}$$

where $m = n/2$ *if* n *is even and* $m = (n+1)/2$ *if* n *is odd.*

Here it is tacitly understood that a bona fide variable change has a nonvanishing Jacobian, since the inversion of the variable change is necessary to produce an integral of $\omega = 0$. That is, if $y_i = g_i(x_1, \ldots, x_n)$, $i = 1, \ldots, n$, denotes the inverse of the variable transformation posited in Pfaff's theorem, and if $C_1, \ldots, C_m$ are constants, then the m simultaneous equations $g_i(x_1, \ldots, x_n) = C_i$, $i = 1, \ldots, m$, represent an integral of $\omega = 0$ because these equations state that $y_i = C_i$ for $i = 1, \ldots, m$, and so imply that $dy_i = 0$ for $i = 1, \ldots, m$, and so by (6.5) that $\omega = 0$ for the values of $(x_1, \ldots, x_n)$ satisfying these equations. This solution can be thought of geometrically as the integral manifold formed by the intersection of the m hypersurfaces $g_i(x_1, \ldots, x_n) = C_i$ and hence "in general" of dimension $d = n - m$. In the case of ordinary space ($n = 3$), $m = 2$ and so $d = 1$, i.e., the solution to the generic equation $\omega = 0$ in this case is a curve—the sort of solution envisioned by Monge.

Pfaff pointed out [471, p. 7] that exceptional cases exist for which the number m of terms in (6.5) could be less than $n/2$ or $(n+1)/2$, respectively. Indeed, as explained below, when $n = 3$ in (6.2) the case envisioned by Euler as the sole meaningful one, namely $\omega = Nd\Phi$, means that a variable change, namely $y_1 = \Phi(x_1, x_2, x_3)$, $y_2 = x_2$, $y_3 = x_3$, exists such that $\omega = N(y_1, y_2, y_3)dy_1$ and thus $m = 1 < (3+1)/2$, whereas Monge had argued for the legitimacy of the case $m = 2 = (3+1)/2$, i.e., what turns out in Pfaff's theorem to be the "general" case. Although Pfaff recognized exceptions to his theorem, he restricted his attention to the generic case stated therein. He worked out detailed, successive proofs for the generic cases of $n = 4, \ldots, 10$ variables in (6.2) and then stated the general generic theorem along with a brief proof sketch reflecting the approach detailed in the worked-out cases $n = 4, \ldots, 10$ [471, §16].

In an important and influential paper of 1827 [309, p. 28], Jacobi expressed the necessary condition that Euler's relation $\omega = Nd\Phi$ hold in the following elegant form:

$$a_1 a_{23} + a_2 a_{31} + a_3 a_{12} = 0, \quad \text{where} \quad a_{ij} = \frac{\partial a_i}{\partial x_j} - \frac{\partial a_j}{\partial x_i}. \tag{6.6}$$

The above expressions a_{ij} were introduced by Jacobi with the notation (i, j) for a_{ij} and were defined for Pfaffian equations $\omega = 0$ in any number n of variables. Perhaps consideration of the a_{ij} was motivated initially by the fact that when ω is exact, all $a_{ij} = 0$ necessarily. Then, more generally, (6.6) gives the necessary condition for Euler's relation to hold. Thus in general, the a_{ij}, along with the a_i, seem to contain the information needed to decide about the nature of the integrals of $\omega = 0$.

The main object of Jacobi's paper [309] was not (6.6), however, but a general proof, using the elegant n-variable notation he had introduced into analysis, of Pfaff's theorem in the case of $n = 2m$ variables—the case that was relevant to its application to partial differential equations. In Pfaff's method of proof, the reduction from $n = 2m$ to $n-1$ differential terms was attained by a variable change provided by the complete solution to a system of ordinary differential equations, and in Jacobi's rendition he introduced the coefficient system

$$A = (a_{ij}), \quad a_{ij} = \frac{\partial a_i}{\partial x_j} - \frac{\partial a_j}{\partial x_i}, \tag{6.7}$$

to write down and manipulate the system of differential equations. Since $a_{ji} = -a_{ij}$, the $n \times n$ matrix A is what is now called skew-symmetric. In what is to follow I will refer to it as *Jacobi's skew-symmetric matrix.*

With the skew-symmetry of A evidently in mind, Jacobi remarked that the system (6.7) shows "great analogy" with the symmetric linear systems that had arisen in diverse analytic applications [309, p. 28],[8] and at first he apparently believed that skew-symmetric systems were new in applications of analysis. Although that turned out to be not quite correct,[9] Jacobi's influential paper certainly served to bring skew-symmetric matrices to the attention of mathematicians, Frobenius included, for Jacobi showed that these matrices have many interesting algebraic properties. Implicit in his formulas for the solution of $Ax = b$, $A^t = -A$, A being $2m \times 2m$ [309, pp. 25–27], is the fact that $\det A$ is the square of a homogeneous degree-m polynomial in the coefficients of A, a polynomial now called a Pfaffian.[10] In fact, Jacobi stressed that it was the introduction of the skew symmetric system (6.7) and its application to Pfaff's procedure that constituted the original contribution of his paper. Since he was proceeding on the generic level of Pfaff's Theorem 6.1, he took it for granted that $\det A \neq 0$, and he realized that this was permissible because n was assumed even and that when n is odd, the skew-symmetry of A forces $\det A$ to vanish identically, i.e., no matter what values are given to the a_{ij}.[11] He did briefly discuss what happens when n is odd, so that $\det A = 0$, and again it was on the generic level. Expressed in more modern terms, Jacobi's tacit generic assumption was that A has maximal rank $n-1$ when n is odd [309, p. 28].

---

[8]Although Jacobi implied that there were many such applications, the only one he explicitly mentioned was the method of least squares, presumably because of the symmetric normal equations $M^tMx = M^tb$ used to obtain a least squares solution to the linear system $Mx = b$. A better-known example would have been the symmetric systems in Lagrange's analysis of a finite mechanical system near a point of equilibrium as discussed above in Section 4.2.1.

[9]At the conclusion of his paper, Jacobi made a point of noting that after he had written it, he discovered that Lagrange and Poisson had already introduced skew-symmetric systems in their work on the calculus of variations.

[10]This result seems to have been first explicitly stated and proved by Cayley (who knew Jacobi's paper) in 1849 [81]. It became a standard theorem in the early treatises on determinants by Brioschi [40, p. 57] and Baltzer [12, p. 29].

[11]As Jacobi pointed out, since $A^t = -A$, $\det A = \det A^t = \det(-A) = (-1)^n \det A$ [309, p. 26]. Thus when n is odd, $\det A = -\det A$, and $\det A = 0$ follows.

Pfaff's method of integrating (6.1), namely by constructing the variable change leading to the normal form (6.5), reduced the integration of $\omega = 0$ to the integration of systems of ordinary differential equations, which was also the generally accepted goal in the theory of partial differential equations, as he noted. The variable change of Theorem 6.1 was obtained by a succession of $n-m$ variable changes, the kth of which transformed ω from an expression involving $n-k$ differentials into one involving $n-k-1$ differentials, where $k = 0,\ldots,n-(m+1)$. Each such variable change required completely integrating a system of first-order ordinary differential equations.

For $n \gg 2$, Pfaff's method required the complete solution of a large number of systems of ordinary differential equations, but it was not until after Cauchy and Jacobi had discovered direct methods for solving nonlinear first-order partial differential equations (6.1) in any number of variables that the inefficiency of Pfaff's method was made evident by the new methods. After Jacobi published his method in 1837, the goal in the theory of first-order partial differential equations became to devise methods for integrating them that were as efficient as possible, i.e., that reduced the number of systems of ordinary differential equations that needed to be considered, as well as their size and the number of solutions to such a system that is required by the method. In this connection, Jacobi devised a new method that was published posthumously in 1862 and which gave impetus to the quest for more efficient methods in the late 1860s and early 1870s.[12]

Although the original motivation for Pfaff's theory—the integration of first-order partial differential equations—had lost its special significance due to the later direct methods of Cauchy and Jacobi, during the 1860s and 1870s the interest in partial differential equations extended to Pfaffian equations as well, since the papers by Pfaff and Jacobi raised several questions. First of all, there was the matter of the admitted exceptions to Pfaff's generic theorem. Given a specific Pfaffian equation $\omega = 0$, how can one tell whether Pfaff's theorem applies, and if it does not, what can one say of the integrals of $\omega = 0$? For example, what is the minimal number k of equations $f_i(x_1,\ldots,x_n) = C_i$, $i = 1,\ldots,k$, that are needed to specify an integral? If we express this question in geometric terms—something rarely done in print in this period—the question becomes, what is the maximal dimension $d = n-k$ of an integral manifold for $\omega = 0$? Secondly, there was the problem of determining more efficient ways to integrate a Pfaffian equation, in the generic case and also the nongeneric ones. These two questions constituted what was known as the "problem of Pfaff" in the period leading up to the work of Frobenius.

Nowadays, the problem of Pfaff is characterized more narrowly as the problem of determining for a given Pfaffian equation $\omega = 0$ in n variables the maximal dimension d of its integral manifolds [307, p. 1623] or equivalently, since $d = n-k$, the minimal value possible for k. Thus in the modern conception of the problem, the second "efficiency" question is ignored, although in the nineteenth century, most mathematicians interested in partial differential and Pfaffian equations deemed it of paramount importance. As we shall see, Frobenius, more of an algebraist at

[12] On Jacobi's two methods and their influence see [273, §§2–3] or (with less detail) [276, §2.1].

heart, ignored it and was criticized for doing so (Section 6.5). Instead, he focused on the first problem and, utilizing the work of his predecessors—above all that of Clebsch—and spurred on by the disciplinary ideals of the Berlin school, he definitively and rigorously solved it.[13]

As we shall see in more detail in what follows, Frobenius showed that the integration of $\omega = 0$ depends on an integer invariant p associated to ω and determined in a simple manner from the coefficient functions $a_i(x)$ defining ω. By way of a preview of Frobenius' results, let us consider the case $n = 5$, so $\omega = a_1(x)dx_1 + \cdots + a_5(x)dx_5$. Pfaff's Theorem 6.1 gives the generic value of $k = m = 3$ differential terms in (6.5) and so $m = 3$ equations defining an integral of $\omega = 0$ and thus an integral manifold of dimension $d = 5 - 3 = 2$. By contrast, Frobenius showed that a variable change $x_i = f_i(y_1, \ldots, y_5)$ exists[14] such that in the z-variables one has, depending on the value of p,

$$p = 1 \Rightarrow \omega = dz_5, \qquad p = 4 \Rightarrow \omega = z_3\,dz_1 + z_4\,dz_2,$$
$$p = 2 \Rightarrow \omega = z_2\,dz_1, \qquad p = 5 \Rightarrow \omega = dz_5 + z_3\,dz_1 + z_4\,dz_2.$$
$$p = 3 \Rightarrow \omega = dz_5 + z_3\,dz_2,$$

Let $z_i = g_i(x_1, \ldots, x_5)$ denote the (local) inverse of $x_i = f_i(y_1, \ldots, y_5)$. Then in the case $p = 1$, $\omega = dg_5$ is exact, and the integral manifold is given by $z_5 = g_5(x_1, \ldots, x_5) = C$ and has dimension $d = 5 - 1 = 4$. This case and the case $p = 2$ cover the Pfaffian equations integrable in the sense of Euler, since for $p = 2$, $\omega = N(x)d\Phi$ with $N(x) = z_2 = g_2(x)$ and $\Phi = z_1 = g_1(x)$. Here also $k = 1$ and $d = 4$. When $p = 3$, then $k = 2$, and an integral manifold is defined by the simultaneous equations $z_2 = g_2(x) = C_1, z_5 = g_5(x) = C_2$ and so has dimension $d = 3$. Likewise, when $p = 4$, we have $k = 2$, and the integral manifold is $z_1 = g_1(x) = C_1$, $z_2 = g_2(x) = C_2$, and has dimension $d = 3$. Finally, when $p = 5$, $k = m = 3$, and we are in the generic case of Pfaff's theorem with integral manifold defined by three equations and so of dimension $d = 5 - 3 = 2$. Thus Pfaff's theorem covers just one of the five possibilities distinguished by Frobenius' results, as expressed below in Theorem 6.7, specialized to the case of $n = 5$ variables.

6.3 The Contributions of Clebsch

After Jacobi, significant advances on Pfaff's problem were not made until the early 1860s, when two mathematicians, Clebsch and Natani, independently and almost simultaneously took up the matter.[15] Natani's paper appeared first, but it was the

---

[13]Rigorously, in the sense indicated in Section 6.1, which for the nineteenth century was indeed exceptionally rigorous.

[14]Strictly speaking, such a transformation exists in a neighborhood of any generic point, as explained in the discussion of Frobenius' Theorem 6.7.

[15]What follows does not represent a full account of work on the problem of Pfaff up to that of Frobenius. For more details see [165,575].

work of Clebsch that made the greatest impression on Frobenius and so will occupy our attention here. Some aspects of Natani's work will be discussed briefly in Section 6.4.3.

Alfred Clebsch (1833–1873) had obtained his doctorate in mathematics in 1854 from Königsberg in the post-Jacobi era. His teachers were Franz Neumann, Friedrich Richelot, and Otto Hesse (who had been Jacobi's student). At Königsberg, Clebsch received a broad and thorough training in mathematical physics, which included on his part a detailed study of the publications of Euler and Jacobi. He was known personally to the mathematicians in Berlin, where Jacobi had ended his career, since during 1854–1858, he had taught in various high schools there as well as (briefly) at the university. Apparently, Borchardt, as editor of *Crelle's Journal*, asked him to edit the manuscript by Jacobi that contained his new method of integrating partial differential equations so that it might be published in the journal, where it did in fact appear in 1862 [315].[16]

The study of Jacobi's new method led Clebsch to ponder the possibility of its extension to the integration of a Pfaffian equation

$$\omega = a_1(x)dx_1 + \cdots + a_n(x)dx_n = 0. \tag{6.8}$$

Such an extension would greatly increase the efficiency of integrating a Pfaffian equation in the sense explained above. Clebsch satisfied himself that he could do this and in fact do it in complete generality—not just for the case of an even number of variables with a nonsingular Jacobian matrix $A = (a_{ij})$, $a_{ij} = \partial a_i/\partial x_j - \partial a_j/\partial x_i$. Thus he wrote that, "The extension of this method [of Jacobi] to the problem of Pfaff in complete generality and in all possible cases is the subject of the following work" [96, p. 193].

Clebsch's entire treatment of the problem was based on a distinction he did not adequately justify. Let m denote the minimum number of differential terms into which (at a generic point in the sense of Section 6.1) ω can be transformed by a variable change $x = g(y)$, so that ω takes the form $\omega = b_1(y)dy_1 + \cdots + b_m(y)dy_m$. If $y_i = f_i(x)$, $i = 1, \ldots, n$, denotes the inverse variable change, then, following the notation of Clebsch, let $F_i(x) = b_i(f_1(x), \ldots, f_n(x))$. Then since $dy_i = df_i$, ω is expressible as

$$\omega = F_1 df_1 + \cdots + F_m df_m, \tag{6.9}$$

which was Clebsch's notation. Now consider the $2m \times n$ Jacobian

$$\partial(f_1, \ldots, f_m, F_1, \ldots, F_m)/\partial(x_1, \ldots, x_n).$$

[16]For further details about Clebsch's career, see the anonymous memorial essay [3] and [563, pp. 7ff.].

Then (to use modern terms) either (I) this Jacobian matrix has full rank or (II) it does not have full rank. In case (I), the $2m$ functions $f_1,\ldots,F_m$ are by definition functionally independent, and Clebsch referred to case (I) as the "determinate" case. In the "indeterminate" case (II), they are not independent. Clebsch went on to claim without adequate justification that it is always possible in case (II) to take $F_1(x) \equiv 1$ with the remaining functions independent [96, pp. 217–220]. For convenience of reference, I will refer to the proposition implicit in Clebsch's remarks as Clebsch's theorem. With a change of notation in case (II) it may be stated in the following form.

Theorem 6.2 (Clebsch's theorem). *Let m denote the minimal number of differential terms into which ω can be transformed. Then either (I) $2m$ independent functions $f_1,\ldots,f_m,F_1,\ldots,F_m$ exist such that*

$$\omega = F_1(x)df_1 + \cdots + F_m(x)df_m; \tag{6.10}$$

or (II) $2m-1$ independent functions $f_0,f_1,\ldots,f_{m-1},F_1,\ldots,F_{m-1}$ exist such that

$$\omega = df_0 + F_1(x)df_1 + \cdots + F_{m-1}(x)df_{m-1}. \tag{6.11}$$

Clebsch stressed that (I) and (II) represented two general and quite distinct classes into which Pfaffians ω are divided, and he was apparently the first to emphasize this distinction [96, p. 217].

Clebsch pointed out that by starting from the assumption of Theorem 6.2, "one is spared the trouble of carrying out direct proofs that lead to very complicated algebraic considerations, which, to be sure, are of interest in their own right ..." [96, p. 194]. This led him later to refer to this approach as his *indirect method.* Theorem 6.2 is equivalent to asserting that a change of variables $x_i = \sigma_i(z_1,\ldots,z_n)$, $i = 1,\ldots,n$, is possible that in the z-variables,

$$\begin{aligned} \mathrm{I}:\ \omega &= z_{m+1}\,dz_1 + \cdots + z_{2m}\,dz_m, \\ \mathrm{II}:\ \omega &= dz_0 + z_m\,dz_1 + \cdots + z_{2m-1}\,dz_{m-1}. \end{aligned}$$

Clebsch thought that he could distinguish the determinate and indeterminate cases by means of the $n \times n$ Jacobian skew-symmetric matrix $A = (a_{ij})$ associated to ω by (6.7). According to him, the determinate case (I) occurred when (in modern parlance) the rank of A is $2m$ [96, p. 208], that is—as articulated in Clebsch's time—when all $k \times k$ minors of A with $k > 2m$ vanish, but some of the $2m \times 2m$ minors do not vanish. In his own paper on the problem of Pfaff [179], Frobenius also used similar cumbersome language, but in a brief sequel submitted seven months later (June 1877), he introduced the notion of the rank of a matrix, so that (by his definition of rank) when, e.g., the minors of the matrix A have the above property, it is said to have rank $2m$ [183, p. 435]. For concision I will use Frobenius' now-familiar rank terminology in what follows, even though it is slightly anachronistic. As for the indeterminate case (II), Clebsch claimed that it corresponded to the case in which the rank of A is $2m-1$ [96, p. 218], although this turns out to be impossible, since (as Frobenius was to prove) the rank of any skew-symmetric matrix is always even.

In case (I) of Clebsch's Theorem 6.2, the system of integrals $f_i(x) = C_i$, $i = 1,\ldots,m$, defines a solution manifold for $\omega = 0$, and in case (II), the manifold is given by $f_i(x) = C_i$ for $i = 0,\ldots,m-1$. Taking these implications of Theorem 6.2 as his starting point, Clebsch turned to the real problem of interest to him, namely to determine the functions f_i by means of solutions to systems of first-order ordinary differential equations—and in a way, based on Jacobi's new method, that was more efficient than Pfaff's original method. Clebsch's idea in case (I) was first to determine one of the functions f_i, e.g., f_m. Then the equation $f_m(x_1,\ldots,x_n) = C_m$ is used to eliminate, e.g., x_n, thereby diminishing both the number of variables and the number m of differential terms by one unit. This new, reduced case could then be handled in the same manner to get f_{m-1}, and so on until all m functions f_i were determined [96, p. 204].

In his indirect method, the function f_m was a solution to a system of linear homogeneous partial differential equations, i.e., equations of the form (in Jacobi's notation)

$$A_i(f) = 0,\ i = 1,\ldots,s, \quad \text{where} \quad A_i(f) = \sum_{j=1}^{n} \alpha_{ij}(x)\frac{\partial f}{\partial x_j}. \tag{6.12}$$

Whereas a single such equation $A(f) = 0$ was known already in the eighteenth century to be equivalent to a system of first-order ordinary differential equations—so that a solution was known to exist—a simultaneous system need not have a solution. Such systems had already occurred in Jacobi's new method,[17] and Jacobi showed that when the s equations in (6.12) are linearly independent,[18] then $n-s$ independent solutions exist, provided the system satisfies the integrability condition

$$A_j[A_k(f)] - A_k[A_j(f)] \equiv 0 \quad \text{for all} \quad j,k. \tag{6.13}$$

For systems satisfying this condition, Jacobi sketched a method for finding the solutions via integration of systems of ordinary differential equations. Clebsch argued that the system $A_i(f) = 0$ he had arrived at also satisfied Jacobi's integrability condition, and so a solution f_m could be obtained by Jacobi's method of integration, which Jacobi had already proved to be "efficient" for integrating nonlinear partial differential equations. However, the reasoning leading to these equations in the nongeneric cases was vague and sketchy and reflected an incorrect understanding of the algebraic implications of cases (I) and (II) of Theorem 6.2, implications that Frobenius first correctly determined (Theorems 6.5 and 6.6). Thus Clebsch's

---

[17] See [273, p. 209] or [276, §2.1] for details.

[18] By "linearly independent" I mean that the $s \times n$ matrix (α_{ij}) has full rank r.

proof that the partial differential equations at which one arrives satisfy Jacobi's integrability condition was not well founded.

In a paper of 1863, Clebsch himself expressed dissatisfaction with his indirect method because it assumed the forms (I) and (II) of Theorem 6.2 rather than revealing how they are obtained. "In the present essay I will therefore directly derive these defining equations of the problem of Pfaff" [97, p. 146]. One key point he failed to mention in his introductory remarks was that he had succeeded in providing a direct derivation of these defining equations only in the case in which n is even and $\det A \neq 0$, i.e., the generic case dealt with by Jacobi in 1827. Within that limited framework, however, his direct approach was indeed far more satisfying.

In Jacobi's new method, the systems $A_i(f) = 0$ of (6.13) had been defined in terms of the Poisson bracket operation [276, p. 47]. Clebsch introduced two bracket operations as analogues of the Poisson bracket.[19] Using these, he defined Jacobi-like differential operators $B_i(f) = \sum_{i=1}^{n} \beta_{ij}(x)\partial f/\partial x_j$, $i = 1,\ldots,s$, $s < n$, with the property that if $B_i(f) = 0$, $i = 1,\ldots,s$, has $n-s$ functionally independent solutions, then repeated application of this result establishes the existence of $2m$ functions $f_1,\ldots,f_m,F_1,\ldots,F_m$ such that $\omega = F_1 df_1 + \cdots + F_m df_m$. However, Clebsch's operators did not satisfy Jacobi's condition (6.13) but rather the more general condition

$$B_i(B_j(f)) - B_j(B_i(f)) = \sum_{k=1}^{s} c_{ijk}(x)B_k(f), \quad i,j = 1,\ldots,s, \tag{6.14}$$

where not all coefficients c_{ijk} vanish identically. Clebsch realized, however, that the independent operators $B_i(f)$ could be replaced by linear combinations $A_i(f)$ of them, so that the $A_i(f)$ satisfied Jacobi's condition (6.13), and so Jacobi's theorem could be applied to establish the following extension of Jacobi's result.[20]

Theorem 6.3 (Jacobi–Clebsch). *A system of $s < n$ independent first-order partial differential equations $B_i(f) = 0$ in n variables is complete in the sense that $n-s$ functionally independent solutions $f_1,\ldots,f_{n-s}$ to the system exist if and only if the integrability condition (6.14) is satisfied.*

Clebsch's new method was thus direct in the sense that it avoided reliance on his earlier Theorem 6.2. Furthermore, the successive systems of linear partial differential equations satisfying the above theorem that need to be integrated are precisely described, and consequently, the integrability conditions guaranteeing an adequate supply of solutions were satisfied—in contrast to the situation in the indirect method, where only the first system is written down. But the direct method was limited to the even-generic case, $n = 2m$ and $\det A \neq 0$. Inspired by Clebsch's efforts, Frobenius sought to deal with the completely general problem of Pfaff by a direct method, i.e., one that did not start from Theorem 6.2. Evidently, the

---

[19] A more detailed discussion of this part of Clebsch's work is given in [277, pp. 394ff.].

[20] In this generality, Clebsch first stated his result in 1866 [99, pp. 260–261].

challenge of the "very complicated algebraic considerations" predicted by Clebsch (in the nongeneric cases) when that theorem is avoided did not deter Frobenius and probably appealed to him, for he sought to deal with the challenge Berlin-style by first seeking to determine the intrinsic invariant-based grounds for Clebsch's Theorem 6.2.

6.4 Frobenius' Solution to the Problem of Pfaff

The opening paragraphs of Frobenius' paper on the problem of Pfaff, which appeared in 1877, make it clear that Clebsch's work had provided the principal source of motivation. Thus Frobenius wrote [179, pp. 249–250]:

> After the preliminary work by Jacobi ... the problem of Pfaff was made the subject of detailed investigations primarily by Messrs. Natani ... and Clebsch .... In his first work, Clebsch reduces the solution of the problem to the integration of many systems of homogeneous linear partial differential equations by means of an indirect method, which he himself later said was not completely suitable for presenting the nature of the relevant equations in the proper light. For this reason in the second work he attacked the problem in another, direct, manner but treated only such differential equations ... $[\omega = 0]$ ... for which the determinant of the magnitudes $a_{\alpha\beta}$ ... differs from zero.
>
> It seems desirable to me to deal with the more general case ... by means of a similar direct method, especially since from the cited works I cannot convince myself that the methods developed for integrating the Pfaffian differential equation in this case actually attain this goal .... Under the above-mentioned assumption,[21] in the very first step toward the solution one arrives at a system of many homogeneous linear partial differential equations, rather than a single one. Such a system must satisfy certain integrability conditions if it is to have a nonconstant integral.... I fail to see, on the part of either author,[22] a rigorous proof for the compatibility of the partial differential equations to be integrated in the case in which the determinant $|a_{\alpha\beta}|$ vanishes.
>
> Clebsch distinguishes two cases in the problem of Pfaff, which he calls *determinate* and *indeterminate*.... However, the criterion for distinguishing the two cases has not been correctly understood by Clebsch.... Were the distinction specified by Clebsch correct, the indeterminate case would never be able to occur.
>
> For the purposes of integration, the left side of a first-order linear differential equation $[\omega = 0]$ is reduced by Clebsch to a canonical form that is characterized by great formal simplicity. It was while seeking to derive the posited canonical form on intrinsic grounds (cf. Kronecker, Berl. Monatsberichte 1874, January ...) that I arrived at a new way of formulating the problem of Pfaff, which I now wish to explicate.

Frobenius' above-quoted words not only indicate the many ways in which Clebsch's work motivated his own, they also reveal how he hit on the approach that he sets forth in his paper, namely by seeking to derive the canonical forms I–II of Clebsch's Theorem 6.2 on "intrinsic grounds" (*innere Gründen*) in the sense

[21] Namely, that $A = (a_{\alpha\beta})$ does not have full rank.

[22] Meaning Clebsch in his papers [96, 97] and Natani in his paper [451].

of Kronecker. Here Frobenius was citing the passage containing Kronecker's first disciplinary ideal as discussed in Section 5.6.2. What Frobenius meant was that the intrinsic grounds for a canonical form would be an invariant or set of invariants (analogous to Weierstrass' elementary divisors or Kronecker's set of invariants for singular families), determined from the coefficients $a_i(x)$ of $\omega = a_1dx_1 + \cdots + a_ndx_n$, that would indicate whether two Pfaffian expressions $\omega(x) = \sum_{i=1}^n a_i(x)dx_i$ and $\omega'(x') = \sum_{i=1}^n a_i'(x')dx_i'$ are equivalent in the analytic sense that a nonsingular transformation $x = \Phi(x') = \left(\varphi_1(x) \cdots \varphi_n(x)\right)^t$ exists such that under this transformation and the concomitant transformation of differentials $dx = J(\Phi)dx'$, where $J(\Phi)$ is the Jacobian matrix of Φ, $\omega(x)$ is transformed into $\omega'(x')$. His goal was to see if Clebsch's normal forms characterize these equivalence classes by finding the "intrinsic grounds"—the invariant(s)—underlying this phenomenon and determining analytic equivalence. And of course, by dealing with all the nongeneric cases that Clebsch had excluded from his preferred direct method, Frobenius was also adhering to Kronecker's second disciplinary ideal (Section 5.6.3), which rejects generic reasoning and declares faith in the possibility of dealing with the plethora of special cases in a uniform manner, as both Weierstrass and Kronecker had done in their work on the transformation of quadratic and bilinear forms.

For this sort of a situation at the interface of analysis and algebra, Frobenius had a paradigm conveniently at hand in the papers on the transformation of differential forms by Christoffel and Lipschitz, who had independently developed the mathematics hinted at in Riemann's 1854 lecture "On the Hypotheses at the Basis of Geometry," which had been published posthumously in 1868 [496]. As we shall see, using ideas gleaned from their work, Frobenius confirmed Kronecker's above declaration that "algebra itself can also supply sufficient inducement for positing canonical forms." In papers published back to back in *Crelle's Journal* in 1869, Christoffel [94] and Lipschitz [417] concerned themselves, among other things, with the problem of determining the conditions under which two nonsingular quadratic differential forms $\sum_{i,j=1}^n g_{ij}(x)dx_idx_j$ and $\sum_{i,j=1}^n g_{ij}'(x')dx_i'dx_j'$ can be transformed into one another by means of general (presumably analytic) transformations $x = \Phi(x')$, $dx = J(\Phi)dx'$.

Of particular interest was the question of when $\sum_{i,j=1}^n g_{ij}(x)dx_idx_j$ could be transformed into a sum of squares $\sum_{i=1}^n (dx_i')^2$ so as to define (when $n = 3$) Euclidean geometry.[23] For the discussion of Lipschitz's paper below it is helpful to note that if a transformation $x = \Phi(x')$ exists for which $\sum_{i,j=1}^n g_{ij}(x)dx_idx_j = \sum_{i,j=1}^n c_{ij}dx_idx_j$, where the c_{ij} are constants, then when (g_{ij}) is symmetric and positive definite as in Riemann's lecture, a further linear transformation may be made so that the original quadratic form becomes a sum of squares $\sum_{i=1}^n (dx_i'')^2$.

As we shall see, Frobenius extracted mathematical ideas from each author's paper, ideas that enabled him to formulate a path to the Kroneckerian intrinsic grounds behind Clebsch's canonical forms. Let us consider first what he found in Lipschitz's paper. The approach of Lipschitz was somewhat more general than that

[23] Keep in mind that there is at this time no sensitivity to a distinction between local and global results and that the actual mathematics is being done on a strictly local level.

of Christoffel in that he considered homogeneous functions $f(dx)$ of the differentials dx of any fixed degree k—the analytic analogue of the homogeneous polynomials of algebra. With those polynomials in mind, he suggested that two such functions $f(dx)$ and $f'(dx')$ of the same degree should be regarded as belonging to the same class if there existed a nonsingular variable transformation $x = \Phi(x')$, $dx = J(\Phi)dx'$ such that $f(dx) = f'(dx')$. Of particular interest was a class containing a function $f'(dx')$ with constant coefficients for the reason indicated above.

By way of illustrative example, Lipschitz considered the case $k = 1$, so that $f(dx) = a_1(x)dx_1 + \cdots + a_n(x)dx_n$ [417, pp. 72–73]. He used this example to explain his interest in what he regarded as the analytic counterpart to a covariant in the algebraic theory of invariants. As an example, he gave the bilinear form in variables $dx_1, \ldots, dx_n$ and $\delta x_1, \ldots, \delta x_n$

$$\Omega(dx, \delta x) = \sum_{i,j=1}^{n} \left(\frac{\partial a_i}{\partial x_j} - \frac{\partial a_j}{\partial x_i} \right) dx_i \delta x_j. \tag{6.15}$$

The coefficient matrix of Ω is of course Jacobi's skew-symmetric matrix of the theory of Pfaff's problem, although Lipschitz made no allusion to that theory. However, in what follows it will be convenient to express (6.15) with the notation $\Omega = \sum_{i,j=1}^{n} a_{ij} dx_i \delta x_j$ with a_{ij} as in (6.15) and so defining Jacobi's skew-symmetric matrix associated to $a_1, \ldots, a_n$.

With the aid of results from Lagrange's treatment of the calculus of variations in *Mécanique Analytique*, he proved the following result [417, pp. 75–77].

Theorem 6.4 (Lipschitz). *Ω is a covariant of $\omega = f(dx)$ in the sense that if $f(dx) = f'(dx')$ under $x = \Phi(x')$ and the concomitant linear transformation of differentials $dx = J(\Phi)dx'$, then one has as well*

$$\Omega(dx, \delta x) = \sum_{i,j=1}^{n} a_{ij} dx_i \delta x_j = \sum_{i,j=1}^{n} a'_{ij} dx'_i \delta x'_j = \Omega'(dx', \delta x'), \tag{6.16}$$

where also $\delta x = J(\Phi)\delta x'$.

Lipschitz emphasized that Ω is the source of the condition that $f(dx) = \sum_{i=1}^{n} a_n(x) dx_i$ be transformable into $\sum_{i=1}^{n} c_i dx'_i$, namely that this bilinear form vanish identically, i.e.,

$$a_{ij} = \frac{\partial a_i}{\partial x_j} - \frac{\partial a_j}{\partial x_i} \equiv 0. \tag{6.17}$$

As Lipschitz observed, this was the well-known condition that the differential $\omega = \sum_{i=1}^{n} a_n(x) dx_i$ be exact, from which the above transformation property followed. All this served to motivate the case of degree $k = 2$, namely quadratic differential forms ds^2, where the same process leads to a quadrilinear "covariant" form in four variables that defines what later became known as the Riemann curvature tensor, so that the vanishing of its coefficients gives the condition that ds^2 may be transformed into a sum of squares.

Of primary interest to Frobenius, however, was Lipschitz's result in the motivational case $k = 1$—his Theorem 6.4. In keeping with the work on quadratic and bilinear forms within the Berlin school, as well as the above-mentioned work of Christoffel and Lipschitz, Frobenius focused on the question of when two Pfaffian expressions $\omega = \sum_{i=1}^{n} a_i(x)dx_i$ and $\omega' = \sum_{i=1}^{n} a'_i(x')dx'_i$ are analytically equivalent in the sense that a nonsingular transformation $x = \Phi(x')$, $dx = J(\Phi)dx'$, exists such that $\omega(x) = \omega'(x')$. His goal was to see whether Clebsch's normal forms characterized these equivalence classes and, if so, to find the "intrinsic grounds"—the invariants—responsible for this phenomenon. What Lipschitz's Theorem 6.4 showed him was that the analytic equivalence of ω and ω' brought with it the analytic equivalence of their associated "bilinear covariants" Ω and Ω' as in (6.16)—as Frobenius called them in keeping with Lipschitz's theorem and terminology. Today the bilinear covariant Ω associated to ω is understood within the framework of the theory of differential forms initiated by Élie Cartan, where $\Omega = -d\omega$. In Section 6.6, the influence of Frobenius' work on Cartan's development of his theory of differential forms will be considered.

It was in seeking to use Lipschitz's Theorem 6.4 that Frobenius drew inspiration from Christoffel's paper. To determine necessary and sufficient conditions for the analytic equivalence of two quadratic differential forms, Christoffel had determined purely algebraic conditions involving a quadrilinear form that were necessary for equivalence, and he then asked whether they were sufficient for the analytic equivalence or whether additional analytic conditions needed to be imposed. He characterized this question as "the crux of the entire transformation problem" [94, p. 60].

Applying this strategy to the problem at hand, Frobenius began by giving an algebraic proof of Lipschitz's Theorem 6.4 [179, pp. 252–253] that was elegant, clear, and simple, and, in particular, did not rely on results from the calculus of variations. His proof makes it clear to present-day readers that Lipschitz's theorem follows from a relatively straightforward calculation in the tangent space of the manifold of generic points where Ω and Ω' have maximal rank.[24]

Frobenius could readily see that a necessary consequence of the analytic equivalence of ω and ω' is the *algebraic equivalence* of the form-pairs (ω,Ω) and (ω',Ω') in the following sense: Fix x at x_0 and x' at $x'_0 = \Phi(x_0)$.[25] Then we have two form-pairs $(\omega,\Omega)_{x_0}$, $(\omega',\Omega')_{x'_0}$ with constant coefficients that are equivalent in the sense that $\omega_{x_0}(u) = \omega'_{x'_0}(u')$ and $\Omega_{x_0}(u,v) = \Omega'_{x'_0}(u',v')$ by means of a nonsingular linear transformation $u = Pu'$, $v = Pv'$, where $P = J(\Phi(x_0))$ and $u,v \in \mathbb{C}^n$. The question that Frobenius posed to himself was whether the algebraic equivalence

[24] See [277, p. 401] for a sketch of Frobenius' proof.

[25] As indicated in Section 6.1, strictly speaking, x_0 and x'_0 should be generic points, i.e., points at which the bilinear forms Ω, Ω' have maximal rank.

of the form-pairs was sufficient to guarantee the (local) analytic equivalence of ω and ω'. Thus it was first necessary to study the algebraic equivalence of form-pairs (w,W) under a nonsingular linear transformation $u = Pu'$, $v = Pv'$, where $w(u) = \sum_{i=1}^{n} a_i u_i$ is a linear form and $W(u,v) = \sum_{i,j=1}^{n} a_{ij} u_i v_j$ is an alternating bilinear form ($a_{ji} = -a_{ij}$) and all coefficients a_i, a_{ij} are constant. The hope was that the algebraic analogue of Clebsch's two canonical forms (6.10) and (6.11) of cases (I) and (II) of Theorem 6.2 would yield the distinct equivalence classes for (w,W). As we shall see in the following section, Frobenius confirmed that this was the case. Indeed, his proof provided a paradigm that led him by analogy to a proof of Clebsch's Theorem 6.2—including a correct way to algebraically distinguish the two cases—and thereby to the conclusion that the algebraic equivalence of (ω,Ω) and (ω',Ω') at generic points x_0 and x'_0 (in the sense of Section 6.1) implies the analytic equivalence of ω and ω' at those points. In this manner, he found what he perceived to be the true "intrinsic grounds" in Kronecker's sense for Clebsch's canonical forms.

Clebsch had shied away from a direct approach in the nongeneric case in favor of his indirect approach because "In this way, one is spared the trouble of carrying out direct proofs that lead to very complicated algebraic considerations, which, to be sure, are of interest in their own right ..." [96, p. 194]. It was just these sorts of "complicated algebraic considerations" that attended nongeneric reasoning in linear algebra and that Weierstrass and Kronecker had shown could be successfully transformed into a satisfying theory, and the paradigm of their work clearly encouraged Frobenius to deal in the above-described manner with the theory of Pfaffian equations. Indeed, Frobenius realized that his friend Ludwig Stickelberger, a fellow student at Berlin and a colleague at the Zurich Polytechnic when Frobenius wrote [179], had already considered the simultaneous transformation of a bilinear or quadratic form together with one or more linear forms in his 1874 Berlin doctoral thesis [179, p. 264n].

6.4.1 The algebraic classification theorem

In discussing form-pairs (w,W), I will use more familiar matrix notation and write $W = u^t Av$, where A is skew-symmetric ($A^t = -A$), and $w = a^t u$, with a, u, and v here being regarded as $n \times 1$ column matrices, e.g., $a = \begin{pmatrix} a_1 & \cdots & a_n \end{pmatrix}^t$. Frobenius began to develop such notation himself shortly after his work on Pfaffian equations, as indicated in the next chapter.

The first algebraic question that Frobenius considered concerned the rank of A, i.e., the rank of a skew symmetric matrix or, as he called it, an alternating system [179, pp. 255–261]. I would guess that this was also one of the first questions related to Pfaffian equations he investigated, since the resultant answer plays a fundamental role in the ensuing theory. For ease of reference, I will name it the *even rank theorem*.

Theorem 6.5 (Even-rank theorem). *If A is skew-symmetric, then its rank r must be even.*

Following Frobenius, let us say that a *principal minor* is one obtained from A by deleting the like-numbered rows and columns, e.g., the $(n-3)\times(n-3)$ determinant of the matrix obtained by deletion of rows $1,3,5$ and columns $1,3,5$ of A. It is easy to give examples of matrices of rank r for which all the principal minors of degree r vanish, but Frobenius showed that when A is symmetric or skew-symmetric of rank r, then there is always a principal minor of degree r (i.e., $r\times r$) that does not vanish. From this result Theorem 6.5 follows directly, since the matrix of a principal minor of A is also skew-symmetric, and as we saw in Section 6.2, Jacobi had already observed that skew-symmetric determinants of odd degree must vanish. Thus the degree r of the nonvanishing principal minor must be even.

With this theorem in mind, let us consider the two canonical forms of cases (I) and (II) of Clebsch's Theorem 6.2. In Clebsch's statement of the theorem, the integer m in (I) and (II) denoted the number of terms in the normal forms (I) and (II). However, this is not necessary, and following Frobenius, the two types of normal form for which he sought the intrinsic grounds will be denoted by

$$\begin{aligned} \mathrm{I}: \omega &= z_{m+1}dz_1+\cdots+z_{2m}dz_m, \\ \mathrm{II}: \omega &= dz_0+z_{m+1}dz_1+\cdots+z_{2m}dz_m. \end{aligned} \tag{6.18}$$

In (I), the number of variables x_i before transformation to normal form is $n=2m+q$, $q\geq 0$. In this case, $a=\begin{pmatrix} z_{m+1} & \cdots & z_{2m} & 0 & \cdots & 0\end{pmatrix}^t$, and so in block matrix form, the Jacobi skew-symmetric matrix A is, by calculation,

$$A_{\mathrm{I}}=\begin{pmatrix} 0 & I_m & 0 \\ -I_m & 0 & 0 \\ 0 & 0 & 0 \end{pmatrix}, \tag{6.19}$$

where I_m denotes the $m\times m$ identity matrix. The associated bilinear covariant is thus

$$\Omega=(dz_1\delta z_{m+1}-dz_{m+1}\delta z_1)+\cdots+(dz_m\delta z_{2m}-dz_{2m}\delta z_m).$$

The corresponding pair of algebraic forms (w,W) is obtained by setting the variables z_i equal to constants and the differentials dz_i, δz_i equal to variables u_i, v_i, respectively, to obtain $w=c_{m+1}u_1+\cdots+c_{2m}u_m$ and $W=(u_1v_{m+1}-u_{m+1}v_1)$ $+\cdots+(u_mv_{2m}-u_{2m}v_m)$. Thus $W=u^tAv$, where A is as in (6.19) above, and $w=a^tu$, $a=\begin{pmatrix} 0 & \cdots & 0 & c_{m+1} & \cdots & c_{2m}\end{pmatrix}^t$.

For case (II) in (6.18), $a=\begin{pmatrix} 1 & z_{m+1} & \cdots & z_{2m} & 0 & \cdots & 0\end{pmatrix}^t$, $n=2m+q$, $q\geq 1$, and so

$$A_{\mathrm{II}}=\begin{pmatrix} 0 & 0 & 0 & 0 \\ 0 & 0 & I_m & 0 \\ 0 & -I_m & 0 & 0 \\ 0 & 0 & 0 & 0 \end{pmatrix}. \tag{6.20}$$

The corresponding algebraic form-pair is $w = a^t u$,

$$a = \begin{pmatrix} 1 \; c_{m+1} \; \cdots \; c_{2m} \; 0 \cdots 0 \end{pmatrix}^t,$$

and $W = u^t A v$ with A as in (6.20). As we saw, Clebsch had erroneously thought that the parity of Jacobi's skew-symmetric matrix would distinguish (I)—even rank—from (II)—odd rank, but the Frobenius even-rank theorem shows that this characterization is impossible. In fact, in both cases (I) and (II) in (6.18), the rank of A is the same even number ($2m$).

On the algebraic level considered here, where the link between ω and Ω no longer exists, one could say that this is because we are seeking an invariant of a form-pair $(w, W) = (a^t u, u^t A v)$ under linear transformations but looking only at A and ignoring a. In his Berlin doctoral thesis of 1874 [549, §2], Stickelberger had already introduced an appropriate invariant for a system consisting of a bilinear form and several linear forms. Stickelberger's thesis was well known to Frobenius, who along with Wilhelm Killing[26] was one of the three appointed "adversaries" at Stickelberger's thesis defense. Consider, for example, $\Gamma = u^t C v$, $\gamma_1 = c^t u$, $\gamma_2 = d^t u$, where C is any $n \times n$ matrix and u, v, c, and d are $n \times 1$ column matrices. From Weierstrass' paper on elementary divisors it was well known that by virtue of a theorem on minor determinants due to Cauchy, the rank of the bilinear form is invariant under nonsingular linear transformations $u = P\bar{u}$, $v = Q\bar{v}$, i.e., if

$$u^t C v = \bar{u}^t (P^t C Q) \bar{v} \equiv \bar{u}^t \bar{C} \bar{v},$$

then rank $(\bar{C}) =$ rank C. Since the linear forms transform by

$$c^t u = u^t c = \bar{u}^t (P^t c) \equiv \bar{c}^t \bar{u},$$

Stickelberger observed that if one introduces the bilinear form $\hat{\Gamma}$ in $n+1$ variables with $n+1 \times n+1$ coefficient matrix

$$\hat{C} = \begin{pmatrix} C & c \\ d^t & 0 \end{pmatrix},$$

which amalgamates the original bilinear and linear forms into a single bilinear form, then under the linear transformations $u = \hat{P}\bar{u}$, $v = \hat{Q}\bar{v}$, where

$$\hat{P} = \begin{pmatrix} P & 0 \\ 0 & 1 \end{pmatrix}, \quad \hat{Q} = \begin{pmatrix} Q & 0 \\ 0 & 1 \end{pmatrix},$$

---

[26] Killing was to go on to make contributions of fundamental importance to the theory of semisimple Lie algebras [276, Part II].

one has

$$\hat{C} \to \hat{P}^t \hat{C} \hat{Q} = \begin{pmatrix} \bar{C} & \bar{c} \\ \bar{d}^t & 0 \end{pmatrix}.$$

This shows that the rank of $\hat{C}$ is an invariant of the system $(\Gamma, \gamma_1, \gamma_2)$.

To apply this to the pair $(w,W) = (a^t u, u^t A v)$ with A skew-symmetric, Frobenius introduced the analogous alternating form $\hat{W}$ with coefficient matrix

$$\hat{A} = \begin{pmatrix} A & a \\ -a^t & 0 \end{pmatrix}. \tag{6.21}$$

Then it follows that the rank of $\hat{A}$ is an invariant of the system (w,W). Since $\hat{A}$ is skew-symmetric, it follows readily from the even rank theorem that either $\operatorname{rank} \hat{A} = \operatorname{rank} A$ or $\operatorname{rank} \hat{A} = \operatorname{rank} A + 2$. Thus if, following Frobenius, we set $p = \frac{1}{2}(\operatorname{rank} A + \operatorname{rank} \hat{A})$, p can be odd or even.

Returning to the pairs (w,W) obtained above corresponding to cases (I) and (II) of (6.18) with A given by (6.19) and (6.20), respectively, it follows that we have, respectively,

$$\hat{A}_{\mathrm{I}} = \begin{pmatrix} 0 & I_m & 0 & c \\ -I_m & 0 & 0 & 0 \\ 0 & 0 & 0 & 0 \\ -c^t & 0 & 0 & 0 \end{pmatrix}, \quad \hat{A}_{\mathrm{II}} = \begin{pmatrix} 0 & 0 & 0 & 0 & 1 \\ 0 & 0 & I_m & 0 & c \\ 0 & -I_m & 0 & 0 & 0 \\ 0 & 0 & 0 & 0 & 0 \\ -1 & -c^t & 0 & 0 & 0 \end{pmatrix},$$

where $c = \begin{pmatrix} c_{m+1} & \cdots & c_{2m} \end{pmatrix}^t$. In $\hat{A}_{\mathrm{I}}$, the $2m$ rows containing I_m and $-I_m$ are linearly independent, but the last row is a linear combination of the rows involving $-I_m$, so that $\operatorname{rank} \hat{A}_{\mathrm{I}} = 2m = \operatorname{rank} A_{\mathrm{I}}$ and $p = 2m$ is even. In $\hat{A}_{\mathrm{II}}$, the first row, the rows involving I_m, the rows involving $-I_m$, and the last row are linearly independent so that $\operatorname{rank} \hat{A}_{\mathrm{II}} = 2m+2$. Since $\operatorname{rank} A_{\mathrm{II}} = 2m$, this means that $p = 2m+1$ is odd. Thus the parity of p distinguishes the two normal forms. Of course, it remained to prove that conversely, depending on the parity p of a form-pair (w,W) it could be transformed into the requisite normal form-pair of the same parity. Frobenius did this and so established the following algebraic theorem.

Theorem 6.6 (Algebraic classification theorem). *Let $w = a^t u$ be a linear form and $W = u^t A v$ an alternating bilinear form. Then $p = (\operatorname{rank} A + \operatorname{rank} \hat{A})/2$ is an integer and is invariant with respect to nonsingular linear transformations of the pair (w,W), which is thus said to be of* class p. *When* $\operatorname{rank} \hat{A} = \operatorname{rank} A = 2m$, *$p = 2m$ is even; and when* $\operatorname{rank} \hat{A} = \operatorname{rank} A + 2 = 2m+2$, *$p = 2m+1$ is odd. If (w,W) is of class $p = 2m$, then there exists a nonsingular linear transformation $u = P\bar{u}$, $v = P\bar{v}$ such that $w = c_{m+1}\bar{u}_1 + \cdots + c_{2m}\bar{u}_m$, and $W = \bar{u}^t \bar{A} \bar{v}$ with $\bar{A}$ as in (6.19). When $p = 2m+1$, a nonsingular linear transformation exists such that w takes the form $w = \bar{u}_0 + c_{m+1}\bar{u}_1 + \cdots + c_{2m}\bar{u}_m$ and $W = \bar{u}^t \bar{A} \bar{v}$ with $\bar{A}$ as in (6.20). Consequently, two form-pairs (w,W) and (w',W') are equivalent if and only if they are of the same class p.*

The developments leading up to this theorem as well as collateral results regarding Pfaffian determinants were presented in Sections 6–11 of Frobenius' paper and totaled 21 pages. Turning next to the analytic theory of the equivalence of differential forms $\omega = \sum_{i=1}^{n} a_i(x)dx_i$, however, Frobenius explained that in effect, his Theorem 6.6 would not serve as the mathematical foundation of the analytic theory but rather as a guide. In developing the analytic theory, "I will rely on the developments of §§. 6–11 as little as possible, and utilize them more by analogy than as a foundation" [179, p. 309]. No doubt he took this approach to encourage analysts not enamored (as he was) by algebra to read the largely self-contained analytic part. Comparison of the two parts shows that the algebraic part, however, provided the blueprint for the analysis, for the reasoning closely parallels the line of reasoning leading to the algebraic classification theorem. Indeed, as we shall see in Section 6.5, it proved too algebraic for the tastes of many mathematicians primarily interested in the integration of differential equations, i.e., in the efficiency problem described toward the end of Section 6.2.

The fact that Frobenius deemed the purely algebraic results sufficiently noteworthy in their own right to present them carefully worked out in Sections 6–11 is indicative of his appreciation for Berlin-style linear algebra, and indeed the theory presented in Sections 6–11 and culminating in Theorem 6.6 was the first but hardly the last instance of Frobenius' creative involvement with linear algebra. For example, undoubtedly inspired by the theory of Sections 6–11, during 1878–1880 Frobenius published several highly important and influential memoirs on further new and important aspects of the theory of bilinear forms [181, 182, 185], which are discussed in Chapters 7 and 8.

6.4.2 The analytic classification theorem

The main theorem in the analytic part of his paper is the result at which Frobenius arrived by developing the analogue of the reasoning leading to the above algebraic classification theorem. Thus given a Pfaffian form $\omega = \sum_{i=1}^{n} a_i(x)dx_i \equiv a(x)^t u$, he considered the form-pair (ω, Ω) determined by ω and its bilinear covariant

$$\Omega(u,v) = \sum_{i,j=1}^{n} a_{ij}(x)u_i v_j, \quad a_{ij} = \frac{\partial a_i}{\partial x_j} - \frac{\partial a_j}{\partial x_i}. \tag{6.22}$$

Corresponding to this form-pair we have by analogy with (6.21) the augmented skew-symmetric matrix

$$\hat{A}(x) = \begin{pmatrix} A(x) & a(x) \\ -a^t(x) & 0 \end{pmatrix}, \quad A = (a_{ij}(x)). \tag{6.23}$$

Since the ranks of A and $\hat{A}$ figure prominently in what is to follow, recall from Section 6.1 that by the rank of, e.g., $A = A(x)$, Frobenius meant the maximal rank of $A(x)$. What Frobenius tacitly showed [179, p. 309] was that the maximal rank of $\hat{A}(x)$ is also invariant with respect to nonsingular transformations $x = \Phi(x')$ at the generic points corresponding to $\hat{A}(x)$. Hence $p = \frac{1}{2}[\operatorname{rank} A(x) + \operatorname{rank} \hat{A}(x)]$ is also an invariant in this sense on the (open and dense) set of generic points x where both $A(x)$ and $\hat{A}(x)$ attain their maximal ranks, and p so defined can therefore be used to define the class of ω. The main goal of his paper was to establish the following analytic analogue of the algebraic classification theorem, which should be interpreted as a theorem about the existence of local transformations at the above-mentioned generic points.

Theorem 6.7 (Analytic classification theorem). *Let ω be of class p. If $p = 2m$, there exists a transformation $x = \Phi(z)$ such that*

$$\text{(I)} \quad \omega = z_{m+1}dz_1 + \cdots + z_{2m}dz_m;$$

and if $p = 2m + 1$, then a transformation $x = \Phi(z)$ exists such that

$$\text{(II)} \quad \omega = dz_0 + z_{m+1}dz_1 + \cdots + z_{2m}dz_m.$$

Consequently, ω and ω' are equivalent if and only if they are of the same class p.

This theorem implies Clebsch's Theorem 6.2 and, in addition, provides through the notion of the class of ω a correct algebraic criterion distinguishing cases (I) and (II). The theorem also shows that the algebraic equivalence of two form-pairs (ω, Ω) and (ω', Ω') is sufficient for the analytic equivalence of ω and ω'. That is, if x and x' are fixed generic points with respect to both ω and ω', and if they are algebraically equivalent for the fixed values x and x', then by the algebraic classification theorem, $(\omega, \Omega)_x$ and $(\omega', \Omega')_{x'}$ must be of the same class p. But this then means that ω and ω' are of the same class p, and so by the analytic classification theorem, each of ω and ω' can be locally transformed into the same canonical form (I) or (II) and hence each into the other, i.e., they are analytically equivalent.

The proofs of both the algebraic and analytic classification theorems are quite similar in most respects, with the analytic version evidently inspired by the algebraic version. Both are lengthy, but the analytic version is longer by virtue of a complication attending the analytic analogue of one point in the algebraic proof. It is this complication that led Frobenius to formulate his integrability theorem for systems of Pfaffian equations, which is discussed in the next section. The remainder of this section is devoted to a brief summary of the complication.[27]

In proving the algebraic classification theorem, Frobenius at one point was faced with the following situation, which I describe in the more familiar vector space

[27] Frobenius' proof is spread out over pp. 309–331 of [179] and also draws on results from other parts of the paper.

terms into which his reasoning readily translates. Given k linearly independent vectors $w_1,\dots,w_k$ in $\mathbb{C}^n$, determine a vector w such that w is linearly independent of $w_1,\dots,w_k$ and also w is orthogonal to a certain subspace $\mathcal{V}$, with basis vectors $v_1,\dots,v_d$. Thus w needs to be picked such that

$$w \cdot v_i = 0, \quad i = 1,\dots,d. \tag{6.24}$$

This means that w is to be picked from $\mathcal{V}^\perp$, the orthogonal complement of $\mathcal{V}$. The problem is complicated by the fact that the vectors $w_1,\dots,w_k$ also lie in $\mathcal{V}^\perp$, and w must be linearly independent of them. However, it is known that the dimension d of $\mathcal{V}$ satisfies the inequality $d \le n+1+k-2m$, where m is a given fixed integer such that $k < m$. Since the dimension of $\mathcal{V}^\perp$ is $n-d \ge n-(n+1+k-2m) = 2m-(k+1) \ge m > k$, it is possible to pick w to be linearly independent of $w_1,\dots,w_k$ as desired.

In the analytic version of this situation that arises en route to the analytic classification theorem, k functions $f_1,\dots,f_k$ of $x_1,\dots,x_n$ are given that are functionally independent, i.e., their gradient vectors $\nabla f_1,\dots,\nabla f_k$ are linearly independent in the sense that the $k \times n$ matrix with the ∇f_i as its rows has full rank k (in a neighborhood of the point under consideration). In lieu of the d vectors $v_1,\dots,v_d$ of the algebraic proof, there are now d vector-valued functions $v_i = (b_{i1}(x),\dots,b_{in}(x))$, where again $d \le n+1+k-2m$ and m is a given fixed integer such that $k < m$. The problem now is to determine a function f such that $\nabla f_1,\dots,\nabla f_k,\nabla f$ are linearly independent and in lieu of (6.24), f must satisfy

$$\nabla f \cdot v_i = 0, \quad \text{i.e.,} \quad B_i(f) \stackrel{\text{def}}{=} \sum_{j=1}^{n} b_{ij}(x)\frac{\partial f}{\partial x_j} = 0, \quad i = 1,\dots,d. \tag{6.25}$$

By analogy with the algebraic proof, the situation is complicated by the fact that the functions $f_1,\dots,f_k$ are also solutions to this system of partial differential equations. Thus for the desired f to exist, the system (6.25) must have at least $k+1$ functionally independent solutions. This is the type of system considered in the Jacobi–Clebsch Theorem 6.3. If the system satisfies the integrability condition (6.14) for completeness, $[B_i,B_j] = \sum_{l=1}^{n} c_{ijl}B_l$, it will have $n-d$ independent solutions, and since (as we already saw) $n-d \ge m > k$, the existence of the desired function f will then follow. The system, however, was not explicitly given (see [179, pp. 312–313])—as was also the case with the systems in Clebsch's indirect method. It was consequently uncertain whether Clebsch's integrability condition was satisfied. As the long quotation given at the beginning of the section shows, Frobenius had criticized Clebsch for glossing over a similar lack of certainty in his indirect method. He was certainly not about to fall into the same trap himself! But how to salvage the proof? To this end, he turned to a general duality between systems such as $B_i(f) = 0$ and systems of Pfaffian equations. As we shall see in the next section, this duality had come to light in reaction to the work of Clebsch and Natani, but Frobenius developed it in a more elegant and general form than his predecessors.

6.4.3 The integrability theorem

In his paper of 1861 on Pfaff's problem [451], Leopold Natani, who was unfamiliar with the still unpublished new method of Jacobi that had inspired Clebsch, did not seek to determine the functions $f_1,\ldots,f_m$ in $\omega = \sum_{i=1}^{2m} a_i dx_i = \sum_{i=1}^{m} F_i df_i$ by means of linear first-order partial differential equations. Instead, he used successive systems of special Pfaffian equations. With the representation $\omega = F_1 df_1 + \cdots + F_m df_m$ as the goal, Natani first constructed a system of Pfaffian equations out of the coefficients a_i, a_{ij} that yielded f_1 as a solution. Then he constructed a second system using as well f_1 to obtain f_2, and so on.[28]

The contrasting treatments of Pfaff's problem by Natani and Clebsch turned the attention of some mathematicians to the connections between (1) systems of linear homogeneous partial differential equations, such as Clebsch's systems, and (2) systems of Pfaffian equations, such as Natani's. It turns out that associated to a system of type (1) is a "dual" system of type (2) with the property that the independent solutions f to (1) are precisely the independent integrals $f = C$ to (2). This general duality was apparently not common knowledge in 1861, since, judging by Clebsch's remarks on Natani's work [97, p. 146n], he failed to realize that Natani's successive systems of Pfaffian equations were the duals in the above sense to the systems of partial differential equations in his direct method. This was pointed out by Hamburger in a paper of 1877 [261] to be discussed below.

Apparently the first mathematician to call attention to the general existence of a dual relation between systems of Pfaffian equations and systems of linear homogeneous partial differential equations was Mayer (1839–1908) in a paper of 1872 [438]. Consideration of Mayer's way of looking at and establishing this reciprocity is of interest by way of comparison with the approach to the matter taken by Frobenius. We will see that on the local level where both tacitly reasoned, Mayer's approach lacked the complete generality and strikingly modern algebraic elegance achieved by Frobenius.

Ever since the late eighteenth century, mathematicians had realized that the integration of a single linear homogeneous partial differential equation

$$\xi_1 \frac{\partial z}{\partial x_1} + \cdots + \xi_n \frac{\partial z}{\partial x_n} = 0 \tag{6.26}$$

was equivalent to the integration of a system of first-order ordinary differential equations, namely the system that—with x_n picked as independent variable—can be written as

$$\frac{dx_1}{dx_n} = \frac{\xi_1}{\xi_n}, \ldots, \frac{dx_{n-1}}{dx_n} = \frac{\xi_{n-1}}{\xi_n}. \tag{6.27}$$

[28] For a clear exposition of the details of Natani's method, see Hamburger's paper [261].

Jacobi, for example, gave an elegant treatment of this equivalence in a paper of 1827 on partial differential equations [308].[29] Mayer began by noting the above-described equivalence between (6.26) and (6.27), which evidently inspired his observation that "in an entirely similar way" it is easy to establish a reciprocal connection between systems of linear homogeneous partial differential equations and Pfaffian systems, which "in particular cases ... has already been observed and utilized many times" [438, p. 448].

Mayer began with a system of m independent linear partial differential equations $A_i(f) = 0$ in $n > m$ variables. Normally, the $A_i(f)$ would be written in general form $A_i(f) = \sum_{j=1}^{n} \alpha_{ij}(x)\partial f/\partial x_j$, but Mayer assumed that they were written in the special form

$$A_i(f) = \frac{\partial f}{\partial x_i} + \sum_{k=m+1}^{n} a_{ik}(x)\frac{\partial f}{\partial x_k}, \quad i = 1,\ldots,m. \tag{6.28}$$

In other words, he assumed that the $m \times n$ matrix $M(x) = (\alpha_{ij}(x))$ can be put in the reduced echelon form $\begin{pmatrix} I_m & A \end{pmatrix}$. It was well known that any particular equation $A_i(f) = 0$ of the system could be replaced by a suitable linear combination of the equations. In other words, elementary row operations may be performed on the matrix $M(x)$. In this manner, for a *fixed* value of x, $M(x)$ can be transformed into its reduced echelon form, and then by permuting columns, which corresponds to reindexing the variables $x_1,\ldots,x_n$, the form $\begin{pmatrix} I_m & A \end{pmatrix}$ can be obtained. However, Mayer apparently did not realize that this cannot be done analytically, i.e., for all x in a neighborhood of a fixed point x_0.

For example, with $m = 2$ and $n = 3$, consider

$$M(x,y,z) = \begin{pmatrix} x & y & 1 \\ z & -1+x+\frac{yz}{x} & 3+\frac{z}{x} \end{pmatrix}$$

in a neighborhood of $(x,y,z) = (1,1,1)$. There $M(x,y,z)$ has full rank, which means that the corresponding partial differential equations are linearly independent, as Mayer evidently assumed. The reduced echelon forms of $M(1,y,z)$ and of $M(x,y,z)$ with $x \neq 1$, for all (x,y,z) close to $(1,1,1)$, are, respectively,

$$\begin{pmatrix} 1 & y & 1 \\ 0 & 0 & 1 \end{pmatrix} \quad \text{and} \quad \begin{pmatrix} 1 & 0 & \frac{x-3y-1}{x(x-1)} \\ 0 & 1 & \frac{3}{x-1} \end{pmatrix},$$

so that it is impossible to bring $M(x,y,z)$ into the form $\begin{pmatrix} I_2 & A \end{pmatrix}$ for all (x,y,z) in a neighborhood of $(1,1,1)$. Hence his treatment of duality on the customary local level was not completely general. That being said, let us consider his introduction of the dual to system (6.28).

[29] For an exposition of the equivalence presented in the spirit of Jacobi see [273, pp. 201ff.].

Since solutions to the system (6.28) are precisely the solutions to

$$A_\lambda(f) = \lambda_1(x)A_1(f) + \cdots + \lambda_m(x)A_m(f) = 0,$$

for any choice of functions $\lambda_i(x)$, the generic equation $A_\lambda(f) = 0$ is equivalent to the system. Now this is a single equation and so corresponds in the above-described manner to a system of ordinary differential equations, namely

$$(1) \quad \frac{dx_i}{dx_n} = \lambda_i(x), \quad i \leq m, \qquad (2) \quad \frac{dx_k}{dx_n} = \sum_{i=1}^{m} \lambda_i a_{ik}(x), \quad k \geq m+1.$$

Substituting (1) into (2) and multiplying through by dx_n then yields the system of $n-m$ Pfaffian equations

$$dx_k = \sum_{i=1}^{m} a_{ik}(x)dx_i, \quad k = m+1,\ldots,n. \tag{6.29}$$

Thus (6.29) is the Pfaffian system that corresponds to (6.28).

Mayer defined a Pfaffian system in the form (6.29) and thus consisting of $n-m$ equations to form a completely integrable system (*unbeschränkt integrables System*) [438, p. 451] if $n-m$ independent integrals $f_k(x_1,\ldots,x_n) = C_k$, $k = m+1,\ldots,n$, exist in the sense that if this integral system is solved for the variables $x_{m+1},\ldots,x_n$ to get the functions $x_k = \varphi_k(x_1,\ldots,x_m,C_{m+1},\ldots,C_n)$, $k = m+1,\ldots,n$, then (6.29) is satisfied identically if these functions are substituted, i.e., if x_k is replaced by $\varphi_k(x_1,\ldots,x_m,C_{m+1},\ldots,C_n)$ and dx_k is replaced by $\sum_{i=1}^{m}(\partial\varphi_k/\partial x_i)dx_i$ for $k = m+1,\ldots,n$.

It follows that if (6.29) is completely integrable in this sense and the above-mentioned substitutions are made, then by comparing the two sides of the resulting equation, we get

$$\frac{\partial\varphi_k}{\partial x_i} = a_{ik}(x_1,\ldots,x_m,C_{m+1},\ldots,C_n).$$

From this and the chain rule, Mayer then obtained the condition

$$0 = \frac{\partial^2\varphi_k}{\partial x_j \partial x_i} - \frac{\partial^2\varphi_k}{\partial x_i \partial x_j} = \frac{\partial a_{ik}}{\partial x_j} - \frac{\partial a_{jk}}{\partial x_i} + \sum_{l=m+1}^{n} a_{jl}\frac{\partial a_{ik}}{\partial x_l} - a_{il}\frac{\partial a_{jk}}{\partial x_l},$$

which in Jacobi operator notation is

$$A_j(a_{ik}) - A_i(a_{jk}) = 0, \quad i,j = 1,\ldots,m, \quad k = m+1,\ldots,n.$$

Since this condition must hold for all $x_1,\ldots,x_m$ and all values of the constants $C_{m+1},\ldots,C_n$, it must hold identically in $x_1,\ldots,x_n$ and so is equivalent to the condition that $A_i(A_j(f)) - A_j(A_i(f)) = 0$ for all f and all $i, j = 1, \ldots, m-1$, which is the Jacobi integrability condition (6.13). This then is a necessary condition of the complete integrability of the Pfaffian system (6.29): the corresponding system (6.28) of partial differential equations must satisfy Jacobi's integrability condition. According to Mayer, implicit in his discussion of the integration of (6.29) was a proof that this condition on (6.29) is also sufficient.

Mayer's interest in the above duality was motivated by his interest in the goal of integration efficiency mentioned in Section 6.2, which was of interest to many analysts at this time. His idea was to start with a system of partial differential equations satisfying Jacobi's integrability condition—the type of system involved in the integration of nonlinear partial differential equations in accordance with Jacobi's new method and its extensions—and then go over to the dual system of Pfaffian equations and integrate it to see whether it yielded a more efficient method. He showed that it did: that the number of integrations needed could be reduced by almost 50% over what the latest theories offered.

Although Mayer had indeed established a reciprocity or duality between systems of linear homogeneous partial differential equations and systems of Pfaffian equations, he had done so by assuming the systems in a special form so that the duality could be obtained from the well-known equivalence of a single linear homogeneous partial differential equation and a system of ordinary differential equations. As we shall see, by taking a more algebraic and elegant approach and by virtue of a powerful new construct—the bilinear covariant—Frobenius, who never mentioned Mayer's paper, was able to establish the reciprocity without assuming the systems in a special generic form. This enabled him to formulate a criterion for complete integrability that was directly applicable to any system of Pfaffian equations.

Given the special form of the systems required for duality in Mayer's sense and the focus of his attention on efficiency matters, it is not clear that he himself realized that the systems of partial differential equations of Clebsch's direct method and the Pfaffian systems of Natani were in fact duals of one another. This was pointed out by Meyer Hamburger (1838–1903) in a paper [261] submitted for publication a few months after Frobenius had submitted his own paper on the problem of Pfaff. Hamburger's method for establishing the duality between Natani's systems of Pfaffian equations and Clebsch's systems of partial differential equations in [261] was probably known to Frobenius, because Hamburger had presented it in the context of a different problem in a paper of 1876 [260, p. 252] in *Crelle's Journal* that Frobenius did cite in his paper. Hamburger's method is more algebraic than Mayer's and does not require that the systems be put in a special generic form. It may have encouraged Frobenius' own simpler algebraic method.

In accordance with the approach of Hamburger but especially that of Frobenius [179, §13], who unlike Hamburger presented everything with elegant simplicity and algebraic clarity, let us consider a system of $r < n$ independent Pfaffian equations

$$\omega_i = a_{i1}(x)dx_1 + \cdots + a_{in}(x)dx_n = 0, \quad i = 1, \ldots, r. \tag{6.30}$$

This means that if $a_i = \begin{pmatrix} a_{i1} & \cdots & a_{in} \end{pmatrix}$ and A is the matrix whose rows are $a_1, \ldots, a_r$, then $\operatorname{rank} A = r$. A Pfaffian equation $\varpi = \sum_{i=1}^{n} b_i(x)dx_i = 0$ is said to belong to the system (6.30) if $\varpi = \sum_{i=1}^{r} c_i(x)\omega_i$, which is equivalent to saying that $b(x) = \begin{pmatrix} b_1 & \cdots & b_n \end{pmatrix} = \sum_{i=1}^{r} c_i(x)a_i(x)$, or, in more modern terms, that $b(x) \in \operatorname{Row}[A(x)]$, where $\operatorname{Row}[A(x)]$ denotes the row space of $A(x)$. In effect, Frobenius identified the system (6.30) with $\operatorname{Row}[A(x)]$, because he said that this system could be replaced by any system $\varpi_i = 0$, $i = 1, \ldots, r$, of independent Pfaffian equations that belong to the system (6.30) in the above sense.

Frobenius defined an equation $f(x) = C$ to be an integral of the Pfaffian system (6.30) if its differential $df = \sum_{i=1}^{n}(\partial f/\partial x_i)dx_i$ belongs to the system in the above-defined sense. (This is equivalent to saying that ∇f, the gradient of f, is in $\operatorname{Row}[A(x)]$.) Now $f(x) = C$ defines an integral manifold $\mathcal{M}$ for the Pfaffian equation $df = 0$, but if $r > 1$, $\mathcal{M}$ is not an integral manifold for the entire system (6.30) in a sense consistent with that of Section 6.2, which would require that every vector dx in the tangent space $\mathcal{T}$ to $\mathcal{M}$ at a point of $\mathcal{M}$ be orthogonal to all of $a_1, \ldots, a_r$, i.e., that $\mathcal{T} \subset \operatorname{Row}[A(x)]^{\perp}$. This is impossible unless $r = 1$, because $\mathcal{T} \subset \operatorname{Row}[A(x)]^{\perp}$ means that $n - 1 = \dim \mathcal{M} = \dim \mathcal{T} \leq \dim \operatorname{Row}[A(x)]^{\perp} = n - r$. Likewise, if the system (6.30) has $\rho < r$ independent integrals $f_i(x) = C_i$, $i = 1, \ldots, \rho$, in Frobenius' sense, this means that the manifold defined by the intersection of the ρ hypersurfaces $f_i(x) = C_i$ is an integral manifold for the system of ρ Pfaffian equations $df_i = 0$ but not for the entire system (6.30). It is only when $\rho = r$ that the manifold defined by the r equations $f_i(x) = C_i$ is an integral manifold for the entire system (6.30), since in that case, every ω_i is necessarily a linear combination of $df_1, \ldots, df_r$. This, then, is the motivation behind Frobenius' definition of when the system (6.30) is said to be complete [179, p. 286].

Definition 6.8. A system of r Pfaffian equations

$$\omega_i = a_{i1}(x)dx_1 + \cdots + a_{in}(x)dx_n = 0, \quad i = 1, \ldots, r,$$

that is independent in the sense that $\operatorname{rank}\left[(a_{ij})(x)\right] = r$ is said to be *complete* if r independent functions $f_1, \ldots, f_r$ exist such that the r equations $f_i(x) = C_i$ are integrals of this system.

By virtue of what was said above, the Pfaffian system $\omega_i = 0$ is complete in Frobenius' sense precisely when it is integrable, i.e., when the manifold defined by $f_i(x) = C_i$, $i = 1, \ldots, r$, is an integral manifold for the system.

In his approach to duality, Hamburger tacitly assumed for convenience that since $A = (a_{ij})$ has rank r, the $r \times r$ minor matrix A_r of A defined by the first r rows and columns of A has nonzero determinant. Suppose $f(x_1, \ldots, x_n) = C$ is an integral of the Pfaffian system (6.30), so that ∇f is a linear combination of the rows of A.[30]

[30] Hamburger's presentation is not as clear as I am indicating. He never defined what it means for $f = C$ to be an integral, but concluded that the equations (6.30) imply that $df = 0$ and that this in turn implies the $n - r$ vanishing determinants indicated below.

Then the $(r+1)\times n$ matrix

$$A^* = \begin{pmatrix} A \\ \nabla f \end{pmatrix} = \begin{pmatrix} a_1^{(1)} & \cdots & a_n^{(1)} \\ \vdots & \ddots & \vdots \\ a_1^{(r)} & \cdots & a_n^{(r)} \\ \frac{\partial f}{\partial x_1} & \cdots & \frac{\partial f}{\partial x_n} \end{pmatrix}$$

has rank r, i.e., every $(r+1)\times(r+1)$ minor matrix of A^* has vanishing determinant. These determinants set equal to 0 yield a system of linear homogeneous partial differential equations having f as solution. Not all of these equations are independent, however, and Hamburger singled out the $n-r$ systems that arise by forming minors using the first r columns of A^* plus one of the remaining $n-r$ columns. Hamburger's approach was thus more algebraic and general than Mayer's, and that is undoubtedly why Frobenius mentioned it. Hamburger, however, never justified that his equations were independent. Frobenius avoided the need for this by proceeding somewhat differently after mentioning Hamburger's approach.

Frobenius considered the linear homogeneous system of equations associated to the Pfaffian system (6.30), namely,

$$a_{i1}u_1 + \cdots + a_{in}u_n = 0, \quad i = 1,\ldots,r, \tag{6.31}$$

or in more familiar notation, $Au = 0$, where $A = (a_{ij})$ is the $r\times n$ matrix of the coefficients and $u = \begin{pmatrix} u_1 & \cdots & u_n \end{pmatrix}^t$. It was well known at this time that the system $Au = 0$—with A a matrix of constants and of rank r—has $n-r$ linearly independent solutions.

For example, in the 1870 edition of Baltzer's text on determinants, which is often cited by Frobenius, there is a theorem due to Kronecker that gives a formula for the general solution to $Au = 0$ [13, pp. 66–67].

Theorem 6.9 (Kronecker's theorem). *Given an $m\times n$ matrix $A = (a_{ij})$ of rank r, suppose for specificity that the minor determinant of A formed from its first r rows and columns is nonzero. Then the solutions to the homogeneous system $Au = 0$ are given as follows. Consider the $(r+1)\times(r+1)$ matrix*

$$\begin{pmatrix} a_{11} & \cdots & a_{1r} & a_{1r+1}u_{r+1} + \cdots + a_{1n}u_n \\ \vdots & \ddots & \vdots & \vdots \\ a_{r1} & \cdots & a_{rr} & a_{rr+1}u_{r+1} + \cdots + a_{rn}u_n \\ * & \cdots & * & * \end{pmatrix},$$

where the last row can be anything. Then the general solution to $Au = 0$ is $u = \begin{pmatrix} C_1/C_{r+1} & \cdots & C_r/C_{r+1}, u_{r+1} & \cdots & u_n \end{pmatrix}^t$, where $C_1,\ldots,C_{r+1}$ denote the cofactors along the (starred) last row (so $C_{r+1} \neq 0$).

The general solution involves $n-r$ "free variables" $u_{r+1},\dots,u_n$, and by successively setting one of these equal to 1 and the rest 0, we obtain (in the now-familiar way) $n-r$ solutions that can be shown to be linearly independent.

Frobenius surely knew Kronecker's above theorem, but in the purely algebraic part of his paper, he presented a more elegant, perfectly general way of establishing the $n-r$ independent solutions, which seems to have originated with him.[31] It is of historical interest because it shows how he was compelled by a penchant for algebraic elegance and generality to a type of linear algebra with strikingly modern overtones despite the continued reliance on determinants.

Frobenius proceeded as follows [179, pp. 255ff.]. Consider an $r\times n$ system of equations $Au=0$ $(r<n)$ with the coefficients of A assumed constant and $r=\operatorname{rank}A$. (Keep in mind that $A=A(x)$ with x fixed.) Then we may pick $n-r$ n-tuples $w_k=(w_{k1},\dots,w_{kn})$ such that the $n\times n$ matrix D with rows consisting of the r rows $a_i=(a_{i1},\dots,a_{in})$ of A followed by the $n-r$ rows w_k, viz.,

$$D=\begin{pmatrix} A \\ w_1 \\ \vdots \\ w_{n-r} \end{pmatrix},$$

satisfies $\det D\neq 0$. Then let b_{kj} denote the cofactor of D corresponding to the w_{kj} coefficient. If D is modified to D^* by replacing row w_k by row a_i, then $\det D^*=0$, since it has two rows equal. Expanding $\det D^*$ by cofactors along the changed row yields the relation

$$\sum_{j=1}^{n} a_{ij}b_{kj}=0, \quad i=1,\dots,r,\ k=1,\dots,n-r, \tag{6.32}$$

where b_{kj} denotes the cofactor corresponding to a_{kj}. I will express these relations as the dot products $a_i\cdot b_k=0$ for $i=1,\dots,r$ and $k=1,\dots,n-r$, where $b_k=(b_{k1},\dots,b_{kn})$. Using well-known determinant-theoretic results that go back to Jacobi, Frobenius easily concluded that the $(n-r)\times n$ matrix B with the b_k as its rows has full rank $n-r$, so that the b_k represent $n-r$ linearly independent solutions to the homogeneous system $Au=0$. He also showed that no more than $n-r$ independent solutions of $Au=0$ can exist. However, he did not stop with this result.

Frobenius defined the above two coefficient systems $A=(a_{ij})$, which is $r\times n$, and $B=(b_{ij})$, which is $(n-r)\times n$, to be "associated" or "adjoined"; likewise, the two systems of equations $Au=0$ and $Bv=0$ are said to be adjoined. The reason for this terminology was the following immediate consequence of (6.32): $a\cdot b=0$, where a is any linear combination of the rows of A and b is any linear combination

[31] Judging by his remarks some thirty years later [226, pp. 349ff.].

of the rows of B. In other words, as we could now say it, Row A and Row B are orthogonal. In fact, given the ranks of A and B, Frobenius realized that any a such that $a \cdot b = 0$ for all $b \in \text{Row } B$ must belong to Row A and vice versa, i.e., he realized what would now be expressed by

$$\text{Row } A = [\text{Row } B]^{\perp} \quad \text{and} \quad \text{Row } B = [\text{Row } A]^{\perp}. \tag{6.33}$$

Furthermore, adjoinedness meant that "the coefficients of the one system of equations are the solutions of the other" [179, p. 257], i.e., in modern terms,

$$\text{Row } A = \text{Null } B \quad \text{and} \quad \text{Row } B = \text{Null } A, \tag{6.34}$$

where, e.g., Null A denotes the null space of A. This was an immediate consequence of the obvious fact, frequently used by Frobenius, that u is a solution to a system of equations $Au = 0$ if and only if $u \cdot a = 0$ for any a that is a row of A or a linear combination thereof, i.e., what would now be expressed by

$$\text{Null } A = [\text{Row } A]^{\perp}. \tag{6.35}$$

Returning now from this excursion into Frobenius-style linear algebra, let us consider how he applied the key relation (6.33) as it relates to the matrix $A = A(x)$ associated to the Pfaffian system (6.30) to obtain its dual. By his definition, $f = C$ is a solution to the Pfaffian system (6.30) if and only if ∇f is a linear combination of the rows of $A(x)$, and by (6.33), this is equivalent to saying that ∇f is orthogonal to the rows of the matrix $B = B(x)$ adjoined to $A(x)$, i.e.,

$$X_k(f) \stackrel{\text{def}}{=} b_{k1}\frac{\partial f}{\partial x_1} + \cdots + b_{kn}\frac{\partial f}{\partial x_n} = 0, \quad k = 1, \ldots, n-r. \tag{6.36}$$

The above system is then defined to be the system of linear partial differential equations adjoined to the Pfaffian system (6.30)—the dual system as I will call it. Likewise, as Frobenius showed, if one starts with a system $X_k(f) = 0$, $k = 1, \ldots, n-r$, the above reasoning can be reversed to define the associated Pfaffian system $\omega_i = 0$, $i = 1, \ldots, r$.

Frobenius used the correspondence between the systems $X_k(f) = 0$ of (6.36) and $\omega_i = 0$ of (6.30) to translate Clebsch's integrability condition (6.14) for the system $X_k(f) = 0$ into an integrability condition for the system $\omega_i = 0$ [179, §14]. His reasoning utilized his version of (6.33)–(6.34). With this fact in mind, note the following implications of the duality correspondence $A \leftrightarrow B$ or, equivalently, $X_k(f) = 0 (k = 1, \ldots, n-r) \leftrightarrow \omega_i = 0 (i = 1, \ldots, r)$. First of all, $\omega = \sum_{i=1}^{n} a_i(x)dx_i$ belongs to the system $\omega_i = 0$ if and only if $a = (a_1, \ldots, a_n) \in \text{Row } A = [\text{Row } B]^{\perp} = \text{Null } B$. Secondly, $X(f) = \sum_{i=1}^{n} b_i(x)\partial f/\partial x_i = 0$ belongs to the system $X_k(f) = 0$, i.e., $X(f)$ is a linear combination of the $X_k(f)$ if and only if $b = (b_1, \ldots, b_n)$ is a linear combination of the b_k, i.e., $b \in \text{Row } B = [\text{Null } B]^{\perp}$.

Let us now consider, along with Frobenius, Clebsch's integrability condition (6.14). It implies that if the equations

$$X(f) = \sum_{i=1}^{n} b_i(x)\partial f/\partial x_i = 0 \quad \text{and} \quad Y(f) = \sum_{i=1}^{n} c_i(x)\partial f/\partial x_i = 0$$

belong to the system $X_k(f) = 0$, then so does $X(Y(f)) - Y(X(f)) = 0$, where

$$X(Y(f)) - Y(X(f)) = \sum_{i=1}^{n}\sum_{j=1}^{n}\left(b_j\frac{\partial c_i}{\partial x_j} - c_j\frac{\partial b_i}{\partial x_j}\right)\frac{\partial f}{\partial x_i}. \tag{6.37}$$

In view of the above preliminary remarks, the integrability condition may be stated as follows. Let $b = (b_1,\ldots,b_n)$ and $c = (c_1,\ldots,c_n)$ and let $[b,c]$ denote the coefficient n-tuple of $X(Y(f)) - Y(X(f))$ as given in (6.37), i.e.,

$$[b,c]_i = \sum_{j=1}^{n}\left(b_j\frac{\partial c_i}{\partial x_j} - c_j\frac{\partial b_i}{\partial x_j}\right), \quad i = 1,\ldots,n.$$

Thus Clebsch's integrability condition may be stated as follows: if $b \in [\text{Null } B]^{\perp}$ and $c \in [\text{Null } B]^{\perp}$, then also $[b,c] \in [\text{Null } B]^{\perp}$. Since $a \in \text{Row } A = \text{Null } B$ by (6.34), this says that if $b \cdot a = 0$ and $c \cdot a = 0$ for all $a \in \text{Null } B$, then $[b,c] \cdot a = 0$ for all $a \in \text{Null } B$. In terms of the Pfaffian expression $\omega = \sum_{i=1}^{n} a_i(x)dx_i$ defined by a, the criterion takes the following form. Let $\omega(b) = \sum_{i=1}^{n} a_i(x)b_i = a \cdot b$. The above considerations may then be summarized in the following form.

Lemma 6.10. *If $\omega(b) = 0$ and $\omega(c) = 0$ for all $\omega = \sum_{i=1}^{n} a_i(x)dx_i$ belonging to the system of Pfaffians (6.30) in the sense that $a(x) = \begin{pmatrix} a_1(x) & \cdots & a_n(x)\end{pmatrix} \in \text{Row}[(a_{ij}(x))]$, then $\omega([b,c]) = 0$.*

Now

$$\omega([b,c]) = \sum_{i=1}^{n}\sum_{j=1}^{n}\left(b_j\frac{\partial c_i}{\partial x_j} - c_j\frac{\partial b_i}{\partial x_j}\right)a_i. \tag{6.38}$$

Frobenius observed that he could rewrite this expression by utilizing the relations $\omega(b) = 0$ and $\omega(c) = 0$. For example, differentiating the first equation with respect to x_j yields

$$0 = \frac{\partial}{\partial x_j}\left(\sum_{i=1}^{n} a_i b_i\right) = \sum_{i=1}^{n}\frac{\partial a_i}{\partial x_j}b_i + \sum_{i=1}^{n} a_i\frac{\partial b_i}{\partial x_j},$$

which gives $\sum_{i=1}^{n} a_i\partial b_i/\partial x_j = -\sum_{i=1}^{n} b_i\partial a_i/\partial x_j$. Doing the same to $\omega(c) = 0$ likewise shows that $\sum_{i=1}^{n} a_i\partial c_i/\partial x_j = -\sum_{i=1}^{n} c_i\partial a_i/\partial x_j$. If these expressions are substituted in (6.38), the result is that

$$\omega([b,c]) = \sum_{i,j=1}^{n} \left(\frac{\partial a_i}{\partial x_j} - \frac{\partial a_j}{\partial x_i} \right) b_i c_j.$$

Thus $\omega([b,c]) = \Omega(b,c)$, where Ω is the bilinear covariant associated to ω, and so Lemma 6.10 can be reformulated in terms of bilinear covariants. In this manner, Frobenius showed that the Jacobi–Clebsch Theorem 6.3, which asserts that the system $X_k(f) = 0$, $k = 1, \ldots, n-r$, is complete in the sense of having $n - (n-r) = r$ independent solutions if and only if it satisfies the Clebsch integrability condition (6.14), translates into the following completeness theorem for the system $\omega_i = 0$ [179, p. 290].

Theorem 6.11 (Frobenius integrability theorem). *Given a system of r independent Pfaffian equations $\omega_i = \sum_{j=1}^{n} a_{ij}(x)dx_j = 0$, $i = 1, \ldots, r$, it is complete if and only if the following integrability condition holds: whenever $\omega_i(b) = 0$ and $\omega_i(c) = 0$ for all i, it follows that $\Omega_i(b,c) = 0$ for all i, where Ω_i is the bilinear covariant associated to ω_i.*

Theorem 6.11 is the source of the appellation "Frobenius' theorem," which is now commonplace. The first to so name it appears to have been Élie Cartan in his 1922 book on invariant integrals.[32] Singling out Frobenius' name to attach to this theorem is a bit unfair to Jacobi and Clebsch, since the above theorem is just the dual of the Jacobi–Clebsch Theorem 6.3. Furthermore, although Frobenius also gave a proof of the theorem that is independent of the Jacobi–Clebsch theorem and the consideration of partial differential equations, that proof was, as he explained [179, p. 291], simply a more algebraic and "symmetric" version of one by Heinrich Deahna (1815–1844) that he had discovered in *Crelle's Journal* for 1840 [112].[33]

There are thus bona fide historical reasons for renaming Frobenius' Theorem 6.11 the Jacobi–Clebsch–Deahna–Frobenius Theorem. That Cartan should focus on Frobenius is nonetheless also understandable on historical grounds. First of all, it was Frobenius who discovered that the "joint" result could be stated in the elegant form of Theorem 6.11. Also, as we shall see in Section 6.6, from Cartan's perspective it was Frobenius who had first revealed the important role the bilinear covariant can play in the theory of Pfaffian equations. Such a role is exemplified by the above theorem; and, inspired by such applications, Cartan sought to apply the bilinear covariant—the derivative of a 1-form in his calculus—to a wider range of problems. In the concluding section, I will offer some preliminary arguments to the effect that the phenomenon of the Frobenius integrability theorem is paradigmatic of

[32]Chapter X is on completely integrable Pfaffian systems, and the first section is entitled "Le théorème de Frobenius" [64, pp. 99ff.]. Immediately after stating Frobenius' theorem (as above), Cartan pointed out that the integrability condition could be formulated differently using his calculus of differential forms. This formulation, which he first published in 1901, is given in (6.44) below. In his monograph of 1945 on differential forms [66, p. 49], Cartan gave necessary and sufficient condition for completeness in his own form (6.44), and no mention is made of Frobenius.

[33]See [277, pp. 416–417] for a bit more on Deahna's paper.

one of the principal ways in which the work of Frobenius has affected the emergence of present-day mathematics.

Let us now briefly return to the reason Frobenius had established his integrability theorem. As we saw at the end of Section 6.4.2, Frobenius' proof of the analytic classification theorem, Theorem 6.7, required establishing that a certain system of partial differential equations $B_i(f) = 0$, $i = 1, \ldots, d$, has $n - d$ independent solutions, but because these equations were not given explicitly it was not possible to show that Clebsch's integrability condition (6.14) held. It was to show that the requisite number of solutions existed that Frobenius had proved his integrability theorem. The application of the theorem to this end was complicated in its details, but the basic lemma was the following.

Lemma 6.12. *Let* $\omega = \sum_{i=1}^{n} a_i(x)dx_i$ *and let* $\Omega(u,v) = \sum_{i,j=1}^{n} a_{ij}u_iv_j$ *be the associated binary covariant. Consider the following system of n Pfaffian equations derived from* Ω: $\omega_i = \sum_{j=1}^{n} a_{ij}dx_j$, $i = 1, \ldots, n$. *Let* $r = \text{rank}\,[(a_{ij})]$. *Then the above system contains a subsystem of r independent equations* $\omega_{i_k} = 0$, $k = 1, \ldots, r$. *This system satisfies the integrability condition of Theorem 6.11 and so possesses r independent integrals* $f_i(x) = C_i$, $i = 1, \ldots, r$.

Frobenius proved this using an identity Clebsch had established in the course of his direct method. This lemma is applied first to the differential form ω hypothesized in the analytic classification theorem in the case $p = 2m$, and then the case $p = 2m + 1$ is reduced to the previous case by a clever application of the above lemma.[34]

In addition to providing the motivation for Frobenius' integrability theorem, the analytic classification theorem itself has been used to develop Carathéodory's approach to thermodynamics.[35]

6.5 Initial Reactions

Compared to the work on Pfaff's problem by his predecessors, Frobenius' contribution was unique in two fundamental respects. It was the first clear and systematic attempt to deal with the problem in complete algebraic generality and by methods—dominated by rank considerations—well suited to such generality. It was a Berlin-style solution to the problem of Pfaff. The other unique feature was the introduction of the bilinear covariant of a Pfaffian as a key theoretical tool. Of course, Jacobi's skew-symmetric matrix had been central to the theory since its introduction in 1827, but by thinking of it as defining a bilinear form Ω associated to a linear form ω and by establishing the importance to the theory of its invariance

[34] For more details, see [277, pp. 417–418].

[35] This has been done by Bamberg and Sternberg [14], who use the "normal form theorem" (viz., Theorem 6.7) to establish a theorem due to Carathéodory that is fundamental to his approach to thermodynamics. See Ch. 22, especially pp. 771ff., and its appendix (pp. 838ff.).

under variable changes as in the analytic classification theorem and the integrability theorem (Theorem 6.11), Frobenius had added a new dimension to the theory that was eventually explored more deeply and broadly by Élie Cartan starting in 1899, as will be seen in the next section. Here I consider briefly the reaction to Frobenius' paper in the intervening years.

The appearance of Frobenius' paper prompted two mathematicians, Sophus Lie (1842–1899) and Gaston Darboux (1842–1917), to publish papers containing some analogous results that had been discovered independently of his work. Lie's interest in the problem of Pfaff was a natural part of his interest during the early 1870s in the theory of first-order partial differential equations. Indeed, the theory of contact transformations that he developed in this connection was directly related to Pfaffian equations, since by 1873, he was characterizing contact transformations as transformations in $2n+1$ variables $z, x_1, \dots, x_n, p_1, \dots, p_n$ that leave the Pfaffian equation $dz - \sum_{i=1}^{n} p_i dx_i = 0$ invariant.[36] In this period, Lie claimed that all his work on partial differential equations and contact transformations could be extended to the general problem of Pfaff, but the only work he published concerned an efficient method of integrating a Pfaffian equation in an even number of variables in the generic case. It was not until Klein called his attention to Frobenius' paper [179] on the general problem that Lie composed his own paper on the subject, which he published in 1877 in the Norwegian journal that he edited [415].

Invoking without proof theorems "from the theory of partial differential equations" that were actually based more specifically on his (largely unpublished) theory of contact transformations, Lie quickly arrived at Clebsch's Theorem 6.2 as his Theorem I: either (I) $2m$ independent functions F_i, f_i exist such that $\omega = \sum_{i=1}^{m} F_i df_i$ or (II) $2m+1$ independent functions Φ_i, φ_i exist such that $\omega = d\varphi_0 + \sum_{i=1}^{m} \Phi_i d\varphi_i$. Lie called these expressions normal forms of ω. He focused on the number of functions in a normal form. This number is, of course, precisely the Frobenius class number p, although Lie made no reference to Frobenius' paper, even though he knew through Klein of its existence.[37] Whether Lie ever looked at Frobenius' paper is unclear, but he gave his own proof that p is the sole invariant of ω, i.e., that any two normal forms of ω have the same p and two Pfaffians ω and ω' with the same p can be transformed into one another [415, §2]. Lie also described a procedure for determining p for a given Pfaffian ω that was a development of observations made by Jacobi [313, §22] and Natani [451, §8] in case (I) [415, §4]. Like the formulations of his predecessors, Lie's was cumbersome. In the light of Frobenius' paper, it is easily seen that if $A = (a_{ij})$ is Jacobi's skew-symmetric matrix associated

[36] A discussion of the evolution of Lie's theory of contact transformations and its relation to the origins of his theory of transformation groups is given in Chapters 1–2 of my book [276].

[37] In a letter to Mayer in March 1873 he wrote, "Frobenius' work is probably very good? Since I have little time, I have not been able to bring myself to read it" [155, p. 713]. In his reply [155, p. 714], Mayer pointed out that it was rather strange that Frobenius did not mention work of either of them in his paper, and this may have prompted Lie's decision not to cite Frobenius' paper.

to $\omega = \sum_{i=1}^{n} a_i dx_i$, and if $a = \begin{pmatrix} a_1 & \cdots & a_n \end{pmatrix}^t$, then one always has $p = \operatorname{rank} \begin{pmatrix} A \\ a^t \end{pmatrix}$ (as Frobenius surely realized),[38] and this is what Lie was getting at by his procedure, albeit without seeming to fully realize it.

The title of Lie's paper, "Theory of the Pfaffian Problem I," suggested a sequel, and Lie had indeed entertained the idea of a sizable second part containing an extension to the general problem of Pfaff of his theory of contact transformations, first-order partial differential equations, and transformation groups. This part never materialized. In 1883, Lie explained to Mayer that "Because of Frobenius I have lost interest in the problem of Pfaff .... I have already written too much that goes unread" [155, p. 714]. No doubt, to Lie, Frobenius' paper was representative of the "analytic" mathematical style of the Berlin school, which he and Klein opposed in favor of a more intuitive, geometric or "synthetic" approach. Since in the 1870s the Berlin school was one of the most prestigious and influential centers for mathematics, Lie's remarks probably reflect the seeming hopelessness of competing with the Berlin treatment of the problem of Pfaff presented by Frobenius.

While Frobenius was working on the problem in 1876, so was Gaston Darboux, although what he wrote up was not immediately submitted for publication. Instead, he gave his notes to Joseph Bertrand, who wished to incorporate them into his lectures at the Collège de France during January 1877. As Darboux explained in 1882, when he eventually published his work [111, p. 15n],

> Shortly thereafter a beautiful memoir by Mr. Frobenius appeared ... bearing a date earlier than that of January 1877 (September 1876) and there this learned geometer proceeded in a manner somewhat analogous to what I had communicated to Mr. Bertrand in the sense that it was based on the use of invariants and of the bilinear covariant of Mr. Lipschitz. Upon returning recently to my work, it seemed to me that my exposition was more calculation-free and, in view of the importance the method of Pfaff has assumed, that it would be of interest to make it known.

This passage makes it fairly certain that it was by virtue of Frobenius' "beautiful" (but calculation-laden) treatment of the problem of Pfaff that Darboux was now publishing his own approach. By considering the problem in such great generality and detail, I suspect that Frobenius had contributed greatly to the perception of "the importance the method of Pfaff has assumed," and it was because Darboux's own approach was more calculation-free than Frobenius' that he now thought it worthwhile to publish it. It would seem from Darboux's remarks that the first part

---

[38]Frobenius apparently never explicitly gave the above expression for p, but it is an immediate consequence of his remark directly following his even rank theorem (Theorem 6.5), namely that Theorem 6.5 implies that $\hat{A} = \begin{pmatrix} A & a \\ -a^t & 0 \end{pmatrix}$ has the same rank $2m$ as A (whence $p = 2m$) if and only if this is true of the rank of $\begin{pmatrix} A \\ -a^t \end{pmatrix}$ [179, p. 263]. Thus the latter matrix has rank $2m+1$ precisely when $p = 2m+1$. At this point in his paper, Frobenius had not yet introduced the class number p and so was not in a position to explain this point to his readers.

of his memoir, discussed below, represented what he had written in 1876 and given to Bertrand.

Like Frobenius, Darboux began by establishing the "fundamental formula" that shows that $\sum_{i,j=1}^{n} a_{ij}dx_i\delta_j$ is a bilinear covariant, i.e., Lipschitz's Theorem 6.4. However, he did not make Frobenius' bilinear covariant $\Omega(u,v)$ the conceptual basis for his theory but rather focused on the system of linear differential equations

$$a_{i1}(x)dx_1 + \cdots + a_{in}(x)dx_n = \lambda a_i(x)dt, \quad i = 1,\ldots,n,$$

where t denotes an auxiliary variable (so the x_i can be regarded as arbitrary functions of t) and λ is "a quantity that could be chosen arbitrarily as 0, a constant, or a function of t, according to the case" [111, p. 19]. For brevity I will denote this system in matrix notation by $A\,dx = \lambda a\,dt$. Given two Pfaffians ω and ω' in variables x_i and x_i', respectively, and with respective systems $A\,dx = \lambda a\,dt$ and $A'\,dx' = \lambda a'\,dt$, Lipschitz's Theorem 6.4 then implied that if $\omega = \omega'$ by virtue of $x = \Phi(x')$ and $dx = J(\Phi)\,dx'$, then the corresponding systems are likewise equivalent in the sense that $A\,dx = \lambda a\,dt$ transforms into $A'\,dx' = \lambda a'\,dt$.

Like Lie, Darboux first quickly established Clebsch's Theorem 6.2. His proof was based on a theorem (stated at the end of Section IV) that he proved carefully in the two generic cases, namely n is even and $\det A \neq 0$, and n is odd (so $\det A = 0$) and $\operatorname{rank} A = n-1$ [111, §III]. He considered briefly only the nongeneric case with n even [111, §IV], and in this connection his vague argument seems to have taken for granted that a system of linear partial differential equations $X_j(f) = 0$, $i = 1,\ldots,r$, will have $n-r$ independent solutions even though it is not clear that the system satisfies Clebsch's integrability condition (6.14).[39] This is precisely the same sort of problem that had compelled Frobenius into a lengthy detour that involved among other things duality considerations and his integrability theorem, Theorem 6.11 (as indicated at the end of Section 6.4.2).

Although Darboux's penchant for generic reasoning had prevented him from giving a valid proof of Clebsch's Theorem 6.2, he showed how to use that theorem to determine a condition distinguishing the two cases and to determine the number of terms in the appropriate normal form for any given system $A\,dx = \lambda a\,dt$. His results may be summed up as follows.

Theorem 6.13 (Darboux). *If the system of differential equations $A\,dx = \lambda a\,dt$ associated to the Pfaffian equation $\omega = 0$ has solutions only when $\lambda = 0$, then a variable change $x \to y, z$ is possible such that*

$$\text{(II)} \quad \omega = dy - [z_1 dy_1 + \cdots + z_m dy_m],$$

and $2m$ is the number of linearly independent equations to which $Adx = \lambda a dt$ reduces, i.e., $2m$ is the number of linearly independent rows of the matrix $\begin{pmatrix} A & \lambda a \end{pmatrix} =$

[39] See [277, p. 421] for more details.

$\begin{pmatrix} A & 0 \end{pmatrix}$.[40] *If* $A\,dx = \lambda a\,dt$ *has solutions for* $\lambda \neq 0$, *then* ω *may be put in the form*

$$\text{(I)} \quad \omega = z_1 dy_1 + \cdots z_m dy_m,$$

and $2m$ *is the number of linearly independent equations to which* $A\,dx = \lambda a\,dt$ *reduces, i.e.,* $2m$ *is the number of linearly independent rows of* $\begin{pmatrix} A & \lambda a \end{pmatrix}$.

Darboux's proof proceeded as follows. Suppose by Clebsch's theorem that a variable $x \to y, z$ change puts ω in its normal form. If that normal form is (II), then the associated system $A\,dx = \lambda a\,dt$ is transformed into

$$dy_i = 0, \quad dz_i = -\lambda dz_i, \; i = 1, \ldots, m, \quad \text{and} \quad 0 = \lambda dt.^{41} \tag{6.39}$$

Thus a solution can exist only when $\lambda = 0$ in case (II), and the system $A\,dx = \lambda a\,dt$ reduces to a completely integrable system of $2m$ equations $dy_i = 0$, $dz_i = 0, i = 1, \ldots m$. In this case, (6.39) consists of the $2m$ independent equations $dy_i = 0$, $dz_i = 0$, i.e., $dy_i/dt = 0$, $dz_i/dt = 0$, $i = 1, \ldots, m$, and so this must be true, by virtue of the covariance of A, of the original system $A\,dx = \lambda a\,dt$ as well, i.e., the matrix $\begin{pmatrix} A & 0 \end{pmatrix}$, and so the matrix A, has $2m$ linearly independent rows.

Next suppose $x \to y, z$ puts ω in the normal form (I). Then $A\,dx = \lambda a\,dt$ transforms into

$$dy_i = 0, \quad dz_i = \lambda z_i dt, \quad i = 1, \ldots, m, \tag{6.40}$$

which has solutions with $\lambda = \text{const} \neq 0$, namely $y_i = C_i$, $i = 1, \ldots, m$, and $z_i = C_i' e^{\lambda t}$, or, to eliminate t, $z_i/z_1 = D_i$, $i = 2, \ldots, m$ [111, pp. 28–29]. Thus in this case, there are $2m$ independent equations (e.g., taking $\lambda = 1$, as we may), and so $2m$ is the number of linearly independent rows in $\begin{pmatrix} A & a \end{pmatrix}$.

In principle, Darboux's theorem gives a criterion, based on consideration of $A\,dx = \lambda a\,dt$, of determining the appropriate normal form for ω, although the task of determining when, for a given system $A\,dx = \lambda a\,dt$, solutions exist only when $\lambda = 0$ does not seem easy in general. Darboux, however, did show how to apply his theorem in "the most general case," by which he meant the generic case [111, pp. 29–30]—the sole case for which Darboux's Theorem 6.13 was proved rigorously. That is, he supposed first that (a) n is even and that $\det A \neq 0$ (the case dealt with by Pfaff and Jacobi). Then "one can solve the equations ... $[A\,dx = \lambda a\,dt]$... for the differentials dx_i." That is, $dx/dt = \lambda A^{-1} a$, and so by the theory of ordinary differential equations, a solution with $\lambda \neq 0$ exists and case (I) obtains. Hence by Darboux's Theorem 6.13, $2m = n$, and so $m = n/2$. Next he supposed that (b) n is odd, so that necessarily the skew-symmetry of A forces $\det A = 0$. "[B]ut

---

[40] Darboux spoke of the number of "distinct" equations.

[41] This follows because the transformed system is $A'dx' = \lambda a'dt$, where $a' = (1 - z_1 \cdots - z_m 0 \cdots 0)^t$ and A' is as in (6.20).

its minors of the first order are not zero in general. As we have seen, it is thus necessary, save for an exceptional case, that $\lambda = 0$ and thus that the equations reduce to $n-1$ distinct ones ... ." This statement is full of the ambiguity that attends generic reasoning (which, as we have seen, Darboux tended to favor). The "exceptional case" could be interpreted as the case in which $\operatorname{rank} A < n-1$, but that is not what one means by "the most general case." Darboux must have meant that assuming $\operatorname{rank} A = n-1$, then (as we would put the matter) "in general," a will not lie in the $(n-1)$-dimensional column space of A, i.e., "in general," $\operatorname{rank} \begin{pmatrix} A & a \end{pmatrix} > \operatorname{rank} A$. The exceptional case ignored by Darboux would then be $\operatorname{rank} \begin{pmatrix} A & a \end{pmatrix} = \operatorname{rank} A = n-1$. (This exception occurs, e.g., for $\omega = 2(x_1dx_1 - x_1dx_2 + x_1dx_3)$.) If, following Darboux, we ignore this exceptional case, then the case $\lambda \neq 0$ could not hold. For if it did, then by Darboux's Theorem 6.13 we would have $2m = \operatorname{rank} \begin{pmatrix} A & a \end{pmatrix} = n$, which is impossible since n is odd. Thus we are in the $\lambda = 0$ case, as Darboux concluded.

The part of Darboux's paper we have been discussing was written before he knew of Frobenius' paper on the subject. As we have seen, Darboux was one of many mathematicians who felt it was acceptable to focus primarily on the generic cases, whereas in his paper Frobenius showed that completely general and formally elegant arguments could be given to deal simultaneously with all cases. However, Frobenius needed many such arguments, including the detour of his integrability theorem, to achieve his end. As we shall see in the following section, Darboux's paper may have suggested to Cartan the idea that if a normal form for ω could be quickly (albeit rigorously) established, then many results of the theory could follow easily in the normal form and yet be true in general by covariance. Cartan did this by means of a clever definition of Frobenius' class number p articulated within the framework of his differential calculus of 1-forms, a calculus inspired by the bilinear covariant, especially in the form given to it by Frobenius.

Although analysts appear to have been generally impressed by the masterful command of algebra manifested in Frobenius' lengthy paper, some found his entire approach too algebraic and regretted his focus on the equivalence theory of Pfaffian forms to the neglect of the issue of efficient methods for integrating Pfaffian equations. Mayer's review of Frobenius' paper in the 1880 volume of *Fortschritte der Mathematik* [439] was along these lines. Although admitting that Frobenius' paper was "substantial in both scope and content," he proceeded to contrast, somewhat unfavorably, Frobenius' approach with the "entirely different one" of Lie. Lie "at the very outset with the help of a few simple theorems on partial differential equations" quickly established Clebsch's Theorem 6.2 and the necessary and sufficient conditions for equivalence of two Pfaffians and showed how one can algebraically determine the number p of variables in the normal form of ω. These matters thus quickly dispatched, he then "focused his attention primarily on the reduction to normal form by the smallest possible number of integrations ... ." But Frobenius "conceives of the problem in purely algebraic terms and, as a consequence of this, while the algebraic side of the problem is more deeply grounded, leaves the question of how to best integrate each successively occurring complete system entirely untouched" [439, p. 250].

Mayer's attitude seems to have been fairly typical of most analysts of the period who were primarily interested in the integration of differential equations. Thus A.R. Forsyth, in the historical remarks to the 1890 volume of his *Theory of Differential Equations*, which was devoted to Pfaffian equations, wrote that

> Lie's results constitute a distinct addition to the theory ... About the time of publication of the memoir by Lie just referred to, Frobenius had ... completed his memoir dealing with Pfaff's problem. He discusses the theory of the normal form rather than the integration of the equation; and the analysis is more algebraic than differential [165, p. 87].

Not surprisingly, only a very slim chapter—Chapter X—was devoted to "Frobenius' Method," and it began with a justification for its brevity: "The investigations of Frobenius ... deal rather with the general theory of the reduction of the [Pfaffian] expression to a normal form than with any process for the integration of equations which occur in the reduction ... [165, p. 272].

Forsyth's book was more a compendium of diverse methods for treating Pfaff's problem than a synthesis of those methods into a coherent theory. The first attempt at such a synthesis was made in 1900 by von Weber [575], an instructor (*Privatdozent*) at the University of Munich. In von Weber's book, Frobenius' work was given a more fundamental role to play, but in reality, von Weber's whole approach was destined for obsolescence because in 1899, Élie Cartan (1869–1951) had begun to develop an entirely new approach to the theory of Pfaffian equations.[42]

6.6 Cartan's Calculus of Differential Forms

Cartan had obtained his doctorate in 1894 with a brilliant thesis that provided a rigorous foundation for Killing's groundbreaking theory of finite-dimensional complex semisimple Lie algebras and its most impressive consequence—a complete classification of simple Lie algebras.[43] During the following four years, most of his attention was focused on developing applications of the ideas and results of his thesis, but he was also on the lookout for new areas of research. One such area he seems to have been considering at the time was the application of Lie's theory of groups to Poincaré's theory of invariant integrals. This comes out in a little paper of 1896 on Darboux-style differential geometry [59], which is especially relevant to the present account because in it we see that Cartan already realized that the variable change formulas in multiple integrals could be derived by submitting the differentials involved to certain rules of calculation. In a footnote, he made the following observation.[44] Consider an oriented surface integral in n-dimensional space such as

[42] A more detailed discussion of von Weber's work has been given by Cogliati [101, §5].

[43] Cartan's thesis work is the subject of Chapter 6 of [276].

[44] See [59, p. 143n]. For expository reasons, I have changed Cartan's notation and expanded his brief remarks. On earlier work, e.g., by Poincaré, on the transformation of multiple integrals and

$\iint_\Sigma dx_i dx_j$, where Σ denotes a 2-manifold. Then if new variables $x_i = f_i(y_1,\ldots,y_n)$ are introduced, the multiple integral transformation formula may be derived by formally multiplying out $dx_i dx_j = [\sum_{k=1}^n (\partial x_i/\partial y_k)dy_k][\sum_{l=1}^n (\partial x_j/\partial y_l)dy_l]$ using the rules

$$dy_k dy_k = 0 \quad \text{and} \quad dy_l dy_k = -dy_k dy_l \ (k \neq l) \tag{6.41}$$

to obtain for $dx_i dx_j$ the expression

$$\sum_{k,l=1}^{n} \frac{\partial x_i}{\partial y_k}\frac{\partial x_j}{\partial y_l} dy_k dy_l = \sum_{k<l}\left(\frac{\partial x_i}{\partial y_k}\frac{\partial x_j}{\partial y_l} - \frac{\partial x_i}{\partial y_l}\frac{\partial x_j}{\partial y_k}\right) dy_k dy_l = \sum_{k<l} \frac{\partial(x_i,x_j)}{\partial(y_k,y_l)} dy_k dy_l.$$

Since this formalism is to be applied to an integral over Σ, which is represented parametrically by $y_k = \varphi_k(s,t)$, $k = 1,\ldots,n$, Cartan added another rule:

$$dy_k dy_l = \frac{\partial(y_k,y_l)}{\partial(s,t)}\, ds\, dt, \tag{6.42}$$

so that

$$\iint_\Sigma dx_i dx_j = \iint_D \sum_{k<l} \frac{\partial(x_i,x_j)}{\partial(y_k,y_l)}\frac{\partial(y_k,y_l)}{\partial(s,t)}\, ds\, dt,$$

where D is the parameter domain in the st-plane.

Thus already in 1896, we see that Cartan was in possession of the fundamental rules on which the now-familiar exterior algebra of differential forms is built. But it was not until 1899 that he found in the problem of Pfaff a sufficiently good reason to systematically develop a differential calculus based on those rules when they are combined with the key new notion of exterior differentiation. As he explained in the introductory remarks to an 1899 paper on the problem [61, pp. 241–242],

> The present work constitutes an exposition of the problem of Pfaff based on the consideration of certain differential expressions ... [which] ... are subject to the usual rules of calculation as long as the order of differentials in a product is not permuted. In sum, the calculus of these quantities is that of the differential expressions that are placed under the multiple integral sign. This calculus has many analogies with that of Grassmann; in addition, it is identical to the geometric calculus used by Mr. Burali-Forti in a recent book.

The similarity of his differential algebra to the "Grassmann algebra" underlying Grassmann's theory of extension (*Ausdehnungslehre*) was thus recognized by him by 1899, although the above quotation together with Cartan's observations of 1896 make it clear that Cartan's calculus was not suggested by Grassmann's work but by the consideration of multiple integrals. In fact, like most nineteenth-century

---

the relation of this work to the generalized Stokes theorem of the exterior calculus of differential forms, see [336, 337].

mathematicians, Cartan may not have attempted to penetrate the unusual notation of the *Ausdehnungslehre* (1862) so as to digest it contents. His knowledge of the latter may have been indirectly obtained through the above-mentioned Burali-Forti book, *Introduction à la géométrie différentielle suivant la méthode de H. Grassmann*, which appeared in 1897 [45].

It is nonetheless possible that Grassmann's work helped encourage Cartan to apply his calculus of differential forms to the problem of Pfaff, because through his familiarity with Forsyth's 1890 volume on the problem he would have known that Grassmann had considered the Pfaff problem in the 1862 version of his *Ausdehnungslehre* [253, §§500–527].[45] Noting the "remarkable formal conciseness" that Grassmann had achieved with his algebraic apparatus [165, p. 84], Forsyth included a brief chapter on Grassmann's work despite his suspicion that it would be unintelligible to most readers because "adequate knowledge of the analytic method of the Ausdehnungslehre" was "probably not common at present" [165, p. 120n]. The possibility that his algebra of differential forms might also lend itself to a concise treatment of the problem of Pfaff may thus have been suggested by Grassmann's work. But Cartan's overall approach drew more definite inspiration from Frobenius, who, Cartan wrote, "in his beautiful memoir in *Crelle's Journal* employs an entirely new method. It is based on consideration of what he calls the *bilinear covariant* of a Pfaffian expression" [61, p. 241].

Frobenius' bilinear covariant prompted Cartan to introduce something not found in Grassmann or Burali-Forti: the idea of what is now called the exterior derivative of a differential form. Indeed, if the rules (6.41) Cartan gave in 1896 are applied provisionally in the spirit of the above multiple integral calculations to formally calculate a derivative $d\omega$ of the Pfaffian $\omega = \sum_{i=1}^n a_i dx_i$, the result is

$$\begin{aligned} d\omega &= \textstyle\sum_{i=1}^n d(a_i dx_i) = \sum_{i=1}^n d(a_i)dx_i + \sum_{i=1}^n a_i d(dx_i) \\ &= \textstyle\sum_{i=1}^n \left\{\sum_{j=1}^n \frac{\partial a_i}{\partial x_j} dx_j\right\} dx_i + \sum_{i=1}^n a_i d(dx_i) \\ &= -\textstyle\sum_{i<j}^n \left(\frac{\partial a_i}{\partial x_j} - \frac{\partial a_j}{\partial x_i}\right) dx_i dx_j + \sum_{i=1}^n a_i d(dx_i). \end{aligned}$$

This suggests defining $d(dx_i) = 0$ in order that

$$d\omega \equiv \sum_{i=1}^n d(a_i)dx_i = -\sum_{i<j}^n a_{ij} dx_i dx_j, \tag{6.43}$$

where $A = (a_{ij})$ is the Jacobi skew-symmetric matrix associated to ω. For Cartan, an expression such as that in (6.43) was regarded as symbolical. To evaluate it, the rule (6.42) is applied to the $dx_i dx_j$ to obtain

[45] Forsyth's book [165] is cited by Cartan [61, p. 239n1] as a source of information on the problem of Pfaff.

$$dx_i dx_j = \frac{\partial(x_i,x_j)}{\partial(s,t)}dsdt = \left(\frac{\partial x_i}{\partial s}\frac{\partial x_j}{\partial t} - \frac{\partial x_i}{\partial t}\frac{\partial x_j}{\partial s}\right)dsdt = u_i v_j - u_j v_i,$$

where $u_k = (\partial x_k/\partial s)ds$ and $v_k = (\partial x_k/\partial t)dt$.[46] In effect $dx_i dx_j$ was for Cartan an alternating bilinear function of u,v (as is $dx_i \wedge dx_j$ for us today); and the identification of $d\omega$ as given in (6.43) with $-\Omega(u,v)$ would thus have been evident to Cartan—as well as the all-important fact that this sort of differentiation is covariant: if $\omega(x,dx) = \tilde{\omega}(\tilde{x},d\tilde{x})$, where $x = \Phi(\tilde{x})$, then also $d\omega = d\tilde{\omega}$ by virtue of the covariance of Ω [61, pp. 251–252].

In view of the above considerations it is not surprising to find Cartan defining the derivative of a Pfaffian expression, or 1-form as we shall now say, as $d\omega = \sum_{i=1}^{n} d(a_i)dx_i$.[47] Indeed, immediately after presenting this definition, he remarked that "consideration of ... $[d\omega]$ ... or, what amounts to the same thing, of the bilinear covariant of ω, forms the basis of the beautiful investigations on the problem of Pfaff by Frobenius and Darboux" [61, p. 252n].

By viewing Lipschitz's bilinear covariant as an alternating bilinear form Ω, Frobenius certainly made the connection between ω and the idea of its derivative $d\omega$ more readily apparent than did Darboux with his focus on the system $A\,dx = \lambda a\,dt$. But as noted at the end of Section 6.5, Darboux's paper may have suggested to Cartan the possibility of quickly establishing the existence of a normal form so as to take advantage of its simplicity to establish results easily for them that would then hold in any coordinate system. But how to do this rigorously, i.e., nongenerically, as per Frobenius rather than Darboux?

Cartan's clever idea was to define the class number p of a Pfaffian ω as the smallest integer p for which a coordinate change $x = \Phi(y)$ exists such that $\omega = \sum_{i=1}^{p} B_i(y_1,\dots,y_p)dy_i$, i.e., the B_i depend only on the first p variables $y_1,\dots,y_p$ of y. Thus a quasinormal form is factored into Cartan's definition of p. His rendition of the problem of Pfaff then consisted in proving that ω can be transformed into the form (I) or (II) of Frobenius' analytic classification theorem according to whether p is even or odd, respectively. Not only did this establish the equivalence of his definition of p with that of Frobenius, it also established the goal of Frobenius' memoir, his analytic classification theorem; and the route there was far more calculation-free due to Cartan's use of the above quasinormal form and covariance.

Here is an outline of that route. For a given 1-form ω, $d\omega$ is a 2-form, which is covariant. Cartan saw that he could use ω and $d\omega \stackrel{\text{sdef}}{=} \omega'$ to create k-forms for all $k > 2$ that are also covariant. He set $\omega'' = \omega \wedge d\omega$, which is a 3-form, and $\omega''' = \frac{1}{2}(d\omega \wedge d\omega)$, which is a 4-form. In general, $\omega^{(2m-1)} = \frac{1}{m!}(\omega^{(m)} \wedge \omega^{(m)})$ is a $2m$-form and $\omega^{(2m)} = \omega \wedge \omega^{(2m-1)}$ is a $(2m+1)$-form. It followed from the

[46] The evaluation of differential expressions, a natural consequence of their origin under multiple integral signs, became essential to Cartan's method of establishing relations among differential expressions.

[47] Cartan's notation for $d\omega$ in 1899 was ω'. In his 1945 monograph [66] he followed the lead of Kähler [332] and replaced ω' with $d\omega$.

covariance of $d\omega = -\Omega(u,v)$ that likewise if $\omega = \varpi$ under $x = \Phi(y)$, then also $\omega^{(k)} = \varpi^{(k)}$. From this it is easily seen that if ω is of class p in Cartan's sense and if $\varpi = \sum_{i=1}^{p} B_i(y_1,\ldots,y_p)dy_i$ is the associated quasinormal form in his definition, then $\varpi^{(p)} = 0$, and so $\omega^{(p)} = 0$. Cartan proved that ω is of class p if and only if p is the smallest integer such that $\omega^{(p)} = 0$ [61, p. 255], thereby demonstrating the succinctness possible with his notational apparatus.

As with the method of Frobenius, Cartan's route to the analytic classification theorem also required the solutions to successive systems of linear partial differential equations $A_k(f) = 0$, $k = 1,\ldots,l$ [61, p. 258]. We saw that because Frobenius was unable to show that Clebsch's integrability condition (6.14) was satisfied by his systems, he was forced to make a lengthy (albeit consequential) detour into duality considerations leading to his integrability theorem for systems of 1-forms. Not so with Cartan. For example, the first such system arises from the equation $\omega^{(p-2)} \wedge df = 0$, where ω is assumed to be a 1-form of class p. Since $\omega^{(p-2)}$ is a $(p-1)$-form, $\omega^{(p-2)} \wedge df$ is a p-form, and if its coefficients are all set equal to zero, the system $A_k(f) = 0$ results. To see that this system satisfies Clebsch's integrability condition, Cartan let ω, being of class p, be transformed into its associated quasinormal form $\varpi = \sum_{i=1}^{p} B(y_1,\ldots,y_p)dx_i$. Then $\omega^{(p-2)} \wedge df = 0$ takes the form $\varpi^{(p-2)} \wedge dg = 0$, $g(y) = f(\Phi(y))$, and by virtue of the simple nature of ϖ, it is easily seen that the system corresponding to $A_k(f) = 0$ in the y-variables satisfies Clebsch's integrability condition—and in fact in the special case of Jacobi's condition (6.13), viz. $[A_j,A_k] = 0$.

Although Cartan had no need of Frobenius' integrability theorem, Theorem 6.11, in his proof of the analytic classification theorem, he incorporated the integrability condition of that theorem—if $\omega(b) = 0$ and $\omega(c) = 0$, then $\Omega(b,c) = 0$—into his own notion of the integrability of an incomplete system of Pfaffian equations.[48] Cartan found a use for Frobenius' integrability theorem per se in his study of incomplete systems of Pfaffians with "characteristic elements" passing through any point x, thereby affording a simplified integration process. The characteristic elements were determined by a system of Pfaffian equations, which I will call the characteristic equations. The fundamental theorem of Cartan's theory of characteristics is that the characteristic equations form a complete system in the sense of Frobenius. Cartan's initial proof of this theorem [62, p. 304] was brief and did not use Frobenius' integrability theorem; but shortly thereafter, he presented an exposition of his results within the framework of his calculus of differential forms [63, Chs. 1–2], and then he used Frobenius' integrability theorem to establish the completeness of the characteristic equations. To do this, however, he first translated the integrability condition of Frobenius' theorem into a form more congenial to his calculus of differential forms.

He did this as follows [63, p. 496]. If $\omega_i = 0$, $i = 1,\ldots,r$, is a given system of linearly independent 1-forms, then $n-r$ additional 1-forms ϖ_j, $j = 1,\ldots,n-r$, may

[48] For a sketch of how he did this, see [277, p. 428].

be added so as to obtain n linearly independent 1-forms ω_i, ϖ_j. From the calculus of these forms it followed that any 2-form could be expressed as a linear combination of twofold products of these n 1-forms, namely $\varpi_j \wedge \varpi_k$, $\varpi_j \wedge \omega_i$, and $\omega_i \wedge \omega_j$. This applies in particular to the 2-forms $d\omega_i$, and so

$$d\omega_i = \sum_{j<k} b_{ijk} \varpi_j \wedge \varpi_k + \sum_{j=1}^{r} \theta_{ij} \wedge \omega_j,$$

where the θ_{ij} are the 1-forms obtained by factoring out ω_j from the terms of $d\omega_i$ involving $\varpi_k \wedge \omega_j$ or $\omega_i \wedge \omega_j$. According to Frobenius' integrability condition, $d\omega_i(b,c) = 0$ whenever $\omega_i(b) = 0$ and $\omega_i(c) = 0$ for all i. Since the ϖ_j are independent of the ω_i, the only way this can happen is if all the coefficients b_{ijk} are identically zero. The integrability condition in Cartan's rendition of Frobenius' theorem thus becomes the condition that

$$d\omega_i = \sum_{j=1}^{r} \theta_{ij} \wedge \omega_j, \quad i = 1, \ldots, r, \tag{6.44}$$

or, as he expressed it, $d\omega_i \equiv 0 \pmod{\omega_1, \ldots, \omega_r}$.

Although the problem of Pfaff had supplied the initial impetus for Cartan to develop his calculus of differential forms, his contribution to that problem pales in magnitude and significance with the subsequent applications he made of his calculus. Even his paper of 1899 contains considerably more than his new derivation of Frobenius' analytic classification theorem—the theorem that had formed the goal of Frobenius' own paper. From the outset, Cartan seems to have been interested in Pfaffian equations as a new means of dealing with partial differential equations of any order, and his 1899 paper contains much along these lines. Likewise, Cartan's papers of 1901 on incomplete systems of Pfaffian equations (discussed briefly above) provided a new and geometrically informed approach to systems of partial differential equations, which in 1934 was extended by Kähler [332] to include systems involving k-forms with $k > 1$.[49] Cartan also developed his theory of Pfaffian systems into the basis for his theory of the structure of infinite-dimensional Lie transformation groups, which concluded in 1909 with a classification of the simple types.[50] As he wrote retrospectively in 1931, "This notion of the bilinear covariant is of major importance, and it plays a dominant role in all my work on systems of partial differential equations and on infinite continuous groups" [65, p. 39].

In all of this work by Cartan, the consideration of bilinear covariants—derivatives of 1-forms—played a key role, and the idea that they should play such a role was something that Cartan evidently took away from the papers of Frobenius and Darboux. The following passage from Cartan's first paper of 1901 on incomplete systems is typical of the many such remarks in his papers that signal his debt to

[49] In 1945, Cartan gave his own exposition of his and Kähler's results in his monograph on the exterior calculus of differential forms and its geometric applications [66]. On the genesis of Cartan–Kähler theory, see [101].

[50] See Sections 8.1–8.2 of my book [276].

them. Before Cartan, incomplete systems had been considered by Otto Biermann in 1885 [23] but only in the generic case. Since then, Cartan wrote [62, p. 241],

> nothing has been done except to demonstrate the same results in another form but without ever achieving perfect rigor, and almost nothing has been done on the case in which the coefficients of the system are not generic.[51]
>
> Precise and general results can be achieved by taking into consideration the bilinear covariants of the right-hand sides of the equations of the system, whose introduction by Frobenius and Darboux has proved to be so fertile in the theory of a single Pfaffian equation.

The second paragraph of this passage indicates that Cartan had come to see the bilinear covariant $d\omega$ as a key mathematical tool. It also suggests that in seeking to assess the influence that Frobenius' paper had on Cartan, it is impossible to fully extricate the influence of Frobenius from that of Darboux, because invariably both men are mentioned together when a reference to the introduction of the bilinear covariant is made. Granted that caveat, the contents of the two papers in question are sufficiently different that it seems reasonable to assume that they impressed Cartan in different ways. From Darboux's paper, Cartan may have obtained the important idea of avoiding complicated calculations by invoking the covariance of his forms and their derivatives under variable change to normal forms—as in his above-sketched proof of Frobenius' analytic classification theorem. That constitutes the principal contribution that could have been made exclusively by Darboux. And what would have impressed Cartan when reading Frobenius' paper? It is there that the bilinear covariant as alternating bilinear form Ω actually occurs as a central concept (as opposed to Darboux's systems $A\,dx = \lambda a\,dt$), thereby making the idea of introducing the derivative of a Pfaffian more palpable. Furthermore, through his integrability theorem and the applications he made of it, Frobenius did far more than Darboux to suggest the idea that the bilinear covariant is the key to the study of Pfaffian equations. Indeed, as we have seen, Frobenius' manner of characterizing integrability—if $\omega_i(b) = 0$ and $\omega_i(c) = 0$, then $\Omega_i(b,c) = 0$—is central to the constructs by means of which Cartan obtained "precise and general results ... by taking into consideration the bilinear covariants" in his work on incomplete systems.

Frobenius' treatment of the problem of Pfaff, although encumbered more by calculations, was also more carefully worked out and self-contained than Darboux's. Frobenius was the first and only mathematician prior to Cartan to correctly understand and systematically analyze the completely general case of the problem of Pfaff, i.e., completely general in the algebraic sense of Section 6.1; and this precedent-setting work must have impressed Cartan, who saw himself faced with the same sort of challenge with regard to the theory of incomplete Pfaffian systems, as indicated in the preceding quotation. Whereas Frobenius had systematically applied Berlin-style linear algebra, Cartan drew on the multilinear algebra behind his calculus of differential forms—as well as his fertile geometric imagination—to deal with the more formidable problems he faced.

---

[51] I have translated Cartan's *quelconque* as generic, since that is his meaning.

6.7 Paradigmatic Aspects of Frobenius' Paper

As we shall see in the ensuing chapters, Frobenius' paper on the problem of Pfaff is paradigmatic of one of the principal ways in which his work has left its mark on present-day mathematics. That work was initiated, rather typically, by a specific problem, a problem of the sort encouraged by his experiences at Berlin: to deal successfully with the problem of Pfaff on the nongeneric level, something that Clebsch had first attempted but without complete success. Another feature of his work on the problem of Pfaff that turned out to be characteristic of much of his later work was that it was a problem formulated within the context of a body of earlier mathematical results on, in this case, the problem of Pfaff—the work of Pfaff, Jacobi, Clebsch, and Natani being the most significant—and Frobenius read the literature as a scholar, carefully and thoroughly. The novel approach to the problem he posed was also very much in harmony with his Berlin schooling: by means of Lipschitz's passing observations on the bilinear covariant of a 1-form (Theorem 6.4), Frobenius was able to view his problem within the context of the simultaneous transformation of a 1-form and its associated binary covariant and so to use the construct his friend and fellow student Stickelberger had introduced in his Berlin doctoral dissertation. Then in accordance with the procedure used by Christoffel and the disciplinary ideals articulated by Kronecker, he sought to reduce the problem to an algebraic one involving the canonical forms of a pair consisting of a linear and a skew-symmetric bilinear form. Having thus formulated an appropriate way to go about resolving the problem he had posed, Frobenius then applied his creative talent to develop the new approach systematically, so that the end result—his paper on the problem of Pfaff—was essentially a carefully worked out and original "monograph" on the problem.

Like Weierstrass' cycle of lectures on analysis, Frobenius' publications convey the conviction that the clear and systematic presentation of results in the proper manner, i.e., developed rigorously from the proper unifying mathematical viewpoint, was just as important as the discovery of new results by whatever means.[52] This penchant on Frobenius' part is clearly set forth in a paper of 1875 (thus predating his Pfaff paper) on the application of the theory of determinants—the foundation of Berlin-style linear algebra—to metric geometry [176], an outgrowth of his work as a second-year student in the prize problem competition discussed in Section 1.1. Frobenius' own account of why he published this paper is worth quoting in full:

> In 1868, I was induced by a prize question set by the philosophical faculty of the University of Berlin to concern myself with the application of the theory of determinants to metric geometry, and at that time, I wrote some works on this subject, which have hitherto been held back from publication by other work. In the meantime, the work of Mr. Darboux [110]

[52]As Weierstrass wrote to Sonya Kovalevskaya, "What I require of a scientific work, however, is unity of method, consequent pursuit of a definite plan, an attendant working out of the details, and that it bear the stamp of independent research." Letter dated 1 January 1875 and published in [28]. Also quoted by Mittag-Leffler [441, p. 155]. The above translation is mine.

> ... came to my attention, just as I was putting together a brief sketch of my investigations. In it I found a large part of the metric relations and geometric constructions treated by me developed in a very elegant and original manner. Whereas with Mr. Darboux there is no connection between the metric relations and the geometric constructions, a main objective of my work is to show how the solution to complicated geometric problems can be read off simply and easily from a few metric relations.

> Partly on these grounds and partly on account of the difficulty of presenting those of my results that are not found in that work [by Darboux] separated from the rest, I wish to communicate here in abbreviated form my developments.

Thus despite the elegant approach of Darboux and the fact that his memoir contained many of the same results, Frobenius felt compelled and justified in presenting those same results by means of his own approach, which he clearly deemed superior for the reason given, even though it required sixty-two pages to do this.

Frobenius had a genuinely scholarly approach to his mathematics in the sense that once interested in a particular problem, he made a thorough search of the literature, which he then creatively viewed and developed from the unifying approach that he deemed the proper one for the subject at hand. In the case of his paper on metric geometry, it was the idea of a small set of geometric relations from which, with the aid of the theory of determinants, a multitude of complicated geometric constructions could be immediately obtained. In the case of the problem of Pfaff, the approach was that of the transformational equivalence of Pfaffian expressions (or 1-forms), and the key unifying concept was the bilinear covariant. In developing his own approach to the material he of course drew as needed on his broad-based knowledge of the mathematical literature. Thus he borrowed the key notion of the bilinear covariant from Lipschitz, which thanks to Jacobi's introduction of his skew-symmetric matrix into the problem of Pfaff, Frobenius could see as central to the problem of Pfaff. And of course, from Clebsch's work he obtained not only the mathematical problem that motivated his work, but also the two normal forms I and II of his equivalence theory. Likewise, as noted in Section 6.4.3, his complete integrability theorem drew on the results of Jacobi, Clebsch, and Deahna, as well as the general realization in the literature of a duality between systems of linear homogeneous differential equations and systems of Pfaffian equations.

Frobenius' genius was to combine these elements in a clear, systematic manner unified by a central concept, that of the bilinear covariant. He had a considerable talent for clear mathematical exposition—even more so than his mentor Weierstrass. His contemporaries, including Cartan, regarded his work as exceptionally thorough and rigorous. The resulting "monograph" that Frobenius produced did not appeal to everyone, as we saw in Section 6.5. Yet top mathematicians such as Darboux and Cartan were impressed by Frobenius' "beautiful" essay on the problem of Pfaff, and they paid him the compliment by seeking to improve and build on it.

Thus Darboux pointed out the value of quickly establishing and then utilizing the covariance of the bilinear covariant as a means of avoiding some of Frobenius' extensive algebraic calculations; and Cartan, besides finding a way to carry out Darboux's suggestion rigorously, saw that the central role played in Frobenius' work by the bilinear covariant could be incorporated into his still fragmentary, incomplete, and seemingly insignificant calculus of differential forms, thereby providing it with the central notion of the exterior derivative of a 1-form so as to obtain the associated bilinear covariant and make his calculus applicable to the problem of Pfaff. Not only had Frobenius formulated the results of Jacobi, Clebsch, and Deahna in terms of the bilinear covariant as the integrability theorem, he also made several applications of it, thereby suggesting its further utility; and Cartan, after reformulating it within the context of his now-complete calculus of differential forms, went on to demonstrate its utility within the framework of the problems of interest to him in the theory of partial differential equations.

Of course, as indicated in Section 6.6, Cartan went on to apply his calculus of differential forms to more than just the problem of Pfaff, thereby establishing that calculus as a basic tool in the repertory of present-day mathematics and advancing the theory of partial differential equations, Lie group theory, and algebraic topology in brilliant ways far removed from the work and interests of Frobenius. Yet it seems to me that Frobenius' "monograph" on the problem of Pfaff, which by nineteenth-century standards was remarkably clear and conceptually coherent, was, for the above reasons, as well as by virtue of Frobenius' attempt (like Cartan's) to deal with problems on an algebraically nongeneric level, a major influence on Cartan as he initiated the theory and application of his calculus of differential forms. That Frobenius' name alone has remained attached to his complete integrability theorem, although not at all historically accurate, has a kind of historical validity in the sense that it is a reflection of the way Frobenius' work influenced Cartan and, through him, present-day mathematics. Cartan's work rendered Frobenius' "monograph" obsolete, but some of its central ideas were subsumed in the work of Cartan and thereby eventually into contemporary mathematics. The association of Frobenius' name with the modern integrability theorem is a telltale of this phenomenon. Many of Frobenius' other monographic essays on various mathematical problems also influenced the development of mathematics in similar ways, as we shall see in the following chapters.

Chapter 7
The Cayley–Hermite Problem and Matrix Algebra

Less than 2 years after Frobenius submitted his "monograph" on the problem of Pfaff, he submitted another monograph [181], in which he showed that bilinear forms in n variables (or equivalently, their coefficient matrices) form a linear associative algebra that can be represented symbolically and used to great advantage in solving linear algebraic problems. He was motivated to do so by a problem, called here the Cayley–Hermite problem, which had hitherto been treated on the generic level.[1] As we shall see, Frobenius was not the first to introduce symbols for matrices or to add or multiply them—such considerations can be traced all the way back to Gauss' *Disquisitiones Arithmeticae* of 1801. In fact, in 1858, in connection with the Cayley–Hermite problem, Cayley had already observed that $n \times n$ matrices form, in effect, an associative algebra. But Cayley's treatment of the problem and of matrix algebra itself remained on the generic level, whereas Frobenius fused the symbolic algebra of forms with the theory of forms developed by Weierstrass and Kronecker in 1868–1874 (Chapter 5) and demonstrated the resultant power of this newly forged tool by using it to definitively solve several problems in addition to the Cayley–Hermite problem. As we shall see in the concluding section, this is why Frobenius rather than his predecessors was the principal force behind the development of the modern theory of matrices.

7.1 The *Disquisitiones Arithmeticae* of Gauss

Matrix algebra is frequently regarded as an outgrowth of the development of the theory of determinants in the nineteenth century, but it would be more accurate to regard both as, to a large extent, the outgrowth of the treatment of the arithmetic

[1]Much of this chapter is based on my paper [270], where I first called attention to the historical importance of this problem.

T. Hawkins, *The Mathematics of Frobenius in Context*, Sources and Studies in the History of Mathematics and Physical Sciences, DOI 10.1007/978-1-4614-6333-7_7,

theory of quadratic forms as contained in Gauss' *Disquisitiones Arithmeticae* [244]. Although Gauss restricted himself to binary forms and to certain parts of the theory of ternary forms needed to complete the treatment of the binary case, he made it perfectly clear that his own work was "only one section of a general treatise concerning rational algebraic functions that are integral and homogeneous in many unknowns and many dimensions.... It is sufficient to draw this broad field to the attention of geometers. There is ample material for the exercise of their genius" [244, Art. 266]. These remarks, coming as they did from a mathematician of unquestioned greatness, were a source of inspiration and challenge to subsequent generations of mathematicians. Moreover, as they responded to Gauss' challenge, they developed symbolic notation that facilitated the extension to a larger number of variables. In this respect they followed the lead of Gauss himself.

Even in his incomplete study of ternary forms, Gauss began to introduce new notational conventions. Thus for linear substitutions in three variables,

$$\begin{aligned} x &= \alpha y + \beta' y' + \gamma' y'', \\ x' &= \alpha' y + \beta' y' + \gamma' y'', \\ x'' &= \alpha'' y + \beta'' y' + \gamma'' y'', \end{aligned}$$

he decided, "for brevity," to refer to them by their coefficients [244, Art. 268]:

$$\begin{matrix} \alpha, \beta, \gamma \\ \alpha', \beta', \gamma' \\ \alpha'', \beta'', \gamma'' \end{matrix} \tag{7.1}$$

When dealing with numerical examples, he modified the notation in (7.1) by adding brackets [244, Art. 273]:

$$\left\{ \begin{matrix} 1, & 0, & 0 \\ 0, & 2, & -7 \\ 0, & -3, & 11 \end{matrix} \right\}$$

Also, linear substitutions were denoted by single letters, such as S, when reference to the coefficients was unnecessary.

The composition of linear substitutions also played an important part in the arithmetic theory of forms, although this statement is in need of further clarification. As in the binary case, Gauss made the observation that if the substitution defined by (7.1) transforms a form f into f' and if another substitution (also indicated by its coefficient matrix) takes f' into f'', then there exists a substitution taking f directly into f''. Gauss wrote down the coefficient matrix of this substitution, which is, of course, the composite or product of the two matrices as we would put it, but he did not explicitly designate this process as a kind of multiplication of objects that are not numbers. It was not that such ideas were too abstract for him. Indeed, he was the first to introduce such ideas, albeit in a different context, namely his profound theory of the composition of certain equivalence classes of binary forms (discussed in Section 9.1.1). Gauss' idea of composing objects that were not ordinary numbers led others to apply it to linear substitutions or, in fact, to their coefficient matrices.

One of the first mathematicians to rise to the challenge of the general research program sketched by Gauss in Article 266 was Cauchy. As we saw in Section 4.3, in his memoir on the theory of determinants [68], which was inspired by Gauss' use of the formula $\det AB = \det A \cdot \det B$ for 2×2 and 3×3 coefficient systems, Cauchy introduced the notion of a "symmetric system" [68, p. 115]

$$\begin{cases} a_{1,1}, \; a_{1,2}, \; a_{1,3}, \; \ldots, \; a_{1,n}, \\ a_{2,1}, \; a_{2,2}, \; a_{2,3}, \; \ldots, \; a_{2,n}, \\ \ldots \quad \ldots \quad \ldots \quad \ldots \quad \ldots \\ a_{n,1}, \; a_{n,2}, \; a_{n,3}, \; \ldots, \; a_{n,n}. \end{cases}$$

which he also represented in the more abbreviated form $(a_{1,n})$. The determinant, although previously defined, is then characterized as "belonging to" the above symmetric system [68, p. 116]. The reader should not be confused by Cauchy's terminology; the matrix represented by the above system is not assumed to be symmetric, i.e., it is not assumed that $a_{ji} = a_{ij}$.

In Cauchy's theory of determinants, his notion of a symmetric system is thus actually more fundamental: the determinant is a number associated with this system. Furthermore, in his formulation of the product theorem for determinants (established for two and three variables by Gauss [244, Art. 159 & 268], Cauchy explicitly recognized the idea of composing two systems $(\alpha_{1,n})$ and $(a_{1,n})$ to form a third $(m_{1,n})$, where $m_{i,j} = \sum_{k=1}^{n} \alpha_{i,k} a_{j,k}$. After proving that the determinant of $(m_{1,n})$ is the product of those of $(\alpha_{1,n})$ and $(a_{1,n})$, he added, "I shall say that ... the system $(m_{1,n})$ results from the composition of the two systems $(a_{1,n})$ and $(\alpha_{1,n})$" [68, p. 143]. There is little doubt that the phrasing of this remark reflects Cauchy's familiarity with Gauss' composition of binary forms.

Cauchy clearly distinguished between a symmetric system and its determinant. Nevertheless, it was the determinant that chiefly interested him, and in this connection there seemed no reason to further pursue the idea of the composition of systems. In fact, examination of the early expositions of Cauchy's work by Lebesgue [404] and Jacobi [311] reveals that although the distinction between a system and its determinant was maintained by Lebesgue, it was completely ignored by Jacobi. For the further development of the composition of systems, we must return to the theory of numbers.

7.2 Eisenstein and Hermite

While still a gymnasium student, Gotthold Eisenstein (1823–1852) had made a careful study of the *Disquisitiones Arithmeticae*, and much of the work he published during his short career was inspired by it. In particular, Gauss' comments in Article

266 stimulated him in 1847 to investigate the theory of quadratic forms in three and more variables [153, p. 117]. One consequence of his work was the further development of Gauss' symbolism. In his "General Investigations of Forms of Third Degree ..." (1844), he introduced the notation $S \times T$ for "the substitution composed from S and T" [151, p. 324] and $\frac{1}{S}$ for "the inverse system of the system S" [151, p. 328]. Such symbolism was primarily employed by Eisenstein to state results succinctly, but he occasionally used it to facilitate the mathematical argument. (See especially Section 8 of [151].)

Although the object of Eisenstein's paper was the arithmetic study of ternary forms, he realized the possibility of an analogous symbolic algebra corresponding to the n-variable case [151, p. 354]:

> Incidentally, an algorithm for calculation can be based on this; it consists in applying the usual rules for the operations of multiplication, division, and exponentiation to symbolic equations between linear systems; correct symbolic equations are always obtained, the sole consideration being that the order of the factors, i.e., the order of the composing systems, may not be altered.

In a paper published the year of his untimely death, Eisenstein reiterated and amplified the above observations [154, p. 354]:

> Here some considerations on the composition of linear transformations should be indicated, which will be applied frequently in what follows. If, after the introduction of a linear substitution, the variables are then replaced by linear combinations of the same, the substitution so obtained is said to be composed out of the two applied one after the other; and its determinant is equal to the product of the determinants of the two components. It is known that the composition of transformations has the greatest analogy with the multiplication of numbers and should also be denoted by the sign for this operation when the substitution systems themselves enter into the calculation as if they were independent quantities,* except that here the order of the factors may not be reversed.

The asterisk refers to a footnote in which Eisenstein added: "An addition of systems can also be introduced to advantage, by taking the sum of two systems to be the one in which every coefficient consists of the sum of the corresponding coefficients in the given systems."

These quotations make it clear that Eisenstein realized that linear substitutions, considered as entities, can be added and multiplied as if they were ordinary numbers, except that the multiplication is not commutative. That he regarded the analogy between the composition of substitutions and ordinary multiplication as something generally known is also not surprising, since it is suggested by the product theorem for determinants, especially in Cauchy's formulation. Furthermore, Eisenstein undoubtedly had in mind Gauss' similar observations about the composition of binary forms: just as Gauss had employed a symbolism for the composition of forms that reflected the analogy with ordinary arithmetic, so Eisenstein employed it for linear substitutions.

Eisenstein's symbolism was adopted by his contemporary Charles Hermite (1822–1901), who had met Eisenstein in Berlin and maintained a correspondence with him on mathematical matters. Hermite's use of Eisenstein's symbolism is especially interesting because it is connected with the Cayley–Hermite problem discussed in the next section. In a paper of 1854 [288], Hermite studied the problem of determining the linear substitutions with integer coefficients that transform a given indefinite ternary form into itself. To solve this arithmetic problem, he first solved the corresponding algebraic problem in which no restriction is put on the nature of the coefficients. The algebraic solution is then used as the starting point for the study of the arithmetic solutions, and at this stage Hermite found Eisenstein's symbolism helpful: all arithmetic solutions, he showed, are expressible in the form $S^{-1}\Sigma S$, where S and Σ are given explicitly.

More extensive use of such symbolism is found in Hermite's next memoir the same year [289]. The reason for employing it was essentially the same. In order to express his results on the nature of the substitutions leaving some form invariant, it was necessary to consider the composition of substitutions, and to do this succinctly, symbolic representations were very useful. A good illustration is afforded by the theorem that Hermite stated in the introduction. He considered the arithmetic problem of determining the linear substitutions that leave a certain nth-degree form in n variables invariant. The theorem is that all such substitutions "will be given symbolically by the formula $S_1^{m_1} S_2^{m_2} \cdots S_\omega^{m_\omega}$," where the S_i are fixed substitutions and the m_i are arbitrary integers. Any two such substitutions S and T satisfy the relation $ST = TS$, which is also a property suitable for symbolic representation. One of the objectives of Hermite's paper [289] is to show that the situation for an indefinite ternary form is essentially different: $ST = TS$ if and only if S and T are powers of the same substitution [289, pp. 113ff.]. Hermite, like Eisenstein, employed the symbolism chiefly as a means of succinctly summarizing results with occasional use of symbolic algebra as a mode of reasoning. (See pp. 214 and 231 of [289].)

Matrices, or "systems" as they were called, were also employed by Eisenstein [152] and Hermite [290] in the theory of elliptic and abelian functions. In these works, equations such as

$$\left\{\begin{matrix} a_0 & a_1 & a_2 & a_3 \\ b_0 & b_1 & b_2 & b_3 \\ c_0 & c_1 & c_2 & c_3 \\ d_0 & d_1 & d_2 & d_3 \end{matrix}\right\} \times \left\{\begin{matrix} \alpha_0 & \alpha_1 & \alpha_2 & \alpha_3 \\ \beta_0 & \beta_1 & \beta_2 & \beta_3 \\ \gamma_0 & \gamma_1 & \gamma_2 & \gamma_3 \\ \delta_0 & \delta_1 & \delta_2 & \delta_3 \end{matrix}\right\} = \left\{\begin{matrix} A_0 & A_1 & A_2 & A_3 \\ B_0 & B_1 & B_2 & B_3 \\ C_0 & C_1 & C_2 & C_3 \\ D_0 & D_1 & D_2 & D_3 \end{matrix}\right\}$$

are found to "express the relation of composition" (Hermite [291, p. 447]). Eisenstein even used the now-familiar curved brackets for his linear systems.

Eisenstein and Hermite thus clearly recognized the fact that linear substitutions or the "system" of their coefficients could be regarded as quantities and, as such, multiplied, added, and manipulated by the ordinary symbolic rules of arithmetic, except where noncommutativity is relevant. Their study of the symbolic algebra of systems was, however, not developed any more than was needed in their work on the theory of numbers and abelian functions. In particular, in spite of Eisenstein's claim that the addition of systems could be introduced to good advantage, neither he nor Hermite employed it. Perhaps Eisenstein would have, had he not died the year he made the above claim; but Hermite apparently saw no use for addition in his own work. We now turn to a problem in which addition, as well as multiplication, of linear substitutions became relevant.

7.3 The Cayley–Hermite Problem

The Cayley–Hermite problem (defined below) is extremely important for an understanding of the historical development of the symbolic algebra of matrices. All three mathematicians who carried the investigation further than Eisenstein and Hermite (namely Cayley, Laguerre, and Frobenius) were concerned with this problem. At least in the cases of Cayley and Frobenius, it was the primary motivation behind their study of the symbolic algebra of matrices.

The history of the Cayley–Hermite problem begins with an 1846 paper, "On certain properties of skew determinants" [80], which Cayley published (in French) in *Crelle's Journal*. It appeared during the period when Cayley was familiarizing himself with Cauchy's theory of determinants and reflects his appreciation and mastery of the new mode of algebraic reasoning. He had read an 1840 paper by Olinde Rodrigues in the *Journal de Mathématiques Pures et Appliquées* [498] in which the latter showed that the nine cosines of the angles between two rectangular coordinate systems in space could be expressed as rational functions of three parameters. The derivation relied heavily on trigonometry. Cayley presented an unoriginal reworking of Rodrigues' result in 1843 [79], but in the 1846 paper [80], he showed that the result could be established by a simpler method that was at the same time much more general. The key was the replacement of manipulations involving trigonometric relations by considerations based on the theory of determinants.[2]

Cayley's starting point was the notion of a "skew system" (*système gauche*), a system in which the coefficients λ_{rs}, $r,s = 1,2,\ldots,n$, satisfy the relations $\lambda_{sr} = -\lambda_{rs}$, $r \neq s$, and $\lambda_{rr} = 1$. Thus in modern notation the coefficient matrix $L = (\lambda_{rs})$ is of the form $L = I + T$, where T is skew-symmetric ($T^t = -T$). Let $\Lambda = \det L$ and let Λ_{rs} denote the cofactor corresponding to λ_{rs}. Cayley had discovered that if n^2 quantities α_{rs} are defined by

$$\begin{aligned} \alpha_{rs} &= 2\Lambda_{rs}/\Lambda \quad (r \neq s), \\ \alpha_{rr} &= (2\Lambda_{rr}/\Lambda) - 1, \end{aligned} \tag{7.2}$$

then they satisfy the orthogonality relations

$$\sum_{r=1}^{n} \alpha_{rs}\alpha_{rs'} = 0 \ \ (s \neq s'), \quad \sum_{r=1}^{n} \alpha_{rs}^2 = 1. \tag{7.3}$$

[2]In this respect, Cayley was astutely following the lead of Lagrange, whose treatment of the principal axes theorem 1775 (Section 4.4.1) involved such a replacement, which Cauchy then generalized to n variables in 1829 (Section 4.4.2).

Cayley never indicated whether his parameters λ_{rs} are assumed to be real or are allowed to be complex. Such ambiguity was commonplace in the period. His reasoning is valid even if the λ_{rs} are complex. When they are assumed real, the coefficient system $U = (\alpha_{rs})$ describes a real orthogonal transformation such as Cauchy had discussed in his 1829 paper on the transformation of quadratic forms (Section 4.4.2), which for $n = 3$ describes the sort of relation between two rectangular coordinate systems that Rodrigues had described using cosines. Equation (7.2) shows that the α_{rs} are rational functions of $(n-1)n/2$ parameters λ_{rs}, $r > s$, and so in particular when $n = 3$, the coefficients α_{rs} depend rationally on three parameters. Thus Rodriques' result is replaced by a more general one about orthogonal transformations of n coordinates.

Cayley's result employed the "adjoint" formula for the inverse of a given coefficient system, which was well known at the time, thanks to Cauchy's memoir on determinants.[3] It can be translated into modern symbolism as follows. If $U = (\alpha_{rs})$, then (7.2) can be expressed in the form $U = 2[L^t]^{-1} - I = 2(I-T)^{-1}$. Using matrix algebra and the fact that $I - T$ and $(I+T)^{-1}$ commute, we may also write without ambiguity

$$U = \frac{I+T}{I-T}. \tag{7.4}$$

(Nowadays, (7.4) is called the Cayley transform.) No such notation as in (7.4) was introduced by Cayley; and although it does provide a concise way to define the α_{rs}, equation (7.2) is itself sufficiently concise, thanks to the use of determinant notation. It is thus uncertain that Cayley would have pursued his problem or the matter of notation any further had it not been for the intervention of Hermite.

Hermite was the first to formulate what I have referred to as the *Cayley–Hermite problem: determine linear substitutions of the variables of a given quadratic form of nonzero determinant that leave the form invariant.* As we have seen, the problem interested him because of its relevance to his work on the arithmetic theory of ternary forms, which is the context within which he first formulated it. He was familiar with Cayley's paper [80], which, from Hermite's point of view, could be regarded as a solution to the above problem for the form $f = x_1^2 + x_2^2 + \cdots + x_n^2$. Hermite realized that his method of solving the above problem for indefinite ternary forms [288, p. 309] could be generalized to quadratic forms in any number of variables by utilizing the skew-symmetric coefficient systems introduced by Cayley. The generalization was published in the *Journal of the Cambridge and Dublin Mathematical Society* [287].

Hermite considered a linear substitution that takes the variables $x_1, x_2, \ldots, x_n$ into $X_1, X_2, \ldots, X_n$ and has the property that

[3] For any invertible matrix M, $M^{-1} = 1/(\det M) \cdot [\mathrm{Cof}(M)]^t$, where $\mathrm{Cof}(M)$ denotes the matrix of cofactors, i.e., the (i,j) entry of $\mathrm{Cof}(M)$ is the cofactor corresponding to m_{ij}. See the discussion following (4.15).

$$f(x_1,x_2,\ldots,x_n) = f(X_1,X_2,\ldots,X_n), \tag{7.5}$$

where $f = \sum_{i,j=1}^{n} a_{ij}x_ix_j$, $a_{ji} = a_{ij}$ and $\det(a_{ij}) \neq 0$. Define auxiliary variables ξ_r by

$$x_r + X_r = 2\xi_r. \tag{7.6}$$

By expressing the relation (7.5) in terms of the ξ_r, Hermite concluded that "the most general way" in which the resulting relation could hold would be if

$$x_r = \xi_r + \frac{1}{2}\sum_{s=1}^{n} \lambda_{r,s}\frac{\partial f(\xi_1,\ldots,\xi_n)}{\partial \xi_s}, \qquad r = 1,2,\ldots,n, \tag{7.7}$$

where $\lambda_{r,s} = -\lambda_{s,r}$ and $\lambda_{r,r} = 0$.

Equation (7.7) is a system of linear equations in the unknowns ξ_r. Once the system is solved for the ξ_r and (7.6) applied, one has the equations for the linear substitution taking the x_i into the X_i or vice versa. Some modern notation is again instructive. If $A = (a_{rs})$ and $T = (\lambda_{rs})$, so that $A^t = A$, $T^t = -T$, then (7.7) translates immediately into

$$x = (I + TA)q, \tag{7.8}$$

where q denotes the column matrix with components $\xi_1,\ldots,\xi_n$. Since $q = \frac{1}{2}(x+X)$, substitution of this relation into (7.8) yields upon rearrangement $(I - TA)x = (I + TA)X$ and hence

$$x = (I - TA)^{-1}(I + TA)X, \tag{7.9}$$

assuming that the above inverse exists.

Equation (7.9) shows that the linear substitutions satisfying (7.5) can be expressed succinctly using the symbolic algebra of linear substitutions. (Actually, not all solutions are given by (7.9), a point to be discussed further in Section 7.5.) In the previous section, we saw that Hermite had used the symbolic representation of substitutions to study the arithmetic solutions in the ternary case, but he apparently failed to see its value in the present context. Thus he left the solution in the form (7.7), without discussing, or symbolically representing, the solution to that system of equations. Equation (7.7) was actually all Hermite needed to give a simple proof that (7.5) is satisfied. It was left to Cayley to point out the value of some symbolic notation in order to remove the "incompleteness" of Hermite's (7.7).

Cayley's observations were made in two notes published in *Crelle's Journal* in 1855 [82, 83]. In the first, he introduced the notation

$$\left|\begin{matrix} \alpha, & \beta, & \gamma, & \ldots \\ \alpha', & \beta', & \gamma', & \ldots \\ \alpha'', & \beta'', & \gamma'', & \ldots \\ \ldots & \ldots & \ldots & \ldots \end{matrix}\right|$$

"to represent what I call a *matrix*; that is, a *system* of quantities arranged in the form of a square..." [82, p. 185]. (According to Cayley [87, v. 2, p. 188], the manuscript of the papers had actually used curved brackets.) He introduced the notation because it

> seemed to me very convenient for the theory of linear equations; for example, I write
>
> $$(\xi,\eta,\xi,\ldots)=\begin{vmatrix}\alpha, & \beta, & \gamma, & \ldots\\ \alpha', & \beta', & \gamma', & \ldots\\ \alpha'', & \beta'', & \gamma'', & \ldots\\ \ldots & \ldots & \ldots & \ldots\end{vmatrix}(x,y,z,\ldots)$$
>
> to represent the system of equations
>
> $$\begin{aligned}\xi &= \alpha x+\beta y+\gamma z+\cdots,\\ \eta &= \alpha' x+\beta' y+\gamma' z+\cdots,\\ \xi &= \alpha'' x+\beta'' y+\gamma'' z+\cdots,\\ &\ldots\ldots\ldots\ldots\ldots\ldots\ldots\ldots\end{aligned}$$

The solution of the above system in terms of $x,y,z\ldots$ could then be expressed using the notation

$$\begin{vmatrix}\alpha, & \beta, & \gamma, & \ldots\\ \alpha', & \beta', & \gamma', & \ldots\\ \alpha'', & \beta'', & \gamma'', & \ldots\\ \ldots & \ldots & \ldots & \ldots\end{vmatrix}^{-1}.$$

And if the $(x,y,z,\ldots)$ are themselves given linearly in terms of other variables, one thus arrives at the idea of a *composite matrix*, for example

$$\begin{vmatrix}\alpha, & \beta, & \gamma, & \ldots\\ \alpha' & \beta' & \gamma' & \ldots\\ \alpha'', & \beta'', & \gamma'', & \ldots\\ \cdot & & & \\ \cdot & & & \\ \cdot & & & \end{vmatrix}\ \begin{vmatrix}a, & b, & c, & \ldots\\ a', & b', & c', & \ldots\\ a'', & b'', & c'', & \ldots\\ \cdot & & & \\ \cdot & & & \\ \cdot & & & \end{vmatrix}.$$

The material presented in the previous sections makes it clear that these observations by Cayley were neither profound nor unprecedented. The second note, however, did contain something that was new. In presenting his notation for representing systems of linear equations in the first note, Cayley had a definite system in mind, namely the system (7.7) of Hermite's paper, which translates into (7.8) in modern notation. Thus if the coefficients of the quadratic form f are denoted by the matrix

$$\begin{vmatrix} a, & h, & g, & \dots \\ h, & b, & f, & \dots \\ g, & f, & c, & \dots \\ \dots & \dots & \dots & \dots \end{vmatrix}, \tag{7.10}$$

then "the solution given by M. Hermite can be summed up in the single equation" that in more abbreviated modern notation—necessary to fit it onto this page—is $X = A^{-1}(A-T)(A+T)^{-1}Ax$, where A is expressed by Cayley as in (7.10) above, X and x are written as $(\mathrm{x},\mathrm{y},\mathrm{z},\dots)$ and $(x,y,z,\dots)$, respectively, and, for example, $A+T$ is expressed by Cayley as

$$\begin{vmatrix} a, & h+\nu, & g+\mu, & \dots \\ h-\nu, & b, & f+\lambda, & \dots \\ g-\mu, & f-\lambda, & c, & \dots \\ \dots & \dots & \dots & \dots \end{vmatrix},$$

"where $\lambda,\mu,\nu,\dots$ are arbitrary quantities" [83, p. 192]. It should be noted that the "arbitrary constants" λ,μ,ν forming T are signed so as to make T skew-symmetric. Given the expansive form of Cayley's actual equation relating $(\mathrm{x},\mathrm{y},\mathrm{z},\dots)$ and $(x,y,z,\dots)$, it is surprising that he did not sense the need at this point for single letter notation for his coefficient systems. This would have meant defining symbolic addition so as to write $A+T$ and $A-T$, which is consequently also missing in the 1855 notes. These missing steps were taken by Cayley a few years later.

7.4 Cayley's Memoir of 1858

On 10 December 1857, Cayley presented two memoirs [84, 85] for publication in the *Philosophical Transactions* of the Royal Society. The first was his oft-cited memoir on the theory of matrices, in which single-letter notation for matrices and matrix addition are introduced along with matrix and scalar multiplication. No attention had been given by historians to the second memoir, but from our viewpoint, it is equally important, because it documents Cayley's continuing interest in the Cayley–Hermite problem. In [85], he studied the generalized problem obtained by considering the linear substitutions that leave invariant a bilinear form instead of a quadratic form. Here it suffices to observe that he made use of the notational innovations contained in the first memoir. For example, the solution to the original Cayley–Hermite problem is now expressed in the form suggested by his 1855 note:

$$(x,y,z) = (\Omega^{-1}(\Omega-\gamma)(\Omega+\gamma)^{-1}\Omega)(x_1,y_1,z_1),$$

where Ω is the matrix corresponding to the quadratic form and γ is a skew-symmetric matrix [85, p. 502]. The two memoirs of 1858 thus complement one another just as the 1855 notes had, and it seems very likely that the notational innovations in [84] were initially motivated by the Cayley–Hermite problem.

Underlying these notational innovations is the idea that "matrices … comport themselves as single entities; they may be added, multiplied or compounded together, &c: the law of addition of matrices is precisely that for addition of ordinary quantities; as regards their multiplication (or composition), there is the peculiarity that matrices are in general not convertible …" [84, p. 476]. Exactly the same idea had been expressed in very similar terms by Eisenstein in 1852, as a glance at the quotations in Section 7.2 indicates. But Cayley explored in print some of the consequences of this idea, whereas Eisenstein and Hermite had not.

Cayley's interest in matrix algebra for its own sake seems to derive chiefly from his discovery of the Cayley–Hamilton theorem, which he communicated to his friend Sylvester in a note dated 19 November 1857.[4] It begins:

> Dear Sylvester,
> I have just obtained a theorem which appears to me *very remarkable*. You know what the composition of matrices means, e.g., if
>
> $$M = \begin{pmatrix} a, & b \\ c, & d \end{pmatrix} \text{ then } M^2 = \begin{pmatrix} a^2+bc, & b(a+d) \\ c(a+d), & d^2+bc \end{pmatrix}$$
>
> and I define as the addition of matrices
>
> $$\begin{pmatrix} a, & b \\ c, & d \end{pmatrix} + \begin{pmatrix} a', & b' \\ c', & d' \end{pmatrix} = \begin{pmatrix} a+a', & b+b' \\ c+c', & d+d' \end{pmatrix}.$$
>
> Suppose now that M is the matrix $\begin{pmatrix} a, & b \\ c, & d \end{pmatrix}$ and form the determinant $\begin{vmatrix} a-M, & b \\ c, & d-M \end{vmatrix}$. I say that will be equal to the matrix $\begin{pmatrix} 0, & 0 \\ 0, & 0 \end{pmatrix}$, viz. expanding, the determinant is
> $(ad-bc)M^0 - (a+d)M^1 + M^2$.

Presumably the justification then followed by computing $(ad-bc)M^0 - (a+d)M^1 + M^2$ and confirming that the result is a matrix of zeros. A similar computation with

$$M = \begin{pmatrix} sa & b & c \\ d & e & f \\ g & h & i \end{pmatrix} \quad \text{and} \quad \begin{vmatrix} a-M & b & c \\ d & e-M & f \\ g & h & i-M \end{vmatrix}$$

---

[4] The letter is located in the library of St. John's College, Cambridge, along with others that passed between Cayley and Sylvester. I am grateful to Dr. I. Grattan-Guinness for calling my attention to the existence of these letters, to Prof. E. Koppelman for informing me of their content, and to the Masters and Fellows of St. John's College for permission to make the above quotation.

yields for the symbolic determinant $(-ceg+bfg+cdh-afh-bdi+aei)M^0 + (bd-ae+cg+fh-ai-ei)M^1+(a+e+i)M^2-M^3$, which upon substituting for the powers of M simplifies to the zero matrix. This calculation was carried out by Cayley (see below). As we shall see, these computations convinced him of the validity of the general result, namely what is customarily called the Cayley–Hamilton theorem, although the inclusion of Hamilton seems dubious, as indicated in the footnote below. Thus Cayley posited the truth of the following assertion.

Theorem 7.1 (Cayley–Hamilton). *If M is any $n \times n$ matrix and if $\varphi(x) = \det(M - xI)$, then $\varphi(M) = 0$, i.e., any square matrix satisfies its own characteristic equation.*

As the note indicates, this theorem presupposes the single letter notation for matrices and matrix addition—precisely those notational innovations suggested by the Cayley–Hermite problem. Indeed, the theorem itself is formally suggested by the symbolism, i.e., $\varphi(M) = \det(M - M \cdot I) = \det(M - M) = 0$. It thus seems likely that the notational innovations suggested by the Cayley–Hermite problem led Cayley to his discovery of the possible truth of the Cayley–Hamilton Theorem, which, once verified, led in turn to his interest in matrix algebra for its own sake as is done in his 1858 paper [84].[5]

Forming as it does the centerpiece of Cayley's memoir, the Cayley–Hamilton theorem provides an appropriate focal point for an appraisal of Cayley's work with matrices. In the first place, he did not prove the theorem, even by mid-nineteenth-century standards. He presented a computational verification for two-by-two matrices, assured his readers that he had also verified the theorem for three-by-three matrices, and added, "I have not thought it necessary to undertake the labor of a formal proof of the theorem in the general case of a matrix of any degree" [84, p. 483]. This statement may reflect more than Cayley's limited interest in proofs where inductive evidence seemed convincing. A proof in the general case may have appeared laborious to him because he envisioned it along the lines of his computational verifications for small-sized matrices, computations in the course of which matrices as entities play no role; matrix algebra is reduced to ordinary algebra. If so, then Cayley failed to realize that his matrix calculus, when sufficiently developed, makes it possible to give a simple proof in the $n \times n$ case, as Frobenius, who was unaware of Cayley's papers of 1858 [84,85], showed in 1878 (Section 7.5).

The significance of the Cayley–Hamilton theorem was, in Cayley's view, that it showed that any "algebraical function" of a matrix M is equal to a polynomial in M of degree less than the order of M. For polynomials or rational functions of M, Cayley felt that the reasoning was too evident to require proofs. If $L = p(M)$, where $p(x)$ is a polynomial, the proof is indeed straightforward. For if $c(x)$ denotes

[5]Special cases of the theorem were presented by Hamilton in the context of his theory of quaternions [262, p. 567], but that theory does not appear to have been a major influence upon Cayley. Cayley himself denied any such influence in a polemical exchange with Tait [349, pp. 153, 164]. The absence of any mention of quaternions in his letter of 19 November 1857 supports his denial, as does our discussion of the role of the Cayley–Hermite problem.

the characteristic polynomial of M, we may write $p(x) = q(x) \cdot c(x) + r(x)$, where $\deg r < \deg c = n$. Thus $L = p(M) = q(M) \cdot c(M) + r(M) = r(M)$. The proof when $f(x) = p(x)/q(x)$ is less obvious, although Cayley's conclusion is correct, as Frobenius showed (Section 7.5).

Also, if A commutes with B, then it commutes with B^{-1} (assuming this exists). With this in mind, consider a rational function $f(x) = p(x)/q(x)$, so that $L = f(M)$ means $L = p(M)[q(M)]^{-1}$. Since any polynomial in M evidently commutes with M, both $p(M)$ and $q(M)$ commute with M. Hence also $[q(M)]^{-1}$ commutes with M and so then does $L = p(M) \cdot [q(M)]^{-1}$. Now, as we shall see below, Cayley believed that every L that commutes with M must be a polynomial in M and so, by the above, a polynomial of degree less than the degree n of M. This must then be true of $L = f(M)$.

But what about algebraic functions involving irrationalities, e.g., $f(x) = \sqrt{x}$? Cayley never seemed to doubt that $L = f(M)$ exists, for he wrote,

> If we only had the equation satisfied by the matrix itself, such extension [to irrational functions of the matrix] could not be made; but we have besides the equation of the same order satisfied by the irrational function of the matrix, and by means of these two equations, and the equation by which the irrational function of the matrix is determined, we may express the irrational function as a rational and integral function of the matrix, of an order at most equal to that of the matrix, less unity; such expression will, however, involve *the coefficients of the equation satisfied by the irrational function* ... [84, p. 483].

Accompanying this vague argument is an illustration. Cayley considered a matrix M of "order 2," i.e., a 2×2 matrix, and the irrational function $L = \sqrt{M}$, which is presumed to exist. Since L satisfies its characteristic equation, $L^2 + aL + b = 0$ for some numbers a, b; but $L^2 = M$, so that $M + aL + b = 0$, or $L = -(1/a)(M + b)$. In other words, Cayley took it for granted that $L = f(M)$ exists for any algebraic function $f(x)$ and so satisfies its characteristic equation. In addition, one has the impression that Cayley's proof was simply an inductive generalization of special cases such as this. It was not so much a proof as an expression of confidence that what he had done in special cases could be done in general—under the tacit assumption that the matrix $L = f(M)$ exists for any algebraic function.

Furthermore, Cayley's "proof" was based on generic reasoning. Suppose, e.g., in the argument for the existence of $L = \sqrt{M}$ one had a characteristic equation with $a = b = 0$, i.e., $L^2 = 0$. Then the above argument fails, since $L = -(1/a)(M + b)$ becomes meaningless. In fact, Cayley's claim for irrational algebraic functions is false. Consider, e.g., $M = \begin{pmatrix} 0 & 1 \\ 0 & 0 \end{pmatrix}$. Clearly, $M^2 = 0$, from which it follows that $L^4 = M^2 = 0$. This implies by reasoning first supplied by Frobenius (Section 7.5) that the characteristic equation of L must be $L^2 = 0$, i.e., this is the nongeneric case in which $a = b = 0$. Not only does Cayley's proof fail, but in fact, $L = \sqrt{M}$ cannot exist.[6] In Section 16.1, we will see how Frobenius later (1896) rigorously

[6] If it did, then since all its characteristic roots are 0, we would have $L = SMS^{-1}$, where $\det S \neq 0$, since M is the Jordan canonical form for this situation. But then $L^2 = SM^2S^{-1} = 0 \neq M$.

established the existence of $\sqrt{M}$ for certain types of $n \times n$ matrices M and also indicated how in general, $f(M)$ could be defined for matrices M and algebraic or transcendental functions $f(x)$ that are analytic in neighborhoods of the characteristic roots of M.

As an application of his incorrect, "algebraic function theorem," Cayley considered the problem of determining the matrices L that commute with a given matrix M. Again he began with some vague generalities: since $LM = ML$ can be regarded as a system of n^2 equations in n^2 unknowns (the coefficients of L), he apparently assumed that L will be some sort of a function of M:

> But whatever the form of the function is, it may be reduced to a rational and integral function of an order equal to that of M, less unity, and we have thus the general expression for the matrices convertible with a given matrix, viz. any such matrix is a rational and integral function ... of the given matrix, the order being that of the given matrix less unity. In particular, the general form of the matrix L convertible with a given matrix M of order 2, is $L = \alpha M + \beta$, or what is the same thing, the matrices
>
> $$\begin{pmatrix} a, & b \\ c, & d \end{pmatrix}, \quad \begin{pmatrix} a', & b' \\ c', & d' \end{pmatrix}$$
>
> will be convertible if $a' - d' : b' : c'' = a - d : b : c$. [84, p. 488].

Cayley's characterization of the solutions to $LM = ML$ is generally false. A theorem due to Frobenius (discussed in Section 7.5) shows that it is only when the minimal polynomial of M coincides with its characteristic polynomial that all the L's are polynomials in M. It is instructive to see how Cayley's conclusion follows from his mode of reasoning.

Once again, the general remarks appear to be essentially an expression of confidence that the results deduced in special cases, such as matrices of order 2, can be extended to matrices of any order. Furthermore, examination of Cayley's discussion of that special case makes it clear how he arrived at the conclusion he did. The generic solution to the system of equations implied by $LM = ML$, where $M = \begin{pmatrix} a & b \\ c & d \end{pmatrix}$ and $L = \begin{pmatrix} w & x \\ y & z \end{pmatrix}$, does indeed imply that $L = \alpha M + \beta I$, with $\alpha = (y/c)$ and $\beta = z - (dy/c)$. Here again, Cayley's reasoning is tacitly generic; the reasoning leading to the expression for L requires that $a - d$, b, and c be all nonzero, and indeed, Cayley's proportion $a - d : b : c$ becomes meaningless when these quantities vanish. As in his proof of the above incorrect "algebraic function theorem," his reasoning is also based on incomplete induction, with general conclusions extrapolated from the (generic) treatment of matrices of small size.

The number of linearly independent solutions L to $LM = ML$ and their nature depends on the Jordan–Weierstrass canonical form of M. Indeed, as we saw in Section 5.5, it was precisely the problem of determining the solutions to $LM = ML$ (with L, M interpreted as linear substitutions) that led Camille Jordan to introduce it (Section 5.5). Cayley's matrix algebra had thus prompted him to pose a question that could have led him, as it later led Jordan, to make a fundamental contribution to the theory of matrices. But the naive, generic level on which he dealt with the problem made such a contribution impossible.[7]

7.5 Frobenius' Memoir on Matrix Algebra

Perhaps because they were published in the *Philosophical Transactions* of the Royal Society of London, Cayley's memoirs of 1858 [84, 85] do not seem to have been known outside of England. The ideas they contained were, however, rather inevitable consequences of the symbolic conventions of Eisenstein and Hermite and the Cayley–Hermite problem. The work of Laguerre [393] and Frobenius [181] supports such a view.

Edmond Nicolas Laguerre (1834–1886) was tutor at the École Polytechnique when he submitted his memoir [393] through Hermite.[8] The choice of Hermite was especially fitting, because what Laguerre did was to continue the work of Eisenstein and Hermite on the application of the symbolic algebra of linear systems to problems in the arithmetic theory of forms and the theory of abelian functions. Accordingly, Laguerre's memoir has two parts. The first is devoted to the development of the symbolic algebra of linear systems along the lines of Cayley's 1858 memoir [84]. The second part contains applications to the above-mentioned fields as well as to the Cayley–Hermite problem. Laguerre's work suffered a fate similar to that of Cayley's 1858 memoirs. It was published in the journal of his alma mater, the École Polytechnique. As in Cayley's case, the chosen journal was not widely read outside of France.

Laguerre's work is nonetheless historically significant for several reasons. In the first place, it shows that the further development of the symbolic algebra of matrices was a natural outgrowth of the work of Eisenstein and Hermite. In fact, it appears that Laguerre was unfamiliar with Cayley's 1855 notes as well as his memoirs of 1858. Secondly, although Laguerre, like Cayley, developed matrix algebra exclusively on the generic level, on that level he was able to use the matrix symbolism to generalize an important paper of Hermite's on abelian functions in two variables to n variables. Frobenius' first reference to Laguerre's paper [393] was in 1883, when he worked on a problem in the theory of abelian and theta functions, and found Laguerre's symbolic treatment of Hermite's work especially enlightening in relation to the problem he was seeking to solve (as will be seen in Section 10.6).

Unaware of Laguerre's memoir in 1877, Frobenius also investigated the Cayley–Hermite problem and developed the symbolic algebra of matrices (in the guise of bilinear forms) in a paper submitted to *Crelle's Journal* in May 1877 and published in 1878 [181]. It bore the title "On linear substitutions and bilinear forms" and,

---

[7]The same is true of Sylvester's subsequent treatment of this problem, as I have shown in [270, §6].

[8]Nowadays, Laguerre is remembered primarily because of the polynomials that bear his name. For a broader perspective on his life and work, see [16, 482].

as we will see, represented a fundamental contribution to the development of linear algebra. Frobenius cited all of Cayley's papers of 1846 and 1855 that were discussed in the previous section, namely [80, 84, 85], but not the papers of 1858 [84, 85], and it is doubtful he was aware of them.

By the time Frobenius began his career as a mathematician, the Cayley–Hermite problem had long since been forgotten; but attention was focused on it again in 1873 by Paul Bachmann (1837–1920) [10]. Bachmann, who had received his doctorate under Kummer at Berlin in 1862 and in 1873 was an assistant professor at the University of Breslau, was led through his research on the arithmetic theory of forms to examine Hermite's paper of 1854 [288] containing his solution to the problem for ternary forms. (No one in Germany appears to have known of Hermite's generalization of his solution to quadratic forms in any number of variables, for it was published in the *Cambridge and Dublin Mathematics Journal.*) He discovered that Hermite's method failed to yield all solutions to the problem, and so he devised an alternative method of obtaining them [10, p. 338]. Bachmann used no matrix symbolism and simply applied determinant-theoretic reasoning in conjunction with symbol manipulation. He determined proper transformations $x = U(p)y$ in three variables that for $p > 0$, gave the solutions given by Hermite's generic formula and for $p = 0$ gave those that Hermite had omitted.

Bachmann's paper induced Hermite in 1874 [293] to reconsider his work of 20 years ago. Restricting himself to ternary forms (as had Bachmann), he examined the cases not covered by his generic proof and sought to show that the several exceptional cases "followed," in various ways, from the generic solution. One exceptional case followed, for example, from the generic formula by letting some of the parameters become infinite. Incidentally, in this paper Hermite finally introduced a bit more of his own symbolism for linear substitutions. Denoting the above-mentioned limiting-case solution by S, he observed that "$S^{-1} = S$, or $S^2 = 1$" [293, p. 189]. As we shall see in more detail below, the complete solutions to the Cayley–Hermite problem for ternary forms that were given by Bachmann and Hermite did not generalize to the case $n > 3$, as Frobenius realized.

Bachmann's interest in the Cayley–Hermite problem was shared by his colleague at the University of Breslau, Jacob Rosanes (1842–1922),[9] who investigated the n-variable case from a different point of view in 1875 [506]. Rosanes' paper was motivated by the observations of Hermite and Cayley to the effect that the invertible transformations $x = PX$ discovered by them with the property that a nonsingular quadratic form $F(x) = x^t Sx$ is left invariant had characteristic polynomials that were "reciprocal" in the sense that if $r = a$ was a root of $p(r) = \det(rI - P)$, then so was its reciprocal, $r = 1/a$. This suggested to Rosanes the possibility

[9] Rosanes was later M. Born's teacher at Breslau. According to Born, it was by recalling Rosanes' lectures of 1903 on algebra and analytic geometry that he recognized the connection between Heisenberg's new approach to quantum mechanics and matrix algebra. See Jammer [318, p. 204], who, however, incorrectly states that Rosanes was Frobenius' student, although, as the following discussion indicates, Rosanes undoubtedly read, and was taught by, Frobenius' 1878 paper.

of a characterization of such transformations that would be independent of the particular quadratic form left invariant, e.g., something along the lines of that an invertible transformation $x = PX$ leaves some nonsingular quadratic form invariant if and only if it has a reciprocal characteristic equation. Rosanes also considered an analogous characterization of the more general invertible transformations that leave a nonsingular bilinear form invariant. He referred to Kronecker's 1866 paper on families of bilinear forms of the type $rA - A^t$ [353], where it was shown that the characteristic polynomial of $rA - A^t$ is reciprocal, assuming that the characteristic roots are distinct. (Kronecker's paper, the subject of Section 5.3, was written before Kronecker was acquainted with Weierstrass' nongeneric techniques for dealing with families of bilinear forms.) Rosanes, who employed no matrix symbolism, realized that except for a constant factor, this is the characteristic polynomial of the transformation defined by the system of equations corresponding to $Ax = \varepsilon A^t y$, with $\varepsilon = \pm 1$, i.e., in matrix notation, $P = \varepsilon A^{-1}A^t$, and that leaves the bilinear form $F(x,y) = x^t Ay$ invariant.[10] More generally, $Q = L^{-1}PL$ leaves $G(x,y) = x(L^t AL)y$ invariant.

Rosanes, like Kronecker (in 1866), sought to deal only with the generic case of distinct multiple roots, an assumption he made explicit. Within the context of that assumption, he focused on the invertible transformations of the form $x = L^{-1}(\varepsilon A^{-1}A)LX$. Kronecker's theorem implied that the characteristic polynomial of such a transformation is reciprocal. By way of a converse, Rosanes showed that if an invertible transformation $x = PX$ has a reciprocal characteristic polynomial that not only has distinct roots but also does not have *both* $r = 1$ and $r = -1$ as roots, then P is of the form $P = L^{-1}\varepsilon A^{-1}AL$, and so leaves a nonsingular bilinear form $G(x,y) = x(L^t AL)y$ invariant.

Actually, given Rosanes' hypothesis of distinct roots, his additional assumption about the roots $r = \pm 1$ turns out to be superfluous. As Frobenius was to point out, Kronecker, in a paper of 1874 [359] (discussed in Section 5.6.4), had given a nongeneric analysis of the equivalence classes of bilinear families $uA + vA^t$ under congruence, and his results implied that if both $r = 1$ and $r = -1$ are characteristic roots of $\varepsilon A^{-1}A^t$, then one of them must be a multiple root.[11] Thus Rosanes' assumption of distinct roots obviates the need for his assumption about $r = \pm 1$. It does not seem that Rosanes was aware of Kronecker's recent paper, but even if he had known of it, it is doubtful that he would have mastered its contents and applied them. Indeed, he did refer to Weierstrass' paper on the theory of elementary divisors [588], but he made no attempt to utilize its contents to go beyond the generic case. By contrast, to Frobenius, a Berlin-trained mathematician well versed in Weierstrass' theory of elementary divisors as well as Kronecker's

---

[10]That $P^t AP = A$ follows by elementary matrix algebra, which Rosanes, regrettably, lacked.

[11]This follows from the more general result given by Frobenius that for a family $uA + vA^t$, any elementary divisors of the form $(u+v)^{2\kappa}$ or $(u-v)^{2\kappa+1}$ occur doubled [181, p. 364, I]. (There is a typographical error in Frobenius' paper (p. 22) that is only partially corrected in the collected works (p. 364).)

more general theory, the obvious nongeneric problem suggested by Rosanes' paper was to characterize those invertible transformations P leaving some nonsingular bilinear form invariant in terms of the structure of the characteristic polynomial of P as a product of the elementary divisors into which it necessarily factors. As we shall see (Section 7.5.3), Frobenius investigated—and completely solved—this problem as well, and it provided some helpful insights into how to deal with the Cayley–Hermite problem.

The renewed discussion of the Cayley–Hermite problem in the 1870s thus generated some good problems from the viewpoint of the second disciplinary ideal of the Berlin school articulated by Kronecker (Section 5.6.3). It was in the spirit of that ideal that Frobenius approached the Cayley–Hermite problem. As he explained in the introduction to his paper [181]:

> Investigations on the transformation of quadratic forms into themselves have so far been limited to consideration of the general [= generic] case, while exceptions to which the results are subject in certain special cases have been exhaustively treated only for ternary forms ... I have thus attempted to fill in the gaps that occur in the proofs of the formulas that ... Cayley ... and ... Hermite ... have given, as well as those, in the reflections that ... Rosanes ... has made about the character of the transformation.

Frobenius' paper ran for 63 of the large pages of *Crelle's Journal* and runs for 71 pages in his collected works. The reason for its length is significant. Frobenius realized that in order to fill in the gaps in a manner that was both rigorous and elegant, it would be essential to fuse the theory of bilinear forms of Weierstrass and Kronecker (Chapter 5) with the symbolic algebra of linear substitutions, to compose, in effect, a monograph on the theory of matrices. The objects in Frobenius' calculus are actually bilinear forms that are denoted by $A = \sum_{i,j=1}^{n} a_{ij}x_iy_j$, $B = \sum_{i,j=1}^{n} b_{ij}x_iy_j$. The addition and scalar multiplication of such forms was already used in the work of Weierstrass and Kronecker. To these operations Frobenius added that of multiplication, which can be defined succinctly when forms are the symbolic objects

$$AB = \sum_{k=1}^{n} \frac{\partial A}{\partial y_k}\frac{\partial B}{\partial x_k}.$$

(The reader can check that the (i,j) coefficient of AB is indeed $\sum_{k=1}^{n} a_{ik}b_{kj}$.) The transformation of the form A by the substitutions $x_i = \sum_j p_{ij}X_j$, $y_i = \sum_j q_{ij}Y_j$ yields the form P^tAQ if the above substitutions are identified with the forms $P = \sum_{i,j} q_{ij}x_iy_j$ and $Q = \sum_{i,j} p_{ij}x_iy_j$.[12] The underlying idea is, of course, to make no distinction between forms and the associated linear substitutions of the variables, just as in ordinary arithmetic, the distinction between multiplier and multiplicand vanishes once the multiplication has been performed. Although the word "matrix" was not yet a part of Frobenius' mathematical vocabulary, his algebra of forms and matrix

---

[12]Frobenius used the notation P' to denote the transpose of P.

algebra differ only in name, and I will frequently employ matrix terminology in what follows, even though in 1877 Frobenius did not.

According to Frobenius, such considerations "led me to treat the composition of linear substitutions instead of the transformation of bilinear forms" [181, p. 344]. Thus the original Cayley–Hermite problem becomes in Frobenius' notation that of determining all nonsingular P such that $P^tAP = A$, where $A^t = A$ and $\det A \neq 0$. Such considerations, without however the single-letter notation, are already implicit in Cayley's 1855 note [83], and it seems likely that when Frobenius became interested in the Cayley–Hermite problem, he read Cayley's two notes of 1855 [82, 83] and was influenced by them. Frobenius was also familiar with the symbolic algebra of linear substitutions of Eisenstein and Hermite. As he explained in a note announcing his results to Hermite, "This operation of form multiplication is nothing other than the composition of substitutions which you have often employed" [180, p. 340]. Furthermore, the algebra of linear substitutions under composition was fundamental to another work known to Frobenius: Camille Jordan's *Traité des substitutions* of 1870 [322], where linear substitutions (interpreted as congruences with respect to a prime number) are employed to represent finite groups of permutations.

Thus by the 1870s, the symbolic algebra of linear substitutions under composition was an established practice in the theory of numbers and the theory of groups, whereas the consideration of the addition and scalar multiplication of bilinear forms was customary in the theory of the transformation of bilinear forms as developed by Weierstrass and Kronecker. The idea of identifying the operations of the transformation of bilinear forms with the composition of linear substitutions, which arises naturally from the Cayley–Hermite problem, also leads naturally to the consideration of bilinear forms as constituting what is now called a linear associative algebra. That simple yet crucial idea was implicit in Cayley's 1855 note [83] on the Cayley–Hermite problem and is the likeliest source of Frobenius' inspiration.

Even in his later memoirs of 1858 and 1866 [84–86], however, Cayley never fully exploited his symbolic algebra of matrices as a mode of reasoning, whereas Frobenius, in characteristic fashion, composed in effect a monograph on matrix algebra that combined the results of the Weierstrass–Kronecker theory of equivalence of bilinear forms with the symbolic algebra of forms. Much of it reads like a clear exposition of matrix algebra such as would be found in a modern sophomore-level linear algebra textbook, but Frobenius not only systematically established such basics as, e.g.,

$$(AB)^{-1} = B^{-1}A^{-1}, \quad (AB)^t = B^tA^t, \quad (AB)C = A(BC),$$

he also pushed more deeply into the implications of the algebra of forms. The result was a rigorously established and powerful theory, as he demonstrated by several applications, including a complete resolution of the general Cayley–Hermite problem. In what follows, I will first sketch some of the new results of the theory and then proceed to consider the applications he gave. Particularly impressive later applications of matrix algebra by Frobenius will be given in Sections 10.5 and 16.1.2.

7.5.1 The minimal polynomial

With the elementary consequences of his matrix algebra established, Frobenius turned to the characteristic polynomial $\varphi(r) = \det(rI - A)$.[13] As remarked in discussing Cayley's 1858 paper on matrix algebra, formal considerations suggest but do not prove that $\varphi(A) = 0$. This equality is easily seen to be true in the generic case of distinct characteristic roots, in which case, by Weierstrass' elementary divisor theory, A can be assumed without loss of generality to be a diagonal matrix with distinct diagonal entries $a_1,\ldots,a_n$. Since $\varphi(A) = \prod_{i=1}^n (A - a_i I)$ is a product of diagonal matrices each with a 0 in a different diagonal position, the product is clearly 0. In terms of Weierstrass' theory (Section 5.4), the generic case occurs when $\varphi(r) = E_n(r)$, where $E_n(r) = D_n(r)/D_{n-1}(r)$. If, e.g., A is again a diagonal matrix but with $a_1 = a_2$ and all other a_i distinct from each other and from a_2, it follows readily that $D_n(r) = \varphi(r) = (r - a_2)^2 \prod_{i=3}^n (r - a_i)$, and $D_{n-1}(r)$, the greatest common divisor of all the $(n-1)\times(n-1)$ minors of $rI - A$, equals $r - a_2$, so that $E_n(r) = D_n(r)/D_{n-1}(r) = \prod_{i=2}^n (r - a_i)$. In this nongeneric case, $E_n(r)$ is thus a proper divisor of $\varphi(r)$, although clearly $E_n(A) = \prod_{i=2}^n (A - a_i I) = 0$. Probably such considerations led Frobenius to suspect that $E_n(A) = 0$ for any A. Since $E_n(r)$ always divides $D_n(r) = \varphi(r)$, this would imply $\varphi(A) = 0$ as well. But how to prove $E_n(A) = 0$?

Frobenius began by observing that since a form A has n^2 coefficients, any linearly independent set of forms can have at most n^2 forms in it [181, p. 353]. Thus given any form A, there will exist a smallest integer p such that the forms $I = A^0, A^1, \ldots, A^{p-1}$ are linearly independent but $A^0, A^1, \ldots, A^p$ are linearly dependent. This dependency relation may be expressed in the form $a_0 A^0 + a_1 A^1 + \cdots + a_p A^p = 0$, where $a_p \neq 0$. The corresponding polynomial

$$\psi(r) = a_0 + a_1 r + \cdots + a_p r^p \tag{7.11}$$

plays a fundamental role in Frobenius' paper, although he did not give it a special name. Of course, it is only unique up to constant multiples as defined by him, but by multiplying (7.11) through by $1/a_p$, it can be assumed that the coefficient of r^p is 1. In accordance with present-day terminology, I will refer to this choice of ψ as *the minimal polynomial of* A. Frobenius' goal was to show that $\psi(r)$ and $E_n(r)$ are the same modulo a constant factor, which means that *the* minimal polynomial coincides with $E_n(r)$, because both are monic.

To relate $\psi(r)$ and $E_n(r)$, Frobenius considered $(rI - A)^{-1}$, which exists for all r with $|r|$ sufficiently large. On the one hand, there is the classical identity $(rI - A)\,\mathrm{Adj}\,(rI - A) = \det(rI - A)I$, which may now be written as $(rI - A)^{-1} = \mathrm{Adj}\,(rI - A)/\varphi(r)$. Since the entries of $\mathrm{Adj}\,(rI - A)$ are signed degree-$(n-1)$ minors of $rI - A$, it follows from the definition of $D_{n-1}(r)$ that it is the polynomial greatest common

[13] Frobenius denoted the identity matrix by the letter E (for *Einheit*).

divisor of all the polynomials that form the coefficients of Adj $(rI - A)$, so Adj $(rI - A) = D_{n-1}(r)M(r)$, and the polynomial coefficients of the matrix $M(r)$ have no common nonconstant factor. Also $\varphi(r) = D_n(r) = D_{n-1}(r)E_n(r)$, and so

$$(rI - A)^{-1} = M(r)/E_n(r), \tag{7.12}$$

and the right-hand side is in "lowest terms," i.e., there is no nontrivial factor of $E_n(r)$ that divides all the polynomial coefficients of $M(r)$. The relation (7.12) is thus a simple consequence of Weierstrass' theory.

To bring $\psi(r)$ into the picture, Frobenius observed that $(rI - A)^{-1} = S$, where $S = \sum_{i=1}^{\infty} A^{i-1}/r^i$. This matrix Laurent series converges absolutely for all $|r|$ sufficiently large, and matrix algebra then shows that $(rI - A)S = I$. Now consider $\psi(r)S = (a_0 + a_1 r + \cdots + a_p r^p)S$. In multiplying $a_k r^k$ by the terms of S, consider first the products of $a_k r^k$ with $A^k/r^{k+1}, A^{k+1}/r^{k+2}, \ldots$. These products, added for all k and suitably rearranged, equal $\psi(A)/r + A\psi(A)/r^2 + A^2\psi(A)/r^3 + \cdots = 0$, since $\psi(A) = 0$. The remaining products of $a_k r^k$ with I/r, A/r^2, $\ldots$, A^{k-1}/r^k, for $k = 0, 1, \ldots, p$, yield a matrix with polynomial coefficients in r of degree at most $p - 1$, which Frobenius denoted by $G(r)$. Thus $\psi(r)S = G(r)$ and $\deg G = p - 1$ in the sense that the diagonal entries are $a_p r^{p-1}$ plus terms involving lower powers of r. Thus $(rI - A)^{-1} = G(r)/\psi(r)$. In view of (7.12), this means that $M(r)/E_n(r) = G(r)/\psi(r)$. If the fraction on the right-hand side is also in lowest terms, then $\psi(r)$ and $E_n(r)$ will differ by at most a constant factor, the proof of which was the desired goal.

To show that $\psi(r)$ and $G(r)$ have no common factor, Frobenius observed that if $\varphi(r) = \lambda(r)\chi(r)$ and $G(r) = \lambda(r)H(r)$, so $G/\psi = H/\chi$ with $\deg \chi < \deg \psi$, then

$$\chi(r)S = K(r) + \chi(A)/r + A\chi(A)/r^2 + A^2\chi(A)/r^3 + \cdots \tag{7.13}$$

by the same reasoning used to show

$$\psi(A)S = G(r) + \varphi(A)/r + A\varphi(A)/r^2 + A^2\varphi(A)/r^3 + \cdots = G(r).$$

If this last equation is divided through by $\lambda(r)$, the result is $\chi(r)S = H(r)$. This would say that the matrix Laurent series (7.13) equals $H(r)$ and so has no negative powers of r, which means by uniqueness of Laurent series that $\chi(A) = 0$, contrary to the definition of ψ.

In sum, what Frobenius had established was the following fundamental theorem of matrix algebra.

Theorem 7.2 (Minimal polynomial theorem). *Given any $n \times n$ matrix A, let $\varphi(r)$ and $\psi(r)$ denote respectively the characteristic and minimal polynomials of A, and let $D_{n-1}(r)$ denote the greatest common divisor of the $(n-1) \times (n-1)$ minors of $rI - A$. Then $\psi(r) = \varphi(r)/D_{n-1}(r) = E_n(r)$. Hence (i) $\deg \psi = n - \deg D_{n-1} \le n$; (ii) $\varphi(A) = 0$; (iii) $\psi(r)$ has the same distinct roots as $\varphi(r)$; (iv) If $f(r)$ is any polynomial such that $f(A) = 0$, then $\psi(r)$ divides $f(r)$.*

From the factorization $\varphi(r) = D_{n-1}(r)\psi(r)$, (i) and (ii) follow immediately. Part (ii) is the Cayley–Hamilton theorem, which Frobenius was the first to prove (without knowing of Cayley's assertion of it). Part (iii) follows from Weierstrass' theory of elementary divisors, which shows that the invariant factor $E_n(r)$ has the same distinct roots as $\varphi(r)$. Part (iv) is an easy consequence of the definition of ψ, for if $f(A) = 0$, the minimality of ψ means that $\deg f \geq \deg \psi$, and so $f(r) = \mu(r)\psi(r) + \rho(r)$, where $\deg \rho < \deg \psi$. Hence $0 = f(A) = \mu(A)\cdot 0 = \rho(A)$ and $\rho(A) = 0$. Given the minimality of ψ and the degree of ρ, we must have $\rho = 0$, i.e., $\psi(r)$ divides $f(r)$.

Because Frobenius left ψ determined only up to a constant factor, he did not explicitly write down $\varphi(r) = D_{n-1}(r)\psi(r)$. He also did not explicitly point out the fact that both $\varphi(r)$ and $D_{n-1}(r)$ can be calculated by established algorithms—carried out nowadays preferably on a computer—and so the same is true of $\psi(r) = \varphi(r)/D_{n-1}(r)$.

An easy application of (iv) made by Frobenius is to matrices A such that $A^k = 0$ for some integer $k > 0$—nilpotent matrices as we would say today [181, p. 357, VI]. Then if $p(r) = r^k$, $p(A) = 0$, and so by (iv), ψ divides r^k. If $p = \deg \psi$, then the minimality of ψ implies $\psi(r) = r^p$ and that p is the smallest integer such that $A^p = 0$. Also (ii) implies that $\varphi(r) = r^n$, and so all the characteristic roots of A are 0. In fact, clearly A is nilpotent in the above sense if and only if $\varphi(r) = r^n$. Similar considerations imply [181, p. 358, VIII]:

Corollary 7.3 (Frobenius). *If $A^m = I$, then all the characteristic roots are mth roots of unity and all the elementary divisors are linear (i.e., A can be diagonalized).*

Frobenius also used his new symbolic calculus to obtain new results about determinants. For example, if A is arbitrary and B is a nilpotent matrix that commutes with A, then $\det(A+B) = \det A$ [181, p. 357, VII].

Another easy consequence of Theorem 7.2 is a rigorous proof of a sharper version of one of Cayley's claims (Section 7.4):

Corollary 7.4. *Suppose $f(r) = p(r)/q(r)$ is a rational function such that $q(A)$ is invertible so that $f(A) = p(A)[q(A)]^{-1} = [q(A)]^{-1}p(A)$ is defined.*[14] *Then there is a polynomial $\rho(r)$ of degree at most $p-1$, $p = \deg \psi$, such that $f(A) = \rho(A)$.*

Frobenius proof goes as follows. If $r_1, \ldots, r_n$ are the characteristic roots of A, i.e., the roots of $\varphi(r)$, each counted as often as its multiplicity, then the characteristic roots of $q(A)$ are $q(r_1), \ldots, q(r_n)$ [181, p. 353], and so $\det[q(A)] = q(r_1)\cdots q(r_n)$. Since $\det[g(A)] \neq 0$, it follows that none of the r_k are roots of $q(r)$, and so the characteristic polynomial $\varphi(r)$ and $q(r)$ are relatively prime, and by Theorem 7.2, the same is therefore true of the minimal polynomial ψ and $q(r)$. Thus polynomials P_1 and P_2 exist such that $1 = P_1(r)\psi(r) + P_2(r)q(r)$, and so $p(r) = p(r)P_1(r)\psi(r) + p(r)P_2(r)q(r)$. But then, since $\psi(A) = 0$, $p(A) = p(A)P_2(A)q(A)$ or $f(A) = p(A)[q(A)]^{-1} = p(A)P_2(A)$. This shows that $f(A)$ equals a polynomial in A.

[14] As Frobenius showed [181, §§1–2], $p(A)$ and $q(A)$ commute, and so $p(A)$ and $[q(A)]^{-1}$ commute.

Of course, if $p(r)p_2(r)$ is of degree p or higher, then $p(r)P_2(r) = \psi(r)\mu(r) + \rho(r)$ with $\deg\rho \le p-1$. Thus $f(A) = p(A)P_2(A) = \rho(A)$, since $\psi(A) = 0$.

Another problem Frobenius considered as an application of the minimal polynomial was this: *For a given A, determine all B such that AB = BA.* This was the problem Cayley had considered in his 1858 paper [84] as indicated above in Section 7.4, although he was apparently unaware of this paper. I will summarize his results on this problem in the following single theorem [181, pp. 370–371].

Theorem 7.5 (Centralizer theorem). *Let A be a given $n \times n$ matrix. (i) If A is such that $D_{n-1}(r) = 1$, i.e., if $\psi(r) = \varphi(r)$, then the only B that commute with A are polynomials in A. (ii) If A is such that the roots $r_1, \ldots, r_p$ of $\psi(r)$ are all distinct, then all B that commute with A are expressible in the following form: set $\psi_k(r) = \psi(r)/(r - r_k)$, $k = 1, \ldots, p$; then $B = \sum_{k=1}^{p} \psi_k(A) C_k \psi_k(A)$, where $C_1, \ldots, C_p$ denote any p matrices. (iii) For any A, if $n_k = \deg D_{n-k}(r)$, where $D_{n-k}(r)$ denotes the greatest common divisor of the $n-k \times n-k$ minors of $rI - A$, then the number of linearly independent B that commute with A is*

$$n + 2(n_1 + n_2 + \cdots). \tag{7.14}$$

Part (i) shows that Cayley's solution to the problem was indeed the generic one, which occurs when the roots of $\varphi(r)$ are all distinct, so that by part (ii) of Theorem 7.2, we must have $\psi(r) = \varphi(r)$, i.e., $D_{n-1}(r) = 1$. Parts (ii)–(iii) of course provide a Berlin-style solution to the problem, according to which (in Kronecker's words)[15] one dares to go beyond the case of distinct characteristic roots. These parts show just how far off the mark Cayley's generic solution could be for certain choices of A. Indeed, as Frobenius pointed out, if we take $A = I$, then $n_k = n - k$, and so (7.14) equals $n + 2[(n-1) + (n-2) + \cdots] = n^2$.

Frobenius did not supply a proof of his centralizer theorem, just a hint as to how to prove it [181, pp. 370–371]. Proofs were later supplied by Maurer in his 1887 doctoral dissertation [437] and then by Voss [576]), who actually used Frobenius' hint.

7.5.2 *Fusion with Weierstrass–Kronecker theory*

Frobenius devoted over ten pages of his paper to interpreting and extending the results of Weierstrass and Kronecker on bilinear forms (Chapter 5) within the framework of his symbolic algebra of matrices [181, pp. 359–371]. What follows are a few important highlights.

One matter Frobenius considered was the various types of equivalence of families of forms considered by Weierstrass and Kronecker. Recall that two families $rA - B$ and $rC - D$ are equivalent in the sense of Weierstrass if the one can be transformed

[15] Quoted in Section 5.6.3.

into the other by means of transformations $x = HX$, $y = KY$. As Frobenius expressed it, this means that $P(rA - B)Q = rC - D$, where $P = H^t$ and $Q = K$, and this is equivalent to $PAQ = C$ and $PBQ = D$. Frobenius said that A and C (and likewise B and D) were *equivalent*. More generally, when $PAQ = C$ for invertible P, Q, Frobenius said that A and C were equivalent. A special case of equivalence occurs when the two families are of the form $rI - A$ and $rI - B$, a form that (as we have already seen) is most common with Frobenius. In that case, the equivalence of $rI - A$ and $rI - B$ means that $PQ = I$ and $PAQ = B$, and so $Q = P^{-1}$ and thus $PAP^{-1} = B$ (or $Q^{-1}AQ = B$). In this case, Frobenius said that A and B were *similar*, as is still done nowadays. For Kronecker, "equivalence" of $rA - B$ and $rC - D$ usually meant that the one family could be transformed into the other by means of $x = PX$ and $y = PY$, which means that $P^tAP = C$ and $P^tBP = D$. Following Kronecker's terminology, Frobenius said that A and C (and likewise B and D) were *congruent*. The above terminology will be used in what follows.

Recall that Kronecker had emphasized in his papers of 1874 the idea of the decomposition of forms into a "direct sum" of more elementary forms, where I use "direct sum" to indicate that each elementary form had its own exclusive set of variables (Section 5.6.1). Frobenius denoted such a direct sum decomposition simply by $A = \sum_k A_k$, but I will use the notation $A = \bigoplus_k A_k$ or $A = A_1 \oplus A_2 \oplus \cdots$ to remind the reader that the sum is direct in the above sense. It then follows that if B is another form with the same type of direct sum decomposition $B = \bigoplus_k B_k$ (meaning that the variable sets for A_k and B_k are the same, then since $A_iB_j = 0$ for $i \neq j$, one has $AB = \bigoplus_k A_kB_k$. Continuing this line of reasoning, Frobenius quickly arrived at the fact that if $p(r)$ is any polynomial, then $p(\bigoplus_k A_k) = \bigoplus_k p(A_k)$, and if $f(r) = p(r)/q(r)$ is any rational function such that $f(A) = p(A)[q(A)]^{-1}$ is defined, then since $f(A) = p(A)$ for some polynomial $p(r)$ by Theorem 7.4, it follows that $f(\bigoplus_k A_k) = \bigoplus_k f(A_k)$.

Suppose now that

$$\varphi(r) = \det[rI - A] = (r-a)^{\alpha}(r-b)^{\beta}\cdots \tag{7.15}$$

denotes the factorization of the characteristic polynomial of A into its elementary divisors in accordance with Weierstrass' theory (Section 5.4). Frobenius used the fact that this means that a nonsingular P exists such that $P^{-1}AP = J$, where $J = J_{\alpha}(a) \oplus J_{\beta}(b) \oplus \cdots$ is the bilinear form version of the Jordan canonical form, so that, e.g.,

$$J_{\alpha}(a) = a(x_1y_1 + \cdots + x_{\alpha}y_{\alpha}) + x_1y_2 + \cdots + x_{\alpha-1}y_{\alpha}$$

is the bilinear form with coefficient matrix equal to the $\alpha \times \alpha$ Jordan block for $r = a$. (Frobenius evidently preferred Jordan's canonical form to the similar one that Weierstrass had introduced (see (5.17)). In other words, specifying the factorization of the characteristic polynomial into its elementary divisors, as in (7.15), is tantamount to specifying (succinctly) the associated Jordan canonical form. The coefficient matrix of the form J is thus

$$\begin{pmatrix} J_\alpha(a) & & \\ & J_\beta(b) & \\ & & \ddots \end{pmatrix}, \quad \text{where now} \quad J_\alpha(a) = \begin{pmatrix} a & 1 & & \\ & a & 1 & \\ & & \ddots & 1 \\ & & & a \end{pmatrix}$$

denotes the $\alpha \times \alpha$ Jordan block for the root $r = a$, and so on. This connection between the structure of the Jordan canonical form in terms of its Jordan blocks and the factorization of the characteristic polynomial into its elementary divisors should be keep in mind in what follows.

Given the factorization (7.15) for $\varphi(r) = \det(rI - A)$, if $f(r)$ is any rational function such that $f(A)$ is defined, then since $f(A)$ equals a polynomial in A by Theorem 7.4, and since $P^{-1}A^kP = (P^{-1}AP)^k$, it follows that

$$P^{-1}f(A)P = f(P^{-1}AP) = f(J_\alpha \oplus J_\beta \oplus \cdots) = f(J_\alpha) \oplus f(J_\beta) \oplus \cdots.$$

Since the characteristic roots of, e.g., $f(J_\alpha)$, when regarded as a form in just α variables, are what f does to the characteristic roots $a,\ldots,a$ of J_α [181, p. 353], it follows that $\det[rI_\alpha - f(J_\alpha)] = (r - f(a))^\alpha$. Hence if $\Phi(r) = \det[rI - f(A)]$ is the characteristic polynomial of $f(A)$, it follows that $\Phi(r)$ equals (with $|M| = \det M$)

$$|rI - f(J)| = |rI_\alpha - f(J_\alpha)| \cdot |rI_\beta - f(J_\beta)| \cdots = (r - f(a))^\alpha (r - f(b))^\beta \cdots.$$

It would be tempting to assume that the above is actually the factorization of $f(A)$ into its elementary divisors. However, this is not the case. For example, if A corresponds to the Jordan block $\begin{pmatrix} 2 & 1 & 0 \\ 0 & 2 & 1 \\ 0 & 0 & 2 \end{pmatrix}$ and $f(r) = (r-2)^2$, then the factorization of $|rI - A|$ into its elementary divisors is $(r-2)^3$, but $(r - f(2))^3 = r^3$ is not the factorization of $\det[rI - f(A)]$ into its elementary divisors, because $f(A) = (A - 2I)^2$ has coefficient matrix $\begin{pmatrix} 0 & 0 & 1 \\ 0 & 0 & 0 \\ 0 & 0 & 0 \end{pmatrix}$, and its elementary divisors are r^2, r, that is, $\det[rI - f(A)] = r \cdot r^2$ is the factorization into elementary divisors, not r^3. To put the matter another way: if J_α is a Jordan block, $f(J_\alpha)$ need not be. This problem occurs when $f'(a) = 0$, where $(r-a)^\alpha$ is a nonlinear elementary divisor, i.e., $\alpha > 1$. Frobenius was able to prove the following result.

Theorem 7.6. *Let $\varphi(r) = \det[rI - A] = (r-a)^\alpha (r-b)^\beta \cdots$ denote the factorization of the characteristic polynomial of A into its elementary divisors. Then if $f(r)$ is any rational function such that $f(A)$ is defined and for which $f'(r)$ does not vanish for any root of a nonlinear elementary divisor, it follows that $\det[rI - f(A)] = (r - f(a))^\alpha (r - f(b))^\beta \cdots$ gives the factorization of the characteristic polynomial of $f(A)$ into its elementary divisors.*

An important special case of this theorem occurs when $f(r) = 1/r$ and A is invertible. For then $f(A) = A^{-1}$ and $f'(r) = -r^{-2}$ never vanishes. Thus one has the following corollary.

Corollary 7.7. *If A is invertible and* $\det[rI - A] = (r-a)^{\alpha}(r-b)^{\beta}\cdots$ *gives the factorization of the characteristic polynomial into elementary divisors, then*

$$\det[rI - A^{-1}] = \left(r - \frac{1}{a}\right)^{\alpha}\left(r - \frac{1}{b}\right)^{\beta}\cdots$$

gives the factorization of the characteristic polynomial of A^{-1} into its elementary divisors.

7.5.3 The problem of Rosanes

Frobenius applied Corollary 7.7 to the problem suggested by Rosanes' paper, as indicated above: (1) *characterize the characteristic polynomials of those invertible P that transform some nonsingular bilinear form $F(x,y) = x^tAy$ into itself, i.e., such that $P^tAP = A$ for some A with* $\det A \neq 0$. Frobenius realized that the same ideas apply to the following more general problem: (2) *characterize the characteristic polynomials of those pairs of invertible matrices P, Q such that $PAQ = A$ for some A with* $\det A \neq 0$. Frobenius first solved (2) and then specialized the result to solve (1) [181, pp. 372–376].

For brevity, let us say that a pair (P,Q) of invertible matrices is a *Rosanes pair* if there is a nonsingular matrix A such that $PAQ = A$. For such a pair, it follows from $PAQ = A$ that

$$(rI - P)AQ = rAQ - PAQ = rAQ - A = A(rQ - I).$$

This means that $rI - P$ and $rQ - I$ are equivalent. Suppose, conversely, that P and Q are such that $rI - P$ and $rQ - I$ are equivalent. Then invertible L and M exist such that

$$L(rQ - I)M = rI - P, \quad \text{or} \quad LM = P, \quad LQM = I.$$

Thus

$$(PLQ)M = P(LQM) = P = LM,$$

and so multiplication by M^{-1} on the right yields $PLQ = L$, i.e., (P,Q) is a Rosanes pair (with respect to L). The conclusion is that *(P,Q) is a Rosanes pair if and only if the families $rI - P$ and $rQ - I$ are equivalent.*

Frobenius pushed this result further. Since $Q^{-1}(rQ-I) = rI - Q^{-1}$, it follows from the above italicized result that (P,Q) is a Rosanes pair if and only if $rI - P$ and $rI - Q^{-1}$ are equivalent. From Weierstrass' theory of elementary divisors, it followed that if

$$\det(rI - P) = (r-a)^{\alpha}(r-b)^{\beta}\cdots \tag{7.16}$$

denotes the factorization of the characteristic polynomial of P into its elementary divisors, then the factorization of the characteristic polynomial of Q^{-1} into its elementary divisors is the same. If this observation is combined with Corollary 7.7, the result is the following.

Theorem 7.8. *(P,Q) is a Rosanes pair if and only if when the factorization of the characteristic polynomial of P into its elementary divisors is given by (7.16), then the factorization of the characteristic polynomial of Q into its elementary divisors is given by*

$$\det(rI - Q) = \left(r - \frac{1}{a}\right)^{\alpha}\left(r - \frac{1}{b}\right)^{\beta}\cdots. \tag{7.17}$$

Although Frobenius did not mention it, he surely realized that this theorem makes it possible to construct all Rosanes pairs. That is, let J denote the Jordan canonical form that has the factors on the right-hand side of (7.16) as its elementary divisors, viz., $J = J_{\alpha}(a) \oplus J_{\beta}(b) \oplus \cdots$, and let $K = J^{-1} = J_{\alpha}(a)^{-1} \oplus J_{\beta}(b)^{-1} \oplus \cdots$. Then K is not a Jordan canonical form, since, e.g., $J_{\alpha}(a)^{-1} \neq J_{\alpha}(1/a)$, but its characteristic polynomial has the elementary divisor factorization given in (7.17), and so it is similar to $J_{\alpha}(1/a) \oplus J_{\beta}(1/b) \oplus \cdots$. Then all Rosanes pairs are of the form $P = LJL^{-1}$, $Q = MKM^{-1}$ for any invertible L, M, and P,Q is a Rosanes pair for the bilinear form $F(x,y) = x^t(LM^{-1})y$. Thus problem (2) is completely resolved.

Frobenius then turned to problem (1)—the problem actually posed in Rosanes' paper. Let us say that P *is a Rosanes transformation* if it is invertible and satisfies

$$P^tAP = A \tag{7.18}$$

for some invertible A. Then (P^t, P) is a Rosanes pair. This means that the characteristic polynomials of P^t and P have the elementary divisor factorizations specified in Theorem 7.8. But since $rI - P^t = (rI - P)^t$, it follows that the characteristic polynomial of P^t has the same W-series and so the same invariant factors and elementary divisors as those of P. Combining these two facts, Frobenius arrived at his solution to problem (1), namely the following theorem [181, III, p. 376].

Theorem 7.9. *P is a Rosanes transformation if and only if corresponding to every elementary divisor $(r-a)^{\alpha}$, $a \neq \pm 1$, of the characteristic polynomial of P, there is an elementary divisor $(r-1/a)^{\alpha}$.*

Once again, the Jordan canonical form can be used to construct all possible Rosanes transformations.[16]

In the course of working out his solution to the Cayley–Hermite problem, Frobenius realized that some of the considerations behind his theory of Rosanes pairs could be modified to prove a lemma that would facilitate his solution to the Cayley–Hermite problem in the cases not covered by Hermite's solution.

Lemma 7.10. *Let A be a $p\times q$ matrix of rank m, and suppose that square invertible matrices P,Q exist such that $PAQ = A$. Then the characteristic polynomial of Q has m (not necessarily distinct) roots that are reciprocals of roots of the characteristic polynomial of P. Hence if no root of Q is the reciprocal of a root of P, an equation of the form $PAQ = A$ can hold only if $A = 0$.*

Frobenius' proof of this lemma [181, p. 374] readily translates into the language of block multiplication of partitioned matrices, and since it is easiest to follow in this form, I will use it. (As we shall see in Section 10.6, it was only 5 years later that Frobenius discovered the virtues of block multiplication.)

The case in which A has full rank was already covered by Theorem 7.8, and so we can assume that $m < \min\{p,q\}$. Then Gaussian elimination teaches that invertible matrices H,K exist such that $HAK = E_1$, where $E_1 = \begin{pmatrix} I_m & 0 \\ 0 & 0 \end{pmatrix}$, I_m denoting the $m\times m$ identity matrix. From $PAQ = A$, it then follows that $P_0E_1Q_0 = E_1$, where $P_0 = HPH^{-1}$ and $Q_0 = K^{-1}QK$. Thus P_0 and P have the same characteristic polynomial and likewise so do Q_0 and Q. The lemma can therefore be established for P_0 and Q_0. If we write $P_0 = \begin{pmatrix} P_{11} & P_{12} \\ P_{21} & P_{22} \end{pmatrix}$, $Q_0 = \begin{pmatrix} Q_{11} & Q_{12} \\ Q_{21} & Q_{22} \end{pmatrix}$, then $P_0E_1Q_0 = E_1$ gives by block multiplication

$$\begin{pmatrix} P_{11}Q_{11} & P_{11}Q_{12} \\ P_{21}Q_{11} & P_{21}Q_{12} \end{pmatrix} = \begin{pmatrix} I_m & 0 \\ 0 & 0 \end{pmatrix}.$$

Comparing $(1,1)$ entries, we see that P_{11} and Q_{11} are nonsingular and are inverses of one another, so that every characteristic root of Q_{11} is the reciprocal of a root of P_{11}. Looking at the $(1,2)$ and $(2,1)$ entries in the light of the invertibility of P_{11} and Q_{11} shows that $P_{21} = Q_{12} = 0$. Thus P_0 and Q_0 are, respectively, lower block triangular and upper block triangular. This means that the characteristic polynomial of P_0 has that of P_{11} as a factor, and the characteristic polynomial of Q_0 has that of Q_{11} as a factor, thereby establishing the lemma.

---

[16]For example, the canonical form $J = J_3(2)\oplus J_3(1/2)\oplus J_2(-1)$ has elementary divisors in accord with Theorem 7.9. Hence all Rosanes transformations with elementary divisors $(r-2)^3$, $(r-1/2)^3$, and $(r+1)^2$, are given by $P = LJL^{-1}$, where L is any invertible 7×7 matrix. To determine a bilinear form left invariant by P, use the fact that in general, P^t is similar to P, and by Theorem 7.9, P is similar to P^{-1}, which is similar to J^{-1}. Thus P^t is similar to J^{-1}, and so $P^t = MJ^{-1}M$ for some invertible M. It then follows that P leaves invariant the form $F(x,y) = x^tAy$, $A = ML^{-1}$.

7.5.4 The Cayley–Hermite problem

Let us now turn to Frobenius' solution to the Cayley–Hermite problem: *given a quadratic form* $F(x) = x^t Sx$, $S = S^t$, *and* $\det S \neq 0$, *determine all the invertible transformations* $x = UX$ *that are proper* ($\det U = 1$) *and leave* $F(x)$ *invariant, i.e., such that*

$$U^t SU = S. \tag{7.19}$$

Expressed in Frobenius' notation and with his conventions, what Hermite had done was to show that "the most general" way to solve the problem was to take

$$U = (S+T)^{-1}(S-T), \tag{7.20}$$

where T is skew-symmetric.[17] As the quoted phrase indicates, Hermite realized that he was offering the generic solution, and of course it is tacitly assumed that $\det(S+T) \neq 0$, so that his solution is meaningful.

To overcome the problems afforded by the n-variable case without resorting to generic reasoning was, of course, the challenge of Kronecker's second disciplinary ideal (Section 5.6.3), and Frobenius set himself the task of dealing with this challenge in the case of the Cayley–Hermite problem. As we shall see, he accomplished it by means of extensive matrix algebra—a new arena for application of his extraordinary calculational skills—combined with the matrix form of Weierstrass' theory of elementary divisors (as discussed in Section 7.5.2).

Frobenius began by clarifying Hermite's generic solution in the n-variable case. Since in Hermite's formula (7.20), $S-T$ is the transpose of $A = S+T$, it is indeed the case that $\det U = \det(A^{-1}A) = +1$, so that U defines a *proper* transformation in the sense of Bachmann and Frobenius. Furthermore, matrix algebra shows that with U as in (7.20), $(S+T)(I+U) = 2S$, so that $\det(I+U) \neq 0$, because $\det S \neq 0$. In other words, in Hermite's generic solution, $r = -1$ is not a characteristic root of U. Frobenius established the following converse, which shows that all solutions to $U^t SU = S$ satisfying $\det(I+U) \neq 0$ are given by Hermite's formula [181, p. 346, III].

Theorem 7.11. *If* U *is any solution to* $U^t SU = S$ *(*$\det S \neq 0$*) that satisfies* $\det(I+U) \neq 0$, *then there is a unique skew-symmetric matrix* T *such that* $U = (S+T)^{-1}(S-T)$, *where* $T = S(I-U)(I+U)^{-1}$. *(Hence* U *is necessarily proper.)*

The proof goes like this. Straightforward matrix algebra shows that if T is defined as in the theorem, then $(I+U)(S+T) = 2S$, and since $I+U$ and S have nonzero

[17] The above quotation is from Hermite's paper in *Crelle's Journal* [288, p. 309], which is restricted to ternary forms. German mathematicians, including Frobenius, were not familiar with Hermite's solution for quadratic forms in n variables [287], published as it was in an obscure British journal. The n-variable solution contains nothing that was not already implied by the solution in the ternary case.

determinants, $\det(S+T) \neq 0$. Further simple matrix algebra then shows that $U = (S+T)^{-1}(S-T)$.[18] It remains to show that T is actually skew-symmetric. To see this, Frobenius considered $T_0 = (I+U)^t T(I+U)$, which is congruent to T, so that if $T_0^t = -T_0$, the same will be true of T. A straightforward calculation shows that $T_0 = U^t S - SU$, from which $T_0^t = -T_0$ follows. Although the proof of this theorem is straightforward, this is because matrix algebra makes it so. Without matrix algebra it is not so easy to see.

Theorem 7.11 showed that with his generic formula $U = (S+T)^{-1}(S-T)$, Hermite had in fact obtained all solutions to $U^t SU = S$, $\det S \neq 0$, that are proper ($\det U = 1$) and do not have $r = -1$ as a characteristic root ($\det(I+U) \neq 0$). What remained was to go beyond this generic case and determine all proper U such that $U^t SU = S$ ($\det S \neq 0$) and $\det(I+U) = 0$. This turned out to be a much more difficult problem, but Frobenius solved it for arbitrary n and without the need to consider myriad special cases.

Before describing Frobenius' solution, a few preliminary remarks need to be made. Consider, first of all, Bachmann's solution $x = U(p)y$ for a ternary quadratic form $F(x) = x^t Sx$ in the case $p = 0$—the case that provides the solutions to the problem for which $\det(I+U) = 0$, i.e., for which $r = -1$ is a root of the characteristic equation. In matrix form with $U(0) = U_0$,

$$U_0 = -I_3 + 2(\text{Adj } S)Q,$$

where $\text{Adj } S = (\det S)S^{-1}$ and $Q = (q_i q_j)$, q_1, q_2, q_3 being parameters chosen such that, setting $q = \begin{pmatrix} q_1 & q_2 & q_3 \end{pmatrix}^t$, $q^t(\text{Adj } S)q = 1$. Unlike $U(p)$ for $p > 0$, this formula clearly generalizes to $n > 3$ and yields $n \times n$ matrices U_0 such that $U_0^t SU_0 = S$, although the U_0 are improper for even values of n. That these U_0 with n odd cannot yield all solutions when $r = -1$ follows from the fact that (for any n) $U_0^2 = I$.[19] By Frobenius' minimal polynomial theorem, it follows that either $U_0 = -I$ or that the minimal polynomial is $\psi(r) = r^2 - 1$, which has distinct roots $r = \pm 1$ and so implies that all the elementary divisors of U_0 are linear (U_0 can be diagonalized). This, however, is not true of transformations $U^t SU = S$, be they proper or improper, that have $r = -1$ as a characteristic root. This is because Frobenius was able to strengthen his Theorem 7.9 characterizing Rosanes transformations when the nondegenerate bilinear form is actually symmetric to the following result [181, p. 383, V].

Theorem 7.12. *$U^t SU = S$, where S is symmetric and nondegenerate, if and only if, except for elementary divisors of the form $(r \pm 1)^{2\kappa+1}$, corresponding to every elementary divisor $(r-a)^\alpha$ of the characteristic polynomial of U, there is an elementary divisor $(r-1/a)^\alpha$.*

---

[18]The formula for T says that $S(I-U) = T(I+U)$, which may be rewritten as $SU + TU = S - T$, i.e., $U = (S+T)^{-1}(S-T)$.

[19]This follows by writing Q in block form as $Q = \begin{pmatrix} q_1 q & \cdots & q_n q \end{pmatrix}$, from which it follows from $q^t(\text{Adj } S)q = 1$ that $Q(\text{Adj } S)Q = Q$ and so $U_0^2 = I_n$.

The additional information provided by Theorem 7.12 vis à vis Theorem 7.9 is that if the characteristic polynomial of U has an elementary divisor of the form $(r-1)^{2\kappa}$, then it has one of the form $(r-1/1)^{2\kappa}$, i.e., the number of elementary divisors of the form $(r-1)^{2\kappa}$ will be even. Likewise, the number of elementary divisors of the form $(r+1)^{2\kappa}$ will be even (possibly zero). The theorem shows that for proper transformations, when $n=3$ if $r=-1$ is a characteristic root, then since $\det U=+1$, $r=-1$ must be a double root, but the elementary divisors must be linear, because if $(r+1)^2$ were an elementary divisor, then by the above theorem, there would be at least two such, which with $n=3$ is impossible. This is why Bachmann's formula for U_0 gives all the solutions with $\det(I+U)=0$ in the ternary case. On the other hand, for $n>3$, the theorem asserts that U's exist with $U^tSU=S$ and with, e.g., $(r+1)^2$ as a (repeated) elementary divisor. They cannot be given by Bachmann's formula generalized to $n>3$, however, since, as we saw, U_0 has only linear elementary divisors because $U_0^2=I$. The generalization to $n>3$ of Bachmann's formula for U_0 does not capture all the solutions to the Cayley–Hermite problem with $r=-1$ as a root. Likewise, Hermite's solutions U obtained by letting some parameters become infinite will not generalize, since again, $U^2=I$ means that the elementary divisors must be linear. In sum: *The generalizations of the solutions of Bachmann and Hermite cannot yield all solutions to the problem when* $\det(I+U)=0$ *because they produce only solutions U with linear elementary divisors.*[20]

How then to deal with the solutions U to the Cayley–Hermite problem for which $\det(I+U)=0$? Frobenius motivated his approach as follows [181, pp. 384–385]. Consider for a fixed nonsingular symmetric matrix S the equation

$$U^tSU=S. \tag{7.21}$$

This represents a system of quadratic equations in the n^2 unknowns u_{ij} that are the coefficients of U, and since transposition does not alter the system, the quadratic equation corresponding to the (i,j) entry of (7.21) is the same as the one corresponding to the (j,i) entry, and so the number of equations is $m=n(n+1)/2<n^2$ (for $n>1$). Thus there will generally be many solutions to this system. (In general, the solutions will form a submanifold of dimension n^2-m within $\mathbb{C}^{n^2}$.) One such solution is Hermite's: $U=(S+T)^{-1}(S-T)$, where T is given as in Theorem 7.11, from which it follows (as we saw) that

$$(I+U)(S+T)=2S. \tag{7.22}$$

[20]Thus, e.g., $U=\begin{pmatrix}-1&1&0&0\\0&-1&0&0\\0&0&-1&1\\0&0&0&-1\end{pmatrix}$, which has $(r+1)^2$ as a repeated elementary divisor and satisfies $U^tSU=S$ for $S=\begin{pmatrix}0&0&0&1\\0&4&-1&2\\0&-1&0&0\\1&2&0&3\end{pmatrix}$, is a solution not covered by generalizing the formulas of Bachmann and Hermite.

In a Hermite solution

$$U = (S+T)^{-1}(S-T) = [\mathrm{Adj}\,(S+T)\cdot(S-T)]/\det(S+T),$$

the u_{ij} are rational functions of the coefficients t_{ij} of T.

Suppose now that T_h is a 1-parameter family of such T's, so that $U_h = (S+T_h)^{-1}(S-T_h)$ is a family of Hermite solutions. Then if $U_0 = \lim_{h\to 0} U_h$ exists, it follows that (7.21) will hold in the limit, and so will yield a solution to the problem. Now if $T_0 = \lim_{h\to 0} T_h$ also exists, we will get another Hermite solution if $\det(S+T_0) \neq 0$. If, however, $\det(S+T_0) = 0$, (7.22) implies $\det(I+U_h)\cdot\det(S+T_h) = 2^n \det S$, and since $\det S \neq 0$, it must be that $\det(I+U_h)$ becomes infinite as $h \to 0$, so that $U_0 = \lim_{h\to h} U_h$ does not exist. On the other hand, if some of the coefficients of T_h become infinite as $h \to 0$, so that $\det(S+T_h)$ becomes infinite as $h \to 0$, then (7.22) implies, again taking determinants, that $\det(I+U_h) \to 0$. Thus if $U_0 = \lim_{h\to 0} U_h$ exists, it will satisfy $U_0^t S U_0 = S$ and $\det(I+U_0) = 0$, i.e., it will be a solution to the Cayley–Hermite problem not included in Hermite's generic solution.[21] Frobenius was able to show that in fact, all solutions U to the Cayley–Hermite problem satisfying $\det(U+I) = 0$ are obtained in this way! That is, he proved the following result.

Theorem 7.13. *Let S be symmetric with $\det S \neq 0$, and suppose that U is a proper transformation (viz., $\det U = +1$) such that $U^t S U = S$ and $\det(I+U) = 0$. Let h denote a (complex) parameter. Then for all $h \neq 0$ sufficiently small, skew-symmetric matrices T_h can be defined with coefficients that are rational functions of h such that $S+T_h$ is invertible and*

$$U = \lim_{h\to 0} (S+T_h)^{-1}(S-T_h).$$

Thus every proper transformation U satisfies either $\det(I+U) \neq 0$, so that U is given by Hermite's formula (by Frobenius' Theorem 7.11), or $\det(I+U) = 0$, so that U is the limit of a 1-parameter family of proper transformations U_h, each of which is given by Hermite's formula by virtue of Frobenius' Theorem 7.13. Frobenius' proof of Theorem 7.13 is worth sketching, because it displays how he utilized his extraordinary talent for complex algebraic manipulations (now applied to matrix algebra) in tandem with Weierstrass' theory of elementary divisors and the concomitant theory of the Jordan canonical form.

The starting point of his proof seems to have been his observation that when U is one of Hermite's solutions, so that $U = (S+T)^{-1}(S-T)$, then $T = S(I-U)(I+U)^{-1}$. If T were invertible, then $K = T^{-1} = (I+U)(I-U)^{-1}S^{-1}$ would also be skew-symmetric. Notice that $K = (I+U)(I-U)^{-1}S^{-1}$ is well defined for

---

[21] Although Frobenius did not mention it, since $\det(I+U_h) \to 0$ implies that the coefficients of U will be bounded for $|h| \leq \varepsilon$, the Bolzano–Weierstrass theorem, which he surely knew, would imply that a sequence $h_n \to 0$ exists such that $U_0 = \lim_{n\to\infty} U_{h_n}$ exists.

every U satisfying $U^tSU = S$ and $\det(I+U) = 0$, provided that $\det(I-U) \neq 0$, i.e., provided $r = 1$ is not also a characteristic root. Frobenius realized that if he could prove the theorem in the special case in which $r = 1$ is not a characteristic root of U, then he could use the Jordan canonical form of U to reduce the general case to this case (as indicated below). So, following Frobenius, let us begin by assuming that U is a proper transformation satisfying $S^tUS = S$ and such that $\det(I+U) = 0$, but $\det(I-U) \neq 0$. The first thing to consider is that $1 = \det U$ is the product of all the characteristic roots of U. Since U is a Rosanes transformation, Theorem 7.9 says that except for the root $r = -1$, all other roots can be paired as $a, 1/a$, with each root in the pair having the same multiplicity. Thus (i) the total number m of roots $r \neq -1$ is even and (ii) their product equals $+1$. From (i), it follows that $q+m = n$, where q is the multiplicity of $r = -1$. From (ii), it follows that $1 = \det U = (-1)^q$, and so q must be even, which means that $n = q+m$ must be even. This means that $n \times n$ skew-symmetric matrices H exist for which $\det H \neq 0$. With this in mind, we return to the matrix

$$K = (I+U)(I-U)^{-1}S^{-1},$$

which is well defined.

It is not immediately clear that K is skew-symmetric, but Frobenius showed that it is, in a manner similar to how he showed that the matrix T in Theorem 7.11 is skew-symmetric: it suffices to show that $K_0 = [S(I-U)]^tK[S(I-U)]$, which is congruent to K, is skew-symmetric. Since $SKS = S(I+U)(I-U)^{-1}$ and $U^tSU = S$, K_0 can be expanded and simplified to $K_0 = SU - U^tS$, from which $K_0^t = -K_0$ follows. Of course, $\det K = 0$, since $\det(I+U) = 0$, and so K cannot be inverted to give a skew-symmetric T such that $U = (S+T)^{-1}(S-T)$. However, we may consider $K_h = K + 2hH$, where H is any skew-symmetric matrix with $\det H \neq 0$. The K_h converge to K as $h \to 0$, and they are invertible for $0 < |h| \ll 1$. This is because although the polynomial $p(h) = \det(K_h) = \det(K+2hH)$ vanishes for $h = 0$, it cannot be identically zero, because $|p(h)| = |h|^n|\det[(1/h)K+2H]| \to \infty \cdot |\det(2H)| = \infty$, and so $p(h) \not\equiv 0$. Thus $p(h)$ has only a finite number of zeros, and so $p(h) \neq 0$ in a deleted neighborhood of $h = 0$, say $0 < |h| < \varepsilon$, and K_h is invertible for these h. For these h we may then define $T_h = K_h^{-1}$. The T_h are inverses of skew-symmetric matrices, and so skew-symmetric. Furthermore, by means of clever matrix algebra, one has [181, pp. 386–387]

$$K_h(S+T_h) = TS + 2hHS + I = 2(I-U)^{-1}[I + h(I-U)HS],$$
$$K_h(S-T_h) = TS + 2hHS - I = 2(I-U)^{-1}[U + h(I-U)HS].$$

Since the term in square brackets in the first equation approaches I as $h \to 0$, taking determinants in the first equation shows that $\det(S+T_h) \neq 0$ for all $h \neq 0$ sufficiently small. Taking the inverse of the first equation thus gives

$$(S+T_h)^{-1}K_h^{-1} = \frac{1}{2}[I + h(I-U)HS]^{-1}(I-U),$$

and so the product of this equation with the second of the two above gives

$$(S+T_h)^{-1}(S-T_h) = [I+h(I-U)HS]^{-1}[U+h(I-U))HS].$$

Thus $\lim_{h\to 0}(S+T_h)^{-1}(S-T_h) = U$, and Theorem 7.13 is proved in the special case in which $r=1$ is not a characteristic root of U.

To deal with the case in which $r=1$ is a characteristic root of U, Frobenius proceeded as follows. In this case, both $r=1$ and $r=-1$ are roots of U, the latter being a root because $\det(I+U)=0$ by hypothesis. The Jordan canonical form J of U can therefore be expressed as $J=\begin{pmatrix} J_1 & 0 \\ 0 & J_2 \end{pmatrix}$, where J_1 is the block diagonal matrix consisting of all the Jordan blocks of J for the root $r=-1$ and J_2 is the block diagonal matrix with all the other Jordan blocks of J. This means that $r=1$ is not a characteristic root of J_1 and that the sole root $r=-1$ of J_1 is not the reciprocal of any root of J_2, since J_2 contains no blocks for $r=-1$. Let G denote the invertible matrix such that $GUG^{-1}=J$ and set $\tilde{S}=(G^{-1})^t SG^{-1}$, where S is the nonsingular symmetric matrix for which $U^tSU=S$. Since $\tilde{S}$ is congruent to S, it is also a symmetric nonsingular matrix and $J^t\tilde{S}J=\tilde{S}$. If we write $\tilde{S}=\begin{pmatrix} \tilde{S}_{11} & \tilde{S}_{12} \\ \tilde{S}_{21} & \tilde{S}_{22} \end{pmatrix}$, then $J^t\tilde{S}J=\tilde{S}$ expressed using block multiplication shows that[22]

$$\begin{aligned} J_1^t\tilde{S}_{11}J_1 &= \tilde{S}_{11}, \quad J_1^t\tilde{S}_{12}J_2 = 0, \\ J_2^t\tilde{S}_{21}J_1 &= 0, \qquad J_2^t\tilde{S}_{22}J_2 = \tilde{S}_{22}. \end{aligned}$$

Consider the $(2,1)$ and $(1,2)$ equations above in the light of Lemma 7.10. The only root of J_1 is $r=-1$, which is not the reciprocal of any root of J_2, and so by the lemma, it follows that $\tilde{S}_{21}=0$ and $\tilde{S}_{12}=0$, and so $\tilde{S}=\begin{pmatrix} \tilde{S}_{11} & 0 \\ 0 & \tilde{S}_{22} \end{pmatrix}$.

The $(1,1)$ equation above says that J_1 is a proper transformation that leaves invariant the quadratic form corresponding to the nonsingular symmetric matrix $\tilde{S}_{11}$. Also, $r=1$ is not a characteristic root of J_1, so that $\tilde{S}_{11}$ falls under the case of Theorem 7.13 already established by Frobenius. Consequently, skew-symmetric matrices $\tilde{T}_{11}(h)$ exist such that

$$J_1 = \lim_{h\to 0}[\tilde{S}_{11}+\tilde{T}_{11}(h)]^{-1}[\tilde{S}_{11}-\tilde{T}_{11}(h)]. \tag{7.23}$$

Now consider the $(2,2)$ equation above. It says that J_2 is a proper transformation that leaves invariant the quadratic form corresponding to the nonsingular symmetric matrix $\tilde{S}_{22}$. Also, $r=-1$ is not a root of J_2, so $\det(I+J_2)\neq 0$. This means that J_2

[22] As noted earlier, Frobenius did not use block multiplication of matrices. He knew that his symbols represented linear substitutions and coefficient matrices as well as bilinear forms, but the form interpretation was favored and so he worked with direct sums of forms to achieve the same results.

is covered by Frobenius' Theorem 7.11, and so a skew-symmetric matrix $\tilde{T}_{22}$ exists such that

$$J_2 = [\tilde{S}_{22} + \tilde{T}_{22}]^{-1}[\tilde{S}_{22} - \tilde{T}_{22}]. \tag{7.24}$$

Together, (7.23) and (7.24) show that $J = \lim_{h\to 0}(\tilde{S} + \tilde{T}_h)^{-1}(\tilde{S} - \tilde{T}_h)$, where $\tilde{T}_h$ is the skew-symmetric matrix

$$\tilde{T}_h = \begin{pmatrix} \tilde{T}_{11}(h) & 0 \\ 0 & \tilde{T}_{22} \end{pmatrix}.$$

Since $U = G^{-1}JG$, the above result can be pulled back to give $U = \lim_{h\to 0}(S + T_h)^{-1}(S - T_h)$, where $T_h = G^t \tilde{T}_h G$ inherits the skew-symmetry of $\tilde{T}_h$.[23]

This completes my sketch of Frobenius' remarkable proof of his Theorem 7.13. Looking at the proof from the standpoint of the present, it appears to be a modern proof, but in 1878, it was a pioneering proof, because such proofs are not to be found in the literature of linear algebra circa 1878. The papers of Hermite, Bachmann, and Rosanes are more typical of the period. Frobenius was creating linear algebra as we know it today.

7.5.5 Orthogonal transformations

Frobenius devoted a section of his paper to "orthogonal forms," by which he meant matrices R such that $R^{-1} = R^t$ [181, pp. 390–400]. These are, of course, precisely the (real or complex) orthogonal matrices of present-day linear algebra. As we saw in Chapter 4, real orthogonal matrices had been an integral part of the transformation of real quadratic forms for a long time, as Frobenius certainly realized. His primary reason for including them in his paper of 1878, however, seems to have been as a response to the work relating to the Cayley–Hermite problem. We saw in Section 7.3 that Cayley had showed that if T is any skew-symmetric matrix such that $\det(I+T) \neq 0$, then $R = (I+T)^{-1}(I-T)$ is orthogonal.[24] Not all orthogonal matrices are obtained by Cayley's formula, however, since the formula implies that $I + R = 2(I+T)^{-1}$, so that $\det(I+R) \neq 0$, i.e., $r = -1$ is not a characteristic root of R. Also, $\det(I-T) = \det(I-T)^t = \det(I+T)$, so $\det R = [\det(I+T)]^{-1}\det(I-T) = 1$. Examples of real orthogonal transformations R with -1 as a characteristic root and $\det R = -1$ are easy to give. Thus not every real (or complex) orthogonal transformation is given by Cayley's formula. Furthermore, it was not even clear

---

[23] In detail: $U = G^{-1}JG = \lim_{h\to 0}\{G^{-1}(\tilde{S} + \tilde{T}_h)^{-1}[(G^t)^{-1}G^t](\tilde{S} - \tilde{T}_h)G\} = [G^t(\tilde{S} + \tilde{T}_h)G]^{-1} \times G^t(\tilde{S} - \tilde{T}_h)G = (S + T_h)^{-1}(S - T_h)$.

[24] This is (7.4), except T has been replaced by $-T$ because Frobenius dealt with it in that (equivalent) form.

from Cayley's work whether all orthogonal R such that $\det(I+R) \neq 0$ are given by his formula. This raised the question as to how general Cayley's formula was.

Another question was raised by a paper written by the Italian mathematician Francesco Brioschi (1824–1897), who published a paper on orthogonal transformations in 1854 [41]. His main result was that if $x = Ry$ is a real orthogonal transformation, then all the characteristic roots of R have absolute value 1. This is, of course, a correct result, but Brioschi had assumed that R could be represented by Cayley's formula—and also that the characteristic roots were all distinct. Thus it was a typical generic proof using a generic formula, namely Cayley's. To a critical thinker like Frobenius, Brioschi's paper thus raised two questions: (1) To what extent is Cayley's formula for orthogonal transformations valid? (2) For an arbitrary real orthogonal matrix R, what can be said about its characteristic roots and also about its elementary divisors? Frobenius' answer to question (1) is not surprising, since it is the special case of his Theorem 7.11 with $S = I$:

Theorem 7.14. *Every orthogonal matrix R for which $|R+I| \neq 0$ has a unique representation of the form $R = (I+T)^{-1}(I-T)$, with T skew-symmetric. The matrix T is given by $T = (I+R)^{-1}(I-R)$, and it follows that* $\det R = 1$.

The second question was answered by the following result.

Theorem 7.15. *If R is any real orthogonal matrix, then (i) the characteristic roots of R all lie on the unit circle $|z| = 1$; (ii) the elementary divisors of $\varphi(r) = \det(rI - R)$ are all linear, i.e., R can be diagonalized.*

By virtue of (i) and the fact that the coefficients of $\varphi(r) = \det(rI - R)$ are real, it follows immediately that if a is a root, then so is $1/a = \bar{a}$. Nowadays, (i) is proved by a short and simple inner product argument,[25] and even though Christoffel had shown the value of inner-product-like reasoning in his 1864 paper on Hermitian symmetric forms (Section 5.2), Frobenius did not realize that he could use it to prove (i). In fact, it was not until his work on nonnegative matrices in the early 1900s that he finally made use of inner-product-like reasoning to establish properties of characteristic roots (see Section 17.2.1). Instead, his proof of (i) utilized properties of elementary divisors and featured a matrix Laurent expansion, a technique that he indicated [181, p. 395] was inspired by Weierstrass' use of Laurent expansions in his influential paper of 1858 (see Weierstrass' proof of Lemma 4.8).

The same technique was used by Frobenius to prove (ii). Even today, proofs of this result are nontrivial. Here is how Frobenius fused matrix algebra with Weierstrass' ideas to give a relatively simple proof. Let $r = a$ be a characteristic root of R, so that by (i), $|a| = 1$. Matrix algebra enabled Frobenius to express Weierstrass' Laurent expansion idea in the succinct form of a matrix Laurent expansion,

$$(rI - R)^{-1} = A(r-a)^{-k} + \text{higher powers of } (r-a), \tag{7.25}$$

[25]For every x, $\|Rx\| = \sqrt{Rx \cdot Rx} = \sqrt{x \cdot R^t Rx} = \sqrt{x \cdot x} = \|x\|$. If a is a characteristic root of R and $x \neq 0$ is such that $Rx = ax$, then the above implies $|a| \cdot \|x\| = \|Ux\| = \|x\|$, and so $|a| = 1$.

valid in some region $0 < |r-a| < \varepsilon$ and with $A \neq 0$.[26] Starting with (7.25), differentiation with respect to r gives

$$(rI-R)^{-2} = kA(r-a)^{-k-1} + \text{higher powers of } (r-a), \tag{7.26}$$

whereas squaring both sides of (7.25) gives

$$(rI-U)^{-2} = A^2(r-a)^{-2k} + \text{higher powers of } (r-a). \tag{7.27}$$

Thus if $A^2 \neq 0$, comparison of (7.26) and (7.27) implies by uniqueness of Laurent expansions that $(rI-U)^{-2}$ has a pole of order $2k = k+1$ at $r = a$, whence $k = 1$ in (7.25). To anyone familiar with elementary divisor theory, the fact that $(rI-R)^{-1}$ has a pole of order $k = 1$ at every characteristic root $r = a$ meant that the elementary divisors of R are all linear.[27]

It remained to show that $A^2 \neq 0$. To show that this is the case for the Laurent coefficient A, Frobenius manipulated (7.25), using matrix transposition and complex conjugation and the facts that $R^t = R^{-1}$, $\bar{R} = R$, and, since a lies on the unit circle $|z| = 1$, that $a^{-1} = \bar{a}$. In addition, he assumed $r \in \Gamma$, where Γ denotes the arc of the unit circle (minus point a) that is inside $|r-a| < \varepsilon$. The resultant manipulations yielded for $r \in \Gamma$,

$$(rI-R)^{-1} = c\bar{A}^t R^t (r-a)^{-k} + \text{higher powers of } (r-a), \tag{7.28}$$

where $c = (-1)^{k-1}a^{2k-1} \neq 0$.[28] Comparison of (7.28) with (7.25) implied (by the identity theorem of complex analysis) that $A = c\bar{A}^t R^t$, and so $A^2 = A(c\bar{A}^t R^t) =$

---

[26]Recall from the discussion surrounding (4.36) that k is the largest order of a pole at $r = a$ of a coefficient $\varphi_{ij}(r)/\varphi(r)$ of $\mathrm{Adj}(rI-R)/\det(rI-R) = (rI-R)^{-1}$.

[27]Since $(rI-R)^{-1} = \mathrm{Adj}(rI-R)/\det(rI-R) \stackrel{\text{def}}{=} \varphi_{ij}(r)/\varphi(r)$, all the coefficients $\varphi_{ij}(r)/\varphi(r)$ have poles of order at most 1 at $r = a$, i.e., if $(r-a)^q$ divides $\varphi(r)$, then $(r-a)^{q-1}$ divides $\varphi_{ij}(r)$ for all i, j. This means that $D_{n-1}(r)$, the polynomial greatest common divisor of all the $\varphi_{ij}(\lambda)$, is also divisible by $(r-a)^{q-1}$; and so $\varphi(r)/D_{n-1}(r) = E_n(r) = \psi(r)$, the minimal polynomial of R, is the product of distinct linear factors. But from Weierstrass' definition of elementary divisors, $E_n(r) = \prod_{i=1}^{d}(r-a_i)^{e_i}$, where $a_1, \ldots, a_d$ are the distinct characteristic roots of R and $(r-a)^{e_i}$ is the elementary divisor for a_i of maximal exponent. Hence all e_i equal 1, and all elementary divisors are linear.

[28]Frobenius used his calculational skill to derive (7.28) from (7.25). First be multiplied both sides of (7.25) by $rR = [r^{-1}R^{-1}]^{-1}$ to get

$$(R^{-1} - r^{-1}I)^{-1} = (R^t - r^{-1})^{-1} = rRA(r-a)^{-k} + \cdots.$$

Now take the transpose of this equation to get

$$(R - r^{-1})^{-1} = rA^t R^t (r-a)^{-k} + \cdots.$$

So far, the assumption that $r \in \Gamma$ has not been used, but now the complex conjugate of the above equation gives, since on Γ, $r^{-1} = \bar{r}$ and $(R - r^{-1})^{-1} = -(r^{-1}I - R) = -(\bar{r}I - R)$,

$c(A\bar{A}^t)R^t$. Now, the diagonal coefficients of $A\bar{A}^t$ are $[A\bar{A}^t]_{jj} = \sum_{k=1}^n |a_{jk}|^2$ and so are not 0 for every j, since $A \neq 0$. The fact that $A\bar{A}^t \neq 0$ and that R^t is invertible then implies $A^2 \neq 0$.

The above proof illustrates a characteristic feature of Frobenius' brand of linear algebra: combine matrix algebra with complex analysis, especially Laurent expansions, and Weierstrass' theory of elementary divisors to achieve general, rigorously established theorems. His proof of the minimal polynomial theorem (Theorem 7.2) had also used power series and Laurent expansions to great effect. Another example is given by his proof of a square root theorem for matrices (Theorem 16.5), which enabled him to give a simple solution to problems raised by the work of Weierstrass and Kronecker.

7.5.6 A theorem on division algebras

In the concluding section of his paper, Frobenius turned to what was apparently a natural consideration: the relationship between his symbolic calculus and quaternions. The quaternions $a \cdot \mathbf{1} + b \cdot \mathbf{i} + c \cdot \mathbf{j} + d \cdot \mathbf{k}$, where a, b, c, d are real numbers, had been introduced by Hamilton in the early 1840s as a result of his efforts to introduce a three-dimensional analogue of the complex numbers. This turned out to be impossible, but he discovered along the way that in four dimensions it was almost possible. That is, with the multiplication defined by the rules

$$\mathbf{i}^2 = \mathbf{j}^2 = \mathbf{k}^2 = -\mathbf{1}, \tag{7.30}$$

$$\begin{aligned} &\mathbf{i} \cdot \mathbf{j} = \mathbf{k}, \quad && \mathbf{j} \cdot \mathbf{k} = \mathbf{i}, \quad && \mathbf{k} \cdot \mathbf{i} = \mathbf{j}, \\ &\mathbf{j} \cdot \mathbf{i} = -\mathbf{k}, \quad && \mathbf{k} \cdot \mathbf{j} = -\mathbf{i}, \quad && \mathbf{i} \cdot \mathbf{k} = -\mathbf{j}, \end{aligned} \tag{7.31}$$

the expressions $\mathbf{q} = a \cdot \mathbf{1} + b \cdot \mathbf{i} + c \cdot \mathbf{j} + d \cdot \mathbf{k}$ shared most of the properties of the ordinary complex numbers. In particular, every $\mathbf{q} \neq \mathbf{0}$ has an inverse $\mathbf{q}^{-1} = \bar{\mathbf{q}}/N(\mathbf{q})$, where $\bar{\mathbf{q}} = a \cdot \mathbf{1} - b \cdot \mathbf{i} - c \cdot \mathbf{j} - d \cdot \mathbf{k}$ is the conjugate of $\mathbf{q}$ and $N(\mathbf{q}) = a^2 + b^2 + c^2 + d^2$ is the norm of $\mathbf{q}$. Of course as (7.31) indicates, multiplication is not commutative, although $\mathbf{q}_1 \cdot \mathbf{q}_2 = -\mathbf{q}_2 \cdot \mathbf{q}_1$.

In his paper of 1858 on matrix algebra, Cayley had observed that 2×2 matrices can be selected that multiply like the quaternions **i**, **j**, and **k** [84, p. 491].

$$\begin{aligned} (rI - R)^{-1} &= -r^{-1}\bar{A}^t R^t (r^{-1} - a^{-1})^{-k} + \cdots \\ &= -r^{-1}\bar{A}^t R^t r^k r^{-k} (r^{-1} - a^{-1})^{-k} a^{-k} a^k + \cdots \\ &= (-1)^{k-1} r^{k-1} a^k \bar{A}^t R^t (r - a)^{-k} + \cdots . \end{aligned} \tag{7.29}$$

Now expand $f(r) = r^{k-1}$ in a power series about $r = a$ to get $r^{k-1} = a^{k-1} + (k-1)a^{k-2}(r-a) + \cdots$. If this expansion is substituted in (7.29), the result is (7.28), and Frobenius' proof is completed.

And Laguerre in his paper of 1867 had regarded his calculus of linear systems as providing a "simple and, so to speak, arithmetic interpretation" of quaternions as well as of ordinary complex numbers, Cauchy's "clefs algébriques," and Galois' "imaginaires congruentielles" (Galois fields). According to Laguerre, his calculus of linear systems provided the sufficiently general standpoint necessary for a deeper study of these number systems. It was Frobenius, however, who made this point with a concrete example: an interesting theorem.

Frobenius considered a set of $m+1$ linearly independent forms or $n \times n$ matrices that included the identity, viz., $I = E_0, E_1, \ldots, E_m$, with the property that the product of any two of them is a linear combination of $I, E_1, \ldots, E_m$. It then follows that the system of all matrices expressible as linear combinations of $E_0, E_1, \ldots, E_m$, viz., $A = \sum_{k=0}^{m} a_i E_i$, is closed under multiplication. Thus such a system of forms is an example of a linear associative algebra. "Especially noteworthy" to Frobenius were systems of *real* matrices such that $\det A \neq 0$ for all $A \neq 0$. He called such A's "complex numbers" with "units" $I, E_1, \ldots, E_m$ and with $\det A$ as the "norm" of A. Although Frobenius did not mention it explicitly, it follows immediately that for such a system of "complex numbers," for every $A \neq 0$ in the system, A^{-1} also belongs to the system. This is because $A^{-1} = f(A)$, where $f(r) = 1/r$, and so by Theorem 7.4, A^{-1} is actually a polynomial in A and so belongs to the system. In particular, this means that for any A, B in the system, if $AB = 0$, then one of the factors must be 0, i.e., the system has no divisors of zero. Such systems of "complex numbers" are examples of what are now called real division algebras. Frobenius observed that "it is easy to construct all possible systems of complex numbers," and proceeded to prove the following theorem.

Theorem 7.16 (Frobenius' theorem on real division algebras). *Aside from the real numbers ($m = 0$), the complex numbers ($m = 2$) and the quaternions ($m = 3$), there are no other systems of complex numbers in the above-defined sense.*

The key to Frobenius' proof of this theorem was the minimal polynomial $\psi(r)$ of a matrix A belonging to such a system. Since for such A, $\psi(r)$ is a polynomial with real coefficients, it factors into linear or irreducible quadratic factors with real coefficients. Frobenius made the telling observation that $\psi(r)$ cannot have more than one such factor; for if $\psi(r) = p_1(r)p_2(r)$, where $p_1(r)$ and $p_2(r)$ have real coefficients, then $0 = \psi(A) = p_1(A) \cdot p_2(A)$. Since this factorization of $\psi(A)$ exists in the given system, one of the polynomial factors, say $p_1(A)$, must be 0 in accordance with the characteristic feature of the systems under consideration. Since $\deg p_1(r) < \deg \psi(r)$, the relation $p_1(A) = 0$ contradicts the minimality of ψ. Thus the real factorization of ψ is either (1) $\psi(r) = r - a$ or (2) $\psi(r) = r^2 - 2pr + (p^2 + q^2)$, where a, p, q are real and $q \neq 0$, so that $\psi(r)$ is irreducible over $\mathbb{R}$.

If (1) holds for all A in the system, then $\psi(A) = 0$ for all A means that $A = aI$ for all A, i.e., every A in the system is some multiple a of I, and so $m = 0$, and the system can be identified with $\mathbb{R}$. If $m > 0$, then (2) must hold for every unit E_i, $i > 0$. But $\psi(E_i) = E_i^2 - 2p_iE_i + (p_i^2 + q_i^2)I = 0$ means that E_i can be replaced by $J_i = (1/q_i)(E_i - p_i I)$ to obtain $m+1$ linearly independent units $I, J_1, \ldots, J_m$ with $J_i^2 = -I$ for $i = 1, \ldots, m$. Since $J_1^2 = -I$ implies $(\det J_1)^2 = (-1)^n$, it follows that n must be even. If $m = 1$, so the units are I and J_1, then since I and J_1 commute, the system $aI + bJ_1$ can be identified with the complex numbers $\mathbb{C}$. A concrete example is obtained by taking $n = 2$ with $I = I_2$ and $J_1 = \begin{pmatrix} 0 & 1 \\ -1 & 0 \end{pmatrix}$.

Now consider the possibility that $m = 2$. Then

$$J_1 J_2 = aI + bJ_1 + cJ_2.$$

If this equation is multiplied through on the right by J_2, the result is

$$-J_1 = aJ_2 + bJ_1 J_2 - cI.$$

If $J_1 J_2$ is eliminated from these two equations, the result is

$$(ab - c)I + (b^2 + 1)J_1 + (bc + a)J_2 = 0.$$

The linear independence of I, J_1, J_2 would then imply that $b^2 + 1 = 0$, which is impossible, since b is real. Hence $m = 2$ is impossible.

Now suppose that $m > 2$. Then for $k > 0$ and $l > 0$ with $k \neq l$, since $J_k \pm J_l$ is not a multiple of I, the minimal polynomials of these matrices must be irreducible quadratics. Let the equations $\psi(A_\pm) = 0$ for $A_\pm = J_k \pm J_l$ be written as

$$\begin{aligned} (J_k + J_l)^2 + a(J_k + J_l) + bI &= 0, \\ (J_k - J_l)^2 + a'(J_k - J_l) + b'I &= 0. \end{aligned} \tag{7.32}$$

If these equations are expanded (using $J_k^2 = J_l^2 = -I$) and added, the result is

$$(a + a')J_k + (a - a')J_l + (b + b' - 4)I = 0,$$

and the linear independence of I, J_k, J_l implies $a = a' = 0$. Accordingly, the first equation in (7.32) reduces to $(J_k + J_l)^2 + bI = 0$, or, if one sets $2 - b = 2s_{kl}$, to

$$J_k J_l + J_l J_k = 2s_{kl} I.$$

In this equation, clearly $s_{lk} = s_{kl}$, and if we set $s_{kk} = -1$, it remains valid for $k = l$. Consider now any element of the system of the form $U = \sum_{k=1}^{m} u_k J_k$. Then

$$U^2 = \sum_{k,l} u_k u_l J_k J_l = \left(\sum_{k,l} s_{kl} u_k u_l \right) I$$

cannot vanish unless $U = 0$, i.e., unless all $u_i = 0$. This means that the quadratic form $\sum_{k,l} s_{kl} u_k u_l$ is strictly definite, and since $s_{kk} = -1$, it is negative definite. Hence there is a real nonsingular linear transformation

$$u_k = \sum_{l=1}^{m} a_{kl} v_l$$

that takes the form into a sum of m negative squares:

$$\sum_{k,l} s_{kl} u_k u_l = -\sum_j v_j^2.$$

Then if the J_k are replaced by J_k^*, where

$$J_k^* = \sum_k a_{kl} J_l, \tag{7.33}$$

it follows that $U = \sum_k u_k J_k = \sum_l v_l J_l^*$, and so

$$U^2 = \sum_{k,l} v_k v_l J_k^* J_l^* = \left(-\sum_l v_l^2\right) I,$$

which implies, for the choices $U = J_k$ and $U = J_k + J_l$, that respectively,

$$(J_k^*)^2 = -I, \quad J_k^* J_l^* = -J_l^* J_k^*. \tag{7.34}$$

As Frobenius said, using (7.33)–(7.34), "it is easy to show that m cannot be greater than 3" [181, p. 404]. That is, consider J_1^*, J_2^*, J_k^*, where $k \neq 1, 2$. Then

$$\begin{aligned}(J_1^* J_2^* J_k^*)^2 &= J_1^* J_2^* (J_k^* J_1^*) J_2^* J_k^* = -J_1^* J_2^* (J_1^* J_k^*) J_2^* J_k^* = +(J_1^*)^2 J_2^* J_k^* J_2^* J_k^* \\ &= -J_2^* J_k^* J_2^* J_k^* = (J_2^*)^2 (J_k^*)^2 = I.\end{aligned}$$

This means that

$$(J_1^* J_2^* J^* J_k + I)(J_1^* J_2^* J^* J_k - I) = 0,$$

and so, since one of these two factors must be 0, $J_1^* J_2^* J_k^* = \pm I$, and so, multiplying by J_k^* on the right, $J_1^* J_2^* = \mp J_k^*$. In these calculations k was any index satisfying $3 \leq k \leq m$. The fact that $J_k^* = \mp J_1^* J_2^*$ shows that there can be at most one such index, i.e., that $m > 3$ is impossible.

Of course, for $m = 3$, such a system is possible, since, as Frobenius noted, if we take $n = 4$, the units I_4, J_1^*, J_2^*, J_3^*, where

$$J_1^* = \begin{pmatrix} 0 & 1 & 0 & 0 \\ -1 & 0 & 0 & 0 \\ 0 & 0 & 0 & 1 \\ 0 & 0 & -1 & 0 \end{pmatrix}, \quad J_2^* = \begin{pmatrix} 0 & 0 & 1 & 0 \\ 0 & 0 & 0 & -1 \\ -1 & 0 & 0 & 0 \\ 0 & 1 & 0 & 0 \end{pmatrix}, \quad J_3^* = \begin{pmatrix} 0 & 0 & 0 & 1 \\ 0 & 0 & 1 & 0 \\ 0 & -1 & 0 & 0 \\ -1 & 0 & 0 & 0 \end{pmatrix},$$

satisfy (7.34), which is the same as (7.30)–(7.31), and so define the quaternions. Frobenius also pointed out that since the J_k^* are skew-symmetric, if $A = a_0 I_4 + \sum_{k=1}^3 a_k J_k^*$, then $A^t = a_0 I_4 - \sum a_{k=1}^3 J_k^*$ is the quaternion conjugate of A and

$$(\det A)^2 = \det(AA^t) = \det\left[\left(\sum_{k=0}^{3} a_k^2\right) \cdot I\right] = \left[\sum_{k=0}^{3} a_k^2\right]^4.$$

Thus calling $\det A$ the norm of A, as Frobenius did, makes sense.

A result analogous to Frobenius' Theorem 7.16 was independently discovered in 1881 by Peirce [464] and was based on the computations of all linear associative algebras of small dimension done earlier by his father, Benjamin Peirce.[29] Peirce's result was published in the *Proceedings of the American Academy of Arts and Sciences*. Frobenius not only published his result first, but in a widely read journal—that of Crelle—and provided a more satisfactory proof. It is not surprising, therefore, to find that it was Frobenius' Theorem 7.16 and proof that was influential. Frobenius' theorem has been generalized to division rings that are algebraic over $\mathbb{R}$, and in that form still usually bears his name. Indeed, his proof idea, namely to utilize the fact that minimal polynomials of elements are either linear or quadratic over $\mathbb{R}$, is still found in proofs of the theorem in this general form [294, §7.3].

[29] See in this connection [267, pp. 245–248].

Chapter 8
Arithmetic Investigations: Linear Algebra

From the time of his dissertation (1870) through his work on the problem of Pfaff in 1876 (Chapter 6), Frobenius' penchant for working on algebraic problems had been pursued within the framework of analysis, especially differential equations. As we saw, his work on the problem of Pfaff—ostensibly a problem within the field of differential equations—had engaged him more fully with linear algebra. Frobenius then had the opportunity to further develop his interests in this area through his work on the purely algebraic Cayley–Hermite problem (1877, Chapter 7), his first major work that did not involve differential equations. The following year, he began to turn to the theory of numbers as a source for problems.

Starting with the publication in 1801 of Gauss' *Disquisitiones Arithmeticae*, the theory of numbers had become an area of mathematics that posed many deep and intriguing problems and suggested many far-reaching generalizations to nineteenth-century mathematicians. During the 1830s and 1840s, Dirichlet made many fundamental contributions to the research program implied by Gauss' *Disquisitiones*. Gauss' publications in 1831–1832 on the law of biquadratic reciprocity, which led him to introduce the Gaussian integers, provided further inspiration for arithmetic investigations, among which was Kummer's work on cyclotomic fields that led him to introduce the radically new idea of ideal complex numbers in order to restore, for the cyclotomic integers he was considering, the properties of prime numbers that held for ordinary and Gaussian integers. Kronecker even claimed to be able to extend Kummer's theory to any finite extension of the rational field, although he never published the details of his theory. In the meantime, Dedekind, who began attending Dirichlet's lectures on number theory after the latter's move to Göttingen in 1855, worked those lectures into a publishable form in 1863 and then continued to supplement Dirichlet's work with his own results in subsequent editions of the lectures.[1] Most notable among these supplements was Dedekind's theory of ideals in algebraic number fields, which first appeared in the second edition (1871) of

[1]More detailed information about Dedekind can be found at the beginning of Chapter 2.

T. Hawkins, *The Mathematics of Frobenius in Context*, Sources and Studies in the History of Mathematics and Physical Sciences, DOI 10.1007/978-1-4614-6333-7_8,

Dirichlet's lectures and then in a different, improved form in a monograph published in French in 1876.

As a former student at Berlin, where both Kummer and Kronecker were professors, Frobenius was well grounded in the theory of numbers. His interest in this area, however, seems to have quickened in the later 1870s and seems to have been due in part to the publications of Dedekind—his theories of modules and ideals and, more generally, his emphasis on the importance of an abstract theory of groups that would embrace not only Galois theory, which was an early interest of Frobenius (Section 1.2) and had since been systematically developed by Camille Jordan in his *Traité des substitutions* (1870), but also developments within number theory that, as Kronecker had noted in 1870, could be stated abstractly in a form now recognizable as a part of the fundamental theorem of finite abelian groups.

Frobenius' early arithmetic work relating to groups is treated in the next chapter. Here the focus is on his work toward an arithmetic theory of bilinear forms, which was inspired by his study of Gauss' *Disquisitiones Arithmeticae*. Frobenius took up Gauss' challenge to extend his theory of binary quadratic forms to more general forms. Given his background, it is not surprising that he sought to develop certain aspects of an arithmetic theory of bilinear forms in any number of variables. This he succeeded in doing by supplementing ideas he took from Gauss with critically important twists of his own devising. He discovered that some of his results had been anticipated in a less general and less elegant form 17 years earlier by H.J.S. Smith, but he had found many more important applications for his results than had Smith. In particular, under the influence of Weierstrass' theory of elementary divisors and Dedekind's theory of algebraic number fields, he applied his arithmetic results to develop the first rational theory of elementary divisors, which implied that Weierstrass' theory was valid over any field (finite or infinite) and ultimately led at the hands of a later generation of mathematicians to the development of Weierstrass' theory for abstract fields as an application of the fundamental theorem of finitely generated modules (Sections 16.2–16.3).

8.1 Two Gaussian Problems for Bilinear Forms

A substantial portion of Gauss' *Disquisitiones Arithmeticae*, namely Section V, was devoted to the arithmetic theory of binary quadratic forms $F(x) = ax_1^2 + 2bx_1x_2 + cx_2^2$, where a, b, and c are integers, i.e., quadratic forms $F(x) = x^tAx$ with A a 2×2 integral and symmetric matrix. In order to treat certain aspects of the arithmetic theory of these forms most naturally, Gauss had also developed parts of the theory of ternary quadratic forms, i.e., quadratic forms corresponding to an integral symmetric 3×3 matrix. Gauss realized that the arithmetic theory of binary and ternary forms was but a small part of a vast theory, and so by way of introducing the reader to ternary forms, he wrote [244, Art. 266]:

> Thus far we have restricted our discussion to functions of the second degree with *two* unknowns .... But manifestly this topic is only a very special part of a more general study of *rational integral algebraic functions which are homogeneous of some degree in several unknowns*. Such functions can be conveniently divided according to their degree as *binary, ternary, quaternary, etc. forms* .... We have devoted the present section to the treatment of binary forms of the second degree. But there are many beautiful truths concerning these forms whose true source is to be found in the theory of ternary forms of the second degree. We will therefore make a brief digression into this theory and will especially treat of those elements which are necessary to complete the theory of binary forms .... We must, however, reserve a more exact treatment of this important subject for another occasion .... At this time we will completely exclude from the discussion quaternary, quinary, etc. forms .... It is sufficient to draw this broad field to the attention of geometers. There is ample material for the exercise of their genius, and higher Arithmetic will surely benefit by their efforts.[2]

Frobenius certainly read this passage, and the arithmetic problems he posed to himself seem to constitute his contribution to the vast theory alluded to by Gauss.

The two problems that primarily concerned him involved nontrivial generalizations of some elementary notions in Gauss' theory of binary quadratic forms. (A discussion of some of the deeper parts of Gauss' theory are considered in Section 9.1.1.) Let $F(x) = x^tAx$ denote a binary quadratic form. Gauss ruled out the possibility that $\det A = 0$ [244, Art. 156]. A basic concern of Gauss was the question of which integers n could be represented by F in the sense that $F(x_0) = n$ for some $x_0 \in \mathbb{Z}^2$. In this connection he introduced the notion of a second form $G = y^tBy$ being *contained* in F [244, Art. 157]. This is the case if there is a nonsingular linear transformation $x = Py$ with integer coefficients such that $F(x) = G(y)$. This means that if n is represented by G, so that $G(y_0) = n$ for some $y_0 \in \mathbb{Z}^2$, then it is represented by F as well, since $x_0 = Py_0 \in \mathbb{Z}^2$ and $F(x_0) = G(y_0) = n$. Thus the totality of integers represented by G is contained within the totality of integers represented by F; but it does not follow (as is easily seen by examples) that every integer represented by F is representable by G. This led Gauss to the notion of the *equivalence* of forms.

Clearly if G is contained in F and F is contained in G, they represent the same integers. Suppose F and G are contained in each other. In matrix notation, the fact that G is contained in F means that $B = P^tAP$, and so $\det B = (\det P)^2 \det A$. Likewise, since F is contained in G, $A = Q^tBQ$, and so $\det A = (\det Q)^2 \det B$. These equalities show that $(\det P)^2 = (\det Q)^2 = 1$ and that $\det A = \det B$. In fact, since $\det P = \pm 1$, P^{-1} is also integral,[3] and so we may take $Q = P^{-1}$. Such a transformation P is called *unimodular*, and Gauss defined two forms F, G to be *equivalent* if there is a unimodular transformation $x = Py$ such that F is transformed into G. As we have seen, this means not only that G is contained in F but also that

[2] Clark's translation has "transcendental Arithmetic," but "higher Arithmetic" is closer to the original Latin.

[3] Recall that $P^{-1} = (\det P)^{-1}\,\mathrm{Adj}\,P$, where $\mathrm{Adj}\,P$ is the transposed matrix of cofactors of P. Thus $\det P = \pm 1$ means that $P^{-1} = \pm\,\mathrm{Adj}\,P$ is integral.

F is contained in G, since we may take $Q = P^{-1}$. Thus G is contained in F and vice versa if and only if F and G are equivalent.

Let us now consider how Frobenius generalized these notions and the problems he then formulated. Let

$$F(x,y) = \sum_{i=1}^{m} \sum_{j=1}^{n} a_{ij} x_i y_j$$

denote a bilinear form with integer coefficients a_{ij}. If $A = (a_{ij})$ denotes the $m \times n$ coefficient matrix of $F(x,y)$, then $F(x,y)$ may be expressed in more compact notation (not used by Frobenius) as

$$F(x,y) = x^t Ay.$$

Then if P and Q denote, respectively, $m \times m$ and $n \times n$ matrices with integer coefficients, and if (following Frobenius) we write the corresponding variable changes in the form $x^t = u^t P$, $y = Qv$, then $F(x,y)$ is transformed into $G(u,v) = u^t Bv$, where $B = PAQ$. In this case, Frobenius defined G with coefficient matrix $B = PAQ$ to be *contained* in F. This would seem to be a straightforward generalization of Gauss' notion, but *Frobenius did not assume that P and Q are nonsingular in his definition of containment*, although it still follows that the integers representable by G are contained within those representable by F. Likewise, of course, A being rectangular in general does not have a determinant to be nonzero, but Frobenius also made no assumption about the rank of A. He defined two forms F, G with corresponding matrices A, B to be *equivalent* if integral matrices P, Q exist that are unimodular ($\det P = \pm 1$, $\det Q = \pm 1$) and such that $B = PAQ$. This is, of course, a straightforward extension of Gauss' definition, and as with Gauss', it is an equivalence relation. However, due to Frobenius' weaker requirements for the containment of one form in another, it no longer follows readily that if G is contained in F and vice versa, then F and G are equivalent.

Frobenius proposed to consider the following arithmetic problem:

Problem 8.1. (I): *Determine necessary and sufficient conditions that two integral bilinear forms be equivalent.* (II): *Determine necessary and sufficient conditions that one such form be contained in another.*

Frobenius solved problem (I) in his 1879 paper [182, §§1–7]. The solution to problem (II), which was based in part on the results obtained in (I), was submitted 6 months after he submitted [182] and was published in 1880 [185]. As we shall see, Frobenius' solution to Problem 8.1 implied that if G is contained in F (in his weaker sense) and if F is contained in G, then they are equivalent precisely as in Gauss' theory, although now this is far from the trivial consequence it was in Gauss' theory. To prove it, Frobenius had to establish what I call his "containment theorem" (Theorem 8.16).

8.2 Solution to Problem (I): Invariant Factors

Problem (I) is of course analogous to the problem posed by Weierstrass in his theory of elementary divisors (Section 5.4), and so it is not surprising that Frobenius transferred relevant notions from that theory to the context of problem (I). This transference is possible because of the analogy between polynomials in one variable λ with complex coefficients and ordinary integers. For example, for both there is a notion of greatest common divisor, and if $p(\lambda)$ and $q(\lambda)$ are polynomials, we can write $p(\lambda) = s(\lambda)q(\lambda) + r(\lambda)$, where $\deg r < \deg q$, which makes possible an analogue of the Euclidean algorithm for polynomials, thereby yielding the greatest common divisor $\delta(\lambda)$ of two polynomials $p(\lambda)$ and $q(\lambda)$ in the form $\delta(\lambda) = a(\lambda)p(\lambda) + b(\lambda)q(\lambda)$. As we would express the matter nowadays, both $\mathbb{Z}$ and $\mathbb{C}[\lambda]$ are examples of Euclidean domains and so principal ideal domains.

If F is the above form with $m \times n$ matrix A, we may define the W-series $\{d_m, \ldots, d_1, d_0\}$ of Weierstrass and Kronecker in analogous fashion: $d_0 = 1$ and for $i > 0$, d_i is the (positive) greatest common divisor of the integers representing the $i \times i$ minor determinants of A. Of course, for every $i > r = \operatorname{rank} A$, $d_i = 0$, and so attention will be restricted to d_i for $i = 0, \ldots, r$. It follows from the properties of determinants, as in Weierstrass' theory, that d_{i-1} divides d_i, and so

$$e_i = d_i / d_{i-1}, \quad i = 1, \ldots, r, \tag{8.1}$$

is a positive integer. Frobenius called the e_i the elementary divisors of A [182, p. 484] even though they were not the analogues of Weierstrass' elementary divisors. Adhering to current terminology, I will refer to the e_i as the *invariant factors* of A (or of the associated form F). If the factorization of e_i into primes is $e_i = p_i^{a_i} q_i^{b_i} \cdots$, then (as Frobenius realized) the prime powers $p_i^{a_i}$, $q_i^{b_i}, \ldots$, $i = 1, \ldots, m$, are the analogues of Weierstrass' elementary divisors. (Frobenius called them the simple elementary divisors.)

The same properties of determinants used in Weierstrass' theory of elementary divisors show that if arithmetic forms F and G are equivalent, then they must have the same invariant factors. Stated in matrix form, the result is that if A and B are $m \times n$ integral matrices such that $B = PAQ$, where P, Q are unimodular, then A and B have the same invariant factors. The question facing Frobenius was whether when A and B have the same invariant factors $e_1, \ldots, e_r$, unimodular P, Q may be determined such that $B = PAQ$. To establish the existence of P, Q, Frobenius proceeded much as Weierstrass had in the sense that he also followed a line of reasoning to the effect that if N were a simple "canonical form" matrix N with invariant factors $e_1, \ldots, e_r$, it would suffice to show that P, Q could be determined such that $PAQ = N$ for any $m \times n$ matrix with $e_1, \ldots, e_r$ as invariant factors. As for N, the obvious choice was the matrix

$$N = \begin{pmatrix} e_1 & & & & & & \\ & e_2 & & & & & \\ & & \ddots & & & & \\ & & & e_r & & & \\ & & & & 0 & & \\ & & & & & \ddots & \\ & & & & & & 0 \end{pmatrix}, \tag{8.2}$$

where $r = \operatorname{rank} A$.

To establish the existence of P, Q such that $PAQ = N$, Frobenius was guided by two results. The first was a known result due to Hermite [284]:

Lemma 8.2 (Hermite). *Given a $1 \times n$ integral row vector such that the greatest common divisor of its components is 1, there exists an $n \times n$ unimodular matrix A whose first row is the given row vector.*

Frobenius may have learned of Hermite's lemma from a paper of Jacobi's that was published posthumously in 1868 [316]; in it, Jacobi gave several proofs of the lemma. But Frobenius, in characteristic fashion, devised his own proof, which was inspired by some ideas he drew from Article 279 of Gauss' *Disquisitiones Arithmeticae* and was simpler than the proofs of Hermite and Jacobi. (As we will see in Section 10.4, this proof was to serve him well, since it was readily adaptable to deal with an analogous problem suggested by Hermite's work on the transformation of abelian functions.)

What Frobenius proved was the following [182, pp. 486ff.]:

Lemma 8.3 (Frobenius). *Let $m < n$ and suppose A and B are integral matrices that are, respectively, $m \times n$ and $n \times m$, and satisfy $AB = I_m$, where I_m denotes the $m \times m$ identity matrix. Then a $1 \times n$ integral row matrix a and an $n \times 1$ integral column matrix b can be determined such that $A^* = \begin{pmatrix} A \\ a \end{pmatrix}$ and $B^* = \begin{pmatrix} B & b \end{pmatrix}$ satisfy $A^*B^* = I_{m+1}$. Hence $n-m$ integral rows may be added to A and $n-m$ integral columns to B so that the resulting $n \times n$ matrices A^{**} and B^{**} satisfy $A^{**}B^{**} = I_n$.*

Since $\det A^{**} \det B^{**} = \det I_n = 1$, it follows that A^{**} is unimodular with B^{**} as its (unimodular) inverse. Frobenius' proof of Lemma 8.3 was quite simple.[4]

---

[4]Let $a_1, \ldots, a_m$ denote the $(1 \times n)$ rows of A and $b_1, \ldots, b_m$ the $(1 \times n)$ columns of B. Then $AB = I_m$ means that $a_i b_j = \delta_{ij}$. Thus for $A^*B^* = I_{m+1}$ to hold, it is necessary that $ab_j = 0$ for all j, i.e., that $aB = 0$. Since $m < n$, we can certainly determine an integral $a \neq 0$ such that $aB = 0$. Furthermore, we can take a such that $\gcd a = 1$. It then follows that an integral column matrix b may be determined such that $ab = \sum a_j b_j = 1$, where b is the $n \times 1$ column matrix with the b_j as entries. Thus the $(m+1, m+1)$ entry of A^*B^* will be 1, as required. However, for $k < m+1$, the $(m+1, k)$ entries must all be zero, i.e., we must have $a_i b = 0$ for all j, or equivalently, $Ab = 0$. The chosen b need not satisfy this condition, and so we replace b with $b - c$. To have $a(b-c) = 1$,e we need that $ac = 0$. To this end, we may choose $c = Bx$ for any x, for then $ac = a(Bx) = (aB)x = 0x = 0$. We also need that $A(b-c) = 0$. But $A(b-c) = Ab - ABx = Ab - x$, since $AB = I_m$. Thus we need to

Notice that if we take $m = 1$ in the above lemma, we have the following corollary.

Corollary 8.4 (Frobenius). *If p and q are integral $n \times 1$ column matrices such that $p^t q = 1$, then it is possible to determine a unimodular matrix P with p^t as its first row and a unimodular matrix Q with q as its first column.*

This follows by taking $m = 1$, $A = \begin{pmatrix} p^t \end{pmatrix}$, and $B = \begin{pmatrix} q \end{pmatrix}$ in Lemma 8.3. Then $P = A^{**}$ and $Q = B^{**}$. Hermite's lemma is easily seen to be equivalent to Corollary 8.4,[5] but as we shall see a bit further on, it is in the form of Corollary 8.4 that Frobenius put it to good use.

The second result was entirely Frobenius' own discovery and is the key to his overall approach:

Lemma 8.5 (Frobenius). *If $F(x,y) = x^t Ay$, where A is an $m \times n$ integral matrix, and if f is the greatest common divisor of its coefficients, then integral column vectors a and b can be determined such that $F(a,b) = a^t Ab = f$.*

It follows readily from Gauss' theorems that this lemma does not hold for binary quadratic forms.[6] To see why it probably should hold for bilinear forms, note that *if* $A = PNQ$, where N is the normal form (8.2), and if we write $e_i = f_1 \cdots f_i$, where the f_i are integers (which is possible because $e_i \mid e_{i+1}$), then $\gcd A = \gcd N = f_1$. Obviously, integral u, v exist such that $u^t N v = f_1$. We can then choose a, b such that $P^t a = u$ and $Qb = v$, and so $a^t Ab = f_1$. The lemma is therefore true if the desired result, the arithmetic equivalence of A and N, is true. What about the converse? Frobenius saw a way to deduce the arithmetic equivalence of A and N from Lemma 8.5 in conjunction with his version of Hermite's lemma, namely Corollary 8.4. That is, using this lemma and corollary, he could establish the following result:

Lemma 8.6 (Reduction lemma). *If $A \neq 0$ and if $f_1 = \gcd A$, then unimodular P_1, Q_1 exist such that*

$$P_1 A Q_1 = \begin{pmatrix} f_1 & 0 \\ 0 & f_1 A_1 \end{pmatrix}.$$

---

set $x = Ab$, and so $c = BAb$. The desired matrices are $A^* = \begin{pmatrix} A \\ a \end{pmatrix}$ and $B^* = \begin{pmatrix} B & c - BAc \end{pmatrix}$:

$$A^* B^* = \begin{pmatrix} AB & Ab - (AB)Ab \\ aB & ab - (aB)Ab \end{pmatrix} = \begin{pmatrix} I_m & 0 \\ 0 & 1 \end{pmatrix},$$

since $AB = I_m$ by hypothesis and $aB = 0$ by construction.

[5]If p^t is the given row vector of Hermite's lemma, since $\gcd p^t = 1$, integers $q_1, \ldots, q_n$ may be determined such that $\sum_{i=1}^{n} p_i q_i = 1$. Then if q denotes the column vector with entries q_i, $p^t q = 1$, and Frobenius' corollary applies and yields Hermite's lemma. To show that Hermite's lemma implies Frobenius' corollary, observe that $p^t q = 1$ implies $\gcd p = \gcd q = 1$.

[6]For example, if $F(x) = 2x_1^2 + 6x_1x_2 + 7x_2^2$, so that $f = \gcd A = 1$, Gauss' Theorem I [244, Art. 229] implies that $F(x) = 1$ is impossible, since $F(0,1) = 7 \equiv 3 \pmod 4$, and so all odd numbers representable by F must be congruent to 3 mod 4.

Frobenius' proof of this lemma proceeds as follows. Apply Lemma 8.5 to obtain integral column matrices p,q such that $p^tAq = f_1$, $f_1 = \gcd A$. Then if $q_1 = (1/f_1)Aq$, we have $p^tq_1 = 1$, and so by Corollary 8.4, we know that a unimodular matrix U may be constructed with p^t as its first row. Likewise, if we take $p_1 = (1/f_1)A^tp$, then $(p_1)^tq = 1$, and Corollary 8.4 yields a unimodular matrix V with q as its first column. In block form,

$$U = \begin{pmatrix} p^t \\ U_1 \end{pmatrix} \quad \text{and} \quad V = \begin{pmatrix} q \ V_1 \end{pmatrix},$$

and block multiplication gives

$$B = UAV = \begin{pmatrix} p^tAq & p^tAV_1 \\ U_1Aq & U_1AV_1 \end{pmatrix}.$$

Since $f_1 = \gcd A$, all the entries of all four blocks above are divisible by f_1, with, of course, $p^tAq = f_1$. Thus we may write

$$B = UAV = \begin{pmatrix} f_1 & f_1R \\ f_1S & f_1T \end{pmatrix},$$

where R,S,T, are integral matrices. Finally, since

$$X = \begin{pmatrix} 1 & 0 \\ -S & I_{m-1} \end{pmatrix} \quad \text{and} \quad Y = \begin{pmatrix} 1 & -R \\ 0 & I_{n-1} \end{pmatrix} \tag{8.3}$$

are unimodular and (by block multiplication)

$$XBY = \begin{pmatrix} f_1 & 0 \\ 0 & f_1A_1 \end{pmatrix}, \quad A_1 = T - SR,$$

the reduction lemma follows with $P_1 = XU$ and $Q_1 = VY$.

It is not difficult to see that continued application of the reduction lemma yields the following reduction theorem.[7]

[7] If $A_1 = 0$, there is nothing to prove. If $A_1 \neq 0$, i.e., if $r = \operatorname{rank} A > 1$, then the reduction lemma may be applied to A_1 with $f_2 = \gcd A_1$. The conclusion is that unimodular P', Q' can be determined such that $P'A_1Q' = \begin{pmatrix} f_2 & 0 \\ 0 & f_2A_2 \end{pmatrix}$. Then

$$P_2 = \begin{pmatrix} 1 & 0 \\ 0 & P' \end{pmatrix} \quad \text{and} \quad Q_2 = \begin{pmatrix} 1 & 0 \\ 0 & Q' \end{pmatrix}$$

are unimodular, and block multiplication gives

Theorem 8.7 (Frobenius' reduction theorem). *If A is an integral $m \times n$ matrix of rank r, then unimodular matrices P,Q can be determined such that*

$$PAQ = \begin{pmatrix} f_1 & & & & & & \\ & f_1 f_2 & & & & & \\ & & \ddots & & & & \\ & & & f_1 f_2 \cdots f_r & & & \\ & & & & 0 & & \\ & & & & & \ddots & \\ & & & & & & 0 \end{pmatrix}. \tag{8.4}$$

As an immediate corollary we have the complete solution to part (I) of Problem 8.1:

Theorem 8.8 (Frobenius' normal form theorem). *If A is an integral $m \times n$ matrix of rank r with invariant factors $e_1, \ldots, e_r$, then unimodular matrices P and Q can be determined such that $PAQ = N$, where N is the normal form matrix (8.2). Thus two $m \times n$ integral matrices are equivalent if and only if they have the same invariant factors.*

Frobenius referred to N in (8.2) as providing "the normal form of F," viz., $F(x,y) = \sum_{i=1}^{r} e_i x_i' y_i'$, $x = P^t x'$, $y = Qy'$. Nowadays, the normal form (8.2) is known as the Smith normal form. The reason is that unbeknownst to Frobenius, H.J.S. Smith had discovered Frobenius' above theorem for square nonsingular matrices in 1861, although his extension of the normal form theorem to nonsquare matrices (of full rank) was not achieved via unimodular matrices. Smith's work will be discussed and compared with that of Frobenius in Section 8.5.

The above shows that the solution to problem (I) will follow once it has been proved that the equation $x^t Ay = f$, $f = \gcd A$, always has integral solutions, i.e., once Lemma 8.5 has been proved. According to Frobenius, his proof of this lemma was "modeled on the procedure taught by Gauss" in Article 236 of *Disquisitiones Arithmeticae* [182, p. 490]. Article 236 was concerned with the construction of the composite of two given quadratic forms,[8] and Frobenius perceived in some of the ideas in that complicated Gaussian construction procedure the initial steps in his method of finding integral solutions to $F(x,y) = f$ and so establishing Lemma 8.5. But as the interested reader can see from Section 8.2.1, the key steps,

---

$$P_2(P_1AQ_1)Q_2 = \begin{pmatrix} f_1 & 0 \\ 0 & f_1 P' A_1 Q' \end{pmatrix} = \begin{pmatrix} f_1 & 0 & 0 \\ 0 & f_1 f_2 & 0 \\ 0 & 0 & f_1 f_2 A_2 \end{pmatrix}.$$

If $A_2 \neq 0$, i.e., if $r > 2$, we may apply the reduction lemma to A_2, and so on, until we end up with unimodular matrices P_i, Q_i, $i = 1, \ldots, r$, such that PAQ, with $P = P_r \cdots P_1$ and $Q = Q_1 \cdots Q_r$, has the form in (8.4). In other words, the reduction theorem follows from the reduction lemma.

[8] See Section 9.1.1 for the definition of form composition.

and particularly the inductive part of Frobenius' proof, were original to him, as was the end result, namely Lemma 8.5. (Other readers may proceed to Section 8.3.)

8.2.1 Frobenius' proof of Lemma 8.5

Let us now consider Frobenius' proof of Lemma 8.5, which says that the equation $x^t Ay = f$, $f = \gcd A$, always has a solution with x, y integral. As we shall see, his method of solving $F(x,y) = f$ leads readily to his reduction theorem (Theorem 8.7), the key to his solution to both parts (I) and (II) of Problem 8.1 as well as to his rational version of Weierstrass' theory.

Let A denote an $m \times n$ matrix of integers and $F(x,y) = x^t Ay$ the associated bilinear form with f the greatest common divisor of all entries in A.

Preliminary step. Choose any integral column vector q with $\gcd q = 1$ and such that $Aq \neq 0$, and let h be the greatest common divisor of the components of Aq, so that we may write

$$Aq = hb, \tag{8.5}$$

where $\gcd b = 1$. This means that integers $p_1, \ldots, p_m$ exist such that $\sum p_i b_i = 1$, i.e., $p^t b = 1$, where p is the column vector with the p_i as components. (Keep in mind that $\gcd p = 1$ by virtue of $p^t b = 1$.) Thus

$$p^t Aq = p^t hb = h. \tag{8.6}$$

Since f divides $p^t Aq$, (8.6) implies that f divides h. If $f = h$, we are done: $F(p,q) = f$. If not, then $f < h$. In this case, let h' denote the greatest common divisor of the components of $p^t A$, so that we may write

$$p^t A = h' c^t, \tag{8.7}$$

where $\gcd c = 1$.

Relations (8.5)–(8.7) were inspired by Gauss, who applied them to a 4×4 skew-symmetric coefficient system that arises from the data of the given forms from which the composite form is to be constructed [244, Art. 236, eqn. (I)]. That is really all the inspiration Frobenius seems to have obtained from Gauss' construction. What follows after (8.7) and leads to the critically important inductive step is all due to Frobenius.

Notice from (8.6)–(8.7) that since $h = p^t Aq = h' c^t q$, h' divides h, and so $h' \leq h$. Now since $\gcd c = 1$, we know that an integral column vector q' exists for which $c^t q' = 1$. Thus $p^t Aq' = h' c^t q = h'$. Since we know that f divides $p^t Aq'$, we see that f divides h'. If $h' = f$, we are done: $F(p,q') = f$. Otherwise, let h'' denote the greatest common divisor of the components of Aq' and continue the process.

Since $h \geq h' \geq h'' \geq \cdots \geq h^{(i)} \geq \cdots > 0$ and f divides each $h^{(i)}$, there must be an i such that $h^{(i)} = h^{(i+1)}$. If $h^{(i)} = f$, we are done. If this is not the case, we proceed as follows.

Inductive step. To illustrate the procedure, let us assume for the sake of simplicity that $h = h'$ with $f < h$. Then, since $h' = h$, equations (8.5)–(8.7) take the form

$$(1)\quad Aq = hb, \qquad (2)\quad p^t Aq = h, \qquad (3)\quad p^t A = hc^t. \tag{8.8}$$

The significance of (8.8) is most readily seen using block multiplication of partitioned matrices.[9] Since $\gcd p = 1$, Hermite's Lemma 8.2, in the form of Frobenius' Corollary 8.4, guarantees that a unimodular matrix U may be constructed that has p^t as its first row. Likewise, since $\gcd q = 1$, there is a unimodular matrix with q^t as its first row. If V denotes the transpose of this matrix, then V is unimodular, and in block partitioned form we have

$$U = \begin{pmatrix} p^t \\ U_1 \end{pmatrix} \quad \text{and} \quad V = \begin{pmatrix} q & V_1 \end{pmatrix}. \tag{8.9}$$

Using block multiplication and (8.8), we obtain

$$B = UAV = \begin{pmatrix} p^t Aq & p^t AV_1 \\ U_1 Aq & U_1 AV_1 \end{pmatrix} = \begin{pmatrix} h & hR \\ hS & T \end{pmatrix}, \tag{8.10}$$

where $R = cV_1$ and $S = U_1 b$. If we introduce the unimodular matrices

$$X = \begin{pmatrix} 1 & 0 \\ -S & I_{m-1} \end{pmatrix} \quad \text{and} \quad Y = \begin{pmatrix} 1 & -R \\ 0 & I_{n-1} \end{pmatrix}, \tag{8.11}$$

then

$$XBY = \begin{pmatrix} h & 0 \\ 0 & A_1 \end{pmatrix}, \quad \text{where } A_1 = T - hSR. \tag{8.12}$$

Thus if $P = XU$ and $Q = VY$, P and Q are unimodular and

$$C = PAQ = XBY = \begin{pmatrix} h & 0 \\ 0 & A_1 \end{pmatrix}. \tag{8.13}$$

It is easily seen that because A and C are equivalent, the greatest common divisor of each matrix is the same.[10] Assuming that we are proceeding by induction on $N = m + n$, either $A_1 = 0$, in which case $h = f$, and we are done by virtue of (8.6);

---

[9] Frobenius did not use this mode of expression, although he began using it a few years later as a result of work on a problem involving abelian functions. See Section 10.6.

[10] $C = PAQ$ implies that every coefficient of C is an integral linear combination of coefficients of A; and $A = P^{-1}CQ^{-1}$ implies that every coefficient of A is an integral linear combination of coefficients of C.

or $A_1 \neq 0$, and by the induction hypothesis we know that column vectors $u_1 \in \mathbb{Z}^{m-1}$ and $v_1 \in \mathbb{Z}^{n-1}$ exist such that $u_1^t A_1 v_1 = g$, where g is the greatest common divisor of the coefficients of A_1. Now by (8.13), $f = \gcd(h,g)$, and so integers w and z exist such that

$$f = zh + wg = zh + wu_1^t A_1 v_1. \tag{8.14}$$

Consider any integer factorizations of z and of w, say $z = xy$ and $w = uv$. Then (8.14) takes the form

$$f = hxy + (uu_1)^t A_1 (vv_1) = u^t C v,$$

with $u = \begin{pmatrix} x_1 & uu_1 \end{pmatrix}^t$ and $v = \begin{pmatrix} y_1 & vv_1 \end{pmatrix}^t$. In other words, $x = u$, $y = v$ yields an integral solution to $x^t C y = f$; and since $C = PAQ$, $(x')^t A y' = f$ for $x' = P^t u$, $y' = Qv$. This completes my sketch of Frobenius' proof of his Lemma 8.5.

8.3 Applications

Frobenius showed that the above-developed theory had many applications. The most far-reaching was the rational theory of elementary divisors that he created. It is the subject of Section 8.6. Another application was his solution to Problem 8.1, part (II) (Section 8.4). Here I consider three other applications. The first is the obvious application of the normal form to solving linear systems of equations and congruences—the sole application considered as well by Smith. The remaining two later proved to be useful in Frobenius' work on theta functions, as we shall see in Section 11.3.

8.3.1 Linear systems of equations and congruences

Frobenius realized that Theorem 8.8 could be applied to linear systems of equations $Ax = b$ in which A and b are integral and only integral solutions x are sought. That is, if N is the normal form matrix (8.2), then $PAQ = N$, and so $Ax = b$ can be written as $Ny = c$, where $y = Q^{-1}x$ and $c = Pb$. Due to the simple nature of the system $Ny = c$, it is easy to determine criteria for integral solutions to exist, and these can then be translated into criteria for $Ax = b$. The resulting theorem is easy to state if we introduce some *Mathematica*-like notation:

Definition 8.9. For an integral matrix A, let GCD[Minors[A,k]] denote the greatest common divisor of the integers representing all $k \times k$ minor determinants of A.

Frobenius' result can then be stated as follows [182, IV, p. 507].

Theorem 8.10. *The system of equations $Ax = b$, where A and b are integral, has integral solutions if and only if the following two conditions hold: (1) If $r = \text{rank}\, A$, then r is also the rank of the augmented matrix $(A|b)$; (2)* $\text{GCD}[\text{Minors}[(A|b),r] = \text{GCD}[\text{Minors}[A,r]]$.

Frobenius also used the normal form of A to obtain results on linear systems of congruences. His main results are summarized in the theorem below. Before stating it, the meaning of Frobenius' symbol (A,k) needs to be indicated. Let A be any $m \times n$ integral matrix and let k be a positive integer. Then (A,k) denotes the number of expressions Ax, $x \in \mathbb{Z}^n$, that are "incongruent (mod k)," i.e., (A,k) is the number of equivalence classes of $(\mathbb{Z}/k\mathbb{Z})^m$ containing an element of the form Ax, with $x \in \mathbb{Z}^n$. Thus (A,k) *gives the number of incongruent $b \in \mathbb{Z}^m$ such that $Ax \equiv b \pmod{k}$ has solutions*. Using his normal form theorem (Theorem 8.8), Frobenius derived the following formula[11]:

$$(A,k) = k^m / \text{GCD}[\text{Minors}\,[(A|K),m]], \quad K = kI_m. \tag{8.15}$$

Here $(A|K)$ represents the matrix A augmented by the columns defining K. As the *Mathematica*-like notation suggests, (A,k) is easy to calculate using a present-day computer. Frobenius' above-mentioned theorem can now be stated as follows.

Theorem 8.11. *Consider the system of congruences $Ax \equiv b \pmod{k}$ with A and b integral, and A $m \times n$. (1) Integral solutions exist if and only if $(A,k) = ((A|b),k)$. (2) When this condition is satisfied, the number of incongruent solutions mod k is the same as the number of incongruent solutions mod k to the homogeneous system $Ax \equiv 0 \pmod{k}$ and equals $k^n/(A,k)$. (3) The number of incongruent b for which $Ax \equiv b \pmod{k}$ has integral solutions is (A,k).*[12]

Thus all the information about $Ax \equiv b \pmod{k}$ is readily obtainable from the value of (A,k).

As an illustration of this theorem, let $k = 6$ and consider the 3×4 matrix

$$B = \begin{pmatrix} 5 & 9 & 4 & 1 \\ 21 & 22 & 1 & 20 \\ 8 & 17 & 9 & -1 \end{pmatrix},$$

which has rank 2. By (8.15) we have (with the assistance of a computer) $(A,6) = 6^3/6 = 36$. Thus by part (3) there are 36 $b \in (\mathbb{Z}/6\mathbb{Z})^3$ (out of a total of $6^3 = 216$) for which $Ax \equiv b \pmod{6}$ has a solution. If $b = \begin{pmatrix} 3 & 8 & -5 \end{pmatrix}^t$, then by (8.15), $((A|b),6) = 36$, and so by (1), $Ax \equiv b \pmod{6}$ has solutions $x \in (\mathbb{Z}/6\mathbb{Z})^4$, and by part (3), the number of incongruent solutions is $(A,6) = 36$. Finally, if $b = \begin{pmatrix} 3 & 4 & 3 \end{pmatrix}^t$, then by (8.15), $((A|b),6) = 6^3/2 = 108 \neq (A,6)$, and so $Ax \equiv b \pmod{6}$ has no solutions.

---

[11] This is the formula given verbally in [182, III, p. 513] in the special case $K = kI_m$.

[12] Part (1) is the special case of [182, III, p. 519] when $K = kI_m$. Part (2) is informally stated [182, p. 520]. Part (3) is just Frobenius' definition of (A,k).

8.3.2 Alternating forms

We saw in Chapter 6 that Frobenius' interest in the problem of Pfaff had brought with it an interest in skew-symmetric matrices. He knew from Jacobi that a $2m \times 2m$ skew-symmetric matrix $A = (a_{ij})$ with complex coefficients a_{ij} regarded as variables had the remarkable property that $\det A$, a homogeneous polynomial in the a_{ij}, is the square of another polynomial, which is now called the Pfaffian and will be denoted here by $\mathrm{Pf}(a_{ij}) = \mathrm{Pf}(A)$. Since $\det A = \mathrm{Pf}(A)^2$, $\mathrm{Pf}(A)$ is determined only up to a factor of ± 1, and Jacobi and Frobenius took the coefficient of the term $a_{12}a_{34}\cdots a_{2m-12m}$ as $+1$, which is equivalent to taking $\mathrm{Pf}(L) = 1$, where L has the block diagonal form $L = J_2 \oplus \cdots \oplus J_2$, with $J_2 = \begin{pmatrix} 0 & 1 \\ -1 & 0 \end{pmatrix}$. When working on the problem of Pfaff, Frobenius had also proved that every skew-symmetric matrix has even rank.

Such properties of skew-symmetric integral matrices are not evident from the normal form N of (8.2) of Theorem 8.8, and Frobenius asked whether, when A is integral and skew-symmetric, there might not be a more revealing normal form for A. For example, if A is 2×2, then $A = \begin{pmatrix} 0 & a_{12} \\ -a_{12} & 0 \end{pmatrix}$. From this "normal form," it is immediate that A has even rank (0 or 2), that (for rank 2) the invariant factors are $e_1 = e_2 = |a_{12}|$, and that $\det A = a_{12}^2$ is the square of a polynomial in the a_{ij}. Frobenius saw how the normal form of his Theorem 8.8 and the results leading to it could be used to obtain a skew-symmetric normal form for $n \times n$ skew-symmetric A that generalized the above 2×2 case [182, §7]. What he showed was that a unimodular matrix P and positive integers $f_1, \ldots, f_l$ can be determined such that

$$P^t AP = f_1 J_1 \oplus + f_1 f_2 J_2 \oplus \cdots \oplus f_1 f_2 \cdots f_l J_2. \tag{8.16}$$

Frobenius' interesting proof of (8.16) is simple in conception but not obvious. For those interested, it is sketched below at the end of this subsection.

It follows from (8.16) that the rank of A is $2l$, and if, in a slight abuse of notation, we set $e_\lambda = f_1 \cdots f_\lambda$, then the $2l$ invariant factors of A are easily seen to be $e_1, e_1, \ldots, e_l, e_l$. Frobenius' skew-symmetric normal form for A is thus

$$G = P^t AP = \begin{pmatrix} 0 & e_1 & & & & & \\ -e_1 & 0 & & & & & \\ & & \ddots & & & & \\ & & & 0 & e_l & & \\ & & & -e_l & 0 & & \\ & & & & & \ddots & \\ & & & & & & 0 \end{pmatrix}. \tag{8.17}$$

From this normal form it follows that the W-series $d_n,\ldots,d_{2l}$ has the property that $d_{2i}=(e_1e_2\cdots e_i)^2$, $i=1,\ldots,l$, and, in particular, that when A is nonsingular, so that $n=2l$, $d_n=\det A=(e_1\cdots e_l)^2=[\mathrm{Pf}(A)]^2$.

It easy to see how to permute the rows of the matrix in (8.17) so that if the same permutation is also applied to the columns, one gets the matrix

$$J^*=\begin{pmatrix} 0 & D_l & 0 \\ -D_l & 0 & 0 \\ 0 & 0 & 0 \end{pmatrix}, \quad D_l=\begin{pmatrix} e_1 & & \\ & \ddots & \\ & & e_l \end{pmatrix}. \tag{8.18}$$

This means that if P' is the (unimodular) permutation matrix corresponding to the row permutations, then $P'G(P')^t=J^*$. A few years later, in an important work on theta functions (the subject of Chapter 11), Frobenius preferred the normal form (8.18).

For future reference, I will summarize Frobenius' results as the following theorem.

Theorem 8.12 (Symplectic basis theorem). *If A is an $n\times n$ skew-symmetric integral matrix of rank $2l$, then a unimodular matrix P may be determined such that $P^tAP=J^*$, where J^* is given by (8.18) and $e_1,e_1,\ldots,e_l,e_l$ are the invariant factors of A such that $e_i\mid e_{i+1}$ for all $i=1,\ldots,l-1$.*

In the special case in which $\det A\neq 0$, so that $n=2l$, if $\varepsilon_1,\ldots,\varepsilon_{2m}$ is the standard basis for $\mathbb{Z}^{2m}$, then the basis $\sigma_1,\ldots,\sigma_{2m}=P\varepsilon_1,\ldots,P\varepsilon_{2m}$, with P as in Theorem 8.12, is sometimes called a symplectic basis with respect to the alternating form $F(x,y)=x^tAy$ because $F(\sigma_i,\sigma_j)$ gives the (i,j) entry of $J^*=P^tAP$ [508, p. 90].

8.3.2.1 Frobenius' proof of (8.16)

Frobenius began by observing that if $f_1=\gcd A$, then by Lemma 8.5, integral row matrices

$$m_1=\begin{pmatrix} m_{11} & m_{12} & \cdots & m_{1n} \end{pmatrix} \quad \text{and} \quad m_2=\begin{pmatrix} m_{21} & m_{22} & \cdots & m_{2n} \end{pmatrix}$$

can be determined such that $m_1Am_2^t=f_1$, and so $m_1(f_1^{-1}A)m_2^t=1$. Because $a_{\beta\alpha}=-a_{\alpha\beta}$, the latter equation can be written out in the form

$$\sum_{\alpha>\beta}(a_{\alpha\beta}/f_1)(m_{1\alpha}m_{2\beta}-m_{1\beta}m_{2\alpha})=1.$$

This shows that the determinants $\begin{vmatrix} m_{1\alpha} & m_{1\beta} \\ m_{2\alpha} & m_{2\beta} \end{vmatrix}=m_{1\alpha}m_{2\beta}-m_{1\beta}m_{2\alpha}$ can have no common divisor for $\alpha>\beta$, and so also for $\alpha<\beta$, since these are just the negatives of those with $\alpha>\beta$.

The above consequence of Lemma 8.5 can be thought of as follows. If M is the $2 \times n$ matrix whose rows are m_1 and m_2, then d_2, the greatest common divisor of all its 2×2 minor determinants, equals 1. This means that $d_1 = 1$ also and that the two invariant factors of M are $e_1 = e_2 = 1$. Consequently, the normal form of M is $N = \begin{pmatrix} I_2 & 0 \end{pmatrix}$. Theorem 8.8 then says that $M = PNQ$, where P, Q are unimodular, P is 2×2 and Q is $n \times n$. Block multiplication applied to $M = PNQ$, with Q partitioned as $\begin{pmatrix} Q_2 \\ Q_{n-2} \end{pmatrix}$, where Q_2 is $2 \times n$ and Q_{n-2} is $n-2 \times n$, shows that $M = PQ_2$ and so

$$\tilde{M} \stackrel{\text{def}}{=} \begin{pmatrix} M \\ Q_{n-2} \end{pmatrix} = \begin{pmatrix} P & 0 \\ 0 & I_{n-2} \end{pmatrix} \begin{pmatrix} Q_2 \\ Q_{n-2} \end{pmatrix}.$$

Thus $\det \tilde{M} = \det P \det Q = \pm 1$, and $\tilde{M}$ is unimodular. In sum, we have added $n-2$ integral rows to the $2 \times n$ matrix M so as to create the $n \times n$ unimodular matrix $\tilde{M}$.

Going back to the skew-symmetric matrix A, consider $B = \tilde{M} A \tilde{M}^t$. From the fact that $m_1 A m_2^t = f_1$, it follows that $MAM^t = f_1 J_2$, and so calculation of B by block multiplication using the definition of $\tilde{M}$ gives

$$B = \begin{pmatrix} MAM^t & MAQ_{n-2}^t \\ Q_{n-2}AM^t & Q_{n-2}AQ_{n-2}^t \end{pmatrix} = \begin{pmatrix} f_1 J_2 & R \\ -R^t & T \end{pmatrix},$$

where T is also skew-symmetric. This is reminiscent of the first step in the proof of the reduction lemma, and by analogy with the second step—see (8.3)—if we take $X = \begin{pmatrix} I_2 & 0 \\ -R^t J_2 & I_{n-2} \end{pmatrix}$, then X is unimodular, and by block multiplication

$$XBX^t = f_1 \begin{pmatrix} J_2 & 0 \\ 0 & A_1 \end{pmatrix}, \quad A_1 = T - R^t J_2 R.$$

Since A_1 is skew-symmetric, the same reasoning can be applied to A_1, $f_2 = \gcd A_1$, and so on until $f_1, \ldots, f_l$ and P satisfying (8.16) are determined.

8.3.3 Modules

Under the influence of Dedekind, Frobenius actually established Theorem 8.11 in a more general form, which involved utilizing Dedekind's notion of a module as presented in his 1877 expository monograph on his theory of algebraic numbers and ideals [113].

In the first section of his monograph, Dedekind defined a system $\mathfrak{a}$ of real or complex numbers to be a module when $\alpha \pm \beta \in \mathfrak{a}$ whenever α, β belong to $\mathfrak{a}$. He pointed out that if $\alpha + \cdots + \alpha$ (n summands) is denoted by $n\alpha$ and likewise $-\alpha - \cdots - \alpha$ by $-n\alpha$, then a module $\mathfrak{a}$ has the property that $n\alpha \in \mathfrak{a}$ for all $n \in \mathbb{Z}$ and all

$\alpha \in \mathfrak{a}$. The modules that primarily concerned Dedekind in his essay were ideals of algebraic integers, but after developing those properties of modules he would need for his theory of ideals, he made the following observation [113, p. 82]:

> The researches in this first chapter have been expounded in a special form suited to our goal, but it is clear that they do not cease to be true when the Greek letters denote not only numbers, but any objects of study, any two of which α, β produce a determinate third element $\gamma = \alpha + \beta$ of the same type, under a commutative and uniformly invertible operation (composition), taking the place of addition. The module $\mathfrak{a}$ becomes a *group* of elements, the composites of which all belong to the same group. The rational integer coefficients indicate how many times an element contributes to the generation of another.

Inspired by Dedekind's results on modules and by this observation, Frobenius devoted a section of his 1879 paper to "modules" [182, §9], which he defined in the following manner [182, p. 510]:

> If in a system of m independent linear forms
>
> $$K_\alpha = k_{\alpha 1}y_1 + k_{\alpha 2}y_2 + \cdots + k_{\alpha p}y_p \quad (\alpha = 1, 2, \ldots, m)$$
>
> the greatest common divisor of the determinants of mth degree is greater than 1, then the system cannot represent all integers for integral values of the variables. The totality of the [integral] value systems $[K_1, \ldots, K_m]$ that can be represented by these forms is called a module* ....

Frobenius' footnote * was to the first section of Dedekind's monograph [113], although—encouraged by Dedekind's above-quoted observation—he is considering additive subgroups of $\mathbb{Z}^m$ as modules.

Frobenius used the symbol K to denote the above system of linear forms; I will let K represent the associated $m \times p$ coefficient matrix $K = (k_{\alpha j})$. Thus Frobenius' module is the system of all elements $m \in \mathbb{Z}^m$ of the form $m = Kn$ for some $n \in \mathbb{Z}^p$. I will denote this system by $\mathcal{M}_K$. It is, of course, an example of a module in the above-quoted sense of Dedekind, and a $\mathbb{Z}$-module in the modern sense. Frobenius wrote $m_1 \equiv m_2 \pmod{K}$ when $m_1 - m_2 \in \mathcal{M}_K$. When $K = kI_m$ (so $p = m$), then $m_1 \equiv m_2 \pmod{K}$ means simply that $m_1 \equiv m_2 \pmod{k}$. By developing properties of his modules, Frobenius was able to formulate Theorem 8.11 with "mod k" replaced by "mod K." One such property is worth discussing for two reasons: to indicate how the derivation utilizes the normal form theorem (Theorem 8.8) and because Frobenius was to make use of it in his theory of generalized theta functions (Section 11.3).

The property in question is the solution to the problem of determining the number of distinct congruence classes mod K, i.e., the number of elements in $\mathbb{Z}^m/\mathcal{M}_K$. Let N denote the normal form of K, so that $K = PNQ$, where P and Q are unimodular matrices of the requisite dimensions. Frobenius observed that for any $m \times p$ integral matrix B, $\mathcal{M}_{BQ} = \mathcal{M}_B$. Also, if $m \to m' = Pm$ for all $m \in \mathbb{Z}^m$, then $m_1 \equiv m_2 \pmod{N}$ if and only $m_1' \equiv m_2' \pmod{PN}$. Thus (in modern terms) $\mathbb{Z}^m/\mathcal{M}_N \cong \mathbb{Z}^m/\mathcal{M}_{PN} = \mathbb{Z}^m/\mathcal{M}_{PNQ} = \mathbb{Z}^m/\mathcal{M}_K$, so the number of congruence classes mod K is the same as the number mod N. The latter number, however, is easy to calculate. If

$$N = \begin{pmatrix} e_1 & & & & & \\ & \ddots & & & & \\ & & e_m & & & \\ & & & 0 & & \\ & & & & \ddots & \\ & & & & & 0 \end{pmatrix},$$

then representatives of the distinct congruence classes are given by

$$\left(s_1 \cdots s_m \, 0 \cdots 0\right)^t, \quad 0 \leq s_j < e_j,$$

and so the total number is $\ell = e_1 e_2 \cdots e_m = d_m$, where d_m is the greatest common divisor of the $m \times m$ minors of N. Thus $\ell = d_m$ is also the greatest common divisor of the $m \times m$ minors of $K = PNQ$. This is summarized in the following theorem [182, I, p. 511]:

Theorem 8.13. *If K is an $m \times p$ integral matrix of full rank m, then the number ℓ of distinct congruence classes* $\bmod K$ *is* $\ell = \mathrm{GCD}[\mathrm{Minors}[K,m]]$. *In particular, when K is $m \times m$ and nonsingular,* $\ell = |\det K|$.

Frobenius' theory of modules was but one of many manifestations of Dedekind's influence on him. As we shall see in the following chapter, Dedekind's views, along with those of Kronecker, encouraged Frobenius to systematically develop the theory of finite abstract abelian groups. In fact, his paper on the subject, written in collaboration with his friend and colleague Stickelberger, was published before Frobenius' paper with the solution to part (II) of Problem 8.1 (discussed below). As we shall see in the next chapter, Frobenius' interest in abstract finite abelian groups soon led to an interest in general finite abstract groups; and as will be seen in Chapters 12–13, it was once again under the influence of Dedekind that Frobenius was led to create his theory of group characters and representations.

8.4 Solution to Problem (II): The Containment Theorem

The above results, and many more, are in Frobenius' paper of 1879. His second paper, which he described as a continuation of the first, was submitted 6 months later, in January 1879, and was published in 1880 [185]. It ran for 21 pages in *Crelle's Journal* and was devoted exclusively to part (II) of Problem 8.1—to determine necessary and sufficient conditions for one form to be contained in another. Recall that form G with matrix B is contained in form F with matrix A if P,Q exist such that $B = PAQ$. Here P,Q are integral but are not necessarily unimodular or even nonsingular. The symbolic matrix notation then suggests saying that B *is a multiple of* A, the latter, more easily remembered terminology being adopted by Frobenius' successors.[13]

It is easy to determine a sufficient condition for B to be a multiple of A. That is, let A and B (still assumed to be $m \times n$) have respective normal forms

$$N = \begin{pmatrix} e_1 & & & \\ & e_2 & & \\ & & \ddots & \\ & & & e_l \end{pmatrix}, \quad M = \begin{pmatrix} \varepsilon_1 & & & \\ & \varepsilon_2 & & \\ & & \ddots & \\ & & & \varepsilon_l \end{pmatrix},$$

where $l = \min\{m,n\}$, $e_1,\ldots,e_r$, $r = \operatorname{rank} A$, and $\varepsilon_1,\ldots,\varepsilon_\rho$, $\rho = \operatorname{rank} B$, are the invariant factors of A and B, respectively, while the remaining e_i and ε_i are zero. Thus $e_{i-1} \mid e_i$ and $\varepsilon_{i-1} \mid \varepsilon_i$ for all $i = 1,\ldots,l$. Now suppose that each invariant factor of B is a multiple of the corresponding invariant factor of A. This means that for all $i = 1,\ldots,l$, we have $\varepsilon_i = m_i e_i$, where m_i is an integer, including possibly $m_i = 0$. Since B is a multiple of A if and only if B^t is a multiple of A^t, we may assume without loss of generality that $l = m$. Thus $M = DN$, where D is the $l \times l$ diagonal matrix with $m_1,\ldots,m_l$ on the diagonal. Now, by Frobenius' normal form theorem (Theorem 8.8), unimodular matrices R, R_1, S, S_1 exist such that $RAR_1 = N$ and $SBS_1 = M$. This means that $SBS_1 = M = DN = DRAR_1$, and so $B = S^{-1}DRAR_1S_1^{-1}$, or $B = PAQ$, where $P = S^{-1}DR$ and $Q = R_1S_1^{-1}$ are integral matrices. Summing up:

Proposition 8.14. *Let A, B be $m \times n$ integral matrices with the property that for every i, the ith invariant factor of B is a multiple of the ith invariant factor of A. Then B is a multiple of A, i.e., integral matrices P, Q exist such that $B = PAQ$. (In Frobenius' language, the form $G = (u)^t Bv$ is contained in the form $F = x^t Ay$.)*[14]

But is the converse of this proposition true? Frobenius was able to prove that the answer was affirmative—thereby solving problem (II)—but this turned out to be far more difficult for him to prove than the above proposition. To push through a proof he was forced to develop a "mod-k" analogue of his reduction theorem and its corollary on transformation to normal form for any positive integer k, and this involved him with subtle considerations involving the definition of the mod-k rank of an integral form $F(x,y) = x^t Ay$ [185, §15] and will not be discussed here.

Here it will suffice to say that after many pages, Frobenius established the following result.

---

[13]It was introduced by Hensel in 1895 [282] and adhered to by Bachmann in his book on the arithmetic of quadratic (and bilinear) forms [11, p. 299].

[14]Frobenius, usually an excellent expositor, failed to inform the reader of this easy consequence of his normal form theorem at the beginning of his 1880 paper. It was only after the difficult task of proving the necessity of the above condition was completed that its sufficiency was quickly deduced as above (but without matrix notation).

Theorem 8.15 (Frobenius' mod-k reduction theorem). *Let $k > 0$ be a fixed integer. If $F(x,y) = x^t Ay$ with A $m \times n$ and integral, then there is a change of coordinates $x = Pu$, $y = Qv$, where P, Q are integral and have determinants relatively prime to k, such that F is transformed into $N(u,v) \stackrel{\text{def}}{=} F(Pu, Qv)$, where*

$$N \equiv g_1 u_1 v_1 + g_2 u_2 v_2 + \cdots + g_\rho u_\rho v_\rho \pmod{k}, \tag{8.19}$$

and $g_{i-1} \mid g_i$ for $i = 1, \ldots, \rho$. Furthermore, the reduced form on the right of (8.19) is unique.

The actual reduction $F \to N$ of (8.19) was obtained via an analogue of Lemma 8.5: if f is the greatest common divisor of k and all the coefficients a_{ij} of A, then integral x and y exist such that $x^t Ay \equiv f \pmod{k}$ [185, §14]. This result was then used to obtain (8.19) in much the same way as Lemma 8.5 was used to obtain the normal form of Theorem 8.8 as indicated in Section 8.2. It is the uniqueness part of Theorem 8.15 that is not so straightforward [185, §§16–17]. By virtue of that uniqueness, it followed that ρ is the mod-k rank of F as originally formulated by Frobenius. I will denote the mod-k rank of F by $\text{rank}_k\, F$.

Eventually, by means of several additional propositions, Frobenius educed from Theorem 8.15 his solution (II), which may be summed up in the following theorem [185, p. 609].

Theorem 8.16 (Frobenius' containment theorem). *Given forms F and G with respective invariant factors $e_1, \ldots, e_r$ and $\varepsilon_1, \ldots, \varepsilon_\rho$, G is contained in F if and only if (a) $\text{rank}\, G \leq \text{rank}\, F$ and (b) $e_i \mid \varepsilon_i$ for all $i \leq \rho$. Stated in terms of the matrices A, B associated to F, G, respectively, B is a multiple of A in the sense that $B = PAQ$, where P, Q are integral, if and only if ε_i is an integral (possibly zero) multiple of e_i for every i.*

Frobenius must have been quite pleased with his solution Problem 8.1, for it shows that the answers to the questions posed by parts (I) and (II) can be read off immediately from the invariant factors $e_1, \ldots, e_r$ and $\varepsilon_1, \ldots, \varepsilon_\rho$ of the two given forms F and G. It also shows that if we define A and B to be "equivalent" if A is a multiple of B and B is a multiple of A, then by the containment theorem, $e_i(A)$ is a multiple of $e_i(B)$ for all i and vice versa, which means that $e_i(A) = e_i(B)$ for all i, so by Theorem 8.8, A and B are equivalent in the earlier sense that unimodular P and Q exist such that $PAQ = B$.

For readers interested in seeing in outline how Frobenius managed to use Theorem 8.15 to arrive at Theorem 8.16, read on. Others may proceed to Section 8.5 without any loss of continuity.

8.4.1 Outline of Frobenius' proof of Theorem 8.16

The fact that P and Q in Theorem 8.15 have determinants relatively prime to k means that the customary formula for inverses, e.g., $P^{-1} = (\det P)^{-1}\, \text{Adj}\, P$, makes sense mod k (i.e., over $\mathbb{Z}/k\mathbb{Z}$ as we would now say) and $P^{-1}P \equiv I \pmod{k}$. For example, if $k = 10$ and $P = \begin{pmatrix} 2 & 3 \\ 5 & 6 \end{pmatrix}$, then $\det P = -3 \equiv 7 \pmod{10}$, which is relatively prime to $k = 10$. Since $7 \cdot 3 = 21 \equiv 1 \pmod{10}$, the mod 10 inverse of 7 is 3, and so

$$P^{-1} = 3\,\mathrm{Adj}\,P = 3\begin{pmatrix} 6 & -3 \\ -5 & 2 \end{pmatrix} \equiv \begin{pmatrix} 8 & 1 \\ 5 & 6 \end{pmatrix} \pmod{10}.$$

Suppose now that $G = (u)^t Bv$ is contained in $F = x^t Ay$, so that integral matrices P, Q exist for which $x^t = Pu^t$, $y = Qv$ transforms F into G, whence $PAQ = B$. I will denote the containment relation by $G \prec F$ or by $B \prec A$. Frobenius defined G to be contained in F mod k if, integral P, Q exist such that $PAQ \equiv B \pmod{k}$. I will denote this equivalence relation by $G \prec F \pmod{k}$ or by $B \prec A \pmod{k}$. Clearly, if $G \prec F$, then $G \prec F \pmod{k}$ for all $k > 0$, a fact that Frobenius used in solving problem (II). Also used in the solution of problem (II) is the fact that $B \prec A$ implies $\operatorname{rank} B \leq \operatorname{rank} A$ and the following mod-k analogue [185, II, p. 596]:

Proposition 8.17. *If* $B \prec A \pmod{k}$, *then* $\operatorname{rank}_k B \leq \operatorname{rank}_k A$.

Frobenius defined equivalence mod k by analogy with ordinary equivalence over $\mathbb{Z}$. Thus G and F are said to be equivalent mod k if $G \prec F \pmod{k}$ and $F \prec G \pmod{k}$. He showed that this means that integral P, Q exist with determinants relatively prime to k such that $PAQ \equiv B \pmod{k}$, and of course, it follows that $(P^{-1})BQ^{-1} \equiv A \pmod{k}$, where P^{-1} and Q^{-1} denote the mod-k inverses of P and Q. Thus Theorem 8.15 states that a form F is uniquely equivalent mod k to a normal form (8.19), and so the integers $g_1, \ldots, g_\rho$ are defined to be the mod-k invariants of F [185, p. 601]. Frobenius showed how the mod-k rank and invariants $g_1, \ldots, g_\rho$ of F can be easily computed from the invariant factors $e_1, \ldots, e_r$ of F [185, I–II, p. 605]:

Proposition 8.18. *If F has invariant factors $e_1, \ldots, e_r$, then (i)* $\operatorname{rank}_k F$ *is the largest i such that k e_i; and (ii) the mod-k invariant factors are given by* $g_i = \gcd(e_i, k)$, $1 \leq i \leq \operatorname{rank}_k F$.

Thus, for example, if $F = x^t Ay$ with A 4×4, and if its invariant factors are computed to be $e_1 = 2$, $e_2 = 2^2 \cdot 3$, and $e_3 = 2^2 \cdot 3 \cdot 5 \cdot 7$, so that $\operatorname{rank} A = 3$, then (1) if $k = 6$, then $\operatorname{rank}_k A = 1$, and $g_1 = 2$; (2) if $k = 15$, then $\operatorname{rank}_k A = 2$, and $g_1 = 1$, $g_2 = 3$; (3) If $k = 8$, $\operatorname{rank}_k A = 3$, and $g_1 = 2$, $g_2 = 4$, $g_3 = 4$.

Using Propositions 8.17–8.18, Frobenius solved problem (II) in the following manner [185, §21]. First suppose that $G \prec F$ and that $\varepsilon_1, \ldots, \varepsilon_\rho$ and $e_1, \ldots, e_r$ are the invariant factors of G and F, respectively. Then (a) $\operatorname{rank} G \leq \operatorname{rank} F$. Now $G \prec F$ implies $G \prec F \pmod{k}$ for every k, and so by Proposition 8.17, we also have $\operatorname{rank}_k G \leq \operatorname{rank}_k F$ for every k. In particular, for $k = e_i$, it follows that $\operatorname{rank}_{e_i} G \leq \operatorname{rank}_{e_i} F$. Since for any fixed $i \leq r$, $e_i \mid e_i$, part (i) of Proposition 8.18 says that $\operatorname{rank}_{e_i} F < i$, which implies that $\operatorname{rank}_{e_i} G < i$ as well. But by part (i) of Proposition 8.18, $\operatorname{rank}_{e_i} G < i$ means that (b) $e_i \mid \varepsilon_i$. In sum, if $G \prec F$, then

(a) rank $G \leq$ rank F and (b) $e_i \mid \varepsilon_i$ for all $i \leq \rho$. Conditions (a) and (b) are thus necessary in order that $G \prec F$. We already saw at the beginning of Section 8.4, that (a) and (b) are sufficient for $G \prec F$, and so the proof outline of Theorem 8.16 is now complete.

8.5 The Work of H. J. S. Smith

At the beginning of his first paper, after briefly alluding to its contents, namely part (I) of Problem 8.1 on the arithmetic equivalence of forms and its application to linear systems of equations and congruences, Frobenius wrote in a footnote that "The works of Mr. Smith ... [537,539,540] ... first came to my attention after completion of this work. In the following the first of these will be cited as *Sm.* [182, p. 483n**]." The second and third papers cited by Frobenius were published in the *Proceedings of the London Mathematical Society* in 1873 and were mostly restatements of results from a lengthy memoir Smith had published in the *Philosophical Transactions of the Royal Society of London* in 1861—the memoir Frobenius referred to by *Sm.* Since Frobenius does not seem to have been in the habit of scanning the pages of the *Philosophical Transactions* for mathematics of interest,[15] it was perhaps Smith's notes of 1873 that he discovered first, and then, of course, he turned to their source, the memoir of 1861.

What Frobenius discovered was that in several of his key results, he had been anticipated 18 years earlier by Smith (1826–1883), who since 1860 had been Savilian professor of geometry at Oxford. Frobenius nonetheless published his work as he had developed it, presumably because his results, where they overlapped with Smith's, were more general, since Smith worked under the implicit hypothesis, common at that time, that the matrices A under consideration were generic, i.e., of full rank. Smith had also restricted his attention to $m \times n$ matrices with $m \leq n$ and so assumed rank $A = m$. In addition, many of Frobenius' results, such as the containment theorem, the applications in Sections 8.3.2 and 8.3.3, and his application of the solution of problem (I) to Weierstrass' theory of elementary divisors (the subject of Section 8.6 below), were not anticipated by Smith. Another consideration that I suspect had considerable weight with Frobenius was the fact that his two papers [182, 185] constituted in his mind an original, systematically developed treatise on the arithmetic theory of bilinear forms and as such had a value that transcended that of the individual results it contained. Leaving aside the possible merits of Frobenius' lucid and rigorously developed "treatise," in what follows I will briefly consider Smith's principal results and their relation to those of Frobenius.

Smith's memoir of 1861 bore the title "On systems of linear indeterminate equations and congruences," which suggests that, unlike Frobenius, who was

---

[15] For example, Frobenius never cited Cayley's 1858 "A Memoir on Matrices," which was also published in the *Philosophical Transactions*, even though he did cite Cayley's earlier work, in *Crelle's Journal*, on the Cayley–Hermite problem (Chapter 7).

interested in Problem 8.1 on the arithmetic theory of bilinear forms and, thanks to his knowledge of Weierstrass' theory of elementary divisors, took the invariant factors of a matrix as the starting point, Smith was primarily motivated by the theory of linear systems of equations and congruences. Indeed, in 1859, in his "Report on the Theory of Numbers" for the British Association for the Advancement of Science, after noting that Gauss' treatment of linear systems of congruences in *Disquisitiones Arithmeticae* [244, Art. 37] was "imperfect" [536, p. 43], Smith had observed how "by means of a few subsidiary propositions relating to determinants, we may, in every case, obtain directly all possible solutions to any proposed system; and (what is frequently of more importance) we can decide a priori whether a given system is resoluble or not, and if it be resoluble we can assign the number of its solutions" [536, pp. 43–44]. In his memoir of 1861, Smith showed how both the resolubility question and the number of solutions of $Ax \equiv b \pmod{k}$ could be determined in a simpler, case-free manner by means of a factorization of A he had discovered.

Smith's factorization involved what are now called the invariant factors of A. That is, he considered an $m \times n$ matrix of integers, which he denoted by $\parallel a \parallel$ (he was familiar with Cayley's work on matrices).[16] Because Smith (like Cayley) operated on the generic level, he assumed explicitly that $m \leq n$ and implicitly that at least one $m \times m$ minor determinant of $A = \parallel a \parallel$ is not zero, i.e., in the language later introduced by Frobenius, A is assumed to have full rank m. For such matrices he introduced, apparently for the first time in the history of number theory, the series $d_m, \ldots, d_1, d_0$, where $d_0 = 1$ and d_i is the integer greatest common divisor of all $i \times i$ minor determinants of A and, realizing that d_{i-1} divides d_i, the associated series of quotients $e_i = d_i/d_{i-1}$, $i = 1, \ldots, m$. Smith's main theorem [537, pp. 391–392] was that when such a generic A is square ($m = n$), then unimodular matrices P, Q can be determined such that

$$A = PNQ, \quad N = \text{Diag. Matrix}(e_n, \ldots, e_1). \tag{8.20}$$

With this fundamental theorem Smith thus anticipated Frobenius' normal form theorem (Theorem 8.8) in the case of a square nonsingular matrix. He even proved that e_{i-1} divides e_i [537, p. 396].

To apply his fundamental theorem to an $m \times n$ matrix A with $m < n$ (and of course full rank m), Smith proceeded as follows [537, pp. 396ff.]. First he proved that A can always be factored as $A = SA'$, where S is $m \times m$, and, in the notation of Definition 8.9, $\det S = \text{GCD}[\text{Minors}[A,m]]$ and $\text{GCD}[\text{Minors}[A',m]] = 1$ [537, pp. 370–371]. It then followed from a property of minor determinants that was used later by Weierstrass in formulating his theory of elementary divisors that if, as in the case of a square matrix, we set $d_i = \text{GCD}[\text{Minors}[A,m]]$, then $d_i =$

[16] Smith was familiar with Cayley's 1858 memoir on matrices [84] (discussed in Section 7.4); he referred to it in his report on the theory of numbers for the British Association for the Advancement of Science [538, p. 167].

GCD[Minors[S,m]]. Thus, as we would now say, A and S have the same invariant factors $e_i = d_i/d_{i-1}$. Then by Smith's fundamental theorem, unimodular matrices P and Q may be determined such that $S = PNQ$, with N the diagonal matrix with diagonal entries $e_n,\ldots,e_1$, left to right. In this manner, Smith obtained the factorization

$$A = PNQA' = PNR, \quad R = QA'. \tag{8.21}$$

In Frobenius' factorization in Theorem 8.8, of course R is $n\times n$ (rather than $m\times n$) and unimodular.

Smith's above matrix R had a generalized unimodular property in the sense that $\delta_n = \mathrm{GCD}[\mathrm{Minors}[R,m]] = 1$. This property of Smith's R enabled him to apply the factorization $A = PNR$ to the theory of systems of linear congruences $Ax \equiv b(\mathrm{mod}\, k)$ in much the same way as Frobenius had used his. In order to state Smith's results in a clear and concise form, I will utilize the notion of the rank of a matrix, although Smith did not.[17] Let A be an $m\times n$ matrix of integers of full rank r. Hence $r = m$ if $m \leq n$ and $r = n$ if $m > n$. The integers $e_i = d_i/d_{i-1}$, $i = 1,\ldots,r$, where d_i denotes the greatest common divisor of all $i\times i$ minors of A (and $d_0 = 1$) introduced by Smith, will be called the invariant factors of A. Smith's results are summed up in the following theorem [537, pp. 399–404].

Theorem 8.19. *Let A be an $m\times n$ matrix of integers (of full rank) and consider the linear system of congruences*

$$Ax \equiv b\,(\mathrm{mod}\,k). \tag{8.22}$$

Let $e_1,\ldots,e_r$ and $e'_1,\ldots,e'_{r'}$ denote the invariant factors of A and $(A|b)$, respectively, and set $g_i = \gcd(e_i,k)$, $i = 1,\ldots,r$, $g'_i = \gcd(e'_i,k)$, $i = 1,\ldots,r$, and $g = \prod_{i=1}^{r} g_i$, $g' = \prod_{i=1}^{r} g'_i$. (1) If $m = n$ (so $r = r' = n$), then (8.22) has solutions if and only if $g = g'$. If this is the case, then the number of incongruent solutions x is g. (2) If $m < n$ (so $r = r' = m$), then (8.22) has solutions if and only if $g = g'$, and the number of incongruent solutions is $g\cdot k^{n-m}$. (3) If $m > n$ (so $r = n$ and $r' = n+1$), then (8.22) has solutions if and only if $g = g'$ and in addition, $e'_{n+1} \equiv 0\,(\mathrm{mod}\,k)$. In this case, the number of incongruent solutions is g.

Smith's theorem thus provided answers to the same questions about the system of congruences $Ax \equiv b\,(\mathrm{mod}\,k)$ as did Frobenius' Theorem 8.11, albeit only for matrices of full rank by virtue of the generic level on which he reasoned. Thus Smith's theorem does not apply to the matrix A introduced immediately after Frobenius' Theorem 8.11. The fact that Smith did not specify the number of incongruent b for which (8.22) has solutions—the number denoted by (A,k) by Frobenius—is not significant, since this number is readily deducible from the sort of

[17] Recall from Section 6.3 that the determinant-theoretic definition of rank was first introduced by Frobenius in 1877, although the rank property was used earlier.

equivalence relation counting found in Gauss' *Disquisitiones Arithmeticae*, which had been studied carefully by Smith.[18] Frobenius' answers to the questions are, however, not only nongeneric but simpler, since the form of his answers is the same whatever the dimensions of A, and easier to compute, since only (A,k) need be computed, not all the invariant factors of A and $(A|b)$ as in Smith's theorem.

By virtue of Theorem 8.19 above and the normal form factorizations (8.20)–(8.21) on which it was based, Smith had anticipated, on the generic level, several of Frobenius' main theorems. Their approaches also had points in common. Both were by induction and both made use of Hermite's Lemma 8.2: a unimodular matrix may be constructed with a given row of relatively prime integers as its first row. In the case of Frobenius, the other key result he used was his Lemma 8.5 that if $f = \gcd(A)$, then $x \in \mathbb{Z}^m$ and $y \in \mathbb{Z}^n$ can be determined for which $x^t Ay = f$. Actually, Smith also employed a lemma, which is easily seen to be equivalent to Frobenius' Lemma, although of course Smith proved it only for A $m \times n$ with rank $A = m$: if $\gcd(A) = 1$, then $x \in \mathbb{Z}^n$ can be determined so that $u = Ax$ and $\gcd(u) = 1$ [537, p. 392].[19]

We saw that in Frobenius' case, his proof of Lemma 8.5 contained within it the ideas that he used to prove the reduction theorem (Theorem 8.7) from which his normal form theorem (Theorem 8.8) followed. In the case of Smith, whose generic proof of the above lemma was relatively simple, two additional theorems were needed to produce his normal form factorization of A in (8.20). The first was the generic version of Frobenius' Theorem 8.10 giving necessary and sufficient conditions that $Ax = b$ (A $m \times n$) have integral solutions. In the generic version, Frobenius' condition (1), viz., rank A = rank $[A|b]$, is automatically satisfied, since rank $A = m$, and so Frobenius' condition (2) with $r = m$, GCD[Minors[$[A|b]$, m] = GCD[Minors[A, m], is the sole condition in Smith's theorem [537, p. 387]. Here too, then, Smith had anticipated a theorem of Frobenius, albeit only in the generic case. Of course, Frobenius deduced his theorem from his normal form theorem, whereas Smith used his generic version in the proof of his generic normal form theorem for square matrices.[20] The other major component of Smith's proof was another theorem going back to Hermite [285, pp. 166ff.]:

Theorem 8.20. *(1) If A is any $n \times n$ matrix such that* $\det A \neq 0$, *then a unimodular matrix Q exists such that $K = AQ$ is a nonnegative upper triangular matrix with the*

---

[18] That is say that $x_1, x_2 \in \mathbb{Z}^n$ are equivalent if $Ax_1 \equiv Ax_2 \pmod{k}$. This is an equivalence relation. The number of incongruent x in an equivalence class equals the number P of solutions to $Ax \equiv 0 \pmod{k}$, which is given by Smith's Theorem 8.19, so if Q is the number of equivalence classes, then $PQ = k^n$ (the number of elements in $(\mathbb{Z}/k\mathbb{Z})^n$) and Q equals the number of incongruent b such that (8.22) has a solution.

[19] Granted Smith's lemma, if $f = \gcd(A)$, apply Smith's lemma to $A' = (1/f)A$ to obtain y such that $u = A'y$ and $\gcd(u) = 1$. Thus $fu = Ay$ and $v = fu$ has $\gcd(v) = f$, so that integers $x_1, \ldots, x_m$ exist for which $\sum_{l=1}^{m} x_i w_i = f$. Thus if $x = \begin{pmatrix} x_1 & \cdots & x_m \end{pmatrix}^t$, we have $x^t Ay = y^t w = f$, which is Frobenius' conclusion. Conversely, if Frobenius' Lemma 8.5 is assumed and $\gcd(A) = 1$, then $x^t Ay = 1$. This says that if $u = Ay$, then $x^t u = 1$, which means that $\gcd(u) = 1$.

[20] Incidentally, just before publication of his memoir, Smith discovered that his theorem on integral solutions to $Ax = b$ had been proved earlier (in 1858 [281]) by Ignaz Heger [537, p. 387n].

property that each diagonal entry is strictly greater than the entries to the right of it in the same row. (2) Likewise, a unimodular matrix P exists such that $H = PA$ is a nonnegative lower triangular matrix such that each diagonal entry is strictly greater than the entries above it in the same column. (3) The matrices H and K are unique with respect to their respective properties.

Hermite simply stated part (1) of the theorem. Smith used Hermite's lemma to supply a proof by induction. He also showed that K is unique. Of course, (2) then follows immediately by matrix transposition. By combining the reasoning behind (1) and (2), Smith obtained his normal form $N = PAQ$ [537, Art.14*]. His proof thus depended critically on the fact that $\det A \neq 0$.

Both Smith and Frobenius were talented mathematicians, but Frobenius had the good fortune to have been trained within the Berlin school of Weierstrass and Kronecker. The Berlin disciplinary ideals, which he enthusiastically accepted, were, by virtue of his exceptional talent, adhered to and creatively applied in his work, thereby producing results that were not only nongeneric but usually presented in a clear and simple fashion. These hallmarks of his work are apparent in his papers on arithmetic linear algebra of 1879–1880 [182, 185] and stand out all the more by comparison with the admirable, highly original work of Smith. Thanks to a felicitous combination of educational background and extraordinary talent, Frobenius was able to obtain completely general results on matters where Smith had attained generic results. Furthermore, he was able to push the consequences of the normal form further. Thus, for example, he developed the theory of the containment of bilinear forms, which culminated in his Theorem 8.16, as well as a normal form theorem for skew-symmetric integral matrices (Section 8.3.2) that proved of considerable value in the theory of theta functions and abelian varieties. And because of his familiarity with Weierstrass' elementary divisor theory—something that did not exist when Smith did his work—Frobenius made what was probably the most influential application of the Smith–Frobenius normal form (Theorem 8.8): the development of a rational theory of elementary divisors, the subject of the next section.

8.6 A Rational Theory of Elementary Divisors

8.6.1 The rationality paradox

As Frobenius explained at the beginning of his 1879 paper [182, p. 483], long before he had formulated and solved Problem 8.1, he had noted what he regarded as an unsatisfying aspect of Weierstrass' theory of elementary divisors; but it was not until the work on part (I) of Problem 8.1 was successfully concluded that he saw the way to avoid it. Here is what Frobenius found unsatisfactory. Consider two families of bilinear forms $\lambda A_1 + A_2$ and $\lambda B_1 + B_2$ that are nonsingular, so that $\det A_1 \neq 0$ and $\det B_1 \neq 0$. Then Weierstrass' theory of elementary divisors (Section 5.4) provided a means of deciding by rational operations whether the two families are equivalent in the sense that nonsingular transformations P and Q exist such that $P(\lambda A_1 + A_2)Q = \lambda B_1 + B_2$, or equivalently, such that

$$PA_1Q = B_1 \quad \text{and} \quad PA_2Q = B_2. \tag{8.23}$$

Here by "rational operations" Frobenius meant the operations of addition, subtraction, multiplication, and division as applied to rational numbers and the coefficients of the given forms. It followed from Weierstrass' theory that the two nonsingular families are equivalent if and only if they have the same W-series. (The W-series for $\lambda A_1 + A_2$ is $\{d_n, d_{n-1}, \ldots, d_1\}$, where $d_i = d_i(\lambda)$ is the polynomial greatest common divisor of the $i \times i$ minor determinants of $\lambda A_1 + A_2$.) All the $i \times i$ minors can be determined by rational operations, and since each such minor is a polynomial in λ, the greatest common divisor $d_i(\lambda)$ of these polynomials can also be determined by rational operations—those that form the analogue of the Euclidean algorithm for polynomials in λ. That algorithm is based on the fact that if $f(\lambda)$ and $g(\lambda)$ are polynomials with coefficients in a given field $\mathbb{F}$, then polynomials $h(\lambda)$ and $k(\lambda)$ with coefficients in $\mathbb{F}$ can be determined such that $f(\lambda) = h(\lambda)g(\lambda) + k(\lambda)$, where $\deg k < \deg g$. In sum, whether $\lambda A_1 + A_2$ and $\lambda B_1 + B_2$ are equivalent can be determined by rational operations.

But what about the *proof* of Weierstrass' elementary divisor theorem (Theorem 5.8)? As Frobenius observed, Weierstrass' proof of 1868, as well as all the subsequent proofs he knew, namely that by Hamburger (1873) and those by Kronecker, Jordan, and Darboux—all given in 1874—required irrational operations. This was because in order to establish the existence of P and Q for two families with (in effect) identical W-series, it was necessary to transform a family $\lambda A_1 + A_2$ with given W-series into a family in a normal or canonical form with the same W-series. The problem was that this canonical form (essentially the Jordan canonical form in the case of Weierstrass' approach) involved the roots of the characteristic polynomial $\varphi(\lambda) = \det(\lambda A_1 + A_2)$ and thus in general involved irrational operations, namely those required to obtain the roots $\varphi(\lambda)$. These notions may be expressed in terms of Dedekind's notion of a field (*Körper*) as Frobenius realized (see below). Thus if $\mathbb{F}$ denotes the field generated by the adjunction of the coefficients of the forms A_i, B_i, $i = 1,2$, to the rational number field $\mathbb{Q}$, one can perform operations within $\mathbb{F}$ to determine whether $\lambda A_1 + A_2$ and $\lambda B_1 + B_2$ have the same W-series. But the known proofs that when the W-series coincide, then P, Q exist such that $P(\lambda A_1 + A_2)Q = \lambda B_1 + B_2$, used the existence of a canonical form with the same elementary divisors, a canonical form that required for its definition elements of $\mathbb{C}$, namely the roots of $\varphi(\lambda) = \det(rA_1 + A_2)$ that were generally not in $\mathbb{F}$.

Frobenius deemed this to be unsatisfactory for the following reason [182, pp. 482–483]. Suppose that the above two families have the same W-series and so are equivalent by virtue of Weierstrass' theory. Thus nonsingular P and Q exist such that (8.23) holds. Now in Weierstrass' theory, the determination of P and Q requires irrational operations, but actually P and Q can be determined by rational operations alone. To see this, note that (8.23) can be written as

$$PA_1 = B_1S \quad \text{and} \quad PA_2 = B_2S, \tag{8.24}$$

where $S = Q^{-1}$. Now (8.24) represents a linear homogeneous system of $2n^2$ equations in the $2n^2$ unknowns p_{ij}, s_{ij} that are the coefficients of P and S. If this system is denoted by $Mx = 0$, then we know—if we assume Weierstrass' theorem established—that the requisite P and Q exist, and so nontrivial solutions x exist, i.e., $\det M = 0$. By the rational operations of Gaussian elimination we can express all solutions to $Mx = 0$ in terms of d parameters $t_1, \ldots, t_d$, where $d = 2n^2 - r$ and $r = \operatorname{rank} M$. Then we need only pick values of the parameters that make the determinants of $(p_{ij}(t_1, \ldots, t_d))$ and $(s_{ij}(t_1, \ldots, t_d))$ both nonzero.[21] Having done that, nonsingular P and S are obtained satisfying (8.24). Furthermore, $Q = S^{-1} = (\det S)^{-1} \operatorname{Adj} S$ can also be determined by rational operations. In other words, by means of rational operations alone, not only can it be decided when two families are equivalent, but also the equivalence transformations P, Q can be determined. Should it not then be possible to give a proof of Weierstrass' elementary divisor theorem (stated using W-series rather than elementary divisors) that involved only rational operations? Once he had solved part (I) of Problem 8.1, Frobenius saw that his normal form theorem (Theorem 8.8) provided a path to such a proof.

8.6.2 *Frobenius' approach and its scope*

Since Frobenius was interested in the rationality problem of Weierstrass' theory, the relevant matrices were of the form $A = rA_1 + A_2$, where $\det A_1 \neq 0$. The coefficients of A are polynomials in r (of degree at most one). Frobenius realized that polynomials in one variable of any degree also possessed the requisite integer-like properties, i.e., that they form what we would now call a Euclidean ring or domain, so that analogous results, such as an analogue of the normal form theorem, could be established for them. As he put it [182, p. 538]:

> The theorems that I have developed for systems A whose elements $a_{\alpha\beta}$ are whole numbers also hold for systems whose elements are polynomial functions of a parameter r. Two bilinear forms $A = \sum a_{\alpha\beta} x_\alpha y_\beta$, $B = \sum b_{\alpha\beta} x_\alpha y_\beta$ are said to be equivalent when they can be transformed into one another by [linear] substitutions whose coefficients are polynomial functions of r and whose determinants are independent of r and nonzero.

Although Frobenius was vague about the domain of the coefficients of the polynomials in the above quotation, it would have been clear to him that in order for the polynomial division algorithm to work, the domain had to be closed under the operations of addition, subtraction, multiplication, and division, i.e., that the coefficient domain had to form what Dedekind had called a *field* in his 1877 essay [113, §15].

---

[21] The set of parameters $(t_1, \ldots, t_d)$ such that these determinants do not vanish is open and dense in $\mathbb{C}^d$, but Frobenius did not indicate rational operations to determine such a choice of $(t_1, \ldots, t_d)$.

For the sake of conciseness and clarity, I will use the anachronistic notation $\mathbb{F}[\lambda]$ for the ring of polynomials in λ (rather than r, which will be reserved for denoting the rank of a matrix) with coefficients in the field $\mathbb{F}$, although it must be kept in mind that no such notation was common at the time. In this notation, what Frobenius had noted in the above quotation about the meaning of equivalence was that the analogue of unimodular substitutions or matrices in ordinary arithmetic are matrices P with coefficients from $\mathbb{F}[\lambda]$ such that $\det P$ is a unit of $\mathbb{F}[\lambda]$, i.e., $\det P \neq 0$ and $\det P \in \mathbb{F}$. For the sake of conciseness I will refer to such matrices P as $\mathbb{F}[\lambda]$-*unimodular*. Thus Frobenius' arithmetic proof of his normal form theorem yielded mutatis mutandis a proof of the following theorem.

Theorem 8.21. *Two matrices A and B with coefficients from $\mathbb{F}[\lambda]$ are equivalent by means of $\mathbb{F}[\lambda]$-unimodular matrices if and only if they have the same invariant factors $e_1(\lambda),\ldots,e_r(\lambda)$, where r denotes their common rank.*

Theorem 8.21 applies in particular when A and B are nonsingular families in the sense of Weierstrass' theory, i.e., of the special form $A = \lambda A_1 + A_2$, $B = \lambda B_1 + B_2$, where the A_i and B_i are square matrices with coefficients from $\mathbb{F}$ and A_1 and B_1 have nonzero determinants. However, the notion of equivalence in Theorem 8.21 is not the same as in Weierstrass' theory, since in the latter, equivalence meant that $PAQ = B$, where P and Q are nonsingular matrices with coefficients in $\mathbb{F}$, not just in $\mathbb{F}[\lambda]$. Frobenius showed that this discrepancy was easy to remedy [182, pp. 539–540]. For future reference, I will state his result as a lemma.

Lemma 8.22. *Let $A = \lambda A_1 + A_2$ and $B = \lambda B_1 + B_2$ denote two families of matrices with entries in $\mathbb{F}$, and $\det B_1 \neq 0$. Then if $\mathbb{F}[\lambda]$-unimodular matrices P,Q exist such that $PAQ = B$, it is possible to determine matrices P_0, Q_0 with entries in $\mathbb{F}$ and nonzero determinants such that $P_0AQ_0 = B$.*

The idea of the proof is quite simple. It depends on the fact that matrices A with entries from $\mathbb{F}[\lambda]$ can be expressed in the form of a "matrix polynomial" $A = A_0 + A_1\lambda + \cdots + A_n\lambda^n$, where the matrices A_i have entries from $\mathbb{F}$ and $A_n \neq 0$, so that $\deg A = n$ by definition. Frobenius realized that the same proof that gives the division algorithm for the polynomials in $\mathbb{F}[\lambda]$ works for polynomials with matrix coefficients.[22] The only difference is that matrix algebra rather than ordinary algebra is used, so the noncommutativity of the former yields two division algorithms for matrix polynomials: if A and B are such that $\deg B \leq \deg A$ and the leading coefficient B_n of B is invertible, then a unique Q with $\deg Q = \deg A - \deg B$ and a

[22] If $a(\lambda) = \sum_{i=0}^{\alpha} a_i\lambda^i, b(\lambda) = \sum_{i=0}^{\beta} b_i\lambda^i \in \mathbb{F}[\lambda]$ have degrees α and $\beta \leq \alpha$, then one wishes to determine the "quotient" $q(\lambda) = \sum_{i=0}^{\alpha-\beta} c_i\lambda^i$ such that the "remainder" $r(\lambda) = a(\lambda) - b(\lambda)q(\lambda)$ has degree less than β. If $a(\lambda) - b(\lambda)q(\lambda)$ is expanded in powers of λ, then the coefficients c_i of $q(\lambda)$ must be determined so that the coefficients of λ^i in the expansion vanish for all $i \geq \beta$. This yields a system of $\alpha - \beta + 1$ equations in the $\alpha - \beta + 1$ unknown coefficients c_i of $q(\lambda)$ that since $b_\beta \neq 0$, yield, successively, unique values of $c_{\alpha-\beta}, c_{\alpha-\beta-1}, \ldots, c_0$. For example, setting the coefficient of λ^α equal to zero yields the equation $a_\alpha - b_\beta c_{\alpha-\beta} = 0$, so $c_{\alpha-\beta} = a_\alpha / b_\beta$.

unique C with $\deg C < \deg B$ exist such that $A = BQ + C$. Likewise, unique Q_1, C_1 exist with the same degree properties as Q, C, respectively, such that $A = Q_1 B + C_1$ [182, I, p. 539]. The determination of Q and Q_1 (and hence also $C = A - BQ$ and $C_1 = A - Q_1 B$) involves operations within $\mathbb{F}$, as in the division algorithm for ordinary polynomials.

Granted the above division algorithms, suppose that A and B are as in the lemma and so have degree 1. Since P, Q have nonzero determinants in $\mathbb{F}$, it follows that

$$R = P^{-1} = (\det P)^{-1} \operatorname{Adj} P \quad \text{and} \quad S = Q^{-1} = (\det Q)^{-1} \operatorname{Adj} Q$$

exist over $\mathbb{F}[\lambda]$ and are also $\mathbb{F}[\lambda]$-unimodular. Thus $PAQ = B$ may be expressed in the form

$$PA = BS, \quad AQ = RB. \tag{8.25}$$

Then by the above division algorithms $P_1, P_0, \ldots, S_1, S_0$ exist such that

$$\begin{aligned} P &= BP_1 + P_0, & Q &= Q_1 B + Q_0, \\ R &= AR_1 + R_0, & S &= S_1 A + S_0, \end{aligned} \tag{8.26}$$

where the "remainders" P_0, Q_0, R_0, S_0 are all of degree zero, i.e., have coefficients in $\mathbb{F}$. Using equations (8.25)–(8.26) and matrix algebra, Frobenius then deduced that P_0 and Q_0 are invertible and that $P_0 A Q_0 = B$ [182, pp. 540–541].

Theorem 8.21 combined with Lemma 8.22 thus yielded his main theorem on elementary divisor theory.

Theorem 8.23. *Let $A = \lambda A_1 + A_2$ and $B = \lambda B_1 + B_2$ be two nonsingular families of matrices with coefficients in a field $\mathbb{F}$. Then A and B are equivalent in the sense that $PAQ = B$ for nonsingular matrices with coefficients in $\mathbb{F}$ if and only if A and B have the same invariant factors.*

The proof of this theorem consists of the above-sketched proof of Lemma 8.22 together with the proof of Frobenius' normal form theorem—albeit with elements of $\mathbb{F}[\lambda]$ playing the role played by elements of $\mathbb{Z}$ in the proof as given. In particular, the desired "rational" demonstration of Weierstrass' theory consists of this proof of Theorem 8.23 when $\mathbb{F} = \mathbb{Q}\left(a_{ij}^{(1)}, a_{ij}^{(2)}, b_{ij}^{(1)}, b_{ij}^{(2)}\right)$, is the field obtained from $\mathbb{Q}$ by adjoining all the coefficients of the matrices A_i, B_i of two given nonsingular families $A = \lambda A_1 + A_2$, $B = \lambda B_1 + B_2$.

In the above exposition of Frobenius' results, I took the liberty of couching them in terms of a field $\mathbb{F}$, although as I said, Frobenius did not. He was certainly acquainted with Dedekind's fairly abstract notion of a field, but such a notion, with attendant notation, was not in general use in 1878 when he wrote his paper. I suspect that Frobenius did not want to scare away his potential readership by introducing an unaccustomed degree of abstractness. For example, in Section 9.2 we will see that when, 3 months after submitting the paper under discussion,

Frobenius submitted a paper with his friend Ludwig Stickelberger, in which they expounded the theory of abstract finite abelian groups, they instructed the reader to regard the symbols $A, B, C, \ldots$ for group elements as representing the congruence classes of integers relatively prime to a fixed integer "in order to be able to portray the abstract development conveniently and intelligibly" [235, p. 546]. The same attitude is apparent in Frobenius' presentation of his rational development of elementary divisor theory. The analogy between ordinary arithmetic and the properties of polynomials in one variable (with no precise mention of the domain of the coefficients) is simply pointed to as yielding the equivalence transformations P, Q of Theorem 8.21 by means of "rational operations from the coefficients of the forms" [182, p. 538]. This is as close as Frobenius comes to mentioning the field $\mathbb{F}$ obtained from $\mathbb{Q}$ by adjoining the coefficients of the forms, namely $\mathbb{F} = \mathbb{Q}\left(a_{ij}^{(1)}, a_{ij}^{(2)}, b_{ij}^{(1)}, b_{ij}^{(2)}\right)$ as discussed above.

Nonetheless, Frobenius did throw out enough remarks to make it clear that he was thinking—concretely if not abstractly—in terms of fields, and not just the field $\mathbb{F} = \mathbb{Q}\left(a_{ij}^{(1)}, a_{ij}^{(2)}, b_{ij}^{(1)}, b_{ij}^{(2)}\right)$. Thus at one point he remarked, regarding the transfer mutatis mutandis of the proof of his normal form theorem (Theorem 8.8) to that of Theorem 8.21 that [182, p. 538]:

> The coefficients of the substitutions that transform two equivalent forms [A and B] into one another are found by rational operations from the coefficients of the forms. Hence if the coefficients of the polynomials $a_{\alpha\beta}$ and $b_{\alpha\beta}$ in r are algebraic numbers of a certain [algebraic number] field, then A can be transformed into B by substitutions whose coefficients belong to the same field.

In other words, if $a_{\alpha\beta}$ and $b_{\alpha\beta}$ are polynomials in $\mathbb{F}[r]$, where $\mathbb{F}$ is an algebraic number field in the sense of Dedekind, then $PAQ = B$, where $P = (p_{\alpha\beta})$, $Q = (q_{\alpha\beta})$ are such that the $p_{\alpha\beta}$ and $q_{\alpha\beta}$ are also in $\mathbb{F}[r]$. (Of course, when $A = rA_1 + A_2$ and $B = rB_1 + B_2$, where the A_i and B_i have coefficients in $\mathbb{F}$ and A_1, B_1 have nonzero determinants, then P and Q can be taken with coefficients in $\mathbb{F}$, but the above quotation occurs before Frobenius had proved Theorem 8.23.)

Further on, Frobenius pointed out that

> The developed principles also remain applicable in the case in which the elements $a_{\alpha\beta}$ of the system A are polynomial functions of r with integer coefficients, and two such functions are not regarded as distinct when their respective coefficients are congruent modulo the prime number p [182, p. 543].

In other words, Theorem 8.23, he is pointing out, also remains valid for matrices with entries from $\mathbb{F}_p[r]$, where $\mathbb{F}_p = \mathbb{Z}/p\mathbb{Z}$, the finite field of the integers modulo a prime p. Indeed, he went on to point out that $\mathbb{F}$ could also consist of the "complex numbers introduced by Galois" [182, p. 543], by which he meant what are now called the Galois fields $\mathrm{GF}(p^\nu)$, $\nu \geq 2$.[23]

---

[23]In his *Traité des substitutions* of 1870 [322, Livre I,§III], Jordan defined these fields as follows. Let $f(x)$ be an irreducible polynomial of degree $\nu > 1$ over the integers mod p. (He proved that such polynomials always exist by proving that $x^{p^\nu} - x$ always has irreducible factors of degree ν over the integers mod p.) Then "nothing prevents us" from introducing the "imaginary symbol" i,

By virtue of these observations, Frobenius made it clear that he realized that Theorems 8.21 and 8.23, although not articulated as abstractly as above, were valid for any field, finite or infinite, that had hitherto arisen in algebraic or number-theoretic investigations. As we shall see in Section 16.2, Frobenius' results were well known, and they initiated the development of linear algebra over an arbitrary field. His "rational" proofs of Theorems 8.21 and 8.23 also initiated another trend. Because they were based on the analogy with his arithmetic proofs of his reduction theorem, they were essentially modern algebraic proofs and, as such, were completely devoid of all the formal transformations based on determinant-theoretic constructs that characterized Weierstrass' development of elementary divisor theory. Frobenius was an admirer and master of the theory of determinants. Indeed, as we shall see in Section 16.1.1, when an error was discovered in Weierstrass' theory as it applied to symmetric matrices, it was Frobenius who succeeded in filling in the gap by a subtle determinant-theoretic argument. Nonetheless, Frobenius' rational theory of elementary divisors had the inadvertent effect of greatly diminishing the role of determinant-based formulas in linear algebra. This will become even clearer in Sections 16.2–16.3, where the role of Frobenius' work on the emergence of the module-theoretic approach to elementary divisor theory is considered. Furthermore, Frobenius also showed that his above-mentioned subtle determinant-theoretic argument could be avoided by solving the problem in Weierstrass' theory by means of a relatively short and simple matrix-algebraic argument (Section 16.1).

8.6.3 A rational canonical form

"The question arises," Frobenius remarked upon proving Theorem 8.23, "as to whether forms of the first degree exist that possess prescribed elementary divisors" [182, p. 541]. Recall that Frobenius' elementary divisors of a nonsingular family $A = \lambda A_1 + A_2$ are what I have termed the invariant factors $e_i(\lambda) = d_i(\lambda)/d_{i-1}(\lambda) \in \mathbb{F}[\lambda]$, where $d_i(\lambda)$ is the polynomial greatest common divisor of all $i \times i$ minors of $\lambda A_1 + A_2$. If A is $n \times n$, let

$$e_n(\lambda) = [\varphi_1(\lambda)]^{\varepsilon_1} \cdots [\varphi_m(\lambda)]^{\varepsilon_m}$$

---

which is subject to the condition that $f(i) \equiv 0 \pmod{p}$. (This i is not to be confused with $\sqrt{-1}$.) Then for any polynomial $F(x)$ with integer coefficients, $F(x) \equiv f(x)q(x) + r(x) \pmod{p}$, where $\deg r(x) < \nu$. Since $f(i) \equiv 0 \pmod{p}$, we have $F(i) \equiv r(i) \pmod{p}$. The remainder expressions $r(i)$ are called "complex integers," and they can be partitioned into p^ν congruence classes mod p, which are the elements of GF(p^ν). (In effect Jordan's complex integers are the elements of $\mathbb{F}_p[x]/(f(x))$ with i taken as the equivalence class of x.) Jordan also showed that $i, i^p, \ldots, i^{p^{\nu-1}}$ are the ν (distinct) roots of $f(x) \equiv 0 \pmod{p}$. Of course, i^{p^k} is congruent mod p to a complex integer, i.e., is an element of GF(p^ν).

denote the factorization of $e_n(\lambda)$ into distinct irreducible factors $[\varphi_i(\lambda)]$ over $\mathbb{F}$. Since $e_{n-1}(\lambda)$ divides $e_n(\lambda)$, it follows that $e_{n-1}(\lambda) = [\varphi_1(\lambda)]^{\varepsilon_1'} \cdots [\varphi_m(\lambda)]^{\varepsilon_m'}$, where $\varepsilon_i' \leq \varepsilon_i$ for all i. Likewise, $e_{n-2}(\lambda) = [\varphi_1(\lambda)]^{\varepsilon_1''} \cdots [\varphi_m(\lambda)]^{\varepsilon_m''}$ with $\varepsilon_i'' \leq \varepsilon_i'$ for all i, and so on. By analogy with Weierstrass' terminology, the polynomial powers $[\varphi_i(\lambda)]^{\varepsilon_i}, [\varphi_i(\lambda)]^{\varepsilon_i'}, [\varphi_i(\lambda)]^{\varepsilon_i''}, \ldots$ corresponding to nonzero powers of the $\varphi_i(\lambda)$ are the elementary divisors of $\lambda A_1 + A_2$ with respect to $\mathbb{F}$. (Frobenius called them the *simple* elementary divisors.) Since invariant factors are determined only up to units, i.e., up to multiplication by a nonzero element of $\mathbb{F}$, it can be assumed without loss of generality that the $e_i(\lambda)$ have leading coefficients equal to 1, i.e., are monic polynomials, and so the irreducible factors $\varphi_i(\lambda)$ can also be taken as monic polynomials.

In response to his question, Frobenius observed that if $\varphi(\lambda) = \lambda^\alpha + a_1\lambda^{\alpha-1} + \cdots + a_\alpha$ is an irreducible polynomial over $\mathbb{F}$, then $\varphi(\lambda) = \det F[\varphi]$, where

$$F[\varphi] = \begin{pmatrix} \lambda + a_1 & -1 & 0 & 0 & \cdots & 0 \\ a_2 & \lambda & -1 & 0 & \cdots & 0 \\ a_3 & 0 & \lambda & -1 & \cdots & 0 \\ a_4 & 0 & 0 & \lambda & \cdots & 0 \\ \cdot & \cdot & \cdot & \cdot & \cdots & -1 \\ a_\alpha & 0 & 0 & 0 & \cdots & \lambda \end{pmatrix}, \tag{8.27}$$

as is easily seen by cofactor expansion down the first column.[24] The $\alpha \times \alpha$ matrix $F[\varphi]$ is the matrix of the nonsingular family $\lambda I + A$, where A is $F[\varphi]$ with $\lambda = 0$, i.e., the matrix with $a_1, \ldots, a_\alpha$ down the first column and -1's along the superdiagonal.[25] Also $\varphi(r)$ is the sole nontrivial invariant factor of this family.[26] Likewise, $F[\varphi^\nu]$, $\nu > 1$, corresponds to the family $\lambda I + B$, where if $[\varphi(\lambda)]^\nu = \lambda^{\alpha\nu} + b_1\lambda^{\alpha\nu-1} + \cdots + b_{\alpha\nu}$, then B is the $\alpha\nu \times \alpha\nu$ matrix with $b_1, \ldots, b_{\alpha\nu}$ down its first column and -1's on the superdiagonal. It follows that $e_{\alpha\nu}(\lambda) = [\varphi(\lambda)]^\nu$ and $e_i(\lambda) = 1$ for all $i < \alpha\nu$.

Frobenius used the symbol $+$ for what we could now call a direct sum. For example, $F[\varphi^\nu] + F[\varphi]$ denotes the matrix

$$\begin{pmatrix} F[\varphi^\nu] & 0 \\ 0 & F[\varphi] \end{pmatrix},$$

---

[24]The notation $F[\varphi]$ is mine, not Frobenius'. The matrix $F[\varphi]$ is related to what A. Loewy later called the companion matrix (*Begleitmatrix*) of φ, as we will see in Section 16.3.1. I will refer to $F[\varphi]$ as the *Frobenius companion matrix of* φ. As I have already explained, Frobenius himself never spoke of fields $\mathbb{F}$ in general. In introducing $F[\varphi]$, he assumed for the sake of concreteness that the coefficients of φ are algebraic numbers "in a certain field" [182, p. 542].

[25]It should be noted than every nonsingular family $\lambda A_1 + A_2$ is equivalent to one in the form $\lambda I + B$, since $A_1^{-1}(\lambda A_1 + A_2)I = \lambda I + B$, $B = A_1^{-1}A_2$; and so in discussing invariant factors, only families $\lambda I + B$ need be considered.

[26]The $(\alpha, 1)$ minor of $F[\varphi]$ is $(-1)^\alpha$, making $d_{\alpha-1}(r) = 1$ and $e_\alpha(r) = d_\alpha(r) = \varphi(r)$, while $e_i(\lambda) = 1$, for all $i < \alpha$.

which is $N \times N$ with $N = \alpha\nu + \alpha$. This matrix has two nontrivial invariant factors, $e_N = [\varphi(\lambda)]^\nu$ and $e_{N-1} = \varphi(\lambda)$. Since φ is irreducible, φ^ν and φ are also the elementary divisors of $F[\varphi^\nu] + F[\varphi]$. More generally, if

$$e_i = [\varphi_1(\lambda)]^{\varepsilon_1^{(i)}} \cdots [\varphi_m(\lambda)]^{\varepsilon_m^{(i)}}, \quad i = 1,\ldots,k,$$

where the φ_i are irreducible over $\mathbb{F}$ and $0 \le \varepsilon_j^{(i-1)} \le \varepsilon_j^{(i)}$, then the nonsingular family corresponding to

$$F\left[\varphi_1^{\varepsilon_1^{(k)}}\right] + \cdots + F\left[\varphi_m^{\varepsilon_m^{(k)}}\right] + \cdots + F\left[\varphi_1^{\varepsilon_1^{(1)}}\right] + \cdots + F\left[\varphi_m^{\varepsilon_1^{(1)}}\right] \tag{8.28}$$

has precisely the $e_i(\lambda)$ as nontrivial invariant factors, or equivalently, (8.28) has precisely the set $[\varphi_j(\lambda)]^{\varepsilon_j^{(i)}}$, $i = 1,\ldots,k$, $j = 1,\ldots,m$, as its elementary divisors. Of course, as Frobenius also pointed out [182, p. 541], in the context of Weierstrass' theory (where polynomials may have any complex coefficients) the elementary divisors take the form $[\varphi_j(\lambda)]^{\varepsilon_j^{(i)}} = (\lambda - a_j)^{\varepsilon_j^{(i)}}$, and the Jordan canonical form may be used in lieu of Frobenius' rational canonical form (8.28).

Frobenius' rational canonical form (8.28) was thus a replacement for the canonical form of Weierstrass and Jordan. Like Weierstrass, Frobenius proved his rational elementary divisor theorem (Theorem 8.23) by proving that a given nonsingular family could be transformed into a family corresponding to a canonical form, but it was the Smith–Frobenius normal form with the invariant factors along the diagonal, not the rational canonical form. Frobenius introduced the latter simply to show that any set of invariant factors corresponded to an actual family, but a consequence of his Theorem 8.23 was that any nonsingular family could be transformed by matrices P and Q into the rational canonical form (8.28) with the same elementary divisors. As we shall see in Section 16.3, other mathematicians looked for a proof of Frobenius' Theorem 8.23 that was based on a line of reasoning that showed directly how a given family could be transformed into its rational canonical form, and this line of thought eventually led to what became the approach to elementary divisor theory via the fundamental theorem of finitely generated modules.

One final observation is in order in preparation for Section 16.3. A year earlier, in his paper [181] on matrix algebra and the Cayley–Hermite problem (Chapter 7), Frobenius had introduced the now-familiar notion of similar square matrices and observed that A and B are similar ($S^{-1}AS = B$ for some nonsingular S) if and only if $S^{-1}(\lambda I - A)S = \lambda I - B$ and so if and only if the characteristic polynomials of A and B have the same elementary divisors [181, p. 363]. In the context of Theorem 8.23, this implies the following.

Theorem 8.24. *If A and B are square matrices with entries from a given field $\mathbb{F}$, then they are similar if and only if their characteristic polynomials have the same invariant factors and hence if and only if A and B have the same rational canonical form (8.28) (with $\lambda = 0$).*

This is the consequence of Frobenius' work that became the focal point of some later renditions of elementary divisor theory. Of course, as I already explained, in 1878 when he submitted [182] Frobenius himself would not have articulated Theorem 8.24 so explicitly in terms of an arbitrary field $\mathbb{F}$, although he certainly realized that Theorem 8.24 held for any specific field that had occurred in mathematical research, such as the algebraic number fields of Kummer and Dedekind and the finite fields $\mathbb{F}_p$ and $\mathrm{GF}(p^{\nu})$ of Galois theory.

As we shall see in Section 16.3, Frobenius' rational approach to Weierstrass' theory had a profound effect on the way that elementary divisor theory was conceived and developed in the twentieth century.

Chapter 9
Arithmetic Investigations: Groups

Frobenius' interest in arithmetic problems during the late 1870s was not limited to bilinear forms. The groundbreaking work of Gauss on the composition of binary quadratic forms had brought with it a line of thinking that would now be characterized as group-theoretic. This same line of thought resurfaced in Kummer's revolutionary work on his theory of ideal numbers and prompted Ernst Schering to develop it into what would now be interpreted as the existence part of the fundamental theorem of finite abelian groups, already expressed ambiguously by Schering so as to encompass the finite abelian groups that were implicit in Gauss' theory of composition of binary quadratic forms as well as those in Kummer's theory of ideal numbers (ideal class groups). Soon thereafter, both Kronecker and Dedekind expressed Schering's result explicitly in abstract terms, with Dedekind expressly making the connection with Galois' notion of a group.

Frobenius, who had great respect for the work of both Kronecker and Dedekind, accepted the appropriateness of the abstract notion of a group, and within that context considered, in collaboration with his colleague Ludwig Stickelberger, some uniqueness problems related to the abstract version of Schering's theorem. The result was a typical Frobenian work: a clear, but lengthy, systematic "treatise" on the theory of finite abelian groups with the fundamental theorem as its centerpiece. It represented the first such treatise and was Frobenius' first publication on the theory of groups since his youthful 1871 complex-analytic exposition of Galois groups (Section 1.2). Frobenius' paper with Stickelberger is the subject of Section 9.2. It is preceded by a section devoted to the work of Gauss, Kummer, Schering, Kronecker, and Dedekind that formed the background and motivation for the Frobenius–Stickelberger paper. The section is lengthy, perhaps inordinately so, but I wanted to convey to the reader the exciting and novel developments within the theory of numbers that Frobenius absorbed as a student and that led to his interest in the theory of groups, which he always considered as a part of arithmetic. These remarkable developments form part of the context of Frobenius' mathematics. In particular, they form the background to the evolution of the notion of a character from Gauss

T. Hawkins, *The Mathematics of Frobenius in Context*, Sources and Studies in the History of Mathematics and Physical Sciences, DOI 10.1007/978-1-4614-6333-7_9,

to Dedekind (Section 12.2) and so provide the context within which Frobenius was led to generalize Dedekind's characters.

Unlike Kronecker, Dedekind and Frobenius expressly viewed the mathematics related to Schering's theorem as a branch of the theory of groups, a term introduced by Galois, the other main branch being the Galois theory of groups. A work by Kronecker involving analytic or Dirichlet densities led Frobenius to reinterpret and extend Kronecker's work by regarding it within the context of the Galois theory of groups. This is the subject of Section 9.3, and it shows that Frobenius had mastered both Galois theory and Dedekind's theory of ideals. It was in this work that several density theorems due to Frobenius originated, as well as a conjectured density theorem that was eventually proved by Chebotarev. It was also by virtue of this work that Frobenius began corresponding with Dedekind, a correspondence that continued sporadically and eventually led to an exchange that started Frobenius down the path to his creation of the theory of group characters and representations (Chapters 12 and 13). In his work on densities, Frobenius was confronted for the first time with the need to develop the means for dealing with noncommutative groups. In Section 9.4, we will see that during this period he developed several group-theoretic techniques that became a staple in his mathematical tool box as his interest in noncommutative groups increased and began to be pursued for its own sake, i.e., independently of connections with problems involving Galois theory.

9.1 Origins of the Fundamental Theorem of Finite Abelian Groups

9.1.1 Gauss

What we would now characterize as finite abelian groups arose in arithmetic investigations as equivalence classes of mathematical objects with a commutative multiplication defined on the classes. It was in Gauss' influential *Disquisitiones Arithmeticae* of 1801 [244] that we find the first two examples.

The first such example arose from the study of congruence relations, which began in the eighteenth century and received a masterful treatment by Gauss in the first four sections of his *Disquisitiones*. For example, if congruence is with respect to a prime number p, so that (in Gauss' notation) $a \equiv b$ modulo p when $a - b$ is divisible by p, the corresponding equivalence relation is $a \sim b$ when $a \equiv b$ mod p. Thus the equivalence class of a consists of all "residues of a modulo p," namely all integers of the form $a + kp$, k any positive or negative integer or zero. In the theory of congruences, Gauss did not explicitly introduce the notion of equivalence classes, choosing to speak, as he could, in more familiar terms. In later sections of *Disquisitiones Arithmeticae* dealing with binary quadratic forms, their equivalence and their composition, however, equivalence classes and their multiplication were of necessity introduced, as we shall see. In studying the equivalence classes of binary quadratic forms, Gauss frequently pointed to the analogy with results from that part of the theory of congruences dealing with residues of powers [244, §III], which concerns the residues of powers of $1, 2, \ldots, p-1$ modulo p. In the light of the treatment of binary forms, readers of *Disquisitiones Arithmeticae* could see that the residues of $1, 2, \ldots, p-1$ modulo p were the equivalence classes with multiplication inherited from the ordinary multiplication of the integers. And of course, since $ab \equiv 0 \bmod p$ implies that $a \equiv 0 \bmod p$ or $b \equiv 0 \bmod p$, it was clear that if a and b were any two of the residue classes $1, 2, \ldots, p-1$, then so was ab. From the later viewpoint of the abstract theory of groups, the residue classes $1, 2, \ldots, p-1$, i.e., the equivalence classes modulo p represented by $1, 2, \ldots, p-1$, form a finite group of commuting elements, which I will denote in what follows with the modern notation $(\mathbb{Z}/p\mathbb{Z})^{\times}$.

In his study of residues of powers [244, §III], Gauss proved many propositions that are now regarded as staples of group theory. For example, he showed that for any $a \not\equiv 0 \bmod p$ there are powers $e > 0$ such that $a^e \equiv 1 \bmod p$ [244, Art. 45]. The smallest of these corresponds to what is now called the *order* of the residue class of a in $(\mathbb{Z}/p\mathbb{Z})^{\times}$. Since Gauss introduced no term for "order," I will use this term to describe Gauss' work more succinctly. If t is the order of a, then the system of powers a^0, a, a^2, a^{t-1} is called the *period* of a [244, Art. 46]; in $(\mathbb{Z}/p\mathbb{Z})^{\times}$ it corresponds to the cyclic subgroup generated by the residue class of a. Gauss proved the theorem that the order of a always divides $p-1$ [244, Art. 49], which corresponds nowadays to the fact that the order of an element of $(\mathbb{Z}/p\mathbb{Z})^{\times}$ always divides the order of $(\mathbb{Z}/p\mathbb{Z})^{\times}$.[1] An immediate consequence of Gauss' theorem was one of Fermat's: $a^{p-1} \equiv 1 \bmod p$ for any $a \not\equiv 0 \bmod p$ [244, Art. 50]. Gauss also gave a rigorous demonstration of a result believed to be true by eighteenth-century mathematicians, namely that for any p there is always an integer with order $p-1$ [244, Art. 55]. Such an integer he called, following Euler, a primitive root. In the group-theoretic language of $(\mathbb{Z}/p\mathbb{Z})^{\times}$, the corresponding result is that $\mathbb{Z}_p^{\times}$ is cyclic for any prime p. Gauss used the existence of a primitive root a to index the $p-1$ residue classes by associating to each such class the power $0 \leq g \leq p-1$ to which a must be raised to obtain a number of that class. Since the index of ab is congruent mod $p-1$ to the sum of the indices of a and b, this logarithmic-type property made indexing residue classes a useful way to solve various problems, as Gauss showed.

Gauss also considered briefly the case of a nonprime modulus m and, by analogy with the case $m = p$, the residue classes of integers relatively prime to m [244, Art. 82ff.]. The number of residue classes is of course equal to the number of positive integers that are less than m and relatively prime to it. Gauss denoted this

---

[1]Gauss' proof was based on partitioning $(\mathbb{Z}/p\mathbb{Z})^{\times}$ into cosets modulo the cyclic subgroup generated by a, an idea already introduced by Lagrange in the early 1770s that ultimately led to the proposition that the order of a subgroup always divides the order of the containing group.

number, already studied by Euler [599, p. 192], by ϕm; I will write it as $\phi(m)$.[2] The abelian group implicit in these considerations will be denoted by $(\mathbb{Z}/m\mathbb{Z})^{\times}$. Gauss pointed out that his earlier proof that the order of any a divides $p-1=\phi(p)$ shows, mutatis mutandis, that the order of any a relatively prime to m divides $\phi(m)$. However "the most elegant property," namely the existence of primitive roots, was another matter. Gauss showed they exist when $m = p^n$, p an odd prime [244, Art. 84ff.], but not when $m = 2^n$, $n > 1$. For m that are not powers of a prime, he noted in passing that primitive roots exist only when $m = 2p^n$, p an odd prime. Stated in group-theoretic terms, Gauss' claim was that $(\mathbb{Z}/m\mathbb{Z})^{\times}$ is cyclic only when m is a prime, a power of an odd prime, or twice an odd prime power. The general noncyclic nature of $(\mathbb{Z}/m\mathbb{Z})^{\times}$ meant that there was no index theory in these cases for facilitating the solution of congruence equations, but Gauss explained why this was not a serious deficiency thanks to the Chinese remainder theorem [244, Art. 92].

Within the context of a different "equivalence class group," however, Gauss showed more interest in the question of a suitable indexing. This occurred in his profound theory of the composition of forms, a theory that "[t]hus far no one has considered" [244, Art. 234ff.], although it was inspired by his predecessors. For example, Lagrange in 1783 had proved some conjectures of Fermat and Euler by observing, inter alia, that the product of two integers of the form $2x^2+2xy+3y^2$ is an integer of the form x^2+5y^2, all x's and y's being integers [551, pp. 17ff.]. This followed from the algebraic identity

$$(2x^2+2xy+3y^2)(2x'^2+2x'y'+3y'^2) = X^2+5Y^2,$$

where

$$X = 2xx'+xy'+yx'-2yy', \quad Y = xy'+yx'+yy'. \tag{9.1}$$

Other such identities were known, and in fact, Legendre showed that

$$(ax^2+2bxy+cy^2)(a'x'^2+2b'x'y'+c'y'^2) = (AX^2+2BY^2+CY^2),$$

where $a,b,c,a',b'c',A,B,C$ are all integers with the first six subject to certain conditions and, analogous to (9.1), X and Y are bilinear functions of x,y,x',y'. It is uncertain whether Gauss was familiar with Legendre's results when he devised his own rigorously developed theory of the composition of binary quadratic forms.[3]

Gauss' theory began with considerations analogous to those of his predecessors. A form $F = AX^2+2BXY+CY^2$ with integer coefficients A,B,C is said to be transformable into the product of two forms with integer coefficients

[2]In Article 38, Gauss had proved $\phi(m) = m\frac{p-1}{p}\frac{q-1}{q}\cdots$, where $p,q,\ldots$ are the distinct prime divisors of m, a formula first stated by Euler.

[3]Regarding Legendre's work and whether it could have been known to Gauss see [599, pp. 332–334].

$$f = ax^2 + 2bxy + cy^2 \quad \text{and} \quad f' = a'x'^2 + 2b'x'y' + c'y'^2 \tag{9.2}$$

provided integers $p^{(i)}, q^{(i)}$, $i = 1, \ldots, 4$, exist such that the substitutions

$$\begin{aligned} X &= p^{(i)}xx' + p^{(i)}xy'' + p^{(i)}yx' + p^{(i)}yy', \\ Y &= q^{(i)}xx' + q^{(i)}xy'' + q^{(i)}yx' + q^{(i)}yy', \end{aligned}$$

make $F(X,Y) = f(x,y) \cdot f'(x',y')$. Gauss called F a composite of f and f' if not all 2×2 minors of the matrix

$$\begin{pmatrix} p^{(0)} & p^{(1)} & p^{(2)} & p^{(3)} \\ q^{(0)} & q^{(1)} & q^{(2)} & q^{(3)} \end{pmatrix} \tag{9.3}$$

vanish [244, Art. 235], i.e., the matrix has full rank 2. It is clear that the ordering of f and f' is irrelevant to the composite F. Gauss spoke of F as *a* composite of f and f' rather than *the* composite because he realized that other forms can be composites of f and f'. He made a careful study of the relations among what he called the determinants of f, f', F, namely $d = b^2 - ac, d' = b'^2 - a'c', D = B^2 - AC$.[4] Based on these complicated relations, he was able to show, among other things, that when f and f' have the same determinant (so $d = d'$), then a composite form F , and its determinant D is by necessity the same as that of f and f' [244, Art. 236].

In terms of rigor and generality, Gauss' above results surpassed those of his predecessors, but these results were but a prelude to the real substance of his theory of composition. Realizing that even for forms f, f' of the same determinant D, a composite is not unique, he considered the equivalence classes formed by the forms of a fixed determinant D under proper equivalence.[5] He showed that the number of such classes is finite [244, Art. 223] and that two composites F and G of f and f' are always properly equivalent and, more generally, that if the pairs F, G, f, g, f', g' are respectively equivalent, then F is a composite of f and f' if and only if G is a composite of g and g' [244, Art. 238]. Thus one can speak of *the* composite class of two equivalence classes. Moreover, as Gauss showed, the resulting operation of composition is associative [244, Art. 240].

Following Gauss, I will denote the forms f, f' in (9.2) by $f = (a,b,c)$ and $f' = (a',b',c')$. By a masterly analysis of the many relations that must hold when $F = (A,B,C)$ is a composite of $f = (a,b,c)$ and $f' = (a',b',c')$, Gauss showed that when the representative forms of two equivalence classes are chosen so that the coefficients a,b,c,a',b',c' satisfy certain simple conditions, which is always possible, then the coefficients A,B,C of a form F in the composite class can be

---

[4]Thus Gauss' determinants are the negatives of the determinants (in the present sense) of the matrices associated to the forms. See Section 4.3 for the important role of Gauss' work in the development of the theory of determinants.

[5]Recall that the two forms in (9.2) are properly equivalent when f can be transformed into f' by a linear transformation of variables $(x,y) \to (x',y')$ with integer coefficients and determinant $+1$.

determined directly from a, b, c, a', b', c'.[6] Using this procedure, he showed that for forms of determinant D, the class $\mathcal{E}$ containing the form $e = (1, 0, -D)$ has the property that if $\mathcal{C}$ is any class, then the composite of the two classes is just $\mathcal{C}$ [244, Art. 243]. Thus the class $\mathcal{E}$, which Gauss named the *principal class*, plays the role of the identity element. Finally, corresponding to each class $\mathcal{C}$ is a *reciprocal class* $\mathcal{C}'$ such that their composite is the principal class $\mathcal{E}$. In other words, Gauss had shown that the equivalence classes of forms, under the operation of composition, form what would later be called a finite abelian group. Gauss denoted the composite of two classes $\mathcal{C}, \mathcal{C}'$ by $\mathcal{C} + \mathcal{C}'$, but I will use the multiplicative notation $\mathcal{C}\mathcal{C}'$ in what follows for consistency with later developments.

Two subgroups of the above group particularly interested Gauss [244, Art. 227]. They involved what he called *properly primitive forms*, meaning forms $f = (a, b, c)$ for which $\gcd(a, 2b, c) = 1$. If f is properly primitive, then so is every form equivalent to f, and so the entire equivalence class of f contains properly primitive forms [244, Art. 226]. Thus we may speak of properly primitive form-classes. Also, the composite of two properly primitive forms is properly primitive, and so if $\mathfrak{F}_1(D)$ denotes the totality of all properly primitive classes for determinant D, it is closed under composition and so forms a finite abelian group. Gauss discovered that the classes in $\mathfrak{F}_1(D)$ could be classified by virtue of certain characteristics of the odd numbers they could represent. That is, he determined a few characteristics, which he called characters, and put form-classes with the same characters into the same genus. (Here it suffices to leave the discussion on this vague level; these matters will be discussed in more specific terms in Section 12.2.) The genus containing the class $\mathcal{E}$ of $e = (1, 0, -D)$ he termed the principal genus. I will denote the principal genus of the properly primitive form-classes by $\mathfrak{F}_0(D)$.

Gauss showed that $\mathfrak{F}_0(D)$ had the following special property with respect to class composition: if $\mathcal{A}$ and $\mathcal{B}$ are any two properly primitive classes of determinant D belonging to the same genus, then the composite class $\mathcal{A}\mathcal{B}$ always belongs to $\mathfrak{F}_0(D)$ [244, Art. 247]. This means in particular that the composite of any two classes in $\mathfrak{F}_0(D)$ is again in $\mathfrak{F}_0(D)$, i.e., in more familiar terms, $\mathfrak{F}_0(D)$ is a subgroup of $\mathfrak{F}_1(D)$. As we shall see, Gauss was aware of an analogy between $\mathfrak{F}_0(D)$ and the residue classes of integers modulo a prime p, viz., $(\mathbb{Z}/p\mathbb{Z})^\times$.[7] The class $\mathcal{E} \in \mathfrak{F}_0(D)$ is analogous to $1 \in (\mathbb{Z}/p\mathbb{Z})^\times$ and $\mathfrak{F}_0(D)$ is closed under composition, just as $(\mathbb{Z}/p\mathbb{Z})^\times$ is closed under multiplication of residue classes. The same could be said of $\mathfrak{F}_1(D)$, as Gauss realized. Gauss was particularly interested in the class number $h_1 = |\mathfrak{F}_1(D)|$. He knew that $h_1 = 2^m h_0$, where $h_0 = |\mathfrak{F}_0(D)|$ and $m \geq 0$ can be determined from the known value of D [244, Art. 301]. This reduced the problem of determining h_1 as a function of D to the same problem for h_0, a problem that intrigued Gauss, although he could not solve it.

---

[6] See [244, Art. 242–243]. A variation of Gauss' procedure, which is easier to apply, was later developed by Dirichlet and Dedekind (§146 of [137–139].

[7] Actually, the analogy with $(\mathbb{Z}/m\mathbb{Z})^\times$ would be closer, since as we shall see, $\mathfrak{F}_0(D)$ need not be cyclic; but Gauss focused on $(\mathbb{Z}/p\mathbb{Z})^\times$ as he had in Section III.

The concluding sections of his theory of form composition are devoted to matters related to this problem [244, Art. 305–306]. He began by proving the analogue of a theorem already proved for $(\mathbb{Z}/p\mathbb{Z})^\times$, namely that (in present-day terms) the order of $\mathcal{C} \in \mathfrak{F}_0(D)$ divides the order of $\mathfrak{F}_0(D)$. He did this to make the following point:

> The demonstration of the preceding theorem is quite analogous to the demonstrations in articles 45, 49 and, in fact, the theory of the ... [exponentiation][8] ... of classes has a great affinity in every way with the subject treated in Section III. But the limits of this work do not permit us to pursue this theory though it is worthy of greater development; we will add only a few observations here, suppress those demonstrations that require too much detail, and reserve a more complete discussion to another occasion [244, Art. 306].

Gauss then proceeded to list observations, denoted by I–X, pertaining to $\mathfrak{F}_0(D)$. The last posed the formidable problem of determining the "general connection" between the class number $h_0 = |\mathfrak{F}_0(D)|$ and D. (As we shall see in Section 12.2, one of Dirichlet's great achievements was to solve this problem.) The previous nine observations had to do with the properties of $\mathfrak{F}_0(D)$ as a group, i.e., as the analogue of $(\mathbb{Z}/p\mathbb{Z})^\times$, and most were straightforward. The most interesting are V–IX, which have to do with the question whether $\mathfrak{F}_0(D)$ is cyclic. Thus V observes that if $\mathfrak{F}_0(D)$ is cyclic of order n, then "the best way to handle" $\mathfrak{F}_0(D)$ is to take a class $\mathcal{C}$ of order n as base and to index each element $\mathcal{K} \in \mathfrak{F}_0(D)$ by g, where $\mathcal{K} = \mathcal{C}^g$, $0 \leq g \leq n-1$. Then the index of the composite of two classes will just be the sum of the indices of the two classes modulo n. In VI, Gauss observed that

> Although the analogy with Section III [Gauss is thinking here of $(\mathbb{Z}/p\mathbb{Z})^\times$ and not the briefly considered $(\mathbb{Z}/m\mathbb{Z})^\times$] and an induction on more than 200 negative determinants and many more positive nonquadratic determinants would seem to give the highest probability that the supposition [of V] is true for *all* determinants, nevertheless such a conclusion would be false, and would be disproved by the continuation of the table of classifications. For the sake of brevity we will call *regular* those determinants for which the whole principal genus [$\mathfrak{F}_0(D)$] can be included in one period, and *irregular* those for which this is not true. We can illustrate with only a few observations this argument which depends on the most profound mysteries of higher Arithmetic and involves the most difficult investigation.[9]

Thus D is regular if and only if $\mathfrak{F}_0(D)$ is cyclic.

In VII, Gauss indicated a procedure by which to determine a class in $\mathfrak{F}_0(D)$ of maximal order. As a preliminary to VIII–IX he then defined the *exponent of irregularity* of D (or of $\mathfrak{F}_0(D)$) as the order of $\mathfrak{F}_0(D)$ divided by the maximal order of a class in $\mathfrak{F}_0(D)$. Thus when D is regular, i.e., when $\mathfrak{F}_0(D)$ is cyclic, the exponent of irregularity is 1; otherwise, it is greater than 1. In VIII, he then made two observations: (A) Consider the number of classes $\mathcal{C} \in \mathfrak{F}_0(D)$ such that $\mathcal{C}^2 = \mathcal{E}$. If this number is greater than two, then the exponent of irregularity must be even, which

---

[8]The text has "multiplication" where I have written "exponentiation" because Gauss used the additive notation $\mathcal{C} + \mathcal{K}$ for the composition of classes. Thus the multiplication of a class refers to expressions such as the m-fold sum of $\mathcal{C}$, i.e., $m\mathcal{C}$, which means $\mathcal{C}^m$ when multiplicative notation is used to denote the operation of composition.

[9]Clark's translation has "transcendental Arithmetic," but "higher Arithmetic" is closer to the original Latin.

means that D is irregular. (In other words, if $\mathfrak{F}_0(D)$ has more than two idempotent elements, it cannot be cyclic.) (B) If the exponent of irregularity is divisible by a prime p, then n [the order of $\mathfrak{F}_0(D)$] will be divisible by p^2. It follows that if n has no square divisors, D must be regular. (That is, $\mathfrak{F}_0(D)$ is cyclic whenever its order has no square divisor.) Finally, in IX, Gauss addressed briefly the issue of how best to deal with $\mathfrak{F}_0(D)$ when it is not cyclic:

> For brevity's sake we cannot treat here the most useful disposition of the system of classes contained in a principal genus with an irregular determinant [so $\mathfrak{F}_0(D)$ is not cyclic]; we observe only that since one base is not sufficient, we must take two or more classes and from their ... [exponentiation][10] ... and composition produce all the rest. Thus we will have *double* or *multiple indices* that will perform the same task that simple ones do for regular determinants. But we will treat this subject more fully another time.

Gauss never did come back in print to justify the observations in I–IX, but he did leave behind a fragmentary manuscript, probably written in 1801, that relates to the unproved observations in VIII–IX. It was written in French and published posthumously in the second volume (1863) of his collected works [247]. The manuscript contains two theorems. The first concerns the totality of properly primitive form-classes of a fixed determinant D, namely $\mathfrak{F}_1(D)$. As we saw, Gauss had shown that $\mathfrak{F}_1(D)$ has all the properties of a finite abelian group, and, of course $\mathfrak{F}_0(D)$ is a subgroup. His theorem may be stated as follows.

Theorem 9.1. *For a fixed prime p, let $\mathfrak{G}_p$ denote the collection of all classes $\mathcal{C} \in \mathfrak{F}_1(D)$ with the property that $\mathcal{C}^m = \mathcal{E}$, where m is some power of p. Then the number of classes in $\mathfrak{G}_p$ is a power p^f, where $f \geq 0$.*

Clearly $\mathfrak{G}_p$ is closed under class composition. Nowadays, it is sometimes referred to as the p-primary component.

Gauss' manuscript contains a complete, correct proof of this theorem. The idea of the proof contains a procedure that is worth mentioning. If $\mathcal{A} \neq \mathcal{E}$ is in $\mathfrak{G}_p$, consider the cyclic subgroup generated by $\mathcal{A}$, which I will denote by $(\mathcal{A})$. Let p^α be the order of $(\mathcal{A})$. If $(\mathcal{A}) = \mathfrak{G}_p$, we are done. If not, let $\mathcal{B} \in \mathfrak{H}_p \sim (\mathcal{A})$, and let b denote the smallest positive power such that $\mathcal{B}^b \in (\mathcal{A})$. Then it is not difficult to show that b is a power of p, say $b = p^\beta$. Also, all classes $\mathcal{A}^j\mathcal{B}^k$ with $0 \leq j < p^\alpha$ and $0 \leq k < p^\beta$ are distinct (as Gauss proves), so there are $p^\alpha p^\beta = p^{\alpha+\beta}$ classes. If there are still further classes $\mathcal{C}$ not among these, let c be the smallest positive integer such that $\mathcal{C}^c = \mathcal{A}^j\mathcal{B}^k$ with j,k as above. Then $c = p^\gamma$, and all the elements $\mathcal{A}^j\mathcal{B}^k\mathcal{C}^l$, with j,k as before and $0 \leq l < p^\gamma$ will be distinct. And so on until $\mathfrak{G}_p$ is exhausted. Thus $\mathfrak{G}_p$ has $p^{\alpha+\beta+\gamma+\cdots}$ elements, which was to be proved.

In effect, Gauss had proved that every class $\mathcal{K} \in \mathfrak{G}_p$ has a unique representation in the form

$$\mathcal{K} = \mathcal{A}^j\mathcal{B}^k\mathcal{C}^l \cdots, \quad 0 \leq j < a, \ 0 \leq k < b, \ 0 \leq l < c, \ \ldots, \tag{9.4}$$

[10]See Footnote 8.

where a is the order of $\mathcal{A}$ and $b, c, \ldots$ are as defined in the proof. Of course, in the proof all of $a, b, c, \ldots$ are powers of p, but the same argument clearly works in any finite abelian group when a is the order of $\mathcal{A}$ and $b, c, \ldots$ are defined as in the proof. It should be noted that although (9.4) implies that $\mathfrak{G}_p$ is a product of the cyclic subgroups $(\mathcal{A}), (\mathcal{B}), (\mathcal{C}), \ldots$, this product is not direct, i.e., pairs of these cyclic subgroups do not generally have $(\mathcal{E})$ as their intersection.[11] As this sketch of Gauss' proof suggests, his proof applies equally well to $\mathfrak{F}_0(D)$.

Gauss' second theorem was stated for $\mathfrak{F}_0(D)$, although it too could have been stated for $\mathfrak{F}_1(D)$.

Theorem 9.2. *If the number of classes in the principal genus $\mathfrak{F}_0(D)$ is $p^\alpha q^\beta \cdots$, where $p, q, \ldots$ are distinct primes, then $\mathfrak{G}_p, \mathfrak{G}_q, \ldots$ (as defined in Theorem 9.1 but now relative to $\mathfrak{F}_0(D)$) contain, respectively, $p^\alpha, q^\beta, \ldots$ elements [and every class $\mathcal{K} \in \mathfrak{F}_0(D)$ has a unique representation in the form $\mathcal{K} = \mathcal{A}\mathcal{B}\cdots$, where $\mathcal{A} \in \mathfrak{G}_p$, $\mathcal{B} \in \mathfrak{G}_q, \ldots$].*

The bracketed part of the above theorem was not part of the theorem as stated by Gauss but is an obvious implication of his proof, as we shall see. Gauss' proof proceeded as follows. First he showed that if $\mathcal{A}, \mathcal{A}'$ are in $\mathfrak{G}_p$, $\mathcal{B}, \mathcal{B}'$ are in $\mathfrak{G}_q$, and so on, then $\mathcal{A}\mathcal{B}\cdots = \mathcal{A}'\mathcal{B}'\cdots$ only if $\mathcal{A} = \mathcal{A}'$, $\mathcal{B} = \mathcal{B}'$, and so on. Then he considered the totality $\mathfrak{P}$ of all such products $\mathcal{A}\mathcal{B}\cdots$. Evidently, $\mathfrak{P}$ is contained in $\mathfrak{F}_0(D)$. The last complete line of the proof asserts that $\mathfrak{P} = \mathfrak{F}_0(D)$. Gauss had apparently intended to supply a proof of this assertion, since he wrote "Let" but then no more. Certainly he must have known how to continue, because the proof that $\mathfrak{P} = \mathfrak{F}_0(D)$ follows easily from what he has already done.[12] Gauss' proof thus shows that every element in $\mathfrak{F}_0(D)$ can be expressed uniquely as a product of elements from $\mathfrak{G}_p$, $\mathfrak{G}_q$, ... In effect, he had shown that $\mathfrak{F}_0(D)$ is the direct product of its primary component groups $\mathfrak{G}_p$, $\mathfrak{G}_q$, ...

It seems likely that the results in this manuscript were the basis for Gauss' unproved mathematical claims in VIII above, for both assertions (A) and (B) of VIII can be easily deduced from Theorem 9.2. As for Gauss' remarks in IX about the use of "multiple indices" when D is irregular, it would seem he had in mind choosing several base classes $\mathcal{A}, \mathcal{B}, \mathcal{C}, \ldots$ such that every element of $\mathfrak{F}_0(D)$ could be represented uniquely as $\mathcal{A}^\alpha \mathcal{B}^\beta \mathcal{C}^\gamma \cdots$ with $0 \leq \alpha < a$, $0 \leq \beta < b$, $0 \leq \gamma < c, \ldots$.

---

[11] As a simple illustration of this fact, let $\mathfrak{C}_9 = (\mathcal{C})$ denote a cyclic group of order 9 with generator $\mathcal{C}$. If we take $\mathcal{A} = \mathcal{C}^3$, then $(\mathcal{A}) = \{\mathcal{E}, \mathcal{C}^3, \mathcal{C}^6\}$. Thus $\mathcal{B} = \mathcal{C}^2 \notin (\mathcal{A})$, and the smallest power of $\mathcal{B}$ residing in $(\mathcal{A})$ is the third: $\mathcal{B}^3 = \mathcal{C}^6 \in (\mathcal{A})$. It follows that every element of $\mathfrak{C}_9$ can be uniquely represented in the form $\mathcal{A}^i\mathcal{B}^j = \mathcal{C}^{3i}\mathcal{C}^{2j}$ for $0 \leq i \leq 2$ and $0 \leq j \leq 2$. Since $(\mathcal{B}) = \mathfrak{C}_9$, it follows that $(\mathcal{A})(\mathcal{B}) = \mathfrak{C}_9$, but $(\mathcal{A}) \cap (\mathcal{B}) = (\mathcal{A}) \neq (\mathcal{E})$.

[12] Let $\mathcal{K} \in \mathfrak{F}_0(D)$. Since the order k of $\mathcal{K}$ must divide $p^\alpha q^\beta \cdots$ (the order of $\mathfrak{F}_0(D)$), it must be of the form $k = p^{\alpha'} q^{\beta'} \cdots$. Set $\mathcal{A} = \mathcal{K}^{k/p^{\alpha'}}$, $\mathcal{B} = \mathcal{K}^{k/q^{\beta'}}$, and so on. Then $\mathcal{A}$ has order $p^{\alpha'}$ and so is in $\mathfrak{G}_p$, $\mathcal{B}$ has order $q^{\beta'}$ and so is in $\mathfrak{G}_q$, and so on. The numbers $k/p^{\alpha'}, k/q^{\beta'}, \ldots$ are relatively prime, and so integers $s, t, \ldots$ exist for which $1 = s(k/p^{\alpha'}) + t(k/q^{\beta'}) + \cdots$. Thus $\mathcal{K} = \mathcal{K}^{s(k/p^{\alpha'})+t(k/q^{\beta'})+\cdots} = \mathcal{A}^s\mathcal{B}^t\cdots$ is in $\mathfrak{P}$.

Then $(\alpha,\beta,\gamma,\ldots)$ would be the corresponding multiple indices. In fact, such a representation is implied by combining the proof of Theorem 9.2 ($\mathfrak{F}_0(D)$ is the direct product of its primary subgroups) with the product representation of the primary subgroups given in (9.4) from his proof of Theorem 9.1. Presumably, however, Gauss had in mind a set of multiple indices that would have the critically important additive property of a single index: the product of two elements with indices $(\alpha,\beta,\gamma,\ldots)$ and $(\alpha',\beta',\gamma',\ldots)$ would have indices $(\alpha'',\beta'',\gamma'',\ldots)$, where $\alpha'' \equiv \alpha+\alpha' \pmod{a}$, $\beta'' \equiv \beta+\beta' \pmod{b}$, and so on. The construction leading to (9.4), however, does not provide this additive property. As we will see below, it is possible to modify Gauss' construction so as to obtain the requisite additivity—so that $\mathfrak{F}_0(D)$ is, in effect, a *direct* product of cyclic subgroups. Given his remarks in VIII about an analogous theory of multiple indices, I suspect that Gauss had the additivity property in mind and probably realized how to modify the construction in the manuscript to accomplish it; but he left behind no evidence to confirm this likelihood.

9.1.2 Kummer

Among the many ways in which Gauss had advanced the progress of "higher arithmetic" in his *Disquisitiones Arithmeticae* was in his treatment of what has become known as the law of quadratic reciprocity. In the eighteenth century, Euler had discovered a version of the law, and Legendre had ventured an admittedly incomplete proof [244, Art. 151]. Gauss' treatment of the law, which he referred to as the "fundamental theorem" [244, Art. 131], included a rigorous proof [244, Art. 135ff.]. The law has to do with quadratic residues. An integer a is said to be a quadratic residue of a prime p if $x^2 \equiv a \pmod{p}$ has a solution; otherwise, it is a nonresidue. To state the law, it is helpful to utilize the symbol Legendre had introduced and let $\left(\frac{n}{p}\right)$ equal 1 when n is a quadratic residue of p and -1 when it is not. Then the law of quadratic reciprocity states that if p and q are odd primes, one has

$$\left(\frac{q}{p}\right) = (-1)^{\frac{p-1}{2}}(-1)^{\frac{q-1}{2}}\left(\frac{p}{q}\right). \tag{9.5}$$

Thus, for example, if both p and q are congruent to 1 (mod 4), then $\frac{p-1}{2}$ and $\frac{q-1}{2}$ are both even, and so (9.5) says that p is a quadratic residue of q if and only if q is a quadratic residue of p.

Gauss deemed this "fundamental theorem" so important that he gave a second proof of it in *Disquisitiones Arithmeticae* using his theory of forms [244, Art. 296ff.], and by 1818, he had published six further proofs. He also looked for higher-order reciprocities. That is, quadratic reciprocity involves a relation between the solvability of $x^2 \equiv q \pmod{p}$ and $x^2 \equiv p \pmod{q}$, and such a relation may be sought for $x^n \equiv q \pmod{p}$ and $x^n \equiv p \pmod{q}$ for $n > 2$. In 1805, while considering such matters in the biquadratic case $n = 4$, Gauss conceived of an idea that was to have a major impact on the development of nineteenth-century number theory. His idea, which he finally published in 1832, was the outcome of the conviction that "the natural source" for a theory of biquadratic residues was to be sought "in an extension of the field of arithmetic," i.e., the results he had achieved regarding biquadratic residues would first appear in their "complete simplicity and natural beauty" when complex integers of the form $a+bi$, $i=\sqrt{-1}$, $a,b \in \mathbb{Z}$, became the object of arithmetic study [246, p. 540]. With these words, Gauss introduced the now familiar ring of Gaussian integers $\mathbb{Z}[i]$ and developed its arithmetic properties. He determined all $\alpha = a+ib$ that are prime in the sense that it is impossible to factor α as $\alpha = \beta\gamma$ where $\beta,\gamma \in \mathbb{Z}[i]$ are not units, i.e., have norms greater than 1.[13] For example, he showed that a positive prime $p \in \mathbb{Z}$ is prime in $\mathbb{Z}[i]$ if $p \equiv 3 \pmod 4$ and that $\alpha = a+bi$ with $ab \neq 0$ is prime in $\mathbb{Z}[i]$ if and only if its norm $N(\alpha) = a^2+b^2$ is congruent to $1 \pmod 4$ [246, pp. 543–544]. Early on in *Disquisitiones Arithmeticae*, Gauss had observed that "It is clear from elementary considerations that any composite number can be resolved into prime factors, but it is often wrongly taken for granted that this cannot be done in several different ways" [244, Art. 16]. He then proceeded to prove the unique factorization theorem for $\mathbb{Z}$. Thus it is not surprising that Gauss realized the need to prove unique factorization in $\mathbb{Z}[i]$ as well, and he gave a valid proof [246, p. 547].

After a detailed development of the arithmetic of $\mathbb{Z}[i]$ and the computation of many examples, Gauss finally presented his biquadratic reciprocity law [246, p. 576], although he declined to prove it, since despite the simplicity of the law, "its proof belongs to the most hidden secrets of arithmetic," involving subtleties that could not be fitted into the framework of a paper.[14] Here then was a formidable challenge to aspiring number theorists: to obtain a relatively simple proof of Gauss' biquadratic law. Another challenge was contained in Gauss' remark, upon introducing the complex integers $\mathbb{Z}[i]$, that "the theory of cubic residues must be based in similar fashion upon the consideration of numbers of the form $a+bh$, where h is an imaginary root of the equation $h^3-1=0$, say $h = -\frac{1}{2}+\sqrt{\frac{3}{4}}\cdot i$, and likewise the theory of residues for higher powers requires other imaginary quantities" [246, p. 541n]. As Gauss' remarks make clear, if one wished to study nth-degree residues and the possibility of a corresponding reciprocity law, then such a study, according to Gauss, should be carried out within the context of the elements in $\mathbb{Z}[\omega_n]$, where ω_n is a primitive nth root of unity.

The first challenge—a proof of the law of biquadratic reciprocity—was taken up and resolved by Jacobi in his Königsberg lectures of 1837 [414, p. vii] and (independently) in 1844 by Eisenstein, then 21 years old. The second challenge was

[13] The norm of $\beta = c+di$ is defined as $N(\beta) = \beta\bar{\beta} = c^2+d^2$. The elements of $\mathbb{Z}[i]$ of norm 1, namely ± 1 and $\pm i$, Gauss called units; they have inverses in $\mathbb{Z}[i]$ [246, p. 541].

[14] Although Gauss did not do so, the law of biquadratic reciprocity can be stated in a form analogous to the statement of the law of quadratic reciprocity as given in (9.5). See [306, p. 123], [414, pp. vi–vii].

taken up by Kummer, then a professor at the University in Breslau.[15] Kummer focused on the arithmetic properties of $\mathbb{Z}[\omega_p]$ when ω_p is a primitive pth root of unity for some odd prime p, e.g., $\omega_p = e^{2\pi i/p}$. Thus the elements of $\mathbb{Z}[\omega_p]$ are of the form $f(\omega) = a_0 + a_1\omega_p + \cdots + a_{p-1}\omega_p^{p-1}$, $a_i \in \mathbb{Z}$, and are sometimes referred to as cyclotomic integers. They are precisely the algebraic integers in $\mathbb{Q}(\omega_p)$, but this notion was first introduced in 1871 by Dedekind (see Section 9.1.5). The analogue of the Gaussian integer norm is $N(f(\omega_p)) = \prod_{k=1}^{p-1} f(\omega_p^k)$, which is always a positive integer for $f(\omega_p) \neq 0$. As in $\mathbb{Z}[i]$, the units in $\mathbb{Z}[\omega_p]$ are those $f(\omega_p)$ with $N(f(\omega_p)) = 1$, so that $[f(\omega_p)]^{-1}$ is also in $\mathbb{Z}[\omega_p]$. Also as in $\mathbb{Z}[i]$, $N(f(\omega_p)g(\omega_p)) = N(f(\omega_p))N(g(\omega_p))$.

Kummer's extensive computations with cyclotomic integers, which were under way by 1844 [378], combined with a blunder caught by Jacobi, apparently led Kummer to the discovery that for $p = 23$ the cyclotomic integers lack critically important features of the Gaussian integers.[16] He had come to realize that (with $p = 23$) the prime 47 cannot be expressed as the norm of a cyclotomic integer. This apparently led him to the realization that the customary definition of a prime number is problematic for $\mathbb{Z}[\omega_p]$ when $p = 23$. For example, it turns out that $\alpha = f(\omega_{23}) = 1 - \omega_{23} + (\omega_{23})^{21}$ has norm $N(\alpha) = 47 \cdot 139$. This means that α cannot be factored as $\alpha = \beta\gamma$, with β, γ nonunits; for such a factorization would imply

$$47 \cdot 139 = N(\alpha) = N(\beta\gamma) = N(\beta)N(\gamma),$$

so that one of β, γ would be forced to have norm 47, which, as noted above, is impossible. The irreducibility of α would make it a prime in the sense that this term was used by Gauss in the context of the Gaussian integers. However, α fails to have a property characteristic of primes in $\mathbb{Z}$ or in $\mathbb{Z}[i]$: if a prime divides a product of nonunits, then it must divide one of the factors. That is, since $\alpha = f(\omega_{23})$ is one of the 22 factors of $N(\alpha) = \prod_{i=1}^{22} f(\omega_{23}^i) = 47 \cdot 139$, α divides the product $47 \cdot 139$; but it cannot divide either 47 or 139. If, e.g., α divided 139, $N(\alpha) = 47 \cdot 139$ would divide $N(139) = 139^{22}$, and so 47 would divide 139^{22}, which is absurd, since 139 is also prime. The failure of α to have this characteristic property of primes in $\mathbb{Z}$ and $\mathbb{Z}[i]$ also shows that if the traditional definition of primality is retained so that α is a prime, then unique factorization into prime factors cannot hold in $\mathbb{Z}[\omega_p]$ for $p = 23$.[17]

---

[15] Regarding the rivalry between Eisenstein and Kummer, see [147].

[16] The following is based on the detailed analysis of Kummer's work by Edwards, especially [144, 145] but see also [142, 143].

[17] Since α divides $47 \cdot 139$, $\beta \in \mathbb{Z}[\omega]$ exists such that $\alpha\beta = 47 \cdot 139$. If each of $\beta, 47, 139$ is factored into prime factors, then if α were deemed to be a bona fide prime, uniqueness of the prime factorization of $\alpha\beta = 47 \cdot 139$ would require that α be one of the prime factors of either 47 or 139; but then α would divide either 47 or 139, which, as we saw, is impossible. In the eighteenth century, it was well known that, e.g., $x^2 + 5y^2 = (x - \sqrt{-5}y)(x - \sqrt{-5}y)$, which for $x = y = 1$ gives $2 \cdot 3 =$

Kummer's response to the problems posed by the traditional notion of primality in $\mathbb{Z}[\omega_p]$ for primes such as $p = 23$ was announced to the Berlin Academy in 1846 [379] and developed in full in the pages of *Crelle's Journal* in 1847 [380]. Kummer deemed cyclotomic integers such as $\alpha = 1 - \omega_{23} + (\omega_{23})^{21}$ not to be true primes, since although irreducible, they failed to have the characteristic property of primes in $\mathbb{Z}$ or in $\mathbb{Z}[i]$, namely that if a prime divides a product, it must divide one of the factors. Kummer thus defined a cyclotomic integer to be prime if it is irreducible and also has the above characteristic property. Of course, this means that then α is neither a prime nor factorable into prime factors; it has no prime divisors. To remedy this situation, Kummer showed that it was possible to introduce ideal prime divisors, so that any nonunit in $\mathbb{Z}[\omega_p]$ is divisible by ideal prime divisors and has a unique factorization into a product of a finite number of such prime divisors.[18] He likened the introduction of ideal prime divisors to the introduction of "imaginary formulas in algebra and analysis" [379, p. 319]. In other words, since an equation such as $x^2 - 4x + 13 = 0$ has no actual solutions, mathematicians invented the imaginary ones $x = 2 \pm 3i$ via the formula for quadratic equations that do possess actual roots. Likewise, since $\alpha = 1 - \omega_{23} + (\omega_{23})^{21}$ has no actual prime factors, Kummer devised a theory in which α has ideal prime divisors that possess the customary properties of primes.

The details of Kummer's groundbreaking and incredibly influential theory are nontrivial and reflect his considerable talent as a mathematician.[19] It earned him a full professorship at Berlin. Here it is impossible to do justice to his creation, and I will simply summarize some of the salient features of the theory necessary for what is to follow. A prime divisor P in Kummer's sense represents a specific algorithm depending on certain parameters by means of which it can be determined whether a given cyclotomic integer $\alpha = f(\omega_p)$ is "divisible by P." If the totality of all $\alpha \in \mathbb{Z}[\omega_p]$ that are divisible by P in accordance with the algorithm turns out to be identical with the totality of $\alpha \in \mathbb{Z}[\omega_p]$ divisible by some fixed $\beta \in \mathbb{Z}[\omega_p]$, then P can be identified with the actual cyclotomic integer β. Likewise, there is an algorithm for determining whether P divides a given cyclotomic integer α exactly m times, i.e., with multiplicity m. As one would hope, any $\alpha \in \mathbb{Z}[\omega_p]$ has only a finite number of distinct prime divisors $P_1, \ldots, P_r$.[20] If $m_1, \ldots, m_r$, are the corresponding

$6 = (1+\sqrt{-5})(1-\sqrt{-5})$. Today these factorizations show that uniqueness of prime factorization and all the customary arithmetic divisibility properties that depend on it fail in $\mathbb{Z}[\sqrt{-5}]$, but prior to Kummer, no such observations were made in print (see [145, p. 323]). Indeed, as late as 1847, Lamé presented a "proof" of Fermat's last theorem to the Paris Academy of Sciences that took for granted that the factorization $x^n + y^n = (x+y)(x+\rho y)(x+\rho^2 y)\cdots(x+\rho^{n-1}y)$, where $\rho = e^{2\pi i/n}$, could be treated by the usual rules of ordinary arithmetic [143, pp. 76–80].

[18] Kummer spoke of ideal complex numbers or factors rather than divisors, but as Edwards' exposition [143, Ch. 4] makes clear, what Kummer actually did is most accurately expressed in the language of divisors.

[19] Edwards has given an insightful overview of the theory [145, pp. 324–328] as well as a full exposition of it [143, Chs. 4–6]. These works are highly recommended to the interested reader.

[20] The number can be exactly determined from the rational prime factorization of the norm $Nf(\omega)$ [143, p. 139, Exer. 2].

multiplicities, then the list of these primes together with their multiplicities gives a new divisibility algorithm defining a new divisor, which I will denote by $D[\alpha]$. It is called *the* divisor of α. It will be expressed as $D[\alpha] = P_1^{m_1} \cdots P_r^{m_r}$ and regarded as the product of m_1 factors of P_1, m_2 factors of P_2, and so on. A fundamental theorem is that two cyclotomic integers $\alpha, \beta \in \mathbb{Z}[\omega_p]$ are such that $D[\alpha] = D[\beta]$ if and only if $\alpha = (\text{unit})\beta$. This is the unique factorization theorem of Kummer's theory. Thus the cyclotomic integer $\alpha = 1 - \omega_{23} + (\omega_{23})^2$ now has a factorization into ideal prime divisors, which turns out to be of the form $\alpha = P_1P_2$, where P_1 is an ideal prime divisor of 47 and P_2 is an ideal prime divisor of 139 [143, p. 139, Exer. 2].

$D[\alpha]$ is called the divisor of $\alpha \in \mathbb{Z}[\omega_p]$, but more generally, any expression $A = P_1^{e_1} \cdots P_s^{e_s}$ with the e_i positive rational integers is called a divisor, even though there may be no $\alpha \in \mathbb{Z}[\omega_p]$ such that $D[\alpha] = A$. Still, it is clear what it means for A to be a divisor of some $\beta \in \mathbb{Z}[\omega_p]$, namely that each prime divisor P_i of A divides β with multiplicity e_i in accordance with Kummer's algorithms. In order to relate Kummer's theory to the modern theory of ideals that Dedekind later created, it should be noted that if for any such divisor A we set

$$\mathfrak{i}(A) = \{\beta \in \mathbb{Z}[\omega_p] : A \text{ divides } \beta\}, \tag{9.6}$$

then $\mathfrak{i}(A)$ is an ideal.[21] In particular, when $A = P$, a prime divisor, $\mathfrak{i}(P)$ is a prime ideal; and if $\mathfrak{i}(A)$ is the principal ideal (γ), then A can be identified with γ, since A is a divisor of exactly the same cyclotomic integers as are divided by γ. Also, if $\mathfrak{i} \subset \mathbb{Z}[\omega_p]$ is an ideal, then a divisor A exists such that $\mathfrak{i} = \mathfrak{i}(A)$.

Any two divisors A and B can also be multiplied. If they are written in the form $A = P_1^{e_1} \cdots P_k^{e_k}$, $B = P_1^{f_1} \cdots P_k^{f_k}$, where now $e_i \geq 0$ and $f_i \geq 0$, then $AB = P_1^{e_1+f_1} \cdots P_k^{e_k+f_k}$. Kummer realized that his theory provided many examples of ideal divisors A with the property that another ideal divisor M exists such that AM is "actual," i.e., may be identified with an element in $\mathbb{Z}[\omega_p]$. For example, it followed from his theory that the rational prime 47 is the product of 22 distinct ideal prime divisors $P_1, \ldots, P_{22}$. Thus $A = P_1$ is an ideal divisor, which is not actual; but if $M = P_2 \cdots P_{22}$, then AM can be identified with 47 and so is actual. For a given ideal divisor A, Kummer investigated the existence of "multipliers" M such that AM is actual [380, §8]. He showed not only that every A has such a multiplier, but more importantly, that a finite number of multipliers $M_1, \ldots, M_h$ could be determined such that for any ideal divisor A, AM_i is actual for some i, and no subset had the above property. Thus the multipliers $M_1, \ldots, M_h$ are both necessary and sufficient for transforming all ideal divisors via multiplication into actual ones [380, p. 352].

The existence of the multipliers $M_1, \ldots, M_h$ led Kummer to introduce the following classification of divisors [380, §9]: $A \sim B$ if a divisor M exists such

[21] It turns out that $\mathfrak{i}(A)$ is never empty. By virtue of the way prime divisors are defined, a prime divisor is always a divisor of some rational prime q (including the special case $q = p$). Thus if P is any ideal prime divisor, $\mathfrak{i}(P)$ will contain the associated prime q. If $A = P_1P_2 \cdots P_r$ is any divisor, then $\mathfrak{i}(A)$ is the product of the ideals $\mathfrak{i}(P_k)$ and hence nonempty.

that AM and BM are both actual. Kummer showed that $\sim$ has the properties that now define an equivalence relation. It follows from the existence of the multipliers $M_1,\ldots,M_h$ that there are at most h equivalence classes. Furthermore, to prove that $\sim$ is transitive, he had shown that if AM is actual, then BM is actual for every $B \sim A$ [380, p. 353]. This implies that the number of equivalence classes is exactly equal to h.

With Gauss' equivalence classes of forms in mind, Kummer showed that his equivalence classes had similar properties, e.g., if $A \sim C$ and $B \sim D$, then $AB \sim CD$. He realized that this meant that the resulting equivalence classes have a well-defined multiplication with $\mathrm{Cl}(A) \cdot \mathrm{Cl}(B)$ taken as $\mathrm{Cl}(AB)$. Kummer had shown that the product of an ideal divisor by a divisor that is actual must always be ideal, and this implies, as he observed, that all the actual divisors form one of the equivalence classes, which, following Gauss' terminology for form-classes, he called the *principal class*. As with Gauss' principal class, if we let $\mathcal{E}$ denote Kummer's principal class, then $\mathcal{E} \cdot \mathcal{C} = \mathcal{C}$. What he was doing was, he realized, completely analogous to what Gauss had done, e.g., in his discussion of $\mathfrak{F}_1(D)$ and $\mathfrak{F}_0(D)$. Thus he showed that for every class $\mathcal{C}$ there is a class $\mathcal{C}'$ such that $\mathcal{C} \cdot \mathcal{C}' = \mathcal{E}$. Kummer used Gauss' Latin term for idempotents, *classes ancipitas*, for $\mathcal{C}$ such that $\mathcal{C}' = \mathcal{C}$, so that $\mathcal{C}^2 = \mathcal{E}$. He also observed that corresponding to each class $\mathcal{C}$ is a least power k such that $\mathcal{C}^k = \mathcal{E}$ and that k must divide h. And like Gauss, he noted that if there is a class $\mathcal{C}$ for which $k = h$ so that every class is represented by a power of $\mathcal{C}$, then the totality of classes may be profitably indexed (or ordered as Kummer says) by those powers. When such a class does not exist, "then by these principles one can only divide the classes into groups and then order the classes within a group; for the ordering of the groups among one another, however, another principle is then necessary" [380, p. 356]. It is not clear whether Kummer was thinking of the multiple indexing Gauss alluded to in point IX of Article 306 of *Disquisitiones* (quoted and discussed in Section 9.1.1), but his words suggest that the "new principle" required was not one he had worked out.

I will denote the set of equivalence classes of divisors in $\mathbb{Z}[\omega_p]$ by $\mathfrak{H}$. From the vantage point of the present, Kummer had shown that $\mathfrak{H}$ is a finite abelian group, and he had repeated essentially the same group-theoretic arguments used by Gauss to establish analogous results. Also, from the vantage point of the present, we can see that $\mathfrak{H}$ can be identified with the ideal class group of $\mathbb{Z}[\omega_p]$.[22] That $\mathfrak{H}$ and particularly the number h of classes in it—the class number—is of theoretical importance is immediately clear. For example, if $h = 1$, so that $\mathfrak{H}$ reduces to the principal class $\mathcal{E}$, then every divisor is actual;the divisors can be identified with the

---

[22]Indeed, if a divisor C is actual, this means that the cyclotomic integers β divisible by C are precisely those divisible by some fixed $\gamma \in \mathbb{Z}[\omega_p]$, i.e., $C = D[\gamma]$ in the notation used earlier. Thus the associated ideal $\mathfrak{i}(C)$ is the principal ideal (β). Thus to say that $A \sim B$, namely that a divisor M exists such that both AM and BM are actual, is to say that the ideal $\mathfrak{i}(M)$ is such that $\mathfrak{i}(A) \cdot \mathfrak{i}(M) = \mathfrak{i}(AM)$ and $\mathfrak{i}(B) \cdot \mathfrak{i}(M) = \mathfrak{i}(BM)$ are both principal. This means that $\mathfrak{i}(A)$ and $\mathfrak{i}(B)$ are in the same class of the ideal class group, which is thus isomorphic to $\mathfrak{H}$ under $\mathrm{Cl}(A) \to \mathrm{Cl}(\mathfrak{i}(A))$.

elements of $\mathbb{Z}[\omega_p]$ (up to units), and Kummer's theorem on uniqueness of ideal prime factorization becomes a unique factorization theorem for the actual elements of $\mathbb{Z}[\omega_p]$.

It was also possible for Kummer to see a connection between the class number and a possible proof of Fermat's last theorem, which asserts that $x^n + y^n = z^n$ has no nontrivial rational integral solutions x, y, z for any exponent $n \geq 3$.[23] It suffices to prove the theorem for all odd rational primes p, although in 1847, valid proofs were known only for 4 and for $p = 3, 5$. For a fixed odd prime p, Kummer devised a two-step proof. The first step involved the class number h of the ideal class group $\mathfrak{H}$ associated to $\mathbb{Z}[\omega_p]$ and showed that if p does not divide h, the first step, and hence the entire proof, succeeds. Primes p with this property are now called *regular primes*, and so Kummer ended up proving that Fermat's last theorem holds for all regular primes.[24]

Kummer returned to Fermat's last theorem in a paper of 1857 [383], where he sought conditions for irregular primes p under which the theorem would be true. The first condition was that $p^2 - h$. He used consequences of this condition to obtain a new criterion for when $\mathcal{A}^p = \mathcal{E}$ implies $\mathcal{A} = \mathcal{E}$, so that $\mathcal{A}$ consists of actual cyclotomic integers (a critical part of his proof of Fermat's theorem for regular primes). As he explained, "this criterion lies somewhat deeper than the above propositions and to establish it completely requires some theorems that are not to be found in earlier works on the theory of complex integers" [383, p. 50]. The new theorems would now be described as group-theoretic in nature, and they involved the ideal class group $\mathfrak{H}$. By a line of reasoning similar to that used by Gauss to prove (9.4) in his manuscript of 1801, which was still unpublished when Kummer wrote his paper, Kummer proved that classes $\mathcal{A}, \mathcal{B}, \mathcal{C}, \ldots$ can be chosen in $\mathfrak{H}$ such that every class $\mathcal{K}$ is expressible uniquely in the form

$$\mathcal{K} = \mathcal{A}^j \mathcal{B}^k \mathcal{C}^l \cdots, \quad 0 \leq j < a, \ 0 \leq k < b, \ 0 \leq l < c, \ \ldots, \tag{9.7}$$

where $a > 1$ is the order of some $\mathcal{A} \in \mathfrak{H}$, $b > 1$ is the order of $\mathcal{B}$ modulo the cyclic subgroup $(\mathcal{A})$, i.e., b is the smallest power k such that $\mathcal{B}^k \in (A)$, $c > 1$ is the order of $\mathcal{C}$ modulo $(\mathcal{A})(\mathcal{B})$, and so on [383, pp. 51–52]. From (9.7) it follows that $h = abc \cdots$, and since (by Kummer's first condition) p divides h but p^2 doesn't, it follows that p divides *exactly* one of $a, b, c, \ldots$. If, e.g., $p \mid b$, it follows that the order of $\mathcal{B}$ must be divisible by p. This means there is an element of order p. (Kummer had thus established Cauchy's theorem for $\mathfrak{H}$.) Next, Kummer observed that in the argument leading to (9.7), the choice of $\mathcal{A}$ was arbitrary, and he could thus have assumed

---

[23] See [143, pp. 165–173] for a clear, detailed account.

[24] It was not known how to compute h, but Kummer also found a necessary and sufficient condition for the regularity of p that was amenable to computation [143, p. 182]. By 1849, he had used this condition to show that of the 11 primes between 5 and 43, all are regular except $p = 37$ [382, p. 138]. This confirmed Fermat's last theorem for the first time for $p = 7, 11, 13, 17, 19, 23, 31, 41, 43$—a great advance at the time. It was not until the 1990s that Fermat's theorem was finally confirmed for all odd primes [159, 535].

that $\mathcal{A}$ was an element of order p, i.e., that $a = p$ in (9.7). Since none of $b, c, \ldots$ are divisible by p, it follows that the only elements of order p are in $\mathcal{E}, \mathcal{A}, \ldots, \mathcal{A}^{p-1}$ [383, p. 53].

Kummer regarded his work on Fermat's theorem as a minor episode en route to his goal of reciprocity laws beyond the known quadratic, cubic, and biquadratic cases. In a lengthy (144-page) paper published in 1859, he developed the mathematical principles he felt would prove viable to this end.[25] This involved replacing the ring $\mathfrak{R} = \mathbb{Z}[\omega_p]$ by $\mathfrak{R}' = \mathbb{Z}[\omega_p, w]$, where w is a pth root of $D \in \mathbb{Q}[\omega_p]$, and introducing a corresponding system of ideal divisor classes analogous to the introduction of $\mathfrak{H}$ [384, §6], i.e., in effect the ideal class group associated to $\mathfrak{R}'$.

9.1.3 Schering

Kummer's factorization (9.7) implied, in modern terms, that $\mathfrak{H}$ is the product of cyclic subgroups $(\mathcal{A}), (\mathcal{B}), (\mathcal{C}), \ldots$. As was the case with Gauss' analogous factorization, the resulting product is not direct, since, e.g., in general, $(\mathcal{A}) \cap (\mathcal{B}) \neq (\mathcal{E})$; $\mathcal{B}$ was selected such that $\mathcal{B}^b \in (A)$, but in general, $\mathcal{B}^b \neq \mathcal{E}$. Likewise, $\mathcal{C}^c \in (\mathcal{A}) \cdot (\mathcal{B})$, but in general, $\mathcal{C}^c \neq \mathcal{E}$, and so on. In 1869, Ernst Schering (1833–1897), a mathematician who was in charge of editing Gauss' collected works for publication, published a proof of a factorization that yielded a direct product [512].[26] Schering realized that his factorization theorem applied as well to Kummer's ideal classes $\mathfrak{H}$ and to the more general ideal classes introduced by Kummer in 1859 [384, §6] (described in the previous paragraph), and would presumably apply as well to the ideal classes of Kronecker's general theory of ideal divisors, which Kummer had said "will appear shortly" [384, p. 57].[27] Schering therefore formulated his theorem for a finite system of unspecified classes, which I will denote by $\mathfrak{S}$ and which could be the classes of $\mathfrak{F}_0(D)$ or $\mathfrak{F}_1(D)$ or those of one of the above systems of ideal classes $\mathfrak{H}$ introduced by Kummer [512, pp. 8ff.].

Theorem 9.3 (Schering's theorem). *Classes $\mathcal{A}, \mathcal{B}, \mathcal{C}, \ldots$ may be chosen in $\mathfrak{S}$ such that every class $\mathcal{K} \in \mathfrak{S}$ is uniquely expressible in the form*

$$\mathcal{K} = \mathcal{A}^j \mathcal{B}^k \mathcal{C}^l \cdots, \quad 0 \le j < a, \; 0 \le k < b, \; 0 \le l < c, \; \ldots, \tag{9.8}$$

[25] See Weil's informative account of this paper from the mathematical perspective of present-day number theory [386, pp. 8–11].

[26] According to Schering [512, pp. 4–5], he had discovered his proof in 1855, and the proof method was independent of the methods used by Gauss in his manuscript of 1801 [247], i.e., the methods leading to Gauss' Theorems 9.1 and 9.2. Gauss died in February 1855, and it is unclear whether Schering knew of Gauss' manuscript when he devised his own method.

[27] Kronecker finally did publish a sketch of his theory, but not until 1882 [363], and even Dedekind could not understand it. A plausible reconstruction of what Kronecker had in mind has been given by Edwards [150].

where $a, b, c, \ldots$ are the orders of $\mathcal{A}, \mathcal{B}, \mathcal{C}, \ldots$ and, in addition, (i) a is the maximal order of any class in $\mathfrak{G}$, b is the maximal order of all classes $\mathcal{B}'$ such that $(\mathcal{A}) \cap (\mathcal{B}') = (\mathcal{E})$, c is the maximal order of all classes $\mathcal{C}'$ such that $(\mathcal{A}) \cdot (\mathcal{B}) \cap (\mathcal{C}') = (\mathcal{E}) \ldots$; (ii) b divides a, c divides b, ...[28]

As we shall now see, some number-theoretic results obtained by Kummer and communicated to Kronecker suggested to the latter the value of an explicitly abstract formulation of Schering's theorem.

9.1.4 Kronecker

In 1870, Kummer presented to the Berlin Academy the results [385] of his investigation of the real units r in $\mathbb{Z}[\omega_p]$, $\omega_p = e^{2\pi i/p}$, p an odd prime, with the property that r and all its conjugates are positive. Let us call this property (P). If r is the square of a unit, then it has property (P), and the question Kummer investigated was whether every real unit with property (P) was necessarily a square. He discovered a condition on p that guaranteed that this was the case. In his proof of Fermat's theorem for regular primes, Kummer had introduced a special factorization of the class number $h = h_1 h_2$, which was valid for all primes p. He now discovered that the only real units in $\mathbb{Z}[\omega_p]$ with property (P) are squares, provided that p is such that 2 does not divide h_1 [385, (II.), p. 861]. This result was also of interest to Kummer because he was able to use it to obtain information about the second factor h_2 [385, (III.), p. 861]:

Theorem 9.4. *If $2 \mid h_2$, then $2 \mid h_1$, and so $4 \mid h$.*

Kummer pointed out that the above theorem was analogous to his earlier discovery (1847), en route to verifying Fermat's theorem for regular primes, that if $p \mid h_2$, then $p \mid h_1$, and so $p^2 \mid h$ [381, p. 317].

Kummer communicated his results to Kronecker before they were presented to the academy. By the time they were presented, Kronecker had discovered an entirely different approach to the matter, as he explained in a paper [355] immediately following Kummer's. The new approach derived from Kronecker's realization that Kummer's second factor could also be interpreted as class number: h_2 is the ideal class number associated to $\mathbb{Z}(\rho_1, \ldots, \rho_\mu)$, where $\rho_1, \ldots, \rho_\mu$ are the $\mu = (p-1)/2$ periods of length two [355, pp. 273–274]. Once he saw this, Kronecker discovered not only that he could give a completely different proof of Kummer's Theorem 9.4, but that he could use the proof idea to generalize Kummer's theorem to the following.

[28] I have stated Schering's theorem using multiplicative notation for composition, but Schering followed Gauss and used additive notation.

Theorem 9.5. *Let h denote the ideal class number associated to $\mathbb{Z}[\omega_p]$ as above. Then if $p-1=\mu\cdot m$ is any nontrivial factorization of $p-1$, there is a corresponding factorization $h=h_1^*h_2^*$ with the property that for any prime divisor q of m, $q \mid h_2^*$ implies that $q \mid h_1^*$, and so $q^2 \mid h$.*

Kronecker's theorem was a generalization of Kummer's because when $m=2$ (so $q=2$ is the only possibility), $h_2^*=h_2$ by Kronecker's above-mentioned observation.

The new approach was based on utilizing an easy consequence of Schering's Theorem 9.3 as it applies to the class system $\mathfrak{S}=\mathfrak{H}$. As we saw, Schering realized that his theorem was valid for many class systems $\mathfrak{S}$, including $\mathfrak{H}$, but he did not attempt to characterize them. By contrast, Kronecker observed that since the "simple principles" underlying the theorems of Gauss and Schering also apply to "the most elementary parts of the theory of numbers"—presumably a reference to the theory of congruences (and especially to $(\mathbb{Z}/m\mathbb{Z})^\times$)—

> it is easy to convince oneself that the above principles belong to a more general, more abstract, sphere of ideas. Hence it seems appropriate to free their development from all inessential limitations so that the same line of reasoning need not be repeated in each instance of application. The advantage of doing this is evident from the development itself, and the presentation, if it is given in the most general way possible, attains at the same time simplicity and—through the distinct emergence of what alone is essential—lucidity [355, pp. 274–275].

With these words, Kronecker clearly set forth the case for the abstract viewpoint in mathematics, although, of course, his words were meant to apply exclusively to arithmetic considerations. At the time (1870), he indicated no connection with the theory of permutation groups, although later, in 1877 [361, III], he found it useful to apply the abstract version of Schering's theorem (stated below) to obtain, in effect, a multiple indexing (in the manner Gauss predicted and Schering established) for the Galois group of an abelian extension [361, III].

To carry out the desideratum of the above quotation, Kronecker posited a finite number of "elements" $\theta_1,\theta_2,\theta_3,\ldots$ of such a nature that from any two of them "by means of a definite procedure a third is determined" [355, p. 275]. He used the notation $\theta_3=\mathfrak{f}(\theta_1,\theta_2)$ to denote the element θ_3 determined by the given procedure from θ_1 and θ_2. Kronecker allowed $\mathfrak{f}(\theta_1,\theta_2)$ to be regarded as multiplication and denoted by $\theta_1\cdot\theta_2$, but instead of writing $\theta_3=\theta_1\cdot\theta_2$, he wrote $\theta_3\sim\theta_1\cdot\theta_2$ because in the specific arithmetic applications, it was always a matter of some kind of equivalence of elements rather than strict equality. (Kronecker did not think of his elements $\theta_1,\theta_2,\theta_3,\ldots$ as equivalence classes, which are sets with an infinite number of elements, but rather as fixed representatives of these equivalence classes.) Kronecker assumed that his multiplication was commutative and associative and that in addition, the following cancellation law held: If $\theta_2\not\sim\theta_3$, then $\theta_1\cdot\theta_2\not\sim\theta_1\cdot\theta_3$. The usual cancellation law follows from Kronecker's version. It turns out that Kronecker's conditions on his elements implies the existence of an identity element. In typical fashion, Kronecker, whose papers are frequently difficult to follow, never mentioned this fact, but in stating (without proof) the first of five basic consequences of his abstract setup, he asserted that corresponding to any element θ is a smallest positive integer k_1 with the property that θ^k is equivalent to "unity" (*die Einheit*), which is expressed with the notation $\theta^{k_1} \sim 1$ [355, p. 276].

Using the five unproven propositions about the multiplicative properties of the elements $\theta_1, \theta_2, \theta_3, \ldots$, Kronecker sketched an abstract version of Schering's proof so as to obtain an abstract version of Schering's Theorem 9.3, which would now be regarded as a part of the fundamental theorem of abelian groups. Kronecker's version of Schering's theorem is based on the following definition. Say that a finite set of elements $\theta_1, \ldots, \theta_r$ of orders $N_1, \ldots, N_r$ forms a *fundamental system* if (1) N_1 is the smallest integer k such that $\theta^k \sim 1$ for all θ, and for $i > 1$, N_i is the smallest integer k such that $\theta^k \sim 1$ modulo all elements of the form $\theta_1^{a_1} \cdots \theta_{i-1}^{a_{i-1}}$; (2) every element θ is expressible in the form

$$\theta \sim \theta_1^{h_1} \cdots \theta_r^{h_r}, \quad 0 \leq h_i < N_i.$$

It follows from this definition that $N_i \mid N_{i+1}$ for all i and that the above representation of θ is unique, and so the total number N of elements θ is $N = N_1 N_2 \cdots N_r$. Kronecker's abstract version of Schering's theorem is now simply that given any finite system of elements θ satisfying his assumptions (commutativity, associativity, cancellation law), a fundamental system of elements $\theta_1, \ldots, \theta_r$ exists. This implies in present-day terms that the finite abelian group $\mathfrak{G}$ comprising all the elements θ is the direct product of the cyclic subgroups generated, respectively, by $\theta_1, \ldots, \theta_r$, i.e., $\mathfrak{G} = (\theta_1) \times \cdots \times (\theta_r)$.

It follows immediately from this theorem that a prime q divides $N = N_1 \cdots N_r$ if and only if it divides N_1. Using this consequence of the above abstract version of Schering's theorem, Kronecker readily deduced a corollary, which he then applied to obtain Theorem 9.5.[29]

9.1.5 *Dedekind*

When Frobenius turned to the study of arithmetic problems with his work on the arithmetic theory of forms (Chapter 8), he was well versed in nineteenth-century number theory, including the work of Gauss, Kummer, Schering, and Kronecker sketched above. Another standard arithmetic work of the period was Dedekind's editions of Dirichlet's lectures on the theory of numbers, which began appearing in 1863. (See the beginning of Chapter 2 for more details.) The second edition of 1871 [137] contained as Supplement X the first presentation of his theory of ideals.

Actually, Supplement X was entitled "The Composition of Forms," and it was only after he had presented his own development of Gauss' theory that Dedekind

[29] I must confess that I am unable to understand how Kronecker applied the corollary to obtain Theorem 9.5.

turned, in Section 159, to his theory of ideals with the remark that "The theory of binary quadratic forms is but a special case of the theory of those nth-degree homogeneous forms with n variables that can be factored into linear factors with algebraic coefficients" [137, pp. 423–424].

As this remark suggests, for inclusion as an appendix to Dirichlet's lectures, Dedekind did not present his theory of ideals in the now-familiar manner, but more or less forced it into the Procrustean framework dictated by his above remark; and he was evidently not satisfied with this first effort to communicate his new ideas, which he had apparently already developed, more or less, in the now-familiar manner [137, p. viii].

The earlier parts of Supplement X, on the composition of binary forms, also contained many new viewpoints and demonstrations. Of particular interest here is that Dedekind was the first to stress that Gauss' theory of the composition of forms was a special case of a theory of groups that included as well the groups of Galois theory (where the term "group" originated). This occurred in his discussion of the Gaussian period associated to a class A of properly primitive forms, i.e., the system of powers $A^0 = E, A, \ldots, A^{a-1}$, where a is the order of A. Dedekind emphasized that:

> Such a class period is only a special case of the following new concept, which is of the utmost importance for the laws of composition: A system $\mathfrak{A}$ of primitive classes of the first kind will be called a *group** if the composition of two classes of the system $\mathfrak{A}$ always produces another class of the same system ... [137, p. 388] .... The easiest groups to survey are the above-mentioned periods [cyclic groups] ... every irregular group [noncyclic group] can be represented as the smallest multiple of certain regular groups, of which any two have only the principal class [identity class] in common. However, here we will not go any further into this representation and the associated theorems of Gauss,** the proofs of which can be easily based upon what has preceded [137, p. 388].

In the footnote $*$ he added:

> I intentionally chose this nomenclature that Galois introduced into algebra, since his theory and the above, which corresponds to the so-called Abelian equations, are both contained within the more general theory of composition, in which $(KK')K'' = K(K'K'')$ and, in addition, from $KK' = KK''$, as well as from $K'K = K''K$, $K' = K''$ always follows ....

And in footnote $**$ Dedekind cited the articles of *Disquisitiones Arithmeticae* in which Gauss stated what Dedekind could see were group-theoretic properties of $\mathfrak{F}_0(D)$.

This passage and accompanying footnotes make it clear that Dedekind realized that the theory of the composition of forms, in so far as it involved, e.g., $\mathfrak{F}_1(D)$ or $\mathfrak{F}_0(D)$ (in the notation of Section 9.1.1), should be regarded as a special case of an abstract theory of composition of elements $K, K', K'', \ldots$ forming a finite system $\mathfrak{K}$ with a composition (or multiplication) of elements defined such that the associative law holds, as well as the left-hand and right-hand cancellation laws. When $\mathfrak{K}$ is finite, as Dedekind seems to have assumed in the quotation, these laws do imply that $\mathfrak{K}$ is a group. The theory of these abstract finite groups would also include the permutation groups of Galois theory as a special case. Dedekind did not cite

Kronecker's paper [355], and it is doubtful whether he had seen it yet.[30] Thus it would seem that Dedekind had independently arrived at the view that the theorems of Gauss and Schering, which he cited [137, p. 389n**], should be extricated from their particular contexts and viewed abstractly; but unlike Kronecker, he expressly envisioned the abstract theory within the broader context of finite abstract groups, which do not necessarily commute and which also include as special cases the theory of permutation groups. As we shall see, this broader understanding of the theorems of Gauss and Schering was accepted by Frobenius and Stickelberger and seems to have formed a good deal of the motivation behind their paper on those theorems and to have been a factor in turning Frobenius into one of the foremost cultivators of the theory of finite groups.

Although Dedekind's presentation of his theory of ideals in Supplement X (1871) was not to his liking, all the basic notions were there: the definitions of algebraic numbers [137, p. 427], algebraic integers [137, p. 437], modules [137, p. 442], and ideals in an algebraic number field [137, p. 452]. As we saw in Section 9.1.2, Kummer's ideal divisors could be identified with ideals in the modern sense, i.e., the sense in which Dedekind conceived of them; but Kummer's theory was restricted to ideal divisors of the "complex integers" in the cyclotomic fields $\mathbb{Q}(\omega_p)$, where ω_p is a primitive pth root of unity for some prime p. Kummer's complex integers, the elements in $\mathbb{Z}[\omega_p]$, were an immediate generalization of the Gaussian integers, the elements in $\mathbb{Z}[i]$. Dedekind's definitions of algebraic integers and algebraic numbers represented a different, more general, approach to what "complex integer" should mean. For Dedekind, any complex number α such that α is the root of a monic polynomial with coefficients from $\mathbb{Q}$ is called an *algebraic number*; α is an *algebraic integer* if it is the root of a monic polynomial with coefficients from $\mathbb{Z}$. It follows from these definitions that if α is an algebraic number, then $\mathbb{Q}(\alpha)$ consists entirely of algebraic numbers—it is now called an *algebraic number field*—and the algebraic integers in $\mathbb{Q}(\alpha)$ play the role of Kummer's complex integers. However, the identification of algebraic numbers with $\mathbb{Z}[\alpha]$ is no longer generally the case.[31] By means of his definitions, Dedekind was able to extend his ideal-based version of Kummer's theory far beyond the above special cyclotomic fields. He was able to develop a theory of ideals of algebraic integers in $\mathbb{Q}(\alpha)$, α an algebraic number, and to establish that any such ideal could be factored uniquely into prime ideals. Kummer's theory, when translated into Dedekind's conceptual framework, became just a very special case of a far more extensive theory.

[30]Kronecker's paper was read at the 10 December 1870 session of the Berlin Academy. Dedekind wrote the preface of the 1871 edition of Dirichlet's lectures in March 1871. In the 1879 edition, Kronecker's paper is cited along with Schering's [138, p. 397n].

[31]Clearly, when α is an algebraic number but not an algebraic integer, not all elements in $\mathbb{Z}[\alpha]$ (starting with α itself) will be algebraic integers; and if α is an algebraic integer, then although all the elements of $\mathbb{Z}[\alpha]$ are then algebraic integers, in some cases (depending on α) there are also algebraic integers that are not in $\mathbb{Z}[\alpha]$ (see [551, pp. 39ff.]).

With the encouragement and assistance of Lipschitz, Dedekind was given the opportunity to present a new, more satisfactory exposition of his theory for publication in Darboux's *Bulletin des sciences mathématiques et astronomiques*.[32] The resulting memoir "On the theory of algebraic integers" was published in two installments in volumes of the *Bulletin* and then as a separate work in 1877 [113]. One of the features of Dedekind's new exposition was that he built it up more systematically, basing it on the properties of modules, to which the first chapter of [113] is devoted.

There Dedekind wrote that a "system" $\mathfrak{a}$ of real or complex numbers formed a module if sums and differences of elements from $\mathfrak{a}$ are also in $\mathfrak{a}$, so that (in the obvious sense) $n\alpha \in \mathfrak{a}$ for all $n \in \mathbb{Z}$ and all $\alpha \in \mathfrak{a}$ [113, Ch. 1, §1]. The most important examples of modules from the point of view of Dedekind's memoir were ideals in algebraic number fields, and he limited his attention in his memoir to those properties of modules that would be used in developing his theory of algebraic numbers and ideals [113, Ch. 2–4]. At the end of his chapter on modules, however, Dedekind added the following remark about the notion of a module[33]:

> The researches in the first chapter have been expounded in a special form suited to our goal, but it is clear that they do not cease to be true when the Greek letters denote not only [complex] numbers, but any objects of study, any two of which α, β produce a determinate third element $\gamma = \alpha + \beta$ of the same type, under a commutative and uniformly invertible operation (composition), taking the place of addition. The module $\mathfrak{a}$ becomes a *group* of elements, the composites of which all belong to the same group. The rational integer coefficients [e.g., in $\beta = m_1\alpha_1 + \cdots + m_n\alpha_n$] indicate how many times an element contributes to the generation of another.

As the quotations above from Supplement X (1871) show, Dedekind was consciously borrowing the term "group" from Galois theory because he believed in the importance of an abstract theory of groups. His abstract modules comprised the commutative abstract groups and could be infinite, as they are when they represent ideals of algebraic integers in an algebraic number field. However, although he deemed these abstract notions important, he himself dealt with groups only as they related to his work on the theory of numbers.[34] In that connection, Dedekind extended Kummer's notion of the ideal class group $\mathfrak{H}$ (Section 9.1.2) to that of an ideal class group $\mathfrak{H}$ associated to any fixed algebraic number field [113, Ch. 4, §28].

Dedekind's penchant for abstraction was an aspect of his emphasis on conceptual thinking as opposed to reasoning based on computations with explicit formulas and explicit forms of mathematical representation. As Dedekind wrote to Lipschitz in 1876,

---

[32]Dedekind wrote his exposition in German, and it was translated into French by the French mathematician J. Hoüel. For further details on how Dedekind came to write his memoir, see [551, pp. 44ff.].

[33]I have followed the translation by Stillwell [120, p. 82].

[34]As we will see in Section 12.3, an exception occurred later (1895).

> My efforts in number theory have been directed toward basing the work not on arbitrary representations or expressions but on simple foundational concepts and thereby ... to achieve in number theory something analogous to what Riemann achieved in function theory, in which connection I cannot suppress the passing remark that Riemann's principles are not being adhered to in a significant way by most writers. Almost always they mar the purity of the theory by unnecessarily bringing in forms of representation which should be results, not tools, of the theory.[35]

The sort of conceptual approach to mathematics that Dedekind aspired to, although commonplace now, was not at all typical in 1876. In particular, it was quite different from nineteenth-century Berlin-style mathematics, which was largely based on explicit representations, such as finite algebraic expressions, power series, and Laurent expansions.

It was this type of mathematics that Frobenius had absorbed in Berlin and had practiced in his own work, as can be seen from the preceding chapters. From the Berlin point of view (as expressed by Frobenius and Stickelberger in 1878), Dedekind's theory of modules was identical to "the theory of linear forms with integer coefficients" [235, p. 546]. That is, with his ideals in mind, Dedekind had focused his attention on "finite"modules $\mathfrak{a} \subset \mathbb{C}$ [113, Ch.1, §§3–4] by which he meant what would now be called finitely generated free modules. For these modules he showed that elements $\alpha_1, \ldots, \alpha_n$ in $\mathfrak{a}$ can be determined that are linearly independent over $\mathbb{Z}$ and such that for every $\alpha \in \mathfrak{a}$ there is a unique set of integers $m_1, \ldots, m_n$ for which $\alpha = m_1\alpha_1 + \cdots + m_n\alpha_n$. From the present-day point of view, Dedekind had shown that $\mathfrak{a}$ is a free $\mathbb{Z}$-module of rank n for some n, and what Frobenius and Stickelberger were observing, in effect, was that $\mathfrak{a}$ is isomorphic to the system of all linear forms $f = m_1x_1 + \cdots + m_nx_n$ in n indeterminates $x_1, \ldots, x_n$. As Berlin-trained mathematicians, they preferred the more formal, algebraic viewpoint of linear forms—a concrete "form of representation" (to use Dedekind's above words)— but such a preference did not prevent them from appreciating and being influenced by Dedekind's work.[36]

9.2 The Frobenius–Stickelberger Paper

Dedekind's emphasis on the importance of an abstract theory of groups that would embrace both the groups of Galois theory and the commutative groups that were arising in the theory of numbers seems to have inspired Frobenius and Stickelberger to develop, in a paper submitted to *Crelle's Journal* in July 1878, the commutative

---

[35] I have followed the translation of Edwards [146, p. 11], who discusses at greater length Dedekind's approach to mathematics and its possible sources. The original German text is on pp. 468–469 of Volume 3 of Dedekind's *Werke*.

[36] In his *Moderne Algebra* [568, §106], van der Waerden defined a free module over a principal ideal ring $\mathfrak{R}$ in the same formal, linear form manner of Frobenius and Stickelberger but with $\mathbb{Z}$ replaced by $\mathfrak{R}$. See Section 16.3.3.

group part of that envisioned theory, with the theorems of Gauss and Schering (Theorems 9.2 and 9.3) as a focal point. This viewpoint can be seen in their opening remark:

> The theory of finite groups of commuting elements was founded, on the one hand, by Euler and Gauss, and, on the other hand, by Lagrange and Abel, the former in their number-theoretic work on residues of powers, the latter in their algebraic work on the resolution of equations. After these foundational works, Gauss and Schering developed the theory further. Gauss ... showed how to decompose a group into primary groups, whose orders are relatively prime .... Schering its decomposition into cyclic groups[37] whose orders are such that each is divisible by those that follow ... [235, p. 545].

In other words, the theorems of Gauss and Schering were to be viewed as part of a more general theory of groups that included as well the Galois groups associated to what are now called abelian field extensions, to which now the theorems of Gauss and Schering could be applied.

The specific objective of their paper had to do with the decompositions into cyclic groups implied by the theorems of Gauss and Schering, i.e., the decomposition in Schering's theorem, as well as the decomposition obtained by applying Schering's theorem to the primary group factors of Gauss' decomposition theorem so that a factorization of the group into a direct product of cyclic subgroups of prime power orders results. Their concern is with uniqueness properties of the decompositions and seems to have been inspired by Dedekind's theorem that the factorization into distinct prime ideal powers is unique. Such uniqueness is easily seen to be out of the question for the above group decompositions, but the question remains as to what extent these decompositions are nonetheless unique. As they put it:

> These decompositions are completely determined, but they can be brought about in different ways. This observation formed the point of departure of our investigation in the sense that it led to the question as to whether there existed common properties of all these decompositions. We realized first of all that the orders of the cyclic groups into which Mr. Schering decomposes the entire group are constant numbers, independent of the choice of the subgroups. By then combining the Gaussian decomposition with Schering's to forge ahead to the irreducible components of the group, we succeeded, with the help of a sharper formulation of the concept of a primitive root (§3), in determining to what extent the irreducible factors of a group are independent of one another and to what extent they are dependent [235, p. 545].

The latter decomposition—into cyclic subgroups of prime power orders—is most analogous to Dedekind's ideal factorization, and the related questions of uniqueness seem to have been the main concern of the authors. Readers must keep in mind that in 1879, the notion of "uniqueness up to isomorphism" was not yet a standard part of mathematicians' conceptual framework, and, as we shall see, Frobenius and Stickelberger approached the uniqueness question more in the spirit of Kronecker and Weierstrass, who sought complete systems of invariants for equivalence classes

---

[37] The authors actually spoke of "elementary groups" [*elementare Gruppen*], but I will use the more familiar term cyclic group in discussing their work.

of families of bilinear forms (Sections 5.4 and 5.6). However, their results do immediately imply uniqueness up to isomorphism in the modern sense.

The influence of both Kronecker and Dedekind is manifest in the fact that the authors will proceed abstractly, whereas shortly before, as we saw in Section 8.6.2, Frobenius had been content to observe that "by analogy" his normal form theorem could be established with the integers replaced by the polynomial ring of various fields. He did not attempt an abstract formulation of his rational elementary divisor theory, for that would have meant introducing something entirely new, the notion of an abstract field, and Frobenius was not inclined toward that sort of innovation.[38] In the case of groups, however, the way had been prepared already not only by Dedekind but also by a Berliner–Kronecker, as we saw in Section 9.1.4. In proposing to develop the theory abstractly, Frobenius and Stickelberger apparently still felt that they were treading on dangerous ground in the sense that prospective readers of their paper might be turned away by the abstraction. In anticipation of such a reaction, they explained that "in order to present the abstract development as conveniently and intelligibly as possible, we link it to the investigation of the classes of integers that, with respect to a given modulus, are incongruent and relatively prime to the same, but without making use of any special properties of these elements" [235, p. 546]. And so the abstract development began with the statement that "The elements of our investigation are the $\varphi(M)$ classes of (real) whole numbers that are incongruent and relatively prime with respect to a modulus M.[39] Two elements are said to be equal, $A = B$, when they are represented by numbers congruent (mod. M). A number of these elements form a (finite) group when the product of any two of them is also included among them." [235, p. 546]. Thus the reader can think of a group as subgroup of $\mathbb{Z}_M^\times$. In this way, the authors avoided an axiomatic definition of a finite group (along the lines of the definitions of Kronecker and Dedekind). But from that point on, the talk is only of elements $A, B, C, \ldots$ and groups $\mathfrak{A}, \mathfrak{B}, \mathfrak{C}, \ldots$, with no reference whatsoever to $\mathbb{Z}_M^\times$. As I will indicate below (following Theorem 9.11), the authors in a sense justified thinking of an abstract finite commutative group $\mathfrak{A}$ as a subgroup of $\mathbb{Z}_M^\times$ by showing that for any given $\mathfrak{A}$, an integer M can be determined such that $\mathfrak{A}$ can be identified with a subgroup of $\mathbb{Z}_M^\times$.

Frobenius' paper with Stickelberger resembles the type of paper we have come to see as characteristic of Frobenius. Within a developed tradition, namely the

[38]The concept of an abstract field seems to have been explicitly formulated for the first time in 1893 by Heinrich Weber, who was strongly influenced by Dedekind's approach to mathematics, which underlay their joint work on a purely algebraic, ideal-based reformulation (in 1880) of Riemann's theory of algebraic functions [121]. On the development of the notion of an abstract field, see [491,492].

[39]By regarding the elements of $(\mathbb{Z}/M\mathbb{Z})^\times$ as the equivalence classes themselves, rather than as fixed representatives of those classes, the authors were tacitly breaking with tradition and following the lead of Dedekind, for whom sets were regarded as basic building blocks of mathematics. They also pointed out that the totality of all complex numbers α that are roots of unity ($\alpha^n = 1$, for some integer n) forms a "system" with "infinitely many elements" (*unzählig vielen Elementen*) from which finite groups can be formed [235, p. 547n].

arithmetic tradition sketched in Section 9.1, a problem is posed, namely uniqueness questions related to Schering's theorem, as well as the theorem that results from combining it with Gauss' theorem to obtain a decomposition of an abelian group into cyclic subgroups of prime power orders. The problem having been posed, the theory of finite abelian groups is then developed systematically in accordance with what is deemed the proper approach to the material. In this case, the proper approach involved, among other things, the abstract approach—carefully disguised for the sake of the reader—of Dedekind's theory of modules. The result in this case was the first "treatise" on the theory of commutative abstract finite groups with a focus on what has since become known as the fundamental theorem of finite abelian groups, including uniqueness up to isomorphism. The authors, I believe, regarded it as a contribution to the abstract theory of finite groups, which Dedekind deemed of considerable importance.

Frobenius and Stickelberger commenced their treatise by introducing several basic notions and terms from Dedekind's theory of modules, albeit with Dedekind's additive notation replaced by a multiplicative one. Following Dedekind, a *basis* for a finite abelian group $\mathfrak{G}$ is defined as a finite collection $A_1,\ldots,A_n$ of elements from $\mathfrak{G}$ such that every element in $\mathfrak{G}$ is expressible in the form $A_1^{m_1}A_2^{m_2}\cdots A_n^{m_n}$. The *rank* r of $\mathfrak{G}$ is the smallest integer n for which a basis of n elements exists [235, p. 547]. Of course, since Dedekind had focused on modules that were ideals in algebraic number fields, he expressly limited development of the theory to that end. For their purposes, Frobenius and Stickelberger introduced the notion of what would now be described as a direct product decomposition. Thus $\mathfrak{G}$ is said to be decomposable into subgroup factors $\mathfrak{H}_1,\ldots,\mathfrak{H}_n$ if $\mathfrak{G}=\mathfrak{H}_1\mathfrak{H}_2\cdots\mathfrak{H}_n$, i.e., if very element in $\mathfrak{G}$ is expressible as $H_1H_2\cdots H_n$ with $H_i\in\mathfrak{H}_i$, and if, in addition, the $\mathfrak{H}_i$ are "divisor-free" in the sense that the "greatest common divisor" of each pair $\mathfrak{H}_i,\mathfrak{H}_j$, viz., $\mathfrak{H}_i\cap\mathfrak{H}_j$, is the group consisting solely of the identity element E [235, pp. 548–549].[40] In what follows I will denote such a direct product decomposition by $\mathfrak{G}=\mathfrak{H}_1\times\mathfrak{H}_2\times\cdots\times\mathfrak{H}_n$. Because the factor groups are divisor-free, no $\mathfrak{H}_i$ is equal to $\{E\}$. The authors of course realized that when $\mathfrak{G}=\mathfrak{H}_1\times\mathfrak{H}_2\times\cdots\times\mathfrak{H}_n$, each $G\in\mathfrak{G}$ has a unique representation of the form $G=H_1\times\cdots\times H_n$ with $H_i\in\mathfrak{H}_i$.

With Dedekind's theory of factorization of ideals apparently in mind, a group $\mathfrak{G}$ is defined to be irreducible if it cannot be factored as $\mathfrak{G}=\mathfrak{H}\times\mathfrak{K}$. It then follows that every group that is not irreducible has a (direct product) factorization into irreducible subgroups [235, I, p. 549]. In the case of ideals, one obtains a unique factorization, the irreducible factors being the prime ideals. In the present situation, the authors continue, the factorization is not unique, although the number of irreducible factors is the same in any two such factorizations and the ordering of the factors in the two

[40]It should be noted that Frobenius and Stickelberger *reversed* the meaning that divisibility had in Dedekind's theory of modules and ideals. Whereas Dedekind said $\mathfrak{a}$ divides $\mathfrak{b}$ when $\mathfrak{b}\subset\mathfrak{a}$ [113, Ch. 1, §1]—and defined least common multiples and greatest common divisors accordingly, as was necessary for his theory of ideals—Frobenius and Stickelberger said that a group $\mathfrak{A}$ divides a group $\mathfrak{B}$ if $\mathfrak{A}\subset\mathfrak{B}$.

factorizations can be chosen so that corresponding factors have the same order. "The proof of this assertion (§. 8) as well as a precise characterization of the irreducible factors ... (§. 9) is the main subject of the following investigation" [235, p. 549]. And so from this point of view, Frobenius and Stickelberger commenced to develop the machinery that would yield not only Schering's theorem but also a factorization into prime power cyclic subgroups (the above irreducible groups) and in both cases an investigation of the degree to which these factorizations were unique.

A central construct in developing the theory was that of the group of all kth roots of unity. That is, if k is any positive integer, let

$$\mathfrak{R}_k = \{X \in \mathfrak{G} : X^k = E\}. \tag{9.9}$$

If the order of $\mathfrak{G}$, a term introduced by Frobenius and Stickelberger [235, p. 547], is g ($g = (\mathfrak{G} : 1)$ in the notation I will use),[41] and if p^e is the largest power of a prime p dividing g, then, e.g., $\mathfrak{R}_{p^e}$ is Gauss' group $\mathfrak{G}_p$ of all elements of order p^m for some $m \geq 0$ (see Gauss' Theorem 9.1), which was perhaps the motivation for introducing the more general groups $\mathfrak{R}_k$ and studying their properties.[42] An abstract version of Gauss' Theorems 9.1 and 9.2 and the ideas implicit in his proofs is summed up as follows [235, IV, p. 559].

Theorem 9.6 (Gauss). *If* $(\mathfrak{G} : 1) = p^a q^b \cdots$ *gives the prime factorization of the order of* $\mathfrak{G}$*, then there is one and only one way to express* $\mathfrak{G}$ *as the direct product of groups of respective orders* $p^a, q^b, \ldots$*, namely* $\mathfrak{G} = \mathfrak{R}_{p^a} \times \mathfrak{R}_{q^b} \times \cdots$.

Frobenius and Stickelberger introduced the term "primary group" for an abelian group of prime power order [235, p. 547]. I will refer to Theorem 9.6 as *Gauss' primary factorization theorem.*

After establishing Gauss' theorem in Section 4 of their paper, the authors proceeded to develop the properties of the rank of a group in Section 5 in preparation for proving Schering's theorem. For example, they proved that if $\mathfrak{G}$ is the direct product of groups $\mathfrak{H}_i$, $i = 1, \ldots, n$, whose orders are pairwise relatively prime, then

$$\text{rank}\, \mathfrak{G} = \max_{i=1,\ldots,n} (\text{rank}\, \mathfrak{H}_i). \tag{9.10}$$

Schering's Theorem 9.3 is then stated abstractly in Section 6 as follows [235, II, p. 564]:

Theorem 9.7 (Schering). *Every group that is not cyclic*[43] *can be decomposed into a direct product of cyclic groups whose orders can be arranged so that the order of each is divisible by the orders that follow.*

---

[41]This notation was later introduced by Frobenius; see the discussion of his paper [194] in Section 9.4.

[42]The authors denote the order of $\mathfrak{R}_k$ by $|k, \mathfrak{G}|$, a notation that is used extensively in their paper.

[43]The term "cyclic group" was not introduced by Frobenius and Stickelberger. They called such groups "elementary groups."

Since Schering had given a valid and straightforward proof of this theorem and Kronecker had then repeated it in an abstract setting, there was no need to give a proof, although further on in their paper, the authors did give an interesting proof of Schering's theorem as an application of Frobenius' theory of invariant factors (see Theorem 9.11). At this point in their paper, however, the authors focused on the uniqueness problem implied by Schering's theorem.

According to that theorem,

$$\mathfrak{G} = \mathfrak{C}_{e_1} \times \mathfrak{C}_{e_2} \times \cdots \times \mathfrak{C}_{e_r}, \tag{9.11}$$

where $\mathfrak{C}_{e_i}$ denotes a cyclic subgroup of order e_i and $e_{i+1} \mid e_i$ for all i. The first problem was thus to show that the integers $r, e_1, \ldots, e_r$ are all invariants of $\mathfrak{G}$, i.e., are independent of the specific factorization in (9.11), which is easily seen to be nonunique. Of course, it was already clear from Schering's proof that e_1 was an invariant of $\mathfrak{G}$, since it was the smallest value of k for which $A^k = E$ for all $A \in \mathfrak{G}$, but it remained to characterize the other e_i intrinsically. To this end, they made extensive use of the orders of the groups $\mathfrak{R}_k$ defined by (9.9). Their solution may be summed up in the following theorem.

Theorem 9.8 (Frobenius–Stickelberger). *In the factorization (9.11) posited by Schering's theorem, the integers r, e_1, e_2, …, e_r are all invariants of $\mathfrak{G}$. That is, (1) $r = \operatorname{rank} \mathfrak{G}$; (2) e_r is the largest integer k such that $(\mathfrak{R}_k : 1) = k^r$, and in general, e_i is the largest integer k such that $e_{i+1} \mid k$ and $(\mathfrak{R}_k : 1) = e_r e_{r-1} \cdots e_{i+1} k^i$.*

An example is helpful to make sense of the above formulas for the e_i. Let $\mathfrak{G} = (\mathbb{Z}/675\mathbb{Z})^\times$, which has order $(\mathfrak{G} : 1) = \phi(675) = 360 = 2^3 \cdot 5 \cdot 3^2$. As we shall see in Section 9.2.2, $\mathfrak{G} = (G_2) \times (G_4) \times (G_5) \times (G_9)$, where G_2, G_4, G_5, G_9 are elements of $\mathfrak{G}$ of respective orders 2,4,5,9. Since $\gcd(4,5,9) = 1$, it follows that $A = G_4 G_5 G_9$ has order $4 \cdot 5 \cdot 9 = 180$, and that $\{G_2, A\}$ is a basis for $\mathfrak{G}$. Since $\mathfrak{G}$ cannot be cyclic, the above basis has minimal size, i.e., $r = \operatorname{rank} \mathfrak{G} = 2$. Now $e_r = e_2$ is the largest integer k such that $(\mathfrak{R}_k : 1) = k^r = k^2$. Since $\mathfrak{R}_k = \{X \in \mathfrak{G} : X^k = E\}$ is a subgroup of $\mathfrak{G}$, $(\mathfrak{R}_k : 1) = k^2$ must divide $(\mathfrak{G} : 1) = 2^3 \cdot 5 \cdot 3^2$. Thus k must divide $2 \cdot 3$, i.e., the only possibilities for k are $k = 2, 3, 6$. Consider $k = 2$. From the direct product representation of $\mathfrak{G}$, it is easily seen that $X^2 = E$ has precisely the four solutions $X = G_2^a G_4^b$, $a = 0,1$, $b = 0,2$, and so $(\mathfrak{R}_2 : 1) = k^2$ for $k = 2$. For the remaining (larger) possibilities for k, it is easily seen that $(\mathfrak{R}_k : 1) \neq k^2$. (If, e.g., $k = 3$, then $\mathfrak{R}_3 = \{E, G_9^3, G_9^6\}$, so $(\mathfrak{R}_3 : 1) = 3 \neq 3^2$.) Thus $e_2 = 2$. Now consider e_1. It is the largest k such that $e_2 = 2$ divides k (so k is even) and $(\mathfrak{R}_k : 1) = e_2 k = 2k$. Since $(\mathfrak{R}_k : 1) = 2k$ divides $(\mathfrak{G} : 1) = 2^3 \cdot 3^2 \cdot 5$, k must divide $2^4 \cdot 3^2 \cdot 5 = 180$. Clearly the maximal such k is $k = 180$, since $X^{180} = E$ for all $X \in \mathfrak{G}$, and so $\mathfrak{R}_{180} = \mathfrak{G}$ and $[\mathfrak{R}_{180} : 1] = 360 = 2 \cdot 180$. In other words, $e_1 = 180$.

The abstract version of Schering's theorem as in Theorem 9.7, when combined with the uniqueness properties posited in Theorem 9.8, is sometimes referred to as the fundamental theorem of finite abelian groups. Of course, nowadays, uniqueness is expressed in terms of the notion of isomorphism, but Frobenius and Stickelberger did not introduce this notion, even though it had been used in the context of permutation groups in Jordan's 1870 *Traité* [322, p. 317]. Theorem 9.8 does more than prove uniqueness up to isomorphism in the sense that it gives recursive formulas in terms of the groups $\mathfrak{R}_k$ for the e_i. This sort of thing appealed to Frobenius, as can be seen from his other publications, and it seems to have pleased the authors to give a fundamental role to the groups $\mathfrak{R}_k$ throughout their paper.[44] However, it should be pointed out that within the framework of Schering's original proof of his theorem, a fairly short and simple proof of the invariance of the orders $e_1,\ldots,e_r$ is possible, as Kronecker's student Eugen Netto showed in his 1882 book [453, pp. 146–147], where he credited Frobenius and Stickelberger with the result. The lasting contribution of Theorem 9.8 has proved to be the fact that Schering's factorization is unique up to isomorphism.

In the first section of their paper, with Dedekind's factorization of ideals apparently in mind, Frobenius and Stickelberger had defined a group of order greater than 1 to be *irreducible* if it could not be factored as the direct product of nontrivial subgroups. It was then easy to prove the following proposition [235, pp. 549, 551].

Proposition 9.9. *If $\mathfrak{G}$ is a nontrivial group, then $\mathfrak{G}$ can be factored into the direct product of irreducible groups, and the order of $\mathfrak{G}$ is the product of the orders of the irreducible factors.*

Although this proposition was easy to prove, no one had previously considered it because no one before Frobenius and Stickelberger had attempted to develop a general theory of finite abelian groups.

Once they had established Gauss' primary factorization theorem (Theorem 9.6), it followed from Proposition 9.9 that $\mathfrak{G}$ is irreducible if and only if $\mathfrak{G} = \mathfrak{R}_{p^a}$ for some prime p and rank $\mathfrak{R}_p = 1$. In other words, the only irreducible groups are cyclic groups of prime power orders [235, pp. 568ff.]. Thus Proposition 9.9 implied that any nontrivial group could be factored into cyclic groups of prime power orders. This was analogous to Dedekind's factorization of an ideal into powers of prime ideals. A fundamental theorem in Dedekind's theory was that his factorization is unique. The above group factorization is, of course, not unique, but Dedekind's uniqueness theorem seems to have inspired the authors to determine to what extent such a factorization is unique. Their answer is contained in the following theorem (also sometimes referred to as the fundamental theorem of abelian groups) [235, II, p. 570].

Theorem 9.10 (Frobenius–Stickelberger). *If $\mathfrak{G}$ has order $g = p^a q^b \cdots$ $(p,q,\ldots$ distinct primes), then $\mathfrak{G}$ has a factorization into irreducible prime power cyclic subgroups*

$$\mathfrak{G} = \mathfrak{C}_{p^{\alpha_1}} \times \cdots \times \mathfrak{C}_{p^{\alpha_s}} \times \mathfrak{C}_{q^{\beta_1}} \times \cdots \times \mathfrak{C}_{q^{\beta_t}} \times \cdots, \tag{9.12}$$

[44] Another characteristically Frobenian touch seems to have been the introduction of the "power groups" $\mathfrak{P}_k = \{G \in \mathfrak{G} : G = H^k, H \in \mathfrak{G}\}$, which, as they showed are a sort of dual of the groups $\mathfrak{R}_k$. In particular $(\mathfrak{G} : 1) = (\mathfrak{P}_k : 1)(\mathfrak{R}_k : 1)$ [235, V, p. 555], which enables results involving $(\mathfrak{R}_k : 1)$, such as Theorem 9.8, to be translated into results involving $(\mathfrak{P}_k : 1)$ [235, p. 566].

where $\alpha_i \geq \alpha_{i+1}$, $\beta_j \geq \beta_{j+1}, \ldots$. In any two such factorizations of $\mathfrak{G}$, the numbers $s,t,\ldots$ and the numbers $\alpha_i, \beta_j, \ldots$ are the same.

The authors called the numbers $p^{\alpha_i}, q^{\beta_j}, \ldots$ the *primary invariants* of $\mathfrak{G}$.

The proof of Theorem 9.10 [235, pp. 369–370] combined Gauss' primary factorization theorem (Theorem 9.6) with an application of their Theorem 9.8 on the uniqueness properties of the factorization in Schering's theorem (as applied to the primary subgroups of $\mathfrak{G}$). My guess is that *all* the uniqueness questions that motivated their paper (as indicated by the quotation given at the beginning of the section) were inspired by Dedekind's uniqueness theorem for ideals. Frobenius, in particular, was drawn to Dedekind's theory. His mastery of that theory is evident from an example included in his paper with Stickelberger [235, §12] and even more so in his work on density theorems, the subject of Section 9.3.

As will be seen in Section 16.3.2, Theorem 9.10 later inspired Frobenius' student Robert Remak to establish an analogue for finite nonabelian groups, and Remak's proof helped Felix Krull prove his version of the Krull–Schmidt theorem, which in turn paved the way for van der Waerden's development of Frobenius' rational theory of elementary divisors (Section 8.6) as an application of Schering's theorem generalized to finitely generated modules over a principal ideal domain (Section 16.3.3). As we shall now see, the close connection between the reasoning underlying Frobenius' version of elementary divisor theory and Schering's theorem was already perceived by Frobenius and Stickelberger, who used the former to derive the latter.

9.2.1 Schering's theorem via the Smith–Frobenius normal form

After a detailed treatment of primitive roots [235, §9] based on their refined definition of this notion [235, §5], the authors turned to the connection between the invariants $e_1, \ldots, e_r$ of a group and Frobenius' arithmetic theory of bilinear forms (Section 8.2), which he had submitted for publication just 3 months before submission of their joint paper. In establishing the properties of the rank of a group [235, §5], they had utilized the following construction, which is reminiscent of those behind Gauss' (9.4) and Kummer's (9.7). Let $\mathcal{A} = \{A_1, \ldots, A_n\}$ be some basis for $\mathfrak{G}$. (Recall that this meant for the authors that every element of $\mathfrak{G}$ is representable in the form $A_1^{m_1} A_2^{m_2} \cdots A_n^{m_n}$ but not necessarily uniquely.) Define subgroups $\mathfrak{H}_i$ by $\mathfrak{H}_1 = \{E\}$, $\mathfrak{H}_2 = (A_1)$, and in general, $\mathfrak{H}_i$ is the group with basis $A_1, \ldots, A_{i-1}$. Then for $1 \leq i \leq n$, let m_{ii} denote the order of A_i mod $\mathfrak{H}_i$, i.e., m_{ii} is the smallest positive integer such that $A_i^{m_{ii}} \in \mathfrak{H}_i$. Thus integers m_{ij}, $j = 1, \ldots, i-1$, exist such that $A_i^{m_{ii}} = \prod_{j=1}^{i-1} A_j^{-m_{ij}}$, which we may express in the form

$$A_1^{\nu_1} A_2^{\nu_2} \cdots A_n^{\nu_n} = E, \tag{9.13}$$

where $(\nu_1, \ldots, \nu_n) = (m_{i1}, \ldots, m_{ii}, 0, \ldots, 0)$ for any fixed i between 1 and n.

If $M_{\mathcal{A}} = (m_{ij})$, $M_{\mathcal{A}}$ is a lower triangular nonsingular $n \times n$ matrix, and any row $v = \begin{pmatrix} v_1 & v_2 & \cdots & v_n \end{pmatrix}$ of $M_{\mathcal{A}}$ satisfies (9.13). From this it follows that if

$$v_i = \sum_{j=1}^{n} z_j m_{ji}, \quad i = 1, \ldots, n, \text{ with } z_j \in \mathbb{Z}, \tag{9.14}$$

then (9.13) holds. Conversely, if $v_1, \ldots, v_n$ are any integers satisfying (9.13), the above definition of the m_{ij} implies that (9.14) holds. In other words, the matrix $M_{\mathcal{A}}$ has the property that (9.13) holds if and only if (9.14) holds.

With this in mind, it is helpful to consider the previous example $\mathfrak{G} = (\mathbb{Z}/675\mathbb{Z})^{\times}$. As we will see in Section 9.2.2, $\mathfrak{G} = (G_2) \times (G_4) \times (G_5) \times (G_9)$, where G_i is an element of $\mathfrak{G}$ of order i. As we already noted, $\mathfrak{G}$ has rank $r = 2$. Also, $\mathfrak{G}$ has $\mathcal{A} = \{G_2, G_4, G_5, G_9\}$ as a basis. Because these basis elements are the generators of a direct product, it is easily seen that $M_{\mathcal{A}}$ is a diagonal matrix with diagonal entries $m_{11}, m_{22}, m_{33}, m_{44} = 2,4,5,9$. If $\varepsilon_i = d_i/d_{i-1}$, $i = 1,2,3,4$, denote the invariant factors of $M_{\mathcal{A}}$, it follows readily that $\varepsilon_4 = 180$, $\varepsilon_3 = 2$, and $\varepsilon_2 = \varepsilon_1 = 1$. Thus $r = \operatorname{rank} \mathfrak{G}$ equals the number of invariant factors $\varepsilon_i > 1$, and the two group invariants e_1, e_2 of Schering's theorem are $\varepsilon_4, \varepsilon_3$.

Frobenius and Stickelberger realized that this example could be generalized so as to use any basis of a group to determine, by means of the Smith–Frobenius normal form (Theorem 8.8), its invariants $e_1, \ldots, e_r$ and a corresponding basis $\mathcal{B} = \{B_1, \ldots, B_r\}$, $r = \operatorname{rank} \mathfrak{G}$, giving the factorization $\mathfrak{G} = \mathfrak{C}_{e_1} \times \cdots \times \mathfrak{C}_{e_r}$ of Schering's Theorem 9.7 [235, §10]. To see this, let $\mathcal{A} = \{A_1, \ldots, A_n\}$ denote some basis for a given group $\mathfrak{G}$. Then *any (not necessarily triangular) $n \times n$ matrix $M_{\mathcal{A}} = (m_{ij})$ is called a matrix associated to the basis $\mathcal{A}$ of $\mathfrak{G}$ if (9.13) holds when and only when (9.14) holds.* From the above discussion, it follows that such matrices always exist and may be constructed for any given basis in the manner described.

By choosing special values for the z_j in (9.14), it is easy to see that $v = \begin{pmatrix} v_1 & v_2 & \cdots & v_n \end{pmatrix}$ can be made equal to any row of $M_{\mathcal{A}}$ and that $M_{\mathcal{A}}$ is nonsingular [235, I., p. 578]. Although the authors did not do so, it is helpful to express (9.14) in matrix form as

$$v = zM_{\mathcal{A}}, \tag{9.15}$$

where z is the row vector $z = \begin{pmatrix} z_1 & \cdots & z_n \end{pmatrix}$.[45] Thus an $n \times n$ nonsingular matrix $M_{\mathcal{A}}$ is a matrix associated to the basis $\mathcal{A}$ of $\mathfrak{G}$ if and only if every solution to (9.13) is of the form (9.15).

Suppose now that $M_{\mathcal{A}}$ is any matrix associated to the basis $\mathcal{A} = \{A_1, \ldots, A_n\}$. Then by Frobenius' Theorem 8.8 on the Smith–Frobenius normal form, unimodular matrices P and Q may be determined such that $PM_{\mathcal{A}}Q = N$, where N is the diagonal matrix with invariant factors $e_1, \ldots, e_n$ along its diagonal *but the notation is now*

---

[45] In fact, the authors do not use any matrix notation. They speak of the n linear forms $\sum_{\alpha=1}^{n} m_{\alpha i} x_\alpha$, $i = 1, \ldots, n$, associated to $\mathfrak{G}$ via the basis $A_1, \ldots, A_n$ rather than the matrix $M_{\mathcal{A}}$ or the corresponding bilinear form.

chosen so that e_1 *is the largest, i.e.,* $e_{i+1} \mid e_i$ *for all* i. Let $Q = (q_{ij})$ and $Q^{-1} = (s_{ij})$. Define $B_1, \ldots, B_n$ in $\mathfrak{G}$ by

$$B_i = \prod_{j=1}^{n} A_j^{s_{ij}}. \tag{9.16}$$

Then $\mathcal{B} = \{B_1, \ldots, B_n\}$ is also a basis for $\mathfrak{G}$. This follows if each A_k can be generated by $\mathcal{B}$. Straightforward computation shows that

$$A_k = \prod_{\ell=1}^{n} B_\ell^{[Q]_{k\ell}} \quad \text{and} \quad B_k = \prod_{\ell=1}^{n} A_\ell^{[Q^{-1}]_{k\ell}}, \tag{9.17}$$

and so we can think of Q^{-1} as taking $\mathcal{A}$ to $\mathcal{B}$ and Q as taking $\mathcal{B}$ back to $\mathcal{A}$.[46]

The next point to observe is that the normal form N is a matrix associated to the basis $\mathcal{B}$. This means showing that all solutions to $\prod_{i=1}^{n} B_i^{v_i'} = E$ are of the form (9.15), viz., $v' = z'N$, with z' running through all elements of $\mathbb{Z}^n$. By virtue of the relation (9.17), a computation of the sort that produced it shows that

$$\prod_{i=1}^{n} B_i^{v_i'} = E \text{ if and only if } \prod_{i=1}^{n} A_i^{v_i} = E, \text{ where } v' = vQ.$$

This means that a necessary and sufficient condition for $\prod_{i=1}^{n} B_i^{v_i'} = E$ is that $v = zM_{\mathcal{A}}$ for all $z \in \mathbb{Z}^n$, where $v = v'Q^{-1}$. This condition can be written as $v' = zM_{\mathcal{A}}Q$ for all $z \in \mathbb{Z}^n$. Since P is unimodular, if we set $z' = zP^{-1}$, then $z \in \mathbb{Z}^n$ if and only if $z' \in \mathbb{Z}^n$, and as z runs through $\mathbb{Z}^n$, so does z'. This means that the necessary and sufficient condition for $\prod_{i=1}^{n} B_i^{v_i'} = E$ can be expressed as $v' = z'PM_{\mathcal{A}}Q = z'N$ for all $z' \in \mathbb{Z}^n$. In other words, N is a matrix associated to the basis $\mathcal{B}$.

Since N is diagonal with invariant factors $e_1, \ldots, e_n$ running left to right down the diagonal and satisfying $e_{i+1} \mid e_i$, we know from $v' = z'N$ that all solutions to $\prod_{i=1}^{n} B_i^{v_i'} = E$ are of the form $\prod_{i=1}^{n} B_i^{z_i' e_i} = E$. This implies in particular that $B_i^{v_i} = E$ if and only if $v_i = z_i' e_i$, and so e_i is the order of B_i. If $e_\rho > 1$ and $e_{\rho+1} = 1$, then $e_{i+1} \mid e_i$ means that $e_i = 1$ for all $i > \rho$. Thus $B_i = E$ for all $i > \rho$, and $B_1, \ldots, B_\rho$ form a basis for $\mathfrak{G}$. And of course $\prod_{i=1}^{\rho} B_i^{v_i'} = E$ if and only if $v_i' = z_i' e_i$ means that $B_i^{v_i'} = E$ for all i, i.e., that $\mathfrak{G} = \mathfrak{C}_{e_1} \times \cdots \times \mathfrak{C}_{e_\rho}$, where $\mathfrak{C}_{e_i}$ is the cyclic subgroup generated by B_i. Thus "the fundamental theorem of Mr. Schering is hereby proved anew" [235, p. 581]. Of course, by virtue of the corresponding uniqueness theorem (Theorem 9.8), $\rho = r$, the rank of $\mathfrak{G}$, and the invariant factors $e_1, \ldots, e_r$ of N are precisely the invariants of $\mathfrak{G}$ in Theorem 9.8.

---

[46] Apply (9.16) to get

$$\prod_{\ell=1}^{n} B_\ell^{q_{k\ell}} = \prod_{\ell=1}^{n} \left[\prod_{j=1}^{n} A_j^{s_{\ell j}} \right]^{q_{k\ell}} = \prod_{j=1}^{n} A_j^{\sum_\ell q_{k\ell} s_{\ell j}} = \prod_{j=1}^{n} A_j^{[QQ^{-1}]_{kj}} = A_k.$$

The realization of a close connection between the theory of the Smith–Frobenius normal form and Schering's version of the fundamental theorem of abelian groups, a realization that is fundamental to the module-theoretic approach to Frobenius' rational theory of elementary divisors (Section 16.3.3), was thus something Frobenius himself had worked out, albeit in a form that made sense in terms of notions at hand in 1878. That form, though now quaint, can be summarized in the following theorem.

Theorem 9.11. *Let* $\mathfrak{G}$ *be a finite abelian group and* $\mathcal{A} = \{A_1,\ldots,A_n\}$ *any set of generators for* $\mathfrak{G}$*. Then* $n \times n$ *matrices* $M_{\mathcal{A}}$ *associated to* $\mathcal{A}$ *in the sense defined above at (9.13) and (9.14) always exist, e.g., the matrix* $M_{\mathcal{A}}$ *defined by the procedure surrounding (9.13) is such a matrix. If the invariant factors* $e_1,\ldots,e_n$ *of any such matrix* $M_{\mathcal{A}}$ *are ordered as a decreasing sequence* $e_1,\ldots,e_n$*, then* $r = \max\{i : e_i > 1\}$ *is the rank of* $\mathfrak{G}$ *and* $e_1,\ldots,e_r$ *are precisely the group invariants* $\mathfrak{G}$ *as described in Theorems 9.7 and 9.8. In addition, if* P *and* Q *are the unimodular matrices such that* $N = PM_{\mathcal{A}}Q$ *is the Smith–Frobenius normal form of* $M_{\mathcal{A}}$ *(with decreasing diagonal entries), then the basis* $\mathcal{B} = \{B_1,\ldots,B_n\}$ *obtained from* $\mathcal{A}$ *by means of* Q *as in (9.17) gives the type of factorization posited in Schering's theorem, namely,* $\mathfrak{G} = \mathfrak{C}_{e_1} \times \cdots \times \mathfrak{C}_{e_r}$*, where* $\mathfrak{C}_{e_i} = (B_i)$ *for* $i = 1,\ldots,r$*.*

9.2.2 *Cyclic factorization of* $(\mathbb{Z}/M\mathbb{Z})^\times$

In the concluding two sections of their paper, the authors considered two examples. The first was $\mathfrak{G} = (\mathbb{Z}/M\mathbb{Z})^\times$ for a composite modulus

$$M = 2^\mu p_1^{\alpha_1} p_2^{\alpha_2} \cdots p_\pi^{\alpha_\pi}, \tag{9.18}$$

where the p_i are the distinct odd primes in the factorization of M [235, §11]. Except for Gauss' claim in *Disquisitiones Arithmeticae* that (in effect) for nonprime M, $(\mathbb{Z}/M\mathbb{Z})^\times$ is cyclic if and only if $M = 2_1^\mu p_1^{\alpha_1}$, where $\mu = 0, 1$ and p_1 is an odd prime (Section 9.1.1), nothing more seems to have been done along these lines. Frobenius and Stickelberger completely settled the matter of the factorization of $(\mathbb{Z}/M\mathbb{Z})^\times$ into cyclic subgroups.

They distinguished three cases: (1) $\mu = 0, 1$ in (9.18), (2) $\mu = 2$, and (3) $\mu \geq 3$. They showed that if $n = \text{rank}\ (\mathbb{Z}/M\mathbb{Z})^\times$, then $n = \pi$ in case (1), $n = \pi + 1$ in case (2), and $n = \pi + 2$ in case (3). Furthermore, for each case they determined n elements $G_1,\ldots,G_n$ in $(\mathbb{Z}/M\mathbb{Z})^\times$ such that $(\mathbb{Z}/M\mathbb{Z})^\times$ is the direct product of the cyclic subgroups generated, respectively, by these elements. If we let $\mathfrak{C}_m \subset (\mathbb{Z}/M\mathbb{Z})^\times$ denote a cyclic subgroup of order m, their results for cases (1)–(3), respectively, may be summed up as follows[47]:

[47]For a present-day version of this result, using isomorphisms and external direct products, see e.g., [141, p. 315].

$$(\mathbb{Z}/M\mathbb{Z})^\times = \begin{cases} \mathfrak{C}_{\phi(p_1^{\alpha_1})} \times \cdots \times \mathfrak{C}_{\phi(p_\pi^{\alpha_\pi})}; \\ \mathfrak{C}_2 \times \mathfrak{C}_{\phi(p_1^{\alpha_1})} \times \cdots \times \mathfrak{C}_{\phi(p_\pi^{\alpha_\pi})}; \\ \mathfrak{C}_2 \times \mathfrak{C}_{2\mu-2} \times \mathfrak{C}_{\phi(p_1^{\alpha_1})} \times \cdots \times \mathfrak{C}_{\phi(p_\pi^{\alpha_\pi})}. \end{cases} \tag{9.19}$$

They also showed that if $g_1,\ldots,g_n$ are the orders of $G_1,\ldots,G_n$, they form a "complete set of invariants" for $(\mathbb{Z}/M\mathbb{Z})^\times$, meaning that the prime power factors of the g_i are the primary invariants of $(\mathbb{Z}/M\mathbb{Z})^\times$ in the sense of Theorem 9.10, so that $(\mathbb{Z}/M\mathbb{Z})^\times$ decomposes into a direct product of cyclic subgroups with these prime powers as their orders. Thus for $M = 675 = 3^3 \cdot 5^2$, the first case of (9.19) applies with $g_1 = \phi(3^3) = 18 = 2\cdot 3^2$ and $g_2 = \phi(5^2) = 20 = 2^2 \cdot 5$, so the primary invariants are $2, 2^2, 3^2, 5$, and $(\mathbb{Z}/675\mathbb{Z})^\times = \mathfrak{C}_2 \times \mathfrak{C}_4 \times \mathfrak{C}_5 \times \mathfrak{C}_9$ gives the decomposition of Theorem 9.10. Of course, it then follows that $(\mathbb{Z}/675\mathbb{Z})^\times = \mathfrak{C}_{180} \times \mathfrak{C}_2$ gives a factorization as in Schering's Theorem 9.7, where $\mathfrak{C}_{180}$ denotes the cyclic subgroup generated by the product of the generators of $\mathfrak{C}_4, \mathfrak{C}_5$, and $\mathfrak{C}_9$.

Frobenius and Stickelberger also showed that any set of integers $e_1,\ldots,e_n$ with all $e_i > 1$ and such that $e_{i+1} \mid e_i$ for all i are the invariants (in the sense of Theorem 9.7) of a subgroup of rank n of $(\mathbb{Z}/M\mathbb{Z})^\times$ for a suitable choice of M [235, p. 584]. (Consequently, any finite abelian group can be realized as a subgroup of $(\mathbb{Z}/M\mathbb{Z})^\times$ for some M.) To do this, they applied Dirichlet's theorem that any arithmetic progression $a, a+k, a+2k, \ldots$ such that a and k are relatively prime contains infinitely many primes. It follows that primes p_i can be chosen in the respective progressions corresponding to $a = 1$ and $k = e_i$ that are odd and such that for $i' \neq i$, $p_{i'} \neq p_i$. Then $M = p_1 p_2 \cdots p_n$ falls under case (1) above with $\alpha_1 = \cdots = \alpha_\pi = 1$, so $\pi = n$. If $G_1,\ldots,G_n$ are the above-mentioned generating elements for $(\mathbb{Z}/M\mathbb{Z})^\times$ determined by Frobenius and Stickelberger as in (9.19), then G_i has order $\phi(p_i) = p_i - 1$. Now since p_i is in the progression $1, 1+e_i, 1+2e_i, \ldots$, $p_i - 1$ is a multiple of e_i. Thus $(p_i - 1)/e_i$ is an integer, $H_i = G_i^{(p_i-1)/(e_i)}$, $i = 1,\ldots,n$, has order e_i, and $H_1,\ldots,H_n$ form a "Schering" basis for a subgroup with the invariants $e_1,\ldots,e_n$.

9.2.3 Summary

Frobenius and Stickelberger had, in effect, written the first monograph on the theory of abstract finite abelian groups and so were exploring the terrain systematically in ways dictated by their mathematical tastes. Although not all their discoveries and proof techniques have remained a part of the basic theory, their work became a standard reference. Thus Weber in his treatment of abelian groups in his influential *Lehrbuch der Algebra* listed the Frobenius–Stickelberger paper among the basic references on the subject [583, p. 65n], and Burnside did the same in both editions of his influential book *Theory of Groups of Finite Order* [48, p. 71n], [56, p. 99n]. The reason is clear: their paper [235] provided a lucid systematic account of the theory of finite abelian groups and included several original contributions of lasting significance to the theory, such as Theorems 9.8, 9.10, and 9.11.

9.3 Analytic Densities and Galois Groups

Frobenius' paper with Stickelberger was the first indication of a serious interest on his part in the theory of finite abstract groups, a theory that he continued to regard as a part of higher arithmetic in the sense of Gauss. His interest in the connection between finite groups and arithmetic problems continued with his reformulation and generalization (under the spell of Dedekind) of a problem on the density of certain types of primes suggested by a paper Kronecker published in 1880.

9.3.1 *A challenging paper by Kronecker*

To mark the occasion of Kummer's 70th birthday, Kronecker dedicated an 1880 paper [362] to him that presented a new criterion for the irreducibility of a polynomial with integer coefficients, a subject that had interested Kronecker for many years. Suppose that $\Phi(x) \in \mathbb{Z}[x]$ and that the discriminant Δ of Φ is not zero.[48] For a fixed prime p, let ν_p denote the number of solutions to $\Phi(x) \equiv 0 \pmod{p}$. Neither Kronecker nor Frobenius specified whether ν_p meant (1) the number of *distinct* solutions mod p or (2) the number of solutions (viz., roots) *each counted as often as its multiplicity*. Probably this was because it does not matter which way ν_p is defined, since the only primes p for which $\Phi(x) \equiv 0 \pmod{p}$ could possibly have a multiple root are those p dividing the discriminant of $\Phi(x)$, i.e., at most a finite number of primes p. Consequently, either definition of ν_p gives the same result in Kronecker's formula (9.20) below and in the densities of sets of primes that are based on it. In what is to follow I will assume that ν_p is defined in accordance with (1), so that in speaking of the number of roots with a certain property, multiplicities are never considered. If we regard $\Phi(x)$ as an element of $\mathbb{F}_p[x]$, where $\mathbb{F}_p$ denotes the finite field of integers mod p, i.e., $\mathbb{F}_p = \mathbb{Z}/p\mathbb{Z}$, then ν_p denotes the number of distinct roots of $\Phi(x)$ lying in $\mathbb{F}_p$. If n is the degree of $\Phi(x)$, then $0 \le \nu_p \le n$.

The starting point of Kronecker's paper was the following theorem, which he presented without proof.[49]

---

[48]This seems to be a tacit assumption on the part of both Kronecker and Frobenius [210, p. 721].

[49]To my knowledge, Kronecker never published a proof of this highly nontrivial theorem. It turns out that Kronecker's theorem (Theorem 9.12) is equivalent to Frobenius' density theorem (Theorem 9.20). As we shall see below, Frobenius used Kronecker's theorem to prove his density theorem. Conversely, it can be shown using modern results that Frobenius' density theorem implies Kronecker's theorem. In fact, the equivalence of these two theorems also holds with natural densities (defined in the next footnote) replacing the analytic densities used by Kronecker and Frobenius. I am grateful to Michael Rosen for informing me of these facts. See in this connection his paper [509]. J.-P. Serre has pointed out (in an email to me) that Frobenius could have based his results on those of Dedekind on his zeta function (1871) instead of on Kronecker's theorem. Frobenius was probably familiar with Dedekind's results, but to utilize them along the lines suggested by Serre, Frobenius would have had to develop, among other things, something like

Theorem 9.12. *For any fixed polynomial $\Phi(x)$ with integer coefficients, if m is the number of irreducible factors into which $\Phi(x)$ decomposes over $\mathbb{Z}$, then*

$$m = \lim_{x \to 0+} \frac{\sum_p \nu_p / p^{1+x}}{\log(1/x)}. \tag{9.20}$$

The summation in (9.20) is over all primes p, and when the limiting value is $m = 1$, the theorem implies that $\Phi(x)$ is irreducible.

Kronecker's idea was to recast this theorem in terms of what he regarded as a way to measure the density of certain types of primes. As was often the case with his papers, he did this in a confusing, hard-to-follow manner. It will therefore be helpful to provide some introduction to Kronecker's discussion of densities. To this end, let A denote some set of primes. (Admittedly, the introduction of sets that in what follows are usually infinite is not something Kronecker would have done explicitly, but these sets are implicit in his remarks, and making them explicit facilitates understanding what Kronecker is doing, as well as the way Frobenius generalized his ideas.) The density of the set A of primes relative to the set of all primes can be defined as the limiting value

$$D_A = \lim_{x \to 0+} \frac{\sum_{p \in A} 1/p^{1+x}}{\log(1/x)}, \tag{9.21}$$

when this limit exists; D_A is what is now usually called an *analytic (or Dirichlet) density*, although Kronecker may have been the first to regard the above limit as a density measure. If $A = P$, the set of all primes, then, as Kronecker realized, $D_P = 1$. Thus for any set of primes A, $0 \le D_A \le D_P = 1$, and D_A can be regarded, Kronecker observed [362, p. 86], as a measure of the density of the primes in A relative to all primes.[50] It also follows from (9.21) that if A is the union of disjoint sets $A_1, \ldots, A_n$ such that D_{A_i} exists for all $i = 1, \ldots, n$, then $D_A = D_{A_1} + \cdots + D_{A_n}$. In particular, if A is a finite set of primes $p_1, \ldots, p_n$, then $D_A = D_{\{p_1\}} + \cdots + D_{\{p_n\}} = 0$, since evidently the definition (9.21) implies $D_{\{p_i\}} = 0$.

Kronecker defined D_A for the following specific sets A. For the fixed polynomial $\Phi(x) \in \mathbb{Z}[x]$ under consideration, say that a prime number p is of *type k* if $\Phi(x)$, regarded as an element of $\mathbb{F}_p[x]$, has k (distinct) roots in $\mathbb{F}_p$. Thus $0 \le k \le n$, where n is the degree of Φ. For each such k, let A_k denote the set of all primes of type k. It was the density D_{A_k} that Kronecker introduced with the notation D_k. Since ν_p in (9.20) is the number of distinct roots of $\Phi(x)$ lying in $\mathbb{F}_p$, we see that $\nu_p = k$ if and only if $p \in A_k$. Thus

---

the formalism of induced characters, which he did not do until 1898 (Section 15.1), after he had created his theory of group characters and representations.

[50]There is also another, more natural definition of density, which, when it exists, agrees with D_A. It is given by $\lim_{n \to \infty}(a_n/\pi_n)$, where a_n is the number of primes $\le n$ in A and π_n is the number of primes $\le n$ [529, pp. 73–76], [548, p. 31].

$$\sum_{p\in P} \nu_p/p^{1+x} = \sum_{k=1}^{n} k\Big\{\sum_{p\in A_k} 1/p^{1+x}\Big\},$$

and if both sides of this equality are divided by $\log(1/x)$ and then the limit is taken as $x \to 0^+$, Theorem 9.12 implies that $\sum_{k=0}^{n} kD_k = m$ and, in particular, gives Kronecker's criterion for the irreducibility of $\Phi(x)$:

Corollary 9.13. *The polynomial $\Phi(x)$ is irreducible if and only if $\sum_{k=1}^{n} kD_k = 1$.*

Kronecker's paper was primarily devoted to the discussion of several nontrivial examples of types of polynomials $\Phi(x)$ for which the D_k could be determined and (in some cases) irreducibility could be concluded by the above corollary. It was essentially a research announcement, and Kronecker described his results with little or no proof or explanation. In Zurich, Frobenius read Kronecker's paper soon after its publication.[51] One of Kronecker's examples involved an irreducible polynomial of prime degree that is solvable by radicals. Without any explanation, Kronecker gave the values of the densities D_k for such a polynomial, and it was in the course of trying to derive these values on his own that in 1880, Frobenius hit upon his group-theoretic approach to densities.[52]

9.3.2 *First density theorem and conjecture*

Frobenius' inspiration seems to have come from the following two-part consideration. For the first part, let $\varphi(x) \in \mathbb{Z}[x]$ be of degree n with nonvanishing discriminant. For a prime number p, let

$$\varphi(x) \equiv \varphi_1(x)\varphi_2(x)\cdots\varphi_r(x) \pmod{p} \tag{9.22}$$

denote the mod p factorization of $\varphi(x)$ into irreducible factors with integer coefficients. If f_i is the degree of $\varphi_i(x)$, then the r positive integers $f_1,\dots,f_r$ satisfy

$$f_1 + f_2 + \cdots + f_r = n. \tag{9.23}$$

Frobenius defined a prime number p to be of *class* $\lambda = \{f_1,\dots,f_r\}$ if (9.22) holds with f_i being the degree of $\varphi_i(x)$ [210, p. 720]. One may then consider the

[51] Kronecker presented his paper to the Berlin Academy on 2 February 1880; Frobenius' results on densities (described below) were obtained, according to him [210, p. 719], the following November, although they were first published (essentially as written down in 1880) in 1896 (in [210]). The main results of [210, §§1–3] are also contained in unpublished letters to Dedekind written in 1882 (see below).

[52] Frobenius explained this in 1886, at the beginning of his paper [194]. (The last line of p. 157 of Kronecker's paper to which Frobenius refers is the last line of p. 87 of the paper as it appears in Volume 2 of Kronecker's *Werke*).

density D_λ of the set A_λ of all primes p of class λ. The connection of the D_λ with Kronecker's density D_k is given by the fact that for almost all primes p,[53] $\varphi(x) \equiv 0 \pmod{p}$ has exactly k solutions in $\mathbb{F}_p$ if and only if exactly k of the factors $\varphi_i(x)$ in (9.22) have degree $f_i = 1$. Thus p is a prime of type k in Kronecker's sense precisely when it is of some class $\lambda^* = \{f_1^*, \dots, f_{r^*}^*\}$ with exactly k of the f_i^* equal to 1. This means that $A_k = \cup_{\lambda^*} A_{\lambda^*}$ and $D_k = \sum_{\lambda^*} D_{\lambda^*}$, provided the densities D_{λ^*} exist.

The second, and critical, part of the inspirational consideration was this. By virtue of (9.23), the set of numbers $\lambda = \{f_1, \dots, f_r\}$ determines a conjugacy class of the symmetric group $\mathfrak{S}_n$, namely the class $\mathcal{S}_\lambda$ of all permutations expressible as a product of r disjoint cycles of respective lengths $f_1, \dots, f_r$. As we saw in Section 1.2, already by 1872, Frobenius had mastered the classical works on Galois' theory by Abel and Galois; furthermore, by 1880 he had also familiarized himself with Camille Jordan's comprehensive treatment of the subject in his *Traité des substitutions* of 1870 [322]. Given the interest in groups and group-theoretic reasoning manifest in his paper with Stickelberger, it is not a complete surprise to find that he now sought to relate the density D_λ to the Galois group of $\varphi(x)$, $\mathfrak{H} = \mathrm{Gal}(\mathbb{L}/\mathbb{Q})$, where $\mathbb{L}$ is a splitting field for φ. Since by the discriminant hypothesis, the roots of φ are distinct, and so n in number, $\mathfrak{H}$ may be regarded as a subgroup of the symmetric group $\mathfrak{S}_n$. With this interpretation of $\mathfrak{H}$ in mind, Frobenius considered the number h_λ of $H \in \mathfrak{H}$ that are in the class $\mathcal{S}_\lambda$ of $\mathfrak{S}_n$, i.e., in the notation I will use,

$$h_\lambda = |\mathfrak{H} \cap \mathcal{S}_\lambda|. \tag{9.24}$$

Clearly $\sum_\lambda h_\lambda = h$, where h is the order of $\mathfrak{H}$. Thus $\sum_\lambda (h_\lambda/h) = 1$. Also, since $\cup_\lambda A_\lambda = P$ (P again the set of all primes), the finite additivity of densities gives $\sum_\lambda D_\lambda = 1$. It would be natural to wonder whether, in fact, $D_\lambda = h_\lambda/h$.

Frobenius' idea was to prove that $D_\lambda = h_\lambda/h$ by utilizing equation (9.20) of Kronecker's Theorem 9.12. (See Section 9.3.5 below for an outline of his proof.) This he managed to do, and so established the following theorem:

Theorem 9.14. *Let $\varphi(x) \in \mathbb{Z}[x]$ be of degree n with nonzero discriminant. Then if $\mathfrak{H}$ is the Galois group of φ regarded as a subgroup of the symmetric group $\mathfrak{S}_n$, the density D_λ of the primes p of class λ is $D_\lambda = |\mathfrak{H} \cap \mathcal{S}_\lambda|/|\mathfrak{H}|$.*

The sets $\mathcal{S}_\lambda \cap \mathfrak{H}$ form a partition of $\mathfrak{H} \subset \mathfrak{S}_n$ into equivalence classes determined by saying that $H_1 \sim H_2$ if there is an $S \in \mathfrak{S}_n$ such that $SH_1S^{-1} = H_2$, i.e., that they are conjugate within $\mathfrak{S}_n$ (although not necessarily within $\mathfrak{H}$ itself). Frobenius' Theorem 9.14 thus says that each such equivalence class, i.e., each nonempty set of the form $\mathcal{S}_\lambda \cap \mathfrak{H}$, has associated to it an infinite number primes: $\mathcal{S}_\lambda \cap \mathfrak{H} \neq \emptyset$ means that $h_\lambda = |\mathcal{S}_\lambda \cap \mathfrak{H}| > 0$, and so $D_{A_\lambda} = h_\lambda/h > 0$, which means that A_λ cannot be finite ($D_A = 0$ for finite A).

[53] That is, all primes except possibly a finite number.

To see what Theorem 9.14 says about Kronecker's density D_k, recall that $D_k = \sum_{\lambda^*} D_{\lambda^*}$, where $\lambda^* = \{f_1^*, \dots, f_r^*\}$ is a partition of n with exactly k of the f_j^* equal to 1. Thus Frobenius' above theorem now gives $D_k = (1/h)\sum_{\lambda^*} h_{\lambda^*}$. Now the $\sum_{\lambda^*} h_{\lambda^*}$ permutations of $\bigcup_{\lambda^*}(\mathfrak{H} \cap \mathcal{S}_{\lambda^*})$ are precisely those that have a factorization into disjoint cycles involving k 1-cycles. In other words, the permutations of $\mathfrak{H}$ in these classes are precisely those that fix k roots of $\varphi(x)$. Theorem 9.14 therefore implies the following characterization of D_k [210, III, p. 726]:

Theorem 9.15. *The Kronecker density D_k of the primes p for which $\varphi(x) \equiv 0 \pmod{p}$ has exactly k incongruent integral solutions mod p equals the fraction of elements of the Galois group of φ that fix exactly k of its roots.*

Thanks to his familiarity with Dedekind's publications, Frobenius realized that there was a connection between the mod p factorization of polynomials and the ideal factorization of p. That is, suppose that $\mathbb{K}$ is an extension of $\mathbb{Q}$ of degree n, and let $\mathfrak{o}_{\mathbb{K}}$ denote the algebraic integers of $\mathbb{K}$. The fundamental theorem of Dedekind's theory asserts that the principal ideal generated by p, namely $p\mathfrak{o}_{\mathbb{K}}$, factors uniquely into (not necessarily distinct) prime ideals, viz.,

$$p\mathfrak{o}_{\mathbb{K}} = \mathfrak{p}_1\mathfrak{p}_2\cdots\mathfrak{p}_r. \tag{9.25}$$

Furthermore, if f_i is the degree of $\mathfrak{p}_i$, i.e., $N(\mathfrak{p}_i) = p^{f_i}$, then $n = f_1 + \cdots + f_r$. This is of course the relation (9.23) associated to the mod p factorization of a polynomial as in (9.22). Indeed, in his monograph of 1877 [113, §26.3], which Frobenius had studied carefully, Dedekind pointed out that $\theta \in \mathfrak{o}_{\mathbb{K}}$ can be chosen so that it is a root of an irreducible monic polynomial $\varphi(x) \in \mathbb{Z}[x]$ of degree n (whence $\mathbb{K} = \mathbb{Q}(\theta)$) and that as a consequence, for all but at most a finite number of primes p, one has

$$\varphi(x) \equiv \varphi_1(x)\varphi_2(x)\cdots\varphi_r(x) \pmod{p}, \tag{9.26}$$

where r is the same as in (9.25), and in addition, if f_i is the degree of $\varphi_i(x)$, then f_i is also the degree of the prime ideal $\mathfrak{p}_i$ in (9.27). In 1878, Dedekind published a paper [114] in which he proved the above results and established further theorems related to exceptional primes p for which the factorization (9.26) cannot be asserted.[54]

This result of Dedekind shows that Frobenius' Theorem 9.14 can be applied to the ideal factorization of (9.25).[55] That is, by analogy with the case of polynomial factorization modulo p, let us say that a prime p is of the *Dedekind class* $\lambda = \{f_1, \dots, f_r\}$ if the prime factorization (9.25) holds with f_i being the degree of $\mathfrak{p}_i$; and let A_λ^* denote the set of all primes p of Dedekind class λ. Then by virtue of Dedekind's above result, this set differs from the set A_λ of all primes p of Frobenius' class $\lambda = \{f_1, \dots, f_r\}$ with respect to the factorization in (9.26) by at most a finite set

[54] In [210, p. 726], Frobenius cites only Dedekind's paper [114]. The excluded primes are those dividing what Dedekind referred to as the index k of θ [114, p. 204].

[55] None of the reasoning leading to Theorem 9.16 below is explicitly given by Frobenius.

of primes p. Thus $D_{A_\lambda^*}$ is the same as Frobenius' density $D_\lambda = D_{A_\lambda}$ and so is given by Frobenius' Theorem 9.14. In other words, if $\mathbb{L}$ is the splitting field of Dedekind's polynomial $\varphi(x)$ and $\mathfrak{H} = \mathrm{Gal}(\mathbb{L}/\mathbb{Q})$ is regarded as a subgroup of $\mathfrak{S}_n$, then $D_{A_\lambda^*}$ is the number of $H \in \mathfrak{H}$ expressible as a product of r disjoint cycles of respective lengths f_i divided by the order of $\mathfrak{H}$ [210, II, p. 726]. Frobenius could thus translate his Theorem 9.14 into the following.

Theorem 9.16. *Let $\mathbb{K}$ be an extension of $\mathbb{Q}$ of degree n, and let $\mathfrak{o}_{\mathbb{K}}$ denote the algebraic integers in $\mathbb{K}$. Let $\varphi(x) \in \mathbb{Q}[x]$ be an irreducible polynomial of degree n with a root $\alpha \in \mathbb{K}$ such that $\mathbb{K} = Q(\alpha)$. Then if $\mathbb{L}$ denotes a splitting field of φ and $\mathfrak{H} = \mathrm{Gal}(\mathbb{L}/\mathbb{Q})$, regarded as a subgroup of the symmetric group $\mathfrak{S}_n$, then $D_{\mathbb{A}_\lambda^*} = |\mathfrak{H} \cap \mathcal{S}_\lambda|/|\mathfrak{H}|$.*[56]

Theorem 9.16 implies that when $\mathfrak{H}$ contains an element H that is a product of r disjoint cycles of lengths f_i, then $D_{A_\lambda^*} > 0$ for $\lambda = \{f_1,\ldots,f_r\}$, and so there are infinitely many primes p of Dedekind class λ, since the density of finite sets is zero. On the other hand, for sets A_λ^* with $D_{A_\lambda^*} = 0$, and so $\mathfrak{H} \cap \mathcal{S}_\lambda = \emptyset$, Frobenius was able to prove that there is at most a finite number of primes p in A_λ^*.[57] He conjectured that in fact there are *no* primes in A_λ^* when $\mathfrak{H} \cap \mathcal{S}_\lambda = \emptyset$. He formulated this conjecture in the following contrapositive form.

Conjecture 9.17 (Frobenius' first conjecture). Let $\mathbb{K}$ and $\mathfrak{H}$ be as in Theorem 9.16. Then if a prime p has the factorization $p\mathfrak{o}_{\mathbb{K}} = \mathfrak{p}_1\mathfrak{p}_2\cdots\mathfrak{p}_r$, with $deg\, \mathfrak{p}_i = f_i$, there exists a permutation $H \in \mathfrak{H}$ that is a product of r disjoint cycles of respective lengths f_i.

Frobenius was able to confirm his conjecture for all but the finite number of primes p that divide the discriminant of $\mathbb{K}$.[58] The exceptional primes, he realized, are those for which the factorization of $p\mathfrak{o}_{\mathbb{K}}$ involves repeated prime ideal factors, i.e., what are now called *ramified primes*.

9.3.3 Correspondence with Dedekind

The above theorems and question, as well as another theorem and a related "wish conjecture" mentioned below, were formulated and (in the case of the theorems) proved by Frobenius in November 1880 [210, pp. 719–720].[59] Although he was not able to establish the truth of his conjecture, he had reason to think that Dedekind

[56] This theorem is stated in Frobenius' letter to Dedekind of 3 June 1882 (see below) and in [210, §2]. Here $\alpha \in \mathfrak{o}_{\mathbb{K}}$ is not necessary, since α is an algebraic number, and so an integral multiple $m\alpha = \theta$ has the properties described in Dedekind's result.

[57] This according to his letter to Dedekind dated 3 June 1882.

[58] This according to his letter to Dedekind dated 12 June 1882.

[59] Frobenius' paper of 1896 [210] contains Theorems 9.14 and 9.15, and according to Frobenius [210, p. 720] they are stated and proved there as they were in 1880.

would be in a position to deal with it conclusively by virtue of the latter's remark, in his 1877 monograph, that "From the very general researches I am going to publish shortly, the ideals of a normal field ... immediately allow us to find the ideals of an arbitrary subfield" [113, §27, p. 142]. Frobenius seems to have set aside his work with densities in the hope that Dedekind's promised publication would provide him with the means of resolving his conjecture. By mid-1882, however, Dedekind had still not published the promised work, and so Frobenius decided to write to him and ask him about the conjecture. In 1880, he had met Dedekind when the latter paid a visit to Zurich, where he had been a professor at the Polytechnic during 1858–1862, just as Frobenius was in 1880. Dedekind met Frobenius' wife and son as well and retained fond memories of the occasion.[60] The personal contact would have made it easier for Frobenius to write to Dedekind about his conjecture, which he did in a letter of 3 June 1882, thereby initiating a correspondence that continued sporadically for 20 years and was to provide Frobenius with the most consequential mathematical problem of his career—the problem of factoring Dedekind's "group determinant," the problem that led Frobenius to create his theory of group characters and representations (Chapter 12).

In the letter of 3 June 1882, after stating Theorem 9.16, Frobenius presented his conjecture and suggested that Dedekind's still unpublished work alluded to in his 1877 monograph would provide the means to answer it. To give Dedekind an idea of the approach he had taken in investigating the question—and which had presumably proved useful in answering it in the case of unramified primes—Frobenius explained that it was based on a theorem that I formulate here in modern notation and terminology as follows:

Theorem 9.18. *Let $\mathbb{L}$ be a normal extension of $\mathbb{Q}$, p a rational prime, and $\mathfrak{p}$ one of the prime ideals in the factorization of $p\mathfrak{o}_{\mathbb{L}}$. If f is the degree of $\mathfrak{p}$, then there exists an element $F_{\mathfrak{p}} \in \mathrm{Gal}(\mathbb{L}/\mathbb{Q})$ of order f such that for every $\omega \in \mathfrak{o}_{\mathbb{L}}$, $F_{\mathfrak{p}}(\omega) \equiv \omega^p \pmod{\mathfrak{p}}$. Moreover, $F_{\mathfrak{p}}$ is unique if p is unramified in $\mathfrak{o}_{\mathbb{L}}$.*[61]

The element $F_{\mathfrak{p}}$ became known as a *Frobenius substitution or automorphism*; it became a fundamental tool in the theory of numbers. As we shall see below, Dedekind had discovered this theorem as well, and Hilbert independently discovered it later.

After presenting Theorem 9.18 in his letter to Dedekind, Frobenius added, "I am probably right in my assumption that you previously followed the approach of this proposition but then eventually abandoned it and replaced it by a better one." No doubt this was a cryptic allusion to the fact that Theorem 9.18 had not enabled him to answer his question for ramified primes. Dedekind's reply on 8 June 1882 showed that he was half correct: Dedekind had indeed discovered Theorem 9.18, but had not abandoned its approach, because it was, he felt, the best approach to

---

[60]This information is contained in Dedekind's letters to Frobenius dated 8 June 1882 and 8 February 1895. The present location of the Dedekind–Frobenius correspondence is indicated in the preface.

[61]Frobenius simply stated Theorem 9.18 in his letter. His proof is in [210, pp. 728–729].

describing the structure of ideals in the ring $\mathfrak{o}_{\mathbb{K}}$ of algebraic integers of $\mathbb{K}$. Thus the Frobenius automorphism had been discovered by Dedekind as well—and apparently earlier, since Theorem 9.18 is contained in a manuscript entitled "Some theorems from the investigation of the relations between the ideals of different fields," which was enclosed with Dedekind's letter and clearly represents an outline of some of the "very general researches" alluded to by Dedekind in his 1877 memoir (as quoted above).[62]

In the enclosed manuscript, which was an outline, Dedekind indicated how to use properties of $\mathfrak{H} = \mathrm{Gal}\,(\mathbb{L}/\mathbb{Q})$ to describe the factorization

$$\mathfrak{o}_{\mathbb{K}} = \mathfrak{p}_1^{a_1} \cdots \mathfrak{p}_s^{a_s}, \quad \mathfrak{p}_1, \ldots, \mathfrak{p}_s \text{ distinct.} \tag{9.27}$$

To this end, he introduced what is now called (following Hilbert [295]) the decomposition group $\mathfrak{D}$ at p and its subgroup, the inertia group. He showed how to determine the number s of distinct primes, the powers a_i to which they occur, and their degrees. In particular, he showed that s equals the number of double cosets of $\mathfrak{H}$ modulo the two subgroups $\mathfrak{D}$ and $\mathfrak{G} = \mathrm{Aut}(\mathbb{L}/\mathbb{K})$. Frobenius replied that he had discovered all the concepts and results of Dedekind's outline but only for unramified primes.[63] As for double cosets, as mentioned above, they had also been involved in Frobenius' (undisclosed) proof of Theorem 9.14, as noted above. Thus Dedekind and Frobenius also independently introduced the notion of a double coset, although only Frobenius went on to develop the theory of double cosets within the context of abstract finite groups (as indicated in Section 9.4 below).

In his letter of 8 June, Dedekind indicated how Theorem 9.18 can be used to show that the answer to Frobenius' question is affirmative, although he also considered only the unramified case. This was probably due to a confusion regarding notation. Dedekind wrote the prime factorization of p as in (9.27), whereas Frobenius wrote it in the form (9.25), with the $\mathfrak{p}_i$ not necessarily distinct. Thus when Frobenius wrote the factorization (9.25) in his letter, Dedekind probably assumed that Frobenius meant the factorization (9.27) with all $a_i = 1$. When Frobenius explained that his concern was with the ramified case, Dedekind responded hurriedly, due to the press of school-related duties, and wrote that the approach via Theorem 9.18 extended to the general situation (letter dated 14 June 1882), although in a subsequent postcard dated 18 June 1882, he announced that there was an error in his proof-extension and that consequently he was not sure whether the answer was affirmative in all

---

[62]Dedekind did not prove Theorem 9.18 in the enclosed manuscript, but in his letter of 14 June 1882 to Frobenius, he sketched his proof, which differed from Frobenius'. This portion of Dedekind's letter was published by E. Noether in Vol. II of Dedekind's *Mathematische Werke*, pp. 415–416. (Cf. Frobenius' remarks contrasting Dedekind's proof and his own, which was more in line with Kronecker's approach to ideal numbers [210, pp. 729–730]. Dedekind's outline was finally published, with only a few minor notational changes, in 1894 [117]. According to Miyake [442, p. 347], Hasse introduced the term "Frobenius substitution" in [263], apparently unaware that Dedekind had discovered it independently and probably earlier.

[63]This according to his letter to Dedekind dated 12 June 1882.

cases. It turns out that it is not always true when p is ramified. (A counterexample, kindly provided by J.-P. Serre, is given below.) When Frobenius finally published his above-described work of November 1880 in 1896, he simply repeated the conjecture and gave Dedekind's proof for the unramified case [210, pp. 726–27].[64]

9.3.4 Counterexample to the first conjecture

Recall the background to Frobenius' question: $\mathbb{K}$ is an extension of $\mathbb{Q}$ of degree n; $\mathfrak{o}_{\mathbb{K}}$ denotes the algebraic integers in $\mathbb{K}$; $\varphi(x) \in \mathbb{Q}[x]$ is an irreducible polynomial of degree n with a root $\alpha \in \mathbb{K}$ such that $\mathbb{K} = \mathbb{Q}(\alpha)$; $\mathbb{L}$ is the splitting field of φ, and since the elements of $\mathfrak{H} = \text{Gal}\,(\mathbb{L}/\mathbb{Q})$ permute the n roots of φ, $\mathfrak{H}$ will be regarded as a subgroup of the symmetric group $\mathfrak{S}_n$. Thus if $\lambda = \{f_1, \ldots, f_r\}$ denotes a partition of n, the set of all permutations in $\mathfrak{S}_n$ that are expressible as a product of r disjoint cycles of respective lengths f_i is a conjugacy class of $\mathfrak{S}_n$, which I will denote by $\mathcal{S}_\lambda$. A rational prime p will be said to be of class λ if the ideal prime factorization of p in $\mathfrak{o}_{\mathbb{K}}$ is $p\mathfrak{o}_{\mathbb{K}} = \mathfrak{p}_1 \cdots \mathfrak{p}_r$, where the degree of $\mathfrak{p}_i$ is f_i and the $\mathfrak{p}_i$ are not necessarily distinct. Frobenius' conjecture was that if p is of class λ, then a $\sigma \in \mathfrak{H}$ always exists that is in the conjugacy class $\mathcal{S}_\lambda$.

As we saw, Frobenius and Dedekind were both able to prove the theorem that the answer is affirmative for all primes p except the finite number that divide the discriminant of $\mathbb{K}$, i.e., except for the ramified primes. This occurred during the period 1880–1882, after which Frobenius turned to other areas of research and published nothing more in algebraic number theory—except finally to publish in 1896 as [210] the theorems and conjectures he had arrived at in November 1880, as well as to present Dedekind's 1882 proof of the above theorem for unramified primes. In the decades following 1896, there seems to have been no discussion in print of Frobenius' above conjecture, which is not entirely surprising, since—given the proof for unramified primes—the ramified case probably did not seem of much arithmetic interest.[65]

In 1996, and so 100 years after Frobenius published his theorems and conjectures, while I was doing research on the Dedekind–Frobenius correspondence, I asked J.-P. Serre's opinion on Frobenius' above conjecture for ramified primes. He graciously responded by sending me a counterexample, which involved a field $\mathbb{K}$ of degree $n = 10$. Four years later, when I asked him whether I might publish his counterexample in a historical work about Frobenius, he sent me a simpler counterexample (with $n =$

[64] I am grateful to Peter Roquette for calling my attention to the fact that this same result was stated and proved by Artin in 1923 [6, *Hilfsatz*, p. 156] without any reference to Frobenius or Dedekind.

[65] I am grateful to Professors Peter Roquette and Franz Lemmermeyer for sharing their expert knowledge of the literature on algebraic number theory during the first 30 years of the twentieth century.

6), which he later modified slightly so as to explicitly compute the polynomial φ. In what follows, I present the counterexample with $n = 6$ as modified for computational purposes.

Let $\psi(x) \in \mathbb{Q}[x]$ be irreducible of degree 4. Let $\mathbb{L}$ be a splitting field for ψ and let $\mathfrak{H} = \mathrm{Gal}\,(\mathbb{L}/\mathbb{Q})$ denote its Galois group. Choose ψ such that $\mathfrak{H}$, regarded as a subgroup of the symmetric group $\mathfrak{S}_4$, coincides with $\mathfrak{S}_4$. Assume in addition that there is a rational prime p with the following property:

Property 9.19. (a) The decomposition group $\mathfrak{D}$ of p is cyclic of order 4 [one may represent a generator of it by the cyclic permutation (1234)]; (b) The inertia subgroup $\mathfrak{I}$ of $\mathfrak{D}$ is of order 2 [its nontrivial element may be represented by $(1234)^2 = (13)(24)$].

That such a p exists will be proved further on.

There is a natural embedding of $\mathfrak{H}$ into the alternating group $\mathfrak{A}_6$, obtained by making $\mathfrak{H}$ act on the six subsets of $\{1,\ldots,4\}$ with two elements. The image is a transitive subgroup of the alternating group $\mathfrak{A}_6$, which also will be denoted by $\mathfrak{H}$, but now $\mathfrak{H}$ is regarded as a subgroup of $\mathfrak{A}_6$. With the six sets ordered as

$$\{1,2\},\ \{1,3\},\ \{1,4\},\ \{2,3\},\ \{2,4\},\ \{3,4\}, \tag{9.28}$$

let $\beta_1 = \alpha_1 + \alpha_2, \ldots, \beta_6 = \alpha_3 + \alpha_4$ denote the corresponding sums of pairs of the roots $\alpha_1, \ldots, \alpha_4$. Then the polynomial

$$\varphi(x) = \prod_{i=1}^{6} (x - \beta_i) \tag{9.29}$$

is in $\mathbb{Q}[x]$ and is irreducible (since $\mathfrak{H}$ acts transitively). Thus, e.g., $\mathbb{K} = \mathbb{Q}(\beta_1)$ is a subfield of $\mathbb{L}$ of degree 6. The Galois group of φ is $\mathfrak{H}$, which we regard as a subgroup of $\mathfrak{A}_6$. Thus $\mathfrak{H}$ permutes the roots β_i of φ in accordance with the above embedding.

With this interpretation of $\mathfrak{H}$, the above-mentioned generator of $\mathfrak{D}$ is represented in $\mathfrak{A}_6$ by the permutation $S = (25)(1463)$. Likewise, the generator of the inertia subgroup $\mathfrak{I}$ is represented by $S^2 = (16)(34)$. This tells us the decomposition of p in $\mathbb{K}$. It splits as

$$p\mathfrak{o}_{\mathbb{K}} = \mathfrak{p}_1^2 \cdot \mathfrak{p}_2,$$

where $\mathfrak{p}_1, \mathfrak{p}_2$ are each of degree 2. This means that p is of class $\lambda = \{2,2,2\}$.[66] Hence Frobenius' conjecture, if true, would imply the existence in $\mathfrak{H}$ (regarded as a subgroup of $\mathfrak{S}_6$) of a product of three disjoint transpositions. This is impossible, because a disjoint product of three transpositions has signature -1, whereas $\mathfrak{H}$ is

---

[66]Given the above information about $\mathfrak{D}$ and $\mathfrak{I}$, the fact that p is of class $\lambda = \{2,2,2\}$ can be deduced using the theorems in the outline Dedekind sent to Frobenius in 1882 and finally published as [117].

actually a subgroup of the alternating group $\mathfrak{A}_6$. Thus φ and $\mathbb{K} = \mathbb{Q}(\beta_1)$ would provide a counterexample to Frobenius' conjecture, provided ψ can be chosen so as to satisfy (a) and (b) of Property 9.19.

According to Serre, "the easiest way (short of writing a specific equation ...)" is to take a fourth degree polynomial in $\mathbb{Q}[x]$ with Galois group equal to $\mathfrak{C}_4$, a cyclic group of order 4, and such that the decomposition group $\mathfrak{D}$ is equal to $\mathfrak{C}_4$ and the inertia subgroup $\mathfrak{J}$ is of order 2. For example, let $p = 3$ and consider the irreducible polynomial

$$\psi_0(x) = x^4 - x^3 - 4x^2 + 4x + 1. \tag{9.30}$$

The roots of ψ_0 are $\rho_k = 2\cos(2^{k-1}\theta)$, where $\theta = 2\pi/15$ and $k = 1,2,3,4$. Then choose for ψ a polynomial that is 3-adically close enough to ψ_0 so that the decomposition and inertia groups are the same, but random enough so that the Galois group $\mathfrak{H}$ is now the full symmetric group $\mathfrak{S}_4$: "Hilbert's irreducibility theorem tells us we can do that."[67]

With the help of PARI, a software package for computer-aided number theory, Serre provided me with an explicit polynomial ψ and the corresponding polynomial φ of (9.29). Start by adding a "3" to ψ_0 as given in (9.30) to get the polynomial

$$\psi(x) = x^4 - x^3 - 4x^2 + 4x + 4,$$

and check that its Galois group is now the full symmetric group $\mathfrak{S}_4$, and that the decomposition and inertia groups at $p = 3$ have not changed. The proof uses the value of the discriminant of ψ, which is $-2^4 \cdot 3^2 \cdot 67$; the discriminant of the field is smaller: $-2^2 \cdot 3^2 \cdot 67$; an essential fact is that both discriminants are of the form $9u$, with $u \equiv -1 \pmod{3}$. The polynomial φ of (9.29) corresponding to this choice of ψ is

$$\varphi(x) = x^6 - 3x^5 - 5x^4 + 15x^3 - 12x^2 + 4x - 4.$$

To summarize: The Galois group of φ, regarded as a subgroup of $\mathfrak{S}_6$, is a transitive subgroup consisting of permutations of signature $+1$. As a consequence, corresponding to the prime $p = 3$, which is of class $\lambda = \{2,2,2\}$, the Galois group cannot contain an element in the conjugacy class $\mathcal{S}_\lambda$, since $\mathcal{S}_\lambda$ consists of products of three disjoint transpositions, which have signature -1.

[67] Hilbert's irreducibility theorem and its implications were not available to Dedekind and Frobenius in 1882.

9.3.5 An outline of Frobenius' proof of Theorem 9.14

Readers not interested in how Frobenius deduced Theorem 9.14 from Kronecker's Theorem 9.12 can proceed to Section 9.3.6 with little loss of continuity.

For those interested, here is the gist of the proof. Recall that Theorem 9.14 asserts that D_λ, the density of the set A_λ of primes p such that a given polynomial $\varphi(x) \in \mathbb{Z}[x]$ with nonvanishing discriminant has the prime factorization $\varphi(x) \equiv \varphi_1(x)\varphi_2(x)\cdots\varphi_r(x) \pmod p$, is given by $D_\lambda = |\mathfrak{H} \cap \mathcal{S}_\lambda|/h$, where $\mathfrak{H}$ is the Galois group of φ regarded as a subgroup of the symmetric group $\mathfrak{S}_n$, h is the order of $\mathfrak{H}$, and $\mathcal{S}_\lambda$ is the conjugacy class of $\mathfrak{S}_n$ consisting of all products of disjoint cycles of respective lengths $f_1,\ldots,f_r$.

The starting point of Frobenius' proof was the starting point of Kronecker's paper, namely equation (9.20) of Theorem 9.12. To apply Kronecker's theorem, Frobenius expressed (9.20) in the form

$$\sum_{p\in P} \nu_p p^{-(1+x)} = m\log(1/x) + \mathfrak{P}(x), \tag{9.31}$$

where $\mathfrak{P}(x)$ denotes a power series in x that converges in a neighborhood of $x = 0$.

Frobenius' idea was to take an arbitrary subgroup $\mathfrak{G} \subset \mathfrak{S}_n$ and to use a construction of Abel's to create a polynomial $\Phi_{\mathfrak{G}}(x) \in \mathbb{Z}[x]$ of degree $s = n!$ for which he could compute the constants ν_p and m of Kronecker's formula (9.20) in terms of constants associated to $\mathfrak{S}_n$, $\mathfrak{G}$, and $\mathfrak{H}$ [210, pp. 721–723], namely, their respective orders

$$s = (\mathfrak{S}_n : 1) = n!, \quad g = (\mathfrak{G} : 1), \quad \text{and} \quad h = (\mathfrak{H} : 1),$$

and the numbers

$$s_\lambda = |\mathcal{S}_\lambda|, \quad g_\lambda = |\mathfrak{G} \cap \mathcal{S}_\lambda|, \quad h_\lambda = |\mathfrak{H} \cap \mathcal{S}_\lambda|.$$

For expository purposes, it will sometimes be helpful to think of the λ as indexed in some way as $\lambda_1,\ldots,\lambda_l$, where l is the number of conjugacy classes $\mathcal{S}_\lambda$ of $\mathfrak{S}_n$. For the polynomial $\Phi_{\mathfrak{G}}$ Frobenius showed that [210, pp. 723–724, (3.), (5.)]

$$\text{for } p \in A_\lambda, \quad \nu_p = \frac{g_\lambda s}{s_\lambda}. \tag{9.32}$$

To calculate m, the number of irreducible factors of $\Phi_{\mathfrak{G}}(x)$ over $\mathbb{Z}$, in a way that would bring the group $\mathfrak{H}$ into the picture was more difficult. To this end, he invoked a theorem in Jordan's *Traité*, which when applied to $\Phi_{\mathfrak{G}}(x)$ enabled him to show the following: For $S \in \mathfrak{S}_n$, let $d_\lambda^S = |\mathfrak{H} \cap (S^{-1}\mathfrak{G}S) \cap \mathcal{S}_\lambda|$. In other words, d_λ^S is the number of elements of $\mathfrak{H}$ of class λ that are in the group $S^{-1}\mathfrak{G}S$. The result was that

$$\sum_{i=1}^{l}\Big\{\sum_{S\in\mathfrak{S}_n} d^S_{\lambda_i}\Big\} = mh. \tag{9.33}$$

Frobenius considered the meaning of the sum on the left-hand side of (9.33) in a way that was to become a characteristic part of his group-theoretic toolbox, as we shall see in Section 9.4. Let $G_1,\ldots,G_{g_\lambda}$ denote the elements of $\mathfrak{G}\cap\mathcal{S}_\lambda$ and consider the following $g_\lambda \times s$ array of group elements:

$$\begin{array}{ccc} S_1^{-1}G_1S_1 & \cdots & S_s^{-1}G_1S_s \\ \vdots & \vdots & \vdots \\ S_1^{-1}G_{g_\lambda}S_1 & \cdots & S_s^{-1}G_{g_\lambda}S_s \end{array}.$$

The elements in the Sth column are all distinct, and the number of them that are from $\mathfrak{H}$ is by definition d^S_λ. Thus $\sum_{S\in\mathfrak{S}_n} d^S_\lambda$ *equals the total number of places in the above array that are occupied by (not necessarily distinct) elements of* $\mathfrak{H}$.

This number can also be calculated by first considering a row of the array. Consider, e.g., the first row. Its length is $s = n!$ and it contains all s_λ elements of the conjugate class $\mathcal{S}_\lambda$, obviously with repetitions. Frobenius realized that each element is repeated the same number n_1 of times, so that $n_1 s_\lambda = s$, and so the number of repetitions is $n_1 = s/s_\lambda$ and is independent of the row chosen. (Here n_1 is what would now be called the order of the centralizer of G_1 in $\mathfrak{S}_n$. The above way to count the distinct elements in a group list became fundamental to Frobenius' subsequent work on noncommutative groups; see Proposition 9.22.) Now, row one contains all h_λ of the elements from $\mathfrak{H}$ of class λ, and each one is repeated s/s_λ times, so that a total of $h_\lambda(s/s_\lambda)$ places in the first row are occupied by elements of $\mathfrak{H}$, and since this number is independent of the row, $g_\lambda h_\lambda(s/s_\lambda)$ places in the array are occupied by elements from $\mathfrak{H}$. Thus $\sum_{S\in\mathfrak{S}_n} d^S_\lambda = (sg_\lambda h_\lambda)/s_\lambda$, and so (9.33) becomes

$$mh = \sum_{i=1}^{l}\Big\{\sum_{S\in\mathfrak{S}_n} d^S_{\lambda_i}\Big\} = s\sum_{i=1}^{l}\frac{g_{\lambda_i}h_{\lambda_i}}{s_{\lambda_i}}. \tag{9.34}$$

With (9.34), Frobenius had arrived at a suitable expression for m. If it is substituted in (9.31), along with the expression for v_p in (9.32), the result is

$$s\sum_{i=1}^{l}\frac{g_{\lambda_i}}{s_{\lambda_i}}\Big\{\sum_{p\in A_{\lambda_i}} p^{-(1+x)}\Big\} = \frac{s}{h}\Big[\sum_{i=1}^{l}\frac{g_{\lambda_i}h_{\lambda_i}}{s_{\lambda_i}}\Big]\log(1/x)+\mathfrak{P}(x). \tag{9.35}$$

To gain an idea of what might be gleaned from this equation, let us take $\mathfrak{G} = \{E\}$, the group consisting of only the identity permutation E. Then since one of the conjugate classes is $\mathcal{S}_{\lambda_e} = \{E\}$, where $\lambda_e = \{1,1,\ldots,1\}$, we have $g = 1$, $g_{\lambda_e} = h_{\lambda_e} = s_{\lambda_e} = 1$, and $g_{\lambda_i} = 0$ for all $i \neq e$. Thus (9.35) reduces to $\sum_{p\in A_{\lambda_e}} p^{-(1+x)} = (1/h)\log(1/x) + (1/s)\mathfrak{P}(x)$. Dividing both sides of this equation by $\log(1/x)$ and letting $x \to 0^+$, we get, as the density of A_{λ_e}, $D_{\lambda_e} = 1/h$: the density of the primes of class λ_e is the reciprocal of the order of the Galois group of φ. Class λ_e primes are those for which $\varphi(x)$ factors mod p completely into linear factors; and so as the order of the Galois group increases, the primes for which the given polynomial $\varphi(x)$ has a mod p linear factorization become more and more rare in the sense of density.

Frobenius' next idea was to introduce suitably chosen subgroups $\mathfrak{G}$ so that the system of equations that would result from (9.35) would yield $D_{\lambda_i} = h_{\lambda_i}/h$ for all $i = 1, \dots, l$. To this end, he defined $\mathfrak{P}_{\lambda_i}(x)$ by

$$\sum_{p \in A_{\lambda_i}} p^{-(1+x)} = \frac{h_{\lambda_i}}{h} \log(1/x) + \mathfrak{P}_{\lambda_i}(x). \tag{9.36}$$

The functions $\mathfrak{P}_{\lambda_i}(x)$ are thus defined for all sufficiently small values of $x \neq 0$, but despite the suggestive notation, it is not clear that they are power series in x and that they converge for all $|x|$ sufficiently small. Clearly *if* they can be shown to be convergent power series for all $|x|$ sufficiently small, *then* $D_{\lambda_i} = h_{\lambda_i}/h$ is equivalent to (9.36).

Let us consider how Frobenius showed that the $\mathfrak{P}_{\lambda_i}(x)$ are actually convergent power series in x. Substitution of the identities (9.36) in (9.35) gives

$$\sum_{i=1}^{l} \frac{g_{\lambda_i}}{s_{\lambda_i}} \mathfrak{P}_{\lambda_i}(x) = \mathfrak{P}(x). \tag{9.37}$$

For the groups $\mathfrak{G}$, Frobenius then considered the following succession of groups. For each $j = 1, \dots, l$, pick an element $G_j \in \mathcal{S}_{\lambda_j}$ and let $\mathfrak{G}^{(j)}$ be the cyclic group generated by G_j. (Thus one of these groups is $G^{(e)} = \{E\}$, which gave us the density of A_{λ_e}.) For each choice $\mathfrak{G} = \mathfrak{G}^{(j)}$, let $g_{\lambda_i} = g_{\lambda_i}^{(j)}$. Substituting these choices in (9.37) then yields a system of l equations in the l unknowns $\mathfrak{P}_{\lambda_i}$, which in matrix notation (not used by Frobenius) is

$$Mx = \mathfrak{P}(x)e, \quad M = (g_{\lambda_i}^{(j)}/s_{\lambda_i}), \tag{9.38}$$

where $x = \begin{pmatrix}\mathfrak{P}_{\lambda_1} & \cdots & \mathfrak{P}_{\lambda_l}\end{pmatrix}^t$ and $e = \begin{pmatrix}1 & \cdots & 1\end{pmatrix}^t$.

If M is invertible, $x = \mathfrak{P}(x)M^{-1}e$, and so taking ith components of both sides, we see that $\mathfrak{P}_{\lambda_i}(x) = [M^{-1}e]_i\mathfrak{P}(x)$ is a constant multiple of $\mathfrak{P}(x)$, and so, like $\mathfrak{P}(x)$, is a power series in x that converges in a neighborhood of $x = 0$. Thus (9.36) would imply that $D_{\lambda_i} = h_{\lambda_i}/h$. Frobenius introduced partitions of $\mathfrak{S}_n$ into what are now called double cosets with respect to $\mathfrak{H}$ and $\mathfrak{G}^{(j)}$ in order to explicitly calculate $\mathfrak{P}_{\lambda_i}(x) = (M^{-1}e)_i\mathfrak{P}(x)$.[68]

[68] Frobenius never published these calculations per se, but analogous calculations to solve an analogous equation involving the matrix M were published in 1887 [194, pp. 310–311, (8.)–(10)].

By proving that (9.36) holds with the $\mathfrak{P}_{\lambda_i}$ being convergent power series in a neighborhood of $x = 0$, Frobenius had attained his goal and established Theorem 9.14.

9.3.6 Second density theorem and conjecture

Frobenius told Dedekind that he had sent off "a little work" featuring Theorem 9.16 to *Crelle's Journal* in 1881.[69] That work, however, never appeared in that journal nor anywhere else. It would seem that the journal, most likely in the person of Kronecker, was dragging its heels on publication, because in his letter of 12 June 1882 to Dedekind, Frobenius wrote sarcastically that he still hoped to live to see the appearance of this work.[70] Then in his letter of 3 January 1883, Frobenius wrote to Dedekind that during the previous fall, Kronecker had promised him that his little work would soon be published. That it never appeared suggests that Frobenius withdrew his paper with the idea of publishing it in a reworked form after Dedekind had published a full account of his remarkable outlined results.[71] Indeed, when Frobenius finally did publish his 1880 results in 1896, it was after Dedekind had finally published his manuscript in 1894.[72] And so in 1896, Frobenius explained about his correspondence with Dedekind in 1882 and about Dedekind's outline and added that "I had always wished that this draft be published before my own, and this was the reason why I have now finally decided on publication" [210, pp. 719–720].

When Frobenius finally did publish his work on densities in 1896 (for reasons discussed below), he omitted these calculations, as well as any mention of double cosets—except for one telling citation to the 1887 paper [210, p. 725])—and simply introduced a special indexing (a natural ordering) of the λ [210, p. 720] so that the above matrix M is lower triangular with nonzero diagonal entries [210, p. 725] and so the nonsingularity of M is immediate. Frobenius' letter to Dedekind dated 12 June 1882 (quoted in the next section) makes it clear that his original, more complicated, proof of Theorem 9.14 relied on the theory of double cosets. Frobenius' subsequent work on double cosets is discussed in Section 9.4.

[69] Letter dated 3 July 1882.

[70] In his letter of 12 June 1882 to Dedekind, Frobenius referred to this work "deren Erscheinen ich noch zu erleben hoffe." When he submitted his paper in 1881, Borchardt had passed away and Kronecker and Weierstrass were the editors of the journal; and Kronecker, the obvious candidate to review the paper, may have been unhappy with Frobenius' articulation of his density theorem in terms of Dedekind's theory of ideals (as Theorem 9.16 above), a philosophically unsavory rival to his own (announced but still unpublished) theory of ideal numbers.

[71] In 1896, Frobenius said he had originally hoped to rework his results in light of Dedekind's published outline [210, p. 720].

[72] Dedekind simply published his outline of 1882 with a few minor notational changes [117] and did so only because Hilbert [295] was rediscovering the results of his outline, including the existence of Frobenius automorphisms [295, p. 13]. (Thus Dedekind, Frobenius, and Hilbert, in that order, independently introduced Frobenius automorphisms.)

That may have been the main reason behind the decision to publish, but there was another as well. The publication of Dedekind's outline in 1894 had enabled Adolf Hurwitz to discover one of Frobenius' unpublished theorems—Theorem 9.20, stated below and usually known as Frobenius' density theorem—as Frobenius learned from a letter Hurwitz sent him dated 2 January 1896.[73] According to Frobenius [210, p. 720], this letter prompted him to publish the above-described theorems and a related question and conjecture as he had obtained them and written them down in 1880—except for a few abbreviations (*Kürzungen*), presumably such as the replacement of the "double coset" part of the proof of Theorem 9.14 by the above-mentioned shorter and simpler argument, and the use of Dedekind's proof of Theorem 9.18 on Frobenius automorphisms. It is not entirely clear that Frobenius would have gotten around to publishing these old results without the fillip provided by Hurwitz's letter. Judging by his publications between 1880 and 1893, his interest in investigating problems in algebraic number theory had been overridden by the demands of research in other areas. It should come as no surprise that one such area was the theory of finite groups, since group-theoretic considerations underlay all the arithmetic work by Frobenius described in this chapter and reflect his increasing fascination with the properties of groups. As we shall see in the next section, it was not long before he was exploring the theory of finite groups independently of a motivating number-theoretic context.

In November 1880, Frobenius had actually explored the connections between densities and Galois groups further than he had revealed in his letters to Dedekind. This further work was also published in his paper of 1896 and is worth describing here. The "Frobenius automorphism theorem" (Theorem 9.18) suggested to Frobenius another way to divide primes p into classes [210, p. 730]. That is, suppose that $\mathbb{L}$ is a normal extension of $\mathbb{Q}$ of degree n and $\mathfrak{H} = \mathrm{Gal}(\mathbb{L}/\mathbb{Q})$. Consider a rational prime p that is unramified, so that by Theorem 9.18 there is a unique Frobenius automorphism, $F_{\mathfrak{p}} \in \mathfrak{H}$, associated to the prime ideal divisor $\mathfrak{p}$ of p. It was known that every other prime divisor $\mathfrak{p}'$ of p is conjugate to $\mathfrak{p}$ in the sense that $H \in \mathfrak{H}$ exists such that $\mathfrak{p}' = H[\mathfrak{p}] \stackrel{\text{def}}{=} \mathfrak{p}_H$. The Frobenius automorphism associated to $\mathfrak{p}_H$ is $F_{\mathfrak{p}_H} = HF_{\mathfrak{p}}H^{-1}$ (reading the composite automorphism from right to left). Thus $F_{\mathfrak{p}_H}$ is conjugate to $F_{\mathfrak{p}}$ with respect to $\mathfrak{H}$, and the conjugacy class of $\mathfrak{H}$ determined by $F_{\mathfrak{p}}$ contains all the Frobenius automorphisms $F_{\mathfrak{p}'}$ associated to prime ideal factors $\mathfrak{p}'$ of $p\mathfrak{o}_{\mathbb{L}}$. These considerations suggested to Frobenius associating an unramified prime p to the unique conjugacy class $\mathfrak{C}_\mu$ of $\mathfrak{H}$ containing all Frobenius automorphisms $F_{\mathfrak{p}_i}$ associated to a prime $\mathfrak{p}_i$ in the factorization of $p\mathfrak{o}_{\mathbb{L}}$. He then posed the problem of determining, for a fixed conjugacy class $\mathfrak{C}_\mu$ of $\mathfrak{H}$, the density D_μ of the set of all unramified primes p associated to $\mathfrak{C}_\mu$ in the above sense. A similar problem had led to his first density theorem, Theorem 9.14.

Let A_μ denote the set of all unramified primes p associated to the class $\mathfrak{C}_\mu$. By analogy with Theorem 9.14, the hope was to be able to prove that $D_\mu = D_{A_\mu}$ equals

[73]For more information on this encounter with Hurwitz, see [109, p. 47]. Hurwitz did not publish his results, but they were published posthumously in 1926 [305].

$|\mathfrak{C}_\mu|/|\mathfrak{H}|$. As in his proof of Theorem 9.14, Frobenius wanted to use Kronecker's Theorem 9.12, which relates via (9.20) the number m of irreducible factors of a polynomial $\Phi(x) \in \mathbb{Z}[x]$ with the integers ν_p, which give the number of solutions to $\Phi(x) \equiv 0 \,(\mathrm{mod}\, p)$. In this case, he showed that given a subgroup $\mathfrak{G} \subset \mathfrak{H} = \mathrm{Gal}\,(\mathbb{L}/\mathbb{Q})$ of order g, there is a natural candidate for Φ corresponding to $\mathfrak{G}$. Let ξ denote an algebraic integer in the fixed field of $\mathfrak{G}$, and let $H_1,\ldots,H_k$ denote the $k = h/g$ coset representatives of $\mathfrak{H}$ modulo $\mathfrak{G}$. Then $\psi(x) = \prod_{i=1}^{k}(x - H_i(\xi)) \in \mathbb{Z}[x]$ is irreducible and $\Phi(x) = \prod_{H\in\mathfrak{H}}(x - H(\xi)) = [\psi(x)]^g$. Thus the number m of irreducible factors of Φ is $m = g = |\mathfrak{G}| = \sum_\mu g_\mu$, where $g_\mu = |\mathfrak{G} \cap \mathfrak{C}_\mu|$.

Frobenius also showed that for Φ, one has $\nu_p = hg_\mu/h_\mu$ for $p \in A_\mu$, where $h_\mu = |\mathfrak{C}_\mu|$. This meant that the reasoning behind the proof of Theorem 9.14 could be used, and would result in the analogue of (9.37). However, as Frobenius explained [210, p. 731], in this case, the use of cyclic subgroups did not lead to a system of equations analogous to (9.38) that had sufficed to solve for the $\mathfrak{P}_i(x)$ and so prove that $D_\mu = h_\mu/h$. Recall that in that theorem, $h_\lambda = |\mathfrak{H} \cap \mathcal{S}_\lambda|$, i.e., $\mathfrak{H}$ was partitioned into the sets $\mathfrak{H} \cap \mathcal{S}_\lambda$. This was a coarser partition than the partition into conjugacy classes C_μ of $\mathfrak{H}$ behind the failed attempt to prove $D_\mu = h_\mu/h$, since in general, $\mathcal{S}_\lambda$ can contain several conjugate classes $\mathfrak{C}_\mu, \mathfrak{C}_{\mu'}, \ldots$ of $\mathfrak{H}$, i.e., $H_1, H_2 \in \mathfrak{H}$ can be conjugate within $\mathfrak{S}_n$ without being conjugate within $\mathfrak{H}$. In particular, if $\mathfrak{C}_\mu$ is a conjugacy class of $\mathfrak{H}$, and if $F \in \mathfrak{C}_\mu$ has order f, then F^r is always conjugate to F within $\mathfrak{S}_n$ when r and f are relatively prime; but F^r need not be in $\mathfrak{C}_\mu$.

These considerations forced Frobenius to consider a partition of $\mathfrak{H}$ that was less refined than the partition into its conjugacy classes—a partition into what he called *divisions* (*Abteilungen*). The division $\mathfrak{D}_\nu$ of $\mathfrak{H}$ determined by an element $F \in \mathfrak{H}$ consists of the conjugacy class $\mathfrak{C}_\mu$ of F together with the conjugacy classes of all the elements F^r as r runs through the $\phi(f)$ positive integers $r < f$ that are relatively prime to f, the order of F. Two divisions are either identical or disjoint so that the divisions provide a partition of $\mathfrak{H}$. Let $\mathfrak{H} = \mathfrak{D}_{\nu_1} \cup \cdots \cup \mathfrak{D}_{\nu_d}$ denote this partition. Frobenius showed that for all $\mathfrak{C}_\mu, \mathfrak{C}_{\mu'} \subset \mathfrak{D}_\nu$, one has $h_\mu = h_{\mu'}$ and $g_\mu = g_{\mu'}$. These relations enabled him to obtain a $d \times d$ system of equations [210, p. 732] that could be solved so as to establish the following result, which is usually called the Frobenius density theorem.

Theorem 9.20 (Frobenius density theorem). *Let $\mathbb{L}$ be a normal extension of $\mathbb{Q}$ and $\mathfrak{H} = \mathrm{Gal}\,(\mathbb{L}/\mathbb{Q})$. Then if D_ν denotes the density of the set of all unramified primes associated to some conjugate class $\mathfrak{C}_\mu \subset \mathfrak{D}_\nu$, one has $D_\nu = h_\nu k_\nu/h$, where k_ν is the number of conjugacy classes $C_\mu \subset \mathfrak{D}_\nu$, h_ν is the common cardinality of these classes, and $h = |\mathfrak{H}|$. (In other words, $D_\nu = |\mathfrak{D}_\nu|/|\mathfrak{H}|$.)*

Frobenius was clearly disappointed that he had to resort to divisions to get the proof idea of his first density theorem to remain viable within the context of a classification of rational primes based on the conjugacy classes of their Frobenius automorphisms. After stating the above theorem, he pointed out that *had* he been able to imitate the proof of his first density theorem, "then the simple expression $[D_\mu = h_\mu/h]\ldots$ would have resulted" [210, p. 732]. Indeed, Frobenius conjectured that this was the case:

Conjecture 9.21 (Frobenius' second conjecture). "To every [conjugacy] class of substitutions of the group $\mathfrak{H}$ correspond infinitely many rational prime numbers. Their density is proportional to the number of distinct substitutions of the class" [210, p. 732].

Frobenius had used the word "conjecture" (*Vermutung*) when speaking of his first conjecture (Conjecture 9.17), and in 1896, when he finally published his results, after mentioning Conjecture 9.17 and the fact that he had posed it to Dedekind, he simply gave Dedekind's reply (and concomitant proof) that the conjecture was true for unramified primes [210, p. 727]. He no longer mentioned the still unresolved case of ramified primes and had either lost interest in the question or now doubted whether the conjecture might hold in that case. With regard to Conjecture 9.21, however, Frobenius never used the word "conjecture"; rather, he expressed it as Theorem V [210, p. 732], Theorem IV being Theorem 9.20 above, which strongly suggests that he was convinced of its veracity although unable to prove it. One of the readers of Frobenius' above-quoted words was the Russian number theorist N. Chebotarev (1894–1947), who in 1922 succeeded in proving Theorem V, which is now known as the Chebotarev density theorem.[74] As we shall see in Section 15.6.3, Chebotarev's proof was used by E. Artin, in a historically interesting and indirect way, to justify the generalization of Dirichlet's L-series that he had obtained using the Frobenius automorphism and Frobenius' theory of group characters and representations.[75]

9.4 Group Lists and Group Equations

In the previous two sections we have seen how arithmetic work had led Frobenius to the study of groups. In Section 9.1, the arithmetic context was provided inter alia by Gauss' theory of composition of forms and the theories of ideal numbers (Kummer) and ideal sets (Dedekind), and it led to Frobenius' systematic study with Stickelberger of the properties of finite abelian groups. In Section 9.3, the motivation came from Kronecker's study of Dirichlet densities and Dedekind's theory of ideals. One common thread we find running through the group-theoretic parts of these investigations by Frobenius had to do with what I would describe as the problem of relating the length of a listing of group elements, which need not be distinct, to the number of solutions to a group equation so as to determine the number of distinct elements in the list.

---

[74]For further information on Chebotarev, his density theorem, and its significance, see [548].

[75]When in 1896, Frobenius finally published his theorems and conjectures from 1880, he was in the process of creating his theory of group characters and representations, but there is no evidence to suggest that he ever considered using his characters to generalize L-series or that he suspected a connection with Theorem V.

The simplest example occurred early in Frobenius' paper with Stickelberger [235, §2]. If $\mathfrak{A}, \mathfrak{B}, \mathfrak{C}, \ldots$ are subgroups of some abelian group $\mathfrak{G}$, consider the group $\mathfrak{H} = \mathfrak{A}\mathfrak{B}\mathfrak{C}\cdots$, which is the list of all elements of the form $ABC\cdots$ that are generated as A runs through $\mathfrak{A}$, B through $\mathfrak{B}$, C through $\mathfrak{C}$, and so on. If $a,b,c,\ldots$ are the orders of $\mathfrak{A}, \mathfrak{B}, \mathfrak{C}, \ldots$, then the length of the list is $L = abc\cdots$. The order h of the group product $\mathfrak{H}$ is the number of distinct elements in this list. The numbers L and h are easily seen so be related to the number e of solutions $(A,B,C,\ldots)$ to the group equation $ABC\cdots = E$, where E is the identity element of $\mathfrak{G}$. That is, $L = he$ or $abc\cdots = he$ [235, p. 551]. From this result, the authors obtained Cauchy's theorem for abelian groups: if a prime p divides the order of a finite abelian group $\mathfrak{H}$, then $\mathfrak{H}$ contains an element of order p. The proof is simple. If $A,B,C,\ldots$. form any basis for $\mathfrak{H}$, then $\mathfrak{H} = \mathfrak{A}\mathfrak{B}\mathfrak{C}\cdots$, where $\mathfrak{A}, \mathfrak{B}, \mathfrak{C}, \ldots$ are now the cyclic subgroups generated, respectively, by $A,B,C,\ldots$. Thus $he = abc\cdots$, where $a,b,c,\ldots$ are the orders of the basis elements $A,B,C,\ldots$. Thus $p \mid h$ means $p \mid he = abc\cdots$ and so p divides one of the factors, e.g., $p \mid a$. Then $a = pa'$ and $P = A^{a'}$ is an element of $\mathfrak{H}$ of order p.

In Frobenius' work on Dirichlet densities, the groups considered were subgroups $\mathfrak{H}$ of the symmetric group $\mathfrak{S}_n$ and so not generally abelian, but he discovered that the same sort of reasoning with lists and equations could frequently be put to good use. One of the simplest illustrations occurs early on with $\mathfrak{H} = \mathfrak{S}_n$ [210, §1] and then again with $\mathfrak{H}$ a subgroup of $\mathfrak{S}_n$ [210, §4], which is eventually taken as the Galois group of a normal extension of $\mathbb{Q}$. Fix an element $F \in \mathfrak{H}$ and consider the totality of elements conjugate to F, which is given by the list of all elements $H^{-1}FH$ as H runs through $\mathfrak{H}$. Frobenius actually denoted the distinct elements of $\mathfrak{H}$ by $H_1,\ldots,H_h$ and wrote out the above list as an actual list [210, (10.), p. 728]:

$$H_1^{-1}FH_1, \quad H_2^{-1}FH_2, \quad \ldots, \quad H_h^{-1}FH_h.$$

The elements in this list he called a *class* of $\mathfrak{H}$, and of course it is what is now called a *conjugacy class* of $\mathfrak{H}$. The length of this list is clearly h, the order of $\mathfrak{H}$. Again, simple considerations show that each distinct element in the list is repeated exactly e times, where now e is the number of solutions H to the equation $HF = FH$, or as would now be said, e is the order of the centralizer of F. In other words:

Proposition 9.22. *If $\mathfrak{H}$ is a group of order h and if $\mathfrak{C}_\lambda$ is a conjugacy class of $\mathfrak{H}$ containing h_λ elements, then in the list of elements $H^{-1}FH$ as H runs through $\mathfrak{H}$ and $F \in \mathfrak{C}_\lambda$ is fixed, every element of $\mathfrak{C}_\lambda$ is repeated exactly e_λ times, where e_λ is the number of solutions H to $HF = FH$. Hence $h = h_\lambda e_\lambda$.*

As we have seen, Frobenius' work on analytic densities was apparently set aside, awaiting a thoroughgoing reworking that would be based on the paper Dedekind would eventually publish using the outline he had sent in 1882. The work on densities was actually a research sideline for Frobenius, who was busily exploring various algebraic aspects of the theory of elliptic and abelian functions. During the period 1880–1889 he published 18 papers, and 13 of them were on some aspect of that theory. (Some of this work is discussed in Chapters 10 and 11 and in

Section 12.4.) Nonetheless, he did find the time to write up for publication many of his group-theoretic results, all presented within the context of the theory of finite abstract groups.

9.4.1 *An abstract proof of Sylow's first theorem*

In March 1884, Frobenius submitted a short paper [193] presenting a new proof of Sylow's theorem of 1872 that if a prime power p^ν divides the order of a permutation group $\mathfrak{H}$, then it contains a subgroup of order p^ν.[76] This was Frobenius' first paper dealing with the theory of finite groups without any connection whatsoever to arithmetic applications.[77] Sylow's theorem was a generalization of Cauchy's theorem that if p divides the order of a permutation group $\mathfrak{H} \subset \mathfrak{S}_n$, then $\mathfrak{H}$ contains an element of order p. Cauchy had in fact demonstrated his theorem by means of a lemma, which amounts to Sylow's theorem for the full symmetric group $\mathfrak{S}_n$. Since Sylow had assumed Cauchy's theorem in his own proof, this circumstance led Kronecker's student Eugen Netto to give a new proof of Sylow's theorem in 1877 [452] that was based directly on Cauchy's lemma.

No doubt it was Netto's paper that provided the fillip for Frobenius' search for a different proof, since all previous proofs, including Netto's, "drag the symmetric group" into the argument, even though it is "completely foreign to the content of Sylow's theorem" [193, p. 301]. And so Frobenius, following the lead of Kronecker and Dedekind and continuing the precedent of his paper with Stickelberger on abelian groups, gave a completely abstract group-theoretic formulation and proof of Sylow's theorem.[78] As Frobenius showed, by removing Sylow's theorem from the extraneous setting of permutation groups, it is possible to give a simpler proof. In retrospect, it was a pioneering effort toward establishing

---

[76] Sylow first proved that if a permutation group $\mathfrak{G}$ has order $p^\sigma q$, where p is prime and $p - q$, then it contains a subgroup of order p^σ [558, p. 586, Thm. I]. Then he proved that a group $\mathfrak{H}$ of order p^σ has a composition series in which the factor groups all have order p, which implies that $\mathfrak{H}$ (and so $\mathfrak{G}$) contains subgroups of order p^ν for $1 \leq \nu \leq \sigma$ [558, p. 588, Thm. III].

[77] For unknown reasons, this paper did not appear in print until 1887. Usually Frobenius' papers in *Crelle's Journal* appeared less than a year after being submitted. Another exception to this generality was the paper on densities and Galois groups, which was submitted in 1881 but was never published. In both cases, Kronecker was the editor of the journal who would have been responsible for judging Frobenius' papers. As already noted in the previous section, Kronecker may not have been completely satisfied with the content of the first paper, because it utilized Dedekind's theory of ideals. The paper submitted in 1884 may also have raised objections. It was motivated by the proof of Sylow's theorem given by Kronecker's student Eugen Netto, and in effect criticized that proof (as indicated below).

[78] Frobenius prefaced his definition of an abstract finite group by citing Kronecker's 1870 paper (Section 9.1.4) and an 1882 paper by Dedekind's collaborator Heinrich Weber [578], who systematically developed Dedekind's ideas about abstract groups.

the value of an abstract approach to many group-theoretic results that had been obtained within the context of Galois' theory of equations, where Galois groups were regarded as groups of permutations of the roots of the associated polynomial. It also represents another example in which Frobenius' originality was primarily in his method of dealing with an already established result. His proof, with its elegant application of his list-equation techniques, has become one of the standard proofs.[79] I will sketch it to show the extent to which it turns on the use of these techniques.

Let $\mathfrak{H}$ denote a finite group whose order h is divisible by the prime power p^ν. Frobenius' proof is by induction on h, and so it is assumed that if $\mathfrak{G}$ is any group of order $g < h$ such that $p^\mu \mid g$, then $\mathfrak{G}$ contains a subgroup of order p^μ. Frobenius then considered the subgroup $\mathfrak{G}$ of all elements of $\mathfrak{H}$ that commute with all elements, i.e., what is now called the *center* of $\mathfrak{H}$. Let g denote the order of $\mathfrak{G}$. There are two cases to consider: (1) $p \mid g$ and (2) $p - g$. If $p \mid g$, then by Cauchy's theorem for abelian groups, $\mathfrak{G}$ contains an element P of order p. Frobenius could have invoked this result from his paper with Stickelberger, but instead he described the list-equation reasoning (given above) that proves it. The proof then follows readily in case (1), since the quotient group $\mathfrak{H}/(P)$, where (P) denotes the cyclic subgroup generated by P, has order $h/p < h$, which is divisible by $p^{\nu-1}$. The induction hypothesis then implies the existence of a subgroup $\mathfrak{K} \subset \mathfrak{H}$ such that $\mathfrak{K}/(P)$ has order $p^{\nu-1}$. It then follows that (since P is in the center of $\mathfrak{H}$) the product $\mathfrak{K} \cdot (P)$ is a subgroup of $\mathfrak{H}$ with order p^ν.

To deal with case (2), Frobenius also utilized a list-equation argument borrowed from his work on densities (and described above). He divided $\mathfrak{H}$ into conjugacy classes and observed that each element in the center $\mathfrak{G}$ constitutes the sole member of its conjugacy class. Since these classes partition $\mathfrak{H}$, this means that

$$h = g + h_1 + \cdots + h_m, \tag{9.39}$$

where $h_1, \ldots, h_m$ are the numbers of elements in the conjugacy classes $\mathfrak{C}_1, \ldots, \mathfrak{C}_m$ determined by elements not in the center. (Nowadays, this is sometimes called the *class equation.*) A list-equation argument then follows and shows that $h = h_i e_i$, where e_i denotes the number of solutions to $HH_i = H_i H$ for $H_i \in \mathfrak{C}_i$. (This is, of course, a repetition of the proof of Proposition 9.22.) These solutions, Frobenius now observed, form a subgroup $\mathfrak{G}_i$ of order e_i (the centralizer of H_i). Now p^ν divides $h = h_i e_i$, but it cannot be that $p^\nu \mid h_i$ for all i, for then (9.39) would imply that $p^\nu \mid g$, contrary to the $p - g$ hypothesis of case (2). Thus for some i, it must be that p^ν divides the order e_i of the group $\mathfrak{G}_i$. Since $h_i > 1$, it follows from $h = h_i e_i$ that $e_i < h$, and so the induction hypothesis implies that $\mathfrak{G}_i$, and hence $\mathfrak{H}$, contains a subgroup of order p^ν.

---

[79] See the account by Curtis [109, p. 41], who also outlines Frobenius' proof using current terminology.

9.4.2 *Double cosets*

Frobenius continued to publish work dealing with algebraic aspects of the theory of elliptic and abelian functions, but he found enough time to submit, in December 1886, a substantial paper dealing with further consequences of list-equation considerations in the theory of finite groups [194]. This paper concerned what are now called double cosets. If $\mathfrak{U}$ is a group and $\mathfrak{G},\mathfrak{H}$ are two subgroups, write $A \sim B \pmod{\mathfrak{G},\mathfrak{H}}$ if $GAH = B$ for some $G \in \mathfrak{G}$ and $H \in \mathfrak{H}$. This is an equivalence relation and so partitions $\mathfrak{U}$ into equivalence classes, which are now called double cosets. As I mentioned in Section 9.3.2, Frobenius had considered a double coset decomposition of the symmetric group $\mathfrak{S}_n$ in order to find the solution to the system of equations (9.38) that brought his proof of his first density theorem (Theorem 9.14) to a successful conclusion. Indeed, referring to his ill-fated paper submitted to *Crelle's Journal* in 1881, Frobenius wrote to Dedekind in 1882 that "The method of decomposing a permutation group Φ by means of two subgroups Ψ and Φ' contained therein also forms the essential foundation of my generalization of Kronecker's theorem, and is analyzed at considerable length in my work."[80] Since the paper submitted in 1881 was not to appear, Frobenius had evidently decided to extract the purely group-theoretic part, translate it into abstract form, and publish it in [194].

As Frobenius pointed out in [194], the equivalence $A \sim B \pmod{\mathfrak{G},\mathfrak{H}}$ is analogous to the more familiar one: $A \sim B \pmod{\mathfrak{G}}$ if $GA = B$ for some $G \in \mathfrak{G}$, or equivalently, if $AB^{-1} \in \mathfrak{G}$. The familiar partition into equivalence classes (now called cosets) was, however, much simpler. Each coset has the same number of elements in it, and so if (following Frobenius) we denote the number of cosets by $(\mathfrak{U} : \mathfrak{G})$ and the orders of $\mathfrak{U}$ and $\mathfrak{G}$ by $(\mathfrak{U} : 1)$ and $(\mathfrak{G} : 1)$, respectively, then clearly $(\mathfrak{U} : 1) = (\mathfrak{U} : \mathfrak{G})(\mathfrak{G} : 1)$. For double cosets, no such simple formula for their number is possible, because double cosets do not all have the same cardinality. Thus the problem of determining the number of double cosets, which Frobenius denoted by $(\mathfrak{U} : \mathfrak{G},\mathfrak{H})$, is far more challenging. Frobenius' solution in his paper [194, pp. 307–310], given below, must have been worked out in 1880, since he used the solution to solve the system of equations (9.38) behind his first density theorem.[81]

Here is the idea of his proof. Consider a fixed double coset $\mathfrak{G}U_1\mathfrak{H}$ of $\mathfrak{U}$. Its elements are all contained in the list of elements GU_1H, as G and H run through $\mathfrak{G}$ and $\mathfrak{H}$, respectively. The length of this list is clearly gh, $g = (G : 1)$, and $h = (H : 1)$. If $U_1, \ldots, U_c$ are the distinct elements of $\mathfrak{G}U_1\mathfrak{H}$, it is easily seen that each element U_i is repeated in the list e_i times, where e_i is the number of solutions (G,H) to

---

[80] Letter of 12 June 1882. The quoted passage is this: "Die Methode der Zerlegung einer Substitutionsgruppe Φ nach zwei darin enthaltenden Untergruppen Ψ und Φ' bildet auch für meine Verallgemeinerung des Kroneckerschen Satzes die wesentliche Grundlage und ist in meiner Arbeit ... ziemlich breit auseinandergesetzt."

[81] See [194, pp. 310–311].

$GU_1H = U_i$. Furthermore, it is easy to see that $e_1 = e_2 = \cdots = e_c$. This means that every element in the list GU_1H is repeated e_1 times, so that $c = gh/e_1$, i.e.,

$$ce_1 = gh. \tag{9.40}$$

By virtue of (9.40), Frobenius could see that the number of solutions to $GUH = U$ with $G \in \mathfrak{G}$, $H \in \mathfrak{H}$, and $U \in \mathfrak{G}U_1\mathfrak{H}$, is $e_1 + \cdots + e_c = ce_1 = gh$. This number is independent of the particular double coset, and so he had proved the following:

Proposition 9.23. *If $m = (\mathfrak{U} : \mathfrak{G}, \mathfrak{H})$, then the number e of solutions (G,U,H) to the equation*

$$GUH = U, \quad \text{with} \quad G \in \mathfrak{G},\ U \in \mathfrak{U},\ H \in \mathfrak{H}, \tag{9.41}$$

is $e = mgh$.

Proposition 9.23 does not provide a formula for $m = (\mathfrak{U} : \mathfrak{G}, \mathfrak{H})$ in terms of known group constants, since e is equally unknown, but Frobenius' idea was to find an alternative characterization of e, which might be combined with the one above so as to eliminate e. The springboard for the idea was the fact that (9.41) can be written as $U^{-1}GU = H^{-1}$, which clearly has the same number of solutions as

$$U^{-1}GU = H. \tag{9.42}$$

Now (9.42) says that G and H are in the same conjugacy class of $\mathfrak{U}$, so let $\mathfrak{C}_\lambda$, $\lambda = 1, \ldots, l$, denote the conjugacy classes of $\mathfrak{U}$, and set

$$u_\lambda = |C_\lambda|, \quad g_\lambda = |\mathfrak{G} \cap \mathfrak{C}_\lambda|, \quad h_\lambda = |\mathfrak{H} \cap \mathfrak{C}_\lambda|.$$

First consider the number of solutions to (9.42) with $G = G_\lambda \in \mathfrak{G} \cap \mathfrak{C}_\lambda$ fixed, $H \in \mathfrak{U}$ fixed, and U running through $\mathfrak{C}_\lambda$. The list of elements $U^{-1}G_\lambda U$, with U running through $\mathfrak{U}$, has length $u = (\mathfrak{U} : 1)$ with u_λ distinct elements. By Proposition 9.22, each element is repeated the same number $e_\lambda = u/u_\lambda$ of times. Thus in (9.42), with G fixed in $\mathfrak{G} \cap \mathfrak{C}_\lambda$, we get a particular $H \in \mathfrak{U}$ exactly $e_\lambda = u/u_\lambda$ times, i.e., the number of solutions to (9.42) with U running through $\mathfrak{C}_\lambda$ and G_λ and H fixed is u/u_λ. If we also let G_λ run through $\mathfrak{G} \cap \mathfrak{C}_\lambda$, we then get $g_\lambda u/u_\lambda$ solutions to (9.42). Next, if we also let H run through $\mathfrak{H}$, (9.42) implies that it is constrained to run through $\mathfrak{H} \cap \mathfrak{C}_\lambda$, and so now the number of solutions will be $h_\lambda g_\lambda u/u_\lambda$. This is the number of solutions to (9.42) with U, G, and H in $\mathfrak{C}_\lambda$, $\mathfrak{G} \cap \mathfrak{C}_\lambda$, and $\mathfrak{H} \cap \mathfrak{C}_\lambda$, respectively. Thus the total number e of solutions to (9.42) will be $e = \sum_{\lambda=1}^{l} h_\lambda g_\lambda u/u_\lambda$. Comparison of this expression for e with the one in Proposition 9.23 then yields the desired formula for $m = (\mathfrak{U} : \mathfrak{G}, \mathfrak{H})$:

$$(\mathfrak{U} : \mathfrak{G}, \mathfrak{H}) = \sum_{\lambda=1}^{l} \frac{g_\lambda h_\lambda u}{ghu_\lambda}. \tag{9.43}$$

9.4.3 Double cosets and Sylow's three theorems

All the above-described results were certainly developed by Frobenius in 1880—at least with $\mathfrak{U} = \mathfrak{S}_n$—since (9.43) with $\mathfrak{U} = \mathfrak{S}_n$ was used to solve the system of equations (9.38), which enabled Frobenius to prove his first density theorem (Theorem 9.14) in November 1880 (cf. [194, pp. 310–312]). There are many other applications of double cosets to group theory in Frobenius' 27-page paper, although some may have been discovered after 1880. Of particular interest are his abstract proofs of other theorems contained in Sylow's paper of 1872. All of Sylow's results were presented within the context of permutation groups, but as Frobenius was the first to show, several of them can be extricated from that context and, as a result, be given simpler proofs. The three main results were summarized by Frobenius as follows:

Theorem 9.24 (Sylow). *(I) If $\mathfrak{U}$ has order $u = p^\sigma q$, where p is prime and $p - q$, then $\mathfrak{U}$ contains a subgroup of order p^ν for all $1 \leq \nu \leq \sigma$. (II) With $\mathfrak{U}$ as in (I), if $\mathfrak{G}$ is a subgroup of order p^σ, then every subgroup $\mathfrak{H} \subset \mathfrak{U}$ of order p^ν, $1 \leq \nu \leq \sigma$, is conjugate to a subgroup of $\mathfrak{G}$. In particular, every subgroup $\mathfrak{H}$ of order p^σ is conjugate to $\mathfrak{G}$. (III) The number N of subgroups of $\mathfrak{U}$ of order p^σ is such that $N \equiv 1 \pmod{p}$.*

Frobenius had of course already proved (I) in [193], but in [194, p. 313, II], he gave another proof using double cosets. It may be that after discovering his first proof of (I) he discovered that all of Sylow's results (I)–(III) could be proved abstractly using his theory of double cosets. Below, I will sketch his proofs of (II) and (III).

Consider first (II), which Sylow proved only for subgroups of order p^σ [558, p. 588, Thm. II]. Recall that en route to proving Proposition 9.23, Frobenius had proved (9.40), i.e., that if the distinct double cosets of $\mathfrak{U}$ with respect to two subgroups $\mathfrak{G}, \mathfrak{H}$ are

$$\mathfrak{G}U_1\mathfrak{H}, \ldots, \mathfrak{G}U_\lambda\mathfrak{H}, \ldots, \mathfrak{G}U_m\mathfrak{H}, \quad m = (\mathfrak{S} : \mathfrak{G}, \mathfrak{H}), \tag{9.44}$$

then

$$c_\lambda e_\lambda = gh, \text{ where } g = (\mathfrak{G} : 1),\ h = (\mathfrak{H} : 1),\ c_\lambda = |\mathfrak{G}U_\lambda\mathfrak{H}|, \tag{9.45}$$

and e_λ is the number of solutions $(G, H) \in \mathfrak{G} \times \mathfrak{H}$ to $GU_\lambda H = U_\lambda$. This is the same as saying that e_λ is the number of $G \in \mathfrak{G}$ such that $U_\lambda^{-1} G U_\lambda \in \mathfrak{H}$, which implies that

$$e_\lambda = \text{ order of } (U_\lambda^{-1}\mathfrak{G}U_\lambda) \cap \mathfrak{H}. \tag{9.46}$$

From this characterization of e_λ it follows that e_λ divides the orders of both $U^{-1}\mathfrak{G}U_\lambda$ and $\mathfrak{H}$, i.e., e_λ is a common divisor of g and h. From (9.45) it then follows that c_λ is a common multiple of g and h. These facts were used by Frobenius in the following manner. By (9.45), $c_\lambda = g(h/e_\lambda) = gf_\lambda$, where f_λ is an integer, since e_λ divides h. Frobenius discovered that the relations

$$c_\lambda = gf_\lambda, \quad c_\lambda = |\mathfrak{G}U_\lambda\mathfrak{H}|, \quad g = (\mathfrak{G} : 1), \tag{9.47}$$

were very useful for proving Sylow's theorems. Since $c_\lambda e_\lambda = gh$ by virtue of (9.45) and $(\mathfrak{U} : 1) = u = \sum_{\lambda=1}^{m} c_\lambda = \sum_{\lambda=1} gf_\lambda$, we also have the following two consequences of (9.47):

$$(i)\ h = e_\lambda f_\lambda; \quad \text{and} \quad (ii)\ \frac{u}{g} = \sum_{\lambda=1}^{m} f_\lambda. \tag{9.48}$$

The above general setup will apply to Frobenius' proof of both (II) and (III).

Focusing first on (II), he let $\mathfrak{G}$ be a subgroup of $\mathfrak{U}$ of order p^σ [by (I)] and $\mathfrak{H}$ a subgroup of $\mathfrak{U}$ of order p^ν. To prove that $\mathfrak{H}$ is conjugate to a subgroup of $\mathfrak{G}$, he proceeded as follows. Since now $g = p^\sigma$ and $u = p^\sigma q$, $p - q$, equation (9.48) (ii) implies that $q = (u/g) = \sum_{\lambda=1}^{m} f_\lambda$, and since $p - q$, it must be that $p - f_\lambda$ for some $\lambda = \lambda_0$. Then by (i) of (9.48), $p^\nu = h = e_{\lambda_0} f_{\lambda_0}$, and so $p - f_{\lambda_0}$ means that $f_{\lambda_0} = 1$ and $e_{\lambda_0} = p^\nu$. The latter equality means that $h = (\mathfrak{H} : 1) = e_{\lambda_0}$, and since by (9.46), e_{λ_0} is also the order of a subgroup of $\mathfrak{H}$, it must be that $(U_{\lambda_0}^{-1}\mathfrak{G}U_{\lambda_0}) \cap \mathfrak{H} = \mathfrak{H}$, i.e., that $\mathfrak{H} \subset U_{\lambda_0}^{-1}\mathfrak{G}U_{\lambda_0}$, and so $U_{\lambda_0}\mathfrak{H}U_{\lambda_0}^{-1} \subset \mathfrak{G}$. This shows that $\mathfrak{H}$ is conjugate to a subgroup of $\mathfrak{G}$, and (II) is proved.

Now consider (III), which concerns the number N of subgroups of $\mathfrak{U}$ of order p^σ. By (I)–(II), we know that there is at least one such subgroup, here denoted by $\mathfrak{H}$, and that every other such subgroup is conjugate to $\mathfrak{H}$. All possible conjugate subgroups of $\mathfrak{H}$ are thus contained in the list of subgroups $U^{-1}\mathfrak{H}U$ as U runs through $\mathfrak{U}$. The length of this list is $u = (\mathfrak{U} : 1)$. Consider the set $\mathfrak{G}$ of all $U \in \mathfrak{U}$ such that $U^{-1}\mathfrak{H}U = \mathfrak{H}$. Then $\mathfrak{G}$ is a subgroup of $\mathfrak{U}$, and it was already introduced by Sylow in his proofs. It is now known as the *normalizer of* $\mathfrak{H}$, and clearly $\mathfrak{H} \subset \mathfrak{G}$. It then follows that if g is the order of $\mathfrak{G}$ (so g is the number of solutions U of the equation $U^{-1}\mathfrak{H}U = \mathfrak{H}$), then the number of distinct conjugates in the list is u/g. Thus $N = u/g$.

Now let (9.44) denote the double coset decomposition of $\mathfrak{U}$ with respect to the above two subgroups, $\mathfrak{H}$ and its normalizer $\mathfrak{G}$. Assume that the notation has been chosen so that the identity element is in the first double coset, $\mathfrak{G}U_1\mathfrak{H}$. This means that we could use $U_1 = E$ as double coset representative, and so the coset is $\mathfrak{G}\mathfrak{H} = \mathfrak{G}$. Thus $c_1 = |\mathfrak{G}U_1\mathfrak{H}| = |\mathfrak{G}| = g$, and since by (9.47), $c_1 = gf_1$, we see that $f_1 = 1$. Now since $N = u/g$, by part (ii) of (9.48), $N = u/g = \sum_{\lambda=1}^{n} f_\lambda = 1 + \sum_{\lambda>1} f_\lambda$, and so $N \equiv 1 \pmod{p}$ will follow provided $p \mid f_\lambda$ for all $\lambda > 1$. Actually, since $p^\sigma = h = e_\lambda f_\lambda$, it suffices to show that $f_\lambda \neq 1$. Suppose that $f_\lambda = 1$, so $e_\lambda = p^\sigma$ gives the order of $\mathfrak{H}$. Then (as before) (9.46) implies that $\mathfrak{H}$ is contained in $U_\lambda^{-1}\mathfrak{G}U_\lambda$, or equivalently, $\mathfrak{H}' = U_\lambda\mathfrak{H}U_\lambda^{-1}$ is contained in $\mathfrak{G}$. Now $\mathfrak{H}'$ has order p^σ also. Application of (II) with $\mathfrak{U} = \mathfrak{G}$ means that its two subgroups $\mathfrak{H}, \mathfrak{H}'$ of order p^σ must be conjugate *within* $\mathfrak{G}$, i.e., $\mathfrak{H}' = G^{-1}\mathfrak{H}G$ with $G \in \mathfrak{G}$. Since $\mathfrak{G}$ is the normalizer of $\mathfrak{H}$, this means that $\mathfrak{H}' = \mathfrak{H}$. Going back to the definition of $\mathfrak{H}'$, we see that $U_\lambda \in \mathfrak{G}$, the first double coset, and so $\lambda = 1$, contradicting the assumption that $\lambda > 1$. Thus $N \equiv 1 \pmod{p}$ is proved. Frobenius did not mention it, but obviously since $N = u/g$, and $p^\sigma = h$ divides g, it follows that N divides (u/p^σ), i.e., N divides q, where $u = [\mathfrak{U}:1] = p^\sigma q$, a fact already implicit in one of Sylow's theorems [558, Thm. I, p. 586].

Besides giving the first abstract proofs of Sylow's theorems, Frobenius also applied his theory of double cosets to permutation groups and obtained new, simpler proofs of many theorems first discovered by others [194, §§6–8], displaying in the process a remarkably extensive grasp of the theory of finite groups as it stood in the early 1880s. Although he published nothing on finite groups during the 5-year period 1888–1892, focusing his attention instead on more established areas of mathematics, such as algebraic geometry [196], elliptic and abelian functions [195, 197], invariant theory [198], and the theory of surfaces [199], once Kronecker passed away and Frobenius was called to Berlin as his replacement, the keen interest in group theory evidenced in the papers described in this chapter was allowed free rein (Section 12.1) and, thanks to the timely intervention of Dedekind, eventually combined in a remarkably fortuitous manner with Frobenius' fondness for linear-algebraic problems to produce his greatest application of linear algebra: his theory of group characters and matrix representations, the subject of Chapters 12, 13, and 15.

Chapter 10
Abelian Functions: Problems of Hermite and Kronecker

During the 1880s, Frobenius published several papers investigating diverse aspects of the theory of abelian and theta functions. In this and the following chapter, three of these works from 1880 to 1884 will be discussed in detail. (Some other papers from this era dealing with theta functions with half-integer characteristics are considered more tangentially in Chapter 12.)

When Frobenius was a student at the University of Berlin, the theory of abelian functions was an area of active research, especially in Berlin, where Weierstrass was one of the leading investigators. As we noted in Section 1.1, on his oral doctoral examination in 1870, Frobenius had impressed Weierstrass with his complete familiarity with this difficult theory. The theory represented a nontrivial generalization of the theory of elliptic functions that had been founded in the 1820s by the work of Abel and Jacobi.

10.1 Abelian Functions and the Jacobi Inversion Problem

Abelian functions are examples of what were known in the nineteenth century as functions of g complex variables $z_1,\ldots,z_g$ that at each finite point possess the "character of a rational function." This terminology had been introduced by Weierstrass [592, pp. 127–129]. What it meant was that at each point $a \in \mathbb{C}^g$, a neighborhood $|z_j - a_j| < r$, $j = 1,\ldots,g$, exists within which the function $f(z_1,\ldots,z_g) = f(z)$ can be represented as the quotient of two absolutely convergent power series in $(z_1 - a_1),\ldots,(z_g - a_g)$. It can be assumed that the two power series are relatively prime in the sense that they do not have a common factor that vanishes at a. Such functions are now called *meromorphic*. The points a at which the power series in the denominator does not vanish are called regular points, and $f(z)$ can be represented locally as a power series. The points a at which the series in the denominator vanishes are the nonessential singular points of f. If at such a point a, the series in the numerator does not vanish, then a is called a pole of f, and f

T. Hawkins, *The Mathematics of Frobenius in Context*, Sources and Studies in the History of Mathematics and Physical Sciences, DOI 10.1007/978-1-4614-6333-7_10,

can be assigned the value ∞ at a. At points a where both power series vanish, the value of f is indeterminate. When points of indeterminacy exist, they form a set of codimension 1 in $\mathbb{C}^g$.

A meromorphic function $f(z)$ of g complex variables is an *abelian function* if (1) $2g$ g-tuples of complex numbers $\omega_\alpha = \begin{pmatrix}\omega_{1\alpha} & \cdots & \omega_{g\alpha}\end{pmatrix}^t$, $\alpha = 1,\ldots,2g$, exist that are linearly independent over $\mathbb{R}$ and satisfy $f(z+\omega_\alpha) = f(z)$, $\alpha = 1,\ldots,2g$; (2) f is not expressible as a function of $h < g$ variables.[1] No abelian function in this sense can have more than $2g$ periods that are linearly independent over $\mathbb{R}$. Clearly an integral linear combination $\omega = \sum_{\alpha=1}^{2g} n_\alpha \omega_\alpha$ satisfies $f(z+\omega) = f(z)$ and so is also a period of f. If every period of f is so expressible, then $\omega_1,\ldots,\omega_{2g}$ is said to be a *primitive* system of periods. In particular, an abelian function of $g = 1$ complex variables is an *elliptic function*. Thus if $f(z)$ is an elliptic function it is doubly periodic, which means that complex numbers ω_1, ω_2 with a nonreal ratio exist so that $f(z+\omega_1) = f(z+\omega_2) = f(z)$.

The theories of elliptic and abelian functions owed their existence to the solution to an inversion problem that showed, by virtue of its solution, that such functions actually exist. The general inversion problem, which is often referred to as the Jacobi inversion problem, may be explained briefly (and loosely) as follows. Let $F(z,w)$ denote an irreducible polynomial (over $\mathbb{C}$) in the two complex variables z and w, and consider the complex curve $F(z,w) = 0$. By the time of Frobenius' student years in Berlin this curve had been studied in connection with the inversion problem from two different points of view by, respectively, Riemann and Weierstrass. Riemann interpreted $F(z,w) = 0$ geometrically as a Riemann surface $\mathfrak{R}$, a notion he had introduced in his doctoral dissertation of 1851 and then used as the context within which to solve the inversion problem in 1857. Weierstrass called $F(z,w) = 0$ an "algebraic structure" (*algebraische Gebilde*) and extensively developed relevant properties by purely algebraic and analytic means as a preliminary to solving the inversion problem. In the following description of the inversion problem, Riemann's viewpoint will be adopted.

Associated to $F(z,w) = 0$ is an integer g, which represents the genus of $\mathfrak{R}$. A rational function $R(z,w)$ may be interpreted as a function on $\mathfrak{R}$, and if $\int_\gamma R(z,w)dz$ is finite for all curves $\gamma \subset \mathfrak{R}$, it is called an abelian integral of the first kind. It turns out that there are g linearly independent abelian integrals of this kind, say $\int_\gamma R_\alpha(z,w)dz$, $\alpha = 1,\ldots,g$, by means of which all such abelian integrals can be expressed as linear combinations. For example, when $F(z,w) = w^2 - p_m(z)$, where $p_m(z)$ is a polynomial of degree $m \geq 3$ with no multiple roots, then using the standard greatest integer notation, $g = [(m-1)/2]$ and $R_\alpha(z,w) = z^{\alpha-1}/w = z^{\alpha-1}/\sqrt{p_m(z)}$, $\alpha = 1,\ldots,g$, gives the requisite independent integrals. When $m = 3,4$, so $g = 1$, the abelian integrals are called *elliptic integrals*. For $m \geq 5$ (so $g \geq 2$), the integrals are called *hyperelliptic integrals*.

[1] Functions satisfying (1) but not (2) are sometimes called trivial or degenerate abelian functions. They include constant functions ($h = 0$) and have the property that for every $\varepsilon > 0$, they have a period ω with $\| \omega \| < \varepsilon$.

If $\mathfrak{c} = (a,b)$ and $\mathfrak{w} = (z,w)$ are two points of $\mathfrak{R}$, and if γ, γ' are any two curves on $\mathfrak{R}$ from $\mathfrak{c}$ to $\mathfrak{w}$, then corresponding to any abelian integral of the first kind there are $2g$ complex numbers $\omega_1, \ldots, \omega_{2g}$ (called periods) such that

$$(\gamma') \int_{\mathfrak{c}}^{\mathfrak{w}} R(u,v)du = (\gamma) \int_{\mathfrak{c}}^{\mathfrak{w}} R(u,v)du + \sum_{\alpha=1}^{2g} n_\alpha \omega_\alpha, \tag{10.1}$$

where the n_α are integers. This relation is fundamental to the inversion problem. For example, in the case of elliptic integrals, the inversion problem centers on the equation

$$z = \int_{\mathfrak{c}}^{\mathfrak{w}} R(u,v)du. \tag{10.2}$$

Here z is a multivalued function of $\mathfrak{w}$ because it depends not only on $\mathfrak{w}$ but also on the curve γ from $\mathfrak{c}$ to $\mathfrak{w}$ along which the integral is taken. In view of (10.1) with $g = 1$, any two values z, z' of the integral in (10.2) are related by $z' = z + n_1\omega_1 + n_2\omega_2$, for some choice of n_1, n_2. The inversion problem is to show that $\mathfrak{w}$, and so w, can be expressed as a single-valued function of z, $w = f(z)$. This means that $w = f(z) = f(z') = f(z + n_1\omega_1 + n_2\omega_2)$, and so f is doubly periodic. Of course, the complete solution to the inversion problem in this case involves showing that f is meromorphic and that ω_1, ω_2 are independent over $\mathbb{R}$.

The more formidable general inversion problem involved the system of equations

$$z_\alpha = \int_{\mathfrak{c}_1}^{\mathfrak{w}_1} R_\alpha + \cdots + \int_{\mathfrak{c}_g}^{\mathfrak{w}_g} R_\alpha, \quad \alpha = 1, \ldots, g. \tag{10.3}$$

As in the elliptic case, the z_α are multivalued functions of $\mathfrak{w}_1, \ldots, \mathfrak{w}_g$ and hence of $w_1, \ldots, w_g$, but for $g > 1$, the problem is more difficult, since it cannot be shown that the w_α are the requisite single-valued functions of $z_1, \ldots, z_g$ but only that symmetric polynomials $\sigma = \sigma(w_1, \ldots, w_g)$ in $w_1, \ldots, w_g$ can be expressed as single-valued meromorphic functions of $z_1, \ldots, z_g$. If $\int R_\alpha(u,v)du$, $\alpha = 1, \ldots, g$, is a set of independent integrals of the first kind, and if $\omega_{\alpha\beta}$, $\beta = 1, \ldots, 2g$, are the periods of $\int R_\alpha(u,v)du$, then $\sigma = f(z_1, \ldots, z_g)$ is meromorphic with the $2g$ periods $\omega_\beta = \left(\omega_{1\beta} \cdots \omega_{g\beta}\right)^t$, $\beta = 1, \ldots, 2g$.

The inversion problem for nonelliptic integrals was first solved in the hyperelliptic case with $g = 2$ in 1851 by Jacobi's student J.G. Rosenhain, as well as by A. Göpel. Before he knew of the work of Rosenhain and Göpel, Weierstrass had discovered how to solve the inversion problem in the general hyperelliptic case, i.e., for hyperelliptic curves of any genus g.[2] His results were published in *Crelle's Journal* in 1854 [586] and eventually served to gain him a professorship at the University of Berlin. According to Weierstrass, by the summer of 1857 he had also

[2] This according to Weierstrass [594, p. 9].

solved the inversion problem for general abelian integrals and had submitted an extensive memoir on the subject to the Berlin Academy [594, pp. 9–10]. Before it appeared, however, Riemann's solution to the same problem (using Riemann surface techniques) was published in *Crelle's Journal* [495]. This caused Weierstrass to withdraw his own manuscript, in order to relate Riemann's different approach to the problem to his own so as to be able to compare their respective results. After having done this, Weierstrass felt that his own approach was in need of a thorough revision. Because of other work and personal constraints it was not until 1869 that a first draft of his revised theory was completed. It was accomplished within the context of his lectures on abelian integrals and functions at the university and so although known to Weierstrass' colleagues and students, including Frobenius, Weierstrass' results, especially in their details, were not generally known.[3]

In 1855, Hermite published an important paper [290] that was motivated by the above-mentioned work of Göpel and Rosenhain. They had worked with a system (10.3) in the $g = 2$ hyperelliptic case, using two particular independent integrals, but Hermite put forth another pair of independent integrals and posed the problem of determining when the hyperelliptic functions obtained by inversion from the latter pair of integrals were rational functions of those obtained from the former pair. This problem led Hermite to entirely new results about the transformation of hyperelliptic functions in $g = 2$ variables that soon suggested the possibility of an extension to abelian functions in any number of variables.

Hermite's work became of interest to Frobenius because Kronecker, at Weierstrass' request, looked into the possibility of its extension to $g > 2$ variables; and then Heinrich Weber pursued several aspects of Kronecker's work in a paper of 1878. Implicit in the work of Hermite, Kronecker, and Weber were two problems that attracted Frobenius' attention. I have called the first *Hermite's abelian matrix problem*. This problem, the treatments of it by Kronecker and Weber, and Frobenius' definitive solution to it by means of ideas and results from his arithmetic theory of bilinear forms are the subjects of Sections 10.2–10.4. The second problem forms the subject of Section 10.5. Hermite's work as generalized to g variables suggested to Kronecker an appropriate context within which to generalize to abelian functions the notion of an elliptic function admitting complex multiplication. Kronecker's generalization and concomitant observations posed a general problem, which I have called *Kronecker's complex multiplication problem*. As with Hermite's classification problem, it was Frobenius who provided Kronecker's problem with a completely general, definitive solution (Section 10.6). In this case, he made critical use of the symbolic algebra of matrices that he had developed a few years earlier (Chapter 7), and it was here that the notion and properties of unitary matrices were first introduced. As we shall see in Section 10.7, Frobenius' work on complex multiplication proved useful in the development of the theory of abelian varieties with complex multiplication.

[3]A version of Weierstrass' lectures was finally published in 1902, five years after his death [594], as part of his collected works.

10.2 Hermite's Abelian Matrix Problem

In order to state Hermite's problem, as well as the complex multiplication problem of Kronecker (Section 10.5), some definitions are prerequisite.

10.2.1 Abelian matrices

Throughout this chapter, a $g \times 2g$ matrix $\Omega = \begin{pmatrix}\omega_1 & \cdots & \omega_{2g}\end{pmatrix}$ with the property that its $2g$ columns $\omega_1, \ldots, \omega_{2g}$ are linearly independent over $\mathbb{R}$ will be called a *period matrix*. In order for abelian functions with periods $\omega_1, \ldots, \omega_{2g}$ to actually exist, Ω must satisfy additional conditions that go back to Riemann and Weierstrass.[4] For the purposes of Sections 10.2–10.5, it will suffice to express them in a special form equivalent to how Riemann expressed them (for special abelian functions) in his solution to the Jacobi inversion problem:

$$\text{(I) } \Omega J \Omega^t = 0; \quad \text{(II) } C = (i/2)\Omega J \Omega^h \succ 0. \tag{10.4}$$

Here $\Omega^h = \overline{\Omega}^t$ denotes the Hermitian transpose of Ω, $C \succ 0$ means that C is a positive definite Hermitian matrix, and

$$J = \begin{pmatrix} 0 & I_g \\ -I_g & 0 \end{pmatrix}, \tag{10.5}$$

where I_g denotes the $g \times g$ identity matrix.[5] If we partition Ω into two $g \times g$ matrices Ω_i, they are invertible, and we may write

$$\Omega = \begin{pmatrix}\Omega_1 & \Omega_2\end{pmatrix}, \quad T = \Omega_1^{-1}\Omega_2 \in \mathfrak{H}_g. \tag{10.6}$$

Here $\mathfrak{H}_g$ denotes what is now usually called the *Siegel half-space*; it consists of all $g \times g$ complex symmetric matrices with positive definite imaginary part. By virtue of Riemann's condition (I), $T = \Phi + i\Psi$ is symmetric; and by virtue of (II), its imaginary part Ψ is positive definite.[6]

Hermite's problem was related to that aspect of the theory dealing with the transformation of abelian and theta functions by means of a linear variable change

---

[4]Riemann's work is discussed in Section 11.4. Weierstrass' more general conditions (now commonly known as Riemann's conditions) are discussed in Section 11.2.

[5]The material under discussion here is now part of the theory of principally polarized abelian manifolds. For a clear modern account, including a simple example of a period matrix with no abelian functions, see Rosen's expository article [508, pp. 96ff.].

[6]Since $C = \frac{i}{2}(\Omega_1\Omega_2^h - \Omega_2\Omega_1^h) = \frac{i}{2}\Omega_1(\overline{T} - T)\Omega_1^h = \Omega_1\Psi\Omega_1^h$, C is positive definite if and only if Ψ is.

$z = Mu$, where $\det M \neq 0$. The objective was to choose the $g \times g$ matrix M so that if $f(z)$ is an abelian function with a primitive period matrix $\Omega = (\Omega_1\ \Omega_2)$, where $T = \Omega_1^{-1}\Omega_2 \in \mathfrak{H}_g$, then $g(u) = f(Mu)$ should have a period matrix of the form $\Omega' = (\Omega_1'\ \Omega_2')$, where $T' = (\Omega_1')^{-1}\Omega_2' \in \mathfrak{H}_g$. This means that if ω_j' denotes the jth column of Ω', $1 \leq j \leq 2g$, then $g(u+\omega_j') = g(u)$, or equivalently, $f(z+M\omega_j') = f(z)$; and so $M\omega_j'$ is a period of f and hence a $\mathbb{Z}$-linear combination of the $2g$ periods of f given by the $2g$ columns of $\Omega = (\Omega_1\ \Omega_2)$. In other words, M takes the lattice of periods associated to $\Omega' = (\Omega_1'\ \Omega_2')$ into the corresponding lattice associated to Ω. Thus M must transform each column of Ω' into some integral linear combination of the columns of Ω. Since what M does to each column of Ω' is given by the columns of $M\Omega' = (M\Omega_1'\ M\Omega_2')$, the above integrality condition says that each column of $M\Omega_1'$ and $M\Omega_2'$ is an integral linear combination of the columns of Ω_1 and Ω_2. This may be stated in the following form:

$$\begin{aligned} M\Omega_1' &= \Omega_1\mathrm{A} + \Omega_2\Gamma = \Omega_1(\mathrm{A}+T\Gamma), \\ M\Omega_2' &= \Omega_1\mathrm{B} + \Omega_2\Delta = \Omega_1(\mathrm{B}+T\Delta), \end{aligned} \tag{10.7}$$

where the capital Greek letters $\mathrm{A},\mathrm{B},\Gamma,\Delta$ stand for $g \times g$ matrices with integer coefficients.[7] Since $M = \Omega_1(\mathrm{A}+T\Gamma)(\Omega_1')^{-1}$ is assumed invertible and $T' = (\Omega_1')^{-1}\Omega_2'$, (10.7) implies that

$$T' = (\mathrm{A}+T\Gamma)^{-1}(\mathrm{B}+T\Delta). \tag{10.8}$$

The $g \times g$ blocks of integers were combined into a $2g \times 2g$ array, which may be represented by the block-partitioned matrix

$$\tilde{A} = \begin{pmatrix} \mathrm{A} & \mathrm{B} \\ \Gamma & \Delta \end{pmatrix}. \tag{10.9}$$

The equations (10.7) can then be combined into the single equation

$$M\Omega' = \Omega\tilde{A}. \tag{10.10}$$

The matrix $\tilde{A}$ is assumed to have an additional property related to the fact that when Ω satisfies Riemann's conditions (10.4), so should Ω'. Consider condition (I), which is satisfied by Ω, so that $\Omega J \Omega^t = 0$. By virtue of (10.10), we may write $\Omega' = M^{-1}\Omega\tilde{A}$, and so $\Omega' J (\Omega')^t = M^{-1}\Omega(\tilde{A}J\tilde{A}^t)\Omega^t(M^{-1})^t$. Thus *if* we assume that $\tilde{A}J\tilde{A}^t = nJ$, then

[7] For example, if $[M\Omega_1']_\beta$ denotes the βth column of $M\Omega_1'$, then to say that $[M\Omega_1']_\beta$ is a $\mathbb{Z}$-linear combination of the columns of $\Omega = (\Omega_1\ \Omega_2)$ means that integers $a_{\alpha\beta}$ and $\gamma_{\alpha\beta}$, $\alpha = 1,\ldots,g$, exist such that $[M\Omega_1']_\beta = \Sigma_{\alpha=1}^{g}(a_{\alpha\beta}[\Omega_1]_\alpha + \gamma_{\alpha\beta}[\Omega_2]_\alpha)$ for $\beta = 1,\ldots,g$. These relations are equivalent to $M\Omega_1' = \Omega_1\mathrm{A} + \Omega_2\Delta$, where $\mathrm{A} = (a_{\alpha\beta})$ and $\Gamma = (\gamma_{\alpha\beta})$.

$$\Omega' J \Omega'^t = nM^{-1}(\Omega J \Omega^t)(M^{-1})^t = 0,$$

and so Ω' also satisfies condition (I). The factor n in $\tilde{A}J\tilde{A}^t = nJ$ must be an integer, since J and $\tilde{A}$ are integral. In fact, it must be a positive integer if Ω' is to satisfy condition (II).[8] Thus the integral matrices $\tilde{A}$ of (10.10) were assumed to have the additional property that

$$\tilde{A}^t J \tilde{A} = n\tilde{J}, \quad n \in \mathbb{Z}^+. \tag{10.11}$$

The $2g \times 2g$ integral matrices $\tilde{A}$ satisfying (10.11) and with $g = 2$ were introduced by Hermite in his 1855 paper. Since the term "Hermitian matrix" already has an established meaning, I will follow Laguerre [393, p. 260], who extended Hermite's results to $g > 2$ variables in 1867, and call $\tilde{A}$ an *abelian matrix*.[9] Following Frobenius [188, §3], the integer n in (10.11) will be called the *order of* $\tilde{A}$ and denoted by $\operatorname{ord}\tilde{A}$. It should be noted that (10.11) is equivalent to $\tilde{A}J\tilde{A}^t = nJ$, because $J^{-1} = -J$. Also, if $\tilde{A}$ is any abelian matrix, then with $\tilde{A}$ block-partitioned as in (10.9), $\tilde{A}$ may be identified with the transformation $\tilde{A} : T \mapsto T'$ given by (10.8). This transformation takes $\mathfrak{H}_g$ into itself, since when $T \in \mathfrak{H}_g$, it follows that $T' \in \mathfrak{H}_g$.[10] (For this reason, Frobenius referred to $\tilde{A}$ as a principal transformation.)

10.2.2 Hermite's problem

In his paper of 1855, Hermite had developed the above-described approach to the transformation of abelian and theta functions for the case $g = 2$, and among many other things, he briefly summarized his results on the problem of describing all 4×4 abelian matrices [290, §III].[11] He observed that if $\tilde{A}$ has order m and $\tilde{B}$ has order n, then $(\tilde{A}\tilde{B})J(\tilde{A}\tilde{B})^t = mnJ$, which shows that $\tilde{A}\tilde{B}$ is abelian of order mn. Thus the class of abelian matrices is closed under matrix multiplication and their orders satisfy the relation

$$\operatorname{ord}(\tilde{A}\tilde{B}) = \operatorname{ord}(\tilde{B}\tilde{A}) = (\operatorname{ord}\tilde{A})(\operatorname{ord}\tilde{B}). \tag{10.12}$$

[8] That is, using $\Omega' = M^{-1}\Omega\tilde{A}$ again, the Hermitian symmetric matrix C' associated to Ω' is $C' = (i/2)\Omega' J \Omega'^h = M^{-1}(i/2)\Omega(\tilde{A}J\tilde{A}^t)\Omega^h(M^{-1})^h = n(M^{-1})C(M^{-1})^h$, and so C' is positive definite when C is, provided that $n > 0$.

[9] Laguerre's paper [393] and its influence on Frobenius' solution to Kronecker's complex multiplication problem are discussed in Section 10.6. However, to my knowledge, neither Frobenius nor anyone else adopted Laguerre's term "abelian matrix."

[10] The nonsingularity of $\mathrm{A} + T\Gamma$ for every abelian $\tilde{A}$ follows from the fact that $T \in \mathfrak{H}_g$, as does the fact that $T' \in \mathfrak{H}_g$ [188, p. 105].

[11] Hermite presented his results in the *Comptes rendus* of the Paris Academy, and although the three-page limit was not yet in force, his substantial results were presented in outline.

The last property implies in particular that if $\tilde{A}$ and $\tilde{B}$ have order one, then so does $\tilde{A}\tilde{B}$, so that the class of first-order abelian matrices is closed under matrix multiplication. Furthermore, Hermite realized that if $\tilde{A}$ is abelian of order n, then $\det \tilde{A} = n^g = n^2$, so that abelian matrices $\tilde{A}$ of order $n = 1$ have $\det \tilde{A} = 1$, and so are properly unimodular, and in fact, it follows readily by taking inverses in (10.11) with $n = 1$ that $\tilde{A}^{-1}$ is also abelian of order one. From a modern perspective and for any g, the abelian matrices of order one are the elements of the symplectic group $\mathrm{Sp}(2g, \mathbb{Z})$ defined with respect to the skew-symmetric form $\langle x, y\rangle = x^t Jy$, since (10.11) with $n = 1$ implies that $\langle \tilde{A}x, \tilde{A}y\rangle = \langle x, y\rangle$ for every $x, y \in \mathbb{Z}^{2g}$.

From the multiplicative property (10.12), it follows that if $\tilde{A}$ has order n and if $\tilde{Q}$ has order 1, then $\tilde{B} = \tilde{A}\tilde{Q}$ also has order n. With this in mind, Hermite defined two abelian matrices $\tilde{A}, \tilde{B}$ of order $n > 1$ to be equivalent if $\tilde{B} = \tilde{A}\tilde{Q}$ for some abelian $\tilde{Q}$ of order one. Hermite evidently knew how to generate all abelian matrices of order one in the case $g = 2$, although he did not pause to disclose his method. (It was probably similar to Kronecker's method discussed below.) Using this knowledge, he determined all the equivalence classes for matrices of a prime order p: Every abelian matrix of order p is equivalent in the above sense to a matrix of one of four types. The first type was represented by the sole diagonal matrix $\tilde{T}_1 = \mathrm{Diag.Matrix}(1, 1, p, p)$, whereas the other types involved integer parameters a, b, c running between 0 and $p - 1$:

$$\tilde{T}_2 = \begin{pmatrix} 1 & 0 & 0 & 0 \\ 0 & p & a & 0 \\ 0 & 0 & 1 & 0 \\ 0 & 0 & 0 & p \end{pmatrix}, \quad \tilde{T}_3 = \begin{pmatrix} p & a & 0 & b \\ 0 & 1 & 0 & 0 \\ 0 & 0 & p & -a \\ 0 & 0 & 0 & 1 \end{pmatrix}, \quad \tilde{T}_4 = \begin{pmatrix} p & 0 & a & b \\ 0 & p & c & a \\ 0 & 0 & 1 & 0 \\ 0 & 0 & 0 & 1 \end{pmatrix}.$$

Within a given type, different parameter values give representatives of different equivalence classes of that type. Thus, for example, the type $\tilde{T}_3$ gives representatives of p^2 different equivalence classes, each corresponding to a specific choice of a and b. By counting the number of possibilities for each type, Hermite arrived at a total of $1 + p + p^2 + p^3$ equivalence classes of abelian matrices of prime order p.

The above shows that every abelian matrix $\tilde{A}$ of prime order p can be generated from a specific matrix $\tilde{T}_k$, $k = 1, \ldots, 4$, by postmultiplication by a first-order matrix $\tilde{Q}$: $\tilde{A} = \tilde{T}_k\tilde{Q}$. Hermite also realized that by multiplying any $\tilde{T}_k$ on both the right and left by first-order matrices $\tilde{R}, \tilde{S}$, he could bring $\tilde{T}_k$ into a diagonal form, viz., $\tilde{R}\tilde{T}_k\tilde{S} = \mathrm{Diag.Matrix}(p, p, 1, 1)$. This means that $\tilde{A} = \tilde{T}_k\tilde{Q} = \tilde{R}[\mathrm{Diag.Matrix}(p, p, 1, 1)](\tilde{S}\tilde{Q})$, i.e., that every abelian matrix of order p can be generated from $\mathrm{Diag.Matrix}(p, p, 1, 1)$ by pre- and postmultiplication by first-order abelian matrices. The choice of $\mathrm{Diag.Matrix}(p, p, 1, 1)$ is somewhat arbitrary. As Hermite certainly realized, there are simple first-order abelian matrices $\tilde{P}', \tilde{Q}'$ such that $\tilde{P}'\,\mathrm{Diag.Matrix}(p, p, 1, 1)\tilde{Q}' = \mathrm{Diag.Matrix}(1, 1, p, p)$. For future reference, I will summarize Hermite's result as follows: *if $\tilde{A}$ is any abelian matrix of prime order p, then first-order abelian matrices $\tilde{P}$ and $\tilde{Q}$ can be determined such that*

$$\tilde{P}\tilde{A}\tilde{Q} = \text{Diag.Matrix}(1,1,p,p). \tag{10.13}$$

Hermite thus had two ways to describe all abelian matrices of any prime order p, namely in terms of the "postmultiplication" equivalence classes described by $\tilde{T}_1,\ldots,\tilde{T}_4$ above or by means of (10.13). But he never mentioned how these two methods relate to the general problem of describing all abelian matrices of any order n, prime or not. Any attentive reader of his paper would have realized what was obvious to Hermite, namely that by virtue of the multiplicative property (10.12) of orders, if $n > 0$ has the prime factorization $n = p_1 \cdots p_k$, where the p_i are not necessarily distinct, then knowing how to generate all abelian matrices $\tilde{A}_i$ of order p_i, we may generate an unlimited number of abelian matrices of order n by forming the products $\tilde{A}_1 \cdots \tilde{A}_k$. The obvious question, however, is whether *all* order n abelian matrices are thereby generated. That is, *is every abelian matrix of order* $n = p_1 \cdots p_k$ *expressible as a product of abelian matrices of orders* p_i? Hermite never broached this question; he probably believed that the answer was affirmative.

Hermite's work thus suggested the following problem.

Problem 10.1 (Hermite's abelian matrix problem). *For every* $g \geq 2$, *determine a method for generating all* $2g \times 2g$ *abelian matrices of any given order* n.

His discussion of the case $g = 2$ suggested two approaches to its solution, each of which constitutes a problem in its own right.

Problem 10.2 (Approach 1). *(a) Solve the above problem for* $2g \times 2g$ *abelian matrices of order one. (b) For* p *prime, determine the equivalence-class types that generalize* $\tilde{T}_1,\ldots,\tilde{T}_4$. *(c) Prove that every abelian matrix of order* $n = p_1 \cdots p_k$, p_i *prime, is a product of abelian matrices of orders* p_i.

Problem 10.3 (Approach 2). *Solve problem (a) and then solve (d): Show that if* $\tilde{A}$ *is an abelian matrix of prime order* p, *then first-order abelian matrices* $\tilde{P},\tilde{Q}$ *may be determined such that*

$$\tilde{P}\tilde{A}\tilde{Q} = \text{Diag.Matrix}(1,\ldots,1,p,\ldots,p). \tag{10.14}$$

As we shall see, Kronecker took the first approach but did not completely solve Problem 10.2, whereas Frobenius took the second approach, albeit in a modified, more illuminating form that enabled him to use the results of his arithmetic study of bilinear forms (Chapter 8) to completely and definitively solve Problem 10.3.

10.3 Kronecker and Weber on Hermite's Problem

In 1858, Weierstrass asked Kronecker to investigate Hermite's problem. He wished to include this topic in a planned work on abelian functions. Kronecker obliged; he wrote up his results the following year and gave them to Weierstrass. This manuscript does not seem to have survived, but in 1866 [353, pp. 158–162] Kronecker briefly described its contents.[12] His most notable achievement was to have solved (a) of Problem 10.2. Expressed in the language of matrices (not used by Kronecker), what he did was to determine $g+2$ first-order elementary matrices $\tilde{E}_i$ with 0's and 1's as coefficients and possessing the following property. Given any abelian matrix $\tilde{A}$ of order one, by means of an algorithm (not fully described in [353]), a succession $\tilde{E}_{i_1}, E_{i_2}, \ldots, \tilde{E}_{i_N}$ of the $\tilde{E}_i$ (each corresponding to an elementary row operation) could be determined such that $\tilde{E}_{i_N} \cdots \tilde{E}_{i_1} \tilde{A} = I_{2g}$. This meant that $\tilde{A} = \tilde{E}_{i_1}^{-1} \cdots \tilde{E}_{i_N}^{-1}$. The $\tilde{E}_i^{-1}$, $i = 1, \ldots, g+2$, thus form a set of generators for the abelian matrices of order one, i.e., for $\mathrm{Sp}(2g, \mathbb{Z})$, and represent a satisfying solution to subproblem (a) of Hermite's problem.[13]

As for Hermite's problem itself, Kronecker had the idea of using his elementary abelian matrices E_i (now operating on the right as column operations) and his algorithm to reduce an abelian matrix $\tilde{A}$ of any order n—not just $n = p$, a prime—to something analogous to the four types $\tilde{T}_1, \ldots, \tilde{T}_4$ specified by Hermite. Thus he wrote that when $\operatorname{ord} \tilde{A} \neq 1$,

> in the specified reduction procedure diagonal terms arise that are divisors of the determinant $[n^g]$—rather than 1—and it is no longer possible to make all the off-diagonal terms zero. But by this procedure all the inequivalent systems are obtained, and the corresponding results represent generalizations of those that Mr. Hermite gave for the case [g]=2 ... [353, p. 161].

Although Kronecker's above-described reduction procedure $\tilde{A} \to \tilde{N}$ for $\operatorname{ord} \tilde{A} = n > 1$ is lacking in details, it is clear that it led to a matrix $\tilde{N} = \tilde{A}\tilde{Q}$, $\operatorname{ord} \tilde{Q} = 1$, that in turn facilitated a description of equivalence-class types analogous to the four types of Hermite. To what extent his procedure constituted a solution to Hermite's problem, however, is a question that cannot be answered without a closer look at the details of the procedure. An idea of what these details involve is suggested by a paper of 1878 by Heinrich Weber [577], in which he fleshed out Kronecker's algorithm, developed its consequences along the lines suggested by Kronecker, and sought to apply the results to deal with Hermite's problem.

Weber showed that by means of elementary abelian column operations, any $2g \times 2g$ abelian matrix $\tilde{A}$ of order n could be reduced to the following normal form [577, p. 135]:

---

[12]The reason Kronecker made these results known in 1866 was that Clebsch and Gordan had published a book on abelian functions that year in which they used similar elementary matrices to reduce a first-degree abelian matrix, not to I_{2g} but to several simple canonical forms [100, §86]. They also attributed the idea of using elementary matrices to reduce unimodular matrices to Kronecker [100, p. 308n], but evidently Kronecker wanted the world to know that he had actually applied his ideas to first-order abelian matrices and had obtained a better reduction than that of Clebsch and Gordan.

[13]In the case $g = 2$, Kronecker's four matrices correspond to the elementary row operations (1) $-$row 1 $\to$ row 3 and row 3 $\to$ row 1; (2) add row 1 to row 3; (3) switch rows 1 and 2 and rows 3 and 4; (4) add row 2 to row 4. For more on generators for degree-one abelian matrices, see [350, pp. 148ff.]. It turns out that the minimal number of generators is 2 for $g = 2$ and 3 for $g \geq 3$.

$$\tilde{A} \to \tilde{A}\tilde{Q} = \tilde{N} = \begin{pmatrix} L & V+W \\ 0 & U \end{pmatrix}, \quad \operatorname{ord} Q = 1, \tag{10.15}$$

where L is lower triangular, U is nonnegative and upper triangular, V is strictly lower triangular, W is nonnegative and upper triangular, and the coefficients of these matrix blocks are subject to the following restrictions: (1) the diagonal entries of L and U are strictly positive and satisfy $l_{ii}u_{ii} = n$ for all $i = 1, \ldots, g$, so that all diagonal entries of $\tilde{N}$ are divisors of n; (2) the entries in row i of U to the right of the diagonal entry u_{ii} are all strictly less than u_{ii}; (3) the entries in row i of the upper triangular matrix W are all strictly less than l_{ii}.

By virtue of the equation $\tilde{N}J\tilde{N}^t = nJ$, it turns out that all coefficients of $\tilde{N}$ are determined by the coefficients of U and W. (For example, $L = n(U^t)^{-1}$.) Thus for any choice of U and W satisfying (1)–(3), a unique $\tilde{N}$ is determined that satisfies $\tilde{N}J\tilde{N}^t = nJ$. However, $\tilde{N}$ need not be abelian, because it need not be integral. For example, if U is chosen in accordance with (1) and (2), it can happen that $L = n(U^t)^{-1}$ has some fractional coefficients. It can also happen that a choice of U for which L is integral will produce, for certain choices of W satisfying (3), a matrix $\tilde{N}$ with some fractional coefficients in V. Thus a case-by-case analysis is required to determine the choices of U and W that actually produce an abelian matrix $\tilde{N}$ of order n.

The number of cases to be considered is dictated by restriction (1) above, which says that the diagonal entries of $\tilde{N}$ are $n/u_{11}, \ldots, n/u_{gg}, u_{11}, \ldots, u_{gg}$, where each u_{ii} is a positive divisor of n. The number of cases to be considered thus depends on the nature of the prime factorization of n. If, e.g., $n = p$, a prime, then each u_{ii} is either 1 or p, and so there are 2^g ways to choose the u_{ii} and hence 2^g cases to consider. For each case, it is necessary to determine which choices for the remaining coefficients of U and for the coefficients of W produce an integral $\tilde{N}$. If $g = 2$, there are $2^2 = 4$ cases to consider in this manner, and they lead to the four equivalence types $\tilde{T}_1, \ldots, \tilde{T}_4$ of Hermite's paper given toward the end of Section 10.2. For $g = 3$, there are $2^3 = 8$ cases to consider, and Weber showed that each case leads to an equivalence type [577, p. 139]. Clearly, even if only orders $n = p$ are considered, the fact that there are 2^g cases to analyze shows that it is necessary to establish a pattern to the nature of the resulting types, but Weber did not do this, and it is unclear whether it can be done. Even if g is kept small, the number of cases to consider is unlimited, for if the prime factorization of n is $n = \prod_{i=1}^{k} p_i^{m_i}$, each coefficient u_{ii} may be chosen in $N = \prod_{i=1}^{k}(m_i + 1)$ ways, for a total of N^g cases to be considered. Thus for $n = 2^4 3^3 = 432$, $N = 20$, and so 20^g cases need to be considered, e.g., 400 cases for $g = 2$ and $8{,}000$ cases for $n = 3$.

Based on these considerations, it is possible to characterize what Kronecker had probably done in response to Weierstrass' request (as expressed by Kronecker) "to represent all integral systems $[\tilde{A}]$" for any g [353, p. 162]. First of all, he had shown how the abelian matrices of order one that are behind the scenes in Hermite's paper may be generated, for any g, by means of a few elementary transformations corresponding to elementary row and column operations. He had also shown how these same operations may be used to reduce an abelian $\tilde{A}$ of order $n > 1$ to a normal form akin to $\tilde{N}$, from which in principle, N^g equivalence types may be determined in the manner described above. Kronecker had certainly provided Weierstrass with the mathematical underpinning of Hermite's treatment of abelian matrices in the case $g = 2$ in a form that extends to any g and to any n, not just $n = p$. For any specific values of g and n, Kronecker's methods indicated how, granted adequate computational capabilities, the N^g cases could be analyzed—albeit by a procedure left on an ad hoc basis—to determine which actually correspond to equivalence class types. Thus he had provided an explanation of how Hermite's solution could be extended to any g and n but not a general solution to Hermite's problem, since his method did not provide a general algorithm, valid for all g and n, for deciding which $\tilde{N}$ are actually abelian. His methods failed to provide an overview of all abelian matrices of order greater than 1.

Evidently Weber had the same opinion, for in considering Hermite's problem he focused on the case $g = 3$. He realized that, at least in this case, Hermite's problem would be resolved if, in addition to specifying the eight types of equivalence classes for abelian matrices of order p, he could show that every abelian matrix $\tilde{A}$ of degree $n = p_1 \cdots p_k$, with the primes p_i not necessarily distinct, is expressible as $\tilde{A} = \tilde{A}_1 \cdots \tilde{A}_k$, where ord $\tilde{A}_i = p_i$ (subproblem (c) of Problem 10.2). Starting with the normal form $\tilde{N}$ of a given $\tilde{A}$ of degree n, he showed that by means of elementary abelian row operations, $\tilde{N}$ could be reduced to a diagonal form, *provided n* has no quadratic divisors, i.e., provided n factors into distinct primes. In this way, he established the following result [577, p. 138].

Proposition 10.4. *Let $\tilde{A}$ be a 6×6 abelian matrix of order n, where n is a product of distinct primes. Then (1) first-order abelian matrices $\tilde{P}$ and $\tilde{Q}$ may be determined such that*

$$\tilde{P}\tilde{A}\tilde{Q} = \text{Diag. Matrix}(m_1, m_2, m_3, n/m_1, n/m_2, n/m_3), \tag{10.16}$$

where the integers m_i are positive divisors of n; (2) $\tilde{A}$ is a product of abelian matrices of prime orders.

To see why (2) follows from (1), note first of all that any diagonal matrix of the form given in (10.16) is abelian of order n. Now suppose, e.g., that $n = 12 = 2^2 \cdot 3$, and consider the diagonal matrix corresponding to $m_1 = 2$, $m_2 = 2^2$, and $m_3 = 2 \cdot 3$ in (10.16), namely, $\tilde{D} = \text{Diag. Matrix}(2,4,6,6,3,2)$. Then the prime factorization of the diagonal elements of $\tilde{D}$ brings with it a factorization of $\tilde{D}$ into the product of two diagonal abelian matrices of orders 3 and 4, respectively, namely, Diag. Matrix$(1,1,3,3,3,1)$ and Diag. Matrix$(2,4,2,2,1,2)$. The latter may be further factored into abelian matrices of order 2, namely Diag. Matrix$(2,2,2,1,1,1)$ and Diag. Matrix$(1,2,1,2,1,2)$. And so the original diagonal abelian matrix $\tilde{D}$ is the product of three diagonal abelian matrices of prime orders.

Weber implied that the above proposition was true when $\tilde{A}$ is $2g \times 2g$ for any g and that he had stated it only in the case $g = 3$ "for simplicity of expression" [577,

p. 136], which suggests the difficulty of using the normal form (10.15) to give a proof valid for any g. Another difficulty was of course that to arrive at part (1) of his proposition, Weber had to assume that n is the product of distinct primes. As a result, he was unable to solve Hermite's problem even for $g = 3$, although he opined that part (2) was most likely true for all n, not just those that are square-free [577, p. 154].

Weber also observed that Hermite's diagonalization (10.13) extends to the case $g = 3$: if $\operatorname{ord}\tilde{A} = p$, then abelian matrices $\tilde{P}, \tilde{Q}$ of order one may be determined such that $\tilde{P}\tilde{A}\tilde{Q} = \text{Diag.Matrix}(1,1,1,p,p,p)$. Imitating what Hermite had done in the case $g = 2$, he deduced this diagonal form from the eight types of equivalence classes with the assistance of some pre- and postmultiplications by elementary abelian matrices. For $g = 3$, this provides a solution to Hermite's problem by the approach of Problem 10.3—but again only for abelian matrices with square-free orders: every prime-order abelian matrix is of the form $\tilde{R}\,\text{Diag.Matrix}(1,1,1,p,p,p)\tilde{S}$, and by (2) of Proposition 10.4, every abelian matrix of a square-free order is the product of abelian matrices of prime orders.

In sum, Weber had provided two partial solutions to Hermite's problem, both being limited to $g = 3$ and to square-free orders.

10.4 Frobenius' Solution to Hermite's Problem

It may have been Frobenius' general interest in the theory of abelian functions that prompted him to take a look at Weber's paper, which appeared at the time he was working on his arithmetic theory of bilinear forms (the subject of Chapter 8). Frobenius' interest in Weber's paper must have quickened when he realized that it dealt, albeit without much success, with abelian matrices and Hermite's problem. After all, abelian matrices were integral matrices, and abelian matrices of degree one were, in particular, unimodular. For an abelian matrix $\tilde{A}$ of degree n, it thus followed that unimodular matrices P, Q may be determined such that $P\tilde{A}Q$ is a diagonal matrix—the Smith–Frobenius normal form (8.2) of his first paper on his arithmetic theory of bilinear forms, written in 1878 [182]. Of course, P and Q need not be abelian, but perhaps, he must have wondered, the fact that $\tilde{A}$ is abelian could be used to extract from the methods in [182] a way to make P and Q abelian of degree one. If so, the resulting normal form would generalize (10.14) of Approach 2 to abelian matrices of any degree n and would constitute a completely general solution to Hermite's problem. One can imagine Frobenius' increasing delight as he reconsidered the methods of his paper [182] in the light of the above question and discovered that his hopes could be realized, as he showed in a paper submitted in 1879 [186]. Let us consider what was involved.

We saw in Section 8.2 that Frobenius' reduction theorem, from which his normal form theorem followed, was based on the following reduction lemma (Lemma 8.6): *If A is an integral matrix with $f_1 = \gcd A$, then unimodular matrices P_1, Q_1 may be determined such that*

$$P_1AQ_1 = \begin{pmatrix} f_1 & 0 \\ 0 & f_1A_1 \end{pmatrix}, \text{ where } A_1 \text{ is integral.} \tag{10.17}$$

Clearly $\gcd(f_1A_1)$ is divisible by f_1, and so $\gcd(f_1A_1) = f_1f_2$ for some integer f_2. The original reduction lemma was then applied to $f_1f_2A_1$ to transform it into the block form $\begin{pmatrix} f_1f_2 & 0 \\ 0 & f_1f_2A_2 \end{pmatrix}$, and so on until the process ends with A transformed into a diagonal matrix, which is the Smith–Frobenius normal form of A. The question now facing Frobenius was whether his reduction lemma could be modified so that when A is abelian, the successive unimodular matrices $P_1, Q_1, \ldots$ could be taken as first-order abelian matrices.

Now, the original reduction lemma had been based on two propositions (Corollary 8.4 and Lemma 8.5): *(1) If a and b are integral row matrices satisfying $ab^t = 1$, then unimodular matrices P, Q may be determined such that a is the first row of P and b^t is the first column of Q. (2) If $f = \gcd A$, then integral row matrices p and q exist such that $pAq^t = f$.*

Evidently, the first question was, can the proof of (1) be modified so that P and Q are first-order abelian when A is abelian? It was easy to show that the answer is affirmative:

Lemma 10.5. *(1') If a and b are integral $1 \times 2g$ row matrices satisfying $aJb^t = 1$, where J is as in (10.11), then a $2g \times 2g$ abelian matrix $\tilde{P}$ of order one can be determined such that a and b are rows 1 and $g+1$, respectively, of $\tilde{P}$.*

Frobenius next considered whether $(1')$ and (2) could be used, in much the same way as he had earlier used (1) and (2), to obtain an abelian matrix version of his original reduction lemma. Suppose that $\tilde{A}$ is a $2g \times 2g$ abelian matrix of order n [186, §2]. Imitating the first step in the proof of the original reduction lemma (Lemma 8.6), he invoked (2) to obtain row matrices p_1 and q_1 such that

$$p_1\tilde{A}q_1^t = f_1, \quad f_1 = \gcd \tilde{A}. \tag{10.18}$$

Imitating the second step, he sought to apply $(1')$ by picking p_{g+1} with the aid of (10.18) such that $p_1Jp_{g+1}^t = 1$. It is easily seen that (since $J^t = -J$)

$$p_{g+1} = (1/f_1)q_1\tilde{A}^tJ \tag{10.19}$$

is the right choice, i.e., $p_1Jp_{g+1}^t = 1$. Then by $(1')$ with $a = p_1$ and $b = p_{g+1}$, there is an abelian matrix $\tilde{P}_1$ of order one with p_1, p_{g+1} as rows 1 and $g+1$. Likewise, if

$$q_{g+1} = (1/f_1)p_1\tilde{A}J, \tag{10.20}$$

then $q_1Jq_{g+1}^t = 1$, and so by $(1')$, an abelian matrix $\tilde{Q}_1$ of order one exists that has q_1, q_{g+1} as rows 1 and $g+1$. Thus the order-one abelian matrix $\tilde{Q}_1^t$ has q_1^t, q_{g+1}^t as columns 1 and $g+1$.

As in the proof of the original reduction lemma, the $(1,1)$ entry of

$$\tilde{B} = P_1 \tilde{A} \tilde{Q}_1^t \tag{10.21}$$

is now seen to be $p_1 \tilde{A} q_1^t = f_1$. In order to get the reduced form (10.17) in the original reduction lemma, a further transformation by additional unimodular matrices had been necessary. In the present context, however, as Frobenius observed, because $\tilde{P}_1$ and $\tilde{Q}_1$ are first-order abelian and so satisfy $\tilde{P}_1 J \tilde{P}_1^t = J$ and $\tilde{Q}_1 J \tilde{Q}_1^t = J$, one obtains via (10.19)–(10.21) further information about the coefficients $[\tilde{B}]_{\gamma 1}$ and $[\tilde{B}]_{1\delta}$ of $\tilde{B}$,[14] which gives the analogue of (10.17), viz.,

$$\tilde{B} = \begin{pmatrix} f_1 & 0 \\ 0 & C \end{pmatrix}. \tag{10.22}$$

But that is not all. Because $\tilde{A}$ is not simply integral but abelian of order n, the same is true of $\tilde{B}$, and so the additional relations

$$\tilde{B} J \tilde{B}^t = nJ \quad \text{and} \quad \tilde{B}^t J \tilde{B} = J \tag{10.23}$$

are at hand. They imply that more is actually known about C, so that (10.22) can be sharpened to

$$\tilde{B} = \tilde{P}_1 \tilde{A} \tilde{Q}_1^t = \begin{pmatrix} f_1 & 0 & 0 & 0 \\ 0 & C_{11} & 0 & C_{12} \\ 0 & 0 & (n/f_1) & 0 \\ 0 & C_{21} & 0 & C_{22} \end{pmatrix}, \tag{10.24}$$

where $\tilde{C} = \begin{pmatrix} C_{11} & C_{12} \\ C_{21} & C_{22} \end{pmatrix}$ is $2(g-1) \times 2(g-1)$ and abelian of order n [186, p. 5].[15]

Since $f_1 = \gcd \tilde{A}$ evidently divides all coefficients of $\tilde{B}$ and so of $\tilde{C}$, we know that $\gcd \tilde{C} = f_1 f_2$ for some integer f_2. The above reasoning may thus be applied to $\tilde{C}$ to deduce a $2(g-1) \times 2(g-1)$ analogue of (10.24) with $f_1 f_2$ playing the role of f_1. And so on. The end result of this serendipitous conjunction of Hermite's theory and problem with the methods of Frobenius' arithmetic theory of forms was thus, in his skilled hands, an "abelian" analogue of the reduction and normal form theorems.

---

[14] $[\tilde{B}]_{\gamma,1} \overset{(10.21)}{=} (p_\gamma \tilde{A}) q_1^t = p_\gamma \tilde{A} (q_1^t) \overset{(10.19)}{=} f_1 (p_\gamma J p_{g+1}^t) = f_1 [\tilde{P}_1 J \tilde{P}_1^t]_{\gamma, g+1} = f_1 [J]_{\gamma, g+1} = f_1 \delta_{\gamma,1}$. Likewise, $[\tilde{B}]_{1,\delta} = (p_1 \tilde{A}) q_\delta^t = p_1 (\tilde{A} q_\delta^t) \overset{(10.20)}{=} -f_1 q_{g+1} J q_\delta^t = -f_1 [\tilde{Q}_1 J \tilde{Q}_1^t]_{g+1,\delta} = -f_1 [J]_{g+1,\delta} = f_1 \delta_{\delta,1}$.

[15] For example, the first relation in (10.23) means that $[\tilde{B} J \tilde{B}^t]_{1,\alpha} = n[J]_{1,\alpha} = n\delta_{\alpha,g+1}$, whereas if $[\tilde{B} J \tilde{B}^t]_{1\alpha}$ is computed using (10.22), one gets $[\tilde{B} J \tilde{B}^t]_{1,\alpha} = \sum_\mu [\tilde{B}]_{1,\mu} [J \tilde{B}^t]_{\mu,\alpha} = f_1 [J \tilde{B}^t]_{1,\alpha} = f_1 [\tilde{B}]_{\alpha, g+1}$. Comparison of the two expressions for $[\tilde{B} J \tilde{B}^t]_{1,\alpha}$ implies that $[\tilde{B}]_{\alpha,g+1} = (n/f_1)\delta_{\alpha,g+1}$. In similar fashion, if the second relations in (10.23) and (10.22) are used to compute $[\tilde{B}^t J \tilde{B}]_{\beta,1}$ in two ways, the result is $[\tilde{B}]_{g+1,\beta} = (n/f_1)\delta_{\beta,g+1}$. Thus we have (10.24).

The former says that first-order abelian matrices $\tilde{P}, \tilde{Q}$ can be determined such that $\tilde{P}\tilde{A}\tilde{Q}$ is the diagonal matrix

$$\tilde{F} = \text{Diag. Matrix}(f_1, (f_1 f_2), \ldots, (f_1 \cdots f_g), n/f_1, n/(f_1 f_2), \ldots, n/(f_1 \cdots f_g)).$$

The latter follows from the above. In stating it, "$\tilde{A}$ is equivalent to $\tilde{B}$" will no longer have Hermite's meaning ($\tilde{A} = \tilde{B}\tilde{Q}$, $\text{ord}\,\tilde{Q} = 1$) but will mean $\tilde{A} = \tilde{P}\tilde{B}\tilde{Q}$, with $\text{ord}\,\tilde{P} = \text{ord}\,\tilde{Q} = 1$.

Theorem 10.6 (Frobenius). *If $\tilde{A}$ is any $2g \times 2g$ abelian matrix of order n, then abelian matrices $\tilde{P}$ and $\tilde{Q}$ of order one can be determined such that $\tilde{P}\tilde{A}\tilde{Q} = \tilde{F}$, where $\tilde{F}$ is the diagonal matrix*

$$\tilde{F} = \text{Diag. Matrix}(e_1, \ldots, e_g, n/e_1, \ldots, n/e_g) \tag{10.25}$$

and $e_1, \ldots, e_g, n/e_g, \ldots, n/e_1$ are (in that order) the invariant factors of $\tilde{A}$. Hence two abelian matrices of order n are equivalent if and only if they have the same abelian normal form $\tilde{F}$, or equivalently, if and only if they have the same invariant factors $e_1, \ldots, e_g$.

This theorem provides a completely general solution to Hermite's problem (stated at the end of Section 10.2.2) when combined with Kronecker's solution to the problem of determining all abelian matrices of order 1: all abelian matrices are generated by choosing integers $e_1, \ldots, e_g$ such that $e_i \mid e_{i+1}$; this then determines an abelian normal form $\tilde{F}$, which, on pre- and postmultiplication by abelian matrices of order 1, yields all abelian matrices of the type determined by $\tilde{F}$. Frobenius' solution is especially satisfying because it gives an overview of all the equivalence classes of abelian matrices of a given order, each class being determined by the corresponding normal form $\tilde{F}$, which in turn is determined by the invariant factors of $\tilde{A}$. As Adolf Kräzer wrote in his treatise on abelian and theta functions of 1903 [350, p. 137], in his paper [186], "Frobenius shows how to form all [abelian matrices of order n]."

Note that since the invariant factor e_g must divide its successor, n/e_g, it follows that e_g^2 must divide n. This limits the possibilities for e_g, and so for $e_1, \ldots, e_{g-1}$ as well. In particular, when n is square-free (as Weber had to assume), it must be that $e_g = 1$, and so $e_i = 1$ for all $i \leq g$, and there is only one abelian normal form, viz., $\tilde{F} = \text{Diag. Matrix}(1, \ldots, 1, n, \ldots, n)$. This means that Weber's diagonal forms (10.16), extended to any g, are all equivalent to $\text{Diag. Matrix}(1, \ldots, 1, n, \ldots, n)$. And when $n = p$, a prime, Hermite's diagonal form (10.14) of Problem 10.3, part (d) follows. As for part (c) of Problem 10.2, prove that every abelian matrix of degree $n = p_1 \cdots p_k$ is the product of abelian matrices of degrees p_i, it is irrelevant to Frobenius' solution to Hermite's problem; but as Frobenius noted in passing [186, p. 5n], its proof follows immediately from Theorem 10.6 because the proof is thereby reduced to the consideration of diagonal matrices $\tilde{F}$ and, as Weber had already observed, the proof is easy for diagonal abelian matrices.

10.5 Kronecker's Complex Multiplication Problem

The study of Weber's paper [577] did more than acquaint Frobenius with Hermite's problem. It also acquainted (or reacquainted) him with a problem that Kronecker had posed in 1866 but never resolved. In that year Kronecker, having become familiar with Hermite's theory of abelian matrices in 1858–1859 as a result of Weierstrass' request, considered using abelian matrices as the context for generalizing to abelian functions a notion that had arisen in the study of elliptic functions, namely that of an elliptic function admitting a complex multiplication. (Kronecker's 1866 paper was already discussed in Section 5.3, albeit from the point of view of the theory of the transformation of bilinear forms.)

10.5.1 Elliptic functions with complex multiplication

The phenomenon that came to be known as elliptic functions admitting complex multiplication had been noted with interest by both Jacobi and Abel [350, p. 213] and then further studied by other mathematicians, primarily because of deep arithmetic connections. Kronecker's first paper on the subject was published in 1857 [351],[16] and he became a leader in the development of the arithmetic aspects of elliptic functions with complex multiplication.[17] Before describing Kronecker's generalization of complex multiplication to abelian functions, it will be helpful to first describe the simpler elliptic case.

Let $f(z)$ be an elliptic function with periods 1 and τ, and so (in the terminology and notation of Section 10.2.1) with period matrix $\Omega = \begin{pmatrix} 1 & \tau \end{pmatrix}$, where now $T = (\tau) \in \mathfrak{H}_g$ simply means that the complex number τ lies in the upper half-plane. Then if m is any integer, it follows that $g(z) = f(mz)$ has the same periods as f and so was known to be algebraically related to f. This phenomenon was described by saying that f admits ordinary (or real) multiplication by m. Thus no matter what τ is, ordinary multiplication by any $m \in \mathbb{Z}$ is admitted by f. However, there are certain periods $1, \tau_0$ such that elliptic functions with those periods admit a "complex multiplication" as well. To see this, and the relation to abelian matrices, consider $\tilde{A}_1 = \begin{pmatrix} 1 & 6 \\ -8 & 5 \end{pmatrix}$, which is abelian of order $n = 53$. Thus, in the sense explained in Section 10.2.1, $\tilde{A}_1$ induces a transformation of periods $\begin{pmatrix} 1 & \tau \end{pmatrix} \mapsto \begin{pmatrix} 1 & \tau' \end{pmatrix}$, where τ' is given by (10.8), which for $g = 1$ becomes $\tau' = (1 - 8\tau)(6 + 5\tau)^{-1}$; and we can ask whether this equation has a suitable solution with $\tau' = \tau$, i.e., whether $\tau = (6 + 5\tau)^{-1}(1 - 8\tau)$ has a solution in the upper half-plane. By solving the quadratic equation involved, we see that $\tau_0 = (-1 + \sqrt{11}\, i)/4$ is such a solution. This means that the corresponding transformation

[16] For a description of Kronecker's paper and its relation to his *Jugendtraum*, see [572, pp. 66ff.].

[17] See in this connection the comments of Weber [581, pp. vi–vii], whose book [581] of 1891 and its second edition of 1908 expounded the theory as it had developed in the nineteenth century.

$z = Mu$ given by (10.7) is in this case $z = Mu$ with $M = 1 - 8\tau_0 = 3 - 2\sqrt{11}\,i$. Thus $g(u) = f(Mu)$ has the same periods as f, and so f and g are algebraically related. In this case, f is said to admit a complex multiplication by $M = 3 - 2\sqrt{11}\,i$. Unlike the above-defined real multiplication, however, not every abelian matrix gives rise to a complex multiplication. For example $\tilde{A}_2 = \begin{pmatrix} 9 & -3 \\ -4 & 3 \end{pmatrix}$ is abelian of order $n = 15$. However, in this case, (10.8) with $T' = T$ is $\tau = (-3 + 3\tau)(9 - 4\tau)^{-1}$, and the corresponding quadratic equation $4\tau^2 - 6\tau - 3 = 0$ has only real solutions. Thus for $\tilde{A}_2$ there is no solution to (10.8) with $\tau' = \tau$ for τ in the upper half-plane. Hence $\tilde{A}_2$ does not give rise to a complex multiplication.

10.5.2 Kronecker's problem

Kronecker's familiarity with Hermite's theory of abelian matrices led him, with encouragement from Weierstrass,[18] to view elliptic functions with complex multiplication in terms of abelian matrices as I have done above and to consider in this manner its analogue for abelian functions in $g > 1$ variables. Thus he posed the problem of determining which abelian matrices $\tilde{A} = \begin{pmatrix} \mathrm{A} & \mathrm{B} \\ \Gamma & \Delta \end{pmatrix}$ have the property that (10.8) has a solution with $T' = T$, i.e.,

$$T = (\mathrm{A} + T\Gamma)^{-1}(\mathrm{B} + T\Delta) \quad \text{for some } T \in \mathfrak{H}_g, \tag{10.26}$$

where $T \in \mathfrak{H}_g$ means that T is symmetric with $\Psi = \operatorname{Im} T$ positive definite. When an abelian matrix $\tilde{A}$ has this property, a T satisfying (10.26) yields a period matrix $\Omega = \begin{pmatrix} I_g & T \end{pmatrix}$ with the property that the lattice Λ generated by the columns of Ω is taken into itself by $z = Mu$, where $M = \mathrm{A} + T\Gamma$. If there were an abelian function f with period matrix Ω, then f would admit the complex multiplication $z = Mu$. Actually, since $T \in \mathfrak{H}_g$, Ω satisfies Riemann's conditions (10.4), which implies that such f actually exist, but as will be clear from Section 11.4.1, Riemann's conditions were not widely known in 1866, when Kronecker wrote his paper.[19]

---

[18] In his memorial essay on Kronecker in 1893, Frobenius, who had recently spoken with the ailing Weierstrass about the happy years of mathematical give and take Weierstrass had enjoyed with Kronecker, wrote, "Since the investigation of elliptic functions with singular modules had led Kronecker to such extraordinarily interesting results, Weierstrass encouraged him to extend his researches to the complex multiplication of theta functions in several variables" [202, p. 719].

[19] On the eventual proof of sufficiency of Riemann's conditions, see Section 11.4. It turns out that Frobenius' work on generalized theta functions was involved in the first proof (by Wirtinger).

I will adopt the terminology introduced later by Frobenius and call an abelian matrix $\tilde{A}$ *principal* if (10.26) has a solution. Likewise, if $\tilde{A}$ is principal, a solution $T = (\tau_{ij})$ to (10.26) will be called a *singular parameter system* for $\tilde{A}$.

In order to state the result Kronecker obtained by investigating this problem, it is necessary to observe that if T is a singular parameter system for a principal $\tilde{A}$ and so satisfies (10.26), then after multiplying through by $\mathrm{A} + T\Gamma$, (10.26) may be rewritten as

$$\mathrm{B} + T\Delta - \mathrm{A}T - T\Gamma T = 0. \tag{10.27}$$

Kronecker's theorem, which he published in 1866 [353], may be stated in the following form using matrix symbolism.

Theorem 10.7. *Let $\tilde{A}$ be an abelian matrix, and set $\tilde{B} = -J\tilde{A}$, where J is defined as in (10.11). Then if $\varphi(\lambda) = \det(\lambda\tilde{B} - \tilde{B}^t)$ has no multiple roots, there exists a complex symmetric matrix T satisfying (10.27).*

As we saw in Section 5.3, Kronecker proved this theorem by deducing it from his Theorem 5.5 on the congruent transformation of certain families of bilinear forms. Theorem 10.7, however, does not provide even a partial solution to the complex multiplication problem in the sense of providing a sufficient condition that $\tilde{A}$ be principal. This is because the theorem does not provide any information about whether $T \in \mathfrak{H}_g$, i.e., whether the imaginary part of T has the critically important property of being positive definite. If the T posited by the theorem happens to be positive definite, then $\mathrm{A} + T\Gamma$ is invertible, and so (10.27) implies (10.26) and T is a singular parameter system. However, because Theorem 10.7 does not establish that the T it posits has a positive definite imaginary part, it does not provide a sufficient condition for $\tilde{A}$ to be principal.

Kronecker, whose interest in the problem had become secondary to the study of the transformation of the pencils of bilinear forms to which it had led him, was well aware of this fact and made two important observations pertaining to it: (1) abelian $\tilde{A}$ exist such that no T satisfying (10.27) has positive imaginary part; (2) abelian $\tilde{A}$ exist such that $\varphi(\lambda)$ has multiple roots and "the numbers τ remain partially undetermined, i.e., in this case there exist certain functions of one or more variables that if set equal to the τ_{ik} solve the problem" [353, p. 157]. Kronecker's first observation is easy to see in the elliptic case $g = 1$. (Take the solutions to (10.27) corresponding to the example $\tilde{A}_2$ considered above.) His observation (2) must have been based on examples with $g > 1$. What he meant by "solve the problem" is, however, not entirely clear. Did he simply mean that he knew of examples in which $\varphi(\lambda)$ has multiple roots and (10.27) has infinitely many solutions, or did he know of examples in which $\varphi(\lambda)$ has multiple roots and infinitely many T exist that satisfy (10.27) *and* are positive definite? Whatever he meant, his remarks raised the question whether an abelian $\tilde{A}$ such that $\varphi(\lambda)$ has multiple roots can be principal, and if so, whether $\tilde{A}$ can have infinitely many singular parameter systems associated to it.

Thus Kronecker's original problem, when combined with his remarks, suggests the following elaboration.

Problem 10.8 (Kronecker's complex multiplication problem). *Without imposing any generic preconditions, determine which abelian $\tilde{A}$ are principal, and for principal $\tilde{A}$, determine exactly when the associated singular parameter system $T = (\tau_{ij})$ is unique.*

It was for the elaborated Problem 10.8 that Frobenius supplied a definitive solution.

10.6 Frobenius' Solution to Kronecker's Problem

In his paper of 1878 on abelian matrices, Weber also considered Kronecker's problem, and it was probably Weber's remarks that induced Frobenius to consider it as well. In treating Kronecker's problem, Weber assumed that $\tilde{A}$ was a principal abelian matrix with singular parameter system T, and he sought to deduce properties of $\tilde{A}$, i.e., necessary conditions that an abelian matrix be principal. He did not practice the generic mode of reasoning in linear algebra, but his linear-algebraic tools at the time were the traditional ones, and before long, he was forced to posit the assumption that the characteristic roots of $\tilde{A}$ are all distinct.[20] The conclusions he reached under this assumption can be summarized as follows.

Proposition 10.9 (Weber). *Let $\tilde{A}$ be a principal abelian matrix with the property that the characteristic polynomial of $\tilde{A}$ has no multiple roots. Then (1) $\tilde{A}$ can have no real characteristic roots and (2) there is only one associated singular parameter system T.*

Part (1) of Weber's proposition generalized what is easily seen to be true in the elliptic case $g = 1$.[21] Part (2) confirmed that the phenomenon discovered by Kronecker, namely that when a principal $\tilde{A}$ has multiple roots, the number of singular parameter systems can be infinite, is indeed limited to the multiple root case. Of course, it remained moot whether a principal $\tilde{A}$ with multiple roots *necessarily* gives rise to more than one singular parameter system T. (It need not, as Frobenius was to show.)

[20] Incidentally, Weber's assumption is weaker than Kronecker's assumption that the roots of $\varphi(\lambda) = \det(\lambda\tilde{B} - \tilde{B}^t)$ are distinct, i.e., $\varphi(\lambda)$ can have multiple roots when $\det(\lambda I - \tilde{A})$ does not, but not conversely.

[21] In that case $\tilde{A} = \begin{pmatrix} a & b \\ c & d \end{pmatrix}$ is principal if and only if (10.26) holds, which for $g = 1$ (as noted above) means that the quadratic equation $c\tau^2 + (a-d)^2 - b = 0$ must have nonreal solutions (one of which will be in the upper half-plane), i.e., its discriminant must be negative. The condition for that may be written as $(\mathrm{tr}\tilde{A})^2 - 4\det\tilde{A} < 0$, whereas the condition that $\tilde{A}$ have real distinct characteristic roots is that $(\mathrm{tr}\tilde{A})^2 - 4\det\tilde{A} > 0$.

As with Weber's treatment of Hermite's abelian matrix problem (Section 10.3), his treatment of Kronecker's complex multiplication problem raised more questions than it answered, and Frobenius most likely took note of this in 1879, when he resolved Hermite's problem using ideas and results from his arithmetic theory of forms. In the case of Kronecker's complex multiplication problem, however, the appropriate mathematical tools needed to resolve it were not so readily manifest. The need for new ideas may have induced Frobenius to scour the mathematical literature for work relating to Hermite's theory of abelian matrices. In any case, it seems to have been at this time that he came across Laguerre's paper of 1867 [393], in which the symbolic algebra of matrices was developed independently of Cayley's paper of 1858 (Section 7.5). One of the principal applications Laguerre made of matrix algebra was to Hermite's theory. He not only translated Hermite's results into matrix symbolism, but, aided by the resulting symbolic form, extended them from $g = 2$ to any $g \geq 1$ [393, §V]. Although Laguerre developed matrix algebra exclusively on the formal, generic level, what he did on that level in reformulating some of Hermite's results was substantial and, as we shall see, provided Frobenius with the key to resolving Kronecker's problem.

In his pioneering memoir of 1855 establishing the theory of abelian matrices, Hermite needed to prove that if $\tilde{A}$ is abelian and if $T \to T'$ by virtue of $\tilde{A}$ in the sense of (10.8), then when $T = \Phi + i\Psi \in \mathfrak{H}_g$, the same is true of $T' = \Phi' + i\Psi'$. The proof that T' is symmetric was relatively easy, but proving that Ψ' is positive definite when Ψ is was more difficult [290, pp. 456–458]. To that end, Hermite used the coefficients of Φ and Ψ to define a real quadratic form $f(x) = x^t Hx$ in four ($= 2g$) variables, which he showed to be positive definite and to have special properties under transformations of the form $x = \tilde{A}y$, properties he had singled out earlier in his paper [290, pp. 450–452]. Hermite used these properties to prove that Ψ' is positive definite.

Hermite had developed the theory of abelian matrices based on the identity $\tilde{A}^t J_h \tilde{A} = nJ_h$, where $J_h = \begin{pmatrix} 0 & J_2 \\ -J_2 & 0 \end{pmatrix}$ with $J_2 = \begin{pmatrix} 0 & 1 \\ 1 & 0 \end{pmatrix}$, rather than $J = \begin{pmatrix} 0 & I_2 \\ -I_2 & 0 \end{pmatrix}$. Laguerre chose to work with J instead of J_h, and so Hermite's form $f(x) = x^t Hx$ was replaced by $f(x) = x^t Lx$, where L is now $2g \times 2g$ and has a particularly enlightening form as a block-partitioned matrix, viz.,

$$L = \begin{pmatrix} \Psi_0 & \Psi_0\Phi \\ \Phi\Psi_0 & \Phi\Psi_0\Phi + (\det\Psi)\Psi \end{pmatrix}, \tag{10.28}$$

where Ψ_0 is Laguerre's notation for Adj Ψ, the transposed matrix of cofactors of Ψ [393, pp. 263, 265].[22] Laguerre seems to have been the first to exploit the block

[22] Of course this means that $\Psi_0 = (\det\Psi)\Psi^{-1}$, but surprisingly, Laguerre introduced no symbolic notation for an inverse. In the theory of determinants, attention was focused on the adjoined system Adj Ψ, and Laguerre apparently adhered to custom. In fact, he called the transpose of a matrix its inverse [393, pp. 223–224].

multiplication of partitioned matrices, which, of course, enters naturally into the theory of abelian matrices, since from the outset they were regarded as partitioned into the four $g \times g$ blocks, namely $\mathrm{A}, \mathrm{B}, \Gamma, \Delta$, of (10.9). Replacing Hermite's form H by the matrix L enabled him to make effective use of block multiplication.

To show that $\Psi' = \operatorname{Im} T'$ is also positive definite, Hermite had obtained a remarkable relationship [290, p. 457], which in Laguerre's rendition became the following [393, p. 264, eqn. (10)]. Let L' denote the matrix of (10.28) corresponding to $T' = \Phi' + i\Psi'$. Then

$$L' = \mu \tilde{A}^t L \tilde{A}, \quad \mu = (\det \Psi')/(n \det \Psi). \tag{10.29}$$

Following Hermite's line of thought, Laguerre used (10.29) to show that Ψ' is positive definite when Ψ is [393, pp. 265–256].

Frobenius read Laguerre's paper with an interest in abelian matrices that are principal. When $\tilde{A}$ is principal, it has a singular parameter system $T = \Phi + i\Psi$ for which $T' = T$ in (10.8). Thus when $\tilde{A}$ is principal, $\Phi' = \Phi$, $\Psi' = \Psi$, $L' = L$, and μ simplifies to $1/n$, so that Laguerre's relation (10.29) becomes $L = (1/n)\tilde{A}^t L \tilde{A}$, which we may write as

$$P^t L P = L, \quad P = n^{-1/2}\tilde{A}. \tag{10.30}$$

Thus Laguerre's relation states that the linear transformation given by $P = n^{-1/2}\tilde{A}$ takes the positive definite real quadratic form defined by L into itself. This makes P and therefore $\tilde{A}$ quite special, as Frobenius realized. For example, when $L = I$, (10.30) becomes $P^t P = I$, which says that $P = n^{-1/2}\tilde{A}$ is real and orthogonal. More generally, since L is positive definite, the principal axes theorem implies that an orthogonal transformation $x \to y$ exists such that $x^t L x = \lambda_1 y_1^2 + \cdots + \lambda_{2g} y_{2g}^2$, and since all $\lambda_i > 0$, the further transformation $y_i = \lambda_i^{-1/2} z_i$ gives $x^t L x = z_1^2 + \cdots + z_{2g}^2$. Expressed in terms of matrix algebra, this says that K exists such that $K^t L K = I$, or equivalently, $L = Q^t Q$, where $Q = K^{-1}$. Substituting this expression for L in (10.30), we obtain $P^t Q^t Q P = Q^t Q$, and if this equation is multiplied on the left by $(Q^t)^{-1} = (Q^{-1})^t$ and on the right by Q^{-1}, the result may be expressed in the form

$$\left(QPQ^{-1}\right)^t \left(QPQ^{-1}\right) = I, \tag{10.31}$$

which implies that $S = QPQ^{-1}$ satisfies $S^t S = I$ and so defines a real orthogonal transformation. Thus Frobenius could see from his Theorem 7.15 on orthogonal matrices that since $P = n^{-1/2}\tilde{A}$ is similar to a real orthogonal matrix S, it inherits the two properties of orthogonal matrices given in that theorem: every characteristic root has absolute value 1 and all the elementary divisors are linear, i.e., $P = n^{-1/2}\tilde{A}$ can be diagonalized. Since $\tilde{A} = \sqrt{n}P$, it then follows that the characteristic roots of $\tilde{A}$ all have absolute value $\sqrt{n}$ and that its elementary divisors must also be linear.

According to Frobenius, once it is realized that these two properties of a principal abelian matrix are necessary, it is easy to show that they are also sufficient [188,

p. 111]. Of course, this was easy for Frobenius, because he was a master of the algebraic aspects of the theory of abelian and theta functions. The point to be made here is that it was through a fertile combination of matrix algebra and Weierstrass' elementary divisor theory that he was led, with some invaluable assistance from Hermite and Laguerre, to the discovery of the following theorem, which solves Kronecker's problem as originally formulated.

Theorem 10.10 (Principal matrix theorem I). *If $\tilde{A}$ is an abelian matrix of order n, then $\tilde{A}$ is principal if and only if (1) all characteristic roots of $\tilde{A}$ have absolute value $\sqrt{n}$ and (2) the elementary divisors of $\tilde{A}$ are all linear ($\tilde{A}$ can be diagonalized).*

What had made Kronecker's problem seem so intractable was the lack of realization of property (1), which is not generally true for abelian matrices. As for property (2), it indicated the extent to which the ad hoc generic assumptions of Kronecker and Weber were justified. That is, by part (iv) of Frobenius' minimal polynomial theorem (Theorem 7.2), (2) is equivalent to saying that it is the minimal polynomial of $\tilde{A}$, rather than its characteristic polynomial, that must have distinct roots when $\tilde{A}$ is principal.

Frobenius was able to obtain a significant corollary to Theorem 10.10 by combining it with a theorem due to Kronecker. In 1857, the latter had proved that if $\varphi(\lambda)$ is a monic polynomial with integer coefficients with the property that all its roots α_i have $|\alpha_i| = 1$, then the α_i must actually be roots of unity, i.e., $\alpha_i^{n_i} = 1$ for some $n_i \in \mathbb{Z}^+$ [352, I, p. 103]. By virtue of Theorem 10.10, the characteristic polynomial of an abelian matrix of order $n = 1$ is just the sort to which Kronecker's theorem applies, and so Frobenius obtained the following corollary to his theorem [188, VII, p. 115]:

Corollary 10.11. *If $\tilde{A}$ is a principal abelian matrix of order $n = 1$, then all its characteristic roots are roots of unity.*

As we shall see in Section 10.7, Hurwitz found a geometric application for this corollary, which was then further developed by Scorza.

The above-described line of reasoning leading from Laguerre's equation (10.29) via (10.30) to properties (1) and (2) of Theorem 10.10 is my reconstruction of how Frobenius discovered them based on his remarks [188, p. 98]. Once having discovered them, however, he proceeded in characteristic fashion to develop the theory ab initio along the lines he deemed most suitable for publication. The new approach led him to introduce the notion of a unitary matrix—apparently for the first time in the history of mathematics[23]—and to establish the main properties of

[23]In the 1850s and 1860s, when Hermitian symmetric matrices and forms were introduced and studied as analogues of real symmetric matrices and quadratic forms, the focus was exclusively on the reality of the characteristic roots. No interest was shown in generalizing the principal axes theorem, which would have led naturally to the notion of a unitary matrix as the "Hermitian" analogue of a real orthogonal matrix. That Frobenius was the first to have found a use for unitary matrices in his 1883 paper [188] on principal transformations is suggested by a remark by Hurwitz in 1897. In his seminal paper on invariant integrals on Lie groups, Hurwitz had occasion to

such matrices. The new approach also brought to light further properties of principal abelian matrices that readily generalized to the context of a much more general, geometrically motivated, conception of complex multiplication, as will be seen in Section 10.7. A brief digression is thus in order.

In the reasoning leading to his Proposition 10.9, Weber had shown that when $\tilde{A}$ is principal and all $2g$ of its characteristic roots are distinct, then exactly half of them, say $\alpha_1, \ldots, \alpha_g$, are characteristic roots of the $g \times g$ matrix M defined by (10.7), viz., $M = \mathrm{A} + T\Gamma$, and the other half are all of the form $\alpha_1', \ldots, \alpha_g'$, where $\alpha_j \alpha_j' = n$ [577, pp. 141–142]. In view of property (1) of Theorem 10.10, namely $\alpha\bar{\alpha} = n$, Frobenius could see that $\alpha_j' = \bar{\alpha}_j$ when $\tilde{A}$ is principal. Thus, at least in the generic case, $\tilde{A}$ and $\tilde{M} = \begin{pmatrix} M & 0 \\ 0 & \overline{M} \end{pmatrix}$ have the same characteristic roots, viz., $\alpha_1, \ldots, \alpha_g, \bar{\alpha}_1, \ldots, \bar{\alpha}_g$. Since the generic case considered by Weber is not far removed from the case of linear elementary divisors guaranteed by property (2), it was perhaps natural for Frobenius to consider showing that $\tilde{A}$ and $\tilde{M}$ are always similar when $\tilde{A}$ is principal. This he did using matrix algebra, including the expansion of $\tilde{A}^t J \tilde{A} = nJ$ by the block multiplication of partitioned matrices, to show that

$$\tilde{A} = \tilde{P}^{-1} \tilde{M} \tilde{P}, \quad \text{where} \quad \tilde{P} = \begin{pmatrix} I & T \\ I & \overline{T} \end{pmatrix} \tag{10.32}$$

and T is a singular parameter system for $\tilde{A}$ [188, p. 105]. This meant that in order to establish the necessity of conditions (1) and (2), it sufficed to establish them for $\tilde{M}$, and given how $\tilde{M}$ is related to M, it sufficed to show that (1) and (2) hold for M. Before proceeding further, it should be noted that (10.32) brings with it the following result about the characteristic polynomial $\varphi(\lambda)$ of $\tilde{A}$:

$$\varphi(\lambda) = \det(\lambda I - \tilde{A}) = \det(\lambda I - M)\det(\lambda I - \overline{M}), \tag{10.33}$$

which shows that the characteristic polynomial of M is a factor of φ and that the roots of φ are precisely the roots of M and their conjugates.

Further and more complicated matrix algebra enabled Frobenius to establish a relation analogous to Laguerre's (10.29) but for M rather than $\tilde{A}$: if $\tilde{A}$ is abelian and $\tilde{A} : T \mapsto T'$ in the sense (10.8), then

$$M \Psi' \overline{M}^t = n \Psi, \tag{10.34}$$

introduce unitary transformations (and the special unitary group) in order to perform what Weyl later called the "unitarian trick" (see [276, p. 393]). Apparently because unitary transformations (or substitutions) were still sufficiently novel in 1897, Hurwitz pointed out to his readers that "these substitutions also come into consideration in other investigations" [304, p. 556, n. 1] and referred them to Frobenius' paper [188].

where Ψ, Ψ' are the imaginary parts of T, T', respectively [188, p. 105]. Frobenius was well acquainted with the theory of Hermitian forms, where conjugate transpositions such as $\overline{M}^t$ replace ordinary transposition so as to define Hermitian symmetry. Although he did not do so, I will use the notation $M^h = \overline{M}^t$ to denote the Hermitian transpose of a matrix M. Thus (10.34) is a Hermitian transpose analogue of Laguerre's equation (10.29), and when $T' = T$, so that $\Psi' = \Psi$, it yields an analogue of (10.30), namely

$$S^h \Psi S = \Psi, \quad S = n^{-1/2} M^h. \tag{10.35}$$

Reasoning completely analogous to that leading from (10.30) to (10.31)—but with Hermitian transposes replacing ordinary transposition—then shows that

$$(QSQ^{-1})^h (QSQ^{-1}) = I \tag{10.36}$$

and so implies that S is similar to $R = QSQ^{-1}$, where R satisfies $R^h R = I$.

Frobenius realized that this is the Hermitian conjugate analogue of the defining relation for a real orthogonal transformation, and he observed that the proof of his Theorem 7.15 on real orthogonal matrices "can be carried over to the more general systems R considered here without the least change" [188, p. 100]. Over two decades later Frobenius and his student I. Schur named the more general systems R "unitary" (*unitär*) [233, p. 356]—presumably because of the important role they had come to play in the representation theory of finite groups that Frobenius had begun creating in 1896 (Chapters 12–15). By virtue of his work on the Cayley–Hermite problem, Frobenius had thus provided, mutatis mutandis, a proof that if R is unitary, then all its characteristic roots have absolute value one and its elementary divisors are linear, so that R is similar to a diagonal matrix. Thus M and therefore also $\tilde{M}$ and $\tilde{A}$ have properties (1) and (2). I should point out that Frobenius deduced these conclusions by applying the reasoning leading from (10.35) to (10.36) to prove a still more general theorem, which he characterized as the "fundamental theorem" underlying his paper: If H is any positive definite Hermitian symmetric matrix and if S is such that $S^h HS = H$, then all the characteristic roots of S have absolute value one and the elementary divisors of S are all linear [188, p. 100].

We saw that Kronecker's remarks about his Theorem 10.7 tacitly suggested the further problem of characterize the principal $\tilde{A}$ for which the associated singular parameter system T is unique. Frobenius completely solved this problem as well. Again, ideas in Weber's proof of Proposition 10.9 formed the starting point. Let α denote a characteristic root of the principal abelian matrix $\tilde{A}$. Then it is a characteristic root of $\tilde{A}^t$ as well, and so $x \neq 0$ exists such that $\tilde{A}^t x = \alpha x$. Since $\tilde{A} J \tilde{A}^t = nJ$ follows from (10.11), we have $nJx = (\tilde{A}J)\tilde{A}^t x = \alpha(\tilde{A}J)x$, or $\tilde{A}(Jx) = (n/\alpha)x$. So much was in effect implicit in Weber's proof. By virtue of property (1) of Frobenius' principal matrix theorem I (Theorem 10.10), however, $n/\alpha = \bar{\alpha}$, and so $\tilde{A}(Jx) = \bar{\alpha}(Jx)$. Taking complex conjugates then gives $\tilde{A}(J\bar{x}) = \alpha(J\bar{x})$, since $\tilde{A}$ and J are real. Summing up: if $x = \begin{pmatrix} y & z \end{pmatrix}^t$ is a characteristic vector for α with respect to $\tilde{A}^t$, then

$x^* = J\bar{x} = (\bar{z}\ -\bar{y})^t$ is a characteristic vector for α with respect to $\tilde{A}$. Now suppose α is a characteristic root of multiplicity m for $\tilde{A}$ (and so for $\tilde{A}^t$ as well). Then by virtue of property (2) of Theorem 10.10, there are m linearly independent characteristic vectors for α as a characteristic root of $\tilde{A}^t$. If these are denoted by $x_j = (y_j\ z_j)^t$, $j = 1,\ldots,m$, then $x_j^* = (\bar{z}_j\ -\bar{y}_j)^t$ are linearly independent characteristic vectors for α with respect to $\tilde{A}$. Frobenius introduced the matrix $Z_\alpha = (c_{jk})$ defined by the dot product[24] $c_{jk} = (1/2i)(x_j \cdot x_k^*) = (1/2i)(y_j \cdot \bar{z}_k - z_j \cdot \bar{y}_k)$ [188, §4]. Then Z_α is an $m \times m$ Hermitian symmetric matrix of full rank m, and Frobenius discovered that the properties of the Z_α were the key to determining when the singular parameter system T of $\tilde{A}$ is unique: T is unique precisely when for every characteristic root α of $\tilde{A}$, Z_α is positive or negative definite [188, p. 114, Satz V].

The reasoning leading to this result did not utilize matrix algebra, but in order to express his uniqueness condition directly in terms of constructs coming from the coefficients of $\tilde{A}$, Frobenius turned to matrix algebra [188, §8]. He showed that Z_α could be replaced by a $2g \times 2g$ Hermitian symmetric matrix $\tilde{Z}_\alpha$ of rank m (the multiplicity of α) and related to Z_α by $\tilde{Z}_\alpha = R^h \begin{pmatrix} Z_\alpha & 0 \\ 0 & 0 \end{pmatrix} R$, where the unitary matrix R is chosen such that $\tilde{Z}_\alpha$ has the following simple symbolic form. Let $\psi(\lambda)$ denote the minimal polynomial of $\tilde{A}$. Then since (2) of Theorem 10.10 states that the elementary divisors of $\tilde{A}$ are all linear, Frobenius' Theorem 7.2 on minimal polynomials implies that $\psi(\lambda) = \prod_{j=1}^{d}(\lambda - \alpha_j)$, where $\alpha_1,\ldots,\alpha_d$ denote the distinct characteristic roots of $\tilde{A}$. Set $\psi_k(\lambda) = \psi(\lambda)/(\lambda - \alpha_k) = \prod_{j \neq k}(\lambda - \alpha_j)$. Then

$$\tilde{Z}_{\alpha_k} = i\psi'(\bar{\alpha}_k)\psi_k(\tilde{A})J, \quad k = 1,\ldots,d. \tag{10.37}$$

The second part of Frobenius' solution to Kronecker's problem may now be stated in the following form.

Theorem 10.12 (Principal matrix theorem II). *Let $\tilde{A}$ be a principal abelian matrix with distinct characteristic roots $\alpha_1,\ldots,\alpha_d$ and let $\tilde{Z}_{\alpha_k}$ be as in (10.37). Then $\tilde{Z}_{\alpha_k}$ is Hermitian symmetric of rank equal to the multiplicity of α_k, and there is a unique singular parameter system T associated to $\tilde{A}$ if and only if for each k, $\tilde{Z}_{\alpha_k}$ is definite, i.e., either nonnegative or nonpositive definite.*

In the above criterion for uniqueness of T, $\tilde{Z}_{\alpha_k}$ nonnegative definite, e.g., means that all its characteristic roots are nonnegative. Also if, e.g., $\tilde{Z}_{\alpha_k}$ is nonnegative definite, then for $l \neq k$, $\tilde{Z}_{\alpha_l}$ may be either nonnegative definite or nonnegative positive. Part (2) of Weber's Proposition 10.9 is an immediate consequence of Theorem 10.12, since when all the roots of $\tilde{A}$ are distinct, as Weber assumed, then each Z_{α_k} has rank 1 (the multiplicity of α_k) and so has exactly one nonzero root. Thus Z_{α_k} is

[24] Frobenius did not utilize the notion of a dot product. I have used it for succinctness of expression.

nonnegative or nonpositive definite, depending on the sign of the nonzero root, and Theorem 10.12 implies that T is unique.

Frobenius' above theorem suggests, but does not prove, that principal abelian matrices $\tilde{A}$ with multiple roots can exist that nonetheless have a unique singular parameter system because all the $\tilde{Z}_{\alpha_k}$ are definite. Frobenius, with his penchant for thoroughness, naturally considered this matter. He hypothesized the existence of a principal abelian $\tilde{A} = \begin{pmatrix} \mathrm{A} & \mathrm{B} \\ \Gamma & \Delta \end{pmatrix}$ of order n for which the associated $g \times g$ matrix $M = \mathrm{A} + T\Gamma$ of (10.7) is μI_g, where $\mu = \frac{1}{2}(p + iq)$ and $q \neq 0$. In this case, since $\tilde{A}$ is similar to $\tilde{M} = M \oplus \overline{M}$ by (10.32), $\tilde{A}$ would have two distinct characteristic roots, μ and $\bar{\mu}$, each with multiplicity $g > 1$. By (2) of the principal matrix theorem I (Theorem 10.10), $\tilde{A}$ has linear elementary divisors, so that [by Frobenius' Theorem 7.2 on minimal polynomials, part (iv)] the minimal polynomial $\psi(\lambda)$ of $\tilde{A}$ must be $\psi(\lambda) = (\lambda - \mu)(\lambda - \bar{\mu})$, which equals $\lambda^2 - p\lambda + n$, since $\mu\bar{\mu} = |\mu|^2 = n$ by (1) of Theorem 10.10. Starting with the equation $M = \mathrm{A} + T\Gamma$ with $M = \mu I_g$, and its complex conjugate equation, Frobenius deduced by matrix algebra that $\mathrm{A} - \bar{\mu}I_g$ is invertible and that $T = (\mathrm{A} - \bar{\mu}I_g)^{-1}\mathrm{B}$ [188, p. 125], which suggests that T would be unique if such an $\tilde{A}$ were to exist.

To construct such an $\tilde{A}$, Frobenius applied his formula (10.37) for $\tilde{Z}_\mu$ to his hypothetical $\tilde{A}$ to deduce that because $\psi(\lambda) = \lambda^2 - p\lambda + n$, one has by (10.37),

$$\tilde{Z}_\mu = q(\tilde{A} - \bar{\mu}I_{2g})J = \begin{pmatrix} -\mathrm{B} & \mathrm{A} - \bar{\mu}I_g \\ -\Delta + \bar{\mu}I_g & \Gamma \end{pmatrix}.$$

Since Theorem 10.12 says that $\tilde{Z}_\mu$ is Hermitian symmetric, Frobenius used block multiplication on $\tilde{Z}_\mu^h - \tilde{Z}_\mu = 0$ to deduce that the blocks constituting $\tilde{A}$ must satisfy

$$\mathrm{B}^t = \mathrm{B}, \quad \Gamma^t = \Gamma, \quad \mathrm{A} = pI_g - \Delta^t. \tag{10.38}$$

To satisfy these conditions, which, as he showed, also imply $\psi(\tilde{A}) = 0$, Frobenius considered

$$\mathrm{B} = \begin{pmatrix} a & b \\ b & c \end{pmatrix}, \quad \Gamma = \begin{pmatrix} -c/k & b/k \\ b/k & -a/k \end{pmatrix}, \quad \Delta = 0, \quad \mathrm{A} = pI_g, \tag{10.39}$$

where a, b, c, k, p are all integers, and in order that $\tilde{A}$ be integral, (1) k is a common divisor of a, b, c. Thus the conditions (10.38) are satisfied. In addition, direct computation shows that $\tilde{A}J\tilde{A}^t = nJ$, where $n = (ac - b^2)/k$. By (1), n is an integer, but to make $\tilde{A}$ abelian of order n, we need to assume (2) $(ac - b^2)/k > 0$. Computation of the roots of the minimal polynomial $\psi(\lambda) = \lambda^2 - p\lambda + n$ with $n = (ac - b^2)/k$ shows that in order for the roots to be nonreal, it is necessary to assume that (3) $p^2 < 4n$. Then $q = \pm\sqrt{4n - p^2}$. Finally, since

$$T = (\mathrm{A} - \bar{\mu}I_g)^{-1}\mathrm{B} = \mu^{-1}\mathrm{B},$$

it follows that

$$\Psi = \operatorname{Im} T = \frac{-2q}{p^2+q^2}\mathrm{B}$$

should be positive definite, which means that $q\mathrm{B}$ should be negative definite. If we take $q < 0$ (as we may), then the roots of B must be positive, and calculation shows that this occurs, provided $k > 0$.

By such considerations Frobenius was led to the principal matrices $\tilde{A}$ with blocks defined as in (10.39) and subject to the conditions (1)–(3). He showed that when $k > 0$, there is a unique singular parameter system $T = (\tau_{ij})$, but that when $k < 0$, there are infinitely many singular T's. In the latter case, he derived elegant formulas for the τ_{ij} as rational functions of a parameter $\lambda \in D$, where the domain $D \subset \mathbb{C}$ is explicitly given [188, pp. 126–127]. Frobenius also gave an example (with $g = 2$) to show that a principal $\tilde{A}$ can exist with all real characteristic roots, which by (1) of the principal matrix theorem I must all be equal to $\pm\sqrt{n}$ and so have multiplicities greater than 1 [188, pp. 127–128]. Thus (1) of Weber's Proposition 10.9, which asserts that a principal $\tilde{A}$ cannot have real characteristic roots when the latter are distinct, does not extend to the case of multiple roots.

In 1883, while Frobenius' paper [188] containing the above-described results was in press, a paper [607] appeared in *Mathematische Annalen* that also dealt with Kronecker's remarks in 1866 about complex multiplication in the case $g > 1$. The author was Eduard Wiltheiss (1855–1900). In 1879, he had written his Berlin doctoral dissertation under Weierstrass' direction on a topic in the theory of abelian functions. Two years later, as his *Habilitationsschrift* at the university in Halle, Germany, he turned to the matter of complex multiplication in the case $g = 2$ as Kronecker had discussed it in his paper of 1866 [353], which Weierstrass may have called to his attention. (Wiltheiss made no reference to Weber's paper.) His *Annalen* paper apparently represented the fruits of the habilitation work. In it, Wiltheiss proposed a different approach to determining the singular parameter systems T associated to principal abelian matrices $\tilde{A}$, one that drew more fully on the analytic theory of the theta functions defined by the parameter systems $T = (\tau_{ij})$ and $T' = (\tau'_{ij})$ of (10.8) than had Kronecker's more purely algebraic approach [607, p. 385]. (Frobenius had also developed such an approach in the more general setting of his own paper; he used it to prove the sufficiency of conditions (1) and (2) of Theorem 10.10 and to deduce Theorem 10.12 [188, §§4–5].)

Although Wiltheiss praised Kronecker's approach, he clearly intended to use his own to get a clearer overview than Kronecker had provided—at least in the case $g = 2$—of when T is unique and when it is not, and to relate this to the properties of the "multipliers," i.e., the characteristic roots μ_i of M. (Recall that Frobenius proved that $\tilde{A}$ is similar to $\tilde{M} = M \oplus \overline{M}$, so that the multipliers μ_i and their conjugates are the characteristic roots of $\tilde{A}$.) Thus Wiltheiss set out to explore the very issues that Frobenius' work had settled definitively, and so it is of interest to ask to what extent he anticipated Frobenius' g-variable conclusions. Do we have here yet another example of multiple discovery involving Frobenius? The answer is no; Wiltheiss did not anticipate Frobenius' discoveries. For example, Wiltheiss realized that the multipliers μ_i were among the characteristic roots of $\tilde{A}$ and that α is a characteristic root of $\tilde{A}$ if and only if $\alpha' = n/\alpha$ is also a root, where $n = \deg\tilde{A}$; but like Weber, he never realized that when $T' = T$, i.e., when $\tilde{A}$ is principal, one has $|\alpha| = \sqrt{n}$, so that $\alpha' = \bar{\alpha}$. Furthermore, his computational case-by-case approach [607, pp. 391–397] implied that the multipliers are always distinct and that whenever a principal $\tilde{A}$ has multiple characteristic roots, there are always infinitely many singular parameter systems T, implications shown to be false by Frobenius' above-discussed example.

In 1903, Frobenius' Theorem 10.10 was given a detailed exposition in Adolf Krazer's treatise on abelian and theta functions [350, pp. 214–234]. Because Frobenius' results on matrix algebra could not be taken for granted as common knowledge even in 1903, Krazer included an exposition of the basics of matrix algebra and related theorems, such as Frobenius' Theorem 7.15 on real orthogonal matrices. In this manner, matrix algebra, Frobenius-style, and its mathematical advantages were called to the attention of the many mathematicians of the period with an interest in abelian and theta functions. As we shall see in the following section, one such mathematician seems to have been S. Lefschetz.

10.7 Geometric Applications of Frobenius' Results

The focus of Frobenius' paper [188] on complex multiplication had been algebraic, as was the case with the papers of Kronecker, Weber, and Wiltheiss. By definitively solving Kronecker's problem, Frobenius' paper brought this algebraic direction to a close. However, as we shall see in this section, mathematicians with a more geometric viewpoint primarily in mind found Frobenius' results useful.

10.7.1 Hurwitz

The first such seems to have been Adolf Hurwitz (1859–1919), who in a paper of 1888 [302], posed the problem of determining all Riemann surfaces $\mathcal{S}$ defined by an irreducible polynomial equation $P(s,z) = 0$ with the property that $\mathcal{S}$ is mapped into itself by a birational transformation.[25] His solution consisted of three main theorems, the first of which showed that any such transformation T must be periodic: $T^k = I$ for some integer k. Hurwitz gave a proof of this [302, §3] but then observed that Frobenius' Theorem 10.10 could be used to establish a more general theorem about algebraic correspondences that implied the periodicity result as a special case [302, §4].

[25] On Hurwitz's paper and related work by him, see [411, pp. 332–333, 344–345].

Briefly and loosely described, a correspondence of type (m,n) on $\mathcal{S}$ may be thought of as a subset $\mathcal{C}$ of $\mathcal{S}\times\mathcal{S}$ with the property that for every $(p,q)\in\mathcal{C}$, there are precisely m points $p_i\in\mathcal{S}$ such that $(p_i,q)\in\mathcal{C}$ and precisely n points $q_i\in\mathcal{S}$ such that $(p,q_i)\in\mathcal{C}$. Thus the correspondence associates to each point q exactly m points $p_1,\ldots,p_m$ and to each point p exactly n points $q_1,\ldots,q_n$. If $\mathcal{S}$ has genus g, then there are g independent integrals of the first kind with $g\times 2g$ period matrix Ω, and the existence of the correspondence implies relations among the integrals that in turn imply that

$$M\Omega=\Omega A, \tag{10.40}$$

where M is $g\times g$ and A is a $2g\times 2g$ integral matrix. As in the discussion leading to (10.10), viz., $M\Omega'=\Omega\tilde{A}$ when $\tilde{A}$ is abelian, (10.40) simply says that $z'=Mz$ takes the columns of Ω into integral linear combinations of those same columns, the integer coefficients being supplied by the matrix A in (10.40).

A relation of the sort (10.40) is satisfied by every principal abelian matrix $\tilde{A}$, since in that case, $T'=T$ means that $\Omega'=\begin{pmatrix}I & T'\end{pmatrix}=\begin{pmatrix}I & T\end{pmatrix}=\Omega$, and so $M\Omega'=\Omega\tilde{A}$ becomes $M\Omega=\Omega\tilde{A}$. But for A in (10.40) to be principal, it must first of all be *abelian*, i.e., it must satisfy $AJA^t=nJ$ for some positive integer n with J as in (10.5). Hurwitz set out to determine the conditions for this to occur. He determined necessary and sufficient conditions that $AJA^t=nJ$, and he observed that these conditions are always fulfilled when the correspondence is of the type $(1,n)$, in which case $AJA^t=nJ$ [302, pp. 300–301]. In particular, when $n=1$, so that the correspondence is one-to-one, A is abelian of order one. From Frobenius' Corollary 10.11, it then followed that all the characteristic roots of A are roots of unity. Since Frobenius had also shown that A and M can be diagonalized and that A is similar to $M\oplus\overline{M}$, it followed that for some positive integer k, $A^k=I_{2g}$, and so also $M^k=I_g$. Using these relations, Hurwitz was then able to conclude that the one-to-one correspondence associated to $M\Omega=\Omega A$ is periodic [302, §6].

10.7.2 Humbert

Hurwitz's relation $M\Omega=\Omega A$ also arose in the geometric study of what would now be called abelian varieties. Suppose that $\Omega=\begin{pmatrix}\omega_1 & \cdots & \omega_{2g}\end{pmatrix}$ is a period matrix satisfying Riemann's conditions (10.4), so that abelian functions exist with $\omega_1,\ldots,\omega_{2g}$ as periods. It was known that abelian functions $f_1(z),\ldots,f_{g+1}(z)$ with period matrix Ω can be determined such that any other abelian function $g(z)$ with the same period matrix Ω is a rational function of $f_1(z),\ldots,f_{g+1}(z)$. Let Π_{2g} denote the parallelotope of points $z=\sum_{i=1}^{2g}t_i\omega_i\in\mathbb{C}^g$, $0\le t_i<1$, and consider the equations

$$x_i=f_i(z),\quad z\in\Pi_{2g},\quad i=1,\ldots,g+1. \tag{10.41}$$

The fact that every abelian function is a rational function of $f_1,\ldots,f_{g+1}$ implies that if $z,z' \in \Pi_{2g}$ and $z \neq z'$, then $f_i(z) \neq f_i(z')$ for at least one i, and so the correspondence $(z_1,\ldots,z_g) \to (x_1,\ldots,x_{g+1})$ is one-to-one on Π_{2g} [410, pp. 109–110]. The equation (10.41) are the parametric equations of a g-dimensional "hyperelliptic" surface later called an abelian variety [410, p. 409] and denoted here by $\mathcal{V}_g$. If Λ denotes the lattice of all $\mathbb{Z}$-linear combinations of the periods $\omega_1,\ldots,\omega_{2g}$, then when Ω satisfies Riemann's conditions (10.4), $\mathcal{V}_g$ can be identified with the g-dimensional torus $\mathbb{C}^g/\Lambda$, with coordinate charts given by (10.41) but with Π_{2g} replaced by various open sets.

Suppose now that for $\mathcal{V}_g = \mathbb{C}^g/\Lambda$, a $g\times g$ matrix M and a $2g\times 2g$ integral matrix A exist such that $M\Omega = \Omega A$. This relation says that $w = Mz$ takes Λ into itself and so determines a well-defined transformation $\hat{M}$ on $\mathbb{C}^g/\Lambda$, viz., $\hat{M}(z+\Lambda) = Mz+\Lambda$, with the property that $\hat{M}(\hat{z}_1+\hat{z}_2) = \hat{M}\hat{z}_1+\hat{M}\hat{z}_2$ for any $\hat{z}_1,\hat{z}_2 \in \mathbb{C}^g/\Lambda$. (In modern parlance, $\hat{M}$ is an endomorphism of the abelian group $\mathbb{C}^g/\Lambda$.) In the coordinate system given by (10.41), $\hat{M}$ transforms the point with coordinates $x_i = f_i(z)$ into the point with coordinates $x_i' = f_i(Mz)$. Now $g_i(z) = f_i(Mz)$ is also an abelian function with period matrix Ω, and so $g_i(z) = R_i(f_1(z),\ldots,f_{g+1}(z))$, where $R_i(w_1,\ldots,w_{g+1})$ is a rational function. Thus the transformation $\hat{M}$ induced by M is rational: $x_i' = R_i(x_1,\ldots,x_{g+1})$, $i = 1,\ldots,g+1$.

Such transformations of abelian varieties with $g > 1$ became of interest to algebraic geometers starting with the work of Georges Humbert (1859–1921) in Paris.[26] In seminal papers of 1899–1900 [299,300], Humbert generalized Hermite's theory of transformations of abelian functions (described above in Section 10.2.1) in a way that dropped Hermite's assumption that the integral matrix $\tilde{A}$ in the relation (10.10), viz., $M\Omega' = \Omega\tilde{A}$, had to be abelian ($\tilde{A}J\tilde{A}^t = nJ$) and replaced it instead with "singularity conditions" imposed upon Ω. (Expressed in matrix form, a singularity condition is an equation $\Omega K \Omega^t = 0$, where $K \neq \lambda J$ is skew-symmetric.) This meant, in particular, that the notion of complex multiplication introduced by Kronecker and investigated definitively by Frobenius in his 1883 paper [188] was replaced by the vastly more general one described above: the abelian variety $\mathcal{V}_g$ admits complex multiplications if invertible matrices M,A exist such that $M\Omega = \Omega A$, where A is integral and (to rule out ordinary multiplication) $M \neq mI_g$ for $m \in \mathbb{Z}$. Accordingly, a complex multiplication (M,A) in Humbert's sense is a complex multiplication in the sense of Kronecker and Frobenius only when $A \neq mI_{2g}$ is a principal abelian matrix, which requires that (1) $AJA^t = nJ$ for some integer $n > 0$; (2) there be a symmetric matrix T with $\Psi = \operatorname{Im} T \succ 0$ that satisfies the equation $T = (\mathrm{A}+T\Gamma)^{-1}(\mathrm{B}+T\Delta)$, where $A = \begin{pmatrix} \mathrm{A} & \mathrm{B} \\ \Gamma & \Delta \end{pmatrix}$. Humbert, who worked in the case $g = 2$, claimed that the most geometrically interesting examples of varieties with complex multiplication occurred when A was not a principal abelian matrix [300, pp. 327–328]. Using the theory he had developed, he described many classes of complex multiplications

---

[26] I am grateful to J.-P. Serre for calling my attention to the work of Humbert.

(M,A) depending on integer parameters, and he pointed out that within such a class of pairs (M,A), A is principal abelian only for exceptional values of the parameters involved [300, pp. 332ff.]. Based on Humbert's paper, one would be tempted to conclude that Frobenius' work, restricted as it was to geometrically uninteresting principal abelian matrices, would not play a significant role in the study of abelian varieties with complex multiplication; but that turns out to be incorrect, as we shall see.

10.7.3 *Scorza*

Further work exploring the properties of abelian varieties with complex multiplications was done in the case $g = 3$ by Humbert and Paul Lévy and then by the Italian algebraic geometers Carlo Rosati (1879–1929) and Gaetano Scorza (1876–1939).[27] It was Scorza's work during 1914–1916 that led to the more general notion of a multiplication on $\mathcal{V}_g$ being defined by any pair (M,A) satisfying $M\Omega = \Omega A$, where A has coefficients from $\mathbb{Q}$, and A and M are allowed to be singular [411, p. 380]. And it was Scorza who showed that some of the ideas in Frobenius' paper [188] remained viable within the more general context of the Humbert–Scorza notion of complex multiplication.

In a lengthy memoir published in 1916 [528], Scorza expounded his general notion of complex multiplication within the context of his definition of a *Riemann matrix*. By a Riemann matrix he meant any $g \times 2g$ matrix $\Omega = \begin{pmatrix}\omega_1 & \cdots & \omega_{2g}\end{pmatrix}$ with the following properties: there is an alternating form $H(x,y) = x^t Hy$, H a $2g \times 2g$ skew-symmetric matrix with coefficients from $\mathbb{Q}$ such that (I) $\Omega H \Omega^t = 0$, and (II) $i\Omega H \Omega^h \succ 0$. Scorza called $H(x,y)$ a principal Riemannian form associated to Ω, as is still done nowadays. Riemann's conditions (10.4) correspond to the special case in which $H = J$. The more general conditions (I) and (II), which apply to any period matrix for which abelian functions exist, go back to Weierstrass and became well known after 1883.[28] It is, of course, always possible to multiply a principal Riemannian form by a suitably chosen integer so as to obtain a principal form with integral coefficient matrix. It was known that abelian functions exist when the period matrix Ω is a Riemann matrix,[29] and so $\mathcal{V}_g = \mathbb{C}^g/\Lambda$ becomes an abelian variety as indicated above.

Scorza was familiar with Frobenius' work on complex multiplication, and he showed that many of Frobenius' key results (with proofs modeled on those by Frobenius) extend to the context of multiplications (M,A) in his sense [528, §4]:

---

[27] For detailed references to the literature, see [411, p. 380].

[28] See Section 11.2 for a discussion of why Weierstrass introduced conditions (I)–(II) in unpublished work. In Section 11.4, the circumstances surrounding the publication of (I)–(II) are described.

[29] In particular, this result follows from Frobenius' theory of Jacobian functions, as noted in Section 11.4 in the discussion of Wirtinger's work.

Theorem 10.13 (Frobenius–Scorza). *(1) If (M,A) is any multiplication, then A satisfies $\tilde{\Omega}A\tilde{\Omega}^{-1} = \tilde{M}$, where*

$$\tilde{\Omega} = \begin{pmatrix} \Omega \\ \overline{\Omega} \end{pmatrix} \quad \text{and} \quad \tilde{M} = \begin{pmatrix} M & 0 \\ 0 & \overline{M} \end{pmatrix}.$$

Hence the characteristic polynomial $F(\alpha)$ of A factors as

$$F(\alpha) = \det(A - \alpha I_{2g}) = \det(M - \alpha I_g) \cdot \det(\overline{M} - \alpha I_g),$$

and so the characteristic roots of A are precisely the characteristic roots of M (the multipliers) and their complex conjugates and $\det A = |\det M|^2 \geq 0$. (2) If A is principal in the sense that $AHA^t = qH$, where H is the principal skew-symmetric matrix of Ω and $q \in \mathbb{Q}$, then $q > 0$, A can be diagonalized, and all its characteristic roots have absolute value $\sqrt{q}$.

Scorza's proof of part (1) of this theorem was a generalization of Frobenius' proof that all the roots of a principal abelian matrix $\tilde{A}$ of order n have absolute value $\sqrt{n}$. Recall that this had followed from Laguerre's results but that Frobenius had then devised his own, preferred, route to the end result. It involved showing that $\tilde{P}\tilde{A}\tilde{P}^{-1} = \tilde{M}$, where $\tilde{P} = \begin{pmatrix} P \\ \overline{P} \end{pmatrix}$, $P = \begin{pmatrix} I_g & T \end{pmatrix}$, and $\tilde{M} = \begin{pmatrix} M & 0 \\ 0 & \overline{M} \end{pmatrix}$ [see (10.32)]. Here $P = \begin{pmatrix} I_g & T \end{pmatrix}$ is just a normalized period matrix with principal skew-symmetric Riemannian form $H = J$. Thus in the more general context of Scorza's multiplications (M,A), it was natural to consider the analogous matrix $\tilde{\Omega} = \begin{pmatrix} \Omega \\ \overline{\Omega} \end{pmatrix}$ in lieu of $\tilde{P}$. In fact, Frobenius had himself introduced $\tilde{\Omega}$ in his 1884 paper on Jacobian functions. (See Proposition 11.4.) Then, since A is real, $\overline{\Omega A} = \overline{\Omega}A$, and so

$$\tilde{\Omega}A = \begin{pmatrix} \Omega A \\ \overline{\Omega A} \end{pmatrix} = \begin{pmatrix} M\Omega \\ \overline{M\Omega} \end{pmatrix} = \tilde{M}\tilde{\Omega}, \quad \text{i.e.,} \quad \tilde{\Omega}A\tilde{\Omega}^{-1} = \tilde{M}.$$

The other parts of Frobenius' proof also translated readily into the proof of part (2).

As a simple application of Theorem 10.13, consider the rational transformation $x_i' = R_i(x_1, \ldots, x_{g+1})$ associated to (M,A) as indicated following (10.41). Since A is assumed integral, if $\det A = 1$, it follows that A^{-1} is integral and that M is invertible, since from part (1), $|\det M|^2 = \det A = 1$. Multiplication of $M\Omega = \Omega A$ on the left by M^{-1} and then on the right by A^{-1} shows that $M^{-1}\Omega = A^{-1}\Omega$, and so (M^{-1}, A^{-1}) also defines a multiplication. This means that the rational transformation corresponding to (M,A) is actually birational. If, in addition, A is principal with $q = 1$, and so a principal abelian matrix of order 1 in Frobenius' terminology, Frobenius' Corollary 10.11 says that the characteristic roots of M are roots of unity, and so $M^n = I_g$ for some $n \in \mathbb{Z}$, and the corresponding birational transformation is periodic. Scorza proved that conversely, when the birational transformation is periodic, A is a principal abelian matrix of order 1 [528, p. 289].

I mentioned that Humbert's extension of the notion of a complex multiplication was based on his consideration of normalized Riemann matrices that satisfied certain singularity conditions. Scorza placed Humbert's approach in a general framework. Humbert's singularity condition on Ω can be expressed in the form $\Omega K \Omega^t = 0$, where K is an integral skew-symmetric matrix, which is not a multiple of the normalized Riemannian form J of (10.5). In the case $g = 2$, Humbert had systematically considered the cases in which there are, respectively, one, two, or three linearly independent K's. With this in mind, for a general Riemann matrix Ω with associated principal form H, Scorza considered other alternating "Riemannian forms" $K(x,y) = x^t K y$ satisfying $\Omega K \Omega^t = 0$ (and now allowed to have rational coefficients) and considered the number of linearly independent Riemannian forms K. Clearly, this number is at least one, because a Riemann matrix always has a principal alternating form H. Echoing Humbert's terminology, Scorza wrote the number of linearly independent alternating Riemannian forms as $1+k$, said that Ω was singular when $k > 0$, and called k the singularity index of Ω.

Scorza had slightly generalized Humbert's notion of a multiplication (M,A) by allowing A to have rational coefficients and allowing M and A to be singular. As a result, they now had the following algebraic properties. Given two multiplications defined by (M_1,A_1) and (M_2,A_2), respectively, it follows that for any $q_1,q_2 \in \mathbb{Q}$, the pairs $(q_1M_1 + q_2M_2, q_1A_1 + q_2A_2)$ and (M_1M_2, A_1A_2) also define multiplications.[30] Thus the multiplications (M,A) form an algebra $\mathcal{M}$ over $\mathbb{Q}$, which was called the *multiplication algebra* associated to Ω. Since the mappings $(M,A) \to M$ and $(M,A) \to A$ define algebra isomorphisms, $\mathcal{M}$ can be identified with the multipliers M or with the corresponding matrices A. The complex multiplications are those $(M,A) \in \mathcal{M}$ with $M \neq qI_g$. Scorza made a detailed study of $\mathcal{M}$ in 1921 [408, p. 381]. Thus $\mathcal{M}$ is the precursor of the modern endomorphism ring of an abelian variety with complex multiplication.

Already in his 1916 paper, Scorza also considered the number $1+h$ of linearly independent forms $L(x,y) = x^t L y$, L rational, such that $\Omega L \Omega^t = 0$. These forms are not necessarily alternating, so clearly $h \geq k$. If (M,A) is any multiplication, then $L = AH$ is such a form, because $\Omega L \Omega^t = (\Omega A) H \Omega^t = M(\Omega H \Omega^t) = 0$. Scorza called h the multiplicability index, because $1+h$ is the dimension of the multiplication algebra $\mathcal{M}$ [528, p. 283]. Thus when $h > 0$, $\mathcal{M}$ contains elements $M \neq qI_g$, and so associated to these M are complex multiplications (M,A). Since $h \geq k$, it follows that when Ω is singular $(k > 0)$, Ω admits complex multiplications, as Humbert had shown in the case $g = 2$. The indices h,k are also isomorphism invariants, i.e., are the same for two Riemann matrices Ω, Ω' that are isomorphic in the sense that nonsingular matrices R,S exist, S with coefficients from $\mathbb{Q}$, such that $\Omega' = R\Omega S$ [528, pp. 271–272].[31] Later, in 1921, Scorza showed that when Ω is nonsingular

[30]For example, $\Omega(A_1A_2) = (\Omega A_1)A_2 = (M_1\Omega)A_2 = M_1(\Omega A_2) = (M_1M_2)\Omega$, which shows that (M_1M_2, A_1A_2) is a multiplication.

[31]The corresponding abelian varieties are then in algebraic correspondence [408, pp. 79ff.].

($k = 0$), then all complex multiplications (M,A) are principal in the sense of part (2) of the Frobenius–Scorza Theorem 10.13 [411, p. 383].

10.7.4 Lefschetz

In his 1916 paper, Scorza sought relations among the integers g, h, and k for a given Riemann matrix Ω. His results were not definitive. He showed, for example, that for any Riemann matrix, $0 \leq k \leq g^2 - 1$ and $0 \leq k \leq h \leq 2g^2 - 1$ [528, p. 276]. Equations among g, h, and k (rather than inequalities) were obtained in many cases by Solomon Lefschetz (1884–1972). Since 1913, Lefschetz had been working at the University of Kansas, first as instructor and then, starting in 1916, as assistant professor.[32] As Lefschetz later recounted, during his years in Kansas he had worked in isolation, freely developing his own ideas as he saw fit. His own ideas were mainly new algebraic-topological ones that he applied within the context of algebraic geometry. He described his early achievements with a striking metaphor: "it was my lot to plant the harpoon of algebraic topology into the body of the whale of algebraic geometry" [412, p. 13]. In the course of planting that harpoon, Lefschetz drew on the work of mathematicians such as Picard and Poincaré—he had received his undergraduate education in engineering and mathematics in Paris—but he also drew on the work of Frobenius, whose mathematics he evidently admired, as we shall see. Although the role of Frobenius' work, unlike that of Poincaré or Picard, was not fundamental to Lefschetz's overall research program, it did play a significant role in his major, prize-winning memoir of 1921, "On certain numerical invariants of algebraic varieties with applications to abelian varieties" [408],[33] a work that, together with his 1924 monograph *L'Analysis situs et la géométrie algébrique* [410], are usually cited as the basis for Lefschetz's above-quoted remark.

The first part of Lefschetz's memoir [408] represented a pioneering development of new algebraic-topological methods within the context of algebraic varieties. In the second part, the developed machinery was applied to abelian varieties, and it was

---

[32]Lefschetz, who was born in Moscow, had spent most of the first two decades of his life living in Paris, where his parents (Turkish citizens) resided. (For further details on Lefschetz's life and work see [564] and the references cited therein.) He was educated at the École Centrale in Paris and was graduated in 1905 as *ingénieur des arts et manufactures*. That same year, he emigrated to the United States to pursue a career as an engineer, but in 1907, an accident that resulted in the loss of both hands caused him to turn instead to a career in mathematics. (In Paris, he had studied mathematics under the instruction of Picard and Appell, both of whom had done important work in the theory of abelian functions.) He spent 1910–1911 at Clark University, in Worcester, Massachusetts, where he earned a doctoral degree with a dissertation on a topic in algebraic geometry. After two years teaching at the University of Nebraska, he moved to the above-mentioned positions at the University of Kansas.

[33]Lefschetz's memoir was an English translation (with minor modifications) of an essay in French that was awarded the *Prix Bordin* of the Paris Academy of Sciences for the year 1919. In 1923, the English version was awarded the Bôcher Memorial Prize of the American Mathematical Society.

here that the work of Scorza became relevant. Lefschetz focused on the topological significance of Scorza's invariants h and k. His most general result was that if Ω is a Riemann matrix and $\mathcal{V}_g$ the associated abelian variety, then $1+k=\rho$, where ρ is the Picard number associated to $\mathcal{V}_g$, a notion he had generalized to algebraic varieties from Picard's work on surfaces [408, pp. 42–44].

The more difficult task of relating both h and k to g required focusing on the more tractable abelian varieties with complex multiplications, and in that connection, Frobenius' paper [188] on Kronecker's complex multiplication problem, which Lefschetz had evidently read,[34] proved important. As Lefschetz explained in his prefatory remarks,

> The consideration of Abelian varieties possessing certain complex multiplications leads us to the determination of h, k, and ρ in a wide range of cases. We touch here in many points investigations of Scorza, Rosati, and Frobenius, our methods being more nearly related to the last-named author [408, p. 44].

Judging by the part of his memoir dealing with the relation of h and k to g [408, Ch. II, pp. 105–151], what Lefschetz meant was that his methods included utilizing matrix algebra in a manner that was essential to some of his reasoning (as was Frobenius' use of matrix algebra in his solution to Kronecker's complex multiplication problem).

A good illustration of this mode of reasoning is provided by Lefschetz's proof that if (M,A) is a multiplication such that A is nonsingular and has a minimal polynomial $f(\alpha)$ that is irreducible over $\mathbb{Q}$, then $m=\deg f$ divides $1+h=\dim\mathcal{M}$ [408, pp. 109–110].[35] As noted above, the multiplication algebra $\mathcal{M}$ can be identified with the matrices B such that for some M', (M',B) is a multiplication. Granted this identification, it follows that $I,A,\ldots,A^{m-1}$ form a linearly independent subset of $\mathcal{M}$, and if there is no $B\in\mathcal{M}$ that is not spanned by the powers of A, then $1+h=m$. If there is a B not spanned by the powers of A, then, Lefschetz claimed, $I,A,\ldots,A^{m-1},B,BA,\ldots BA^{m-1}$ must be linearly independent. Otherwise, there would be a dependency relation expressible in the form $\phi(A)+B\psi(A)=0$, where $\phi,\psi\in\mathbb{Q}[\alpha]$ have degrees less than $m=\deg f$. Now because $\deg\psi<m$, it follows that $\psi(A)$ is nonsingular: if $\alpha_1,\ldots,\alpha_{2g}$ are the characteristic roots of A, then the characteristic roots of $\psi(A)$ are $\psi(\alpha_j)$, $j=1,\ldots,2g$, and we cannot have $\psi(\alpha_j)=0$, because ψ cannot share a root with the irreducible polynomial f. The nonsingularity of $\psi(A)$ means that $\chi\in\mathbb{Q}[\alpha]$ may be determined such that $\psi(A)\chi(A)=nI$, where n is an integer.

> Indeed the computation of the coefficients ... [of χ] is formally the same as that which presents itself when given an algebraic number α such that $\psi(\alpha)\neq 0$, it is proposed to put $1/\psi(\alpha)$ in the form ... $[(1/n)\chi(\alpha)]$, a problem which is easily solved [408, p. 109].

---

[34] Lefschetz refers to results in Frobenius' paper that are not treated by Scorza [408, p. 133].

[35] Following terminology introduced in 1916 by Rosati, Lefschetz referred to $f(\alpha)=0$ as the "minimum equation of Frobenius" [408, p. 109n].

Multiplication of $\phi(A)+B\psi(A)=0$ by $\chi(A)$ then gives $\phi(A)\chi(A)+nB=0$, contrary to the hypothesis that B is not in the span of $I,A,\ldots,A^{m-1}$. The claim that the $2m$ matrices $I,A,\ldots,A^{m-1},B,BA,\ldots,BA^{m-1}$ are linearly independent is thus verified. If they span $\mathcal{M}$, then $1+h=2m$. If not, there is a $C\in\mathcal{M}$ that is not spanned by the above matrices, and the same line of reasoning shows that the $3m$ matrices A^s,BA^s,CA^s, $s=0,\ldots,m-1$, are linearly independent. If they span $\mathcal{M}$, then $1+h=3m$. If not, the above reasoning may be repeated but eventually must terminate with the conclusion that $1+h$ is some multiple of m.

Lefschetz also employed matrix algebra within the context of the characteristic polynomial $F(\alpha)=\det(\alpha I-A)$ of a multiplication (M,A) and, in particular, in conjunction with the Frobenius–Scorza Theorem 10.13. It was with this theorem in mind that he wrote that

> The characteristic equation whose properties from the point of view that interests us here, have been thoroughly studied by Frobenius, will play a fundamental part in the sequence. The problem which we propose to consider is the following: *Given a certain equation with rational coefficients, to find the most general abelian varieties possessing a complex multiplication of which this equation is the characteristic equation* [408, p. 106].

The rationale for solving this problem was that for the abelian varieties so determined, the relations among g, h, and k would be forthcoming.

Although the italicized problem seems exceedingly general, it was known that the general theory could be reduced to the case of "pure" Riemann matrices.[36] When Ω is pure, the multiplication algebra $\mathcal{M}$ is a division algebra. Furthermore, as Lefschetz had shown, for all multiplications (M,A) such that A is a "general projectivity" in the sense that it has a full complement of $2g$ linearly independent characteristic vectors, the associated characteristic polynomial takes the special form $F(\alpha)=\det(A-\alpha I_{2g})=[f(\alpha)]^r$, where $f(\alpha)$ is the minimal polynomial of A and is irreducible (over $\mathbb{Q}$) [408, p. 110].

In order to make headway on the above italicized problem for pure Ω with $F(\alpha)=[f(\alpha)]^r$, Lefschetz had to restrict his attention to the case $r=1$. Before doing that, however, he pointed out that "according to Frobenius" [408, p. 113] if the multiplication (M,A) corresponding to $F(\alpha)=[f(\alpha)]^r$ is principal ($AHA^t=qH$), then all the roots of $F(\alpha)$ have the same absolute value $\beta\,(=\sqrt{q})$. Thus when $s=\deg f(\alpha)$ is odd, $f(\alpha)$ has a real root, which must be $\pm\beta$. Since the coefficients of $f(\alpha)$ are rational, it follows that the product of its s roots and hence the absolute value of that product, namely β^s, is rational. This means that the multiplication (M^s,A^s) is such that A^s has a rational characteristic root $\ell=\pm\beta^s$. Since A^s is also a general projectivity, its characteristic polynomial is also a power of its irreducible minimal polynomial ψ; and because ψ has the rational number ℓ as a root, it must be that $\psi(\alpha)=\alpha-\ell$. Thus $\psi(A)=A^s-\ell I_{2g}=0$, and so the minimal polynomial of A is$f(\alpha)=\alpha^s-\ell$. Thus when Ω is pure, if a

[36]The Riemann matrix Ω is pure if it is not isomorphic in Scorza's above-defined sense to a Riemann matrix Ω' that is a direct sum of two Riemann matrices: $\Omega'=\Omega_1'\oplus\Omega_2'$.

multiplication (M,A) exists for which A principal and its minimal polynomial is of odd degree s, then the associated characteristic polynomial takes the simple form $F(\alpha) = (\alpha^s - \ell)^r$.

When Lefschetz referred to Frobenius, he cited Frobenius' paper on complex multiplication [188], even though he was really applying part (2) of the Frobenius–Scorza theorem (Theorem 10.13), Scorza's generalization of Frobenius' results. Lefschetz was apparently paying homage to Frobenius because he felt that it was Frobenius' work in [188] and more generally his development of the symbolic algebra of matrices—Lefschetz explicitly mentions his use of "the notation of Frobenius" [408, p. 80]—including the notion and properties of the minimal polynomial, that facilitated his work on the italicized problem for pure Riemann matrices.

Let us now consider Lefschetz's solution to the above italicized problem for a pure Riemann matrix Ω such that $r = 1$, so that $F(\alpha) = \det(A - \alpha I_{2g})$ is irreducible over $\mathbb{Q}$ of degree $2g$ and so has no multiple roots. The starting point and basis for his solution was a theorem due to Scorza. Since Scorza derived it using the Frobenius–Scorza Theorem 10.13, I will sketch his proof. It was another manifestation of the viability of Frobenius' ideas.

Unlike Frobenius, Scorza thought about linear-algebraic matters in terms of projective geometry. He thought of the rows of a Riemann matrix Ω as the homogeneous coordinates of points in the complex projective space $\mathbb{P}^{2g-1}$. Associated to Ω is the $(g-1)$-dimensional projective space τ consisting of all points of $\mathbb{P}^{2g-1}$ whose homogeneous coordinates are linear combinations of the rows of Ω. Thus the homogeneous coordinates $x^t = \begin{pmatrix} x_1 & \cdots & x_g \end{pmatrix}$, $x \neq 0$, of the points in τ are all of the form $x^t = z^t\Omega$ for some $z \neq 0$. Likewise, the conjugate space $\bar{\tau}$ consists of all points $x^t = z\overline{\Omega}$. If Ω admits the multiplication (M,A), then one can consider the projective transformation on $\mathbb{P}^{2g-1}$ given in homogeneous coordinates by $y^t = x^tA$. (Scorza called it a Riemannian homography.) It takes the spaces τ and $\bar{\tau}$ into themselves by virtue of $M\Omega = \Omega A$.[37]

Suppose that the homography A is nonsingular with no multiple roots. Then the Frobenius–Scorza theorem states that $\tilde{\Omega}A\tilde{\Omega}^{-1} = M \oplus \overline{M}$, $\tilde{\Omega} = \begin{pmatrix} \Omega \\ \overline{\Omega} \end{pmatrix}$, and so all the characteristic roots of A are nonreal and may be written as $\alpha_1,\ldots,\alpha_g,\bar{\alpha}_1,\ldots,\bar{\alpha}_g$, where $\alpha_1,\ldots,\alpha_g$ are the characteristic roots of M (the multipliers). For any multiplier α_j, if x_j^t is the corresponding fixed point of the homography, so that $\alpha_j x_j^t = x_j^t A$, then it follows from $\tilde{\Omega}A\tilde{\Omega}^{-1} = M \oplus \overline{M}$ that x_j^t is in τ and so of the form

$$x^t = z_j^t\Omega.^{38} \tag{10.42}$$

---

[37] For example, for x^t in τ, $y^t = x^tA = z^t\Omega A = (z^tM)\Omega = z_1^t\Omega \in \tau$.

[38] Make the coordinate change $x^t = u^t\tilde{\Omega}$. In these coordinates, the homography is given by $v^t = u^t(M \oplus \overline{M})$, and the fixed point satisfies $\alpha_j u_j^t = u_j^t(M \oplus \overline{M})$, which by block multiplication with $u_j^t = \begin{pmatrix} u_1^t & u_2^t \end{pmatrix}$ gives $\alpha_j u_1^t = u_1^t M$ and $\alpha_j u_2^t = u_2^t\overline{M}$ and so $u_2^t = 0$. Thus $u_j^t = \begin{pmatrix} u_1^t & 0 \end{pmatrix}$, and it follows

On the other hand, from the fact that x^t is a solution to $x^t(A - I_{2g}) = 0$ Scorza assumed that

$$x_j^t = \left(p_1(\alpha_j) \cdots p_{2g}(\alpha_j)\right), \quad j = 1, \ldots, 2g, \tag{10.43}$$

where

$$p_k(t) = q_{k1} + q_{k2}t + \cdots + q_{k2g}t^{2g-1} \in \mathbb{Q}[t]. \tag{10.44}$$

This is certainly the case if the characteristic polynomial of A is assumed to be irreducible.[39] In matrix notation, (10.42) and (10.43) combine to give $Z\Omega = \Omega^* Q^t$, where Z is the $g \times g$ matrix with the z_j as rows, $Q = (q_{kl})$, and Ω^* is

$$\begin{pmatrix} 1 & \alpha_1 & \alpha_1^2 & \cdots & \alpha_1^{2g-1} \\ \cdots & \cdots & \cdots & \cdots & \cdots \\ 1 & \alpha_g & \alpha_g^2 & \cdots & \alpha_g^{2g-1} \end{pmatrix}. \tag{10.45}$$

From the nonsingularity of Z and Q and the rationality of Q, it follows that Ω is isomorphic to Ω^*. Scorza had thus established the following result.

Theorem 10.14 (Scorza). *If (M,A) defines a multiplication with A having an irreducible characteristic polynomial, then Ω is isomorphic to the Riemann matrix (10.45), where $\alpha_1, \ldots, \alpha_g$ are the characteristic roots of M.*

In view of Scorza's theorem and the fact that h, k are the same for isomorphic Riemann matrices, for any given polynomial $F(\alpha)$ irreducible over $\mathbb{Q}$, Lefschetz focused on determining necessary and sufficient conditions that a matrix Ω of the form (10.45), with $\alpha_1, \ldots, \alpha_g$ being a subset of the roots of F, be a pure Riemann matrix. In view of the Frobenius–Scorza theorem, an obvious necessary condition is that F have no real roots,[40] and so Lefschetz assumed that F not only is irreducible but has no real roots.

Lefschetz broke the problem facing him into three parts as follows. (1) Determine necessary and sufficient conditions that Ω in (10.45) be a Riemann matrix, i.e., that there be a principal Riemannian form $H(x,y)$ associated to Ω in the sense

---

that x_j is in τ because in the new coordinate system, τ is characterized by $\tilde{\Omega}u = z^t\Omega$ and $\tilde{\Omega}u_j^t = u_1^t\Omega$ is of the requisite form.

[39] In that case, it follows from Gaussian elimination, since the fields $\mathbb{Q}(\alpha_j)$, $j = 1, \ldots, g$, are isomorphic. It also follows by algebraic-topological considerations, as Lefschetz showed [408, pp. 111–112]. Whether it is true when the minimal polynomial of A is not assumed irreducible is unclear, although Scorza stated Theorem 10.14 below without any explicit assumption on the minimal polynomial of A.

[40] If F is to be the characteristic polynomial associated to a multiplication (M,A), then since $\tilde{\Omega}A\tilde{\Omega}^{-1} = M \oplus \overline{M}$, any real roots would be multiple roots, contrary to the assumption that F is irreducible.

described above. (2) When Ω in (10.45) is a Riemann matrix, determine necessary and sufficient conditions that Ω be pure. (3) For matrices (10.45) that are pure Riemann matrices, determine the relation of h, k, and g.

In order to solve problems (1) and (2), Lefschetz turned to the Galois group $\mathfrak{G}$ of $F(\alpha)$ over $\mathbb{Q}$. He showed that Ω is a Riemann matrix if and only if $\mathfrak{G}$ has the following property [408, p. 116]:

Property 10.15. There is no $\sigma \in \mathfrak{G}$ such that for every pair (α_i, α_j), $1 \leq i < j \leq g$, the image pair $(\sigma(\alpha_i), \sigma(\alpha_j))$ consists of conjugates.

If Ω has this property, and so is a Riemann matrix, it need not be pure, but when Ω is pure, then, Lefschetz showed, $2(1+k) = 1+h = 2g$ [408, p. 117]. (Since $g \geq 1$, this shows that $h > 0$ and so Ω admits complex multiplications.) He also established necessary and sufficient conditions on $\mathfrak{G}$ that Ω be pure,[41] thereby solving problems (2) and (3). When $\mathfrak{G}$ is abelian, he was able to simplify these results as follows. Let $\mathfrak{G}_1$ denote the subgroup of $\mathfrak{G}$ that leaves the set $\{\alpha_1, \ldots, \alpha_g\}$ invariant, and let n denote its order. Then Ω is pure if and only if $n = 1$ [408, p. 121]. In that case, of course, $2(1+k) = 1+h = 2g$, but because $\mathfrak{G}$ is assumed abelian, he was also able to establish the more general result that for any Ω of the form (10.45) satisfying Property 10.15, one has $2(1+k) = 1+h = 2ng$ [408, p. 122].

Lefschetz's lengthy paper contains many more results establishing relations between h and k in various special contexts, but the above discussion suffices to indicate the role played by Frobenius' work. In addition to Frobenius-style matrix algebra, part (1) of the Frobenius–Scorza Theorem 10.13 together with Scorza's Theorem 10.14 formed the basis of Lefschetz's study of the relation among g, h, and k. The combined work of Frobenius and Scorza helped Lefschetz to plant the harpoon of topology in the whale of algebraic geometry.

10.7.5 Postscript: A. A. Albert

The tool of matrix algebra had also been used by Lefschetz to study the structure of multiplication algebras $\mathcal{M}$ that are commutative [408, pp. 13–14]. In retrospect, it can be seen as the first step toward a characterization of those division algebras that can be realized as the multiplication algebra of a pure Riemann matrix Ω. In 1924, Lefschetz moved to Princeton University, where in 1928 he met A. Adrian Albert (1905–1972), a visiting postdoctoral student. Albert had been one of L. E. Dickson's students at the University of Chicago and was working on the theory of division algebras. After hearing Albert talk on his work, Lefschetz took him aside, explained to him how division algebras arise as multiplication algebras associated to Riemann matrices, and encouraged him to work on the problem of characterizing the non-commutative division algebras that occur as the multiplication algebra of a pure Riemann matrix. Aided by an important 1929 paper by Rosati on the structure of

[41] Conditions (a) and (b) on p. 117 of [408].

multiplication algebras [507], Albert immediately began to work on the problem in tandem with research on the structure of division algebras. His complete solution was presented in three papers published in 1934–1935.[42] In this work, Albert continued Lefschetz's practice of utilizing matrix algebra, but there are no longer any references to Frobenius in this connection. By Albert's time, Frobenius' work on matrix algebra had become an anonymous part of basic mathematics.

[42]See papers 33, 37, and 42 of [2]. Earlier papers in [2] document Albert's progress toward the solution. See also Jacobson's account on pp. lvi–lvii and lii–lv of [2]. The modern theory of abelian varieties with complex multiplication was initiated in 1955 (Weil, Shimura, Taniyama), and was motivated by arithmetic considerations originating in the elliptical case $g = 1$ (Kronecker, Deuring, Hasse) [530, pp. x–xi]. The modern theory involves restrictions on $\mathcal{M}$ [530, pp. x, 35].

Chapter 11
Frobenius' Generalized Theory of Theta Functions

This chapter is devoted to Frobenius' theory of generalized theta functions, which he called "Jacobian functions" in honor of Jacobi, who had pointed out the fundamental role that can be played by theta functions in establishing the theory of elliptic functions and solving the inversion problem.[1] (The inversion problem and related matters that will be helpful for an appreciation of the following chapter have been discussed in Section 10.1.) In Frobenius' day, and especially among Weierstrass' students, a theta function in g variables $z = \begin{pmatrix} z_1 & \cdots & z_g \end{pmatrix}^t$ was an entire function defined by means of an infinite series of the form

$$\theta(z) = \sum_{n \in \mathbb{Z}^g} e^{G(z,n)}, \tag{11.1}$$

where $G(z,n)$ is a polynomial of degree two in $z_1, \ldots, z_g, n_1, \ldots, n_g$, and so, using matrix notation with $a = \begin{pmatrix} a_1 & \cdots & a_g \end{pmatrix}^t$, $b = \begin{pmatrix} b_1 & \cdots & b_g \end{pmatrix}^t$, and $c \in \mathbb{C}$, we may write

$$G(z,n) = \pi i[z^t Rz + z^t Sn + n^t Tn + a^t z + b^t n + c], \tag{11.2}$$

where the matrices R and T are symmetric, $\det S \neq 0$ (so that $\theta(z)$ is not a function of fewer than g variables), and, in order to ensure uniform convergence on compact subsets, $T = \Phi + i\Psi$ must be such that Ψ is positive definite, so that the real part of $\pi i[n^t Tn]$ is negative for all n.

Generalizing what Jacobi had done in the elliptic case, Weierstrass showed in his Berlin lectures that the abelian functions arising in the solution to the Jacobi inversion problem could be expressed in terms of special theta functions. These theta functions had coefficients R, S, T in (11.2) that were related to the periods $\omega_{\alpha\beta}$ of the abelian integrals of the first kind, as well as to the periods $\eta_{\alpha\beta}$ of associated

[1] The honorific term "Jacobian functions" was used in a similar, albeit less general, manner by Weierstrass [595, p. 55] and Klein [342, p. 324].

T. Hawkins, *The Mathematics of Frobenius in Context*, Sources and Studies in the History of Mathematics and Physical Sciences, DOI 10.1007/978-1-4614-6333-7_11,

abelian integrals of the second kind. For both types, $\alpha = 1,\ldots,g$ and $\beta = 1,\ldots,2g$. As in the previous chapter, it is convenient to introduce the $g \times 1$ column matrices $\omega_\alpha = (\omega_{1\alpha} \cdots \omega_{g\alpha})^t$ and $\eta_\alpha = (\eta_{1\alpha} \cdots \eta_{g\alpha})^t$, $\alpha = 1,\ldots,2g$, and to refer to the ω_α and η_α as periods as well. As in the elliptic case these periods satisfied characteristic functional equations, which now take the form

$$\theta(z+\omega_\alpha)=e^{2\pi i L_\alpha(z)}\theta(z), \quad L_\alpha(z)=\eta_\alpha^t z+b_\alpha,\ b_\alpha \in \mathbb{C}, \quad \alpha=1,\ldots,2g. \tag{11.3}$$

In what follows, I will refer to (11.3) as a system of *equations of quasiperiodicity for* $\theta(z)$. Repeated application of these equations to $\theta(z+\omega)$, where $\omega = \sum_{\alpha=1}^{2g} n_\alpha \omega_\alpha$ and the n_α are integers, yields an analogous quasiperiodic equation $\theta(z+\omega) = e^{2\pi i L(z,\omega)}\theta(z)$, which is frequently the starting point in modern expositions.[2]

By 1869, Weierstrass realized that the special abelian functions generated by solving the Jacobi inversion problem did not produce all possible abelian functions. This led him to consider the more general theta functions that arise when the polynomial $G(z,n)$ in (11.2) is arbitrary except for the above conditions on S and T. He also considered more general entire functions constructed from general theta functions, which he called theta functions of order r. These functions all satisfied equations of quasiperiodicity of the same general form as in (11.3). It is important to realize that in the quasiperiodic equations of Weierstrass' theta functions, the coefficients occurring in the linear functions L_α of (11.3) were rather special, so that not all theta functions in the customary modern sense are theta functions in the sense of Weierstrass and his contemporaries. *In what follows, I will use the term "ordinary theta function" to denote theta functions in the nineteenth-century sense of Weierstrass (as defined precisely in Section 11.1) in order to distinguish such functions from the theta functions (in the modern sense), which are more general and were first introduced by Frobenius under the name of "Jacobian functions."*

Although Frobenius had learned about ordinary theta functions in the sense of Weierstrass while a student in Berlin, his understanding of Weierstrass' most coherent and definitive rendition of the theory was facilitated by the publication in 1880 of a book on ordinary theta functions by Weierstrass' student Friedrich Schottky, who had become Frobenius' colleague at the Zurich Polytechnic in 1882. As we will see, Schottky's up-to-date presentation of Weierstrass' theory also diverged in subtle ways from that of Weierstrass, ways that seem to have encouraged Frobenius to consider the possibility of placing Weierstrass' theory within a more general and algebraically elegant framework, a possibility also encouraged, I would suggest, by a problem suggested by the publication by Hurwitz of Weierstrass' hitherto unpublished theorem (Theorem 11.3 below) giving the modern formulation

[2]The explicit formula for $L(z,\omega)$ is given by Frobenius [189, p. 174, (5)–(7)]. It shows that L is also linear in the z_j, and if we write $L(z,\omega) = H(z,\omega) + J(\omega)$, where H is homogeneous in the z_i, so $H(z+z',\omega) = H(z,\omega) + H(z'\omega)$, then $J(\omega+\omega') \equiv J(\omega) + J(\omega')$ mod $\mathbb{Z}$, and $H(z,\omega+\omega') = H(z,\omega) + H(z,\omega')$. These properties of $L(z,\omega)$ can be taken as the starting point of a more abstract development of the theory of theta functions; see, e.g., [508, p. 88].

of the so-called Riemann conditions on a period matrix necessary for the existence of nontrivial abelian functions with those periods. The result was his theory of Jacobian functions (1884), the subject of Section 11.3. As will be seen in Section 11.4, Frobenius' theory, in tandem with a theorem from the 1890s due to Appell and Poincaré, eventually formed the foundation for the theory of abelian functions and varieties.

11.1 Weierstrass' Lectures on Theta Functions

As noted in the introductory remarks to Chapter 10, Weierstrass published very little of his theory of abelian and theta functions at the time he developed it, choosing instead to present his results in his lectures on abelian integrals and functions, which he usually gave once every two years. Frobenius most likely attended these lectures as a student during the summer semester of 1869.[3] As we saw in Section 1.1, on his oral doctoral examination the following year, he had impressed Weierstrass with his extensive knowledge of the theory of abelian integrals and functions.

The 1869 version of Weierstrass' lectures was the first to adopt the general form of presentation that he subsequently adhered to, although how much of what Frobenius had learned then was at his fingertips in the fall of 1883, when he worked on his paper, is uncertain. Fortunately for him, and for the mathematical public at large, in 1880, Weierstrass' student Friedrich Schottky (1851–1935) published, with Weierstrass' encouragement, a little book entitled *Sketch of a Theory of Abelian Functions in Three Variables* [519]. Although Schottky focused on the three-variable case, in the first three sections of the book he presented, as he explained, Weierstrass' theory of theta functions in any number of variables. Not only was Frobenius familiar with the contents of Schottky's book, he also had its author as colleague at the Polytechnic in Zurich from 1882 until 1892, when both left Zurich, Frobenius headed for Berlin and Schottky for Marburg.

It was not until 1889 that Weierstrass began to think that he should publish his lectures. It was decided to use primarily his lectures during the winter term 1875–1876, since this version of the lectures was deemed the most coherent. This was most likely the term that Schottky attended the lectures.[4] The edited version of the lectures, undertaken by G. Hettner and J. Knoblauch, finally appeared in 1902 [594], 5 years after Weierstrass' death. Except for minor—albeit suggestive—differences, Schottky's presentation in his book agrees with what is found in Weierstrass' lectures as published. For this reason, I will begin by indicating Weierstrass' treatment of ordinary theta functions, followed by an indication of the minor differences introduced by Schottky that are relevant to Frobenius' work.

[3]According to the schedule of Weierstrass' lectures (*Werke* 3, pp. 355ff.), 1869 was the first time since Frobenius had matriculated at Berlin in 1867 that the lectures had been given.

[4]Since Schottky was in Berlin from 1874 to 1879, he could have attended Weierstrass' lectures on the subject as given in either the winter term 1875–1876 or the winter term 1877–1878.

Weierstrass' lectures began with, and were mainly concerned with, the theory of abelian integrals and the solution to the associated Jacobi inversion problem. As indicated in the introductory remarks, the theory of ordinary theta functions was an important component of the solution to the inversion problem. The most general theta functions involved two sets of parameters—or characteristics as Schottky and later authors termed them—$\mu = (\mu_1 \cdots \mu_g)^t$ and $\nu = (\nu_1 \cdots \nu_g)^t$, and here I will use the notation $\theta(z;\mu,\nu)$ for these functions, where $z = (z_1 \cdots z_n)^t$. (Weierstrass' own notation was quite similar.) From their origin via abelian integrals, the functions $\theta(z;\mu,\nu)$ had associated with them two sets of $2g$ periods corresponding to abelian integrals of the first and second kinds. Here these sets will be denoted, respectively, by the $g \times 2g$ matrices

$$\Omega = (\omega_1 \cdots \omega_{2g}) \quad \text{and} \quad \mathrm{H} = (\eta_1 \cdots \eta_{2g}), \tag{11.4}$$

so that the ω_α and η_β are $g \times 1$ column matrices. By virtue of their origins in the theory of abelian integrals, these periods satisfied many relations, and Weierstrass singled out the following two, which are most readily stated in matrix notation with $\Omega = (\Omega_1\ \Omega_2)$ and $\mathrm{H} = (\mathrm{H}_1\ \mathrm{H}_2)$ denoting partitions into $g \times g$ matrices:

$$\Omega_2\Omega_1^t - \Omega_1\Omega_2^t = 0 \quad \text{and} \quad \mathrm{H}_2\mathrm{H}_1^t - \mathrm{H}_1\mathrm{H}_2^t = 0. \tag{11.5}$$

A key feature of these theta functions was that, as with those that arise in the case $g = 1$ of elliptic integrals, they satisfied equations of quasiperiodicity, namely

$$\begin{aligned} \theta(z+\omega_\alpha;\mu,\nu) &= e^{L_\alpha(z)}\theta(z;\mu,\nu), \\ L_\alpha(z) &= \eta_\alpha^t z + \tfrac{1}{2}(\eta_\alpha^t\omega_\alpha) + \begin{cases} \pi i\nu_\alpha, & \alpha \le g, \\ -\pi i\mu_\alpha, & \alpha > g. \end{cases} \end{aligned} \tag{11.6}$$

(These are a special case of the general equations (11.3), but keep in mind that no one prior to Frobenius had any occasion to consider such general equations.) Using his equations and the theorem that $\det\Omega_1 \neq 0$, Weierstrass showed that $\theta(z;\mu,\nu)$ could be represented by an everywhere convergent series of the form

$$\theta(z;\mu,\nu) = \sum_{n\in\mathbb{Z}^g} \exp[G(z,n)], \tag{11.7}$$

where $G(z,n)$ is a polynomial in the z_j and n_k of degree two and generally inhomogeneous. In matrix–vector notation like that used in (11.2),

$$G(z,n) = z^t Rz + z^t Sn + n^t Tn + a^t z + b^t \cdot n + c, \tag{11.8}$$

where the $g \times g$ matrices R and T are symmetric and $c \in \mathbb{C}$.

The theory of abelian integrals, of which the above was a part, constituted the bulk of Weierstrass' lectures. In the published version, this material covered Chapters 1–29 and 31–33 for a total of 591 pages. The remaining two chapters (30 and 34) were devoted to what Weierstrass called "general theta functions." That is, the functions $\theta(z;\mu,\nu)$ considered above were fabricated from data originating in Weierstrass' theory of abelian integrals. The series representation (11.7), however, suggested the idea of *defining* a class of functions $\theta(z;\mu,\nu)$ by means of (11.7), where $G(z,n)$ is now a degree-two polynomial that is completely arbitrary, except for the conditions necessary to ensure that (1) the series in (11.7) converges absolutely and uniformly when z is restricted to any compact subset of $\mathbb{C}^g$ and (2) the series (11.7) defines a function that cannot be transformed by a linear variable change into a function of fewer than g variables. The condition for (1), he showed, was that the real part of T be negative definite,[5] and the condition for (2) was that $\det S \neq 0$. Under these assumptions on $G(z,n)$, Weierstrass showed that the function $\theta(z;\mu,\nu)$ defined a priori by the series in (11.7) has associated to it two $g \times 2g$ systems of numbers Ω and H as in (11.4) and parameters $\mu = \begin{pmatrix}\mu_1 & \cdots & \mu_g\end{pmatrix}^t$, $\nu = \begin{pmatrix}\nu_1 & \cdots & \nu_g\end{pmatrix}^t$ such that the equations of quasiperiodicity (11.6) are satisfied. Weierstrass' formulas [594, pp. 569–570] are easy to express in the matrix notation of (11.8). Let $\Omega = \begin{pmatrix}\Omega_1 & \Omega_2\end{pmatrix}$ and $\mathrm{H} = \begin{pmatrix}\mathrm{H}_1 & \mathrm{H}_2\end{pmatrix}$, where the blocks Ω_j and H_j are $g \times g$. Then Weierstrass' formulas translate into

$$\begin{aligned} \Omega_1 &= 2\pi i(S^{-1})^t, & \Omega_2 &= 2(S^{-1})^t T, \\ \mathrm{H}_1 &= 4\pi i R(S^{-1})^t, & \mathrm{H}_2 &= 4R(S^{-1})^t T - S, \\ \mu &= \pi i(2TS^{-1}a - b), & \nu &= 2S^{-1}a. \end{aligned} \tag{11.9}$$

Thus the period matrices Ω, H are completely determined by the quadratic terms of $G(z,n)$, whereas the characteristics μ,ν depend as well on the linear terms. The linear terms of $G(z,n)$ may be chosen so as to make μ,ν equal any two vectors in $\mathbb{C}^g$. As Weierstrass showed, the period matrices defined by (11.9) satisfy the relations

$$\begin{aligned} \mathrm{H}_1^t\Omega_1 - \Omega_1^t\mathrm{H}_1 &= 0, \\ \mathrm{H}_2^t\Omega_2 - \Omega_2^t\mathrm{H}_2 &= 0, \\ \mathrm{H}_1^t\Omega_2 - \Omega_1^t\mathrm{H}_2 &= 2\pi i I_g. \end{aligned} \tag{11.10}$$

The relations in (11.5) can also be deduced from (11.9).

Weierstrass referred to the theta functions defined a priori by the series in (11.7) as *general theta functions* because for those arising from the theory of abelian integrals on a complex algebraic curve, the coefficients τ_{jk} of T in (11.8) are

[5] As I remarked in the previous section, Frobenius was familiar with all this when he composed his paper on complex multiplication. He preferred to write $G(z,n) = \pi i[u^t Ru + u^t Sn + n^t Tn + \cdots]$—as I did in (11.2)—and so the convergence condition for him was that the imaginary part of T be positive definite.

determined by the coefficients of the equation of the curve, and so, at least for $g \geq 4$, are subject to more constraints than $\mathrm{Re}\, T \prec 0$. "Accordingly, for theta functions the remarkable fact presents itself that by solution of the Jacobi inversion problem, one does not arrive at the most general functions with the same characteristic properties" [594, p. 581]. It was Weierstrass' theory of these general theta functions that Schottky presented in the beginning sections of his book. I will now briefly indicate this theory and then the slight but significant differences that are found in Schottky's exposition.

One of Weierstrass' results [594, p. 577], which was to figure prominently in Frobenius' theory, shows that the general theta function defined by the series in (11.7) is essentially determined by its equations of quasiperiodicity (11.6):

Proposition 11.1. *Let $\theta = \theta(z;\mu,\nu)$ denote the general theta function defined by the series in (11.7). Then if $\psi(z)$ is any entire function satisfying the same equations of quasiperiodicity (11.6) as θ, it must be that ψ is a constant multiple of θ.*

In his lectures, Weierstrass' discussion of general theta functions [594, Ch. 34] was focused on the fact that when the characteristics μ,ν have integer components, these functions are always even or odd and have remarkable properties. (Characteristics with integer components in Weierstrass' notation correspond to characteristics with half-integer components in present-day notation.) He concluded his lectures with a remarkable addition theorem for the 2^{2g} functions $\theta(z;\mu,\nu)$ obtained by restricting the g components of μ to 0 and 1 and those of ν to 0 and -1. In order to obtain it, he began by considering an arbitrary product of r theta functions with arbitrary, not necessarily integral, characteristics

$$\Pi(z) = \prod_{\alpha=1}^{r} \theta\left(z;\mu^{(\alpha)},\nu^{(\alpha)}\right). \tag{11.11}$$

He showed [594, p. 612] that $\Pi(z)$ satisfies equations of quasiperiodicity similar to those in (11.6), i.e., if $\mu^* = \sum_{\alpha=1}^{r}\mu^{(\alpha)}$ and $\nu^* = \sum_{\alpha=1}^{r}\nu^{(\alpha)}$, then $\Pi(z+\omega_\alpha) = e^{L_\alpha(z)}\Pi(z)$, but where now

$$L_\alpha = r\left[\eta_\alpha^t z + \frac{1}{2}(\eta_\alpha^t\omega_\alpha)\right] + \begin{cases} \pi i\nu_\alpha^*, & \alpha \leq g, \\ -\pi i\mu_\alpha^*, & \alpha > g. \end{cases} \tag{11.12}$$

The above equations of quasiperiodicity have the same form as those in (11.6) with μ^*,ν^* playing the role played there by μ,ν *except* for the occurrence of the factor $r > 1$ in L_α. With this in mind, Weierstrass considered the theta functions with periods $\Omega' = \left(r^{-1}\Omega_1\ \Omega_2\right)$ and $\mathrm{H}' = \left(\mathrm{H}_1\ r\mathrm{H}_2\right)$ and characteristics μ^*, ν^*/r. These he denoted by $\theta^r(z;\mu^*,\nu^*/r)$ and called "transformed theta functions" [594, p. 614].[6] By virtue of their definition as theta functions, they satisfy the equations

[6] If $G(z,n)$, with coefficients given by R,S,T,a,b as in (11.8), defines via the series (11.7) the theta function with periods Ω,H and characteristics μ^*, ν^*, then the series with $G'(z,n)$ defined by $R' = rR$, $S' = rS$, $T' = rT$, $a' = a$, $b' = b$, gives $\theta^r(z;\mu^*,\nu^*/r)$.

of quasiperiodicity (11.6) but with $\Omega', \mathrm{H}', \mu^*$, ν^*/r playing the roles played there by Ω, H, μ, ν. Since for $1 \leq \alpha \leq g$, $z+\omega_\alpha = z' + \omega'_\alpha$, where $z' = z + [(r-1)/r]\omega_\alpha$ and $\omega'_\alpha = (1/r)\omega_\alpha$ is the αth column of Ω', repeated application of these equations of quasiperiodicity shows that $\theta^r(z) \stackrel{\text{def}}{=} \theta^r(z;\mu^*,\nu^*/r)$ also satisfies $\theta^r(z+\omega_\alpha) = e^{L_\alpha(z)}\theta^r(z)$ with L_α the same as in (11.12). In other words, Π and θ^r satisfy the same equations of quasiperiodicity (11.12) with respect to Ω. Because the functions L_α in (11.12) differ from those in (11.6) due to the term with the $r > 1$ factor, Weierstrass' Proposition 11.1 does not apply, and Π is not necessarily a constant multiple of θ^r. However, Weierstrass used the fact that Π satisfies the quasiperiodic equations (11.12) to show that Π is a linear combination of transformed theta functions such as θ^r [594, p. 617]:

$$\Pi(z) = \frac{1}{r^g} \sum_{n\in\mathbb{Z}^g}{}' C_n \theta^r\left(z, \mu^*, \frac{nu^* + 2n}{r}\right), \tag{11.13}$$

where $\sum'$ indicates that the sum is restricted to those $n \in \mathbb{Z}^g$ with components between 0 and $r-1$, and so there are r^g summands. All the functions θ^r on the right-hand side of (11.13) satisfy the quasiperiodic equations $\theta^r(z+\omega_\alpha) = e^{L_\alpha(z)}\theta^r(z)$ with L_α as in (11.12), because for any $n = \begin{pmatrix} n_1 & \cdots & n_g \end{pmatrix}^t$ in (11.13), $\exp[\pi i(\nu^*_\alpha + 2n_\alpha)] = \exp[\pi i \nu_\alpha]$.

Weierstrass observed that (11.13) could be "generalized a bit more" [594, p. 18] by considering a product of the form

$$\Pi^*(z) = \prod_{k=1}^{r} \theta\left(z + \sigma^{(k)}; \mu^{(k)}, \nu^{(k)}\right), \tag{11.14}$$

where $\sigma^{(k)} = \begin{pmatrix} \sigma_1^{(k)} & \cdots & \sigma_g^{(k)} \end{pmatrix}^t$ is subject to the sole condition that $\sum_{\alpha=1}^{r} \sigma_\alpha^{(k)} = 0$. As he showed, the reason a formula similar to (11.13) could be obtained for $\Pi^*(z)$ was that it satisfies the same equations of quasiperiodicity satisfied by $\Pi(z)$. It was, in fact, the analogue of (11.13) for $\Pi^*(z)$ in the case $r = 2$ that eventually led him to a remarkable addition theorem [594, p. 624] for the 2^{2g} theta functions with the integral characteristics described above. That theorem served as the conclusion to his lectures.

In the first three sections of his book, Schottky presented the above-described material up to Weierstrass' addition theorem. Although he did not present the addition theorem, he made an extensive study of complicated relations among theta functions with integer characteristics, which is the reason why most of his book is devoted to the case $g = 3$. What is particularly notable about the first three sections is the fact that Schottky stressed that the considerations leading to Weierstrass' formula (11.13) could be expressed in more general terms. After introducing Weierstrass' product function $\Pi(z)$ as in (11.11), he defined any entire function $\varphi(z)$ that satisfies the same equations of quasiperiodicity as Weierstrass' product function $\Pi(z)$, namely, $\varphi(z+\omega_\alpha) = e^{L_\alpha(z)}\varphi(z)$, $\alpha = 1, \ldots, 2g$, with the L_α

given by (11.12), to be a *theta function of order r with characteristics* (μ^*, ν^*) [519, pp. 5–6]. Thus Weierstrass' two product functions Π and Π^* of (11.14) and also Weierstrass' transformed theta functions θ^r are all examples of Schottky's theta functions of order r. This definition involved a tacit extension of the notion of a theta function of order r. For Weierstrass, this term was restricted to the functions θ^r; for Schottky, the term covered any entire function with the same quasiperiodic equations with respect to Ω as the functions θ^r.

Schottky's notion of theta functions $\varphi(z)$ of order r was simply a reflection of his realization that by virtue of the common underlying equations of quasiperiodicity satisfied by any theta function of order r, Weierstrass' derivation of his key formula (11.13) could be extended to this larger class of functions [519, pp. 6–9]:

Theorem 11.2 (Weierstrass–Schottky). *Any theta function $\varphi(z)$ of order r with period matrices Ω, H and characteristics μ^*, ν^* is expressible in the form*

$$\varphi(z) = \frac{1}{r^g} \sum_{n\in\mathbb{Z}^g}{}' C_n \theta^r\Big(z, \mu^*, \frac{nu^* + 2n}{r}\Big), \tag{11.15}$$

where $\sum'$ indicates that the sum is restricted to those $n \in \mathbb{Z}^g$ with components between 0 *and $r-1$, and so there are r^g summands. Also, the functions $\theta^r\Big(z, \mu^*, \frac{\nu^*+2n}{r}\Big)$ are linearly independent.*

Schottky's proof was essentially the one given by Weierstrass with Weierstrass' product function $\Pi(z)$ replaced by any theta function of order r in Schottky's sense. (The linear independence part follows by taking $\varphi(z) \equiv 0$ and observing that the expressions giving the C_n all vanish.) Although it is not certain whether Weierstrass had imagined this more general theorem, it certainly would have come as no surprise.

11.2 Weierstrass on General Abelian Functions

Weierstrass' theory of general theta functions as expounded by Schottky formed a large part of the motivation and setting for Frobenius' paper, but there was another motivational source as well. It involved a theorem of Weierstrass' on abelian functions that was included neither in his lectures [594] nor in Schottky's book [519] but was revealed in a paper by Hurwitz published early in 1883 [301] and known to Frobenius. Thanks to a manuscript left behind by Weierstrass and included in his collected works [595], it is possible to indicate the considerations that led to Weierstrass' theorem.

We have seen that Weierstrass had stressed that the ordinary theta functions defined a priori by any series of the form (11.7) were more general than those originating in his theory of abelian integrals. Likewise, one could imagine meromorphic functions of g variables with $2g$ independent periods that did not come via inversion of abelian integrals—functions we shall call *general abelian functions* as opposed to the *special abelian functions* that arise by virtue of the solution to the Jacobi inversion problem for abelian integrals—a terminology that goes back to Poincaré. The latter functions are all expressible in terms of the special theta functions that arise from the theory of abelian integrals, and in December 1869, Weierstrass presented a brief paper to the Berlin Academy [589] in which he pointed out that such functions are not the most general abelian functions. In order to generate all abelian functions, Weierstrass had devised a more general inversion process, which he claimed achieved this end, although there were still some "algebraic difficulties" to resolve [589, p. 46]. What these were he did not say, but he hinted at them in his cryptic concluding remark that he was taking under advisement the question whether general abelian functions exist that are not expressible in terms of theta functions.

What Weierstrass meant by his remark is suggested by the above-mentioned manuscript [595, pp. 66–67]. Let $\theta(z)$ be a general theta function with period matrices $\hat{\Omega}, \hat{\mathrm{H}}$ and zero characteristics, i.e., $\mu = \nu = 0$ in (11.7), and consider the function

$$\wp(z) = \frac{\prod_{k=1}^{m} \theta(z + a_k)}{\prod_{k=1}^{m} \theta(z + b_k)}, \tag{11.16}$$

where a_k and b_k, $k = 1, \ldots, m$, are any two sets of m vectors in $\mathbb{C}^g$ with the property that $\sum_{k=1}^{m}(a_k - b_k) = 0$. Then computation of $\wp(z + \hat{\omega}_k)$ using the equations of quasiperiodicity for θ shows that $\wp$ is an abelian function with period matrix $\hat{\Omega}$. In his lectures, Weierstrass had shown that every special abelian function, i.e., every abelian function arising by virtue of the inversion of abelian integrals (as sketched in the introductory remarks to the previous chapter), is expressible as a $\wp$-function associated to a special theta function [594, pp. 604–607]. This generalized what Jacobi had shown in the case $g = 1$. Once Weierstrass realized that abelian functions exist that are not special, it was natural to ask whether they, too, were expressible as $\wp$-functions associated to some general theta function θ as in (11.16). Weierstrass conjectured that every abelian function f with period matrix $\hat{\Omega}$ could be expressed as a rational function of $g + 1$ $\wp$-functions, all of which are defined in terms of the same theta function θ but with different choices for the a_k, b_k in (11.16).

The algebraic sticking point was that if the conjecture were true, it would imply that every period matrix Ω^* of every abelian function f is subject to constraints. To see this, let $G(z,n) = z^t\hat{R}z + z^t\hat{S}n + n^t\hat{T}n$ be the homogeneous polynomial (in the notation of (11.8)) that produces $\theta(z)$ via the series (11.7). Then from (11.9), it follows that $\hat{T} = \pi i(\hat{\Omega}_1)^{-1}\hat{\Omega}_2$, and so if we set $T = (\hat{\Omega}_1)^{-1}\hat{\Omega}_2$, the fact that $\hat{T} = \pi i T$ is symmetric with negative definite real part means that T is symmetric with positive definite imaginary part. The symmetry of T implies, as Weierstrass observed (albeit without the aid of matrix algebra), that

$$\hat{\Omega}_2^t(\hat{\Omega}_1^t)^{-1} - (\hat{\Omega}_1)^{-1}\hat{\Omega}_2 = \hat{\Omega}_1\left(T^t - T\right)\hat{\Omega}_1^t = 0,$$

which is one of the basic relations in (11.5) and which for later reference I will state in the equivalent form

$$\hat{\Omega} J \hat{\Omega}^t = 0, \quad J = \begin{pmatrix} 0 & I_g \\ -I_g & 0 \end{pmatrix}. \tag{11.17}$$

Likewise, the fact that the imaginary part of T is positive definite means that the Hermitian symmetric matrix $i\hat{\Omega} J \hat{\Omega}^h$ is positive definite[7]:

$$i\hat{\Omega} J \hat{\Omega}^h \succ 0. \tag{11.18}$$

Weierstrass realized all this, although he did not express his observations in terms of matrices.

Now suppose Ω is any primitive period matrix for f. Then the periods $\hat{\omega}_j$ that make up $\hat{\Omega}$ must be integral linear combinations of the periods constituting Ω, which means in matrix notation that $\hat{\Omega} = \Omega P$, where P is a $2g \times 2g$ nonsingular matrix of integers. Thus by (11.17), $0 = \hat{\Omega} J \hat{\Omega}^t = \Omega P J P^t \Omega^t = \Omega L \Omega^t$, where $L = PJP^t$ is a nonsingular integral skew-symmetric matrix. Likewise, setting $\hat{\Omega} = \Omega P$ in (11.18) shows that $i\Omega L \Omega^h \succ 0$. In other words, if Ω is any primitive period matrix for an abelian function that is expressible rationally in terms of $g+1$ functions of the form of $\wp(z)$ in (11.16), then there exists a nonsingular skew-symmetric integral matrix L such that (I) $\Omega L \Omega^t = 0$ and (II) $i\Omega L \Omega^h \succ 0$. The same is true for every period matrix Ω^* of f. That is, since $\Omega Q = \Omega^*$ for some nonsingular integral matrix Q, it follows readily that (I) and (II) for Ω and L imply (I) and (II) for Ω^* and $L^* = MLM^t$, where $M = \operatorname{Adj} Q$.

But does every period matrix of a general abelian function necessarily have properties (I) and (II) with respect to some nonsingular integral skew-symmetric matrix? If not, then the envisioned representation of f in terms of the $\wp$-functions of (11.16) would not be possible. That was Weierstrass' sticking point, and according to the manuscript [595, pp. 66–67], in February 1870, he announced to the Berlin Academy that the answer to the above question was affirmative, i.e., he had proved the following theorem.

Theorem 11.3 (Weierstrass' conditions). *If $f(z)$ is an abelian function with period matrix Ω, then there exists a nonsingular skew-symmetric integral matrix L such that*

$$\text{(I)}\ \Omega L \Omega^t = 0; \qquad \text{(II)}\ i\Omega L \Omega^h \succ 0. \tag{11.19}$$

According to Weierstrass, once he had established this theorem, "the remainder of the investigation presented no substantive difficulties" and led to the conclusion that

[7] If $T = \Phi + i\Psi$, then $i\hat{\Omega} J \hat{\Omega}^h = i[\hat{\Omega}_1 (T^h - T)\hat{\Omega}_1^h] = i[-2i(\hat{\Omega}_1 \Psi \hat{\Omega}_1^h)] = 2\hat{\Omega}_1 \Psi \hat{\Omega}_1^h$ is positive definite because $\Psi = \operatorname{Im} T$ is and $\det \hat{\Omega}_1 \neq 0$.

every abelian function is expressible as a rational function of $g+1$ $\wp$-functions associated to the same theta function θ as in (11.16) [595, p. 67]. The main body of the manuscript is devoted to proving the theorem and this conclusion. It should be mentioned that despite Weierstrass' above-quoted remark, the proof is rather long and nontrivial, which may account for the fact that it was not part of his lectures [594]. It should also be noted that in the published proceedings of the Berlin Academy for February 1870, it was simply stated that Weierstrass had presented further results related to his note of 1869; neither Theorem 11.3 nor anything else of a specific nature was revealed.[8]

Conditions (I) and (II) of the above theorem are customarily called Riemann's conditions on Ω. In his 1857 paper presenting his solution to the Jacobi inversion problem for abelian integrals, Riemann showed that the $g\times 2g$ period matrix of the g abelian integrals involved had the following important properties. Let $\varpi_j = (\omega_{j1} \cdots \omega_{j2g})$ denote the jth *row* of Ω. (Thus the entries of ϖ_j give the $2g$ periods of the jth independent abelian integral used in the inversion process.) Then for any j and k, $\varpi_j J \varpi_k^t = 0$ [495, §20]. This says that $[\Omega J \Omega^t]_{jk} = 0$ and so is equivalent to $\Omega J \Omega^t = 0$. Riemann also showed that for any row ϖ_j one has $(i/2)\varpi J \varpi^h > 0$ [495, §21]. From this it follows (as Riemann realized) that if $\varpi = \sum_{j=1}^{g} m_j \varpi_j$, where the m_j are any integers that are not all zero, then

$$\frac{i}{2}\sum_{j,k} \varpi_j J \varpi_k^h m_j m_k > 0. \tag{11.20}$$

Since $[(i/2)\Omega J \Omega^h]_{jk} = (i/2)\varpi_j J \varpi_k^h$, the above inequality is equivalent to the assertion that $(i/2)\Omega J \Omega^h \succ 0$. Thus Riemann had in effect shown that if f is a special abelian function, it necessarily has a primitive period matrix Ω satisfying (I) and (II) of Theorem 11.3 with $L = J$. It seems that by 1860, Riemann was asserting that something like this was also true for general abelian functions. These matters are discussed below in Section 11.4. Given that Weierstrass' conditions (I)–(II) are more general than Riemann's and that he most likely was unaware of Riemann's views circa 1860, which were not well known, I will refer to (I)–(II) of Weierstrass' Theorem 11.3 as the *Riemann–Weierstrass conditions on* Ω.

Weierstrass' Theorem 11.3 was revealed to the mathematical public through a paper by Adolf Hurwitz that appeared in the first issue of *Crelle's Journal* for 1883 [301]. Although Hurwitz was one of Felix Klein's students, he had spent a considerable amount of time learning mathematics in Berlin. In addition to spending

[8]Weierstrass said that he had announced Theorem 11.3 at the session of the Academy on 14 February, but he must have meant the session of 17 February 1870. In the minutes from that session we read, "Hr. Weierstrass machte sodann—im Anschluss an die am 2. December v. J. gelesene Notiz—eine weitere Mittheilung aus seinen Untersuchungen über die $2n$fach periodischen Funktionen" (*Monatsberichte der Königlich Preussischen Akademie der Wissenschaften zu Berlin. Aus dem Jahre 1870* (Berlin, 1871), p. 139). This uninformative statement was the extent of published information about what Weierstrass said at the 17 February 1870 session.

three semesters in Berlin while working on his doctorate (under Klein's direction from Leipzig), after obtaining it in 1881, he also spent 1881–1882 in Berlin. Hurwitz began his paper by thanking "my highly esteemed teacher" Weierstrass for encouraging the research that he was now communicating. The geometrically oriented research involved, when expressed analytically, the study of the properties of an abelian function $f(z)$ that is real-valued for real $z \in \mathbb{C}^g$, and it would appear that as part of his encouragement, Weierstrass told Hurwitz about his Theorem 11.3. Actually, in Weierstrass' manuscript [595, pp. 65ff.] once conditions (I) and (II) were obtained, they were transformed into an equivalent form derived by expressing the period matrix Ω in terms of its real and imaginary parts. Thus if we write $\Omega = \Sigma + i\Delta$ and set

$$A = \Sigma L \Sigma^t, \quad B = \Sigma L \Delta^t, \quad C = \Delta \Sigma \Delta^t,$$

it is not difficult to see that (I) and (II) are equivalent to (1) $A = C$; (2) $B^t = B$; (3) $B + iA \succ 0$.[9] Judging by the manuscript, Weierstrass preferred this form of his conditions, and it was in this form that Hurwitz stated (and presumably received) Weierstrass' Theorem 11.3 [301, pp. 8–9], which he used to prove the lemma that if the periods of an abelian function are such that r of them are real and $s = 2g - r$ are imaginary, then necessarily $r = s = g$.

11.3 Frobenius' Theory of Jacobian Functions

In December 1883, Frobenius submitted a paper "On the Foundations of the Theory of Jacobian Functions" to *Crelle's Journal* [189]. It represented a generalization of the Weierstrass–Schottky theory of ordinary theta functions and what I have called Schottky functions toward the end of Section 11.1. What motivated him to do it? He does not tell us in so many words, but I believe that there was a problem behind this work as well, as I will now attempt to argue.

In 1883 Hurwitz published a paper [301] that presented Weierstrass' hitherto unpublished Theorem 11.3 giving necessary conditions on a matrix Ω in order for abelian functions with period matrix Ω to exist. Weierstrass' theorem, which Frobenius does not seem to have known beforehand, must have proved intriguing to him in the following sense. Recall from Section 11.2 that Weierstrass had proved in his lectures that every special abelian function $f(z)$ can be expressed as a $\wp$-function associated to an ordinary theta function θ as in (11.16). This means that $f(z) = \varphi_a(z)/\varphi_b(z)$, where $\varphi_a(z) = \prod_{k=1}^{m} \theta(z + a_k)$, $\varphi_b(z)$ is defined similarly with respect to $b_1, \ldots, b_m$, and $\sum_{k=1}^{m} a_k = \sum_{k=1}^{m} b_k$. These functions must also have the same periodicity matrix Ω as $f(z)$, and so by Weierstrass' Theorem 11.3, the

[9] Hurwitz described the necessary conditions (1)–(3) on a period matrix as "zuerst von Weierstrass gefundenen" [301, p. 8].

periodicity matrices of functions like φ_a and φ_b must satisfy Weierstrass' conditions (I) and (II) with respect to some integral matrix L. In fact, there are reasons to believe that Frobenius mistakenly thought that Weierstrass had proved (but neither in print nor in his lectures) a similar result for general abelian functions, as indicated below following Frobenius' Theorem 11.11. The entire functions analogous to φ_a and φ_b occurring in the quotient representation of general abelian functions would thus also have period matrices satisfying Weierstrass' conditions.

Now φ_a and φ_b are not necessarily ordinary theta functions or Schottky functions, but it is easily seen by calculation that they nonetheless satisfy quasiperiodicity relations of the same general form as Schottky functions, viz.,

$$\varphi(z+\omega_k) = \mathrm{e}^{2\pi i L_k(z)}\varphi(z), \quad L_k(z) = \eta_k^t z + b_k, \tag{11.21}$$

where $\omega_1,\ldots,\omega_{2g}$ are the periods of $\varphi(z)$, η_k is a $g\times 1$ column matrix, and b_k is a constant, for $k=1,\ldots,2g$. Frobenius made a simple but inspirational observation about functions φ satisfying these generalized quasiperiodicity relations. He observed that if $\Omega = \begin{pmatrix}\omega_1 & \cdots & \omega_{2g}\end{pmatrix}$ is the periodicity matrix for any such function, then evidently, for every j and k between 1 and $2g$ one has

$$\varphi([z+\omega_j]+\omega_k) = \varphi([z+\omega_k]+\omega_j).$$

If the left- and right-hand sides of this equality are calculated by applying the equations of quasiperiodicity (11.21) twice to each side, the equation that results after cancellation of common terms reduces to $\exp[2\pi i(\omega_k^t\eta_j - \eta_k^t\omega_j)] = 1$, and so $\omega_k^t\eta_j - \eta_k^t\omega_j$ must be an integer. Since $\omega_k^t\eta_j - \eta_k^t\omega_j$ is the (j,k) entry of the $2g\times 2g$ skew-symmetric matrix

$$K = \Omega^t \mathrm{H} - \mathrm{H}^t\Omega, \tag{11.22}$$

where $\mathrm{H} = \begin{pmatrix}\eta_1 & \cdots & \eta_{2g}\end{pmatrix}$, we see that K must have integer entries. Could it be, Frobenius might have wondered, that this integral matrix is related to the integral matrix L of Weierstrass' conditions?

This, then, is the problem that I believe motivated Frobenius' theory of Jacobian functions: Seek to distinguish a class of functions $\varphi(z)$ that has the following properties: (a) each function of the class satisfies quasiperiodicity relations of the form (11.21); (b) all the Schottky functions, i.e., all the functions considered by Weierstrass and Schottky, are included; (c) necessary and sufficient conditions on a period matrix Ω can be determined that are analogous to those of Weierstrass' Theorem 11.3 for abelian functions; (d) there is a suitably generalized version of the Weierstrass–Schottky Theorem 11.2 for the functions of the class. The obvious candidates for the integral matrix L implicit in (c) would be either K or its adjoint $\operatorname{Adj} K$, both of which are necessarily integral by virtue of (a). The following exposition of Frobenius' paper [189] makes, I suggest, a compelling case for the above problem as the principal motivating force behind the remarkable general theory that Frobenius presented in its pages.

The starting point of Weierstrass' theory of general theta functions had been a convergent theta series $\theta = \sum_n e^{G(z,n)}$ establishing the existence of θ. The period matrices Ω and H were then determined by the coefficients of $G(z,n)$ via (11.9), and then, lastly, the equations of quasiperiodicity—(11.6) for θ and (11.12) for θ^r, Π, Π^*—were established. By contrast, Frobenius' starting point was more general and abstract. He began by introducing arbitrary $g \times 2g$ matrices $\Omega = \begin{pmatrix}\omega_1 \cdots \omega_{2g}\end{pmatrix}$ and $\mathrm{H} = \begin{pmatrix}\eta_1 \cdots \eta_{2g}\end{pmatrix}$ subject to the sole condition that the columns ω_α of Ω be linearly independent over $\mathbb{Z}$, and he considered entire functions $\varphi(z)$ satisfying the above equations of quasiperiodicity (11.21). In keeping with the terminology of Weierstrass and Schottky, he called the ω_k and η_k periods of the first and second kinds. Also in keeping with Weierstrass and Schottky, he introduced constants $c_1, \ldots, c_{2g}$, which correspond to Weierstrass' characteristics $\mu_1, \ldots, \mu_g, \nu_1, \ldots, \nu_g$. They are defined by $c_k = b_k - \frac{1}{2}(\omega_k^t \eta_k)$, $j = 1, \ldots, 2g$, and enable the equations (11.21) to be written in a form analogous to Weierstrass' quasiperiodicity equations (11.6), viz.,

$$L_k(z) = \eta_j^t \left(z + \frac{1}{2}\omega_j \right) + c_j. \tag{11.23}$$

If $p(z)$ is any second-degree polynomial in the complex variables z_j, it is easily seen by calculation that $\varphi(z) = e^{p(z)}$ satisfies quasiperiodic equations with L_k as in (11.23) for any choice of Ω, with H and c then determined by the choice of Ω. Frobenius called such functions Jacobian functions of order zero. (Today, they would be called trivial theta functions.) These functions have what was known as infinitely small periods ω, meaning that for every $\varepsilon > 0$ there is an ω satisfying quasiperiodic equations of the form (11.21) with $\| \omega \| < \varepsilon$. From theorems due to Riemann [497] and Weierstrass [590], it followed that entire functions φ with infinitely small periods were all products of a Jacobian function of order zero and a "degenerate function" $\varphi(z)$, meaning that the z_i are actually linear functions of fewer than g variables [189, p. 172].

Functions $\varphi(z)$ satisfying (11.21), or equivalently (11.23), with the $2g$ periods ω_j linearly independent over $\mathbb{Z}$ and without infinitely small periods were called *Jacobian functions of rank g* by Frobenius. The algebraic study of these functions was the primary concern of his paper. I will refer to these functions as *Jacobian functions of type* (Ω, H, c), where $c = \begin{pmatrix}c_1 \cdots c_{2g}\end{pmatrix}^t$, or simply as *Jacobian functions*. Jacobian functions are now simply called theta functions, and so Frobenius' paper was in effect the first study of theta functions in the modern sense.[10] All the ordinary theta functions and theta functions of order r of Weierstrass and Schottky functions are examples of Jacobian functions. For if Ω_w and H_w are the period matrices for a Weierstrass theta function θ, we may take

[10]The extension of the term "theta function" to include Jacobian functions may have been initiated by Weil in 1949; see Section 11.4.

$$\Omega = \Omega_w, \quad \mathrm{H} = \frac{1}{2\pi i}\mathrm{H}_w, \quad c = \frac{1}{2}(\nu - \mu)$$

to see from Weierstrass' quasiperiodicity equations (11.6) that θ is a Jacobian function of type (Ω, H, c). Likewise, Weierstrass' functions θ^r, Π, and Π^* and, more generally, Schottky's theta functions of order r—*Schottky functions* for brevity in what follows—are, by virtue of Weierstrass' quasiperiodicity equations (11.12), all examples of Jacobian functions of type $(\Omega, r\mathrm{H}, c^*)$ with $c^* = \frac{1}{2}(\nu^* - \mu^*)$. (Thus Weierstrass' integer characteristics correspond to half-integer characteristics in Frobenius' notation, as is still the case today.) Of course, at this point, it is unclear for what more general types (Ω, H, c), Jacobian functions actually exist. The problem of clarifying this point was obviously fundamental to Frobenius' prospective theory of Jacobian functions.

11.3.1 A fundamental existence theorem

To get started, Frobenius used the fact that Jacobian functions by definition have no infinitely small periods to establish the following result [189, I, p. 173].

Proposition 11.4. *If φ is of type (Ω, H, c), then the $2g \times 2g$ matrix $\tilde{\Omega} = \begin{pmatrix} \Omega \\ \overline{\Omega} \end{pmatrix}$ has full rank $2g$ and so is invertible.*

This result implies that the columns of Ω are actually linearly independent over $\mathbb{R}$, so that, in the terminology of Section 10.2.1, Ω is a period matrix. Proposition 11.4 played a key role in Frobenius' efforts to establish necessary and sufficient conditions for the existence of Jacobian functions of type (Ω, H, c) in terms of properties of $K = \Omega^t \mathrm{H} - \mathrm{H}^t \Omega$.

One necessary condition is of course that K must be integral. For a second such condition, Frobenius considered iK, which is Hermitian symmetric by virtue of the skew-symmetry of K. This means that $i(\overline{w}^t K w)$ is real for all $w \in \mathbb{C}^{2g}$, and since iK can never be positive or negative definite,[11] the sign of $i(\overline{w}^t K w)$ will vary with the choice of w. It is instructive to see how the sign of $i(\overline{w}^t K w)$ behaves in the special case in which Schottky functions of type (Ω, H, c) exist. Recall that the Weierstrass–Schottky theorem shows that all Schottky functions are linear combinations of Weierstrass functions θ^r, which as Jacobian functions are all of the same type. Thus the existence of Schottky functions depends on the existence of these Weierstrass functions θ^r. As we saw in Section 11.1, the latter are defined by theta series $\sum_n e^{G(z,n)}$ with respect to the Weierstrass period matrix $\Omega_w' = \left(\frac{1}{r}\Omega_1 \ \Omega_2\right)$, where $\Omega = (\Omega_1 \ \Omega_2)$. As Weierstrass showed, in order for the series to converge and define a function of g complex variables that cannot be linearly transformed into

[11] The skew-symmetry of K implies that if ρ is a characteristic root of K, then so is $-\rho$.

a function of fewer variables, the coefficients defining $G(z,n)$ must have certain properties—$\det S \neq 0$, $T = \Phi + i\Psi$, with $\Phi \prec 0$ in the notation of (11.8)—that by Weierstrass' equations (11.9) translate into the conditions that $\det \Omega_1 \neq 0$ and that $T = \pi r i \Omega_1^{-1}\Omega_2 = \Phi + i\Psi$ with $\Phi \prec 0$. Since $K = -rJ$ for Schottky functions by virtue of Weierstrass' equations (11.10), if we write $w = \begin{pmatrix} w_1 \\ w_2 \end{pmatrix}$, $w_k \in \mathbb{C}^g$, then

$$i\left(\overline{w}^t K w\right) = -ir\left(\overline{w}^t J w\right) = -ir\left(\overline{w}_1^t w_2 - \overline{w}_2^t w_1\right). \tag{11.24}$$

To get $\Phi = \frac{1}{2}(T + \overline{T})$ into the picture, suppose w_1 and w_2 are related by $w_1 = -\Omega_1^{-1}\Omega_2 w_2 = -\frac{1}{r\pi i}T w_2$. Substitution of this expression for w_1 in (11.24) shows that

$$i\left(\overline{w}^t K w\right) = -\frac{1}{\pi}\overline{w}_2^t\left(T + \overline{T}\right)w_2 = -\frac{2}{\pi}\left(\overline{w}_2^t \Phi w_2\right) > 0,$$

because $\Phi \prec 0$. In sum, when Schottky functions exist, it is necessary that $i(\overline{w}^t K w) > 0$ for all $w \neq 0$ for which $w_1 = -\Omega_1^{-1}\Omega_2 w_2$, or equivalently, for which $\Omega w = \Omega \omega_1 + \Omega_2 \omega_2 = 0$.

The above sort of considerations were, I believe, familiar to Frobenius and most likely led him to conjecture the following generalization to Jacobian functions [189, p. 177, B].

Theorem 11.5 (Existence theorem I). *In order for Jacobian functions of type* (Ω, H, c) *to exist, it is necessary and sufficient that (A) the skew-symmetric matrix* $K = \Omega^t \mathrm{H} - \mathrm{H}^t \Omega$ *have integer coefficients; and (B)* $i(\overline{w}^t K w) > 0$ *for all* $w \neq 0$ *such that* $\Omega w = 0$.

For reasons indicated below, Frobenius seems to have had in the back of his mind the idea that to prove that (B) holds when Jacobian functions φ of a type (Ω, H, c) exist, he should seek to construct out of φ a nonconstant entire function that would be bounded unless $i(\overline{w}^t K w) > 0$ for all $w \neq 0$ such that $\Omega w = 0$. Liouville's well-known theorem then ensured that such an entire function cannot be bounded, and so (B) must necessarily hold. The idea is simple enough, but it required considerable ingenuity on Frobenius' part to bring it to fruition. A sketch of how he did it is given below in Section 11.3.1.1 for those interested. Others may proceed without loss of continuity to Section 11.3.2.

11.3.1.1 Sketch of Frobenius' proof that condition (B) is necessary

Suppose that a Jacobian function φ of type (Ω, H, c) exists. For a fixed (as yet unspecified) $v \in \mathbb{C}^g$, set $\psi(z) = \varphi(v + z)$. Then ψ is a Jacobian function of type $(\Omega, \mathrm{H}, c^*)$, where $c^* = c + v^t \mathrm{H}$. In what follows, I will employ Frobenius' notation and define $E[u] = e^{2\pi i u}$, a type of notation still in common use. Define a complex-valued function L of $2g$ real variables $\xi_1, \ldots, \xi_{2g}$ by

$$L(\xi) = E\left[-\frac{1}{2}\xi^t \Omega^t \mathrm{H}\xi - c^*\xi\right]\psi(\Omega\xi), \quad \xi \in \mathbb{R}^{2g}. \tag{11.25}$$

Since ψ is an entire function, L is continuous on $\mathbb{R}^{2g}$ and so bounded on the unit hypercube $\mathfrak{C}$ consisting of all ξ with $0 \le \xi_k \le 1$ for all $k = 1,\ldots,2g$, i.e.,

$$|L(\xi)| \le G \quad \text{for all} \quad \xi \in \mathfrak{C},$$

where G is independent of ξ. If e_k denotes the kth column of I_{2g}, then since $\Omega e_k = \omega_k$, it follows that $\psi(\Omega(\xi + e_k)) = \psi(\Omega\xi + \omega_k)$, and this can be computed using the equations of quasiperiodicity for ψ. This computation, on simplification, shows that

$$L(\xi + e_k) = E\left[-\frac{1}{2}[K\xi]_k\right]L(\xi), \quad [K\xi]_k = k\text{th component of } K\xi.$$

Since $[K\xi]_k$ is real, the exponential factor has absolute value 1, which means that L has the remarkable property that

$$|L(\xi + e_k)| = |L(\xi)| \quad \text{for all} \quad k = 1,\ldots,2g.$$

This shows that $\xi \to |L(\xi)|$ is periodic with the $2g$ e_k as periods, and so its values in the hypercube $\mathfrak{C}$ are repeated periodically throughout $\mathbb{R}^{2g}$, whence

$$|L(\xi)| \le G \quad \text{for all} \quad \xi \in \mathbb{R}^{2g}. \tag{11.26}$$

Next, Frobenius considered a nonzero vector $w \in \mathbb{C}^{2g}$ such that $\Omega w = 0$. His goal was to use the above results to deduce that $i(\overline{w}^t K w) > 0$. To this end, he used Proposition 11.4: since $w \ne 0$, $\tilde{\Omega} w \ne 0$, which means that $\overline{\Omega} w \ne 0$, and so by conjugation, $r \stackrel{\text{def}}{=} \Omega\overline{w} \ne 0$. For ξ he took $\xi_w = w + \overline{w}$. Using $\xi = \xi_w$ in (11.25), together with $\Omega w = 0$ and $\Omega\xi_w = \Omega\overline{w} = r$, he was able to rewrite (11.25) as

$$E[-\alpha - \beta]\psi(r) = e^p F, \tag{11.27}$$

where

$$\begin{aligned} \alpha &= (1/2)(\overline{w}^t \Omega^t \mathrm{H}\overline{w}), & \beta &= (c^* - \overline{c^*})\overline{w}, \\ p &= \pi i(\overline{w}^t K w), & F &= E[c^* w + \overline{c^* w}]L(\xi_w). \end{aligned}$$

Since $c^* w + \overline{c^* w}$ is real, the exponential factor in F has absolute value 1, and so $|F| = |L(\xi_w)| \le G$ by (11.26).

Now consider $z \in \mathbb{C}$ and observe that if we replace w by the scalar multiple $w' = \bar{z}w$, then $\Omega w' = \bar{z}\Omega w = 0$; and so all the above calculations go through with w replaced by w'. This implies that (11.27) still holds but with α, β, p, F replaced by

α', β', p', F', where $\alpha' = \alpha z^2$, $\beta' = \beta z$, $p' = p|z|^2$, and one still has $|F'| \leq G$. Thus by replacing w with $w' = \bar{z}w$, and so $r = \Omega\overline{w}$ with $r' = \Omega(\overline{\bar{z}w}) = zr$, equation (11.27) becomes

$$E\left[-\alpha z^2 - \beta z\right] \psi(zr) = e^{p|z|^2} F'. \tag{11.28}$$

Now $\chi(z) \stackrel{\text{def}}{=} E[-\alpha z^2 - \beta z]\psi(zr)$ is an entire function of the single complex variable z. The definition of $\chi(z)$ depends on v because $\psi(zr) = \varphi(v + zr)$, and it can be shown that χ cannot be a constant function for every choice of v, because if it were, then φ would have infinitely small periods, contrary to the definition of φ [189, p. 176]. So assume that v is chosen such that $\chi(z)$ is a nonconstant function. Then Liouville's theorem asserts that $\chi(z)$ cannot be bounded for all $z \in \mathbb{C}$. But (11.28) implies that $|\chi(z)| \leq e^{p|z|^2} G$ for all $z \in \mathbb{C}$, so that χ would be bounded if $p \leq 0$. In other words, it must be that $p = \pi i(\overline{w}^t K w) > 0$, and (B) is proved!

In a footnote to his proof, Frobenius wrote, "In his lectures Mr. Weierstrass applied this theorem [Liouville's theorem] in a similar manner in order to derive the conditions for the convergence of theta series" [189, p. 177n]. This remark by Frobenius is puzzling. None of Weierstrass' published treatments of the conditions for the convergence of a theta series make any use of Liouville's theorem, and in fact, it is difficult to imagine a reasonable approach to the conditions based on that theorem.[12] Perhaps Frobenius was confusing Weierstrass' treatment of theta series convergence with his treatment of some other matter. Be that as it may, Frobenius' remark does suggest that in devising his remarkable proof, he had in the back of his mind the idea that the fact that $p = \pi i(\overline{w}^t K w)$ is positive when $\Omega w = 0$ should be obtained by creating out of φ an entire nonconstant function that would be bounded unless $p > 0$. Even if he absorbed this general idea from Weierstrass' lectures, his application of it, as described above, is remarkably clever and reflects his extraordinary talent for dealing effectively with computational complexities to educe important consequences.

11.3.2 Connection with the Riemann–Weierstrass conditions (I)–(II) on a period matrix

Frobenius' proof of the sufficiency of (A) and (B) will be considered further on. He first developed the implications of the fact that conditions (A) and (B) are necessary with an eye toward possible connections between K and the L of Weierstrass'

[12] For Weierstrass' treatment of the convergence of theta series in one variable, see [597, pp. 567ff.], and for several variables, see [596]. In the published version of his lectures on abelian integrals and functions [594, p. 568], the convergence condition is simply stated without proof and with a footnote reference to [596].

Theorem 11.3.[13] He proceeded as follows: If Jacobian functions of type (Ω, H, c) exist, so that (A) and (B) hold, then the $2g \times 2g$ matrix

$$M = \begin{pmatrix} \Omega \\ \mathrm{H} \end{pmatrix}$$

must be nonsingular, i.e., if $Mw = 0$, then $w = 0$. For if $Mw = 0$, then both $\Omega w = 0$ and $\mathrm{H}w = 0$, but then $Kw = \Omega^t(\mathrm{H}w) - \mathrm{H}^t(\Omega w) = 0$, and so $i(\overline{w}^t Kw) = 0$, which means by (B) that $w = 0$, and so M is nonsingular. Frobenius used the nonsingularity of M to show that if functions of type (Ω, H, c) exist, then so do functions of type (Ω, H, c') for every $c' \in \mathbb{C}^{2g}$. This explains why (A) and (B) involve only Ω and H. Another consequence of the nonsingularity of M is the nonsingularity of K, since an easy calculation shows that

$$M^t JM = \Omega^t \mathrm{H} - \mathrm{H}^t \Omega = K, \tag{11.29}$$

where as usual, $J = \begin{pmatrix} 0 & I_g \\ -I_g & 0 \end{pmatrix}$. Since $\det J = 1$, $\det K = (\det M)^2 > 0$. Indeed, since K is skew-symmetric, Frobenius knew from his work on the problem of Pfaff that its determinant is the square of its Pfaffian $\mathrm{Pf}[K]$ (see Section 8.3.2), and so

$$\det K = \ell^2, \quad \ell \in \mathbb{Z}^+, \quad \ell = \pm \mathrm{Pf}[K]. \tag{11.30}$$

Frobenius called ℓ the *order of* K. If φ is a Schottky function, so that $K = -rJ$, then, since $\det(-rJ) = r^{2g}$, $\ell = r^g$. Thus by the Weierstrass–Schottky theorem, in this case ℓ gives the number of linearly independent Schottky functions of (Schottky) order r, a result that Frobenius will generalize to Jacobian functions of a given type.

The necessary part of Theorem 11.5 is vaguely reminiscent of Weierstrass' Theorem 11.3, although the resemblance is not close, because the integral skew-symmetric matrix L of Weierstrass' theorem satisfies $\Omega L \Omega^t = 0$ and $i\Omega L \Omega^h \succ 0$. There is, however, another integral skew-symmetric matrix naturally associated to K, namely $L = \mathrm{Adj}\, K$, the transpose of the matrix of cofactors of K, which satisfies $KL = (\det K) I_{2g} = \ell^2 I_{2g}$ [189, pp. 187–188]. Of course, $L = \ell^2 K^{-1}$, but in this paper, Frobenius avoided all use of the still novel matrix algebra that had been so essential in his paper on complex multiplication a year earlier (Section 10.6), preferring here to use instead the Weierstrass-type notation to which his readers were accustomed. Hence K^{-1} was never mentioned in connection with L, and in general, the symbolism of matrix algebra was avoided.

If (11.29), viz., $M^t JM = K$, is rewritten in the form $MLM^t = \ell^2 J^{-1} = \ell^2(-J)$, and if this equation is then expanded by block multiplication, it becomes

[13] This is done in [189, §3], albeit under the more general hypothesis that Ω and H are $g \times \sigma$, with $\sigma \leq 2g$, and so the functions involved are not what I have called Jacobian functions. In what follows I consider only the case $\sigma = 2g$.

$$\begin{pmatrix} \Omega L\Omega^t & \Omega L\mathrm{H}^t \\ \mathrm{H}L\Omega^t & \mathrm{H}L\mathrm{H}^t \end{pmatrix} = \ell^2 \begin{pmatrix} 0 & -I_g \\ I_g & 0 \end{pmatrix},$$

and so implies Frobenius' equations (A′) [189, p. 188], namely

$$(\mathrm{A}') \quad \begin{matrix} \Omega L\Omega^t = 0, & \mathrm{H}L\Omega^t = \ell^2 I_g, \\ \mathrm{H}L\mathrm{H}^t = 0, & \det L = \ell^{4g-2}. \end{matrix} \tag{11.31}$$

Thus, in particular, $L = \mathrm{Adj}\, K$ satisfies $\Omega L\Omega^t = 0$, which is the first condition of Weierstrass' Theorem 11.3. Frobenius naturally sought to relate condition (B) of Theorem 11.5 to Weierstrass' second condition ($i\Omega L\Omega^h \succ 0$).

Condition (B) says that $i(\overline{w}^t Kw) > 0$ for all $w \in \mathbb{C}^{2g}$ satisfying $\Omega w = 0$. To bring L into the picture, Frobenius observed that if $y = Kw$, then $LK = \ell^2 I_{2g}$ implies that $w = I_{2g}w = (1/\ell^2)LKw = (1/\ell^2)Ly$, and so, taking the conjugate transpose, $\overline{w}^t = (1/\ell^2)\overline{y}^t L^t$, whence

$$0 < i(\overline{w}^t Kw) = (i/\ell^2)(\overline{y}^t L^t y). \tag{11.32}$$

Now L is in the picture. To bring Ω into it as well, Frobenius turned to $M = \begin{pmatrix} \Omega \\ \mathrm{H} \end{pmatrix}$. Since M is nonsingular, for any $x \in \mathbb{C}^g$ the equation $Mw = \begin{pmatrix} \Omega w \\ \mathrm{H}w \end{pmatrix} = \begin{pmatrix} 0 \\ x \end{pmatrix}$ has a unique solution w; in this manner, he obtained a one-to-one correspondence between all $x \neq 0$ in $\mathbb{C}^g$ and the $w \in \mathbb{C}^{2g}$ for which $\Omega w = 0$ and $i(\overline{w}^t Kw) > 0$. Since $\mathrm{H}w = x$ and $\Omega w = 0$, this means that $y = Kw = \Omega^t \mathrm{H}w - \mathrm{H}^t \Omega w = \Omega^t x$. Thus for all $x \neq 0$ in $\mathbb{C}^g$, we have Frobenius' equation (B′) [189, p.188]:

$$(\mathrm{B}') \quad 0 < i(\overline{y}^t L^t y) = i(\overline{x}^t \overline{\Omega} L^t \Omega^t x), \tag{11.33}$$

which means that the Hermitian symmetric matrix $i(\overline{\Omega} L^t \Omega^t) = -i(\overline{\Omega} L \Omega^t)$ is positive definite. Since this is also true of its complex conjugate $i\Omega L\Omega^h$, we see that $i\Omega L\Omega^h \succ 0$, and so L satisfies *both* the conditions in Weierstrass' Theorem 11.3. For future reference I will state this result as a theorem:

Theorem 11.6. *If Jacobian functions of type* (Ω, H, c) *exist, then* $L = \mathrm{Adj}\, K$ *is a nonsingular integral skew-symmetric matrix such that (I)* $\Omega L\Omega^t = 0$ *and (II)* $i(\Omega L\Omega^h) \succ 0$.

At the conclusion of his proof that conditions (A) and (B) of Theorem 11.5 imply those of (11.31) and (11.33), i.e., Frobenius' conditions (A′) and (B′), Frobenius claimed that (A′) and (B′) were "completely equivalent" to (A) and (B) [189, p. 189]. He continued, as justification of his claim, by showing that (A′) and (B′) imply that $\det L \neq 0$. (What he showed, in fact, was that every skew-symmetric L satisfying conditions (I) and (II) of Theorem 11.6 must have a nonzero determinant.) Since by (A′), $\det L = \ell^{4g-2}$, we may assume that $\ell > 0$. The remaining relations in (A′) are equivalent to $MLM^t = \ell^2(-J)$, where as above, $M = \begin{pmatrix} \Omega \\ \mathrm{H} \end{pmatrix}$. Taking determinants in this relation shows that $\det M = \pm\ell \neq 0$. After showing this, Frobenius continued by saying, "Now one need only proceed in reverse order though the developments of these paragraphs in order to obtain the conditions (A) and (B) from (A′) and (B′) [189, p. 189]. What, precisely, did Frobenius mean by these words? As we shall see in the discussion of the work of Wirtinger and Castelnuovo in Section 11.4 below, the answer is of some historical significance and so worthy of attention. My interpretation, justified in Section 11.3.2.1 below, is that Frobenius meant the following theorem.

Theorem 11.7 (Existence theorem II). *Let Ω be a $g \times 2g$ matrix with columns linearly independent over $\mathbb{R}$. Then Jacobian functions of some type $(\Omega, m\mathrm{H}, \ldots)$ exist, $m \in \mathbb{Z}$, if and only if an integral, skew-symmetric matrix L exists that satisfies conditions (A′) and (B′), viz., (11.31) and (11.33).*

We shall see in Section 11.4 that although Frobenius apparently did not realize it, this theorem remains valid if all the conditions of (A′) and (B′) are replaced by the subset consisting of conditions (I) and (II) of Theorem 11.6, i.e., the two conditions of Weierstrass' Theorem 11.3. This improved version of Frobenius' Theorem 11.7 eventually became a foundation stone of the theory of abelian functions.

11.3.2.1 Justification of Theorem 11.7 (optional)

Here is the rationale for interpreting Frobenius' above-quoted remarks as Theorem 11.7. Suppose that a skew-symmetric integral matrix L exists satisfying (A′) and (B′), i.e., (11.31) and (11.33), with $\ell \geq 0$. Then, as already noted, these conditions imply that $\det L \neq 0$, that $\ell > 0$, and that M is invertible with $\det M = \pm\ell$. The invertibility of M means that $MLM^t = \ell^2(-J) = \ell^2 J^{-1}$ may be rewritten as $M^t = M^{-1}\ell^2 L^{-1} J^{-1}$. If we *define* K by $K = \ell^2 L^{-1}$, then $M^t = M^{-1}KJ^{-1}$, which may be rewritten as $MJM^t = K$. Taking determinants in this relation, we see that since $\det L = \ell^{4g-2}$ by (A′), $\det K = \ell^2$. Furthermore, since by definition of K, $KL = \ell^2 I_{2g} = (\det K) I_{2g}$, we see that $L = \mathrm{Adj}\, K$. On the other hand, since $K = MJM^t$, computation of the right-hand side shows that $K = \Omega^t \mathrm{H} - \mathrm{H}^t \Omega$, i.e., K satisfies the defining relationship (11.22) of condition (A) in Theorem 11.5, although it does not follow that K is integral. The relation $K = MJM^t$ was the key relation (11.29) in the reasoning leading from (A) and (B) to (A′) and (B′), and the reasoning can now be reversed, as Frobenius said, to establish condition (B) of Theorem 11.5.

In short, what Frobenius meant by his above-quoted remark may be summed up as follows. Given Ω and H, if L is an integral skew-symmetric matrix satisfying (A′) and (B′), then L is nonsingular, $\ell > 0$, $L = \mathrm{Adj}\, K$, where K satisfies (B) and a weakened form of (A), namely,

$$(\mathrm{A}^*) \qquad K = \Omega^t \mathrm{H} - \mathrm{H}^t \Omega.$$

Since $K = \ell^2 L^{-1}$, its coefficients are rational numbers, but they are not necessarily integers. It is not difficult to construct examples of integral L satisfying (A′) and (B′) for which K is not integral, and so no Jacobian functions of types $(\Omega, \mathrm{H}, \ldots)$ can exist.[14] Frobenius certainly must have realized this; what he intended by his above-quoted remarks was that (A′) and (B′) are completely equivalent to (A*) and (B)—with K, L, and ℓ simply assumed to be rational. He surely also realized that even if (A′) and (B′) are assumed and the resulting K is not integral, being rational, an integer m can be chosen such that $K_1 = mK$ is integral. Since $K_1 = \Omega^t \mathrm{H}_1 - \mathrm{H}_1^t \Omega$ with $\mathrm{H}_1 = m\mathrm{H}$, Jacobian functions of type $(\Omega, \mathrm{H}_1, \ldots)$ exist when (A′) and (B′) are assumed. In other words, implicit in Frobenius' remarks about the "complete equivalence" of conditions (A)–(B) and (A′)–(B′) was Theorem 11.7 above.

11.3.3 A formula for the number of independent Jacobian functions of a given type

I will now sketch the reasoning by which Frobenius established the sufficiency of his conditions (A) and (B). As we shall see, he utilized suitably generalized results and techniques from Weierstrass' theory of ordinary theta functions and theta functions of order r as expounded by Schottky, but he also drew upon results from his work on arithmetic linear algebra, such as his containment theorem (Theorem 8.16) and his Theorem 8.13 on modules. The reasoning behind his proof of the sufficiency of (A) and (B) also segued into his proof that the number of Jacobian functions of a given type (Ω, H, c) satisfying (A) and (B) is $\ell = \sqrt{\det K}$ (Theorem 11.10 below). Readers wishing to skip these admittedly nontrivial technicalities should proceed to Section 11.3.4, which contains an interesting application of Theorem 11.10 that leads to the question whether Frobenius thought that Weierstrass had in effect proved that every abelian function is the quotient of two Jacobian functions.

As we noted earlier in this section, Weierstrass' functions θ^r are Jacobian functions of Frobenian type $(\Omega, r\mathrm{H}, \ldots)$. However, as we saw, the functions θ^r were created from Weierstrassian theta functions with Weierstrass periods $(r^{-1}\Omega_1 \quad \Omega_2)$ and $(\mathrm{H}_1 \quad r\mathrm{H}_2)$. Thus θ^r is also a Jacobian function of type $(\hat{\Omega}, \hat{\mathrm{H}}, \ldots)$, where $\hat{\Omega} = (r^{-1}\Omega_1 \quad \Omega_2)$ and $\hat{\mathrm{H}} = \frac{1}{2\pi i}(\mathrm{H}_1 \quad r\mathrm{H}_2)$. When θ^r is so regarded, the associated skew-symmetric matrix is $\hat{K} = \frac{1}{2\pi i}(\Omega_1^t \mathrm{H}_2 - \mathrm{H}_1^t \Omega_2) = -J$. *In sum,* θ^r *is both a*

---

[14] Let D and T be $g \times g$ and symmetric. Assume that D has rational coefficients and is invertible. Assume $\Psi = \operatorname{Im} T \succ 0$. Take $\Omega = (D \quad -T)$ and $\mathrm{H} = (0 \quad I_g)$. Then $K = \Omega^t \mathrm{H} - \mathrm{H}^t \Omega = \begin{pmatrix} 0 & D \\ -D & 0 \end{pmatrix}$. Assume $\ell = \det D > 0$, so $\det K = \ell^2$. Finally, set $L = \operatorname{Adj} K = \ell^2 K^{-1} = \ell^2 \begin{pmatrix} 0 & -D^{-1} \\ D^{-1} & 0 \end{pmatrix}$. Then L satisfies (A′) and (B′), and D can be chosen such that L is integral but K is not. Take, e.g., $D = \begin{pmatrix} 4 & 0 \\ 0 & 1/2 \end{pmatrix}$, so $\ell = 2$.

Jacobian function of type $(\Omega,\mathrm{H},\ldots)$ *with order* $\ell = r^g$ *and a Jacobian function of type* $(\hat{\Omega},\hat{\mathrm{H}},\ldots)$ *with order* $\ell = 1$. Note that the integral nonsingular matrix $G = \begin{pmatrix} rI_g & 0 \\ 0 & I_g \end{pmatrix}$ transforms the latter period pair into the former: $\hat{\Omega}G = \Omega$ and $\hat{\mathrm{H}}G = \mathrm{H}$.

The above interpretation of a theta function as a Jacobian function of two types, suitably generalized, played a fundamental role in Frobenius' reasoning leading to a proof of the sufficiency of conditions (A) and (B), as well as to his generalization of the Weierstrass–Schottky theorem. For ease of reference, the first step of the generalization process is summarized in the following lemma.

Lemma 11.8. *If* φ *is a Jacobian function of type* $(\Omega_a,\mathrm{H}_a,c_a)$ *and if* $G = (g_{\alpha\beta})$ *is a* $2g \times 2g$ *nonsingular integral matrix, then* φ *is also a Jacobian function of type* $(\Omega_b,\mathrm{H}_b,c_b)$*, where*

$$\Omega_b = \Omega_a G, \quad \mathrm{H}_b = \mathrm{H}_a G, \quad c_b = c_a G + \frac{1}{2} n(K_a, G), \tag{11.34}$$

and $n(K_a,G)$ *is the row matrix with components*

$$[n(K_a,G)]_\lambda = \sum_{\alpha<\beta} k^a_{\alpha\beta} g_{\alpha\lambda} g_{\beta\lambda}, \quad \lambda = 1,\ldots,2g, \tag{11.35}$$

where $K_a = (k^a_{\alpha\beta}) = \Omega_a^t \mathrm{H}_a - \mathrm{H}_a^t \Omega_a$.

The equation $\Omega_b = \Omega_a G$ says that every column ω_b of Ω_b is an integral linear combination of the columns ω_a of Ω_a. Repeated application of the quasiperiodic equations for type $(\Omega_a,\mathrm{H}_a,c_a)$ to $\varphi(z+\omega_b)$ then yields quasiperiodic equations for the type $(\Omega_b,\mathrm{H}_b,c_b)$ and establishes the lemma [189, pp. 174, 186].

Lemma 11.8 says that a Jacobian function with period matrix Ω_a can be regarded as a Jacobian function with period matrix $\Omega_b = \Omega_a G$ for any choice of a nonsingular integral matrix G. It follows from (11.34) that the skew-symmetric matrices K_a, K_b associated to (Ω_a,H_a) and (Ω_b,H_b) satisfy $K_b = G^t K_a G$. In the language of Frobenius' arithmetic theory of bilinear forms (Section 8.4), this means that K_b is contained in K_a. Recall that a major result of that theory was the containment theorem (Theorem 8.16), which says that an integral matrix B is contained in A ($B = GAH$, where G,H are integral) if and only if rank $B \le$ rank A and the invariant factors of B are all multiples of those of A. Frobenius realized that the proof of the "if" part of this theorem could be specialized to skew-symmetric $2g \times 2g$ matrices so as to yield the following version: *If* A *and* B *are nonsingular integral skew-symmetric* $2g \times 2g$ *matrices, and if all the invariant factors of* B *are integral multiples of the corresponding invariant factors of* A*, then* B *is contained in* A *in the sense that an integral matrix* G *exists such that* $B = G^t AG$.[15] Let us now see how he used this result to establish the sufficiency of conditions (A) and (B).

---

[15]Frobenius invoked the above result without providing any proof, but it is easy to see that the proof of the "if" part of his containment theorem indicated at the beginning of Section 8.4 remains viable if modified as follows: (1) N,N' are now the skew-symmetric normal forms (8.18) of A

Suppose that (Ω,H,c) is such that $K=\Omega^t\mathrm{H}-\mathrm{H}^t\Omega$ satisfies (A) and (B) of Theorem 11.5. Since the invariant factors of $-J$ are all $+1$, the invariant factors of K, being positive integers, are all multiples of those of $-J$, and so by the above result, an integral matrix G exists such that $K=G^t(-J)G$. Taking determinants, we see that $\ell^2=(\det G)^2$, and so

$$|\det G|=\ell, \tag{11.36}$$

where (as we saw) $\ell>0$ because K satisfies (B) and so is nonsingular. Set $\hat{\Omega}=\Omega G^{-1}$ and $\hat{\mathrm{H}}=\mathrm{H}G^{-1}$. Then $\Omega=\hat{\Omega}G$, $\mathrm{H}=\hat{\mathrm{H}}G$, and so by Lemma 11.8 with $\Omega_a=\hat{\Omega}$ and $\Omega_b=\Omega$, etc., we know that if Jacobian functions of type $(\hat{\Omega},\hat{\mathrm{H}},c_a)$ exist, then they are also Jacobian functions of type (Ω,H,c_b), i.e., this latter type exists as well. In particular if we take $c_a=\hat{c}$, where

$$\hat{c}=\left[c-\frac{1}{2}n(-J,G)\right]G^{-1}, \tag{11.37}$$

then by (11.34), $c_b=c$. In sum, Jacobian functions of type (Ω,H,c) exist, provided Jacobian functions of type $(\hat{\Omega},\hat{\mathrm{H}},\hat{c})$ exist.

Now $\hat{K}=\hat{\Omega}^t\hat{\mathrm{H}}-\hat{\mathrm{H}}^t\hat{\Omega}=-J$. This means that $\hat{K}$ is integral and so satisfies condition (A). Also, since K is assumed to satisfy (B) with respect to Ω, it follows readily that $\hat{K}$ satisfies (B) with respect to $\hat{\Omega}$.[16] By virtue of the equivalence of conditions (A)–(B) with (A′)–(B′) as in Theorem 11.7, we know that $\hat{L}=\operatorname{Adj}\hat{K}=\operatorname{Adj}(-J)=J$ satisfies (I) $\hat{\Omega}J\hat{\Omega}^t=0$ and (II) $i\hat{\Omega}J\hat{\Omega}^h\succ 0$. It then follows from Weierstrass' theory of theta functions that a theta function $\theta(z)=\theta(z;\mu,\nu)$ exists with Weierstrass period matrices $\Omega_w=\hat{\Omega}$ and $\mathrm{H}_w=2\pi i\hat{\mathrm{H}}$ and parameters μ,ν given by $(\nu \quad -\mu)=2\hat{c}$. That is, from the data $\Omega_w,\mathrm{H}_w,\mu,\nu$, Weierstrass' equations (11.9) uniquely determine the coefficients R,S,T,a,b of the quadratic polynomial $G(z,n)$ of (11.8)[17]; and (I)–(II) imply that T is symmetric with negative definite real part, which means that $\theta(z)=\sum_{n\in\mathbb{Z}^g}e^{G(z,n)}$ converges and defines the requisite theta function. This function is a Jacobian function of type $(\hat{\Omega},\hat{\mathrm{H}},\hat{c})$ and so by Lemma 11.8 is also a Jacobian function of type (Ω,H,c). Thus conditions (A) and (B) are indeed sufficient, and Theorem 11.5 is now completely proved.

The line of reasoning given in the above paragraph remains valid for any choice of the parameter vector $\hat{c}$, not just the one specified by (11.37), and so implies the following slightly more general result, which will be used in what follows.

---

and B so $R^tAR=N$ and $S^tBS=N'$. (2) With $e_i'=m_ie_i$ for $i=1,\ldots,g$, write $N'=M^tNM$, where $M=\begin{pmatrix}\Delta & 0\\ 0 & I_g\end{pmatrix}$ and Δ is the diagonal matrix with $m_1,\ldots,m_g$ as diagonal entries.

[16] Condition (B) is satisfied because $i(\overline{w}^tKw)=i(\overline{u}^t\hat{K}u)$, where $u=Gw$, and so $\Omega w=0$ if and only if $\hat{\Omega}u=0$. Thus (B) follows.

[17] $R=\frac{1}{2}\mathrm{H}_1\Omega_1^{-1}$, $S=2\pi i\Omega_1^{-1}$, $T=\pi i\Omega_1^{-1}\Omega_2$, $a=\pi i(\Omega_1^t)^{-1}\nu$, $b=\pi i\Omega_1^{-1}\Omega_2\nu-\frac{1}{\pi i}\mu$.

Lemma 11.9. *If $K = \Omega^t \mathrm{H} - \mathrm{H}^t \Omega$ satisfies (A)–(B) of Theorem 11.5, then a nonsingular integral G exists such that $K = G^t(-J)G$ and Weierstrass' theta functions of all Frobenius types $(\hat{\Omega}, \hat{\mathrm{H}}, \ldots)$ exist, where $\hat{\Omega} = \Omega G^{-1}$ and $\hat{\mathrm{H}} = \mathrm{H}G^{-1}$.*

Frobenius realized that he could apply this lemma to generalize the Weierstrass–Schottky theorem (Theorem 11.2) as follows.

Theorem 11.10 (Weierstrass–Schottky–Frobenius). *Let (Ω, H, c) be a type satisfying conditions (A) and (B), so that Jacobian functions of that type and order $\ell = \sqrt{\det K}$ exist. Then ℓ linearly independent Jacobian functions $\varphi_1, \ldots, \varphi_\ell$ of this type may be determined such that every Jacobian function φ of this type is given by $\varphi = \sum_{k=1}^{\ell} C_k \varphi_k$.*

This theorem is usually referred to as Frobenius' theorem, although it would be more just to call it the Weierstrass–Schottky–Frobenius theorem. In fact, in a footnote [189, p. 194n], Frobenius explained that the reasoning leading to his Theorem 11.10 was modeled after that of Weierstrass as presented in Schottky's monograph [519]. However, Frobenius combined the type of reasoning employed by Schottky and Weierstrass with results from his work on arithmetic linear algebra. What follows is a brief sketch of his proof [189, pp. 193–197] that focuses on how he applied arithmetic linear algebra.

In order to construct functions analogous to the basis functions θ^r of Theorem 11.2 of Weierstrass and Schottky, Frobenius proceeded as follows. Let φ denote a fixed Jacobian function of type (Ω, H, c), and let $\hat{\Omega}, \hat{\mathrm{H}}$ be as in Lemma 11.9. For each $p \in \mathbb{Z}^{2g}$, set $\omega_p = \hat{\Omega} p$, $\eta_p = \hat{\mathrm{H}} p$, and $c_p = \hat{c} p + \frac{1}{2} n(-J, G)$, where $\hat{c}$ is defined by (11.37). Then define $\varphi(z; p) = \mathrm{e}^{-L(z,p)} \varphi(z + \omega_p)$, where $L(z, p) = \eta^t(z + \frac{1}{2}\omega_p) + c_p$. Some nontrival calculations show that (i) $\varphi(z; p)$ is also of type (Ω, H, c) and (ii) if $p' \equiv p \,(\mathrm{mod}\, G)$, then $\varphi(z; p') = \varphi(z; p)$ [189, p. 195]. As we saw in Section 8.3.3, congruence mod G was a notion that Frobenius had introduced in 1879 under the influence of Dedekind's theory of modules. Thus $p' \equiv p \,(\mathrm{mod}\, G)$ means that $p' - p$ belongs to the $\mathbb{Z}$-module $\mathcal{M}_G$ consisting of all $n \in \mathbb{Z}^{2g}$ expressible as $n = Gn$ for some $n \in \mathbb{Z}^{2g}$. A result from the 1879 paper that he used here is Theorem 8.13: *The number of distinct congruence classes* mod G *is* $|\det G$. In the present context, we have $|\det G| = \ell$ by (11.36), so there are ℓ congruence classes.

Let $p_1, \ldots, p_\ell$ denote representatives of these classes, with the indexing chosen such that $p_1 \equiv 0 \,(\mathrm{mod}\, G)$, i.e., p_1 is the representative of the class $\mathcal{M}_G$. By (ii), this means that $\varphi(z, p_1) = \varphi(z)$. Likewise, let $n_1, \ldots, n_\ell$ denote representatives of the ℓ congruence classes mod G^t, and define the rational column matrix $q_k = (G^t)^{-1} n_k$ so that

$$n_k = G^t q_k, \quad k = 1, \ldots, \ell. \tag{11.38}$$

Then if $\psi_{n_k} = \sum_{j=1}^{\ell} \mathrm{e}^{2\pi i(-p_j^t q_k)} \varphi(z; p_j)$, calculation shows that $\psi_{n_k}(z)$ is a Jacobian function of type $(\hat{\Omega}, \hat{\mathrm{H}}, \hat{c} + q_k^t)$. By Lemma 11.9, there is a Weierstrass theta function of this type, and so by Weierstrass' Proposition 11.1, $\psi_{n_k}(z)$ must be a constant multiple of the Weierstrass theta function $\theta(z;\hat{c}+q_k^t)$ associated to this type. If we write the constant multiple in the form ℓC_k, we have

$$\psi_{n_k} = \sum_{j=1}^{\ell} \mathrm{e}^{2\pi i(-p_j^t q_k)} \varphi(z;p_j) = \ell C_k \theta(z;\hat{c}+q_k^t).$$

Summing these equalities over all $k = 1,\ldots,\ell$, we have (reversing summation order)

$$\sum_{j=1}^{\ell} \left\{ \sum_{k=1}^{\ell} \mathrm{e}^{2\pi i(-p_j^t q_k)} \right\} \varphi(z;p_j) = \sum_{k=1}^{\ell} \ell C_k \theta(z;\hat{c}+q_k^t). \tag{11.39}$$

Calculation shows that the term $\{\cdots\}$ in (11.39) equals ℓ for $j=1$ and vanishes for $j>1$ [189, p. 196]. As a result, the left-hand side of (11.39) reduces to the $j=1$ term, and (11.39) becomes

$$\ell\varphi(z) = \ell\varphi(z;p_1) = \ell \sum_{k=1}^{\ell} C_k \theta(z;\hat{c}+q_k^t). \tag{11.40}$$

Equation (11.40) shows that the Jacobian function φ of type (Ω,H,c) is a linear combination of ℓ Weierstrass theta functions, all of which have the same Frobenian primary and secondary periods $\hat{\Omega}$ and $\hat{\mathrm{H}}$ but different parameters $\hat{c}+q_k^t$. Although the functions $\varphi_k(z) \stackrel{\text{def}}{=} \theta(z;\hat{c}+q_k^t)$ were obtained by a line of reasoning starting from a given φ, they are independent of the choice of φ, since they depend only on Ω,H,G and constructs related to them, viz., $\hat{\Omega},\hat{\mathrm{H}},\hat{c},p_k,n_k,q_k$. Since φ_k is a Jacobian function of type $(\hat{\Omega},\hat{\mathrm{H}},\hat{c}+q_k^t)$, by Lemma 11.8 with $\Omega_a=\hat{\Omega},\mathrm{H}_a=\hat{H},c_a=\hat{c}+q_k^t$, it is also a Jacobian function of type (Ω,H,c_b), where from the equation for c_b at (11.34) together with the formula (11.37) defining $\hat{c}$, we have $c_b=c+q_k^t G=c+n_k^t$, the last equality due to (11.38). Since n_k^t is integral, the types $(\Omega,\mathrm{H},c+n_k^t)$ and (Ω,H,c) have identical equations of quasiperiodicity, and so are identical.[18] In this way, Frobenius showed that any Jacobian function of type (Ω,H,c) is a linear combination of the ℓ Jacobian functions φ_k, which are all of type (Ω,H,c). The linear independence of $\varphi_1,\ldots,\varphi_\ell$ was shown by Schottky's argument: The above reasoning can be applied to $\varphi(z)\equiv 0$ (even though it is not a proper Jacobian function), and it shows that the C_k are all 0 in that case. Frobenius' basis functions φ_k are analogous to Weierstrass' functions θ^r. Both are theta functions of a type $\hat{\Omega},\hat{\mathrm{H}}$, where $\hat{\Omega}=\Omega G^{-1}$, $\hat{\mathrm{H}}G^{-1}$, and G is such that $K=G^t(-J)G$. (In the case of the θ^r, G is given just before Lemma 11.8.)

In this manner Frobenius proved Theorem 11.10. Two months later, he submitted a "Part II" to his paper on Jacobian functions [190]. It was inspired by an 1883

[18] If L'_α and L_α denote the respective linear functions (11.23) in the equations of quasiperiodicity of the two types, then $c+n_k^t \equiv c \bmod \mathbb{Z}$ means $L'_\alpha \equiv L_\alpha \bmod \mathbb{Z}$, and so $\mathrm{e}^{2\pi i L'_\alpha} = \mathrm{e}^{2\pi i L_\alpha}$.

paper by Kronecker (on complex multiplication in the elliptic case $g = 1$) and led Frobenius to a new proof of Theorem 11.10 that, as he emphasized, did not depend very much on arithmetic linear algebra [190, pp. 205–206]. It seems clear, however, that it was the link with his work on arithmetic linear algebra that originally helped inspire his theory of Jacobian functions.

11.3.4 An application of Theorem 11.10

Frobenius illustrated the value of his version of the Weierstrass–Schottky theorem for his theory of Jacobian functions by making an application of it that was quite different from the sort made by Weierstrass and Schottky. He observed that if $\varphi_1,\ldots,\varphi_r$ are all Jacobian functions of a common type (Ω,H,c) and order ℓ, then it is easily seen from the common quasiperiodic equations satisfied by these functions that the product $\varphi^* = \prod_{j=1}^r (\varphi_j)^{n_j}$ will be a Jacobian function of type $(\Omega, n\mathrm{H},\ldots)$, where $n = \sum_{j=1}^r n_j$. Since the associated Frobenius form (11.22) is $K^* = nK$, $\ell^* = \sqrt{\det K^*} = \sqrt{n^{2g}\ell^2}$, and so φ^* has order $\ell^* = n^g\ell$. With this in mind, Frobenius considered the general homogeneous polynomial in variables $u_1,\ldots,u_r$ of degree n with undetermined coefficients. I will denote this polynomial by $H(u_1,\ldots,u_r)$. It has $\binom{n+r-1}{r-1}$ terms. Thus $H(\varphi_1,\ldots,\varphi_r)$ is a linear combination of $N = \binom{n+r-1}{r-1}$ Jacobian functions φ^* of the same type and with order $\ell^* = n^g\ell$. If $N > \ell^*$, then by Theorem 11.10, these N functions must be linearly dependent, and so values that are not all zero may be assigned to the coefficients of H, so that $H(\varphi_1,\ldots,\varphi_r) \equiv 0$. The hypothetical inequality $N > \ell^*$ simplifies to

$$\left(1+\frac{1}{n}\right)\cdots\left(1+\frac{r-1}{n}\right) > (r-1)!\ell n^{g+1-r}.$$

As $n \to \infty$, the left-hand side approaches the limit 1, while for $r > g+1$, the right-hand side approaches the limit 0. This means that when $r = g+2$, sufficiently large values of n exist such that the above inequality holds, and so Frobenius had proved the following result:

Theorem 11.11 (Frobenius). *Any $g+2$ Jacobian functions of the same type satisfy a homogeneous polynomial equation.*

Frobenius referred to the above theorem as "the theorem posited by Mr. Weierstrass (Berlin *Monatsberichte* 1869)" [189, p. 197]. This is the paper mentioned in Section 11.2 in which Weierstrass called attention to the fact that abelian functions exist that are not the special abelian functions generated by solving the Jacobi inversion problem for abelian integrals. There, Weierstrass stated (without proof) the following theorem [589, p. 46].

Theorem 11.12 (Weierstrass). *If $f(z)$ is any abelian function with period matrix Ω, then g additional abelian functions $f_1,\ldots,f_g$ that have the same period matrix*

Ω and are functionally independent can be determined such that the $g+1$ functions $f, f_1, \ldots, f_g$ are algebraically dependent, i.e., satisfy a polynomial equation $P(f(z), f_1(z), \ldots, f_g(z)) \equiv 0$.

This was one of several theorems about general abelian functions that Weierstrass established with the ultimate goal of showing that such a function (like the special abelian functions) is expressible in terms of an ordinary theta function. In the 1869 paper, the functions f_j were the partial derivatives $f_j = \partial f/\partial z_j$, but in a brief paper in *Crelle's Journal* in 1880 [592], Weierstrass showed that there were many other ways to determine $f_1, \ldots, f_g$.

What could Frobenius have meant when he identified his Theorem 11.11 with Weierstrass' theorem? The only straightforward answer to this question seems to involve the following considerations. Let $f(z)$ be an abelian function with period matrix Ω. Then if $f(z) = \varphi_1(z)/\varphi_0(z)$, where φ_1, φ_0 are Jacobian functions of the same type (Ω, H, c), and if $\varphi_2, \ldots, \varphi_{g+1}$ are g more Jacobian functions of this type, then by Theorem 11.11, there is a homogeneous polynomial in $g+2$ variables, $H(u_0, u_1, \ldots, u_{g+1})$, such that $H(\varphi_0, \varphi_1, \ldots, \varphi_{g+1}) \equiv 0$. If H has degree n, then

$$H(\varphi_0, \varphi_1, \ldots, \varphi_{g+1}) = \varphi_0^n H(1, \varphi_1/\varphi_0, \ldots, \varphi_{g+1}/\varphi_0).$$

Thus if $P(u_1, \ldots, u_{g+1}) = H(1, u_1, \ldots, u_{g+1})$, then $P(f, f_1, \ldots, f_g) \equiv 0$, where $f_k = \varphi_k/\varphi_0$ is abelian with period matrix Ω. This conclusion is reminiscent of Weierstrass' stated theorem, but of course requires that the given function f be a quotient of Jacobian functions of the same type. In fact, if it is assumed that every abelian function with period matrix Ω is expressible as the quotient of two Jacobian functions, then the same sort of application of Frobenius' Theorem 11.11 shows that any $g+1$ abelian functions $f_1, \ldots, f_{g+1}$ must satisfy a polynomial equation $P(f_1, \ldots, f_{g+1}) \equiv 0$ and so are algebraically dependent.[19]

It would seem that Frobenius' identification of his Theorem 11.11 with Weierstrass' Theorem 11.12 from 1869 was based on the assumption that every abelian function is expressible as the quotient of Jacobian functions. As we shall see in the following section, in the 1890s, this assumption was proved to be true by Appell (for $g = 2$ variables) and by Poincaré (for any number of variables); but what could Frobenius have been thinking in 1883 when he wrote his paper? Perhaps something along the following lines.

In his lectures, Weierstrass had proved that the above quotient representation was true for the special abelian functions that arise from the Jacobi inversion process. That is, as we saw in Section 11.2, he had proved that every special abelian function

[19]Let $f_k = \varphi_k/\psi_k$, where φ_k and ψ_k are Jacobian of the same type. Then we also have $f_k = \Phi_k/\Phi_0$, where $\Phi_0 = \prod_{j=1}^{g+1} \psi_j$ and $\Phi_k = \varphi_j \cdot \prod_{j \neq k} \psi_j$. All the functions Φ_k, $k = 0, 1, \ldots, g+1$, are products of Jacobian functions with period matrix Ω, and so are also Jacobian functions with period matrix Ω. In fact, they must all be of the same type, because $f_k = \Phi_k/\Phi_0$ is periodic with Ω as period matrix, and so Φ_k and Φ_0 must be of the same type. Now Frobenius' Theorem 11.11 may be applied to $\Phi_0, \Phi_1, \ldots, \Phi_{g+1}$ to deduce $P(f_1, \ldots, f_{g+1}) \equiv 0$ in the same manner as above.

f can be expressed as a $\wp$-function associated to an ordinary theta function θ as in (11.16). This means that $f(z) = \varphi_a(z)/\varphi_b(z)$, where $\varphi_a(z) = \prod_{k=1}^{m} \theta(z+a_k)$, $\varphi_b(z)$ is defined similarly with respect to $b_1,\ldots,b_m$, and $\sum_{k=1}^{m} a_k = \sum_{k=1}^{m} b_k$. It is easily seen that φ_a and φ_b are Jacobian functions of the same type $(\Omega, m\mathrm{H},\ldots)$, where θ is of type $(\Omega,\mathrm{H},\ldots)$. However, as we saw in Section 11.2, for a general abelian function f, Weierstrass proved only that f can be expressed as a rational function of $g+1$ $\wp$-functions associated to the same theta function, and this result does not imply that a general abelian function is a quotient of Jacobian functions. During his lifetime, Weierstrass never published a proof or even a precise statement of his theorem. The closest he came was in the above-mentioned paper of 1880 (originally a letter to Borchardt), where he explained that his Theorem 11.12 was part of the research that had enabled him to attain his end goal, namely the theorem that every abelian function in r arguments $z_1,\ldots,z_r$ "can be expressed by means of a Θ-function of r arguments" [592, p. 133]. Given this imprecise statement, it is conceivable that Frobenius had incorrectly interpreted it to mean the same thing it had meant for special abelian functions and so hastily concluded that every abelian function is expressible as the quotient of Jacobian functions. If so, although unjustified at the time, the conclusion turned out to be correct and, as will be seen in the next section, eventually led mathematicians to utilize Frobenius' theory of Jacobian functions as a foundation stone in a systematic development of the theory of abelian functions and varieties that was independent of the theory of abelian integrals.

Before proceeding to the next section, it should be noted that Frobenius' paper contains more than has been indicated here. Much of its contents, including the fundamental Theorem 11.5, was developed within the broader framework of entire functions without infinitely small periods that satisfy the equations of quasiperiodicity (11.21) for $\sigma \leq 2g$ primary and secondary periods, so that Ω and H are $g \times \sigma$. Within that framework, he also developed an algorithm for constructing all matrix pairs (Ω,H) satisfying conditions (A) and (B) of Theorem 11.5. He also proved that when Jacobian functions of some type (Ω,H,c) exist, there is always a Jacobian function of that type for which Ω is a primitive set of periods [189, §9]. As we have seen, this sort of generality and thoroughness was characteristic of Frobenius' mathematics.

11.4 Assimilation into the Mainstream

With the notable exception of Hermite's memoir of 1855 [290], during the 1850s, 1860s and 1870s, the theory of abelian integrals and functions had been advanced primarily by German mathematicians, most notably by Riemann and Weierstrass. However, whereas the advances in the theory of abelian integrals were well documented by Riemann's paper of 1857 solving the Jacobi inversion problem, the same was not true regarding the theory of general abelian functions (as opposed to the theory of the special abelian functions that arise from the

Jacobi inversion problem). Both Riemann and Weierstrass had stated fundamental theorems regarding general abelian functions without providing any proofs.

11.4.1 Developments in France

In France, these claims were taken as challenges by the new generation of talented mathematicians that had emerged by 1880. Their ranks included Henri Poincaré, Paul Appell, and Emile Picard, who in 1880 were 26, 25, and 24 years old, respectively. For example, in 1884, Poincaré published a paper in the *Bulletin* of the Mathematical Society of France [478], in which he proved generalized versions of two theorems attributed to Weierstrass in an 1874 paper by Sofia Kovalevskaya. As Poincaré pointed out [478, p. 125], the generalizations were probably known to Weierstrass, but the point was that he had never published proofs of the theorems even in the communicated form; they had simply been informally stated in letters to students. The theorems need not concern us here, except to say that they had to do with the properties of the ordinary theta functions associated to a system of abelian integrals with certain singular properties, which are then reflected in the period matrices of the theta functions. In 1886, Poincaré presented these results, along with some related supplementary material, in the pages of the *American Journal of Mathematics* [480].

Included in the supplementary material was a discussion of what Poincaré called "intermediary functions" [480, §IV].[20] These were by definition entire functions $\Phi(z)$ of g complex variables that satisfied quasiperiodicity equations differing from Frobenius' quasiperiodicity equations (11.21) only in notation. Poincaré, who was unaware of Frobenius' 1884 papers on Jacobian functions [189, 190], did not exclude the possibility of infinitely small periods, because he wished to include the functions $\Phi(z) = \mathrm{e}^{P(z)}$, where $P(z)$ is some second-degree polynomial in the variables z_j, an example of what Frobenius had called a Jacobian function of order zero. Poincaré, like Frobenius, was apparently led to consider the notion of an intermediary function as a natural generalization of that of an ordinary theta function; and, like Frobenius, he realized the implication of the identity $\Phi([z+\omega_j]+\omega_k) = \Phi([z+\omega_k]+\omega_j)$, namely that, in the notation of the previous section, the skew-symmetric matrix $K = \Omega^t \mathrm{H} - \mathrm{H}^t \Omega$ has integer coefficients. As we have seen, this was the starting point of Frobenius' theory of Jacobian functions. Here, then, is another example of a multiple discovery involving Frobenius. Let us now consider to what extent and in what directions Poincaré proceeded from the same starting point.

Immediately after defining an intermediary function, Poincaré restricted his attention to the case of $g = 2$ variables "in order to fix the ideas" [480, p. 349]

[20]Poincaré took the term *fonctions intermédiaires* from Briot and Bouquet, who used it their book on elliptic functions [44, p. 236].

and further assumed that the period matrix Ω of the intermediary functions Φ to be considered is normalized, meaning that $\Omega J \Omega^t = 0$. It is easy to see that under these assumptions, Jacobian functions exist with

$$K = \begin{pmatrix} 0 & D \\ -D & 0 \end{pmatrix}, \quad D = \begin{pmatrix} e_1 & 0 \\ 0 & e_2 \end{pmatrix}$$

for any choice of e_1, e_2 with $e_1 \mid e_2$, i.e., with K equal to any nonsingular Frobenian skew-normal form (8.18).[21] However, because he introduced intermediary functions to deal with his generalizations of the theorems of Weierstrass, Poincaré ruled out pairs (Ω, H) that did not correspond to the context of Weierstrass' theorems. For the pairs (Ω, H) not excluded, it followed (in the above notation) that $D = mI_2$, i.e., $K = mJ$ [480, pp. 350–351]. By imposing a further normalizing condition on Ω, Poincaré arranged that m must be positive. The intermediary functions with $K = mJ$ were called *intermediary functions of order m* by Poincaré. (The Frobenian order is $\ell = m^2$.) Poincaré's intermediary functions of order m are thus a very special type of Jacobian function, namely Schottky's theta functions of order m (Section 11.1).

Thus although Poincaré had introduced the equivalent of the notion of a Jacobian function and had discovered the integrality property of the Frobenius form K of (11.22), the property that had proved inspirational to Frobenius, Poincaré's preoccupation with the task of proving a version of Weierstrass' theorem focused his attention narrowly on a very special type of Jacobian function. As a result, we find in Poincaré's lengthy discussion of intermediary functions [480, pp. 352–362] no theorems about intermediary functions in general, such as Frobenius' Theorems 11.5 and 11.11. However, as we shall now see, Poincaré's interest in proving another theorem stated without proof—this one due to Riemann—ultimately led to the discovery that more general intermediary functions play an important role in the representation of abelian functions in terms of ordinary theta functions.

In the spring of 1860, Riemann had spent a month in Paris, where he was warmly received by the leading mathematicians, including Hermite, to whom he communicated some of his discoveries regarding general abelian functions. Although Riemann never published these discoveries, Hermite included a description of them in—of all places!—his supplementary note to the 1862 edition of Lacroix's treatise on the calculus [292, pp. 388–394]. According to Hermite, Riemann had explained that if $f(z)$ is any abelian function, then its period matrix is not entirely arbitrary but subject to constraints. That is, for a period matrix $\Omega = \begin{pmatrix} \Omega_1 & \Omega_2 \end{pmatrix}$ for f that is "suitably chosen" [292, p. 388],[22] Ω is subject to conditions that are best indicated by considering $g(z) = f(\Omega_1 z)$, which has period matrix $\hat{\Omega} =$

---

[21] Let $\Omega = \begin{pmatrix} I_2 & iI_2 \end{pmatrix}$ and $H = \begin{pmatrix} 0 & D \end{pmatrix}$. Then K is as above, and Frobenius' Theorem 11.5 implies the existence of Jacobian functions for this choice of Ω and H.

[22] The meaning of "suitably chosen," which was not clarified by Hermite, is discussed below in the context of Wirtinger's normalized period matrix.

$\Omega_1^{-1}\Omega = (I_g \quad \Omega_1^{-1}\Omega_2)$. The remarkable condition was that $T = \Omega_1^{-1}\Omega_2$ must be symmetric. As we have seen (Section 11.2), this symmetry condition is equivalent to condition (I) of Weierstrass' (unpublished) Theorem 11.3 with $L = J$, viz., $\Omega J \Omega^t = 0$. Although Hermite focused on the fact that $T = \Omega_1^{-1}\Omega_2$ is symmetric, judging by Riemann's 1857 paper [495, §17], Riemann certainly realized that $\operatorname{Im} T$ is positive definite, the equivalent of (II) of Weierstrass' Theorem 11.3 with $L = J$, viz., $i\Omega J \Omega^h \succ 0$.

The assumption that $\operatorname{Im} T$ is positive definite is in fact evidently necessary for what Hermite next reported about his conversation with Riemann [292, p. 392]. Fix a positive integer k, and consider the series

$$\Phi(z) = \sum_{m \in \mathbb{Z}^g} a_m \exp\left[2\pi i(m^t z) + \frac{\pi i}{k}(m^t T m)\right],$$

where the coefficients a_m are chosen such that if $m' \equiv m \pmod{k}$, then $a_{m'} = a_m$. Thus there are k^g ways to choose the coefficients, and so k^g functions Φ. For the series to converge (as Hermite took for granted), $\operatorname{Im} T$ must be positive definite. Hermite pointed out that Φ satisfies quasiperiodicity equations that make it what Frobenius later called a Jacobian function of type $(\hat{\Omega}, \hat{\mathrm{H}}, \hat{c})$, where $\hat{\Omega} = (I_g \quad T)$, $\hat{\mathrm{H}} = (0 \quad -kT)$, and $\hat{c}$ depends only on k and T and so is independent of the choice of the coefficients a_m in the definition of Φ. As Hermite pointed out, this means that if Φ and Φ' are two such functions corresponding to different choices of coefficients, then (assuming that $\Phi' \neq C\Phi$) $f = \Phi'/\Phi$ will be an abelian function with period matrix $\hat{\Omega}$, from which an abelian function with period matrix $\Omega = (\Omega_1 \quad \Omega_2)$ is obtained by a linear variable change. Presumably it was in this manner that Riemann showed Hermite that his conditions—$T = \Omega_1^{-1}\Omega_2$ is symmetric and $\operatorname{Im} T$ is positive definite—are also sufficient for the existence of abelian functions with period matrix $\Omega = (\Omega_1 \quad \Omega_2)$.

What Riemann apparently realized in 1860 when he met Hermite (as indicated above) will be summarized in the following theorem.

Theorem 11.13 (Riemann). *(a) If f is an abelian function, then there is a period matrix $\Omega = (\Omega_1 \quad \Omega_2)$ for f with the following properties: (i) $T = \Omega_1^{-1}\Omega_2$ is symmetric and (ii) $\operatorname{Im} T$ is positive definite. (b) If $\Omega = (\Omega_1 \quad \Omega_2)$ is $g \times 2g$ with columns linearly independent over $\mathbb{R}$, then abelian functions with period matrix Ω exist if Ω satisfies (i) and (ii) of part (a).*

It would seem that Riemann did not indicate to Hermite how part (a) is to be proved, since in 1883 Hermite communicated to the Paris Academy a note by Poincaré and Picard [474] filling this gap.

To establish (a), Poincaré and Picard proceeded as follows. Given an abelian function f with period matrix Ω, they invoked Weierstrass' Theorem 11.12 from 1869, which posits the existence of functionally independent abelian functions $f_1, \ldots, f_g$ such that $f, f_1, \ldots, f_g$ satisfy a polynomial equation. This enabled them to construct a Riemann surface on which they could proceed much as Riemann had in his 1857 paper. (See the discussion of Riemann's work surrounding (11.20).) In this way they determined that if ϖ_j is the jth row of Ω (and so represents the period system of an integral on the Riemann surface), then integers $c_{jk} = -c_{kj}$ exist such that $C = (c_{jk})$ satisfies $\varpi_j C \varpi_k^t = 0$, which is equivalent to (I) $\Omega C \Omega^t = 0$. To show that $\det C \neq 0$, they used the fact, again following Riemann's lead, that $i\varpi_j C \varpi_k^h > 0$. Thus, although they did not emphasize it, they had in effect proved that (II) $i\Omega C \Omega^h \succ 0$. Finally, to deduce Riemann's symmetry condition, they observed (presumably by consideration of integrals on the Riemann surface)[23] that a nonsingular integral matrix M may be determined such that the period matrix $\hat{\Omega} = \Omega M$ satisfies $\hat{\Omega} J \hat{\Omega}^t = 0$, which is equivalent to $T = \hat{\Omega}_1^{-1}\hat{\Omega}_2$ being symmetric.

Thus en route to proving part (a) of Riemann's Theorem 11.13, Picard and Poincaré had unwittingly rediscovered Weierstrass' Theorem 11.3 (with $C = L$): In order that abelian functions with period matrix Ω exist, it is necessary that Ω satisfy the following conditions:

$$\begin{array}{l} \textit{An integral, skew-symmetric nonsingular matrix } L \textit{ exists} \\ \textit{such that } (I)\ \Omega L \Omega^t = 0 \textit{ and } (II)\ i\Omega L \Omega^h \succ 0. \end{array} \tag{11.41}$$

As we saw in Section 11.2, Weierstrass himself had never published Theorem 11.3 but had communicated it to Hurwitz, who presented it (without any proof) in *Crelle's Journal* at about the same time Picard and Poincaré published their paper. Thus by 1883, thanks to the combined work of Riemann, Weierstrass, Picard, Poincaré, and Hurwitz, the now familiar Riemann–Weierstrass conditions on a period matrix, namely (I) and (II) of (11.41), became generally known, although a complete published proof of the necessity of these conditions was still lacking, because the only published proof was that by Picard and Poincaré, and their proof was based on Weierstrass' unproved Theorem 11.12 from 1869. Also, the question of sufficiency of these conditions had not been addressed.

In 1891, Paul Appell (1855–1930) saw in the Picard–Poincaré paper hope for an affirmative answer to the following question [5, pp. 157–158]: It is known that every elliptic function of a complex variable z is rationally expressible in terms of ordinary theta functions; is this also true for abelian functions of $g > 1$ variables? (This is, of course, the question that Weierstrass had thought about, and resolved, over twenty years earlier, albeit without publishing his result—see Section 11.2.) At first, Appell noted, such a representation might seem doubtful, because it was known that the periods of theta functions are not arbitrary but are subject to relations. But, he continued, Riemann's theorem, proved by Picard and Poincaré in their 1883 paper, showed that the period matrices of abelian functions are also not arbitrary but subject to conditions, thereby making it more plausible that all abelian functions of g variables are rationally expressible in terms of ordinary theta functions. Appell's idea was to use a theorem Poincaré had published the same year as his note with Picard.

---

[23] See the exposition of their proof by Krazer [350, pp. 117–120], where this is the approach taken.

In [477], Poincaré had shown that a well-known theorem due to Weierstrass, viz., that any meromorphic function of one complex variable is expressible as the quotient of two entire functions, could be extended to meromorphic functions of two complex variables. In what follows, I will refer to this highly nontrivial extension as *Poincaré's quotient theorem for meromorphic functions*.[24] Since abelian functions are meromorphic, Appell's idea was to start from Poincaré's above quotient theorem, which meant limiting his considerations to abelian functions of two complex variables. If $f(z)$ is such a function, then by Poincaré's theorem, we have $f(z) = g_1(z)/g_2(z)$, where g_1, g_2 are entire. Since f is also quadruply periodic, Appell was able to use the periodicity of f to show that the above quotient representation could be replaced by $f(z) = \Phi(z)/\Psi(z)$, where Φ, Ψ are also entire but satisfy certain relations with regard to the periods of f. These relations, namely equations (29) and (32) on pages 191 and 195 of [5], respectively, when interpreted in terms of Frobenius' paper [189], say that Φ and Ψ are Jacobian functions of the same type (Ω, H, c).

Appell's functions Φ and Ψ are of course also intermediary functions in the sense of Poincaré, although not necessarily the very special intermediary functions of order m on which Poincaré had focused in 1886. When he wrote his paper, Appell was apparently unfamiliar with the relevant papers of Frobenius and Poincaré. But he realized that Φ and Ψ, by virtue of the quasiperiodicity equations (32) they satisfied, were not far removed from ordinary theta functions. The remainder of his paper was devoted to showing that Φ and Ψ could be expressed in terms of such functions, thereby obtaining the desired representation of $f(z)$ in terms of ordinary theta functions. It must be emphasized that Appell's main theorem was that any abelian function f in two variables is expressible in terms of ordinary theta functions. The above representation $f = \Phi/\Psi$ was simply the means to establish the main theorem and was not regarded as especially significant in its own right.

Neither Poincaré nor Appell remarked upon the possible validity of their respective theorems for functions of any number of variables, but in a dissertation done in consultation with them and published in *Acta Mathematica* in 1895 [108], Pierre Cousin (1867–1933) presented an entirely different proof of Poincaré's quotient theorem for meromorphic functions, valid for any number of variables. Apparently, the papers of Appell and Cousin got Poincaré thinking, and by June 1897, he realized that the ideas behind his 1883 proof of the quotient theorem for meromorphic functions in two variables, which had employed the theory of potential, could be modified so as to directly establish Appell's representation $f = \Phi/\Psi$ [481, p. 71] and that, more importantly, by generalizing the potential theory techniques to any number of variables, he could obtain Appell's representation $f = \Phi/\Psi$ for abelian functions in any number of variables:

[24]The extension from $g = 1$ to $g = 2$ variables involved considerations of an entirely different nature, which, through the work of Cousin mentioned below, had a considerable influence on the development of the theory of functions of several complex variables [89].

Theorem 11.14 (Poincaré quotient theorem for abelian functions). *Any abelian function of g variables is expressible as the quotient of intermediary functions.*

The detailed proof leading to the above theorem was published in *Acta Mathematica* in 1899 [483]. The potential-theoretic results required to set the stage for proving Theorem 11.14 filled more than fifty pages.[25]

Although I will refer to Theorem 11.14 as Poincaré's quotient theorem for abelian functions, it must be emphasized that Poincaré did not single out this result and designate it as a theorem. As with Appell, it was simply a step, admittedly the major one, in the proof of the main theorem, namely that every abelian function can be expressed in terms of the more special nineteenth-century theta functions, i.e., what I have termed ordinary theta functions. Thus in an influential 1902 paper expounding his many contributions to the theory of abelian and theta functions in the pages of *Acta Mathematica* (at the request of its editor, Mittag-Leffler), it was this result that Poincaré presented as a fundamental theorem ([484, p. 486, Theorem B]); it is only in the course of the ensuing proof that the conclusion is reached that the given abelian function is the quotient of two intermediary functions [484, p. 509]. Since intermediary functions could be expressed in terms of ordinary theta functions, Theorem B followed.

11.4.2 The contributions of Wirtinger

Although Frobenius' papers on Jacobian functions still seem to have been unknown to Poincaré in 1899, his g-variable extension of Appell's main theorem by extending his representation $f = \Phi/\Psi$ to g variables, namely Theorem 11.14, would have made it easy for anyone well acquainted with both Poincaré's paper of 1899 and Frobenius' paper [189] on Jacobian functions to see that Frobenius' theory could be used in conjunction with Poincaré's Theorem 11.14 to establish the foundational theorems of the theory of general abelian functions. For example, one such foundational theorem was the theorem that the Riemann–Weierstrass conditions (11.41) were necessary for the existence of abelian functions. As we saw, the proof implicit in the joint 1883 paper by Picard and Poincaré used a theorem of Weierstrass' from 1869 that was still lacking a proof, so that their proof, in turn, was incomplete. A complete proof follows readily from Theorem 11.14 together with Frobenius' Theorem 11.6. For if Ω is such that abelian functions f exist, then by the former theorem, $f = \varphi_1/\varphi_2$, where the φ_i are Jacobian functions of some common type (Ω, H, c), and so by the latter theorem, $L = \operatorname{Adj} K$, $K = \Omega^t \mathrm{H} - \mathrm{H}^t \Omega$, satisfies the Riemann–Weierstrass conditions (I)–(II) of (11.41).

---

[25] In this page count I have excluded the pages devoted to sections IV and V of the paper, which, according to Poincaré [483, p. 164], were not necessary for the end goal of the paper. In 1902, Poincaré sketched shorter proofs that combined the ideas of his original proof with those of Cousin's proof of the g-variable quotient theorem for meromorphic functions [484, pp. 486–509].

The first to point this out in print was the Austrian mathematician Wilhelm Wirtinger (1865–1945), who was an adherent of Felix Klein's Riemann-surface-based approach to complex function theory. In a two-part paper on abelian functions published in 1895–1896 [609, 610], he pointed out that Picard and Poincaré's 1883 proof of Riemann's theorem was incomplete, because besides ignoring certain singular cases, it used Weierstrass' unproved 1869 Theorem 11.12. Wirtinger's paper was written several years before Poincaré published his quotient theorem for abelian functions, but Wirtinger knew Appell's 2-variable version of the theorem, and taking for granted that the theorem could be extended to g variables, he pointed out that the g-variable version (in effect Poincaré's Theorem 11.14) when combined with Frobenius' theorems on Jacobian functions would provide a rigorous proof of the necessity of the Riemann–Weierstrass conditions (I)–(II) for the existence of abelian functions with Ω as period matrix [609, pp. 69–70].

Wirtinger's remark was made in passing, which is probably why he took the liberty of assuming that Appell's proof could be extended to $g > 2$ variables.[26] His paper focused on his discovery that if abelian functions with period matrix Ω exist, then—without invoking Weierstrass' 1869 theorem—it is possible to construct a Riemann surface on which the type of reasoning employed by Picard and Poincaré in their 1883 paper can be employed to show that the Riemann–Weierstrass conditions hold. Once that was accomplished, Wirtinger proceeded to use Riemann surface techniques to establish other basic theorems such as Weierstrass' 1869 theorem.

In this way, Wirtinger became the first to publish a complete proof of the necessity of the Riemann–Weierstrass conditions. He was apparently also the first mathematician to give a proof that these conditions are also sufficient. This he did by invoking Frobenius' theory of Jacobian functions. His brief proof [609, §8, p. 83], which involved a non sequitur, consisted of the following two observations. (1) If Ω is a period matrix that satisfies the Riemann–Weierstrass conditions (11.41), then "Frobenius showed that" Jacobian functions of some type (Ω, H, c) exist. (2) If φ denotes such a Jacobian function then the second logarithmic derivatives of φ, viz., $f_j = \frac{\partial}{\partial z_j}\left[\left[\frac{\partial \varphi}{\partial z_j}\right]/\varphi\right]$, are abelian functions with Ω as period matrix.[27]

The non sequitur in Wirtinger's proof was the claim that Frobenius had showed that if Ω satisfies the conditions (I) and (II) of (11.41), then Jacobian functions of some type (Ω, H, c) exist. Wirtinger was no doubt thinking of that part of Frobenius'

---

[26] As we shall see below, such extensions were first made over forty years later and were not as routine as Wirtinger had imagined.

[27] If $\varphi(z+\omega_\alpha) = \exp 2\pi i[\eta_\alpha \cdot z + b_\alpha]\varphi(z)$, $\alpha = 1,\ldots,2g$, denote the quasiperiodicity equations for φ, then taking logarithms,

$$\log[\varphi(z+\omega_\alpha)] = 2\pi i(\eta_\alpha \cdot z + b_\alpha) + \log[\varphi(z)],$$

and twofold differentiation with respect to z_j eliminates the linear terms, so that $\partial^2\left[\log[\varphi(z+\omega_\alpha)]\right]/\partial z_j^2 = \partial^2\left[\log[\varphi(z)]\right]/\partial z_j^2$. In other words, if $D_j = \frac{\partial}{\partial z_j}$, then the functions $f_j = D_j\left[D_j[\varphi]/\varphi\right]$, $j = 1,\ldots,g$, are abelian with period matrix Ω.

paper containing Theorem 11.7. However, that theorem imposes assumptions in addition to (I) and (II) for the requisite Jacobian functions to exist. Strictly speaking, part (1) of Wirtinger's proof is not correct. However, it is correct to say that the ideas and results in Frobenius' paper can be combined in such a manner as to establish the existence of the requisite Jacobian functions, even though Frobenius did not make this observation. Whether this is what Wirtinger really meant is uncertain, but it is of interest to note that a useful development of one of Frobenius' ideas—Wirtinger's normal form for a period matrix satisfying the Riemann–Weierstrass conditions [609, pp. 95ff.]—can be used to this end. Since this normal form became a standard part of the theory of abelian and theta functions, I will describe it and then show how it can be used to easily establish the theorem posited by Wirtinger, namely, that the Riemann–Weierstrass conditions are sufficient for the existence of Jacobian functions with period matrix Ω and so for the existence of abelian functions with period matrix Ω.

Frobenius had observed [189, pp. 189ff.] that to obtain simple forms for the necessary conditions for Jacobian functions of type $(\Omega, \mathrm{H}, \ldots)$ to exist, it was useful to invoke his symplectic basis theorem (Theorem 8.12) and replace Ω by $\hat{\Omega} = \Omega P$ and H by $\hat{\mathrm{H}} = \mathrm{H}P$, where P is a unimodular matrix such that $P^t K P = J^*$, J^* being the skew-symmetric normal form (8.18) of K, viz.,

$$J^* = \begin{pmatrix} 0 & D \\ -D & 0 \end{pmatrix}, \quad D = \begin{pmatrix} e_1 & & \\ & \ddots & \\ & & e_g \end{pmatrix}, \tag{11.42}$$

where $e_1, e_1, \ldots, e_g, e_g$ are the invariant factors of K, and so $e_i \mid e_{i+1}$ for all $i = 1, \ldots, g$. It then follows that $\hat{K} = \hat{\Omega}^t \hat{\mathrm{H}} - \hat{\mathrm{H}}^t \hat{\Omega} = J^*$, and as a result, many relations appear in a simpler form, such as the conditions (A′) of (11.31).

Wirtinger was familiar with the above observations by Frobenius, but in his paper he added a twist [609, §15]: If a period matrix Ω satisfies the Riemann–Weierstrass conditions (11.41), then Frobenius' idea can be applied to the matrix L of (11.41). He thus replaced Ω by $\hat{\Omega} = \Omega P$, where now the unimodular matrix P is such that $PLP^t = J^*$, i.e., J^* of (11.42) is now the skew-symmetric normal form of L rather than K.[28] Since P is unimodular, Ω is primitive if and only if $\hat{\Omega}$ is. Then conditions (I) and (II) for L imply that $\hat{\Omega} J^* \hat{\Omega}^t = 0$ and that $i(\hat{\Omega} J^* \hat{\Omega}^h) \succ 0$. Now consider the period matrix

$$\tilde{\Omega} = D^{-1} \hat{\Omega}_1^{-1} \hat{\Omega} = \left(D^{-1} \; D^{-1} \hat{\Omega}_1^{-1} \hat{\Omega}_2\right) \stackrel{\text{def}}{=} \left(D^{-1} \; T\right). \tag{11.43}$$

Later, $\tilde{\Omega}$ became known as the normal (or normalized) form of the period matrix Ω [350, pp. 120ff.]; I will refer to it as the *Wirtinger normal form of* Ω *with*

[28] Since e_1 divides all the e_j, $L' = e_1^{-1} L$ is also integral and satisfies (I)–(II). Wirtinger worked with L', i.e., he assumed $e_1 = 1$ in J^* [609, p. 95].

respect to L. Of course, we have changed to a period matrix for $g(z) = f(\hat{\Omega}_1 Dz)$. It follows from $\hat{\Omega}J^*\hat{\Omega}^t = 0$ and $i\hat{\Omega}J^*\hat{\Omega}^h \succ 0$ that $T = D^{-1}\hat{\Omega}_1^{-1}\hat{\Omega}_2$ is symmetric and has positive definite imaginary part.[29] Furthermore, this is a primitive period matrix for $g(z)$ if Ω was a primitive period matrix for $f(z)$.[30] Incidentally, these considerations can be used to specify Riemann's "suitably chosen" period matrix $\tilde{\Omega} = (\tilde{\Omega}_1 \ \tilde{\Omega}_2)$ such that $\tilde{T} = \tilde{\Omega}_1^{-1}\tilde{\Omega}_2$ is symmetric with positive definite imaginary part, as in part (a) of Theorem 11.13.[31]

Wirtinger's normal form (11.43) can be used in conjunction with Frobenius' theorems on Jacobian functions to easily prove the theorem Wirtinger attributed to Frobenius:

Theorem 11.15. *If a period matrix Ω satisfies the Riemann–Weierstrass conditions (11.41), then Jacobian functions of type (Ω, H, c) exist for some* H *and all c.*

To see this, let Ω satisfy (I)–(II) of (11.41) and let $\tilde{\Omega} = (D^{-1} \quad T)$ denote the Wirtinger normal form (11.43) associated to Ω and L. With notation as in (11.42), let $\tilde{\mathrm{H}} = (0 \quad -e_g I_g)$. Then

$$\tilde{K} = \tilde{\Omega}^t\tilde{\mathrm{H}} - \tilde{\mathrm{H}}^t\tilde{\Omega} = \begin{pmatrix} 0 & -e_g D^{-1} \\ e_g D^{-1} & 0 \end{pmatrix}$$

is integral because $e_j \mid e_g$ for all j. Thus condition (A) of Frobenius' Theorem 11.5 is satisfied for $\tilde{\Omega}, \tilde{\mathrm{H}}$. Condition (B) follows from the fact that $\Psi = \operatorname{Im} T \succ 0$.[32] Thus Jacobian functions of types $(\tilde{\Omega}, \tilde{\mathrm{H}}, \tilde{c})$ exist by Frobenius' Theorem 11.5. Using the inverse of the variable change associated to $\hat{\Omega} \to \tilde{\Omega}$, it then follows that Jacobian functions of a type (Ω, H, c) exist,[33] which concludes the proof of Theorem 11.15.

---

[29]If we write $\hat{\Omega} = (\hat{\Omega}_1 \ \hat{\Omega}_2)$, then $\hat{\Omega}J^*\hat{\Omega}^t = 0$ becomes $(\hat{\Omega}_1 D)[-T + T^t](\hat{\Omega}_1 D)^t = 0$ and so $T^t = T$, since $\hat{\Omega}_1 D$ is invertible. Likewise, if $T = \Phi + i\Psi$, then

$$i\hat{\Omega}J^*\hat{\Omega}^h = i(\hat{\Omega}_1 D)[-T + \overline{T}](\hat{\Omega}_1 D)^h = 2(\hat{\Omega}_1 D)\Psi(\hat{\Omega}_1 D)^h,$$

and so $i\hat{\Omega}J^*\hat{\Omega}^h \succ 0$ if and only if $\Psi \succ 0$.

[30]To see this, let $\tilde{\omega}_\alpha = [\tilde{\Omega}]_\alpha$, i.e., $\tilde{\omega}_\alpha$ is the αth column of $\tilde{\Omega}$. Let $\tilde{\omega}$ be any period of $g(z)$. Thus $g(z + \tilde{\omega}) = g(z)$, i.e., $f(\hat{\Omega}_1 Dz + \hat{\Omega}_1 D\tilde{\omega}) = f(\hat{\Omega}_1 Dz)$. This says that $\hat{\Omega}_1 D\tilde{\omega}$ is a period of f, and so assuming that Ω and thus $\hat{\Omega}$ is primitive, it follows that $\hat{\Omega}_1 D\tilde{\omega} = \sum_{j=1}^{2g} n_j\hat{\omega}_j$, so $\tilde{\omega} = \sum_{j=1}^{2g} n_j D^{-1}\hat{\Omega}_1^{-1}\hat{\omega}_j = \sum_{j=1}^{2g} n_j\tilde{\omega}_j$, which implies that $\tilde{\Omega}$ of (11.43) is primitive for $g(z)$.

[31]When $J^* = J$, the choice is Wirtinger's $\hat{\Omega}$. If $J^* \neq J$, then $J^* = M^t JM$, where $M = \begin{pmatrix} D & 0 \\ -I & 0 \end{pmatrix}$ and $\tilde{\Omega} = \hat{\Omega}M$. When $J^* \neq J$, $\det M = \det D > 1$, so M is not unimodular, and consequently $\tilde{\Omega}$ will not be primitive, even though Ω and $\hat{\Omega}$ are.

[32]If $\tilde{\Omega}w = 0$ for $w = (w_1 \quad w_2)^t \neq 0$, then $w = (-DTw_2 \quad w_2)$. A straightforward computation then shows that $i(w^h\tilde{K}w) = 2e_g(w_2^h\Psi w_2) > 0$, since $\Psi \succ 0$.

[33]It follows from routine calculation that if $\tilde{\varphi}$ is a Jacobian function of type $(\tilde{\Omega}, \tilde{\mathrm{H}}, \ldots)$, then for any invertible $g \times g$ matrix M, $\hat{\varphi}(z) = \tilde{\varphi}(Mz)$ is Jacobian of type $(M^{-1}\tilde{\Omega}, M^t\tilde{\mathrm{H}}, \ldots)$. Taking $M = (\hat{\Omega}_1 D)^{-1}$, we obtain functions $\hat{\varphi}$ of type $(\hat{\Omega}, M^t\tilde{\mathrm{H}}, \ldots)$. Since every column ω_α of Ω is an integral

It is perhaps worth mentioning that a trivial modification of the above reasoning directly establishes the sufficiency of the Riemann–Weierstrass conditions (I)–(II) for the existence of abelian functions, i.e., without the need to introduce Wirtinger's second logarithmic derivative functions, as in part (2) of his proof. The reason is that according to Frobenius' Theorem 11.10, there are $\ell = \sqrt{\det \tilde{K}}$ linearly independent functions of type $(\tilde{\Omega}, \tilde{\mathrm{H}}, \tilde{c})$. Thus, if $\sqrt{\det \tilde{K}} \geq 2$, two linearly independent Jacobian functions $\tilde{\varphi}_1, \tilde{\varphi}_2$ of type $(\tilde{\Omega}, \tilde{\mathrm{H}}, \tilde{c})$ exist, and so $\tilde{f} = \tilde{\varphi}_1 / \tilde{\varphi}_2$ is an abelian function with period matrix $\tilde{\Omega}$, which means (by the above-mentioned variable change) that abelian functions with period matrix Ω exist. To ensure that $\ell \geq 2$ in the above reasoning, we need only take a function φ of type $(\tilde{\Omega}, \tilde{\mathrm{H}}, \hat{c})$ and consider φ^2, which is a Jacobian function of a type $(\tilde{\Omega}, 2\tilde{\mathrm{H}}, \ldots)$. Since the corresponding Frobenius form is $\tilde{K}' = 2\tilde{K}$, it follows that there are $\ell' = 2^g \ell \geq 2$ linearly independent functions of this type.

Wirtinger's second paper was ostensibly devoted to Frobenius' theory of Jacobian functions. In the opening paragraph he wrote [610, p. 1]:

> The subject of the following investigation is in large part the so-called theory of Jacobian functions. Although the fundamental theorems of this theory have been developed by Frobenius ... in a most general and elegant form, I have taken the liberty, in the initial sections, of deriving anew the most important of these in a manner more akin to the methods used by Hermite in the theory of ordinary theta functions ....

What Wirtinger did was to start with Jacobian functions of general type (Ω, H, c) but then go over to the normal form of Ω as in (11.43), viz., $\tilde{\Omega} = (D^{-1} \quad T)$, and then by means of a variable change and choice of $\tilde{\mathrm{H}} = (0 \quad RI_g)$ for a certain R and $c = 0$, to end up with Jacobian functions satisfying quasiperiodic equations formally analogous to the quasiperiodic equations for theta functions of order R and zero characteristic, viz., (11.12) with $\mu^* = \nu^* = 0$ [610, p. 6]. Thus e.g., the number of linearly independent Jacobian functions of type $(\tilde{\Omega}, \tilde{\mathrm{H}}, 0)$ is given by $R^g / \prod_{j=1}^{g} e_j$ [610, p. 7], which is analogous to the formula for theta functions of order R in the Weierstrass–Schottky theorem (Theorem 11.2) and reduces to it when all $e_j = 1$, i.e., when $J^* = J$. In bringing out the formal analogy with theta functions of order R, Wirtinger had regrettably removed from Frobenius' theory its "most general and elegant form," and it was essentially in this specialized, inelegant form that Frobenius' theory was presented in Krazer's *Lehrbuch der Thetafunktionen* (1903).[34]

11.4.3 *New foundations for the theories of abelian functions and varieties*

We saw that in 1899, Poincaré was still not familiar with Frobenius' theory of Jacobian functions. By 1902, possibly because of Wirtinger's papers, Poincaré

linear combination of those of $\hat{\Omega}$ ($\Omega = \hat{\Omega} P^{-1}$, P unimodular), it follows from calculation of $\hat{\varphi}(z + \omega_\alpha)$ that $\hat{\varphi}$ is also Jacobian of some type $(\Omega, \mathrm{H}, \ldots)$.

[34] See [350, pp. 126ff.]. Krazer allowed parameters $c \neq 0$.

had learned of Frobenius' work. In his above-mentioned comprehensive expository paper for *Acta mathematica*, Poincaré included a section on intermediary functions. As we saw, in 1886 Poincaré had developed that theory in a very special case suited to the application he had in mind. In his 1902 paper, noting that Frobenius had developed a theory of intermediary functions "from a different point of view" [484, p. 510], Poincaré did not reproduce his earlier specialized work on intermediary functions but simply referred his readers to the 1886 paper containing it. Instead, he focused on a geometric–analytic interpretation of the integers k_{ij} of the skew-symmetric matrix $K = \Omega^t\mathrm{H} - \mathrm{H}^t\Omega$ and its relation to Frobenius' Theorem 11.5 giving the necessary and sufficient conditions (A) and (B) on K for Jacobian functions to exist.

Poincaré began by observing that any two periods constituting Ω, say ω_α and ω_β, form a parallelogram $\Pi_{\alpha\beta}$ of points $z = t_1\omega_\alpha + t_2\omega_\beta$, $0 \leq t_j \leq 1$, $j = 1,2$. If $\gamma_{\alpha\beta}$ denotes the boundary of $\Pi_{\alpha\beta}$ and if φ is an intermediary (i.e., Jacobian) function of type $(\Omega, \mathrm{H}, \ldots)$, then, he showed, $\int_{\gamma_{\alpha\beta}} d\log\varphi = 2\pi i k_{\alpha\beta}$ [484, p. 512]. Since the integrand is a total differential, this integral, and so $k_{\alpha\beta}$, will be zero unless φ has a zero in the interior of the parallelogram, so that $\log\varphi$ has a singularity there. This led Poincaré to study the intersection of the variety of zeros of φ—an important geometric object of study—with the interiors of the parallelograms $\Pi_{\alpha\beta}$. All this was carried out by means of geometrically informed analytic considerations. As a corollary to these considerations, he obtained, as he noted, a proof that Frobenius' condition (B) of Theorem 11.5 is satisfied: when intermediary functions φ exist, $i(\overline{w}^t Kw) > 0$ for all $w \neq 0$ such that $\Omega w = 0$ [484, p. 519]. Poincaré emphasized that his proof of the necessity of condition (B) was quite different from Frobenius'; and he evidently preferred it to the latter's nongeometric one, for he added that he believed that "the preceding considerations are of a nature better suited to understanding the significance of the numbers M_{kj} [$= 2\pi i k_{kj}$] and the relations of these numbers to the distribution of the zeros of the [intermediary] functions θ" [484, p. 519].

The above-described section on intermediary functions immediately followed the section with Poincaré's second proof of Theorem B, which had ended with the sentence, "Our function F is thus the quotient of two intermediary functions" [484, p. 509]. This was a perfect lead-in to the section on intermediary functions and suggests a good reason why Poincaré focused on Frobenius' work on intermediary functions rather than his own, namely because abelian functions are quotients of intermediary functions, and Frobenius had been concerned with various necessary and sufficient conditions for the existence of intermediary functions for which Theorem 11.5 was the foundation stone. There is little doubt in my mind that Poincaré, like Wirtinger, realized that his quotient theorem for abelian functions could be combined with Frobenius' results to obtain necessary and sufficient conditions for the existence of abelian functions, as well as other fundamental theorems about abelian functions. Nonetheless, he did not see fit to make this point, and the reason seems fairly clear. We have seen that Frobenius' essentially algebraic development of his theory did not appeal to Poincaré. He had simply paused in the exposition of his own work to suggest what he regarded as a more enlightening analytic-geometric approach to Frobenius' theory.

The 1902 volume of *Acta Mathematica* celebrated the 100th anniversary of the birth of Abel, which is why Poincaré had been asked by Mittag-Leffler to contribute an article expounding his work on abelian functions. An article was requested from Wirtinger for the same reason.[35] In it, he began with a review of his earlier work of 1895 (discussed above), in which he had given the first complete proof of the necessity of the Riemann–Weierstrass conditions on a period matrix Ω in order that abelian functions exist. Recall that at that time, he had pointed to Appell's quotient theorem for abelian functions of two variables, if extended to any number of variables, as the basis for another complete proof when combined with Frobenius' theory of Jacobian functions. Realizing that Poincaré had since established the requisite extension of Appell's theorem, he now wrote that "the proofs [of the necessity of the Riemann–Weierstrass conditions] of Appell for two variables and the later one by Poincaré are based in essence on the theory of single-valued functions of several [complex] variables. I myself sought to achieve the proof from the outset by giving precedence to connections with the theory of abelian integrals ... [611, pp. 134–135]. Thus Wirtinger, like Weierstrass, whose lectures on abelian functions and integrals also appeared in 1902, preferred to base the general theory of abelian functions on the theory that had spawned it—the theory of abelian integrals—rather than on more general principles of multivariable complex function theory. By opting to utilize the well-developed theory of abelian integrals, he was able to arrive more quickly at the desired result. Although Wirtinger's attitude might now seem backward-looking, it should be kept in mind that Poincaré's proof of the g-variable version of Appell's theorem was long and complicated. Furthermore, Wirtinger continued to see a role for Frobenius' theory. He again pointed out that the sufficiency of the Riemann–Weierstrass conditions "follows from the investigations of Frobenius on general Jacobian functions" [611, p. 135].

By 1902, the main problems of the theory of abelian functions had been solved. Not only Riemann's solution to the Jacobi inversion problem, but now also Weierstrass', had been published [594]. In addition, the theory of general (as opposed to special) abelian functions had been worked out and the fundamental theorems discussed and established in a variety of ways. Indicative of the mature state of the theory is the appearance in 1903 of Adolf Krazer's *Lehrbuch der Thetafunktionen* [350], which represented the first comprehensive treatise on the theory of general abelian functions. As the title suggests, Krazer's approach was to use the theory of theta functions in the sense of Weierstrass and Schottky (Section 11.1) to build up the theory of abelian functions. The result was a tedious, unmotivated development of the theory that made no use of Frobenius' theory

[35] Frobenius also contributed a paper [221]. By this time, his interests were focused on finite groups and their representations, and he contributed a theorem about the solvability of certain groups of order $p^a q^b$ that was eventually extended to all such groups by Burnside (see following Theorem 15.4).

of Jacobian functions. In particular, the necessity of the Riemann–Weierstrass conditions (11.41) for the existence of abelian functions with period matrix Ω was established using a proof based on the one given by Picard and Poincaré in 1883 (and discussed above).

With its main problems solved, the theory of abelian functions began to lose its status as a principal area of mathematical research. In his lectures on the development of mathematics circa 1914, Felix Klein (1849–1925) wrote with apparent regret that "When I was a student, abelian functions were, as an effect of the Jacobian tradition, considered the uncontested summit of mathematics, and each of us was ambitious to make progress in this field. And now? The younger generation hardly knows abelian functions."[36] Although Klein's remarks may have reflected the situation in Germany, they are somewhat misleading if one considers what was transpiring in France and Italy. There, research activity involving abelian functions was sustained by a growing interest in the geometric objects naturally associated with abelian functions, namely abelian varieties.

Some idea of the interest in abelian varieties among French and Italian mathematicians can be seen from the work on complex multiplications associated to abelian varieties described in Section 10.7. As we saw there, one of the principal figures was Solomon Lefschetz, who worked in mathematical isolation in Kansas but with an extensive knowledge of the work related to abelian varieties being done in France and Italy. In 1919, he was awarded the *Prix Bordin* of the Paris Academy of Sciences for his essay containing pioneering applications of algebraic topology to the study of algebraic and abelian varieties. This essay was published in 1921 [408], and some of its contents have already been discussed in Section 10.7. There, we saw that Lefschetz had a great appreciation for Frobenius' work on matrix algebra and complex multiplication (both of which Krazer had discussed at length in his book). Frobenius' theory of Jacobian functions was, however, not known to him. He had not even read Poincaré's proof of the necessity of Frobenius' condition (B) [408, p. 91n]. But he knew about intermediary functions, and in his prize-winning essay he established what he regarded as a new and important theorem: if φ is an intermediary function of some type $(\Omega, \mathrm{H}, \ldots)$ and if $K = \Omega^t\mathrm{H} - \mathrm{H}^t\Omega$, then $C = \pm K^{-1}$ satisfies (I) $\Omega C \Omega^t = 0$ and (II) $i\Omega C \Omega^h \succ 0$ [408, p. 97], so that Ω is a Riemann matrix in the sense of Scorza (Section 10.7). He did not realize that Frobenius had already proved this result by showing that $L = \mathrm{Adj}\, K = \pm(\det K)C$ satisfies (I)–(II) (Theorem 11.6).

Before Lefschetz's memoir appeared, his oversight was pointed out by the Italian geometer Guido Castelnuovo (1865–1952), who was familiar with Frobenius' work on Jacobian functions. Castelnuovo admired the innovative algebraic topological methods Lefschetz was introducing into the study of abelian varieties, and for this reason arranged for Lefschetz's proof to appear in the January 1921 proceedings of

[36] The above quotation is on page 312 of Klein's published lectures [345]. I have followed the translation by Ackerman [347, p. 294].

the *Accademia dei Lincei* in Rome [409], followed by his own proof of a stronger version of Lefschetz's result [67]:

Theorem 11.16 (Frobenius–Castelnuovo). *Let Ω be a $g \times 2g$ matrix with columns linearly independent over $\mathbb{R}$. Then intermediary (viz., Jacobian) functions of some type $(\Omega, H, \ldots)$ exist if and only if Ω is a Riemann matrix, i.e., if and only if an integral skew-symmetric matrix L exists such that (I) $\Omega L \Omega^t = 0$ and (II) $i\Omega L \Omega^h \succ 0$.*

Castelnuovo explained that his approach to the above theorem "was, to tell the truth, already indicated by Frobenius" [67, p. 313]. The "only if" part of his theorem is, as noted, Frobenius' Theorem 11.6, and as we saw in discussing Wirtinger's work, the "if" part follows readily from Wirtinger's normal form for a period matrix in conjunction with Frobenius' Theorem 11.5. Thus in both directions the proof depends on Frobenius' fundamental Theorem 11.5. Castelnuovo's idea was to establish Theorem 11.16 without invoking Frobenius' fundamental Theorem 11.5. In its stead, he used the classical theorem (implicit, e.g., in Weierstrass' theory of theta functions) that theta functions of Frobenian types $(\Omega^*, \mathrm{H}^*, \ldots)$, with $\Omega^* = (I_g \quad T^*)$, $\mathrm{H}^* = (0 \quad I_g)$, and T^* symmetric exist if and only if $\Psi = \operatorname{Im} T \succ 0$.[37]

Thanks to Castelnuovo's observations, Lefschetz's appreciation of Frobenius' work was extended to include his theory of Jacobian functions. This appreciation is reflected in his 1924 book *L'analysis situs et la géométrie algébrique*, where in the chapter on abelian varieties, Frobenius' papers on Jacobian functions are cited, along with Wirtinger's 1895 paper [609], Poincaré's 1902 paper [484], and his own 1921 memoir [408], as the basic references for the chapter [410, Ch. VI, n. 1]. Then in a 1928 report on developments in the theory of abelian varieties, Lefschetz in a sense threw out a challenge, in the following passage [411, p. 354]:

> In our presentation [of basic theorems on intermediary and abelian functions in p variables] we have followed the same lines as most authors. Appell [5] proceeded in a distinctly different manner and while the details are only given for $p = 1, 2$, it is very likely that his method can be extended to any p. Starting directly with a multiply periodic function attached to ... $[\Omega]$ ... he shows that it may be written as the quotient of two intermediary functions with the same periods. In connection with the work of Frobenius [189, 190], Appell's method contains all the elements of a very elegant exposition of the subject.

The challenge then was to extend to g variables Appell's method of proving that abelian functions can be expressed as quotients of intermediary functions, which was quite different from the method Poincaré had introduced to achieve this end (by applying potential theory). Assuming that Lefschetz was aware

---

[37] Assume that Ω, H are chosen such that $K = J^*$, and take $L = -(J^*)^{-1}$. By the sort of reasoning behind the Wirtinger normal period matrix, if intermediary functions with Ω as period matrix exist, then they exist for period matrix $\tilde{\Omega} = D\Omega_1^{-1}\Omega = (D \quad T)$, where $T = D\Omega_1^{-1}\Omega_2$ is symmetric with $\operatorname{Im} T \succ 0$. It is then possible to transform an intermediary function with period matrix $\tilde{\Omega}$ into a theta function of the above-described type, and this process can be reversed, thereby establishing the theorem.

of Poincaré's method,[38] he evidently did not see it as affording the "very elegant exposition" he envisioned by extending Appell's method. Although Lefschetz did not mention it, the prospect of extending Appell's method had been made more feasible by Cousin's 1895 extension to any number of variables of Poincaré's quotient theorem for meromorphic functions, the starting point of Appell's method.

The first mathematician to respond to Lefschetz's challenge was Fabio Conforto (1909–1954). In his lectures on "abelian functions and Riemann matrices" at the University of Rome in 1940–1941, he sought to develop the theory based on two principles: (1) one should begin with the general definition of an abelian function as any meromorphic function of g variables with $2g$ linearly independent periods over $\mathbb{R}$ and directly establish their representation as quotients of intermediary (viz., Jacobian) functions; (2) the fact that the Riemann–Weierstrass conditions (I)–(II) on Ω are necessary and sufficient for the existence of abelian functions with period matrix Ω should follow as a natural consequence of the theorem that these conditions are necessary and sufficient for the existence of Jacobian functions with Ω as period matrix [104, p. vii]. The latter theorem is of course the Frobenius–Castelnuovo Theorem 11.16, which he knew from Castelnuovo's paper, and so for the first time, Frobenius' theory of Jacobian functions became an integral part of a systematic development of the theory of abelian functions.

Conforto's lectures were published in 1942 in a photomechanically reproduced form [103]. How widely they were known is uncertain, but one mathematician who knew about them was Carl Ludwig Siegel (1896–1981), who was Frobenius' "mathematical grandson," having obtained his doctorate at Göttingen in 1920 under the direction of Frobenius' student E. Landau.[39] In his 1948–1949 lectures on functions of several complex variables at the Institute for Advanced Study in Princeton, Siegel developed the theory of abelian functions along the lines laid down by Conforto [531, Ch. I–IX].[40] Siegel's lectures were published in a photomechanically reproduced form in 1949. In 1956, Conforto's friend the Austrian mathematician Wolfgang Gröbner (1899–1980) published with Springer-Verlag a revised version of Conforto's lectures as *Abelsche Funktionen und algebraische Geometrie* [104], and in 1966, Siegel published a revised version of his lectures [532], with an English translation appearing in 1973 [534]. Through the work of Conforto and Siegel, Frobenius' theory of Jacobian functions had become a basic element in what would now be described as the classical treatment of abelian functions and varieties.

---

[38]In his report, Lefschetz actually never mentioned that Poincaré had extended Appell's theorem, although not Appell's proof method, to any number of variables.

[39]In 1915, as a beginning university student at Berlin, Siegel had attended Frobenius' lectures on number theory and later reminisced about how those lectures had turned him into a mathematician [533]. Siegel's career plans were interrupted by the First World War, and by the time he was able to return to mathematics, Frobenius was dead.

[40]Siegel's "Short Bibliography on Abelian Functions" [531, p. 123] includes Conforto's lectures [103], as well as Frobenius' papers on Jacobian functions [189, 190].

Independently of the work of Conforto and Siegel, Frobenius' theory also became fundamental to the modern approach to abelian varieties thanks to a paper by André Weil (1906–1998). In 1949, Weil presented a *Seminaire Bourbaki* paper on "fundamental theorems in the theory of theta functions (après Poincaré and Frobenius)" [598]. Here he used the term "theta function" to include intermediary or Jacobian functions, a practice that has continued to the present.[41] Although Weil was apparently unfamiliar with the lectures of Conforto (1942) or Siegel (1949), he too had in mind the foundational role that could be played by Frobenius' theory of Jacobian functions in conjunction with the theorem that every abelian function is the quotient of Jacobian functions. Weil was familiar with Poincaré's 1899 paper [483] in which this theorem was proved using potential theory. The most remarkable part of Weil's two-part paper was the first, which dealt with Poincaré's theorem. I have already mentioned the length and complexity of Poincaré's proof. The proofs given by Conforto and by Siegel, although in accordance with Appell's original approach, were themselves quite long ([104, pp. 26–55], [531, pp. 29–54]). Weil showed that by applying ideas underlying recent proofs of de Rham's theorems, a significantly shorter proof of Poincaré's theorem could be obtained.

As Weil later explained in his commentary on [598], once Poincaré's theorem is in hand, "the theory of meromorphic functions on the complex torus can be developed in an essentially algebraic manner; as I saw in 1949, it is just a matter of following Frobenius, about whom I knew nothing before then" [600, pp. 570–571]. Thus the second part of Weil's paper began, "The algebraic study of theta functions has been made in detail by Frobenius in two little-known but very interesting memoirs .... What follows reproduces some of his results, obtained by means of a slightly different viewpoint" [598, p. 417]. Weil evidently found Frobenius' essentially algebraic approach congenial. His "slightly different viewpoint" consisted in a more abstract algebraic formulation with Frobenius' quasiperiodic equation for $\omega = \sum_{\alpha=1}^{2g} n_\alpha \omega_\alpha$ expressed in terms of matrices associated to the real vector space E spanned over $\mathbb{R}$ by a primitive period system $\omega_1, \ldots, \omega_{2g}$. He focused on Frobenius' main results, viz., Theorem 11.5, giving necessary and sufficient conditions (A) and (B) for the existence of Jacobian functions, and Theorem 11.10, specifying the number of linearly independent Jacobian functions of a given type. These results and the attendant proof ideas translated readily into the language and notation of Weil's viewpoint.

Weil concluded his paper by observing that "The majority of known results on abelian functions and varieties (in the 'classical case' where the field of constants is the complex field) can be deduced very easily from the preceding and knowledge of the cohomology ring of the torus" [598, p. 421]. The approach envisioned by Weil has now become preponderant, and so Frobenius' algebraic theory of Jacobian functions, coupled with Poincaré's theorem (with Weil's proof), also lives on as well in the modern treatment of complex abelian varieties.[42]

---

[41] Authors in the classical tradition, such as Conforto, Siegel, and, more recently, Markushevich [433], however, continued to speak of Jacobian functions.

[42] See, e.g., [557, Ch. II], [395, Chs. VI,VIII], [508, §3].

Chapter 12
The Group Determinant Problem

This and the following three chapters are devoted to Frobenius' greatest mathematical achievement, his theory of group characters and representations. The first two chapters consider how he was led to create the basic theory. Then a chapter is devoted to other lines of mathematical thought that led other mathematicians to independently discover at least a part of Frobenius' results—yet another example of multiple discovery involving Frobenius. The fourth chapter discusses further work by Frobenius on the theory and application of representation theory as well as the contributions made by his best student, I. Schur, and by Schur's student R. Brauer.

As with most of Frobenius' mathematical achievements, his theory of group characters and representations was motivated by his efforts to solve a specific mathematical problem, this one on the cusp of two of his favorite subjects at the time (1896), the theory of determinants and the theory of finite groups. The problem, which was posed to Frobenius by Dedekind, had to do with what Dedekind called the group determinant of a finite group $\mathfrak{H}$. It was a notion of Dedekind's own imagination, not something that was well known at the time. Nor is it a familiar notion nowadays. Before commencing a historical discussion of the manner in which the problem emerged and how it led to Frobenius' theory of characters and representations, I will begin with a brief mathematical exposition of the group determinant and how the solution of the associated problem of its factorization relates to modern group representation theory. The reader will as a consequence be better able to appreciate the historical developments traced in the ensuing sections.

Let $\mathfrak{H} = \{E, A, B, C, \ldots\}$ be a finite group of order h, with E denoting the identity element of $\mathfrak{H}$, and associate to each element of $\mathfrak{H}$ an independent variable, $x_E, x_A, x_B, \ldots$. The group determinant of $\mathfrak{H}$ is then defined as follows. Consider the $h \times h$ matrix whose rows and columns are ordered by the given ordering $E, A, B, C, \ldots$ of the elements of $\mathfrak{H}$ and whose (P, Q) entry is the variable $x_{PQ^{-1}}$. Then the group determinant $\Theta(x_A, x_B, x_C, \ldots) = \Theta(x)$ is the determinant of this matrix, i.e.,

T. Hawkins, *The Mathematics of Frobenius in Context*, Sources and Studies in the History of Mathematics and Physical Sciences, DOI 10.1007/978-1-4614-6333-7_12,

$$\Theta = \det \begin{pmatrix} x_E & x_{A^{-1}} & x_{B^{-1}} & x_{C^{-1}} & \cdots \\ x_A & x_E & x_{AB^{-1}} & x_{AC^{-1}} & \cdots \\ x_B & x_{BA^{-1}} & x_E & x_{BC^{-1}} & \cdots \\ x_C & x_{CA^{-1}} & x_{CB^{-1}} & x_E & \cdots \\ \vdots & \vdots & \vdots & \vdots & \vdots \\ \cdots & \cdots & \cdots & \cdots & x_E \end{pmatrix}. \tag{12.1}$$

It is easily seen that Θ is a homogeneous polynomial of degree h in the variables $x_E, x_A, x_B, x_C, \ldots$; and the problem of the group determinant, as formulated by Frobenius (Dedekind had formulated it differently, as we shall see), was to consider the factorization of Θ over $\mathbb{C}$ into irreducible homogeneous polynomial factors, viz., $\Theta = \prod_{\lambda=1}^{k} \Phi_\lambda^{e_\lambda}$. Since Θ was determined by the multiplication table of $\mathfrak{H}$, the problem was to see how the nature of its factorization reflected the structure of $\mathfrak{H}$. In particular, how is k, the number of distinct irreducible factors, related to $\mathfrak{H}$? Also, if $f_\lambda = \deg \Phi_\lambda$, how are e_λ and f_λ related to $\mathfrak{H}$?

If Frobenius' problem is viewed anachronistically in the light of present-day group representation theory, it is not difficult to see that it translates into some of the basic questions posed—and answered—by the modern theory. To begin with, there is a close connection between Θ and the regular representation of $\mathfrak{H}$. Consider the vector space over the field $\mathbb{C}$ of complex numbers that consists of all formal sums $\sum_{R\in\mathfrak{H}} \alpha_R R$, where $\alpha_R \in \mathbb{C}$. Addition and scalar multiplication are defined in the obvious way, and if the same is done for multiplication, we obtain the group algebra $\mathbb{C}\mathfrak{H}$. Let T_R denote the linear transformation that acts on this algebra as left multiplication by $R \in \mathfrak{H}$:

$$T_R\left(\sum_{Q\in\mathfrak{H}} \alpha_Q Q\right) = \sum_{Q\in\mathfrak{H}} \alpha_Q RQ.$$

The representation $R \to T_R$ is the left regular representation of $\mathfrak{H}$. If $\sigma(R)$ denotes the $h \times h$ matrix of T_R with respect to the basis $E, A, B, C, \ldots$, then $(x_{PQ^{-1}}) = \sum_{R\in\mathfrak{H}} x_R \sigma(R)$,[1] and so

$$\Theta = \det\left(\sum_{R\in\mathfrak{H}} x_R \sigma(R)\right).$$

Suppose now that σ decomposes into two representations μ and ν of degrees p and q, respectively. That is, suppose M is a nonsingular $h \times h$ matrix over $\mathbb{C}$ such that for all $R \in \mathfrak{H}$,

$$M\sigma(R)M^{-1} = \begin{pmatrix} \mu(R) & 0 \\ 0 & \nu(R) \end{pmatrix},$$

[1] The matrix $\sigma(R)$ is a permutation matrix. It has a 1 in its (P,Q) entry if and only if $T_R(Q) = P$, i.e., if and only if $RQ = P$. Thus $R = PQ^{-1}$, and so $x_R = x_{PQ^{-1}}$, which means that $\sum x_R \sigma(R) = (x_{PQ^{-1}})$.

where $\mu(R)$ and $\nu(R)$ are $p \times p$ and $q \times q$ matrices, respectively. Then $\Theta(x) = \Phi(x)\Psi(x)$, where

$$\Phi(x) = \det\left(\sum_{R\in\mathfrak{H}} x_R\mu(R)\right) \quad \text{and} \quad \Psi(x) = \det\left(\sum_{R\in\mathfrak{H}} x_R\nu(R)\right)$$

are homogeneous polynomials in the x_R with complex coefficients of degrees p and q respectively. By virtue of the complete reducibility theorem, we know that σ decomposes into a finite number of irreducible representations. Corresponding to this decomposition is a factorization of the group determinant $\Theta(x)$ into a product of factors that are, in fact, irreducible as polynomials over $\mathbb{C}$. Frobenius' problem thus translates into the following. How many distinct irreducible representations are contained in the regular representation, and how are their degrees f_λ and the number e_λ of times they occur related to $\mathfrak{H}$?

This problem, given the time of its formulation, did not seem to have anything to do with matrix representations, i.e., with homomorphisms from $\mathfrak{H}$ to groups of invertible matrices, and indeed, when certain complex-valued functions $\psi(R)$ emerged from Frobenius' study of the factorization of Θ as central to that factorization, it was because of analogies with the special problem of factoring Θ when $\mathfrak{H}$ is abelian, which Dedekind had already solved, that Frobenius tentatively called them "characters" That is, ever since the time of Gauss' *Disquisitiones Arithmeticae* there was a notion of a character in the theory of numbers, a notion connected with some of the deepest results of that field of mathematics, a notion that, at the hands of Dedekind and Weber, had been articulated within the context of a finite abstract abelian group. This evolution of the notion of a character within number theory is the subject of Section 12.2. Indeed, as we shall also see in that section, the idea of the group determinant itself was suggested to Dedekind by analogy with the discriminant in a normal algebraic number field. Frobenius was thus indebted to Dedekind in particular, and to number theory in general, for the problem that led to his generalization of the arithmetic notion of a group character. Although Frobenius himself developed his theory of group characters from the perspective of linear algebra with any eye toward applications to the theory of finite groups, its arithmetic aspects were developed extensively by Issai Schur (Section 15.5), and ultimately the theory Frobenius created paid back its debt to the theory of numbers through the work of Artin (Section 15.6).

Let us now consider the sequence of events that led Frobenius to take on the problem of the group determinant.

12.1 The Fountain of Youth

We saw in Chapter 3 that the death of Kronecker in 1891 enabled Weierstrass to get Frobenius appointed as Kronecker's successor in 1892. A professorship at the University of Berlin brought with it ordinary membership in the Berlin Academy of

Sciences, and on 29 June 1893, Frobenius gave his inaugural address (*Antrittsrede*) to the academy [201]. After acknowledging his debt to his teachers at Berlin—Kummer, Weierstrass, and Kronecker—which oriented his own research in the disciplines they had pursued, namely algebra, arithmetic, analysis and complex function theory, Frobenius spoke in general terms about the orientation of his own mathematical work [201, p. 574]:

> From the outset, the treatment of algebraic problems was especially appealing to me, and time and again I returned by preference to them when a break was needed after taxing analytic work. Both directions in modern algebra, the theory of equations and the theory of forms, equally captivated me. In the latter I was drawn by preference to the theory of determinants, in the former to the theory of groups. The group concept, introduced into mathematics by Gauss and Galois, has in recent years attained a fundamental significance in all branches of our science, and especially in that part of arithmetic for which Kummer's discovery of ideal numbers laid the foundation. Indeed, a large part of the results that we classify under the name of number theory is nothing but a theory of groups of commuting elements.

This description of his research, with its great emphasis on the theory of groups, is rather inaccurate as a description of his past work, because of the 42 papers he had published through 1892, only three had dealt with group theory—his paper with Stickelberger on abelian groups (1879—Section 9.2), and his two papers on Sylow's theorems and double cosets (Section 9.4). Indeed, in the nontechnical overview of his work that followed the above words in the inaugural address, no mention was made of his work on group theory. The above quotation is really a declaration of his growing interest in the still relatively new field of the abstract theory of groups, a field of mathematics of modest status at the time and described above as serving as a respite from the rigors of more traditional analytic work.

After completing the above-mentioned overview with a discussion of his prodigious output in the area of elliptic, abelian, and theta functions during the 1880s and early 1890s, Frobenius made another veiled reference to his current attraction to group theory [201, pp. 575–576]:

> In the theory of theta functions it is easy to set up an arbitrarily large number of relations, but the difficulty begins when it comes to finding a way out of this labyrinth of formulas.
>
> Consideration of that mass of formulas seems to have a withering effect on the mathematical imagination. Many a distinguished researcher who, through tenacious perseverance, has advanced the theory of theta functions in two, three, or four variables, has, after an outstanding demonstration of brilliant analytic talent, grown silent either for a long time or forever. I have attempted to overcome this paralysis of the mathematical creative powers by time and again seeking renewal at the fountain of youth of arithmetic. I hope it will be granted me to draw further results from this inexhaustible source, results that will make me worthy of the honor bestowed on me by my election to the academy [201, pp. 369–370].

In speaking of the "labyrinth of formulas," Frobenius probably had in mind especially the theory of theta functions with integral characteristics. By "arithmetic" he did not mean the theory of numbers per se but rather the theory of abstract finite, not necessarily abelian, groups. He perhaps came to regard abstract group theory as a part of arithmetic because his above-mentioned work on finite abelian groups as well as on Dirichlet densities and Galois groups (Section 9.3) were linked to problems in number theory, as we have seen. Although he had published only two papers on group theory that were independent of number-theoretic considerations before returning to Berlin, namely his papers on Sylow's theorems and double cosets [193, 194], they had clearly whetted his appetite for the subject, and a Berlin professorship gave him the freedom to pursue whatever mathematics he deemed important and interesting and to publish it quickly in the proceedings of the Berlin Academy. From 1893 onward, all but four of his 58 mathematical publications were in the proceedings, and the first, presented at the same session of the academy in which he gave his inaugural address, was on solvable groups [200].

The motivation behind [200] came again from Sylow's important paper of 1872. Sylow had proved that every permutation group of order p^σ is solvable [558, p. 588]. The order of such a group is the product of σ identical primes. As a sort of counterpoint to Sylow's theorem, Frobenius now proved that if a group is a product of σ distinct primes, it is solvable. The proceedings of the academy for 1895 contained three further papers by Frobenius on solvable groups and on other aspects of the theory of abstract finite groups.[2] The "fountain of youth of arithmetic" had become synonymous with "group theory" for Frobenius.

One of the group theory papers of 1895 is especially noteworthy. Entitled simply "On finite groups" [204], it reflects Frobenius' penchant for reworking a subject systematically from a new point of view. Here the new viewpoint was supplied by his concept of a complex. Frobenius considered a (possibly infinite) universal system $\mathfrak{U}$ of elements $A, B, C, \ldots$ on which a binary operation $(P, Q) \to PQ$ is defined and has the following additional properties: (1) the operation is associative; (2) if $PQ = P'Q'$, then $P = P'$ if and only if $Q = Q'$; (3) for every finite subset $\mathfrak{A}$ of $\mathfrak{U}$, if $\mathfrak{A}^*$ denotes the set of all $U \in \mathfrak{U}$ that are (finite) products of elements from $\mathfrak{A}$, then $\mathfrak{A}^*$ is also finite. He noted that these properties on $\mathfrak{U}$ imply the existence of a unique element $E \in \mathfrak{U}$ such that $E^2 = E$ and for every $U \in \mathfrak{U}$, there is a unique element $U^{-1} \in \mathfrak{U}$ satisfying $UU^{-1} = U^{-1}U = E$. Every subset of $\mathfrak{U}$ is then by definition a *complex*. Thus, in particular, $\mathfrak{U}$ is a (possibly infinite) complex. In this paper, Frobenius proposed to limit himself to finite complexes, and all the complexes $\mathfrak{A}, \mathfrak{B}, \mathfrak{C}, \ldots$ occurring in his theorems are assumed to be subsets of the complex $\mathfrak{U}$.

In 1895, this notion of a complex must have appeared very "abstract" in a somewhat pejorative sense to most mathematicians, one notable exception being Dedekind (and his collaborator Weber). Indeed, Frobenius told his readers that he was following Dedekind's notation in writing $\mathfrak{A} = A + B + C + \cdots$ to denote the complex consisting of elements $A, B, C, \ldots$ and likewise $\mathfrak{H} = \mathfrak{A} + \mathfrak{B} + \mathfrak{C} + \cdots$ to denote the union of the complexes $\mathfrak{A}, \mathfrak{B}, \mathfrak{C}, \ldots$. Here Dedekind's symbol $+$ is in essence being used as we would use the set union symbol $\cup$, and this is a reflection of the fact that Dedekind and, following him here, Frobenius, were in effect seeking to provide a set-theoretic foundation for certain arithmetic considerations. Today, we take such a foundation for granted, but in 1895, a systematic set-theoretic approach to mathematics was not at all common practice. Cantor and Dedekind were the

[2]See Frobenius, *Abhandlungen* **2**, 565–573, 632–694.

earliest exponents of this approach, but Cantor's interest in set theory had led him away from mainstream mathematics into the realm of cardinal and ordinal numbers, whereas Dedekind applied set-theoretic ideas to the theory of algebraic numbers—an area of considerable interest to many mathematicians. And now Frobenius was seeking to do the same for the theory of finite groups. Since this interesting paper will be seen below to play an important circumstantial role in the events that led Frobenius to create his theory of group characters and representations, and since it is about as far into the realm of abstraction that Frobenius ever ventured, I will attempt to briefly convey some idea of its contents.

A complex $\mathfrak{B}$ is said to be divisible by a complex $\mathfrak{A}$ if (in modern notation) $\mathfrak{A} \subset \mathfrak{B}$. If $\mathfrak{A} = A_1 + \cdots + A_r$ and $\mathfrak{B} = B_1 + \cdots + B_s$, then $\mathfrak{A}\mathfrak{B}$ denotes the complex of all elements AB with A running through $\mathfrak{A}$ and B running through $\mathfrak{B}$. (This is analogous to Dedekind's definition of the product of two ideals [113, S12, p. 98].) As Frobenius pointed out, this multiplication of complexes satisfies the associative law. If $\mathfrak{B}$ consists of a single element B, then $\mathfrak{A}\mathfrak{B}$ is also written as $\mathfrak{A}B$. A group is then defined as a complex $\mathfrak{G}$ with the property that $\mathfrak{G}$ is divisible by $\mathfrak{G}^2$, whence $\mathfrak{G}^2 = \mathfrak{G}$ follows. Other important examples of complexes are what are now called cosets $\mathfrak{H}G$ and $G\mathfrak{H}$ with respect to a subgroup $\mathfrak{H}$ of a finite group $\mathfrak{G}$. The double cosets $H\mathfrak{G}K$ with respect to subgroups $\mathfrak{H}, \mathfrak{K}$ of a group $\mathfrak{G}$ that Frobenius had introduced in 1880 and presented abstractly in his paper of 1887 [194] are also complexes. The notion of a complex also afforded Frobenius an elegant and strikingly modern way to define the factor group $\mathfrak{G}/\mathfrak{H}$ corresponding to a normal subgroup $\mathfrak{H}$ of a group $\mathfrak{G}$, since for the complexes $\mathfrak{H}A$, $\mathfrak{H}B$, now regarded as elements, a multiplication—that of two complexes—was already in place, which by virtue of the normality of $\mathfrak{H}$ and the associativity of complex multiplication satisfied $(\mathfrak{H}A)(\mathfrak{H}B) = \mathfrak{H}(A\mathfrak{H})B = \mathfrak{H}(\mathfrak{H}A)B = \mathfrak{H}AB$ and defined thereby a group $\mathfrak{G}/\mathfrak{H}$.

Starting with some general theorems about complexes, Frobenius proceeded to establish an incredibly large number of theorems about finite groups, some of which were known and some that were new—all presented from the rather abstract unifying viewpoint of his theory of complexes. Here are a few examples. First some general theorems: (1) Let $\mathfrak{A}$ and $\mathfrak{B}$ denote groups of orders $a = (\mathfrak{A} : 1)$ and $b = (\mathfrak{B} : 1)$, and let $d = (\mathfrak{D} : 1)$, where $\mathfrak{D} = \mathfrak{A} \cap \mathfrak{B}$ is, by definition, the greatest common divisor of $\mathfrak{A}$ and $\mathfrak{B}$. Also define the least common multiple of $\mathfrak{A}$ and $\mathfrak{B}$ to be the smallest group $\mathfrak{C}$ containing $\mathfrak{A} \cup \mathfrak{B}$ and set $c = (\mathfrak{C} : 1)$. Then the complex $\mathfrak{A}\mathfrak{B}$ contains exactly ab/d distinct elements, and it is a group if and only if $\mathfrak{A}\mathfrak{B} = \mathfrak{B}\mathfrak{A}$. In general, $c \geq ab/d$ and $c = ab/d$ if and only if $\mathfrak{A}\mathfrak{B}$ is a group, in which case $\mathfrak{A}\mathfrak{B} = \mathfrak{C}$ [204, p. 635]. (2) Suppose $\mathfrak{A}$ and $\mathfrak{B}$ are groups and $\mathfrak{A}B = B\mathfrak{A}$ for all $B \in \mathfrak{B}$. This implies the weaker condition that $\mathfrak{A}\mathfrak{B} = \mathfrak{B}\mathfrak{A}$, and so by (1), $\mathfrak{A}\mathfrak{B} = \mathfrak{C}$, the least common multiple of $\mathfrak{A}$ and $\mathfrak{B}$. Frobenius' result is that the greatest common divisor $\mathfrak{D} = \mathfrak{A} \cap \mathfrak{B}$ is a normal subgroup of $\mathfrak{B}$ and that the factor groups $\mathfrak{C}/\mathfrak{A}$ and $\mathfrak{B}/\mathfrak{D}$ are isomorphic [204, p. 638]. Here the implicit fact that $\mathfrak{A}$ is normal in $\mathfrak{C}$ follows from the hypothesis, since $C\mathfrak{A} = (AB)\mathfrak{A} = A(B\mathfrak{A}) = A(\mathfrak{A}B) = (A\mathfrak{A})B = (\mathfrak{A}A)B = \mathfrak{A}(AB) = \mathfrak{A}C$. As Frobenius pointed out, the special case of (2) when $\mathfrak{A}$ and $\mathfrak{B}$ are both normal subgroups of $\mathfrak{C}$ had formed the foundation of Hölder's theorem of 1889 on the invariance properties of composition series of finite groups [297].

General results such as (1) and (2) formed the content of the first section of Frobenius' paper. He then used them in the second section to obtain results relating to subgroups of prime power order. As he pointed out in another paper of 1895, Theorems V and VIII of that section were just what was needed to avoid "the language of the theory of permutations" that Sylow had employed to prove his theorem [205, p. 670]. More generally, the results of his paper "On finite groups" [204] enabled Frobenius—as he explained—to view matters relating to Sylow's theorem in a more satisfying manner. In fact, the main theorem of the sequel [205] was a generalization of Sylow's theorem. As noted earlier, in his 1872 paper, Sylow had in effect proved that if the prime power p^{ν} divides the order of a (permutation) group $\mathfrak{H}$, then $\mathfrak{H}$ has a subgroup of order p^{ν}. Sylow also proved that when $\nu = e$ is the largest exponent such that p^{ν} divides $(\mathfrak{H} : 1)$, then the number of subgroups of this order is congruent to $1 \bmod p$, but his proof did not work for exponents $\nu < e$. Using results from "On finite groups," Frobenius proved in [205] that for exponents $\nu < e$ the number of subgroups is still congruent to $1 \bmod p$.

In 1895, Frobenius published a third paper on finite groups, entitled "On solvable groups II" [206]. Using the results of "On finite groups" [204], he was able to generalize theorems and prove conjectures contained in his 1893 paper "On solvable groups" [200] mentioned earlier. At this time, there was a growing interest among some mathematicians (e.g., Hölder, Burnside, Cole) in two problems: determining what groups are solvable and what groups are simple. Solvable groups were of interest in Galois theory, since the polynomials with solvable Galois groups were the ones solvable by radicals. Simple groups were of interest because the factor groups in a composition series of any finite group are simple, and thus the problem of classifying all simple groups was regarded as a significant step toward the classification of all finite groups.

Among the results Frobenius obtained in [206] relating to these problems I will mention two: (1) If $p < q < r$ are primes with $(p,q,r) \neq (2,3,5)$, and if $(\mathfrak{G} : 1) = p^2qr^c$, then $\mathfrak{G}$ is solvable [206, p. 692]; (2) (conjectured by Frobenius in [200]) among all groups whose order is a product of five (not necessarily distinct) primes, there are only three nonabelian groups that are simple, namely the groups of proper linear fractional transformations mod p for $p = 7, 11, 13$. These are now called the projective special linear groups $\mathrm{PSL}_2(p)$. As Frobenius and his contemporaries conceived of them, for a fixed prime p, $\mathrm{PSL}_2(p)$ consists of all linear fractional (or projective) transformations

$$w \equiv \frac{az+b}{cz+d} \ (\mathrm{mod}\, p), \quad ad - bc \equiv 1 \ (\mathrm{mod}\, p), \tag{12.2}$$

where a, b, c, d are integers. It was known that for groups whose order is a product of at most four primes, the only nonabelian simple group is $\mathrm{PSL}_2(5)$, which has order 60 and was known to be isomorphic to the alternating group $\mathfrak{A}_5$ and to Klein's icosahedral group. Thus Frobenius' result (2) was an extension of what was known at the time about nonabelian simple groups.

In January 1895, while Frobenius was readying his above described paper "On finite groups" [204] for presentation to the Berlin Academy, he received a letter from Dedekind dated 19 January 1895.[3] Dedekind had written to Frobenius about Kronecker. As we have seen throughout this book, Frobenius was well acquainted with some of Kronecker's mathematical work and had been influenced by it in many ways. He had also written a thoughtful memorial essay (*Gedächtnisrede*) on Kronecker for the Berlin Academy in 1893 [202]. Dedekind had written to Frobenius to suggest that a letter he had received from Kronecker in 1880 was of sufficient mathematical interest to warrant publication in the proceedings of the Academy.[4] On 24 January, Frobenius responded to Dedekind with a long, friendly letter. Besides expressing his agreement with Dedekind's suggestion, he touched on many matters of common interest: a quarrel Dedekind had with Hilbert, Weierstrass' failing health, the reactions of the Frobenius family to their new surroundings in Berlin, and so on. One passing remark turned out to be especially consequential. As we have seen (Section 9.1.5), in the supplements to Dirichlet's lectures, Dedekind had suggested the importance of an abstract theory of finite groups, embracing thereby both the abelian groups that arose in number theory and the generally nonabelian ones that arose in Galois theory. Since Frobenius was now fully occupied with that theory, it was natural for him to wonder whether Dedekind, who—as Frobenius knew by first-hand experience—left many significant results unpublished, had done some work on the theory of finite groups. And so he wrote, "I am curious what you will say about my work on the theory of groups that I will present next to the Academy. I mean, I know you have concerned yourself with the subject, but I do not know how far you have gone into it." Frobenius was referring to his paper "On finite groups," which, as we saw, was very Dedekind-like in its general, abstract approach through complexes.

Dedekind responded to this display of curiosity on Frobenius' part by writing in his reply (8 February 1895):

> I am very excited about your work on groups, since I was pleased with the simplicity of your methods, among others your proof that in a group whose order is divisible by the prime number p there is always an element of order p;[5] in the first years of my studies on groups (1855–1858), I arrived at it in a much more involved way. Later, I pursued certain questions about groups only in so far as the motivation arose from other quarters; therefore, if it should happen that I at some point already considered the subject of your work, I would certainly not have advanced as far as you. For good measure, let me ask: do hypercomplex numbers with noncommutative multiplication also intrude in your research? But I do not wish to trouble you for an answer, which I will best obtain from your work.

---

[3] The present location of the Dedekind–Frobenius correspondence is given in the preface.

[4] This is the letter containing Kronecker's "Jugendtraum theorem" on abelian extensions of imaginary quadratic fields. See [149, p. 30].

[5] Probably Dedekind was referring to Frobenius' simple (and abstract) proof [193] of the more general result that if p^{ν} divides the order of a group, then it contains a subgroup of order p^{ν}. Frobenius' proof is sketched in Section 9.4.

Before we proceed to Frobenius' response to this cryptic question, we need to understand what Dedekind meant by it.

12.2 Dedekind Characters and Group Determinants

Dedekind's penchant for abstraction, for seeking out general principles and concepts that underlay many seemingly disparate lines of arithmetic investigation, led him to the concept of a character on a finite abelian group. If $\mathfrak{H}$ is such a group, and its order is h, then any nonzero function $\chi : \mathfrak{H} \to \mathbb{C}$ such that

$$\chi(RS) = \chi(R)\chi(S) \quad \text{for all } R, S \in \mathfrak{H} \tag{12.3}$$

is by definition a character on $\mathfrak{H}$. The arithmetic nature of characters is revealed by the abstract version of Schering's theorem (Theorem 9.7 in the Frobenius–Stickelberger version), which implies the weaker result that $\mathfrak{H}$ can be represented as a direct product of cyclic subgroups, e.g., $\mathfrak{H} = (R_1) \times (R_2) \times \cdots \times (R_k)$, where (R_i) denotes a cyclic subgroup with generator R_i and order n_i. Thus for any element $H \in \mathfrak{H}$, we have the unique representation

$$H = R_1^{a_1} \cdots R_k^{a_k}, \quad 0 \leq a_i < n_i,$$

and so by the multiplicative property (12.3), $\chi(H) = [\chi(R_1)]^{a_1} \cdots [\chi(R_k)]^{a_k}$. Since $R_i^{n_i} = E$, that same property implies that $\chi(R_i)^{n_i} = \chi(E) = 1$. Thus $\chi(R_i) = \varepsilon_i$, an n_ith root of unity, and so

$$\chi(H) = \varepsilon_1^{m_1} \varepsilon_2^{m_2} \cdots \varepsilon_k^{m_r}. \tag{12.4}$$

Since any function defined by (12.4) with any choice of n_ith roots of unity ε_i will be multiplicative in the sense of (12.3), it follows that there are exactly $h = n_1 \cdots n_k$ characters defined on $\mathfrak{H}$, including the 1-character, $\chi_1(H) = 1$ for all $H \in \mathfrak{H}$, which arises by taking all $\varepsilon_i = 1$. To get some sense of the mathematical gravitas underlying Dedekind's definition of a character, I will now briefly indicate the profound arithmetic investigations that motivated his abstract formulation.

The term "character" originated with Gauss' classification of equivalence classes of binary forms into genera (Section 9.1.1). Stated generally, his idea was that form-classes that represented integers with the same characteristic properties should be regarded as belonging to the same genus. The characteristic properties he had in mind were supplied by the important discoveries summarized in the theorem below [244, §§229ff.]. Before stating it, recall that the Legendre symbol $\left(\frac{n}{p}\right)$, where p is a prime and n is not divisible by p, is equal to $+1$ if n is a quadratic residue of 1 mod p ($n^2 \equiv 1 \pmod{p}$) and equals -1 if n is not a quadratic residue of 1 mod p.

Theorem 12.1. *Let $F = ax^2 + 2bxy + cy^2$ be a primitive form ($\gcd(a,b,c) = 1$) of Gaussian determinant*[6] *D, defined by $D = b^2 - ac$, and let n denote any integer representable by F in the sense that $n = F(x_n, y_n)$ for some integers x_n, y_n. Then:*

I. *If p is an odd prime divisor of D, then $\left(\frac{n}{p}\right)$ remains constant for all such n not divisible by p.*
II. *If $D \equiv 3 \pmod 4$ (so $D \equiv 3 \pmod 8$ or $D \equiv 7 \pmod 8$), then $(-1)^{\frac{1}{2}(n-1)}$ remains constant for all such odd n.*
III. *If $D \equiv 2 \pmod 8$, then $(-1)^{\frac{1}{8}(n^2-1)}$ remains constant for all such odd n.*
IV. *If $D \equiv 6 \pmod 8$, then $(-1)^{\frac{1}{2}(n-1)+\frac{1}{8}(n^2-1)}$ remains constant for all such odd n.*
V. *If $D \equiv 4 \pmod 8$, then $(-1)^{\frac{1}{2}(n-1)}$ remains constant for all such odd n.*
VI. *If $D \equiv 0 \pmod 8$, then each of $(-1)^{\frac{1}{2}(n-1)}$ and $(-1)^{\frac{1}{8}(n^2-1)}$ remains constant for all such odd n.*

The manner of expressing I–VI in the above theorem so that the properties of the integers n representable by F are characterized by certain functions of n taking the value $+1$ or -1 came from Dirichlet's groundbreaking memoirs (1839–1840) on the analytic theory of numbers [132, pp. 335–336] and not from Gauss. For example, Part II in *Disquisitiones Arithmeticae* [244, Art. 229, I] reads as follows: "When ... $D \equiv 3 \pmod 4$, odd numbers representable by the form F will be all $\equiv 1$, or all $\equiv 3 \pmod 4$." It is easily seen that an odd integer n satisfies $n \equiv 1 \pmod 4$ if and only if $(-1)^{\frac{1}{2}(n-1)} = +1$ and $n \equiv 3 \pmod 4$ if and only if $(-1)^{\frac{1}{2}(n-1)} = -1$. Gauss and Dirichlet were saying the same thing, but Dirichlet was expressing the character as a numerically valued expression that could be used in analytic formulas.

For example, Gauss indicated the characters determined by the form

$$F = 11x^2 + 4xy + 15y^2, \quad D = -161 \equiv 3 \pmod 4, \tag{12.5}$$

with the notation: 3, 4; R7 N23, where 3, 4 means that all the odd numbers n representable by F are congruent to 3 modulo 4. Also, since $D = -7 \cdot 23$, 7 and 23 are the odd prime divisors of D, and R7 N23 means that n is a quadratic residue mod 7 and a nonresidue mod 23. In terms of Dirichlet's notation, these properties translate respectively into $(-1)^{\frac{1}{2}(n-1)} = -1$, $\left(\frac{n}{7}\right) = 1$, and $\left(\frac{n}{23}\right) = -1$. Thus for Dedekind, Gauss' three characters for forms with $D \equiv 3 \pmod 4$ become three functions of n, namely $C_1(n) = (-1)^{\frac{1}{2}(n-1)}$, $C_2(n) = \left(\frac{n}{7}\right)$, and $C_3(n) = \left(\frac{n}{23}\right)$. Since all forms in the equivalence class of the above form F represent the same integers n, it follows (as Gauss realized) that characters are properties of the equivalence class $\mathcal{A}$ determined by F. Thus $\mathcal{A}$ has Gaussian characters 3, 4: R7 N23, or, as Dedekind would later express the Dirichlet approach, $C_1(\mathcal{A}) = -1$, $C_2(\mathcal{A}) = +1$, and $C_3(\mathcal{A}) = -1$.

[6]In modern versions of the theory, Gauss' determinant D is replaced by the discriminant of F, which is equal to $4D$.

Although Gauss' theorem is valid for primitive forms, for his classification into genera he focused on the subclass of primitive forms that are properly primitive in the sense that $\gcd(a, 2b, c) = 1$. As in Section 9.1.1, let $\mathfrak{F}_1(D)$ denote the set of (proper) equivalence classes of properly primitive forms F of fixed Gaussian determinant D. Recall that Gauss showed, in effect, that under his definition of form composition, the form-classes form a finite abelian group. He now divided the classes of $\mathfrak{F}_1(D)$ into *genera*, where by definition, all form-classes with the same character values belong to the same genus. Thus, for example, when $D = -161$, since $D \equiv 3 \pmod 4$, there are three characters C_1, C_2, C_3 (defined above). The genus characterized by the values $C_1 = -1, C_2 = +1, C_3 = -1$ thus contains the class $\mathcal{A}$ determined by F in (12.5). The class $\mathcal{B}$ determined by $G = 14x^2 + 14xy + 15y^2$ differs from $\mathcal{A}$ but, as the reader can easily check, has the same three character values; and so $\mathcal{B}$ is in the same genus as $\mathcal{A}$. On the other hand, as we saw in Section 9.1.1, the form $E = x^2 - Dy^2 = x^2 + 161y^2$ determines the class $\mathcal{E}$ that acts as the identity element in $\mathfrak{F}_1(D)$. Since Dedekind realized that every Gaussian character C satisfies $C(\mathcal{C}_1\mathcal{C}_2) = C(\mathcal{C}_1)C(\mathcal{C}_2)$,[7] the fact that $\mathcal{E} = \mathcal{E}^2$ implies that for any of the three Gaussian characters, $C(\mathcal{E}) = C(\mathcal{E})^2 = +1$. In other words, the genus containing $\mathcal{E}$ is defined by character values $C_i = +1$ for $i = 1, 2, 3$, and so is different from the genus containing $\mathcal{A}$ and $\mathcal{B}$.[8] The genus of $\mathcal{F}_1(D)$ containing $\mathcal{E}$ is called the *principal genus* and denoted by $\mathfrak{F}_0(D)$. It is a subgroup of $\mathfrak{F}_1(D)$, because, as Gauss proved, it is closed under his definition of composition.

From Gauss' Theorem 12.1 it is easy to determine the number λ of Gaussian characters corresponding to a given Gaussian determinant D. Since Gaussian determinants D congruent to 1 mod 4 are not covered by II–VI, being congruent to 1 or 5 mod 8, it follows that $\lambda = m$, where m is the number of distinct odd prime divisors of D, when $D \equiv 1 \pmod 4$. In this case, the m Legendre symbols in Part I define all the Gaussian characters. Also, from VI, it is clear that for $D \equiv 0 \pmod 8$, $\lambda = m + 2$. For all other D, clearly $\lambda = m + 1$. Thus λ is easy to determine, and it follows that there are 2^λ possible sequences of values for $\mathcal{C}_1, \ldots, \mathcal{C}_\lambda$ evaluated on an equivalence class and thus at most 2^λ genera in $\mathfrak{F}(D)$. Gauss proved the following remarkable theorem.

Theorem 12.2. *(1) $\mathfrak{F}_1(D)$ always contains $2^{\lambda-1}$ genera. (2) Each genus has the same number of classes in it.*

It follows immediately that the class numbers $h_1 = (\mathfrak{F}_1(D) : 1)$ and $h_0 = (\mathfrak{F}_0(D); 1)$ are related by $h_1 = 2^{\lambda-1}h_0$. That is why Gauss had proposed to his successors the problem of expressing h_0 as a function of D (Section 9.1.1).

Gauss' successors, however, were faced with the problem of trying to master the methods he had employed in his theory of composition of forms. As Dirichlet

---

[7]Dedekind explicitly states this fact in Supplement X of the 2nd and later editions of Dirichlet's lectures: [137, p. 399], [138, p. 407], [139, p. 408].

[8]Gauss gave a list of the forms representing each of the 16 classes of $\mathfrak{F}_1(D)$ for $D = -161$ and divided them into genera by their characters [244, Art. 231].

pointed out with a quotation from Legendre [132, p. 414], he—and probably Dirichlet as well—found them inscrutable, and hence unusable. Dirichlet's greatest achievement was the development of entirely new methods for dealing with Gauss' theory, methods based, as he said, on the "methods of infinitesimal analysis." These methods were applied to Gauss' theory in several papers published in *Crelle's Journal* in 1838–1840 [131–133], but earlier, in 1837, he first applied these methods to prove a "simple" theorem: *Every arithmetic progression* $a, a+k, a+2k, a+3k, \ldots$, *with a and k relatively prime, must contain infinitely many primes* [129, 130]. According to Dirichlet ([129, pp. 315–316], [131, p. 355–356]), Legendre had used this theorem as a lemma in proving several theorems, including his law of quadratic reciprocity for odd primes, but his attempt to prove the lemma was incomplete; and it was only after Dirichlet had abandoned his efforts to complete it that he hit upon a different, viable approach to a completely rigorous proof.

Expressed from Dedekind's viewpoint, the starting point of Dirichlet's new approach was the realization that the reasoning leading to Euler's identity

$$\sum_{n=1}^{\infty} \frac{1}{n^s} = \prod_p \left(1 - \frac{1}{p^s}\right)^{-1}, \quad s > 1, \tag{12.6}$$

where the product is over all primes p, shows more generally that if $\psi(n)$ is any complex-valued function of $n \in \mathbb{Z}^+$ such that (1) $\psi(nn') = \psi(n)\psi(n')$ and (2) $\sum_{n=1}^{\infty} |\psi(n)| < \infty$, then

$$\sum_{n=1}^{\infty} \psi(n) = \prod_p (1 - \psi(p))^{-1}. \tag{12.7}$$

To prove the arithmetic progression theorem one considers a character χ on $\mathfrak{H} = (\mathbb{Z}/k\mathbb{Z})^{\times}$ and sets

$$\psi(n) = \begin{cases} \chi(\overline{n})/n^s & \text{if } (n,k) = 1, \\ 0 & \text{otherwise}, \end{cases}$$

where $\overline{n}$ denotes the equivalence class of $(\mathbb{Z}/k\mathbb{Z})^+$ containing n. The series in (12.7) then becomes what is now known as a Dirichlet L-series:

$$L(s,\chi) = \sum_{(n,k)=1} \frac{\chi(\overline{n})}{n^s}, \tag{12.8}$$

and (12.7) takes the form $L(s,\chi) = \prod_{(p,k)=1} \left(1 - \frac{\chi(\overline{p})}{p^s}\right)^{-1}$.

It is easily seen by comparison of the above product formula for $L(s,\chi)$ with (12.6) that for $\chi = \chi_1$, the 1-character, $L(s,\chi_1)$ is equal to $F(s)\zeta(s)$, where $\zeta(s) = \sum_{n=1}^{\infty} \frac{1}{n^s}$ is Riemann's zeta function and $F(s) = \prod_{p|k}(1 - \frac{1}{p^s})$ is well behaved in a neighborhood of $s = 1$, so that $\lim_{s \to 1^+} L(s,\chi_1) = +\infty$. A pivotal point in Dirichlet's proof involved showing that for all $\chi \neq \chi_1$, $\lim_{s\to 1^+} L(s,\chi)$ exists as a finite, nonzero number. Since the order of $\mathfrak{H} = (\mathbb{Z}/k\mathbb{Z})^+$ is $h = \phi(k)$, the characters of $\mathfrak{H}$ will be written as $\chi_1,\ldots,\chi_{\phi(k)}$. In Dedekind's somewhat simplified rendition of Dirichlet's proof, a sequence of highly nontrivial reasoning finally leads to the equation

$$\sum_{i=1}^{\phi(k)} \chi_i(\overline{a})^{-1} \log L(s,\chi_i) = \phi(k)\Big[\sum_{p\equiv a\,(\mathrm{mod}\,k)} \frac{1}{p^s} + Q\Big],$$

where Q is a finite positive number. As $s \to 1^+$, the left-hand side approaches $+\infty$ due to the above-mentioned behavior of the L-functions, and so it must be that the series on the right-hand side diverges, whence there must be an infinite number of primes $p \equiv a\,(\mathrm{mod}\,k)$, and the arithmetic progression theorem is proved!

The success of his analytic methods in proving the arithmetic progression theorem encouraged Dirichlet to develop them further and to apply them to problems in Gauss' theory of composition of forms [131, pp. 359–360], starting with the most difficult problem of all: to solve the problem posed by Gauss of expressing the class number $h_0 = (\mathfrak{F}_0(D) : 1)$ in terms of D (Section 9.1.1). Since Gauss had shown that $h_1 = 2^{\lambda-1}h_0$, where $h_1 = (\mathfrak{F}_1(D) : 1)$, Dirichlet sought to express h_1 in terms of D; and after much effort spread out over three papers [131–133], he succeeded [133, pp. 492–493]. It was in the course of these analytic calculations that the numerical interpretation of Gauss' characters in Theorem 12.1 was utilized.

Dirichlet also proved Gauss' Theorem 12.2, which has $h_1 = 2^{\lambda-1}h_0$ as a consequence. In Supplement IV of the second (1871) and third (1879) editions of Dirichlet's lectures, Dedekind gave his own proof of Gauss' Theorem 12.2. It drew, of course, on ideas already in Dirichlet's proof but, among other things, emphasized the role of characters more general than the Gaussian characters $C_1,\ldots,C_\lambda$. Thus he introduced the notation $\chi(n)$ (already used by Dirichlet) for any one of the 2^λ terms of $\prod_{i=1}^{\lambda}(1+C_i(n))$ ([137, p. 320], [138, p. 324]), which is just a succinct way to say that χ is either the 1-character or some product of distinct Gaussian Characters C_i. Thus the χ are characters on $\mathfrak{F}_1(D)$ as in the definition at the beginning of the section. In proving part (2) of Gauss' Theorem 12.2, Dedekind introduced the L-series $L(s,\chi) = \sum_{(n,2D)=1} \frac{\chi(n)}{n^s}$ and made critical use of the fact that except when χ is the 1-character, $\lim_{s\to 1^+} L(s,\chi)$ exists as a finite number ([137, p. 322], [138, p. 326]).

Not only did Dedekind emphasize the role of characters in his rendition of results due to Dirichlet that are found in various supplements to Dirichlet's lectures, he also did so in the final supplement, which contained his theory of ideals in algebraic number fields. If $\mathbb{K}$ is a finite extension of $\mathbb{Q}$, then associated to $\mathbb{K}$ is the ideal class group,[9] which Dedekind denoted by $\mathfrak{H}$ with h denoting its order. One of the problems Dedekind naturally considered was that of expressing h in terms of numerical expressions associated to $\mathbb{K}$, such as its discriminant Δ. In this connection he introduced the analogue of (12.7), namely [138, p. 578]

---

[9]On the introduction of ideal class groups, see Sections 9.1.2 and 9.1.5.

$$(39) \quad \sum_{\mathfrak{a}\subset\mathfrak{o}_{\mathbb{K}}} \varphi(\mathfrak{a}) = \prod_{\mathfrak{p}} (1-\varphi(\mathfrak{p}))^{-1}, \quad \text{where} \quad (40) \quad \varphi(\mathfrak{a}\mathfrak{b}) = \varphi(\mathfrak{a})\varphi(\mathfrak{b}).$$

And a bit further on in the third edition (1879), he felt compelled to add the following general remarks [138, pp. 580–581]:

> Deeper investigations, to which belong, e.g., those on the genera of quadratic forms ... and those on the distribution of prime ideals into the various ideal classes, are connected with the consideration of more general series and products that arise from (39) when one sets
>
> $$\varphi(\mathfrak{a}) = \chi(\mathfrak{a})/N(\mathfrak{a})^s,$$
>
> where, besides (40), the function $\chi(\mathfrak{a})$ also possesses the property that it takes the same value on all ideals $\mathfrak{a}$ belonging to the same class A; this value is therefore appropriately denoted by $\chi(A)$ and is clearly always an hth root of unity. Such functions χ, which in an extended sense can be termed *characters*, always exist; and indeed, it follows easily from the theorems mentioned at the conclusion of §149 that the class number h is also the number of all distinct characters $\chi_1, \chi_2, \ldots, \chi_h$ and that every class A is completely characterized, i.e., is distinguished from all other classes, by the h values $\chi_1(A), \chi_2(A), \ldots, \chi_h(A)$.

The theorems in §149 to which Dedekind referred are, of course, those of Gauss, Schering, and Kronecker (Section 9.1), which imply that $\mathfrak{H}$ can be decomposed into a direct product of cyclic subgroups.

Referring to the role characters had played in number theory (as described above), Dedekind later wrote to Frobenius that "After all this, it was not much to introduce the concept and name of characters for every abelian group, as I did in the third edition of Dirichlet's Zahlentheorie."[10] Of course, Dedekind had limited his above remarks on characters to the subject at hand, ideal class groups, although the general implications were clear. It was actually his friend and collaborator Heinrich Weber to whom Dedekind had dedicated the third edition of Dirichlet's lectures, who presented the theory of characters within the framework of arbitrary abstract finite abelian groups.

Weber did this in a paper of 1882 [578], which was motivated by a brief paper of Dirichlet's, presented at a session of the Berlin Academy in 1840 [134]. Dirichlet had pointed out that his methods for proving the arithmetic progression theorem could, with a few modifications, be applied to prove that every properly primitive form represents infinitely many prime numbers. To illustrate the sort of modifications that would be needed, Dirichlet considered the special case in which $D = -p$, where p is a prime that is congruent to 3 mod 4. In addition, he assumed that D was "regular" in Gauss' sense that the classes of $\mathfrak{F}_1(D)$ may be written as the powers of a single such class, i.e., $\mathfrak{F}_1(D)$ is cyclic. As Weber explained, "one is easily freed from these restrictions by application of the general theorem on abelian groups first proved by Schering but later ... by others in various ways" [578, p. 301]. Among the others Weber naturally included Kronecker as well as Frobenius and

---

[10]Letter to Frobenius dated 7 August 1896. The quoted portion is reproduced in Dedekind's *Werke* 2, p. 434.

Stickelberger. Because of the importance of the subject, however, Weber deemed it appropriate to begin his paper with an exposition of the basic properties of finite abelian groups [578, pp. 302–309]. He included his own proof of a weaker version of Schering's theorem, namely that every such group is the direct product of cyclic subgroups—the theorem stated at the beginning of this section.

An integral part of Weber's exposition of finite abelian groups was the attendant theory of group characters, starting with the definition and properties of characters as given at the beginning of this section [578, pp. 307–309]. In particular, he proved that if $\chi_1,\ldots,\chi_h$ are the characters on the abelian group $\mathfrak{H} = \{H_1,\ldots,H_h\}$, with χ_1 the 1-character and $H_1 = E$, then (in Weber's numbering)

$$\text{(10)–(11)} \qquad \sum_{i=1}^{h} \chi_j(H_i) = \begin{cases} h \text{ if } j = 1, \\ 0 \text{ otherwise,} \end{cases}$$

and

$$\text{(12)–(13)} \qquad \sum_{j=1}^{h} \chi_j(H_i) = \begin{cases} h \text{ if } i = 1, \\ 0 \text{ otherwise.} \end{cases}$$

The reader familiar with the modern theory of group characters will recognize these relations as special cases of the general orthogonality relations

$$\sum_{i=1}^{h} \chi_j(H_i)\overline{\chi_k(H_i)} = h\delta_{jk} \quad \text{and} \quad \sum_{j=1}^{h} \chi_j(H_i)\overline{\chi_k(H_j)} = h\delta_{ik}, \tag{12.9}$$

obtained by setting $k = 1$. Weber, however, did not indicate any awareness of these more general relations. In subsequent papers of 1886–1887 dealing with abelian number fields, Weber again presented the basic concepts and results concerning abelian groups and characters, since he had occasion to use them [579, pp. 200–202, 222–224], [580, pp. 111–116]. They were also included in the second volume of his widely read *Lehrbuch der Algebra* of 1896 [583]. In all these presentations of characters, Weber repeated the relations (10–11) and (12–13) of his 1882 paper, but never realized (12.9). As we shall see below, it would seem that (12.9) first came to light in the more general form it takes in Frobenius' theory of characters on arbitrary finite groups.

For nonabelian groups $\mathfrak{H}$, the Dedekind–Weber definition of a character as any nonzero complex-valued function such that $\chi(RS) = \chi(R)\chi(S)$ still makes sense. Nowadays it is called a linear character of $\mathfrak{H}$, although for historical emphasis I will frequently refer to it as a Dedekind character. The existence of a nontrivial Dedekind character, however, no longer follows. For example, if $\mathfrak{H}$ is simple but not abelian, then it cannot have nontrivial Dedekind characters, since if such a character existed, being a group homomorphism, its kernel would be a nontrivial normal subgroup of $\mathfrak{H}$. Unless the concept of a character is *generalized* in such a way that nontrivial characters always exist, it would not seem to be of great significance in the study of matters related to nonabelian groups. But how should it be generalized and to what purpose? The concept of a character had grown out of problems in number theory involving abelian groups, and there seemed no need for an extension to nonabelian groups. As we shall see, it was the work of Frobenius on the group determinant problem that eventually brought with it the rationale for a generalized notion of a group character, which then in turn was discovered to have arithmetic applications.

The notion of a group determinant was as unfamiliar in 1895 as it is today, being the private property of Dedekind. Like the notion of a character on an abelian group, the notion of a group determinant came to Dedekind through number theory sometime about 1880 or possibly earlier.[11] According to Dedekind, the idea for such a notion suggested itself by analogy with certain discriminants in a normal extension $\mathbb{L}$ of $\mathbb{Q}$. Let $\mathbb{L}$ have degree h over $\mathbb{Q}$ and let $\mathfrak{H} = \mathrm{Gal}\,(\mathbb{L}/\mathbb{Q}) = \{S_1,\ldots,S_h\}$. Then if $\alpha_1,\ldots,\alpha_h$ is any basis for $\mathbb{L}$ over $\mathbb{Q}$, the discriminant $\Delta = \Delta(\alpha_1,\ldots,\alpha_h)$ of the basis is $\Delta = (\det M)^2$, where M is the $h \times h$ matrix whose ith row consists of the h conjugates of α_i, viz., $\alpha_i S_1,\ldots,\alpha_i S_h$, so that $M = (\alpha_i S_j)$. The idea of a group determinant occurred to Dedekind when he was considering the special "very useful" bases that arise by picking an appropriate element $\omega \in \mathbb{L}$ (or $\omega \in \mathfrak{o}_{\mathbb{L}}$) such that the h conjugates of ω, viz., $\omega S_1,\ldots,\omega S_h$, form a basis for $\mathbb{L}$. In this case, the matrix M in the definition of the discriminant takes the form $M = (\omega S_i S_j)$, which suggested to Dedekind the group determinant $\Theta = \det(x_{S_i S_j})$. In fact, in his initial computations of group determinants for nonabelian groups [116, pp. 7r–17r], Dedekind defined the group determinant as $\Theta = \det(x_{PQ})$; later, he changed to $\Theta = \det(x_{PQ^{-1}})$. The two expressions for θ differ only in sign, but (as indicated in the introduction) the latter has the variable x_E down the diagonal, which makes the coefficient of the term x_E^h equal to 1—niceties that probably induced Dedekind to make the change.

About 1880, Dedekind had discovered that when $\mathfrak{H}$ is abelian, the factorization of Θ involves the characters $\chi_1,\chi_2,\ldots,\chi_h$ of $\mathfrak{H}$. In this case, Θ is the product of linear factors with the characters as coefficients:

$$\Theta = \prod_{\lambda=1}^{h}\left(\sum_{R\in\mathfrak{H}} \chi_\lambda(R)x_R\right). \tag{12.10}$$

It was known at this time that a determinant whose rows are cyclic permutations of the first can be factored into linear factors with roots of unity as coefficients.[12] For example, if $\omega \neq 1$ is a cube root of unity, then

$$\begin{vmatrix} a_1 & a_2 & a_3 \\ a_3 & a_1 & a_2 \\ a_2 & a_3 & a_1 \end{vmatrix} = (a_1+a_2+a_3)\left(a_1+\omega a_2+\omega^2 a_3\right)\left(a_1+\omega^2 a_2+\omega a_3\right).$$

[11] This according to Dedekind's letter to Frobenius dated 8 July 1896. This portion of the letter is included in Dedekind's *Werke* 2, 433–434.

[12] See Muir's history [449, v. 2, 401–412, v. 3, 372–392, v. 4, 356–395].

It was probably the Gauss–Schering–Kronecker theorem discussed in Section 9.1 that led Dedekind to see that a similar result obtains when the indices of the rows are permuted according to any group of permutations that is abelian. Indeed, it was this theorem that Burnside used in a paper of 1894 [46] to establish this result without knowing of Dedekind's unpublished results. It should be kept in mind, however, that Burnside did not speak of characters, just roots of unity. The fact that Dedekind did couch this result in terms of characters became significant in connection with Frobenius' work on the group determinant problem. Let us now see how a version of that problem arose in Dedekind's mind and prompted his cryptic allusion, in the above-quoted 1895 letter to Frobenius, to hypercomplex numbers that intrude in group-theoretic considerations.

Dedekind's discovery of the factorization of Θ when $\mathfrak{H}$ is abelian is distinguished from Burnside's in another notable respect: it led him to consider the group determinant for nonabelian groups with regard to the nature of its factorization into irreducible factors. This occurred in February 1886. At this time and for reasons he could not recall ten years later, he decided to compute the group determinant for some nonabelian groups. The first such group he considered was the symmetric group $\mathfrak{S}_3$ with elements $S_1 = 1$, $S_2 = (1\ 2\ 3)$, $S_3 = (1\ 3\ 2)$, $S_4 = (2\ 3)$, $S_5 = (1\ 3)$, $S_6 = (1\ 2)$.[13] Let $x_i = x_{S_i}$ for $i = 1, \ldots, 6$. Dedekind discovered that

$$\Theta(x_1, \ldots, x_6) = (u+v)(u-v)(u_1u_2 - v_1v_2)^2,$$

where, with $\rho \neq 1$ denoting a cube root of unity,

$$\begin{aligned} u &= x_1 + x_2 + x_3, \quad u_1 = x_1 + \rho x_2 + \rho^2 x_3, \quad u_2 = x_1 + \rho^2 x_2 + \rho x_3, \\ v &= x_4 + x_5 + x_6, \quad v_1 = x_4 + \rho x_5 + \rho^2 x_6, \quad v_2 = x_4 + \rho^2 x_5 + \rho x_6. \end{aligned}$$

Thus, as a homogeneous function of the x_i's, Θ factors into linear factors and a repeated irreducible factor $\Phi = u_1u_2 - v_1v_2$ of degree 2. The fact that Θ factors entirely into linear factors when the group is abelian apparently led Dedekind to consider the possibility of a number system, call it $\mathcal{H}$, over which Φ decomposes into linear factors: $\Phi = \alpha\beta$, $\alpha = \sum_{r=1}^{6} \alpha_r x_r$, $\beta = \sum_{r=1}^{6} \beta_r x_r$, $\alpha_r, \beta_r \in \mathcal{H}$, $r = 1, 2, \ldots, 6$. He assumed that $\mathcal{H}$ had the properties of a hypercomplex number system, i.e., a linear associative algebra over the field of complex numbers. The previous year, Dedekind had concerned himself with commutative hypercomplex number systems (see Section 13.3), and this may have prompted him to consider them in relation to group determinants.

For simplicity, Dedekind assumed that $\alpha_1 = \beta_1 = 1$. By comparing the coefficients of $x_r x_s$ in Φ and in $\alpha\beta$, he then obtained 20 conditions that the numbers α_r, β_r must satisfy in order that $\Phi = \alpha\beta$. For example,

[13]The following is based upon Dedekind's computations in the manuscript [116, p. 10r] but is presented in the notation he subsequently adopted. He also communicated these computations to Frobenius in a letter of 13 July 1896. The portion of the letter containing these calculations is contained in Dedekind's *Werke* 2, 437–441.

$$\begin{aligned} \alpha_4\beta_4 &= \alpha_5\beta_5 = \alpha_6\beta_6 = \alpha_2+\beta_2 = \alpha_3+\beta_3 = -1, \\ \alpha_2\beta_2 &= \alpha_3\beta_3 = 1, \\ \alpha_4+\beta_4 &= \alpha_5+\beta_5 = \alpha_6+\beta_6 = 0. \end{aligned} \tag{12.11}$$

Dedekind deduced that $\alpha_4^2 = \alpha_5^2 = \alpha_6^2 = 1$ and that $1+\alpha_2+\alpha_2^2 = 1+\alpha_3+\alpha_3^2 = 0$. The latter equality implies that $\alpha_2^3 = \alpha_3^3 = 1$, although Dedekind did not write it down at this point. It was perhaps these relations that suggested to Dedekind identifying α_r with $S_r \in \mathfrak{S}_3$ and defining β_r in terms of the α_s by means of (12.11). He found that if this is done, then the conditions for $\Phi = \alpha\beta$ are satisfied, provided one sets both $\eta = 1+\alpha_2+\alpha_3$ and $\omega = \alpha_4+\alpha_5+\alpha_6$ equal to zero. Expressed in more familiar terms, Dedekind's discovery was that Φ decomposes into linear factors over $\mathcal{H} = \mathbb{C}\mathfrak{S}_3/\Omega$, where $\mathbb{C}\mathfrak{S}_3$ denotes the group algebra of $\mathfrak{S}_3$ with respect to the complex numbers $\mathbb{C}$, and Ω is the ideal generated by η and ω.

Thus Dedekind obtained a decomposition of Θ into linear factors. "With this result," he later wrote to Frobenius, "I was thoroughly satisfied at the time (3 February 1886), and because it seemed very noteworthy to me, I proceeded directly to other examples ...."[14] Next he considered a nonabelian group of order ten formed by taking a semidirect product of cyclic groups of orders 2 and 5. Again he was able to factor Θ into linear and irreducible second-degree factors. After some inconclusive computations on the conditions for a number system such that Θ decomposes entirely into linear factors, however, he turned to the quaternion group. Here he had more success, because on making a change of variables $(x_1,\ldots,x_8) \to (u_1,\ldots,u_4,v_1,\ldots,v_4)$ similar to the one employed to obtain the factorization of Θ for $\mathfrak{S}_3$, he obtained

$$\Theta = (u_1+u_2+u_3+u_4)(u_1+u_2-u_3-u_4)(u_1-u_2+u_3-u_4)(u_1-u_2-u_3+u_4)\Phi^2,$$

where $\Phi = v_1^2+v_2^2+v_3^2+v_4^2$. The second-degree factor Φ thus represents the norm of the quaternion $v_1+iv_2+jv_3+kv_4$, which means that Φ factors into linear factors if quaternions are allowed as coefficients:

$$v_1^2+v_2^2+v_3^2+v_4^2 = (v_1+iv_2+jv_3+kv_4)(v_1-iv_2-jv_3-kv_4).$$

Hence in this case also, Θ can be factored into linear factors if hypercomplex numbers are permitted as coefficients.

As these examples indicate, Dedekind's interest in the group determinant was focused mainly on the problem of determining, for a given group $\mathfrak{H}$, a hypercomplex number system over which its group determinant would decompose into linear factors, presumably in the hope of finding some interesting relations between its structure and the structure of $\mathfrak{H}$. Hence, his cryptic question to Frobenius

[14] Letter to Frobenius dated 13 July 1896. The quoted portion is reproduced in Dedekind's *Werke* 2, 440.

(from the quotation at the end of Section 12.1): "do hypercomplex numbers with noncommutative multiplication also intrude in your research? But I do not wish to trouble you for an answer, which I will best obtain from your work." That question was posed in Dedekind's letter to Frobenius on 8 February 1895. Let us now return to the correspondence and see how Frobenius reacted.

12.3 Frobenius Learns About Group Determinants

Frobenius had once briefly considered a certain type of hypercomplex number system in the concluding section of his 1878 paper on linear algebra [181], which contains his well-known theorem classifying finite-dimensional division algebras over $\mathbb{R}$ (Section 7.5.6). But as we shall see, he did not find hypercomplex numbers appealing in and of themselves. His classification of division algebras was done as an application of the matrix algebra he had developed in [181] rather than as the beginning of a research program on more general hypercomplex systems. And no such systems had entered into his work on finite groups. Frobenius' lack of enthusiasm for hypercomplex numbers seems reflected in the lack of curiosity about Dedekind's remark implied by his reply on 10 February 1895, for he wrote simply, "My work on groups is now appearing. There is no discussion in it about hypercomplex numbers; previously obtained results are summarized, the methods of Sylow are further developed, and the investigations in my last work are carried further."

Had Frobenius expressed more interest, or at least curiosity, Dedekind most likely would have discussed his work on group determinants in his next letter. Instead, he replied, a bit apologetically,

> [M]y question regarding the use of hypercomplex numbers in the theory of groups was very audacious; it arose from an observation I made in February 1886 but then did not pursue further, although it seemed noteworthy enough to me; perhaps sometime I will venture to present it to you at the risk that it will entirely vanish before your criticism ...."[15]

The occasion for Dedekind's and Frobenius' letters during January and February 1895 had been the business of getting Kronecker's letter published in an expurgated form acceptable to Dedekind. With Dedekind's last-quoted letter, that business was finished,[16] and so there was no reason to continue the correspondence—unless, of course, Frobenius was now sufficiently curious about Dedekind's use of hypercomplex numbers in group theory to write and encourage him to present his ideas. But he was not, and the correspondence broke off. Had Dedekind never returned to the matter, it is doubtful that Frobenius would have played such a

[15]Letter of 12 February 1895. A portion of this letter containing the above quotation was published in Dedekind's *Werke* 2, 420.

[16]The letter was published as [371]. Regarding Dedekind's concern to omit parts of the letter, and more generally his relations with Kronecker, see Edwards' article [145, pp. 370–372].

pioneering role in the creation and development of the theory of group characters and representations, which would have emerged instead from the developments traced in Chapter 14.

Frobenius was thus fortunate that Dedekind did decide to renew the correspondence a year later and to tell him about group determinants. During the fall of 1895, perhaps stimulated by Frobenius' activity in the theory of groups, Dedekind decided to pursue some group-theoretic research of his own. In February 1886, the period when he had studied the group determinant, Dedekind had also studied normal extensions of the rational field that have the property that all subfields are normal. The Galois group of such a normal extension then has the property that all of its subgroups are normal. That work now (in 1895) suggested to him the purely group-theoretic problem of characterizing abstract finite groups with the property that all subgroups are normal—"Hamiltonian groups" as he called them. To his surprise, he discovered the answer was relatively simple, and he communicated it to his close friend Heinrich Weber. Weber was an editor of *Mathematische Annalen* and urged his friend to publish his result there.[17] Dedekind, however, did not believe in rushing into print. He wanted to be certain the result was new. Perhaps he also wanted to be certain it was significant, i.e., not a simple consequence of known results. Who would know better than Frobenius?

Dedekind therefore wrote to him on 19 March 1896:

> Some time ago I had intended to write to you and first of all to express my thanks for your works, through which the African darkness of the theory of groups is brightened. I also wanted to communicate some studies on groups and fields to you which, in Weber's opinion, contain new results but which do not touch upon the same areas as yours. I am, however, reluctant to engage you in mathematical conversation right now and prefer to wait until you feel more inclined toward it.

The reason for Dedekind's hesitancy was that he had heard from Frobenius' colleague at Berlin Kurt Hensel that Frobenius was not feeling well. Frobenius assured him that he was well enough for mathematics and invited him to discuss his work.[18] Dedekind accepted the invitation and quickly sent off his theorem on Hamiltonian groups to Frobenius. After presenting it, he added, "Since I am speaking about groups, I would like to mention another consideration that I came upon in February."[19]

With that introduction, Dedekind proceeded to present the concept of the group determinant and stated the theorem about its factorization for abelian groups ($\Theta = \prod_{\alpha=1}^{h}\{\sum_{R\in\mathfrak{G}}\chi^{(\alpha)}(R)x_R\}$). For nonabelian groups, Dedekind explained, undoubtedly with his computed examples in mind, that Θ always seems to have higher-degree irreducible factors, although these can be factored into linear factors

[17] It was eventually published there as [118].

[18] Letter dated March 22, 1896.

[19] Letter dated 25 March 1896. The portion of the letter containing the above quotation as well as those that follow was published in Dedekind's *Werke* 2, 420–421.

if hypercomplex numbers with noncommutative multiplication are allowed as coefficients. He then continued:

> It would be reasonable to conjecture that the properties of a group $\mathfrak{G}$ that relate to its subgroups would be reflected in its determinant $[\Theta]$. ... Except for one clue, which suggests a connection between the number of ordinary linear factors of $[\Theta]$ ... and those normal divisors $\mathfrak{A}$ of $\mathfrak{G}$ that have the property $\mathfrak{A}RS = \mathfrak{A}SR$; however, I have found nothing at all, and it is possible that for the time being, little will result from the entire matter.

Frobenius quickly and enthusiastically responded to Dedekind's letter with a reply of 18 pages dated 29 March 1896:

> Long ago it surprised me that you had not participated more actively in the development of the abstract theory of groups, even though, by virtue of your disposition, this field must have been especially attractive to you. Now I see that you have concerned yourself with it for ten years and have kept back your extremely beautiful results from your friends and admirers (also, unfortunately, by virtue of your disposition?).[20]

Most of the 18 pages are filled with a technical discussion of Dedekind's theorem on Hamiltonian groups and its relation to Frobenius' own results on groups. But he began the letter with some comments on group determinants (in which s' is used to denote s^{-1}):

> First of all, the nth-degree determinant is $|x_{rs'}| = (-1)^{n-m-1}|x_{rs}|$, where m is the number of elements of order 2. You probably prefer $x_{rs'}$ because the elements on the diagonal are all x_1 and the coefficient of x_1^n is 1. I believe I am well acquainted with the theory of determinants, and I think that the formula $[\Theta = \prod_{\alpha=1}^{h}\{\sum_R \chi^{(\alpha)}(R)x_R\}]$ ... has not been expressed in this generality for abelian groups. For cyclic groups it has been known a long time. ... But I also never thought of this generalization which is so close at hand.

Dedekind's idea of using hypercomplex number systems to factor nonlinear factors of Θ into linear ones, however, did not appeal to Frobenius, who wrote, "I do not know yet whether I will be able to reconcile myself to your hypercomplex numbers." From the outset, he was more interested in the ordinary factorization of Θ and its relation to the group $\mathfrak{G}$. He also did not understand what clue Dedekind had discovered: "the entire subject is so new to me that I cannot see yet how the irreducible factors of the determinant are connected with the (invariant?) subgroups. If you know something about this, please tell me."

An additional reason for Frobenius' evident interest in the factorization of Dedekind's group determinant was that it bore certain similarities to a homogeneous polynomial that had emerged in some of his work on the theory of theta characteristics published in 1884 [192]:

[20]There are hundreds of pages of unpublished manuscripts concerning groups by Dedekind in the archives of the Niedersächsische Staats- und Universitätsbibliothek in Göttingen. In the archives of the library of the Technical University at Braunschweig, Germany, there are 86 pages by Dedekind dealing with groups. Scharlau [511] has analyzed Dedekind's unpublished algebraic work from the period 1855–1858.

> Are you familiar with my work: *Ueber Thetafunctionen mehrerer Variabeln...*? There a polynomial of degree $[2^p]$ ... in 2^{2p} variables is treated, a determinant that for the group of the theta functions is closely related to your group determinant. If some of the variables are set equal to 0, it decomposes into linear factors or into second-degree factors or into fourth-degree factors, according to the relations between the characteristics (syzygetic–asyzygetic).

Frobenius' paper [192] will be discussed below, because his study of the above-mentioned polynomial involved some techniques that he was to put to good use in his study of the group determinant. In response to Frobenius' request for further information, Dedekind specifically stated the conjecture that his above-mentioned clue had suggested to him[21]:

Conjecture 12.3 (Dedekind). The number of linear factors in Θ equals the index of the commutator subgroup $\mathfrak{G}'$ and hence equals the order of the abelian group $\mathfrak{G}/\mathfrak{G}'$.

He also gave Frobenius a hint as to how he had arrived at this conjecture by adding that the linear factors of Θ correspond "in a certain way" to the characters of the abelian group $\mathfrak{G}/\mathfrak{G}'$. Finally, he invited Frobenius to pursue these matters, since "I distinctly feel that I will not achieve anything here." Undoubtedly, he also realized that the study of Θ would take him too far afield from his principal interests in the theory of numbers. At the age of 65, one obviously cannot afford to squander one's time! The analysis of Θ was nevertheless perceived by Dedekind to be a good research problem, especially for someone with Frobenius' interests and talents, and he clearly hoped Frobenius would work on it. By way of additional encouragement, he added the following comment about Frobenius' above-mentioned polynomial in his 1884 paper [192]: "After a glance at your work (Crelle 96, p. 100) it seems to me that there is no doubt whatsoever that your determinant for the characteristic group of the theta functions coincides essentially with my $[\Theta]$ (better conversely), and therefore the full priority for this group determinant belongs to you." It will be clear after we consider Frobenius' paper [192] that the polynomial treated there would never have suggested to Frobenius the idea of a group determinant. Dedekind was being overly generous, and Frobenius never took Dedekind's disclaimer seriously.[22]

12.4 Theta Functions with Integral Characteristics

Frobenius' paper [192] was one in a series of papers that had appeared between 1880 and 1884 and were devoted to the investigation of diverse aspects of the theory of theta functions in any number of variables. Three of Frobenius' papers in this

---

[21]The letter is dated 31 March–3 April 1896. The portion of the letter discussed below was published in Dedekind's *Werke* 2, 421–423.

[22]Although in his publications, Frobenius mentioned the analogy with this polynomial, he also made clear his indebtedness to Dedekind and in particular noted Dedekind's priority in envisioning the importance of group determinants for the theory of groups [212, pp. 38–39].

series [186, 188, 189] have already been discussed in Chapters 10–11. As we saw there, the first two were both motivated in large part by the appearance in 1878 of Weber's paper extending Hermite's theory of the transformation of abelian functions from $g = 2$ to $g = 3$ variables.

Weber's paper also triggered Frobenius' interest in the theory of theta functions with integral characteristics. Theta functions with such characteristics had already been studied by Weierstrass and by Schottky, as we saw in Chapter 11, and that fact undoubtedly also encouraged Frobenius to continue the tradition. For theta functions in g complex variables $z = \begin{pmatrix} z_1 & \cdots & z_g \end{pmatrix}^t$, the level at which Frobenius generally preferred to work, a characteristic is a $2 \times g$ matrix

$$A = \begin{pmatrix} a_1 & \cdots & a_g \\ b_1 & \cdots & b_g \end{pmatrix} \stackrel{\text{def}}{=} \begin{pmatrix} a \\ b \end{pmatrix}, \quad a, b \in \mathbb{Z}^g. \tag{12.12}$$

Characteristics A and B are considered equal when $A \equiv B \pmod 2$, so that there are 2^{2g} distinct integral characteristics, each determined by a $2 \times g$ matrix of 0's and 1's. Corresponding to a characteristic A is the theta function

$$\vartheta[A](z) = \sum_{n \in \mathbb{Z}^g} \exp\left[2\pi i \left(z + \frac{1}{2}a\right) \cdot \left(n + \frac{1}{2}b\right) + \pi i \left(n + \frac{1}{2}b^t\right) T \left(n + \frac{1}{2}b\right)\right],$$

where (as before) $T = \Phi + i\Psi$ is a complex symmetric matrix with an imaginary part Ψ that is positive definite, which guarantees that the series converges absolutely and uniformly on compact subsets. When $A = \begin{pmatrix} 0 \\ 0 \end{pmatrix} \stackrel{\text{def}}{=} 0$, $\vartheta[A](z)$ reduces to a theta function without characteristics, i.e., the sort that figured in Frobenius' papers [186, 188]. Theta functions with characteristics $A \neq 0$ had proved useful in the analytic theory of elliptic and abelian functions and in related applications to algebraic geometry. They had been introduced by Jacobi in the case $g = 1$, by Göpel and Rosenhain for $g = 2$, and by Weierstrass for g arbitrary (see Section 11.1).[23]

Frobenius' first paper on theta characteristics [187] (1880) was motivated by his observation that Weber, in generalizing to $g = 3$ variables a key result for $g = 2$ in Hermite's theory, had weakened it considerably. By developing the theory of characteristics, especially what he called "Göpel systems," Frobenius was able to prove an addition theorem analogous to Hermite's from which he then obtained a sharper version of Hermite's key theorem, and he then went on to apply his addition theorem to reformulate, extend, and make rigorous work done by Hermann Stahl [547] (1880), one of Weierstrass' students.[24]

[23]For an idea of the rich and complicated analytic theory of theta characteristics and its history, including some indications of Frobenius' contributions, see the lengthy (131 pages) seventh chapter of Krazer's treatise [350].

[24]My brief summary of Frobenius' paper is based on his own extensive overview [187, pp. 11–14].

Frobenius' above-described paper [187] was written not long after he and Stickelberger had developed the theory of finite abelian groups. It is clear that the totality of the 2^{2g} characteristics with matrix addition modulo 2 as the group operation form an abelian group, which will be denoted by $\mathfrak{K}_g$. For notational convenience (as we shall see), Frobenius used multiplicative notation for the group operation:

$$AA' \equiv \begin{pmatrix} a+a' \\ b+b' \end{pmatrix} \pmod{2}. \tag{12.13}$$

Clearly, $\mathfrak{K}_g$ has the property that every element is idempotent: $A^2 = O$ for all $A \in \mathfrak{K}_g$. In his paper, Frobenius made use of a subgroup of $\mathfrak{K}_g$ associated to a Göpel system [187, p. 23]. In a paper submitted in 1883, three months after his paper on principal transformations of theta functions [188], he submitted one entitled "On Groups of Theta Characteristics" [191]. Here he considered subgroups of $\mathfrak{K}_g$ classified by their rank and developed their theory so as to obtain thereby "a sharper insight into the essence of the formulas" that Stahl, Max Nöther, and Prym had obtained in various publications [191, p. 130]—a characteristically Frobenian enterprise.

A month after he submitted his paper on groups of theta characteristics, he submitted the paper referred to in his letter to Dedekind, "On Theta Functions of Several Variables" [192]. This paper was different from his earlier papers on theta characteristics. Previously he had developed the theory of characteristics so as to rework known results in a new, more general, coherent, and rigorous fashion. In the paper [192], however, Frobenius used his mastery of the theory of theta functions with characteristics to develop the theory in an entirely new direction. The focus of his paper was a determinant that seems to have been of his own devising.

Some preliminaries are required in order to define Frobenius' determinant and then to discuss its properties. The first preliminary is some of the notation that Frobenius had introduced in his first paper (1880) on theta functions with integral characteristics [187, pp. 14ff.]. For characteristics $A = \begin{pmatrix} a \\ b \end{pmatrix}$, $A' = \begin{pmatrix} a' \\ b' \end{pmatrix}$, ... in $\mathfrak{K}_g$ and with $a \cdot b$ denoting the dot product of a and b, let

$$(A) = (-1)^{a \cdot b}, \quad \begin{pmatrix} A \\ A' \end{pmatrix} = (-1)^{a \cdot b'}, \tag{12.14}$$

and

$$\begin{aligned} |A,A'| &= a' \cdot b - a \cdot b', \\ |A,A',A''| &= |A,A'| + |A',A''| + |A'',A|. \end{aligned} \tag{12.15}$$

Incidentally, Frobenius called (A) the character of A [187, p. 16], presumably because like Dirichlet's version of Gauss' characters, it takes the values ± 1.

However, $(AB) = (A)(B)$ does not hold in general.[25] Thus $A \to (A)$ lacks what Dedekind perceived to be the defining property of a character on a group, and suggests that in 1880, Frobenius, who was not so deeply involved with number theory as Dedekind, had not arrived at Dedekind's precise notion of what "character on an abelian group" should mean.

By way of a second preliminary, Frobenius considered for each characteristic $R \in \mathfrak{K}_g$ and for $z, w \in \mathbb{C}^g$, the following product of two theta functions:

$$\vartheta[R](z,w) = \vartheta[R](z+w) \cdot \vartheta[R](z-w).$$

Then if z_α, w_β, $\alpha, \beta = 1, \ldots, r$, denote $2r$ variable vectors in $\mathbb{C}^g$, corresponding to each $R \in \mathfrak{K}_g$ we may define the $r \times r$ matrix

$$M(R) = \Big(a_{\alpha\beta}(R)\Big) = \Big(\vartheta[R](z_\alpha, w_\beta)\Big). \tag{12.16}$$

The coefficients of $M(R)$ depend on the g^2 vectors $z_\alpha, w_\beta \in \mathbb{C}^g$ as well as on $R \in \mathfrak{K}_g$. Making extensive and masterly use of complicated formulas from his earlier papers, Frobenius showed [192, p. 152] that the matrix $M(R)$ of (12.16) has the remarkable property that for all $S, T \in \mathfrak{K}_g$ [192, p. 152],

$$\det M(S) = \begin{cases} \det M(T) & \text{if } g > 1, \\ -(ST)\det M(T) & \text{if } g = 1. \end{cases} \tag{12.17}$$

Thus for $g > 1$, the determinant of $M(S)$ is independent of the chosen characteristic S, and for $g = 1$, since $-(ST) = \pm 1$, the determinants $\det M(S)$ and $\det M(T)$ can differ at most by a factor of ± 1.

Corresponding to each of the r^2 elements $R \in \mathfrak{K}_g$, Frobenius next introduced an independent complex variable x_R and considered the determinant

$$\Delta = \det\Big[\sum_{R \in \mathfrak{K}_g} M(R) x_R\Big].$$

He showed that the ratio of determinants

$$F = \Delta / \det M(A)$$

is independent of the g^2 vectors z_α, w_β [192, p. 153]. Of course, (12.17) also shows that F does not depend on the choice of A (except for a ± 1 factor in the elliptic case

---

[25]For example, in the case $g = 2$, if $A = \begin{pmatrix} 1 & 0 \\ 1 & 1 \end{pmatrix}$ and $B = \begin{pmatrix} 0 & 1 \\ 0 & 0 \end{pmatrix}$, then $(A)(B) = (-1)(+1) \neq +1 = (AB)$, since by (12.13) $AB = \begin{pmatrix} 1 & 1 \\ 1 & 1 \end{pmatrix}$.

$g = 1$). Thus the quotient F defines a function of the r^2 variables x_R alone, which he denoted by $F[x_R]$. In sum:

$$F[x_R] = \det\Big[\sum_R x_R \vartheta[R](z_\alpha, w_\beta)\Big] / \det M(A) \qquad (12.18)$$

is a homogeneous polynomial of degree $r = 2^g$ in the r^2 group variables $x_A, x_B, x_C, \dots$ with coefficients that do not depend on z_α and w_β.

Having established these results, Frobenius turned to the true object of his paper: the properties of $F[x_R]$. He began by showing that $F[x_R]$ has interesting symmetry properties, e.g., $F[x_R] = F[(R)x_R]$ [192, p. 153]. Since in $\mathfrak{K}_2$, exactly 6 of the 16 characteristics R have $(R) = -1$, it follows that replacement of x_R by $-x_R$ in $F[x_R]$ for these six variables does not change the value of the polynomial, which is thus somewhat special. Another sort of symmetry property he obtained was $F[x_R] = F[x_{AR}]$ for every $A \in \mathfrak{K}_g$ [192, p. 154]. In other words, if the variables are permuted according to $R \to AR$, the polynomial remains unchanged. By virtue of such symmetry properties, he was able to write down explicit expressions for $F[x_R]$ for $g = 1$ and $g = 2$ [192, pp. 154, 156].

One other general property of $F[x_R]$ noted by Frobenius turned out to be especially relevant to his study of Dedekind's group determinant [192, p. 161]:

Theorem 12.4. *If for each $A \in \mathfrak{K}_g$ we set*

$$z_A = \sum_{B \in \mathfrak{K}_g} \binom{B}{A} x_B y_{BA}, \quad \text{then} \quad F[z_R] = F[x_R] \cdot F[y_R]. \qquad (12.19)$$

In other words, if in the polynomial $F[x_R]$ we replace each variable x_A by z_A as given above, so that the resultant polynomial involves variables x_R and y_R, then it factors into $F[x_R]$ times $F[y_R]$. For later reference, note that since $B^2 = 0$ and so $B(BA) = B^2A = A$, we could replace BA by C, where $BC = A$, and express (12.19) in the form

$$z_A = \sum_{BC=A} \binom{B}{BC} x_B y_C. \qquad (12.20)$$

As remarkable as Theorem 12.4 is, Frobenius managed to make only one application of it [192, pp. 161–162].

Most of Frobenius' further results about $F[x_R]$ involved a technique of *variable specialization*, as I shall call it. That is, if $\mathfrak{S}$ is any subset of $\mathfrak{K}_g$, which need not be a subgroup, we may set all variables $x_R = 0$ for $R \notin \mathfrak{S}$ so as to obtain from $F[x_R]$ a new polynomial, which I will denote by $F^*[x_R]$ and refer to as the specialized polynomial associated to $\mathfrak{S}$. For example, if $\mathfrak{S}$ consists of any two characteristics A and B, then $F^*[x_R] = \big(x_A^2 - (AB)x_B^2\big)^{r/2}$, as Frobenius showed [192, p. 155]. One of his main results about specializations $F^*[x_R]$ has to do with the distinction between triples

A,B,C of characteristics that are *syzygetic* ($|A,B,C| \equiv 0 \pmod 2$ in the notation of (12.15)) or *asyzygetic* ($|A,B,C| \equiv 1 \pmod 2$). This distinction was introduced in his 1880 paper in connection with what he called Göpel systems [187, §2] and was further utilized in his paper on groups of characteristics [191, pp. 134ff.]. Now he utilized this distinction to prove the following theorem [192, V, p. 164].

Theorem 12.5. *Let $\mathfrak{S}$ be a subset of $\mathfrak{K}_g$ and $F^*[x_R]$ the corresponding specialization of $F[x_R]$. (I) $F^*[x_R]$ factors completely into linear factors if and only if every triple $A,B,C \in \mathfrak{S}$ is syzygetic. (II) $F^*[x_R]$ is a power of a quadratic polynomial if and only if every triple $A,B,C \in \mathfrak{S}$ is asyzygetic.*

Frobenius wrote down a formula for $F^*[x_R]$ in case (II) [192, p. 163, (3)].

The final two sections of Frobenius' paper, which are chock-full of complicated identities and equations, many of which were taken from his earlier work, especially his paper [191] on groups of theta characteristics, were aimed at proving a more general factorization theorem than Theorem 12.5. Before stating it, a few preliminaries are necessary. If $\mathfrak{H}$ is any subgroup of $\mathfrak{K}_g$ of rank γ, the set $\mathfrak{U}$ of all $U \in \mathfrak{H}$ such that (in the notation of (12.15)) $|U,H| \equiv 0 \pmod 2$ for all $H \in \mathfrak{H}$ forms a subgroup of $\mathfrak{H}$ called the *syzygetic subgroup* [191, p. 131] of $\mathfrak{H}$. If α is the rank of $\mathfrak{U}$, then it is always the case that $\gamma \pm \alpha$ is even [191, p. 133]. Frobenius proved the following factorization theorem.[26]

Theorem 12.6. *Let $\mathfrak{H}$ be a subgroup of $\mathfrak{K}_g$ of rank γ and let α denote the rank of its syzygetic subgroup $\mathfrak{U}$. Consider the specialization $F^*[x_R]$ associated to $\mathfrak{H}$. Then $F^*[x_R] = \left(\prod_{i=1}^{2^\alpha} G_i[x_R]\right)^m$, where $m = 2^{(g-(1/2)(\gamma+\alpha))}$ and each G_i is a homogeneous polynomial of degree $d = 2^{(1/2)(\gamma-\alpha)}$.*

In this theorem, Frobenius was able to relate the factorization of $F^*[x_R]$ to invariants of the associated group $\mathfrak{H}$, namely its rank γ and the rank α of its syzygetic subgroup. As illustrations of the theorem, first let $\mathfrak{H}$ be the subgroup of $\mathfrak{K}_4$ consisting of all characteristics of the form $A = \begin{pmatrix} x & 0 & y & 0 \\ 0 & z & 0 & 0 \end{pmatrix}$, where the variables x,y,z can take the values 0 and 1. Then $\mathfrak{H}$ has order 2^3 and rank $\gamma = 3$. The syzygetic subgroup coincides with $\mathfrak{H}$, so that $\alpha = 3$. Thus $m = 2^1 = 2$, $d = 2^0 = 1$, and $2^\alpha = 2^3 = 8$, so $F^*[x_R]$ is the square of a product of eight linear factors. Next let $\mathfrak{H}$ be the subgroup of $\mathfrak{K}_4$ consisting of all characteristics of the form $A = \begin{pmatrix} x & y & 0 & z \\ w & 0 & u & v \end{pmatrix}$, where the variables $x,y,\ldots,v$ can take the values 0 and 1. Thus $\mathfrak{H}$ is a subgroup of order 2^6 and rank $\gamma = 6$. The syzygetic subgroup of $\mathfrak{H}$ has order 4, and rank $\alpha = 2$. According to the theorem, this means that $m = 1$, $d = 4$, and $2^\alpha = 4$, so that $F^*[x_R]$ is a product of four quartic factors G_i. If we take $\mathfrak{H} = \mathfrak{K}_g$, so that $F[x_R]$ is not specialized, then $\gamma = 2g$ and the syzygetic subgroup consists only of 0, so that $\alpha = 0$ [191, p. 133]. This

[26]Frobenius [192, p. 171] stated the theorem more generally for a system $\mathcal{S}$ of characteristics containing $\gamma+1$ "essentially independent" characteristics as defined earlier [187, p. 15]. When $\mathcal{S}$ is a subgroup $\mathfrak{H}$ of $\mathfrak{K}_g$, it takes the form given in Theorem 12.6.

means that $m = 2^0 = 1$ and $d = 2^g = r$, the degree of $F[x_R]$. In other words, when $F[x_R]$ is not specialized, Theorem 12.6 yields no information about its factorization.

Now it should be clear why, when Dedekind explained his notion of a group determinant $\Theta = \Theta(x_R)$ to Frobenius in his letter of 31 March–3 April, Frobenius immediately thought of his old study of the polynomial $F[x_R]$ and its factorization (when appropriately specialized). Certainly $F[x_R]$ is not the group determinant of $\mathfrak{K}_g$, nor is $F^*[x_R]$ the group determinant of the group $\mathfrak{H}$ associated to the specialization in Theorem 12.6. But given Frobenius' evident fascination with $F[x_R]$ in 1884, it is not surprising that he jumped at the chance to explore the somewhat analogous group determinant Θ. Indeed, whereas his study of $F[x_R]$ had involved him, by virtue of its definition, with a myriad of formulas and relations attendant on the theory of theta characteristics, the study of the group determinant, which is linked in a simpler, more straightforward fashion to a finite group, no doubt struck Frobenius as a welcome respite from the arduous task of dealing with theta function relations, i.e., as an opportunity to renew himself in an especially appealing way at the "fountain of youth of arithmetic."

Chapter 13
Group Characters and Representations 1896–1897

Having now established the great appeal to Frobenius of Dedekind's suggestion that he study Θ and its factorization, let us consider, with the aid of his correspondence with Dedekind, how he progressed. His first progress report to Dedekind came in a letter dated 12 April 1896, just nine days after Dedekind had finished writing his letter to Frobenius.

13.1 Frobenius' Letter of 12 April

Dedekind's irresistible invitation had come at a propitious time for Frobenius, since it arrived during the break between the winter and summer semesters when he had more free time. His letter to Dedekind indicates that he had spent that time probing into the mysteries of Θ from every possible angle.

The first matter treated by Frobenius in his letter was most likely the first thing he considered after receiving Dedekind's invitation, namely the latter's conjectured theorem on the number of linear factors in Θ. Following Dedekind's generous hint that there is a connection between the linear factors in the group determinant of a nonabelian group $\mathfrak{H}$ and the characters on the abelian group $\mathfrak{H}/\mathfrak{H}'$, where $\mathfrak{H}'$ is the commutator subgroup of $\mathfrak{H}$, Frobenius observed that if $\bar{\chi}$ is a character on $\mathfrak{H}/\mathfrak{H}'$, then it determines a complex-valued function χ on $\mathfrak{H}$, namely $\chi(A) = \bar{\chi}(\mathfrak{H}'A)$ for all $A \in \mathfrak{H}$, which has the property that

$$\chi(AB) = \chi(A)\chi(B) \quad \text{for all} \quad A, B \in \mathfrak{H}. \tag{13.1}$$

Frobenius referred to complex-valued functions $\chi \neq 0$ satisfying (13.1) as "characters" in accordance with Dedekind's terminology. To distinguish between these characters and the generalized characters that Frobenius was eventually to introduce, the former will be referred to as Dedekind (or linear) characters whenever there is any chance of confusion. Frobenius next observed that conversely, every Dedekind

T. Hawkins, *The Mathematics of Frobenius in Context*, Sources and Studies in the History of Mathematics and Physical Sciences, DOI 10.1007/978-1-4614-6333-7_13,

character χ on $\mathfrak{H}$ determines a corresponding character $\bar{\chi}$ on $\mathfrak{H}/\mathfrak{H}'$, because the set $\mathfrak{K}$ of all $A \in \mathfrak{H}$ such that $\chi(A) = 1$—the kernel of χ as we would now say—is a normal subgroup of $\mathfrak{H}$ containing $\mathfrak{H}'$. Thus $\mathfrak{K}AB = \mathfrak{K}BA$, in accordance with Dedekind's earlier, more cryptic hint in his letter of 25 March quoted above; and it follows that χ is constant on each coset $\mathfrak{H}'A$, so that $\bar{\chi}(\mathfrak{H}'A) = \chi(A)$ defines a character $\bar{\chi}$ on $\mathfrak{H}/\mathfrak{H}'$.

Judging by Dedekind's hints to Frobenius, the former also realized all of the above. In addition, he must have realized that for every character χ on $\mathfrak{H}$, $\Phi = \sum_{R\in\mathfrak{H}} \chi(R)x_R$ is a linear factor of the group determinant $\Theta = \det(x_{AB^{-1}})$. Frobenius proved this in his letter to Dedekind.[1] It is likely that Dedekind had anticipated Frobenius' line of argument. Whether Dedekind had proved that conversely, every linear factor $\Phi = \sum_{R\in\mathfrak{H}} c_R x_R$ of Θ is such that $\chi(R) = c_R$ defines a Dedekind character on $\mathfrak{H}$ is less certain. In his letter (as quoted in the previous section), he referred to a "clue" that "suggests" that the number of linear factors equals the number of Dedekind characters on $\mathfrak{H}/\mathfrak{H}'$, so he may not have attempted to prove the converse. The way Frobenius proved it in his letter involved a line of thought that became fundamental to his study of the group determinant. It may have been encouraged by his study of the polynomial $F[x_R]$ in his 1884 paper [192].

Let $x_E, x_A, x_B, \ldots$ and $y_E, y_A, y_B, \ldots$ denote two independent sets of group variables and consider the corresponding group determinants $\Theta(x) = \det(x_{AB^{-1}})$ and $\Theta(y) = \det(y_{AB^{-1}})$. Then if $z_{P,Q}$ denotes the row P, column Q entry of the matrix product $(x_{AB^{-1}}) \cdot (y_{AB^{-1}})$, it follows from the multiplicative property of determinants that $\det(z_{P,Q}) = \Theta(x)\Theta(y)$. Since by definition,

$$z_{P,Q} = \sum_{R\in\mathfrak{H}} x_{PR^{-1}} y_{RQ^{-1}} = \sum_{AB=PQ^{-1}} x_A y_B,$$

it follows that $z_{P,Q}$ depends only on the group element $C = PQ^{-1}$, so that we may write $z_{P,Q} = z_{PQ^{-1}} = z_C$. In other words $\det(z_{P,Q}) = \Theta(z)$, the group determinant with respect to the group variables z_C, and the multiplicative property of the group determinant takes the following form.

Theorem 13.1. *If* $z_C = \sum_{AB=C} x_A y_B$, *then* $\Theta(z) = \Theta(x) \cdot \Theta(y)$.

---

[1]The proof was as follows. (It is helpful to refer to (12.1).) Multiply column "B" of Θ, the column with entries $x_{AB^{-1}}$ with B fixed, by $\chi(B^{-1})$ and add the resulting column to the first column. Once every column is so multiplied and added to the first, the row "A" entry of column 1 is now

$$x_{AE^{-1}} + \sum_{B\neq E} \chi(B^{-1})x_{AB^{-1}} = \sum_{B\in\mathfrak{H}} \chi(B^{-1})x_{AB^{-1}}. \tag{13.2}$$

With $R = AB^{-1}$ in (13.2), the row A entry of column 1 takes the form

$$\sum_{R\in\mathfrak{H}} \chi(A^{-1}R)x_R = \chi(A^{-1})\Phi, \quad \Phi = \sum_R \chi(R)x_R.$$

Thus every entry in column 1 contains the factor Φ, which means that Φ is a factor Θ.

In connection with this theorem, Frobenius referred in his letter to his analogous Theorem 12.4, for $F[x_R]$—the only difference being the telltale factor $\binom{B}{BC} = \pm 1$. Theorem 12.4, however, was a bona fide theorem: it had been deduced from the complicated properties of theta functions rather than from the elementary multiplicative property of determinants; and despite its remarkable nature, it had found little application in Frobenius' paper [192]. By contrast, Theorem 13.1 was such a simple consequence of the definition of Θ and the multiplicative property of determinants that Frobenius seems to have sensed that it should play a fundamental role in the ensuing theory. Indeed, he began by using it to prove that every linear factor Φ of Θ is given by a Dedekind character χ.

The line of reasoning that Frobenius used to this end can be extended to any irreducible factor Φ of the group determinant (as he realized further on in the letter) to show that the irreducible factors of Θ possess the same multiplicative property as Θ:

Theorem 13.2. *If $\Phi(x)$ is any irreducible factor of $\Theta(x)$, then when $z_C = \sum_{AB=C} x_A y_B$ it follows that $\Phi(z) = \Phi(x) \cdot \Phi(y)$.*

Before indicating Frobenius' proof of this theorem, some notational conventions are in order. Let $\Phi_\lambda(x)$, $\lambda = 1, \ldots, l$, denote the distinct irreducible factors of $\Theta(x)$ and let f_λ denote the degree of Φ_λ. Finally, let

$$\Theta = \prod_{\lambda=1}^{l} \Phi_\lambda^{e_\lambda} \tag{13.3}$$

denote the factorization of Θ into irreducible factors. Since the diagonal entries of $\Theta(x) = \det(x_{AB^{-1}})$ are all x_E, $\Theta = x_E^h + \cdots$, where h is the order of $\mathfrak{H}$. Thus by (13.3), $\Phi_i = c_i x_E^{f_i} + \cdots$, where $\prod_{i=1}^{l} c_i^{e_i} = 1$. Since the distinct irreducible factors are uniquely determined only up to a constant multiple, without loss of generality it may be assumed that each $c_i = 1$. Then if we set $x = e$, where

$$e = (\varepsilon_E, \varepsilon_A, \varepsilon_B, \ldots) \quad \text{and} \quad \varepsilon_R = \begin{cases} 1 & \text{if } R = E, \\ 0 & \text{if } R \neq E, \end{cases} \tag{13.4}$$

it follows that $\Phi(e) = 1$.

To establish Theorem 13.2, Frobenius observed that Theorem 13.1 implies

$$\Phi(z) = M(x)N(y),$$

where M, N are polynomials in the x_R and y_R, respectively. Although he did not indicate the considerations underlying this conclusion, since they seemed so obvious to him, for historical reasons they are worth noting. The prime factorization of $\Theta(x)$ in the unique factorization ring $\mathbb{C}[x_E, x_A, x_B, \ldots]$ is given by (13.3). Although Frobenius did not employ such terminology, the fact that polynomial rings have properties analogous to the ring of integers and rings of algebraic integers in number fields was second nature to him. Kronecker had emphasized the importance of an arithmetic approach to algebra in his lectures in Berlin since the 1860s and in publications in the early 1880s such as [363–365]; and the same idea underlay Dedekind's 1882 paper with Weber on the theory of algebraic functions [121]. Not only was Frobenius well acquainted with the work of Kronecker and Dedekind, as we have seen, he himself had developed an arithmetic approach to Weierstrass' theory of elementary divisors based on the same idea (Section 8.6.2). In studying the group determinant, he always had this arithmetic point of view in mind. In fact, although in his letter he referred to the Φ_i as "irreducible factors" (as did Kronecker), in print he employed the more suggestive term "prime factors" [212]. The relation $\Phi(z) = M(x)N(y)$ follows immediately from such arithmetic considerations. That is,

$$\Theta(z) = \Theta(x)\Theta(y) = \left[\prod_{\lambda=1}^{k} \Phi_\lambda(x)^{e_\lambda}\right] \cdot \left[\prod_{\lambda=1}^{k} \Phi_\lambda(y)^{e_\lambda}\right],$$

and since $\Phi(z)$ divides $\Theta(z)$ and the $\Phi_\lambda(x)$ and $\Phi_\lambda(y)$ are prime factors in the polynomial ring $\mathbb{C}[x_E, x_A, \ldots, y_E, y_A, \ldots]$, it follows that $\Phi(z) = M(x)M(y)$, where $M(x)$ is some product of the $\Phi_\lambda(x)$, and $N(y)$ is some product of the $\Phi_\lambda(y)$.

To deduce Theorem 13.2 from $\Phi(z) = M(x)N(y)$, Frobenius employed the technique of variable specialization that had proved valuable in dealing with the "theta characteristic" polynomial $F[x_R]$. Setting $y = e$ (as defined in (13.4)) implies that $z_C = \sum_{AB=C} x_A y_B = x_C$, so that $\Phi(z) = M(x)N(y)$ becomes $\Phi(x) = M(x)N(e)$. Similarly, $\Phi(y) = M(e)N(y)$. Finally, setting $x = y = e$ in $\Phi(z) = M(x)N(y)$ yields $1 = \Phi(e) = M(e)N(e)$, and so it follows that

$$\Phi(x)\Phi(y) = M(x)N(e)M(e)N(y) = M(x)N(y) = \Phi(z),$$

and therefore Theorem 13.2 holds. (Cf. [212, p. 42].)

Once Theorem 13.2 is established, it follows by direct calculation that when $\Phi = \sum_{R\in\mathfrak{G}} c_R x_R$ is a linear factor of $\Theta(x)$, the relation $\Phi(z) = \Phi(x)\Phi(y)$ entails that the coefficients c_R satisfy $c_A c_B = c_{AB}$ and so define a Dedekind character $\chi(R) = c_R$ on $\mathfrak{H}$. Thus *every* linear factor of $\Theta(x)$ is determined by a Dedekind character on $\mathfrak{H}$ and hence (as we saw above) by a character on the abelian group $\mathfrak{H}/\mathfrak{H}'$. To complete the proof of Dedekind's conjecture, Frobenius had to show that no linear factor of Θ is repeated in the factorization of Θ. Frobenius did this by making clever use of basic properties of determinants. In this manner, Frobenius arrived at a proof of Dedekind's conjectured theorem:

Theorem 13.3. *Every linear factor of $\Theta(x)$ is of the form*

$$\Phi(x) = \sum_{R\in\mathfrak{H}} \chi(R) x_R,$$

where χ is a Dedekind character on $\mathfrak{H}$. Moreover, every linear factor of $\Theta(x)$ occurs only to the first power in the factorization (6.6) of Θ. Hence the number of linear factors in Θ is equal to the order of the abelian group $\mathfrak{H}/\mathfrak{H}'$, where $\mathfrak{H}'$ is the commutator subgroup of $\mathfrak{H}$.

With this result, Frobenius had confirmed what Dedekind had suspected. The result must have been satisfying because of the way it related properties of the factorization of Θ to the structure of the underlying group $\mathfrak{H}$. Of course, both Dedekind and Frobenius hoped that there would be further results of this kind to establish. Indeed, Dedekind's factorization of Θ for the symmetric group $\mathfrak{S}_3$ and the quaternion group (Section 12.2), which he had sent to Frobenius in a brief letter dated 6 April,[2] suggested the following theorems:

Theorem 13.4. *A linear change of variables is possible such that each irreducible factor Φ_λ becomes a function of its own set of v_λ variables.*

Theorem 13.5. $v_\lambda = e_\lambda f_\lambda$.

Frobenius himself factored Θ for the dihedral group of order 8—the only other nonabelian group of order 8 besides the quaternion group (as he noted)—and found these theorems confirmed in this case as well, with Θ factoring over $\mathbb{C}$ into four distinct linear factors and one quadratic factor with multiplicity two.

Nonetheless, after giving the proof of Theorem 13.3 in his letter, he indicated that he did not have much of an idea as to how next to proceed: "Naturally all the irreducible factors $[\Phi]\ldots$ of $[\Theta]\ldots$, and the powers to which they occur, must derive from the group ... $[\mathfrak{H}]$. But I still have no idea how." What follows this passage is a portion of the letter that was apparently written in stages, as Frobenius explored different approaches to the study of Θ and its factorization. At one point, he raised the question whether Theorems 13.4 and 13.5 were generally true. Further on, he finally proved Theorem 13.4 but not, he notes, Theorem 13.5. After some further inconclusive and, in fact, confusing computations relative to Theorem 13.5, he turned to nonmathematical matters; and after writing what was seemingly intended as the concluding paragraph of the letter, he suddenly announced that he had succeeded in proving Theorem 13.5. By taking into consideration the progressive nature of the letter, we obtain a rather good idea of how Frobenius waged his war—to use his metaphor—against the group determinant. In particular, we see how the generalized characters emerged and gradually forced themselves more and more into a central place in his research strategy, thereby setting the stage for the breakthrough that inspired his letter of 17 April, which is discussed in the following section.

Frobenius' initial assaults on Θ were based on the following strategy: "One must now transfer the properties of $[\Theta]\ldots$ to $[\Phi]\ldots$." Theorem 13.2 was one example of a successful execution of this strategy. It had derived from the multiplicative property of determinants, and Frobenius considered other properties of determinants that he

[2]The mathematical portion of this letter is given in Dedekind's *Werke* 2, pp. 423ff.

could apply to Θ and then, hopefully, to the irreducible factors as well. One such property of determinants is that for any $n \times n$ matrix $A = (a_{i,j})$,

$$A \cdot \operatorname{Adj} A = (\det A)I \tag{13.5}$$

where $\operatorname{Adj} A$ is the corresponding adjoint matrix. The (i,j) entry of $\operatorname{Adj} A$, the (j,i) cofactor of A, was frequently expressed in the nineteenth century in the form $[\operatorname{Adj} A]_{i,j} = \partial D/\partial a_{j,i}$, where $D = \det A$. Thus (13.5) implies that the following relation holds for the off-diagonal entries in (13.5):

$$\sum_{j=1}^{n} a_{ij} \frac{\partial D}{\partial a_{kj}} = 0, \quad i \neq k.$$

Applied to the group determinant Θ, this relation becomes, after some simplification,

$$\sum_{R} \frac{\partial \Theta}{\partial x_R} x_{AR} = 0, \quad A \neq E. \tag{13.6}$$

In order to transfer (13.6) to the irreducible factors, Frobenius set $\Theta = \prod_{\lambda=1}^{l} \Phi_\lambda^{e_\lambda}$ in (13.6) and applied the product formula for derivatives, which implies

$$\frac{\partial \Theta}{\partial x_R} = \sum_{\lambda} e_\lambda \frac{\Theta}{\Phi_\lambda} \frac{\partial \Phi_\lambda}{\partial x_R}. \tag{13.7}$$

Using (13.7), the relation (13.6) becomes, after some straightforward manipulations,

$$\sum_{\lambda=1}^{l} e_\lambda \Theta_\lambda \left\{ \sum_{R} \frac{\partial \Phi_\lambda}{\partial x_R} x_{AR} \right\} = 0, \quad A \neq E, \tag{13.8}$$

where

$$\Theta_\lambda = \prod_{\mu \neq \lambda} \Phi_\mu. \tag{13.9}$$

Now since Φ_λ and Θ_λ are relatively prime, whereas Φ_λ divides Θ_ν for $\nu \neq \lambda$, (13.8) implies that Φ_λ must divide $\Delta_\lambda^{(A)} = \sum_R \frac{\partial \Phi_\lambda}{\partial x_R} x_{AR}$. This must be true, in particular, if Φ_λ and $\Delta_\lambda^{(A)}$ are regarded as polynomials in x_E, i.e., as elements in $\mathbb{C}[x_E]$. Suppose, suppressing the subscript λ, we express $\Phi = \Phi_\lambda$ as a polynomial in x_E with the notation

$$\Phi = x_E^f + \left(\sum_{A \neq E} \psi(A) x_A \right) x_E^{f-1} + \cdots, \tag{13.10}$$

so that $\psi(A)$ denotes the coefficient of $x_E^{f-1}x_A$ in Φ. Direct calculation then shows that

$$\Delta^{(A)} = \sum_R \frac{\partial \Phi}{\partial x_R} x_{AR} = \psi(A^{-1})x_E^f + \cdots, \quad A \neq E. \tag{13.11}$$

Since Φ divides $\Delta^{(A)}$, Frobenius could deduce from (13.10) and (13.11) that

$$\Delta^{(A)} = \sum_R \frac{\partial \Phi}{\partial x_R} x_{AR} = \psi(A^{-1})\Phi. \tag{13.12}$$

Since $\Delta^{(E)} = \sum_R \frac{\partial \Phi}{\partial x_R} x_R = f\Phi$, (13.12) will remain valid for $A = E$ if $\psi(E)$ is set equal to f, and so Frobenius did this. In sum, for each irreducible factor Φ_λ of Θ he defined the function $\psi^{(\lambda)}$ on $\mathfrak{H}$ by

$$\psi^{(\lambda)}(A) = \begin{cases} \text{coefficient of } x_E^{f_\lambda - 1}x_A \text{ in } \Phi_\lambda & \text{if } A \neq E, \\ f_\lambda & \text{if } A = E. \end{cases} \tag{13.13}$$

Equation (13.12) represents the property of the irreducible factor Φ that derives from the relation (13.6) for Θ, which, in turn, is simply an expression of the basic relation (13.5) from the theory of determinants as specialized to group determinants. It turns out that the function ψ introduced in this manner is $\psi(A) = \operatorname{tr}\sigma(A)$, i.e., ψ is the character associated to the irreducible representation σ of $\mathfrak{H}$ that corresponds to Φ as indicated in the introduction to Chapter 12.

When Φ is a linear factor, (13.10) with $f = 1$ and $\psi(E) = f$ implies that $\Phi = \sum_A \psi(A)x_A$, whereas, as Frobenius had shown in proving Theorem 13.3, $\Phi = \sum_A \chi(A)x_A$, so that $\psi = \chi$, a Dedekind character. The functions ψ could thus be regarded as a generalization of Dedekind's characters. Of course, Frobenius had not set out purposely to generalize the notion of a character, and nowhere in his letter of 12 April does he stress the idea of the functions ψ as generalized characters. But as his investigation proceeded, as he obtained further relationships by a variety of strategies, the importance of (13.12) and of the function ψ became increasingly apparent. In Frobenius' capable hands, ultimately the entire theory of group characters was to flow from (13.12).

In his letter, after deriving (13.7), Frobenius tried another line of attack, which had as its starting point the multiplicative property of the irreducible factors as expressed in Theorem 13.2. He tried an idea used on $F[x_R]$ as explained in Section 12.4: specialize the variables and see what the multiplication property becomes. If $y_{R^{-1}} = 1$ and $y_H = 0$ for $H \neq R^{-1}$, then $z_A = x_{AR}$, and the multiplicative property becomes

$$\Phi(x_{AR}, x_{BR}, \ldots) = \chi(R^{-1})\Phi(x_A, x_B, \ldots), \tag{13.14}$$

where in general, $\chi(H)$ denotes the coefficient of x_H^f in Φ. Thus $\Phi(y)$, with the above specialization, viz., $y_H = \delta_{H,R^{-1}}$, equals $\chi(R^{-1})$. From (13.14), it follows that $\chi(HK) = \chi(H)\chi(K)$, so that χ is a Dedekind character.

A third line of attack is also reminiscent of the approach used to deal with $F[x_R]$: See how Θ can be factored if all x_R are set equal to 0 except those belonging to some set $\mathfrak{S}$. In keeping with the notation of Section 12.4, the resulting specialized determinant will be denoted by $\Theta^*(x)$. Frobenius tried this with $\mathfrak{S} = \mathfrak{G}$, a subgroup of $\mathfrak{H}$ of order g. By partitioning the elements of $\mathfrak{H}$ into disjoint cosets $\mathfrak{G}H$, he saw that $\Theta^*(x) = \Delta^{h/g}$, where Δ denotes the (unspecialized) group determinant of $\mathfrak{G}$. When $\mathfrak{G}$ is the cyclic group generated by an element R of order r, then by Dedekind's result for the abelian case, Δ factors into linear factors. Consequently, with this choice of $\mathfrak{G}$,

$$\Theta^*(x) = \prod_{\rho} \left(x_E + \rho x_R + \rho^2 x_{R^2} + \cdots + \rho^{r-1} x_{R^{r-1}}\right)^{h/r}, \tag{13.15}$$

where ρ runs through all rth roots of unity. Although Frobenius apparently forgot to mention it until the beginning of his next letter, where he summarized results from the 12 April letter before presenting new results, it follows from (13.15) that $\psi^{(\lambda)}(R)$ is a sum of f_λ rth roots of unity and so is one of Dedekind's algebraic integers.[3] As we will see, in his next letter (17 April) Frobenius deduced further information about the values of the $\psi^{(\lambda)}$ from (13.15).

It is at this point in the letter that Frobenius raised the question of the general validity of Theorems 13.4 and 13.5, and the remainder of the letter is largely devoted to his attempts to prove them. The letter gives the distinct impression, however, that in developing the consequences to the three approaches to Θ as sketched above, Frobenius was simply communicating some potentially useful ideas and relationships to Dedekind. Then he became interested in the possibility of proving Theorems 13.4 and 13.5, and the next stage of the letter contains the first fruits of that interest: a proof of Theorem 13.4. The proof is based on further properties of the function ψ that derive from (13.12). It is not surprising that he should concentrate on ψ and (13.12), for their importance was implied by the results already obtained. For example, (13.14) indicates that the coefficient of $x_A^{f-1} x_B$ in Φ is $\chi(A)\psi(BA^{-1})$ and hence is expressible in terms of χ and ψ. From this fact and (13.12), it can be seen with a bit of effort that all coefficients of terms in Φ are expressible in terms of χ and ψ. Let us now see what further relations involving ψ Frobenius uncovered en route to his proof of Theorem 13.4.

Equation (13.12) can be regarded as the "Hth" equation in a linear system of h equations in the unknowns $\partial\Phi/\partial x_R$, viz.,

$$\sum_R \frac{\partial \Phi}{\partial x_R} x_{HR} = \psi\left(H^{-1}\right)\Phi, \quad H = E, A, B, \ldots.$$

[3]Since Φ_λ and Φ_λ^* both have the same coefficient for the term $x_E^{f-1} x_R$, namely $\psi^{(\lambda)}(R)$ by the definition at (13.13), and since Φ_λ^* consists of f_λ of the factors in (13.15), it follows that the coefficient of $x_E^{f-1} x_R$ in Φ_λ^* is a sum of f_λ of the roots of unity ρ in (13.15).

Solving this system for $\frac{\partial\Phi}{\partial x_R}$ by Cramer's rule yields the relation

$$\frac{h}{\Phi}\frac{\partial\Phi}{\partial x_R} = \sum_H \psi\left(RH^{-1}\right)\frac{1}{\Theta}\frac{\partial\Theta}{\partial x_H}. \tag{13.16}$$

As in deriving (13.12), Frobenius next translated this into a relation involving all the irreducible factors by setting $\Theta = \prod_\lambda \Phi_\lambda^{e_\lambda}$ in (13.16) to obtain via (13.7) the relation

$$\frac{h}{\Phi}\frac{\partial\Phi}{\partial x_R} = \sum_\lambda e_\lambda \left\{\sum_H \psi\left(RH^{-1}\right)\frac{1}{\Phi_\lambda}\frac{\partial\Phi_\lambda}{\partial x_H}\right\}. \tag{13.17}$$

From this equation he then concluded that all bracketed terms on the right-hand side must vanish except the term with $\Phi_\lambda = \Phi$.[4]

The fact that all the bracketed terms on the right hand side of (13.17) vanish, except for the term corresponding to the irreducible factor Φ, can be expressed in the following form:

$$\frac{h}{e}\frac{\partial\Phi}{\partial x_R} = \sum_H \psi\left(RH^{-1}\right)\frac{\partial\Phi}{\partial x_H}$$

and

$$0 = \sum_H \psi\left(RH^{-1}\right)\frac{\partial\Phi'}{\partial x_H} \qquad (\Phi' \neq \Phi).$$

He then compared the coefficient of $x_A^{f-1}x_R$ on both sides of these equations, using the fact that the coefficient of $x_A^{f-1}x_R$ in Φ is, by virtue of (13.14) (as noted above), $\chi(A)\psi(RA^{-1})$. The results of the comparison are

$$\sum_R \psi\left(AR^{-1}\right)\psi\left(RB^{-1}\right) = \frac{h}{e}\psi\left(AB^{-1}\right), \tag{13.18}$$

[4]Although Frobenius gave no justification for this conclusion, he undoubtedly used the same sort of divisibility considerations he had employed to obtain (13.12). (This assumption is supported by Frobenius' published account [212, §5], where such considerations are explicitly employed.) That is, if (13.17) is multiplied through by the product of all the irreducible factors $\prod_\lambda \Phi_\lambda$, and if we set $\Phi = \Phi_{\lambda_0}$, it may be rewritten in the form

$$\Theta_{\lambda_0}\left[h\frac{\partial\Phi_{\lambda_0}}{\partial x_R} - e_{\lambda_0}\Delta_{\lambda_0}\right] = \sum_{\lambda\neq\lambda_0} e_\lambda\Theta_\lambda\Delta_\lambda, \quad \Delta_\lambda = \sum_H \psi\left(RH^{-1}\right)\frac{\partial\Phi_\lambda}{\partial x_H}.$$

Now Φ_{λ_0} does not divide Θ_{λ_0}, nor can it divide $h\frac{\partial\Phi_{\lambda_0}}{\partial x_R} - e_{\lambda_0}\Delta_{\lambda_0}$, which is homogeneous of degree $f_{\lambda_0} - 1$. Since, however, Φ_{λ_0} divides Θ_λ for all $\lambda \neq \lambda_0$, it divides the right-hand side of the above equation, which means that the right-hand side must vanish identically. Since neither e_λ nor Θ_λ is 0, this means that $\Delta_\lambda = 0$ for all $\lambda \neq \lambda_0$.

$$\sum_R \psi\left(AR^{-1}\right)\psi'\left(RB^{-1}\right) = 0 \quad (\Phi' \neq \Phi). \tag{13.19}$$

These relations are a precursor of the now familiar *first orthogonality relations*. That is, if we take $R = E$ and $A = B$ in these equations, since then $\psi\left(AB^{-1}\right) = \psi(E) = f$, they imply

$$\sum_R \psi^{(\lambda)}(A)\psi^{(\mu)}\left(A^{-1}\right) = h(f/e)\delta_{\lambda,\mu}, \tag{13.20}$$

which would be the first orthogonality relations had Frobenius known at this point that $e = f$ and $k = l$. As we shall see in Sections 13.3 and 13.4, that $e = f$ turned out to be very difficult for Frobenius to prove; it was the last part of his theory of characters that he established.

The main significance of (13.18) and (13.19) for Frobenius at the moment was that they enabled him to prove Theorem 13.4. Using them, he showed that if new variables ξ_A are defined by

$$\xi_A = \sum_R \psi\left(AR^{-1}\right)x_R, \tag{13.21}$$

then Φ can be expressed as a function of the ξ_A, whereas if $\Phi' \neq \Phi$, Φ' is independent of the ξ_A. The number of linearly independent ξ_A's is equal to the rank of the matrix $(\psi(AB^{-1}))$. Another consequence of (13.18), which he simply noted in passing, is that e must divide h, since the numbers $\psi(H)$, being sums of roots of unity, are algebraic integers. (At this point in the letter Frobenius made a point of noting that he was making use of Dedekind's results on algebraic integers.)

Although the above orthogonality relations had enabled Frobenius to prove Theorem 13.4, he could not immediately see how to conclude that $v = \text{rank}\left(\psi(AB^{-1})\right)$ must equal ef. The letter was about to be concluded when he announced, "Finally, I think I have found a proof for $v = [ef]$... It does not please me at all and must be greatly simplified." The proof was indeed unsatisfactory, being long and complicated, and its details need not concern us. What is more significant is the result that the proof brought with it: "It is very remarkable that the coefficients of $[\Phi^e]$... depend only on the $\psi(R)$, while in $[\Phi]$ the $\chi(R)$ still occur. Since, however, the extraction of the [eth] root of $[\Phi^e]$... requires only rational operations, the $\chi(R)$ must be rationally expressible in terms of the $\psi(R)$." In other words, the coefficients of Φ are completely determined by the values of the corresponding function ψ.[5] Certainly now there could be no doubt about the value of the functions ψ for the theory of the group determinant. For linear Φ's, ψ was a Dedekind character. Furthermore, Theorems 13.4 and 13.5 and the corollary $h = \sum_\lambda e_\lambda f_\lambda$ were all derived from the properties of ψ, as was the fact that e divides h.

At the same time, Frobenius had some reasons to be a bit disappointed. His theorems were not as satisfying as Dedekind's conjectured Theorem 13.3, which

[5]Frobenius proved this differently when he finally published his results [212, p. 46]. See in this connection the interesting observations and computations by Conrad [106, pp. 380–383].

linked a property of Θ (the number of linear factors) with the structure of $\mathfrak{H}$ (the index of the commutator subgroup). Also, Theorem 13.5 had required a terribly long proof. That proof definitely bothered Frobenius, and after giving it he asked Dedekind for some help: "Through your investigations on numbers [formed] from several units you are certainly completely familiar with the methods of the above research and can specify simplifications. For my conclusions are so complicated that I myself do not rightly know where the main point* of the proof is, and in fact I am still slightly mistrustful of it." Frobenius was referring to a paper on commutative hypercomplex number systems Dedekind had written in 1885.[6] Since his proof of Theorem 13.5 had made use of the fact that the matrices $(\psi(AB^{-1}))$ and $(x_{AB^{-1}})$ commute, he apparently thought Dedekind might have something helpful to say. The fact that the matrices commute follows from the fact that $\psi(AB) = \psi(BA)$. Frobenius had noted this property of ψ and its equivalent, $\psi(B^{-1}AB) = \psi(A)$, when he derived (13.16). Therefore, he added the following footnote corresponding to the asterisk in the above quotation:

> I believe it is contained in the equation $\psi(AB) = \psi(BA)$. For from it follows $\Sigma_R \psi(RA^{-1})x_{RB^{-1}} = \Sigma_R x_{AR^{-1}} \psi(RB^{-1})$ (by replacing R by ARB), i.e., the system $(x_{AB^{-1}})$ is permutable with $(\psi(BA^{-1}))$.

Frobenius thus closed his letter of 12 April meditating on the significance of the relation $\psi(AB) = \psi(BA)$. Five days later, on 17 April, he wrote jubilantly to Dedekind:

> My former colleague Schottky was and is one of the greatest optimists that I know; otherwise he would not have been able to endure my pessimism so well. He used to say: If in an investigation, after difficult mental exertion, the feeling arises that nothing will be achieved on the matter in question, then one can rejoice for he is standing before the solution. Many times I have found this truth confirmed and this time as well. At the end of my last letter I gave up the search and requested your assistance. The next day I saw, if not the entire solution, at least the way to it. My feeling that the equation $\psi(AB) = \psi(BA)$ provided the key did not deceive me. I still have a long way to go but I am certain I have chosen the right path.... Do you know of a good name for the function ψ? Versor? Or should ψ be called the character of Φ (which agrees for linear Φ)?

We shall now consider how the equation $\psi(AB) = \psi(BA)$ indicated new perspectives for Frobenius' research.

13.2 Frobenius' Letter of 17 April

The equation $\psi(AB) = \psi(BA)$ implies that the function ψ remains constant on each conjugacy class of the group $\mathfrak{H}$. That is, if $P, Q \in \mathfrak{H}$ are conjugate, so that $P = R^{-1}QR$ for some $R \in \mathfrak{H}$, then $\psi(P) = \psi(R^{-1}(QR)) = \psi((QR)R^{-1}) = \psi(Q)$. The letter of 17 April makes it clear that it was this link with the structure of the underlying group

[6] As will be seen in Section 13.3, Frobenius later discovered that the results in Dedekind's 1885 paper are intimately related to the functions ψ.

$\mathfrak{H}$ that drew Frobenius' attention. Probably he saw in this connection the possibility of obtaining results as satisfying as Dedekind's conjectured Theorem 13.3. And of course, he succeeded in doing just that by proving that the number l of distinct irreducible factors of Θ is equal to the number k of conjugacy classes of $\mathfrak{H}$, a result comparable in kind to Theorem 13.3 but more far-reaching. Although Frobenius' letter of 17 April is already a more or less polished exposition of his new discoveries, it does suggest how he came to make them. As I shall attempt to show in what follows, the invariance of the functions ψ on the conjugacy classes probably quickly led him to suspect that $k = l$. Then in order to prove it, he further developed the consequences of this invariance in conjunction with his earlier results on the group determinant as presented in his letter of 12 April. These further consequences not only produced the desired proof, but provided Frobenius with other new theorems and unexpected connections with earlier work by him and also by Dedekind.

Since Frobenius had ended his letter of 12 April by using the orthogonality relations (13.16) and (13.17) to prove Theorems 13.4 and 13.5, he probably began by seeing what the relationships satisfied by the functions ψ look like when their invariance on the conjugacy classes is taken into account, i.e., in a notation that reflects this fact. Thus let $(1),\ldots,(\alpha),\ldots,(k)$ denote the k conjugacy classes, where $(1) = \{E\}$.[7] If $P, Q \in (\alpha)$ so that $P = R^{-1}QR$ for some $R \in \mathfrak{H}$, then $P^{-1} = R^{-1}Q^{-1}R$, so the inverses of the elements in (α) form a conjugacy class, which will be denoted by (α'). The number of elements in (α) will be denoted by h_α. It is easily seen that $h_{\alpha'} = h_\alpha$. Finally, let $\psi_\alpha^{(\lambda)}$ denote the value of the character corresponding to Φ_λ on the class (α). In this notation the orthogonality relations (13.20) can be written as

$$\sum_{\kappa=1}^{k} h_\kappa \psi_\kappa^{(\lambda)} \psi_{\kappa'}^{(\mu)} = \frac{h}{e_\lambda} f_\lambda \delta_{\lambda\mu}, \quad \lambda,\mu = 1,\ldots,l. \tag{13.22}$$

This is, in fact, the first thing that Frobenius wrote down in his letter after the necessary preliminaries, including the above notation. The significance of the equation is, as Frobenius observed, that it indicates a relation between the number of conjugacy classes k and the number l of distinct irreducible factors Φ. As he explained, in the language of matrices, the orthogonality relations (13.22) assert that if we introduce the $l \times k$ and $k \times l$ matrices

$$M = (M_{\lambda\kappa}) = \left(\psi_\kappa^{(\lambda)} \frac{e_\lambda}{f_\lambda h}\right) \quad \text{and} \quad N = (N_{\kappa\lambda}) = \left(h_\kappa \psi_{\kappa'}^{(\lambda)}\right), \tag{13.23}$$

then they are "reciprocal," i.e., $MN = I_l$. It therefore follows by elementary linear algebra that $l \leq k$.[8] Probably this inequality was the first thing Frobenius discovered

[7] Here I diverge slightly from Frobenius, who denoted the classes by $(0),\ldots,(k-1)$, with $(0) = \{E\}$.

[8] Frobenius gave no reason. He most likely used the fact that $l = \operatorname{rank} I_l = \operatorname{rank}(MN) \leq \min\{\operatorname{rank} M, \operatorname{rank} N\} \leq \min\{k,l\}$. Thus if it were the case that $l > k$, the preceding inequalities would imply the contradiction that $l \leq k$.

as a consequence of the new way of regarding the characters. It is the first result mentioned in the letter, and it undoubtedly encouraged him to hope that $k = l$ and, with an eye toward proving it, to further develop the consequences of the new notation.

In the letter, after showing that $l \leq k$, Frobenius wrote, "Since I did everything in my head, I cannot at the moment recall how I showed that $k = l$." What then follows in the letter are further consequences of the new notation and viewpoint that were to prove decisive for his development of the theory of the group determinant. No doubt in proceeding to lay out these consequences for Dedekind, Frobenius hoped that they would remind him of how the proof that $k = l$ had gone. If so, his hopes were realized, because near the end of the letter, he suddenly announced, "Now I will indeed finally show that $k = l$." And he proceeded to do so. It turns out that much of the intervening material in the letter was irrelevant to the proof, except for the psychological factor of jogging his memory for the missing pieces of the proof. In what follows, I expound Frobenius' proof without the many detours found in the letter. However, one is worth mentioning before we turn to the proof that $k = l$.

In his letter of 12 April, Frobenius had derived the formula (13.15), which gives the value of Θ when it is specialized by picking an $R \neq E$ of order r in $\mathfrak{H}$ and setting all variables equal to zero except for $x_E, x_R, x_{R^2}, \ldots, x_{R^{r-1}}$. In his letter of 17 April, Frobenius further specialized (13.15) by also setting the variables $x_{R^2}, \ldots, x_{R^{r-2}}$ equal to 0, assuming $r > 2$.[9] Thus (13.15) becomes

$$\Theta^{**} = \prod_{\rho} \left(x_E + \rho x_R + \rho^{-1} x_{R^{-1}}\right)^{h/r}, \tag{13.24}$$

where the product is over all rth roots of unity ρ. Since Φ_λ^{**} divides Θ^{**} and $f_\lambda = \deg \Phi_\lambda$, (13.24) implies that for every λ, the above variable specialization yields Φ_λ^{**} as a product of f_λ of the factors in (13.24). This means that the coefficient of $x_E^{f_\lambda - 1} x_R$ in Φ_λ^{**} is a sum of f_λ rth roots of unity, and the coefficient of $x_E^{f_\lambda - 1} x_{R^{-1}}$ in Φ_λ^{**} is the sum of the reciprocals of those same roots of unity. Since for any root of unity, its inverse is its complex conjugate, it follows from the definitions of $\psi^{(\lambda)}(R)$ and $\psi^{(\lambda)}(R^{-1})$ in (13.13) that they are complex conjugates. As Frobenius remarked in a footnote, when $r = 2$, so that $R^{-1} = R$, $\psi^{(\lambda)}(R)$ is real-valued, as is the case with Dedekind's characters. Thus $\psi^{(\lambda)}(R^{-1}) = \overline{\psi^{(\lambda)}(R)}$ for all $R \in \mathfrak{H}$.[10]

Let us now return to Frobenius' proof that $k = l$. Having introduced the matrices M and N and realized that $MN = I_l$ implies $l \leq k$, Frobenius, whose bailiwick was

---

[9] See following Frobenius' equation (28.).

[10] In his first paper on the theory of the group characters $\psi^{(\lambda)}$, Frobenius gave another nonelementary proof that $\psi^{(\lambda)}(R^{-1}) = \overline{\psi^{(\lambda)}(R)}$ for all $R \in \mathfrak{H}$ [211, p. 11]. Once he realized (in 1897—see (13.44)) that $\psi^{(\lambda)}(R)$ is simply the trace function of an irreducible matrix representation of $\mathfrak{H}$, this result followed from the basic linear algebra he had developed in response to the Cayley–Hermite problem (Section 7.5).

linear algebra, realized that $k = l$ *if and only if* $M^t N^t = I_k$. For if $k = l$, so M and N are square matrices, then $MN = I$ implies that M and N are invertible and $N = M^{-1}$, so that $NM = I$ and therefore, taking transposes, $M^t N^t = I$. On the other hand, suppose k and l are not known to be equal but that $M^t N^t = I_k$. This is equivalent to

$$\sum_{\lambda=1}^{l} \frac{e_\lambda}{f_\lambda} h_\kappa \psi_\kappa^{(\lambda)} \psi_{\nu'}^{(\lambda)} = h\delta_{\kappa\nu}, \quad \kappa,\nu = 1,\dots,k. \tag{13.25}$$

The relations (13.25) are a precursor of the now familiar *second orthogonality relations*, the difference being that Frobenius did not yet know that $k = l$ and that $e_\lambda = f_\lambda$.

Frobenius could see that (13.25), when combined with (13.22), implies $k = l$. For if we take $\mu = \lambda$ in (13.22), we can use the result to write

$$X \stackrel{\text{def}}{=} \sum_{\lambda=1}^{l} \left\{ \sum_{\kappa=1}^{k} \frac{e_\lambda}{f_\lambda} h_\kappa \psi_\kappa^{(\lambda)} \psi_{\kappa'}^{(\lambda)} \right\} = \sum_{\lambda=1}^{l} h\delta_{\lambda\lambda} = hl. \tag{13.26}$$

If we reverse the order of summation in the expression for X, we get

$$X = \sum_{\kappa=1}^{k} \left\{ \sum_{\lambda=1}^{l} \frac{e_\lambda}{f_\lambda} h_\kappa \psi_\kappa^{(\lambda)} \psi_{\kappa'}^{(\lambda)} \right\}.$$

The bracketed expression above is precisely the left-hand side of (13.25) with $\nu = \kappa$, and so (13.25) together with (13.26) implies that

$$hl = X = \sum_{\kappa=1}^{k} h\delta_{\kappa\kappa} = hk.$$

Thus $k = l$ *follows, provided* $M^t N^t = I_k$, *i.e., provided (13.25) holds.*

Although in the modern development of group representation theory the second orthogonality relations follow readily from the first, for Frobenius, struggling to develop a theory of the group determinant from scratch, and working with a precursor of the first orthogonality relations, namely (13.22), the proof that (13.25) holds was not so simple. Having for the moment forgotten how it proceeded, he next showed Dedekind how the equality $k = l$ implies that if $h_{\alpha\beta\gamma}$ denotes the number of the $h_\alpha h_\beta h_\gamma$ products ABC with $A \in (\alpha)$, $B \in (\beta)$, $C \in (\gamma)$ such that $ABC = E$, then

$$h_\alpha h_\beta \psi_\alpha^{(\kappa)} \psi_\beta^{(\kappa)} = f_\kappa \sum_{\gamma=1}^{k} h_{\alpha\beta\gamma'} \psi_\gamma^{(\kappa)}. \tag{13.27}$$

Regarding these equations—denoted by (11.) in Frobenius' letter but with the κ superscripts suppressed—he wrote:

> From these equations the unknowns ... $[\psi_1^{(\kappa)}, \psi_2^{(\kappa)}, \ldots, \psi_k^{(\kappa)}]$ ... may be calculated. They have k solutions ... $[\psi_1^{(\kappa)}, \psi_2^{(\kappa)}, \ldots, \psi_k^{(\kappa)}, \kappa = 1, \ldots, k]$ .... I recommend that you compute some examples.

Frobenius then went on to discuss other interesting discoveries (summarized below) about what happens to Θ and the Φ_λ when the variables are specialized by $x_P = x_Q$ if and only if P and Q are conjugates in $\mathfrak{H}$. He ended up rederiving (13.27) but without assuming $k = l$, and at that point in the letter, he remembered how (13.27) can be used to prove that $k = l$ by proving the second orthogonality relation (13.25). Here is how he did it.

The equations (13.25) are analogous to those constituting the first orthogonality relation (13.22) except that the summation is over the index λ, i.e., over the distinct characters rather than over the conjugacy classes as in (13.22). In his letter of 12 April, he had observed in passing the relation

$$\sum_{\lambda=1}^{l} e_\lambda \psi^{(\lambda)}(A) = 0 \quad (A \neq E).^{11} \tag{13.28}$$

Now (13.28) involves summation over λ. When $A = E$, the left-hand side of (13.28) becomes $\sum_{\lambda=1}^{l} e_\lambda \psi^{(\lambda)}(E) = \sum_{\lambda=1}^{l} e_\lambda f_\lambda = h$, since $h = \deg \Theta = \sum_\lambda \deg[\Phi_\lambda^{e_\lambda}] = \sum_\lambda e_\lambda f_\lambda$. Thus if we express A as RS, (13.28) implies that

$$\sum_{\lambda=1}^{l} e_\lambda \psi^{(\lambda)}(RS) = h\delta_{R,S^{-1}}, \tag{13.29}$$

where $\delta_{R,S^{-1}} = 0$ unless $R = S^{-1}$, i.e., unless $RS = E$. If (13.29) is summed over all $R \in (\alpha)$ and $S \in (\beta)$, it becomes

$$\sum_{\lambda=1}^{l} e_\lambda \left\{ \sum_{\gamma=1}^{k} h_{\alpha\beta\gamma'} \psi_\gamma^{(\lambda)} \right\} = h_\beta h \delta_{\alpha\beta'}.^{12} \tag{13.30}$$

If the left-hand side of (13.27) with κ replaced by λ is substituted for the bracketed expression in (13.30), then (13.30) simplifies to

$$\sum_{\lambda=1}^{l} \frac{e_\lambda}{f_\lambda} h_\alpha \psi_\alpha^{(\lambda)} \psi_\beta^{(\lambda)} = h\delta_{\alpha\beta'},$$

---

[11] The left-hand side of (13.28) represents the coefficient of $x_E^{h-1} x_A$ in $\Theta = \prod_\lambda \Phi_\lambda^{e_\lambda}$, but from the definition of $\Theta = \det(x_{AB^{-1}})$, it can be seen that this coefficient must be 0. I suspect (13.28) had slipped Frobenius' mind when, in the letter, he could not recall how to go from $k \leq l$ to $k = l$. It is introduced immediately after he announced, toward the end of his letter, that he would now prove that $k = l$.

[12] If $\alpha \neq \beta'$, then $\sum_{R\in(\alpha),S\in(\beta)} h\delta_{R,S^{-1}} = 0$. If $\alpha = \beta'$, then the sum on the left-hand side is $\sum_{R\in(\beta')} h\delta_{S^{-1}S^{-1}} = hh_{\beta'} = hh_\beta$.

which is (13.25). This completes Frobenius' proof of the following theorem:

Theorem 13.6. *The number of distinct irreducible factors of the group determinant is equal to the number of conjugacy classes of the group:* $k = l$.

After presenting his proof of Theorem 13.6 to Dedekind, he wrote, "This relation $[k = l]$ is all the more remarkable since there does not seem to be any relation between the individual prime factors Φ and the individual [conjugacy] classes ... ."

Theorem 13.6 was not the only significant discovery Frobenius had made by the time of his letter of 17 April. He had obtained several results about the factorization of the group determinant under the variable specialization associated to conjugacy, which are summarized in the following theorem.

Theorem 13.7. *Let* Θ^* *and* Φ_λ^* *denote* Θ *and* Φ_λ *with the variable specialization* $x_R \to x_\rho$ *for all* $R \in (\rho)$, $\rho = 1, \ldots, k$. *Then*

$$\Phi_\lambda^* = \left(\frac{1}{f_\lambda} \sum_{\rho=1}^{k} h_\rho \psi_{\rho'}^{(\lambda)} x_\rho \right)^{f_\lambda},$$

so that

$$\Theta^* = \prod_{\lambda=1}^{k} \left(\frac{1}{f_\lambda} \sum_{\rho=1}^{k} h_\rho \psi_{\rho'}^{(\lambda)} x_\rho \right)^{e_\lambda f_\lambda}.$$

Moreover, if $A = (a_{\alpha,\beta})$ *is the* $k \times k$ *matrix with* $a_{\alpha,\beta} = \sum_{\gamma=1}^{k} \frac{h_{\alpha\beta\gamma}}{h_\alpha} x_\gamma$, *then*

$$\det A = \prod_{\lambda=1}^{k} \left(\frac{1}{f_\lambda} \sum_{\alpha=1}^{k} h_\alpha \psi_{\alpha'}^{(\lambda)} x_\alpha \right).$$

Frobenius regarded the above equation for Φ^* as "one of the most important formulas." Together, this equation and the expression for Θ^* that follows from it represent satisfying generalizations of Dedekind's factorization of the group determinant of an abelian group, as given by (12.10). Rather than attempt an extension of the coefficient domain peculiar to each group as Dedekind had considered, Frobenius just specialized the variables to obtain the factorization into linear factors, which had been Dedekind's goal. Furthermore, Frobenius' result involved a generalization of Dedekind's characters. These results alone would certainly justify Frobenius' suggestion, at the beginning of his letter of 17 April (quoted here at the end of the previous section), to call the functions $\psi^{(\lambda)}$ the *characters* of the group $\mathfrak{H}$. Of course, when $\mathfrak{H}$ is abelian, so that conjugacy classes consist of single elements, Theorem 13.7 implies Dedekind's factorization using Dedekind characters. Frobenius obtained the above factorization for $\det A$ by consideration of the multiplication property of Theorem 13.2 for Φ in specialized form:

$$\Phi^*(\ldots z_\alpha \ldots) = \Phi^*(\ldots x_\alpha \ldots)\Phi^*(\ldots y_\alpha \ldots), \quad z_\alpha = \sum_{\beta,\gamma} \frac{h_{\alpha\beta\gamma}}{h_\alpha} x_{\beta'} y_{\gamma'}.$$

The formula for $\det A$ says that each distinct linear factor of Θ^* occurs exactly once in $\det A$, a determinant that is defined entirely in terms of the structural constants $h_{\alpha\beta\gamma}, h_\alpha$ of the group $\mathfrak{H}$. As we shall see in the following section, Theorem 13.7 became the keystone of the first published version of Frobenius' theory of group characters. Today this theorem is not even mentioned, since group determinants are no longer a part of the theory of group characters.

13.3 Frobenius' Paper "On Group Characters"

"I am going to break off here," Frobenius wrote in concluding his letter of 17 April.

> Tomorrow and the day after I will not be working. On Thursday my lectures begin. Consequently I will now have less time left over for this work, but I hope that I will succeed in what remains. I am very grateful to you for suggesting this work, which has given me immeasurable joy.

As we shall now see, Frobenius used his limited time for research to write up his results on group characters for publication—and to do it in a way that gave no inkling of the unfinished nature his research on the factorization of the group determinant.

The letters of 12 and 17 April indicate that within a period of ten days, Frobenius had obtained almost all of the basic results of his first two papers on the theory of group characters [211, 212], both of which appeared in 1896. One basic theorem was, however, missing, namely the theorem that each irreducible factor Φ of Θ occurs as often as its degree, so that $e_\lambda = f_\lambda$ in the notation of (13.3). (Formulated in terms of matrix representations, this is the theorem that an irreducible representation occurs in the regular representation as often as its degree.) It took Frobenius over five months before he managed to prove that $e = f$, thereby providing the missing link in the chain of theorems that constituted his theory of the group determinant (Section 13.4). In the meantime, he polished up his new theory of group characters. In his letter of 26 April 1896, he wrote to Dedekind:

> Since I have come no further [proving $e = f$], I have concerned myself with attractively polishing up the results so far obtained. The first [thing I did] was an act of vile ingratitude against the magnificent determinant $\Delta\,[= \Theta]$, the miraculous source from which everything wonderful has flowed: I have attempted to derive the results directly from the theory of groups .... As I write this, your work: *Theory of Complex Quantities Formed from n Units*[13] is lying next to me, and there is no end to my astonishment.

[13] This is Dedekind's 1885 paper [115].

Frobenius presented these polished results in his paper "On Group Characters," which was officially submitted to the Berlin Academy on 30 July 1896, having been already set in type and proofread.[14]

The paper is of interest for several reasons. In it, Frobenius did indeed introduce and develop his new theory of characters in a way that does not depend upon consideration of the group determinant Θ and thus in a way that is totally different from the manner in which he actually discovered the functions ψ in the course of his investigation of Θ. Undoubtedly, this approach appealed to Frobenius, since the characters arise directly from the consideration of the group $\mathfrak{H}$ and its structure and appear more directly to be generalizations of Dedekind's characters. It also had the advantage that it enabled him to temporarily bypass discussing the connection of the characters with the irreducible factors of Θ and the numbers e and f—and the question of their equality. Finally, the paper also brings out more clearly the connections with hypercomplex number systems (linear associative algebras) that are only alluded to in his letter of 17 April.

The ideas concerning hypercomplex numbers that bear upon Frobenius' work were initiated by a question raised by Gauss in his own review, in 1831, of his second essay on biquadratic residues [246], where Gaussian integers are introduced.[15] Gauss had made systematic use of complex numbers in this work, and in his review he raised the question, with them in mind, as to why "the relations between things which represent a multiplicity of more than two dimensions cannot yield other types of quantities permissible in general arithmetic ...." This question was considered, beginning in 1861, by Weierstrass, who gave it a concrete algebraic formulation. Some of his ideas were subsequently communicated to his former student H.A. Schwarz in a letter dated 19–27 June 1883, which was then published the following year as [593].

In his letter, Weierstrass considered the problem of defining an addition and multiplication for a n-dimensional system of hypercomplex numbers $\mathbf{a} = \alpha_1\mathbf{e}_1 + \alpha_2\mathbf{e}_2 + \cdots + \alpha_n\mathbf{e}_n$, $\mathbf{b} = \beta_1\mathbf{e}_1 + \beta_2\mathbf{e}_2 + \cdots + \beta_n\mathbf{e}_n$, where the α_i and β_i are real numbers and the $\mathbf{e}_i$ are the "units" of the system. Addition is defined by $\mathbf{a}+\mathbf{b} = \sum_{i=1}^n(\alpha_i+\beta_i)\mathbf{e}_i$ and multiplication by $\mathbf{ab} = \sum_{j,k=1}^n \alpha_j\beta_k(\mathbf{e}_j\mathbf{e}_k)$, where

$$\mathbf{e}_j\mathbf{e}_k = \sum_{i=1}^n a_{ijk}\mathbf{e}_i. \tag{13.31}$$

Thus the multiplication is completely determined by the constants a_{ijk}. Weierstrass assumed that the a_{ijk} are chosen so that the resulting multiplication is commutative and associative. In terms of the constants a_{ijk}, this means that

$$a_{ijk} = a_{ikj} \quad [\mathbf{e}_j\mathbf{e}_k = \mathbf{e}_k\mathbf{e}_j], \tag{13.32}$$

---

[14] See Dedekind's letter of 20 July 1896.

[15] See the beginning of Section 9.1.2.

$$\sum_{i=1}^{n} a_{irs}a_{kit} = \sum_{i=1}^{n} a_{irt}a_{kis} \quad [(\mathbf{e}_r\mathbf{e}_s)\mathbf{e}_t = (\mathbf{e}_r\mathbf{e}_t)\mathbf{e}_s]. \tag{13.33}$$

The conditions (13.32) and (13.33) make the system into what would now be called a commutative ring. To study the question of the existence of inverses, Weierstrass considered the hypercomplex equation $\mathbf{ax} = \mathbf{b}$. Written out in terms of the "coordinates" α_i, β_j, x_k of **a**, **b**, and **x**, this becomes a system of n equations in n unknowns x_k, and Weierstrass observed that for $n > 2$ and for any choice of the a_{ijk}, there always exist elements $\mathbf{a} \neq \mathbf{0}$ such that $\mathbf{ax} = \mathbf{0}$ has solutions $\mathbf{x} \neq \mathbf{0}$. In other words, systems of dimension greater than 2 always contain divisors of zero, which was Weierstrass' term for numbers **a** with the above property. Weierstrass seemed to suggest that it was the existence of divisors of zero in systems of dimension greater than 2 that Gauss had in mind when he made his remark.

The publication of Weierstrass' letter evoked a quick response from Dedekind, who perceived therein many similarities with his development of the theory of algebraic number fields in Supplement X of the second edition (1871) of Dirichlet's *Vorlesungen über Zahlentheorie* [137], which was abandoned in subsequent editions in favor of a simpler approach.[16] Thus, when Weierstrass' paper appeared in 1884, Dedekind was in a position to respond to it with a paper of his own [115] in which many of his results from Supplement X were carried over to the hypercomplex number systems studied by Weierstrass and, more generally, to systems in which the coefficients α_i, β_i take on complex values. The motivation behind Dedekind's paper was also partly polemical: by emphasizing the existence of divisors of zero, Weierstrass, Dedekind felt, had misinterpreted Gauss' remark. According to Dedekind, what Gauss had meant was that higher-dimensional commutative hypercomplex number systems do not represent anything really new; they can be understood in terms of ordinary complex numbers. Dedekind's main result indicates the mathematical basis for this interpretation of Gauss:

Theorem 13.8. *Suppose the n^3 complex numbers a_{ijk} satisfy Weierstrass' commutativity and associativity conditions (13.32) and (13.33). Let $P = (p_{ij})$ be defined by $p_{ij} = \sum_{r,s=1}^{n} a_{rsr}a_{sij}$. Then if $\det P \neq 0$, there exist complex numbers $e_i^{(s)}$, $i,s = 1,2,\ldots,n$, such that the n-tuples $(e_1^{(s)}, e_2^{(s)}, \ldots, e_n^{(s)})$, $s = 1,\ldots,n$, are linearly independent and satisfy the multiplication condition (13.31) in the sense that*

$$e_j^{(s)} e_k^{(s)} = \sum_{i=1}^{n} a_{ijk} e_i^{(s)}. \tag{13.34}$$

The condition $\det P \neq 0$—the equivalent of the three conditions assumed by Weierstrass in his above-described discussion of polynomials with hypercomplex

[16]A discussion of the relevant results in Supplement X and their relation to the results in Dedekind's paper [115] can be found in my paper [266, pp. 156–157].

coefficients—is equivalent to the condition that the radical of the hypercomplex system defined by the a_{ijk} be zero, i.e., that the system be semisimple. Although Dedekind did not use the notion of a direct sum decomposition, he showed, in effect (using Theorem 13.8), that the commutative systems satisfying $\det P \neq 0$ can be represented as a direct sum of n copies of the complex numbers, thereby further justifying his claim that these systems are not really new.[17] Concerning the numbers $e_i^{(s)}$, Dedekind also obtained the analogue for complex-valued x_i's of an arithmetic result in Supplement X, namely that if $A(x) = (A_{ij}(x))$ denotes the matrix with $A_{ij}(x) = \left(\sum_{k=1}^{n} a_{ijk}x_k\right)$, then

$$\det A(x) = \prod_{s=1}^{n} \left(\sum_{i=1}^{n} e_i^{(s)} \, x_i \right). \tag{13.35}$$

Dedekind presented (13.35) with the remark, "At the same time one obtains herewith the result, in which, conversely, everything else is contained, that ... [det $A(x)$] is a product of n linear factors" [115, p. 157].

The similarity between the relationship in Theorem 13.8 and the character relation (13.27) did not go unnoticed by Frobenius. Indeed, with a change of notation, (13.27) can be written in the form

$$h_\beta h_\gamma \psi_\beta^{(\lambda)} \psi_\gamma^{(\lambda)} = f_\lambda \sum_{\alpha=1}^{k} h_{\alpha'\beta\gamma} \psi_\alpha^{(\lambda)}. \tag{13.36}$$

As we noted, already in his letter of 17 April, Frobenius had realized that the above equations could be used to *compute* the numbers $\psi_\alpha^{(\lambda)}$, so that they could be determined using the group constants $h_{\alpha'\beta\gamma}, h_\alpha$ rather than via their definition as certain coefficients of the Φ_λ. But now Frobenius had the idea of using the equations (13.36) to *define* the $\psi_\alpha^{(\lambda)}$. That is, he realized that the numbers

$$a_{\alpha\beta\gamma} = \frac{h_{\alpha'\beta\gamma}}{h_\alpha} \tag{13.37}$$

satisfy all the conditions of Dedekind's Theorem 13.8. This means that they define a commutative hypercomplex number system, although at this time, Frobenius did not like to think in such terms. He preferred to express everything in terms of group-related notions, matrices, and determinants.[18] Nonetheless, the conditions of

---

[17] See especially equation (23) [115, p. 146]. The so-called Wedderburn theorem for semisimple associative algebras was later obtained (for the complex field) independently by T. Molien (1893) [443] and E. Cartan (1898) [60] (see Section 14.2).

[18] The numbers $a_{\alpha\beta\gamma} = \frac{h_{\alpha'\beta\gamma}}{h_\alpha}$ define the center of the group algebra associated to the group $\mathfrak{H}$, as Frobenius pointed out in 1897 [213]. The group elements $E, A, B, \ldots$ form a basis for the group algebra, and if $\mathbf{e}_\alpha = \sum_{A\in(\alpha)} A$, then the $\mathbf{e}_\alpha$ form a basis for the center, and $\mathbf{e}_\beta \mathbf{e}_\gamma = \sum_{\alpha=1}^{k} \frac{h_{\alpha'\beta\gamma}}{h_\alpha} \mathbf{e}_\alpha$.

Dedekind's Theorem 13.8 being satisfied when $a_{\alpha\beta\gamma} = h_{\alpha'\beta\gamma}/h_\alpha$, it followed that k^2 complex numbers $e_\alpha^{(\lambda)}$ exist (k being the number of conjugacy classes of $\mathfrak{H}$) such that (13.34) holds: $e_\beta^{(\lambda)} e_\gamma^{(\lambda)} = \sum_{\alpha=1}^{k} \frac{h_{\alpha'\beta\gamma}}{h_\alpha} e_\alpha^{(\lambda)}$. Comparison with (13.36) indicates that

$$\psi_\alpha^{(\lambda)} = \left(\frac{f_\lambda}{h_\alpha}\right) e_\alpha^{(\lambda)}. \tag{13.38}$$

This relation became the keystone of Frobenius' 1896 paper "On Group Characters" [211]; it allowed him to define the characters in terms of the constants $h_{\alpha\beta\gamma}$, h_α which relate directly to the group $\mathfrak{H}$. However, he did not find it necessary to rely in this connection on Dedekind's Theorem 13.8, since he realized that Theorem 13.8 followed from a theorem on commuting matrices that he had known since 1877, a theorem that showed that the numbers $e_\alpha^{(\lambda)}$ were simply the characteristic roots of certain matrices.

Frobenius' theorem from 1877, which is already stated in his letter of 17 April[19] and was published soon thereafter [209] as a preliminary to his paper on group characters, may be stated as follows [209, §7 III]:

Theorem 13.9. *Given m commuting $n \times n$ matrices $A, B, \ldots,$ there exists an ordering,*

$$a_1, a_2, \ldots, a_n,\ b_1, b_2, \ldots, b_n, \ldots,$$

of the characteristic roots of $A, B, \ldots$ such that for any rational function $f(u, v, \ldots)$ in m variables $u, v, \ldots,$ the characteristic roots of $f(A, B, \ldots)$ are $f(a_1, b_1, \ldots)$, $f(a_2, b_2, \ldots)$, $\ldots$, $f(a_n, b_n, \ldots)$.

Of course, when $f = p/q$ with $\deg q > 0$, it is tacitly assumed that $q(A, B, \ldots)$ is invertible, so that $f(A, B, \ldots)$ exists. Frobenius had published special cases of Theorem 13.9 in his paper on the Cayley–Hermite problem: when $m = 1$, so $f(A, B, \ldots) = f(A)$ [181, §3, III] and when $m = 2$ and $f(u, v) = uv$. However, as he explained [209, pp. 707–708], he had previously refrained from giving a proof of Theorem 13.9 because he believed it would follow trivially from the already proved case $m = 1$ when combined with a theorem about finite sets of commuting matrices that he hoped to prove. Since the proof of the latter theorem had eluded him, he never got around to proving Theorem 13.9 until 1896. In that year, of course, he had good reason to provide a proof of Theorem 13.9, and so he did.[20]

---

[19] In the *Excurs* following equation (13.13.).

[20] Nowadays, Theorem 13.9 is a consequence of the theorem that commuting matrices have a common characteristic vector and so can be simultaneously brought into triangular form. Frobenius' proof, however, was not along these lines; it was based completely on clever applications of the symbolic algebra of matrices.

Theorem 13.9 relates to Dedekind's Theorem 13.8 as follows. Given a hypercomplex number system with structure constants a_{ijk} satisfying the conditions of Theorem 13.8, define the matrices A_k by

$$A_k = (a_{ijk}). \tag{13.39}$$

Then these matrices commute by virtue of (13.32) and (13.33), and the matrix $A(x)$ of (13.35) is given by

$$A(x) = \sum_{k=1}^{n} x_k A_k. \tag{13.40}$$

Since $A(x) = f(A_1,\ldots,A_n)$, where $f(u_1,\ldots,u_n) = x_1u_1 + x_2u_2 + \cdots + x_nu_n$, Theorem 13.9 implies that if the characteristic roots of A_k are $r_k^{(1)},\ldots,r_k^{(n)}$, then the characteristic roots of A are the n linear functions of the x_i's, $r_1^{(s)}x_1 + r_2^{(s)}x_2 + \cdots + r_n^{(s)}x_n$, $s = 1,2,\ldots,n$. Then since the determinant of a matrix equals the product of its characteristic roots, it follows that $\det A(x) = \prod_{s=1}^{n}\bigl(\sum_{i=1}^{n} r_i^{(s)}x_i\bigr)$, which is analogous to Dedekind's factorization (13.35). Furthermore, since the a_{ijk} satisfy (13.32) as well as (13.33), it follows that

$$A_jA_k = \sum_{i=1}^{n} a_{ijk}A_i.$$

According to Frobenius' Theorem 13.9, the left-hand side of the above equation has $r_j^{(s)}r_k^{(s)}$ for its sth characteristic root, and the right-hand side has $\sum_{i=1}^{n} a_{ijk}r_i^{(s)}$ as its sth characteristic root. In other words, $r_j^{(s)}r_k^{(s)} = \sum_{i=1}^{n} a_{ijk}r_i^{(s)}$, which is Dedekind's Theorem 13.8. By such considerations as these, Frobenius re-proved Dedekind's theorem and even generalized it somewhat.

Thus in "On Group Characters" [211, pp. 7–8], after defining the numbers $h_{\alpha\beta\gamma}$ and h_α, Frobenius invoked his matrix version of Dedekind's Theorem 13.8 with $a_{\alpha\beta\gamma} = h_{\alpha'\beta\gamma}/h_\alpha$ as in (13.37) to obtain the characteristic roots $r_\alpha^{(\lambda)}$ and then defined numbers $\chi_\alpha^{(\lambda)}$ by $\chi_\alpha^{(\lambda)} = (f_\lambda/h_\alpha)r_\alpha^{(\lambda)}$, since as we have seen from (13.38), he knew this was true for the "real" characters $\psi_\alpha^{(\lambda)}$, although now f_λ is simply a proportionality factor "temporarily left undetermined" [211, p. 7] rather than the degree of Φ_λ. Frobenius explained rather cryptically that he would "later" have something definite to say about the k factors f_λ and that he would "then call the quantities ... $[\chi_\alpha^{(\lambda)}]$ ... the k characters of the group $\mathfrak{H}$." For the time being, however, he just called them "characters." He then went on to prove that these characters possess many of the properties he had already discovered for the functions $\psi^{(\lambda)}$. In particular, he showed that the $\chi^{(\lambda)}$ satisfy the orthogonality relations (13.22) and (13.25), where, however, the numbers e_λ that occur there are now simply "new proportionality factors" that turn out to have the property that $\sum_\lambda e_\lambda f_\lambda = h$, h being the order of $\mathfrak{H}$.

Although Frobenius introduced the group determinant $\Theta = |x_{PQ^{-1}}|$ in "On Group Characters," he did not consider the factorization $\Theta = \prod_\lambda \Phi_\lambda^{e_\lambda}$ but instead showed that the specialized determinant Θ^* factors into linear factors with the characters involved in the coefficients:

$$\Theta^*(x_1,\ldots,x_k) = \prod_{\lambda=1}^{k} \left(\frac{1}{f_\lambda} \sum_{\alpha=1}^{k} h_\alpha \chi_\alpha^{(\lambda)} x_\alpha \right)^{g_\lambda}.$$

It follows that the distinct linear factors of Θ^* are precisely the factors of the determinant of the matrix A given by (13.39) and (13.40), a result that Frobenius felt was "probably one of the most noteworthy results of the developed theory" [211, p. 22]. Of course, he knew that $g_\lambda = e_\lambda f_\lambda$ by virtue of $\Theta = \prod_\lambda \Phi_\lambda^{e_\lambda}$, but he could not establish this with his approach in "On Group Characters." He showed instead that the integers g_λ have the following interpretation. If the characters $\chi^{(\lambda)}$ are extended to $\mathfrak{H}$ by setting $\chi^{(\lambda)}(A) = \chi_\alpha^{(\lambda)}$ for all $A \in (\alpha)$, then g_λ is the rank of the $h \times h$ matrix $(\chi(PQ^{-1}))$.

Frobenius thus managed to rederive, without considering the factorization of Θ and the functions ψ associated to it, most of the content of his Theorem 13.7. He had provided what to him was a satisfying, purely group-theoretic introduction of the group characters and had shown them to represent a genuine generalization of Dedekind's characters by virtue of the linear factorization of $\Theta^*(x)$, which generalizes Dedekind's factorization of Θ when $\mathfrak{H}$ is abelian. He had also brought out interesting connections between the new theory of characters, the theory of commuting matrices, and Dedekind's work on commuting hypercomplex number systems. However, nowhere in "On Group Characters" is it mentioned that $\chi_\alpha^{(\lambda)}$, e_λ, and f_λ all have definite meanings in terms of the factorization $\Theta = \prod_\lambda \Phi_\lambda^{e_\lambda}$. As Frobenius had written, he would be more specific about his proportionality factors e_λ, f_λ "later." Presumably, he meant as soon as he had managed to supply the missing link in the theory of the group determinant Θ, namely a proof that $e_\lambda = f_\lambda$.

13.4 The Missing Link

Near the conclusion of his letter of 17 April, after proving that $k = l$, Frobenius had remarked, "The numbers e and f are therefore not yet determined." What he apparently meant was that although he knew that f represented the degree of an irreducible factor Φ of the group determinant Θ and that e represented the power to which Φ occurs in the factorization of Θ, he did not know how to relate them to the structure of the group $\mathfrak{H}$, as he had related the number l of distinct factors to the number of conjugacy classes of $\mathfrak{H}$. Nor did he know whether e and f are always equal, although there was some evidence for this. As we saw in Section 12.2, Dedekind had factored Θ for the symmetric and quaternion groups, and Frobenius had done the same for the dihedral group of order 8. In all of these examples there was a second-degree irreducible factor, and it occurred to the second power, thereby suggesting the possibility that perhaps $e = f$ in general. (Recall that $e = f$ means in modern terms that each irreducible representation occurs in the regular representation as many times as its degree.) These examples were not convincing evidence, but they were surely suggestive. Also, in proving Theorem 13.3, he had in effect established that $f = 1$ implies $e = 1$.

Sufficient evidence was thus at hand to hint at the possibility that $e = f$ in general, and Frobenius began to investigate this matter after sending off his letter of April 17 to Dedekind. His next letter, dated 26 April 1896, contained the initial results of that investigation. He had succeeded in proving $f = 2$ implies $e = 2$, but he was not pleased with the proof: "I have been able neither to extend this complicated proof to $f = 3$ nor to simplify it, i.e., to understand its essence." Nevertheless: "I am convinced that a prime function of degree f necessarily must depend on f^2 variables .... Now subjective convictions are indeed better than proofs, but the cruel world is not satisfied with such." There is no doubt that Frobenius wanted to believe that $e = f$. As he admitted, "It would be wonderful if $e = f$. For then my theory would supply everything needful for the determination of the prime factors Φ—the degrees $\sqrt{ef}$ and the characters $\psi(R)$, out of which the coefficients of Φ are composed in simple ways." What Frobenius apparently meant by this last remark is that he had ways (discussed below) to compute the products $g = ef$, so that if $e = f$, then $e = f = \sqrt{g}$. Also, once f is known, the corresponding character can be determined by computations comparable in magnitude to those required to obtain the integers g, and once the values $\psi(R)$ are known, it is possible to construct therefrom the corresponding irreducible factor Φ.[21]

The integers g may be computed as follows. The fundamental multiplication relation (13.36) for the characters can be transformed, using the orthogonality relations, into

$$\frac{h}{e_\lambda f_\lambda} h_\alpha \psi_\alpha^{(\lambda)} = \sum_{\beta=1}^{k} f_{\alpha\beta'} \psi_\beta, \quad \text{where} \quad f_{\alpha\beta'}^{(\lambda)} = \sum_{\gamma,\lambda=1}^{k} \frac{h_{\gamma\lambda\alpha}}{h_\gamma} \frac{h_{\gamma\lambda\beta}}{h_\lambda}.$$

These equations state that if $F^{(\lambda)} = (f_{\alpha\beta'}^{(\lambda)})$ and D is the diagonal matrix with $h_1, \dots, h_k$ down the diagonal and $v_\lambda = \left(\psi_1^{(\lambda)} \cdots \psi_k^{(\lambda)}\right)^t$, then $(e_\lambda f_\lambda F^{(\lambda)} - hD)v_\lambda = 0$, which shows that the integer $g_\lambda = e_\lambda f_\lambda$ can be computed as a root x of the kth-degree equation

$$\det\left(xF^{(\lambda)} - hD\right) = 0. \tag{13.41}$$

[21]The formulas that show how to construct Φ from ψ were presented in his letters to Dedekind dated 26 April and 4 September 1896. They were finally published, after he proved $e = f$, in his 1896 paper on the factorization of the group determinant [212, §3].

Since k can be significantly smaller than the order h of the group, it is feasible to compute the roots of (13.41) in examples where the computation of Θ and its factorization is not feasible. In his letter of 26 April, Frobenius indicated that he had done this for five groups. First, he considered the groups of transformations

$$y = \frac{\gamma x + \delta}{\alpha x + \beta},$$

where $\alpha, \beta, \gamma, \delta$ are integers (mod p) and $\alpha\delta - \beta\gamma \equiv 1$ (mod p). These are the projective unimodular groups over the integers mod p. For $p = 3, 5, 7$, these groups are the tetrahedral group, of order 12, the icosahedral group, of order 60, and a group of order 168. Besides these three groups, he also worked with two symmetric groups: the group $\mathfrak{S}_4$, of order $4! = 24$ (octahedral group), and $\mathfrak{S}_5$, of order $5! = 120$.

Direct computation of the factorization of Θ for these groups is impractical except for the tetrahedral group (see below), but using "the methods developed in my last letter," Frobenius computed the integers $g_\lambda = e_\lambda f_\lambda$. Probably, he used (13.41). In the five examples, $4 \leq k \leq 7$. Hence the computations involved would be manageable for someone with Frobenius' exceptional computational skills. The $e = f$ hypothesis was strengthened by these computations, which showed that in each case, the g_λ are squares. Under the assumption that $e = f = \sqrt{g}$, he then derived character tables for each of the five groups.[22] He wrote to Dedekind:

> These tables are very thought provoking. Remember that they supply with certainty only the ratios of the ψ. Always $ef = \psi_0^2$. The equation $h = \sum ef$ yields
>
> $$\begin{aligned} 12 &= 3 \cdot 1^2 + 3^2 \\ 60 &= 1^2 + 2 \cdot 3^2 + 4^2 + 5^2 \\ 168 &= 1^2 + 2 \cdot 3^2 + 6^2 + 7^2 + 8^2 \\ 24 &= 2 \cdot 1^2 + 2^2 + 2 \cdot 3^2 \\ 120 &= 2 \cdot 1^2 + 2 \cdot 4^2 + 2 \cdot 5^2 + 6^2 \end{aligned}$$
>
> thus a representation of the order h as a sum of squares, which is as highly remarkable as it is incomprehensible. For the tetrahedron I computed $\Delta\, [= \Theta]$, you too, probably, and found $e = f = 3$.

The above quotation suggests that although there was a fair amount of evidence supporting the hypothesis $e = f$, and although Frobenius believed $e = f$ to be true in general, he was still a bit baffled by why, mathematically, it was so. Hence there was still some room for doubt, and he concluded his letter of 26 April with the plea, "Should you have an example where $e \neq f$, please write to me as soon as possible so that I do not go astray." The following day, Dedekind wrote to Frobenius.[23] His letter contained some further examples, but they provided no new evidence that

---

[22] These tables are presented in his "On Group Characters" paper of 1896 [211, §8].

[23] The relevant mathematical portions of the letter were published in Dedekind's *Werke* 2, 425–433.

$e = f$. In one example (a semidirect product of an arbitrary abelian group with the cyclic group of order 2), $f \leq 2$; in the other, a recent computation done on 18 April 1896, Θ was not factored entirely. Unable to make any further theoretical headway, Frobenius turned "in despair" to further computation, as he put it, perhaps only half in jest, in a letter dated 7 May 1896. This time, he proposed to work with the general projective unimodular group of order $h = \frac{1}{2}p(p^2 - 1)$. He determined the conjugacy classes, the number k, the numbers h_α and $h_{\alpha\beta\gamma}$, and the characters. Of course, he had to guess at the actual value of $\psi^{(\lambda)}(E) = f_\lambda$ in order to write down the table.[24] But, as he noted, with his choice for the f_λ, $h = \sum_\lambda f_\lambda^2$. More circumstantial evidence!

For almost a month, Dedekind received no further word from Frobenius, who was still involved with the responsibilities of the summer semester. When a letter finally came (dated 4 June 1896), it was much more high-spirited than the previous ones.

He had not yet managed to prove that $e = f$, but he felt he had at least acquired some insight into why it should be true. Moreover, the reasoning behind the new insight implied that f divides e and would prove $e = f$ if a certain fact could be established. Frobenius eventually proved that $e = f$ by a different line of reasoning. At the time, however, he had reason to believe that he was on the brink of success. In that happy frame of mind, he jokingly shared the secret of his success with Dedekind:

> I quickly realized I would not attain the goal with my puny methods ... and I decided to seek the "great method." I call it the "Principle of the Horse Trade." You ... know how a horse is bought (or a diamond or a house). At the market, the desired horse is ignored as much as possible and at last is allowed to be formally acknowledged.
>
> It can also be called, in more elegant language, the "Principle of the Pout." Therefore, in order to find $e = f$, I first of all went to the trade exhibition with my wife, then to the picture exhibition. At home I read *Effie Briest*[25] and rid my fruit trees of caterpillars ....
>
> I gather from many places in your writings that my "Method of the Horse Trade" is probably known to you, albeit by a more civilized name. I hope that you will not give away the trade secret to anyone. My great work *On the Methods of Mathematical Research* (with an appendix on caterpillar catching), which makes use of it, will appear after my death.

Frobenius' "Principle of the Horse Trade" did indeed provide him with the proof that $e = f$. Adhering to that principle, he busied himself with other matters. He published his first two papers relating to group characters [209, 211], meditated on the possible significance of Dedekind's hypercomplex factorization of Θ, and finally published his results from 1880 on the density of primes [210], as discussed in Section 9.3. Then, the summer semester having ended, he traveled to Juist, one

---

[24] The table and much of the supporting computation were presented in his "On Group Characters" paper [211, §§9–10].

[25] A novel by Theodor Fontane, whose title is actually *Effi Briest*. It was published as a book in 1896 after appearing in serialized form during 1894–1895. Frobenius seems to have read most of Fontane's novels. Fontane is generally considered to be one of the greatest German literary figures between Goethe and Thomas Mann.

of the East Frisian Islands, for a vacation and visited with Dedekind and his sister on the return trip to Berlin.

Shortly after his return to Berlin, Frobenius hit on still another approach to proving that $f = 2$ implies $e = 2$.[26] Unlike his earlier proof of this result, the new one generalized to the case of arbitrary f, although in the general case it required proving that a certain expression is not identically zero, and that was by no means obvious. Two days later, however, he finally could announce a complete, general proof for $e = f$. His success depended on his "Principle of the Horse Trade," since the complete break with his research caused by his trip to Juist, combined with his own disorderliness, was instrumental in leading him to the new approach that had proved viable. As he explained to Dedekind (letter dated 6 September 1896):

> I will ... attempt to gather together the entire theory of the group determinant ... out of my highly scattered and disorganized papers. To *some* extent, however, such disorder is useful. That is, after my return home, I could no longer find the proof that I wrote to you long ago: If $f = 2$, then $e = 2$ also. After much torment I arrived at the new form of this proof and recognized here the possibility of generalization which I had completely despaired of in connection with the first proof.

Having finally proved that $e = f$ in general, Frobenius was able to compose his paper on the factorization of the group determinant [212], thereby solving the problem that Dedekind had posed to him six months earlier. His solution is summarized in the following theorem.

Theorem 13.10. *The determinant $\Theta = \det(x_{PQ^{-1}})$ associated to a finite group $\mathfrak{H}$ has the prime factorization $\Theta = \prod_{\lambda=1}^{k} \Phi_\lambda^{f_\lambda}$, where k, the number of distinct prime factors Φ_λ, equals the number of conjugacy classes of $\mathfrak{H}$, and f_λ is equal to the degree of Φ_λ, so that each prime factor occurs as often as its degree. Moreover, the number of linear factors of Θ is equal to the order of $\mathfrak{H}/\mathfrak{H}'$, where $\mathfrak{H}'$ denotes the commutator subgroup of $\mathfrak{H}$.*

Of course, Theorem 13.10 does not tell the whole story of Frobenius' discoveries, for it says nothing about his discovery that the notion of a character on an abelian group is capable of a nontrivial, interesting, and (as we will see) remarkably useful generalization to nonabelian groups. Frobenius' characters had first emerged as coefficients of certain terms of the prime factors Φ but were then seen to have a surprising connection with the theory commutative hypercomplex number systems through Dedekind's Theorem 13.8. It was Frobenius' theory of generalized characters that had provided him with the means for proving Theorem 13.10, and he clearly realized that these generalized characters possessed an interest and importance that transcended their application to the problem of factoring the group determinant. In fact, within six months of his discovery that $e = f$, Frobenius had hit upon another characterization of his characters—as traces of irreducible

---

[26] According to his letter to Dedekind dated 4 September 1896.

matrix representations of $\mathfrak{H}$—and had begun to develop the consequences of the new characterization. Once again Dedekind seems to have provided some of the inspiration for Frobenius' new line of research.

13.5 Matrix Representations

When Dedekind computed factorizations of group determinants in February 1886, he also considered the more general example of a nonabelian group $\mathfrak{H}$ of order $2m$ formed by taking what would now be called a semidirect product of an abelian group $\mathfrak{A}$ of order m and the cyclic group of order 2. This example was communicated to Frobenius (as *Beispiel 3*) in Dedekind's letter of 27 April 1896, when Frobenius was attempting to prove the $e = f$ hypothesis.[27] Using the m characters $\chi^{(\mu)}$ of $\mathfrak{A}$, Dedekind introduced new variables $u_\mu, v_\mu, u'_\mu, v'_\mu$ and showed that $\Theta = \prod_{\mu=1}^{m}(u_\mu u'_\mu - v_\mu v'_\mu)$. Some of these second-degree factors then split into two linear factors, while others are irreducible and occur in pairs (depending on the type of character $\chi^{(\mu)}$ used to define $u_\mu, v_\mu, u'_\mu, v'_\mu$). Then in his letter of 13 July 1896, Dedekind made a point of noting that these considerations could be presented "more completely and attractively" using matrices.[28] That is, if $M(x) = (x_{PQ^{-1}})$ denotes the "group matrix" (so $\Theta = \det[M(x)]$), then it is possible to define a nonsingular matrix L using the values of the characters $\chi^{(\mu)}$ such that

$$N(x) = L^{-1}M(x)L = \begin{pmatrix} u_1 & v'_1 & \cdots & 0 & 0 \\ v_1 & u'_1 & \cdots & 0 & 0 \\ \vdots & \vdots & \ddots & \vdots & \vdots \\ 0 & 0 & \cdots & u_m & v'_m \\ 0 & 0 & \cdots & v_m & u'_m \end{pmatrix}.$$

Although, in characteristic fashion, Dedekind quickly added, "you must have seen through that long ago," his remark indicates that Frobenius had not communicated anything along these lines to him. Eventually, however, Frobenius did consider the implications of Dedekind's approach. He communicated his discoveries to Dedekind in April 1897 (possibly during a visit), and they were published in the proceedings of the Berlin Academy the following November under the title "On the Representation of Finite Groups by Linear Substitutions" [213]. There, after summarizing his main results, Frobenius wrote, "In April of this year I communicated the most significant of these results to Dedekind, to whom I owe the stimulation for these investigations." Whether the stimulation consisted of more than the above remarks by Dedekind remains unclear, although it is

[27] This portion of Dedekind's letter can be found in *Werke* 2, 425–428.

[28] Dedekind *Werkes* 2, 437.

clear (as we shall see) that they would have sufficed to set Frobenius thinking along the lines that resulted in his "Representation" paper. In this section I will summarize the main results of Frobenius' paper and show how they led him to a new way of conceiving of his group characters and the fundamental "multiplicative relation" (13.36) by which they were defined in his "On Group Characters" paper.

In imitation of Dedekind's example, Frobenius proved in Section 13.3 of his paper the following result.

Theorem 13.11. *If $\mathfrak{H}$ is any group of order h, and $\Theta = \det[M(x)] = \prod_{\lambda=1}^{k} \Phi_\lambda^{f_\lambda}$ the factorization of the group determinant into its irreducible factors, then the k characters $\chi^{(\lambda)}$ of $\mathfrak{H}$ may be used to define a matrix L such that*

$$L^{-1}M(x)L = \begin{pmatrix} N_1(x) & 0 & \cdots & 0 \\ 0 & N_2(x) & \cdots & 0 \\ \vdots & \vdots & \ddots & \vdots \\ 0 & 0 & \cdots & N_k(x) \end{pmatrix}, \tag{13.42}$$

where $\det[N_\lambda(x)] = \Phi_\lambda(x)^{f_\lambda}$.

Pointing out that this was as far as one could go using the values of the irreducible characters to define the matrix L, Frobenius showed in Section 13.5 that by utilizing "higher irrationalities," one could define a matrix L that carried the decomposition of the group matrix further:

Theorem 13.12. *If $\mathfrak{H}$ is any group of order h, then a matrix L exists such that (13.42) holds, except that now the $f_\lambda^2 \times f_\lambda^2$ matrix $N_\lambda(x)$ has the form*

$$N_\lambda(x) = \begin{pmatrix} (x_{ij}^{(\lambda)}) & 0 & \cdots & 0 \\ 0 & (x_{ij}^{(\lambda)}) & \cdots & 0 \\ \vdots & \vdots & \ddots & \vdots \\ 0 & 0 & \cdots & (x_{ij}^{(\lambda)}) \end{pmatrix},$$

where $(x_{ij}^{(\lambda)})$ is an $f_\lambda \times f_\lambda$ matrix whose f_λ^2 entries $x_{ij}^{(\lambda)}$ are linearly independent linear homogeneous functions of the group variables $x_E, x_A, x_B, \ldots$ and $\det(x_{ij}^{(\lambda)}) = \Phi_\lambda(x)$.

Theorem 13.12 implies the complete reducibility theorem for the regular representation of $\mathfrak{H}$. As we shall see in Section 15.2, two years later, Frobenius returned to Theorem 13.12 and showed that there was a greater analogy with Theorem 13.11 if in addition to the characters $\chi^{(\lambda)}(R)$, more general functions $a_R^{(\lambda)}$ on the group $\mathfrak{H}$ are introduced to play the role in Theorem 13.12 played by the characters in Theorem 13.11. Frobenius called them characteristic units, and they define what are now called primitive idempotents $e^{(\lambda)} = \sum_R a_R^{(\lambda)} R$ of the group algebra of $\mathfrak{H}$.

In addition to establishing these illuminating generalizations of Dedekind's example, Frobenius began, in Section 13.2, to introduce the now-familiar terminology of matrix representations. Thus if $\mathfrak{H}$ is an abstract group and $\sigma : \mathfrak{H} \to \mathfrak{K}$ is a surjective group homomorphism, where $\mathfrak{K}$ is a group of invertible $m \times m$ matrices $\sigma(R)$, $R \in \mathfrak{H}$, the matrices $\sigma(R)$ are said to *represent* (*darstellen*) the group $\mathfrak{H}$, and σ is called a *representation* (*Darstellung*) of $\mathfrak{H}$. Two representations σ and τ are said to be *equivalent* if an invertible matrix P exists for which $\tau(R) = P^{-1}\sigma(R)P$ for all $R \in \mathfrak{H}$. Corresponding to any representation σ of $\mathfrak{H}$, Frobenius introduced the associated "group matrix" $\sigma(x) = \sum_{R \in \mathfrak{H}} \sigma(R)x_R$ and its determinant $F(x) = \det \sigma(x)$. Thus if $\rho(x)$ is the regular representation of $\mathfrak{H}$, namely $\rho(x) = \sum_{R \in \mathfrak{H}} \sigma(R)x_R = (x_{PQ^{-1}})$, then $F(x) = \Theta(x)$.

Using the properties of the group determinant that he had established, Frobenius was able to show that for every representation σ, the corresponding determinant has a prime factorization $F(x) = \prod_{\lambda=1}^{k} \Phi_\lambda^{s_i}$, where $s_\lambda \geq 0$, so that only the prime factors of the group determinant $\Theta(x)$ are needed for the factorization of $F(x)$. A representation σ is then said to be primitive if $F(x)$ cannot be factored, so that $F(x)$ must equal one of the prime factors $\Phi_i(x)$ of $\Theta(x)$. Thus Theorem 13.12 says that the regular representation ρ decomposes into the primitive representations defined by the matrices $(x_{kl}^{(i)})$; it is the complete reducibility theorem for ρ, although Frobenius' definition of a primitive representation was still dominated by the determinant-theoretic viewpoint that informed most of his proofs. In 1899 [216, p. 130], he referred to the above definition of a primitive representation as "provisional" and replaced it with the equivalent one that $R \to \sigma(R)$ is primitive, or *irreducible* (as he now also said), if σ is not equivalent to a representation of the form

$$\begin{pmatrix} \tau_1(R) & 0 \\ 0 & \tau_2(R) \end{pmatrix}.$$

By virtue of the complete reducibility theorem, this definition is equivalent to the more customary one today that σ is irreducible if it is not equivalent to a representation of the form

$$\begin{pmatrix} \tau_1(R) & 0 \\ * & \tau_2(R) \end{pmatrix}.$$

In what follows, I will use the term "irreducible representation" in the now customary sense since the two notions diverge in meaning when the representations are over fields of characteristic p dividing the order of the group (as will be seen in Section 15.6).

It follows from Theorem 13.12 that for every λ, $\sigma_\lambda(x) = (x_{kl}^{(\lambda)})$ defines an irreducible representation, which is associated with a prime factor Φ_λ of the group determinant $\Theta(x)$ and hence with its character. In Section 4, Frobenius explored the connection between σ_λ and the associated character $\chi^{(\lambda)}$. Thus let $\sigma_\lambda(x) = \sum_R x_R \sigma_\lambda(R)$ where $\det \sigma_\lambda(x) = \Phi_\lambda(x)$ and $f_\lambda = \deg \Phi_\lambda$. Then since $\sigma_\lambda(\alpha x + \beta y) = \alpha\sigma_\lambda(x) + \beta\sigma_\lambda(y)$, where $x = (x_E, x_A, \ldots)$, $y = (y_E, y_A, \ldots)$, it follows that the characteristic polynomial of $\sigma_\lambda(x)$ is

$$\begin{aligned}\det[\sigma_\lambda(x) - uI] &= \det[\sigma_\lambda(x) - u\sigma_\lambda(E)] = \det[\sigma_\lambda(x_E - u, x_A, \ldots)] \\ &= \Phi_\lambda(x_E - u, x_A, \ldots).\end{aligned} \tag{13.43}$$

Recall from Frobenius' original definition (13.10) of $\chi^{(\lambda)}(A)$ as the coefficient of $x_E^{f_\lambda - 1} x_A$ in Φ_λ that

$$\Phi_\lambda(x_E - u, x_A, \ldots) = (x_E - u)^{f_\lambda} + \sum_{A \neq E} \chi^{(\lambda)}(A) x_A (x_E - u)^{f_\lambda - 1} + \cdots.$$

The coefficient of $(-u)^{f_\lambda - 1}$ in this expression is thus

$$f_\lambda x_E + \sum_{A \neq E} \chi^{(\lambda)}(A) x_A = \sum_{A \in \mathfrak{H}} \chi^{(\lambda)}(A) x_A.$$

On the other hand, by (13.43), $\Phi_\lambda(x_E - u, x_A, \ldots)$ equals

$$\det[\sigma_\lambda(x) - uI] = (-u)^{f_\lambda} + \mathrm{tr}[\sigma_\lambda(x)](-u)^{f_\lambda - 1} + \cdots,$$

so comparison of the coefficients of $(-u)^{f_\lambda - 1}$ shows that

$$\mathrm{tr}\, \sigma_\lambda(x) = \sum_{A \in \mathfrak{H}} \chi^{(\lambda)}(A) x_A.$$

Setting $x_R = 1$ and $x_A = 0$ for $A \neq R$ in this equation, Frobenius obtained the now familiar relation

$$\chi^{(\lambda)}(R) = \mathrm{tr}\, \sigma_\lambda(R). \tag{13.44}$$

Thus Frobenius' characters, which had forced themselves into the spotlight of his attention in a rather fortuitous way, had finally lost their air of mystery: they were simply the trace functions of the irreducible representations of the group, and so of course of considerable importance.

According to Frobenius, the trace equation (13.44) "leads to a deeper insight into the significance of the equations which I developed for the calculation of the characters" [213, p. 96]. What he apparently meant was that (13.44) makes possible another derivation of the relations

$$h_\alpha h_\beta \chi_\alpha^{(\lambda)} \chi_\beta^{(\lambda)} = f^{(\lambda)} \sum_\gamma h_{\alpha'\beta\gamma} \chi_\gamma^{(\lambda)}, \tag{13.45}$$

which had formed the foundation stone for his development of the theory of characters in "On Group Characters." Now, as Frobenius will show, it is the trace equation (13.44) rather than (13.45) that is truly fundamental to the theory of characters; the latter is a consequence of the former.

Frobenius' new derivation of (13.45) by means of the trace characterization of his characters is worth sketching, because it illustrates the type of techniques he was in the process of developing, techniques that his student Issai Schur developed further. For an element $A \in \mathfrak{H}$ and an irreducible representation $R \to \sigma(R)$, he considered the matrix

$$M_A = \sum_R \sigma(R^{-1}AR). \tag{13.46}$$

Then $\sigma(S)M_A = M_A\sigma(S)$ for all $S \in \mathfrak{H}$. This follows from a type of calculation that was already fundamental to Frobenius' approach to group determinants. Since

$$\sigma(S)M_A = \sigma(S)\sum_R \sigma(R^{-1}AR) = \sum_R \sigma(SR^{-1}AR),$$

and since as R runs over $\mathfrak{H}$, so does $T = RS^{-1}$, it follows that

$$\sum_R \sigma(SR^{-1}AR) = \sum_T \sigma(T^{-1}ATS) = \sum_T \sigma(T^{-1}AT)\sigma(S) = M_A\sigma(S).$$

This invariance of summations over the group with respect to "translations" such as $T = RS^{-1}$ was also to prove fundamental to the extension of Frobenius' theory to continuous groups initiated by Schur and extended in remarkable ways to a Fourier analysis on such groups by Hermann Weyl (Section 15.5).

Because M_A commutes with $\sigma(S)$, it also commutes with $\sigma(x) = \sum_S \sigma(S)x_S$. But since the f^2 coefficients x_{ij} of $\sigma(x)$ are linearly independent functions of the variables x_R (by Theorem 13.12), $\sigma(x)$ can equal any $f \times f$ matrix for a suitable choice of the variables x_R. This means that M_A commutes with all $f \times f$ matrices and so must be a scalar multiple of the identity matrix $I = \sigma(E)$. Here we see Frobenius using (and proving) what amounts to one of the cases in Schur's lemma (Theorem 15.5). Schur, however, was to prove his lemma by means of elementary considerations—taking Frobenius' technique of summation over the group even further—so that it could then be used to develop the theory of characters and representations, whereas Frobenius' reasoning drew upon some of the principal results of his theory.

Let us now continue Frobenius' rederivation of (13.45). It reflects, as does the first part of the proof, both his calculational skill and his mastery of new techniques for dealing with matrix representations and trace functions rather than determinants. Having shown that $M_A = \lambda\sigma(E)$, Frobenius next determined λ. To this end, he used the group list and equation techniques learned in the 1880s (Section 9.4) to transform the definition of M_A in (13.46) into

$$M_A = \sum_{S \in (\beta)} (h/h_\beta)\sigma(S) \quad \text{if } A \in (\beta). \tag{13.47}$$

Using (13.47) combined with $M_A = \lambda \sigma(E)$, we get by taking traces

$$\lambda f = \sum_{S \in (\beta)} (h/h_\beta) \operatorname{tr} \sigma(S) = h\chi_\beta,$$

and so $\lambda = (h/f)\chi_\beta$ and $fM_A = h\chi_\beta$ for $A \in (\beta)$. This last expression can be rewritten using (13.47) as $f\sum_{S\in(\beta)} \sigma(S) = h_\beta \chi_\beta \sigma(E)$, and if both sides are multiplied on the left by $\sigma(A)$, then for $A \in (\alpha)$ we get $f\sum_{S\in(\beta)} \sigma(AS) = h_\beta \chi_\beta \sigma(A)$. Taking the trace of both sides of this equation then gives

$$f \sum_{S \in (\beta)} \chi(AS) = h_\alpha h_\beta \chi_\alpha \chi_\beta, \tag{13.48}$$

and Frobenius' standard group list and group equation considerations show that since $A \in (\alpha)$,

$$f \sum_{S \in (\beta)} \chi(AS) = \sum_\gamma (h_{\alpha\beta\gamma'}/h_\alpha)\chi_{\gamma}. \tag{13.49}$$

If (13.49) is substituted in (13.48), the result is the fundamental equation (13.45), viz., $h_\alpha h_\beta \chi_\alpha^{(\lambda)} \chi_\beta^{(\lambda)} = f \sum_\gamma h_{\alpha'\beta\gamma} \chi_\gamma^{(\lambda)}$.

The historical significance of Frobenius' new derivation of (13.45) is that even though the main results of his "Representation" paper were obtained by the combination of determinant techniques and character relations that he had devised to prove Theorem 13.10, the new derivation of (13.45) suggested the possibility that the theory of the matrix representation of groups—a theory to which Frobenius himself ascribed far greater importance than to that of the group determinant[29]—could be developed using the theory of matrices rather than determinants. That this could be done was shown by Schur in 1905 (Section 15.5).

After establishing Theorem 13.12, which represents in matrix form the fundamental theorem of Frobenius' theory of group characters and determinants, he noted that in fact, it had been independently discovered by a young Estonian mathematician, Theodor Molien, who had presented it in a slightly more general form in two papers that had appeared shortly before Frobenius' own "Representation" paper. Moreover, Molien's papers were based on work published by him in 1893. As we shall see in the following chapter, Molien's work was but one example, albeit the most impressive one, of the fact that other developments in late-nineteenth-century mathematics were leading by quite different routes to the results of Frobenius' theory.

[29] See Frobenius' remark to this effect at the beginning of his "Representation" paper of 1897 [213].

Chapter 14
Alternative Routes to Representation Theory

The correspondence between Dedekind and Frobenius makes it clear that if Dedekind had not decided to introduce and study group determinants—a subject with no established tradition and really outside his main interests in algebraic number theory—or if he had decided not to communicate his ideas on group determinants to Frobenius, especially given Frobenius' complete lack of curiosity about Dedekind's allusion to a connection between hypercomplex numbers and groups, it is unlikely that Frobenius would be known as the creator of the theory of group characters and representations. This is not to say that the theory would have remained undiscovered for a long time. On the contrary, three lines of mathematical investigation were leading to essentially the same theory that Frobenius had begun to explore: (1) the theory of noncommutative hypercomplex number systems; (2) Lie's theory of continuous groups; and (3) Felix Klein's research program on a generalized Galois theory. The main purpose of this chapter is to briefly indicate how these lines of investigation were leading—or in some cases did lead—to the results of Frobenius' theory.[1] In addition, Frobenius' letters to Dedekind provide us with some interesting commentary on two of the leading figures in these developments, T. Molien and W. Burnside.

14.1 Hypercomplex Numbers and Lie Groups

As in Section 13.3, expressions $\mathbf{a} = \sum_{i=1}^{n} \alpha_i \mathbf{e}_i$, $\mathbf{b} = \sum_{i=1}^{n} \beta_i \mathbf{e}_i$ with α_i and β_i complex numbers will be said to form an n-dimensional system $\mathcal{H}$ of hypercomplex numbers with addition and multiplication given by $\mathbf{a} + \mathbf{b} = \sum_{i=1}^{n} (\alpha_i + \beta_i)\mathbf{e}_i$ and $\mathbf{ab} = \sum_{i,j=1}^{n} \alpha_i \beta_j (\mathbf{e}_i \mathbf{e}_j)$, where

[1] See my paper [267] for further details.

T. Hawkins, *The Mathematics of Frobenius in Context*, Sources and Studies in the History of Mathematics and Physical Sciences, DOI 10.1007/978-1-4614-6333-7_14,

$$\mathbf{e}_i\mathbf{e}_j = \sum_{k=1}^{n} a_{ijk}\mathbf{e}_k. \tag{14.1}$$

It is assumed that the multiplication constants a_{ijk} are such that multiplication is associative but not necessarily commutative. Thus $\mathcal{H}$ is a ring, or if scalar multiplication is introduced in the natural way by defining $\alpha\mathbf{a} = \sum_{i=1}^{n} \alpha\alpha_i\mathbf{e}_i$, then $\mathcal{H}$ is a linear associative algebra. In what follows, $\mathcal{H}$ will be said to be a complete matrix algebra if it is isomorphic to the linear associative algebra of all $m \times m$ matrices for some m. In that case, $\mathcal{H}$ has a basis of m^2 units $\mathbf{e}_{ij}$ with multiplication given by $\mathbf{e}_{ij}\mathbf{e}_{kl} = \delta_{jk}\mathbf{e}_{il}$.

Interest on the part of mathematicians in hypercomplex number systems can be traced back to Hamilton's introduction of quaternions, although it was not until the 1880s that the theory of such systems became of interest to more than a few isolated individuals. The broader interest in hypercomplex numbers was due in large part to the publication in 1884 of papers on hypercomplex numbers by two prominent mathematicians, Weierstrass and Poincaré. Weierstrass' paper [593] dealt with commutative hypercomplex numbers and has already been discussed in Section 13.3. Weierstrass' paper and Dedekind's thoughtful response to it generated much interest—and several additional papers by other mathematicians—in the theory of commutative hypercomplex number systems. Poincaré's paper [479] was concerned with systems that are not necessarily commutative but contain a two-sided identity element.[2] It was prompted by two notes in the Paris *Comptes rendus* by J.J. Sylvester, which called attention to the fact that hypercomplex number systems could be defined by matrices and that some, namely the quaternions over the field of complex numbers (Hamilton's biquaternions) and Sylvester's own "nonions" were complete matrix algebras. Poincaré responded to Sylvester's notes by asserting that the problem of determining all hypercomplex number systems with identity "easily reduced to the following: *To find all the continuous groups of linear substitutions in n variables, the coefficients of which are linear functions of n arbitrary parameters*" [479, p. 740].

What Poincaré had in mind was that every element $\mathbf{u} = \sum_{i=1}^{n} u_i\mathbf{e}_i \in \mathcal{H}$ determines a linear transformation $\mathbf{u}_R : \mathbf{x} \to \mathbf{x}' = \mathbf{x}\mathbf{u}$. In view of (14.1), $\mathbf{u}_R$ is the linear transformation with coordinate equations

$$x'_k = \sum_{i=1}^{n} b_{ki}(u)x_i, \text{ where } \quad u = (u_1, \ldots, u_n) \text{ and } b_{ki}(u) = \sum_{i=1}^{n}\left(\sum_{j=1}^{n} a_{ijk}u_j\right), \tag{14.2}$$

so that they are linear and homogeneous in both variables x_i and parameters u_j. In this way, Poincaré forged a link between hypercomplex number systems and Sophus Lie's theory of continuous transformation groups. The transformations $\mathbf{u}_R$ form a

[2]Actually, Poincaré did not explicitly assume the existence of a two-sided identity, but he—and those who followed him—made an equivalent assumption.

group $\mathfrak{G}$ in the sense that $\mathfrak{G}$ is closed under composition and contains the identity transformation $\mathbf{u}_R^0$. This was how Lie originally conceived of his groups. It follows that for all $\mathbf{u}_R$ sufficiently close to $\mathbf{u}_R^0$, $\mathbf{u}_R^{-1}$ exists, and so $\mathfrak{G}$ defines a Lie group germ. The corresponding Lie algebra $\mathfrak{g}$ may be identified with $\mathcal{H}$ as a vector space with "Lie bracket" given by $[\mathbf{x},\mathbf{y}] = \mathbf{xy} - \mathbf{yx}$.

Lie, who was a professor at the University of Oslo in 1884, had been developing his theory more or less in isolation since 1874.[3] After 1886, when he accepted a professorship at the University of Leipzig in order to found a school devoted to the development and application of his theory, he encouraged his students and associates to develop the connections between hypercomplex number systems and continuous groups along the lines suggested by Poincaré, who became absorbed in other mathematical work and did not develop his ideas. Among those who responded were G. Scheffers, E. Study, T. Molien, and É. Cartan. Lie himself suggested the following problem in 1889 [416, p. 326]: Let $\mathfrak{g}$ denote the Lie algebra associated with the hypercomplex system $\mathcal{H}$. Then $\mathfrak{g}$ is never simple, because the identity of $\mathcal{H}$ generates a proper Lie algebra ideal. In this sense, $\mathfrak{g}$ is like the Lie algebra $\mathrm{gl}(m,\mathbb{C})$ of the general linear group $\mathrm{GL}(m,\mathbb{C})$. Indeed, as Study had shown, it is always possible to choose a basis $\mathbf{e}_0, \mathbf{e}_1, \ldots, \mathbf{e}_{n-1}$ for $\mathfrak{g}$ such that $\mathfrak{g}_{n-1} = \mathrm{span}\{\mathbf{e}_1, \ldots, \mathbf{e}_{n-1}\}$ forms a Lie algebra with the property that $\mathrm{tr}\,\mathbf{u}_R = 0$ for all $\mathbf{u} \in \mathfrak{g}_{n-1}$. Thus $\mathfrak{g}_{n-1}$ resembles the Lie algebra $\mathrm{sl}(m,\mathbb{C})$ of the special linear group $\mathrm{SL}(m,\mathbb{C})$, which is simple. Probably with this in mind, Lie posed the problem of determining those systems $\mathcal{H}$ such that $\mathfrak{g}_{n-1}$ is simple, i.e., contains no nontrivial ideals. Lie pointed out that the complex quaternions are an example of such an $\mathcal{H}$ and, based on his calculations of transformation groups in a small number of variables, that no such $\mathcal{H}$ existed for $n = 5, 6, 7, 8$. Of course, for $n = 9$ and more generally for $n = m^2$, the complete matrix algebra of $m \times m$ matrices has Lie's property. It turns out that if $\mathcal{H}$ has Lie's property, then $\mathcal{H}$ is simple as a hypercomplex system in the sense that $\mathcal{H}$ contains no proper two-sided ideals. The work of Molien and Cartan was to show that the only simple hypercomplex systems are the complete matrix algebras, so that they are the only systems satisfying Lie's condition.

The work of Molien and Cartan differed from that of the other mathematicians seeking to relate the study of hypercomplex systems with Lie's theory in that it was inspired and guided by the groundbreaking work of Wilhelm Killing (1847–1923) on the structure of Lie algebras, which he published between 1888 and 1890.[4] Killing, a professor at the Lyceum Hosianum in Braunsberg, East Prussia (now Braniewo, Poland), had been led through his work on the foundations of geometry to Lie's notion of an infinitesimal group or Lie algebra. Even before Lie's friend Felix Klein had called Killing's attention to Lie's theory of groups, Killing had posed to himself the problem of investigating what amounts to the structure of Lie algebras. When he learned of Lie's work, Killing gained some much-needed focus. He still

---

[3] On Lie's early work on continuous groups and its historical background, see Chapters 1–3 of my book [276] on the history of Lie groups.

[4] On Killing's work and its background, see Chapters 4 and 5 of my book [276].

wished to classify all Lie algebras, but he accorded special attention to *simple* Lie algebras because of their envisioned importance in applications of Lie's theory to differential equations. Central to Killing's approach was the linear transformation $\mathrm{ad}\,\mathbf{u} : \mathbf{x} \to [\mathbf{x},\mathbf{u}]$ and its characteristic polynomial $k_{\mathbf{u}}(\omega) = |\mathrm{ad}\,\mathbf{u} - \omega I|$, which has the form

$$k_{\mathbf{u}}(\omega) = (-1)^n \left[\omega^n - \psi_1(u)\omega^{n-1} + \cdots \pm \psi_{n-1}(u)\omega\right] \tag{14.3}$$

for $\mathbf{u} = \sum_{i=1}^{n} u_i \mathbf{e}_i$, $\mathbf{x} = \sum_{i=1}^{n} x_i \mathbf{e}_i \in \mathfrak{g}$. The linear transformation $\mathrm{ad}\,\mathbf{u}$ is analogous to the linear transformation $\mathbf{u}_R$ introduced by Poincaré, but Killing was not familiar with the literature on hypercomplex numbers and so was not guided by an analogy with associative algebras. Quite the contrary, it was the analogy between Killing's theory for the nonassociative algebra $\mathfrak{g}$ and that of the associative algebras $\mathcal{H}$ that inspired and guided the work of both Molien and Cartan on the structure of hypercomplex number systems. Using the characteristic equation and the remarkable properties of its roots, Killing classified all simple Lie algebras, showing that in addition to the four general types indicated by Lie, there were only five other possibilities. Killing also introduced the term "semisimple" (*halbeinfach*) for a Lie group with Lie algebra $\mathfrak{g}$ that is a direct sum of simple ideals.

14.2 T. Molien and É. Cartan

Theodor Molien (1861–1941) was born in Riga, Latvia, and educated at the University of Dorpat (now Tartu) in Estonia.[5] After completing his formal education there in 1883, he spent several semesters at the University of Leipzig, where Felix Klein was then a professor. Molien came to Leipzig having studied and worked primarily in astronomy and with the intention of studying celestial mechanics. While at Leipzig, under Klein's influence, he became interested in pure mathematics and ended up publishing a paper on elliptic functions that resulted from a problem Klein had posed to him. After returning to Dorpat as an instructor (*Dozent*) in 1885, Molien frequently returned to Leipzig during his vacation periods. There he became acquainted with the work and students of Klein's successor, Sophus Lie. Back in Dorpat, in 1888, Molien was joined by Friedrich Schur, who had become head of Molien's department. Schur had come from Leipzig where he too had participated in the research activities of Lie's school. It is thus not surprising that Molien knew about the work of Lie's students on hypercomplex numbers and about Killing's work on Lie algebras as well. The theory of hypercomplex numbers particularly intrigued him, and it became the subject of his doctoral dissertation, which was begun in 1888, completed in the fall of 1891, and eventually published in *Mathematische Annalen* in 1893 [443].

---

[5]In what follows, I have drawn upon N.F. Kanounov's biography of Molien [335, 14–36].

The first main result of Molien's dissertation was that a hypercomplex system is simple if and only if it is a complete matrix algebra.[6] The ultimate goal of his thesis was to obtain a "normal form" for hypercomplex systems that would make their structure more evident. His approach to finding such a normal form is of historical interest because it amounts to a study of the representations of a hypercomplex system. Let me explain. As we have seen, Poincaré had declared the study of hypercomplex systems with identity to be equivalent to the study of the group of *bilinear* transformations $\mathbf{u}_R$ defined by (14.2). The group is bilinear, since coefficients $b_{ki}(u)$ of $\mathbf{u}_R$ are linear homogeneous functions of the u_j. However, these groups are also simply transitive, so that, as Study had shown, the investigation of hypercomplex systems with identity was equivalent to the study of simply transitive bilinear groups.[7] Molien, however, proposed to drop the assumption of simple transitivity and study bilinear groups of transformations

$$T_{\mathbf{u}} : x_k' = \sum_{i=1}^{m} b_{ki}(u)x_i, \quad k = 1,\ldots,m, \tag{14.4}$$

where $\mathbf{u} = \sum_{i=1}^{n} u_i\mathbf{e}_i \in \mathcal{H}$, the $b_{ki}(u)$ are linear homogeneous functions of $u_1,\ldots,u_n$, and $T_{\mathbf{uv}} = T_{\mathbf{v}} \circ T_{\mathbf{u}}$. From the assumption that the transformations (14.4) define a bilinear n-parameter transformation group in the sense of Lie, it follows that $T_{\alpha\mathbf{u}+\beta\mathbf{v}} = \alpha T_{\mathbf{u}} + \beta T_{\mathbf{v}}$, that $T_{\mathbf{u}}^0 = I$ ($\mathbf{u}^0 =$ the identity element of $\mathcal{H}$), and that the correspondence $\mathbf{u} \to T_{\mathbf{u}}$ is one-to-one. Thus $\mathbf{u} \to T_{\mathbf{u}}$ is a faithful representation of $\mathcal{H}$ of degree m. A nonfaithful representation $\mathbf{u} \to T_{\mathbf{u}}$ can be considered a group belonging to what Molien called an "accompanying hypercomplex system" $\mathcal{H}^*$, which would now be identified with the quotient algebra $\mathcal{H}/\mathcal{I}$, where $\mathcal{I}$ is the kernel of $\mathbf{u} \to T_{\mathbf{u}}$.

Molien's objective was to obtain a normal form for the equations (14.4) for every bilinear group $\mathfrak{G} = \{T_{\mathbf{u}}\}$ associated to $\mathcal{H}$. In particular, the normal form for the group of transformations $T_{\mathbf{u}} = \mathbf{u}_R$ would then yield the normal form for $\mathcal{H}$ itself. In the process of obtaining the normal form for $\mathfrak{G}$, Molien discovered many of the basic theorems on the representation of hypercomplex systems. The interested reader can consult my paper [267, pp. 262–264]. Here I will simply note one particularly important implication of the resulting normal form as it applies to the system $\mathcal{H}$. One of the central ideas behind Molien's approach to the study of $\mathcal{H}$ derived from his observation that if $f(\mathbf{u}) = \sum_{i=1}^{n} \alpha_i u_i$ is any linear form, then $M(\mathbf{u},\mathbf{v}) = f(\mathbf{uv}) = \sum_{i=1}^{n} \alpha_i(\mathbf{uv})_i$, where $\mathbf{uv} = \sum_{i=1}^{n}(\mathbf{uv})_i\mathbf{e}_i$ is a bilinear form in the variables u_i, v_j with the property that $M(\mathbf{uv},\mathbf{w}) = M(\mathbf{u},\mathbf{vw})$. If, in addition f is such that M is symmetric—$M(\mathbf{u},\mathbf{v}) = M(\mathbf{v},\mathbf{u})$—then, Molien realized, it determines an

[6] For a discussion of Molien's actual approach to hypercomplex systems in general and, in particular, an indication of how he proved his theorem on simple systems by means of analogies with Killing's work, see Section 3 of my paper [267].

[7] Here "simply transitive" means simply transitive on an open dense subset of $\mathbb{C}^n$. This follows directly from the fact that $\mathbf{u}_R$ is invertible for all $(u_1,\ldots,u_n)$ in an open dense set.

accompanying hypercomplex system $\mathcal{H}^* \cong \mathcal{H}/\mathcal{I}$, where $\mathcal{I}$ is the two-sided ideal consisting of all $\mathbf{u} \in \mathcal{H}$ for which $M(\mathbf{u},\mathbf{x}) = 0$ for all $\mathbf{x} \in \mathcal{H}$. Molien observed that every system $\mathcal{H}$ has a linear form f such that the bilinear form $f(\mathbf{uv})$ is symmetric, for if

$$p_{\mathbf{u}}(\omega) = |\mathbf{u}_R - \omega I| = (-1)^n \left[\omega^n - \varphi_1(u)\omega^{n-1} + \cdots \pm \varphi_n(u)\right] \tag{14.5}$$

denotes the characteristic polynomial of $\mathbf{u}_R$, then $\varphi_1(u)$ has this property since $\varphi_1(u) = \operatorname{tr}\mathbf{u}_R$. Since the term "trace" was not a common part of a mathematician's vocabulary at the end of the nineteenth century, Molien always spoke of $\varphi_1(u)$ as a coefficient of a characteristic equation. In what follows, however, we shall speak of it as a trace, a term (*Spur* in German) that in 1899 Frobenius began to popularize in the context of matrices "following the lead of Dedekind" [215, p. 119].

The symmetric bilinear form $M(\mathbf{u},\mathbf{v}) = \varphi(\mathbf{uv}) = \operatorname{tr}(\mathbf{uv})_R = \operatorname{tr}(\mathbf{v}_R \circ \mathbf{u}_R)$ is of course the associative analogue of the Killing form for a Lie algebra, $K(\mathbf{u},\mathbf{v}) = \operatorname{tr}(\operatorname{ad}\mathbf{u} \circ \operatorname{ad}\mathbf{v})$, but Molien was not borrowing from Killing, because there is actually no Killing form in Killing's work. In fact, Killing had difficulty establishing the properties of simple Lie algebras from which his classification then flows, and it was Cartan in his doctoral thesis of 1894 [57] who made Killing's classification rigorous by introducing the quadratic form $\psi_2(\mathbf{u})$ associated to the characteristic polynomial (14.3) of $\operatorname{ad}\mathbf{u}$.[8] The bilinear form $\varphi_1(\mathbf{uv})$ played a role in Molien's thesis similar to the role played shortly thereafter by $\psi_2(\mathbf{u})$ in the thesis of Cartan, who did not know of Molien's work. The only difference was that whereas Cartan was interested primarily in the semisimple case, Molien's emphasis was on the classification of all hypercomplex systems and so on the general normal form for such a system. Thus, even though Molien did not state it explicitly, the theorem that $\mathcal{H}$ is a direct sum of complete matrix algebras if and only if the bilinear form $M(\mathbf{u},\mathbf{v}) = \varphi_1(\mathbf{uv})$ is nonsingular was an immediate consequence of the normal form. As we shall see, in the special case in which $\mathcal{H}$ is the group algebra for a finite group $\mathfrak{H}$, this theorem laid the foundation for Molien's approach to the representation theory of $\mathfrak{H}$ by implying the complete reducibility of the regular representation of $\mathfrak{H}$ into irreducible representations. This is because if $\mathfrak{H} = \{E,A,B,C,\ldots\}$ then the h elements $\mathbf{e}_1 = E$, $\mathbf{e}_2 = A, \ldots$ form a basis for a hypercomplex system of dimension h; and a straightforward calculation shows that the determinant of the form $M = \varphi_1(\mathbf{uv})$ equals $\pm h^h$, so that M is nonsingular and the group algebra is semisimple.

When Cartan was busy reworking and improving on Killing's results for his doctoral thesis, he was aware of the connection between bilinear Lie groups and hypercomplex systems to which Poincaré had called attention and that had been

[8]The Killing form of a Lie algebra is related to the quadratic form $\psi_2(\mathbf{u})$ by $K(\mathbf{u},\mathbf{u}) = [\psi_1(\mathbf{u})]^2 - 2\psi_2(\mathbf{u})$, which reduces to $K(\mathbf{u},\mathbf{u}) = -2\psi_2(\mathbf{u})$ for any Lie algebra satisfying $\mathfrak{g}' = \mathfrak{g}$ and so in particular for semisimple Lie algebras. For a comparative discussion of the contributions of Killing and Cartan to the structure of Lie algebras, see Chapters 5 and 6 of my book [276].

further developed by Study. In 1895, the year after his thesis appeared, he applied the new results on the structure of Lie algebras to the bilinear group $\mathfrak{G}$ defined by a hypercomplex system $\mathcal{H}$. Recall that the Lie algebra $\mathfrak{g}$ of $\mathfrak{G}$ is simply $\mathcal{H}$ with $[\mathbf{x},\mathbf{y}] = \mathbf{xy} - \mathbf{yx}$. From the general theory of Lie algebras developed by Cartan, it followed that if $\mathfrak{g}$ is not solvable, it contains a maximal solvable ideal $\mathfrak{r}$ such that $\mathfrak{g}/\mathfrak{r}$ is semisimple. (The ideal $\mathfrak{r}$ is now called the radical of $\mathfrak{g}$—a term introduced by Frobenius in 1903 [224, p. 318].) Thus $\mathfrak{g}/\mathfrak{r}$ is a sum of simple ideals, and Cartan discovered that when $\mathfrak{g}$ is defined by a hypercomplex system $\mathcal{H}$, these simple ideals are all isomorphic to the Lie algebras of special linear groups in $n_1, n_2, \ldots$ variables. Applying this result to the group $\mathfrak{G}$ itself, Cartan concluded that "one can find n_1^2 variables that are exchanged among themselves in the manner of the parameters of the general linear group in n_1 variables, likewise n_2^2 other variables that are exchanged in a similar manner and so on" [58, p. 546]. As Cartan pointed out, this means that the hypercomplex system $\mathcal{H}$ contains a subalgebra that is a sum of complete matrix algebras.

Cartan did not pause to consider when $\mathcal{H}$ itself is a direct sum of complete matrix algebras, but shortly thereafter, he decided to adopt a more direct approach to the theory of hypercomplex systems. Rather than applying the results of Lie algebra theory, he developed the analogous ideas for hypercomplex systems. His results were announced in 1897—the year Frobenius' "Representation" paper [213] appeared. Among his results was the theorem that a hypercomplex system $\mathcal{H}$ is simple (in the sense that it contains no proper two-sided ideals) if and only if it is a complete matrix algebra and that $\mathcal{H}$ is semisimple—a sum of simple ideals that are consequently complete matrix algebras—if and only if the quadratic form $\varphi_2(\mathbf{u})$ associated to the characteristic polynomial (14.5) of $\mathcal{H}$ is nonsingular. Once again, if $\mathcal{H}$ is the group algebra of a finite group $\mathfrak{H}$ of order h, a straightforward calculation shows that the determinant of $\varphi_2(u)$ equals $(-1)^h[\frac{1}{2}(h-1)h]^h$. Thus Cartan's results also implied that group algebras are direct sums of complete matrix algebras. Cartan, however, did not think to apply them to the group algebra of a finite group, but if Frobenius had not created his theory, it seems likely that someone (possibly Poincaré himself)[9] would have applied Cartan's results to group algebras. Indeed, as we shall now see, Molien was led to consider the implications of his equivalent results for group algebras in 1897.

It was probably the work of Klein that suggested to Molien the line of thought that led him to apply his results on hypercomplex systems to group algebras.

No one did more to promote the idea that group-related ideas are fundamental to all of mathematics than Klein. Among other things, his work during the 1870s and 1880s focused attention on finite groups of linear transformations with complex coefficients (which might represent projective transformations). One of his projects

[9]Poincaré was familiar with Frobenius' theory, which in 1912 he deemed the most important advance in the theory of finite groups in many years [486, p. 141], and he realized its connections with the work of Cartan. In 1903 [485, p. 106], he pointed out that "les théories de ces deux savants mathématiciens s'éclairent mutuellement."

was a research program that would generalize Galois' theory of equations. His views were widely circulated in his lectures on the icosahedron, which were published in 1884 [341] and so appeared at the time Molien was studying with Klein. In his lectures, Klein showed how the general quintic equation could be reduced to that of the icosahedral equation because the Galois group of the former is (after adjunction of the square root of the discriminant) the alternating group $\mathfrak{A}_5$, which is isomorphic to the group of 60 linear homogeneous transformations associated to the latter. The quintic equation and its solution through the icosahedral equation served as the paradigm for Klein's more general research program, which centered on the "form-problem" (*Formenproblem*) associated to a finite group of linear transformations. If $\mathfrak{G}$ is a finite group of linear transformations in the variables $x_1, \ldots, x_n$, then the associated form-problem is that of calculating the x_i from the forms (or polynomials) that are left invariant by the transformations of $\mathfrak{G}$. A central problem of Galois' original theory—to determine the roots $x_1, \ldots, x_n$ of a polynomial $f(x)$ by radical extensions of the field of numbers left invariant by the Galois group of $p(x)$—was thus a special type of form-problem.

Closely connected with the form-problem was another, which involved the reduction of the form-problem for $\mathfrak{G}$ to that of another group $\mathfrak{H}$ that is a homomorphic image of $\mathfrak{G}$. As Klein explained:

> The formulation of this problem has a certain importance, because we obtain therewith a general program for the further development of the theory of equations. Among form-problems ... with isomorphic groups we already designated above as the simplest the one which possesses the smallest number of variables. Given an equation $f(x) = 0$, we first seek to determine the smallest number of variables for which the group of linear substitutions can be constructed that is isomorphic to the Galois group of $f(x) = 0$. Then we would set up ... the form-problem belonging to this group and seek to reduce the solution of $f(x) = 0$ to this form-problem. [341, pp. 125–126]

Since it was customary to use the word "isomorphic" in the modern sense of "homomorphic," Klein was probably thinking of homomorphisms, although we cannot be certain. In any case, this "normal problem" (as he later termed it) had first been formulated by him, together with the form-problem, in a paper of 1879. It clearly focused attention on the possible representations of a finite group as a group of linear transformations and invited investigations of the properties of such representations, particularly as to their degrees. In his *Lectures on Mathematics* (1894) [343, p. 74], Klein specifically singled out this part of his general research program as worthy of further research: "A first problem I wish to propose is as follows. In recent years many groups of permutations of $6, 7, 8, 9, \ldots$ letters have been made known. The problem would be to determine the minimum number of variables with which isomorphic groups of linear substitutions can be formed."

Klein himself and the mathematicians directly associated with him in the execution of his program were concerned with representations by groups of projective transformations (expressed via linear transformations modulo scalar multiplication) and dealt with specific groups rather than with a general theory applicable to all finite groups.[10] But it is not difficult to imagine how suggestive this work would have been, especially to mathematicians aware of connections being developed between Lie groups and hypercomplex systems. Indeed, in a paper of 1894 (to be discussed in Section 14.3) in which William Burnside established the basics of the representation theory of finite groups from the starting point of the bilinear group associated to a group algebra, he pointed out that his work "obviously has a bearing on the question of the smallest number of variables in which [a finite group] g can be represented as a group of linear substitutions, i.e., on what Prof. Klein calls the degree of the normal problem connected with g" [50, p. 547]. It seems likely that Molien also had Klein's normal problem in mind when, starting in 1895, he sought to apply his results on hypercomplex systems to finite groups. In 1897, he published two papers [444, 445] that shed light on the normal problem by in effect developing the representation theory of finite groups by linear transformations. In fact, in the second paper, Molien focused on the question of the number of variables in an irreducible representation.

Molien's work on group representations was submitted to the scientific society of the University of Dorpat and acknowledged at the meetings of 24 April and 25 September 1897. At that time, he was still an instructor at the university and had published nothing since his outstanding thesis on hypercomplex systems, five years earlier. In the first note, he explained that he wished to communicate "some general theorems relating to the representability of a given discrete group in the form of a homogeneous linear substitution group," which are derived from the theory of hypercomplex numbers.

Molien began with the observation that a finite group $\mathfrak{G}$ with elements $S_1,\ldots,S_h$ determines a system $\mathcal{H}$ of hypercomplex numbers $\mathbf{x} = x_1S_1 + \cdots + x_hS_h$ with multiplication determined by $\mathfrak{G}$:

$$S_\alpha S_\beta = \sum_{\gamma=1}^{h} a^{\gamma}_{\alpha\beta} S_\gamma, \tag{14.6}$$

where $a^{\gamma}_{\alpha\beta} = 1$ if $S_\alpha S_\beta = S_\gamma$ and $a^{\gamma}_{\alpha\beta} = 0$ otherwise.[11] The system $\mathcal{H}$ is now called the group algebra of $\mathfrak{G}$. It was in this paper that Molien stated the criterion for semisimplicity implicit in his thesis—that $\mathcal{H}$ is semisimple if and only if the bilinear form $M(\mathbf{u},\mathbf{v}) = \mathrm{tr}(\mathbf{uv})_R = \mathrm{tr}\,\mathbf{v}_R \circ \mathbf{u}_R$ is nonsingular. In the case of the group algebra $\mathcal{H}$ defined by (14.6), $M(\mathbf{u},\mathbf{v})$ is easy to compute and has determinant $\pm h^h$. He was thus able to conclude, in effect, that $\mathcal{H}$ is a direct sum of complete matrix algebras and that the product equations for $\mathbf{y} = \mathbf{ux}$ can (by choosing a suitable basis) be put

[10] For references to the literature on the normal problem see [343, 608]. It should also be noted that Frobenius' student I. Schur applied Frobenius' theory of group representations to develop a general theory of projective representations. See Section 15.5.

[11] In presenting Molien's results, his notation has been slightly modified to bring it into line with that of Frobenius. In particular, the letters h, k, l have been given the same significance here as with Frobenius.

in the form

$$y_{\alpha\beta}^{(\lambda)} = \sum_{\gamma=1}^{n_\lambda} u_{\alpha\gamma}^{(\lambda)} x_{\gamma\beta}^{(\lambda)}, \quad \alpha, \beta = 1, \ldots, n_\lambda, \quad \lambda = 1, \ldots, l, \tag{14.7}$$

where $h = \sum_{\lambda=1}^{l} n_\lambda^2$, and $x_{\gamma\beta}^{(\lambda)}$, $u_{\alpha\gamma}^{(\lambda)}$, $y_{\alpha\beta}^{(\lambda)}$ denote the coefficients of $\mathbf{x}$, $\mathbf{u}$, $\mathbf{y}$ with respect to this basis.

If $u_1, \ldots, u_h$ denote the coefficients of $\mathbf{u} \in \mathcal{H}$ with respect to the original basis, so that $\mathbf{u} = u_1 S_1 + \cdots + u_h S_h$, then (14.7) can be written as

$$y_{\alpha\beta}^{(\lambda)} = \sum_{\gamma=1}^{n_\lambda} b_{\alpha\gamma}^{(\lambda)}(u) x_{\gamma\beta}, \quad i, k = 1, \ldots, n_\lambda, \quad \lambda = 1, \ldots, l, \tag{14.8}$$

where the $b_{\alpha\gamma}^{(\lambda)}(u)$ are linear homogeneous functions of $u_1, \ldots, u_h$. Then, since the index β in these equations "has no influence on the coefficients" $b_{\alpha\gamma}^{(\lambda)}(u)$ [444, p. 268], Molien considered the system of equations in $n_1 + \cdots + n_l$ variables

$$y_{\alpha}^{(\lambda)} = \sum_{\gamma=1}^{n_\lambda} b_{\alpha\gamma}^{(\lambda)}(\mathbf{u}) x_{\gamma}^{(\lambda)}, \quad \alpha = 1, \ldots, n_\lambda, \quad \lambda = 1, \ldots, l. \tag{14.9}$$

He observed that each subsystem in (14.9) yields a finite group of linear transformations $T_\beta^{(\lambda)} \leftrightarrow \left(b_{\alpha\gamma}^{(\lambda)}(S_\beta)\right)$, $\beta = 1, \ldots, h$, and that $S_\beta \to T_\beta^{(\lambda)}$ is a group homomorphism. We can recognize these as the irreducible representations of the group $\mathfrak{G}$.

The main point of Molien's first communication concerned the following question [444, p. 270]: "If a discrete group is already given in the form of a linear substitution group, what is its relation to the systems of equations [(14.9)] considered by us?" That is, suppose the group $\mathfrak{G} = \{S_1, \ldots, S_h\}$ is itself a group of linear transformations. How is the group of linear transformations S_β related to the groups defined by the subsystems in (14.9)? To answer this question, he introduced the associated continuous group of transformations $S_u = \sum_{\beta=1}^{h} u_\beta S_\beta$. The answer that Molien gave was already implicit in his thesis [443, Satz 40]: by a linear change of variables, the matrix of equations defining S_u can be put in the form

$$\begin{pmatrix} B_1(u) & 0 & \cdots & 0 \\ 0 & B_2(u) & \cdots & \\ \vdots & \vdots & \ddots & \vdots \\ 0 & 0 & \cdots & B_m(u) \end{pmatrix}, \tag{14.10}$$

where each of $B_1(u)$, $B_2(u), \ldots$ represents one of the matrices $b_{\alpha\gamma}^{(\lambda)}(u)$ from (14.9). In particular, setting $u_\alpha = \delta_{\alpha\beta}$ in (14.10), so that $S_u = S_\beta$, Molien obtained the analogous decomposition of the linear transformations S_β. In effect, he had established the complete reducibility theorem for groups of linear transformations, which, if applied to a representation of an abstract finite group, yields immediately the usual formulation of the complete reducibility theorem. As he himself put the matter, "From the given composition of a discontinuous finite group, all linear groups of substitutions with the same structure can be obtained" [444, p. 276]. That is, from the multiplication table of the group, one obtains the hypercomplex system $\mathcal{H}$ and therefrom the associated irreducible groups defined by the subsystems in (14.9); and he had now proved that every linear group with the same multiplication table is built up out of these groups.

It was in the second note that Molien went beyond what was, more or less, an immediate consequence of the results in his thesis. Having shown in his first note "how a given substitution group can be decomposed into its irreducible components," he proposed in the second to consider "only the properties of the irreducible groups ..." [445, p. 277]. Molien was primarily interested in what could be said about the numbers n_λ, the number of variables occurring in the irreducible groups of (14.9)—the degrees of the irreducible representations, as we would say. As was noted earlier, this problem had a direct bearing on Klein's "normal problem." Of course, it followed immediately from the fact that the group algebra of $\mathfrak{G}$ is a direct sum of complete matrix algebras that $h = \sum_{\lambda=1}^{l} n_\lambda^2$. The main result of his second paper was that the n_λ divide h. In the process of proving this result, Molien obtained further basic theorems of group representation theory, including what amounts to the orthogonality relations for characters.

As noted in Section 13.5, Frobenius realized by the time of his 1897 "Representation" paper that his generalized characters were simply the trace functions of the irreducible representations of the underlying group. Since trace functions had already played a central role in Molien's 1893 dissertation, where much of the theory of hypercomplex systems is based on consideration of the linear function $\varphi_1(\mathbf{u}) = \operatorname{tr}\mathbf{u}_R$ and the associated symmetric bilinear form $M(\mathbf{u},\mathbf{v}) = \varphi_1(\mathbf{uv})$, it is not surprising that characters in the guise of trace functions played a central role in his papers of 1897 as well. When $\mathcal{H}$ is the group algebra of $\mathfrak{G}$, the decomposition of $\mathcal{H}$ into a direct sum of complete matrix algebras implies that the associated trace function $\varphi_1(\mathbf{u}) = \operatorname{tr}\mathbf{u}_R$ has the decomposition $\varphi_1(\mathbf{u}) = \sum_{\alpha=1}^{l} \varphi_\alpha^{(\lambda)}(\mathbf{u})$, where $\varphi_1^{(\lambda)}(\mathbf{u}) = \sum_{\beta=1}^{n_\lambda} b_{\beta\beta}^{(\lambda)}(\mathbf{u})$ is the trace of the λth group in (14.9). Since $\varphi_1^{(\lambda)}(\mathbf{u})$ is a linear function of $u_1,\ldots,u_h$, where $\mathbf{u} = \sum_{\alpha=1}^{h} u_\alpha S_\alpha \in \mathcal{H}$ and $S_1 = E$, the identity element of $\mathfrak{G}$, following Molien, we express $\varphi_1^{(\lambda)}$ in the form

$$\varphi_1^{(\lambda)}(\mathbf{u}) = n_{\lambda 1}u_1 + \cdots + n_{\lambda h}u_h.$$

Then $\varphi_1^{(\lambda)}(S_\beta) = n_{\lambda\beta}$ represents the trace of the linear transformation $S_\beta^{(\lambda)}$ obtained by setting $u_\alpha = \delta_{\alpha\beta}$ in the system of equations defining the λth group in (14.9). (In the more familiar notation of Frobenius, $n_{\lambda\beta} = \chi^{(\lambda)}(S_\beta)$, where $\chi^{(\lambda)}$ denotes the λth irreducible character of the group $\mathfrak{G}$.) As Molien realized, it follows immediately from the fact that $\varphi_1^{(\lambda)}$ is a trace function that $n_{\lambda 1} = \operatorname{tr} S_1^{(\lambda)} = n_\lambda$, the number of variables involved in the λth group in (14.9) and the chief object of his investigation in the second note [445].

The key to Molien's investigation was provided by considering the trace function $\Omega(\mathbf{u},\mathbf{v})$ of the group of transformations $\mathbf{x} \to \mathbf{vxu}$, which Study had introduced in his study of hypercomplex systems. Thus $\Omega(\mathbf{u},\mathbf{v}) = \mathrm{tr}\,\mathbf{u}_R \circ \mathbf{v}_L$ is another symmetric bilinear form defined on any hypercomplex system $\mathcal{H}$. The advantage of Ω over the previously considered form $M(\mathbf{u},\mathbf{v}) = \mathrm{tr}(\mathbf{uv})_R = \mathrm{tr}(\mathbf{v}_R \circ \mathbf{u}_R)$ is that if $\mathcal{H}$ is a complete matrix algebra, then a straightforward calculation shows that $\Omega(\mathbf{u},\mathbf{v}) = \varphi_1(\mathbf{u})\varphi_1(\mathbf{v})$. Since the group algebra is a direct sum of complete matrix algebras, it follows that in this case,

$$\Omega = \sum_{\lambda=1}^{l} \varphi_1^{(\lambda)}(\mathbf{v})\varphi_1^{(\lambda)}(\mathbf{u}). \tag{14.11}$$

On the other hand, Ω can be expressed in terms of the coordinates corresponding to the basis $S_1,\ldots,S_h$ of $\mathcal{H}$. The result is

$$\Omega = \sum_{\alpha,\beta,\gamma,\delta} a^{\alpha}_{\gamma\beta} a^{\beta}_{\alpha\delta} v_\gamma u_\delta,$$

where the multiplication constants $a^{\gamma}_{\alpha\beta}$ are defined following (14.6). By computing Ω in this manner, Molien obtained the following result [445, p. 268]: Let the group elements be arranged so that $S_1 = E$, and $S_1,\ldots S_k$ are representatives of the k conjugacy classes $(1),(2),\ldots,(k)$ of the group $\mathfrak{G}$, and let n_α denote the number of $S \in \mathfrak{G}$ such that $S^{-1}S_\alpha S = S_\alpha$. (In other words, n_α is the order of the normalizer of S_α.) Then if we set $C'_\alpha(\mathbf{v}) = \sum_{S_\beta \in (\alpha')} v_\beta$ and $C_\alpha(\mathbf{u}) = \sum_{S_\beta \in (\alpha)} u_\beta$, where as usual (α') denotes the conjugacy class of inverses of elements in (α), Ω is given by

$$\Omega = \sum_{\alpha=1}^{k} n_\alpha C'_\alpha(v) C_\alpha(u). \tag{14.12}$$

The first consequence Molien derived from equating (14.11) and (14.12) was that $l = k$: the number of irreducible groups in (14.9) is equal to the number of conjugacy classes of $\mathfrak{G}$. This is analogous to Frobenius' theorem that the number of conjugacy classes equals the number of distinct irreducible factors of the group determinant (Theorem 13.6). Although Molien's proof that $l = k$ is typically obscure, the result is a consequence of his approach: the l linear forms $\varphi_1^{(\lambda)}(\mathbf{u})$ are linearly independent, as are the k linear forms $C_\alpha(u)$. By appropriately specializing the values of v in (14.11) and (14.12), it can be shown that each $\varphi_1^{(\lambda)}(\mathbf{u})$ is a linear combination of the $C_\alpha(u)$'s and that conversely, each $C_\alpha(u)$ is a linear combination of the $\varphi_1^{(\lambda)}(\mathbf{u})$'s, so that $l = k$.

Whereas Frobenius had used the two fundamental orthogonality relations (13.22) and (13.25) to deduce the equivalent of $l = k$, Molien used the equality of l and k to obtain what amounts to Frobenius' orthogonality relations. Using the fact that the trace functions $\varphi_1^{(\lambda)}$ are invariant on the conjugacy classes of $\mathfrak{G}$, he wrote

$$\varphi_1^{(\lambda)}(\mathbf{u}) = \sum_{\alpha=1}^{k} n_{\lambda\alpha} C_\alpha(\mathbf{u}), \quad \varphi_1^{(\lambda)}(\mathbf{v}) = \sum_{\alpha=1}^{k} n'_{\lambda\alpha} C'_\alpha(\mathbf{v}), \tag{14.13}$$

where $n'_{\lambda\alpha} = \varphi_1^{(\lambda)}(S_\alpha^{-1})$. Making the substitutions (14.13) in (14.11) and equating the result with (14.12), he obtained the following relations on the coefficients $n'_{\lambda\alpha}$ and $n_{\lambda\beta}$:

$$n_\alpha \sum_{\lambda=1}^{k} n'_{\lambda\alpha} n_{\lambda\beta} = h\delta_{\alpha\beta}. \tag{14.14}$$

In matrix form, this equation is equivalent to an equation of the form $AB = I$. Since $l = k$ the matrices are square, so that $BA = I$ follows. Tacitly using this fact of linear algebra, Molien obtained "the further important formula"

$$\sum_{\alpha=1}^{k} h_\alpha n'_{\lambda\alpha} n_{\mu\alpha} = h\delta_{\lambda\mu}, \tag{14.15}$$

where $h_\alpha = h/n_\alpha$. From the definition of n_α, h_α is easily seen to equal the number of elements in the conjugacy class (α). Equations (14.14) and (14.13) are, respectively, the fundamental orthogonality relations (13.25) and (13.22) of Frobenius' theory.

From the orthogonality relation (14.13), Molien obtained the main result of his note [445], namely that the number of variables in every irreducible group in (14.9) must divide the order of $\mathfrak{G}$: $n_\lambda \mid h$ for every λ. Earlier, Frobenius had used (14.5) in the form (13.18) to prove the analogous result $e \mid h$.[12] Like Frobenius, Molien realized that if $S^{(\lambda)}$ is the λth irreducible representation of $\mathfrak{G}$ as in (14.9), then the numbers $n_{\lambda\alpha} = \operatorname{tr} S^{(\lambda)}(S_\alpha)$ are algebraic integers, since they are a sum of roots of unity by the "known theorem" that since the linear transformations $S^{(\lambda)}(S_\alpha), \alpha = 1,\ldots,h$, form a finite group, the characteristic roots of $S^{(\lambda)}(S_\alpha)$ are roots on unity. Indeed, the above-mentioned theorem was well known to mathematicians interested in finite groups of linear transformations, such as Klein and his students. In one of his many memoirs from the 1870s and early 1880s on finite groups of linear transformations, Camille Jordan had proved that such linear transformations have a diagonal canonical form with roots of unity on the diagonal. Since Molien had been Klein's student, his knowledge of this fact would not be surprising. As will be seen in Section 14.4, Jordan's theorem played a role in the independent discovery

[12] See the remarks following (13.21).

by Maschke, another of Klein's students, of the complete reducibility theorem. As for Molien, by virtue of the "known theorem" and (14.15), he realized that he could prove $n_\lambda \mid h$ by showing that $n_\lambda \mid h_\alpha n_{\lambda\alpha}$ for every α; and he succeeded in doing this, although his proof [445, §5] is rather involved.

Shortly before Frobenius wrote up his "Representation" paper [213] in 1897, he learned of Molien's work from Study [213, p. 92], and so he wrote to Molien and sent him copies of his papers. In his reply of 10 December 1897, Molien explained that he had been unaware of Frobenius' work.[13] Stimulated by the communication from Frobenius, Molien wrote up a bit more of his results, which he asked Frobenius to present in the proceedings of the Berlin Academy. This was done by Frobenius on 16 December [446]. Incidentally, Molien's paper contains the theorem that if $\mathfrak{G}$ is a finite group of linear transformations and if $\mathfrak{G}^p$ denotes its p-fold Kronecker (or tensor) product, then each irreducible representation of $\mathfrak{G}$ is contained in the decomposition of $\mathfrak{G}^p$ for some p.[14] Another noteworthy result of the paper is discussed in Section 15.5.

Two months after submitting Molien's paper to the academy, Frobenius wrote to Dedekind about him:

> You will have noticed that a young mathematician, Theodor Molien in Dorpat, has considered the group determinant independently of me. In volume 41 of the *Mathematische Annalen* he published a very beautiful but difficult work, "Über Systeme höherer complexer Zahlen," in which he has investigated noncommutative multiplication and obtained important general results of which the properties of the group determinant are special cases. Since he was entirely unknown to me, I have made some inquiries regarding his personal circumstances. Details are still lacking. This much I have already learned: that he is still an instructor in Dorpat; that his position there is uncertain and he has not advanced as far as he would have desired in view of his undoubtedly strong mathematical talent. I would very much like to interest you in this talented man; here and there you are virtually privy councilor; if an opportunity presents itself, please think of Herr Molien, and if you have time, look at his work.[15]

Whether Dedekind ever attempted to further Molien's career is uncertain. Frobenius, however, wrote several letters of reference for Molien in the hope of securing him a position in Russia, where he chose to remain. Although he strongly praised Molien's accomplishments and talent in these letters as well as in his own publications—see in this connection the discussion of his reworking of Molien's results in Section 15.3—the political situation in Russia nullified Frobenius' efforts to boost Molien's career.[16] He was refused a vacant professorship at Dorpat, and after a

---

[13]Molien's letter was published (in Russian translation) by Kanounov [334, p. 57]. The rest of his brief correspondence with Frobenius (in Russian translation) may be found in Kanounov's biography [335].

[14] Molien's proof is similar to that given later by Burnside [56, p. 299], who states the theorem in terms of a faithful representation $\mu : \mathfrak{H} \to \mathfrak{G}$ of an abstract group $\mathfrak{H}$.

[15]Letter to Dedekind dated 24 February 1898.

[16]According to Kanounov [333], Molien was refused the professorship at Dorpat as a consequence of the czarist regime's Russification policy.

year studying medieval and Renaissance mathematical manuscripts in the Vatican Library, he decided to become, in 1900, the first mathematics professor in Siberia by accepting a position at the technical institute in Tomsk. At Tomsk, without even a research library at his disposal, Molien devoted his energy to developing and teaching a mathematics curriculum. He published numerous sets of lecture notes but no research papers, thus leaving it to Frobenius to continue the development of the new theory. This Frobenius did, with a steady stream of papers (1898–1903) on the new theory and its applications to the study of hypercomplex systems and finite groups, as will be seen in Sections 15.1–15.4.

14.3 W. Burnside

In the mid-1890s, when Frobenius began to focus on the theory of abstract finite groups, the British mathematician William Burnside (1852–1927) began to do the same as well. In fact, Burnside published several theorems on the theory only to discover, as he explained [47, pp. 191–192], that he had been completely anticipated by Frobenius. After Frobenius learned about group determinants from Dedekind, he discovered that in a paper of 1893, Burnside had considered what amounts to the group determinant of an abelian group and had established Dedekind's unpublished theorem on its factorization into linear factors (12.10), albeit without connecting it with the notion of a group character. When Frobenius informed Dedekind of this fact, he added that

> This is the same Herr Burnside who annoyed me several years ago by quickly rediscovering all the theorems I had published on the theory of groups, in the same order and without exception: first my proof of Sylow's theorem, then the theorem on groups with square-free orders, on groups of order $p^{\alpha}q$, on groups whose order is a product of four or five prime numbers, etc., etc. In any case, a very remarkable and amazing example of intellectual harmony, probably as is possible only in England and perhaps America.[17]

I do not think Frobenius was implying that Burnside was plagiarizing his results, but simply that he was one of the many British and American mathematicians who paid little attention to mathematical developments on the Continent. (Actually, this stereotype does not really apply to Burnside.) Little did Frobenius then realize that yet another instance of "intellectual harmony" was shortly to occur, for in a paper of 1898 [50], Burnside derived many of Frobenius' results about group determinants.

As was the case with Molien, Burnside's work was suggested by the work relating Lie's theory of groups and the theory of hypercomplex numbers. In a paper submitted in January 1898 [49], and thus shortly after the appearance of Molien's papers of 1897 and Frobenius' 1897 "Representation" paper (none of which were known to him), Burnside considered a finite group $\mathfrak{G}$ and the associated continuous group of transformations $T_y : \mathbf{x} \to \mathbf{xy}$ defined by the group algebra

[17] Letter to Dedekind dated 7 May 1896.

$\mathcal{H}$ of $\mathfrak{G}$. As we have seen, the Lie algebra $\mathfrak{g}$ of this group may be identified with $\mathcal{H}$ as a vector space with $[\mathbf{x},\mathbf{y}] = \mathbf{xy} - \mathbf{yx}$. Making this identification, Burnside investigated the Lie group by investigating its Lie algebra $\mathfrak{g}$. He showed that $\mathbf{z} \in \mathfrak{g}$ is "self-conjugate" in the sense that $[\mathbf{z},\mathbf{x}] = 0$ for all $\mathbf{x} \in \mathfrak{g}$ if and only if $\mathbf{z} = \sum_{\alpha=1}^{h} z_\alpha S_\alpha$ is such that $z_\alpha = z_\beta$ whenever S_α and S_β are conjugate in the finite group $\mathfrak{G}$, so that the totality $\mathfrak{z}$ of all self-conjugate elements forms a k-dimensional ideal in $\mathfrak{g}$, where k denotes as usual the number of conjugacy classes of $\mathfrak{G}$. Similar considerations showed that the derived algebra $\mathfrak{g}' = [\mathfrak{g},\mathfrak{g}]$ has dimension $h-k$ and that $\mathfrak{g} = \mathfrak{z} \oplus \mathfrak{g}'$. Burnside was familiar with Cartan's 1894 thesis [57], and he applied Cartan's criterion for semisimplicity to conclude that $\mathfrak{g}'$ is semisimple. To fully understand the structure of $\mathfrak{g}$, it would be necessary to solve the problem of determining the type of simple ideals into which $\mathfrak{g}'$ decomposes. As mentioned earlier in this section, Cartan had done just that (for any hypercomplex system) in his 1895 note [58], but evidently Burnside was not familiar with it and did not provide a solution in his paper. By the end of 1898, however, he had solved the problem and shown that the continuous group with Lie algebra $\mathfrak{g}$ is a direct product of k groups each of which is isomorphic to a general linear group [50]. Furthermore, he had applied his results to rederive many of Frobenius' theorems on the group characters and the group determinant.

When Burnside published his paper of January 1898 [49], he had looked at Frobenius' 1896 papers: the paper on group characters [211] (discussed in Section 13.3) and the paper on the factorization of the group determinant [212] (summed up in Theorem 13.10), and by the time he composed his second paper of 1898 [50], he knew of Molien's 1893 doctoral dissertation on hypercomplex numbers [443] as well. It would seem that this knowledge was a factor in his own discoveries, so that although he rederived many of Frobenius' results by means of a completely different approach, he did not independently rediscover them in the same way that Molien had done.[18] Nonetheless, it would seem that it was the work on hypercomplex systems and Lie groups that had initiated Burnside's investigation of the connections between the finite group $\mathfrak{G}$ and the continuous group with Lie algebra $\mathfrak{g}$ rather than Frobenius' work. Thus it is conceivable that had Frobenius not been induced by Dedekind to study the group determinant, Burnside might have shared the honors with Molien as coinventor of the representation theory of finite groups.

14.4 H. Maschke

Once the group determinant was considered, it was natural to investigate its factorization into irreducible factors (as Frobenius proceeded to do). By contrast, given a finite group of linear transformations, the analogous question did not present

[18]See pp. 273–279 of my paper [267] for a discussion of Burnside's work in general and this point in particular.

itself due to the absence of a complete reducibility theorem to parallel the prime factorization theorem for polynomials. Given a complete reducibility theorem, it would have been only a matter of time before the theory of group representations would have emerged. We have seen how the complete reducibility theorem was emerging from research on the connections between hypercomplex systems and Lie's theory as contained in the work of Cartan, Molien, and Burnside. To conclude the discussion of alternative routes to Frobenius' theory, I will consider how another line of research led Heinrich Maschke (1853–1908) to the complete reducibility theorem independently of the work of Frobenius, Cartan, Molien, and Burnside.

Like Molien, Maschke was a former student of Klein's, and as such, interested in research on finite groups of linear transformations and their invariants, a line of research that, as we noted in Section 14.2, was encouraged by Klein's envisioned generalization of Galois' theory. In 1892, Maschke joined his friend Oskar Bolza, another of Klein's students, and the American E.H. Moore to form the Mathematics Department of the recently established University of Chicago. In 1896, a simple but consequential observation was made independently by Moore and the German mathematician Alfred Loewy:

Theorem 14.1. *If $\mathfrak{G}$ is any finite group of linear transformations of complex variables $x = (x_1,\ldots,x_n)$, then a positive definite Hermitian form $\varphi(x) = \sum_{i,j=1}^{n} a_{ij}x_i\bar{x}_j$ exists that is invariant under the transformations of $\mathfrak{G}$, so that $\varphi(Tx) = \varphi(x)$ for all $T \in \mathfrak{G}$.*[19]

Moore was led to formulate Theorem 14.1 as a consequence of reading Klein's geometric approach to the determination of all finite subgroups of the group of all projective transformations of the plane $\mathrm{PGL}(3,\mathbb{C})$ [340]. Loewy considered Theorem 14.1 because Picard had proved it for finite subgroups of $\mathrm{GL}(2,\mathbb{C})$ and for all but one type of finite subgroup of $\mathrm{GL}(3,\mathbb{C})$ [473].[20] Both of them proved this theorem by observing that if $\psi(x)$ is any positive definite Hermitian form, then $\varphi(x) = \sum_{T\in\mathfrak{G}} \psi(Tx)$ is the desired form since for every $T \in \mathfrak{G}$,

$$\varphi(Tx) = \sum_{S\in\mathfrak{G}} \psi(STx) = \sum_{U\in\mathfrak{G}} \psi(Ux) = \varphi(x). \tag{14.16}$$

Of course, this technique of summing over a finite group and the attendant invariance of such sums under "translations" $T \to ST$ as used in (14.16) was being utilized that very same year by Frobenius to create his theory of characters and group determinants. In fact, as we have seen, Frobenius had already used the technique in his work on integral theta characteristics in the 1880s (Section 12.4), but neither

---

[19]See in this connection Loewy's papers [419] and [420, p. 561n], Moore's paper [448], and Klein's announcement [344].

[20]Picard failed to see that Poincaré's technique of summing over a group to generate invariants (see below), which he himself had extended to certain countably infinite subgroups of $\mathrm{PGL}(3,\mathbb{C})$ [472], would yield a completely general proof of Theorem 14.1.

Moore nor Loewy seemed aware of this fact. It is more likely that they may have been aware that essentially the same technique had been introduced for certain countably infinite groups by Poincaré in his theory of theta Fuchsian functions in the early 1880s.[21] Indeed, it seems to have been this work on automorphic functions that led Hurwitz at about the same time to observe that all the invariants of a finite linear group may be generated by the technique of summation over the group [304, p. 71].[22]

Maschke knew of Theorem 14.1 through his contact with Moore. Moore had applied Theorem 14.1 to give a new proof of Jordan's theorem that if $T \in \mathrm{GL}(n, \mathbb{C})$ has finite order, so $T^k = I$ for some integer k, then it has a diagonal canonical form with kth roots of unity on the diagonal—a theorem independently derived by Frobenius as a consequence of his minimal polynomial theorem (Theorem 7.2). Jordan's result may have induced Maschke to seek to use Theorem 14.1 to prove that for all finite subgroups $\mathfrak{G} \subset \mathrm{GL}(n, \mathbb{C})$, a linear change of variables is possible such that the matrices of the $T \in \mathfrak{G}$ have coefficients that are all elements in the field generated over $\mathbb{Q}$ by the gth roots of unity, g being the order of $\mathfrak{G}$. This turns out to be a very difficult problem, and after many talented mathematicians had obtained partial results (Section 15.5), it was finally solved by R. Brauer in 1945 (Section 15.6). In 1898, Maschke established the result for groups $\mathfrak{G}$ with the property that some $T \in \mathfrak{G}$ has distinct characteristic roots [434]. His proof involved the following lemma [434, p. 497, Satz VII]:

Theorem 14.2. *Suppose $\mathfrak{G} \subset \mathrm{GL}(n, \mathbb{C})$ is a finite group with the property that the (i, k) coefficient of each transformation vanishes for a fixed i, k with $i \neq k$. Then if some $T \in \mathfrak{G}$ has distinct characteristic roots, it follows that $\mathfrak{G}$ is "intransitive," i.e., by a change of variables, the matrices of the $T \in \mathfrak{G}$ are all of the form*

$$\begin{pmatrix} Q_1 & 0 \\ 0 & Q_2 \end{pmatrix}.$$

In what follows, I will refer to the first hypothesis in this theorem as the *vanishing coefficient property* and the second as the *hypothesis of a generic $T \in \mathfrak{G}$*.

In the course of his proof, Maschke showed, using his hypothesis of the existence of a generic $T \in \mathfrak{G}$, that a variable change exists such that the coefficient matrices of all $T \in \mathfrak{G}$ are of the form

$$\begin{pmatrix} Q_1 & 0 \\ R & Q_2 \end{pmatrix}. \tag{14.17}$$

---

[21] See, e.g., Poincaré's papers [476, p. 97], [475, p. 182] or the discussion in Gray's book [255].

[22] Hurwitz's paper was mainly concerned with the extension of the technique to continuous groups. His paper played an important role in the extension of Frobenius' theory of characters and representations to continuous groups as indicated below in Section 15.5.

Using Moore's Theorem 14.1 together with the existence of the generic T, he then showed that the variable change could be chosen such that $R = 0$ for all $T \in \mathfrak{G}$. This then enabled him to put the matrices of all $T \in \mathfrak{G}$ in the form

$$\begin{pmatrix} Q_1 & 0 & \cdots & 0 \\ 0 & Q_2 & \cdots & 0 \\ \vdots & \vdots & \ddots & \vdots \\ 0 & 0 & \cdots & Q_s \end{pmatrix}, \tag{14.18}$$

where none of the Q_i have Maschke's vanishing coefficient property. It then remained to show that groups that do not have the vanishing coefficient property (but do possess a generic transformation) can be transformed so as to have cyclotomic numbers as coefficients.

Shortly after establishing this result, Maschke realized that he could prove Theorem 14.2 without the hypothesis of the generic $T \in \mathfrak{G}$, and so for all finite groups of linear transformations [435]. He had thus established a necessary and sufficient condition that a finite group of linear transformations be "intransitive" in the sense described above. Much of Maschke's proof [434, §§1–2] involved showing that if $\mathfrak{G}$ has the vanishing coefficient property, then a linear change of variables is possible such that the matrices of the $T \in \mathfrak{G}$ have the form in (14.17). Then he used Moore's Theorem 14.1 to show that for any finite group of linear transformations, (14.17) implies the matrix form of Theorem 14.2, i.e., that $R = 0$ in (14.17) [434, §3]. This part of his proof thus amounts to a proof of the complete reducibility theorem. Maschke's proof idea for this result was to show that a change of variables is possible such that the $T \in \mathfrak{G}$ continue to have a matrix representation of the form (14.17), while the invariant Hermitian form posited in Theorem 14.1 takes the form $\varphi = \sum_{i=1}^{n} x_i \bar{x}_i$. In the now familiar language introduced by Frobenius, this means that the matrices $M(T)$ in (14.17) are unitary ($\overline{M(T)}^t M(T) = I$), which implies readily that $R = 0$ in (14.17). In the course of his proof, Maschke had in effect established that every representation of a finite group is equivalent to a unitary representation.

Although Maschke had given a proof of what has since become known as the complete reducibility theorem, it should be kept in mind that he himself did not stress the importance of this result, which is not even highlighted as a theorem in his paper. Of primary importance to him was the improved version of Theorem 14.2 made possible by complete reducibility. Still, it seems likely that eventually Maschke himself or another mathematician would have considered the implications of the complete reducibility property for the study of finite linear groups and, in particular, for the resolution of Klein's normal problem.

Maschke's discovery of complete reducibility further illustrates the fact that the creation of the theory of group characters and representations was related to a broad spectrum of late-nineteenth-century mathematics. Molien's discovery and application of the orthogonality relations for group characters shows that even character theory was not exclusively the product of Frobenius' genius and the arithmetic tradition (Gauss, Dirichlet, Dedekind) out of which the notion of a character had evolved. The developments traced in this chapter are indicative of the multifarious links that Frobenius' theory had with other branches of mathematics. By virtue of such links, it has become an integral part of present-day mathematics.

Of all the mathematicians who discovered some aspect of the theory of group characters and representations—Frobenius, Molien, Burnside, and Maschke—it was by far Frobenius who developed the theory and its applications most extensively and rigorously. This can be seen already in Chapter 12 and, even more so, in the next chapter.

Chapter 15
Characters and Representations After 1897

Frobenius' papers of 1896–1897 marked the beginning of a new theory, a theory that continued to evolve in various directions for over a half-century. Frobenius himself, along with Burnside, made significant contributions to the theory after 1897, and many new ideas, viewpoints, and directions were introduced by Frobenius' student Issai Schur (1875–1941), and then by Schur's student Richard Brauer (1901–1977). In this chapter, these later developments will be sketched, with particular emphasis on matters that relate to the presentation in the previous sections.

15.1 Frobenius' Theory of Induced Characters

From the outset of his theory, Frobenius had realized that the irreducible factors of the group determinant, or equivalently, the irreducible representations of a group, are uniquely determined by the associated irreducible characters. Thus the first step in the determination of the irreducible factors or representations for a given group was to determine the irreducible characters. In order to facilitate the computation of these characters for a given group, he devised two general methods. One method, published in 1899 [215], was based on the fact already observed by Molien in [446] that if σ, σ' are two representations of $\mathfrak{H}$ with characters χ, χ', then their product $\chi\chi'$ is the character of the tensor product representation $\sigma \otimes \sigma'$. Thus if $\chi^{(\alpha)}$, $\alpha = 1, \ldots, k$ are the irreducible characters of $\mathfrak{H}$, then integers $f_{\alpha\beta\gamma}$ exist such that

$$\chi^{(\alpha)}\chi^{(\beta)} = \sum_{\gamma=1}^{k} f_{\alpha\beta\gamma}\chi^{(\gamma)}.$$

By utilizing properties of the integers $f_{\alpha\beta\gamma}$, Frobenius used the above formula and partial knowledge of the characters to obtain further information about their values. This method of "composition of characters" was used by him in conjunction with another method of greater historical significance, since it inspired, as we shall

T. Hawkins, *The Mathematics of Frobenius in Context*, Sources and Studies in the History of Mathematics and Physical Sciences, DOI 10.1007/978-1-4614-6333-7_15,

see, the theory of what are today called induced characters and representations. Frobenius' other method was presented in 1898 [214] and resulted from the problem of seeking relations among the irreducible characters of a group $\mathfrak{H}$ and those of a subgroup $\mathfrak{G}$, the idea being that if the irreducible characters of the subgroup $\mathfrak{G}$ are already known, then this information should help determine the irreducible characters of the larger group $\mathfrak{H}$.

To explain the considerations involved in this method of Frobenius and how they may have suggested the concept of an induced representation, it will be helpful to begin with some notation. Let $\mathfrak{H}$ be a group of order h with a subgroup $\mathfrak{G}$ of order g, and $n = (\mathfrak{H} : \mathfrak{G})$ the index of $\mathfrak{G}$ in $\mathfrak{H}$. Corresponding to $\mathfrak{H}$ are the group variables x_R with $R = E, A, B, \ldots$ running through $\mathfrak{H}$, the group matrix $\hat{\rho}(x) = (x_{RS^{-1}})$, and the associated group determinant $\hat{\Theta}(x) = |x_{RS^{-1}}|$. (Recall that $\hat{\rho}(x)$ with $x_R = 1$ and $x_S = 0$ for all $S \neq R$ yields the regular representation $R \to \hat{\rho}(R)$ of $\mathfrak{H}$.) Finally, with $\hat{k}$ denoting the number of conjugacy classes of $\mathfrak{H}$, and (λ), $\lambda = 0, 1, \ldots, \hat{k} - 1$, denoting the classes themselves with $(0) = \{E\}$, let $\hat{\Theta}(x) = \prod_{\lambda=0}^{\hat{k}-1} \Phi_{\lambda}^{\hat{f}_{\lambda}}$ denote the factorization of $\hat{\Theta}$ into irreducible factors and $\chi^{(\lambda)}$ the irreducible character associated to Φ_{λ}. Likewise, let $\Theta(x) = |x_{PQ^{-1}}|$ denote the group determinant of $\mathfrak{G}$, so that Θ is a homogeneous polynomial of degree g in the g variables x_P, as P ranges over $\mathfrak{G}$. Then, with k denoting the number of conjugacy classes of $\mathfrak{G}$, let $\Theta(x) = \prod_{\kappa=0}^{k-1} \Psi_{\kappa}(x)^{f_{\kappa}}$ be the factorization of Θ into its irreducible factors, and let $\psi^{(\kappa)}$ denote the irreducible character of $\mathfrak{G}$ associated with $\Psi_{\kappa}(x)$.

In order to see how the characters $\chi^{(\lambda)}$ are related to the characters $\psi^{(\kappa)}$, Frobenius partitioned $\mathfrak{H}$ into disjoint cosets with respect to $\mathfrak{G}$. Let $A_1 = E, A_2, \ldots, A_n$ denote elements of $\mathfrak{H}$ effecting such a partition into left cosets, so that $\mathfrak{H} = \cup_{i=1}^{n} A_i\mathfrak{G}$. Then it is easy to see that we have as well a partition of $\mathfrak{H}$ into right cosets, namely $\mathfrak{H} = \cup_{i=1}^{n} \mathfrak{G}A_j^{-1}$. If the elements of $\mathfrak{G}$ are ordered in some manner, starting with E, then these partitions, Frobenius discovered, impose an ordering on the elements of $\mathfrak{H}$ that sheds light on the connections between the group determinants and associated matrices of $\mathfrak{G}$ and $\mathfrak{H}$.

I will consider first (as did Frobenius) the ordering $\mathfrak{G}, \mathfrak{G}A_2^{-1}, \ldots, \mathfrak{G}A_n^{-1}$ corresponding to the right coset partition. The partition orders the group $\mathfrak{H}$ and the group variables x_R in a manner that naturally partitions the group matrix $\hat{\rho}(x) = (x_{RS^{-1}})$ into n^2 $g \times g$ blocks, the (i, j)th block being, with P, Q running through $\mathfrak{G}$,

$$\rho_{ij}(x) = \left(x_{PA_i^{-1}(QA_j^{-1})^{-1}}\right) = \left(x_{PA_i^{-1}A_jQ^{-1}}\right).$$

Since $x_{(PA_i^{-1})(A_iQ^{-1})} = x_{PQ^{-1}}$, each of the diagonal blocks is the group matrix of $\mathfrak{G}$, viz., $\rho_{ii} = (x_{PQ^{-1}})$, $P, Q \in \mathfrak{G}$. Let x^* denote the specialization of the $h = (\mathfrak{H} : 1)$ variables $x_E, x_A, \ldots, x_R, \ldots$, with $x_R = 0$ for all $R \notin \mathfrak{G}$. Then $\rho_{ii} = (x_{PQ^{-1}}) = \rho_{ii}(x^*)$, whereas $\rho_{ij}(x^*) = 0$ for $i \neq j$ because $PA_i^{-1}A_jQ^{-1} \notin \mathfrak{G}$. Thus $\hat{\rho}(x^*)$ consists of n diagonal blocks $\rho_{ii} = (x_{PQ^{-1}})$, and so $\hat{\Theta}(x^*) = \Theta(x^*)^n$, $n = (\mathfrak{H} : \mathfrak{G})$. (Since x denotes the vector of variables for $\mathfrak{H}$, then in this notation, $\Theta = \det(x_{PQ^{-1}}) = \Theta(x^*)$, since Θ is a polynomial in the variables x_R such that $R \in \mathfrak{G}$.) From $\hat{\Theta}(x^*) = \Theta(x^*)^n$, it follows that each specialized factor $\Phi_\lambda(x^*)$ divides $\Theta(x^*)^n$, and so nonnegative integers $r_{\kappa\lambda}$ exist such that

$$\Phi_\lambda(x^*) = \prod_{\kappa=0}^{k} \Psi_\kappa(x^*)^{r_{\kappa\lambda}}. \tag{15.1}$$

From (15.1) and Frobenius' original characterization of irreducible characters as certain coefficients of a corresponding irreducible factor of the group determinant (as in (13.10)), he obtained the relation $\sum_\kappa \psi^{(\kappa)}(P) = \chi^{(\lambda)}(P)$ for all $P \in \mathfrak{G}$. With the help of the orthogonality relations, he then derived the important relation

$$\sum_{\lambda=0}^{\hat{k}-1} r_{\kappa\lambda}\chi_\rho^{(\lambda)} = \frac{h}{gh_\rho} \sum_{P\in(\rho)\cap\mathfrak{G}} \psi^{(\kappa)}(P). \tag{15.2}$$

Frobenius explained that formula (15.2) was especially well suited for obtaining the values of the characters of $\mathfrak{H}$ from those of a subgroup $\mathfrak{G}$ and that he had used it to determine the characters of the symmetric group $\mathfrak{S}_n$ [214, p. 115]. His solution to this formidable problem, which he published in 1900 [217] with a sequel on the alternating group in 1901 [218], will be discussed below.

The above derivation of (15.2) is fairly straightforward, but Frobenius also obtained it by another, less routine, line of thought. We saw in his letters to Dedekind that Frobenius, when initially exploring the group determinant, had probed its properties from every angle he could imagine. With the basics of a theory of group determinants, characters, and representations now established, he evidently continued to employ the same "shotgun" strategy, albeit now with all the above basics at hand. The derivation of (15.2) described above was in effect based on starting from the regular representation $\hat{\rho}$ of $\mathfrak{H}$, i.e., from the group matrix $\hat{\rho}(x) = \sum_{T\in\mathfrak{H}} \hat{\rho}(T)x_T = (x_{RS^{-1}})$, and proceeding to its specialization $\hat{\rho}(x^*)$, i.e., to the restriction of $\hat{\rho}$ to $\mathfrak{G}$. This apparently suggested to Frobenius the possibility of a reverse approach: Start with a representation σ of $\mathfrak{G}$; is it then possible, guided by the above considerations, to construct from σ a representation of $\mathfrak{H}$? In answering this question he was led to what is now called the representation of $\mathfrak{H}$ induced by σ. Let us consider how he constructed the induced representation and what may have suggested the construction to him.

Frobenius preferred to work with the group matrix associated to σ, namely $\sigma(x^*) = \sum_{P\in\mathfrak{G}} \sigma(P)x_P$. The corresponding induced representation $\hat{\sigma}$ is associated to a group matrix $\hat{\sigma}(x)$, which is partitioned into n^2 blocks of $e \times e$ matrices, e being the degree of σ. Such a partitioned matrix is reminiscent of the partitioned group matrix behind Frobenius' above-described derivation of (15.2), and for reasons I will indicate, I suspect it inspired the construction of $\hat{\sigma}(x)$. Going back to the group determinants $\hat{\Theta}(x)$ and $\Theta(x^*)$ and their associated group matrices, suppose we obtain an analogous, but different, partition of $\hat{\rho}(x) = (x_{RS^{-1}})$ using instead the partition of $\mathfrak{H}$ into left cosets: $\mathfrak{H} = \cup_{i=1}^n A_i\mathfrak{G}$. Then the (i,j)th block of $\hat{\rho}(x)$ is

$$\rho_{ij}(x) = \left(x_{A_iP(A_jQ)^{-1}}\right) = \left(x_{A_iPQ^{-1}A_j^{-1}}\right). \tag{15.3}$$

In this case, the diagonal blocks $\rho_{ii}(x) = \left(x_{A_iPQ^{-1}A_i^{-1}}\right)$ are the group matrices for the conjugate groups $A_i\mathfrak{G}A_i^{-1}$. With this in mind, let us consider Frobenius' definition of $\hat{\sigma}(x)$.

Frobenius defined the (i, j) matrix block of $\hat{\sigma}(x)$ as follows:

$$\sigma_{ij}(x) = \sum_{P\in\mathfrak{G}} \sigma(P)x_{A_iPA_j^{-1}},$$

which is the analogue of (15.3). Indeed, as in the situation of (15.3), the diagonal block σ_{ii} is the group matrix for the representation of the conjugate group $A_i\mathfrak{G}A_i^{-1}$ obtained from the representation σ of $\mathfrak{G}$ in the obvious way, viz., $\sigma(A_iPA_i^{-1}) \stackrel{\text{def}}{=} \sigma(P)$, since then,

$$\sigma_{ii}(x) = \sum_{P\in\mathfrak{G}} \sigma(P)x_{A_iPA_i^{-1}} = \sum_{Q\in A_i\mathfrak{G}A_i^{-1}} \sigma(Q)x_Q,$$

which is the group matrix of $A_i\mathfrak{G}A_i^{-1}$. The induced group matrix is then defined as the partitioned matrix

$$\hat{\sigma}(x) = \left(\sigma_{ij}(x)\right),$$

which is thus $ne \times ne$. Of course, it was necessary to show that $\hat{\sigma}(x)$ is a group matrix, namely that $\hat{\sigma}(z) = \hat{\sigma}(x)\hat{\sigma}(y)$, where $z_R = \sum_{ST=R} x_S y_T$, but this is now easy to verify, as Frobenius showed [214, pp. 110–111]. It is equivalent to verifying that $\hat{\sigma}$ is a representation of $\mathfrak{H}$.[1]

Frobenius then proceeded to establish what amounts to his version of the reciprocity theorem that bears his name. It is now usually stated as follows:

Theorem 15.1. *Let τ_λ denote an irreducible representation of $\mathfrak{H}$, and σ_κ an irreducible representation of $\mathfrak{G}$. Then the induced representation $\hat{\sigma}_\kappa$ of $\mathfrak{H}$ contains (in its complete reduction into irreducible representations) τ_λ with the same multiplicity as the restriction of τ_λ to $\mathfrak{G}$ contains σ_κ.*

Let us consider how Frobenius conceived of and established this important theorem. Going back to his derivation of (15.2), you will see that (in the notation above) the integer $r_{\kappa\lambda}$ gives the number of times $\Psi_\kappa(x^*) = \det\sigma_\kappa(x^*)$ occurs in the

[1] Since $\hat{\sigma}(R) = \hat{\sigma}(x)$ with x specialized to $x_R = 1$, $x_S = 0, S \neq R$, it is easy to see that $\hat{\sigma}(z) = \hat{\sigma}(x)\hat{\sigma}(y)$ implies that $\hat{\sigma}(AB) = \hat{\sigma}(A)\hat{\sigma}(B)$; that $\hat{\sigma}(E) = I$ is immediate from the definition. Frobenius' entire definition can be articulated without group matrices as follows: Extend σ to all of $\mathfrak{H}$ by setting $\sigma(R) = 0$ for all $R \notin \mathfrak{G}$. Then it is easy to check that $\hat{\sigma}(S) = \left(\sigma_{ij}(S)\right)$, where $\sigma_{ij}(S) = \sigma(A_i^{-1}SA_j)$ for all $S \in \mathfrak{H}$.

factorization of $\Phi_\lambda(x^*)$. In other words, if $\det\tau_\lambda(x) = \Phi_\lambda(x)$, so that $\det\tau_\lambda(x^*) = \Phi_\lambda(x^*)$, where $\tau_\lambda(x^*)$ is the group matrix associated to the restriction of τ_λ to $\mathfrak{G}$ via the variable specialization $x_R = 0$ for all $R \notin \mathfrak{G}$, then $r_{\kappa\lambda}$ gives the multiplicity of $\det\sigma_\kappa(x^*)$ as a factor of $\det\tau_\lambda(x^*)$. Consider now the induced group matrix $\hat{\sigma}_\kappa(x)$. From results in his "Representation" paper [213, pp. 86–87] Frobenius could conclude that the associated determinant $\det\hat{\sigma}_\kappa(x)$ has the same irreducible factors as $\hat{\Theta}(x)$. (The results being invoked by Frobenius say, when translated out of the language of group matrices and determinants, that every irreducible representation is contained in the regular representation.) Thus we have a factorization of the form

$$\det\hat{\sigma}_\kappa(x) = \prod_\lambda \Phi_\lambda(x)^{s_{\kappa\lambda}}. \tag{15.4}$$

Starting from this factorization for $\det\hat{\sigma}_\kappa(x)$, Frobenius obtained a formula identical to (15.2) but with $s_{\kappa\lambda}$ in place of $r_{\kappa\lambda}$. To show that this was the *same* formula, i.e., that he had derived it in a new way, he observed that (15.2) can be thought of as a matrix equation $R\chi = K$, where $R = (r_{\kappa\lambda})$ is $k \times \hat{k}$ and $\chi = (\chi_\rho^{(\lambda)})$ is $\hat{k} \times \hat{k}$ and invertible. Likewise, he had just derived $S\chi = K$, with $S = (s_{\kappa\lambda})$, and so the invertibility of χ shows that $R = S$, i.e., that $r_{\kappa\lambda} = s_{\kappa\lambda}$. For Frobenius, what was important was the new proof of (15.2) via his method of induced representations. He did not pause to state explicitly the reciprocity theorem implicit in his reasoning: $\det\hat{\sigma}_\kappa(x)$ *contains the factor* $\det\tau_\lambda(x)$ *with the same multiplicity as* $\det\tau_\lambda(x^*)$ *contains* $\det\sigma_\kappa(x^*)$. This is what (15.1) and (15.4) imply in view of the identity $s_{\kappa\lambda} = r_{\kappa\lambda}$.

15.2 Characteristic Units and Young Tableaux

In 1899, Frobenius published a sequel [216] to his 1897 "Representation" paper [213] in which he continued his investigation of the matrix representation of groups. Among other things, he showed that Theorems 13.11 and 13.12 of the earlier paper could be generalized so that the group matrix $M(x) = (x_{PQ^{-1}})$ of the regular representation is replaced by the group matrix $\sigma(x) = \sum_{R\in\mathfrak{H}} x_R\sigma(R)$ corresponding to any representation σ of $\mathfrak{H}$. The generalized theorems implied in particular the complete reducibility theorem for every representation: every representation of $\mathfrak{H}$ is equivalent to a direct sum of irreducible representations (all of which come from the decomposition of the regular representation of $\mathfrak{H}$). Thus Frobenius published a proof of this theorem the same year as did Maschke, although apparently neither knew of the work of the other.

In [216], Frobenius also returned to Theorem 13.12 and showed that there was a greater analogy with Theorem 13.11 than had been apparent when he presented them in 1897. The new approach to Theorem 13.12 taken in [216] was inspired by a theorem in his pioneering paper of 1878 applying elementary divisor theory to matrix algebra (Section 7.5). A special case of a general theorem established there [181, §13] showed that if a matrix K can be diagonalized (i.e., has linear elementary divisors) and if $\rho, \rho', \ldots$ are its distinct characteristic roots, then $(uI - K)^{-1}$ has a Laurent-type expansion in powers of $u - r$, which begins $(uI - K)^{-1} = A(u-\rho)^{-1} + \cdots$. The "residue" matrix A has the property that $A^2 = A$. Also if $(uI - K)^{-1} = A'(u-\rho')^{-1} + \cdots$ is the expansion in powers of $u - \rho'$, then one has not only $(A')^2 = A'$, but also $AA' = 0$. Furthermore, $I = A + A' + \cdots$ and $K = \rho A + \rho' A' + \cdots$, where the sums are over all the distinct roots $\rho, \rho', \ldots$. Frobenius applied this result to the group matrix $M(x) = (x_{RS^{-1}})$ as follows [216, §5]. If $\sigma_1, \ldots, \sigma_k$ denote the irreducible representations of $\mathfrak{H}$, then Theorems 13.11–13.12 show that $M(x)$ is equivalent to the direct sum of f_1 copies of $\sigma_1(x)$, f_2 copies of $\sigma_2(x)$, ..., and f_k copies of $\sigma_k(x)$. Pick $x = c$ such that the characteristic roots $\rho_i^{(\lambda)}$, $i = 1, \ldots, f_\lambda$, $\lambda = 1, \ldots, k$, of all of $\sigma_1(c), \ldots, \sigma_k(c)$ are distinct. This means that $M(c)$ has a total of $\sum_{\lambda=1}^{k} f_\lambda$ distinct roots, with $\rho_i^{(\lambda)}$ of multiplicity f_λ and with all elementary divisors of $M(c)$ linear.[2] Frobenius' 1878 theorem applied to $K = M(c)$ then implied the existence of a total of $\sum_{\lambda=1}^{k} f_\lambda$ matrices $A_i^{(\lambda)}$, $i = 1, \ldots, f_\lambda$, $\lambda = 1, \ldots, k$, such that $I = \sum_{i,\lambda} A_i^{(\lambda)}$, $K = \sum_{i,\lambda} \rho_i^{(\lambda)} A_i^{(\lambda)}$, and $[A_i^{(\lambda)}]^2 = A_i^{(\lambda)}$, $A_i^{(\lambda)} \cdot A_{i'}^{(\lambda')} = 0$ for $(i,\lambda) \neq (i',\lambda')$. Moreover, since K is a group matrix, viz., $K = (c_{PQ^{-1}})$, each $A_i^{(\lambda)}$ is expressible in the form $(a_{PQ^{-1}}^{(\lambda)})$. (The $a_{PQ^{-1}}^{(\lambda)}$ also depend on i, but that dependency is suppressed here.)

Frobenius discovered that the h numbers $a_R^{(\lambda)}$ that define $A_i^{(\lambda)}$ are similar in many respects to the characters $\chi^{(\lambda)}(R)$ of $\mathfrak{H}$. For example, $[A_i^{(\lambda)}]^2 = A_i^{(\lambda)}$ implies $\sum_{PQ=R} a_P^{(\lambda)} a_Q^{(\lambda)} = a_R^{(\lambda)}$, which is analogous to the orthogonality relation (13.18), viz., $\sum_{PQ=R} \chi^{(\lambda)}(P)\chi^{(\lambda)}(Q) = (h/f_\lambda)\chi^{(\lambda)}(R)$; and $A_i^{(\lambda)} \cdot A_{i'}^{(\lambda')} = 0$ for $(i,\lambda) \neq (i',\lambda')$ implies that $\sum_{PQ=R} a_P^{(\lambda)} a_Q^{(\lambda')} = 0$, which is analogous to the orthogonality relation (13.19), viz., $\sum_{PQ=R} \chi^{(\lambda)}(P)\chi^{(\lambda')}(Q) = 0$. Frobenius later termed the $a_R^{(\lambda)}$ primitive characteristic units,[3] and I will use that terminology here. In [216], he also used the properties of the irreducible characters of $\mathfrak{H}$ to further explore the connections between irreducible characters and primitive characteristic units. First of all, if $a_\rho^{(\lambda)} = (1/h_\rho)\sum_{R\in(\rho)} a_R^{(\lambda)}$, so that $a_\rho^{(\lambda)}$ is the average value of $a_R^{(\lambda)}$ over the conjugate class (ρ), then these average values are determined by characters, the exact relation being $a_\rho^{(\lambda)} = (1/h)\chi_{\rho'}^{(\lambda)}$, where as usual (ρ') denotes the conjugate class of the inverses of the elements in (ρ). From this relation it then followed that, conversely, the irreducible characters are completely determined by the primitive characteristic units:

$$\chi^{(\lambda)}(R^{-1}) = \sum_{S\in\mathfrak{H}} a_{S^{-1}RS}^{(\lambda)}. \tag{15.5}$$

[2]In his "Representation" paper, Frobenius had shown that if μ is any representation of $\mathfrak{H}$, then the elementary divisors of $\mu(x)$ are all linear [213, p. 87].

[3]See Frobenius' 1903 paper [222], which is discussed further on.

It follows from the properties of the matrices $A_i^{(\lambda)}$ that the corresponding elements $\mathbf{e}_i^{(\lambda)} = \sum_R a_R^{(\lambda)} R$ of the group algebra of $\mathfrak{H}$ satisfy $[\mathbf{e}_i^{(\lambda)}]^2 = \mathbf{e}_i^{(\lambda)}$ and $\mathbf{e}_i^{(\lambda)} \cdot \mathbf{e}_{i'}^{(\lambda')} = 0$ for $(i,\lambda) \neq (i',\lambda')$ and $\sum_{i,\lambda} \mathbf{e}_i^{(\lambda)} = E$, the identity element in the group algebra. Since Frobenius had been familiar with group algebras ever since Dedekind had introduced him to them in 1896, there is no doubt that he realized that his primitive characteristic units define idempotent elements of $\mathcal{H}$ with the above properties. In terms of the approach to representation theory via modules common today and going back to a 1929 paper [458] by E. Noether, Frobenius' primitive characteristic units define the primitive idempotents of the group algebra, which means that $\mathcal{I} = \mathcal{H}\mathbf{e}_i^{(\lambda)}$ is a minimal left ideal and $x_L : \mathbf{b} \to \mathbf{xb}$ restricted to $\mathcal{I}$ defines an irreducible representation of $\mathfrak{H}$ of degree f_λ. But, as will become clear below, Frobenius' dislike of hypercomplex numbers as a theoretical and conceptual basis for the study of group representations made him disinclined to develop his theory in this direction.

It is noteworthy that whereas Noether's work was inspired to a large extent by Dedekind's theory of ideals, which she formulated in more abstract, axiomatic terms, Frobenius, as much as he admired Dedekind and generally approved his approach to algebra, found certain tendencies of that approach already too abstract. Thus in a letter of 1893 to Heinrich Weber, who was then planning to write a treatise on algebra, Frobenius expressed his relief that Weber rather than Dedekind would be the author. He hoped that Weber would adopt Dedekind's approach, but avoid the "overly abstract nooks" that Dedekind "now so readily seeks out."[4] Frobenius complained in particular that in his additions to Dirichlet's lectures on number theory, Dedekind had pushed "the abstraction" too far, so that, for example, permutations become unnecessarily "incorporeal."

In particular, Frobenius developed the theory of primitive characteristics in a more concrete form [216, §§5–6], which might not be appealing to most present-day mathematicians but which led to a mathematically satisfying result. Recall that Frobenius' 1897 Theorem 13.11 showed that the values of the k irreducible characters $\chi^{(\lambda)}$, $\lambda = 1,\ldots,k$, can be used to construct an $h \times h$ matrix L such that $L^{-1}M(x)L$ is a direct sum of k matrices $N_\lambda(x)$, where $\det N_\lambda(x) = \Phi_\lambda(x)^{f_\lambda}$, $\Theta(x) = \prod_{\lambda=1}^{k} \Phi(x)^{f_\lambda}$ being the prime factorization of the group determinant of $\mathfrak{H}$. Thus $\Phi_\lambda(x)^{f_\lambda}$ is determined by $N_\lambda(x)$, which in turn is determined by the irreducible characters as in Theorem 13.11. Theorem 13.12 then says that each $N_\lambda(x)$ is equivalent to a direct sum of f_λ copies of the corresponding irreducible representation $\sigma_\lambda(x)$. Thus $\Phi_\lambda(x) = \det \sigma_\lambda(x)$, but Theorem 13.12 gives no information on how σ_λ and Φ_λ can be completely determined that would be analogous to what is

[4] Letter dated 23 December 1893, and located in the archives of the Niedersächsiche Staats- und Universitätsbibliothek, Göttingen (Cod. Ms. Philos. 205. Nr. 16). Although Frobenius was critical of Dedekind's penchant for what he regarded as unnecessary abstraction, his admiration for Dedekind is also evident throughout these passages. The book that Weber planned became his classic *Lehrbuch der Algebra* [582,583]. The edition of Dirichlet's *Vorlesungen über Zahlentheorie* to which Frobenius referred was the forthcoming fourth edition of 1894.

in Theorem 13.11. What Frobenius showed in [216] was that σ_λ, and so also $\Phi_\lambda(x) = \det \sigma_\lambda(x)$, $\sigma_\lambda(x) = \sum_{R\in\mathfrak{H}} x_R \sigma_\lambda(R)$, can be calculated from the values of a primitive characteristic unit $a_R^{(\lambda)}$.

To state Frobenius' result precisely, some preliminary notation is required. To avoid an excess of sub- and superscripts, let σ denote an irreducible representation of $\mathfrak{H}$ with $\Phi(x) = \det[\sigma(x)]$ the corresponding irreducible factor of the group determinant $\Theta(x) = \det[M(x)] = \det[(x_{PQ^{-1}})]$ and $f = \deg \sigma = \deg \Phi$. A given primary characteristic unit associated to σ will be denoted by a_R. Let the h elements of $\mathfrak{H}$ be ordered in some manner, e.g., $\mathfrak{H} = \{E, A, B, \ldots\}$, and in what follows, all $h \times h$ matrices will have their rows and columns so ordered. Corresponding to the characteristic unit a_R, let $\bar{A} = (\bar{a}_{P,Q})$ denote the matrix with row P, column Q entry equal to $a_{Q^{-1}P}$, i.e., $\bar{a}_{P,Q} = a_{Q^{-1}P}$. The matrix $\bar{A}$ and the group matrix $M(x) = (x_{PQ^{-1}})$ are both $h \times h$ and they commute. The matrix $\bar{A}$ has rank f. Suppose now that from $\bar{A}$, f rows ($P = R_1, \ldots, R_f$) and f columns ($Q = S_1, \ldots, S_f$) are chosen such that the corresponding $f \times f$ minor determinant does not vanish. Given these choices of rows and columns, introduce the following notation: if L is any $h \times h$ matrix, let L_f denote the $f \times f$ submatrix of L with (i, j) entry equal to the row R_i, column S_j entry of L. Note that $\det[\bar{A}_f]$ is an $f \times f$ minor determinant of $\bar{A}$ and does not vanish. Frobenius' theorem may now be stated as follows [216, p. 145, (4.), p. 147]:

Theorem 15.2. *Given a primitive characteristic unit a_R, the corresponding irreducible representation is given by $\sigma(x) = [M(x)]_f [\bar{A}_f]^{-1}$.*

Frobenius must have been quite pleased with this result, the final one in the sequel to his "Representation" paper, for it shows in a precise and straightforward manner how σ, and so also $\Phi(x) = \det[\sigma(x)]$ and $\chi(R) = \operatorname{tr} \sigma(R)$, are built up from the values of an associated primitive characteristic unit a_R. As we shall see below, he determined the primitive characteristic units for the symmetric group a few years later. Like the equally elegant Theorem 13.7, Theorem 15.2 has disappeared from the theory along with the emphasis on determinants that characterized Frobenius' approach to representation theory.

Frobenius' lack of enthusiasm for approaching the representation theory of finite groups through the theory of the structure of hypercomplex systems comes out explicitly in his important paper of 1903 [222] on the characteristic units of the symmetric group. The catalyst for his paper was provided by two papers by the British mathematician Alfred Young (1873–1940) that appeared in the *Proceedings of the London Mathematical Society* in 1901–1902 [614, 615]. Frobenius' several experiences of "intellectual harmony" with Burnside—further instances of which will be discussed below—undoubtedly made him a faithful reader of the *Proceedings*, where Burnside usually published his work.

Entitled "On Quantitative Substitutional Analysis," Young's papers were motivated by his work on the theory of invariants. It had led him to the problem of determining polynomials $P = P(x_1, \ldots, x_n)$ having what might be called in a broad sense symmetry properties. Young expressed these properties in terms of what he called "substitutional equations" such as

$$(\lambda_1 + \lambda_2 s_2 + \lambda_3 s_3 + \cdots)P = 0, \tag{15.6}$$

where $s_2, s_3, \ldots$ denote the substitutions (or permutations) on n objects (other than the "identical" substitution), and the left-hand side means $P + \lambda_2 s_2 P + \lambda_3 s_3 P + \cdots$, where $s_m P$ is P with its variables permuted according to s_m. In Frobenius' notation, which we will now adopt, (15.6) may be written as $(\sum_{S \in \mathfrak{S}_n} a_S S)P = 0$, where $\mathfrak{S}_n$ denotes the symmetric group. Young showed that all polynomials P satisfying (15.6) may be written in the form $P = \sum_{i=1}^{m} P_i$, where corresponding to each P_i is a "substitutional expression" $\sum_{S \in \mathfrak{S}_n} c_S^{(i)} S$ such that $P_i = (\sum_{S \in \mathfrak{S}_n} c_S^{(i)} S)F_i$, where the polynomial F_i is an arbitrary polynomial in $x_1, \ldots, x_n$. Moreover, each such substitutional expression satisfies the relation

$$\left(\sum_{S \in \mathfrak{S}_n} a_S S\right)\left(\sum_{S \in \mathfrak{S}_n} c_S^{(i)} S\right) = 0, \tag{15.7}$$

where the multiplication on the left-hand side was to be carried out "by virtue of the multiplication table of the group" [614, p. 103]. The above equations could then be used to find the expressions $(\sum_{S \in \mathfrak{S}_n} c_S^{(i)} S)$ and thence to find all P satisfying (15.6).

Frobenius could see immediately that Young's "substitutional expressions" were elements of the group algebra of $\mathfrak{S}_n$ and that the equations (15.7) were equations involving multiplication within this hypercomplex system. And of course, he would have realized that the characters and representations of $\mathfrak{S}_n$ are intimately related to the properties of this hypercomplex system. In a sense, Burnside had made it easier for Frobenius to see these things, for he had been the referee for Young's papers, and it was at his request that the group-theoretic elements were made more prominent [566, p. xxiii]. As we saw in Section 14.3, shortly before Burnside refereed Young's papers, he had become acquainted with the work of Frobenius and Molien. Therefore, he would have realized that since Young's work dealt, in effect, with the group algebra of $\mathfrak{S}_n$, it was relevant to Frobenius' theory. By inducing Young to give more prominence to the group-related notions in his work, Burnside had done Frobenius a favor. Since Frobenius had published a detailed study of the characters of the symmetric group in 1900 [217], what Young had to say would have been of special interest to him. And so Frobenius read on. Actually, the first 36 pages of Young's paper would not have contained anything of interest to him, involving as they did generalizations and specializations of (15.6) and (15.7) and applications to the theory of invariants, a branch of mathematics he held in low esteem (as documented in Section 15.5 below). But then, in the fifteenth section of his first paper, Young proposed to establish a substitutional identity, which Frobenius realized was consequential for the representation theory of $\mathfrak{S}_n$.

In order to state his substitutional identity, Young introduced what later became known as a "Young tableau"—an arrangement $\mathfrak{T}$ of n objects into rows of decreasing length $\alpha_1 \geq \alpha_2 \geq \cdots \geq \alpha_\mu$ as indicated below:

$$\begin{array}{llllllll} a_{1,1} & a_{1,2} & \cdots & \cdot & \cdot & \cdot & \cdot & a_{1,\alpha_1} \\ a_{2,1} & a_{2,2} & \cdots & \cdot & \cdot\, a_{2,\alpha_2} & & & \\ \cdot & \cdot & \cdots & \cdot & & & & \\ a_{\mu,1} & a_{\mu,2} & \cdots & a_{\mu,\alpha_\mu} & & & & \end{array}$$

Let $\mathfrak{R}$ denote the subgroup of $\mathfrak{S}_n$ consisting of all permutations R that permute among themselves the objects in each row of $\mathfrak{T}$, and let $\mathfrak{C}$ denote the subgroup of all those permutations C that permute among themselves the objects in each column of $\mathfrak{T}$. Young introduced "substitutional expressions" such as

$$\mathbf{np}_{\mathfrak{T}} = \Big(\sum_{C\in\mathfrak{C}} (\operatorname{sgn} C)C\Big)\Big(\sum_{R\in\mathfrak{R}} R\Big) = \sum_{R\in\mathfrak{R},C\in\mathfrak{C}} (\operatorname{sgn} C)CR,$$

where $\operatorname{sgn} C = 1$ or -1, depending on whether the permutation C is even or odd, respectively. The hypercomplex number **np** depends on the way the objects $a_1,\ldots,a_n$ are distributed to form $\mathfrak{T}$ as well as on the *shape* of $\mathfrak{T}$, which is determined by the numbers $\alpha_1,\ldots,\alpha_\mu$. Let the class of all tableaux $\mathfrak{T}$ with this fixed shape be denoted by $[\alpha_1,\ldots,\alpha_\mu]$, and consider the hypercomplex number

$$\mathbf{t}_{\alpha_1,\ldots,\alpha_\mu} = \sum_{\mathfrak{T}\in[\alpha_1,\ldots,\alpha_\mu]} \mathbf{np}_{\mathfrak{T}}. \tag{15.8}$$

Young's substitutional identity expresses the identical permutation **1** as a linear combination of the $\mathbf{t}_{\alpha_1,\ldots,\alpha_\mu}$:

$$\mathbf{1} = \sum_{\alpha_1,\ldots,\alpha_\mu} a_{\alpha_1,\ldots,\alpha_\mu}\mathbf{t}_{\alpha_1,\ldots,\alpha_\mu}. \tag{15.9}$$

In his second paper [615], Young showed that

$$a_{\alpha_1,\ldots,\alpha_\mu} = \left[\frac{\prod_{r,s}(\alpha_r-\alpha_s-r+s)}{\prod_r(\alpha_r+\mu-r)!}\right]^2,$$

and he studied the multiplicative properties of the expressions $\mathbf{t},\mathbf{np}_{\mathfrak{T}}$. For example, he showed that

$$(a_{\alpha_1,\ldots,\alpha_\mu})\mathbf{t}^2_{\alpha_1,\ldots,\alpha_\mu} = \mathbf{t}_{\alpha_1,\ldots,\alpha_\mu} \quad \text{and} \quad \mathbf{t}_{\alpha_1,\ldots,\alpha_\mu}\mathbf{t}_{\beta_1,\ldots,\beta_\nu} = \mathbf{0},$$

where $\alpha_1,\ldots,\alpha_\mu \neq \beta_1,\ldots,\beta_\nu$. And if $\mathbf{np}_{\mathfrak{T}}$ is one of the terms of $\mathbf{t}_{\alpha_1,\ldots,\alpha_\mu}$ in (15.9), then

$$(\mathbf{np}_{\mathfrak{T}})^2 = (a_{\alpha_1,\ldots,\alpha_\mu})^{-1/2}(\mathbf{np}_{\mathfrak{T}}), \quad \mathbf{t}_{\alpha_1,\ldots,\alpha_\mu}(\mathbf{np}_{\mathfrak{T}}) = (a_{\alpha_1,\ldots,\alpha_\mu})^{-1}(\mathbf{np}_{\mathfrak{T}}).$$

From relations such as these, Frobenius would have surmised readily that Young's results are intimately connected with the structure of the group algebra $\mathcal{H}$ of $\mathfrak{S}_n$. Although he preferred not to develop his mathematics in terms of hypercomplex systems, he realized (as Molien had explicitly shown) that $\mathcal{H}$ is the sum of k complete matrix algebras, where k is the number of conjugacy classes of $\mathfrak{S}_n$. He knew that k was equal the number of partitions of n, i.e., the number of distinct sequences $\alpha_1 \geq \alpha_2 \geq \cdots$ such that $\alpha_1 + \alpha_2 + \cdots = n$, so that k was also equal to the number of distinct shapes $\alpha_1, \ldots, \alpha_\mu$ of tableaux. Thus there is a one-to-one correspondence between shapes $\alpha_1, \ldots, \alpha_\mu$ and complete matrix algebras and therefore irreducible representations of $\mathfrak{S}_n$. Young's identity could be seen as the representation of the identity of the group algebra as a sum of the k identity elements $(a_{\alpha_1,\ldots,\alpha_\mu})\mathbf{t}_{\alpha_1,\ldots,\alpha_\mu}$ of the complete matrix algebras. In fact, in his paper on the characters of the symmetric group [217], Frobenius had introduced the subgroups $\mathfrak{R}$, albeit for a different reason, namely to bring his theory of induced characters to bear on the problem of determining the irreducible characters.

However, in [217], the connection between a partition of n, $\alpha_1, \ldots, \alpha_\mu$ and an irreducible character $\chi^{(\alpha)}$ had been rather indirect and insignificant. There each irreducible character corresponded to a sequence of n integers $(\kappa) = (\kappa_1, \ldots, \kappa_n)$ satisfying

$$0 \leq \kappa_1 < \kappa_2 < \kappa_3 < \cdots < \kappa_n, \quad \text{and} \quad \kappa_1 + \kappa_2 + \cdots + \kappa_n = \frac{n(n+1)}{2}. \tag{15.10}$$

There are k such sequences, because, as Frobenius showed, each determines a partition of n, $\alpha_1, \alpha_2, \ldots$, where $\alpha_i = \kappa_{n-i+1} - (n-i)$. For example, with $n = 6$ and $(\kappa) = (0,1,3,4,6,7)$, $\alpha_1 = 2, \alpha_2 = 2, \alpha_3 = 1, \alpha_4 = 1$ is the corresponding partition of n. In what follows, the character $\chi^{(\alpha)}$ will be denoted by $\chi^{(\kappa)}$. Frobenius showed that the values of the k characters on the ρth conjugacy class, $\chi_\rho^{(\kappa)}$, are given by certain coefficients of a homogeneous polynomial $F_\rho(x_1, \ldots, x_n)$ of degree $\frac{1}{2}n(n+1)$, which is defined as follows. Set $\Delta = \Delta(x_1, \ldots, x_n) = \prod_{i>j}(x_i - x_j)$ and let $\operatorname{sgn} \Delta = 1, 0, -1$ according to whether $\Delta > 0, = 0, < 0$. Then if the ρth conjugacy class consists of permutations that factor into disjoint cycles consisting of a 1-cycles, b 2-cycles, c 3-cycles, etc., let

$$\begin{aligned} F_\rho &= \Delta(x_1, \ldots, x_n)(x_1 + \cdots + x_n)^a (x_1^2 + \cdots + x_n^2)^b (x_1^3 + \cdots + x_n^3)^c \cdots \\ &= \textstyle\sum_{\kappa_1,\ldots,\kappa_n} C^\rho_{\kappa_1,\ldots,\kappa_n} x_1^{\kappa_1} x_2^{\kappa_2} \cdots x_n^{\kappa_n}, \end{aligned}$$

so that the summation is over all $\kappa_1, \ldots, \kappa_n$ satisfying $\sum \kappa_i = \frac{1}{2}n(n+1)$ with $\kappa_i \neq \kappa_j$ for $i \neq j$, although here it is not required that $\kappa_i < \kappa_{i+1}$. He proved that the character value $\chi_\rho^{(\kappa)}$ is determined by the coefficient $C^\rho_{\kappa_1,\ldots,\kappa_n}$, the exact relation being

$$C^\rho_{\kappa_1,\ldots,\kappa_n} = \chi_\rho^{(\kappa)} \operatorname{sgn} \Delta(\kappa_1, \ldots, \kappa_n). \tag{15.11}$$

He used this remarkable formula to determine values of the characters, such as their degrees $f_\kappa = \chi^{(\kappa)}(E)$. For example, for the symmetric group $\mathfrak{S}_4$ and the irreducible character corresponding to $(\kappa) = (0,1,3,6)$, the conjugacy class of E corresponds to $a = n$ and $b = c = \cdots = 0$ in the formula for F_ρ, so that $F_\rho = \Delta(x_1,\ldots,x_4)(x_1 + \cdots + x_4)^4$. If this expression is expanded, the coefficient of $x_2 x_3^3 x_4^6$ is $C_{0,1,3,6} = +3$, and since $\operatorname{sgn}\Delta(0,1,3,6) = \operatorname{sgn} 540 = +1$, (15.11) implies that $f_\kappa = \chi^{(\kappa)}(E) = 3$.[5]

Thus, although in [217], Frobenius had realized that each irreducible character corresponds to a partition $\alpha_1,\ldots,\alpha_\mu$ of n, that connection was not fundamental to his approach. The connection with partitions of unity was used only to count the sequences $\kappa_1,\ldots,\kappa_n$, i.e., to show that the number is k, the number of conjugacy classes of $\mathfrak{S}_n$, which was already known to equal the number of partitions of n. It was the $\kappa_1,\ldots,\kappa_n$ that were central, because his approach was founded on (15.11). Using this approach, he had not found it easy to establish the relationship between the sequences $\kappa_1,\ldots,\kappa_n$ and $\lambda_1,\ldots,\lambda_n$ corresponding to what he called associated characters $\chi^{(\kappa)}, \chi^{(\lambda)}$. This notion arises as follows. If $\chi^{(1)}(S) = \operatorname{sgn} S = \pm 1$ depending on whether S is even or odd, respectively, then $\chi^{(1)}$ is what we called a Dedekind character of $\mathfrak{S}_n$, which means that it satisfies $\chi^{(1)}(ST) = \chi^{(1)}(S)\chi^{(1)}(T)$ and so defines a 1-dimensional representation of $\mathfrak{S}_n$. Consequently, if $S \to \sigma(S)$ is any representation of $\mathfrak{S}_n$, then $S \to \chi^{(1)}(S)\sigma(S)$ is another. This representation and its character are said to be associates of the original. In particular, corresponding to every irreducible character $\chi^{(\kappa)}$ is an associated character $\chi^{(\lambda)} = \chi^{(1)}\chi^{(\kappa)}$. Frobenius showed how the numbers $\kappa_1,\ldots,\kappa_n$ and $\lambda_1,\ldots,\lambda_n$ are related for associated characters [217, p. 160], but, as he admitted in 1903, the proof (via (15.11)) required "truly involved considerations" [222, p. 244].

Thanks to Young's work, Frobenius was now in a position to see that the expression $(a_{\alpha_1,\ldots,\alpha_\mu})^{1/2}\mathbf{np}_{\mathfrak{T}}$ is a primitive characteristic unit and so by Theorem 15.2 determines the corresponding irreducible representation σ and so the corresponding irreducible character χ. Furthermore, characters $\chi^{(\alpha)}, \chi^{(\beta)}$ are associates precisely when the corresponding tableaux are transposes of one another. Some of Frobenius' results are summed up in the following theorem, which, he felt, "probably contains some of the most noteworthy properties of the symmetric group and its characters" [222, p. 265]:

Theorem 15.3 (Frobenius). *Let $\mathfrak{T}$ have shape $\alpha = (\alpha_1,\ldots,\alpha_\mu)$. Then the equation $CR = R'C'$ with $C, C' \in \mathfrak{C}$ and $R, R' \in \mathfrak{R}$, has more solutions such that $C'C$ is even than solutions with $C'C$ odd. Let the excess of the even solutions over the odd solutions be denoted by $\frac{n!}{f}$, and let $\zeta(S)$ be defined for $S \in \mathfrak{S}_n$ by $\zeta(S) = \operatorname{sgn} C$ if $S = CR$ with $C \in \mathfrak{C}$ and $R \in \mathfrak{R}$ and $\zeta(S) = 0$ otherwise. Then $a_S = \frac{f}{n!}\zeta(S)$ is a primitive characteristic unit of $\mathfrak{S}_n$.*[6]

[5] The value $\chi^{(\kappa)}(E)$ could also have been computed using, e.g., the coefficient $C_{3,1,0,6}$ of $x_1^3 x_2 x_4^6$, which equals -3, for then $\operatorname{sgn}\Delta(3,1,0,6) = \operatorname{sgn} -540 = -1$.

[6] Young interpreted the product PQ of two permutations as Q followed by P, hence reading from right to left, whereas Frobenius adopted the reverse convention. I have followed Young's convention in presenting Frobenius' results.

Once the characteristic unit a_S is known, the corresponding character has a simple expression in terms of a_S as shown in (15.5), but Frobenius also showed that if χ corresponds to a_S, then $\frac{h_\rho}{f}\chi_\rho$ is equal to the difference between the number of even and the number of odd permutations of the form CR, where $C \in \mathfrak{C} \cap (\rho)$ and $R \in \mathfrak{R}$.

In the notation of Theorem 15.3, Young's expression $\mathbf{np}_{\mathfrak{T}}$ takes the form $\mathbf{np}_{\mathfrak{T}} = \sum_{S\in\mathfrak{S}_n} \zeta(S)S$, and, as Frobenius also pointed out, if $\chi^{(\alpha)}$ denotes the character determined by the characteristic unit a_S, then

$$\mathbf{t}_{\alpha_1,\ldots,\alpha_\mu} = \frac{h}{f^{(\alpha)}} \sum_{S\in\mathfrak{S}_n} \chi^{(\alpha)}(S)S.$$

It must have pleased him to show that in lieu of Young's expansion formula for $\mathbf{t}_{\alpha_1,\ldots,\alpha_\mu}$ in (15.8), he could specifically determine the coordinates of $\mathbf{t}_{\alpha_1,\ldots,\alpha_\mu}$ with respect to the group basis for the group algebra $\mathcal{H} = \mathbb{C}\mathfrak{S}_n$ and show that these coordinates are given by the corresponding character. Thus although he described Young's two papers as "very notable," he also stressed the fact that Young "did not recognize the relation of the function $\zeta(R)$ to the character $\chi(R)$ and to the corresponding primitive representation of the symmetric group, and it is first through this that his work attains its proper significance, since otherwise, the numerical coefficients occurring in his formulas remain mostly undetermined" [222, p. 265]. Although Frobenius' comment tends to underplay the importance of the new ideas introduced into representation theory by Young's work, it is true that it is in connection with this theory that Young's work became appreciated. For example, it was Frobenius' reformulation of Young's results in [222] that led Hermann Weyl in 1925 to see how the "Young symmetrizers" $\sum_{S\in\mathfrak{S}_n} \zeta(S)S$ may also be used to determine the irreducible representations of the special linear group.[7]

Frobenius was still not quite finished with the characters of the symmetric group. The following year (1904), he published a paper, "On the characters of multiply transitive permutation groups" [225], that contains several results of interest. A permutation group $\mathfrak{H}$ acting on a set $\mathcal{A} = \{a_1,\ldots,a_n\}$ of n symbols and so a subset of the symmetric group $\mathfrak{S}_n$ is r-fold transitive if for any two subsets of $\mathcal{A}$ of cardinality r, say $\mathcal{X} = \{x_1,\ldots,x_r\}$ and $\mathcal{Y} = \{y_1,\ldots,y_r\}$, there is an $H \in \mathfrak{H}$ such that $H(x_i) = y_i$ for all i. The paper has two methodologically disjoint parts. The first three sections appear to have been inspired by some results in an 1888 paper on permutation groups by Kronecker's protégé Eugen Netto [454]. They led Frobenius to results about characters of $\mathfrak{S}_n$ that are also characters of r-fold transitive subgroups. As indicated above, in his 1,900 paper on the characters of the symmetric group [217] Frobenius had shown that each character $\chi^{(\kappa)}$ of $\mathfrak{S}_n$ is determined by a sequence $(\kappa) = \kappa_1,\ldots,\kappa_n$ of integers as in (15.10). He now defined the dimension d of $\chi^{(\kappa)}$ to be $d = 2n - 1 - \kappa_n$, and he proved that every character $\chi^{(\kappa)}$ of $\mathfrak{S}_n$ of dimension $d \leq r/2$ is a character of every r-fold transitive

[7] On this matter, see Section 11.5 of my book [276].

subgroup of $\mathfrak{S}_n$ [225, I, p. 340]. Using this result, he proved that every twofold transitive subgroup of $\mathfrak{S}_n$ has the character $\chi(R) = \alpha - 1$, α being the number of elements in $\mathcal{A}$ fixed by R—a known result—and that conversely, every transitive group with this character is necessarily twofold transitive. Likewise, he proved that every fourfold transitive subgroup of $\mathfrak{S}_n$ has the characters $\alpha - 1$, $\frac{1}{2}\alpha(\alpha - 3) + \beta$, and $\frac{1}{2}(\alpha - 1)(\alpha - 2) - \beta$, where β is the number of 2-cycles in the representation of R as a product of disjoint cycles, and that conversely, every transitive subgroup of $\mathfrak{S}_n$ that has these three characters is fourfold transitive.

In the second part of his paper, Frobenius communicated a new way to represent the characters of $\mathfrak{S}_n$ that facilitated calculation [225, §4] and then applied it to determine the character tables of the two fivefold transitive Mathieu groups $\mathfrak{M}_{12}$ [225, §5] and $\mathfrak{M}_{24}$ [225, §6], some of the earliest examples of what are now called sporadic simple groups. In dealing with $\mathfrak{M}_{24}$, a subgroup of $\mathfrak{S}_{24}$ of order 244, 823, 040, he showed that $\mathfrak{M}_{24}$ contains a subgroup isomorphic to $\mathfrak{M}_{12}$, something that Mathieu himself had not realized [107, p. viii]. It then followed that $\mathfrak{M}_{24}$ contains isomorphic copies of all of the Mathieu groups. Frobenius' paper is listed in the bibliographies for these two groups in the 1985 *Atlas of Finite Groups* [107, pp. 244, 246].

15.3 Hypercomplex Number Systems à la Frobenius

After establishing Theorem 15.3, Frobenius rederived Young's results by his own methods, which, he emphasized, involved a substantial simplification of the reasoning. Of course, Frobenius' methods did not involve hypercomplex numbers, and it is interesting to read what he had to say on this point: "It is less significant that I abstain from the use of hypercomplex numbers, since, as convenient as they occasionally are, they do not always serve to make the presentation more transparent" [222, p. 266]. Frobenius preferred an approach to algebra based on the consideration of matrices and their determinants. As he put it in a letter to Dedekind (after using matrices and elementary divisor theory to explain a point about hypercomplex numbers), "everything of a hypercomplex nature ultimately reduces in reality to the composition of matrices."[8] This attitude is reflected in his approach to primitive characteristic units, as we have seen. The theory of matrices as developed and utilized by Frobenius depended heavily on determinant-related notions, especially those of Weierstrass' theory of elementary divisors, so that for Frobenius, the theory of determinants was fundamental to algebra. It made possible a rigorous and "basis-free" approach to algebraic problems, including those problems raised by the consideration of hypercomplex systems.This attitude on

[8]Letter dated 15 July 1896.

Frobenius' part comes out in the two papers of 1903 [223, 224] that are devoted to reestablishing the results on hypercomplex number systems due to Molien and Cartan that were discussed in Section 14.2.

Frobenius' first paper on hypercomplex systems was inspired by the 1893 dissertation of Molien [443], which he described in glowing terms as "the most important advance in recent times in that branch of algebra called group theory" [223, p. 284]. Although he praised Molien's brilliance, he also felt that the mathematical tools that he had brought to bear on the subject were "incomplete." And indeed, a key theorem in Molien's theory had been given a proof containing a substantial lacuna that could not readily be filled with the tools Molien had devised. So Frobenius deemed it appropriate to rework Molien's results utilizing his own mathematical tools. These tools were grounded in the theory of determinants, including Weierstrass' theory of elementary divisors and the applications Frobenius had made of it to what would now be called linear algebra, as indicated in Chapters 4–7. As we have seen, at Berlin, linear algebra had been developed with the high standards of rigor typical of the Weierstrass school, to which Frobenius most decidedly belonged. However, it was not simply the rigor of the determinant-based approach that appealed to Frobenius. He also appreciated the fact that determinant-related concepts could be formulated directly without dependence on a particular basis or canonical form. Thus in Frobenius' reformulation of Molien's results, the starting point is the two "group matrices" $S(x)$ and $T(x)$, which correspond, respectively, to left and right multiplication by $\mathbf{x} \in \mathcal{H}$, and their respective determinants.

After he had completed his reformulation of Molien's results, Frobenius learned of Cartan's 1898 study of hypercomplex numbers [60] (discussed briefly in Section 14.2). Although he praised Cartan's study as "outstanding" (*ausgezeichnet*), he was quick to point out that from a methodological point of view, it had nothing in common with what he had done:

> The approach taken by Mr. Cartan has not the least in common with the methods used here .... The transformation of basis, the starting point and the goal of his investigation, is avoided by me as long as possible .... The invariant properties of the [hypercomplex system][9] which are independent of every representation of it and with which I begin, first arise with him at the conclusion by means of a normal form of the [system], obtained through a long series of transformations the goal of which first becomes clear at the end. The distinction between the two methods is thus the same as that between the approaches of Weierstrass and Kronecker in the theory of families of bilinear forms. A particularly noteworthy formula of Mr. Cartan's (§65, (37)), which is not in Molien, I had obtained in the simplest manner by decomposition of the determinant $|S(x)+T(y)|$ into prime factors [223, pp. 285–286].

As these words amply illustrate, Frobenius was invoking Kronecker's first disciplinary ideal (Section 5.6.2), and so was an advocate of a basis-free approach to algebra, but what he practiced differs from the current basis-free emphasis in mathematics in that he did not see abstract structures as providing the foundation

[9] Frobenius used the word "group" instead of "hypercomplex system" and in general used group-related terms to describe properties of hypercomplex systems.

for such an approach but rather the concrete theory of determinants, which gives rise to constructs that are invariant under linear transformations of variables, such as the W-series and related elementary divisors.

Frobenius appreciated the fact that Cartan had gone further than Molien in his study of hypercomplex systems. In addition to the "noteworthy formula" Cartan had discovered (to be discussed below), he had pushed Molien's decomposition of $\mathcal{H}$ into semisimple and nilpotent parts further. Molien had in effect shown that associated to $\mathcal{H}$ is a maximal nilpotent ideal $\mathcal{R}$ such that $\mathcal{H}/\mathcal{R}$ is semisimple. Incidentally, it was Frobenius who first coined the term "radical" for $\mathcal{R}$, although he rejected Killing's term "semisimple" and called such systems instead "Dedekind groups," since (as noted following Theorem 13.8) Dedekind had first introduced this notion implicitly in 1885 within the context of commutative hypercomplex systems and the related debate over the meaning of Gauss' words. Cartan, who was really the first to give prominence to the notion of the radical in his study, had gone beyond Molien's result and proved that $\mathcal{H}$ actually contains a semisimple subsystem $\mathcal{S}$ such that $\mathcal{H} = \mathcal{S} + \mathcal{R}$. Frobenius therefore devoted a second paper [224] to showing how this and other results original with Cartan could also be established by means of the methods and results of his first paper.

The "noteworthy formula" that Cartan had discovered and Frobenius had rederived turned out to be significant in the subsequent development of group characters and representations by Brauer and is therefore worth describing. As Molien had realized, when $\mathcal{H}$ is not semisimple, the left and right representations of $\mathcal{H}$ need not be equivalent. In particular, the left and right group determinants, $\det[S(x)]$ and $\det[T(x)]$, can have different prime factorizations. But Cartan showed [60, p. 60] that these determinants have the same prime factors $\Phi_i(x)$—one for each simple component of the semisimple algebra $\mathcal{S} \subset \mathcal{H}$—and are expressed by formulas of the form

$$\begin{aligned} \det[S(x)] &= \textstyle\prod_{i=1}^{k} \Phi_i(x)^{s_i}, \quad s_i = \sum_{j=1}^{k} c_{ij} r_i, \\ \det[T(x)] &= \textstyle\prod_{i=1}^{k} \Phi_i(x)^{t_i}, \quad t_i = \sum_{j=1}^{k} c_{ji} r_j, \end{aligned} \tag{15.12}$$

where $r_1, \ldots, r_k$ are the degrees of the complete matrix algebras into which $\mathcal{S}$ decomposes, and the c_{ij} are nonnegative integers with $c_{jj} \geq 1$ and c_{ij} possibly distinct from c_{ji}. The numbers c_{ij} for $i \neq j$ and $c_{ii} - 1$ were defined by Cartan in terms of a certain basis.

In [223, p. 308], Frobenius rederived Cartan's formula by showing that $\det[S(x) + T(y)]$, as a polynomial in the $2n$ variables $x_1, \ldots, x_n, y_1, \ldots, y_n$, has the prime factorization

$$\det[S(x) + T(y)] = \prod_{i,j=1}^{k} \Psi_{ij}(x,y)^{c_{ij}},$$

where the integers c_{ij} are as in Cartan's formula (15.12). Thus in Frobenius' approach, the integers c_{ij} arise from the start as invariants of $\mathcal{H}$. Frobenius then derived Cartan's formulas (15.12) by setting $y = 0$ and then $x = 0$ in his own formula. The integers c_{ij} are now called the *Cartan invariants*. They play a fundamental role in Brauer's modular theory of representations, as will be seen in Section 15.6. Incidentally, Frobenius obtained a necessary and sufficient condition that $\det[S(x)]$ and $\det[T(x)]$ have the same prime factorization—expressed, of course, in terms of a determinant [223, p. 290]—and for this reason Brauer later termed such $\mathcal{H}$ *Frobenius algebras* [35, p. 239], a term still in use today.

15.4 Applications to Finite Groups by Burnside and Frobenius

In 1899, in his second paper on the matrix representation of groups, Frobenius wrote that with it he had brought to a conclusion his "general investigations of the group determinant" [216, p. 130]. True to his word, after 1899 he turned his attention to applications of his general theory as well as to other types of mathematics. As we have already seen, he applied his theory of group characters to the nontrivial problem of computing character tables for groups, particularly the symmetric and alternating groups, and he used his theory of the group determinant to place the theory of noncommutative hypercomplex systems on a suitable mathematical basis. He also applied it to the investigation of purely group-theoretic problems, although here he was not the first. That honor belonged to Burnside.

For several years, Burnside had been actively interested in the problem of determining (up to isomorphism) all finite simple groups. As he explained in his book on the theory of finite groups [48, p. 343], solution of this problem was essential for the solution of "the most general problem of pure group-theory," namely that of classifying all finite groups. Burnside felt that a general solution to the simple group problem "is not to be expected," and so he focused on the problem of determining all simple groups with orders not exceeding a given upper bound [48, p. 370]. Building on the work of Hölder (1892) and Cole (1893), in 1895 Burnside pushed this upper bound to 1092. After presenting these results in his book, Burnside remarked:

> No simple group of odd [composite] order is known to exist.[10] An investigation as to the existence or nonexistence of such groups would undoubtedly lead, whatever the conclusion may be, to results of importance; it may be recommended to the reader as well worth his attention [48, p. 379].

Once Burnside became acquainted with Frobenius' theory of group characters, he sought to apply it to the study of groups of odd order. In 1900 [51], he showed that

[10]Burnside of course meant "odd *composite* order," since groups of prime order are simple.

if a group $\mathfrak{H}$ has odd order, then no irreducible character other than the trivial (or principal) character $\chi = 1$ can be entirely real-valued. Using this result and some of its consequences, which are interesting in themselves, he showed that among subgroups of the symmetric group $\mathfrak{S}_n$ for $n \leq 100$ there are no simple groups of odd order. At the end of his paper he wrote:

> The results in this paper, partial as they necessarily are, appear to me to indicate that an answer to the interesting question as to the existence or nonexistence of simple groups of odd composite order may be arrived at by a further study of the theory of group characteristics.

The day that Frobenius picked up the issue of the *Proceedings of the London Mathematical Society* containing Burnside's paper must have been a stimulating one for him, for it also contained A. Young's first paper on "Quantitative Substitutional Analysis" [614]. As we have seen, Frobenius responded to Young's work that same year [222]. As for Burnside's paper, it must have struck Frobenius as yet another instance of the "intellectual harmony" he had spoken about to Dedekind (Section 14.3), for Frobenius, too, had the idea to use character theory to extend theorems relating to solvable groups that he had obtained before creating the theory, although by the time he wrote up his results for publication in 1901 [219], he had Burnside's "interesting" paper in front of him; and he called attention to the fact that it contained the first application of character theory to the study of the properties of a group. Instances of intellectual harmony continued, for soon after publishing [219], Frobenius learned that Burnside had established essentially the same result in the sequel [52] to his above-mentioned paper! Frobenius responded, in another paper from 1901 [220], by using character theory to extend that result and to obtain thereby a proof of a theorem that Burnside had earlier established only under strong restrictive assumptions and without using character theory.

In its original formulation, the theorem concerned permutation groups, so let $\mathfrak{H}$ denote a group of permutations of a set with n elements, $\mathcal{A} = \{a_1, \ldots, a_n\}$. Suppose $\mathfrak{H}$ has order h and that it has the following properties:

(a) $\mathfrak{H}$ is transitive, i.e., for any i and j there is an $R \in \mathfrak{H}$ such that $R(a_i) = a_j$.
(b) No element of $\mathfrak{H}$ fixes two or more elements of $\mathcal{A}$ and there is an $R \neq E$ in $\mathfrak{H}$ that fixes one element of $\mathcal{A}$.

If, e.g., a_1 is fixed by some $H \neq E$, then the set $\mathfrak{G}$ of all $R \in \mathfrak{H}$ such that R fixes a_1 forms a subgroup of $\mathfrak{H}$. If g is the order of $\mathfrak{G}$, the transitivity of $\mathfrak{H}$ implies that $g < h$ and that $H_1, \ldots, H_n$ may be chosen from $\mathfrak{H}$ such that

$$\mathfrak{H} = \bigcup_{i=1}^{n} \mathfrak{G} H_i \tag{15.13}$$

gives a partition of $\mathfrak{H}$ into disjoint cosets. Thus $h = (\mathfrak{H} : \mathfrak{G})(\mathfrak{G} : 1) = ng$. Corresponding to (15.13) are the n conjugate subgroups $H_i^{-1}\mathfrak{G}H_i$, $i = 1, \ldots, n$. The elements of $H_i^{-1}\mathfrak{G}H_i$ fix $H_i^{-1}(a_1)$, and so if $R \in (H_i^{-1}\mathfrak{G}H_i) \cap (H_j^{-1}\mathfrak{G}H_j)$ for $i \neq j$, R fixes

$H_i^{-1}(a_1)$ and $H_j^{-1}(a_1)$, which by (b) means that if $R \neq E$, then $H_i^{-1}(a_1) = H_j^{-1}(a_1)$, so that $H_iH_j^{-1} \in \mathfrak{G}$, which is impossible because $H_1, \ldots, H_n$ determine the coset partition (15.13). Thus $\mathfrak{H}$ has the following property:

(c) For $i \neq j$, $(H_i^{-1}\mathfrak{G}H_i) \cap (H_j^{-1}\mathfrak{G}H_j) = \{E\}$.

Nowadays, a (nontrivial) subgroup $\mathfrak{G} \subset \mathfrak{H}$ with property (c) is said to be *malnormal*, a term I will use in what follows.

All of the above was well known by the time Burnside became interested in permutation groups with properties (a) and (b), and so (c) as well. Now (c) shows that the number of distinct elements in $\bigcup_{i=1}^n H_i^{-1}\mathfrak{G}H_i$ is $ng - (n-1) = h - (n-1)$, and so there are exactly $n-1$ elements of $\mathfrak{H}$ outside $\bigcup_{i=1}^n H_i^{-1}\mathfrak{G}H_i$, and they are precisely the elements of $\mathfrak{H}$ that fix no element of $\mathcal{A}$. Suppose we add the identity element to this set and consider

$$\mathfrak{N} = \left(\mathfrak{H} \sim \bigcup_{i=1}^{n} H_i^{-1}\mathfrak{G}H_i\right) \cup \{E\}. \tag{15.14}$$

If $\mathfrak{N}$ were known to be a subgroup, then it would be a normal subgroup, because the conjugate of a permutation fixing no element of $\mathcal{A}$ has the same property. This would mean that $\mathfrak{H}$ could not be simple. In the 1897 edition of his treatise on finite groups, Burnside showed that $\mathfrak{N}$ is always a subgroup when g is even [48, pp. 141–144]. Then in 1900 [53] he showed that $\mathfrak{N}$ is a subgroup if g is odd, provided that $n < N^2$, where N is the smallest *odd* number such that a simple group of order N exists. At the time, it was known that $N > 9000$ [53, p. 240]. As Burnside noted, the truth of the theorem that $\mathfrak{N}$ is always a subgroup was thus tied up with the question whether simple groups of odd composite order can exist.

Stimulated by Burnside's efforts, Frobenius turned in his 1901 paper [220] to the question whether $\mathfrak{N}$ is always a subgroup. It was well known at the time that the above hypothetical theorem about $\mathfrak{N}$ could be formulated in terms of abstract groups $\mathfrak{H}$ that contain a malnormal subgroup $\mathfrak{G}$ in the sense that (c) holds, and it was in this form that Frobenius considered the question. Burnside had not utilized group character theory in his work on the question, although sums of roots of unity occurred in his arguments. Frobenius discovered that his 1898 theory of induced characters [214] made it possible to prove Burnside's theorem without any restrictions [220, p. 199, V]:

Theorem 15.4 (Frobenius). *Let $\mathfrak{H}$ be an abstract group of order h with a malnormal subgroup $\mathfrak{G}$ of order g, so that $h = ng$, where $n = (\mathfrak{H} : \mathfrak{G})$. Then $\mathfrak{N}$ as given by (15.14) is the unique subgroup of $\mathfrak{H}$ of order n (and so must be a normal subgroup of $\mathfrak{H}$). It consists of the n elements $R \in \mathfrak{H}$ such that $R^n = E$.*

An abstract group $\mathfrak{H}$ satisfying the hypothesis of this theorem is now called a *Frobenius group*, as is a permutation group $\mathfrak{H}$ satisfying (a) and (b) above. Although Frobenius left no trace of his theory of characters in his statement of Theorem 15.4, in the course of his proof he showed that $\mathfrak{N} = \bigcap_{\lambda=1}^{k} \mathfrak{N}_\lambda$, where $\mathfrak{N}_\lambda$ consists of all

$R \in \mathfrak{H}$ such that $\chi^{(\lambda)}(R) = f_\lambda$, $\chi^{(\lambda)}$ being as usual the λth irreducible character of $\mathfrak{H}$ and $f_\lambda = \chi^{(\lambda)}(E)$ its degree [220, p. 198]. Nowadays, $\mathfrak{N}$ is called the Frobenius kernel of the (Frobenius) group $\mathfrak{H}$.

In the 1911 edition of his book, Burnside singled out Frobenius' theorem (stated for permutation groups) as one of two "good examples" of the power of the theory of group representations and characters for establishing properties of finite groups [56, p. 321]. The other example had to do with groups of order $p^a q^b$, where p and q are distinct primes. In 1895, as part of the first instance of "intellectual harmony" between Frobenius and Burnside, they had both proved independently that groups of such an order are solvable if $a \leq m$, where m is the index of p with respect to q.[11] In 1897, Burnside extended the conclusion to groups with $a < 2m$ [48, §243]. Then in 1902, Frobenius extended the conclusion to groups with $a \leq 2m$ [221]. At the conclusion of his paper, Frobenius had pointed out that his proof involved "purely group-theoretic considerations" and made no use of "the theory of substitutions" in the sense that no "representation" of a group as a group of substitutions was used. Here Frobenius was using the word "substitution" in the sense of a permutation, but of course, his proof also did not utilize his new theory of representing groups by "linear substitutions" either. It was left to Burnside to show in 1904 that the theory of group characters yielded a proof that every group of order $p^a q^b$ is solvable [54].[12]

By the time Burnside published the second edition of his treatise on groups in 1911 [56], he had become more convinced that simple groups of odd composite order do not exist. There, in a "Note on Groups of Odd Order," he pointed to his discovery that no nontrivial irreducible character of such a group can have all real values as "the most noticeable property" of these groups [56, p. 503]. Using it, he showed that every irreducible group of linear transformations in three, five, or seven variables must be solvable. This result and the fact that mathematicians had by then shown that no simple group of odd order less than 40,000 exists, "suggests inevitably that simple groups of odd order do not exist." In 1963, Feit and Thompson finally confirmed Burnside's conjecture by proving that every group of odd order is solvable [162]. Their lengthy proof involved many purely group-theoretic results and techniques introduced long after Burnside's time; but, in accordance with Burnside's intuitions, it also made extensive use of the theory of group characters. Burnside's above theorem on the irreducible characters of groups of odd order was used "continually" in their proof [161, p. 969]. The properties of Frobenius groups figure in their proof as well [162, p. 782 (Proposition 3.16)]. In fact, a proof of Frobenius' Theorem 15.4 that does not use characters has never been given; all proofs of Theorem 15.4 are variants of Frobenius' original proof, which is the germ

[11] If r is one of the $\varphi(q-1)$ primitive roots of q, then m is defined by $r^m \equiv p \pmod{q}$.

[12] Using Burnside's $p^a q^b$ theorem, P. Hall generalized it to prove that $\mathfrak{H}$ is solvable if it has the property that for every representation of its order as a product of two relatively prime integers, subgroups of those orders exist. Hall's proof involved the notion of a Hall subgroup, which played a significant role in later work, including Feit and Thompson's "odd-order paper" [162] discussed in the following paragraph.

of most later applications of the theory of characters to the study of the structure of finite groups, such as those found in the Feit–Thompson paper.[13] The Feit–Thompson paper was largely responsible for the swell of research activity aimed at classifying all simple groups, and by 1981, what Burnside had regarded in 1897 as "unexpected" had been achieved.[14]

All of Frobenius' major contributions to the theory of characters and representations had been made by 1903, and his last paper having anything to do with the theory was published in 1907 [227] and involved a generalization of a theorem in the theory of groups made possible using characters. The theory Frobenius had created, however, lived on and continued to flourish at the hands of Frobenius' "mathematical son" Issai Schur (Section 15.5) and his "mathematical grandson" Richard Brauer (Section 15.6).

15.5 I. Schur

During his 24 years as full professor in Berlin, Frobenius had a total of ten doctoral students. Two of them, E. Landau and I. Schur, became well-known mathematicians. Issai Schur (1875–1941), in particular, earned Frobenius' highest respect from the outset. According to Frobenius, Schur posed his thesis problem completely on his own, and his solution so impressed Frobenius that he declared that Schur "with one blow has shown himself to be a master of algebraic research" [22, p. 127]. Schur, who was born in the Russian Empire, had begun attending the University of Berlin in 1894 (initially as an undergraduate). He was thus in attendance during the years that Frobenius was creating his theory of group characters. Seven years after commencing his education in Berlin, he received his doctorate.

15.5.1 Polynomial representations of $\mathrm{GL}(n,\mathbb{C})$

Schur's doctoral dissertation of 1901 [521] was inspired by an 1894 paper on the theory of invariants [303] by one of Klein's best students, Adolf Hurwitz (1859–1919).[15] Hurwitz gave an exposition of invariant theory that emphasized the fact that in the theory, one needs to consider, corresponding to a nonsingular matrix

[13] A proof without characters of Burnside's theorem on groups of order $p^a q^b$ can be gleaned from the Feit–Thompson paper [162]. Relatively short proofs without characters were given in the 1970s by Goldschmidt [251] for p,q odd and by Matsuyama [436] for $p=2$. These proofs utilize some of the modern ideas and results in group theory and are not as elementary as Burnside's proof.

[14] See the accounts by Aschbacher [9] and Gorenstein [252].

[15] A more detailed discussion of Schur's dissertation and its role in the history of the representation theory of Lie groups can be found in my book [276]. See especially Section 3 of Chapter 10.

or linear transformation $A = (a_{ij})$, various other matrices $T(A)$, not necessarily the same size as A, whose coefficients are functions of the coefficients a_{ij} of A. For example, in the traditional approach to the theory of invariants, if

$$f(b;x) = \sum_{e_1+\cdots+e_n=r} b_{e_1,\ldots,e_n} x_1^{e_1} x_2^{e_2} \cdots x_n^{e_n}$$

is the "general" homogeneous polynomial of degree r in $x_1,\ldots,x_n$, then each $A \in \mathrm{GL}(n,\mathbb{C})$ defines a linear transformation $x = Ax'$ of the variables x_i that induces a linear transformation $b' = T(A)b$ of the coefficients $b_{e_1,\ldots,e_n}$ of f, where

$$f(b;x) = f(b;Ax') = f(b';x').$$

The coefficients of the matrix $T(A)$ are homogeneous functions of the coefficients a_{ij} of A of degree r . For example, if $f(x;b) = b_1x_1^2 + b_2x_1x_2 + b_3x_2^2$, so $r = 2$, and if $A = \begin{pmatrix} a_{11} & a_{12} \\ a_{21} & a_{22} \end{pmatrix}$, so $x_1 = a_{11}x_1' + a_{12}x_2'$ and $x_2 = a_{21}x_1' + a_{22}x_2'$, then

$$T(A) = \begin{pmatrix} a_{11}^2 & a_{11}a_{21} & a_{21}^2 \\ 2a_{11}a_{12} & a_{11}a_{22} + a_{12}a_{21} & 2a_{21}a_{22} \\ a_{12}^2 & a_{12}a_{22} & a_{22}^2 \end{pmatrix}.$$

An invariant is then a homogeneous polynomial, $I(b)$, of the coefficients $b_{e_1,\ldots,e_n}$ such that $I(b') = I(b)$.[16] Hurwitz's approach to invariant theory emphasized bypassing the intermediary form $f(b;x)$ and considering directly the requisite transformations $T(A)$. They satisfy $T(AB) = T(A)T(B)$. In addition, their coefficients are homogeneous polynomials of degree r in the coefficients a_{ij} of A [303, §7]. Hurwitz also pointed out that other types of considerations in the theory of invariants involve transformations $T(A)$ with the same properties. Of course, reading Hurwitz's paper today, we see that such transformations $T(A)$ define certain *representations* of the group $\mathrm{GL}(n,\mathbb{C})$.

In his dissertation, Schur proposed a general study of correspondences $A \to T(A)$ such that the coefficients of $T(A)$ are polynomial functions of the coefficients a_{ij} of A and $T(AB) = T(A)T(B)$ for all $A, B \in \mathrm{GL}(n,\mathbb{C})$. The motivation from the theory of invariants is reflected in his terminology, $T(A)$ being called "an invariant form or matrix built out of A." According to Frobenius, Schur posed this problem on his own. For a student of Frobenius, natural questions to ask would be, What are the possibilities for such representations? Can they be described? Does the problem reduce to finding irreducible ones? If so, can one find formulas for the irreducible characters and the corresponding dimensions? These are natural enough questions to ask, but the task of answering them is formidable: one has only to think how

[16]In addition to these "absolute invariants," more general invariants satisfying $I(b') = (\det A)^w I(b)$, where w is a nonnegative integer, were also considered.

nontrivial the analogous problem was, even in Frobenius' hands, for the symmetric group; and the group $\mathrm{GL}(n,\mathbb{C})$ is continuous, not finite, so Frobenius' theory of characters and representations does not directly apply.

In his definition of an invariant form $T(A)$, Schur did not even require that $T(A)$ be invertible; rather, he proved how the theory reduces to the case in which $T(A)$ is invertible and its coefficients are homogeneous polynomials of a fixed degree m, where $m \leq n$. For such homogeneous invariant forms, Schur than established many of the analogues of theorems in Frobenius' theory, e.g., two invariant forms of degree m are equivalent if and only if their "characteristics" $\Phi(A) = \operatorname{tr} T(A)$ are equal; and such forms are completely reducible, so that the problem of their determination reduces to the case of irreducible ones. As Frobenius had done for the symmetric group, Schur determined the irreducible characteristics and the degrees of the corresponding representations.

The key to Schur's success was his discovery that there is a biunique mapping $T \to \tau$ between homogeneous invariant forms $T(A)$ of degree of homogeneity $m \leq n$ and representations τ of the symmetric group $\mathfrak{S}_m$, which brings with it a formula for the characteristic $\Phi = \operatorname{tr} T(A)$ in terms of the character $\chi = \operatorname{tr} \tau$. Understandably, Frobenius was quite impressed by Schur's dissertation. In his evaluation of it, after mentioning some of the main results, he explained that they "are only the most important of the superabundant contents of the work. Between them there are in addition a host of interesting special cases. Some of these less important results would already suffice for the content of a good dissertation."[17] Frobenius went on to say that if Schur had simply shown how his original "beautiful problem"—to determine all (possibly singular) invariant forms—reduced to the problem he eventually studied—that of determining all (nonsingular) invariant forms of order m—the dissertation would have called for the designation "outstanding" (*hervorragend*). "But in addition, he has completely solved the difficult problem and with the simplest means and, so to speak, conformally mapped the theory of invariant forms onto the seemingly totally different theory of the representations of the symmetric group—something that always lends a particular attraction to a mathematical development." With this one work, he concluded, Schur had proved himself to be "a master of algebraic research."

15.5.2 Projective representations and factor sets

Soon after his brilliant doctoral dissertation, Schur presented the *Habilitationsschrift* required to become an instructor at the University. Schur's habilitation officially took place on 5 January 1903. In December 1902, Frobenius wrote his evaluation of the *Habilitationsschrift*, and his praise for what Schur had written was unrestrained:

---

[17]This and the following quotations are my translation of portions of the conclusion of Frobenius' evaluation as transcribed by Biermann [22, p. 127].

"In this substantial work, as earlier in his dissertation, the author demonstrates his outstanding talent for posing, transforming, breaking down, and solving major algebraic problems."[18]

The problem Schur posed in his *Habilitationsschrift*, which appeared in *Crelle's Journal* in 1904 [522], was motivated in large part by the work of Felix Klein and his students. As a part of his program to generalize Galois' theory (see Section 14.2), Klein and his students worked on the problem of determining all finite groups of projective transformations, that is, transformations expressible in inhomogeneous coordinates in the form

$$x_i' = \frac{a_{i,1}x_1 + \cdots + a_{i,n}x_n + a_{i,n+1}}{a_{n+1,1}x_1 + \cdots + a_{n+1,n}x_n + a_{n+1,n+1}},$$

and in homogeneous coordinates in the form

$$\rho y_i' = a_{i,1}y_1 + \cdots + a_{i,n+1}y_{n+1}, \quad i = 1, \ldots, n+1, \tag{15.15}$$

where $|a_{i,j}| \neq 0$ and the factor $\rho \neq 0$ is included because the homogeneous coordinates $\rho y_1', \ldots, \rho y_{n+1}'$ all determine the same point in projective space. Of course, this also means that all matrices $\rho A = (\rho a_{ij})$ determine the same projective transformation. Thus projective transformations are the elements of $\mathrm{PGL}(n+1, \mathbb{C}) = \mathrm{GL}(n+1, \mathbb{C})/(\rho I)$.

Klein had determined all such finite groups for $n = 2$ in 1875 [340], and Valentiner had done the same for $n = 3$ in 1889, but with the rudimentary techniques at their disposal, it was clear that it would be tedious to extend the same exhaustive type classification to values of $n > 3$ and impossible to do it for all n. Some attention was thus focused on determining all finite projective groups in a given number of variables with a specific structure. For example, Maschke in 1898 determined all finite projective groups with $n = 3,4$ that are isomorphic to a symmetric or alternating group. And Wiman in 1899 considered Klein's normal problem (Section 14.2) for projective groups with the structure of the symmetric or alternating groups: determine the projective group with the minimal number of variables that is isomorphic to $\mathfrak{S}_m$, respectively to $\mathfrak{A}_m$. However, only for $m = 6$ did he carry out the construction of these groups. All of this work, however, lacked general methods of sufficient power to make more than a dent in the problem. As Frobenius put it,

[18]Quoted by Biermann [22, p. 135]. Later, in a 1914 memorandum supporting Schur for a position, Frobenius expressed similar sentiments about Schur's work in general: "As only a few other mathematicians do, he practices the Abelian art of correctly formulating problems, suitably transforming them, cleverly dismantling them, and then conquering them one by one" [22, pp. 139, 223]. After his death, Frobenius' words were quoted by Planck when Schur was admitted to membership in the Berlin Academy of Sciences in 1922. See Schur, *Gesammelte Abhandlungen* 2, p. 414.

> F. Klein and his school had computationally investigated the representation of groups by linear fractional substitutions, and the general resolution of this problem appeared as a hopeless undertaking. The difficulties that stood in the way of mastering the problem were first completely overcome through the penetrating acumen and persistent reflection of Schur.[19]

The problem considered by Klein and his students was reformulated by Schur as follows. Given an abstract finite group $\mathfrak{H}$, determine all isomorphisms $\pi : H \to T(H)$, where $T(H) \in \mathrm{PGL}(n,\mathbb{C})$. If $\sigma(H)$ denotes the corresponding matrix of coefficients of $T(H)$ as in (15.15), then matrices $\sigma(H)$ must satisfy the relation

$$\sigma(A)\sigma(B) = r_{A,B}\sigma(AB), \tag{15.16}$$

where $r_{A,B} \in \mathbb{C}$ and $r_{A,B} \neq 0$. Thus the problem is to determine all such mappings $H \to \sigma(H)$. Having so transformed the problem, Schur proceeded to generalize it by posing the problem of finding all homomorphisms of $\mathfrak{H}$ onto a group of projective transformations. This meant determining all inequivalent irreducible mappings $H \to \sigma(H)$ satisfying (15.16). Of course, every ordinary representation of $\mathfrak{H}$ determines a projective representation (with $r_{A,B} = 1$), so that (as Schur put it) the work of Frobenius and Molien could be seen as resolving a special case of the problem. Schur showed that in fact, the general problem could be reduced to a problem in the ordinary representation of a group, and in so doing, he earned the praise that Frobenius had bestowed on his achievement.

Equation (15.16) in conjunction with the associative law for $\mathfrak{H}$ shows that the complex numbers $r_{A,B}$ associated to the projective representation σ satisfy

$$r_{P,Q}r_{PQ,R} = r_{P,QR}r_{Q,R} \quad \text{for all} \quad P,Q,R \in \mathfrak{H}. \tag{15.17}$$

Any system of h^2 complex numbers $r_{A,B} \neq 0$ satisfying this equation will be called a *factor set* in accordance with later terminology. Schur showed that every factor set corresponds to a projective representation σ of degree h (the order of $\mathfrak{H}$): for if the $h \times h$ Frobenius group matrix is modified by setting $X = (r_{P,Q^{-1}}x_{PQ^{-1}})$, and if we write it as $X = \sum_{R\in\mathfrak{H}} \sigma(R)x_R$, then the matrices $\sigma(R)$ satisfy (15.16). Thus σ can be regarded as the regular representation for this factor set.

Suppose now that two projective representations σ and σ' are such that $\sigma(R)$ and $\sigma'(R)$ define the same projective transformation. This means that complex numbers $c_R \neq 0$ exist such that $\sigma'(R) = c_R\sigma(R)$ for all $R \in \mathfrak{H}$. In terms of the corresponding factor sets, the relation is

[19] These words were written in a memorandum supporting Schur for a professorship [22, p. 224], but they are not at all an exaggeration of Schur's accomplishments vis à vis the efforts of the Klein school.

$$r'_{A,B} = \frac{c_A c_B}{c_{AB}} r_{A,B} \tag{15.18}$$

for all $A, B \in \mathfrak{H}$. Schur defined two factor sets to be *associated* if (15.18) holds. Association defines an equivalence relation on the set of all factor sets. He showed that the number of equivalence classes is finite and that if $\mathcal{C}(r_{A,B})$ denotes the class containing factor set $r_{A,B}$, then $\mathcal{C}(r_{A,B}) \circ \mathcal{C}(s_{A,B}) = \mathcal{C}(r_{A,B}s_{A,B})$ defines a multiplication of classes such that they form a finite abelian group $\mathfrak{M}$, which he called the multiplier of $\mathfrak{H}$.[20] The order m of $\mathfrak{M}$ gives the number of nonidentical (but possibly equivalent) projective representations σ of $\mathfrak{H}$.

Schur showed how to define a finite group $\mathfrak{G}$ that contains a subgroup $\mathfrak{N}$ in its center that is isomorphic to $\mathfrak{M}$ and such that $\mathfrak{G}/\mathfrak{N}$ is isomorphic to $\mathfrak{H}$. He called $\mathfrak{G}$ a representing group for $\mathfrak{H}$, because, as he showed, every ordinary representation of $\mathfrak{G}$ determines a projective representation of $\mathfrak{H}$ and vice versa. In particular, the irreducible representations of $\mathfrak{G}$ correspond to the irreducible projective representations of $\mathfrak{H}$. In this way, he solved his problem by breaking it down into two subproblems: the problem of constructing the representing group $\mathfrak{G}$ and the problem of determining all irreducible representations of $\mathfrak{G}$, a problem that in principle, had already been solved by Frobenius. He also studied the question of when every projective representation of $\mathfrak{H}$ is equivalent to an ordinary representation (which means that $\mathfrak{M}$ has order 1). His results implied, for example, that this is the case if the order of $\mathfrak{H}$ is the product of distinct primes or, more generally, if all the Sylow subgroups of $\mathfrak{H}$ are cyclic. Later, in 1911, he solved the formidable problem of determining all irreducible projective representations of the symmetric and alternating groups that are not equivalent to ordinary representations.[21] These results, taken together with Frobenius' results on the ordinary representations of $\mathfrak{S}_n$ and $\mathfrak{A}_n$, thus resolved the problem for these general classes of groups, a problem that had been dealt with only in very special cases by Klein's students.

15.5.3 Schur's Lemma and representations of SO$(n, \mathbb{R})$

We have seen that in Frobenius' formulation of the theory of group characters and representations, considerations based on determinants were fundamental to many of the proofs and thus to the formulations as well, as is illustrated by his reciprocity theorem for induced characters (Section 15.1). In papers of 1904, Burnside set forth a more elementary derivation of the theory. It made use of the fact that an invariant positive definite Hermitian form exists for every finite

[20]The group $\mathfrak{M}$ can be identified with the second cohomology group of $\mathfrak{H}$ over $\mathbb{C}$, but group cohomology did not exist at this time. On this and other anticipations of modern theories by Schur, see [406, p. 101]. See also the discussion in Section 15.6 below of Brauer's work on Schur's index theory and its connections with Galois cohomology.

[21]Schur, *Abhandlungen* 1, 346–441.

group of linear transformations (Theorem 14.1), a fact whose utility Maschke had already discovered in 1899 when he proved his complete reducibility theorem (Section 14.4). In 1905, Schur proposed another reformulation of the theory [523], which, like Burnside's, was elementary—and therefore not reliant on sophisticated determinant-based arguments—but, in addition, avoided the use of Hermitian forms, which Schur found to be extraneous to the theory.

The starting point and foundation for Schur's reformulation was the now well known and remarkably useful "Schur's lemma," which as stated by him runs as follows.

Theorem 15.5 (Schur's lemma). *Let σ and τ be irreducible representations of degrees f and g and let $\sigma(x) = \sum_{R\in\mathfrak{H}} \sigma(R)x_R$ and $\tau(x) = \sum_{R\in\mathfrak{H}} \tau(R)x_R$ be the corresponding group matrices. Suppose there is an $f \times g$ constant matrix P such that $\sigma(x)P = P\tau(x)$ for all x. Then either $P = 0$ or σ and τ are equivalent and P is a square matrix with $|P| \neq 0$. In the latter case, when in particular, σ and τ are actually equal, so that $\sigma(x)P = P\sigma(x)$ for all x, then P must be a scalar matrix, i.e., $P = \lambda I$.*

Of course, in the above theorem, "for all x" could be replaced by "for all $R \in \mathfrak{H}$"; Schur was just following Frobenius' practice. As noted in Section 13.5, Frobenius had already utilized what amounts to this result (with $f = g$) in his 1897 "Representation" paper [213], but his proof was not elementary, since it depended on some of the principal results of the theory and ultimately depended on his theory of the group determinant. Schur's simple proof depended on nothing but the elementary properties of matrices.[22]

By building up the theory of group characters and representations on this lemma, Schur thus completed what Frobenius had begun in his "Representation" papers. In those papers, Frobenius had used matrix algebra in conjunction with his theory of the group determinant to obtain results on characters and matrix representations, as, e.g., in his rederivation of the fundamental character relations (13.45). Schur showed that these results could be established solely by means of matrix algebra and without reference to the theory of the group determinant. As a consequence, the technique of summation over the group and its invariance with respect to translations became even more fundamental than in Frobenius' version of the theory. Among other things, as we shall see momentarily, the new approach helped forge a link between Frobenius' theory and the theory of Lie groups, and in fact, Schur unwittingly took the first steps in this direction with a paper of 1924 on the rotation group $\mathfrak{D}_n = \mathrm{SO}(n,\mathbb{R})$ [526].

As with Schur's initial work dealing with the representation of continuous groups, namely his dissertation [521], his paper [526] was motivated by the theory of invariants and particularly by an 1897 paper by Hurwitz [304]. During the intervening years, Schur had not paid much attention, in print, to the theory of

[22]For an exposition of Schur's proof and its role in deriving Schur's orthogonality relations for irreducible representations, see Curtis' book [109, pp. 140ff.].

invariants. This may have been due to Frobenius' aversion to the theory, which he deemed (in 1902) to be "a subject of meager significance" to which many mathematicians, including Hilbert, had contributed "yeoman's work."[23] Frobenius admired Hilbert's solution to the finiteness problem[24] for the invariants of an n-ary form, because, he felt, for the first time it had brought the concepts of algebra (Hilbert's finite basis theorem) to bear on invariant theory. But he felt that with his solution, Hilbert had also brought the theory to completion. As he put it, with this theorem, Hilbert became "the scientific founder and terminator (*Vollender*) of invariant theory," presumably by resolving its central problem and making it an aspect of algebra. Although Schur's dissertation had been motivated by a problem suggested by invariant theory, the problem itself was not a problem of invariant theory. Furthermore, he had reformulated and treated it in terms of the concepts of algebra, namely the theory of representations. So Frobenius' high opinion of Schur's dissertation is consistent with his attitude toward invariant theory. As late as 1914, three years before his death, Frobenius maintained the view that Hilbert had finished off the theory of invariants, and his attitude may have temporarily restrained Schur's enthusiasm for the subject.

After becoming a chaired full professor in Berlin in 1921, Schur began to lecture on the theory of invariants and to collaborate with A. Ostrowski on invariant-theoretic problems (treated, of course, using the concepts of algebra).[25] In his lectures on invariant theory, which he gave for the first time during the winter semester of 1923–1924,[26] one of the major topics Schur focused on—Cayley's theory of semi-invariants—was precisely an area of Hilbert's early work that Frobenius had dismissed as "yeoman's work." Part of that theory involved Cayley's solution to a certain counting problem related to the invariants of the general binary form, i.e., in modern parlance, the invariants of the homogeneous polynomial representations $A \to T(A)$ of $\mathrm{GL}(n, \mathbb{C})$ associated with the transformation of that form as explained above in conjunction with Hurwitz's paper of 1894 [303].[27] Schur realized that Molien, in the 1898 note [446] he submitted through Frobenius after they had learned of each other's work (as discussed above in Section 14.2), had in effect solved Cayley's counting problem for the invariants of any matrix

[23]Frobenius used the word *Kärrnerarbeit*. See his memoranda of 1902 and 1914 to the Prussian Ministry of Culture regarding the possible appointment of Hilbert to a professorship in Berlin [22, pp. 209–210, 222–223].

[24]The finiteness problem for a given type of invariant is to prove that there is a finite number of such invariants such that any invariant of the given type is expressible as a polynomial function of these.

[25]The joint work is contained in Schur, *Abhandlungen* 2, 334–358, and Ostrowski, *Papers* 2, 127–151, and is discussed in my book on the history of Lie group theory [276, Ch. 10, §4].

[26]I am grateful to the late Mrs. Brauer and to Walter Feit and Jonathan Alperin for kindly making Richard Brauer's notes from these lectures available to me. A later version of Schur's lectures was published by Grunsky [527].

[27]I have discussed Cayley's counting problem and its extensions by Molien and Schur (alluded to below) in [272] and [276, Ch. 7,§4, Ch. 10,§5].

representation of a finite group by using the theory of characters. Schur's familiarity with Hurwitz's 1897 paper [304] made it possible for him to see how to deal with Cayley's problem for the invariants of any matrix representation of the rotation group $\mathfrak{D}_n = \mathrm{SO}(n, \mathbb{R})$.

In his paper [304], Hurwitz had shown how to define an invariant integral on a compact Lie group. This enabled him to extend Hilbert's finite basis theorem to invariants of $\mathfrak{D}_n$, something Hilbert had been unable to do except in the case $n = 3$. This meant in particular that if $R \to \sigma(R)$ is any representation of $\mathfrak{D}_n$, then a finite integral $\int_{\mathfrak{D}_n} f(\sigma(R))\, dm$ is defined for all continuous functions f that is translation invariant:

$$\int_{\mathfrak{D}_n} f(\sigma(RS))\, dm = \int_{\mathfrak{D}_n} f(\sigma(R))\, dm \quad \text{for all } S \in \mathfrak{D}_n.$$

This is just the principle of invariance of "summation" over the group, with summation understood in the sense of integration. In view of his reformulation of Frobenius' theory in [523], Schur could see that an analogue of Frobenius' entire theory could be articulated for the rotation group $\mathfrak{D}_n$: the proofs were direct analogues of those in [523], except that integration over $\mathfrak{D}_n$ replaced summation over the finite group. In this manner, not only could a complete reducibility theorem be established, but in contrast to the case of the polynomial representations of the noncompact $\mathrm{GL}(n, \mathbb{C})$ studied in his doctoral dissertation [521], now the irreducible characters $\chi(R) = \mathrm{tr}\,\sigma(R)$ satisfied orthogonality relations, such as

$$\int_{\mathbf{D}_n} \chi^{(\kappa)}(R)\overline{\chi^{(\lambda)}(R)}\, dm = 0$$

for distinct irreducible characters. The theory of characters and their orthogonality relations had been an essential part of Molien's solution of Cayley's problem for finite groups, and they played a similar role in Schur's solution to the problem for $\mathfrak{D}_n$, which he published in [526].

Although Schur had in effect shown how to extend Frobenius' theory to compact Lie groups, the most important type of Lie groups, semisimple groups, for which a profound theory had been developed by Killing and E. Cartan, were mostly noncompact, examples being the special linear group $\mathrm{SL}(n, \mathbb{C})$, the complex orthogonal groups $\mathrm{SO}(n, \mathbb{C})$ and the symplectic groups $\mathrm{Sp}(2n, \mathbb{C})$. In a brilliant mathematical tour de force, Hermann Weyl discovered in 1924 how to combine the theory of semisimple Lie groups with the "unitarian trick" contained in Hurwitz's 1897 paper [304] to be able to do for semisimple groups what Schur had done for the rotation group $\mathfrak{D}_n$, namely, to develop the analogue of Frobenius' theory of representations and characters for these groups—including even formulas for the irreducible characters and the degrees of the irreducible representations [602, 603]. Weyl's work on Einstein's general theory of relativity and its generalizations had made him interested in the possibility of extending Frobenius' complete reducibility theorem to the special linear group; and the timely appearance of Schur's 1924 paper [526], which included as well a discussion of Hurwitz's unitarian trick, provided the spark that ignited the flame of Weyl's mathematical imagination and culminated in his above-mentioned papers, as well as his paper with Peter [604].[28] Through Weyl's work the ideas and concepts of Frobenius' theory became part of a far broader mathematical framework, including what is sometimes referred to as generalized harmonic analysis.

15.5.4 Index theory

There is one other major new direction in which Schur developed Frobenius' theory. It had to do with what Schur called the arithmetic study of group representations. Maschke, in a sense, initiated the theory with the 1898 paper [434] that inspired his complete reducibility theorem (as indicated above in Section 14.4). Recall that Maschke's paper [434] suggested a problem that I will refer to as Maschke's problem:

Problem 15.6 (Maschke's problem). If $y = A_i x$, $i = 1, \ldots, h$, are the matrix equations of a group $\mathfrak{H}$ of linear transformations, show that a variable change $x = Px'$ exists such that the equations of $\mathfrak{H}$ in the new variables, viz., $y' = P^{-1}A_iPx'$, are such that the coefficients of all $P^{-1}A_iP$ are in $\mathbb{Q}(\varepsilon)$, where ε is a primitive hth root of unity.

Maschke did not solve this problem. In his paper [434], he deduced the conclusion of the problem under the assumption that $\mathfrak{H}$ contains a transformation with distinct characteristic roots. He did not speculate on its validity without this assumption, although his paper clearly raised this question and so implicitly posed the above problem.

Note that in Maschke's problem, $\mathfrak{H}$ can be regarded as a representation σ of an abstract group $\mathfrak{G}$ with the multiplication table of $\mathfrak{H}$, namely $\sigma(G_i) = A_i$, $i = 1, \ldots, h$. When $\mathfrak{H}$ is viewed as a representation, Maschke's problem becomes that of showing that $\mathfrak{H}$ is *equivalent* to a representation with matrix coefficients from $\mathbb{Q}(\varepsilon)$. The language of the representation-theoretic viewpoint will be used in what follows.

The complete reducibility theorem applied to $\mathfrak{H}$ shows that it suffices to solve Maschke's problem when $\mathfrak{H}$ is irreducible. In 1905, Burnside returned to Maschke's problem [55]. Noting that the "determination of the arithmetic nature of the coefficients in an irreducible group of linear substitutions of finite order is a question which has hitherto received little attention despite its undoubted importance," he observed that Maschke's result was the only general theorem on the question known at that time. Burnside was able to show that the conclusion of Maschke's problem follows under a weaker assumption than that made by Maschke, by proving the following theorem.

---

[28] For the full story, see my paper [274], and especially Chapters 11–12 of my book [276].

Theorem 15.7 (Burnside). *Let $\mathfrak{H}$ be an irreducible group of linear transformations of order h, and suppose that $\mathfrak{H}$ has the following property: there is no integer $b > 1$ such that every characteristic root of every $T \in \mathfrak{H}$ has a multiplicity of b or some multiple thereof. Then $\mathfrak{H}$ is equivalent to a group of linear transformations whose coefficients are all in $\mathbb{Q}(\varepsilon)$, where ε is a primitive hth root of unity, and so Maschke's problem is solved for groups with this property.*

Burnside felt that it was "in the highest degree improbable" that a group $\mathfrak{H}$ could exist that did not satisfy the above property, i.e., for which the above-described integer $b > 1$ exists, although he admitted his inability to prove its nonexistence. It may have been the papers of Maschke and Burnside that turned Schur's interest to the arithmetic side of representation theory, for in his first paper (in 1906) on this side of the theory [524], he made further contributions to the solution of Maschke's problem. Unlike Burnside, however, Schur considered the problem within the framework of a new theory of far greater generality.

Suppose $\mathfrak{H}$ is an abstract group, and let $\sigma^{(\lambda)}$, $\lambda = 1,\ldots,k$, denote its irreducible representations over $\mathbb{C}$ with corresponding characters $\chi^{(\lambda)}$. Let $\mathbb{F}$ be an arbitrary subfield of $\mathbb{C}$. Say that $\sigma^{(\lambda)}$ is *rationally representable* in a finite extension $\mathbb{F}(\mu)$ if it is equivalent to a representation with coefficients in $\mathbb{F}(\mu)$. In particular, when $\mathbb{F} = \mathbb{Q}$, it followed from Frobenius' results that $\sigma^{(\lambda)}$ is always rationally representable in an algebraic number field $\mathbb{Q}(\mu)$. However, as will be seen, by replacing $\mathbb{Q}$ by $\mathbb{F}$, Schur gave to his deliberations a generality that proved useful in terms of applications. If $\sigma^{(\lambda)}$ is rationally representable in $\mathbb{F}(\mu)$, then clearly this field must contain $\chi^{(\lambda)}(R) = \operatorname{tr}\sigma^{(\lambda)}(R)$ for every $R \in \mathfrak{H}$. Thus if $\chi_1^{(\lambda)},\ldots,\chi_k^{(\lambda)}$ denote the k values of $\chi^{(\lambda)}$ on the k conjugacy classes of $\mathfrak{H}$, then $\mathbb{F}(\chi_1^{(\lambda)},\ldots,\chi_k^{(\lambda)}) \subset \mathbb{F}(\mu)$. Let $l = \left(\mathbb{F}\left(\chi_1^{(\lambda)},\ldots,\chi_k^{(\lambda)}\right) : \mathbb{F}\right)$. Then $l \mid n$, where $n = (\mathbb{F}(\mu) : \mathbb{F})$ and so $n = lm$. Suppose now n_0 denotes the minimum degree of all extensions $\mathbb{F}(\mu)$ in which $\sigma^{(\lambda)}$ is rationally representable. Then, by the above, $n_0 = l_0 m_0$. Schur called m_0 the index of $\sigma^{(\lambda)}$ or $\chi^{(\lambda)}$ relative to $\mathbb{F}$. It is now called the *Schur index*.

Some of the main results of Schur's index theory are summarized in the theorem below.[29] It involves Schur's notion of conjugate characters, which may be defined as follows. Let $\mathbb{L}$ be a normal extension of $\mathbb{F}$ that contains the values taken by the characters $\chi^{(1)},\ldots,\chi^{(k)}$. Say that two of these characters $\chi^{(\lambda)}$ and $\chi^{(\lambda')}$ are *conjugate with respect to* $\mathbb{F}$ if there is $\pi \in \operatorname{Gal}(\mathbb{L}/\mathbb{F})$ such that $\chi^{(\lambda)}(R) = \pi(\chi^{(\lambda')}(R))$ for all $R \in \mathfrak{H}$. This is an equivalence relation and so partitions the k irreducible characters of $\mathfrak{H}$ into equivalence classes. The significance of the number of these classes comes out in the following result [524, §1, III].

Theorem 15.8. *Let σ denote a representation of $\mathfrak{H}$ that is rationally representable in $\mathbb{F}$ and is also $\mathbb{F}$-irreducible in the sense that there is no matrix M with entries from $\mathbb{F}$ such that $M^{-1}\sigma M = \rho_1 \oplus \rho_2$. Then the following conclusions may be drawn.*

---

[29]For a clear and detailed discussion of the results in Schur's paper [524], including sketches of some of the proofs, sometimes along more modern lines, see Curtis' book [109, pp. 157ff.].

(i) There is an irreducible character $\chi^{(\lambda_1)}$ with index m_0 such that the character $\chi = \operatorname{tr}\sigma$ is expressible as

$$\chi = m_0\left(\chi^{(\lambda_1)} + \cdots + \chi^{(\lambda_{l_0})}\right),$$

where $\chi^{(\lambda_2)},\ldots,\chi^{(\lambda_{l_0})}$ are the conjugates of $\chi^{(\lambda_1)}$ with respect to $\mathbb{F}$. (Hence the number of characters in the equivalence class determined by $\chi^{(\lambda_1)}$ is l_0.) (ii) The number of inequivalent $\mathbb{F}$-rational, $\mathbb{F}$-irreducible representations of $\mathfrak{H}$ is equal to the number of equivalence classes into which the k irreducible characters of $\mathfrak{H}$ are partitioned by the relation of conjugacy with respect to $\mathbb{F}$.

Schur's lemma (Theorem 15.5), in a suitably modified version [524, §1, I], played a key role in the reasoning leading to the conclusions of this remarkable theorem.

Theorem 15.8 included, as a very special case, a theorem that had been obtained in 1903 by Loewy [423]. Recall that Loewy was one of the discoverers of Theorem 14.1, which Maschke had used to establish his complete reducibility theorem. Loewy realized that Maschke's theorem remains true over any field of characteristic zero.[30] If it is applied to the field $\mathbb{R}$ of real numbers, it shows that every finite group of linear transformations is similar over $\mathbb{R}$ to a direct sum of real irreducible groups. In [423], Loewy showed that real irreducible groups either (a) are irreducible over $\mathbb{C}$ or (b) decompose over $\mathbb{C}$ into the sum of two groups that are irreducible over $\mathbb{C}$ and are complex conjugates of each other. He did not consider the related problem of determining whether the conjugate representations are equivalent. As Schur pointed out in the above-mentioned 1906 paper [524, p. 175], Loewy's theorem is an immediate consequence of Theorem 15.8 with $\mathbb{F} = \mathbb{R}$. For if $\chi^{(\lambda_1)},\chi^{(\lambda_2)},\ldots,\chi^{(\lambda_{l_0})}$ are the irreducible characters associated by Theorem 15.8 to a given representation that is irreducible over $\mathbb{R}$, then, since $n_0 = l_0 m_0$ is 1 or 2 when $\mathbb{F} = \mathbb{R}$, there are three possibilities for l_0 and the Schur index m_0 associated to $\chi^{(\lambda_1)},\chi^{(\lambda_2)},\ldots,\chi^{(\lambda_{l_0})}$: (I) $l_0 = m_0 = 1$; (II) $l_0 = 1, m_0 = 2$; (III) $l_0 = 2, m_0 = 1$. By Theorem 15.8, (I) corresponds to case (a) above, and (II) and (III) correspond to the two possibilities that may occur in case (b): either the two conjugate representations are equivalent—case (II) with $\chi = 2\chi^{(\lambda_1)}$—or they are not—case (III) with $\chi = \chi^{(\lambda_1)} + \chi^{(\lambda_2)}$.

These considerations suggested the following question: given a representation σ over $\mathbb{R}$ that is irreducible over $\mathbb{R}$, how can one determine which of cases (I)–(III) applies to σ? In a joint paper also in 1906 [233], Schur and Frobenius provided a very satisfying answer by showing that the desired information can be obtained easily from the character χ of σ. That is, if $c(\sigma)$ is defined by

---

[30]See Section 5 of Loewy's paper [422]. Maschke's theorem is actually valid for any field with characteristic not dividing the order of the group—a fact that is immediately clear from the proof Schur gave using his lemma [523, §3]—but Loewy was not interested in fields of finite characteristic [422, p. 59n].

$$c(\sigma) = \begin{cases} +1 & \text{if } \sigma \text{ falls under case I;} \\ -1 & \text{if } \sigma \text{ falls under case II;} \\ 0 & \text{if } \sigma \text{ falls under case III.} \end{cases}$$

Then, as they showed,

$$c(\sigma) = \frac{1}{h} \sum_{R \in \mathfrak{H}} \chi(R^2).$$

The character expression for $c(\sigma)$ is sometimes called the *Frobenius–Schur indicator*. As they observed, using this expression, it is easy to give examples illustrating that all three cases do occur.

Theorem 15.8 does not exhaust the riches of Schur's index theory. Some other results of the theory that proved useful in applications to Maschke's Problem 15.6 are summarized in the following two theorems [524, §3, IXa, §4, XV]:

Theorem 15.9. *Let χ denote an irreducible character of $\mathfrak{H}$ with associated irreducible representation σ. Let m_0 denote the index of χ with respect to $\mathbb{F}$. Suppose that $\mathfrak{G}$ is a subgroup of $\mathfrak{H}$ with the following properties: (i) $\mathfrak{G}$ has an irreducible representation τ that is rationally representable in $\mathbb{F}$; (ii) the complete reduction of $\sigma|_{\mathfrak{G}}$, the restriction of σ to $\mathfrak{G}$, contains τ $r > 0$ times. Then r must be a multiple of m_0.*

The second theorem is stated in terms of groups of linear transformations so as to apply directly to Burnside's Theorem 15.7.

Theorem 15.10. *Let $\mathfrak{H}$ be an irreducible group of linear transformations. Suppose that $\mathfrak{H}$ contains a normal subgroup $\mathfrak{G}$ of prime index p. Then either $\mathfrak{G}$ is also irreducible or its complete reduction over $\mathbb{C}$ involves p irreducible components of degree p, no two of which are equivalent.*

Schur applied these theorems to rederive Burnside's theorem on Maschke's problem as well as to provide a solution to the problem for solvable groups. In Schur's terminology, Burnside's theorem was this: *If $\mathfrak{H}$ is an irreducible group of linear transformations, then $\mathfrak{H}$ is rationally representable in $\mathbb{F} = \mathbb{Q}(\varepsilon)$, where ε is a primitive hth root of unity, provided there is no integer $b > 0$ such that for every transformation $T \in \mathfrak{H}$, each of its characteristic roots has multiplicity b or some multiple thereof.* To see how Schur dealt with Burnside's theorem, let $\chi(T) = \operatorname{tr} T$, $T \in \mathfrak{H}$, so that χ is an irreducible character of $\mathfrak{H}$, and let m_0 be the Schur index of χ with respect to $\mathbb{F} = \mathbb{Q}(\varepsilon)$. Then a minimal finite extension $\mathbb{F}(\mu)$ exists such that $\mathfrak{H}$ is rationally representable in $\mathbb{F}(\mu)$. Now, in the case under consideration,

$$n_0 = (\mathbb{F}(\mu) : \mathbb{F}) = m_0 l_0 = m_0, \tag{15.19}$$

because $l_0 = (\mathbb{F}(\chi_1, \ldots, \chi_k) : \mathbb{F}) = 1$. This by virtue of the fact that the field $\mathbb{F}(\chi_1, \ldots, \chi_k)$ is just $\mathbb{F} = \mathbb{Q}(\varepsilon)$, since χ_i is a sum of hth roots of unity for every $i = 1, \ldots, k$, and so all χ_i are in $\mathbb{Q}(\varepsilon)$. Thus *(15.19) implies that* $\mathfrak{H}$ *is rationally representable in* $\mathbb{F} = \mathbb{Q}(\varepsilon)$ *precisely when* $m_0 = 1$.

To make use of this connection, Schur observed that he could make use of Theorem 15.9. To this end, consider any transformation $T \in \mathfrak{H}$. Then T generates a cyclic subgroup $\mathfrak{G} \subset \mathfrak{H}$. Since $\mathfrak{G}$ is abelian, all its irreducible representations are one-dimensional and are given by Dedekind characters. (Recall that a Dedekind character ψ on a group $\mathfrak{G}$ is a nonzero complex-valued function on $\mathfrak{G}$ such that $\psi(PQ) = \psi(P)\psi(Q)$ for all $P, Q \in \mathfrak{G}$.) Consider, in particular, the Dedekind character ψ defined as follows. Let ρ denote a characteristic root of T and r its multiplicity. Define ψ on $\mathfrak{G}$ by $\psi(T^i) = \rho^i$. This is a Dedekind character on $\mathfrak{G}$. Since the values of Dedekind characters on $\mathfrak{G}$ are gth roots of unity, g the order of $\mathfrak{G}$, and since h is a multiple of g, they are also hth roots of unity. In other words, ψ, as an irreducible representation of $\mathfrak{G}$, is rationally representable over $\mathbb{F} = \mathbb{Q}(\varepsilon)$ and so satisfies (i) of Theorem 15.9. Also the complete reduction of $\mathfrak{G}$ into one-dimensional representations includes ψ precisely $r > 0$ times, since this reduction simultaneously diagonalizes T and its powers. Thus (ii) of Theorem 15.9 is also satisfied, so that its conclusion holds and r must be a multiple of m_0. Because the same reasoning applies to every root ρ of every $T \in \mathfrak{H}$, if m_0 were greater than 1, we could take $b = m_0$ in the hypothesis of Burnside's theorem. In other words, if the hypothesis of Burnside's theorem is assumed to hold, it must be that $m_0 = 1$, and so $\mathfrak{H}$ is rationally representable in $\mathbb{F} = \mathbb{Q}(\varepsilon)$, thereby establishing Burnside's theorem.

Schur was also able to use his index theory to solve Maschke's Problem 15.6 for any solvable group of linear transformations [524, §4, XIV]. Suppose that not every such group can be rationally represented in $\mathbb{Q}(\varepsilon)$. Let h_0 be the minimum order for which such groups exist and $\mathfrak{H}$ such a group. Without loss of generality, $\mathfrak{H}$ can be assumed irreducible. We will apply Theorem 15.9 with σ the irreducible representation defined by $\mathfrak{H}$, viz., $\sigma(T) = T$ for all $T \in \mathfrak{H}$. Let m_0 denote the index of σ with respect to $\mathbb{F} = \mathbb{Q}(\varepsilon)$, so that $m_0 > 1$ by hypothesis. Now, since $\mathfrak{H}$ is solvable, it contains a normal subgroup $\mathfrak{G}$ with prime index p. Let $\mathfrak{G} \sim \mathfrak{G}_1 \oplus \cdots \oplus \mathfrak{G}_q$ denote the complete reduction of $\mathfrak{G}$ over $\mathbb{C}$. Since $\mathfrak{G}$ is solvable and of order $h_0/p < h_0$, every $\mathfrak{G}_i$ is rationally representable in $\mathbb{Q}(\varepsilon)$. Now Theorem 15.10 says that either $q = 1$ ($\mathfrak{G}$ is irreducible) or $q = p$ and no two $\mathfrak{G}_i$ are equivalent. Thus if τ is the irreducible representation defined by $\mathfrak{G}$ in the first case and by any $\mathfrak{G}_i$ in the second case, then $\sigma|_{\mathfrak{G}}$ contains τ with multiplicity $r = 1$. But according to Theorem 15.9, r is a multiple of $m_0 > 1$, which is impossible. Thus no solvable group of linear transformations can exist that is not rationally representable in $\mathbb{Q}(\varepsilon)$. In this manner, Schur solved Maschke's problem for solvable groups. As we shall see in the following section, his student Richard Brauer completely solved Maschke's problem using his theory of modular characters.

Of Schur's subsequent arithmetic researches on representation theory, mention will be made of only one, a brief paper of 1919 [525].[31] It is of special interest from the historical viewpoint of this and the following section. In [525], Schur showed how the concept of a factor set, which had figured prominently in his theory of projective representations, could be slightly modified so as to apply to arithmetic aspects of representation theory. His student Richard Brauer then went on to make good use of the modified factor sets in his generalization of Schur's index theory (Section 15.6).

Schur's 1919 paper was inspired by a thought-provoking paper that same year by Andreas Speiser (1885–1970) on arithmetic aspects of representation theory [543]. Speiser's paper was based on the following considerations. Suppose σ is an irreducible representation of degree n of some finite group $\mathfrak{H}$. Then by Frobenius' results it is rationally representable in an algebraic number field $\mathbb{K}$, which may be taken as a normal extension of $\mathbb{F} = \mathbb{Q}(\chi_1, \dots, \chi_k)$, where the χ_i are the values of the character $\chi = \operatorname{tr}\sigma$ on the conjugacy classes of $\mathfrak{H}$. The question Speiser considered is this: When is σ rationally representable in $\mathbb{F} = \mathbb{Q}(\chi_1, \dots, \chi_k)$, that is, when is the Schur index m_0 of σ with respect to $\mathbb{Q}$ equal to 1? He observed that for each $S \in \operatorname{Gal}(\mathbb{K}/\mathbb{F})$, the conjugate irreducible representation σ^S (obtained by applying the automorphism S to all coefficients of the matrices of σ) has the same character χ and so is equivalent to σ. This means that for each $S \in \operatorname{Gal}(\mathbb{K}/\mathbb{F})$, a matrix M_S with coefficients in $\mathbb{K}$ exists such that $M_S^{-1}\sigma^S M_S = \sigma$. The matrix M_S is uniquely determined up to a constant factor. It is easily checked that the definition of M_S and M_T implies that $(M_S^T M_T)^{-1}\sigma^{ST}(M_S^T M_T) = \sigma$, where M_S^T denotes the matrix obtained from M_S by applying $T \in \operatorname{Gal}(\mathbb{K}/\mathbb{F})$ to each coefficient of M_S. Since $M_S^T M_T$ is uniquely determined up to a constant factor $r_{S,T} \in \mathbb{K}$, we may write

$$M_S^T M_T = r_{S,T} M_{ST}. \tag{15.20}$$

Suppose now that constants $c_S \in \mathbb{K}$ can be chosen such that the matrices $M'_S = c_S M_S$ satisfy (15.20) with $r_{S,T} = 1$ for all S, T. For example, Speiser showed that this was possible if the degree n of σ is odd and if χ is real-valued. Then in such cases, by virtue of (15.20) with $r_{S,T} = 1$, $\tilde{\sigma}(S) = M'_S$ satisfies

$$\tilde{\sigma}(S)^T \tilde{\sigma}(T) = \tilde{\sigma}(ST) \quad \text{for all} \quad S \in \mathfrak{G} = \operatorname{Gal}(\mathbb{K}/\mathbb{F}). \tag{15.21}$$

Speiser called any mapping $\tilde{\sigma} : \mathfrak{G} \to \mathrm{GL}(n, \mathbb{K})$ satisfying (15.21) a representation of degree n of $\mathfrak{G}$. It is a representation in the ordinary sense only when the coefficients of $\tilde{\sigma}(S)$ lie in $\mathbb{F}$ for all S, so that $\tilde{\sigma}(S)^T = \tilde{\sigma}(S)$. Hence, to avoid possible confusion, I will refer to Speiser's representations as *S-representations*.

[31] For Schur's other arithmetic researches on representation theory, see his *Abhandlungen* 1, 251–265 (1908), 295–311 (1909), 451–463 (1911).

Speiser's main discovery about his representations was that they are all "equivalent," in an appropriate sense of that word, to a sum of trivial representations. That is, he proved the following theorem.

Theorem 15.11. *Every S-representation $\tilde{\sigma}$ of degree n is S-equivalent to the representation $\tilde{\tau}(S) = I_n$, where I_n denotes the $n \times n$ identity matrix, and S-equivalence to I_n means that $P \in \mathrm{GL}(n, \mathbb{K})$ exists such that $P^S \tilde{\sigma}(S) P^{-1} = I_n$ for all $S \in \mathfrak{G}$.*

Applied to the above considerations, Theorem 15.11 shows that if the ordinary irreducible representation σ is such that $\tilde{\sigma}(S) = M_S$ satisfies (15.21), and so defines an S-representation, then a matrix P exists such that $P^S M_S P^{-1} = I_n$. But then $M_S = (P^S)^{-1} P$, and so

$$\sigma = M_S^{-1} \sigma^S M_S = ((P^S)^{-1})P)^{-1} \sigma^S (P^S)^{-1} P = P^{-1} P^S \sigma^S (P^S)^{-1} P,$$

which implies that

$$P\sigma P^{-1} = P^S \sigma^S (P^S)^{-1} = \left(P\sigma P^{-1}\right)^S.$$

In other words, $P\sigma P^{-1}$ is fixed by all $S \in \mathrm{Gal}(\mathbb{K}/\mathbb{F})$ and thus has its coefficients in $\mathbb{F}$. The representation σ is therefore rationally representable in $\mathbb{F} = \mathbb{Q}(\chi_1, \ldots, \chi_k)$.

By means of such considerations, Speiser proved that every irreducible representation of odd degree with a real-valued character is rationally representable over $\mathbb{F} = \mathbb{Q}(\chi_1, \ldots, \chi_k)$. In effect, this solved Maschke's Problem 15.6 for irreducible groups of linear transformations in an odd number of variables and with real-valued traces. He also pointed out in passing that Theorem 15.11, specialized to representations of degree one, yields a quick proof of the well-known Satz 90 of Hilbert's *Zahlbericht* [296, p. 149].

Schur's attention was caught by the relation (15.20), which Speiser had just stated verbally en route to the special case in which $r_{S,T} \equiv 1$. In the notation $\tilde{\sigma}(S) = M_S$, (15.20) takes the form

$$\tilde{\sigma}(S)^T \tilde{\sigma}(T) = r_{S,T} \tilde{\sigma}(ST), \quad \text{for all} \quad S, T \in \mathfrak{G} = \mathrm{Gal}(\mathbb{K}/\mathbb{F}), \tag{15.22}$$

which is analogous to the defining relation (15.16) for a projective representation. In fact, the numbers $r_{S,T}$ in (15.22) satisfy a relation analogous to the defining relation (15.17) for a factor set:

$$r_{S,T}^U r_{ST,U} = r_{S,TU} r_{T,U} \quad \text{for all} \quad S, T, U \in \mathfrak{G} = \mathrm{Gal}(\mathbb{K}/\mathbb{F}). \tag{15.23}$$

Here $r_{S,T}^U$ denotes the image of $r_{S,T}$ under U.

Schur realized that the ideas behind his theory of projective representations and factor sets could be transferred with suitable modifications to the present context by defining $\tilde{\sigma}$ in (15.22) to be a *representation with factor set* $r_{S,T}$. He showed that conversely, if g denotes the order of $\mathfrak{G} = \mathrm{Gal}(\mathbb{K}/\mathbb{F})$, then any g^2 nonzero numbers $r_{S,T} \in \mathbb{K}$ that satisfy (15.23) constitute the factor set for a representation $\tilde{\sigma}$

of degree g. As in the projective case, Schur constructed the regular representation for the factor set. Recall that in the projective case, two representations σ, σ' were considered identical when constants c_S existed such that $\sigma'(S) = c_S\sigma(S)$ for all $S \in \mathfrak{G}$. As we saw above, Speiser had in effect defined his S-representation $\tilde{\sigma}$ by such a multiplication: $\tilde{\sigma}(S) = M'_S = c_S M_S$, $c_S \in \mathbb{K}$. Schur thus declared two representations $\tilde{\sigma}$, $\tilde{\sigma}'$ with factor sets $r_{S,T}$, $r'_{S,T}$ to be essentially the same when $c_S \in \mathbb{K}$ exist such that $\tilde{\sigma}'(S) = c_S\tilde{\sigma}(S)$ for all $S \in \mathfrak{G} = \mathrm{Gal}\,(\mathbb{K}/\mathbb{F})$. The factor sets then satisfy a relation analogous the relation (15.18) for associated factor sets in the projective case, namely

$$r'_{S,T} = \frac{c_S^T c_T}{c_{ST}}.$$

As in the projective case, such factor sets are said to be *associated*. Speiser's S-representations $\tilde{\sigma}$ thus correspond to Schur's representations with a factor set associated to $r_{S,T} \equiv 1$.

Expressed in the language of Schur's factor set representations, what Speiser had proved in Theorem 15.11 was that every factor set representation of degree n associated to the factor set $r_{S,T} \equiv 1$ is S-equivalent to the n-fold direct sum $\tilde{\tau} \oplus \cdots \oplus \tilde{\tau}$, where $\tilde{\tau}$ is the trivial (ordinary or factor set) representation $\tilde{\tau}(S) = 1$ for all $S \in \mathfrak{G} = \mathrm{Gal}\,(\mathbb{K}/\mathbb{F})$. Now $\tilde{\tau}$, being one-dimensional, is evidently irreducible in the sense of S-equivalence. Furthermore, it follows immediately that any two representations with factor sets associated to $r_{S,T} \equiv 1$ and of degree n, being S-equivalent to $\tilde{\tau} \oplus \cdots \oplus \tilde{\tau}$, must be S-equivalent to each other. Schur generalized these results as follows. First of all, any two representations with factor sets and of the same degree n must be S-equivalent. Moreover, corresponding to the class of representations with factor sets associated to a fixed factor set $r_{S,T}$ is an S-irreducible representation $\tilde{\tau}$ of this class such that every representation of the class is S-equivalent to a direct sum $\tilde{\tau} \oplus \cdots \oplus \tilde{\tau}$. This means that the degree m of $\tilde{\tau}$ divides the degree n of every representation of this class. In particular, since every factor set corresponds to a representation of degree g, namely the regular representation for the factor set, m divides as well the order of $\mathfrak{G} = \mathrm{Gal}\,(\mathbb{K}/\mathbb{F})$.

As we shall see in the next section, both Richard Brauer and Emmy Noether made good use of Schur's "arithmetic" theory of factor sets in their work on the theory of algebras. Nowadays, in addition to the use of factor sets to define (following Noether) crossed products of algebras, they are used (suitably generalized) to obtain all extensions $\mathfrak{G}$ of a group $\mathfrak{F}$ by an abelian normal subgroup $\mathfrak{N}$. Each of Schur's factor sets determines an extension of $\mathrm{Gal}\,(\mathbb{K}/\mathbb{F})$ by the group $\mathbb{K}^\times$, the multiplicative group of $\mathbb{K}$.[32]

[32]See *Encyclopedic Dictionary of Mathematics*, 2nd edn., Art. 190.N.

15.6 R. Brauer

During his tenure as full professor in Berlin (1919–1935), Schur had a total of 22 doctoral students, including several who went on to become reputable mathematicians [406, pp. 103–104]. Among the latter, Richard Brauer (1901–1977) was one of the most distinguished, the one who most carried on the tradition of Frobenius and Schur by making fundamental contributions to the representation theory of finite groups.[33] Brauer began to attend Schur's lectures in 1920, and in 1922 began participating in his seminar. In 1924, Schur had begun his study of the representations of the special orthogonal group (Section 15.5.3). This work was followed by several further papers in which Schur also considered the full orthogonal group. Brauer's doctoral dissertation (1925) provided a purely algebraic derivation of the irreducible characters of these groups, that is, one that did not utilize integration.

15.6.1 Generalized index theory

Brauer's next area of research (as instructor at the University of Königsberg) was more significant. It involved applying Schur's index theory to the study of hypercomplex number systems over more general fields than the complex numbers, including in particular finite fields. This approach to hypercomplex systems had been introduced by Wedderburn, who in 1907 had generalized the results of Molien and Cartan to arbitrary fields.[34] The major difference with the theory for $\mathbb{C}$ was contained in Wedderburn's theorem that if $\mathcal{H}$ is simple over the field $\mathbb{F}$, then $\mathcal{H} \cong M_n(\mathcal{D})$, where $M_n(\mathcal{D})$ denotes the complete matrix algebra of all $n \times n$ matrices with coefficients from $\mathcal{D}$, a division algebra (or skew field) over $\mathbb{F}$.

Brauer succeeded in generalizing Schur's theory to algebras as follows. Suppose first that $\sigma^{(\lambda)}$ is an irreducible representation of $\mathfrak{H}$ that is rationally representable in the field $\mathbb{L} = \mathbb{F}(\mu)$. Without loss of generality it is assumed that the values of the corresponding character are already contained in $\mathbb{F}$, so that, in the notation of Section 15.5.4, $l = 1$. Such a field $\mathbb{L}$ was called a *splitting field* for $\sigma^{(\lambda)}$ over $\mathbb{F}$ or for the corresponding group algebra $\mathcal{H} = \mathbb{F}\mathfrak{H}^{(\lambda)}$, where $\mathfrak{H}^{(\lambda)} = \sigma^{(\lambda)}[\mathfrak{H}] \subset M_n(\mathbb{C})$. The algebra $\mathcal{H}$ is simple and "central," that is, its center consists of only the scalar multiples of the identity. The Schur index m_0 of $\sigma^{(\lambda)}$ over $\mathbb{F}$ is then the minimum of $[\mathbb{L} : \mathbb{F}]$ over all such splitting fields $\mathbb{L}$. Now, when $\mathbb{L}$ is a splitting field, it follows

[33]The following presentation of Brauer's work draws heavily on the articles on Brauer by Feit [160] and (especially) Green [256]. Not long after I had written it, Curtis' book *Pioneers of Representation Theory* [109] appeared. The final two chapters provide a much more extensive and mathematically detailed treatment of Brauer's work. In particular, Brauer's theory of blocks and its application to the theory of finite groups [109, Ch. VII, §3] is not covered in my presentation.

[34]Wedderburn's work and its historical background have been treated in detail by Parshall [461].

that

$$\mathbb{L}\mathfrak{H}^{(\lambda)} \cong M_n(\mathbb{L}).$$

Since also

$$\mathbb{L}\mathfrak{H}^{(\lambda)} \cong \mathbb{L} \otimes_{\mathbb{F}} \mathcal{H},$$

these two isomorphisms suggest how to define a splitting field over $\mathbb{F}$ for any central simple algebra $\mathcal{H}$ over $\mathbb{F}$, where $\mathbb{F}$ is an arbitrary field with algebraic closure $\overline{\mathbb{F}}$: Say that $\mathbb{L} \subset \overline{\mathbb{F}}$ is a splitting field for $\mathcal{H}$ if

$$\mathbb{L} \otimes_{\mathbb{F}} \mathcal{H} \cong M_n(\mathbb{L}).$$

Then the Schur index of $\mathcal{H}$ may be defined as before as the minimum of $(\mathbb{L} : \mathbb{F})$ over all splitting fields of $\mathcal{H}$.

In this way, Brauer generalized Schur's index theory and applied it to the study of algebras (as hypercomplex systems gradually came to be called). Not only did he re-prove Wedderburn's results (which brings to mind the way Frobenius re-proved the results of Molien and Cartan with tools drawn from the representation theory of his day), he also showed, for example, that if $\mathcal{H}$ is a central simple algebra, so that $\mathcal{H} \cong M_n(\mathcal{D})$ by Wedderburn's theorem, then $\dim_{\mathbb{F}} \mathcal{D} = m_0^2$, where m_0 is the Schur index of $\mathcal{H}$ with respect to $\mathbb{F}$. And with the assistance of Emmy Noether (1882–1935), who was also studying algebras but using the tools of her theory of modules, they obtained in 1927 the following characterization of splitting fields, which Schur presented to the Berlin Academy [37]: *If* $\mathbb{L}$ *is a splitting field for* $\mathcal{H}$*, so that* $(\mathbb{L} : \mathbb{F}) = m_0 r$ *for some* r*, then* $\mathbb{L}$ *is isomorphic to a maximal subfield of* $M_r(\mathcal{D})$*. Conversely, if* $\mathbb{L}$ *is any maximal subfield of* $M_r(\mathcal{D})$*, then it is isomorphic to a splitting field for* $\mathcal{H}$ *and* $(\mathbb{L} : \mathbb{F}) = m_0 s$*, where* $s \mid r$*.*

In their work on algebras, both Brauer and Noether utilized another idea of Schur's, namely his arithmetic theory of factor sets [525] discussed at the end of Section 15.5. Noether used these factor sets to create her crossed product algebras, while Brauer used a generalized version (to deal with finite separable, but not necessarily normal, extensions) throughout his work relating to Schur's index theory. In particular, just as Schur had used his original factor sets to define the multiplier group $\mathfrak{M}$, so too Brauer used his factor sets to construct a group that will be denoted here by $\mathfrak{H}_{\mathbb{L}}(\mathbb{F})$. Underlying Brauer's work was also another group, which has since become known as the *Brauer group*. This group, here denoted by $\mathfrak{B}(\mathbb{F})$, consists of equivalence classes of algebras over $\mathbb{F}$ that are simple and central. By Wedderburn's theorem, such an algebra is isomorphic to a complete matrix algebra over a division algebra $\mathcal{D}$. Two such algebras $\mathcal{H}$, $\mathcal{H}'$ are in the same equivalence class when $\mathcal{D} \cong \mathcal{D}'$. If $\mathcal{C}(\mathcal{H})$ and $\mathcal{C}(\mathcal{H}')$ denote two such classes, then multiplication is defined by $\mathcal{C}(\mathcal{H}) \circ \mathcal{C}(\mathcal{H}') = \mathcal{C}(\mathcal{H} \otimes_{\mathbb{F}} \mathcal{H}')$, which is well defined and makes $\mathfrak{B}(\mathbb{F})$ into a group with identity element $\mathcal{C}(\mathbb{F})$. A subgroup $\mathfrak{B}_{\mathbb{L}}(\mathbb{F}) \subset \mathfrak{B}(\mathbb{F})$ is formed by all $\mathcal{C}(\mathcal{H})$ for which $\mathbb{L}$ is a splitting field for $\mathcal{H}$. Central to Brauer's application of Schur's index theory to algebras is the isomorphism $\mathfrak{B}_{\mathbb{L}}(\mathbb{F}) \cong \mathfrak{H}_{\mathbb{L}}(\mathbb{F})$.

After the work of Eilenberg, Mac Lane and Steenrod laid the foundation for homological algebra in the mid-1940s, it was possible to express Brauer's isomorphism (in the special case that $\mathbb{L}$ is a Galois extension of $\mathbb{F}$) as $\mathfrak{B}_{\mathbb{L}}(\mathbb{F}) \cong H^2(\mathfrak{G}, \mathbb{L}^\times)$, where $\mathfrak{G} = \mathrm{Gal}\,(\mathbb{L}/\mathbb{F})$ and $\mathbb{L}^\times$ denotes the multiplicative group of $\mathbb{L}$. This isomorphism, together with the isomorphism $H^1(\mathfrak{G}, \mathbb{L}^\times) \cong 0$ (which has its roots in Satz 90 of Hilbert's *Zahlbericht* [296]) became the foundation stones of Galois cohomology, which, particularly through class field theory, has had a great influence on the theory of numbers. In a sense, one could say that representation theory which in the Dedekind–Frobenius tradition owed its inspiration to number theory, was paying back its debt through its role in the creation of Galois cohomology.

15.6.2 Modular representations

In 1933, shortly after he had completed his researches on algebras and Schur's index theory, the anti-Semitic policies of the Nazis led to Brauer's dismissal from his position at Königsberg. Two years later, the same policies forced Schur to resign his professorship at Berlin. Although Schur continued to think about mathematical problems, he was severely depressed as a result of those policies and felt he had no right to publish his discoveries either in- or outside Germany.[35] In 1939 Schur and his family emigrated to Palestine, where he died in 1941 at the age of 66. Brauer emigrated to America, where after a brief stay at the University of Kentucky (1933–1934) and then at the newly founded Institute for Advanced Study in Princeton as Weyl's assistant (1934–1935), he became an assistant professor at the University of Toronto, where he remained until 1948. After a period as professor at the University of Michigan (1948–1952), he moved to Harvard University, where he remained until his retirement in 1971. During his years in America, Brauer made many important contributions to representation theory and its applications. Perhaps the most radical of these had to do with the theory of representations over fields of characteristic p.

L.E. Dickson (1874–1954) had initiated the study of the "modular" theory (as he named it) in 1902 [122, 123]. If $\mathbb{K}_p$ is an algebraically closed field of characteristic p, then, as Dickson showed, the ordinary representation theory over $\mathbb{C}$ may be imitated over $\mathbb{K}_p$ as long as p does not divide the order h of the group $\mathfrak{H}$ being studied [122]. By considering examples of specific groups, Dickson also showed that the case $p \mid h$ is truly exceptional [123]. For example, he showed that when $\mathfrak{H}$ is the symmetric group $\mathfrak{S}_3$ of order 6 and $p = 2$, the regular representation ρ of $\mathfrak{H}$

[35] See the personal reminiscences of Alfred Brauer, Richard Brauer's brother, on p. vii of Schur's *Abhandlungen* 1.

over a suitable field $\mathbb{F}_2$ is equivalent to

$$\begin{pmatrix} \sigma & 0 & 0 \\ 0 & \tau & 0 \\ 0 & 0 & \tau \end{pmatrix},$$

where τ is a second-degree irreducible representation, and $\sigma(R) = \left(\begin{smallmatrix} 1 & 0 \\ a & 1 \end{smallmatrix}\right)$, where $a = 1$ or $a = 0$, depending on whether R is or is not a 2-cycle. Although Dickson did not say it explicitly, it follows that σ is reducible but not completely reducible (since when $a = 1$, $\sigma(R)$ cannot be diagonalized). This simple example illustrates some other exceptions that are related to the failure of the complete reducibility theorem. The representation σ is not equivalent to the representation $\sigma'(R) \equiv I_2$, where I_2 is the 2×2 identity matrix, and yet $\chi(R) = \operatorname{tr}\sigma(R) = \operatorname{tr}\sigma'(R) = \chi'(R)$ for all R, a situation that cannot occur in the ordinary theory, whereby two representations with identical characters are necessarily equivalent.

Five years later, Dickson published two papers [124, 125] in which he attempted to say what he could of a positive nature about modular representations when $p \mid h$. He showed that even in this case, if σ is an irreducible representation, then its group determinant, $\det\sigma(x) = \det(\sum_{R\in\mathfrak{H}} \sigma(R)x_R)$, is still irreducible as a polynomial over $\mathbb{K}_p$. Although he did not state it explicitly, his result has the following implication. Even without the complete reducibility theorem, it follows (as Dickson probably realized) that every representation σ is equivalent to one of the form

$$\begin{pmatrix} \sigma_1 & 0 & \cdots & 0 \\ * & \sigma_2 & \cdots & 0 \\ \vdots & \vdots & \ddots & \vdots \\ * & * & \cdots & \sigma_r \end{pmatrix}, \tag{15.24}$$

where the σ_i are irreducible representations of $\mathfrak{H}$. To distinguish (15.24) from the complete reducibility theorem, the σ_i in (15.24) will be called the *irreducible constituents* of σ rather than the irreducible components. Equation (15.24) implies that if $\sigma(x) = \sum_{R\in\mathfrak{H}} \sigma(R)x_R$ is the group matrix, then $\det\sigma(x) = \prod_{i=1}^{r} \det\sigma_i(x)$. Since by Dickson's theorem, $\det\sigma_i(x)$ is irreducible over $\mathbb{K}_p$ for any p, the above is the prime factorization of the group determinant associated to the representation σ. As in Frobenius' theory, each irreducible factor of a group determinant corresponds to an irreducible representation—one of the irreducible constituents in (15.24).

In Frobenius' theory, however, the character $\chi^\sigma = \operatorname{tr}\sigma$ determines the irreducible components σ_i and hence the prime factors of $\det\sigma(x)$, but this is not so in the modular theory when $p \mid h$: two representations over $\mathbb{K}_p$ can have identical characters yet different irreducible constituents. This is illustrated by one of Dickson's main theorems: If the order of $\mathfrak{H}$ is $h = p^e q$, where p is prime and $(p,q) = 1$, so that p^e is the largest power of p dividing h, and if ρ denotes the regular representation of $\mathfrak{H}$ over $\mathbb{K}_p$, then ρ is equivalent to a matrix of the form

(15.24) in which each block σ_i is the same, so $\sigma_i = \sigma$ for all i, and there are p^e such blocks, so σ is $q \times q$. Here σ is not necessarily irreducible. It follows that for all $R \in \mathfrak{H}$, $\chi^\rho(R) = \operatorname{tr}\rho(R) = p^e \operatorname{tr}\sigma(R) = 0$. If τ is the trivial h-dimensional representation $\tau(R) \equiv I_h$, then $\chi^\tau(R) = \operatorname{tr}\tau(R) = h = p^e q = 0$ for all R, so that χ^ρ and χ^τ are identical. In general, however, ρ and τ will have different irreducible constituents (as is illustrated by the example of $\mathfrak{S}_3$ discussed above). Although Dickson did not explicitly make this observation, he did point out that in the above case, $\Theta(x) = \det\rho(x) = \det[\sigma(x)]^{p^e}$, where $\det\sigma(x)$ has degree q, "in marked contrast to the nonmodular theory, in which every algebraically irreducible factor of $[\Theta]$ ... enters to a power exactly equal to its degree" [125, p. 389].

Whether or not Dickson recognized all the above disappointing features of the modular characters when $p \mid h$, he understandably did not propose abandoning them, since that would make sense only if he had a superior replacement at hand. Indeed, he obtained some relatively encouraging results about linear characters, that is, characters of one-dimensional representations. They satisfy $\chi(AB) = \chi(A)\chi(B)$ and will be called (as in previous sections) *Dedekind characters*. Recall from Section 12.2 that Dedekind had shown that if $\mathfrak{H}$ is abelian of order h, then there are exactly h Dedekind characters and that he had conjectured (and Frobenius had proved) that when $\mathfrak{H}$ is not abelian, the number of Dedekind characters equals the order of $\mathfrak{H}/\mathfrak{H}'$, where $\mathfrak{H}'$ is the commutator subgroup of $\mathfrak{H}$ (Section 12.3 and Theorem 13.3). When $h = p^e q$ (as above), Dickson showed that over $\mathbb{K}_p$, an abelian group $\mathfrak{H}$ has q Dedekind characters, and if $\mathfrak{H}$ is not abelian, then the number of Dedekind characters equals the order of $\mathfrak{H}/\mathfrak{H}'$ divided by p^f, where p^f is the largest power of p that divides the order of $\mathfrak{H}/\mathfrak{H}'$ [124, p. 484]. These results, at least, were reasonable analogues of results in the nonmodular theory.

After Dickson, the modular theory of characters and representations does not seem to have received much attention for almost two decades. Since the overall thrust of Dickson's work was that the case $p \mid h$ was problematic in the sense that many of the theorems of the ordinary theory did not have nice analogues when $p \mid h$, this is not surprising. It should be mentioned, however, that in his 1923 book on finite groups, Andeas Speiser devoted a few pages to modular representations,[36] and for the case $p \mid h$ stated and proved a theorem that is analogous to Dickson's last mentioned theorem. Speiser also proved a theorem about modular representations when $(p, h) = 1$, which will be discussed below.

Brauer's interest in the modular theory was triggered by a remark Schur made to him on one of Brauer's trips to Berlin.[37] Schur surprised him by suggesting that they collaborate on a book covering all aspects of representation theory. A year or so later, Schur decided that he was too busy to undertake such a project, but he encouraged Brauer to do it with the help of a young physicist who could deal with Wigner's groundbreaking application (1931) of representation theory to quantum mechanics [606]. These plans fell through when the Nazis gained power and Brauer

---

[36]See Section 59 of the 1923 edition [544] or Section 69 of the second edition of 1927 [545].

[37]According to Brauer's own recollections, *Papers* 1, p. xviii.

was forced to leave Germany, but in a 15-page monograph of 1935 [32], he began writing about modular representations. From the outset, he had more success than Dickson. Recall from the end of Section 13.1 Frobenius' elation when he was on the brink of proving that the number of irreducible representations of a group equals the number k of its conjugacy classes. Dickson had obtained nothing comparable, but Brauer was able to show that if $\mathfrak{H}$ has order $h = p^e q$—as usual, p is a prime and $(p,q) = 1$—then the number of irreducible representations of $\mathfrak{H}$ over $\mathbb{K}_p$ equals k_p, the number of conjugacy classes of $\mathfrak{H}$ for which the order of the elements is relatively prime to p. He also proved that every irreducible representation over $\mathbb{K}_p$ occurs as an irreducible constituent of the regular representation.

Although Brauer had thus obtained results that, like Dickson's theorems on Dedekind characters, paralleled the results of the ordinary theory, it was still the case, as we have seen, that many of the basic theorems of the ordinary theory do not hold in any analogous sense when $p \mid h$. Nonetheless, Brauer went on to show, in joint work published in 1937 in collaboration with his student C.J. Nesbitt (1912–2001) that a substantial theory could be developed when $p \mid h$ [35, 36]. One of the salient features of this theory is that just as Frobenius had ended up changing the definition of a character in order to deal with noncommutative groups, so also in the new theory, the definition of a modular character is changed in order to deal with the case $p \mid h$.

The key lemma supporting the change in definition is the following [36, p. 5].

Lemma 15.12. *If σ and τ are two representations of $\mathfrak{H}$ over $\mathbb{K}_p$ such that $\sigma(R)$ and $\tau(R)$ have the same characteristic roots for each $R \in \mathfrak{H}$, then σ and τ have the same irreducible constituents.*

Now, over a field of characteristic zero, the hypothesis of this lemma follows from the assumption that $\chi^{\sigma}(R) = \chi^{\tau}(R)$ for all $R \in \mathfrak{H}$. This is because, e.g., for every positive integer l, $\chi^{\sigma}(R^l) = \operatorname{tr}\sigma(R^l) = \operatorname{tr}[\sigma(R)^l]$ is the sum s_l of the lth powers of the characteristic roots of $\sigma(R)$. But if σ is a representation of degree n, the coefficients a_i of the characteristic polynomial $f(\lambda) = \det[\lambda I - \sigma(R)] = \lambda^n + a_1\lambda^{n-1} + \cdots + a_n$, and hence its roots, are completely determined by the sums s_l, $l = 1, \ldots, n$, since by Newton's identities,

$$la_l = -(s_l + a_1 s_{l-1} + a_2 s_{l-2} + \cdots + a_{l-1}s_1), \quad l = 1, \ldots, n.$$

Thus $\chi^{\sigma}(R) = \chi^{\tau}(R)$ implies that the characteristic roots of $\sigma(R)$ and $\tau(R)$ are identical. Of course, over $\mathbb{K}_p$ this argument fails, since $l \equiv 0 \pmod{p}$ is possible in Newton's identities.

With this in mind, however, let as usual the order h of $\mathfrak{H}$ be written as $p^e q$ with $(p,q) = 1$. Then if $R \in \mathfrak{H}$ has order divisible by p, it turns out that R can be written as $R = ST$, where S and T commute and S has order p^a $(1 \leq a \leq e)$ and the order of T is relatively prime to p. Thus if σ is a representation of $\mathfrak{H}$, $\sigma(R) = \sigma(S)\sigma(T)$. Since $\sigma(S)^{p^a} = I$, the characteristic roots of $\sigma(S)$ are p^ath roots of unity and thus equal to 1 in $\mathbb{K}_p$.[38] Since the characteristic roots of $\sigma(R)$ are products of the roots of $\sigma(S)$ and $\sigma(T)$, it follows that in $\mathbb{K}_p$, $\sigma(R)$ has the same characteristic roots as $\sigma(T)$. Consequently, if σ and τ are two representations over $\mathbb{K}_p$ such that $\sigma(T)$ and $\tau(T)$ have the same roots for all elements T with order relatively prime to p, then they have the same roots for all $R \in \mathfrak{H}$, and so Lemma 15.12 implies that they have the same irreducible constituents.

Now, if $R = ST$ as above, then T has an order that divides q, so that $\sigma(T)^q = \sigma(T^q) = I$, which means that the roots of $\sigma(T)$ are qth roots of unity with respect to $\mathbb{K}_p$. The qth roots of unity with respect to $\mathbb{K}_p$ form a cyclic group of order q under multiplication. Likewise, the qth roots of unity in the field $\mathbb{C}$ of complex numbers form a cyclic group of order q. Since these two cyclic groups are isomorphic, let φ denote any isomorphism from the former to the latter group. Then, since the characteristic roots λ_i of $\sigma(T)$ are qth roots of unity with respect to $\mathbb{K}_p$, the numbers $\varphi(\lambda_i) \in \mathbb{C}$ are also qth roots of unity. If we set

$$\psi^\sigma(T) \stackrel{\text{def}}{=} \sum_i \varphi(\lambda_i), \quad \text{where} \quad \chi^\sigma(T) = \operatorname{tr}[\sigma(T)] = \sum_i \lambda_i,$$

then the above remarks involving Newton's identities imply that if $\psi^\sigma(T) = \psi^\tau(T)$ for all $T \in \mathfrak{H}$ of order dividing q, then $\sigma(T)$ and $\tau(T)$ have the same characteristic roots (since their images under φ are the same), and so σ and τ have the same irreducible constituents. By virtue of these considerations, it was deemed "advisable to change the point of view" and to consider the complex-valued function ψ^σ as the modular character of σ rather than χ^σ [36, p. 12]. The characters ψ^σ (which are defined only on the $T \in \mathfrak{H}$ with orders dividing q) are now usually known as the Brauer characters of $\mathfrak{H}$. They play a role in the modular theory for $p \mid h$ akin to the role played by the ordinary characters of Frobenius' theory.

The details of the remarkable modular theory that Brauer with Nesbitt's assistance developed around the new notion of a character are too intricate to describe here, but a few general remarks are in order. As we saw in Section 15.3, Frobenius did not regard the theory of hypercomplex systems as an essential tool in the investigation of group representation theory. Indeed, to develop the theory of such systems, he imported the determinant-theoretic tools and concepts that he had developed to deal with group representations. By contrast, the modular theory of Brauer and Nesbitt was based on the theory of algebras. Since when $p \mid h$, the group algebra $\mathcal{H} = \mathbb{K}_p\mathfrak{H}$ is no longer semisimple, it was the more general theory of nonsemisimple hypercomplex systems (or algebras) that was needed. The reliance on the theory of algebras is connected with another feature of the theory, namely that it emphasized establishing relations between the ordinary characters and representations and their modular counterparts, with an eye toward using the modular theory to gain information about the ordinary characters and more generally about the structure of finite groups.

---

[38] If $\omega^{p^a} = 1$, then the binomial expansion implies that $(\omega - 1)^{p^a} \equiv 0 \pmod{p}$, and so $\omega = 1$ in $\mathbb{K}_p$.

In his book, Speiser had already set the stage for the connection between the ordinary and modular theories. If σ is an ordinary representation of $\mathfrak{H}$, then it is rationally representable in an algebraic number field $\mathbb{K}$, and in fact, the coefficients of the matrices may be taken as members of the ring $\mathfrak{o}_{\mathbb{K}}$ of algebraic integers of $\mathbb{K}$. This follows from Frobenius' results. Given a prime p, if $\mathfrak{p}$ is a prime ideal of $\mathfrak{o}_{\mathbb{K}}$ in the factorization of the ideal $p\mathfrak{o}_{\mathbb{K}}$ generated by p, then $\mathfrak{p}$ is maximal and $\mathbb{K}_p = \mathfrak{o}_{\mathbb{K}}/\mathfrak{p}$ is a finite field of characteristic p. By reducing the representation σ modulo $\mathfrak{p}$, we obtain a representation $\bar{\sigma}$ over $\mathbb{K}_p$. Speiser considered only the nonexceptional case $(p,h) = 1$, and showed that every modular representation arises in this way. For him, the thrust of this result seems to have been that the ordinary theory is the source of, the supplier of, the modular theory (at least when $(p,h) = 1$).[39] Brauer, however, was primarily interested in the case $p \mid h$ and in relations between ordinary and modular representations, which might be used to supply information about the ordinary characters and representations and, more generally, about finite groups.

The theory of algebras and, in particular, the Cartan invariants discussed in Section 15.3 following (15.12) enter into this avenue of research in the following manner. Suppose that $\mathbb{K}$ is a sufficiently large algebraic number field so that every ordinary irreducible representation of $\mathfrak{H}$ has an integral representation (as described above) in $\mathbb{K}$, and denote these irreducible representations by $\sigma_1, \ldots, \sigma_k$. Let $\tau_1, \ldots, \tau_l$ denote the distinct irreducible modular representations over $\overline{\mathbb{K}}$, where $\overline{\mathbb{K}}$ is a minimal algebraically closed field containing the field $\mathbb{K}_p = \mathfrak{o}_{\mathbb{K}}/\mathfrak{p}$. Then over $\overline{\mathbb{K}}$, $\bar{\sigma}_i$ has a decomposition of the form (15.24) with the τ_i on the diagonal. If we let d_{ij} denote the number of times τ_j occurs as an irreducible constituent of $\bar{\sigma}_i$, then $D = (d_{ij})$ is a $k \times l$ matrix. Now $C = (c_{ij})$, the matrix of Cartan invariants for the group algebra $\mathcal{H} = \overline{\mathbb{K}}\mathfrak{H}$, is $l \times l$, and it turns out that $C = D^t D$. The relation $C = D^t D$ is fundamental to Brauer's theory. It shows, in particular, that C is symmetric, so that $c_{ij} = c_{ji}$ for all i and j. In the light of Cartan's formula (15.12), which remains valid for algebras over any algebraically closed field, this shows that the group algebra $\mathcal{H} = \overline{\mathbb{K}}\mathfrak{H}$, while not semisimple, is a Frobenius algebra.

Among the many results about ordinary representations and about the structure of finite groups that Brauer obtained using his modular theory, the one most pertinent to the developments traced in the preceding section is the completely general solution to Maschke's Problem 15.6 that he gave in 1945 [33]. As we have seen, in addition to Maschke, exceptional mathematicians such as Burnside, Schur, and Speiser, all using ordinary characters, had worked on this problem but succeeded in solving it only in special cases. Brauer, using the newly defined modular characters, succeeded

[39] After giving his proof, Speiser wrote "damit ist bewiesen, dass wir in den irreduziblen algebraischen Darstellungen alle irreduziblen Darstellungen im GF(p^n) gefunden haben" [1923:167; 1927:223].

in solving it with no restrictions: *Every group* $\mathfrak{H} \subset \mathrm{GL}(n,\mathbb{C})$ *of order h is rationally representable in* $\mathbb{Q}(\varepsilon)$, *where* ε *is a primitive hth root of unity.*[40]

15.6.3 Artin L-functions and induction theorems

We have seen that the theory of group characters and representations in the tradition of Dedekind and Frobenius had its roots in arithmetic, although it was only the Dedekind characters of abelian groups that had found arithmetic applications. It is thus fitting to conclude the discussion of Frobenius' theory of group characters and its aftermath by noting how that theory was successfully reunited with the theory of numbers. We saw in Section 12.2 that one of the sources of Dedekind and Weber's notion of a character for an abelian group had been the L-series of Dirichlet (12.8). Indeed, as we saw in Section 12.2, when Dedekind introduced the general notion of a character on a finite abelian group $\mathfrak{H}$, he also noted that when $\mathfrak{H}$ is the ideal class group associated to an algebraic number field $\mathbb{K}$, then each character χ on $\mathfrak{H}$ has associated to it an L-function and series, viz.,

$$L(s,\chi) = \prod_{\mathfrak{p}} \left[1 - \frac{\chi(H_{\mathfrak{p}})}{N(\mathfrak{p})^s}\right]^{-1} = \sum_{\mathfrak{p}} \frac{\chi(H_{\mathfrak{p}})}{N(\mathfrak{p})^s}, \tag{15.25}$$

where $H_{\mathfrak{p}} \in \mathfrak{H}$ is the ideal class containing $\mathfrak{p}$.

By 1920, algebraic number theory had developed far beyond Dedekind's version of it, due largely to the influence of Hilbert's groundbreaking work in this area. In particular, various L-functions defined by characters on generalized ideal class groups were studied, in a line of development that later became known as class field theory,[41] to obtain generalizations of Dirichlet's theorem on the infinity of primes in arithmetic progressions (Weber, Hecke). These generalized ideal class groups were also finite abelian groups, and I will refer to the associated L-functions as *abelian* L-functions.

L-functions for nonabelian groups were first introduced in 1923 by Emil Artin (1898–1962), who had just become an instructor at the University of Hamburg. The work of two mathematicians, Teiji Takagi (1875–1960) and Frobenius, seems to have been the principal source of inspiration for what he did. In the case of Takagi, it was his 1920 paper containing what is now called the Takagi existence theorem. In this remarkable paper Takagi showed that every abelian Galois extension $\mathbb{L}$ of an algebraic number field $\mathbb{K}$ is a "class field" in the sense that there is a generalized ideal class group $\mathfrak{H}^*$ associated to $\mathbb{L}/\mathbb{K}$ that is isomorphic to $\mathrm{Gal}(\mathbb{L}/\mathbb{K})$. Takagi

[40]In 1947, Brauer used his induction theorem (discussed below) to show that ε can be taken as a primitive mth root of unity, where m is the least common multiple of the orders of elements in $\mathfrak{H}$. This proof did not use Brauer characters. See Brauer's *Collected Papers* **1**, 553.

[41]For a lucid historical overview of the history of class field theory and references to the primary and secondary literature, see Conrad's exposition [105].

established the isomorphism indirectly, i.e., without providing an explicit mapping between $\mathfrak{H}^*$ and Gal $(\mathbb{L}/\mathbb{K})$; but for reasons indicated below, Artin (1) was convinced that abelian L-functions could be identified with L-functions defined on Gal $(\mathbb{L}/\mathbb{K})$ for $\mathbb{L}/\mathbb{K}$ abelian and (2) realized consequently that the notion of an L-function on Gal $(\mathbb{L}/\mathbb{K})$ could be extended to Galois groups corresponding to nonabelian extensions $\mathbb{L}/\mathbb{K}$. With regard to both (1) and (2), it was the work of Frobenius that proved inspirational.

The first step in the direction of both (1) and (2) was supplied by the Frobenius automorphism theorem (Theorem 9.18), which by this time had been generalized to Galois extensions $\mathbb{L}$ of an algebraic number field $\mathbb{K}$.[42] Artin cited Weber's *Lehrbuch der Algebra* [584, §178] for the extension, which ran as follows. If $\mathfrak{p} \in \mathfrak{o}_{\mathbb{K}}$ and if $\mathfrak{P} \in \mathfrak{o}_{\mathbb{L}}$ is a prime divisor of $\mathfrak{p}$, then there is an automorphism $F_{\mathfrak{P}} \in \text{Gal}(\mathbb{L}/\mathbb{K})$ such that $F_{\mathfrak{P}}(\omega) \equiv \omega^{N(\mathfrak{p})} \pmod{\mathfrak{P}}$ for all $\omega \in \mathfrak{o}_{\mathbb{L}}$. Artin restricted his attention to $\mathfrak{p}$ that are unramified in the sense that $\mathfrak{p}$ does not divide the relative discriminant of $\mathbb{L}/\mathbb{K}$ and showed that in that case (as in Frobenius' original theorem), the automorphism $F_{\mathfrak{P}}$ is unique [6, p. 91].[43]

By 1923, the automorphism nowadays associated with the name of Frobenius was a well-known arithmetic tool that had been independently discovered by Frobenius, Dedekind, and Hilbert,[44] but in Artin's case, his interest in this automorphism was evidently linked to his interest in Frobenius' 1896 paper [210] containing his results from 1880 on density theorems (Section 9.3). In particular, he was interested in Frobenius' conjectured theorem, which is now known as the Chebotarev density theorem.[45] As far as Artin knew in 1923, Frobenius' conjectured theorem had still not been proved or disproved, and he was clearly interested in proving it. As we saw in Section 9.3.6, Frobenius' thoughts on his conjectured theorem were intimately tied up with the existence and uniqueness of Frobenius automorphisms for unramified rational primes p and the fact that each such p determined a conjugacy class $\mathfrak{C}_p$ of the associated Galois group, namely the class containing the Frobenius automorphisms $F_{\mathfrak{p}}, F_{\mathfrak{p}'}, \ldots$ of the prime factors $\mathfrak{p}, \mathfrak{p}', \ldots$ of p.

Artin knew that he could prove Frobenius' conjectured density theorem and also formulate a bona fide generalization of abelian L-functions to the case in which $\mathbb{L}$ is a nonabelian Galois extension of $\mathbb{K}$ by means of a theorem he could prove in many special cases but not in general. It concerns abelian extensions $\mathbb{L}/\mathbb{K}$ with $\mathfrak{H}^* = \{C_1, \ldots, C_n\}$ denoting the generalized ideal class group that is isomorphic to Gal $(\mathbb{L}/\mathbb{K})$ in accordance with Takagi's theorem. In this case, the Frobenius automorphisms $F_{\mathfrak{P}}, F_{\mathfrak{P}'}, \ldots$ of the prime divisors of $\mathfrak{p}$, being conjugate, are actually identical because Gal $(\mathbb{L}/\mathbb{K})$ is abelian; and so we may speak of the Frobenius automorphism $F_{\mathfrak{p}}$. Artin's conjectured theorem ran as follows [6,

---

[42] At this time, Artin still spoke of "substitutions" rather than automorphisms.

[43] Seven years later, in his second paper on generalized L-functions [8], Artin showed how to deal as well with the case of ramified primes $\mathfrak{p}$.

[44] See the footnote to Theorem 9.18.

[45] See following Theorem 9.20.

p. 98, Satz 2]: *(a)* $F_{\mathfrak{p}}$ *depends only on the ideal class* C_i *of* $\mathfrak{H}^*$ *in which* $\mathfrak{p}$ *lies; (b) this correspondence between ideal classes and automorphisms is one-to-one and determines the isomorphism between* Gal $(\mathbb{L}/\mathbb{K})$ *and* $\mathfrak{H}^*$. *Thus the composite of two automorphisms* $F_{\mathfrak{p}}, F_{\mathfrak{p}^*}$ *corresponds to the product of the corresponding ideal classes* C_i, C_{i^*}. As we shall see, this conjectured theorem was finally proved by Artin in 1927; it is now known as *Artin's general reciprocity theorem*.

Assuming this reciprocity theorem, the abelian L-function associated to $\mathfrak{H}^*$ can be thought of as defined with respect to Gal $(\mathbb{L}/\mathbb{K})$ by

$$L(s,\chi,\mathbb{L}/\mathbb{K}) = \prod_{\mathfrak{p}} \left[1 - \frac{\chi(F_{\mathfrak{p}})}{N(\mathfrak{p})^s}\right]^{-1}. \tag{15.26}$$

Furthermore, since χ is just an irreducible representation of the abelian group Gal $(\mathbb{L}/\mathbb{K})$, if we now assume that Gal $(\mathbb{L}/\mathbb{K})$ is not abelian, Frobenius' representations σ : Gal $(\mathbb{L}/\mathbb{K}) \to \mathbb{C}$ and associated characters $\chi = \operatorname{tr}\sigma$ now afford a natural generalization of the notion of an L-function. That is, if $\mathfrak{P}$ is any prime divisor of $\mathfrak{p}$, we may define

$$L(s,\chi,\mathbb{L}/\mathbb{K}) = \prod_{\mathfrak{p}} \left[\det\left(I - \frac{\sigma(F_{\mathfrak{P}})}{N(\mathfrak{p})^s}\right)\right]^{-1}. \tag{15.27}$$

The above definition is well defined, because as Artin pointed out, (15.27) is independent of the choice of the prime divisor $\mathfrak{P}$ of $\mathfrak{p}$. That is, since any two prime divisors $\mathfrak{P}, \mathfrak{P}'$ of $\mathfrak{p}$ have conjugate Frobenius automorphisms $F_{\mathfrak{P}}, F_{\mathfrak{P}'}$, with, e.g., $F_{\mathfrak{P}'} = H^{-1}F_{\mathfrak{P}}H$, $H \in$ Gal $(\mathbb{L}/\mathbb{K})$, it follows that $\sigma(F_{\mathfrak{P}'}) = [\sigma(H)]^{-1}\sigma(F_{\mathfrak{P}})\sigma(H)$ is similar to $\sigma(F_{\mathfrak{P}})$ and so the determinant in (15.27) is unchanged if $F_{\mathfrak{P}'}$ is used rather than $F_{\mathfrak{P}}$, i.e., (15.27) is independent of the choice of prime divisor $\mathfrak{P}$ of $\mathfrak{p}$ and so is well defined. Since the determinant of a complex number is just itself, (15.27) reduces to (15.26) when Gal $(\mathbb{L}/\mathbb{K})$ is abelian and σ is replaced by χ.

Artin proved that his L-functions have the following two properties [6, pp. 92–93].

(I) *If* $\chi = \sum_{i=1}^{n} q_i\chi_i$, *where the* χ_i *are characters of* Gal $(\mathbb{L}/\mathbb{K})$ *and the* q_i *are ordinary rational numbers, then*

$$L(s,\chi,\mathbb{L}/\mathbb{K}) = \prod_{i=1}^{n} L(s,\chi_i,\mathbb{L}/\mathbb{K})^{q_i}.$$

(II) *If* $\hat{\chi}$ *is the character of* Gal $(\mathbb{L}/\mathbb{K})$ *induced from the character* χ *of a subgroup* $\mathfrak{G} \subset$ Gal $(\mathbb{L}/\mathbb{K})$ *in accordance with Frobenius' theory (Section 15.1), then*

$$L(s,\hat{\chi},\mathbb{L}/\mathbb{K}) = L(s,\chi,\mathbb{L}/\mathbb{M}),$$

where $\mathbb{M}$ *is the field fixed by* $\mathfrak{G}$.

Since every representation of Gal $(\mathbb{L}/\mathbb{K})$ is an integral linear combination of irreducible representations, it follows from (I) that every L-function is a product, with possibly repeated factors, of L-functions of irreducible representations.

Propositions (I) and (II) combine to have a further significance by virtue of Artin's proof of what has become known as *Artin's induction theorem* [6, pp. 102ff.]:

(III) *Every character χ on a finite group $\mathfrak{H}$ is expressible in the form $\chi = \sum_{i=1}^{n} q_i \hat{\chi}_i$, $q_i \in \mathbb{Q}$, where each $\hat{\chi}_i$ is induced from a Dedekind (viz., linear) character χ_i on a cyclic subgroup $\mathfrak{G}_i$ of $\mathfrak{H}$.*

Thus, given any L-function $L(s,\chi,\mathbb{L}/\mathbb{K})$, we see that since $\chi = \sum_{i=1}^{n} q_i \hat{\chi}_i$ in accordance with (III) applied to $\mathfrak{H} = \text{Gal}\,(\mathbb{L}/\mathbb{K})$, (I)–(II) imply that

$$L(s,\chi,\mathbb{L}/\mathbb{K}) = \prod_{i=1}^{n} L(s,\chi_i,\mathbb{L}/\mathbb{M}_i)^{q_i}, \tag{15.28}$$

where $\mathbb{M}_i$ is the fixed field of the cyclic subgroup $\mathfrak{G}_i$ of Gal $(\mathbb{L}/\mathbb{K})$. Thus every Artin L-function is expressible as a product of rational powers of abelian L-functions. It followed from work of Hecke [279] that the abelian L-functions $L(s,\chi_i,\mathbb{L}/\mathbb{M}_i)$ can be analytically continued to the entire complex plane as meromorphic functions. If the q_i were integers, this would be true of $L(s,\chi,\mathbb{L}/\mathbb{K})$ as well, but Artin could not prove this.

Artin used his results on L-functions in conjunction with his conjectured general reciprocity theorem to prove Frobenius' conjectured density theorem [6, p. 106, Satz 4]. That same year, however, Chebotarev had independently given a proof of Frobenius' conjectured density theorem that, unlike Artin's, was free from any unproved assumptions. A German version was submitted to *Mathematische Annalen* in 1924 and appeared in 1926 [561]. Artin naturally read Chebotarev's proof and discovered therein the ideas he needed to give a completely general proof of his general reciprocity theorem the following year [7, p. 361].

With the proof of the general reciprocity theorem, Artin's theory of L-functions was established, but Artin believed that one of his results could be improved. He had proved (15.28), which shows that if m is the least common multiple of the denominators of the rational numbers q_i, then $L(s,\chi,\mathbb{L}/\mathbb{K})^m$ can be analytically continued to the entire complex plane as a meromorphic function. Artin believed that this was true of $L(s,\chi,\mathbb{L}/\mathbb{K})$ itself. He shared this conjecture with Brauer as well as his ideas on how it might be proved.[46] Specifically, he conjectured that his above induction theorem (III) could be replaced by a theorem asserting that if $\mathfrak{H}$ is a finite group, then every character χ on $\mathfrak{H}$ is expressible in the form $\chi = \sum_{i=1}^{l} n_i \hat{\chi}_i$, where the n_i are integers and each $\hat{\chi}_i$ is the character induced on $\mathfrak{H}$ from a Dedekind (viz., linear) character on a subgroup $\mathfrak{G}_i$ of $\mathfrak{H}$.

[46]See Brauer's remarks [34, pp. 502–503, 503, n. 3].

That such a theorem would suffice to show that $L(s,\chi,\mathbb{L}/\mathbb{K})$ has a meromorphic continuation to the entire complex plane can be seen as follows. In the first place, the above conjectured theorem together with (I)–(II) shows that in lieu of (15.28), we have

$$L(s,\chi,\mathbb{L}/\mathbb{K}) = \prod_{i=1}^{l} L(s,\chi_i,\mathbb{L}/\mathbb{M}_i)^{n_i}, \tag{15.29}$$

where now $\mathbb{M}_i$ is the field fixed by $\mathfrak{G}_i$, and χ_i is a Dedekind character on the (not necessarily abelian) subgroup $\mathfrak{G}_i$. It does not follow that the extension $\mathbb{L}/\mathbb{M}_i$ is abelian, but Artin also realized that his L-functions had the following property [34, p. 502, III]:

(IV) *Let σ be a representation of $\mathfrak{G} = \mathrm{Gal}\,(\mathbb{L}/\mathbb{K})$ with corresponding character $\chi = \mathrm{tr}\,\sigma$. Then if $\mathfrak{N}$ is the kernel of σ and $\mathbb{N}$ the field fixed by $\mathfrak{N}$, one has $L(s,\chi,\mathbb{L}/\mathbb{K}) = L(s,\chi',\mathbb{N}/\mathbb{K})$, where χ' is the character defined via χ on $\mathfrak{G}/\mathfrak{N} \cong \mathrm{Gal}\,(\mathbb{N}/\mathbb{K})$.*

If (IV) is applied to each factor in (15.29), we have

$$L(s,\chi_i,\mathbb{L}/\mathbb{M}_i) = L(s,\chi_i',\mathbb{N}_i/\mathbb{M}_i).$$

Now, χ_i is a Dedekind character on the subgroup $\mathfrak{G}_i$, which means that $\chi_i = \sigma$ in (IV); and of course, $\mathfrak{H} = \mathrm{Gal}\,(\mathbb{L}/\mathbb{K})$ is replaced by $\mathfrak{H} = \mathrm{Gal}\,(\mathbb{L}/\mathbb{M}_i) \cong \mathfrak{H}/\mathfrak{N}_i$, which is abelian, since the commutator subgroup $\mathfrak{H}'$ of $\mathfrak{H}$ is contained in the kernel $\mathfrak{N}_i$ of χ_i.[47] Thus (15.29) combined with (IV) and the fact that the χ_i are Dedekind characters means that each L-function in (15.29) is an abelian L-function and so has a continuation to the entire complex plane as a meromorphic function. Finally, since the n_i in (15.29) are integers, the same conclusion follows for the general L-function $L(s,\chi,\mathbb{L}/\mathbb{K})$.

Stimulated by Artin's conjectures, in 1946 Brauer [34] established the following more precise version of Artin's conjectured induction theorem.

Theorem 15.13. *Every character χ on a finite group $\mathfrak{H}$ is expressible as $\chi = \sum_{i=1}^{n} n_i\hat{\chi}_i$, where the n_i are ordinary integers and $\hat{\chi}_i$ is induced from a Dedekind character χ_i on an "elementary subgroup" $\mathfrak{G}_i \subset \mathfrak{H}$, that is, a group that is the direct product of a p-primary subgroup[48] of $\mathfrak{H}$ and a cyclic subgroup with order relatively prime to p.*

---

[47] Recall that $\mathfrak{H}'$ is generated by all products of the form $H_1H_2H_1^{-1}H_2^{-1}$, with $H_1,H_2 \in \mathfrak{H}$. Since χ_i is a Dedekind character, $\chi_i(H_1H_2H_1^{-1}H_2^{-1}) = \chi_i(H_1)\chi_i(H_2)\chi_i(H_1)^{-1}\chi_i(H_2)^{-1} = 1$ because complex numbers commute.

[48] A p-primary subgroup is a subgroup every element of which has order a power of p. Since $\mathfrak{H}$ is finite, the p-primary subgroup has order that is a power of p.

Although Brauer's theory of modular characters was not required for his proof of Theorem 15.13, it played a key role in motivating the proof [109, pp. 266–267].

In his paper, Artin actually made another conjecture about the analytic properties of his L-functions, which is now known as Artin's conjecture. Like Artin's conjecture that his L-functions are meromorphic in $\mathbb{C}$, it involves a generalization of a known property of abelian L-functions, and runs as follows: If χ is an irreducible character different from the principal character $\chi = 1$, then $L(s, \chi, \mathbb{L}/\mathbb{K})$ continues to an entire function. He predicted that unlike the other conjecture, this one would "require completely new methods" for its resolution [6, p. 105]. It is still unresolved today, although it has been confirmed when χ is induced from a Dedekind character and in other special cases. It has been subsumed by R. Langlands' program and so is a vital part of a major contemporary research program ultimately concerned with arithmetic questions.[49]

With the work of Artin and Brauer we have in a sense come full circle. As we saw in Section 12.2, it was the theory of numbers that had led to the notion of a Dedekind character, and it was the analogy with arithmetic considerations (discriminants of number fields) that had inspired Dedekind's introduction of the group determinant and the problem of its factorization—the problem that led Frobenius to generalize the concept of a character and to develop the attendant theory of group representations. With the work of Artin and Brauer, we see how Frobenius' theory of group characters was now being used to advance and stimulate the development of number theory: Brauer's work on Schur's index theory provided part of the foundation for Galois cohomology; and the work on generalized L-functions by Artin and Brauer marked an important step toward the generalization of class field theory to nonabelian extensions. It is also of interest that with the induction theorems of Artin and Brauer, we see, by means of Frobenius' theory of induced characters, that Dedekind's characters are more intimately connected to those of Frobenius than either of them probably realized.

---

[49]Gelbart [250] has written an illuminating expository account of the Langlands program that indicates how Artin's conjecture fits into it. See especially pp. 203–204 and 208–209. Langlands' expository article [396] also conveys an idea of the role in the theory of numbers played by representation theory.

Chapter 16
Loose Ends

As its title suggests, this chapter is devoted to tying up several historical loose ends related to the work of Frobenius featured in the previous chapters. The first section focuses on work done by Frobenius in response to the discovery of a gap in Weierstrass' theory of elementary divisors as it applied to families of quadratic forms. Frobenius gave two solutions to the problem of filling the gap. The first drew upon the results and analogical reasoning used in his arithmetic theory of bilinear forms and its application to elementary divisor theory (Chapter 8). In effect, this solution solved the gap problem for matrices over unique factorization domains, although it was lengthy and involved intricate determinant-theoretic considerations. Frobenius' second solution was the one he preferred, for it was both simple and elementary. It involved yet another brilliant application of the symbolic algebra of matrices, one on a par with the applications of matrix algebra to the Cayley–Hermite problem (Chapter 7) and to Kronecker's complex multiplication problem (Sections 10.5 and 10.6).

In the second section, I discuss the gradual assimilation of Frobenius' rational theory of elementary divisors (Section 8.6) into the early decades of the twentieth century. In these developments, the presentation of the theory was along the same lines as outlined by Frobenius, although reasoning based on the analogy between $\mathbb{Z}$ and $\mathbb{F}[\lambda]$, $\mathbb{F}$ a specific known field, was ultimately replaced by reasoning applied to matrices with coefficients from a principal ideal domain. Then, in the final section, I trace the complex developments that culminated in B.L. van der Waerden's development of elementary divisor theory as an application of the fundamental theorem of finitely generated modules over a principal ideal domain. Although van der Waerden's approach was far different from that of Frobenius, we will see that the work of Frobenius and his students on differential equations (Section 1.2), group representation theory (Chapters 13 and 15), and the fundamental theorem of finite abelian groups (Section 9.2), as well as his rational theory of elementary divisors itself, played an important role in the developments leading up to van der Waerden's now familiar rendition of elementary divisor theory.

T. Hawkins, *The Mathematics of Frobenius in Context*, Sources and Studies in the History of Mathematics and Physical Sciences, DOI 10.1007/978-1-4614-6333-7_16,

16.1 Congruence Problems and Matrix Square Roots

The story behind the congruence problems solved by Frobenius begins with the discovery of a gap in the proof of Weierstrass' key theorem on nonsingular families of quadratic forms (Corollary 5.10, part I), which asserts that two families of nonsingular quadratic forms are congruent if and only if they have the same elementary divisors.

16.1.1 A gap in Weierstrass' theory

The fundamental theorem of Weierstrass' theory of elementary divisors was that two families of bilinear forms $F(x,y) = x^t(\lambda A - B)y$ and $G(X,Y) = X^t(\lambda C - D)Y$ that are nonsingular ($\det A \neq 0, \det C \neq 0$) are *equivalent* in the sense that nonsingular transformations $x = HX$, $y = KY$ exist such that $F(x,y) = G(X,Y)$ if and only if they have the same elementary divisors (Theorem 5.8). Expressed in matrix form, equivalence means $H^t(\lambda A - B)K = \lambda C - D$. To prove this theorem Weierstrass applied the Jacobi transformation (5.5) to establish the key part: every nonsingular $F(x,y)$ can be transformed into its Weierstrass normal form $X^t\tilde{W}(\lambda)Y$ (5.18), which depends only on the elementary divisors of $F(x,y)$. Thus two families with the same elementary divisors can be transformed into the same normal form and so into each other.

To apply the generic Jacobi transformation, Weierstrass needed the following preparatory lemma [588, pp. 317, 320ff.].

Lemma 16.1. *Given a nonsingular family of bilinear forms with coefficient matrix $\lambda A - B$, let $p = p(\lambda)$ denote any fixed linear factor of the characteristic polynomial $f(\lambda) = \det(\lambda A - B)$. Then nonsingular linear transformations H_p and K_p can be determined such that the transformed matrix $\lambda A' - B' = H_p^t(\lambda A - B)K_p$ has the following property. For $k = 1,\dots,n-1$, let $f_k(\lambda)$ denote the principal $k \times k$ minor determinant of $sA' - B'$ obtained by striking out the first $n-k$ rows and columns. Then if $p(\lambda)^{a_k}$ is the highest power of $p(\lambda)$ dividing $f_k(\lambda)$, there is no $k \times k$ minor $m(\lambda)$ of $\lambda A' - B'$ such that $p^{a_k}(\lambda)$ does not divide $m(\lambda)$.*

If, as usual, $D_k(\lambda)$ denotes the polynomial greatest common divisor of all $k \times k$ minors of $sA' - B'$, the above property means that $D_k(\lambda) = p(\lambda)^{a_k}q(\lambda)$ where $(p,q) = 1$. In the terminology later introduced by Frobenius [203, p. 578], the minor determinants $f_k(\lambda)$ of $sA' - B'$ are said to be *regular with respect to p*.

Weierstrass proved his lemma and in fact showed that H_p and K_p can be taken as products of elementary matrices [588, pp. 322–324]. He then turned to the special case in which $\lambda A - B$ is the coefficient matrix of a family of quadratic forms such that A and B are symmetric. In this case, $x = y$ and $F(x,x) = x^t(\lambda A - B)x$, and a single transformation $x = HX$ needs to be determined such that $F(x,x) = G(X,X)$. This means that the two families of quadratic forms are *congruent*. In matrix form,

$H^t(\lambda A - B)H = \lambda C - D$. In Weierstrass' construction of matrices H, K such that $H^t(\lambda A - B)K = \tilde{W}(\lambda)$, the part of the construction that related to the characteristic root corresponding to $p(\lambda)$ had used Lemma 16.1. Weierstrass realized that one would have $H = K$ when A and B are symmetric as long as $H_p = K_p$ in Lemma 16.1 [588, p. 324], and he apparently felt that such a choice was clearly possible and in no need of discussion. It turned out, however, that it was not at all clear how to prove Lemma 16.1 with A and B symmetric and $H_p = K_p$. There was thus a gap in Weierstrass' proof of Corollary 5.10.

It is not entirely clear who first realized that there was a gap, but it seems to have been Weierstrass' student Ludwig Stickelberger who first became concerned with the question whether the gap posed an insurmountable obstacle to the tenability of Weierstrass' overall approach to proving Corollary 5.10. According to Stickelberger [550, p. 22], this occurred while he was working on his doctoral dissertation (1874), which involved applications of Weierstrass' theory of elementary divisors. At that time, by means of indirect, case-by-case considerations, he convinced himself that Weierstrass' assumption that $H_p = K_p$ was possible in the symmetric case was valid, i.e., that the following lemma was true:

Lemma 16.2. *If $\lambda A - B$ is a nonsingular family of symmetric matrices and if $p = p(\lambda)$ is any linear factor of* $\det(\lambda A - B)$, *then a nonsingular matrix H_p exists such that for $\lambda A' - B' = H_p^t(\lambda A - B)H_p$, the minors $f_k(\lambda)$ are regular with respect to p.*

However, Stickelberger was unable to give a direct proof. Five years later, as Frobenius' colleague at the Zurich Polytechnic, Stickelberger published a paper in *Crelle's Journal* [550] in which he showed that by a clever rearrangement of the parts of Weierstrass' original proof procedure and without any new results, the above difficulty of proving Lemma 16.2 could be avoided.

Frobenius' first paper on his arithmetic theory of bilinear forms, containing his rational theory of elementary divisors, appeared in the same volume of *Crelle's Journal*, and not surprisingly, Frobenius asked himself whether his arithmetic theory of elementary divisors might yield a direct proof of Lemma 16.2. By the time he published the second part of his theory in 1880, he had found such a proof, as he noted briefly at the end of his paper [185, p. 631]. He did not present all the details since although he had confirmed the veracity of Lemma 16.2, his proof involved his containment theorem (Theorem 8.16) and so all the machinery of his arithmetic approach to elementary divisor theory. Such a proof would not serve any purpose in Weierstrass' theory, where the lemma occurs as a preliminary to his procedure.

After retiring from his professorship in 1892, Weierstrass focused his energy on the publication of his collected works. In that connection, he was faced with the gap in his proof of Corollary 5.10. Stickelberger's rearrangement of his original proof procedure being aesthetically unsatisfying,[1] Weierstrass appealed to Frobenius to find a suitable proof of Lemma 16.2. In his earlier efforts to give a direct proof

[1]This fact and the other information preliminary to Frobenius' discovery of a proof is drawn from Frobenius' account in 1894 [203, pp. 577–578].

of Lemma 16.2, Stickelberger had hoped to find a determinant identity from which it would follow [550, p. 22], but to no avail. Kronecker had established a determinant identity in 1870, which, it was hoped, would serve as the key to proving Lemma 16.2. This was a matter that Kronecker frequently discussed with Frobenius and Stickelberger, who presented objections to Kronecker's efforts to fashion a viable proof out of his identity. It was Frobenius who showed how to do this in a paper of 1894 [203].

The details of how Frobenius accomplished this involved extensive determinant-based reasoning of a subtle nature and will not be considered here. However, it is worth noting that by 1894, he had made the analogical form of reasoning used to establish his rational theory of elementary divisors a bit more openly abstract. In his rational theory, he had developed the mathematics for matrices over $\mathbb{Z}$ and then simply invoked the analogy with $\mathbb{F}[\lambda]$ to derive his rational theory. Now at the very outset, Frobenius explained that the coefficients of the matrices involved in the reasoning to follow could lie in one of several domains [203, p.578]: "Let a system of arbitrarily many rows and columns be given whose elements $a_{\alpha\beta}$ are whole numbers or polynomial functions of one or more variables with arbitrary constant coefficients or integral quantities of any field." In other words, Frobenius realized that his matrices could be assumed to have coefficients from $\mathfrak{R} = \mathbb{Z}$, or from $\mathfrak{R} = \mathbb{C}[\lambda_1, \ldots, \lambda_n]$, or from $\mathfrak{R} = \mathfrak{o}_{\mathbb{K}}$, where $\mathfrak{o}_{\mathbb{K}}$ denotes the ring of algebraic integers in a finite extension $\mathbb{K}$ of $\mathbb{Q}$. His inclusion of the example $\mathfrak{R} = \mathbb{C}[\lambda_1, \ldots, \lambda_n]$ with $n > 1$ reflects his realization that the ensuing proof required only unique factorization into prime factors and not, as in his rational theory of elementary divisors, the additional property that the greatest common divisor d of $a, b \in \mathfrak{R}$ can be expressed as $d = sa + tb$ for $s, t \in \mathfrak{R}$. Expressed in present-day terms, what Frobenius in effect realized was that the reasoning he was about to give was valid for matrices over a unique factorization domain $\mathfrak{R}$ and not just for matrices over principal ideal domains, although of course he expressed himself in terms of specific examples of domains $\mathfrak{R}$.[2] In the case of finite groups, he had adopted the abstract point of view that had already been sanctioned by Dedekind and Kronecker (Sections 9.1.4–9.1.5). It was not until the twentieth century that abstract rings were studied, notably by Emmy Noether. In 1894, in a work done at Weierstrass' request, Frobenius no doubt felt that the above remark was as far as he wished to go in the direction of generality.

The following year, Kronecker's former student Kurt Hensel, who was in Berlin and in communication with Frobenius about his results (see [203, p. 89]), published his own proof of Frobenius' containment theorem (Theorem 8.16). He too stressed the generality of the result by considering matrices whose elements are the "integral quantities of an arbitrary domain of rationality, thus, e.g., integers or polynomials of one variable x" [282, p. 109]. Kronecker had used the term "domain of rationality"

[2] In the case of algebraic integers, some results require excluding p a prime divisor of 2 [203, p. 584] in the definition of "regular with respect to p."

(*Rationalitätsbereich*) for the fields he considered.[3] Hensel presumably excluded polynomials in more than one variable because he realized that the proof of Frobenius' theorem on the Smith–Frobenius normal form (Theorem 8.8), which he took for granted in his proof of the containment theorem, makes repeated use of the fact that the greatest common divisor d of $a, b \in \mathfrak{R}$ is expressible as $d = sa + tb$, for $s, t \in \mathfrak{R}$ and $\mathfrak{R} = \mathbb{F}[x_1, \ldots, x_n]$ fails to have this property for $n > 1$. For example, for $n = 2$, $a = x_1$ and $b = x_2$ are relatively prime, but no polynomials $s = s(x_1, x_2), t = t(x_1, x_2)$ exist for which $1 = sx_1 + tx_2$.

16.1.2 Two matrix congruence problems

Having been drawn into the "gap" issue and then having established Lemma 16.2, albeit by means of a subtle and extensive determinant-based line of reasoning, Frobenius next explored the possibility of a different and simpler route to the resolution of the gap issue, a route that involved matrix algebra rather than determinants. Expressed in Frobenius' language and notation, Weierstrass' inadequately established Corollary 5.10 asserts that two nonsingular families of symmetric matrices, $\lambda A_1 - A_2$ and $\lambda B_1 - B_2$, are congruent, viz.,

$$R^t(\lambda A_1 - A_2)R = \lambda B_1 - B_2, \quad \det R \neq 0,$$

if and only if they have the same elementary divisors. Of course, if they are congruent, then they are equivalent, i.e., $P(\lambda A_1 - A_2)Q = \lambda B_1 - B_2$, with $P = R^t$, $Q = R$, and so they have the same elementary divisors by Weierstrass' Theorem 5.8, which had no gaps in its proof. The converse had been the problem: Suppose $\lambda A_1 - A_2$ and $\lambda B_1 - B_2$ have the same elementary divisors. Then Weierstrass' Theorem 5.8 says that they are equivalent. It was his proof that they are actually congruent that had required patching.

The above considerations suggested the following problem to Frobenius.

Problem 16.3. *Use matrix algebra to show that equivalence plus symmetry implies congruence. That is, if A and B are symmetric matrices such that $PAQ = B$ for nonsingular P, Q, show by matrix algebra that a nonsingular R exists for which $R^tAR = B$.*

If Problem 16.3 could be solved, that would obviate the need for a subtle determinant-theoretic argument to establish Weierstrass' Corollary 5.10, since in Problem 16.3, A and B could be two nonsingular families of symmetric matrices, viz., $A = \lambda A_1 - A_2$ and $B = \lambda B_1 - B_2$.

Frobenius realized that work of Kronecker also suggested Problem 16.3 as well as a related problem (Problem 16.4 below). As we saw in Section 5.6, by 1874,

---

[3]On Kronecker's work with fields, see Purkert's study [491].

Kronecker had determined a set of invariants for a family of possibly singular quadratic (respectively, bilinear) forms $\lambda A_1 - A_2$ such that two such families are congruent (respectively, equivalent) if and only if they have the same set of invariants, e.g., the same W-series and K-series. In 1874, he published only a very brief sketch of the theory in the quadratic case, although by the end of the year, he had written up, but not published, a coherent treatment of the bilinear case. He finally published it in 1890 [367], and then during 1890–1891, he published his theory for families of quadratic forms [368, 369]. The quadratic theory was developed ab initio; he did not attempt to derive it from the bilinear theory, as had been Weierstrass' (problematic) strategy in 1868. From Kronecker's papers of 1890–1891, it now followed that if any two families of quadratic forms, singular or not, are equivalent, then they are congruent; for if $\lambda A_1 - A_2$ and $\lambda B_1 - B_2$ are equivalent, where the A_i and B_i are all symmetric, then they have the same invariants by virtue of the bilinear theory developed in [367], whereas conversely, if $\lambda A_1 - A_2$ and $\lambda B_1 - B_2$ have the same invariants, then they are congruent by virtue of the quadratic theory developed in [368, 369].

It should be noted that *if* for quadratic families it could be proved independently of Kronecker's theories that equivalence implies congruence, e.g., if Problem 16.3 could be solved, then Kronecker's entire quadratic theory could be dispensed with in the sense that his main theorem would now be an immediate consequence of his bilinear theory. That is, since congruence of two symmetric families, $\lambda A_1 - A_2$ and $\lambda B_1 - B_2$, obviously implies equivalence, we would have that $\lambda A_1 - A_2$ and $\lambda B_1 - B_2$ have the same invariants if and only if they are equivalent if and only if—asuming a solution to Problem 16.3—they are congruent.

As we saw in Section 5.6.4, the one theory that Kronecker did fully develop and publish in 1874 had to do with the congruence of the special families of bilinear forms with coefficient matrices $\lambda A - A^t$, the type that had arisen in connection with the complex multiplication problem of Section 5.3. In 1866, when Kronecker considered that problem, he had been content with a generic result, namely Theorem 5.5. In 1874, he was in a position to deal with it on a nongeneric level and without assuming the family of forms to be nonsingular or the number of variables to be even, as in Theorem 5.5. Thus in the course of sixty pages, he showed that for all A and B, $\lambda A - A^t$ and $\lambda B - B^t$ are congruent if and only if they have the same invariants [359]. Of course, in view of the bilinear theory Kronecker published in 1890 [367], the above result would follow as an immediate consequence of the main theorem of that theory *if* it could be shown directly that for families of the special type $\lambda A - A^t$, equivalence implies congruence. That is, *if* the following problem could be solved:

Problem 16.4. *Show by means of matrix algebra that if nonsingular P, Q exist for which $P(\lambda A - A^t)Q = \lambda B - B^t$, then a nonsingular R exists such that $R^t(\lambda A - A^t)R = \lambda B - B^t$.*

Problems 16.3 and 16.4 were both given affirmative answers by Frobenius in a paper published in 1896 [208]. As we shall see, he accomplished this in a matter of a few pages. As he observed in the introduction [208, p. 697]:

> The extremely simple argument presented here provides a complete replacement for the lengthy analysis that Kronecker employed ... [in the above-mentioned papers [359, 368, 369]] ... and also with its help the subtle deliberations in the work of Weierstrass that are required for a precise treatment of ... quadratic ... forms can be avoided.

16.1.3 Frobenius' solution

In order to suggest the line of reasoning that surely led Frobenius to realize how Problems 16.3 and 16.4 could be solved, consider Problem 16.3. By hypothesis, nonsingular P and Q exist such that

$$PAQ = B. \tag{16.1}$$

Taking transposes of both sides and using the symmetry of A and B, we get

$$Q^t AP^t = B. \tag{16.2}$$

Frobenius discovered that equations (16.1) and (16.2) together provide the key to the solution of both Problems 16.3 and 16.4.

Eliminating B from these two equations, we have $PAQ = Q^t AP^t$, which may be rewritten as

$$(Q^t)^{-1}PA = AP^t Q^{-1}. \tag{16.3}$$

Thus if we set $U = (Q^t)^{-1}P$, then $U^t = P^t Q^{-1}$, and so (16.3) becomes

$$UA = AU^t, \quad U = (Q^t)^{-1}P. \tag{16.4}$$

It then follows by repeated application of (16.4) that $U^k A = A(U^t)^k$ for every positive integer k and therefore that for every polynomial $\chi(t) \in \mathbb{C}[t]$,

$$\chi(U)A = A\chi(U^t). \tag{16.5}$$

If $\chi(t)$ is such that $\det[\chi(U)] \neq 0$, then $\chi(U^t) = [\chi(U)]^t$ is invertible, and (16.5) can be written as

$$\chi(U)A[\chi(U^t)]^{-1} = A, \quad U = (Q^t)^{-1}P. \tag{16.6}$$

If we now go back to (16.2) and express A by the left-hand side of (16.6), we get

$$B = Q^t \chi(U)A[\chi(U^t)^{-1}]P^t = RAS, \tag{16.7}$$

with R and S defined by

$$R = Q^t \chi(U) \quad \text{and} \quad S = [\chi(U^t)]^{-1} P^t, \tag{16.8}$$

where $\chi(t)$ is any polynomial with the property that $\det[\chi(U)] \neq 0$ for $U = (Q^t)^{-1}P$.

In effect, (16.7) and (16.8) give an infinite number of equivalence transformations R, S taking A into B. The question is whether it is possible to choose $\chi(t)$ such that $R = S^t$, for then (16.7) asserts that A and B are congruent. From the expressions in (16.8) for R and S, it follows that the condition that $R = S^t$ can be expressed in the form $Q^t \chi(U) = P[\chi(U)]^{-1}$ or as

$$[\chi(U)]^2 = U, \quad U = (Q^t)^{-1}P. \tag{16.9}$$

It was by means of the above sort of matrix-algebraic reasoning (given here essentially as Frobenius presented it in his paper [208, §2]) that Frobenius surely first realized that Problem 16.3 would be solved if he could prove the following square-root theorem:

Theorem 16.5. *If U is any square matrix with* $\det U \neq 0$*, then a polynomial $\chi(z)$ exists such that $[\chi(U)]^2 = U$.*[4]

The relation $[\chi(U)]^2 = U$ of course implies that $\det[\chi(U)] \neq 0$, and so application of this theorem to $U = (Q^t)^{-1}P$ then implies by the above reasoning that S as given in (16.8) satisfies $S^tAS = B$. Thus for A, B symmetric, if they are equivalent, they are congruent.

Frobenius' proof of Theorem 16.5 will be discussed below. First, however, it should be observed, as Frobenius did, that the reasoning leading to (16.9) derived entirely from (16.1) and (16.2), so that for any A and B satisfying these two equations, symmetric or not, the same reasoning as given above would lead to (16.9), so that Theorem 16.5 implies that S as defined in (16.8) satisfies $S^tAS = B$. With that in mind, consider Problem 16.4—show that if $\lambda A - A^t$ and $\lambda B - B^t$ are equivalent, then they are congruent. Let $P(\lambda A - A^t)Q = \lambda B - B^t$. Evidently, the only way this can hold for all λ is if $PAQ = B$ and $PA^tQ = B^t$. The former equality is (16.1), and the latter, after transposition, is (16.2). Thus for this A and B we have (16.9), and so by Theorem 16.5, we may conclude that a nonsingular S exists for which $S^tAS = B$. Transposition of this equality yields $S^tA^tS = B^t$, from which $S^t(\lambda A - A^t)S = \lambda B - B^t$ follows immediately. Thus the equivalence of $\lambda A - A^t$ and $\lambda B - B^t$ does indeed imply their congruence, and Problem 16.4 is also solved.

Before discussing Frobenius' proof of the above square-root theorem, I will give some historical background so that the reader can fully appreciate his achievement. Prior to Frobenius' work on the above congruence problems, both Cayley and Sylvester had considered the matter of matrix square roots. Although it is uncertain whether Frobenius was familiar with what they had to say, their remarks provide a historical and mathematical perspective on his Theorem 16.5 and its proof.

---

[4]Frobenius' theorem actually specifies that $\deg \chi = m - 1$, where m is the degree of the minimal polynomial of U. This additional information can be inferred from Frobenius' proof; see below.

16.1.4 Cayley, Sylvester, and matrix square roots

Cayley had already considered the idea of a square root of a matrix in his paper of 1858 on matrix algebra [84, pp. 483ff.], the subject of Section 7.4. This occurred in his discussion of the implications of the Cayley–Hamilton theorem, namely that $\varphi(A) = 0$, where $\varphi(A) = \det(\lambda I - A)$. It follows immediately from this theorem, Cayley observed, that if $f(t)$ is any polynomial or rational function and M is an $n \times n$ matrix, then $L = f(M)$ is expressible as a polynomial of degree at most $n-1$.[5] "But it is important to consider," Cayley continued, "how far or in what sense the like theorem is true with respect to irrational functions of a matrix" [84, p. 383]. By "irrational functions" of a matrix M, Cayley had in mind expressions such as $L = \sqrt{M}$, which he considered by way of example. In this case, if M is $n \times n$, then $L = \sqrt{M}$ exists precisely when the system of n^2 quadratic equations in the n^2 unknown coefficients of L that corresponds to $L^2 = M$ has a solution. Cayley focused on how to determine L when it exists, presumably with an eye toward determining "how far or in what sense" L is expressible as a polynomial in M. To this end, he showed how the Cayley–Hamilton theorem could be used to facilitate finding L.

Cayley's method applies to matrices of any size, but presumably to avoid complicated notation, he illustrated it in the case $n = 2$. For the purposes at hand, I will illustrate it for 3×3 matrices. If M is 3×3 and $L^2 = M$, then the Cayley–Hamilton theorem implies that constants a, b, c exist such that $L^3 + aL^2 + bL + cI = 0$. Since $L^2 = M$, it follows that $L^3 = ML = LM$, and so the equation for L becomes $LM + aM + bL + cI = 0$ or $L(M + bI) = -(aM + cI)$. Squaring both sides of this last equation and substituting M for L^2, we obtain $M(M + bI)^2 = (aM + cI)^2$. This matrix equation corresponds to $n^2 = 9$ quadratic equations in the $n = 3$ unknowns a, b, c rather than the $n^2 = 9$ unknown coefficients of L in the $n^2 = 9$ equations implied by $L^2 = M$. This reduction of unknowns was the point of Cayley's method.

Cayley's method, however, was ineffective in dealing with the question he had posed, namely, "how far and to what extent" is it the case that irrational functions such as $\sqrt{M}$ can be expressed as polynomials in M. For example, if

$$M = \begin{pmatrix} 0 & 1 & 0 \\ 0 & 0 & 0 \\ 0 & 0 & 0 \end{pmatrix},$$

then since $M^2 = 0$, the equation $M(M + bI)^2 = (aM + cI)^2$ derived above reduces to $c^2 I = (b^2 - 2ac)M$ and implies $b = c = 0$ with a left undetermined. Thus $L = (aM + cI)/(M + bI)$ becomes $L = aM/M$, but since M is not invertible, the method loses its meaning and might seem to suggest that L does not exist, although this

[5]This is indeed correct, as Frobenius showed in his 1878 paper [181, p. 355], assuming that when $f(t) = p(t)/q(t)$, $\det q(A) \neq 0$, so that $[q(A)]^{-1}$ exists.

turns out to be incorrect. The matrix M does have square roots; there are solutions to the system of n^2 equations in n^2 unknowns symbolized by $L^2 = M$, and they are given by

$$L(\alpha,\beta) = \begin{pmatrix} 0 & \alpha & \beta \\ 0 & 0 & 0 \\ 0 & \beta^{-1} & 0 \end{pmatrix},$$

where α and β are arbitrary parameters with $\beta \neq 0$ [240, Vol. 1, p. 239]. Since the minimal polynomial of M is $\psi(t) = t^2$, and so of degree 2, it follows that if $L(\alpha,\beta)$ were expressible as a polynomial in M, then it would be expressible as a polynomial of degree 1. However, it is easily seen that $L(\alpha,\beta) = pM + qI$ is impossible, since it implies $\beta = 0$. This shows that $L = \sqrt{M}$ cannot always be expressed as a polynomial in M. The example $M = \left(\begin{smallmatrix} 0 & 1 \\ 0 & 0 \end{smallmatrix}\right)$ shows that $L = \sqrt{M}$ need not exist.[6] In the two above examples of exceptional M, $\det M = 0$. Frobenius' Theorem 16.5 shows that if $\det M \neq 0$, then M always has a square root, and it is expressible as a polynomial in M.

During the period 1882–1884, Cayley's friend J.J. Sylvester became interested in matrix algebra.[7] Among other things, he considered Cayley's question concerning the determination of irrational functions of a matrix M such as $\sqrt{M}$. Sylvester's general conclusion was that if $f(z)$ is any single- or multiple-valued function of z and if M is $n \times n$ with distinct characteristic roots $\lambda_1, \ldots, \lambda_n$, then $f(M)$ is given by

$$f(M) = \sum_{i=1}^{n} \prod_{j \neq i} \frac{(M - \lambda_j I)}{(\lambda_i - \lambda_j)} f(\lambda_i). \tag{16.10}$$

Formula (16.10) is based on the idea behind the Lagrange interpolation formula and is sometimes called Sylvester's interpolation formula. As Sylvester realized, it applies to the case in which the characteristic roots λ_i are all distinct. When some λ_i are equal, "the formula must be replaced by another obtained from it by the usual method of infinitesimal variation" [559, p. 111]. Perhaps Sylvester could have done this in a specific case, e.g., when M has one double root, but no less an algebraist than Lagrange had been led to false conclusions using "the usual method of infinitesimal variation" because he continued to reason generically.[8] Indeed, when $f(z) = \sqrt{z}$, the examples given above indicate that

---

[6] If $L^2 = M$, then $L^4 = M^2 = 0$. This means that the characteristic roots of L must all be 0 and so (since L is 2×2) $\varphi(t) = t^2$ is the characteristic polynomial of L. The Cayley–Hamilton theorem then implies $L^2 = 0 \neq M$, and so $\sqrt{M}$ does not exist.

[7] For the personal and institutional background to Sylvester's brief flurry of interest in matrix algebra, see [462, pp. 135–138]. A fairly detailed mathematical discussion of Sylvester's work on matrix algebra is given in [270, §6].

[8] I refer to Lagrange's attempt to extend his elegant generic solution to $\ddot{y} + Ay = 0$, $y(0) = y_0$, A $n \times n$, to the case in which $f(\rho) = \det(\rho^2 + A)$ has one root of multiplicity two. See Section 4.2.1.

$f(M)$ need not exist or may exist but not be derivable by infinitesimal considerations applied to (16.10), which would lead to an expression for $f(M)$ as a polynomial in M.

16.1.5 Frobenius' proof of his square-root theorem

Frobenius, by contrast with Sylvester (who is not mentioned by Frobenius), provided a proof of Theorem 16.5 that is nongeneric, yet simple, and completely rigorous by present-day standards [208, §1]. For pedagogical reasons, it would probably not be used today, because it invokes basic results from complex analysis, but for that reason, it provides another good example of how in forging the link between formal matrix algebra and Weierstrass' theory of elementary divisors, Frobenius followed the example of his mentor and employed considerations drawn from Weierstrass-style complex analysis.[9] What follows is an exposition of the resulting proof.

The goal of Theorem 16.5 is to determine a polynomial $\chi(z)$ such that $[\chi(U)]^2 = U$, where U is a given matrix with $\det U \neq 0$. Frobenius turned to the minimal polynomial $\psi(z)$ of U. Since $\psi(U) = 0$ by definition, it suffices to define $\chi(z)$ in such a way that $\psi(z)$ divides $[\chi(z)]^2 - z$, for if $[\chi(z)]^2 - z = \psi(z)\theta(z)$, then $[\chi(U)]^2 - U = \psi(U)\theta(U) = 0$, i.e., $V^2 = U$ for $V = \chi(U)$. Frobenius' proof idea involved utilizing the identity

$$[\chi(z)]^2 - z = \left[\chi(z) - \sqrt{z}\right]\left[\chi(z) + \sqrt{z}\right], \tag{16.11}$$

which is valid for any determination of $\sqrt{z}$.

With this in mind, let $\psi(z) = \prod_{j=1}^{d}(z-\lambda_j)^{e_j}$ denote the factorization of the minimal polynomial of U, so that $\lambda_1, \ldots, \lambda_d$ are all the distinct characteristic roots of U. Since $\det U \neq 0$, each root λ_j differs from 0, and so a branch $f_j(z)$ of $\sqrt{z}$ exists in a sufficiently small neighborhood $\mathcal{N}(\lambda_j)$ of $z = \lambda_j$. Thus $f_j(z)$ is analytic in this neighborhood, and since $f_j(\lambda_j) = \sqrt{\lambda_j} \neq 0$, it follows that $f_j(z)/\psi(z)$ has a pole of order e_j at λ_j, and so we may write

$$\frac{f_j(z)}{\psi(z)} = \frac{a_{-e_j}}{(z-\lambda_j)^{e_j}} + \cdots + \frac{a_{-1}}{z-\lambda_j} + \mathfrak{P}(z-\lambda_j), \tag{16.12}$$

where $\mathfrak{P}(z-\lambda_j)$ is my notation for a power series in $z-\lambda_j$, i.e., a series involving nonnegative integral powers of $z-\lambda_j$. The singular part of the Laurent expansion (16.12) can be expressed as a simple fraction

[9] See in this connection Frobenius' use of power and Laurent series in his proof of his minimal polynomial theorem (Theorem 7.2) and in his proof that a real orthogonal matrix can be diagonalized (Theorem 7.15).

$$\frac{a_{-e_j}}{(z-\lambda_j)^{e_j}}+\cdots+\frac{a_{-1}}{z-\lambda_j}=\frac{A_j(z)}{(z-\lambda_j)^{e_j}}, \tag{16.13}$$

where $A_j(z)$ is a polynomial in z.

With these preliminaries in place, define $\chi(z)$ by

$$\chi(z)=\sum_{j=1}^{d}\frac{A_j(z)\psi(z)}{(z-\lambda_j)^{e_j}}. \tag{16.14}$$

Thus $\chi(z)$ is $\psi(z)$ times the sum of the singular parts of $f_j(z)/\psi(z)$ at each pole $z=\lambda_j$, $j=1,\dots,d$. It is easy to see that $\chi(z)$ is a polynomial, because in the jth term of (16.14), the factor $(z-\lambda_j)^{e_j}$ in the denominator divides $\psi(z)=\prod_{j=1}^{d}(z-\lambda_i)^{e_j}$. To show that $\psi(z)$ divides $\chi(z)^2-z$ is to show that the rational function $([\chi(z)]^2-z)/\psi(z)$ is a polynomial. Frobenius' idea was to use the identity (16.11) to show that the above rational function is analytic at all the roots λ_k of ψ. Since the λ_k are the only possible poles of $([\chi(z)]^2-z)/\psi(z)$, it then follows that it is a polynomial, i.e., that $\psi(z)$ divides $[\chi(z)]^2-z$.

To show that $([\chi(z)]^2-z)/\psi(z)$ is analytic at $z=\lambda_k$, consider, for any fixed k, the difference $\chi(z)-\sqrt{z}=\chi(z)-f_k(z)$ for $z\in\mathcal{N}(\lambda_k)$. From (16.12)–(16.14), it follows that

$$\chi(z)-f_k(z)=\sum_{j\neq k}\frac{A_j(z)\psi(z)}{(z-\lambda_j)^{e_j}}-\psi(z)\mathfrak{P}(z-\lambda_k). \tag{16.15}$$

The presence of $\psi(z)$ as a factor in every term above means that $(z-\lambda_k)^{e_k}$ can be factored from every term, so that (16.15) may be written as

$$\chi(z)-f_k(z)=(z-\lambda_k)^{e_k}\mathfrak{P}^*(z-\lambda_k),$$

which we may assume without loss of generality is also valid in $\mathcal{N}(\lambda_k)$. Since it is clear that $\chi(z)+\sqrt{z}=\chi(z)+f_k(z)=\mathfrak{P}^{**}(z-\lambda_k)$ for $z\in\mathcal{N}(\lambda_k)$, the identity (16.11) becomes

$$[\chi(z)]^2-z=(z-\lambda_k)^{e_k}\mathfrak{P}^*(z-\lambda_k)\mathfrak{P}^{**}(z-\lambda_k)=(z-\lambda_k)^{e_k}\mathfrak{P}^{***}(z-\lambda_k),$$

which shows that $([\chi(z)]^2-z)/(z-\lambda_k)^{e_k}=\mathfrak{P}_k^{***}(z-\lambda_k)$ is analytic in $\mathcal{N}(\lambda_k)$. Consequently,

$$([\chi(z)]^2-z)/\psi(z)=\left[([\chi(z)]^2-z)/(z-\lambda_k)^{e_k})\right]\cdot\left[1/\left(\prod_{j\neq k}(z-\lambda_i)^{e_j}\right)\right]$$

is analytic in a neighborhood of λ_k for any $k = 1, \ldots, d$. This means that the minimal polynomial $\psi(z)$ divides the polynomial $\chi(z)^2 - z$. In view of the above preliminary remarks, Frobenius' proof of Theorem 16.5 is now complete.[10]

16.1.6 The spread of Frobenius-style matrix algebra

By Frobenius-style matrix algebra, I mean not simply the utilization of a symbolic algebra of matrices but also, and more importantly, the utilization of that algebra in conjunction with other rigorously developed mathematical theories and, in particular, with Weierstrass' theory of elementary divisors. The idea of matrix algebra had been introduced independently by several mathematicians besides Frobenius—Cayley, Laguerre, and Sylvester—but he alone developed that idea in the above-described manner, as exemplified by his solutions to the problem of Rosanes and the Cayley–Hermite problem (Sections 7.5.3 and 7.5.4), to Kronecker's complex multiplication problem (Section 10.6), and to the above Problems 16.3 and 16.4. By 1896, when he published his solution to the latter problems, however, his work on matrix algebra was still not widely known or appreciated.

For example, in 1887, Lipschitz published a paper [418] that was prompted by a passing remark that Camille Jordan had made in a lengthy paper on linear differential equations [328, p. 112, no. 36], which had appeared in *Crelle's Journal* 26 pages after Frobenius' 1878 paper [181] on matrix algebra. Jordan observed without proof that if $x_i' = \sum_{i,j=1}^{n} a_{ij} x_j$ is a linear substitution S that belongs to a group of finite order, then S has a diagonal canonical form with roots of unity along the diagonal. Lipschitz realized that Jordan's remark implied that if any linear substitution S composed with itself k times gives the identical substitution, then it has a diagonal form with k roots of unity along the diagonal. Being well versed in Weierstrass' theory of elementary divisors, Lipschitz devoted his paper to a proof of the elementary divisor analogue: if S has the above property, then all its characteristic roots are kth roots of unity and all its elementary divisors are linear. He failed to realize that his proposition was a special case of a more general theorem already proved by Frobenius a few pages earlier in the same issue of *Crelle's Journal* that contained Jordan's paper.[11] Lipschitz's proposition is also an easy consequence of Frobenius' Theorem 7.2 on the minimal polynomial $\psi(t)$ of S, since $S^k = I$ implies by that theorem that $\psi(t)$ divides $f(t) = t^k - 1$. Thus all the roots of $\psi(t)$ are (1) distinct and are (2) kth roots of unity. By part (iv)

---

[10]Frobenius' proof [208, pp. 697ff.] is expressed somewhat more generally than expounded here so as to allow a brief discussion of the problematic case $\det U = 0$ as well as other functions of a matrix.

[11]Frobenius proved that if S is a matrix with the property that the sequence S^j, $j = 0, 1, 2, \ldots$, has only a finite number of distinct terms, then all characteristic roots of S are either 0 or roots of unity, and the elementary divisors corresponding to the roots of unity are all linear [181, Satz VI, p. 357].

of Theorem 7.2, (1) implies that the elementary divisors of A are all linear and by part (iii) of Theorem 7.2, (2) implies that all the characteristic roots of A are kth roots of unity. Lipschitz had clearly overlooked Frobenius' paper! Worse yet, Kronecker responded to Lipschitz's paper by suggesting in a paper of 1890 [366] how Lipschitz's theorem could be deduced (nontrivially) by means of considerations similar to some he had published earlier that year for orthogonal systems. He too seems to have forgotten (or never learned) Frobenius' far simpler proof. As Frobenius said in 1896 regarding his minimal polynomial theorem, "So far, little attention has been paid to this consequential theorem" [209, p. 711]. Not only, he continued, were Lipschitz and Kronecker unfamiliar with the theorem and with his 1878 paper containing it, but also almost all "the English and American algebraists, who have concerned themselves considerably with the theory of matrices" [209, p. 712]. Ironically, the activity in the realm of matrix algebra to which Frobenius referred seems to have been prompted by the need to develop Cayley's and Sylvester's ideas more generally and rigorously, whereas this had already been done by Frobenius.[12]

After 1896, however, Frobenius' accomplishments in the realm of matrix algebra and its applications became better known. As already mentioned, his solution to Kronecker's complex multiplication problem, together with the attendant results on matrix algebra, were highlighted by Adolf Krazer in his treatise on the theory of theta functions, which appeared in 1903 [350, Ch. 6] and, as we saw in Section 10.7.4, had considerable influence on Lefschetz's work on abelian varieties with complex multiplication. Even earlier, in 1899, Peter Muth (1860–1909), who had been one of Moritz Pasch's doctoral students at the University of Giessen, published the first book devoted to a systematic exposition of the theory of quadratic and bilinear forms that had been created by Weierstrass, Kronecker, and Frobenius.[13] Entitled *Theory and Application of Elementary Divisors* [450], Muth's book made matrix algebra, which is developed along the lines set out in Frobenius' 1878 paper [181], central to the theory.[14] The basics were expounded in the second chapter, which closely followed the development as given in Frobenius' 1878 paper, and was then used throughout the book. In particular, Muth stressed the importance of Frobenius' "simple, elegant" solution to the congruence problems (Problems 16.3 and 16.4) [450, pp. xii–xiii, p. 125] and used it to derive Kronecker's theory of singular families of quadratic forms as an easy consequence of his theory for families of bilinear forms [450, pp. 125–128] as well as Kronecker's congruence theory for the special families $x^t(\lambda A + A^t)y$ [450, pp. 142–143]. Muth's book became a standard reference, and Frobenius' matrix-algebraic square-root technique

[12]See in this connection [270, p. 107n.15].

[13]Regarding Muth's life and work, see [463].

[14]In the preface, Muth wrote that he had been encouraged by the fact that "From the outset my undertaking was of special interest to several outstanding experts in the theory of elementary divisors," namely Frobenius, S. Gundelfinger, and K. Hensel (Kronecker's former student) [450, p. iv].

became the standard one for deducing the theory of the congruence of families of quadratic forms from the equivalence theory of pencils of bilinear forms.[15]

Also in 1896, Frobenius began creating the theory of characters and representations of finite groups (Chapters 12–15), which attracted the attention of many mathematicians. One of the principal tools in developing representation theory was linear algebra, and so at the hands of Frobenius and his brilliant student Issai Schur, Frobenius' special brand of linear algebra became a familiar aspect of the theory.

Thanks largely to Frobenius' work, matrix algebra became an additional tool for dealing with problems of a linear-algebraic nature. In the case of the congruence problems (Problems 16.3 and 16.4), matrix algebra rather than the lengthy determinant-based considerations of the papers of Weierstrass and Kronecker had, as Frobenius said, revealed the "proper basis" (*eigentlicher Grund*) for showing why equivalence implies congruence for the families of forms they had considered. This was one of the ways in which Frobenius contributed inadvertently to the decline of the theory of determinants as a principal tool of linear algebra. Another way that was more far-reaching resulted from his "rational" development of elementary divisor theory in 1879, which was based on arithmetic considerations and analogical reasoning rather than heavy use of determinants (Section 8.6). We now turn to the spread and further development of Frobenius' rational approach to the theory.

16.2 Assimilation of Frobenius' Rational Elementary Divisor Theory

In order to discuss the assimilation of Frobenius' rational theory of elementary divisors into the mathematical culture of the twentieth century, it will be helpful as a reference point to outline the main components of his theory as expounded in Section 8.6. For the sake of simplicity, I will limit the outline to $n \times n$ matrices.

Outline 16.6 (Frobenius' rational theory of elementary divisors).

A. Let A be a matrix with coefficients from $\mathfrak{R}$, where $\mathfrak{R}$ is either $\mathbb{Z}$ or $\mathbb{F}[\lambda]$ and $\mathbb{F}$ can be any known field (including the finite Galois fields). Then matrices P, Q with coefficients in $\mathfrak{R}$ and determinants that are units in $\mathfrak{R}$ can be determined such that $PAQ = N$, where N is the diagonal matrix (Smith–Frobenius normal form) with diagonal entries $e_1, \ldots, e_r, 0, \ldots, 0$, where $r = \operatorname{rank} A$, $e_i = d_i/d_{i-1}$ is the ith invariant factor, and $e_{i-1} \mid e_i$. (As usual, $d_1 = 1$, and for $1 < i \leq n$, d_i is the gcd of the $i \times i$ minors of A.)

B. If $A = \lambda A_1 + A_2$, where the A_i have coefficients in $\mathbb{F}$ and $\det A_1 \neq 0$, then matrices P, Q with coefficients in $\mathbb{F}$ (rather than $\mathfrak{R} = \mathbb{F}[\lambda]$) and nonzero determinants can be determined such that $PAQ = N$.

---

[15] See, e.g., [25, pp. 297–301], [127, pp. 120–125], [567, pp. 130–131], [430, pp. 60–61], [240, v. 2, 41–42].

C. Two families of matrices $A = \lambda A_1 + A_2$ and $B = \lambda B_1 + B_2$, where the A_i and B_i have coefficients in $\mathbb{F}$ and A_1 and B_1 have nonzero determinants, are equivalent in the sense that nonsingular P, Q with coefficients in $\mathbb{F}$ exist such that $B = PAQ$ if and only if A and B have the same invariant factors, or equivalently, if and only if they have the same elementary divisors, viz., the prime power factors of the invariant factors. In particular, two matrices A and B with coefficients in $\mathbb{F}$ are similar if and only if $\lambda I - A$ and $\lambda I - B$ have the same invariant factors, respectively elementary divisors.

D. If $e_1, \ldots, e_r$ are monic polynomials in $\mathbb{F}[\lambda]$ such that $e_{i-1} \mid e_i$, then there exists a matrix R of rank r and with coefficients in $\mathbb{F}$ that has the e_i as its invariant factors, i.e., $\lambda I + R = F[\varphi^a] \oplus F[\psi^b] \oplus \cdots$, where $F[\varphi^a], F[\psi^b], \ldots$ are the Frobenius "companion matrices" (8.27) of the prime power factors of the e_i (Section 8.6.3). Call R the rational canonical form associated to $e_1, \ldots, e_r$. (Hence two pencils A, B as in part C above are equivalent if and only if they have the same rational canonical form.)

The first published creative response to Frobenius' theory seems to have come from Kronecker in 1891 (the year of his death) and concerned part A above. In a two-page paper in *Crelle's Journal* [370], Kronecker wrote that Frobenius, in his 1879 paper [182], had given the first method for reducing a matrix A with integer coefficients to a diagonal matrix N with each diagonal entry dividing its successors down the diagonal.[16] Frobenius' proof of part A did indeed imply a method for determining P and Q, but the resulting algorithm was complicated; it was conceived primarily as a proof, not as an efficient method of computing P and Q. The goal of Kronecker's brief note was to sketch out a simpler algorithm for obtaining N by means of "elementary transformations," by which he meant (I) transposition of two rows (or columns) of A; (II) multiplying a row (column) of A by -1; (III) adding an integral multiple of one row (column) of A to a different row (column). Such elementary transformations were well known from the theory of determinants, and Kronecker, in typical fashion, left it to the reader to see that his algorithm was based on repeated use of (I)–(III) as stated above. Each elementary row transformation used defines a unimodular matrix, and their composite is a unimodular matrix P. Likewise, the composite of the elementary column transformations used defines a unimodular matrix Q. Kronecker's method thus produced P and Q such that $PAQ = N$. Kronecker's method of transforming to the normal form N (and the obvious analogue when the entries of A are from $\mathbb{F}[\lambda]$) became the standard approach to part A.[17]

No doubt Kronecker's paper, those by Frobenius and Hensel in 1894–1895 on the containment theorem (Section 16.1.1), and one by Landsberg in 1896, all of them published in *Crelle's Journal*, increased awareness within the mathematical community of Frobenius' rational approach to elementary divisor theory.

---

[16]Kronecker ignored the work of Smith, even though Frobenius referenced it throughout his paper.

[17]For a contemporary rendition, see, e.g., [141, pp. 459ff.].

That awareness was increased much further, however, by the publication in 1899 of Muth's comprehensive book on elementary divisor theory (see Section 16.1.6). Of particular interest here is Muth's exposition of Frobenius' rational approach to the subject, which closely followed that of Frobenius' 1879–1880 papers [182, 185].

Except for the inclusion of Kronecker's "elementary transformation" approach to part A, Muth presented parts A–C essentially as had Frobenius. In particular, the theory of part A was presented for $\mathfrak{R} = \mathbb{Z}$ in great detail and then quickly extended by analogy to the case in which $\mathfrak{R}$ consists of polynomials in one variable. However, Muth tacitly considered only the classical case of polynomials with complex coefficients, i.e., he considered only $\mathbb{F} = \mathbb{C}$ in A–C. Although Frobenius had pointed out that his reasoning was valid for all known fields, Muth emphasized Frobenius' theory because it resolved the rationality paradox of Weierstrass' theory (Section 8.6.1). No doubt because the focus was on the classical case $\mathbb{F} = \mathbb{C}$, part D (the rational canonical form) was ignored. Instead, Muth followed his exposition of Frobenius' theory by a determinant-based alternative development of elementary divisor theory along the lines of Weierstrass' original paper [450, pp. 69ff.]. He then used the results to construct families in Jordan–Weierstrass canonical form with prescribed elementary divisors [450, pp. 85ff.]; in working over $\mathbb{C}$, there is no need of the rational canonical form. Muth's book made Frobenius' rational theory of elementary divisors more readily available to the international community of mathematicians, although its applicability to any known field and the concomitant rational canonical form were features that remained hidden in Muth's more traditional approach. It is quite possible that Frobenius himself approved of this approach for a textbook.

The first textbook on higher algebra to include elementary divisor theory seems to have been written by Maxime Bôcher, a professor of mathematics at Harvard University, who had spent several years at Göttingen with Felix Klein before publishing his book *Introduction to Higher Algebra* in 1907 [25], with a German translation appearing in 1910 [26].[18] In the preface, Bôcher expressed his indebtedness to Frobenius and Kronecker for the form taken by his book. In developing parts A–C of Outline 16.6, Bôcher dropped the analogical approach to part A of Frobenius and Muth and dealt exclusively with $\mathfrak{R} = \mathbb{C}[\lambda]$, whose elements he termed λ-matrices. (Of course, he also utilized the elementary transformations of Kronecker in part A.) Perhaps because his was an introductory text for students, Bôcher, like Muth, limited himself to polynomials with real or complex coefficients [25, p. 1], although in a footnote to the chapter on λ-matrices and their equivalence, he noted that "Various modifications of the point of view here adopted are possible and important" [25, p. 262n]. For example one could consider λ-polynomials with

[18]Weber's *Lehrbuch der Algebra* in its various editions from 1895 onward did not treat elementary divisor theory.

coefficients in “a certain domain of rationality.”[19] As with Muth, the limitation of the text to polynomials over $\mathbb{C}$ obviated the need for part D, and the rational canonical form was not presented. Several later authors of books on matrix theory, e.g., Turnbull and Aitken (1932) [567] and Wedderburn (1934) [585], adhered to Bôcher's approach and developed elementary divisor theory via parts A–C with $\mathfrak{R} = \mathbb{C}[\lambda]$.

Thus in the books of Muth and Bôcher, the implication that Frobenius' theory was applicable to any field was ignored (Muth) or relegated to a passing footnote (Bôcher). A different attitude was taken by Alfred Loewy in his 1910 exposition of elementary divisor theory [425] in the second edition of Pascal's *Repertorium der höheren Mathematik*, a well-known reference work at the turn of the last century. Loewy (1873–1935) was a great admirer of Frobenius' mathematics, including his work with matrices. After receiving his doctorate in 1894 from the University of Munich, Loewy moved to the University of Freiburg, where he became an assistant professor in 1902. Frobenius' old friend and collaborator Ludwig Stickelberger was also at Freiburg, where he had been since 1879, when he left the Zurich Polytechnic and his colleague Frobenius. Undoubtedly, Loewy's association with Stickelberger encouraged his appreciation of Frobenius' mathematics. (Frobenius' multifaceted influence on Loewy will be discussed further in the next section, since it is relevant to the work of Loewy's student Wolfgang Krull.)

In his *Repertorium* article, Loewy, unlike Muth and Bôcher, stressed the wide ranging applicability of Frobenius' results. For example, after beginning (for expository reasons) with matrices A over $\mathbb{Z}$ and introducing the notion of the associated invariant factors $e_i(A) = d_i(A)/d_{i-1}(A)$ and their fundamental properties—namely (1) $e_{i-1}(A) \mid e_i(A)$ and (2) $e_i(AB)$ is a multiple of both $e_i(A)$ and $e_i(B)$—he wrote [425, p. 105]:

> All of these theorems are capable of *far-reaching generalizations*. They are based on the unique decomposition of whole numbers into prime factors and on the existence of the greatest common divisor of two or more whole numbers. Let Ω be a *domain of rationality* or a number field ... [and let] $\rho_1, \ldots, \rho_k$ be k variables. A sum of terms of the form
>
> $$\alpha \rho_1^{r_1} \rho_2^{r_2} \cdots \rho_k^{r_k},$$
>
> where the α are integers from Ω and $r_1, \ldots, r_k$ are positive whole numbers or 0, is called a polynomial in Ω. These can be distinguished as *reducible* (decomposable) or *irreducible* (indecomposable) .... The reducible functions can be decomposed into a finite number of irreducible factors, which are likewise polynomials in Ω; every reducible function uniquely determines its irreducible factors up to multiplicative constants. On this the concept of the *greatest common divisor* of two or more polynomials in Ω is based; it is uniquely determined up to a constant factor ....

[19] Bôcher also mentioned (1) λ-polynomials with integer coefficients and (2) polynomials in several variables; but it is unclear what he thought could be proved in these two cases, neither of which involves a principal ideal domain.

Clearly under the influence of Frobenius' paper of 1894 [203] (Section 16.1.1), what Loewy was excitedly stressing, albeit lacking the requisite vocabulary for succinctness, was that the notion of invariant factors of matrices and the above mentioned properties were valid for $\mathfrak{R} = \Omega[\rho_1,\ldots,\rho_k]$—and not just with $k=1$—because (as Frobenius had shown) all that is needed is uniqueness of factorization into prime factors.

Loewy then pointed out that if A, B are matrices with coefficients in $\mathfrak{R} = \Omega[\rho_1,\ldots,\rho_k]$ and if B is a multiple of A such that $B = PAQ$ for some matrices P, Q with coefficients also from $\mathfrak{R}$, then (by virtue of property (2) above) a necessary condition for $B = PAQ$ is that $e_i(B)$ be a multiple of $e_i(A)$. He pointed out with a simple example with $k = 2$ variables that the condition is not necessary when $k > 1$.[20] Loewy then went on to state Frobenius' containment theorem (Theorem 8.16: B is a multiple of A if and only if $e_i(B)$ is a multiple of $e_i(A)$) for matrices with coefficients in either $\mathbb{Z}$ or $\Omega[\rho]$. (Loewy failed to mention that the proof of the sufficiency part of this theorem requires part A of Outline 16.6 for the two mentioned coefficient domains.) Loewy also sought to get at the basis for the failure of the sufficiency part of the theorem, by attributing it to the fact in $\Omega[\rho_1,\ldots,\rho_k]$ for $k > 1$, "the 'greatest common divisor' does not capture all that is common" [425, p. 108n]. It does not seem from this vague remark that Loewy had put his finger squarely on the problem, but he certainly realized that it was related to the properties of the greatest common divisor.

In Section 8.4, we saw that Frobenius had proved that if A is a multiple of B and vice versa, then in fact, by his containment theorem, A and B are equivalent, i.e., unimodular matrices P and Q exist such that $PAQ = B$. Of course, both Frobenius and Loewy realized that the containment theorem was valid for $\mathfrak{R} = \mathbb{F}[\lambda]$ as well as for $\mathfrak{R} = \mathbb{Z}$. Loewy in effect used this consequence of Frobenius' containment theorem to pass (in effect) from part A directly to part C, which is of course the fundamental theorem of the theory of elementary divisors [425, p. 109]. Specializing to $\Omega = \mathbb{C}$, Loewy then proceeded to expound Weierstrass' theory and to present the Weierstrass–Jordan form as a means of constructing a nonsingular pencil with any given elementary divisors. Thus Loewy also omitted any discussion of the rational canonical form (part D), although as we shall see in the next section, that form later came into prominence due to Loewy's work on differential equations. Loewy concluded his above presentation with some historical remarks. In particular, he emphasized the important role played by Frobenius [425, pp. 115–116]:

> Frobenius laid down the bridge between the investigations of Smith and Weierstrass .... He showed the great generality of the concept of "elementary divisor"; one also owes to him Theorem IV, which is fundamental to the entire theory and holds for matrices with elements that are whole numbers or are polynomials of one or more variables with arbitrary constant coefficients or quantities from a domain of rationality Ω.

---

[20] Let $A = \begin{pmatrix} \rho_1\rho_2 & 0 \\ 0 & 1 \end{pmatrix}$ and $B = \begin{pmatrix} \rho_1 & 0 \\ 0 & \rho_2 \end{pmatrix}$. Then A and B have the same invariant factors but $B = PAQ$ is impossible, because it implies $\det P \det Q = 1$ as well as that $\det P$ depends on the ρ_i and vanishes for $\rho_1 = \rho_2 = 0$ [425, p. 107].

Theorem IV is the theorem asserting the necessity that the invariant factors of B be multiples of those of A if B is to be a multiple of A.

Although subsequent developments of the theory of elementary divisors did not follow Loewy's suggestion and prove the containment theorem so as to make it possible to proceed from part A to part C without needing part B—probably because the proof of the Theorem IV part of the containment theorem was complicated—Loewy's interest in the most general class of mathematical objects for which various parts of the theory could be established presaged the advent of the "modern algebra" movement within mathematics. This movement, for which the universities at Göttingen and Hamburg were centers, was well under way by the 1920s through the work of mathematicians such as Emmy Noether and Emil Artin, and became epitomized in B.L. van der Waerden's two-volume book *Moderne Algebra* (1930–1931), which was based in large part on lectures by Noether and Artin.

Frobenius' theory, of course, transferred readily and without significant modifications to the abstract context of rings and fields. This can be seen in C.C. MacDuffee's book *The Theory of Matrices* [430], which was published in 1933 in Springer-Verlag's series *Ergebnisse der Mathematik und ihrer Grenzgebiete*. Thus part A of Outline 16.6 (reduction to Smith–Frobenius normal form) is established (using Kronecker's elementary transformations) for an abstract principal ideal ring $\mathfrak{R}$ [430, pp. 29ff.]. Then taking $\mathfrak{R} = \mathbb{F}[\lambda]$, with $\mathbb{F}$ an abstract field, part B is established more or less as Frobenius had done it in 1879 [430, pp. 45ff.]. Part C of Frobenius' theory then follows for matrices over $\mathbb{F}$, and then Frobenius' rational canonical form for $\lambda I - A$ (part D) is given, except with Frobenius' blocks $F[\varphi]$ (8.27), $\varphi = \lambda^{\alpha} + a_1\lambda^{\alpha-1} + \cdots + a_{\alpha}$, modified to

$$F^*[\varphi] = \begin{pmatrix} \lambda & -1 & 0 & \cdots & 0 \\ 0 & \lambda & -1 & \cdots & 0 \\ \cdots & \cdots & \cdots & \cdots & \cdots \\ a_{\alpha} & a_{\alpha-1} & a_{\alpha-2} & \cdots & a_1 + \lambda \end{pmatrix}.$$

Like the presentations of Muth and Bôcher, MacDuffee's was in the same mold as outlined by Frobenius. Unlike them, however, he presented a rational canonical form. This was probably due to the fact that it was emphasized by van der Waerden in the second volume of his *Moderne Algebra*, which was familiar to MacDuffee when he wrote his own book.

As we shall see in the next section, van der Waerden's development of a rational approach to elementary divisor theory, like all earlier treatments, had as its starting point part A of Frobenius' theory, but after that, it diverged significantly from Frobenius' approach. Inspired in this connection by work of Loewy's student Wolfgang Krull, van der Waerden introduced the now familiar approach of deducing the rational canonical form of a linear transformation as an application of the fundamental theorem of finitely generated modules over a principal ideal ring. The developments leading up to van der Waerden's approach are sketched in the following section not only because of their relevance to the history of linear algebra but also because these developments were inspired in diverse ways by work of Frobenius and his students. The section thus documents further instances of the role his work played in the emergence of present-day mathematics.

16.3 The Module-Theoretic Approach to Elementary Divisors

The work of Krull that proved inspirational to van der Waerden was in turn motivated by two research programs developed by Loewy. We begin with a discussion of these two programs. Then Krull's work, initially conceived as an abstract theory with applications to both of Loewy's programs, is treated. Finally, van der Waerden's development of Krull's idea within the context of modules over principal ideal rings is considered.

16.3.1 Loewy on differential equations and matrix complexes

Like Frobenius earlier, Loewy was interested in the algebraic aspects of the Fuchsian theory of linear homogeneous ordinary differential equations

$$A(y) = y^{(n)} + a_1 y^{(n-1)} + \cdots + a_{n-1}y' + a_n y = 0, \qquad (16.16)$$

where $y^{(i)} = d^i y/dx^i$ and n is called the *order* of the equation. Loewy worked under the hypothesis that $y = f(x)$ and all the coefficients $a_i = a_i(x)$ belong to a field Σ of functions $f(x)$ that are defined in a fixed region D of the complex plane and are analytic there, except possibly for a finite number of isolated singularities. In addition, Σ is closed under differentiation: if $f(x) \in \Sigma$, then $f'(x) \in \Sigma$. The classical example of Σ was the totality of all rational functions of the complex variable x. Another notable special case arises when Σ contains only constant functions, i.e., Σ is a subfield of $\mathbb{C}$; then $A(y) = 0$ has constant coefficients.

Loewy's program of research can be traced back to Frobenius' paper of 1873 [175], in which the distinction between irreducible and reducible equations $A(y) = 0$ was introduced (see Section 1.2). As we saw, among other things, Frobenius showed that if $A(y) = 0$ is reducible, and so shares a solution with an equation of lesser order, then an irreducible equation $J(y) = 0$ exists, that has all its solutions in common with $A(y) = 0$; furthermore, $B(y) = 0$ exists such that we have a factorization $A = BJ$ in the sense that $A(y) = B(J(y))$ for any $y = f(x)$, i.e., the differential operator A is the composite of J with B. Here, of course, the orders of J and B add up to the order of A, just as in polynomial factorization. It follows from Frobenius' result that an irreducible equation $A(y) = 0$ may be factored as $A = J_1 J_2 \cdots J_g$, where each J_i is irreducible ([280, pp. 191–193], [513, p. 85]). In 1902, Frobenius' student Edmund

Landau, who was then an instructor (*Privatdozent*) at Berlin, pointed out by a simple example that such a factorization is not unique and that in fact, infinitely many such factorizations can exist [394, p. 116]. He went on to show, however, that in any two such factorizations, the number of irreducible factors is the same and the irreducible factors from the two factorizations can be matched up so that matched pairs have the same orders [394, pp. 117ff.].

Landau's result formed the starting point of Loewy's investigations. Using the rationality group of Picard and Vessiot,[21] Loewy showed in a paper of 1903 that in any two factorizations into irreducible factors, the irreducible factors can be matched up so that the matched pairs have the same rationality group, which implies in particular Landau's theorem [421, p. 6]. In a second paper the same year, Loewy further developed his rationality-group approach to factorization [424] by combining it with notions from Frobenius' theory of matrix representations of finite groups (1897–1899; see Chapters 13 and 15). For example, he showed that $A(y)=0$ is irreducible if and only if its rationality group is irreducible in a sense analogous to that used by Frobenius in his theory of matrix representations.

Loewy also added a second improvement to Landau's theorem. That is, given any two factorizations of any A into irreducible factors, $A=J_1\cdots J_g=J_1'\cdots J_{g'}'$, not only does one have $g=g'$, but a permutation σ of $1,\ldots,g$ exists such that the pairs J_i and $J_{\sigma(i)}'$ not only have the same order but also are of the same *type* [424, p. 565]. The notion that two equations $A(y)=0$ and $B(z)=0$ of the same order n are also of the same type had been introduced by Poincaré in the 1880s and perfected by Fuchs. It plays a role in what follows, so here is the definition: $A(y)=0$ is of the same type as $B(y)=0$ if a differential operator P of order $n-1$ and with coefficients from Σ exists with the property that for every solution $y=f(x)$ of $A(y)=0$, there is a solution $z=g(x)$ of $B(z)=0$ such that $f(x)=P(g(x))$. It then follows (as Fuchs showed) that conversely, $B(y)=0$ is also of the same type as $A(y)=0$. In short, $A(y)=0$ and $B(y)=0$ are of the same type.

In 1913 [426], Loewy introduced a new point of view into his work on differential operators. It proved to be motivational for that line of Krull's post-thesis work of interest to us here. If $A(y)=0$ is the nth-order differential equation of (16.16), and if $y_1=f(x)$ is a solution to $A(y)=0$, then $y=\begin{pmatrix}y_1 & \cdots & y_n\end{pmatrix}^t$ is a solution to the linear system of first order equations

$$\frac{dy_1}{dx}=y_2,\ldots,\qquad \frac{dy_{n-1}}{dx}=y_n,\qquad \frac{dy_n}{dx}+a_ny_1+\cdots+a_1y_n=0,$$

since the first $n-1$ equations say that $y_i=d^{i-1}y_1/dx^{i-1}=y_1^{(i-1)}$, and so the last asserts that $y_1^{(n)}+a_1y_1^{(n-1)}+\cdots+a_ny_1=0$, i.e., that $A(y_1)=0$, which is correct, since y_1 is a solution of $A(y)=0$. Conversely, if $y=\begin{pmatrix}y_1 & \cdots & y_n\end{pmatrix}^t$ is a solution to the

[21] On the rationality group and Picard–Vessiot theory, see Gray's historical account [255, pp. 267ff.].

above system, then $A(y_1) = 0$. This connection between the solutions of $A(y) = 0$ and those of the above linear system was not new, but for Loewy it was the starting point for a new approach to his study of $A(y) = 0$.[22]

Expressed in matrix form, the above linear system is $y' + \mathfrak{A}y = 0$, where

$$\mathfrak{A} = \begin{pmatrix} 0 & -1 & 0 & \cdots & 0 \\ 0 & 0 & -1 & \cdots & 0 \\ \cdots & \cdots & \cdots & \cdots & \cdots \\ a_n & a_{n-1} & a_{n-2} & \cdots & a_1 \end{pmatrix}. \tag{16.17}$$

In 1917, Loewy named $\mathfrak{A}$ the *companion matrix* (*Begleitmatrix*) to A [427, p. 255], and, as we shall see, he was aware of the connection with the matrices $F[\varphi]$ (8.27) of Frobenius' rational canonical form [427, p. 262n] and regarded Frobenius' theory as a special case of his own.[23]

Central to the new theory was the translation of the above notion that $A(y) = 0$ and $B(z) = 0$ are of same type into the context of the associated companion matrices. If A and B are two nth-order differential operators with respective companion matrices $\mathfrak{A}$ and $\mathfrak{B}$, then the latter are said to be of the same type if a nonsingular matrix $\mathfrak{P}$ over Σ exists such that $y = \mathfrak{P}z$ transforms $y' + \mathfrak{A}y$ into $z' + \mathfrak{B}z$. If $y = \mathfrak{P}z$ and $y' = \mathfrak{P}z' + \mathfrak{P}'z$ are substituted into $y' + \mathfrak{A}y$, the result (after left multiplication by $\mathfrak{P}^{-1}$) is $z' + (\mathfrak{P}' + \mathfrak{P}^{-1}\mathfrak{A}\mathfrak{P})z$, and so $\mathfrak{A}$ and $\mathfrak{B}$ are of the same type when

$$\mathfrak{B} = \mathfrak{P}' + \mathfrak{P}^{-1}\mathfrak{A}\mathfrak{P}. \tag{16.18}$$

With this definition, A and B are of the same type if and only if $\mathfrak{A}$ and $\mathfrak{B}$ are of the same type [427, p. 256]. As Loewy observed, in the special case in which Σ contains only constant functions, $\mathfrak{P}' = 0$, and so (16.18) becomes the requirement that the matrices $\mathfrak{A}$ and $\mathfrak{B}$ be similar: $\mathfrak{A} \cong \mathfrak{B}$. For brevity, I will therefore denote the more general equivalence relation (16.18) by $\mathfrak{A} \cong_D \mathfrak{B}$. Since Frobenius' distinction between reducible and irreducible differential equations $A(y) = 0$ had played a central role in Loewy's earlier work, it should be noted that he showed that $A(f) = 0$ is reducible if and only if its companion matrix $\mathfrak{A}$ is reducible in a sense analogous to that used by Frobenius in his theory of matrix representations:[24]

$$\mathfrak{A} \cong_D \begin{pmatrix} \mathfrak{P} & 0 \\ \mathfrak{Q} & \mathfrak{R} \end{pmatrix}.$$

[22]Loewy refers to Schlesinger's 1908 book on differential equations [514, pp. 157–158], which may have suggested to him the value of this connection for his own work, as described below.

[23]The main results of Loewy's theory are outlined in [427]. His most complete exposition is contained in a paper published in *Mathematische Zeitschrift* in 1920 and dedicated to Ludwig Stickelberger on the occasion of his 70th birthday [429]. The 1917 paper, however, brings out more fully the connections with Frobenius' theory.

[24]This result is implicit in [427] and explicit in [429, p. 102].

Loewy's first main theorem was as follows ([427, p. 258], [429, p. 81]).

Theorem 16.7. *A differential operator A of order n is expressible as $A = B_1 \cdots B_k$, where B_i has order n_i, if and only if $\mathfrak{A} \cong_D \mathfrak{T}$, where $\mathfrak{T}$ has lower triangular block form with diagonal blocks $\mathfrak{T}_{ii} \cong_D \mathfrak{B}_i$, $\mathfrak{B}_i$ being the companion matrix of B_i. Moreover, $\mathfrak{T} \cong_D \mathfrak{N}$, where $\mathfrak{N}$ is the normal form*

$$\mathfrak{N} = \begin{pmatrix} \mathfrak{B}_1 & 0 & 0 & \cdots & 0 & 0 \\ \mathfrak{Z}_{21} & \mathfrak{B}_2 & 0 & \cdots & 0 & 0 \\ \cdots & \cdots & \cdots & \cdots & \cdots & \cdots \\ 0 & 0 & 0 & \cdots & \mathfrak{Z}_{kk-1} & \mathfrak{B}_k \end{pmatrix}, \tag{16.19}$$

and $\mathfrak{Z}_{ii-1}$ is the $n_i \times n_{i-1}$ matrix

$$\mathfrak{Z}_{ii-1} = \begin{pmatrix} 0 & 0 & \cdots & 0 \\ 0 & 0 & \cdots & 0 \\ \vdots & \vdots & & \vdots \\ -1 & 0 & \cdots & 0 \end{pmatrix}. \tag{16.20}$$

As we have seen, it is always possible to express A as a product of irreducible factors B_i, in which case the corresponding companion matrices $\mathfrak{B}_i$ of the theorem will be irreducible.

Loewy's second main theorem involved operators A that are "completely reducible" in the following sense: the linear transformations T of the rationality group have a matrix representation in block diagonal form with each diagonal block $M_i(T)$ being irreducible. In 1903 [424], Loewy had shown that each block is the rationality group of an irreducible equation $J_i(y) = 0$ and that $A(y)$ can be factored as $A = B_iJ_i$. Thus A can be thought of as a common multiple of all the J_i. Loewy had shown that the B_i had further properties that justified regarding A as the least common multiple of $J_1, \ldots, J_g$ [424, p. 582]. Now, from his new point of view, he was able to prove the following result ([427, p. 258], [429, pp. 98–99]).

Theorem 16.8. *If A is completely reducible and so the least common multiple of irreducible operators $J_1, \ldots, J_g$, then $\mathfrak{A} \cong_D \mathfrak{J}_1 \oplus \cdots \oplus \mathfrak{J}_g$, where $\mathfrak{J}_i$ is the irreducible companion matrix of J_i.*

As we shall see, Theorem 16.8, which shows that $\mathfrak{A}$ has a direct sum decomposition into irreducible components, seems to have provided part of the motivation behind Krull's theory of generalized abelian groups, especially when combined with the uniqueness of such decompositions that followed from Loewy's theory of matrix complexes, to which I now turn.

We have seen that concepts from Frobenius' theory of group representations, such as the distinction between reducible and irreducible representations and the concept of complete reducibility, had influenced Loewy's work on differential equations. Frobenius' theory also turned Loewy's attention, at the same time, to the theory of groups of linear substitutions, i.e., groups of transformations

$$x' = A_\lambda x,$$

where the A_λ are $n \times n$ matrices, indexed by λ and defining a possibly infinite group under matrix multiplication. If the group is finite, then by Frobenius' complete reducibility theorem, it follows that a matrix S exists such that the matrices $S^{-1}A_\lambda S = D_\lambda$ are all in block diagonal form (with block size independent of λ) and the diagonal blocks define irreducible groups. If the group is infinite, however, such a block diagonalization need not be possible. However, if the group $\{A_\lambda\}$ is reducible, it is not difficult to show that a matrix S exists such that for all λ, the matrices $S^{-1}A_\lambda S = L_\lambda$ are in a lower triangular block form with the diagonal blocks $L_\lambda^{(i)}$ defining irreducible groups.

In a paper published in 1903 [422], Loewy showed that the reduction to lower triangular block form is unique in the sense that if $T^{-1}A_\lambda T = M_\lambda$ is also in a lower triangular block form with irreducible diagonal blocks $M_\lambda^{(j)}$, then the number of irreducible blocks of L_λ and M_λ is the same and they can be matched up so that matching pairs $L_\lambda^{(i)}$ and $M_\lambda^{(i')}$ are equivalent representations in Frobenius' sense, i.e., a matrix S_i exists such that $S_i^{-1}L_\lambda^{(i)}S_i = M_\lambda^{(i')}$ for all λ. An immediate corollary is that if the group is actually completely reducible, so that the matrices L_λ are actually in block diagonal form (with irreducible diagonal blocks), then those diagonal blocks are unique up to their ordering along the diagonal.

Frobenius and Schur were able to generalize Loewy's result [234, p. 385]. They assumed that $\{A_\lambda\}$ is a possibly infinite collection of $n \times n$ matrices that is closed under matrix multiplication but need not form a group. Thus A_λ need not be invertible and the identity matrix need not be a A_λ. In other words, $\{A_\lambda\}$ was assumed to be what is now called a semigroup. Under this assumption, they showed that it still follows that S exists such that $S^{-1}A_\lambda S = L_\lambda$ is in lower triangular block form with diagonal blocks $L_\lambda^{(i)}$ that are either irreducible or identically zero. Frobenius and Schur showed that these diagonal blocks are unique up to ordering in the sense of Loewy's above-described result. Of course, it follows that if the semigroup $\{A_\lambda\}$ is completely reducible in the sense that L_λ is actually in block diagonal form, then the diagonal blocks $L_\lambda^{(i)}$ are unique up to ordering.

In a paper of 1917 on "Matrix and differential complexes" [428], Loewy combined his earlier work on differential equations and on groups of linear substitutions to obtain a substantial generalization of the result of Frobenius and Schur. He considered any set (or *complex*, as he called it) of $n \times n$ matrices $\{A_\lambda\}$ rather than a group or semigroup, and he allowed the coefficients of the A_λ to belong to any function field Σ of the sort utilized in his above-described work on differential equations. Such a matrix complex was then said to be reducible if a matrix S over Σ exists such that $S^{-1}A_\lambda S + S' = \begin{pmatrix} B_\lambda & 0 \\ C_\lambda & D_\lambda \end{pmatrix}$. In the notation following (16.18), by virtue of the existence of such an S, one has $A_\lambda \cong_D \begin{pmatrix} B_\lambda & 0 \\ C_\lambda & D_\lambda \end{pmatrix}$ for all λ. In Loewy's terminology, the matrix complex $\mathfrak{C} = \{A_\lambda\}$ and the matrix complex

$$\mathfrak{C}' = \left\{ \begin{pmatrix} B_\lambda & 0 \\ C_\lambda & D_\lambda \end{pmatrix} \right\}$$

are of the same type. I will denote this reduction by $\mathfrak{C} \cong_D \mathfrak{C}'$. Loewy then showed that if $\mathfrak{C}$ is reducible, then the reduction $\mathfrak{C} \cong_D \mathfrak{C}'$ can be carried further, so that $\mathfrak{C} \cong_D \mathfrak{L}$, where the matrices L_λ of $\mathfrak{L}$ are all in block triangular form with block sizes independent of λ and with all the diagonal block complexes $\mathfrak{L}^{(i)} = \{L_\lambda^{(i,i)}\}$ irreducible, i.e., not reducible in Loewy's above-described sense. He also showed that the reduction to lower triangular form $\mathfrak{C} \cong_D \mathfrak{L}$ has the following invariance properties: if $\mathfrak{C} \cong_D \mathfrak{M}$ denotes another reduction to the above-described type of block lower triangular form, then the number of irreducible diagonal block complexes in $\mathfrak{L}$ and in $\mathfrak{M}$ is the same, and there is a pairing $(\mathfrak{L}^{(i)}, \mathfrak{M}^{(i')})$ of the irreducible diagonal block complexes of $\mathfrak{L}$ and $\mathfrak{M}$ such that $\mathfrak{L}^{(i)} \cong_D \mathfrak{M}^{(i')}$ [428, p. 21]. This result, too, applies to the special case in which the complex $\mathfrak{C}$ is completely reducible in the obvious sense. In particular, it applies to the context of Loewy's second main theorem on differential operators (Theorem 16.8) and shows that if $\mathfrak{A}$ is the companion matrix of a completely reducible differential operator A, so that $\mathfrak{A} \cong_D \mathfrak{J}_1 \oplus \cdots \oplus \mathfrak{J}_g$, $\mathfrak{J}_i$ being irreducible, then this representation of $\mathfrak{A}$ is unique in the following sense: if $\mathfrak{A} \cong_D \mathfrak{D}_1 \oplus \cdots \oplus \mathfrak{D}_h$, with $\mathfrak{D}_i$ irreducible, then $h = g$ and there is a permutation $i \to i'$ such that $\mathfrak{D}_i \cong_D \mathfrak{J}_{i'}$ for all i.

16.3.2 Krull's theory of generalized abelian groups

As we shall see, the goal of establishing existence and uniqueness of decompositions of matrices such as Loewy's $\mathfrak{A}$ as an application of a more general and abstract group-theoretic theorem was taken up by Loewy's student Wolfgang Krull a few years after he had completed his doctoral dissertation under Loewy's direction. It is instructive to begin with a few words about the goal of his dissertation, since the experience of seeking to attain that goal put him in a position to later realize that the dissertation goal could be achieved with far greater originality as another application of the same group-theoretic theorem.

The goal of Krull's doctoral dissertation seems to have been inspired by Loewy's reflections on his first main theorem on differential operators, Theorem 16.7. Loewy realized that his theorem was analogous to Frobenius' theorem on the rational canonical form (part D of Outline 16.6 of Frobenius' theory). In fact, as he observed [427, pp. 261ff.], if in the theorem, the function field Σ contains only constant functions, i.e., if Σ is a subfield of $\mathbb{C}$, then Theorem 16.7 asserts that every companion matrix is similar to one in the simple rational canonical form $\mathfrak{N}$ of that theorem, viz., (16.19). To see this clearly, let $\varphi(t) = t^n + a_1 t^{n-1} + \cdots + a_n \in \Sigma[t]$ and set $A_\varphi(y) = y^{(n)} + a_1 y^{(n-1)} + \cdots + a_n y$. Thus A_φ is a differential operator with constant coefficients. Denote the companion matrix of A_φ by $\mathfrak{A}_\varphi$. Then $\varphi(t) = \det[tI + \mathfrak{A}_\varphi]$, and

$\varphi(t)$ is the sole nontrivial invariant factor of $tI + \mathfrak{A}_\varphi$. If $\varphi(t) = \prod_{i=1}^{k} \pi_i(t)^{a_i}$ represents the factorization of $\varphi(t)$ over Σ into distinct irreducible factors $\pi_i(t)$, then the factors $\pi_i^{a_i}$ are the elementary divisors. Now because A_φ has constant coefficients, $A_\varphi = \prod_{i=1}^{k} A_{\pi_i^{a_i}}$. Thus Theorem 16.7 applies with $B_i = A_{\pi_i^{a_i}}$ and implies that $\mathfrak{A}_\varphi \cong \mathfrak{N}$, since $\cong_D$ becomes $\cong$ (similarity) when Σ contains only constants.

The only difference between Loewy's normal form $\mathfrak{N}$ and Frobenius' rational canonical form, which here would give $\mathfrak{A}_\varphi \cong F[\pi_1^{a_1}] \oplus \cdots \oplus F[\pi_k^{a_k}]$, where the $F[\pi_i^{a_i}]$ are the Frobenius "companion matrices" (8.27) for the polynomials $\pi_i^{a_i}$, is that Loewy's rational normal form contains the extra blocks $\mathfrak{Z}_{ii-1}$ of (16.20). (It turns out, as Krull showed in his doctoral dissertation, that the blocks $\mathfrak{Z}_{ii-1}$ are necessary in general but not when the companion matrices are associated to polynomials that are pairwise relatively prime.) Of course, Loewy's theorem shows only that every constant companion matrix is similar to the normal form $\mathfrak{N}$. It does not show the same for any matrix over Σ, and companion matrices are rather special, e.g., they have a single invariant factor. Nonetheless, in the preface to his 1920 exposition of his theory, Loewy pointed to his normal form $\mathfrak{N}$ for companion matrices. Observing that it "agrees with" the one that Frobenius used in his "classical work" of 1879 [182] for matrices with constant coefficients, Loewy declared that by virtue of Theorem 16.7, Frobenius' rational canonical form attained thereby "a much more far-reaching domain of validity" [429, p. 60].

Loewy vowed to go into the significance of companion matrices for an arbitrary matrix later [427, p. 263], but he never did. Instead, he encouraged Krull to write his doctoral dissertation on this subject. The dissertation, entitled "On Companion Matrices and Elementary Divisor Theory," was presented in 1921 [372] and fulfilled Loewy's wish to see companion matrices play a fundamental role in a rational theory of elementary divisors. Thus in the first section, Krull showed that given any matrix A over an arbitrary (abstract) field $\mathbb{F}$, polynomials $\psi_1, \ldots, \psi_m$ from $\mathbb{F}[t]$ may be determined such that (i) $\psi_{i+1} \mid \psi_i$; (ii) if $C[\psi_i]$ denotes the companion matrix of ψ, then for all i, ψ_i is the minimal polynomial of the matrix $C[\psi_i] \oplus \cdots \oplus C[\psi_m]$; (iii) $A \cong K$, where $K = C[\psi_1] \oplus + \cdots \oplus C[\psi_m]$.[25] He then showed that two matrices are similar if and only if they have the same normal form K, i.e., the same $C[\psi_i]$. The polynomials $\psi_i(t)$ turn out to be the invariant factors of $tI - A$ [372, p. 92]. To get a more refined normal form, Krull proved that if $\phi(t)$ and $\theta(t)$ have no common factor over $\mathbb{F}$, then $C[\phi\theta] \cong C[\phi] \oplus C[\theta]$. This means that in the normal form K, if $\psi_i(t) = \prod_{j=1}^{\ell} \pi_j^{a_{ij}}$ gives the prime factorization of $\psi_i(t)$ (the distinct primes are assumed to be monic), then the $\pi_j^{a_{ij}}$ are the elementary divisors corresponding to the ith invariant factor ψ_i and $C[\psi_i] \cong C[\pi_1^{a_{i1}}] \oplus \cdots \oplus C[\pi_\ell^{a_{i\ell}}]$. It then follows from (iii) above that

[25] Krull's companion matrices were defined as the negatives of Loewy's because Krull defined the characteristic polynomial as $\det(tI - A)$, rather than as $\det(tI + A)$, as with Loewy.

$$A \cong \bigoplus_{i=i}^{k} \bigoplus_{j=1}^{\ell} C[\pi_j^{a_{ij}}], \tag{16.21}$$

which is essentially Frobenius' rational canonical form. Krull also proved that if $\pi(t)$ is an irreducible monic polynomial of degree m, then for $a > 1$, there is the further decomposition

$$C[\pi^a] \cong \begin{pmatrix} C[\pi] & 0 & 0 & \cdots & 0 & 0 \\ \mathfrak{Z}_{21} & C[\pi] & 0 & \cdots & 0 & 0 \\ \cdots & \cdots & \cdots & \cdots & \cdots & \cdots \\ 0 & 0 & 0 & \cdots & \mathfrak{Z}_{aa-1} & C[\pi] \end{pmatrix},$$

where $\mathfrak{Z}_{ii-1}$ is an $m \times m$ version of the subdiagonal matrices (16.20) that occur in Loewy's normal form $\mathfrak{N}$ in (16.19).

Except for the prominence given to companion matrices in Krull's version of the theory of elementary divisors, nothing else was radically new, and he never published his results in a mathematical publication; it remained just a Freiburg dissertation. Even as Krull was going through the motions of obtaining his doctorate, he had turned to other mathematical considerations inspired by the work of Emmy Noether. It was only after he had absorbed the abstract way of thinking of Noether that he sought to apply it inter alia to the work of Loewy and elementary divisor theory.

Krull submitted his Freiburg doctoral thesis in October 1921. He had actually spent 1920 and 1921 in Göttingen [494, p. 1], where he became acquainted with Emmy Noether and her work, which had just then turned toward what would now be called abstract algebra. In 1920, she published a joint paper with Werner Schmeidler on ideals in abstract rings of a type that included rings of differential operators such as those considered by Loewy, as well as analogous rings of partial differential operators [459]. The following year, she published a fundamental paper on the theory of ideals in commutative rings [455], which contains her now classical decomposition theorems for ideals in noetherian rings.

While Krull was writing up his doctoral dissertation, inspired by Noether and her work, he was also gearing up for research in the abstract theory of rings and their ideals. His Freiburg *Habilitationsschrift* [373] was submitted for publication in *Mathematische Annalen* and dated "Göttingen 21 January 1922." It was on problems in the new abstract theory of rings and drew upon Noether's results on ideals. Unlike his doctoral dissertation, the *Habilitationsschrift* was published in a mathematics journal and was followed by four more papers on abstract rings and their ideals, all submitted in 1923.[26]

[26]These papers are [4]–[7] in Krull, *Abhandlungen* 1. For an overview of their contents, as well as Krull's subsequent work on the theory of ideals, see P. Ribenboim's essay on Krull [494, pp. 3ff.].

Krull's research on various aspects of the theory of ideals continued for many more years, but in a paper submitted at the end of 1923 [375], he also turned briefly to another line of research that may have been encouraged by Noether's 1920 paper with Schmeidler [459]. Unlike Loewy, who had considered individual differential operators and their factorization into irreducible factors, Noether and Schmeidler focused on ideals in rings of such operators, so that the closest thing to an individual operator was the principal ideal it generated. Loewy's work nonetheless had a considerable influence on theirs. In particular, they generalized the notion of operators of the same type to ideals and obtained several results analogous to those of Loewy. Perhaps their application of abstract algebra to the work of his mentor suggested to Krull the possibility of using abstract algebra to deal with other aspects of Loewy's work, notably the decomposition $\mathfrak{A} \cong_D \mathfrak{J}_1 \oplus \cdots \oplus \mathfrak{J}_g$ of Theorem 16.8 and its uniqueness as in Loewy's theory of matrix complexes. Krull, however, did not seek to follow their ideal-theoretic approach. His work, submitted in December 1923 and published in 1925 [375], was inspired by two decomposition theorems in group theory and a related idea.

The two theorems were contained in the 1911 doctoral dissertation of Frobenius' student Robert Remak (1888–1942), which was of such high quality that it was published in *Crelle's Journal* the same year [493].[27] The first theorem was the Frobenius–Stickelberger version of the fundamental theorem of finite abelian groups (Theorem 9.10), which states, in the mathematical language used by Remak, that every such group can be decomposed into a direct product of "directly indecomposable" subgroups, namely certain cyclic subgroups of prime power orders, and that such a decomposition is unique up to group isomorphisms. Suppose now that $\mathfrak{H}$ is a finite nonabelian group. It is not difficult to see that if $\mathfrak{H}$ is not directly indecomposable, i.e., if it is expressible as a direct product[28] $\mathfrak{H} = \mathfrak{A} \times \mathfrak{B}$, where $\mathfrak{A}$ and $\mathfrak{B}$ are nontrivial subgroups of $\mathfrak{H}$, then $\mathfrak{H} = \mathfrak{G}_1 \times \cdots \times \mathfrak{G}_k$, where the $\mathfrak{G}_i$ are directly indecomposable subgroups of $\mathfrak{H}$. Remak focused on the question of uniqueness of such a decomposition, as had Frobenius and Stickelberger in their paper. He proved that given any two such factorizations $\mathfrak{H} = \mathfrak{G}_1 \times \cdots \times \mathfrak{G}_k = \mathfrak{K}_1 \times \cdots \mathfrak{K}_\ell$, then $k = \ell$ and there is a central automorphism[29] of $\mathfrak{H}$ that takes each $\mathfrak{G}_i$ onto a $\mathfrak{K}_j$. Remak's theorem thus showed that certain parts of the Frobenius–Stickelberger theorem were more generally true.

[27]Despite his considerable mathematical talent, Remak, a Jew, did not have a correspondingly successful career as a mathematician and was eventually deported to Auschwitz, where he perished. For more on Remak's life and work, see [440].

[28]In the terminology and notation of Remak, $\mathfrak{H}$ is the direct product $\mathfrak{H} = \mathfrak{A} \times \mathfrak{B}$ of nontrivial subgroups $\mathfrak{A}$, $\mathfrak{B}$ if (1) every $g \in \mathfrak{G}$ is expressible as $g = ab$, with $a \in \mathfrak{A}$ and $b \in \mathfrak{B}$; (2) $\mathfrak{A} \cap \mathfrak{B} = \{E\}$; (3) every $a \in \mathfrak{A}$ commutes with every $b \in \mathfrak{B}$. When $\mathfrak{H}$ is abelian, this agrees with the definition of $\mathfrak{H} = \mathfrak{A} \times \mathfrak{B}$ given by Frobenius and Stickelberger in 1878. Condition (3) must be added for nonabelian groups.

[29]An automorphism σ of $\mathfrak{H}$ is central if $h^{-1}\sigma(h)$ is in the center of $\mathfrak{H}$ for all $h \in \mathfrak{H}$ [493, p. 293], a notion still in use today.

In conjunction with the theorems of Frobenius–Stickelberger and Remak, Krull developed the idea that an additive, possibly infinite, abelian group $\mathfrak{G}$ may be thought of as a group acted on by a set Ω of operators, namely $\Omega = \mathbb{Z}$, where for $n > 0$, $n \cdot g = g + \cdots + g$ (n summands) and $-n \cdot g = -(n \cdot g)$. Clearly, $n \cdot (g + g') = n \cdot g + n \cdot g'$ for all $g, g' \in \mathfrak{G}$. Dedekind had already made these observations long ago for additive subgroups of $\mathbb{C}$, which he called modules (Section 9.1.5). But Krull pursued the idea further on the abstract level. He considered more generally any additive abelian group $\mathfrak{V}$ with a set of operators $\Omega \supset \mathbb{Z}$ such that (1) $\omega \cdot v \in \mathfrak{V}$ for all $\omega \in \Omega$ and all $v \in \mathfrak{V}$ and that (2) every $\omega \in \Omega$ satisfies $\omega \cdot (v + v') = \omega \cdot v + \omega \cdot v'$ for all $v, v' \in \mathfrak{V}$ [375, §1]. Krull called $\mathfrak{V}$ (with operator domain Ω) a *generalized abelian group*. Likewise, $\mathfrak{W}$ is a (generalized abelian) subgroup of $\mathfrak{V}$ when $\mathfrak{W}$ is a subgroup of the abelian group $\mathfrak{V}$ in the usual sense and in addition, $\omega \cdot \mathfrak{W} \subset \mathfrak{W}$ for all $\omega \in \Omega$. Also, a homomorphism $\phi : \mathfrak{V}_1 \mapsto \mathfrak{V}_2$ of generalized abelian groups with common operator domain Ω must be an ordinary group homomorphism of $\mathfrak{V}_1$ and $\mathfrak{V}_2$ with the following additional property that relates to (1) and (2) above: $\phi(\omega \cdot v) = \omega \cdot \phi(v)$. Henceforth in discussing Krull's work, I will speak of Ω-subgroups and Ω-homomorphisms to avoid ambiguity.

The groups considered by Frobenius and Stickelberger and by Remak were finite, but the applications to Loewy's work (discussed below) involved infinite generalized abelian groups. These particular groups were finite-dimensional as vector spaces, and Krull sought to generalize this sort of finiteness by means of chain conditions with which he had probably become acquainted through his contact with Noether. Thus a "finite" generalized abelian group is a generalized abelian group $(\mathfrak{V}, \Omega)$ with the additional property that both the ascending and descending chain conditions hold for Ω-subgroups.[30] When $\mathfrak{V}$ satisfied these finiteness conditions, Krull called it a *generalized finite abelian group* [375, §2]. It was for these groups that he obtained, by utilizing some of Remak's proof ideas, the following analogue of the theorems of Frobenius–Stickelberger and Remak [375, p. 186].

Theorem 16.9. *If $\mathfrak{V}$ is any generalized finite abelian group with associated operator domain Ω, then $\mathfrak{V}$ has a direct sum decomposition into directly indecomposable[31] Ω-subgroups, and this decomposition is unique up to Ω-isomorphisms.*

In a paper of 1928, the Russian mathematician O. Schmidt showed that Krull's commutativity hypothesis in Theorem 16.9 could be dropped [515], and the

---

[30] The ascending chain condition for ideals had been introduced by Noether in her above-mentioned fundamental paper of 1921 [455, Satz I, p. 30] and goes back to Dedekind, as she noted. All Krull's early papers on ideals cite this paper as a basic reference. Nine months before submitting his paper [375] on generalized abelian groups, Krull had submitted a paper with a descending chain condition on the successive powers of an ideal [374, p. 179, (f)]. As for Noether, her important work on what are now called Dedekind rings [456, 457], involved a descending chain condition. A bit later, Artin used a descending chain condition in his study of what are now called Artinian rings.

[31] Krull spoke simply of indecomposable subgroups, but he meant the analogue of what Remak had called "directly indecomposable" subgroups.

resulting theorem is now called the Krull–Schmidt theorem (or occasionally, the Remak–Krull–Schmidt theorem).

The applications of Theorem 16.9 to Loewy's work involved what Krull called *generalized abelian groups of finite rank* [375, §4]. Expressed in more familiar terms, these groups are finite-dimensional vector spaces $\mathfrak{V}$ over a field $\mathbb{F}$ under vector addition; the dimension n of $\mathfrak{V}$ was called by Krull the rank of $\mathfrak{V}$. The associated operator domain Ω includes the elements $\alpha \in \mathbb{F}$ together with other operators $\Theta, \Theta', \ldots$. These groups satisfy the ascending and descending chain conditions and so are a special type of generalized finite abelian group. Krull remarked that although it is easy to show that these groups decompose into a direct sum of directly indecomposable Ω-subgroups, he did not see any way to substantially simplify the proof of the uniqueness part of Theorem 16.9 when that proof is limited to such groups [375, p. 175]. Such considerations may have encouraged him to develop the theory more generally, as in Theorem 16.9, even though the applications he had in mind involved generalized abelian groups of finite rank.

For the application of Theorem 16.9 to Loewy's work on differential equations, Krull took $\mathfrak{V}$ to be an n-dimensional vector space of infinitely differentiable functions of a complex variable x over Loewy's field Σ (introduced at the beginning of Section 16.3.1) [375, §7]. The operator domain Ω consisted of the elements of Σ together with the operator $\Theta = \frac{d}{dx}$. Then $\mathfrak{V}$ is an abelian group of finite rank provided $\Theta \cdot \mathfrak{V} \subset \mathfrak{V}$. To see what this means, let $y_i = f_i(x)$, $i = 1, \ldots, n$, be a basis for $\mathfrak{V}$. Then $\Theta \cdot y_i \in \mathfrak{V}$ if and only if functions $a_{ij}(x) \in \Sigma$ exist such that $dy_i/dx = \sum_{j=1}^{n} a_{ij} y_j$. In other words, $\mathfrak{V}$ is the solution space of the system of equations $dy/dx + \mathfrak{A}y = 0$, where $y = \begin{pmatrix} y_1 & \cdots & y_n \end{pmatrix}^t$ and $\mathfrak{A} = (-a_{ij}(x))$. The context provided by $\mathfrak{V}$ thus includes Loewy's theory, but does not require that $\mathfrak{A}$ be a companion matrix. If $z_i = g_i(x)$, $i = 1, \ldots, n$, is another basis for $\mathfrak{V}$, then every y_i is a linear combination of the z_j, or, more precisely, $y = \mathfrak{P}z$, where $\mathfrak{P}$ is a nonsingular matrix with coefficients from Σ. As we already saw in discussing Loewy's theory, $y = \mathfrak{P}z$ implies that $dz/dx + \mathfrak{B}z = 0$, where $\mathfrak{B} = \mathfrak{P}^{-1}\mathfrak{P}' + \mathfrak{P}^{-1}\mathfrak{A}\mathfrak{P}$, which is Loewy's relation (16.18) and so $\mathfrak{A} \cong_D \mathfrak{B}$ in the notation used in discussing Loewy's work. Thus the group $\mathfrak{V}$ corresponds to the class of all linear systems of differential equations of the same type, with each basis for $\mathfrak{V}$ corresponding to such a system, and two such groups are Ω-isomorphic if and only if they are associated to the same class of linear systems of differential equations [375, p. 189].

Furthermore, if $\mathfrak{V} = \mathfrak{W}_1 + \mathfrak{W}_2$ is a direct sum decomposition of $\mathfrak{V}$ into Ω-subgroups, and if we choose the basis $\{z_j(x)\}$ for $\mathfrak{V}$ to consist of the union of bases for $\mathfrak{W}_1$ and $\mathfrak{W}_2$, then clearly the system $dz/dx + \mathfrak{B}z = 0$ corresponding to this basis will be such that $\mathfrak{B} = \begin{pmatrix} \mathfrak{B}_1 & 0 \\ 0 & \mathfrak{B}_2 \end{pmatrix} \cong_D \mathfrak{A}$. Such considerations show that Krull's Theorem 16.9 implies that $\mathfrak{A} \cong_D \mathfrak{D}$, where $\mathfrak{D}$ is a block diagonal matrix and the diagonal blocks $\mathfrak{D}_{ii}$ are directly indecomposable in the obvious sense, and furthermore, that in any two such decompositions, the number of blocks is the same and the diagonal blocks $\mathfrak{D}_{ii}$ and $\mathfrak{D}'_{ii}$ of the two decompositions can be paired up so that the pairs are of the same type: $\mathfrak{D}_{ii} \cong_D \mathfrak{D}'_{jj}$ for each such pair $\mathfrak{D}_{ii}, \mathfrak{D}'_{jj}$ [375, Satz 15, p. 190]. Citing Loewy's papers, Krull emphasized that such a result was hitherto known only for $\mathfrak{A}$ corresponding to a completely reducible A.[32]

In order to deal with Loewy's matrix complexes, Krull had to extend his Theorem 16.9 to apply to what he called complexes of generalized abelian groups of finite rank [375, §8], and he had to limit himself to the case of ordinary matrix similarity $\cong$ rather than "same type" similarity $\cong_D$, which meant that the underlying field was not Loewy's field Σ but a field $\mathbb{F}$ of constants. He showed that every complex $\{A_\lambda\}$ of $n \times n$ matrices over $\mathbb{F}$ is similar to a complex of block diagonal matrices with directly indecomposable diagonal blocks and, more importantly, that if $\{A_\lambda\}$ is similar to another such complex, then the number of diagonal blocks is the same, and the blocks can be matched into similar pairs [375, Satz 17, p. 193]. In this way, he obtained, as he pointed out [375, p. 195, 195 n.42], a Loewy-type result, albeit applicable to all matrix complexes over $\mathbb{F}$ and not just completely reducible ones. Incidentally, the generalized abelian group $\mathfrak{V}_\lambda$ of finite rank associated to a matrix A_λ of the complex was an n-dimensional vector space over $\mathbb{F}$ with operator domain Ω consisting of the elements of $\mathbb{F}$ together with $\Theta = T_\lambda$, where T_λ is the linear transformation on $\mathfrak{V}$ with matrix representation A_λ with respect to some basis for $\mathfrak{V}_\lambda$. This idea of thinking of a linear transformation as an operator on a generalized abelian group of finite rank was developed further by Krull, as we shall now see.

In a paper published the following year (1926), Krull explored properties of various composition series for his generalized finite abelian groups (again motivated by results of Loewy on differential and matrix complexes) [376, §§2–3] and provided further examples of such groups, e.g., in the theory of ideals [376, §4]. He also showed how the viewpoint of generalized abelian groups of finite rank operated on by a linear transformation could be used to obtain a highly original derivation of the theory of elementary divisors in which companion matrices arise naturally [376, §§7–8]. To this end, he considered a special kind of generalized abelian group of finite rank $\mathfrak{V}$. If $\mathbb{F}$ is the field underlying $\mathfrak{V}$ as a vector space of dimension n, then the operator domain is $\Omega = \mathbb{F}[\Theta]$, where Θ is an operator on $\mathfrak{V}$, which in addition to satisfying the requisite condition $\Theta \cdot (v_1 + v_2) = \Theta \cdot v_1 + \Theta \cdot v_2$, also satisfies $\Theta(\alpha v) = \alpha(\Theta v)$ for all $\alpha \in \mathbb{F}$ [376, p. 19]. Stated in more familiar terms, Krull's assumption is that Θ is a linear transformation on the vector space $\mathfrak{V}$.[33] Thus Ω is a ring, which is the homomorphic image of the ring $\mathbb{F}[x]$ with kernel consisting of all polynomials $f(x) \in \mathbb{F}[x]$ that are multiples of the minimal polynomial $m(x)$ of Θ.

[32]See [375, p. 191 n.34]. Krull was referring to Loewy's results on matrix complexes (described above) as they apply to the completely reducible operators of Loewy's Theorem 16.8. Of course, it should be kept in mind that although Krull's theorem applies to any $\mathfrak{A}$, the $\mathfrak{D}_{ii}$ are simply directly indecomposable and not necessarily irreducible in Loewy's sense.

[33]Krull did not use the terminology of linear transformations and vector spaces, which was soon brought into linear algebra primarily through the influence of the work of Weyl (see below).

These groups Krull called *elementary divisor groups*, because he realized that the theory of elementary divisors could be derived from the properties of these groups.

Krull did not derive the requisite properties of elementary divisor groups by applying his general decomposition theorem (Theorem 16.9) but by going back to the analogy between the two rings $\mathbb{Z}$ and $\mathbb{F}[x]$ that had first been used by Frobenius to develop his rational theory of elementary divisors (Section 8.6). This analogy brings with it an analogy between the ring of residue classes $\mathbb{Z}/m\mathbb{Z}$ and $\mathbb{F}[x]/(m(x)) \cong \mathbb{F}[\Theta]$. As a result, "The investigation of [elementary divisor groups] hence proceeds almost word for word exactly as for the familiar [finite abelian groups], and a sketch of the proofs will suffice" [376, p. 23]. The sketch consisted mostly of a "dictionary" indicating how familiar concepts from the theory of finite abelian groups were to be translated for application to elementary divisor groups. Thus the order of an element $v \in \mathfrak{V}$ is defined to be the monic polynomial $f(x) \in \mathbb{F}[x]$ of smallest degree such that $f(x) \cdot v \stackrel{\text{def}}{=} f(\Theta) \cdot v = 0$. And $\mathfrak{V}$ is cyclic with generator v_0 if every $v \in \mathfrak{V}$ is of the form $v = f(\Theta) \cdot v_0$ for some $f \in \mathbb{F}[x]$. It then follows that two cyclic groups are Ω-isomorphic if and only if they have the same order.

The proof of the Frobenius–Stickelberger theorem (Theorem 9.10) then translates, Krull declared, into the following theorem [376, Satz 15, pp. 23–24]:

Theorem 16.10. *Every elementary divisor group $\mathfrak{V}$ is a direct sum $\mathfrak{V} = (v_1) + \cdots + (v_s)$, where (v_i) denotes the cyclic group with generator v_i and order $g_i(x)$ and $g_i(x) = p_i(x)^{d_i}$, with $p_i(x)$ irreducible over $\mathbb{F}$. Furthermore, this decomposition is unique up to Ω-isomorphisms.*

Krull called the polynomials

$$g_i(x) = p_i(x)^{d_i} = x^{e_i} - a_1 x^{e_i - 1} - \cdots - a_{e_i}, \tag{16.22}$$

the *polynomial invariants* of $\mathfrak{V}$. To relate Theorem 16.10 to elementary divisor theory, Krull considered the matrix A representing Θ with respect to some arbitrarily chosen basis. Any matrix similar to A is the matrix representation of Θ with respect to some other basis. Theorem 16.10 shows how to pick an especially nice basis. That is, $p^{d_i}(\Theta) \cdot v_i = 0$ means, in view of (16.22), that $\Theta^{e_i} \cdot v_i$ is a linear combination of $v_i, \Theta \cdot v_i, \ldots, \Theta^{e_i - 1} \cdot v_i$, and the fact that $p_i(x)^{d_i}$ is the order of v_i means that

$$\mathcal{B}_i = \left\{ v_i, \Theta \cdot v_i, \ldots, \Theta^{e_i - 1} \cdot v_i \right\}$$

is a set of linearly independent vectors and hence forms a basis for the (Θ-invariant) subspace (v_i). The matrix of Θ with respect to this basis is

$$C_i = \begin{pmatrix} 0 & 0 & \cdots & 0 & a_{e_i} \\ 1 & 0 & \cdots & 0 & a_{e_{i-1}} \\ 0 & 1 & \cdots & 0 & a_{e_{i-2}} \\ \cdots & \cdots & \cdots & \cdots & \cdots \\ 0 & 0 & \cdots & 1 & a_1 \end{pmatrix},$$

which is, of course, a companion matrix associated to the characteristic polynomial $\det(xI - C_i) = p_i(x)^{d_i}$.[34] Thus if we take $\mathcal{B} = \cup_{i=1}^{s} \mathcal{B}_i$ as basis for $\mathfrak{V}$, the corresponding matrix of Θ is the block diagonal matrix $C = C_1 \oplus \cdots \oplus C_s$ with characteristic polynomial $\prod_{i=1}^{s} p_i(x)^{d_i}$, and C is similar to A. Here C is the normal form (16.21) of Krull's doctoral dissertation. Krull's Theorem 16.10 thus implies that two matrices A and B are similar if and only if they have the same polynomial invariants and that this is so if and only if they have the same rational canonical form $C = C_1 \oplus \cdots \oplus C_s$. Although Krull did not mention it, determinant-theoretic considerations applied to the canonical form C show that the elementary divisors of A are Krull's invariant polynomials $g_i(x) = p_i(x)^{d_i}$. (Cf. the discussion of Frobenius' rational canonical form (8.28).)

A Frobenius-like rational theory of elementary divisors thus flows from Krull's Theorem 16.10. Like Frobenius 48 years earlier, Krull invoked a proof by analogy with a rigorously established theorem. In Krull's case it was the Frobenius–Stickelberger version of the fundamental theorem for abelian groups. Evidently, he thought that the proof-by-analogy justification of Theorem 16.10 was so straightforward that a detailed proof based on his general Theorem 16.9 was unwarranted. In 1929, an expanded version of Krull's development of the theory of elementary divisors via Theorem 16.10 was published as an appendix to the second volume of Otto Haupt's *Introduction to Algebra* [264, pp. 617–629], and here as well, the proof of Theorem 16.10 was left to the reader to fill in by analogy with the proof of the Frobenius–Stickelberger version of the fundamental theorem of finite abelian groups. Two years later, in the second volume of his book *Modern Algebra*, B.L. van der Waerden (1903–1996) modified Krull's approach to Theorem 16.10, replacing Krull's proof by analogy with an actual, relatively simple, abstract proof of a generalization of Schering's version of the fundamental theorem (Theorem 9.7), namely what is now usually referred to as the fundamental theorem of finitely generated modules over a principal ideal ring. Let us now consider how this came about and what was involved.

16.3.3 Van der Waerden's Moderne Algebra

Thanks to the recollections about the sources for his *Moderne Algebra* (1930) that van der Waerden made in 1975 [569], we can see the influences that led him to transform Krull's approach to elementary divisor theory into the module-based approach common nowadays. For Chapter 15 (linear algebra) of volume 2 [568], the principal sources were Emmy Noether, A. Châtelet, Otto Schreier, and "the classical papers of Frobenius on elementary divisors" [569, p. 36].

---

[34] Krull's matrix of Θ with respect to a basis is the transpose of the usual definition. Thus he gets the transpose of C_i [376, p. 26], which agrees with how companion matrices are defined in his doctoral dissertation [372, p. 58].

Van der Waerden came to Göttingen in 1924 as a postdoctoral student and digested Noether's work as part of his program to resolve certain foundational questions in algebraic geometry [569, pp. 32–33]. Her influence on him was not limited to his chapter on linear algebra but permeated all of his work. Of particular importance for his overall approach to linear algebra, however, was her abstract notion of a module over a ring. After Dedekind introduced the term "module" in his work on algebraic number theory (Section 9.1.5), this term had been given to other analogous mathematical objects. As we saw, Frobenius had applied Dedekind's definition ($\mathcal{M} \subset \mathbb{C}$ is a module if it forms an abelian group under addition in $\mathbb{C}$) to subsets $\mathcal{M} \subset \mathbb{Z}^n$ (Section 8.3.3). Later, in 1905, Lasker used the term "module" for what we would call an ideal in $\mathbb{C}[x_1,\dots,x_n]$,[35] and Macaulay continued this terminology in his important study of ideal theory in $\mathbb{C}[x_1,\dots,x_n]$, as did Noether and Schmeidler in their abstract study of ideas in rings of partial differential operators (1920) (referred to in Section 16.3.2 above). Of course, in these later uses of the term "module," the ring $\mathfrak{R}$ such that $r \cdot m \in \mathcal{M}$ for $r \in \mathfrak{R}$ was not the ring of integers. Other examples of such modules were subsequently studied by Noether (see, e.g., [455, §9], [283, §1]). Given her penchant for abstraction, it is not surprising that she eventually formulated the modern definition of a module over a ring $\mathfrak{R}$ (or $\mathfrak{R}$-module, as she also termed it). In the context of commutative rings, this can be found in her 1926 paper on what are now called Dedekind rings [457, p. 34] and for noncommutative rings in her Göttingen lectures on algebras and representation theory (1927–1928), which were published in 1929 based on van der Waerden's lecture notes [458, p. 646]. Noether emphasized that $\mathfrak{R}$-modules formed a particular class of the groups with operator domain Ω that had been studied by Krull and O. Schmidt [458, pp. 645–646].

According to van der Waerden [569, p. 36], Noether called attention to a book on the theory of numbers by Albert Châtelet. It was published in 1913 and was based on Châtelet's lectures for the prestigious Cours Peccot at the Collège de France (1911–1912) [88]. What clearly impressed Noether and van der Waerden about the book was the important role given to modules. In 1975, van der Waerden pointed to §106 of his book as influenced by Châtelet [569, p. 36]. Given the content of §106 (indicated below), it is easy to see what van der Waerden found inspiring in Châtelet's treatment of modules.

Châtelet expanded Dedekind's notion of a module along lines suggested by Minkowski's book *Geometrie der Zahlen* (1896). He defined a subset $\mathcal{M}$ of points $(p_1,\dots,p_n)$ of $\mathbb{R}^n$ to be a module when it is closed under the addition of $\mathbb{R}^n$ and so forms an abelian group.[36] For the modules of primary concern, those he called "of type (m,n)," matrix algebra was utilized. A module $\mathcal{M} \subset \mathbb{R}^n$ is said to be of type (m,n) if its points are given by the matrix equation $p = zA$, where $p = \begin{pmatrix} p_1 & \cdots & p_n \end{pmatrix}$, A is $m \times n$, $z = \begin{pmatrix} z_1 & \cdots & z_m \end{pmatrix}$, and the z_i take on all integer values [88, p. 29]. These modules are thus examples of finitely generated free $\mathbb{Z}$-modules of rank m. The

---

[35] Lasker reserved the term "ideal" for an ideal in $Z[x_1,\dots,x_n]$.

[36] Châtelet also extended his notion of module to "smireal" n-dimensional spaces [88, pp. 10, 25].

matrix A he called a base matrix for the module. The m rows of A are what would now be called a basis for the module. He also observed that an $m \times n$ matrix B is another base matrix for $\mathcal{M}$ if and only if $B = PA$, where P is an $m \times m$ unimodular matrix [88, pp. 35ff.]. Then, in discussing submodules $\mathcal{M} \subset \mathbb{Z}^n$ of type (n,n), Châtelet observed that for $n > 1$, "There is an infinity of such [matrix] bases, but one can distinguish one of them, which is particularly remarkable" [88, p. 46]. That is, since $\mathcal{M}$ is a submodule of $\mathbb{Z}^n$ of rank n, its base matrices A are $n \times n$ matrices of integers with $\det A \neq 0$. To these matrices, Hermite's Theorem 8.20 applies: a unimodular matrix P exists such that $H = PA$ is a nonnegative lower triangular matrix with the property that each diagonal entry is strictly greater than the entries below it in the same column; furthermore, the lower triangular matrix H is unique with respect to this property. (As noted in Section 8.5, Hermite's theorem had been used by Smith to obtain his normal form.) The matrix H was for Châtelet the canonical base matrix for the module $\mathcal{M}$. He showed how it could be used to solve problems involving modules.

In sum, from Châtelet's book, the following two ideas, which van der Waerden put to use in §106, are found: (1) unimodular matrices describe basis changes in finitely generated free modules; (2) canonical forms for integral matrices under unimodular transformation can lead to especially useful bases. Rather than modules of type (m,n), van der Waerden considered what he called "linear form modules." Such a module $\mathcal{M}$ consists of all formal expressions $f = r_1u_1 + \cdots + r_mu_m$, where the r_i are elements of a principal ideal domain $\mathfrak{R}$,[37] m is fixed, and the u_i are indeterminates with $r \cdot f$ and $f_1 + f_2$ defined in the obvious way for linear forms. (As noted toward the end of Section 9.1.5, this is the way Frobenius conceived of Dedekind's modules.) In modern parlance, $\mathcal{M}$ is a finitely generated free $\mathfrak{R}$-module of rank m with basis $u_1, \ldots, u_m$. A simple inductive proof shows that if $\mathcal{N}$ is a submodule of $\mathcal{M}$, then $\mathcal{N}$ is a free module of rank $n \leq m$ [568, p. 121]. As I will now show, van der Waerden used ideas (1) and (2) to obtain a canonical pair of bases for $\mathcal{M}$ and $\mathcal{N}$, respectively.

With $\mathcal{M}$ and $\mathcal{N}$ as above, suppose that $u_1, \ldots, u_m$ and $v_1, \ldots, v_n$ are respective bases. Then since $\mathfrak{N} \subset \mathfrak{M}$, every v_i is a linear combination of $u_1, \ldots, u_n$, and so $v_i = \sum_{j=1}^m a_{ji}u_j$, where $a_{ij} \in \mathfrak{R}$ for $1 \leq i \leq n$ and $1 \leq j \leq m$. If we let $A = (a_{ji})$, so that A is $m \times n$ and has rank n, and introduce the row matrices $u = \begin{pmatrix} u_1 & \cdots & u_m \end{pmatrix}$ and $v = \begin{pmatrix} v_1 & \cdots & v_n \end{pmatrix}$, the above linear dependence of the v_i on the u_j may be expressed in the matrix form

$$v = uA. \tag{16.23}$$

By idea (1) extended to $\mathfrak{R}$-modules, two bases $x_1, \ldots, x_m$ and $u_1, \ldots, u_m$ of $\mathfrak{M}$ are related by $u = Ux$, where U is unimodular in the sense that U is a matrix over $\mathfrak{R}$

[37] In §§106–107, van der Waerden also allowed $\mathfrak{R}$ to be what one might call a "noncommutative Euclidean domain" [568, p. 120].

such that $\det U$ is a unit in $\mathfrak{R}$.[38] Likewise, two bases $y_1,\ldots,y_n$ and $v_1,\ldots,v_n$ of $\mathfrak{N}$ are related by $v = yV$, where V is unimodular. Now by (16.23), the relation between the bases $x_1,\ldots,x_n$ and $y_1,\ldots,y_m$ is given by $yV = xUA$, or $y = xUAV^{-1}$. Applying now idea (2), this relation can be made especially simple by applying Frobenius' normal form theorem (Theorem 8.8) as generalized to principal ideal domains: unimodular matrices P, Q exist such that $PAQ = N$, where N is the $m \times n$ "diagonal" matrix of invariant factors e_i,

$$N = \begin{pmatrix} e_1 & & & & \\ & \ddots & & & \\ & & e_n & & \\ & & & \ddots & \\ & & & & 0 \end{pmatrix}, \quad e_i \mid e_{i+1}.$$

If then the bases $x_1,\ldots,x_m$ and $y_1,\ldots,y_n$ are chosen with $U = P$ and $V = P^{-1}$, we have $y = xN$, which yields the following theorem:

Theorem 16.11. *Let $\mathfrak{R}$ be a principal ideal domain. If $\mathcal{M}$ is a free $\mathfrak{R}$-module of rank m and $\mathcal{N}$ a submodule of rank n, then bases $x_1,\ldots,x_m$ and $y_1,\ldots,y_n$ of $\mathcal{M}$ and $\mathcal{N}$ exist such that for all $i = 1,\ldots,n$, $y_i = e_i x_i$, where the elements $e_i \in \mathfrak{R}$ have the property that $e_i \mid e_{i+1}$ for all $i \leq n-1$.*

Van der Waerden called this theorem the "elementary divisor theorem," undoubtedly because when $\mathfrak{R} = \mathbb{Z}$, the e_i were called the elementary divisors of A (rather than the invariant factors of A) by Frobenius and his successors. In *Moderne Algebra*, the Smith–Frobenius normal form theorem for principal ideal rings is nowhere stated as a formal theorem. Van der Waerden had incorporated it and its proof into his proof of his elementary divisor theorem. In effect, this theorem was van der Waerden's version of the Smith–Frobenius normal form theorem.

According to van der Waerden, "Section 107 was influenced by Otto Schreier in Hamburg, who was a specialist in linear algebra and the theory of groups" [569, p. 36]. Before his untimely death in 1929, Schreier (b. 1901) gave lectures at the University of Hamburg on the theory of elementary divisors. A book based on these lectures was published in 1932 by Emmanuel Sperner [520].[39] Based on Schreier's lectures as they appear in that book and on the contents of Section 107 (discussed below), I would suggest that Schreier's influence may have been along the following lines.

---

[38] That is, $\det U$, which belongs to $\mathfrak{R}$, has an inverse in $\mathfrak{R}$. Thus $U^{-1} = (\det U)^{-1}\,\mathrm{Adj}\,U$ is a matrix over $\mathfrak{R}$, and $x = uU^{-1}$.

[39] Sperner explained in the preface that the subject matter of the book was as Schreier intended, although the ordering of the material and some proofs were changed in part to achieve greater simplicity.

Schreier was one of the first mathematicians to follow Weyl's lead and develop the theory of elementary divisors within the context of linear transformations acting on abstract finite-dimensional real or complex vector spaces.[40] Schreier also sought to utilize the new approach to the theory via generalized abelian groups that Krull had introduced in his 1926 paper [376, §§7–8], albeit without using Krull's Theorem 16.10 with its proof by appeal to analogy with the fundamental theorem of finite abelian groups. To do this, he began by following part A of Frobenius' theory (Outline 16.6) to establish that if $A(u)$ is an $n \times n$ matrix with coefficients from $\mathbb{F}[u]$, $\mathbb{F}$ an arbitrary abstract field, then $A(u)$ is equivalent by means of elementary row and column operations to a diagonal matrix with diagonal entries $e_1(u), \ldots, e_r(u), 0, \ldots 0$, where $e_i(u) \mid e_{i+1}(u)$. Following Frobenius' terminology, the polynomials $e_i(u)$ are then defined to be the elementary divisors of $A(u)$ (rather than the invariant factors). Then follows the theorem that $A(u)$ and $B(u)$ are equivalent if and only if they have the same elementary divisors. Next the theory is specialized to the "characteristic matrix of a linear transformation" [520, pp. 86ff.], namely $A(u) = A - uI$, where A is the matrix of a linear transformation T with respect to some basis for the n-dimensional vector space $\mathcal{V}$ under consideration. It follows that if B is another matrix of T, then $PBP^{-1} = A$, where P is over $\mathbb{F}$ with $\det P \neq 0$, and so $P(B - uI)P^{-1} = A - uI$ and $A - uI$ and $B - uI$ are similar, and so any two characteristic matrices of T have the same elementary divisors. (Since $\det(A - uI)$ is not identically zero, there will be n elementary divisors $e_i(u)$ with $e_1(u) = \cdots = e_{n-r}(u) = 1$, where $r = \operatorname{rank} A$.)

It is at this point that Schreier wanted to utilize Krull's results. If instead of appealing to the fundamental theorem of finite abelian groups in the form given to it by Frobenius and Stickelberger, Krull had appealed to it in the form given by Schering, namely that such a group (conceived additively) is a direct sum of cyclic subgroups of orders e_i, where $e_i \mid e_{i+1}$, then his Theorem 16.10 would have taken the form $\mathfrak{V} = (v_1) + \cdots + (v_r)$, where (v_i) is a cyclic subgroup of order $e_{n-r+i}(u)$, $i = 1, \ldots, r$, where the $e_{n-r+i}(u)$ are the nontrivial elementary divisors of $A - uI$. Rather than proving this version of Krull's Theorem 16.10 in detail—the readers could not be presumed to have the proof of the Schering's theorem at their fingertips—theorems about linear transformations were developed to obtain, in effect, this result [520, pp. 88–93]. Then, once again following Krull, Schreier showed that $\mathcal{V} = \mathcal{V}_1 + \cdots + \mathcal{V}_r$, where $\mathcal{V}_i$ is the subspace with basis $v_i, Tv_i, \ldots, T^{a_i - 1}v_i$, where a_i is the degree of $e_{n-r+i}(u)$. This then leads to the normal matrix form for T that consists of the direct sum of companion matrices for each of the polynomials $e_{n-r+i}(u)$, $i = 1, \ldots, r$.

[40] In *Space–Time–Matter*, first published in 1921, Weyl developed the machinery of tensor algebra within the context of abstract vector spaces. In his 1923 monograph *Mathematical Analysis of the Space Problem* [601, Anhang 12], Weyl developed elementary divisor theory over $\mathbb{C}$ within the same context and so made the concept of a linear transformation acting on a vector space fundamental. For more on Weyl and the space problem, see [518, §2.8] and [276, §11.2].

I suspect that Schreier, who was also involved with research on group theory,[41] suggested to van der Waerden the possibility of developing elementary divisor theory by proving a generalization of, say, Schering's theorem for abelian groups with operators Ω that would cover the case of Krull's Theorem 16.10 in which $\Omega = \mathbb{F}[T]$, T a linear transformation on a finite-dimensional vector space $\mathcal{V}$ over $\mathbb{F}$. Such a group is, of course, an $\mathfrak{R}$-module with $\mathfrak{R} = \mathbb{F}[u]$ a principal ideal domain and $f(x) \cdot v = f(T)v$ for all $v \in \mathcal{V}$. In any case, this is essentially what van der Waerden did: in §107, he stated and proved what is nowadays called the fundamental theorem for finitely generated modules over a principal ideal ring. However, he stated the theorem in the language of abelian groups with operators—perhaps reflecting Schreier's influence as well as that of O. Schmidt.[42] Thus the section is titled "The Fundamental Theorem of Abelian Groups," and the theorem is expressed as follows [568, pp. 128–129].

Theorem 16.12. *Let $\mathcal{M}$ denote a finitely generated (additive) abelian group with operator domain a principal ideal domain $\mathfrak{R}$.[43] Then $\mathcal{M}$ is the direct sum of cyclic subgroups, $\mathcal{M} = (h_1) + \cdots + (h_n)$, where the ideal annihilating $(h_i) \subset \mathcal{M}$ is $(e_i) \subset \mathfrak{R}$ with $e_i \mid e_{i+1}$ for all $i < m$, and $(e_i) = (0)$ for $i > m$. Furthermore, the above direct sum decomposition is unique up to group isomorphisms.*

It should come as no surprise that van der Waerden's elementary divisor theorem (Theorem 16.11) played a key role in the existence part of his proof of this theorem, since it is still a standard proof [141, pp. 442ff.]. Here is the idea. (For the proof, van der Waerden reverted to module-theoretic language.) Let $\mathcal{M}$ be an $\mathfrak{R}$-module with n generators $g_1, \ldots, g_n$. Corresponding to $\mathcal{M}$ is the linear form module $\mathcal{M}'$ in n indeterminates $u_1, \ldots, u_n$, i.e., a free $\mathfrak{R}$-module of rank n with basis $u_1, \ldots, u_n$. The mapping $\phi : \sum_{i=1}^n r_i u_i \to \sum_{i=1}^n r_i g_i$ is a module homomorphism of $\mathcal{M}'$ onto $\mathcal{M}$. Thus $\mathcal{M}$ is isomorphic to $\mathcal{M}'/\mathcal{N}$, where $\mathcal{N}$ is the kernel of ϕ. Now apply Theorem 16.11 to $\mathcal{M}'$ and its submodule $\mathcal{N}$ to determine bases $u_1', \ldots, u_n'$ and $v_1, \ldots, v_m$ of $\mathcal{M}'$ and $\mathcal{N}$, respectively, such that $v_i = e_i u_i'$ for $i = 1, \ldots, m$, where $e_i \mid e_{i+1}$ for all $i < m$. Then we may pick $h_1, \ldots, h_n$ in $\mathcal{M}$ such that $\phi(u_i') = h_i$. The existence part of the theorem is concluded by showing that $\mathcal{M} = (h_1) + \cdots + (h_n)$ in accordance with Theorem 16.12.

Van der Waerden thus used the Smith–Frobenius normal form (in the guise of Theorem 16.11) to obtain the "Schering" version of the fundamental theorem of finitely generated modules, just as Frobenius and Stickelberger had used it fifty years earlier to derive Schering's version of the fundamental theorem for finite abelian groups (Section 9.2.1). I do not know whether van der Waerden was familiar

[41] For a discussion of Schreier's important work on continuous groups, see [276, pp. 497ff.].

[42] Under Schmidt's influence, van der Waerden devoted an entire chapter [568, Ch. 6] to groups with operators [569, p. 34].

[43] As with Theorem 16.11, $\mathfrak{R}$ can also be a "noncommutative Euclidean domain." See the citation at that theorem.

with their derivation, but if so, it surely must have helped inspire the overall approach he took to his own fundamental theorem.

Because of the added generality of his fundamental theorem, van der Waerden was able to use it to develop the theory of elementary divisors along the lines introduced by Krull. In particular, he was able to follow Schreier's approach to canonical matrix forms for a linear transformation T on a finite-dimensional vector space $\mathcal{V}$ over a field $\mathbb{F}$—but without the lengthy digressions needed by Schreier to avoid invoking the "Schering"-like version of Krull's Theorem 16.10 mentioned above [568, §109]. That is, if Theorem 16.12 is applied with $\mathfrak{R} = \mathbb{F}[x]$ and $\mathcal{M} = \mathcal{V}$ with $f(x) \cdot v = f(T)v$, it yields the "Schering"-like version of Krull's Theorem 16.10 mentioned above: if T has rank r, then $\mathcal{V} = (v_1) + \cdots + (v_r)$, where the annihilating ideal of (v_i) is the principal ideal generated by the monic polynomial $e_i(x)$, and $e_i(x) \mid e_{i+1}(x)$ for all $i < r$. Use of the basis $v_i, Tv_i, \ldots, T^{a_i - 1}v_i$, where $a_i = \deg e_i(x)$, for (v_i), then leads to the matrix representation of T as the direct sum of companion matrices for the invariant factors $e_i(x)$. As van der Waerden noted in passing [568, p. 137], these companion matrices can be further decomposed as direct sums of companion matrices by expressing the cyclic module (v_i) as a direct sum of (indecomposable) cyclic submodules with orders that are powers of irreducible polynomials. These "prime power" polynomials are of course precisely the elementary divisors of T (in the terminology of Weierstrass vis à vis Frobenius), and the corresponding matrix representation of T is essentially the Frobenius rational canonical form of Section 8.6.3.

Chapter 17
Nonnegative Matrices

This final chapter on Frobenius' mathematics is devoted to the paper he submitted to the Berlin Academy on 23 May 1912 with the title "On matrices with nonnegative elements" [231].[1] This turned out to be his last great mathematical work. He was 62 at the time and in declining health. The paper was inspired by a remarkable theorem on positive matrices that a young mathematician by the name of Oskar Perron (1880–1975) had discovered in the course of studying generalized continued fraction algorithms with periodic coefficients. Perron's work (1905–1907) is discussed in Section 17.1, and Frobenius' creative reaction to it (1908–1912) in Section 17.2. The results on nonnegative matrices obtained by Frobenius in 1912, which are sometimes referred to generically as Perron–Frobenius theory, although motivated by purely mathematical considerations, as was the case as well with the contributions of Perron, later provided the mathematical foundation for a broad spectrum of applications to such diverse fields as probability theory, numerical analysis, economics, dynamic programming, and demography. Section 17.3 is devoted to the first such application of Frobenius' theory, which was to the probabilistic theory of Markov chains. As originally formulated by Markov in 1908, the probabilistic analysis underlying the theory was based on algebraic assumptions regarding the characteristic roots of an associated stochastic probability matrix P. Markov attempted to characterize which $P \geq 0$ had the requisite properties, but his efforts were unclear and inadequate. Consequently, his successors for many years restricted their attention to the case $P > 0$. It was eventually realized that Frobenius' theory provided clear and definitive answers to all the questions left unresolved by Markov and provided the requisite theoretical tools for dealing with Markov chains associated to matrices $P \geq 0$. The theory of Markov chains became one of the earliest developed applications of Frobenius' theory and seems to have

[1] In what follows, such matrices A will be called nonnegative, and this property will be indicated with the notation $A \geq 0$. Similarly, $A > 0$ will mean that all the elements of A are positive.

T. Hawkins, *The Mathematics of Frobenius in Context*, Sources and Studies in the History of Mathematics and Physical Sciences, DOI 10.1007/978-1-4614-6333-7_17,

served to call general attention among mathematicians and mathematically inclined scientists to the existence and utility of the theory.[2]

17.1 The Work of Perron

Perron began his studies at the University of Munich and then, as was commonplace, spent semesters at several other universities—Berlin, Tübingen, and Göttingen in his case—but Munich was his mathematical home base.[3] In 1902, he obtained his doctorate there with a dissertation on a problem involving the rotation of a rigid body. He wrote it under the direction Ferdinand Lindemann, who is remembered nowadays for his proof that π is a transcendental number. Also at the University of Munich was Alfred Pringsheim (1850–1941), who was renowned as a brilliant lecturer and conversationalist.[4] Perron's postdoctoral research interests turned in the direction of Pringsheim's current work, which involved the theory of continued fractions with real or complex coefficients.

A continued fraction is a formal expression of the form

$$a_0 + \frac{a_1}{b_1 + \frac{a_2}{b_2 + \cdots}}, \tag{17.1}$$

where the coefficients $a_0, a_1, a_2, \ldots, b_1, b_2, \ldots$ can be any real or complex numbers. Such expressions with positive integer coefficients are naturally suggested by the Euclidean algorithm.[5] I will use the following notation, which is due to Pringsheim, to express (17.1) in the typographically simpler form

$$a_0 + \frac{a_1|}{|b_1} + \frac{a_2|}{|b_2} + \cdots. \tag{17.2}$$

The continued fraction (17.2) is said to converge if the sequence of partial continued fractions

$$S_n = a_0 + \frac{a_1|}{|b_1} + \frac{a_2|}{|b_2} + \cdots + \frac{a_n|}{|b_n} \tag{17.3}$$

has a finite limit S as $n \to \infty$. In this case, the continued fraction (17.2) is said to converge to S.

---

[2]The present chapter is based upon my paper [278], which contains more details, especially regarding the work of Perron and Markov.

[3]For details on Perron's life and family background see [166].

[4]The information about Pringsheim is based on Perron's memorial essay [470].

[5]For an overview of the history of continued fractions, see [39]. On the connection with the Euclidean algorithm, see also [278, pp. 665ff.].

17.1.1 *Stolz's theorem revisited*

One part of the theory of continued fractions that drew Pringsheim's interest, and so Perron's as well, had to do with continued fractions that are periodic in their coefficients (see below). It was Perron's work in this area that is relevant to the developments traced in this chapter. The starting point for both Pringsheim's and Perron's work in this area was a theorem due to Otto Stolz (1842–1905). In 1885–1886 Stolz, who was a professor at the University of Innsbruck, published a two-volume series of lectures on what he called "general arithmetic from the modern viewpoint." The second volume [552] was devoted to the arithmetic of complex numbers, and its final chapter considered the subject of continued fractions.

In particular, Stolz considered continued fractions $\frac{a_1|}{|b_1} + \frac{a_2|}{|b_2} + \cdots$ with possibly complex but periodic coefficients [552, pp. 299ff.]. Here it will suffice to consider *purely periodic continued fractions*, i.e., those that are periodic from the outset. Thus if the positive integer $\mathbf{p}$ denotes the period, the coefficients of the continued fraction satisfy

$$a_{i\mathbf{p}+j} = a_j,\ b_{i\mathbf{p}+j} = b_j, \quad j = 1,\ldots,\mathbf{p}, \quad i = 1,2,3,\ldots. \tag{17.4}$$

If the continued fraction converges to x, i.e., if $x = \lim_{n\to\infty} S_n$ in the notation of (17.3) (with $a_0 = 0$), then due to the assumed periodicity,

$$x = \frac{a_{\mathbf{p}+1}|}{|b_{\mathbf{p}+1}} + \frac{a_{\mathbf{p}+2}|}{|b_{\mathbf{p}+2}} + \cdots,$$

and so

$$x = \frac{a_1|}{|b_1} + \frac{a_2|}{|b_2} + \cdots + \frac{a_{\mathbf{p}}|}{|b_{\mathbf{p}} + x}. \tag{17.5}$$

For any (not necessarily periodic) continued fraction, Stolz also wrote the νth partial continued fraction S_ν in the form

$$S_\nu = \frac{A_\nu}{B_\nu}, \tag{17.6}$$

where A_ν and B_ν are the numerator and denominator of S_ν when expressed as a simple fraction. Thus, e.g.,

$$S_3 = \frac{a_1|}{|b_1} + \frac{a_2|}{|b_2} + \frac{a_3|}{|b_3} = \frac{a_1a_3 + a_1b_2b_3}{a_3b_1 + a_2b_3 + b_1b_2b_3},$$

and so $A_3 = a_1a_3 + a_1b_2b_3$ and $B_3 = a_3b_1 + a_2b_3 + b_1b_2b_3$. Stolz established the following recurrence relations satisfied by A_ν and B_ν [552, p. 267, eqn. (II)]:

$$\begin{aligned} A_0 &= 0, \quad A_1 = a_1, \quad A_{\nu+2} = a_{\nu+2}A_\nu + b_{\nu+2}A_{\nu+1}, \\ B_0 &= 1, \quad B_1 = b_1, \quad B_{\nu+2} = a_{\nu+2}B_\nu + b_{\nu+2}B_{\nu+1}. \end{aligned} \tag{17.7}$$

Using them [552, p. 299], he was able to express (17.5) in the form

$$x = \frac{A_{\mathbf{p}-1}x + A_{\mathbf{p}}}{B_{\mathbf{p}-1}x + B_{\mathbf{p}}}, \tag{17.8}$$

which can be rewritten as a quadratic equation in x:

$$B_{\mathbf{p}-1}x^2 + (B_{\mathbf{p}} - A_{\mathbf{p}-1})x - A_{\mathbf{p}} = 0. \tag{17.9}$$

Stolz then proceeded to present the first general convergence theorem for periodic continued fractions [552, pp. 300–302]. It may be summed up as follows.

Theorem 17.1 (Stolz's theorem). *Let* $\frac{a_1|}{|b_1} + \frac{a_2|}{|b_2} + \cdots$ *denote a purely periodic continued fraction with period* $\mathbf{p}$*. Then the condition*

$$\text{(A)} \qquad B_{\mathbf{p}-1} \neq 0$$

is a necessary condition for convergence. To obtain necessary and sufficient conditions, two cases must be distinguished. Case I. Suppose the quadratic equation (17.9) has a double root. Then condition (A) is also sufficient, and the continued fraction converges to this root. Case II. Suppose (17.9) has two distinct roots x_0 *and* x_1*. Then it is also necessary that*

$$\text{(B)} \qquad |B_{\mathbf{p}} + x_0 B_{\mathbf{p}-1}| \neq |B_{\mathbf{p}} + x_1 B_{\mathbf{p}-1}|.$$

Assuming (B) holds, let the notation for x_0, x_1 *be chosen such that*

$$\text{(B}'\text{)} \qquad |B_{\mathbf{p}} + x_0 B_{\mathbf{p}-1}| > |B_{\mathbf{p}} + x_1 B_{\mathbf{p}-1}|.$$

Then a further necessary condition is

$$\text{(C)} \qquad A_\nu - x_1 B_\nu \neq 0, \quad \nu = 1, 2, \ldots, \mathbf{p} - 2.$$

In Case II, conditions (A), (B′), and (C) are necessary and sufficient for convergence, and the continued fraction converges to x_0*.*

With this theorem, Stolz had certainly given a definitive answer to the question of the convergence of periodic continued fractions, but his formulation of his results and the concomitant proofs were neither simple nor insightful. Pringsheim's 1900 paper [490] aimed to improve on this aspect of Stolz's work, although it achieved little in the way of clarity or simplicity [278, p. 663]. The source of the dissatisfaction with Stolz's theorem involved Case II, the case of distinct roots in the quadratic equation (17.9).

Stolz's condition (B) of Case II seemed unenlightening because his lengthy proof lacked any intuitive motivation. It was Perron who discovered the underlying reason why condition (B) made sense and this enabled him to state Stolz's theorem more clearly and simply than either Stolz or Pringsheim. He showed how to do this in his second paper of 1905 [466]. As we shall see, the discovery on his part was especially important because it revealed to him a new and promising approach to a generalized theory of continued fraction algorithms inspired by Jacobi's algorithm for determining the greatest common divisor of a set of more than two integers, an approach that he developed in his *Habilitationsschrift* of 1907 (Section 17.1.2) and that in turn led him to his theorem on positive matrices, which was also published in 1907 (Section 17.1.4).

Perron observed that the source of Stolz's quadratic equation (17.9), namely (17.8), can be expressed in the homogeneous form

$$\rho x = A_{\mathbf{p}-1}x + A_{\mathbf{p}}, \quad \rho = B_{\mathbf{p}-1}x + B_{\mathbf{p}}, \tag{17.10}$$

which I will express in the more suggestive form

$$Av = \rho v, \quad A = \begin{pmatrix} A_{\mathbf{p}-1} & A_{\mathbf{p}} \\ B_{\mathbf{p}-1} & B_{\mathbf{p}} \end{pmatrix}, \quad v = \begin{pmatrix} x \\ 1 \end{pmatrix}. \tag{17.11}$$

Although Perron did not use any matrix notation in his paper, he certainly was familiar with it and realized (17.11). He observed [466, p. 497] that elimination of x in (17.10) shows that ρ is a root of the quadratic equation

$$f(\rho) = \begin{vmatrix} A_{\mathbf{p}-1} - \rho & A_{\mathbf{p}} \\ B_{\mathbf{p}-1} & B_{\mathbf{p}} - \rho \end{vmatrix} = 0, \tag{17.12}$$

which is of course the characteristic equation of A, a term he adopted in his *Habilitationsschrift*.

If x_0, x_1 denote the roots of Stolz's quadratic equation (17.9) as specified by condition (B$'$) of his Theorem 17.1, and if ρ_0, ρ_1 are the corresponding values from (17.10), then we see that

$$\rho_i = B_{\mathbf{p}-1}x_i + B_{\mathbf{p}}, \quad i = 0, 1, \tag{17.13}$$

and Stolz's condition (B$'$) is that $|\rho_0| > |\rho_1|$. Furthermore, since $B_{\mathbf{p}-1} \neq 0$ by condition (A), it is easily seen that Case II of Stolz's Theorem 17.1 (Stolz's quadratic equation has distinct roots) corresponds to the roots ρ_i of the characteristic equation (17.12) being distinct, so that in Case II, Stolz's condition (B)/(B$'$) simply says that the characteristic equation must have roots of differing absolute values.

Perron not only reformulated Stolz's condition (B)/(B′) in terms of the roots of the characteristic equation, but he gave an entirely different proof of the necessity of the condition $|\rho_0| > |\rho_1|$ for convergence in the case of distinct roots. He utilized the recurrence relations (17.7) to deduce from the assumed periodicity that more generally,

$$\begin{aligned} A_{i\mathbf{p}+j} &= A_{(i-1)\mathbf{p}+j}A_{\mathbf{p}-1} + B_{(i-1)\mathbf{p}+j}A_j, \\ B_{i\mathbf{p}+j} &= A_{(i-1)\mathbf{p}+j}B_{\mathbf{p}-1} + B_{(i-1)\mathbf{p}+j}B_j. \end{aligned} \tag{17.14}$$

These relations formed the starting point of his proof that of the two roots, one of them, denoted below by ρ_0, is such that

$$L = \lim_{j\to\infty} \left(\frac{\rho_1}{\rho_0}\right)^j \tag{17.15}$$

exists as a finite number [466, §2], from which it follows immediately that $L = 0$ and $|\rho_0| > |\rho_1|$.

Perron was the first to realize the important role of the coefficient matrix A of (17.11) and its characteristic roots in the study of periodic continued fractions. From this new vantage point, Stolz's results assume a far simpler form. For future reference, I will sum them up as a theorem.

Theorem 17.2. *Given a continued fraction that is purely periodic with period* $\mathbf{p}$, *let* A *denote the corresponding matrix as defined in (17.11). Case I: if the characteristic roots* A *are equal, then a necessary and sufficient condition for convergence is that* $B_{\mathbf{p}-1} \neq 0$. *Case II: If the characteristic roots are unequal, then a necessary and sufficient condition for convergence is that they have different absolute values. Furthermore, if* ρ_0 *is the root with the larger absolute value, then the continued fraction converges to* $\alpha = x_1/x_2$, *where* $x = \begin{pmatrix} x_1 & x_2 \end{pmatrix}^t$ *is a nonzero solution to* $Ax = \rho_0 x$.

In Case II, Study's necessary condition (A) that $B_{\mathbf{p}-1} \neq 0$ is automatically fulfilled, because $B_{\mathbf{p}-1} = 0$ would imply by (17.13) that $\rho_0 = \rho_1 = B_{\mathbf{p}}$, contrary to the assumption that $|\rho_0| > |\rho_1|$. Perron's proof of Case I was also based on this new vantage point provided by consideration of A [465, §3].

Although Perron presented his reasoning in print directly in terms of iteration relations such as (17.14), it should be noted, as he surely realized, that if we introduce the 2×1 column matrix

$$C_\nu = (A_\nu \quad B_\nu)^t, \tag{17.16}$$

then the above equations state that $AC_{(i-1)\mathbf{p}+j} = C_{i\mathbf{p}+j}$ for any $i \geq 1$. This relationship, and thus (17.14), follows by iteration from the case $i = 1$, i.e.,

$$AC_j = C_{\mathbf{p}+j}. \tag{17.17}$$

17.1.2 Generalized continued fraction algorithms

Perron pointed out in the beginning of his 1905 paper described above that not only did his new methods "derive the basic formulas of Stolz in a rational manner" but in addition, "As I will show elsewhere, my procedure has the additional advantage that by means of a natural extension of it, the convergence of the general Jacobi continued fraction algorithms can be decided" [466, p. 495]. Perron showed this in his *Habilitationsschrift*, which was published in *Mathematische Annalen* in 1907 [467]. As we shall now see, in this tour de force extension of his work of 1905, he not only established the promised convergence criteria in the far more complicated case of Jacobi's algorithm, but did it in a manner that led naturally to his theorem on positive matrices.

In his *Habilitationsschrift*, Perron showed that Jacobi's original algorithm [317] could be developed and generalized in a manner that brought out the analogy with the case of ordinary continued fractions, particularly his approach to the case of periodic continued fractions [278, pp. 665–669]. Here is how the Jacobi algorithm (as I will call it) was defined by Perron.

Definition 17.3. Let $n \geq 1$ be a fixed integer, and let $a_i^{(\nu)}$ denote real or complex numbers, where the indices i, ν satisfy $1 \leq i \leq n$ and $0 \leq \nu < \infty$. Let the numbers $A_j^{(\nu)}$ be defined by the following $(n+1)$-term recurrence relation:

$$\begin{aligned} A_j^{(\nu)} &= \delta_{j\nu}, \quad 0 \leq j, \nu \leq n, \\ A_j^{(\nu+n+1)} &= A_j^{(\nu)} + \textstyle\sum_{i=1}^{n} a_i^{(\nu)} A_j^{(\nu+j)}, \quad \nu \geq 0. \end{aligned} \tag{17.18}$$

Then the Jacobi algorithm is said to converge and have limiting values $(\alpha_1, \ldots, \alpha_n)$ if $\alpha_j = \lim_{\nu \to \infty} A_j^{(\nu)} / A_0^{(\nu)}$ exists as a finite real or complex number for all $j = 1, \ldots, n$.

The $(n+1)$-term recurrence relation (17.18) is analogous to Stolz's two-term recurrence relation (17.7).

Also by analogy with the case of ordinary continued fractions, Perron defined a Jacobi algorithm to be purely periodic with period $\mathbf{p}$ if the coefficients $a_i^{(\nu)}$ satisfy

$$a_i^{(m\mathbf{p}+j)} = a_i^{(j)}, \quad 1 \leq i \leq n, \quad 0 \leq j \leq \mathbf{p} - 1, \tag{17.19}$$

for any positive integer m. Perron's successful reworking of Stolz's Theorem 17.1 had been based on consideration of the matrix A of (17.11) associated to the two-term recurrence relation (17.7). For purely periodic Jacobi algorithms he introduced the analogous $(n+1) \times (n+1)$ matrix

$$A = \begin{pmatrix} A_0^{(\mathbf{p})} & \cdots & A_0^{(\mathbf{p}+n)} \\ & \ddots & \\ A_n^{(\mathbf{p})} & \cdots & A_n^{(\mathbf{p}+n)} \end{pmatrix}. \tag{17.20}$$

I will refer to A as *the matrix of the* (**p**-*periodic) algorithm*. If we introduce the $(n+1)\times 1$ column matrices

$$\mathbf{C}_\nu = (A_0^{(\nu)} \quad \cdots \quad A_n^{(\nu)})^t, \quad \nu \geq 0, \tag{17.21}$$

then A is the matrix with columns $\mathbf{C}_\mathbf{p}, \ldots, \mathbf{C}_{\mathbf{p}+n}$. The analogue of the fundamental relation (17.17) for ordinary continued fractions is[6]

$$AC_\nu = C_{\mathbf{p}+\nu}. \tag{17.22}$$

As an illustration of these definitions, consider the Jacobi algorithm with $n = 3$ and period $\mathbf{p} = 2$ that is defined by the coefficients $a_i^{(\nu)} = 0$ for $i = 1, 2$ and $\nu = 0, 1$ and $a_3^{(0)} = 1$, $a_3^{(1)} = 2$. All other coefficients $a_i^{(\nu)}$ are then determined by the periodicity relations (17.19), e.g., $a_1^{(3)} = a_1^{(2+1)} = a_1^{(1)} = 0$ and $a_3^{(3)} = a_3^{(1)} = 2$. The $A_j^{(\nu)}$ are then determined via the recurrence relations (17.18). To write down the matrix A of the algorithm as defined in (17.20), it is necessary in this manner to determine the values of $A_j^{(\nu)}$ for $0 \leq j \leq 3$ and $2 \leq \nu \leq 2+3 = 5$. Computation of these values gives for A the matrix

$$A = \begin{pmatrix} 0 & 0 & 1 & 2 \\ 0 & 0 & 0 & 1 \\ 1 & 0 & 0 & 0 \\ 0 & 1 & 1 & 2 \end{pmatrix}. \tag{17.23}$$

We saw that in the case of ordinary continued fractions, Perron had discovered that the properties of the characteristic roots of the matrix A of (17.11) were related to the convergence or nonconvergence of the continued fraction. He found the same to be true in the case of Jacobi algorithms. For such algorithms of any period $\mathbf{p}$, he showed that a necessary condition for convergence was that the characteristic polynomial $f(\rho) = \det(A - \rho I_{n+1})$ of the matrix associated to the algorithm be *regular* in the following sense.

Definition 17.4. A polynomial $f(\rho)$ is regular when it has a root ρ_0 with the property that every other root either has smaller absolute value or, when the absolute values are equal, a smaller multiplicity. The root ρ_0 will be called the principal root (*Hauptwurzel*).

Perron was able to prove that when the period is $\mathbf{p} = 1$, the regularity of the matrix A of the algorithm is also sufficient for convergence, but for periods $\mathbf{p} > 1$, he had to add further assumptions to guarantee convergence [278, pp. 671–673].

---

[6]The relation (17.22) (without the matrix symbolism) is the special case of Perron's formula (9) [467, p. 7] that arises when Perron's λ is taken to be the period $\mathbf{p}$ of a purely periodic algorithm.

We saw that in Perron's reworking of Stolz's theorem as Theorem 17.2, he had also shown that when a periodic continued fraction is known to converge, its limiting value in Case II can be determined from a characteristic vector for the maximal root ρ_0. Perron showed that this result had the following generalization to convergent purely periodic Jacobi algorithms. If $\alpha_j = \lim_{\nu\to\infty} A_j^{(\nu)}/A_0^{(\nu)}$, $j = 1,\dots,n$, are the finite limiting values of the algorithm and if we introduce homogeneous coordinates $(x_0, x_1, \dots, x_n)$, $x_0 \neq 0$, of $(\alpha_1,\dots,\alpha_n)$ so that $\alpha_j = x_j/x_0$ for $j = 1,\dots,n$, then it can be shown that one has as well $\alpha_j = [Ax]_j/[Ax]_0$ for $j = 1,\dots,n$. This means that $([Ax]_0,\dots,[Ax]_n)$ is another set of homogeneous coordinates for $(\alpha_1,\dots,\alpha_n)$, and so a constant $\rho \neq 0$ must exist such that $([Ax]_0,\dots,[Ax]_n) = \rho(x_0, x_1,\dots,x_n)$, i.e., $Ax = \rho x$. This means of course that ρ must be a characteristic root of A and that the limiting values α_j of the algorithm can be determined by a characteristic vector for that characteristic root. But which characteristic root ρ was the right one? Strictly speaking, it could be any one of them ($\rho = 0$ can never be a characteristic root because, as Perron showed, $\det A = (-1)^{n\mathbf{p}} \neq 0$). Since Perron had proved that the characteristic polynomial of A must be regular as defined in Definition 17.4 when the algorithm converges, he no doubt suspected that the desired characteristic root was ρ_0, the principal root in the definition of regularity.

To explore this suspicion further, Perron turned to a class of periodic algorithms for which he had proved convergence. This class was related to the following convergence theorem for Jacobi algorithms that are not necessarily periodic.[7]

Theorem 17.5. *Suppose that the coefficients $a_i^{(\nu)}$ of a Jacobi algorithm are real numbers such that (1) $a_i^{(\nu)} \geq 0$ for all i and ν and (2) $a_n^{(\nu)} > 0$ for all ν. Then if a constant C exists such that for all i and ν,*

$$0 < \frac{1}{a_n^{(\nu)}} \leq C \quad \text{and} \quad 0 \leq \frac{a_i^{(\nu)}}{a_n^{(\nu)}} \leq C, \tag{17.24}$$

it follows that $\alpha_j = \lim_{\nu\to\infty}(A_j^{(\nu)}/A_0^{(\nu)})$ exists as a finite number for all $j = 1,\dots,n$.

The above theorem implies that any purely periodic Jacobi algorithm satisfying (1) and (2) is guaranteed to converge, because (17.24) is always satisfied: If the period is $\mathbf{p}$, the first quotient in (17.24) can assume at most $\mathbf{p}$ distinct values, and the second can assume at most $(n-1)\mathbf{p}$ values. Thus C can be taken as the maximum of this finite set of values.

For periodic Jacobi algorithms that are nonnegative in the sense that they satisfy conditions (1) and (2) of Theorem 17.5, and so converge, Perron investigated the question whether it is the principal root ρ_0 (in the sense of Definition 17.4) that yields the limiting values $\alpha_i = x_i/x_0$ of the algorithm, where $x = (x_0 \quad \cdots \quad x_n)^t$ satisfies $Ax = \rho_0 x$. The nonnegativity conditions (1)–(2) imply that the matrix A of the algorithm has nonnegative entries by virtue of the recurrence relations (17.18)

[7] See Satz II [467, p. 12], which establishes more than I have stated below in Theorem 17.5.

and the definition (17.20) of A. Perron naturally realized this and much more. He showed first of all that the recurrence relations together with the nonnegativity conditions (1)–(2) actually imply that $A_j^{(\nu)} > 0$ for all $j = 1, \ldots, n$, provided that $\nu \geq n$. This fact combined with (17.22) then implies that all coefficients of A^ν are positive numbers as long as $\nu \geq 2n/\mathbf{p}$, a fact that I will express with the now familiar notation $A^\nu > 0$ for all $\nu \geq 2n/\mathbf{p}$. (In the example of the 2-periodic algorithm with algorithm matrix A given by (17.23), which satisfies the nonnegativity conditions (1) and (2) of Theorem 17.5 and so converges, $2n/\mathbf{p} = 6/2 = 3$, and so $A^3 > 0$, as can easily be checked.) Since the characteristic roots of the matrix A^ν are the νth powers of the characteristic roots of A with the same characteristic vectors, information about the characteristic roots of A can be obtained from information about the characteristic roots of the positive matrix $B = A^\nu$, $\nu \geq 2n/\mathbf{p}$, as we shall see below.

17.1.3 Perron's lemma

The positivity of $B = A^\nu$ for a fixed value of ν satisfying $\nu \geq 2n/\mathbf{p}$ apparently induced Perron to ask whether certain properties of the characteristic roots of B germane to the regularity of its characteristic equation and its principal root might hold simply because $B > 0$, i.e., independently of the fact that $B = A^\nu$ with A the matrix (17.20) associated to a periodic Jacobi algorithm. He was able to prove the following remarkable result [467, p. 47, Hilfssatz].

Lemma 17.6 (Perron's lemma). *Let $B = (b_{ij})$ be any $n \times n$ matrix such that $B > 0$. Then B has at least one positive root. The greatest positive root ρ_0 has multiplicity one, and all the cofactors of $\rho_0 I - B$ are positive.*

The positivity of the cofactors of $\rho_0 I - B$ means that $\mathrm{Adj}\,(\rho_0 I - B) > 0$, where as usual, the adjoint matrix $\mathrm{Adj}\,(\rho_0 I - B)$ is the transpose of the matrix of cofactors of $\rho_0 I - B$ (Section 4.3). By the fundamental property of the adjoint matrix, $(\rho_0 I - B)\,\mathrm{Adj}\,(\rho_0 I - B) = \det(\rho_0 I - B) I = 0$. Thus if we express $\mathrm{Adj}\,(\rho_0 I - B)$ in terms of its columns, viz., $\mathrm{Adj}\,(\rho_0 I - B) = (c_1 \quad \cdots \quad c_n)$, then $c_i > 0$ for all i and $0 = (\rho_0 I - B)\,\mathrm{Adj}\,(\rho_0 I - B) = ((\rho_0 I - B)c_1 \quad \cdots \quad (\rho_0 I - B)c_n)$, so that $(\rho_0 I - B)c_i = 0$ for all i. In other words, Perron's lemma implies that every column of the adjoint matrix of $\rho_0 I - B$ is a positive characteristic vector for ρ_0.

Perron's lemma represents the first part of Perron's theorem (Theorem 17.9 below), the second part being that $|\rho| < \rho_0$ for all other roots ρ of B. The truth of Perron's lemma is easy to verify for any 2×2 matrix $B > 0$: the characteristic roots of B are given by the familiar quadratic formula, from which it is easily seen that the roots are real and distinct and that the larger is positive; the positivity of the 1×1 cofactors of $\rho_0 I - B$ also follows readily from the quadratic formula for ρ_0. Such considerations probably suggested to Perron the possibility that Lemma 17.6 might be true for $n \times n$ matrices; but for them, the above sort of straightforward verification is not feasible, since among other things, there is no formula for the roots of the characteristic polynomial. Not surprisingly, Perron sought to establish the general validity of the lemma by induction on n. His proof was entirely correct, although explained poorly.[8]

I now consider how Perron utilized Lemma 17.6 to show how to determine, for a periodic Jacobi algorithm that is nonnegative in the sense that (1) and (2) of Theorem 17.5 hold, which characteristic root of the algorithm matrix A provides the characteristic vectors that yield the limiting values α_j of the algorithm. These deliberations on his part are historically important, because they revealed to him the possibility of extending Lemma 17.6 to include the assertion that $|\rho| < \rho_0$ holds for all roots $\rho \neq \rho_0$ of B.

The nonnegativity conditions (1) and (2) imply, as we saw, that the matrix A associated to the periodic algorithm satisfies $A^\nu > 0$ for all $\nu \geq 2n/\mathbf{p}$, where $\mathbf{p}$ is the period of the algorithm. As we saw, since the algorithm converges Perron knew that its limiting values $\alpha_j = x_j/x_0$ have the property that $x = (x_0 \quad \cdots \quad x_n)^t$ satisfies $Ax = \rho' x$ for some characteristic root ρ' of A. All that remained was to determine which root, and this is where Lemma 17.6 proved useful. Since $A^\nu > 0$ for all $\nu \geq 2n/\mathbf{p}$, Lemma 17.6 as applied to $B = A^\nu > 0$ says that B has a largest positive root σ_0 of multiplicity one. Now, it was well known that if $\rho_0, \rho_1, \ldots, \rho_n$ are the characteristic roots of A, each root being listed as often as its multiplicity, then $\rho_0^\nu, \rho_1^\nu, \ldots, \rho_n^\nu$ is a similar listing of the characteristic roots of A^ν. Suppose by Lemma 17.6 that $\sigma_0 = \rho_0^\nu$. Then it follows that (1) $\rho_0 > 0$, that (2) ρ_0 is the largest positive root of A, and that (3) ρ_0 has multiplicity one.[9] Having established these properties of ρ_0, Perron then used determinant-theoretic relations to show that the positivity of all cofactors of $\sigma_0 I - A^\nu$ (a consequence of Lemma 17.6 applied to $B = A^\nu > 0$) implied the same for the cofactors of $\rho_0 I - A$ [467, pp. 49–50].

Before proceeding to Perron's proof that ρ_0 is in fact the root ρ' whose characteristic vectors $x = (x_0 \quad x_1 \quad \cdots \quad x_n)^t$ yield the limiting values of the algorithm, it will be helpful to point out a fact realized by Perron, namely that the above reasoning establishes the following corollary to Lemma 17.6:

Corollary 17.7 (Corollary to Perron's lemma). *If A is any nonnegative matrix such that $A^\nu > 0$ for some positive integer ν, then A has at least one positive root. The greatest positive root ρ_0 has multiplicity one, and* Adj $(\rho_0 I - A) > 0$.

To show that $\rho' = \rho_0$, Perron used a relation he had derived in his general study of $\mathbf{p}$-periodic algorithms [467, (38), p. 41]: if the algorithm converges to $x = (x_0 \quad x_1 \quad \cdots \quad x_n)^t$, so that $Ax = \rho' x$ for some root ρ' of the characteristic equation, then for any $i, j = 0, 1, \ldots, n$,

$$\rho' = \lim_{\nu \to \infty} [A^{\nu+1}]_{ij} / [A^\nu]_{ij}. \tag{17.25}$$

[8] For an exposition of this historic proof, see [278, Appendix 6.1].

[9] Perron gave only a proof of (1), from which (2) and (3) follow easily. Regarding his proof of (1), see [278, p. 675, n. 15].

From (17.25) it then follows that

$$\rho' = \mathrm{tr}(A^{\nu+1})/\mathrm{tr}(A^{\nu}). \tag{17.26}$$

(See [278, p. 676, n. 16] for details.) Perron did not use trace notation or terminology, but be realized in effect that the trace of a matrix was equal to the sum of its characteristic roots, and so he proceeded to express (17.26) in the form

$$\rho' = \lim_{\nu\to\infty} \frac{\sum_{i=0}^{n} \rho_i^{\nu+1}}{\sum_{i=0}^{n} \rho_i^{\nu}}, \tag{17.27}$$

where the summation is over all roots of A, counted according to multiplicity.

The only roots that contribute to the limiting value in (17.27) are those with absolute value $M = \max_i |\rho_i|$.[10] These roots are all of the form $\rho = Me^{i\theta}$, but Perron showed further that the limit in (17.27) cannot exist if there is a root $\rho = Me^{i\theta}$ with $\theta \not\equiv 0 \pmod{2\pi}$ [467, pp. 50–51]. In other words, since the limit is known to exist, $\rho_0 = M$ is the sole characteristic root with maximal absolute value M, i.e., for any other characteristic root ρ, it must be that $|\rho| < \rho_0$. Furthermore, Corollary 17.7 to Perron's lemma applies and shows that ρ_0 has multiplicity one. Thus formula (17.27) reduces to $\rho' = \lim_{\nu\to\infty} \rho_0^{\nu+1}/\rho_0^{\nu} = \rho_0$. Perron had thus solved his problem: given a purely periodic Jacobi algorithm that satisfies the nonnegativity conditions (1) and (2) of Theorem 17.5 and so converges, it is the maximal positive root ρ_0 of the algorithm matrix A that provides the limiting values of the algorithm in homogeneous form as any characteristic vector $x = (x_0 \quad \cdots \quad x_n)^t$ for ρ_0 [467, p. 51, Satz IX].

For example, if A is the algorithm matrix given in (17.23), so that $A^3 > 0$, it now follows that since a characteristic vector for $\rho_0 = 2.69174\ldots$ is

$$x = (0.861983\ldots \quad 0.371507\ldots \quad 0.320233\ldots \quad 1)^t,$$

[10] Let $M, a, b, \ldots$ be the distinct absolute values of the roots $\rho_0, \rho_1, \ldots, \rho_n$. Then the numerator of (17.27) can be expressed as $M^{\nu+1}\alpha_{\nu+1} + a^{\nu+1}\beta_{\nu+1} + b^{\nu+1}\gamma_{\nu+1} + \cdots$, where each of $\alpha_{\nu+1}, \beta_{\nu+1}, \gamma_{\nu+1}$ is a sum of at most $n+1$ complex numbers of absolute value 1. Thus each of $\alpha_{\nu+1}, \beta_{\nu+1}, \gamma_{\nu+1}$ has absolute value at most $n+1$. Likewise, the denominator of (17.27) is expressible as $M^{\nu}\alpha_{\nu} + a^{\nu}\beta_{\nu} + b^{\nu}\gamma_{\nu} + \cdots$, with each of $\alpha_{\nu}, \beta_{\nu}, \gamma_{\nu}$ having absolute value at most $n+1$. Thus the ratio in (17.27) that is under the limit operation, on division of the numerator and denominator by M^{ν}, is

$$\frac{M\alpha_{\nu+1} + a(a/M)^{\nu}\beta_{\nu+1} + b(b/M)^{\nu}\gamma_{\nu+1} + \cdots}{\alpha_{\nu} + (a/M)^{\nu}\beta_{\nu} + (b/M)^{\nu}\gamma_{\nu} + \cdots}.$$

Clearly, all the terms in the numerator except the first approach 0 as $\nu \to \infty$. Likewise, all the terms in the denominator except the first approach 0 as $\nu \to \infty$. Thus, assuming the limit ρ' exists in (17.27), we see that $\rho' = \lim_{\nu\to\infty} M(\alpha_{\nu+1}/\alpha_{\nu}) = \lim_{\nu\to\infty}(M^{\nu+1}\alpha_{\nu+1})/(M^{\nu}\alpha_{\nu})$, which is to say that only the roots of absolute value M contribute to the limiting value ρ'.

the periodic algorithm associated to A converges to

$$(\alpha_1, \alpha_2, \alpha_3) = (0.861983\ldots,\ 0.371507\ldots,\ 0.320233\ldots).$$

As the above outline of Perron's reasoning suggests, the *only* part of it that used the fact that A is the matrix associated to a periodic Jacobi algorithm satisfying the nonnegativity conditions (1) and (2) of Theorem 17.5 was the reasoning leading to the existence of a characteristic root ρ' satisfying (17.25). In effect, he had proved, as he realized, the following extension of his Corollary 17.7.

Proposition 17.8. *Suppose $A = (a_{ij})$ is any nonnegative matrix that has the following properties: (i) $A^{\nu} > 0$ for some power $\nu > 0$; (ii) there is a characteristic root ρ' of A such that*

$$\lim_{\nu \to \infty} \frac{[A^{(\nu+1)}]_{ij}}{[A^{(\nu)}]_{ij}} = \rho' \quad \text{for all } i, j.$$

Then there is a root $\rho_0 > 0$ of multiplicity 1 such that $|\rho| < \rho_0$ for all other characteristic roots ρ. Furthermore, the cofactors of $\rho_0 I - A$ are all positive, so that every column of Adj $(\rho_0 I - A)$ *provides an $x > 0$ for which $Ax = \rho_0 x$.*

17.1.4 Perron's theorem

Perron thought that he could prove that for *every* matrix $A > 0$ (and so satisfying (i) above), it must also be the case that (ii) follows, so that the conclusions of Proposition 17.8 remain true for all positive matrices. In other words, he was confident that he could prove what we now know as Perron's theorem, although he had not yet fully worked out the proof, which he believed would be rather complicated [278, p. 678].

Within six months of submitting his *Habilitationsschrift* to the *Annalen*, Perron published a paper confirming his claims. The paper was entitled "Towards the theory of matrices" [468], and in it, Perron proposed to show how many of the proof ideas he had developed in his *Habilitationsschrift* could be used to give new and simple proofs of known results about matrices and their characteristic equations as well as to establish some new ones. Among the new ones was the theorem alluded to in his *Habilitationsschrift* that every positive matrix has a regular characteristic equation [467, §5], i.e., what is now usually called Perron's theorem:

Theorem 17.9 (Perron's theorem). *Let A be any square matrix such that $A > 0$. Then A has a characteristic root $\rho_0 > 0$ of multiplicity one such that $\rho_0 > |\rho|$ for all other characteristic roots ρ of A. Moreover, all the cofactors of $\rho_0 I - A$ are positive. [Hence $x > 0$ exists such that $Ax = \rho_0 x$.]*

Perron's proof was dictated by the reasoning that had implied Proposition 17.8. The assumption that $A > 0$ in Theorem 17.9 is a special case of property (i) of Proposition 17.8, and so he now proved that property (ii) holds for every positive matrix. Theorem 17.9 then follows by the same reasoning as that behind Proposition 17.8, as Perron himself explained so as not to repeat it [468, p. 261].

The proof that property (ii) of Proposition 17.8 holds for every positive matrix was achieved by means of the following complicated lemma [468, pp. 259–261].

Lemma 17.10 (Perron's limit lemma). *Let $A = (a_{ij})$ be $n \times n$ with $a_{ij} > 0$ for all i and j. Then (i) $\lim_{\nu\to\infty}[A^\nu]_{ij}/[A^\nu]_{nj}$ exists as a finite number that is independent of j. Denote it by x_i/x_n. Then (ii) $\rho' = \lim_{\nu\to\infty}[A^{\nu+1}]_{ij}/[A^\nu]_{ij}$ exists as a finite positive number that is independent of i and j. (iii) If $x = (x_1 \quad \cdots \quad x_n)^t$ with the x_i as in (i), then $Ax = \rho' x$. Hence ρ' is a positive characteristic root of A.*

Perron's proof of part (i) is indicative of his method of proof. For a fixed value of i, let $\beta_i^{(\nu)}$ and $B_i^{(\nu)}$ denote, respectively, the minimum and the maximum of the n positive numbers $[A^\nu]_{ij}/[A^\nu]_{nj}$, $j = 1,\ldots,n$, so that $\beta_i^{(\nu)} \leq B_i^{(\nu)}$ for all ν. Perron showed that the sequence $\beta_i^{(\nu)}$ increases with ν, and that $B_i^{(\nu)}$ decreases. From "a known theorem," it then followed that $\beta_i = \lim_{\nu\to\infty}\beta_i^{(\nu)}$ and $B_i = \lim_{\nu\to\infty}B_i^{(\nu)}$ exist as finite numbers with $\beta_i \leq B_i$ [467, p. 259].[11] Then a fairly complicated "ε–δ"-type argument was given to show that $\beta_i = B_i$, and since this is true for all $i = 1,\ldots,n$, (i) of Lemma 17.10 follows.

Immediately after stating Theorem 17.9, Perron added two historically consequential comments [467, p. 262]:

> Although this is a purely algebraic theorem, nevertheless I have not succeeded in proving it with the customary tools of algebra. The theorem remains valid, by the way, when the a_{ik} are only partly positive but the rest are zero, *provided only that a certain power of the matrix A exists for which none of the entries are zero.*

The second sentence, of course, reflects Perron's realization that the reasoning in his *Habilitationsschrift* implied Proposition 17.8. Combined with Lemma 17.10, that proposition implies the result noted by Perron. In other words, Perron had also established the following corollary to Theorem 17.9:

Corollary 17.11 (Perron's corollary). *Let $A \geq 0$ be such that $A^\nu > 0$ for some power $\nu > 0$. Then the conclusions of Theorem 17.9 still hold.*

Although Perron simply mentioned the gist of Corollary 17.11 in passing, his remark could be interpreted to mean that there is a substantial class of nonnegative matrices for which the remarkable conclusions of Perron's Theorem 17.9 remain valid. As we shall see, Frobenius eventually pursued this implication to a definitive conclusion in 1912 and in the process created his remarkable theory of nonnegative matrices. It

---

[11] The "known theorem" was presumably that increasing (respectively, decreasing) sequences of real numbers that are bounded above (respectively, below) converge.

was Perron's first sentence above, however, that initially drew Frobenius' interest to the theory of positive and nonnegative matrices.

The first sentence reflects Perron's dissatisfaction with his proof of Theorem 17.9. What he seems to have meant was that his proof depended on the limit considerations of Lemma 17.10 and its proof, and so required more than the "customary tools of algebra," such as the theory of determinants. This is how Frobenius interpreted Perron's remark [228, p. 404], and so he took up the challenge implicit in it: to give a determinant-based proof of Perron's theorem that would avoid the complexities of Perron's limit lemma.[12]

17.2 Frobenius' Theory of Nonnegative Matrices

Although Frobenius no doubt regularly scanned the pages of *Mathematische Annalen*, which had become the journal of the rival Göttingen school of mathematics of Klein, Hilbert, and Minkowski, it seems unlikely to me that he would have paid any attention to Perron's *Habilitationsschrift* when it appeared in the first issue of 1907, due to its subject matter. But when Perron's 16-page paper "Towards the theory of matrices" [468] appeared in a subsequent issue that year, it is not surprising that Frobenius, an expert on the theory of matrices, read it and responded to the challenge set forth by Perron of providing a proof of his Theorem 17.9 that would avoid his limit lemma.

17.2.1 Frobenius' papers of 1908 and 1909

Thus while Perron was writing up a detailed study of the convergence of aperiodic Jacobi algorithms, which appeared in 1908 [469], Frobenius set himself the task of a proof of Perron's theorem that avoided his limit lemma. He succeeded, and his results were published in 1908 [228]. Frobenius proved a slightly stronger version of Perron's Theorem 17.9 [228, §1], which may be stated as follows.

Theorem 17.12 (Frobenius' version of Perron's theorem). *Let $A > 0$ be $n \times n$. Then the following hold. (I) A has a positive characteristic root and hence a*

[12]It does not seem that Perron was seeking a purely algebraic proof of his theorem in the sense of a proof that was completely free of propositions from analysis. For example, he never expressed a similar dissatisfaction with his proof of his seminal Lemma 17.6, despite the fact that it repeatedly invoked basic theorems from analysis such as the intermediate value theorem for continuous functions [278, Appendix 6.1]. The intermediate value theorem was also invoked in Frobenius' proof of Perron's theorem. It should also be noted that although Frobenius took up the challenge of a determinant-based proof of Perron's theorem, as a student of Weierstrass, he was not averse to employing results from complex analysis, notably Laurent expansions, in his proofs of theorems about matrices (as in Sections 7.5.1, 7.5.5, and 16.1.5).

maximal positive root ρ_0. *Furthermore,* ρ_0 *has multiplicity one and* $\mathrm{Adj}\,(\rho I - A) > 0$ *for all* $\rho \geq \rho_0$. *(II) If* ρ *is any other characteristic root of* A*, then* $|\rho| < \rho_0$.

Part (I) of Frobenius' version represents a slightly improved version of Perron's Lemma 17.6, the improvement being that $\mathrm{Adj}\,(\rho I - A)$ is positive not only for $\rho = \rho_0$ but also for all $\rho > \rho_0$. His proof of part (I), like Perron's of his Lemma 17.6, was by induction on n, but it was far shorter and simpler. This was due in part to the fact that Frobenius used the induction hypothesis that $\mathrm{Adj}\,(\rho I - A) > 0$ for all $\rho \geq \rho_0$.[13] This induction hypothesis enabled him to prove quickly via cofactor expansions that A has positive roots and so a maximal positive root ρ_0. It also enabled him to give a quick proof, by further, more subtle, cofactor expansions, that $\mathrm{Adj}\,(\rho I - A) > 0$ for $\rho \geq \rho_0$. The fact that ρ_0 has multiplicity one then followed from an identity not used by Perron. First of all, let $\varphi(\rho) = \det(\rho I - A)$ denote the characteristic polynomial of A.[14] Then

$$\varphi'(\rho) = \sum_{\alpha=1}^{n} \varphi_{\alpha\alpha}(\rho), \tag{17.28}$$

where $\varphi_{\alpha\alpha}(\rho)$ denotes the αth *principal minor determinant of* $\rho I - A$, i.e., the minor determinant obtained from $\rho I - A$ by deleting row α and column α. Equation (17.28) showed that $\varphi'(\rho_0) > 0$. This is because $\alpha + \alpha$ is even and so $\varphi_{\alpha\alpha}(\rho_0) = [\mathrm{Adj}\,(\rho_0 I - A)]_{\alpha\alpha} > 0$.

It was Perron's proof of part (II) of Theorem 17.12 that had required his limit lemma. Frobenius' proof of Part II required less than a full page [228, p. 406] and avoided Perron's Limit Lemma 17.10. Nonetheless, it was somewhat contrived. As we shall see, he soon discovered a proof that was even briefer and yet straightforward, a proof that seems to have provided the fillip for a new and remarkably rewarding line of research that ultimately led to his masterful paper of 1912 on nonnegative matrices.

Besides supplying a far simpler proof of the "noteworthy properties" [228, p. 404] of positive matrices to which Perron had called attention, Frobenius also considered briefly what could be said when A is simply assumed to be nonnegative:

> If the matrix A is only assumed to have elements $a_{\alpha\beta} \geq 0$, then by means of the above methods of proof and continuity considerations it is easy to determine the modifications under which the above theorems remain valid. The greatest root ... $[\rho_0]$ ... is real and ... ≥ 0. It can be a multiple root, but only when all the principal determinants ... $[\varphi_{\alpha\alpha}(\rho_0)]$ ... vanish [228, pp. 408–409].

---

[13] In proving Lemma 17.6, Perron could have used the weaker induction hypothesis $\mathrm{Adj}\,(\rho_0 I - A) > 0$ but did not, thereby unnecessarily complicating his proof; see [278, Appendix 6.1].

[14] In his *Habilitationsschrift*, Perron had defined the characteristic polynomial of A as $\det(A - \rho I)$ [467, p. 30], but in his paper on matrices [468, p. 249], he defined it as $\det(\rho I - A)$, which was more in keeping with the fact that it is $\mathrm{Adj}\,(\rho_0 I - A)$ that is positive in Perron's theorem.

For later reference, I will sum up the above quotation as the following proposition.

Proposition 17.13. *If $A \geq 0$, there is a nonnegative root ρ_0 that is greatest in the sense that $|\rho| \leq \rho_0$ for all characteristic roots ρ of A. It is no longer the case that ρ_0 necessarily has multiplicity one, but in order for it to be a multiple root, it is necessary that $\varphi_{\alpha\alpha}(\rho_0) = 0$ for all α, where $\varphi_{\alpha\alpha}(\rho)$ is defined following (17.28).*

As Frobenius said, Proposition 17.13 was an easy consequence of Perron's theorem, obtained as a limiting case of that theorem.[15] Just how much of Perron's theorem is lost in the limiting case is illustrated by the Jordan–Weierstrass canonical form matrix

$$A = \begin{pmatrix} 0 & 1 & 0 \\ 0 & 0 & 0 \\ 0 & 0 & 0 \end{pmatrix},$$

which has ρ_0 being zero (rather than positive) and with multiplicity three. Proposition 17.13 does not amount to much, but it shows that Frobenius was already wondering about what could be said about nonnegative matrices. The above example shows that some restrictive assumptions would have to be imposed on nonnegative matrices to attain something akin to Perron's theorem. Evidently, Frobenius had not yet determined what they should be.

In 1909, Frobenius published a sequel to his paper on positive matrices, which seems to have been inspired by his discovery that certain properties of characteristic roots could be established quickly and simply by means of what would now be called inner product considerations. As we saw in Section 5.2, this discovery had been made already in the 1860s by Clebsch and Christoffel within the context of Hermitian symmetric matrices, but Frobenius was apparently unaware of their use of the technique; at least he failed to use it in his proof of his Theorem 7.15 on orthogonal matrices, as indicated in Section 7.15. Having now discovered it, he showed how it could be used to give a very short and simple proof of part (II) of Theorem 17.12 [229, pp. 411–412]. He also used the same technique to prove the following proposition, which he regarded as a converse to Perron's theorem. It seems to be the first sign of interest in a line of investigation that was eventually to lead him to a series of remarkable results about nonnegative matrices.

The proposition in question is the following.

Proposition 17.14. *Let $A > 0$. Then if y is a nonnegative characteristic vector for some root ρ of A, it must be that $\rho = \rho_0$ and hence that $y > 0$.*

---

[15]Thus, e.g., in the limit, $\text{Adj}\,(\rho_0 I - A) > 0$ for $A > 0$ becomes $\text{Adj}\,(\rho_0 I - A) \geq 0$ for $A \geq 0$. In particular, the (α, α) entry of $\text{Adj}\,(\rho_0 I - A)$, namely $\varphi_{\alpha\alpha}(\rho_0)$, is nonnegative; and since by (17.28), $\varphi'(\rho_0)$ is the sum of all the $\varphi_{\alpha\alpha}(\rho_0)$, it follows that $\varphi'(\rho_0) \geq 0$ with $\varphi'(\rho_0) = 0$ only if all $\varphi_{\alpha\alpha}(\rho_0)$ vanish, thereby giving the above necessary condition for ρ_0 to be a multiple root.

With the use of modern inner (or dot) product notation $x \cdot y = x_1 y_1 + \cdots + x_n y_n$, Frobenius' proof goes like this [229, p. 410]. First of all, since $A > 0$ and $y \geq 0$, $Ay > 0$, and so the equation $Ay = \rho y$ implies that $\rho > 0$. Now let x be the positive characteristic vector for ρ_0 that exists by virtue of Perron's Theorem 17.9 applied to the transposed matrix A^t, so that $A^t x = \rho_0 x$. Then by hypothesis,

$$\rho(x \cdot y) = (x \cdot \rho y) = (x \cdot Ay) = (A^t x \cdot y) = (\rho_0 x \cdot y) = \rho_0(x \cdot y).$$

Since $x > 0$ and $0 \neq y \geq 0$ means that $x \cdot y > 0$, it follows by canceling $x \cdot y$ in the above equation that $\rho = \rho_0$.

Frobenius' Proposition 17.14 shows that for positive matrices, the only nonnegative characteristic vectors y that exist are those associated to the maximal positive root ρ_0; and because ρ_0 has multiplicity one, y must actually be a multiple of a positive characteristic vector for ρ_0 and so itself positive. From this point of view, Proposition 17.14 suggests the following more general problem:

Problem 17.15. *Given a nonnegative matrix A, determine the characteristic roots of A for which nonnegative characteristic vectors exist.*

Nowadays, in many applications of the theory of nonnegative matrices, the existence of positive or nonnegative characteristic vectors is of great importance; but this was not the case in Frobenius' time.[16] It was evidently as a problem of purely mathematical interest that Frobenius eventually realized that Problem 17.15 could be investigated in an incredibly fruitful manner.

17.2.2 Frobenius' 1912 paper on nonnegative matrices

According to Frobenius, it was Problem 17.15 that led him to his remarkable discoveries about nonnegative matrices. His paper of 1912 presenting these discoveries begins as follows [231, p. 546].

> In my works ... [228, 229] ... I developed the properties of positive matrices and extended them with the necessary modifications to nonnegative [matrices]. The latter, however, require a far more in-depth investigation, to which I was led by the problem treated in §11.

---

[16]Two such present-day applications—Markov chains and input–output-type economic analysis—existed at the time of Frobenius' work on positive and nonnegative matrices (1908–1912). Markov introduced the eponymous chains in 1908 (see Section 17.3), and the mathematician Maurice Potron announced an economic theory analogous to input–output analysis in 1911 [487, 488] with details given in 1913 [489]. Although Frobenius was apparently unaware of these developments, it is of interest to note that neither Markov nor Potron ascribed an importance to nonnegative characteristic vectors in their respective applications. For further general information about Potron, whose work remained unappreciated until recently, see [1, 17, 18]. Note also that Wilfried Parys is working on an annotated Potron bibliography and on historical aspects of Perron–Frobenius theory in economics.

The problem of §11 is Problem 17.15 above. Let us now consider how that problem may have led him to his discoveries.

For $A > 0$, Frobenius' Proposition 17.14 solves Problem 17.15 by showing that the maximal root ρ_0 is the only such root. However, if A and A' are both positive matrices with respective maximal roots ρ_0 and ρ_0', then it is easily seen that the nonnegative matrix $B = \begin{pmatrix} A & 0 \\ 0 & A' \end{pmatrix}$ has nonnegative characteristic vectors for both ρ_0 and ρ_0'. That is, if x, x' denote positive characteristic vectors for A, ρ_0 and A', ρ_0', respectively, then matrix block multiplication shows that $y = \begin{pmatrix} x \\ 0 \end{pmatrix}$ and $y' = \begin{pmatrix} 0 \\ x' \end{pmatrix}$ are nonnegative characteristic vectors of B for ρ_0 and ρ_0', respectively. Frobenius, who was an expert on the application of matrix algebra to linear algebraic problems, had utilized the symbolic algebra of block partitioned matrices on many occasions, especially in his work on principal transformations of theta functions (Section 10.6).

The above observations about the nonnegative matrix B indicate that more generally, Frobenius' Problem 17.15 is trivial to solve for any nonnegative matrix in block diagonal form

$$B = \begin{pmatrix} R_{11} & 0 & \cdots & 0 \\ 0 & R_{22} & \cdots & 0 \\ \cdots & \cdots & \cdots & \cdots \\ 0 & 0 & \cdots & R_{kk} \end{pmatrix}, \quad R_{11} > 0, \quad \ldots, \quad R_{kk} > 0. \tag{17.29}$$

If $\rho_0^{(1)}, \ldots, \rho_0^{(k)}$ are the maximal roots of $R_{11} \ldots, R_{kk}$, then, as in the case of two diagonal blocks, B has a nonnegative characteristic vector $y^{(i)}$ for $\rho_0^{(i)}$, $i = 1, \ldots, m$. More generally, let P_σ denote the $n \times n$ permutation matrix obtained from the identity matrix I_n by permuting its rows according to the permutation $\sigma \in S_n$, and consider the similar matrix $A = P_\sigma B P_\sigma^{-1} = P_\sigma B P_\sigma^t$. (Since P_σ is an orthogonal matrix, $P_\sigma^{-1} = P_\sigma^t$.) Then A is nonnegative, because the similarity transformation $B \to (P_\sigma B) P_\sigma^t$ involves first permuting the rows of B by σ and then permuting the columns of the resulting matrix, $P_\sigma B$, by σ. Furthermore, $z^{(i)} = P_\sigma y^{(i)}$ is also nonnegative and is easily seen to be a characteristic vector of A for $\rho_0^{(i)}$. Thus Frobenius' Problem 17.15 is solved for any nonnegative matrix *permutationally similar* (in the sense described above) to a matrix in the block form (17.29).

Frobenius' problem would be rather trivial if every nonnegative matrix were permutationally similar to a matrix in the block diagonal form (17.29), but this is not the case. For example, $A = \begin{pmatrix} 1 & 0 \\ 1 & 1 \end{pmatrix}$ is not permutationally similar to a matrix in the form (17.29), since the transposition $\sigma = (12)$ is the sole nontrivial permutation of two objects, and $P_\sigma A P_\sigma^t = \begin{pmatrix} 1 & 1 \\ 0 & 1 \end{pmatrix}$. On the other hand, as Frobenius realized [231, p. 555], every nonnegative matrix is permutationally similar to a matrix in a lower triangular block form

$$\begin{pmatrix} R_{11} & 0 & 0 & \cdots & 0 \\ R_{21} & R_{22} & 0 & \cdots & 0 \\ \cdots & \cdots & \cdots & \cdots & \cdots \\ R_{m1} & R_{m2} & R_{m3} & \cdots & R_{mm} \end{pmatrix}, \tag{17.30}$$

where now each diagonal $k_i \times k_i$ block R_{ii} is nonnegative—but not necessarily positive—and assuming that no further reduction is possible, the diagonal blocks R_{ii} have the property that they are not permutationally similar to a matrix in the block form

$$\begin{pmatrix} P & 0 \\ Q & R \end{pmatrix}, \tag{17.31}$$

since if, e.g., $P_\sigma R_{11} P_\sigma^t = \left(\begin{smallmatrix} P & 0 \\ Q & R \end{smallmatrix}\right)$, then the similarity transformation generated by

$$P_{\sigma^*} = \begin{pmatrix} P_\sigma & 0 & \cdots & 0 \\ 0 & I_{k_2} & \cdots & 0 \\ & & \ddots & \\ 0 & 0 & \cdots & I_{k_m} \end{pmatrix}$$

would make (17.30) permutationally similar to a more refined lower triangular block form with the block R_{11} replaced by (17.31).

Thus Frobenius' Problem 17.15 requires dealing with an "irreducible" lower triangular block form (17.30) rather than the diagonal block form (17.29). This means that it is necessary to know, first of all, to what extent the nonnegative matrices R_{ii} that occur on the diagonal in (17.30) possess the properties of positive matrices set forth in Perron's Theorem 17.9. Frobenius called such nonnegative matrices "indecomposable" (*unzerlegbar*). Nowadays, they are said to be *irreducible*, and to avoid confusion, I will use the current terminology. There is, of course, a certain degree of analogy between Frobenius' definition of an indecomposable nonnegative matrix and his definition of an irreducible matrix representation of a finite group; I suspect that Frobenius, who was aware of the analogy, chose his terminology to avoid confusion of the two notions.

Thus a nonnegative matrix A is irreducible if it is not permutationally similar to a matrix of the form (17.31). Nonnegative matrices that are permutationally similar to the block form (17.31) he referred to as "decomposable"; I will use the current term *reducible*. It should be noted that if a nonnegative matrix A is reducible, so that $A = P_\sigma \left(\begin{smallmatrix} P & 0 \\ Q & R \end{smallmatrix}\right) P_\sigma^{-1}$, then $A^\nu = P_\sigma \left(\begin{smallmatrix} P^\nu & 0 \\ Q^* & R^\nu \end{smallmatrix}\right) P_\sigma^{-1}$ can never be positive. Recall that Perron's Corollary 17.11 shows that every nonnegative matrix such that $A^\nu > 0$ for some power ν possesses all the properties of a positive matrix posited in Perron's Theorem 17.9. The above considerations show that the class of irreducible matrices includes all nonnegative matrices satisfying Perron's condition

$A^\nu > 0$. This fact may have raised the hope in Frobenius' mind that the larger class of irreducible matrices might share some of the remarkable properties of those satisfying Perron's condition; if so, the solution of Frobenius' Problem 17.15 would be greatly advanced. The first task, then, would be to investigate the extent to which irreducible matrices satisfy the conclusions of Perron's theorem.

The above characterization of the concept of a reducible matrix—and hence also an irreducible one—is the characterization that Frobenius used in his reasoning and is, as I have suggested, probably the form in which he was led to it by Problem 17.15. It is, however, possible to formulate the concept in a form directly related to the coefficient array of a reducible matrix. That is, an $n \times n$ matrix $A \geq 0$ is reducible if and only if there exist $p > 0$ rows of A and $q = n - p > 0$ *complementary* columns of A such that there are zeros at all the intersections of these rows and columns. This was Frobenius' official definition of reducibility [231, p. 548]. For example, if

$$A = \begin{pmatrix} a_{11} & 0 & a_{13} & a_{14} & 0 \\ a_{21} & a_{22} & a_{23} & a_{24} & a_{25} \\ a_{31} & 0 & a_{33} & a_{34} & 0 \\ a_{41} & 0 & a_{43} & a_{44} & 0 \\ a_{51} & a_{52} & a_{53} & a_{54} & a_{55} \end{pmatrix},$$

where the a_{ij} are positive, then rows 1, 3, and 4 and complementary columns 2 and 5 have zeros at their intersections (so $p = 3$ and $q = 2$). To see that A is reducible in the original sense, consider the transposition $\sigma = (2,4)$ of the columns of A; it puts the two columns with the "intersection zeros" at the far right, and σ applied to the rows of the resulting matrix puts the intersection zeros in the upper right-hand corner, i.e.,

$$P_\sigma A P_\sigma^t = \begin{pmatrix} a_{11} & a_{14} & a_{13} & 0 & 0 \\ a_{41} & a_{44} & a_{43} & 0 & 0 \\ a_{31} & a_{34} & a_{33} & 0 & 0 \\ a_{21} & a_{24} & a_{23} & a_{22} & a_{25} \\ a_{51} & a_{54} & a_{53} & a_{52} & a_{55} \end{pmatrix}, \quad P_\sigma = \begin{pmatrix} 1 & 0 & 0 & 0 & 0 \\ 0 & 0 & 0 & 1 & 0 \\ 0 & 0 & 1 & 0 & 0 \\ 0 & 1 & 0 & 0 & 0 \\ 0 & 0 & 0 & 0 & 1 \end{pmatrix},$$

so that $P_\sigma A P_\sigma^t$ is in the form (17.31) and A is reducible in the first-mentioned sense.

To avoid possible confusion in what is to follow, it should be noted that in order for a matrix $A \geq 0$ to be reducible, it must be at least 2×2. Thus, although Frobenius never mentioned it, it follows that every 1×1 $A \geq 0$ is irreducible, including $A = (0)$. The fact that $A = (0)$ is irreducible is relevant to the solution of Frobenius' Problem 17.15, because it means that some of the irreducible blocks R_{ii} in (17.30) can be (0). (See the discussion following (17.37) below.) Many of Frobenius' theorems about irreducible matrices, however, do not hold for $A = (0)$, and so in discussing them, $A = (0)$ will be excluded by stipulating that the nonnegative matrices A under consideration do not include the zero matrix.

Nowadays, graph-theoretic notions are used with profit in the theory of nonnegative matrices. For example, for an $n \times n$ matrix $A = (a_{ij}) \geq 0$, the directed graph $G(A)$ of A is defined as follows: $G(A)$ has vertices $1,\ldots,n$, and a directed edge $i \to j$ exists when $a_{ij} > 0$. Then it turns out that A is irreducible in Frobenius' sense precisely when $G(A)$ is connected in the following sense: either $G(A)$ has one vertex (so A is 1×1), or $G(A)$ has at least two vertices and for any two vertices $i \neq j$ there is a directed path from i to j.

Judging by the content of Frobenius' 1912 paper, it seems that in exploring the properties of irreducible matrices, he focused on the question whether the maximal root ρ_0 of Proposition 17.13 has multiplicity one. He observed that if A is reducible and so permutationally similar to $\begin{pmatrix} P & 0 \\ Q & R \end{pmatrix}$, then

$$0 = \varphi(\rho_0) = \det(\rho_0 I - A) = \det(\rho_0 I - P) \cdot \det(\rho_0 I - R),$$

and so one of the principal minor determinants,[17] $\det(\rho_0 I - P)$ or $\det(\rho_0 I - R)$, must vanish. By means of determinant-theoretic considerations combined with the matrix algebra of block-partitioned matrices, he was able prove the converse: if some principal minor determinant of $\rho_0 I - A$ vanishes, then A must be reducible. In this way, he proved that a nonnegative $A \neq 0$ is reducible if and only if some principal minor of $\rho_0 I - A$ vanishes. Stated another way, his result was that a nonnegative $A \neq 0$ is irreducible if and only if none of the principal minors of $\rho_0 I - A$ vanish. This meant in particular that when $A \neq 0$ is irreducible, none of the degree-$(n-1)$ principal minors $\varphi_{\alpha\alpha}(\rho_0)$ vanish, i.e., all are positive, and so by the identity (17.28) he had used in his 1908 paper, $\varphi'(\rho_0) = \sum_{\alpha=1}^{n} \varphi_{\alpha\alpha}(\rho_0) > 0$, which means that ρ_0 has multiplicity one. The positivity of the diagonal elements $\varphi_{\alpha\alpha}(\rho_0)$ of $\mathrm{Adj}\,(\rho_0 I - A)$ together with $\mathrm{Adj}\,(\rho_0 I - A) \geq 0$ then implied, by a determinant identity, that $\mathrm{Adj}\,(\rho_0 I - A) > 0$.[18] Finally, the fact that all roots ρ of any $A \geq 0$ satisfy $|\rho| \leq \rho_0$ (Proposition 17.13) shows that $\rho_0 = 0$ is possible for an irreducible A only when $A = (0)$. In this way, Frobenius obtained his first substantial result on nonnegative matrices:

Theorem 17.16 (Irreducible matrix theorem). *If $A \neq 0$ is an irreducible matrix, then ρ_0 is positive, has multiplicity one, and satisfies* $\mathrm{Adj}\,(\rho_0 I - A) > 0$. *Hence there is an $x > 0$ such that $Ax = \rho_0 x$. All other characteristic roots ρ satisfy $|\rho| \leq \rho_0$.*

---

[17] A principal minor determinant of $\rho_0 I - A$ of degree $n-k$ is one obtained by deleting the same k rows and columns of $\rho_0 I - A$, e.g., by deleting the first k rows and the first k columns.

[18] What Frobenius used, without any explanation, was the fact that if B is any matrix and $\mathrm{Adj}\,(B) = (\beta_{ij})$, then $\det B = 0$ implies $\beta_{ij}\beta_{ji} = \beta_{ii}\beta_{jj}$ for all $i \neq j$. Since when $B = \rho_0 I - A$, β_{ii} and β_{jj} are positive (being principal minors), it follows from $\beta_{ij}\beta_{ji} = \beta_{ii}\beta_{jj}$ that β_{ij} and β_{ji} are not just nonnegative but positive. The identity $\beta_{ij}\beta_{ji} = \beta_{ii}\beta_{jj}$ follows from a very special case of a well-known identity due to Jacobi [13, p. 50]. It also follows readily from the more basic identity $B \cdot \mathrm{Adj}\,(B) = (\det B)I$, which when $\det B = 0$ implies $\mathrm{rank}\,\mathrm{Adj}\,(B) \leq 1$, and so all 2×2 minors of $\mathrm{Adj}\,(B)$ must vanish. (Viewed in modern terms, $B \cdot \mathrm{Adj}\,(B) = 0$ means that the range of $\mathrm{Adj}\,(B)$ is contained in the null space of B. When 0 is a simple root of B, this means that $\mathrm{rank}\,\mathrm{Adj}\,B \leq 1$.)

Frobenius was eventually able to strengthen the above theorem by showing that $\mathrm{Adj}(\rho I - A) > 0$ for all $\rho \geq \rho_0$ [231, p. 552], as in his version of Perron's theorem (Theorem 17.12).[19]

Frobenius' Theorem 17.16 showed that nonzero irreducible matrices possessed almost all the properties of the matrices covered by Perron's Corollary 17.11, namely nonnegative matrices satisfying Perron's condition that $A^{\nu} > 0$ for some power ν. The sole difference was that the strict inequality $|\rho| < \rho_0$ of Perron's Theorem 17.9 and Corollary 17.11 is replaced by $|\rho| \leq \rho_0$. Frobenius introduced the irreducible matrices

$$A = \begin{pmatrix} 0 & a_{12} & 0 & \cdots & 0 \\ 0 & 0 & a_{23} & \cdots & 0 \\ \cdots & \cdots & \cdots & \cdots & \cdots \\ 0 & 0 & 0 & \cdots & a_{n-1n} \\ a_{n1} & 0 & 0 & \cdots & 0 \end{pmatrix}, \tag{17.32}$$

where $b = a_{21}a_{23}\cdots a_{n-1n}a_{n1} \neq 0$, to show that $|\rho| \leq \rho_0$ is best possible [231, p. 559].[20] It is easily seen that the characteristic polynomial of A is $\varphi(\rho) = \rho^n - b$, and so the characteristic roots are $\rho_k = \sqrt[n]{|b|} \cdot \varepsilon^k$, where $\varepsilon = e^{(2\pi i)/n}$ and $k = 0, 1, \ldots, n-1$. Thus $\rho_0 = \sqrt[n]{|b|}$, and all roots satisfy $|\rho_k| = \rho_0$.

These considerations led Frobenius to define an irreducible matrix $A \neq 0$ to be *primitive* if $|\rho| < \rho_0$ for every characteristic root $\rho \neq \rho_0$. In other words, those irreducible matrices that possess *all* the properties posited by Perron's Theorem 17.9 are called primitive. The remaining irreducible matrices he termed *imprimitive*. Thus the matrices in (17.32) are imprimitive, whereas every $A \geq 0$ satisfying Perron's condition that $A^{\nu} > 0$ for some ν is primitive by Perron's Corollary 17.11. The obvious question is, are there any irreducible matrices that are primitive besides those satisfying Perron's condition? Frobenius showed that the answer is no [231, p. 553]:

Theorem 17.17 (Primitive matrix theorem). *An irreducible matrix A is primitive if and only if $A^{\nu} > 0$ for some power $\nu > 0$.*

In order to establish Theorem 17.17, it is necessary to prove that a primitive matrix has a power that is positive, since the converse is clear (as noted at the beginning of this section). It is interesting to see where Frobenius got the idea for his proof. As we saw, in his 1908 paper responding to Perron's call for a more satisfactory proof of Theorem 17.9, Frobenius had done just that by giving a simple determinant-based proof that avoided Perron's limit lemma. But Frobenius did not stop there. As a mathematician, he was characteristically thorough

---

[19] Presumably for this reason, Theorem 17.16 is not stated by Frobenius as a formal theorem, although it is alluded to in his prefatory remarks. The proof is given on pp. 549–550 of [231].

[20] In terms of the graph-theoretic characterization of irreducibility given above in a footnote to Frobenius' official definition, $G(A)$ is a directed n-cycle and so connected.

and delighted in exploring mathematical relations from every conceivable angle within the framework of his chosen approach to the subject. Thus even though Perron's limit lemma was no longer needed to establish Perron's theorem, Frobenius could not refrain from considering the possibility of a simpler, more traditionally algebraic, proof. It was based on the following result [228, p. 408]: if a nonnegative $A \neq 0$ has a root ρ_0 of multiplicity one that strictly dominates in absolute value all other characteristic roots, then for any (i, j) entry,

$$\lim_{k\to\infty} \frac{[A^k]_{ij}}{\rho_0^k} = \frac{[\text{Adj}\,(\rho_0 I - A)]_{ij}}{\varphi'(\rho_0)}, \quad \varphi(\rho) = \det(\rho I - A).$$

By virtue of this identity from his 1908 paper, Frobenius saw how to prove the primitive matrix theorem. That is, the identity applies when A is primitive because by definition ρ_0 has the requisite multiplicity and dominance properties. Also, by the irreducible matrix theorem (Theorem 17.16), $\text{Adj}\,(\rho_0 I - A) > 0$, which implies that both numerator and denominator in the above limit are positive, since by (17.28), $\varphi'(\rho_0)$ is the sum of the diagonal terms $\varphi_{\alpha\alpha}(\rho_0)$ of $\text{Adj}\,(\rho_0 I - A) > 0$. Because the limit is positive, it follows that for all sufficiently large values of k, the expressions $[A^k_{ij}]/\rho_0^k$ will be positive for all (i, j). Since $\rho_0 > 0$, it follows that $A^k > 0$ for all sufficiently large k. This then establishes the primitive matrix theorem.

Frobenius also obtained a simple sufficient condition for primitivity as a byproduct of his investigation of the properties of imprimitive matrices:

Theorem 17.18 (Trace theorem). *If A is irreducible and $\text{tr}A > 0$, then A is primitive. Hence all imprimitive A have $\text{tr}A = 0$, and so all diagonal entries must be zero.*

Frobenius' proof of this theorem was a byproduct of the line of reasoning leading to his main theorem on imprimitive matrices (Theorem 17.19); this reasoning is indicated in the sketch of his proof of Theorem 17.19 that is given in Section 17.2.3 below.

Frobenius' primitive matrix theorem showed that Perron's condition $A^\nu > 0$ exactly characterized those irreducible matrices satisfying all the properties posited in Perron's theorem, i.e., the matrices A Frobenius called primitive. The imprimitive matrices could now be seen as the class of nonnegative irreducible matrices that lay outside the province of Perron's investigations: no power of such a matrix could be positive. Frobenius' exploration of their properties yielded his most profound results on nonnegative matrices. These results are summarized in Theorem 17.19 below. For those interested, a broad outline of the main ideas by means of which he established the theorem is given in Section 17.2.3. In stating the theorem, I will use the notation $A \sim_\sigma B$ to mean that A is permutationally similar to B by means of the permutation matrix P_σ, so that $P_\sigma A P_\sigma^{-1} = P_\sigma A P_\sigma^t = B$.

Theorem 17.19 (Imprimitive matrix theorem). *Let $A \neq 0$ be an $n \times n$ imprimitive matrix, and let k denote the number of characteristic roots of A with absolute value equal to ρ_0. Then: (1)*

$$A^k \sim_\sigma \begin{pmatrix} R_{11} & 0 & \cdots & 0 \\ 0 & R_{22} & \cdots & 0 \\ \cdots & \cdots & \cdots & \cdots \\ 0 & 0 & \cdots & R_{kk} \end{pmatrix}, \tag{17.33}$$

where each square block R_{ii} is primitive; (2) if the characteristic polynomial of A is expressed in the notation

$$\varphi(\rho) = \det(\rho I - A) = \rho^n + a_1\rho^{n_1} + a_2\rho^{n_2} + \cdots, \quad a_i \neq 0, \tag{17.34}$$

then k is the greatest common divisor of the differences $n - n_1, n_1 - n_2, \ldots$; (3) if ρ is any characteristic root of A, then so is $\varepsilon^i\rho$, where $\varepsilon = e^{2\pi i/k}$; (4) in particular, the k roots with absolute value ρ_0, viz., $\rho_i = \varepsilon^i\rho_0$, $i = 0, \ldots, k-1$, all have multiplicity one; (5) if the characteristic polynomial of A is expressed in the notation

$$\varphi(\rho) = \rho^n + b_1\rho^{n-k} + b_2\rho^{n-2k} + \cdots + b_m\rho^{n-mk}, \tag{17.35}$$

where $b_m \neq 0$ but $b_i = 0$ for some $i < m$ is possible, and if

$$\psi(\rho) = \rho^m + b_1\rho^{m-1} + b_2\rho^{m-2} + \cdots + b_m, \tag{17.36}$$

then $\psi(\rho)$ has a positive root of multiplicity one that is larger than the absolute value of any other root.

A few comments about this remarkable theorem are in order. First of all, the integer k, which figures so prominently in the theorem, is nowadays usually called the *index of imprimitivity of A*. The definition of k makes sense for $k = 1$ as well and simply defines a primitive matrix. Part (2) gives an easy way to determine k if the characteristic polynomial of A is known. Stated geometrically, part (3) says that the set of characteristic roots of A is invariant under rotations by $2\pi/k$ radians; and (4) says that the k roots of absolute value ρ_0 form a regular k-gon inscribed in the circle $|z| = \rho_0$, with one vertex at $z = \rho_0$. Although Frobenius certainly recognized these simple geometric consequences of his results, he did not mention them explicitly. What fascinated him was the more algebraic part (5), which "shows most palpably the minor modification by means of which the properties of positive matrices are transferred to imprimitive ones, while at the same time they remain entirely unchanged in their validity for primitive ones" [231, p. 558]. Let me explain. The properties of the roots of characteristic polynomials of positive matrices (as given by Perron's theorem) are transferred to imprimitive matrices A in the sense that these properties are inherited by the polynomial $\psi(\rho)$ of (17.36). This polynomial is related to the characteristic polynomial $\varphi(\rho)$ of A by $\varphi(\rho) = \rho^\ell\psi(\rho^k)$, where $\ell = n - mk$. When A is primitive, i.e., when $k = 1$, it follows that $\varphi(\rho) = \rho^\ell\psi(\rho)$, so that in this case, $\varphi(\rho)$ inherits via $\psi(\rho)$ all the properties of positive matrices, viz., a root $\rho_0 > 0$ of multiplicity one of $\varphi(\rho)$ exists and all other roots ρ satisfy $|\rho| < \rho_0$ ($\rho = 0$ being the only possible root of $\varphi(\rho)$ that is not a root of $\psi(\rho)$).

We have now seen how the problem Frobenius posed to himself—that of determining the characteristic roots of a nonnegative matrix A that possess nonnegative characteristic vectors (Problem 17.15)—may have led him, via the lower triangular block forms (17.30), into his penetrating study of irreducible matrices, the distinction between primitive and imprimitive matrices being motivated by Perron's Theorem 17.9 and its Corollary 17.11. Most of Frobenius' paper [231] is concerned with the theory of irreducible matrices. Having worked out that theory, he then turned in the penultimate section of his paper [231, §11] to Problem 17.15, the problem he said was the motivation for him to develop the theory of nonnegative matrices. I will conclude the discussion of Frobenius' paper by indicating his solution.

Recall that Problem 17.15 seems to have been motivated by Proposition 17.14 from his 1909 paper: when $A > 0$, the maximal positive root ρ_0 is the sole characteristic root of A with a nonnegative characteristic vector. This solves Problem 17.15 for positive matrices. Once Frobenius had established the irreducible matrix theorem (Theorem 17.16), the same inner product argument used to prove Proposition 17.14 yields an analogous solution to Problem 17.15 for irreducible A: If $A \neq 0$ is irreducible, then the only characteristic root possessing a nonnegative characteristic vector is ρ_0 [231, pp. 554–555].

Suppose now that A is reducible. Then, as already indicated in (17.30), permutations exist such that A is permutationally similar to a matrix of the form

$$\begin{pmatrix} R_{11} & 0 & 0 & \cdots & 0 \\ R_{21} & R_{22} & 0 & \cdots & 0 \\ \cdots & \cdots & \cdots & \cdots & \cdots \\ R_{m1} & R_{m2} & R_{m3} & \cdots & R_{mm} \end{pmatrix}, \tag{17.37}$$

where the diagonal blocks R_{jj} are irreducible. Nowadays in the theory of nonnegative matrices, (17.37) is called *a Frobenius normal form for A*. Although the irreducible diagonal blocks R_{jj} in (17.37) are uniquely determined by A up to permutational similarity, their ordering on the diagonal depends in general on the chosen normal form (17.37). For example, if A is permutationally similar to the normal form T_1, it is also permutationally similar to the normal form T_2, where

$$T_1 = \begin{pmatrix} R_{11} & 0 & 0 \\ R_{21} & R_{22} & 0 \\ R_{31} & 0 & R_{33} \end{pmatrix} \quad \text{and} \quad T_2 = \begin{pmatrix} R_{11} & 0 & 0 \\ R_{31} & R_{33} & 0 \\ R_{21} & 0 & R_{22} \end{pmatrix},$$

since the block-transposition $\tau = (2,3)$ applied to the rows and columns of T_1 results in T_2.

With this in mind, suppose that (17.37) is some Frobenius normal form for A, and let $\rho_0^{(j)}$ denote the maximal root of the irreducible block R_{jj}, $j = 1,\ldots,m$, in the ordering associated to the chosen normal form (17.37). Using the abovementioned solution to Problem 17.15 for irreducible matrices, Frobenius showed via block multiplication that if ρ is a characteristic root of a reducible A, then ρ can have a nonnegative characteristic vector only if ρ is one of the above maximal roots $\rho_0^{(j)}$, $j = 1,\ldots,m$. Now assume that ρ is one of the $\rho_0^{(j)}$ and consider when it possesses a nonnegative characteristic vector. Since it is possible that $\rho = \rho_0^{(j)}$ for several values of j, let λ denote the *largest* of all indices j for which $\rho_0^{(j)} = \rho$ *in the ordering associated to the given normal form.* Frobenius showed that if $\rho = \rho_0^{(\lambda)}$ is strictly greater than the maximal roots of all blocks further down the diagonal, i.e., if

$$\rho_0^{(\lambda)} > \rho_0^{(\lambda+i)} \quad \text{for all } i = 1,\ldots,m-\lambda, \tag{17.38}$$

then $\rho = \rho_0^{(\lambda)}$ has a nonnegative characteristic vector. His solution to Problem 17.15 then followed by establishing the converse, so as to prove the following.

Theorem 17.20. *If $A \neq 0$ is nonnegative, then a characteristic root ρ of A has a nonnegative characteristic vector if and only if A has a normal form (17.37) such that $\rho = \rho_0^{(\lambda)}$ (with index λ as defined above) for which (17.38) holds.*

As an illustration of this theorem, consider the following matrix [517, p. 168]:

$$A = \begin{pmatrix} 5\,0\,0\,0\,0 \\ 1\,0\,0\,0\,0 \\ 0\,0\,5\,0\,0 \\ 1\,1\,1\,4\,0 \\ 0\,0\,1\,0\,6 \end{pmatrix}.$$

This is in a Frobenius normal form (17.37) with the five diagonal entries representing the five irreducible blocks R_{jj} of A. Thus $\rho = 5,0,5,4,6$ are all possible candidates for having a nonnegative characteristic vector, although the normal form defining A guarantees this, by virtue of the necessary and sufficient condition (17.38) of Theorem 17.20, only for the root $\rho = 6$. Whether other roots have nonnegative characteristic vectors depends on whether other normal forms for A exist with a different ordering of the diagonal blocks so that condition (17.38) applies to characteristic roots $\rho \neq 6$. It turns out that

$$A \sim_\tau \begin{pmatrix} 5\,0\,0\,0\,0 \\ 1\,6\,0\,0\,0 \\ 0\,0\,5\,0\,0 \\ 0\,0\,1\,0\,0 \\ 1\,0\,1\,1\,4 \end{pmatrix}, \quad \tau = \begin{pmatrix} 1\,2\,3\,4\,5 \\ 3\,5\,1\,2\,4 \end{pmatrix}. \tag{17.39}$$

It is clear from (17.38) applied to the normal form in (17.39) that not only $\rho = 6$ but also $\rho = 5$ and $\rho = 4$ have nonnegative characteristic vectors by virtue of Frobenius'

Theorem 17.20. Whether there is a nonnegative characteristic vector for $\rho = 0$ or another linearly independent one for the double root $\rho = 5$ depends on what further normal forms (17.37) are permutationally similar to A.

The above example illustrates that Theorem 17.20, as a solution to Problem 17.15, is not entirely satisfying, because it depends on knowing all possible normal forms for A. Frobenius' proof of Theorem 17.20 actually involved ideas that later proved to be the key to a definitive solution [517, pp. 162, 168]. His own proof methods, however, lacked the graph-theoretic viewpoint that brings the underlying ideas to fruition as in [517, pp. 163–169]. For example, by means of graph-theoretic notions based on (17.38), it follows that $\rho = 4,6,5$ each have one independent nonnegative characteristic vector, whereas $\rho = 0$ has none. The complete solution to Problem 17.15 for the A of Theorem 17.20 can be read off from the information implicit in a graph associated to A (called the reduced graph of A). Graph theory, however, was in its infancy in 1912, and Frobenius was not impressed by what had so far been achieved by applying the theory to linear algebra [516, p. 143].

The following subsection is intended for readers wishing to gain an appreciation of how Frobenius went about proving his extraordinary imprimitive matrix theorem. Others may proceed to Section 17.3 without any loss of continuity.

17.2.3 Outline of Frobenius' proof of Theorem 17.19

A key to the further investigation of the primitive–imprimitive distinction for Frobenius derived from a line of thought that he had used in the past, starting with his 1878 "monograph" on matrix algebra [181, pp. 358ff.]. It provides another illustration, beyond those given in Sections 7.5, 10.6, and 16.1, of the manner in which matrix algebra was an agent of mathematical discovery for Frobenius. Let $\varphi(\rho) = |\rho I - A|$, and set $\varphi(s,t) = (\varphi(t) - \varphi(s))/(t-s)$. Then by the Cayley–Hamilton theorem, which, as we saw (Section 7.5.1), Frobenius had independently discovered via his minimal polynomial theorem, we have $\varphi(A) = 0$, and so for s not a characteristic root, $\varphi(s,A) = \varphi(s)(sI - A)^{-1} = \text{Adj}\,(sI - A)$. The expansion of $\varphi(t)$ in powers of $t - s$ shows that

$$\varphi(s,t) = \frac{\varphi(t) - \varphi(s)}{t - s} = \varphi'(s) + \frac{1}{2}\varphi''(s)(t-s) + \cdots + \frac{1}{n!}\varphi^{(n)}(s)(t-s)^{n-1}.$$

By setting $t = A$ in the above equation, we have

$$\text{Adj}\,(sI - A) = \varphi(s,A) = \varphi'(s)I + \frac{1}{2}\varphi''(s)(A - sI) + \cdots + \frac{1}{n!}\varphi^{(n)}(s)(A - sI)^{n-1}.$$

This equation can be rearranged in the form

$$\text{Adj}\,(sI - A) = \psi_0(s)I + \psi_1(s)A + \cdots + \psi_{n-1}(s)A^{n-1}, \tag{17.40}$$

where the $\psi_i(s)$ are polynomials in s.[21] Since both sides of (17.40) define everywhere continuous (matrix) functions of s, (17.40) remains valid when s is set equal to a characteristic root of A. This implies the following lemma.

Lemma 17.21. *For any $n \times n$ matrix A and any s, $\mathrm{Adj}\,(sI - A)$ is a linear combination of $I, A, \ldots, A^{n-1}$.*

From this lemma and the irreducible matrix theorem (Theorem 17.16), Frobenius readily deduced the following key lemma [231, p. 551].

Lemma 17.22. *If $A \neq 0$ is irreducible, then for any fixed pair of indices (i, j), the n coefficients $[A^m]_{ij}$, $m = 0, \ldots, n-1$, cannot all vanish.*

The proof is as follows. By Lemma 17.21 with $s = \rho_0$, $\mathrm{Adj}\,[\rho_0 I - A]_{ij}$ is a linear combination of the nonnegative numbers $[A^m]_{ij}$, $m = 0, \ldots, n-1$, and by the irreducible matrix theorem, we know that $\mathrm{Adj}\,[\rho_0 I - A]_{ij} > 0$, which means that the coefficients $[A^m]_{ij}$, $m = 0, \ldots, n-1$, cannot all vanish.

A first consequence of Lemma 17.22 is an easy-to-apply sufficient condition for an irreducible matrix to be primitive [231, p. 553], namely the trace theorem (Theorem 17.18), stated already in Section 17.2.2: *If A is irreducible and if* $\mathrm{tr}\, A > 0$, *then A is primitive. Hence all imprimitive A have* $\mathrm{tr}\, A = 0$, *and so all diagonal entries must be zero*. The proof is quite simple. Suppose $\mathrm{tr}\, A > 0$ and that e.g., $a_{11} > 0$. Then $[A^m]_{11} > 0$, since it is a sum of nonnegative terms one of which is $a_{11}^m > 0$. Now by Lemma 17.22 above, for every i there is an $l < n$ for which $[A^l]_{i1} > 0$. Similarly, $m < n$ exists with $[A^m]_{1j} > 0$. Since $[A^{l+m}]_{ij}$ contains the term $[A^l]_{i1}[A^m]_{1j}$, it is positive. In other words, $A^{l+m} > 0$, and so A is primitive by the primitive matrix theorem. Thus $\mathrm{tr}\, A > 0$ is incompatible with the hypothesis of imprimitivity and we must have $\mathrm{tr}\, A = 0$.

The next theorem formed Frobenius' entrée into a deeper understanding of imprimitive matrices [231, p. 554].

Theorem 17.23. *If A is any nonzero $n \times n$ matrix such that the matrix powers $A, A^2, \ldots, A^n$ are all irreducible, then A is primitive. Hence if $A \neq 0$ is imprimitive, there is an integer m, $1 < m \leq n$, such that A^m is reducible.*

Again the proof is easy, given what has gone before. Suppose that A and all its powers up to A^n are irreducible. Then since in particular A is irreducible, the irreducible matrix theorem shows that $B = \mathrm{Adj}\,(\rho_0 I - A) > 0$. This implies that $BA > 0$ as well, since $[BA]_{ij} = 0$ would hold only if the jth column of A were 0, but then A would be reducible, contrary to assumption. Since by Lemma 17.21 B is a linear combination of A^m, $m = 0, \ldots, n-1$, $BA > 0$ is a linear combination of A^m, $m = 1, \ldots, n$, which means that not all of the n quantities $[A^m]_{11}$, $m = 1, \ldots, n$, can vanish. Thus $\mathrm{tr}\, A^{m_0} > 0$ for one of these values of m. Since A^{m_0} is irreducible, the trace theorem implies that it is primitive, and so ρ_0^m strictly dominates the absolute

[21] In his 1878 paper, (17.40) is presented divided through by $\varphi(s)$ so as to give a formula for $(sI - A)^{-1}$ [181, p. 358, (4)].

values of all other characteristic roots ρ^m. Then ρ_0 strictly dominates the absolute values of all other roots ρ of A, and so A is primitive. This establishes the first statement in the theorem, and the second then follows immediately.

With Theorem 17.23 in mind, Frobenius obtained the following result [231, pp. 554–556].

Theorem 17.24. *If $A \neq 0$ is irreducible but A^m is reducible for some $m > 1$, then A^m is completely reducible, in the sense that A^m is permutationally similar to a block diagonal matrix in which the diagonal blocks are all irreducible.*

The starting point of the proof was again the irreducible matrix theorem, specifically the fact that $\mathrm{Adj}\,(\rho_0 I - A) > 0$. As we have seen, this means that both the equations $Ax = \rho_0 x$ and $A^t y = \rho_0 y$ have positive solutions obtained by using a column, respectively row, of $\mathrm{Adj}\,(\rho_0 I - A) > 0$. Frobenius also realized that a result from his 1909 paper, namely Proposition 17.14, remains valid for irreducible A (by the same line of reasoning): *the only nonnegative characteristic vector of A is (up to a positive multiple) the positive characteristic vector x corresponding to the maximal root ρ_0.*

Now, since A^m is reducible and hence permutationally similar to a matrix in lower triangular block form, we can assume without loss of generality that A^m itself is in the block form

$$A^m = \begin{pmatrix} R_{11} & 0 & 0 & \cdots & 0 \\ R_{21} & R_{22} & 0 & \cdots & 0 \\ \cdots & \cdots & \cdots & \cdots & \cdots \\ R_{\ell 1} & R_{\ell 2} & R_{\ell 3} & \cdots & R_{\ell\ell} \end{pmatrix},$$

where the diagonal blocks R_{ii} are irreducible. Using the existence of x and y and the italicized fact given above together with block multiplication, Frobenius deduced that all the nondiagonal blocks R_{ij}, $i \neq j$, must vanish, implying that A^m is indeed completely reducible. Furthermore, the reasoning showed that each irreducible block R_{ii} has ρ_0^m as its maximal root.

Although Frobenius' step-by-step arguments leading up to Theorems 17.23 and 17.24 were fairly simple and straightforward, piecing them together as he did to achieve these theorems was an act of brilliance. Even more brilliant was the way he was able to use these theorems to arrive at his remarkable imprimitive matrix theorem, Theorem 17.19. To do so required reasoning of a more complex nature, and for this reason, the remainder of the outline of Frobenius' proof of Theorem 17.19 will be less complete than what has preceded.

Although Frobenius' proof of Theorem 17.19 is correct, it was not presented with his customary lucidity, possibly due to the more complicated nature of the reasoning. The following lemma, which was not formally stated by Frobenius, represents the guiding idea of the entire proof (see [231, p. 557]).

Lemma 17.25. *Let $A \neq 0$ be imprimitive and let ρ_i, $i = 0, \ldots, k-1$, denote the $k > 1$ characteristic roots of absolute value ρ_0. Then for any positive integer m, A^m*

is completely reducible into primitive blocks R_{ii} if and only if all k roots ρ_i satisfy $\rho_i^m = \rho_0^m$. In that case, the number of diagonal blocks R_{ii} is precisely k.

Lemma 17.25 leaves it unclear whether integers m actually exist for which A^m is completely reducible into primitive parts, but this follows readily. That is, if A is imprimitive, then we know by Theorem 17.23 that there is a power m_0, $1 < m_0 \leq n$, such that A^{m_0} reduces, and so reduces completely by Theorem 17.24. The reasoning behind Lemma 17.25 implies that either all the irreducible blocks R_{ii} of A^{m_0} are primitive or all are imprimitive. In the latter case, we know that a power $R_{ii}^{m_i}$ exists that is completely reducible. Thus if $M = \prod_{i=1}^{\ell} m_i$, then A^M completely reduces into a greater number of irreducible blocks than in A^{m_0}. If these blocks are all imprimitive, we can repeat the above reasoning to get an even larger power of A that reduces into a yet larger number of irreducible parts. Since the total number of irreducible parts cannot exceed the dimension n of A, it follows that this process must come to a stop, i.e., there will be a power $m \leq n$ such that A^m is completely reducible into primitive parts.

Let h denote the smallest power for which A^h is completely reducible into primitive parts. Then by Lemma 17.25, h is the smallest power such that all the k roots ρ_i, $i = 0, \ldots, k$, satisfy $\rho_i^h = \rho_0^h$, i.e., such that all k quotients ρ_i/ρ_0 are hth roots of unity. In particular, it follows that $k \leq h$.

Frobenius then considered the characteristic equation of A:

$$\varphi(\rho) = |\rho I - A| = \rho^n + c_1\rho^{n-1} + \cdots + c_m\rho^{n-m} + \cdots + c_n.$$

Consider the coefficient c_m. If m is not divisible by h, then $m = ph + q$, where p, q are nonnegative integers and $1 \leq q < h$. Thus for every quotient ρ_i/ρ_0 we have $(\rho_i/\rho_0)^m = (\rho_i/\rho_0)^{hp}(\rho_i/\rho_0)^q = (\rho_i/\rho_0)^q$. Hence if all (ρ_i/ρ_0) were mth roots of unity, they would all be qth roots of unity, which is impossible, since $q < h$. This means (by Lemma 17.25) that A^m is not completely reducible into primitive parts, i.e., either A^m is imprimitive or is completely reducible into irreducible blocks R_{jj} that are all imprimitive. Thus in either case, the trace theorem (Theorem 17.18) implies $\operatorname{tr} A^m = 0$, or equivalently, that the sum of the mth powers of all the roots of $\varphi(\rho)$ vanishes. From Newton's identities Frobenius then deduced by induction that $c_m = 0$ for m not divisible by h [231, p. 557].

The fact that $c_m = 0$ whenever m is not divisible by h implies first of all that $h \leq n$. For if $h > n$, then *all* coefficients c_m of φ vanish and $\varphi(\rho) = \rho^n$, which is impossible, since $\rho_0 > 0$ is a root. Thus $h \leq n$ and

$$\varphi(\rho) = \rho^n + a_1\rho^{n-m_1h} + a_2\rho^{n-m_2h} + \cdots, \tag{17.41}$$

where $a_i \neq 0$ for all i and $m_1 < m_2 < \cdots$. From this special form for $\varphi(\rho)$, it follows that if ε is any hth root of unity, then

$$\varphi(\varepsilon\rho) = \varepsilon^n\varphi(\rho), \quad \varphi'(\varepsilon\rho) = \varepsilon^{n-1}\varphi'(\rho).$$

These relations show that if ρ is any root of φ, then so is $\varepsilon\rho$, and if ρ is a root of multiplicity one (so $\varphi'(\rho) \neq 0$), then so is $\varepsilon\rho$. It thus follows that if $\varepsilon = e^{2\pi i/h}$ (a primitive hth root of unity), then the $h \geq k$ roots $\varepsilon^i\rho_0$, $i = 0,\ldots,h-1$, all have absolute value 1, which means that $h = k$ and the much-discussed special roots ρ_i, $i = 0,\ldots,k-1$, are precisely the roots $\varepsilon^i\rho_0$, $i = 0,\ldots,k-1$, and they all have multiplicity one.

From the above proof-sketch, with h everywhere now replaced by k, parts (1)–(4) of the imprimitive matrix theorem follow. Part (5) then follows readily, as indicated following the statement of the theorem in Section 17.2.2.

17.3 Markov Chains 1908–1936

Jacobi's generalization of the Euclidean algorithm had led Perron, in his further generalization of it, to introduce a nonnegative matrix associated to any such algorithm that is periodic. Furthermore, the existence of a characteristic root ρ_0 that possesses certain dominance properties relative to the other characteristic roots was relevant to his primary concern: the convergence of a periodic algorithm and the calculation of its limiting values. It is rather remarkable that at roughly the same time as Perron's work, considerations derived from an entirely difference source, namely the theory of probability, led A.A. Markov (1856–1922) to a type of probabilistic model to which is associated a (stochastic) nonnegative matrix and that furthermore, the existence of a characteristic root (namely $\rho_0 = 1$) with dominance properties relative to the other characteristic roots was critical in carrying out his primary objective, namely the analytic calculation of the associated probabilistic functions so as to show that certain laws of large numbers that had been established by Chebyshev for independent sequences apply as well to many cases of dependent sequences.

Markov's paper was presented to the Academy of Sciences in St. Petersburg on 5 December 1907 and published in its proceedings in 1908 [431]. A German translation was appended to the German edition of his lectures on the theory of probability in 1912 [432]—the same year that Frobenius published his remarkable results on nonnegative matrices. There is no evidence that Frobenius was aware of Markov's paper. Indeed, as we have seen in the previous section, Frobenius' theory of nonnegative matrices was inspired by the work of Perron, and, as we shall now see, by creating his theory, Frobenius unwittingly resolved all the linear-algebraic problems Markov had posed—but did not completely resolve—for stochastic matrices.

Markov's work nonetheless forms part of the historical context of Frobenius' work for two related reasons. (1) As we shall see in Section 17.3.1, to push through the probabilistic analysis of his chains, Markov needed to assume that his stochastic matrices A have the properties that imply, within the context of Frobenius' theory, that A is primitive. In a very sketchy and confusing manner Markov arrived at sufficient (but not necessary) conditions for primitivity, although his proof of their sufficiency when A is nonnegative but not positive ($A \not> 0$) was vague and based in part on an unjustified assumption. Markov's immediate successors understandably assumed $A > 0$ in their renditions of his theory. (2) As we shall see in Section 17.3.2, it was not until the 1930s that an interest in Markov chains became widespread, and it was then by means of Frobenius' theory that it was developed rigorously and in complete generality for nonnegative, rather than just positive, stochastic matrices. In this way, the theory of Markov chains became one of the earliest developed applications of the Perron–Frobenius theory and seems to have served to call general attention among mathematicians and mathematically inclined scientists to the existence and utility of the theory.

17.3.1 Markov's paper of 1908

In his paper, Markov considered a sequence of numbers

$$x_1, x_2, \ldots, x_k, x_{k+1}, \ldots. \tag{17.42}$$

Initially, he assumed that each x_k can assume three values $\alpha = -1$, $\beta = 0$, and $\gamma = +1$, and then he generalized to the case in which each x_k can take a fixed finite number n of distinct values $\alpha, \beta, \ldots, \mu, \ldots \nu, \ldots$. In the general case, he introduced the probability $p_{\mu\nu}$ that (for any k) if $x_k = \mu$, then $x_{k+1} = \nu$. Thus for any μ, we must have

$$p_{\mu,\alpha} + p_{\mu,\beta} + \cdots = 1. \tag{17.43}$$

We see that with these assumptions, (17.42) defines what is now called an n-state Markov chain with transition probability matrix $P = (p_{\mu\nu})$, and in fact, Markov himself occasionally referred to (17.42) as a chain.[22]

Markov wished to calculate the probability distribution of the sum $x_1 + x_2 + \cdots + x_n$ for increasingly large values of n, and in this connection, the properties of the characteristic roots of P, or equivalently, its transpose $A = P^t$, were critical to performing these calculations. Nowadays, A is usually termed a stochastic matrix, i.e., a nonnegative matrix with columns adding to 1, but, especially in older literature, a nonnegative matrix with row sums equaling 1 (e.g., P) is called stochastic [240, v. 2, p. 83]. Markov vacillated between the systems P and $A = P^t$ in his paper. To avoid confusion, I will refer to *column stochastic* and *row stochastic* matrices. Incidentally, Markov made no use of matrix notation in his paper, just as Perron had used none in his *Habilitationsschrift* the preceding year. He expressed

[22]See e.g., pp. 569, 571 and 576 of the English translation by Petelin. In what is to follow, all page references will he to this translation (cited in the bibliographic reference for Markov's 1908 paper [431]).

all his reasoning in terms of determinants and systems of linear equations, written out without any abbreviated matrix notation.

Markov began by treating in considerable detail the case of three numerical states. Then he turned to case of any finite number n of numerical states $\alpha, \beta, \gamma, \ldots$. The mathematical analysis was essentially the same in the more general case, but that analysis (for n states) depended on linear-algebraic assumptions that became more difficult to establish in the n-state case. These assumptions involved the characteristic roots of P, or equivalently, those of $A = P^t$. It is to easy to realize, as Markov did, that $\rho = 1$ is a characteristic root. For example, if $x = (1 \quad 1 \quad \cdots \quad 1)^t$, then $Px = x$ follows from the row-stochastic nature of P. It is also easy to see that because P is row-stochastic, every root $\rho \neq 1$ of $\varphi(r) = \det(rI - P)$ satisfies $|\rho| \leq 1$.[23] The attendant mathematical analysis, however, required assuming that (1) $\rho = 1$ has multiplicity one and that (2) $|\rho| < 1$ for all roots $\rho \neq 1$.

Markov realized that (1) and (2) hold when $P > 0$; but there was no justification for restricting attention to the case $P > 0$ in his theory of chains, and Markov accordingly sought to extend his theory to $P \geq 0$ by imposing conditions on P that would guarantee (1) and (2). To this end, he set forth two conditions articulated somewhat obscurely in terms of determinants.[24] However, these two conditions were never invoked in his proof that they are sufficient for (1) and (2) to hold. In his proof, he tacitly assumed that P had the following two properties that turn out to follow, respectively, from the two determinant-based conditions, although Markov never showed this, nor even mentioned it.[25]

Property 17.26. If C, D is any partition of $\{1, \ldots, n\}$ into nonempty sets, then there are $\mu \in C$ and $\nu \in D$ for which $p_{\mu\nu} \neq 0$.

Property 17.27. For any partition of $\{1, \ldots, n\}$ into nonempty sets E, F there is no corresponding partition of $\{1, \ldots, n\}$ into nonempty sets G, H such that $p_{ij} = 0$ for all $(i, j) \in (G \times F) \cup (H \times E)$.

Markov's Properties 17.26 and 17.27 are never expressly stated but are tacitly assumed in his proofs that (1) and (2) hold [278, pp. 713–715].

Property 17.26 is easily seen to be equivalent to P being irreducible in Frobenius' sense. That is, the failure of Property 17.26 to hold is equivalent to P being reducible in accordance with Frobenius' official definition of this notion. For if Property 17.26

---

[23] If x is a characteristic vector for ρ and $m = \max_i |x_i|$, let i_0 be such that $|x_{i_0}| = m$. Then the i_0th equation of $\rho I = Px$ is $\rho x_{i_0} = \sum_{j=1}^{n} p_{i_0 j} x_j$. Taking absolute values and using the triangle inequality implies $|\rho| m \leq \sum_{j=1}^{n} p_{i_0 j} |x_j| \leq \left(\sum_{j=1}^{n} p_{i_0 j}\right) m = 1 \cdot m$, whence $|\rho| \leq 1$. Markov sought variations on this line of reasoning that would prove $|\rho| < 1$ for all roots $\rho \neq 1$ for P satisfying certain conditions [278, §6.3].

[24] See [278, p. 694], where these conditions are denoted by (ID) and (IID) and Markov's actual words are quoted.

[25] For proofs that the two determinant-based conditions (suitably interpreted) imply Properties 17.26 and 17.27 below (and denoted respectively by (I*) and (II*) in [278]), see [278, pp. 695–697].

fails to hold, then a partition C,D exists such that $p_{ij} = 0$ for all $i \in C$ and all $j \in D$. This means that P is reducible in Frobenius' sense: there are zeros at the intersections of the $p = |C|$ rows of P and the $q = |D|$ complementary columns.

Evidently Markov had tacitly anticipated the equivalent of Frobenius' key notion of an irreducible nonnegative matrix, albeit restricted to the special case of stochastic matrices. Incidentally, in 1911, and thus also before Frobenius, Maurice Potron, a mathematical economist familiar with the work of Perron and Frobenius published during 1907–1909, introduced the equivalent of the notions of reducible and irreducible nonnegative matrices.[26] His strongest results about solutions $x \geq 0$, $y \geq 0$ to $(sI - A)x = By$ with A,B nonnegative and $s \geq \rho_0$ were for A that are not partially reduced. Here we have yet another example of an instance of multiple discovery involving Frobenius. As we have seen, in all the previous instances Frobenius went further in developing the relevant theory, and in most cases with far greater rigor, than any of his fellow discoverers. This is true in particular regarding Markov and the theory of irreducible matrices, as we shall see.

It turns out that Markov's Properties 17.26 and 17.27 together imply that P is primitive in the sense of Frobenius, i.e., that P is irreducible, that the maximal root $\rho_0 = 1$ has multiplicity one, and that $|\rho| < 1$ for all other characteristic roots ρ of P. This can be seen as follows. First of all, Property 17.26 implies (as noted above) that P is irreducible. Frobenius' irreducible matrix theorem (Theorem 17.16) is therefore applicable and implies that $\rho_0 = 1$ has multiplicity one. Secondly, Markov's proof that $|\rho| < 1$ is correct if Properties 17.26 and 17.27 are assumed [278, pp. 714–715]. By definition, P is therefore primitive.

Markov's Properties 17.26 and 17.27, however, do not characterize primitive row-stochastic P; there are such P that do not have Property 17.27. For example, if

$$P = \begin{pmatrix} 0 & p_{12} & p_{13} & 0 \\ p_{21} & 0 & 0 & p_{24} \\ 0 & p_{32} & p_{33} & 0 \\ p_{41} & 0 & 0 & p_{44} \end{pmatrix} \tag{17.44}$$

denotes any row-stochastic matrix with all entries of the form p_{ij} positive, then it is irreducible, i.e., it has Markov's Property 17.26.[27] However, it does not have Property 17.27 by virtue of the partitions $E = \{2,3\}, F = \{1,4\}$ and $G = \{1,3\}, H = \{2,4\}$; all coefficients of P with indices (i,j) in $G \times F$ or in $H \times E$ are zero. Nonetheless, P is primitive by virtue of Frobenius' trace theorem (Theorem 17.18), since $\operatorname{tr} P = p_{33} + p_{44} > 0$. Thus $|\rho| < 1$ for all roots $\rho \neq 1$, even though P fails to have Property 17.27.

Markov had originally obtained his results for row-stochastic matrices P assuming $P > 0$ and then sought to extend them to $P \geq 0$ [431, p. 574n]. His

[26] Potron spoke of "partially reduced" matrices [487, p. 1130], by which he meant the equivalent of reducible matrices.

[27] The directed graph $G(P)$ contains the 4-cycle $1 \to 3 \to 2 \to 4 \to 1$ and so is connected.

proof that $|\rho| < 1$ for all characteristic roots $\rho \neq 1$ certainly applies when $P > 0$, since such P have Properties 17.26 and 17.27. And his proof that $\rho_0 = 1$ is a simple root then follows (as he realized) from a lemma due to Minkowski [278, §6.3.1]. Thus within the more limited context of positive row-stochastic matrices, Markov had independently discovered and proved some of Perron's results about positive matrices. However, as Schneider has pointed out [516, p. 147], there is no mention of Perron's result that $\mathrm{Adj}(I - P) > 0$ and its consequence that $y = 1$ has a positive characteristic vector for $A = P^t$, a consequence that has since become of considerable importance to the theory of Markov chains. Likewise, Perron's Corollary 17.11 implied that row-stochastic $P \geq 0$ possessing a positive power have characteristic roots with the properties Markov wished to establish, but Markov gave no indication in his paper that he realized this. And of course, also missing is the deeper insight, implied by Frobenius' primitive matrix theorem, that a row-stochastic $P \geq 0$ having Markov's Property 17.26 (irreducibility) possesses characteristic roots with the properties Markov needed if and only if it has a positive power.

Although within the context of stochastic matrices $P \geq 0$, Markov seems to have anticipated Frobenius' notion of irreducibility, it was obscured by his emphasis on a determinant-based condition in lieu of a precise and explicit mathematical formulation of irreducibility: Property 17.26 arises only as a tacit assumption in his proofs without any reference to his determinant-based condition. Furthermore, although his main result that irreducible row-stochastic matrices having Property 17.27 are primitive was correct, not only did it fail to characterize primitive row-stochastic matrices, but Markov's proof that $\rho = 1$ has multiplicity one was not at all rigorous for $P \ngtr 0$ due to its dependence on an unproven generalized version of Minkowski's lemma. By contrast, Frobenius' 1912 paper [231] was based on careful definitions, and by means of clear and rigorous proofs he obtained definitive results on irreducible matrices that went far beyond anything found in the paper of Markov, who, after all, was primarily interested in the probabilistic aspects of his chains, which mainly involved him with analytical derivations.

That same probabilistic focus naturally limited Markov's attention to the more amenable class of stochastic matrices, whereas the work of Perron and Frobenius revealed, in retrospect, that the theorems discovered by Markov were more generally true and were but a part of the rich theory of nonnegative matrices. Even the fact that any nonnegative matrix A possesses a root $\rho_0 \geq 0$ with the property that $|\rho| \leq \rho_0$ for all other roots ρ (the limiting case of Perron's theorem) was an unexpected result that came out of Perron's penetrating study of Jacobi's algorithm, whereas the same result is trivial when A is stochastic (as noted above). And of course, from Perron's penetrating study came his even more surprising result that when $A^\nu > 0$, the above inequalities become strict and ρ_0 has multiplicity one. As we have seen, it was these remarkable discoveries by Perron that engaged Frobenius' interest in the theory of nonnegative matrices and ultimately led to his masterly paper of 1912. I will now briefly consider how the theory developed in Frobenius' paper was applied to give a clear and rigorous treatment of Markov's theory of chains for the case $P \ngtr 0$.

17.3.2 *Frobenius' theory and Markov chains*

Frobenius had concluded his paper of 1912 with one application, which was to the theory of determinants [231, §14]. If $X = (x_{ij})$ is a matrix of n^2 independent variables x_{ij}, then it was well known that $\det X$ is an irreducible polynomial in these n^2 variables. From his theory of nonnegative matrices, he now deduced that if $\tilde{X}$ is the matrix obtained from X by setting some $x_{ij} = 0$, then if $\tilde{X}$ is irreducible as a nonnegative matrix in the obvious sense, the polynomial $\det \tilde{X}$ is still irreducible. Thus although Frobenius had written a definitive work on irreducible nonnegative matrices, Markov's theory being unfamiliar, the sole known application was to the theory of determinants. Frobenius' paper thus represented a definitive study of a type of matrix that was not at the time seen to be relevant to many applications or related to the main topics of the linear algebra of the time. For example, the new generation of texts on the theory of matrices that appeared in the early 1930s by Schreier and Sperner [520], Turnbull and Aitkin [567], and Wedderburn [585] make no mention of the Perron–Frobenius theory, being devoted to the main topics in linear algebra, such as canonical matrix forms, properties of symmetric, orthogonal, Hermitian, and unitary matrices, and their applications to quadratic and bilinear forms.

Even though Markov's paper was translated into German and appended to the 1912 German translation of his book on the theory of probability [432], it is uncertain how widely read it was. Apparently, those who did discuss Markov chains in the period 1912–1930 limited their attention to the case $P > 0$ [516, p. 147], perhaps because, as we have seen, when $P > 0$, Markov's proofs are correct and comprehensible. In the late 1920s, there was a renewed interest in Markov chains on the part of a large number of mathematicians, who became more or less simultaneously interested in the subject. Some of them, including J. Hadamard and M. Fréchet, apparently reinvented aspects of the theory without knowing of Markov's pioneering work [258, p. 2083, 2083n.3]. In the early 1930s, in the midst of the revival of interest in Markov chains, two applied mathematicians, R. von Mises and V.I. Romanovsky, independently applied Frobenius' theory of nonnegative matrices in order to deal with chains corresponding to stochastic $P \not> 0$.

17.3.2.1 R. von Mises

In 1920, Richard von Mises (1883–1953) became the first director of the newly formed Institute for Applied Mathematics at the University of Berlin—a type of institute Frobenius would have opposed at Berlin, where for him, mathematics had meant pure mathematics.[28] The arrival of von Mises in fact coincided with a period of renewed vitality and ascendency for mathematics at Berlin, and von

[28] On the founding of the institute, see [22, pp. 148–153].

Mises, with his dynamic personality, was a key player in this revival. In 1921, he became the founder and editor of a journal devoted to applied mathematics, *Zeitschrift für angewandte Mathematik und Mechanik*. In the first issue, he wrote an introductory essay [573] in which he made the point that the line between pure and applied mathematics is constantly shifting with time as mathematical theories find applications [573, p. 456]. Such an area of pure mathematics was constituted by Frobenius' theory of nonnegative matrices. As we saw, it was the pure mathematics of ordinary and generalized continued fractions that motivated Perron's work, which Frobenius further developed solely by virtue of its interesting algebraic content. Von Mises sought to apply this theory to a problem at the foundations of statistical mechanics.

This occurred in his 1931 book *The Calculus of Probabilities and Its Application to Statistics and Theoretical Physics* [574], which formed part of his lectures on applied mathematics. The application to theoretical physics, which constituted the fourth and final section of his book, had to do with the statistical mechanics of gases that had been developed by Maxwell and Boltzmann in the nineteenth century, with alternative statistical models arising in the twentieth century from the work of Planck, Bose, Einstein, and Fermi. All of these physical theories shared a common assumption. Stated in the neutral language of the theory of probability, the assumption was the following. Suppose there are k "states" $S_1,\ldots,S_k$ that a certain "object" can be in. Let p_i denote the probability that the object is in state S_i. Then the assumption is that all states are equally likely, i.e., that $p_i = 1/k$ for $i = 1,\ldots,k$. In Boltzmann's theory, the states represented small cells of equal volume in the 3-dimensional momentum space of an ideal gas molecule (the object) [574, p. 432]. In Planck's quantum theory, the states represented k energy levels $0, h\nu, 2h\nu, \ldots, (k-1)h\nu$ that the ideal gas molecule (the object) may have, where h denotes Planck's constant [574, pp. 439–440]. In the Bose–Einstein–Fermi theory [574, pp. 446–449] the states are the occupancy numbers $0,1,2,\ldots,k-1$ for a cell of volume h^3 and fixed energy in the six-dimensional phase space of an ideal gas molecule. The object in this case is such a cell.

Such a priori assignments of probabilities were anathema to von Mises' approach to probability theory, according to which probabilities were relative frequencies obtained from a repeated experiment, where the experiment could be an empirical one or an Einsteinian "thought experiment." Von Mises believed that he could describe a thought experiment that would provide a sound probabilistic basis for the above assumptions as follows [574, pp. 532ff.]. Imagine k urns $U_1,\ldots,U_k$. Each urn contains k lots, which are numbered from 1 to k. From an arbitrarily chosen urn U_{x_0} a lot is drawn. Let x_1 denote the number of the drawn lot. Proceed to urn U_{x_1} and draw a lot. Let x_2 denote the number drawn. Then proceed to urn U_{x_2} and draw a lot, and so on. Then a sequence $x_0, x_1, x_2, \ldots$ is generated, where each x_i is an integer between 1 and k. This is, of course, an example of what is now called a k-state Markov chain, and von Mises was aware that the mathematics of his thought experiment was "closely connected" to "the problem of Markov chains" [574, p. 562]. He saw in this model a way to justify the a priori "equal probability" assumption underlying the above physical models.

I will use symbolic matrix and vector notation and Markov chain terminology in describing von Mises' work, even though he himself did not. Thus let the components of $v^{(0)} = (p_1^{(0)} \quad \cdots \quad p_k^{(0)})^t$ denote the initial probabilities of being in states $S_1, \ldots, S_k$, respectively. Nothing is assumed about the values of the $p_i^{(0)}$ except of course that they are probabilities, so that $p_i^{(0)} \geq 0$ and $\sum_{i=1}^k p_i^{(0)} = 1$. In other words, $v_0 = (p_1^{(0)} \quad \cdots \quad p_k^{(0)})^t$ is what is now called a *probability vector.* Let as usual $P = (p_{ij})$ denote the matrix of transition probabilities and $A = P^t$ its transpose. Von Mises' own notation was chosen so that the coefficients a_{ij} of A defined his transition probabilities, i.e., he defined a_{ij} as the probability of moving from state j to state i [574, p. 533]. Only the matrix A is considered by von Mises, and A is column stochastic. Since A is column stochastic, it follows that $v_1 = Av_0$ is also a probability vector, and its components give the probabilities of being in states $S_1, \ldots, S_k$ after one iteration of the process. More generally, $v_n = A^n v_0$ is a probability vector with components giving the probabilities of being in states $S_1, \ldots, S_k$ after n iterations of the process. Thus if $v_\infty = \lim_{n\to\infty} A^n v_0$ exists, its components will give the probabilities of being in the various states in the long run. Von Mises' goal was to determine reasonable conditions under which v_∞ exists and to show that the components of v_∞ are all the same, i.e., that $v_\infty = \left(1/k \cdots 1/k\right)^t$, thereby justifying the assumption underlying the above physical models that all states are equally likely.

Von Mises was familiar with Frobenius' three papers on positive and nonnegative matrices, about which he may have learned from Frobenius' former star student Issai Schur, who was also a professor at Berlin. Thus in a footnote [574, p. 536n], von Mises wrote:

> A large part of the propositions that will be derived here and in sections **4** and **5** are closely related to the algebraic theory of matrices with nonnegative elements that was developed in three works ... [228, 229, 231] ... by *G. Frobenius.* A part of the results of course follow only from the special property of our matrices that the column sums have the value 1.

In his work, von Mises utilized Frobenius' notions of reducible and irreducible matrices, as well as the related notion of complete reducibility in the sense of nonnegative matrices [574, pp. 534–536][29]; and he was clearly guided by Frobenius' results, especially those in his paper of 1912. However, von Mises couched everything in probabilistic terms and notation and presented his own proofs rather than appealing to or reproducing Frobenius' own more general and complicated proofs.

Von Mises' key theorem regarding the existence and nature of v_∞ was the following [574, p. 548]:

[29] Complete reducibility for $A \geq 0$ means that A is permutationally similar to a block diagonal matrix in which the diagonal blocks are irreducible. Cf. Theorem 17.24 above.

Theorem 17.28. *If A is (a) irreducible, (b) has* $\operatorname{tr} A > 0$, *and (c) is symmetric, then for any initial state vector $v_0 \neq 0$,*

$$v_\infty = \lim_{n\to\infty} A^n v_0 = \left(\tfrac{1}{k} \; \cdots \; \tfrac{1}{k}\right)^t. \tag{17.45}$$

Although von Mises gave his own proof, he probably first realized that his Theorem 17.28 was an easy consequence of Frobenius' theorems. For example, the assumption that A is irreducible with $\operatorname{tr} A > 0$ means that A is primitive by Frobenius' trace theorem. From the primitivity of A, it follows that $\rho_0 = 1$ has multiplicity one and that $|\rho| < 1$ for all other characteristic roots, which implies that $v_\infty = \lim_{n\to\infty} A^n v_0$ exists.[30] Also $Av_\infty = A \lim_{n\to\infty} A^n v_0 = \lim_{n\to\infty} A^{n+1} v_0 = v_\infty$, so that v_∞ is a characteristic vector for the root $\rho_0 = 1$ of A. Then by the symmetry hypothesis, $A = A^t = P$, so v_∞ is a characteristic vector for the root $\rho_0 = 1$ of P. Since P is row stochastic, another characteristic vector for $\rho_0 = 1$ is $e = \left(1 \; \cdots \; 1\right)^t$. Since $\rho_0 = 1$ has multiplicity one and v_∞ is a probability vector, it follows that $v_\infty = (1/k)e$, and (17.45) is established.

Referring to the considerations culminating in Theorem 17.28, von Mises, in keeping with his frequentist approach to probability theory, declared that

> Our ... deductions are not based on an assumption about probabilities of fixed individual states and also not on the ergodic hypothesis,[31] but rather exclusively on the assumptions a) to c), which concern the transition probabilities and of which only the last is quantitatively decisive. *It is not the assumption that certain states are equally likely, which is hardly physically meaningful, but rather [the assumption] that between these states symmetric ... transition probabilities exist, that forms the proper foundation for the kinetic theory of gas and similar physical-statistical theories* [574, p. 555].

Although von Mises' work on the probabilistic foundations of statistical mechanics was not explicitly about Markov chains, it was known to those working in this area.[32] Incidentally, one of von Mises' students, Lothar Collatz, applied some of Frobenius' results on nonnegative matrices to a problem in numerical analysis [102], thereby suggesting a vast new area for application that proved quite fertile, as can be seen from Varga's 1962 book *Matrix Iterative Analysis* [570].

---

[30] Write $A = S^{-1}JS$, where J is the Jordan canonical form of A. The above-described properties of the characteristic roots of A imply that $\lim_{n\to\infty} J^n = J_\infty = \text{Diag. Matrix}(1, 0, \ldots, 0)$. Thus $\lim_{n\to\infty} A^n v_0 = S^{-1} J_\infty S v_0$ exists. Of course, in von Mises' theorem, A is assumed to be symmetric, so that J is diagonal and $J_\infty = \text{Diag. Matrix}(1, 0, \ldots, 0)$ is easier to see.

[31] For the statement of this hypothesis, see [574, pp. 521–522]. Von Mises joined the ranks of those who criticized invoking it in conjunction with Boltzmann's theory and devoted many pages to critiquing it [574, pp. 526–532].

[32] See, e.g., the paper by Hadamard and Fréchet [258, p. 2083], where von Mises' work is called to the reader's attention and praised. Hadamard and Fréchet also state (on p. 2083) that von Mises (among others mentioned) did his work without knowledge of Markov's paper [431]. Although the basis for this statement is uncertain, it seems to be based on their belief that Markov's work was available only in Russian, whereas, as noted earlier, a German translation had been available since 1912 in the German edition of Markov's book [432].

17.3.2.2 V. I. Romanovsky

V.I. Romanovsky (1879–1954) was born in Vernyi (now Almaty) in Kazakhstan and by 1918 had returned to nearby Tashkent in Uzbekistan as professor of probability and mathematical statistics. During 1900–1908, he had been a student and then docent at the University of Saint Petersburg. In 1904, he completed his doctoral dissertation under the direction of Markov at the university, where Markov had been a professor since 1886.[33]

In 1929, Romanovsky published a paper (in French) in the proceedings of the Academy of Sciences of the USSR, "On Markoff chains" [499]. After giving the basic definitions, he explained that "We call the series of such trials discrete Markoff chains because this eminent geometer was the first to consider them. Here we will expound some new results concerning the general case, which was not considered by Markoff" [499, p. 203]. By the "general case" he meant the generic case in which all the characteristic roots of $A = P^t$ are distinct, where (as in the above discussion of Markov's work) $P = (p_{ij})$ denotes the matrix of transition probabilities of an n-state chain. Of course one of these roots is $\rho_0 = 1$. In addition to assuming no multiple roots, Romanovsky also assumed that $\rho = -1$ was not a root. For $k = 0, 1, 2, \ldots$, he considered the probabilities $q_i^{(k)}$ of being in the ith state after k iterations of the process. Although he did not use any matrix notation—working with systems of linear equations as had Markov—Romanovsky realized the equivalent of $\mathbf{v}^{(k+1)} = A\mathbf{v}^{(k)}$, where $A = P^t$ and for any k $\mathbf{v}^{(k)} = \left(q_1^{(k)} \cdots q_n^{(k)}\right)^t$. He also realized the immediate implication that $\mathbf{v}^{(k)} = A^k\mathbf{v}^{(0)}$. However, he erroneously assumed that since $\rho = -1$ was excluded as a characteristic root, all roots $\rho \neq 1$ satisfy $|\rho| < 1$, so that $\mathbf{v}^{(\infty)} = \lim_{k\to\infty} A^k\mathbf{v}^{(0)}$ exists [499, p. 204].

At this point in time, Romanovsky was not familiar with Frobenius' paper of 1912, which makes it clear that it is only for primitive matrices A (with or without multiple roots) that the above reasoning is valid. In particular, Frobenius' example (17.32) in the special case

$$A = \begin{pmatrix} 0 & 1 & 0 \\ 0 & 0 & 1 \\ 1 & 0 & 0 \end{pmatrix} \tag{17.46}$$

is a stochastic matrix with index of imprimitivity $k = 3$ satisfying all Romanovsky's explicit assumptions but having the three cube roots of unity as characteristic roots, so that $|\rho| = 1$ for all roots ρ and $\mathbf{v}^{(\infty)} = \lim_{k\to\infty} A^k\mathbf{v}^{(0)}$ does not exist. Ignorant of Frobenius' work, Romanovsky repeated his error in two notes in the *Comptes rendus* of the Paris Academy of Sciences in 1930 [500, 501]. A Czech mathematician, J. Kaucký, spotted the error, and in a 1930 note in the *Comptes rendus* [338], he gave as a counterexample the matrix in (17.46), albeit without mentioning Frobenius.

[33]For further information on Romanovsky, see [164].

Kaucký concluded by pointing out that "the classical theory of A. A. Markoff" shows that $\mathbf{v}^{(\infty)} = \lim_{k\to\infty} A^k \mathbf{v}^{(0)}$ exists when $P = (p_{ij}) > 0$. His remark reflects the fact that in the initial phase of the reawakened interest in Markov chains, Markov's "classical" theory was restricted to the case $P > 0$, perhaps because Markov's efforts to extend the theory to some $P \geq 0$ were, as we saw, unclear and partly untenable.

At the session of 19 January 1931 of the Paris Academy, Romanovsky responded to Kaucký's criticism with a note "On the zeros of stochastic matrices" [502].[34] As the title suggests, here he attempted a more careful examination of the possibilities for the characteristic roots of certain stochastic matrices. Some of his propositions (I and II) are valid for any row-stochastic matrix P and not just for those satisfying the additional condition—no zero columns—imposed by him; one of them (Proposition III about a characteristic vector x for $\rho_0 = 1$) is ambiguously stated and, depending on the interpretation, either contains an unnecessary hypothesis (if $x \geq 0$ is asserted) or is false (if $x > 0$ is asserted). The next two (IV–VI) are incorrect.[35] Anyone well versed in the results of Frobenius' 1912 paper would have realized these defects. Romanovsky had clearly not yet studied Frobenius' paper and was probably not yet aware of its existence.[36]

By 1933, Romanovsky had become familiar with Frobenius' paper, for in that year, he published a paper in the *Bulletin de la Société mathématique de France* entitled "A theorem on the zeros of nonnegative matrices" [503], which began by noting that the zeros of such matrices "are profoundly studied by G. Frobenius" in [231]. The theorem of the title was a corrected generalization of his faulty Proposition VI in the note of 19 January 1931. Three years later, in 1936, Romanovsky published a lengthy memoir in *Acta Mathematica* entitled "Investigations on Markoff chains" [504], and by that time he had evidently digested all three of Frobenius' papers on positive and nonnegative matrices [228, 229, 231]. Citing these three papers, he wrote in his introductory remarks:

---

[34]Judging by his remark [502, p. 267], Romanovsky was the first to use the term "stochastic matrix." For him it meant (i) $P \geq 0$ (ii) with row sums equaling 1 and (iii) no zero column. Nowadays, condition (iii) is not usually included in the definition of a stochastic matrix, and I have not included this condition in my references to stochastic matrices.

[35]In the 16 January 1933 session of the Académie, Émile Ostenc gave simple counterexamples to IV–VI [460]. He made no reference to Frobenius' 1912 paper [231].

[36]The most interesting and historically significant part of Romanovsky's paper is the concluding paragraphs, where he responded to Kaucký's criticism by attempting to characterize those P which admit all primitive kth roots of unity, $k \geq 3$, as characteristic roots. These paragraphs are of interest because they involved what turns out to be an alternative characterization of the degree of imprimitivity k of an irreducible matrix, a characterization that has a graph-theoretic interpretation (A is cyclic of index k). Romanovsky himself made no reference to the theory of graphs, and it is doubtful he was thinking in such terms, since his ideas were motivated by the well-known determinant-theoretic formula for the coefficients of the characteristic polynomial $\varphi(r) = |rI - A|$, as is evident from his subsequent, more detailed papers [503, p. 215] and [504, p. 163].

> Since the theory of stochastic matrices and their zeros plays a fundamental role in the theory of Markoff chains and is intimately connected to the theory of nonnegative matrices developed by G. Frobenius, I will begin my memoir with an exposition of the results of G. Frobenius ....

Romanovsky devoted 33 of the 105 pages of his memoir to Frobenius' theory and its application to stochastic matrices, thereby exposing his readers to all of Frobenius' significant results and making clear their relevance to the theory of stochastic matrices and Markov chains. In 1945, he incorporated his exposition of Frobenius' theory into a book on discrete markov chains (in Russian). Citing Romanovsky's book and several of his earlier papers, Felix Gantmacher devoted a chapter to Frobenius' theory of nonnegative matrices in his book (in Russian) on the theory of matrices, which appeared in 1953.[37] Gantmacher's book represented the first genuinely comprehensive treatise on matrix theory and has since become a classic. It was translated into German in 1958 and into English in 1959 and is still in print as [240]. An English translation of Romanovsky's book by E. Seneta was published in 1970 [505].

[37]In 1937, Gantmacher and Krein [241] had already used Perron's Lemma 17.6 as proved by Frobenius in 1908 to develop their theory of strictly positive (respectively nonnegative) matrices—$n \times n$ matrices such that all $k \times k$ minors are positive (respectively, nonnegative) for all $k = 1, \ldots, n$. Such matrices arise in the mechanical analysis of small oscillations. See [242] for a comprehensive account.

Chapter 18
The Mathematics of Frobenius in Retrospect

In terms of their approach to creative work, mathematicians display a spectrum of tendencies. Some focus most of their time and effort on building up a monumental theory. Sophus Lie was such a mathematician, with his focus on his theory of transformation groups. Among Frobenius' mentors, Weierstrass, with his focus on the theory of abelian integrals and functions and the requisite foundations in complex function theory, and Richard Dedekind, with his theory of algebraic numbers and ideals, are further examples of mathematicians who were primarily theory builders. At the other end of the spectrum are mathematicians whose focus was first and foremost on concrete mathematical problems. Of course, many mathematicians fall somewhere between these extremes. A prime example is Hilbert, who created several far-reaching theories, such as his theory of integral equations, but also solved many specific problems, such as the finite basis problem in the theory of invariants, Waring's problem, and Dirichlet's problem; and of course he posed his famous 23 mathematical problems for others to attempt to solve. Frobenius was decidedly at the problem-solver end of the spectrum. Virtually all of his important mathematical achievements were driven by the desire to solve specific mathematical problems, not famous long-standing problems such as Waring's problem, but in general, problems that he perceived in the mathematics of his time.

This view of Frobenius' mathematical orientation is borne out by the preceding chapters. As we saw, it was the problem Frobenius perceived in Clebsch's attempt to handle the problem of Pfaff on a nongeneric level that prompted his own work on the problem of Pfaff. His important work on the symbolic algebra of matrices was motivated by the challenge of solving nongenerically the Cayley–Hermite problem and the related problem he drew from Rosanes' work. His fruitful work on the arithmetic theory of bilinear forms was prompted by a problem of his own devising (Problem 8.1) that was inspired by his reading of Gauss' *Disquisitiones Arithmeticae*. And his rational theory of elementary divisors, which flowed out of that arithmetic work, was motivated by the problem posed by the rationality paradox he perceived in Weierstrass' approach to the theory of elementary divisors and the concomitant problem of how to avoid it.

T. Hawkins, *The Mathematics of Frobenius in Context*, Sources and Studies in the History of Mathematics and Physical Sciences, DOI 10.1007/978-1-4614-6333-7_18,

His work with Stickelberger on abstract finite abelian groups was motivated by the problem of determining what sort of uniqueness could be established for the factorization into cyclic subgroups in Schering's theorem. His important work on density theorems and concomitant conjectures was an outgrowth of the task he set for himself of making sense out of Kronecker's assertions about the analytic densities associated to a certain class of polynomials. It led not only to his own density theorems but to an interest in the theory of finite nonabelian groups, which, in the form of Galois groups, were fundamental to his overall approach (unlike Kronecker's); and the problem of solving his first density theorem led him to develop his theory of double cosets. This interest in the theory of finite nonabelian groups was kept alive by the problem of providing proofs of Sylow's theorems within the context of abstract groups rather than within the context of permutation groups, as had hitherto been the practice. Later work on group theory was focused on two important problems of the time, viz., the problem of determining classes of solvable groups and the problem of determining whether a group is simple.

All of Frobenius' major work on abelian and theta functions was also problem-driven. His work on abelian functions as discussed in Chapter 10 was motivated by two problems, which I have named Hermite's abelian matrix problem and Kronecker's complex multiplication problem. And as I have argued in Section 11.3, his theory of Jacobian functions, viz., theta functions in the modern sense, can also be seen as motivated by a problem, namely the problem of generalizing the Weierstrass–Schottky theory of theta functions in such a way that the Riemann–Weierstrass conditions on the period matrix of an abelian function would play an analogous role in his theory of generalized theta functions.

His single greatest achievement, his theory of group characters and representations, was the outcome of his efforts to solve the group determinant problem. Solving this problem led to his theory of group characters. Dedekind also implicitly posed for Frobenius another problem by showing that his factorization of the group determinant for a special class of nonabelian groups could be presented more attractively and completely in terms of matrices (Section 13.5). This posed for Frobenius the problem as to whether the sort of result Dedekind had achieved in a very special case—namely that for a very particular type of nonabelian group, the associated group matrix $(x_{PQ^{-1}})$ is similar to a block diagonal matrix, with the similarity matrix constructed using the Dedekind characters of the group—could be generalized to all nonabelian groups using his generalized characters. The solution of this problem (in Theorems 13.11, 13.12 and 15.2) led to his development of the theory of matrix representations, including his theory of primitive characteristic units. One of the byproducts of Frobenius' work on group characters and representations was his important contributions to the theory of linear associative algebras over $\mathbb{C}$ (hypercomplex number systems, as they were known then). They were all motivated by problems, first the problem of providing complete and clear proofs of T. Molien's theorems, second the problem of developing the work of Cartan by basis-free methods, viz., the theory of determinants, in accordance with Kronecker's first disciplinary ideal, and third the problem of recasting Alfred Young's work on "quantitative substitutional analysis" in terms of group algebras and Frobenius' notion of primitive characteristic units. In all cases, the solutions of the problems led to new and important results.

Finally, his masterful theory of nonnegative matrices was the result of attempting to solve two successive problems. The first, pointed out by Perron himself, was to give a more algebraic proof of his theorem on positive matrices (Perron's theorem), i.e., one that avoided his limit lemma (Lemma 17.10). Frobenius solved this problem, and the ensuing interest in positive matrices, together with the obvious question of how much of Perron's theory extends to nonnegative matrices, led him to pose a problem (Problem 17.15) regarding the existence of nonnegative characteristic vectors for nonnegative matrices, which turned out to require him to create his beautiful theory of nonnegative matrices, a theory that allowed him to solve Problem 17.15, but turned out to have an importance, especially by virtue of its manifold applications, that far exceeded in depth and significance his solution to that problem.

And so Frobenius was a problem solver, but that characterization of him does not adequately describe the nature of his mathematics and the reason it has endured. It is also necessary to consider the manner in which he went about solving problems. Here his Berlin training, especially what he received from Weierstrass, proved to be especially important. Weierstrass emphasized through his own work the importance of seeking to grapple with the general, i.e., n-variable, case and to do so in a manner that was rigorous and, in particular, nongeneric. Weierstrass also emphasized the importance of expounding mathematical results in a clear and appropriate manner by means of a suitable framework. Indeed, he spent most of his career attempting to do this for the theory of abelian integrals and functions. These Weierstrassian tenets found a resonance with Frobenius, who adopted them in his own work. The previous chapters bear witness to the fact that Frobenius always dealt with his problems in the n-variable case and presented his results with a degree of rigor and clarity remarkable for its time. A good case in point is afforded by Frobenius' rigorous nongeneric solution to the Cayley–Hermite problem, a problem that both Bachmann and Hermite could solve nongenerically only in the ternary case by methods that did not extend to n variables.

In particular, Frobenius took very seriously Weierstrass' emphasis on the presentation of results in an appropriate context. For each of the above-discussed problems that he solved, he developed what he perceived as the appropriate theoretical framework, and his pedagogical skills translated into an ability to present his work with a degree of lucidity that generally exceeded what we find in Weierstrass' own publications. For these reasons, many of his papers resemble carefully conceived and lucidly written monographs. Thus his work on the problem of Pfaff resembled a monograph built on the notion and properties of the bilinear covariant and duality considerations, which led to his integrability theorem. His solution to the Cayley–Hermite problem was woven into a monograph on matrix algebra, a monograph whose results he was able to apply, not only to the problems of Cayley–Hermite and Rosanes, as he did in his monographic paper of 1878, but also later to solve Kronecker's complex multiplication problem and the congruence problems of Weierstrass and Kronecker. Moreover, the theory that Frobenius had developed to solve Kronecker's complex multiplication problem was sufficiently viable that even though geometers subsequently adopted a much broader characterization of complex multiplication than Kronecker's, which Frobenius had focused on, they found in Frobenius' theory results that, in a readily generalizable form, played a key role in the study of abelian varieties with complex multiplication (Section 10.7).

To resolve the question of uniqueness in the fundamental theorem of finite abelian groups, Frobenius and Stickelberger created the first monograph on the theory of abstract finite abelian groups. The arithmetic problems on bilinear forms inspired by his reading of Gauss were solved within the context of an arithmetic theory of his own devising that culminated in his normal form theorem (Theorem 8.8). That theory proved to have many applications, e.g., to the theory of linear systems of equations and congruences and, most notably, to a rational theory of elementary divisors. He also was able to modify his arithmetic theory of the normal form so that the modified theory led to a normal form for abelian matrices (Theorem 10.6) that provided an elegant solution to Hermite's abelian matrix problem. Next, Frobenius took Kronecker's sketchy work on analytic densities and placed it within the theoretical context of Galois theory and its connections with Dedekind's theory of ideals. Dedekind had pioneered this aspect of his theory of ideals, although he had held back many of his results from publication when Frobenius did his work on densities. As a result, Frobenius found it necessary to supplement Dedekind's published theory with his theory of double cosets and his theorem on Frobenius automorphisms. That theorem provided the theoretical context for a different type of density theorem, as exemplified by Frobenius' density theorem and the related conjectured theorem eventually proved by Chebotarev. In addition, his theory of double cosets provided the theoretical foundation for the first abstract theory of all three of Sylow's major theorems.

To solve the problem of having the Riemann-Weierstrass conditions on period matrices play a role in the theory of theta functions analogous to their role in the theory of abelian functions, Frobenius created his theory of Jacobian functions as a generalization of Weierstrass' theory of theta functions. To solve the group determinant problem Frobenius created his most remarkable theory of all, his theory of group characters, and then, with the fillip provided by some observations by Dedekind, transformed that theory into a broader, equally remarkable theory of matrix representations of finite groups. Finally, to solve the problem of what nonnegative matrices possess nonnegative characteristic vectors, he created his masterful theory of nonnegative matrices, the core of present-day Perron–Frobenius theory.

Another feature of Frobenius' mathematics is that despite his predilection for algebraically oriented problems, his ability to master the leading areas of mathematics of his day enabled him to look within all these areas for interesting problems. In this respect, he resembled Kronecker more than Weierstrass or Kummer, as he himself must have realized.[1] At Berlin, Frobenius had received a solid, broad-based mathematical education in the theory of numbers and in complex function theory, including its applications to differential equations and elliptic and abelian functions and integrals and the attendant theory of theta functions, an education that he supplemented by his own extensive readings in the literature. This included the study of classics such as Gauss' *Disquisitiones Arithmeticae* and the work of Galois, as well as more recent literature in all the above-mentioned fields. All of this literature he read with an eye toward finding significant problems, essentially algebraic or formal in nature, suggested by his studies. His resultant work spanned the fields of Galois theory, linear and total differential equations, the theories of determinants and bilinear forms, the theory of matrices, including the special theory for nonnegative ones, algebraic and analytic number theory, abelian and theta functions (with implications for complex abelian varieties), finite group theory, and the theory of linear associative algebras.

Frobenius' broad-based "Kroneckerian" approach to mathematical research was coupled with a penchant for the mathematics of his day. He was not consciously a visionary, seeking to create the mathematics of the future. He was essentially a mathematical conservative, who sought to improve and expand the development of known subjects in what he deemed the best possible manner. He was in this respect a man of his times. As a consequence, he was frequently not the only mathematician to develop ideas based on certain discoveries. Indeed, as can be seen from the index, he was involved in instances of multiple discoveries with sixteen other mathematicians. In all of these instances, except the multiple discovery of the Frobenius automorphism theorem, it was Frobenius who developed the common discovery or idea in the deepest and most far-reaching manner and with the greatest degree of rigor, as can be seen from the preceding chapters. Also, even though Frobenius had a predilection for problems to be solved by formal or algebraic means, he was no purist when it came to solving them, and as a student of Weierstrass, he was quite willing to use real and complex analysis to accomplish his goals whenever he deemed it appropriate. Thus his first postdoctoral work was a complex-analytic rendition of Galois theory, and later he used the uniqueness of Laurent expansions, e.g., in his proofs of his minimal polynomial theorem, his theorem on orthogonal matrices, and his matrix square root theorem. His fundamental existence theorems on Jacobian functions (Theorems 11.5 and 11.7), although established by mostly formal or algebraic reasoning, made critical use of theorems from real and complex analysis in the "sufficiency" parts of the proofs, as indicated in Section 11.3.1.1.

Despite his mathematical conservatism, Frobenius' concern for the mathematics of his own time, by virtue of the clear, rigorous, and far-reaching manner in which he went about solving problems, inadvertently often contributed to the mathematics

[1] See in this connection Frobenius' 1893 memorial essay on Kronecker, in which he contrasted Kummer and Weierstrass, whose fame was based on work in a specific area of mathematics, with Kronecker, whose "far-reaching discoveries" were spread out over many disciplines [202, p. 705].

of the future. Thus his work on the problem of Pfaff was developed by Cartan, whose work in turn led to Cartan–Kähler theory; and his integrability theorem in particular has become foundational in present-day differential topology and calculus on manifolds. His theory of matrix algebra is embedded in present-day mathematics. His rendition of the theory of the Smith normal form and his allied rational theory of elementary divisors, which played a key role in the development of abstract linear algebra in general, and the module-theoretic approach in particular, are part of today's graduate texts on modern algebra. The density theorems that bear the names of Frobenius and Chebotarev are still a fundamental part of number theory. Frobenius' theory of Jacobian functions, or simply theta functions as they are now defined, became a critical part of the foundations of the modern theory of abelian functions and varieties. As already noted, his paper on nonnegative matrices formed the backbone of the "Perron–Frobenius theory" that has found numerous applications in a broad spectrum of present-day science and technology, including, perhaps most recently, an application to Internet search engine ranking algorithms.[2]

Last, but hardly least, his solution to the now old-fashioned group determinant problem, a problem tailor-made to pique Frobenius' interest, produced his theory of group characters and representations, which still plays a fundamental role in mathematics and the sciences (notably theoretical physics). Frobenius' theory of group characters and representations was also significantly expanded by the work of his student Issai Schur and by his student Richard Brauer. Their work, along with that of Frobenius, is at the heart of the representation theory of finite groups today. Furthermore, as I have shown elsewhere, Frobenius' theory and Schur's work on a theory of representations for the general rotation group $\mathrm{SO}(n,\mathbb{R})$ inspired the work of Weyl that in turn inspired present-day research on the representation theory of infinite groups, such as the compact semisimple Lie groups treated by Weyl, as well as a burgeoning theory of harmonic analysis on infinite groups.[3] Frobenius' theory, with an assist from his automorphism theorem, also provided Artin the means to radically generalize the notion of an L-function, which, among other things, later became a part of the ongoing Langlands program (Section 15.6.3).

While Frobenius' theory of group characters and representations was undoubtedly his greatest single mathematical creation, I hope this book will make it clear that his impact on present-day mathematics is hardly limited to his creation of that theory. Within the large number of mathematical fields that drew his attention, he introduced concepts and established theorems that, thanks in part to his habit of presenting his results in a clear, rigorous monographic form, have become a part of the basic mathematics of the present. An Internet browser search for "Frobenius," which brings up a host of concepts, methods, theorems, and constructions bearing the name of Frobenius, although not necessarily explicitly found in Frobenius' own publications, is a good indicator of the viability and inspirational potential of the many ideas and results he introduces into a variety of mathematical fields. Among

---

[2]See [397]. I am grateful to Wilfried Parys for calling this recent application to my attention.

[3]See Chapters 11, 12 and the Afterword of my book [276].

his contemporaries, Frobenius (b. 1849) was neither a creative genius on a par with Poincaré (b. 1854) nor a mathematical visionary such as Lie (b. 1842). Nevertheless, his considerable and highly original talent for seeking out and rigorously and definitively solving a broad spectrum of mathematical problems, after placing them within what he deemed a suitable theoretical framework, produced a body of mathematical work whose sum total has had a cumulative impact on pure and applied mathematics that puts him in the company of those distinguished mathematicians.

References

1. G. Abraham-Frois and E. Lendjel, editors. *Les Oeuvres Économiques de l'Abbé Potron*. L'Harmattan, Paris, 2004.
2. A. A. Albert. *Collected Mathematical Papers Part 1. Associative Algebras and Riemann Matrices*. American Mathematical Society, Providence, RI, 1993.
3. Anonymous. Zum Andenken an Rudolf Friedrich Alfred Clebsch. *Math. Ann.*, 6:197–202, 1873.
4. Anonymous. Georg Frobenius. *Vierteljahrsschrift der Naturforschenden Gesellschaft in Zürich*, page 719, 1917.
5. P. Appell. Sur les fonctions périodiques de deux variables. *Jl. de math. pures et appl.*, (4) 7:157–219, 1891.
6. E. Artin. Über eine neue Art von *L*-Reihen. *Abh. aus d. math. Seminar d. Univ. Hamburg, 1923*, 3:89–108, 1923. Reprinted in *Papers*, 105–124.
7. E. Artin. Beweis des allgemeinen Reziprozitätsgesetzes. *Abh. aus d. math. Seminar d. Univ. Hamburg*, 5:353–363, 1927. Reprinted in *Papers*, 131–141.
8. E. Artin. Zur Theorie der *L*-Reihen mit allgemeinen Gruppencharakteren. *Abh. aus d. math. Seminar d. Univ. Hamburg, 1930*, 8:292–306, 1930. Reprinted in *Papers*, 165–179.
9. M. Aschbacher. The classification of the finite simple groups. *The Mathematical Intelligencer*, 3:59–65, 1981.
10. P. Bachmann. Untersuchungen über quadratische Formen. *Jl. für die reine u. angew. Math.*, 76:331–341, 1873.
11. P. Bachmann. *Die Arithmetik der quadratischen Formen. Erster Abtheilung*. Teubner, Leipzig, 1898.
12. R. Baltzer. *Theorie und Anwendungen der Determinanten*. S. Hirzel, Leipzig, 1857.
13. R. Baltzer. *Theorie und Anwendungen der Determinanten*. S. Hirzel, Leipzig, 3rd edition, 1870.
14. P. Bamberg and S. Sternberg. *A Course in Mathematics for Students of Physics*, volume 2. Cambridge University Press, Cambridge, 1988.
15. H. Begehr. Constantin Carathéodory (1873–1950). In H. Begehr et al., editors, *Mathematics in Berlin*, pages 105–109. Birkhäuser, Berlin, 1998.
16. M. Bernkopf. Laguerre, Edmond Nicolas. In *Dictionary of Scientific Biography*, volume 7, pages 573–576. Charles Scribner's Sons, New York, 1973.
17. C. Bidard, G. Erreygers, and W. Parys. Review of [1]. *European J. of the History of Economic Thought*, 13:163–167, 2006.
18. C. Bidard, G. Erreygers, and W. Parys. Our daily bread: Maurice Potron, from Catholicism to mathematical economics. *European J. of the History of Economic Thought*, 16(1):123–154, 2009.

T. Hawkins, *The Mathematics of Frobenius in Context*, Sources and Studies in the History of Mathematics and Physical Sciences, DOI 10.1007/978-1-4614-6333-7,

19. K.-R. Biermann. Wahlvorschläge zur Wahl von Mathematikern in die Berliner Akademie. *Abhandlungen d. Akad. der Wiss. zu Berlin*, Math-nat. Kl., Nr. 3, 1960.
20. K.-R. Biermann. Karl Weierstrass. Ausgewählte Aspekte seiner Biographie. *Jl. für die reine u. angew. Math.*, 223:193–220, 1966.
21. K.-R. Biermann. Dedekind, (Julius Wilhelm) Richard. *Dictionary of Scientific Biography*, 4:1–5, 1971.
22. K.-R. Biermann. *Die Mathematik und ihre Dozenten an der Berliner Universität 1810–1920*. Akademie-Verlag, Berlin, 1973.
23. O. Biermann. Über n simultane Differentialgleichungen der Form $\sum_{\mu=1}^{n+m} X_\mu dx_\mu$. *Zeitschrift für Mathematik und Physik*, 30:234–244, 1885.
24. J. B. Biot. *Essai de Géométrie analytique appliquée aux courbes et aux surfaces du second ordre*. Bachelier, Paris, 7th ed. edition, 1826.
25. M. Bôcher. *Introduction to Higher Algebra*. Macmillan, New York, 1907. Republished by Dover Publications, New York, 1964. German translation as [26].
26. M. Bôcher. *Einführung in die höhere Algebra*. Teubner, Leipzig, 1910. Translated by H. Beck.
27. M. Bôcher. The published and unpublished work of Charles Sturm on algebraic and differential equations. *Bulletin of the American Mathematical Society*, 18:1–18, 1912.
28. R. Bölling, editor. *Briefwechsel zwischen Karl Weierstrass und Sofja Kowalewskaja*. Akademie Verlag, Berlin, 1993.
29. R. Bölling. Weierstrass and some members of his circle: Kovalevskaia, Fuchs, Schwarz, Schottky. In H. Begehr, editor, *Mathematics in Berlin*, pages 71–82. Birkhäuser, 1998.
30. C. Borchardt. Neue Eigenschaft der Gleichung, mit deren Hülfe man die seculären Störungen der Planeten bestimmt. *Jl. für die reine u. angew. Math.*, 30:38–45, 1846.
31. C. Boyer. *History of Analytic Geometry*. Scripta Mathematica, New York, 1956.
32. R. Brauer. Über die Darstellung von Gruppen in Galoisschen Feldern. *Actualités Sci. Industrielles*, 195:1–15, 1935. Reprinted in *Papers* 1, 323–335.
33. R. Brauer. On the representation of a group of order g in the field of the g-th roots of unity. *American Journal of Math.*, 67:461–471, 1945. Reprinted in *Papers* 1, 461–471.
34. R. Brauer. On Artin's L-series with general group characters. *Annals of Math.*, 48:502–514, 1947. Reprinted in *Papers* 1, 539–551.
35. R. Brauer and C. Nesbitt. On the regular representations of algebras. *Proc. Nat. Acad. Sci. USA*, 23:236–240, 1937. Reprinted in R. Brauer, *Papers* 1, 190–194.
36. R. Brauer and C. Nesbitt. On the modular representations of groups of finite order I. *University of Toronto Studies*, 4:1–21, 1937. Reprinted in R. Brauer, *Papers* 1, 336–354.
37. R. Brauer and E. Noether. Über minimale Zerfällungskörper irreduzibler Darstellungen. *Sitzungsberichte der Akademie der Wiss. zu Berlin 1927 physikalisch-math. Klasse*, 1927. Reprinted in Brauer *Papers* 1, 221–228 and Noether *Abhandlungen*, 552–559.
38. J. Bret. Determination de la longueur des axes principaux dans les surfaces du second ordre qui ont un centre. *Annales de mathématiques (Gergonne)*, 2:33–37, 1812.
39. C. Brezinski. *History of Continued Fractions and Padé Approximants*. Springer-Verlag, Berlin, 1991.
40. F. Brioschi. *La teorica dei determinanti e le sue principali applicazioni*. Bizzoni, Pavia, 1854. German translation, Berlin [1856a]; French translation, Paris [1856b].
41. F. Brioschi. Note sur une théorème reatif aux déterminants gauches. *Jl. de math. pures et appl.*, 19:253–256, 1854.
42. F. Brioschi. *Theorie der Determinanten und ihre hauptsächlichen Anwendungen*. Duncker and Humblot, Berlin, 1856. German translation of [40].
43. F. Brioschi. *Théorie des déterminants et leurs princpales applications*. Mallet–Bachelier, 1856. French translation of [40].
44. C. Briot and J. Bouquet. *Théorie des fonctions elliptiques*. Gauthier-Villars, 2nd edition, 1875.
45. C. Burali-Forti. *Introduction à la géométrie différentielle suivant la méthode de H. Grassmann*. Gauthier–Villars, 1897.
46. W. Burnside. On a property of certain determinants. *Messenger of Mathematics*, 23(2): 112–114, 1894.

47. W. Burnside. Notes on the theory of groups of finite order. *Proceedings London Math. Soc.*, 26:191–214, 1895.
48. W. Burnside. *Theory of Groups of Finite Order*. The Univeristy Press, Cambridge, 1897.
49. W. Burnside. On the continuous group that is defined by any given group of finite order. *Proceedings London Math. Soc.*, 29:207–225, 1898.
50. W. Burnside. On the continuous group that is defined by any given group of finite order (second paper). *Proceedings London Math. Soc.*, 29:546–565, 1898.
51. W. Burnside. On some properties of groups of odd order. *Proceedings London Math. Soc.*, 33:162–185, 1900.
52. W. Burnside. On some properties of groups of odd order (second paper). *Proceedings London Math. Soc.*, 33:257–268, 1900.
53. W. Burnside. On transitive groups of degree n and class $n-1$. *Proceedings London Math. Soc.*, 32:240–246, 1900.
54. W. Burnside. On groups of order $p^{\alpha}q^{\beta}$. *Proceedings London Math. Soc.*, 2(1):388–392, 1904.
55. W. Burnside. On the complete reduction of any transitive permutation group and on the arithmetical nature of the coefficients in its irreducible components. *Proceedings London Math. Soc.*, 3(2):239–252, 1905.
56. W. Burnside. *Theory of Groups of Finite Order*. The University Press, Cambridge, 2nd edition, 1911.
57. É. Cartan. *Sur la structure des groupes de transformations finis et continus*. Nony, Paris, 1894. Reprinted in *Oeuvres* 1, 137–287.
58. É. Cartan. Sur certains groups algébriques. *Comptes Rendus Acad. Sci. Paris*, 120:545–548, 1895. Reprinted in *Oeuvres* I, 289–292.
59. É. Cartan. Le principe de dualité et certaines intégrales multiples de l'espace tangentiel et de l'espace réglé. *Bull. Soc. Math. France*, 24:140–177, 1896. Reprinted in *Oeuvres* II.1, 265–302.
60. É. Cartan. Les groupes bilinéaires et les systèmes de nombres complexes. *Ann. Fac. Sci. Toulouse*, 12:B1–B99, 1898. Reprinted in *Oeuvres* II.1, 7–105.
61. É. Cartan. Sur certaines expressions différentielles et le problème de Pfaff. *Annales scientifiques École Normale Sup. Paris*, 16:239–332, 1899. Reprinted in *Oeuvres* II.1, 303–396.
62. É. Cartan. Sur l'intégration des systèmes d'équations aux différentielles totales. *Annales scientifiques École Normale Sup. Paris*, 18:241–311, 1901. Reprinted in *Oeuvres* II.1, 411–481.
63. É. Cartan. Sur l'intégration de certains systèmes de Pfaff de caractère deux. *Bulletin des sciences mathématiques*, 29:233–302, 1901. Reprinted in *Oeuvres* II.1, 483–553.
64. É. Cartan. *Leçons sur les invariants intégraux*. A. Hermann & fils, Paris, 1922.
65. É. Cartan. Notice sur les travaux scientifiques. In *Selecta. Jubilé scientifique de M. Élie Cartan*. Gauthier–Villars, Paris, 1939. This version of Cartan's notice was drafted in 1931, but apparently first published in 1939. It is reprinted in *Oeuvres* I, 1–98.
66. É. Cartan. *Les systèmes différentiels extérieurs et leurs applications géométriques*. Actualités scientifiques et industrielles, no. 994. Hermann, Paris, 1945.
67. G. Castelnuovo. Sulle funzioni abeliane. I. le funzioni intermediarie. *Rendiconti della R. Accademia dei Lincei*, (5)30:50–55, 1921. Reprinted in *Memorie sceite* (Bologna, 1937), pp. 529–534, and in *Opere* 3, 312–317.
68. A. L. Cauchy. Mémoire sur les fonctions qui ne peuvent obtenir que deux valeurs égales et de signes contraires par suite des transpositions opérées entre les variables qu'elles renferment. *Journal de l'École Polytechnique*, cah.17, t. 10:29ff., 1815. Reprinted in *Oeuvres* (2) 1, 91–169.
69. A. L. Cauchy. *Cours d'analyse de l'École Royale Polytechnique*. Debure, Paris, 1821. Reprinted in *Oeuvres* (2) 3.
70. A. L. Cauchy. Application du calcul des résidus à l'intégration de quelques équations différentielles linéaires et à coefficients variables. *Exercises de mathématiques*, 1, 1826. Reprinted in *Oeuvres* (2) 6, 261–264.

71. A. L. Cauchy. *Leçons sur les applications du calcul infinitésimal à la géométrie*, volume 1. Chez de Bure frères, Paris, 1826. Reprinted in *Oeuvres* (2)5.
72. A. L. Cauchy. Sur l'équation à l'aide de laquelle on determine les inégalités séculaires des mouvements des planètes. *Exer. de math.*, 4, 1829. Reprinted in *Oeuvres* (2) 9, 174–195.
73. A. L. Cauchy. Léquation qui a pour racines les moments d'inertie principaux d'un corps solide, et sur diverse équations du même genre. *Mem. Acad. des Sci.*, 1830. Reprinted in *Oeuvres* (2)1, 79–81.
74. A. L. Cauchy. Méthode générale propre à fournier les équations de condition relatives aux limites des corps dans les problèmes de physique mathématique. *Comptes rendus Acad. Sci. Paris*, 1839. Reprinted in *Oeuvres* (1) 8, 193–227.
75. A. L. Cauchy. Mémoire sur l'intégration des équations linéaires. *Comptes Rendus, Acad. Sci. Paris*, 8, 1839. Reprinted in *Oeuvres* (1) 4, 369–426. Cauchy also republished this as [76].
76. A. L. Cauchy. Mémoire sur l'intégration des équations linéaires. *Exer. d'analyse*, 1, 1840. Reprinted in *Oeuvres* (2) 11, 75–133.
77. A. L. Cauchy. Mémoire sur les arrangements que l'on peut former avec des lettres donneés, et sur les permutations ou substitutions à la aide desquelles on passe d'un arrangement à un autre. *Exercises d'analyse et de physique mathématique*, 3:151–252, 1844. Reprinted in *Oeuvres* (2) 13, 171–282.
78. A. L. Cauchy. Mémoire sur les perturbations produites dans les mouvements vibratoires d'un système de molécules par l'influence d'un autre système. *Comptes rendus Acad. Sc. Paris*, 30, 1850. Reprinted in *Oeuvres* (1) 4, 202–211.
79. A. Cayley. On the motion of rotation of a solid body. *Cambr. Math Journal*, 3:224–232, 1843. Reprinted in *Papers* 1, 28–35.
80. A. Cayley. Sur quelques propriétés des déterminants gauches. *Jl. für die reine u. angew. Math.*, 32:119–123, 1846. Reprinted in *Papers* 1, 332–336.
81. A. Cayley. Sur les déterminants gauches (Suite du Mémoire t. xxxii. p. 119). *Jl. für die reine u. angew. Math.*, 38:93–96, 1849. Reprinted in *Papers* 1, 410–413.
82. A. Cayley. Remarques sur la notation des fonctions algébriques. *Jl. für die reine u. angew. Math.*, 50:282–285, 1855. Reprinted in *Papers* 2, 185–188.
83. A. Cayley. Sur la transformation d'une fonction quadratique en elle-même par des substitutions linéaires. *Jl. für die reine u. angew. Math.*, 50:288–299, 1855. Reprinted in *Papers* 2, 192–201.
84. A. Cayley. A memoir on the theory of matrices. *Phil. Trans. R. Soc. London*, 148:17–37, 1858. Reprinted in *Papers* 2, 475–496.
85. A. Cayley. A memoir on the automorphic linear transformation of a bipartite quadric function. *Phil. Trans. R. Soc. London*, 148:39–46, 1858. Reprinted in *Papers* 2, 497–505.
86. A. Cayley. A supplementary memoir on the theory of matrices. *Phil. Trans. R. Soc. London*, 156:438–48, 1866. Reprinted in *Papers* 5.
87. A. Cayley. *Collected Mathematical Papers*, volumes 1–14. Cambridge University Press, Cambridge, 1889–1898.
88. A. Châtelet. *Leçons sur la théorie des nombres. (Modules. Entiers algébrique. Réduction continuelle.)*. Gauthier–Villars, Paris, 1913.
89. R. Chorlay. From problems to structures: the Cousin problems and the emergence of the sheaf concept. *Archive for History of Exact Sciences*, 64:1–73, 2010.
90. E. B. Christoffel. *De motu permanenti electricitatis in corporibus homogenis. Dissertatio inauguralis*. G. Shade, Berlin, 1856. Reprinted in *Abhandlungen* 1, 1–64.
91. E. B. Christoffel. Verallgemeinerung einiger Theoreme des Herrn Weierstrass. *Jl. für die reine u. angew. Math.*, 63:255–272, 1864. Reprinted in *Abhandlungen* 1, pp. 129–145.
92. E. B. Christoffel. Über die kleinen Schwingungen eines periodisch eingerichteten Systems materieller Punkte. *Jl. für die reine u. angew. Math.*, 63:273–288, 1864. Reprinted in *Abhandlungen* 1, 146–161.
93. E. B. Christoffel. Theorie der bilinearen Functionen. *Jl. für die reine u. angew. Math.*, 68: 253–272, 1868. Reprinted in *Abhandlungen* 1, 277–296.

94. E. B. Christoffel. Über die Transformation der homogenen Differentialausdrücke zweiten Grades. *Jl. für die reine u. angew. Math.*, 70:46–70, 1869. Reprinted in *Abhandlungen* 1, 352–377.
95. A. Clebsch. Theorie der circularpolarisirenden Medien. *Jl. für die reine u. angew. Math.*, 57:319–358, 1860.
96. A. Clebsch. Über das Pfaffsche Problem. *Jl. für die reine u. angew. Math.*, 60:193–251, 1861.
97. A. Clebsch. Über das Pfaffsche Problem. Zweite Abhandlung. *Jl. für die reine u. angew. Math.*, 61:146–179, 1863.
98. A. Clebsch. Über eine Classe von Gleichungen, welche nur reelle Wurzeln besitzen. *Jl. für die reine u. angew. Math.*, 62:232–245, 1863.
99. A. Clebsch. Über die simultane Integration linearer partieller Differentialgleichungen. *Jl. für die reine u. angew. Math.*, 65:257–268, 1866.
100. A. Clebsch and P. Gordan. *Theorie der Abelschen Functionen*. Teubner, Leipzig, 1866.
101. A. Cogliati. The genesis of Cartan–Kähler theory. *Archive for History of Exact Sciences*, 65:397–435, 2011.
102. L. Collatz. Einschliessungssatz für die charakteristischen Zahlen von Matrizen. *Mathematische Zeitschrift*, 48:221–226, 1942.
103. F. Conforto. *Funzioni Abeliane e Matrici di Riemann*. Libreria dell'Università, Rome, 1942.
104. F. Conforto. *Abelsche Funktionen und algebraische Geometrie*, edited from the author's Nachlass by W. Gröbner, A. Andreotti, and M. Rosati. Springer-Verlag, Berlin, 1956.
105. K. Conrad. History of class field theory. This unpublshed essay is available online as a PDF file at www.math.uconn.edu/~kconrad/blurbs/gradnumthy/cfthistory.pdf.
106. K. Conrad. The origin of representation theory. *L'Enseignement mathématique*, 44:361–392, 1998.
107. J. H. Conway et al., editors. *Atlas of Finite Groups. Maximal Subgroups and Ordinary Characters for Simple Groups*. Clarendon Press, Oxford, 1985.
108. P. Cousin. Sur les fonctions de *n* variables complexes. *Acta Mathematica*, 19:1–61, 1895.
109. C. W. Curtis. *Pioneers of Representation Theory: Frobenius, Burnside, Schur and Brauer*. American Mathematical Society, 1999.
110. G. Darboux. Sur les relations entre groupes de points, de cercles et de sphères dans le plan et dans l'espace. *Annales scientifiques École Normale Sup. Paris*, (2) 1:323ff., 1872.
111. G. Darboux. Sur le probème de Pfaff. *Bulletin des sciences mathématiques*, (2)6:14–68, 1882.
112. H. Deahna. Über die Bedingungen der Integrabilität linearer Differentialgleichungen erster Ordnung zwischen einer beliebigen Anzahl veränderlicher Grössen. *Jl. für die reine u. angew. Math.*, 20:340–349, 1840.
113. R. Dedekind. *Sur la théorie des nombres entiers algébriques*. Gauthier–Villars, Paris, 1877. First published in volumes (1) XI and (2) I of *Bulletin des sciences mathématiques*. A partial reprint (that excludes in particular Dedekind's chapter on modules) is given in Dedekind's *Werke* 3, 262–313. An English translation of the entire essay, together with a lengthy historical and expository introduction, is available as [120].
114. R. Dedekind. Über den Zusammenhang zwischen der Theorie der Ideale und der Theorie der höheren Kongruenzen. *Abhandlungen der K. Gesellschaft der Wiss. zu Göttingen*, 23:1–23, 1878. Reprinted in *Werke* 1, 202–230.
115. R. Dedekind. Zur Theorie der aus *n* Haupteinheiten gebildeten komplexen Grössen. *Nachrichten von der Königlichen Gesellschaft der Wissenschaften und der Georg-Augustus-Universität zu Göttingen*, pages 141–159, 1885. Reprinted in *Werke* 2, 1–19.
116. R. Dedekind. Gruppen-Determinante und ihre Zerlegung in wirkliche und übercomplexe Factoren. Niedersächsiche Staats- und Universitätsbibliothek Göttingen, Cod. Ms. R. Dedekind V, 5, 1886.
117. R. Dedekind. Zur Theorie der Ideale. *Nachrichten von der Königlichen Gesellschaft der Wissenschaften und der Georg-Augustus-Universität zu Göttingen*, pages 272–277, 1894. Reprinted in *Werke* 2, 43–48.

118. R. Dedekind. Über Gruppen, deren sämtliche Theiler Normaltheiler sind. *Math. Ann.*, 48: 548–561, 1897. Reprinted in *Werke* 2, 87–101.
119. R. Dedekind. *Gesammelte mathematische Werke. Herausgegeben von Robert Fricke, Emmy Noether, Øystein Ore*, volume 2. Vieweg, 1931. Reprinted by Chelsea Publishing Company, New York, 1969.
120. R. Dedekind. *Theory of Algebraic Integers. Translated and Introduced by John Stillwell.* Cambridge University Press, Cambridge, 1996. English translation of [113], together with an extensive mathematical and historical introduction [551].
121. R. Dedekind and H. Weber. Theorie der algebraischen Funktionen einer Veränderlichen. *Jl. für die reine u. angew. Math.*, 92:181–290, 1882. Dated October 1880. Reprinted in Dedekind's *Werke* 1, 238–349.
122. L. E. Dickson. On the group defined for any given field by the multiplication table of any given finite group. *Trans. American Math. Soc.*, 3:285–301, 1902. Reprinted in *Papers* 2, 75–91.
123. L. E. Dickson. On the groups defined for an arbitrary field by the multiplication tables of certain finite groups. *Proceedings London Math. Soc.*, 35:68–80, 1902. Reprinted in *Papers* 6, 176–188.
124. L. E. Dickson. Modular theory of group characters. *Bull. American Math. Soc.*, 13:477–488, 1907. Reprinted in *Papers* 4, 535–546.
125. L. E. Dickson. Modular theory of group-matrices. *Trans. American Math. Soc.*, 8:389–398, 1907. Reprinted in *Papers* 2, 251–260.
126. L. E. Dickson. *History of the Theory of Numbers*. Carnegie Institution of Washington, Washington, D.C., 1919–1923. 3 volumes.
127. L. E. Dickson. *Modern Algebraic Theories*. Sandborn, Chicago, 1926.
128. L. E. Dickson. Singular case of pairs of bilinear, quadratic, or Hermitian forms. *Transactions of the American Mathematical Society*, 29:239–253, 1927.
129. P. G. Dirichlet. Beweis des Satzes, dass jeder unbegrenzte arithmetische Progression, deren erstes Glied und Differenz ganze Zahlen ohne gemeinschaftlichen Factor sind, unendlich viele Primzahlen enthält. *Abhandlungen d. Akad. der Wiss. zu Berlin*, pages 45–81, 1837. Reprinted in *Werke* 1, 313–342.
130. P. G. Dirichlet. Beweis eines Satzes über die arithmetische Progression. *Berichte über die Verhandlungen der Königl. Preuss. Akademie der Wissenschaften*, pages 108–110, 1837. Reprinted in *Werke* 1, 307–312.
131. P. G. Dirichlet. Sur l'usage des séries infinies dans la théorie des nombres. *Jl. für die reine u. angew. Math.*, 18:259–274, 1838. Reprinted in *Werke* 1, 357–374.
132. P. G. Dirichlet. Recherches sur diverses applications de l'analyse infinitésimale à la théorie des nombres. *Jl. für die reine u. angew. Math.*, 19:324–369, 1839. Reprinted in *Werke* 1, 411–461.
133. P. G. Dirichlet. Recherches sur diverses applications de l'analyse infinitésimale à la théorie des nombres. *Jl. für die reine u. angew. Math.*, 21:1–12, 134–155, 1840. Reprinted in *Werke* 1, 461–496.
134. P. G. Dirichlet. Über eine Eigenschaft der Quadratischen Formen. *Berichte über die Verhandlungen der Königl. Preuss. Akademie der Wissenschaften, Jahrg. 1840*, pages 49–52, 1840. Reprinted in *Jl. für die reine u. angew. Math.* 21 (1840), 98–100 and in *Werke* 1, 497–502.
135. P. G. Dirichlet. Recherches sur les formes quadratiques à coefficients et à indéterminées complexes. *Jl. für die reine u. angew. Math.*, 24:291–371, 1842. Reprinted in *Werke* 1, 533–618.
136. P. G. Dirichlet. Über die Stabilität des Gleichgewichts. *Jl. für die reine u. angew. Math.*, 32:85–88, 1846. Reprinted in *Werke* 2, 5–8.
137. P. G. Dirichlet. *Vorlesungen über Zahlentheorie*. Vieweg, Braunschweig, 2nd edition, 1871. Edited and supplemented by R. Dedekind.
138. P. G. Dirichlet. *Vorlesungen über Zahlentheorie*. Vieweg, Braunschweig, 3rd edition, 1879. Edited and supplemented by R. Dedekind.

139. P. G. Dirichlet. *Vorlesungen über Zahlentheorie*. Vieweg, Braunschweig, 4th edition, 1894. Edited and supplemented by R. Dedekind.
140. P. Dugac. Eléments d'analyse de Karl Weierstrass. *Archive for History of Exact Sciences*, 10:41–176, 1973.
141. D. Dummit and R. Foote. *Abstract Algebra*. Wiley, 2nd edition, 1999.
142. H. Edwards. The background to Kummer's proof of Fermat's last theorem for regular primes. *Archive for History of Exact Sciences*, pages 219–236, 1975.
143. H. Edwards. *Fermat's Last Theorem: A Genetic Introduction to Algebraic Number Theory*. Springer-Verlag, New York, 1977. Russian translation, Moscow, 1980.
144. H. Edwards. Postscript to "The background of Kummer's proof ...." *Archive for History of Exact Sciences*, 17:381–394, 1977.
145. H. Edwards. The genesis of ideal theory. *Archive for History of Exact Sciences*, 23:321–378, 1980.
146. H. Edwards. Dedekind's invention of ideals. *Bulletin of the London Mathematical Society*, 15:8–17, 1983.
147. H. Edwards. Kummer, Eisenstein, and higher reciprocity laws. Number theory related to Fermat's last theorem. *Progress in Mathematics*, 26:31–43, 1983.
148. H. Edwards. *Galois Theory*. Springer-Verlag, New York, 1984.
149. H. Edwards. An appreciation of Kronecker. *The Mathematical Intelligencer*, 9:28–35, 1987.
150. H. Edwards. *Divisor Theory*. Birkhäuser, Boston, 1990.
151. G. Eisenstein. Allgemeine Untersuchungen über die Formen dritten Grades mit drei Variabeln, welche der Kreistheilung ihre Enstehung verdanken. *Jl. für die reine u. angew. Math.*, 28:289–374, 1844.
152. G. Eisenstein. Beiträge zur Theorie der elliptischen Functionen. *Jl. für die reine u. angew. Math.*, 35:137–146, 1847.
153. G. Eisenstein. Neue Theoreme der höheren Arithmetik. *Jl. für die reine u. angew. Math.*, 35:117–136, 1847.
154. G. Eisenstein. Über die Vergleichung von solchen ternären quadratischen Formen, welche verschiedene Determinante haben. *Ber. u. die Verb. der Akad. der Wiss. Berlin 1852*, pages 350–389, 1852.
155. F. Engel. Anmerkungen. In *Sophus Lie Gesammelte Abhundlungen*, volume 3, pages 585–789. Teubner, Leipzig, 1922.
156. L. Euler. *Introductio in analysin infinitorum*. M. M. Bousquet, Lausanne, 1748. Reprinted in *Opera omnia* (1) 9.
157. L. Euler. Recherches sur la connaissance mécanique des corps. *Mémoire de l'académie des sciences de Berlin, 1758*, 1765. Reprinted in *Opera omnia* (2) 8, 178–199.
158. L. Euler. Du mouvement de rotation des corps solides autour d'un axe variable. *Mémoire de l'academe des sciences de Berlin, 1758*, 1765. Reprinted in *Opera omnia* (2) 8, 200–235.
159. G. Faltings. The proof of Fermat's last theorem by R. Taylor and A. Wiles. *Notices of the AMS*, 42:743–746, 1995.
160. W. Feit. Richard D. Brauer. *Bull. American Math. Soc.*, 1(2):1–20, 1979.
161. W. Feit and J. G. Thompson. A solvability criterion for finite groups and some consequences. *Proc. Nat. Acad. Sci. U.S.A.*, 48:968–970, 1962.
162. W. Feit and J. G. Thompson. Solvability of groups of odd order. *Pacific Journal of Math.*, 13:755–1029, 1963.
163. P. Fernández. Review of [397]. *The Mathematical Intelligencer*, 30, 2008.
164. Sh. K. Formanov and R. Mukhamedkhanova. On the origin and development of research in probability theory and mathematical statistics in Uzbekistan up to the middle of the twentieth century (in Russian). *Uzbek. Mat. Zh.*, 4:64–71, 2004.
165. A. R. Forsyth. *Theory of Differential Equations. Part I. Exact Equations and Pfaff's Problem*. Cambridge University Press, 1890.
166. E. Frank. Oskar Perron (1880–1975). *Journal of Number Theory*, 14:281–291, 1982.
167. G. Frei and U. Stammbach. *Hermann Weyl und die Mathematik an der ETH Zürich, 1913–1930*. Birkhäuser, Basel, 1992.

168. G. Frei and U. Stammbach. *Die Mathematiker an den Züricher Hochschulen*. Birkhäuser, Basel, 1994.
169. H. Freudenthal. Riemann, Georg Friedrich Bernhard. *Dictionary of Scientific Biography*, 11:447–456, 1975.
170. H. Freudenthal. Schottky, Friedrich Hermann. *Complete Dictionary of Scientific Biography*, Encyclopedia.com (March 31, 2012), 2008. http://www.encyclopedia.com.
171. G. Frobenius. *De functionum analyticarum unis variabilis per series infinitas repraesentatione. Dissertatio inauguralis mathematica* .... A. W. Schadii, Berlin, 1870.
172. G. Frobenius. Über die Entwicklung analytischer Functionen in Reihen, die nach gegebenen Functionen fortschreiten. *Jl. für die reine u. angew. Math.*, 73:1–30, 1871. Reprinted in *Abhandlungen* 1, 35–64. Essentially a German-language reworking of his Berlin dissertation [171].
173. G. Frobenius. Über die algebraischer Auflösbarkeit der Gleichungen, deren Coefficienten rationale Functionen einer Variablen sind. *Jl. für die reine u. angew. Math.*, 74:254–272, 1872. Reprinted in *Abhandlungen* 1, 65–83.
174. G. Frobenius. Über die Integration der linearen Differentialgleichungen durch Reihen. *Jl. für die reine u. angew. Math.*, 76:214–235, 1873. Reprinted in *Abhandlungen* 1, 84–105.
175. G. Frobenius. Über den Begriff der Irreductibilität in der Theorie der linearen Differentialgleichungen. *Jl. für die reine u. angew. Math.*, 76:236–270, 1873. Reprinted in *Abhandlungen* 1, 106–140.
176. G. Frobenius. Anwendungen der Determinantentheorie auf die Geometrie des Maaßes. *Jl. für die reine u. angew. Math.*, 79:184–247, 1875. Reprinted in *Abhandlungen* 1, 158–220.
177. G. Frobenius. Über algebraisch integrirbare lineare Differentialgleichungen. *Jl. für die reine u. angew. Math.*, 80:183–193, 1875. Reprinted in *Abhandlungen* 1, 221–231.
178. G. Frobenius. Über die regulären Integrale der linearen Differentialgleichungen. *Jl. für die reine u. angew. Math.*, 80:317–333, 1875. Reprinted in *Abhandlungen* 1, 232–248.
179. G. Frobenius. Über das Pfaffsche Problem. *Jl. für die reine u. angew. Math.*, 82:230–315, 1877. Reprinted in *Abhandlungen* 1, 249–334.
180. G. Frobenius. Note sur la théorie des formes quadratiques à un nombre quelconque de variables. *Comptes Rendus, Acad. Sci. Paris*, 85:131–133, 1877. Reprinted in *Abhandlungen* 1, 340–342.
181. G. Frobenius. Über lineare Substitutionen und bilineare Formen. *Jl. für die reine u. angew. Math.*, 84:1–63, 1878. Reprinted in *Abhandlungen* 1, 343–405.
182. G. Frobenius. Theorie der linearen Formen mit ganzen Coefficienten. *Jl. für die reine u. angew. Math.*, 86:146–208, 1879. Reprinted in *Abhandlungen* 1, 482–544.
183. G. Frobenius. Über homogene totale Differentialgleichungen. *Jl. für die reine u. angew. Math.*, 86:1–19, 1879. Reprinted in *Abhandlungen* 1, 435–453.
184. G. Frobenius. Über schiefe Invarianten einer bilinearen oder quadratischen Form. *Jl. für die reine u. angew. Math.*, 86:44–71, 1879. Reprinted in *Abhandlungen* 1, 454–481.
185. G. Frobenius. Theorie der linearen Formen mit ganzen Coefficienten (Forts.). *Jl. für die reine u. angew. Math.*, 88:96–116, 1880. Reprinted in *Abhandlungen* 1, 591–611.
186. G. Frobenius. Zur Theorie der Transformation der Thetafunctionen. *Jl. für die reine u. angew. Math.*, 89:40–46, 1880. Reprinted in *Abhandlungen* 2, 1–7.
187. G. Frobenius. Über das Additionstheorem der Thetafunctionen mehrerer Variabeln. *Jl. für die reine u. angew. Math.*, 89:185–220, 1880. Reprinted in *Abhandlungen* 2, 11–46.
188. G. Frobenius. Über die principale Transformation der Thetafunctionen mehrerer Variableln. *Jl. für die reine u. angew. Math.*, 95:264–296, 1883. Reprinted in *Abhandlungen* 2, 97–129.
189. G. Frobenius. Über die Grundlagen der Theorie der Jacobischen Functionen. *Jl. für die reine u. angew. Math.*, 97:16–48, 1884. Reprinted in *Abhandlungen* 2, 172–204.
190. G. Frobenius. Über die Grundlagen der Theorie der Jacobischen Functionen (Abh. II). *Jl. für die reine u. angew. Math.*, 97:188–223, 1884. Reprinted in *Abhandlungen* 2, 205–240.
191. G. Frobenius. Über Gruppen von Thetacharakteristiken. *Jl. für die reine u. angew. Math.*, 96:81–99, 1884. Reprinted in *Abhandlungen* 2, 130–148.

192. G. Frobenius. Über Thetafunctionen mehrerer Variablen. *Jl. für die reine u. angew. Math.*, 96:100–122, 1884. Reprinted in *Abhandlungen* 2, 141–171.
193. G. Frobenius. Neuer Beweis des Sylowschen Satzes. *Jl. für die reine u. angew. Math.*, 100:179–181, 1887. Reprinted in *Abhandlungen* 2, 301–303.
194. G. Frobenius. Über die Congruenz nach einem aus zwei endlichen Gruppen gebildeten Doppelmodul. *Jl. für die reine u. angew. Math.*, 101:273–299, 1887. Reprinted in *Abhandlungen* 2, 304–330.
195. G. Frobenius. Über das Verschwinden der geraden Thetafunctionen. *Nachrichten von der Königlichen Gesellschaft der Wissenschaften und der Georg-Augustus-Universität zu Göttingen*, 5:67–74, 1888. Reprinted in *Abhandlungen* 2, 376–382.
196. G. Frobenius. Über die Jacobischen covarianten der Systeme von Berührungskegelschnitten einer Curve vierter Ordnung. *Jl. für die reine u. angew. Math.*, 103:139–183, 1888. Reprinted in *Abhandlungen* 2, 331–375.
197. G. Frobenius. Über die Jacobischen Functionen dreier Variabeln. *Jl. für die reine u. angew. Math.*, 105:35–100, 1889. Reprinted in *Abhandlungen* 2, 383–448.
198. G. Frobenius. Theorie der biquadratischen Formen. *Jl. für die reine u. angew. Math.*, 106:125–188, 1890. Reprinted in *Abhandlungen* 2, 449–512.
199. G. Frobenius. Über Potentialfunctionen, deren Hessesche Determinante verschwindet. *Nachrichten von der Königlichen Gesellschaft der Wissenschaften und der Georg-Augustus-Universität zu Göttingen*, 10:323–338, 1891. Reprinted in *Abhandlungen* 2, 513–528.
200. G. Frobenius. Über auflösbare Gruppen. *Sitzungsberichte der Akademie der Wiss. zu Berlin*, pages 337–345, 1893. Reprinted in *Abhandlungen* 2, 565–573.
201. G. Frobenius. Antrittsrede. *Sitzungsberichte der Akademie der Wiss. zu Berlin*, pages 368–370, 1893. Reprinted in *Abhandlungen* 2, 574–576.
202. G. Frobenius. Gedächtnisrede auf Leopold Kronecker. *Abhandlungen d. Akad. der Wiss. zu Berlin*, pages 3–22, 1893. Reprinted in *Abhandlungen* 3, 707–724.
203. G. Frobenius. Über die Elementartheiler der Determinanten. *Sitzungsberichte der Akademie der Wiss. zu Berlin*, pages 7–20, 1894. Reprinted in *Abhandlungen* 2, 577–590.
204. G. Frobenius. Über endliche Gruppen. *Sitzungsberichte der Akademie der Wiss. zu Berlin*, pages 81–112, 1895. Reprinted in *Abhandlungen* 2, 632–663.
205. G. Frobenius. Verallgemeinerung des Sylowschen Satzes. *Sitzungsberichte der Akademie der Wiss. zu Berlin*, pages 981–993, 1895. Reprinted in *Abhandlungen* 2, 664–676.
206. G. Frobenius. Über auflösbare Gruppen II. *Sitzungsberichte der Akademie der Wiss. zu Berlin*, pages 1027–1044, 1895. Reprinted in *Abhandlungen* 2, 677–694.
207. G. Frobenius. Zur Theorie der Scharen bilinearer Formen. *Vierteljahrsschrift der Naturforschenden Gesellschaft in Zürich*, 44:20–23, 1896. This is an excerpt from a letter to Weierstrass dated "November 1881". It is not contained in [232].
208. G. Frobenius. Über die cogredienten Transformationen der bilinearen Formen. *Sitzungsberichte der Akademie der Wiss. zu Berlin*, pages 7–16, 1896. Reprinted in *Abhandlungen* 2, 695–704.
209. G. Frobenius. Über vertauschbare Matrizen. *Sitzungsberichte der Akademie der Wiss. zu Berlin*, pages 601–614, 1896. Reprinted in *Abhandlungen* 2, 705–718.
210. G. Frobenius. Über Beziehungen zwischen den Primidealen eines algebraischen Körpers und den Substitutionen seiner Gruppe. *Sitzungsberichte der Akademie der Wiss. zu Berlin*, pages 689–703, 1896. Reprinted in *Abhandlungen* 2, 719–733.
211. G. Frobenius. Über Gruppencharaktere. *Sitzungsberichte der Akademie der Wiss. zu Berlin*, pages 985–1021, 1896. Reprinted in *Abhandlungen* 3, 1–37.
212. G. Frobenius. Über die Primfaktoren der Gruppendeterminante. *Sitzungsberichte der Akademie der Wiss. zu Berlin*, pages 1343–1382, 1896. Reprinted in *Abhandlungen* 3, 38–77.
213. G. Frobenius. Über die Darstellung der endlichen Gruppen durch lineare Substitutionen. *Sitzungsberichte der Akademie der Wiss. zu Berlin*, pages 994–1015, 1897. Reprinted in *Abhandlungen* 3, 82–103.

214. G. Frobenius. Über Relationen zwischen den Charakteren einer Gruppe und denen ihrer Untergruppen. *Sitzungsberichte der Akademie der Wiss. zu Berlin*, pages 501–515, 1898. Reprinted in *Abhandlungen* 3, 104–118.
215. G. Frobenius. Über die Composition der Charaktere einer Gruppe. *Sitzungsberichte der Akademie der Wiss. zu Berlin*, pages 330–339, 1899. Reprinted in *Abhandlungen* 3, 119–128.
216. G. Frobenius. Über die Darstellung der endlichen Gruppen durch lineare Substitutionen II. *Sitzungsberichte der Akademie der Wiss. zu Berlin*, pages 482–500, 1899. Reprinted in *Abhandlungen* 3, 129–147.
217. G. Frobenius. Über die Charaktere der symmetrischen Gruppe. *Sitzungsberichte der Akademie der Wiss. zu Berlin*, pages 516–534, 1900. Reprinted in *Abhandlungen* 3, 148–166.
218. G. Frobenius. Über die Charactere der alternirenden Gruppe. *Sitzungsberichte der Akademie der Wiss. zu Berlin*, pages 303–315, 1901. Reprinted in *Abhandlungen* 3, 167–179.
219. G. Frobenius. Über auflösbare Gruppen III. *Sitzungsberichte der Akademie der Wiss. zu Berlin*, pages 849–875, 1901. Reprinted in *Abhandlungen* 3, 180–188.
220. G. Frobenius. Über auflösbare Gruppen IV. *Sitzungsberichte der Akademie der Wiss. zu Berlin*, pages 1216–1230, 1901. Reprinted in *Abhandlungen* 3, 189–203.
221. G. Frobenius. Über Gruppen der Ordnung $p^\alpha q^\beta$. *Acta Mathematica*, 26:189–198, 1902. Reprinted in *Abhandlungen* 3, 210–219.
222. G. Frobenius. Über die charakteristischen Einheiten der symmetrischen Gruppe. *Sitzungsberichte der Akademie der Wiss. zu Berlin*, pages 328–358, 1903. Reprinted in *Abhandlungen* 3, 244–274.
223. G. Frobenius. Theorie der hyperkomplexen Grössen. *Sitzungsberichte der Akademie der Wiss. zu Berlin*, pages 504–537, 1903. Reprinted in *Abhandlungen* 3, 284–317.
224. G. Frobenius. Theorie der hyperkomplexen Grössen II. *Sitzungsberichte der Akademie der Wiss. zu Berlin*, pages 634–645, 1903. Reprinted in *Abhandlungen* 3, 318–329.
225. G. Frobenius. Über die Charaktere der mehrfach transitiven Gruppen. *Sitzungsberichte der Akademie der Wiss. zu Berlin*, pages 558–571, 1904. Reprinted in *Abhandlungen* 3, 335–348.
226. G. Frobenius. Zur Theorie der linearen Gleichungen. *Jl. für die reine u. angew. Math.*, 129:175–180, 1905. Reprinted in *Abhandlungen* 3, 349–354.
227. G. Frobenius. Über einen Fundamentalsatz der Gruppentheorie II. *Sitzungsberichte der Akademie der Wiss. zu Berlin*, pages 428–437, 1907. Reprinted in *Abhandlungen* 3, 394–403.
228. G. Frobenius. Über Matrizen aus positiven Elementen. *Sitzungsberichte der Akademie der Wiss. zu Berlin*, pages 471–476, 1908. Reprinted in *Abhandlungen* 3, 404–409.
229. G. Frobenius. Über Matrizen aus positiven Elementen II. *Sitzungsberichte der Akademie der Wiss. zu Berlin*, pages 514–518, 1909. Reprinted in *Abhandlungen* 3, 410–414.
230. G. Frobenius. Gegenseitige Reduktion algebraischer Körper. *Math. Ann.*, 70:457–458, 1911. Extract from a letter to H. Weber dated 19 June 1909. Reprinted in *Abhandlungen* 3, 491–492.
231. G. Frobenius. Über Matrizen aus nicht negativen Elementen. *Sitzungsberichte der Akademie der Wiss. zu Berlin*, pages 456–477, 1912. Reprinted in *Abhandlungen* 3, 546–567.
232. G. Frobenius. *Gesammelte Abhandlungen. Herausgegeben von J.-P. Serre*. Springer-Verlag, Berlin, 1968.
233. G. Frobenius and I. Schur. Über die reellen Darstellungen der endlichen Gruppen. *Sitzungsberichte der Akademie der Wiss. zu Berlin*, pages 186–208, 1906. Reprinted in Frobenius, *Abhandlungen* 3, 355–377.
234. G. Frobenius and I. Schur. Über die äquivalenz der Gruppen linearer Substitutionen. *Sitzungsberichte der Akademie der Wiss. zu Berlin*, pages 209–217, 1906. Reprinted in Frobenius, *Abhandlungen* 3, 378–386.
235. G. Frobenius and L. Stickelberger. Über Gruppen von vertauschbaren Elementen. *Jl. für die reine u. angew. Math.*, 86:217–262, 1879. Reprinted in *Abhandlungen* 1, 545–590.
236. L. Fuchs. Zur Theorie der linearen Differentialgleichungen mit veränderlichen Coefficienten. *Jahresbericht über die städtliche Gewerbeschule zu Berlin, Ostern 1865*, 1865. Reprinted in *Werke* 1, 111–158.

237. L. Fuchs. Zur Theorie der linearen Differentialgleichungen mit veränderlichen Coefficienten. *Jl. für die reine u. angew. Math.*, 66:121–160, 1866. Reprinted in *Werke* 1, 159–202.
238. L. Fuchs. Zur Theorie der linearen Differentialgleichungen mit veränderlichen Coefficienten (Ergänzungen zu der im 66sten Bande dieses Journals enthaltenen Abhandlung). *Jl. für die reine u. angew. Math.*, 68:354–385, 1868. Reprinted in *Werke* 1, 205–240.
239. E. Galois. Oeuvres mathématiques d'Évariste Galois. *Jl. de math. pures et appl.*, 11:381–444, 1846. Reprinted as a book, *Oeuvres mathématiques d'Évariste Galois*, Paris, 1897.
240. F. Gantmacher. *Matrix Theory*. AMS Chelsea Publishing, 2000. This work, published in two volumes, is an English translation of Gantmacher's *Teoriya Matrits* (Moscow, 1953). It first appeared in 1959.
241. F. Gantmacher and M. Krein. Sur les matrices complètement non négative et oscillatoires. *Compositio mathematica*, 4:445–476, 1937.
242. F. Gantmacher and M. Krein. *Ozillationsmatrizen, Ozillationskerne und kleine Schwingungen mechanischer Systeme*. Akademie-Verlag, Berlin, 1960. Originally published in Russian (first edn. 1941). German version of the second edition edited by Alfred Stöhr.
243. F. Gassmann. Bemerkungen zur vorstehenden Arbeit. *Mathematische Zeitschrift*, pages 666–668, 1926. Commentary on [305]. Reprinted in Hurwitz's *Werke* 2, 738–739.
244. C. Gauss. *Disquisitiones arithmeticae*. G. Fleischer, Leipzig, 1801. English translation by A. Clark (Yale University Press, New Haven, 1966). In quotations I have followed Clark's translation unless otherwise noted.
245. C. Gauss. Review of [471]. *Göttingische gelehrte Anzeigen*, pages 1025–1038, 1831.
246. C. Gauss. Theoria residuorum biquadraticorum. Commentatio secunda. *Comm. soc. sci. Gottingenesis*, 7:93–149, 1832. Reprinted in *Werke* 2. Citations are to the German translation on pp. 534–586 of [248].
247. C. Gauss. Démonstration de quelques théorèmes concernant les périodes des classes des formes du second degré. *Werke*, 2:266–288, 1863.
248. C. Gauss. *Carl Friedrich Gauss' Untersuchungen über höhere Arithmetik. Deutsch herausgegeben von H. Maser*. Springer, Berlin, 1889.
249. C. F. Geiser. E. B. Christoffel. In *E. B. Christoffel gesammelte mathematische Abhandlungen*, volume 1, pages v–xv. B. G. Teubner, Leipzig, 1910.
250. S. Gelbart. An elementary introduction to the Langlands program. *Bulletin of the American Mathematical Society*, 10(2):177–219, 1984.
251. D. Goldschmidt. A group theoretic proof of the $p^a q^b$ theorem for odd primes. *Mathematische Zeitschrift*, 13:373–375, 1970.
252. D. Gorenstein. *Finite Simple Groups. An Introduction to Their Classification*. Plenum Press, New York and London, 1982.
253. H. Grassmann. *Die lineale Ausdehnungslehre*, volume 1, Pt. 2. Enslin, Berlin, 1862. Reprinted with extensive notes by F. Engel in [254].
254. H. Grassmann. *Hermann Grassmanns gesammelte mathematische und physikalische Werke ... herausgegeben von Friedrich Engel*, volume I_2. Teubner, 1896.
255. J. Gray. *Linear Differential Equations and Group Theory from Riemann to Poincaré*. Birkhäuser, Boston, 2nd edition, 2000.
256. J. A. Green. Richard Dagobert Brauer. *Bulletin London Math. Soc.*, 10:317–342, 1978. Reprinted in Richard Brauer, *Papers* 1, xxii–xliii.
257. J. N. P. Hachette. *Traité des surfaces du second degré*. Klostermann, Paris, 1813.
258. J. Hadamard and M. Fréchet. Sur les probabilités discontinues des événements en châine. *Zeitschrift für angwandte Mathematik und Mechanik*, 13:92–97, 1933. Reprinted in J. Hadamard *Oeuvres* 4, 2083–2088.
259. M. Hamburger. Review of [172]. *Jahrbuch über die Fortschritte der Mathematik*, 3, 1874.
260. M. Hamburger. Zur Theorie der Integration eines Systems von n linearen partiellen Differentialgleichungen erster Ordnung mit zwei unabhängigen und n abhängigen Veränderlichen. *Jl. für die reine u. angew. Math.*, 81:243–280, 1876.
261. M. Hamburger. Über das Pfaff'sche Problem. *Archiv der Mathematik und Physik*, 60:185–214, 1877.

262. W. R. Hamilton. *Lectures on Quaternions*. Hodges and Smith, Dublin, 1853.
263. H. Hasse. Bericht über neuere Untersuchungen und Probleme aus der Theorie der algebraischen Zahlkörper. *Jahresbericht der Deutschen Mathematiker-Vereinigung*, 35, 36, Ergänzungsband 6:1–55; 233–311; 1–204, 1926, 1927, 1930.
264. O. Haupt. *Einführung in die Algebra*, volume 2. Akademische Verlaggesellschaft, Leipzig, 1929. The theory of generalized abelian groups and its application to elementary divisor theory is presented by Krull in an appendix (pp. 617–629).
265. T. Hawkins. *Lebesgue's Theory of Integration: Its Origins and Development*. Univ. of Wisconsin Press, Madison, 1970. 2nd edition, New York (Chelsea), 1975. Reprint of 2nd edition by the American Mathematical Society, 2001.
266. T. Hawkins. The origins of the theory of group characters. *Arch. Hist. Exact. Sci.*, 7:142–170, 1971.
267. T. Hawkins. Hypercomplex numbers, Lie groups and the creation of group representation theory. *Arch. Hist. Exact Sci.*, 8:243–287, 1972.
268. T. Hawkins. New light on Frobenius' creation of the theory of group characters. *Arch. Hist. Exact Sci.*, 12:217–243, 1974.
269. T. Hawkins. Cauchy and the spectral theory of matrices. *Historia Math.*, 2:1–29, 1975.
270. T. Hawkins. Another look at Cayley and the theory of matrices. *Archives internationales d'histoire des sciences*, 26:82–112, 1977.
271. T. Hawkins. The origins of modern theories of integration. In I. Grattan-Guinness, editor, *From the Calculus to Set Theory: 1630–1910*, chapter 4. Duckworth, 1980. Paperback printing by Princeton University Press, 2000.
272. T. Hawkins. Cayley's counting problem and the representation of Lie algebras. In *Proceedings, International Congress of Mathematicians, Berkeley 1986*. American Mathematical Society, Providence, 1987.
273. T. Hawkins. Jacobi and the birth of Lie's theory of groups. *Archive for History of Exact Sciences*, 42:187–278, 1991.
274. T. Hawkins. From general relativity to group representations. The background to Weyl's papers of 1925–26. In *Matériaux pour l'historie des mathématiques au XXe Siècle. Actes du colloque à la mémoire de Jean Dieudonné. Société Mathématique de France, Séminaires et Congrès No. 3*, pages 69–100. Soc. math. de France, Paris, 1998.
275. T. Hawkins. Weyl and the topology of continuous groups. In I. M. James, editor, *History of Topology*, pages 169–198. North-Holland, 1999.
276. T. Hawkins. *Emergence of the Theory of Lie Groups. An Essay on the History of Mathematics 1869–1926*. Springer, New York, 2000.
277. T. Hawkins. Frobenius, Cartan, and the Problem of Pfaff. *Archive for History of Exact Sciences*, 59:381–436, 2005.
278. T. Hawkins. Continued fractions and the origins of the Perron–Frobenius theorem. *Archive for History of Exact Sciences*, 62, 2008.
279. E. Hecke. Über die *L*-Funktionen und den Dirichletschen Primzahlsatz für einen beliebigen Zahlkörper. *Nachrichten von der Königlichen Gesellschaft der Wissenschaften und der Georg-Augustus-Universität zu Göttingen*, pages 299–318, 1917. Reprinted in *Werke*, 178–197.
280. L. Heffter. *Einleitung in die Theorie der linearen Differentialgleichungen mit einer unabhängigen Variable*. B. G. Teubner, Leipzig, 1894.
281. I. Heger. *Auflösung eines Systems von mehren unbestimmten Gleichungen ersten Grades in ganzen Zahlen*. Gerold's Sohn in Comm., Wien, 1858.
282. K. Hensel. Ueber die Elementheiler componirter Systeme. *Jl. für die reine u. angew. Math.*, 114:109–115, 1895.
283. K. Hentzelt and E. Noether. Bearbeitung von K. Hentzelt: zur Theorie der Polynomideale und Resultanten. *Math. Ann.*, 88:53–79, 1923. Reprinted in E. Noether, *Abhandlungen*, 409–435.
284. C. Hermite. Sur une question relative à la théorie des nombres. *Jl. de math. pures et appl.*, 14, 1849. Reprinted in *Oeuvres* 1, 265–273.

285. C. Hermite. Sur l'introduction des variables continues dans la théorie des nombres. *Jl. für die reine u. angew. Math.*, 41:191–216, 1851. Reprinted in *Oeuvres* 1, 164–192.
286. C. Hermite. Sur la décomposition d'un nombre en quatre carrés. *Comptes Rendus, Acad. Sci. Paris*, 37:133–134, 1853. Reprinted in *Oeuvres* 1, 288–289.
287. C. Hermite. Remarques sur un mémoire de M. Cayley relatif aux déterminants gauches. *Cambr. and Dublin Math. Journal*, 9:63–67, 1854. Reprinted in *Oeuvres* 1, 290–295.
288. C. Hermite. Sur la théorie des formes quadratiques. Premier mémoire. *Jl. für die reine u. angew. Math.*, 47:313–342, 1854. Reprinted in *Oeuvres* 1, 200–233.
289. C. Hermite. Sur la théorie des formes quadratiques. Seconde mémoire. *Jl. für die reine u. angew. Math.*, 47:343–368, 1854. Reprinted in *Oeuvres* 1, 234–263.
290. C. Hermite. Remarque sur un théorème de M. Cauchy. *Comptes Rendus, Acad. Sci. Paris*, 41:181–183, 1855. Reprinted in *Oeuvres* 1, 459–481.
291. C. Hermite. Sur la théorie de la transformation des fonctions abéliennes. *Comptes Rendus, Acad. Sci. Paris*, 40, 1855. Reprinted in *Oeuvres* 1, 444–477.
292. C. Hermite. Note de M. Hermite. In *Traité élémentaire de calcul différentiel et de calcul intégral par S-F. Lacroix*, volume 2, pages 365–491. Mallet–Bachelier, Paris, 6th. edition, 1862.
293. C. Hermite. Extrait d'une lettre de M. Ch. Hermite sur la transformation des formes quadratiques en elles-mêmes. *Jl. für die reine u. angew. Math.*, 78:325–328, 1874. Reprinted in *Oeuvres* 3, 185–189.
294. I. N. Herstein. *Topics in Algebra*. Wiley, New York, 2nd. edition, 1975.
295. D. Hilbert. Grundzüge einer Theorie des Galoisschen Zahlkörpers. *Nachrichten von der Königlichen Gesellschaft der Wissenschaften und der Georg-Augustus-Universität zu Göttingen*, pages 224–236, 1894.
296. D. Hilbert. Die Theorie der algebraischen Zahlkörper. *Jahresbericht der Deutschen Mathematiker-Vereinigung*, 4:175–546, 1897. Reprinted in *Abhandlungen* 1, 63–363.
297. O. Hölder. Zurückführung einer beliebigen algebraischen Gleichung auf eine Kette von Gleichungen. *Math. Ann.*, 34, 1889.
298. R. A. Howard, editor. *Dynamic Probabilistic Systems. Vol I: Markov Models*. Wiley, New York, 1971.
299. G. Humbert. Sur les fonctions abéliennes singulières (Premier Mémoire). *Jl. de math. pures et appl.*, (5)5:233–350, 1899.
300. G. Humbert. Sur les fonctions abéliennes singulières (Deuxième Mémoire). *Jl. de math. pures et appl.*, (5)6:279–386, 1900.
301. A. Hurwitz. Ueber die Perioden solcher eindeutiger, $2n$-fach periodischer Functionen, welche im Endlichen überall den Charakter rationaler Functionen besitzen und reell sind für reelle Werthe ihrer n Argumente. *Jl. für die reine u. angew. Math.*, 4:1–20, 1883. Reprinted in *Werke* 1, 99–118.
302. A. Hurwitz. Ueber diejenigen algebraische Gebilde, welche eindeutige Transformationen in sich zulassen. *Math. Ann.*, 32:290–308, 1888. Reprinted in *Werke* 1, 241–259.
303. A. Hurwitz. Zur Invariantentheorie. *Math. Ann.*, 45:381–404, 1894. Reprinted in *Werke* 2, 508–532.
304. A. Hurwitz. Über die Erzeugung der Invarianten durch Integration. *Nachrichten von der Königlichen Gesellschaft der Wissenschaften und der Georg-Augustus-Universität zu Göttingen*, pages 71–90, 1897. Reprinted in *Werke* 2, 546–564.
305. A. Hurwitz. Über Beziehungen zwischen den Primidealen eines algebraischen Körpers und den Substitutionen seiner Gruppe. *Mathematische Zeitschrift*, 25:661–665, 1926. Published posthumously from Hurwitz's diary by F. Gassmann and accompanied by his notes [243]. Reprinted (with Gassmann's notes) in *Werke* 2, 733–739.
306. K. Ireland and M. Rosen. *A Classical Introduction to Modern Number Theory*. Springer-Verlag, New York, 1982.
307. K. Itô, editor. *Encyclopedic Dictionary of Mathematics*, volume 3. MIT Press, 1987.
308. C. G. J. Jacobi. Über die Integration der partiellen Differentialgleichungen erster Ordnung. *Jl. für die reine u. angew. Math.*, 2:317–329, 1827. Reprinted in *Werke* 4, 1–15.

309. C. G. J. Jacobi. Über die Pfaffsche Methode, eine gewöhnliche lineare Differentialgleichung zwischen $2n$ Variabeln durch ein System von n Gleichungen zu integriren. *Jl. für die reine u. angew. Math.*, 2:347–357, 1827. Reprinted in *Werke* 4, 19–29.
310. C. G. J. Jacobi. De binis quibuslibet functionibus homogeneis secundi ordinis per substitutiones lineares in alias binas transformandis, quae solis quadratis variabilium constant; .... *Jl. für die reine u. angew. Math.*, 12:1–69, 1834. Reprinted in *Werke* 3, 191–268.
311. C. G. J. Jacobi. De formatione et proprietatibus determinantium. *Jl. für die reine u. angew. Math.*, 22:285–318, 1841. Reprinted in *Werke* 3, 255–392.
312. C. G. J. Jacobi. De determinantibus functionalibus. *Jl. für die reine u. angew. Math.*, 22:319–359, 1841. Reprinted in *Werke* 3, 393–438.
313. C. G. J. Jacobi. Theoria novi multiplicatoris systemati aequationum differentialum vulgarium applicandi. *Jl. für die reine u. angew. Math.*, 27 & 29:199–268 & 213–279; 333–376, 1845. Reprinted in *Werke* 4, 317–509.
314. C. G. J. Jacobi. Über eine elementare Transformation eines in Bezug auf jedes von zwei Variablen-Systemen linearen und homogenen Ausdrucks. *Jl. für die reine u. angew. Math.*, 53:265–270, 1857. Reprinted in *Werke* 3, 583–590.
315. C. G. J. Jacobi. Nova methodus, aequationes differentiales partiales primi ordinis inter numerum variabilium quemcumque propositas integrandi. *Jl. für die reine u. angew. Math.*, 60:1–181, 1862. Reprinted in *Werke* 4, 3–189. Translated into German and annoted by G. Kowalewski as *Ostwald's Klassiker der exakten Wissenschaften*, Nr. 156 (Leipzig, 1906).
316. C. G. J. Jacobi. Über die Auflösung der Gleichung $\alpha_1 x_1 + \alpha_2 x_2 + \cdots + \alpha_n x_n = fu$. *Jl. für die reine u. angew. Math.*, 69:1–28, 1868. Published posthumously by E. Heine. Reprinted in *Werke* 6, 355–384.
317. C. G. J. Jacobi. Allgemeine Theorie der kettenbruchähnlichen Algorithmen, in welchen jede Zahl aus drei vorhergehenden gebildet wird. *Jl. für die reine u. angew. Math.*, 69:29–64, 1868. Published posthumously by E. Heine. Reprinted in *Werke* 6, 385–426.
318. M. Jammer. *The Conceptual Development of Quantum Mechanics*. McGraw–Hill, New York, 1966.
319. F.-W. Janssen, editor. *Elwin Bruno Christoffel. Gedenkschrift zur 150. Wiederkehr des Geburtstages*. Kreis Aachen, 1979. Separate printing of *Heitmatblätter des Kreises Aachen*, Heft 3–4 (1978) and Heft 1 (1979).
320. C. Jordan. Mémoire sur la résolution algébrique des équations. *Journal de math. pures et appl.*, 12(2):109–157, 1867. Reprinted in *Oeuvres* 1, 109–157.
321. C. Jordan. Sur la résolution algébrique des équations primitives de degré p^2 (p étant premier impair). *Journal de math. pures et appl.*, 13:111–135, 1868. Reprinted in *Oeuvres* 1, 171–202.
322. C. Jordan. *Traité des substitutions et des équations algébriques*. Gauthier–Villars, Paris, 1870.
323. C. Jordan. Sur la résolution des équations différentielles linéaires. *Comptes Rendus, Acad. Sci. Paris*, 73:787–791, 1871. Reprinted in *Oeuvres* 4, 313–317.
324. C. Jordan. Sur les polynômes bilinéaires. *Comptes Rendus, Acad. Sci. Paris*, 77:1487–1491, 1873. Presented December 22, 1873. Reprinted in *Oeuvres* 3, 7–11.
325. C. Jordan. Sur la réduction des formes bilinéaires. *Comptes Rendus, Acad. Sci. Paris*, 78: 614–617, 1874. Presented March 2, 1874. Reprinted in *Oeuvres* 3, 13–16.
326. C. Jordan. Sur les systèmes de formes quadratiques. *Comptes Rendus, Acad. Sci. Paris*, 78:1763–1767, 1874. Reprinted in *Oeuvres* 3, 17–21.
327. C. Jordan. Mémoire sur les formes bilinéaires. *Jl. de math. pures et appl.*, 19:35–54, 1874. Submitted in August, 1873. Reprinted in *Oeuvres* 3, 23–42.
328. C. Jordan. Mémoire sur les équations différentielles linéaires à intégrale algébrique. *Jl. für die reine u. angew. Math.*, 84:89–215, 1878. Reprinted in *Oeuvres* 2, 13–139.
329. C. Jordan. Observations sur la réduction simultanée de deux formes bilinéaires. *Comptes Rendus, Acad. Sci. Paris*, 92:1437–1438, 1881. Reprinted in *Oeuvres* 1, 189.
330. C. Jordan. *Cours d'analyse de l'École Polytechnique*, volume 3. Gauthier–Villars, Paris, 1887.
331. A. Joseph et al., editors. *Studies in Memory of Issai Schur*. Progress in Mathematics, Vol. 210. Birkhäuser Boston, Springer-Verlag, New York, 2003.

332. E. Kähler. *Einführung in die Theorie der Systeme von Differentialgleichungen*. Teubner, Leipzig, 1934.
333. N.F. Kanounov. O rabotakh F.E. Molina po teorii predstavlenii konechnykh grupp. *Istoriya i metodologiya estestvennykh nauk*, 17:57–88, 1966.
334. N.F. Kanounov. O rabotakh F.E. Molina po teorii predstavlenii konechnykh grupp. *Istoriya i metodologiya estestvennykh nauk*, 11:56–68, 1971.
335. N.F. Kanounov. *Fedor Eduardovich Molin*. Moscow, 1983.
336. V. Katz. The history of differential forms from Clairaut to Poincaré. *Historia Mathematica*, 8:161–188, 1981.
337. V. Katz. Differential forms—Cartan to De Rham. *Archive for History of Exact Sciences*, 33:321–336, 1985.
338. J. Kaucký. Remarques à la note de M. V. Romanovsky. *Comptes Rendus, Acad. Sci. Paris*, 191:919–921, 1930.
339. C. H. Kimberling. Emmy Noether. *American Math. Monthly*, 79:136–149, 1972.
340. F. Klein. Über binäre Formen mit linearen Transformationen in sich selbst. *Math. Ann.*, 9: 183–208, 1875. Reprinted in *Abhandlungen* 2, 275–301.
341. F. Klein. *Vorlesungen über das Ikosaeder*. Teubner, Leipzig, 1884.
342. F. Klein. Über hyperelliptische Sigmafunktionen (Erster Aufsatz). *Math. Ann.*, 27, 1886. Reprinted, with notes, in *Abhandlungen* 3, 323–356.
343. F. Klein. *Lectures on Mathematics*. Macmillan, New York and London, 1894.
344. F. Klein. Über einen Satz aus der Theorie der endlichen (discontinuirlichen) Gruppen linearer Substitutionen beliebig vieler Veränderlicher. *Jahresbericht der Deutschen Mathematiker-Vereinigung*, 5:57, 1897.
345. F. Klein. *Vorlesungen über die Entwicklung der Mathematik im 19. Jahrhundert*, volume I. Springer, Berlin, 1926. Reprinted together with [346] by Chelsea, New York, 1956.
346. F. Klein. *Vorlesungen über die Entwicklung der Mathematik im 19. Jahrhundert*, volume II. Springer, Berlin, 1927. Reprinted together with [345] by Chelsea, New York, 1956.
347. F. Klein. *Development of Mathematics in the 19th Century. Translated by M. Ackerman*. Math Sci Press, Brookline, Mass., 1979. Translation of [345, 346].
348. K. Knopp. *Theory of Functions, Part II*. Dover, New York, 1947. Translated from the 4th German edition.
349. C. G. Knott. *Life and Scientific Work of Peter Guthrie Tait*. Cambridge University Press, Cambridge, 1911.
350. A. Krazer. *Lehrbuch der Thetafunktionen*. Teubner, Leipzig, 1903. Reprinted by Chelsea Publishing Company (New York, 1970).
351. L. Kronecker. Über die elliptischen Functionen, für welche complexe Multiplication stattfindet. *Monatsberichte der Akademie der Wiss. zu Berlin*, pages 455–460, 1857. Reprinted in *Werke* 3, 177–183.
352. L. Kronecker. Zwei Sätze über Gleichungen mit ganzzahligen Coefficienten. *Jl. für die reine u. angew. Math.*, 53:173–175, 1857. Reprinted in *Werke* 1, 103–108.
353. L. Kronecker. Über bilineare Formen. *Monatsberichte der Akademie der Wiss. zu Berlin*, 1:145–162, 1866. Reprinted in *Jl. für die reine u. angew. Math.*, 68:273–285 and in *Werke* 1, 145–162.
354. L. Kronecker. Über Schaaren quadratischer Formen. *Monatsberichte der Akademie der Wiss. zu Berlin*, pages 339–346, 1868. Reprinted in *Werke* 1, 163–174. The above title for the work was added by the editor (K. Hensel). See *Werke* 1, 163n. 1.
355. L. Kronecker. Auseinandersetzung einiger eigenschaften der Klassenzahl idealer complexer Zahlen. *Monatsberichte der Akademie der Wiss. zu Berlin*, pages 881–889, 1870. Reprinted in *Werke* 1, 273–382.
356. L. Kronecker. Über Schaaren von quadratischen und bilinearen Formen. *Monatsberichte der Akademie der Wiss. zu Berlin*, pages 59–76, 1874. Presented Jan 19, 1874. Reprinted in *Werke* 1, 349–372.
357. L. Kronecker. Über Schaaren von quadratischen und bilinearen Formen. *Monatsberichte der Akademie der Wiss. zu Berlin*, pages 149–156, 1874. Presented Feb 16, 1874. Reprinted in *Werke* 1, 373–381.

358. L. Kronecker. Über Schaaren von quadratischen und bilinearen Formen. *Monatsberichte der Akademie der Wiss. zu Berlin*, pages 206–232, 1874. Presented March 16, 1874. Reprinted in *Werke* 1, 382–413.
359. L. Kronecker. Über die congruenten Transformation der bilinearen Formen. *Monatsberichte der Akademie der Wiss. zu Berlin*, pages 397–447, 1874. Presented April 23, 1874. Reprinted in *Werke* 1, 423–483.
360. L. Kronecker. Sur les faisceaux de formes quadratiques et bilinéaires. *Comptes Rendus, Acad. Sci. Paris*, 78:1181–1182, 1874. Presented April 27, 1874. Reprinted in *Werke* 1, 417–419.
361. L. Kronecker. Über Abelsche Gleichungen (Auszug aus der am 16. April 1877 gelesenen Abhandlung). *Monatsberichte der Akademie der Wiss. zu Berlin*, pages 845–851, 1877. Reprinted in *Werke* 4, 63–71.
362. L. Kronecker. Über die Irreducibilität von Gleichungen. *Monatsberichte der Akademie der Wiss. zu Berlin*, pages 155–162, 1880. Reprinted in *Werke* 2, 83–93.
363. L. Kronecker. Grundzüge einer arithmetischen Theorie der allgebraischen Grössen. (Abdruck einer Festschrift zu Herrn E. E. Kummers Doctor-Jubiläum, 10 September 1881.) *Jl. für die reine u. angew. Math.*, 92:1–122, 1882. Reprinted in *Werke* 2, 237–387.
364. L. Kronecker. Zur arithmetischen Theorie der algebraischen Formen. *Jl. für die reine u. angew. Math.*, 93:365–366, 1882. Reprinted in *Werke* 2, 397–401.
365. L. Kronecker. Die Zerlegung der ganzen Grössen eines natürlichen Rationalitäts-Bereichs in ihre irreductibeln Factoren. *Jl. für die reine u. angew. Math.*, 94:344–348, 1883. Reprinted in *Werke* 2, 409–416.
366. L. Kronecker. Über die Composition der Systeme von n^2 Grössen mit sich selbst. *Sitzungsberichte der Akademie der Wiss. zu Berlin*, pages 1081–1088, 1890. Reprinted in *Werke* 3_1, 463–473.
367. L. Kronecker. Algebraische Reduction der Schaaren bilinearer Formen. *Sitzungsberichte der Akademie der Wiss. zu Berlin*, pages 1225–1237, 1890. Reprinted in *Werke* 3_2, 141–155.
368. L. Kronecker. Algebraische Reduction der Schaaren quadratischer Formen. *Sitzungsberichte der Akademie der Wiss. zu Berlin*, pages 1375–1388, 1890. Reprinted in *Werke* 3_2, 159–174.
369. L. Kronecker. Algebraische Reduction der Schaaren quadratischer Formen. *Sitzungsberichte der Akademie der Wiss. zu Berlin*, pages 9–17, 34–44, 1891. Reprinted in *Werke* 3_2, 175–198.
370. L. Kronecker. Reduction der Systeme von n^2 ganzahlige Elementen. *Jl. für die reine u. angew. Math.*, 107:135–136, 1891. Reprinted in *Werke* 4, 123–124.
371. L. Kronecker. Auszug aus einem briefe von L. Kronecker an R. Dedekind. *Sitzungsberichte der Akademie der Wiss. zu Berlin 1895*, pages 115–117, 1895. Reprinted in *Werke* 5, 453–457.
372. W. Krull. Über Begleitmatrizen und Elementartheiler. Inauguraldissertation Universität Freiburg i. Br., 1921. First published in Krull's *Abhandlungen* 1 (1999), 55–95.
373. W. Krull. Algebraische Theorie der Ringe. I. *Math. Ann.*, 88:80–122, 1922. Reprinted in *Abhandlungen* 1, 80–122.
374. W. Krull. Algebraische Theorie der Ringe. II. *Math. Ann.*, 91:1–46, 1924. Reprinted in *Abhandlungen* 1, 166–211.
375. W. Krull. Über verallgemeinerte endliche Abelsche Gruppen. *Mathematische Zeitschrift*, 23:161–196, 1925. Reprinted in *Papers* 1, 263–298.
376. W. Krull. Theorie und Anwendung der verallgemeinten Abelschen Gruppen. *S'ber. Akad. Wiss. Heidelberg, Math.-Natur. Kl.*, 1:1–10, 1926. Reprinted in *Papers* 1, 299–328.
377. M. Kuga. *Galois' Dream: Group Theory and Differenial Equations*. Birkhäuser, Boston, 1993. Translated from the Japanese by Susan Addington and Motohico Mulase.
378. E. E. Kummer. De numeris complexis, qui radicibus unitatis et numeris integris realibus constant. *Gratulationsschrift der Univ. Breslau in Jubelfeier der Univ. Königsberg*, 1844. Reprinted in *Jl. de math. pures et appl.*, 12: 185–212, 1847, and in *Papers* 1, 165–192.
379. E. E. Kummer. Zur Theorie der Complexen Zahlen. *Monatsberichte der Akademie der Wiss. zu Berlin*, pages 87–96, 1846. Reprinted in *Jl. für die reine u. angew. Math.*, 35: 319–326, 1847, and in *Papers* 1, 203–210.

380. E. E. Kummer. Über die Zerlegung der aus Wurzeln der Einheit gebildeten complexen Zahlen in ihre Primfactoren. *Jl. für die reine u. angew. Math.*, 35:327–367, 1847. Reprinted in *Papers* 1, 211–251.
381. E. E. Kummer. Beweis des Fermat'schen Satzes der Unmöglichkeit von $x^\lambda + y^\lambda = z^\lambda$ für eine unendliche Anzahl Primzahlen λ. *Monatsberichte der Akademie der Wiss. zu Berlin*, pages 132–141, 305–319, 1847. Reprinted in *Papers* 1, 274–297.
382. E. E. Kummer. Allgemeiner Beweis des Fermat'schen Satzes, dass die Gleichung $x^\lambda + y^\lambda = z^\lambda$ durch ganze Zahlen unlösbar ist, für alle diejenigen Potenz-Exponenten, welche ungerade Primzahlen sind und in den Zählern der ersten $\frac{1}{2}(\lambda - 3)$ Bernoullischen Zahlen als Factoren nicht vorkommen. *Jl. für die reine u. angew. Math.*, 40:130–138, 1850. Reprinted in *Papers* 1, 336–344.
383. E. E. Kummer. Einige Sätze über die aus Wurzeln der Gleichung $\alpha^\lambda = 1$ gebildeten complexen Zahlen, für den Fall, dass die Klassenzahl durch λ theilbar ist, nebst Anwendung derselben auf einen weiteren Beweis des letzten Fermat'schen Lehrsatzes. *Abhandlungen d. Akad. der Wiss. zu Berlin*, pages 41–74, 1857. Reprinted in *Papers* 1, 639–672.
384. E. E. Kummer. Über die allgemeinen Reciprocitätsgesetze unter den Resten und Nichtresten der Potenzen, deren Grad eine Primzahl ist. *Abhandlungen d. Akad. der Wiss. zu Berlin*, pages 19–159, 1859. Reprinted in *Papers* 1, 699–839.
385. E. E. Kummer. Über eine Eigenschaft der Einheiten der aus den Wurzeln der Gleichung $\alpha^\lambda = 1$ gebildeten complexen Zahlen, und über den zweiten Factor der Klassenzahl. *Monatsberichte der Akademie der Wiss. zu Berlin*, pages 855–880, 1870. Reprinted in *Papers* 1, 919–944.
386. E. E. Kummer. *Collected Papers*, volume 1. Springer-Verlag, Berlin, Heidelberg, New York, 1975. A. Weil, ed.
387. J. L. Lagrange. Solution de différents problèmes de calcul intégral.... *Misc. Taurinensia*, 1766. Reprinted in *Oeuvres* 1, 471–668.
388. J. L. Lagrange. Nouvelle solution du problème du mouvement de rotation d'un corps de figure quelconque qui n'est animé par aucune force accélératrice. *Nouv. Mém. de l'acad. des Sciences de Berlin, 1773*, 1775. Reprinted in *Oeuvres* 3, 577–616.
389. J. L. Lagrange. Recherches sur les équations séculaires des mouvements des noeuds, et des inclinaisons des orbites des planètes. *Hist. de l'Acad. des Sciences, 1774*, 1778. Reprinted in *Oeuvres* 6, 635–709.
390. J. L. Lagrange. Sur différentes questions d'analyse relatives à la théorie des integrales particulières. *Nouv. Mém. Acad. Sci. Berlin*, 1781. Reprinted in *Oeuvres* 4, 585–634.
391. J. L. Lagrange. *Méchanique analitique*. La Veuve Desaint, Paris, 1788. The 4th edition is reprinted in *Oeuvres* 11. Sections V and VI of Part II of the 1st ed. correspond to Sections VI and IX of Part II in later edns.
392. J. L. Lagrange. *Mécanique analytique*, volume 1. Mallet–Bachelier, Paris, 3rd. edition, 1853. The 4th edn. is reprinted in *Oeuvres* 11.
393. E. Laguerre. Sur le calcul des systèmes linéaires. *Journal École Polytechnique*, 62, 1867. Reprinted in *Oeuvres* 1, 221–267.
394. E. Landau. Ein Satz über die Zerlegung homogener linearer Differentialausdrücke in irreducible Factoren. *Jl. für die reine u. angew. Math.*, 124, 1902.
395. S. Lang. *Introduction to Algebraic and Abelian Functions*. Springer, New York, 2nd. edition, 1982.
396. R. Langlands. Representation theory: Its rise and its role in number theory. In *Proceedings of the Gibbs Symposium Yale University, May 15–17, 1989*, pages 181–210. American Mathematical Society, 1990.
397. A. Langville and C. Meyer. *Google's PageRank and Beyond. The Science of Search Engine Rankings*. Princeton University Press, Princeton, 2006. Paperback reprint, 2012. See also the informative book review by Fernández [163].
398. P. S. Laplace. Mémoire sur les solutions particulières des équations différentielles et sur les inégalités séculaires des planètes. *Mém. de l'Acad. des Sciences de Paris 1775*, 1775. Reprinted in *Oeuvres* 8, 325–366.

399. P. S. Laplace. Recherches sur le calcul intégral et sur le système du monde. *Mém. de l'Acad. des Sciences de Paris 1772*, 1776. Reprinted in *Oeuvres* 8, 369–501.
400. P. S. Laplace. Mémoire sur les inégalités séculaires des planètes et des satellites. *Mém. de l'Acad. des Sciences de Paris 1784*, 1787. Reprinted in *Oeuvres* 11, 49–92.
401. P. S. Laplace. Mémoire sur les variations séculaires des orbites des planètes. *Mém. de l'Acad. des Sciences de Paris 1787*, 1789. Reprinted in *Oeuvres* 11, 295–306.
402. P. S. Laplace. *Traité de mécanique céleste*, volume 1. J. B. M. Duprat, Reprinted in *Oeuvres* 1, 1799.
403. P. S. Laplace. *Exposition du système du monde*. Bachelier, 6th. edition, 1835. Reprinted in *Oeuvres* 6.
404. V. A. Lebesgue. Thèses de mécanique et d'astronomie. *Jl. de math. pures et appl.*, 2:337–355, 1837.
405. W. Ledermann. Reduction of singular pencils of matrices. *Proceedings Edinburgh Mathematical Society*, (2) 4:92–105, 1935.
406. W. Ledermann. Issai Schur and his school in Berlin. *Bull. London Math. Soc.*, 15:97–106, 1983.
407. W. Ledermann and P. Neumann. The life of Issai Schur through letters and other documents. In *Studies in Memory of Issai Schur*, Progress in Mathematics, Vol. 210, pages xlv–xc. Birkhäuser Boston, Boston, 2003.
408. S. Lefschetz. On certain numerical invariants of algebraic varieties with application to abelian varieties. *Transactions of the American Mathematical Society*, 22:327–482, 1921. Reprinted in [413], pp. 41–196.
409. S. Lefschetz. Sur le théorème d'existence des fonctions abéliennes. *Rendiconti della R. Accademia dei Lincei*, pages 48–50, 1921.
410. S. Lefschetz. *L'analysis situs et la géométrie algébrique*. Gauthier–Villars, Paris, 1924. Reprinted in [413, pp. 283–439].
411. S. Lefschetz. XV. Transcendental Theory; XVI. Singular Correspondences; XVII. Hyperelliptic Surfaces and Abelian Varieties. In *Selected Topics in Algebraic Geometry. Report of the Committee on Rational Transformations*, volume 63 of *Bulletin of the National Research Council*. Washington, 1928. Reprinted in [542, pp. 310–395].
412. S. Lefschetz. A page of mathematical autobiography. *Bulletin of the American Mathematical Society*, 74:854–879, 1968. Reprinted in [413, pp.13–38].
413. S. Lefschetz. *Selected Papers*. Chelsea, New York, 1971.
414. F. Lemmermeyer. *Reciprocity Laws. From Euler to Eisenstein*. Springer, Berlin, 2000.
415. S. Lie. Theorie des Pfaffschen Problems I. *Archiv for Mathematik*, 2:338–379, 1877. Reprinted in *Abhandlungen* 3, 320–351.
416. S. Lie. Über irreduzible Berührungstransformationsgruppen. *Berichte über d. Verh. d. Sächsischen Gesell. der Wiss. math.-phys. Klasse, 1889*, pages 320–327, 1889. Reprinted in *Abhandlungen* 6, 260–266.
417. R. Lipschitz. Untersuchungen in Betreff der ganzen homogenen Functionen von *n* Differentialen. *Jl. für die reine u. angew. Math.*, 70:71–102, 1869.
418. R. Lipschitz. Beweis eines Satzes aus der Theorie der Substitutionen. *Acta Mathematica*, 10:137–144, 1887.
419. A. Loewy. Sur les formes définies à indéterminées conjuguées de M. Hermite. *Comptes Rendus Acad. Sci. Paris*, 123:168–171, 1896.
420. A. Loewy. Über bilineare Formen mit conjugirt imäginaren Variabeln. *Math. Ann.*, 50:557–576, 1898.
421. A. Loewy. Ueber die irreducible Factoren eines linearen homogenen Differentialausdruckes. *Berichte über d. Verh. d. Sächsischen Gesell. der Wiss., math. -phys. Klasse*, pages 1–13, 1902.
422. A. Loewy. Über die Reducibilität der Gruppen linearer homogener Substitutionen. *Trans. American Math. Soc.*, 4:44–64, 1903.
423. A. Loewy. Über die Reducibilität der reellen Gruppen linearer homogener Substitutionen. *Trans. American Math. Soc.*, 4:171–177, 1903.

424. A. Loewy. Über reduzible lineare homogene Differentialgleichungen. *Math. Ann.*, 56:549–584, 1903.
425. A. Loewy. Kombinatorik, Determinanten und Matrices. In P. Epstein and H. E. Timmerding, editors, *Repertorium der höheren Mathematik*, volume 1, chapter 2. Leipzig and Berlin, 1910.
426. A. Loewy. Über lineare homogene Differentialsysteme und ihre Sequenten. *Sitzungsberichte der Heidelberger Akademie der Wissenschaften, math.–naturwis. Kl., Abt. A*, Abhandlung 17, 1913.
427. A. Loewy. Die Begleitmatrix eines linearen homogenen Differentialausdruckes. *Nachrichten von der Königlichen Gesellschaft der Wissenschaften und der Georg-Augustus-Universität zu Göttingen*, pages 255–263, 1917.
428. A. Loewy. Über Matrizen- und Differentialkomplexe. *Math. Ann.*, 78:1–51, 1917.
429. A. Loewy. Begleitmatrizen und lineare homogene Differentialausdrücke. *Mathematische Zeitschrift*, 7:58–125, 1920.
430. C. C. MacDuffee. *The Theory of Matrices*. Springer, Berlin, 1933.
431. A. A. Markov. Rasprostranenie predel'nykh teorem ischisleniya veroyatnostei na summu velichin svyazannykh v tsep'. *Zap. (Mem.) Imp. Akad Nauk St. Peterb. Fiz.–Mat. Ser. 8, No. 3*, 1908. German translation by H. Liebmann on pp. 272–298 of [432]. English translation by G. Petelin on pp. 552–575 of [298].
432. A. A. Markov. *Warscheinlichkeitsrechnung*. B. G. Teubner, Leipzig, 1912.
433. A. I. Markushevich. *Introduction to the Classical Theory of Abelian Functions*. Translations of Mathematical Monographs. American Mathematical Society, Providence, RI, 1992. Translation of the Russian edition (Moscow, 1979).
434. H. Maschke. Über den arithmetischen Charakter der Coefficienten der Substitutionen endlicher linearer Substitutionsgruppen. *Math. Ann.*, 50:492–498, 1898.
435. H. Maschke. Beweis des Satzes, dass diejenigen endlichen linearen Substitutionsgruppen, in welchen einige durchgehends verschwindende Coefficienten auftreten, intransitiv sind. *Math. Ann.*, 52:363–368, 1899.
436. H. Matsuyama. Solvability of groups of order $2^a p^b$. *Osaka Jl. of Math.*, 10:375–378, 1973.
437. L. Maurer. *Zur Theorie der linearen Substitutionen. Inauguraldissertation Strassburg*. R. Schultz, Strassburg, 1887.
438. A. Mayer. Über unbeschränkt integrable Systeme von linearen totalen Differentialgleichungen und die simultane Integration linearer partieller Differentialgleichungen. *Math. Ann.*, 5:448–470, 1872.
439. A. Mayer. Review of [179]. *Jahrbuch über die Fortschritte der Mathematik*, 9:249–254, 1880.
440. U. Merzbach. Robert Remak and the estimation of units and regulators. In S. S. Demidov et al, editor, *Amphora: Festschrift für Hans Wußing zu seinem 65. Geburtstag*, pages 481–552. Birkhäuser, Berlin, 1992.
441. G. Mittag-Leffler. Weierstrass et Sonja Kowalewsky. *Acta mathematica*, 39:133–198, 1923.
442. K. Miyake. A note on the arithmetical background to Frobenius' theory of group characters. *Expositiones mathematicae*, 7:347–358, 1989.
443. T. Molien. Über Systeme höherer complexer Zahlen. *Math. Ann.*, 41:83–156, 1893.
444. T. Molien. Eine Bemerkung zur Theorie der homogenen Substitutionsgruppen. *Sitzungsberichte der Naturforscher-Gesellschaft b.d. Univ Jurjeff (Dorpat)*, 11:259–274, 1897.
445. T. Molien. Über die Anzahl der Variabeln einer irreductiblen Substitutionsgruppe. *Sitzungsberichte der Naturforscher-Gesellschaft b.d. Univ. Jurjeff (Dorpat)*, 11:277–288, 1897.
446. T. Molien. Über die Invarianten der linearen Substitutionsgruppen. *Sitzungsberichte der Akademie der Wiss. zu Berlin 1897*, pages 1152–1156, 1898.
447. G. Monge and J. N. P. Hachette. Application d'algèbre à la géométrie. *Journal Ec. Poly.*, t. 4, cah. 11:143–169, 1802.
448. E. H. Moore. A universal invariant for finite groups of linear substitutions: with applications to the theory of the canonical form of a linear substitution of finite period. *Math. Ann.*, 50:213–219, 1898.
449. T. Muir. *The Theory of Determinants in the Historical Order of Development*, volume 1–4. Macmillan, London and New York, 2nd. edition, 1911–1923.

450. P. Muth. *Theorie und Anwendung der Elementartheiler*. Teubner, Leipzig, 1899.
451. L. Natani. Über totale und partielle Differentialgleichungen. *Jl. für die reine u. angew. Math.*, 58:301–328, 1861.
452. E. Netto. Neuer Beweis eines Fundamentaltheorems aus der Theorie der Substitutionslehre. *Math. Ann.*, 13:249–250, 1877.
453. E. Netto. *Die Substitutionentheorie und ihrer Anwendung auf die Algebra*. Teubner, Leipzig, 1882.
454. E. Netto. Untersuchungen aus der Theorie der Substitutionen-Gruppen. *Jl. für die reine u. angew. Math.*, 103:321–336, 1888.
455. E. Noether. Idealtheorie in Ringbereichen. *Math. Ann.*, 83:24–66, 1921. Reprinted in *Abhandlungen*, 354–366.
456. E. Noether. Abstrakter Aufbau der Idealtheorie im algebraischen Zahlkörper. *Jahresbericht der Deutschen Mathematiker-Vereinigung*, 33:102, 1924. Reprinted in *Abhandlungen*, p. 102.
457. E. Noether. Abstrakter Aufbau der Idealtheorie in algebraischen Zahl- und Funktionenkörpern. *Math. Ann.*, 96:26–61, 1926–27. Reprinted in *Abhandlungen*, 493–528.
458. E. Noether. Hypercomplexe Grössen und Darstellungstheorie. *Mathematische Zeitschrift*, 30:641–692, 1929. Reprinted in *Abhandlungen*, 563–614.
459. E. Noether and W. Schmeidler. Moduln in nichtkommutativen Bereichen, insbesondere aus Diffzenzausdrücken. *Mathematische Zeitschrift*, 8:1–35, 1920. Reprinted in *Abhandlungen*, 318–352.
460. E. Ostenc. Sur les zéros des matrices stochastiques. *Comptes Rendus, Acad. Sci. Paris*, 196:150–151, 1933.
461. K. H. Parshall. Joseph H.M. Wedderburn and the structure theory of algebras. *Archive for History of Exact Sciences*, 32:223–349, 1985.
462. K. H. Parshall and D. Rowe. *The Emergence of the American Mathematical Research Community, 1876–1900: J. J. Sylvester, Felix Klein, and E. H. Moore*. History of Mathematics, Vol. 8. American Mathematical Society, 1994.
463. M. Pasch. Peter Muth. *Jahresbericht der Deutschen Mathematiker-Vereinigung*, 18:454–456, 1909.
464. C. S. Peirce. On the algebras in which division is unambiguous. *Am. Jl. of Math.*, 4:225–229, 1881.
465. O. Perron. Note über die Konvergenz von Kettenbrüchen mit positiven Gliedern. *Sitzungsberichte der mathematisch–physikalischen Klasse der K. B. Akademie der Wissenschaften zu München 1905*, 35:315–322, 1906.
466. O. Perron. Über die Konvergenz periodischer Kettenbrüche. *Sitzungsberichte der mathematisch–physikalischen Klasse der k. b. Akademie der Wissenschaften zu München 1905*, 35:495–503, 1906.
467. O. Perron. Grundlagen für eine Theorie des Jacobischen Kettenbruchalgorithmus. *Math. Ann.*, 64:1–76, 1907.
468. O. Perron. Zur Theorie der Matrices. *Math. Ann.*, 64:248–263, 1907.
469. O. Perron. Über die Konvergenz der Jacobi-Kettenalgorithmen mit komplexen Elementen. *Sitzungsberichte der mathematisch–physikalischen Klasse der K. B. Akademie der Wissenschaften zu München 1908*, pages 401–481, 1908. Submitted at the 7 December 1907 session.
470. O. Perron. Alfred Pringsheim. *Jahresbericht der Deutschen Mathematiker-Vereinigung*, 56: 1–6, 1953.
471. J. F. Pfaff. Methodus generalis aequationes differentiarum partialium, nec non aequationes differentiales vulgares, utrasque primi ordinis, inter quotcunque variabiles, complete integrandi. *Abhandlungen d. Akad. der Wiss. zu Berlin 1814–15*, pages 76–136, 1818. All references are to the annoted German translation by G. Kowalewski published as *Ostwald's Klassiker der exakten Wissenschaften, Nr. 129*, W. Engelmann, Leipzig, 1902.
472. É. Picard. Sur une classe de groupes discontinus de substitutions linéaires et sur les fonctions de deux variables indépendantes restant invariables par ces substitutions. *Acta Mathematica*, 1:297–320, 1882. Reprinted in *Oeuvres* 1, 311–334.

473. É. Picard. Remarque sur les groupes linéaires d'ordre fini à trois variables. *Bull. Soc. Math. France*, 15:152–155, 1887. Reprinted in *Oeuvres* 1, 597–600.
474. É. Picard and H. Poincaré. Sur un théorème de Riemann relatif aux fonctions de n variables indépendantes admettant $2n$ systèmes de périodes. *Comptes Rendus Acad. Sci. Paris*, 97:1284–1287, 1883. Reprinted in Picard, *Oeuvres* 1, 109–112, and in Poincaré, *Oeuvres* 4, 307–310.
475. H. Poincaré. Sur les fonctions fuchsiennes. *Acta Mathematica*, 1:193–295, 1882. Reprinted in *Oeuvres* 2, 169–257.
476. H. Poincaré. Sur les fonctions uniformes qui se reproduisent par des substitutions linéaires. *Math. Ann.*, 19:553–564, 1882. Reprinted in *Oeuvres* 2, 92–105.
477. H. Poincaré. Sur les fonctions de deux variables. *Acta mathematica*, 2:97–113, 1883. Reprinted in *Oeuvres* 4, 147–161.
478. H. Poincaré. Sur la réduction des intégrales abéliennes. *Bulletin de la société mathématique de France*, 12:124–143, 1884. Reprinted in *Oeuvres* 3, 333–351.
479. H. Poincaré. Sur les nombres complexes. *Comptes Rendus Acad. Sci. Paris*, 99:740–742, 1884. Reprinted in *Oeuvres* 5, 77–79.
480. H. Poincaré. Sur les fonctions abéliennes. *American Journal of Mathematics*, 8:289–342, 1886. Reprinted in *Oeuvres* 4, 318–378.
481. H. Poincaré. Sur les fonctions abéliennes. *Comptes Rendus, Acad. Sci. Paris*, 124:1407–1411, 1897. Reprinted in *Oeuvres* 4, 469–472.
482. H. Poincaré. Préface. In *Oeuvres de Laguerre*, volume 1, pages v–xv. Gauthier–Villars, Paris, 1898.
483. H. Poincare. Sur les propriétés du potentiel et sur les fonctions abéliennes. *Acta Mathematica*, pages 89–178, 1899. Reprinted in *Oeuvres* 4, 162–243.
484. H. Poincaré. Sur les fonctions abéliennes. *Acta mathematica*, 26:43–98, 1902. Reprinted in *Oeuvres* 4, 473–526.
485. H. Poincaré. Sur l'intégration algébrique des équations linéaires et les périodes des intégrales abéliennes. *Jl. des math. pures et appl.*, 9(5):139–212, 1903. Reprinted in *Oeuvres* 3, 106–166.
486. H. Poincaré. Rapport sur les travaux de M. Cartan. fait à la faculté des sciences d'Université de Paris. *Acta Mathematica*, 38:137–145, 1912.
487. M. Potron. Quelques propriétés des substitutiones linéaires à coefficients ≥ 0 et leur application aux problèmes de la production et des salaires. *Comptes Rendus, Acad. Sci. Paris*, 153:1129–1132, 1911.
488. M. Potron. Application aux problèmes de la "production suffisante" et du "salaire vitale" de quelques propriétés des substitutions linéaires à coefficientes ≥ 0. *Comptes Rendus, Acad. Sci. Paris*, 153:1458–1459, 1911.
489. M. Potron. Quleques propriétés des substitutiones linéaires à coefficients ≥ 0 et leur application aux problèmes de la production et des salaires. *Annales scientifiques École Normale Sup. Paris*, (3) 30:53–76, 1913.
490. A. Pringsheim. Ueber die Convergenz periodischer Kettenbrüche. *Sitzungsberichte der mathematisch–physikalischen Klasse der K. B. Akademie der Wissenschaften zu München 1900*, pages 463–488, 1901.
491. W. Purkert. Zur Genesis des abstrakten Körperbegriffs. 1. Teil. *NTM-Schriftenreihe für Geschichte der Naturwiss., Technik, und Med.*, 8:23–37, 1971.
492. W. Purkert. Zur Genesis des abstrakten Körperbegriffs. 2. Teil. *NTM-Schriftenreihe für Geschichte der Naturwiss., Technik, und Med.*, 10:8–20, 1973.
493. R. Remak. Über die Zerlegung der endlichen Gruppen in direkte unzerlegbare Faktoren. *Jl. für die reine u. angew. Math.*, 139:293–308, 1911.
494. P. Ribenboim. Wolfgang Krull—Life, Work and Influence. In *Wolfgang Krull Gesammelte Abhandlingen*, volume 1, pages 1–20. Walter de Gruyter, Berlin, 1999.
495. B. Riemann. Theorie der Abel'schen Functionen. *Jl. für die reine u. angew. Math.*, 54, 1857. Reprinted in *Werke*, pp. 88–144.

496. B. Riemann. Über die Hypothesen, welche der Geometrie zu Grunde liegen. *Abhandlungen der K. Gesellschaft der Wissenschaften zu Göttingen*, 1868. Reprinted in *Werke*, 272–287.
497. B. Riemann. Beweis des Satzes, dass eine einwerthige mehr als $2n$ fach periodische Function von n Veränderlichen unmöglich ist. *Jl. für die reine u. angew. Math.*, 71, 1870. Reprinted in *Werke*, 294–297. Extracted from a letter from Riemann to Weierstrass dated Göttingen 26 October 1859.
498. O. Rodrigues. Des lois géométriques qui régissent les déplacements d'un système solide dans l'espace, et de la variation des coordonnées provenant de ces déplacements considérés indépendent des causes qui peuvent les produire. *Jl. de math. pures et appl.*, 5:380–440, 1840.
499. V. Romanovsky. Sur les chaînes de Markoff. *Doklady Akademii nauk SSSR A*, pages 203–208, 1929.
500. V. Romanovsky. Sur les chaînes discrètes de Markoff. *Comptes Rendus, Acad. Sci. Paris*, 191:450–452, 1930.
501. V. Romanovsky. Sur les chaînes biconnexes continues de Markoff. *Comptes Rendus, Acad. Sci. Paris*, 191:695–697, 1930.
502. V. Romanovsky. Sur les zéros de matrices stocastiques. *Comptes Rendus, Acad. Sci. Paris*, pages 266–269, 1931.
503. V. Romanovsky. Un théorème sur les zéros des matrices non négatives. *Bulletin de la Société mathématique de France*, 61:213–219, 1933.
504. V. Romanovsky. Recherches sur les chaînes de Markoff. *Acta mathematica*, 66:147–251, 1936.
505. V. Romanovsky. *Discrete Markov Chains*. Wolters–Nordhoff, Groningen, 1970. Translated form the Russian edition (1945) by E. Seneta.
506. J. Rosanes. Über die Transformation einer quadratischen Form in sich selbst. *Jl. für die reine u. angew. Math.*, 80:52–72, 1875.
507. C. Rosati. Sulle matrici di Riemann. *Rendiconti Circolo Mat. Palermo*, 53:79–134, 1929.
508. M. Rosen. Abelian varieties over $\mathbb{C}$. In G. Cornell and J. Silverman, editors, *Arithmetic Geometry*, pages 79–101. Springer-Verlag, New York, 1986.
509. M. Rosen. Polynomials mod p and the theory of Galois sets. In M. Lavrauw et al., editors, *Theory and Application of Finite Fields. The 10th International Conference on Finite Fields and Their Applications, July 11–15, 2011, Ghent, Belgium*, pp. 163–178. American Mathematical Society, 2012.
510. D. Rowe. Klein, Hilbert, and the Göttingen mathematical tradition. *Osiris*, (2) 5:186–213, 1989.
511. W. Scharlau. Unveröffentlichte algebraische Arbeiten Richard Dedekinds aus seiner Göttinger Zeit. *Archive for History of Exact Sciences*, 27:335–367, 1982.
512. E. Schering. Die Fundamental-Classen der zusammensetzbaren Formen. *Abhandlungen der K. Gesell. er Wiss. zu Göttingen, Math-Naturwiss. Cl.*, 14:3–13, 1869.
513. L. Schlesinger. *Handbuch der Theorie der linearen Differentialgleichungen*, volume 1. B. G. Teubner, 1895.
514. L. Schlesinger. *Vorlesungen über Differentialgleichungen*. B. G. Teubner, Leipzig, 1908.
515. O. Schmidt. Über unendliche Gruppen mit endlicher Kette. *Jl. für die reine u. angew. Math.*, 29:34–41, 1928–1929.
516. H. Schneider. The concept of irreducibility and full decomposibility of a matrix in the works of Frobenius, König and Markov. *Journal of Linear Algebra and Its Applications*, 1977.
517. H. Schneider. The influence of the marked reduced graph of a nonnegative matrix on the Jordan form and on related properties: A survey. *Journal of Linear Algebra and Its Applications*, 84:161–189, 1986.
518. E. Scholz. Historical aspects of Weyl's Raum–Zeit–Materie. In E. Scholz, editor, *Hermann Weyl's Raum–Zeit–Materie and a General Introduction to His Scientific Work*. Birkhäuser, 2000.
519. F. Schottky. *Abriss einer Theorie der Abel'schen Functionen von drei Variabeln*. Teubner, Leipzig, 1880.

520. O. Schreier and E. Sperner. *Vorlesungen über Matrizen*. B. G. Teubner, Leipzig, 1932.
521. I. Schur. *Über eine Klasse von Matrizen, die sich einer gegebenen Matrix zuordnen lassen*. Dieterich'schen Univ.-Buchdruckerei, Göttingen, 1901. Schur's dissertation. Reprinted in *Abhandlungen* 1, 1–72.
522. I. Schur. Über die Darstellung der endlichen Gruppen durch gebrochene lineare Substitutionen. *Jl. für die reine u. angew. Math.*, 127:20–50, 1904. Reprinted in *Abhandlungen* 1, 86–127.
523. I. Schur. Neue Begründung der Theorie der Gruppencharaktere. *Sitzungsberichte der Akademie der Wiss. zu Berlin, Physikalisch-Math. Kl. 1905*, pages 406–432, 1905. Reprinted in *Abhandlungen* 1, 143–169.
524. I. Schur. Arithmetische Untersuchungen über endliche Gruppen linearer Substitutionen. *Sitzungsberichte der Akademie der Wiss. zu Berlin, Physikalisch-Math. Kl. 1906*, pages 164–184, 1906. Reprinted in *Abhandlungen* 1, 177–197.
525. I. Schur. Einige Bemerkungen zu der vorstehenden Arbeit A. Speiser, Zahlentheoretische Sätze aus der Gruppentheorie. *Mathematische Zeitschrift*, 5:7–10, 1919. Reprinted in *Abhandlungen* 2, 276–279.
526. I. Schur. Neue Anwendungen der Integralrechnung auf Probleme der Invariantentheorie. *Sitzungsberichte der Akademie der Wiss. zu Berlin, Physikalisch-Math. Kl. 1924*, pages 189–208, 1924. Reprinted in *Abhandlungen* 2, 440–459.
527. I. Schur. *Vorlesungen über Invariantentheorie*. Springer-Verlag, Berlin, 1968. H. Grunsky, ed.
528. G. Scorza. Intorno alla teoria generale delle matrici di Riemann e ad alcune applicazione. *Rendiconti Circolo Mat. Palermo*, 41:262–379, 1916.
529. J.-P. Serre. *A Course in Arithmetic*. Springer-Verlag, New York, 1973.
530. G. Shimura. *Abelian Varieties with Complex Multiplication and Modular Functions*. Princeton University Press, 1998.
531. C. L. Siegel. *Analytic Functions of Several Complex Variables. Lectures delivered at the Institute for Advanced Study 1948–1949. Notes by P. T. Bateman*. Institute for Advanced Study, 1949. Reprinted with corrections in March 1962.
532. C. L. Siegel. *Vorlesungen über ausgewählte Kapitel der Funktionentheorie*, volume 3. Mathematisches Insitute, Göttingen, Göttingen, 1966.
533. C. L. Siegel. Erinnerungen an Frobenius. In J.-P. Serre, editor, *Ferdinand Georg Frobenius Gesammelte Abhandlungen*, volume 1, pages iv–vi. Springer-Verlag, 1968.
534. C. L. Siegel. *Topics in complex function theory*, volume 3. Wiley-Interscience, New York, 1973.
535. S. Singh. *Fermat's Enigma: The Epic Quest to Solve the World's Greatest Mathematical Problem*. Anchor Books, New York, 1998. Originally published by Walker and Company (New York, 1997).
536. H. J. S. Smith. Report on the theory of numbers. Part I. *Report of the British Assoc. for the Advancement of Science*, pages 228–267, 1859. Reprinted in *Papers* 1, 368–406, and in [541, p.38ff].
537. H. J. S. Smith. On systems of linear indeterminate equations and congruences. *Phil. Trans. R. Soc. London*, 151:293–326, 1861. Reprinted in *Papers* 1. 368–406.
538. H. J. S. Smith. Report on the theory of numbers [Part III]. *Report of the British Assoc. for the Advancement of Science*, 1861. Reprinted in *Papers* 1, 163–228. The entire report (6 parts) was also reprinted as the book [541].
539. H. J. S. Smith. I. On the arithmetical invariants of a rectangular matrix, of which the constituents are integral numbers. *Proc. London Math. Soc.*, 4:236–240, 1873. Reprinted in *Papers* 2, as Note I of "Arithmetical Notes," pp. 67–85.
540. H. J. S. Smith. II. On systems of linear congruences. *Proc. London Math. Soc.*, 4:241–249, 1873. Reprinted in *Papers* 2, as Note II of "Arithmetical Notes," pp. 67–85.
541. H. J. S. Smith. *Report on the Theory of Numbers*. Chelsea Publishing Co, New York, 1965. A reprint of Smith's six reports "On the Theory of Numbers" to the British Association for the Advancement of Science between 1859 and 1865, together with a biographical sketch by Charles H. Pearson and recollections by B. Jowett.

542. V. Snyder, et al. *Selected Topics in Algebraic Geometry*. Chelsea, New York, 1970.
543. A. Speiser. Zahlentheoretische Sätze aus der Gruppentheorie. *Mathematische Zeitschrift*, 5:1–6, 1919.
544. A. Speiser. *Die Theorie der Gruppen von endlicher Ordnung*. Springer-Verlag, Berlin, 1923.
545. A. Speiser. *Die Theorie der Gruppen von endlicher Ordnung*. Springer-Verlag, Berlin, 2nd edition, 1927.
546. W. Spottiswoode. *Elementary Theorems Relating to Determinants*. Longman, Brown, Green, and Longman Paternoster Row, London, 1851.
547. H. Stahl. Beweis eines Satzes von Riemann über ϑ-Charakteristiken. *Jl. für die reine u. angew. Math.*, 88:273–276, 1880.
548. P. Stevenhagen and H. Lenstra. Chebotarëv and his density theorem. *The Mathematical Intelligencer*, 18(2):26–37, 1996.
549. L. Stickelberger. *De problemate quodam ad duarum bilinearium vel quadraticarium transformationum pertinente*. G. Schade, Berlin, 1874.
550. L. Stickelberger. Ueber Schaaren von bilinearen und quadratischen Formen. *Jl. für die reine u. angew. Math.*, 86:20–43, 1879.
551. J. Stillwell. Translator's introduction. In *Theory of Algebraic Integers by Richard Dedekind*, Cambridge Mathematical Library. Cambridge University Press, Cambridge, 1996.
552. O. Stolz. *Vorlesungen über allgemeine Arithmetik, nach den neueren Ansichten. Zweiter Theil: Arithmetik der complexen Zahlen mit geometrischen Anwendungen*. Teubner, Leipzig, 1886.
553. N. Stuloff. Frobenius: Ferdinand Georg, Mathematiker. In *Neue Deutsche Biographie*, volume 5, page 641. Duncker and Humblot, 1960.
554. C. Sturm. Analyse d'un Mémoire sur la résolution des équations numériques; par M. Ch. Sturm. (Lu à l'Acad. roy. des Scien., le 23 mai 1829. *Bull. Sci. Math.*, 11:419–422, 1829. Reprinted in [556, pp. 323–26].
555. C. Sturm. Extrait d'un mémoire sur l'intégration d'un système d'équations différentielles linéaires, présenté à l'Academie des Sciences le 27 Juillet 1829 par M. Sturm. *Bull. Sci. Math.*, 12:313–322, 1829. Reprinted in [556, pp. 334–42].
556. C. Sturm. *Collected works of Charles François Sturm. Jean-Claude Pont (Editor). In Collaboration with Flavia Padovani*. Birkhäuser, Basel, 2009.
557. H. P. F. Swinnerton-Dyer. *Analytic Theory of Abelian Varieties*. Cambridge University Press, 1974.
558. L. Sylow. Théorèmes sur les groupes des substitutions. *Math. Ann.*, 5:584–594, 1872.
559. J. J. Sylvester. On the equation to the secular inequalities in the planetary theory. *Phil. Mag.*, 16:110–11, 1883. Reprinted in *Papers*, v. 4, 110–111.
560. R. Taton. L'École Polytechnique et le renouveau de la géométrie analytique. In *Mélanges Alexandre Koyré*, volume 1, pages 552–564. Hermann, Paris, 1964.
561. N. Tchebotarev. Die bestimmung der dichtigkeit einer menge von primzahlen, welche zu einer gegebenen substitutionsklasse gehören. *Math. Ann.*, 95:191–228, 1926.
562. L. W. Thomé. Zur Theorie der linearen Differentialgleichungen. *Jl. für die reine u. angew. Math.*, 76:273–302, 1873.
563. R. Tobies and D. Rowe, editor. *Korrespondenz Felix Klein–Adolph Mayer. Auswahl aus den Jahren 1871–1907*. Teubner, Leipzig, 1990.
564. A. W. Tucker and F. Nebeker. Lefschetz, Solomon. In *Dictionary of Scientific Biography*, volume 18, pages 534–539. Charles Scribner's Sons, New York, 1990.
565. H. W. Turnbull. On the reduction of singular matrix pencils. *Proceedings Edinburgh Mathematical Society*, (2) 4:67–76, 1935.
566. H. W. Turnbull. Alfred Young, 1873–1940. *Journal London Math. Soc.*, 16:194–207, 1941. Reprinted in *The Collected Papers of Alfred Young*, Toronto, 1977 pp. xv–xxvii.
567. H. W. Turnbull and A. C. Aitken. *An Introduction to the Theory of Canonical Matrices*. Blackie and Son, London & Glasgow, 1932.
568. B. L. van der Waerden. *Moderne Algebra*, volume 2. Springer, Berlin, 1931.

569. B. L. van der Waerden. On the sources of my book *Moderne Algebra*. *Historia Mathematica*, 2:31–40, 1975.
570. R. Varga. *Matrix Iterative Analysis*. Prentice–Hall, Englewood Cliffs, N. J., 1962.
571. F. Viète. *In artem analyticam isagoge...* Tours, 1591. References are to the English translation in J. Klein *Greek Mathematical Thought and the Origin of Algebra*, Cambridge, Mass., 1968.
572. S. G. Vlăduţ. *Kronecker's Jugendtraum and Modular Functions*. Gordon and Breach, 1991.
573. R. von Mises. Über die Aufgaben und Ziele der angewandten Mathematik. *Zeitschrift für angewandte Mathematik und Mechanik*, 1:1–15, 1921.
574. R. von Mises. *Vorlesungen aus dem Gebiete der angewandten Mathematik. Band I: Warscheinlichkeitsrechnung und ihrer Anwendung in der Statistik und theoretischen Physik*. F. Deuticke, Leipzig, 1931.
575. E. von Weber. *Vorlesungen über das Pfaffsche Problem und die Theorie der partiellen Differentialglichungen erster Ordnung*. Teubner, Leipzig, 1900.
576. A. Voss. Ueber die mit einer bilinearen Form vertauschbaren bilinearen Formen. *Sitzungsberichte der mathematisch–physikalischen Classe der k. b. Akademie der Wissenschaften zu München 1889*, 1890.
577. H. Weber. Ueber die Transformationstheorie der Theta-Functionen, ins Besondere derer von drei Veränderlichen. *Annali di matematica*, (2) 9:126–166, 1878.
578. H. Weber. Beweis des Satzes, dass jede eigentlich primitive quadratische Form unendlich viele Primzahlen darzustellen fähig ist. *Math. Ann.*, 20:301–329, 1882.
579. H. Weber. Theorie der abel'schen Zahlkörper. *Acta Mathematica*, 8:193–263, 1886.
580. H. Weber. Theorie der abel'schen Zahlkörper. *Acta Mathematica*, 9:105–130, 1887.
581. H. Weber. *Elliptische Functionen und algebraische Zahlen*. Vieweg, Braunschweig, 1891. A second edition was published in 1908 under the same title but as the third volume of the second edition of Weber's *Lehrbuch der Algebra*.
582. H. Weber. *Lehrbuch der Algebra*, volume 1. F. Vieweg & Sohn, Braunschweig, 1895. The second ed. (1898), was reprinted with corrections and some new notation as a third edition by Chelsea Publishing Co., New York, 1961.
583. H. Weber. *Lehrbuch der Algebra*, volume 2. F. Vieweg & Sohn, 1896. Second ed. 1899, reprinted with corrections and some new notation as a third edition by Chelsea Publishing Co., New York, 1961.
584. H. Weber. *Lehrbuch der Algebra*, volume 3. Braunschweig, F. Vieweg & Sohn, 2nd edition, 1908. Reprinted with corrections and some new notation as a third edition by Chelsea Publishing Co., New York, 1961.
585. J. H. M. Wedderburn. *Lectures on Matrices*. American Mathematical Society, New York, 1934.
586. K. Weierstrass. Zur Theorie der Abel'schen Functionen. *Jl. für die reine u. angew. Math.*, 47:289–306, 1854. Reprinted in *Werke* 1, 133–152.
587. K. Weierstrass. Über ein die homogenen Functionen zweiten Grades betreffendes Theorem. *Monatsberichte der Akademie der Wiss. zu Berlin*, 1858. Reprinted in *Werke* 1, 233–246.
588. K. Weierstrass. Zur Theorie des quadratischen und bilinearen Formen. *Monatsberichte der Akademie der Wiss. zu Berlin*, pages 311–338, 1868. Reprinted with modifications in *Werke* 2, 19–44.
589. K. Weierstrass. Über die allgemeinsten eindeutigen und $2n$-fach periodischen Functionen von n Veränderlichen. *Monatsberichte der Akademie der Wiss. zu Berlin 1869*, pages 853–857, 1870. Reprinted in *Werke* 2, 45–48.
590. K. Weierstrass. Neuer Beweis eines Hauptsatzes der Theorie der periodischen Functionen von mehreren Veränderlichen. *Monatsberichte der Akademie der Wiss. zu Berlin*, pages 680–693, 1876. Reprinted in *Werke* 2, 55–69.
591. K. Weierstrass. Nachtrag zu der am 4. März ... gelesenen Abhandlung: Über ein die homogenen Functionen zweiten Grades betreffendes Theorem. *Sitzungsberichte der Akademie der Wiss. zu Berlin*, pages 430–439, 1879. Reprinted with footnotes added by Weierstrass in *Werke* 3, 139–148.

592. K. Weierstrass. Untersuchungen über die $2r$-fach periodischen Functionen von r Veränderlichen. *Jl. für die reine u. angew. Math.*, 89:1–8, 1880. Reprinted in *Werke* 2, 125–133.
593. K. Weierstrass. Zur Theorie der aus n haupteinheiten gebildeten complexen Grössen. *Nachrichten von der Königlichen Gesellschaft der Wissenschaften und der Georg-Augustus-Universität zu Göttingen*, pages 395–414, 1884. Reprinted in *Werke* 2, 311–332.
594. K. Weierstrass. Vorlesungen über die Theorie der abelschen Transcendenten. In G. Hettner and J. Knoblauch, editors, *Mathematische Werke von Karl Weierstrass*, volume 4. Mayer and Müller, 1902.
595. K. Weierstrass. Allgemeine Untersuchungen über $2n$-fach periodische Functionen von n Veränderlichen. In J. Knoblauch, editor, *Mathematische Werke von Karl Weierstrass*, volume 3, pages 53–114. Mayer and Müller, Berlin, 1903.
596. K. Weierstrass. Über the Convergenz der θ-Reihen beliebig vieler Argumente. In J. Knoblauch, editor, *Mathematische Werke von Karl Weierstrass*, volume 3, pages 115–122. Mayer and Müller, Berlin, 1903.
597. K. Weierstrass. *Vorlesungen über elliptischen Functionen. Mathematische Werke Vol. 5 (J. Knoblauch, ed.)*. Mayer and Müller, Berlin, 1915.
598. A. Weil. Théorèmes fondamentaux de la théorie des fonctions thêta (d'après des mémoires de Poincaré et Frobenius). *Séminaire Bourbaki*, Exposé 16, 1949. The second, corrected, edition (1959) is reprinted in [600, pp. 414–421].
599. A. Weil. *Number Theory: An Approach Through History from Hammurapi to Legendre*. Birkhäuser, Boston, 1984.
600. A. Weil. *Collected Papers*, volume 1. New York, Springer-Verlag.
601. H. Weyl. *Mathematische Analyse des Raumproblems*. Springer, Berlin, 1923.
602. H. Weyl. Zur Theorie der Darstellung der einfachen kontinuierlichen Gruppen. (Aus einem Schreiben an Herrn I. Schur.) *Sitzungsberichte der Akademie der Wiss. zu Berlin*, pages 338–345, 1924. Reprinted in *Abhandlungen* 2, 453–460.
603. H. Weyl. Theorie der Darstellung kontinuierlicher halbeinfacher Gruppen durch lineare Transformationen. Kap. I–III und Nachtrag. *Mathematische Zeitschrift*, 23–24:271–309 (vol. 23), 328–395, 789–791 (vol. 24), 1925–1926. Reprinted in *Abhandlungen* 2, 543–647.
604. H. Weyl and F. Peter. Die Vollständigkeit der primitiven Darstellungen einter geschlossenen kontinuierlichen Gruppe. *Math. Ann.*, 97:737–755, 1927. Reprinted in Weyl, *Abhandlungen* 3, 58–75.
605. E. Whittaker. *A History of the Theories of Aether and Electricity*, volume 1. E. Nelson, London and New York, 2nd edition, 1951.
606. E. P. Wigner. *Gruppentheorie und ihre Anwendung auf die Quantenmechanik der Atomspektren*. Braunschweig, 1931.
607. E. Wiltheiss. Über die complexe Multiplication hyperelliptischer Functionen zweier Argumente. *Math. Ann.*, 21:385–398, 1883.
608. A. Wiman. Über die Darstellung der symmetrischen und alternirenden Vertauschungsgruppen als Collineationsgruppen von möglichst geringer Dimensionzahl. *Math. Ann.*, 52:243–270, 1899.
609. W. Wirtinger. Zur Theorie der $2n$-fach periodischen Functionen. 1. Abhandlung. *Monatshefte für Mathematik und Physik*, 6:69–98, 1895.
610. W. Wirtinger. Zur Theorie der $2n$-fach periodischen Functionen. (2. Abhandlung). *Monatshefte für Mathematik und Physik*, 7:1–25, 1896.
611. W. Wirtinger. Über einige Probleme in der Theorie der Abel'schen Functionen. *Acta Mathematica*, 26:133–156, 1902.
612. H. Wussing. *Die Genesis des abstrakten Gruppenbegriffes. Ein Betrag zur Entstehungsgeschichte der abstrakten Gruppentheorie*. VEB Deutscher Verlag, Berlin, 1969. English translation as [613].
613. H. Wussing. *The Genesis of the Abstract Group Concept. A Contribution to the History of Abstract Group Theory*. MIT Press, Cambridge, MA, 1984. Abe Shenitzer, transl., 1984. Translation of [612].

614. A. Young. On quantitative substitutional analysis. *Proceedings London Math. Soc.*, 33:97–146, 1901. Reprinted in *Papers*, 42–91.
615. A. Young. On quantitative substitutional analysis. *Proceedings London Math.*, 34:361–397, 1902. Reprinted in *Papers*, 92–128.
616. A. Yvon-Villarceau. Note sur les conditions des petites oscillations d'un corps solide de figure quelconque et la théorie des équations différentielles linéaires. *Comptes rendus Acad. Sci. Paris*, 71:762–766, 1870.

Index

T. Hawkins, *The Mathematics of Frobenius in Context*, Sources and Studies in the History of Mathematics and Physical Sciences, DOI 10.1007/978-1-4614-6333-7,

9781461463320